A230

MULTILINGUAL
THESAURUS
of GEOSCIENCES

Pergamon Books of Related Interest

Gusev et al English-Russian Dictionary of Applied Geophysics
Kuznetsov Russian-English Polytechnical Dictionary
Nickon & Silversmith Organic Chemistry—The Name Game: Modern Coined Terms and Their Origins
Stoliarov Russian-English Oil Field Dictionary
WEC Energy Terminology: A Multilingual Glossary

Pergamon Journals of Related Interest
(Sample copies available on request)

COMPUTERS & GEOSCIENCES
GEOCHIMICA ET COSMOCHIMICA ACTA
INTERNATIONAL JOURNAL OF ROCK MECHANICS AND MINING SCIENCES & GEOMECHANICS ABSTRACTS
JOURNAL OF AFRICAN EARTH SCIENCES
JOURNAL OF SOUTH-EAST ASIAN EARTH SCIENCES
JOURNAL OF STRUCTURAL GEOLOGY
ORGANIC CHEMISTRY
PHYSICS AND CHEMISTRY OF THE EARTH
QUATERNARY SCIENCE REVIEWS

MULTILINGUAL THESAURUS of GEOSCIENCES

Edited by

G.N. Rassam
Chairman, Commission on Geological Documentation (COGEODOC) and
American Geophysical Union (AGU), Washington, D.C., USA

J. Gravesteijn
COGEODOC and Bureau de Recherches Geologiques et Minières (BRGM), Orléans, France

R. Potenza
COGEODOC and Consiglio Nazionale delle Ricerche (CNR), Milan, Italy

Sponsored by the
International Council for Scientific and Technical Information (ICSTI)
and
International Union of Geological Sciences (IUGS)

PERGAMON PRESS

New York · Oxford · Beijing · Frankfurt · São Paulo · Sydney · Tokyo · Toronto

U.S.A.	Pergamon Press, Maxwell House, Fairview Park, Elmsford, New York 10523, U.S.A.
U.K.	Pergamon Press, Headington Hill Hall, Oxford OX3 0BW, England
PEOPLE'S REPUBLIC OF CHINA	Pergamon Press, Room 4037, Qianmen Hotel, Beijing, People's Republic of China
FEDERAL REPUBLIC OF GERMANY	Pergamon Press, Hammerweg 6, D-6242 Kronberg, Federal Republic of Germany
BRAZIL	Pergamon Editora, Rua Eça de Queiros, 346, CEP 04011, Paraiso, São Paulo, Brazil
AUSTRALIA	Pergamon Press Australia, P.O. Box 544, Potts Point, N.S.W. 2011, Australia
JAPAN	Pergamon Press, 8th Floor, Matsuoka Central Building, 1-7-1 Nishishinjuku, Shinjuku-ku, Tokyo 160, Japan
CANADA	Pergamon Press Canada, Suite No. 271, 253 College Street, Toronto, Ontario, Canada M5T 1R5

First printing 1988

Library of Congress Cataloging in Publication Data

Multilingual thesaurus of geosciences.
Bibliography: p.
Includes index.
1. Subject headings—Geology.   2. Subject headings—Mines and Mineral resources.   3. Subject headings—Hydrology.   I. Rassam, Ghassan N.
II. Gravesteijn, J.   III. Potenza, R.   IV. International Council for Scientific and Technical
Information.   V. International Union of Geological Sciences.
Z695.1.G43M84   1987   025.4′9551   86-25353

British Library Cataloguing in Publication Data

Multilingual thesaurus of geosciences
1. Geology. Polyglot dictionaries
I. Rassam, G. N.   II. Gravesteijn, J.
III. Potenza, R.
551′.03
ISBN 0–08–036431–4

Reproduced, printed and bound in Great Britain by
Hazell Watson & Viney Limited
Member of BPCC plc
Aylesbury Bucks

Contents

Preamble: On the Birth of a Thesaurus

"In the beginning was the Word" has had small acceptance by philosophers from Plato to Feyerabend who, through the ages, have wrangled over the meaning of the word, "Word." Was it the sign for an idea of an object, or was it the mark of the object itself? For purposes of translation, epistemologists have argued that, "Schnee ist weiss, if and only if snow is white." This has little relevance to geology, and certainly none to the multilingual thesaurus you have in your hand. What is relevant, however, is that the language of science undergoes strange and complicated transmutations as it crosses linguistic boundaries (e.g., Eisberg in German becomes Iceberg in English, and remains Iceberg in French).

In 1972, at the International Geological Congress in Montreal, the IUGS Commission on Geological Documentation (COGEODOC) presented a modest draft of a multilingual thesaurus of structural geology in Czech, English, French, and German.

For the representatives of the national organizations concerned at that time (American Geological Institute, Bundesanstalt für Geowissenschaften und Rohstoffe, Bureau de Recherches Geologiques et Minières, GEOFOND), the aim was not only to undertake a new task, but to respond to a more and more evident need to organize international cooperation in the field of earth science documentation.

One must bear in mind, that at that epoch—which seems to be so far in the past—computer use had just begun to enter geologic work, and only a few organizations had started to produce bibliographies and indexes by computer methods.

The presentation of this "micro thesaurus" surprised the international community at Montreal and aroused its interest. Several national delegations (among which was the delegation of the Soviet Union) attended the meetings of COGEODOC. The group did not want to confine itself to a pilot project and also did not want to disappoint the new interlocutors at Montreal. Thus, before the end of the Congress, it was decided—maybe with a bit of a happy-go-lucky attitude—to work on a multilingual thesaurus for the earth sciences as a whole.

The general principles for such a thesaurus were defined as follows:

- the work should be organized collectively and should take into account the existing national thesauri and documentary activities;

- any proposed linguistic or national collaboration should be accepted, the initial languages being Czech, English, French, German, Russian, and Spanish;

- the lexical reference material should be the "Glossary of Geology," published by the American Geological Institute; each participant on behalf of any linguistic group should be allowed to propose additional terms and concepts. After discussion in the working group, these terms were to be introduced and translated by all other linguistic groups.

- The final aim of the project is to facilitate the exchange of bibliographic data between the different organizations and linguistic groups through an international network of documentation for the benefit of all geologists.

The project was begun, and the work leading to the present stage took 14 years; a short span for any human enterprise . . . and very short when considered in geological concepts of time! In these 14 years there were many technical problems to be solved, plenty of sessions and discussions, many corrections and—above all—immense good intention and additional participation of the partners from Canada, Czechoslovakia, Finland, France, West Germany, Italy, guests from Denmark, Sweden, Hungary, correspondents from Australia and other countries.

When we took the first steps toward creating the thesaurus each of us had some knowledge of each other's first language, and possessed a working knowledge of a third or fourth and that helped the process along. Borrowed terms (e.g., klippe, graben, molasse, roche moutonnée) created special problems when it became evident that the borrower had altered the original meanings; and common terms with different meaning in different countries (e.g., schist and schiste) created others. Classifications that are troublesome in American usage (e.g., siltstone, claystone, mudstone) became more so when exported. Differences between American English and British English had to be explicated. It was not long before we realized that chalk and a blackboard were indispensable to our deliberations. Can a thrust fault and a reverse fault be explained without a pencil? We were reminded of the hand gestures to be observed in the corridors of any large meeting when geologists are describing the details of a structure to colleagues. There was a time when the formation of a Committee to Standardize Geological Gestures was considered.

In the beginning we took a gamble perhaps and the odds against success were great, but finally we succeeded because our cooperation was founded on the willingness to cooperate, on friendship as well as on modern techniques and on data processing. However, we are aware that nothing would have been accomplished without the support of the International Union of Geological Sciences and also the International Council of Scientific Unions Abstracting Board which has evolved into the International Council for Scientific and Technical Information, and where we have always found scientific and operational help and understanding.

The present multilingual thesaurus is fully in concordance with our initial wish: a collective work of geologists, who often became information specialists by outside demand or by professional interest. Its completion and publication is a technical success. But beyond this—and this seems to us to be essential—during the numerous meetings all over the world, there was the chance of discovery and establishment of friendship and close human relations, and the fact that there are already countries with operational bilateral cooperation has come from this documentary tool and work.

We are pleased that we have this opportunity to offer our congratulations and express our profound admiration for this accomplishment. The geologists who performed the work have gone a long way to respond to the admonition of the great Francis Bacon, who wrote,

"Although we think we govern our words, . . . certain it is that words, as a Tartar's bow, do shoot back upon the understanding of the wisest, and mightily entangle and pervert the judgement. So that it is almost necessary . . . in setting down in the very beginning the definitions of our words and terms, that others may know how we accept and understand them, and whether they concur with us or no. For it cometh to pass, for want of this, that we are sure to end there where we ought to have begun, which is—in questions and differences about words."

L. Delbos, *France*
H. Glashoff, *Federal Republic of Germany*
J.J. Lloyd, *United States*

Introduction

The Multilingual Thesaurus of Geosciences (MT) was developed over the past 14 years by the efforts of two international organizations: International Council for Scientific and Technical Information (ICSTI) and the Commission of Geological Documentation (COGEODOC) of the International Union of Geological Sciences (IUGS). ICSTI, which is a not-for-profit organization like its predecessor ICSU AB (International Council of Scientific Unions Abstracting Board), aims at increasing accessibility to and awareness of scientific and technical information. Its membership includes the leading world database producers in science and technology, as well as representatives of all activity sectors in the information transfer chain. COGEODOC promotes international cooperation among national and international documentation centers covering the earth sciences, and aims at improvement of the availability and the use of geological information by geoscientists and specialists involved in geological activities. Additional and important help was provided by the organizations where the participants worked. For a list of these organizations, see Editorial Group at the end of this introduction.

In 1970 the ICSU AB selected geology as the subject of a pilot project to develop a methodology for the construction of a multilingual thesaurus, which could then be applied to other sciences under the assumption that geology has a more limited vocabulary than other scientific disciplines and that there was less ambiguity in terminology and nomenclature.

ICSU AB was joined in the pilot project by COGEODOC and the language coverage was to include Czech, English, French, and German while the subject field was reduced to tectonics with a subset of 300 terms. The work was completed and presented to the International Geological Congress at Montreal (Canada) in August 1972. At the Congress, IUGS requested that the joint IUGS/ICSU AB working group (WGMT) undertake the expansion of the pilot project to a complete multilingual thesaurus of the vocabulary of geology. After initially examining and experimenting with theoretical approaches, the WGMT switched to a totally pragmatic methodology and used it throughout.

The American Geological Institute's Glossary was taken as a first reference tool. This glossary contains about 33,000 terms and their definitions. It is approximately 80 percent complete without considering archaic nomenclature or slang.

It was quickly shown that the assumption that geology would have a limited vocabulary with few ambiguous terms was overly optimistic. Much time was spent on explanations of the meaning of terms in various languages. It was important to relate concepts rather than terms in a science where cognates may have different meanings (Fault–Falte), and homonyms are sometimes more restricted in one language than in another (schist–schiste).

WHY A MULTILINGUAL THESAURUS IN GEOLOGY?

The treatment of bibliographic information and the creation of manual or automated indexes require the use of documentary methods of classification and organization for bibliographic analysis.

In a manual system, the classification plays a preeminent role. With introduction of automatic methods of retrieval (and computers), the procedures for selection of pertinent documents are refined in the sense that index terms are used for retrieval rather than broad categories. The ensuing dramatic increase in facility and quality of telecommunications has added a new dimension to the dissemination of information: It has shrunk the distance between the laboratory and the database and between the researcher and the documentalist, and as a result, between the scientist as *producer* of data and the scientist as a *user* of data.

It is not surprising therefore that a movement began in various documentation centers to increase the compatibility between their methods and their products.

In 1970, the French, West German and Czech geological surveys cooperated to create and update a joint database. A common indexing vocabulary was used and direct contacts between the documentation centers led to a common indexing practice. On the other hand, in the 1970s, GEOINFORM, the common geoscience information project of the CMEA (Council for Mutual Economic Assistance) countries, started developing a multilingual thesaurus for the analysis of the geoscience literature to be covered by the participating countries.

The major outcome of the ICSTI–IUGS multilingual thesaurus (MT) project is the reinforcement of this international cooperation. Nine European countries (Czechoslovakia, Finland, France, Federal Republic of Germany, Hungary, Italy, Poland, Romania, and Spain) and the United States (represented by the American Geological Institute, AGI) participate in an information network contributing in the updating of the two operational international databases, GeoRef and PASCAL-GEODE and using a common indexing language perfectly compatible with the MT.

The MT thus could become in the near future the switching mechanism for a world network for geological information.

It is planned that the MT project will be a dynamic and ongoing operation that will take into account developments in the international information scene, and which will contribute to a better and easier use of geoscience information.

HOW TO BUILD A MULTILINGUAL THESAURUS?

Thesauri

The creation of thesauri is closely related to the automatic treatment of bibliographic analysis and the search strategy which results therefrom.

The International Standards Organization (ISO) standard for monolingual thesauri defines three criteria:

1. *Controlled language of indexing:* terms selected from the natural language and formally used to represent, in a condensed form, the contents of the document (descriptors).

2. *Establishment of relationships between terms:* these relationships may be hierarchical with "specific" and "generic" terms, or associative with "use" and "use for" or "see also" terms.

3. *Formal organization of relationships:* function of a term (descriptor or non-descriptor) and its relationship with other terms belonging to the same group of apparent concepts.

A multilingual thesaurus is different in that it permits transposition and translation of concepts in many languages. The multilingual thesaurus may be "divided" into a number of distinct monolingual structured thesauri, although it is not a simple sum of them.

Multilingual thesauri may be created *a priori* when different documentation centers decide to construct a common database. The multilingual thesaurus allows each linguistic group to operate in its own language, both for indexing and for retrieval. Such networks have been created by the United Nations, Commission of European Community and the CMEA countries.

In the case of international cooperation between separate existing information systems, the methodologic approach is based on *a posteriori* compatibility between the concepts used for the various systems operating in different languages.

In this case, the comparison of concepts and terms and the analysis of the indexing techniques of the participants take the most important place in the construction of a multilingual thesaurus. The *a posteriori* methodology was chosen for the MT project.

Languages

The Working Group included the following language groups: Czech, English, Finnish, French, German, Italian, Russian, and Spanish. Other languages may be included in the future.

Hungarian and Swedish specialists participated in the work of the group but their linguistic versions have not yet been included in the database. For several years there was also Canadian participation in the project.

Principles

The MT has been conceived as a switching mechanism between documentation centers specializing in geoscience. It involves creating a common indexing vocabulary, compatible with work methods and practical systems in the various centers. In other words, bridges must be created between centers of different languages and with different methodology. Such a task may have seemed impossible but the Working Group started with a phased, pragmatic approach.

Most members of the group had the technical expertise necessary for decision-making. In addition, the cooperation among several members of the group, as it has developed, actually used the methodology and the terminology of the MT thus creating a "proving ground" for the MT concept.

Methodology

- *A posteriori* approach: Selection by each participant of appropriate terms in AGI's *Glossary of Geology*.

- Verification of compatibility of terms with each "national" monolingual thesaurus.

- Discussion in annual meetings of definition of terms, their quality as descriptors or non-descriptors in each linguistic version.

- Introduction or elimination of terms as a function of the following criteria: utility of these terms for indexing or retrieval, and use of them as descriptors by at least one of the systems represented.

- Addition of terms judged necessary for each participant to assure a compatibility between their national thesaurus and the MT.

- Terms are then assigned, by each linguistic group:
  1. Their role in each documentation system (descriptor; non-descriptor)
  2. Establishment of eventual substitute relationships for non-descriptors. Multiple preferred terms may be given. These substitutes may be simple or complex as a function of the specificity of the term or its characteristics, whether post- or pre-coordinated.
  3. Reciprocity of substitute relationships. Preferred usage (use for or *UF*) is indicated for descriptors involved by both unique and multiple preferred terms.

- Some adjectives and general terms used for post-coordinate indexing were then introduced in the MT. These terms are useful only in relation to other descriptors in a syntactic relationship used by certain national systems.

- When a term is used only in one language, its original form is not translated and is repeated in the other languages; for example, EN ECHELON in English.

- Classification of terms by subject or field.

Hierarchical Structure of MT

The hierarchical structure is established at the level of each monolingual national thesaurus. These "national" thesauri make the annexes or the extensions of the multilingual nucleus and are related to it by the MT terms.

Examples of hierarchies in "national" monolingual thesauri follow.

Russian: DVIZHENIE TEKTONICHESKOE (related to MT
term *PLATE TECTONICS*)

V(BT)	TEKTOGENEZ
N(NT)	DVIZHENIE EHSTATICHESKOE
	DREJF KONTINENTOV
	SKLADKOOBRAZOVANIE
	TAFROGENEZ
	EHPEYROGENEZ
a! (RT!)	KOLEBANIE
	SMESHCHENIE
	TEKTONIKA PLIT

English: ANTICLINES (equivalent to MT term)

BT	FOLDS
SA	ANTICLINORIA
SA	ANTIFORM FOLDS
SA	ARCHES
SA	DIAPIRS
SA	DOMES
SA	GEANTICLINES
SA	SYNCLINES

(BT = broader term; NT = narrower term; RT = related term; SA = see also)

CONTENTS

The MT database contains now approximately 5,000 key terms expressed as descriptors or non-descriptors in 8 linguistic versions. The terms are classed in 36 groups or fields, of which 20 correspond to the major subdivisions of geoscience; 11 concern the systematic parts of classification domains such as stratigraphy, rock names, elements, etc.; and 5 distinct fields describe concepts common to all subdivisions (properties, methods, etc.)

For the systematic subdivisions, selection has been made among the terms available. The criteria for selection were the following:

• Stratigraphy: epoch and stage names. Field: STRS

• Paleontology: names of groups (to the family or order level) as a function of the documentation importance of the groups. Field: PALS

• Soils: terms used in American and European classifications. Field: SUSS

• Rocks: names generally accepted, according to the IUGS classifications. Fields: IGNS, IGMS, SEDS

• Minerals: principal groups. Field: MING

• Extraterrestrial bodies: planets and meteorites. Field: EXTS

Geographic and physiographic place names are not included for the time being.

PRESENTATION

Six languages—English, French, German, Italian, Russian and Spanish—were chosen for this printed version of the thesaurus. For other languages, the option of printing a national version of the MT together with one or more of the languages is left open.

The MT includes a main list of key terms as well as several indexes and translations. The main list shows the key terms in alphabetical order carrying a sequential reference number. Each entry is composed of the following:

• A key term in English presented in the initially selected form of the reference concept. The key term is followed by the field of application.

- Six entries corresponding to the different linguistic versions of the MT, exhibiting the documentation nature of the term (Descriptor, Non-descriptor, Adjectival, General or Non-translatable). If the term is in all capital letters, the term is a descriptor in that language and the term is translated. If the designation is in upper and lower case, the term is not a descriptor and is followed by one or two descriptors indicating (in capital letters) the preferred term(s). If two descriptors are assigned, these are separated by a semicolon. Preferred usage is indicated by a *UF* (use for) for each linguistic version. Multiple preferred terms are separated by a semicolon.

- Adjectives are in capital letters for those versions using post-coordinate indexing. For the other linguistic versions, adjectives are followed by (adj.) and are in upper and lower case letters, for example, Abyssal (adj.).

- General terms are treated in a similar manner to adjectives and are followed by (gen.) for those versions not using post-coordinate indexing, for example, Paleo (gen.).

- Non-translated terms are enclosed in brackets, [EN ECHELON].

For further information on the presentation, see the Legend and Preface to Entries of the MT.

In addition, there are for each language a specific linguistic index in alphabetical order referencing the key-term numbers used in the main list.

Those indexes are followed by the field index, in which key terms have been classified by field. A full list of the field codes and their explanations follows.

APPL	Applied geophysics
CHEE	Elements
CHES	Chemical compounds
COMS	Commodities
ECON	Economic geology
ENGI	Engineering geology
ENVI	Environment
EXTR	Extraterrestrial geology
EXTS	Meteorites, planets
GEOC	Geochemistry
GEOH	Hydrology
GEOL	General geology
IGMS	Metamorphic rocks—systematics
IGNE	Petrology
IGNS	Igneous rocks—systematics
INST	Instruments—equipment
ISOT	Isotope geochemistry/Absolute age, geochronology
MARI	Marine geology
MATH	Mathematical geology
METH	Methods
MINE	Mineralogy
MING	Mineral groups
MINI	Mining
MISC	Miscellaneous
PALE	Paleontology
PALS	Paleontology—systematics
PHCH	Physical and chemical properties, processes
SEDI	Sedimentology
SEDS	Sedimentary rocks—systematics
SOLI	Solid Earth geophysics
STRA	Stratigraphy
STRS	Stratigraphy—systematics
STRU	Structural geology
SURF	Geomorphology—Quaternary geology
SUSS	Soils—systematics
TEST	Textures—structures

THE PARTICIPANTS

In ongoing and international projects such as the MT, the number of people who contribute their work and efforts is usually very large and very diverse. The ICSTI–IUGS MT was fortunate to have enthusiastic and energetic participation from many individuals. While these individuals often shared expertise (and languages), they just as often had strong and distinct opinions and expressed those quite forcefully. The result was, at the least, educational.

Work was started by a joint working group of COGEODOC and ICSTI (WGMT). From 1970 to 1976 L. Delbos was the chairman of WGMT, followed by H. Glashoff in 1976 through 1981. The current chairman of WGMT is G. N. Rassam.

The MT database was created at the Bundesanstalt für Geowissenschaften und Rohstoffe (Geological Survey of West Germany) by H. Glashoff. After his resignation in 1981, there was a period of relative inactivity followed by the assumption of database maintenance by the National Research Council of Italy (CNR). Operational processing began again in 1983 and special programming was undertaken by R. Potenza of CNR for the purpose of the publication of the thesaurus.

EDITORIAL GROUP

Czechoslovakia—Geofond-Praha.
> J. Hruska, COGEODOC Vice-Chairman,
> E. Kvetonova, editors, Czech version

Federal Republic of Germany—Bundesanstalt für Geowissenschaften und Rohstoffe.
> H. Glashoff, former COGEODOC Chairman
> G. Niedermeier, editor, German version
> J. Nowak, COGEODOC member

Finland—Geological Survey of Finland.
> C. Kortman, editor, Finnish version, COGEODOC member

France—Bureau de Recherches Géologiques et Minières, ICSTI member.
> P. Calas, editor
> L. Delbos, former COGEODOC Chairman

J. Gravesteijn, COGEODOC Secretary, MT Coordinator, editor, French version
—Centre National de la Recherche Scientifique, ICSTI member.
E. Lagarde

Italy—Consiglio Nazionale delle Ricerche.
> Milan: R. Potenza, coordinator MT database, COGEODOC member
> R. Carimati, B. Testa, editors, Italian version, MT database managers
> Bologna: M. Manzoni, editor, Italian and American (English) version

Spain—Instituto Geologico y Minero de España.
> J. Fernandez Tomas, COGEODOC member
> L. de Lamo, A. Sanchez Lencina, editors, Spanish version

USA—American Geological Institute.
> J. J. Lloyd, editor, American (English) version 1970–1974
> G. N. Rassam, COGEODOC Chairman, IUGS representative at ICSTI, editor, American (English) version

USSR—VSEGEI (All-Union Geological Institute), Leningrad.
> A. N. Oleynikov, COGEODOC member, editor, Russian version
> Z. D. Moskalenko, editor, Russian version
> B. D. Dvorkina, editor, Russian version
> —VNIIMORGEO (All-Union Institute of Marine Geology and Geophysics), Riga.
> G. I. Konshin

Temporary participation:
> Canada—Geological Survey of Canada, K. Gunn
> Hungary—MAFI (Hungarian Geological Survey), J. Pálmai
> Sweden—University of Uppsala, P. Bengtson

BIBLIOGRAPHY

American Geological Institute, 1972, 1980, *Glossary of Geology,* 1st and 2nd editions, Alexandria, Va., USA.

American Geological Institute, Crystal S. Palmer, editor, 1986, GeoRef Thesaurus and Guide to Indexing, 4th edn., Alexandria, Va., USA.

Bureau de Recherches Géologiques et Minières—Centre National de la Recherche Scientifique, Centre du Documentation Scientifique et Technique, 1986, PASCAL-GEODE sciences de la terre lexique Francais Anglais, Paris.

Carimati, R., Potenza, R., Testa, B., 1984, *Lessico internazionale di Scienze della Terra,* edizione italiana, C.N.R., Milan, 215 p.

Glashoff, H., 1980/1981, The multilingual thesaurus in geology, a switching tool between bibliographic databases developed by IUGS and ICSI AB; *Bull. Bur. Rech. Geol. Min.,* Section IV, No. 3, p. 201–206.

Gravesteijn, J., Potenza, R., and Rassam, G. N., 1984. *Le thesaurus multilingue pour les sciences de la terre: chimère ou réalité.* Communication presented at the 27th International Geological Congress, Moscow. Symposium on geological information S20-2.2. Congress, Moscow, 1984.

Moskalenko, Z. D., Oleynikov, A. N., and Dolgopolov, V. M., 1984, Russian national version of Multilingual Thesaurus of Geology: in Mathematical geology and geological information, in the collection: *Proceedings of the 27th International Geological Congress,* V.N.U. Science Press B.V., Utrecht, Netherlands, 20, p. 273–286.

Rassam, G. N., Gravesteijn, J., and Lagarde, E., 1980–1981, A joint international bibliographic data base in geology: an experiment whose time has come: *Bull. Bur. Rech. Geol. Min.,* Section IV, No. 3, p. 215–222.

Vsesoyuznyj Nauchno-Issledovatel'skij Geologicheskij Institut (VSE-GEI) Oleynikov, A. N., editor, 1979, Mnogoyazychnyj tezaurus po geologii (metodika postroeniya anglo-russkoj chasti), Leningrad, USSR, 20 p.

Vsesoyuznyj Nauchno-Issledovatel'skij Geologicheskij Institut (VSE-GEI) Oleynikov, A. N., Moskalenko, Z. D., editors, 1982–1983, Mnogoyazychnyj tezaurus po geologii (Russkij variant) Kniga I, II, Leningrad, USSR., 248 p., 273 p.

Introduction

Le Thésaurus Multilingue des géosciences (MT) fut réalisé au cours des 14 derniéres années grâce aux efforts de deux organisations internationales: l' "International Council of Scientific and Technical Information" (ICSTI)—Conseil International de l'Information Scientifique et Technique—et la Commission de Documentation géologique (COGEODOC) de l'Union Internationale des Sciences Géologiques (UISG). L'ICSTI, d'une manière identique à son prédécesseur l'ICSU AB (International Council of Scientific Unions Abstracting Board/ Bureau des Résumés du Conseil International des Unions Scientifiques), est une organisation à but non lucratif qui a pour objectif de faciliter l'accessibilité et la connaissance de l'information scientifique et technique. Elle regroupe les producteurs de banques de données scientifiques et techniques du monde entier, mais aussi les représentants de tous les secteurs d'activité faisant partie de la chaîne de transfert de l'information. COGEODOC a pour but de promouvoir la coopération internationale entre les centres documentaires en géosciences, nationaux ou internationaux, de faciliter les échanges et l'usage de l'information géologique par les géoscientistes et tous les spécialistes concernés par les activités géologiques. Les organismes au sein desquels ont travaillé tous les participants à cette oeuvre ont apporté quant à eux une aide supplémentaire des plus importantes.

La liste du groupe éditorial figure à la fin de cette introduction.

En 1970, l'ICSU AB avait choisi la géologie pour un projet pilote, dans le but de mettre au point une méthodologie de construction d'un thésaurus multilingue qui pourrait être applicable aux autres sciences. On partait de l'idée, à priori, que la géologie avait un vocabulaire plus limité que les autres disciplines scientifiques et que la terminologie et la nomenclature étaient moins ambigues.

ICSU AB s'associa à COGEODOC pour faire ce projet pilote qui fut réalisé en quatre langues, l'Allemand, l'Anglais, le Français et le Tchèque, alors que le thème choisi fut la Tectonique avec un corpus de 300 termes. Ce travail fut complété et présenté au Congrès géologique international de Montréal (Canada) en Août 1972. A ce congrès, l'UISG demanda que le groupe de travail UISG/ICSU AB (WGMT) élargisse son projet en un véritable thésaurus multilingue du vocabulaire géologique. Après un examen préliminaire et un essai d'approche théorique, le groupe de travail adopta finalement une méthode pragmatique et la conserva jusqu'au bout.

Le glossaire de l'American Geological Institute fut pris comme ouvrage de référence. Il contient 33 000 termes environ avec leurs définitions. Il couvre approximativement 80% du vocabulaire géologique si on fait abstraction des nomenclatures archaïques et des termes "vulgaires."

On s'aperçut rapidement que l'idée selon laquelle la géologie possédait un vocabulaire limité et ne contenait pas de termes ambigus était certainement très optimiste. Dans la réalité, beaucoup de temps fut consacré par les participants à expliquer le sens des termes dans les différentes langues. Il était important de relier entre eux des concepts plutôt que les termes eux-mêmes dans une science où les "termes presque semblables" ont différents sens (Fault–Falte) et où les homonymes ont souvent des définitions plus restrictives dans une langue que dans une autre (schist–schiste).

POURQUOI UN THESAURUS MULTILINGUE EN GEOLOGIE?

Le traitement de l'information bibliographique et la réalisation d'index manuels ou automatisés nécessitent l'emploi de méthodes documentaires de classification et d'organisation pour l'analyse bibliographique.

Dans un système manuel, la classification joue un rôle prédominant. Avec l'introduction des ordinateurs et des méthodes de traitement automatique, les procédures de sélection des documents pertinents deviennent "plus fines," dans le sens que des termes d'indexation sont employés pour la recherche plutôt que les concepts très larges des plans de classement. L'accroissement extraordinaire, tant en facilité qu'en qualité, des télécommunications a ajouté une nouvelle dimension à la diffusion de l'information: les distances entre le laboratoire et la banque de données ont été rétrécies et on observe un rapprochement entre le chercheur et le documentaliste et, par la force des choses, entre le scientifique comme producteur de données et le scientifique comme usager de ces données.

Il n'est donc pas surprenant de voir apparaître pour différents centres documentaires un effort de recherche de compatibilité entre leurs méthodes et leurs produits.

En 1970, les services géologiques de l'Allemagne de l'ouest, de la France et de la Tchécoslovaque ont coopéré pour créer et mettre sur pied une banque de données bibliographiques commune. Ils employèrent un vocabulaire d'indexation commun et, grâce à leurs fréquents contacts, élaborèrent une méthode d'indexation commune. De même, dans les années 70, GEOINFORM, le projet commun en information géologique des pays du COMECON, permet de lancer un thésaurus multilingue servant à l'analyse de la littérature géologique des pays concernés.

Le résultat principal de ce projet de Thésaurus Multilingue (MT) a été le renforcement de la coopération internationale. Neuf pays européens (Espagne, Finlande, France, Hongrie, Italie, Pologne, République Fédérale d'Allemagne, Roumanie et Tchécoslovaquie) et les Etats Unis (représentés par l'American Geological Institute, AGI) participent à un réseau documentaire qui contribue à la mise à jour de deux banques de données internationales opérationnelles, GeoRef et PASCAL–GEODE. De plus, ces banques emploient un vocabulaire commun d'indexation parfaitement compatible avec le thésaurus multilingue. Celui-ci devrait devenir, dans un avenir proche, le moyen adéquat pour un réseau mondial d'information géologique. Il est prévu que ce projet de thésaurus soit une opération dynamique et continue qui tiendra compte de l'evolution internationale du contexte de l'information et contribuera à un meilleur usage de la documentation en géosciences tout en la rendant plus accessible.

COMMENT CONSTRUIRE UN THESAURUS MULTILINGUE?

Les Thésaurus

La réalisation d'un thésaurus est liée étroitement au traitement automatique de l'analyse bibliographique et des procédures de recherches documentaires qui en découlent.

La norme de l'International Standards Organization (ISO) définit trois critères pour les thésaurus monolingues:

1. *Langage contrôlé pour l'indexation:* Termes formalisés choisis à partir du langage naturel et employés pour représenter, sous une forme condensée, le contenu des documents (descripteurs).

2. *Etablissement des relations entre les termes:* Ces relations peuvent être hiérarchiques avec des termes "spécifiques" et "génériques," ou associatives avec des termes du type "employer" et "employés pour" ou "voir aussi."

3. *Formalisation des relations sémantiques:* Statut d'un terme (descripteur ou non-descripteur) et ses relations avec d'autres termes appartenant au même groupe de concepts.

Un thésaurus multilingue se différencie par le fait qu'il permet la transposition et la traduction des concepts dans une ou plusieurs autres langues. Il peut être "divisé" en un certain nombre de thésaurus monolingues, eux-mêmes structurés, bien qu'en fait il ne s'agisse pas simplement de la somme de ces derniers.

Les thésaurus multilingues peuvent être construits *a priori* lorsque différents centres documentaires décident de créer une banque de données commune.

Un thésaurus multilingue permet à chaque groupe linguistique de travailler dans sa propre langue, aussi bien pour l'indexation que pour la recherche. De tels réseaux documentaires ont été mis sur pied par les Nations Unies, la CEE et les pays du COMECON.

Lorsqu'il s'agit d'une coopération internationale entre des centres d'information possédant déjà leurs propres systèmes, l'approche méthodologique est basée sur une compatibilité *a posteriori* entre les concepts employés par les différents centres travaillant dans différentes langues.

La comparaison des concepts et des termes et l'analyse des différentes techniques d'indexation prennent alors une place capitale dans la construction d'un thésaurus multilingue. Le projet du Thésaurus Multilingue (MT) a choisi la méthode *a posteriori*.

Les Langues

Le groupe de travail comprenait les groupes linguistiques suivants: Allemand, Anglais, Espagnol, Finnois, Français, Italien, Russe, Tchèque. D'autres langues peuvent être introduites par la suite. Les spécialistes hongrois et suédois ont participé au travail du groupe, mais leurs versions n'ont pas encore été introduites dans la banque de données. Enfin, pendant quelques années, il y eu une participation canadienne au projet.

Principes

Le Thésaurus Multilingue a été conçu comme un mécanisme d'échange entre des centres documentaires spécialisés en géosciences. Il implique la création d'un vocabulaire d'indexation commun compatible avec les méthodes employées et les systèmes pratiqués dans les différents centres. En d'autres termes, on peut établir des ponts entre ceux-ci, bien qu'ils pratiquent des langues et des méthodologies différentes. Une approche pragmatique, procédant par étapes, fut décidée dès le début par le groupe.

La majorité des membres du groupe, par leurs compétences techniques, avaient pouvoir de décision. De plus, la coopération existant déjà entre eux s'était faite en employant la méthodologie et la terminologie du MT et avait ainsi préparé un "terrain favorable" au concept de ce thésaurus.

Methodologie

- Une approche à postériori: Sélection par chaque participant des termes choisis et retenus, dans le glossaire géologique de l'AGI.

- Vérification de la compatibilité de ces termes avec chaque thésaurus monolingue "national."

- Discussion au cours de réunions annuelles sur la définition des termes, leur qualité de descripteurs ou de non-descripteurs dans chaque version linguistique.

- Introduction ou élimination des termes en fonction des critères suivants: leur utilité pour l'indexation et le traitement, et leur emploi comme descripteurs par au moins un des systèmes documentaires du groupe.

- Adjonction de termes jugés nécessaires par chaque participant afin d'assurer une bonne compatibilité entre leur version nationale et le Thésaurus Multilingue

- Chaque groupe linguistique définit et vérifie alors:
  1. Le statut du terme dans son propre système documentaire (descripteur ou non-descripteur).
  2. Les renvois pour les non-descripteurs à un ou deux descripteurs. En tenant compte du caractère pré- ou post coordonné

du système d'indexation utilisé par les participants, les descripteurs de renvoi seront constitués par des mots généraux ou par un terme composé plus spécifique.

3. La réciprocité des relations entre non-descripteurs et descripteurs. Le code UF est alors généré pour les relations de préférence pour tous les descripteurs concernés en les reliant aux non-descripteurs qui en dépendent.

• Quelques adjectifs (p.ex.ABYSSAL) et termes généraux (p.ex.CYCLE) utilisés dans des systèmes de post-coordination, ont été introduits dans le Thésaurus Multilingue. Ces termes sont toujours employés en relation syntaxique avec d'autres descripteurs dans certains systèmes nationaux.

• Lorsqu'un terme est employé uniquement dans une langue, sa forme originale non traduite est simplement répétée dans les autres langues. Exemple: [EN ECHELON] en Anglais.

• Enfin, l'ensemble des termes est classé par sujet ou par thème.

Structure Hierarchique du Thesaurus Multilingue

La structure hiérarchique est établie au niveau de chaque version linguistique du Thésaurus Multilingue, c'est-à-dire dans les thésaurus "nationaux" monolingues. Ces thésaurus "nationaux" constituent les annexes ou prolongations du noyau multilingue et sont reliés à ce dernier par ses termes de base.

Exemples de hiérarchies dans les thésaurus nationaux monolingues:

Russe: DVIZHENIE TEKTONICHESKOE (relié au TM par le terme *PLATE TECTONICS*)
V (BT)   TEKTOGENEZ
N (NT)   DVIZHENIE EHSTATICHESKOE
          DREJF KONTINENTOV
          SKLADKOOBRAZOVANIE
          TAFROGENEZ
          EHPEYROGENEZ

a! (RT!)   KOLEBANIE
          SMESHCHENIE
          TEKTONIKA PLIT

Américain: ANTICLINES (identique au terme du TM)
BT   FOLDS
SA   ANTICLINORIA
SA   ANTIFORM FOLDS
SA   ARCHES
SA   DIAPIRS
SA   DOMES
SA   GEANTICLINES
SA   SYNCLINES

(BT = Terme générique; NT = Terme spécifique, RT = Terme associé, SA = Voir aussi)

CONTENU DU THESAURUS

Actuellement, la banque de données du Thésaurus Multilingue contient approximativement 5,000 termes clés exprimés sous forme de descripteurs ou nondescripteurs en huit versions linguistiques. Ces termes sont classés en 36 groupes ou champs parmi lesquels 20 correspondent aux grandes subdivisions des géosciences, 11 aux domaines systématiques tels que la stratigraphie, les roches, les éléments chimiques, etc.. Cinq champs distincts sont réservés aux concepts employés dans toutes les subdivisions, par exemple: les propriétés, les méthodes, etc.

Pour les domaines systématiques, le choix des termes a avant tout tenu compte de leur utilisation. Les critéres de ce choix ont été les suivants:

• En Stratigraphie: noms des époques et des étages stratigraphiques. Champ: STRS.

- En Paléontologie: noms des groupes (jusqu'au niveau de la famille ou de l'ordre) en fonction de l'importance documentaire de ces groupes. Champ: PALS.

- Pour les sols: ce sont les termes employés dans les classifications américaine et européenne. Champ: SUSS.

- Pour les roches: la nomenclature généralement acceptée et selon la classification de l'UISG. Champs: IGNS, IGMS, SEDS.

- Minéraux: Principaux groupes. Champ: MING.

- Corps extraterrestres: Planètes et météorites. Champ: EXTS.

Les noms géographiques ne sont pas pris en compte pour le moment.

PRESENTATION

Six langues—Allemand, Anglais, Espagnol, Français, Italien et Russe—ont été choisies pour cette version imprimée du Thésaurus. Pour les autres langues, il est toujours possible d'éditer une version nationale du Thésaurus avec une ou plusieurs autres langues.

Le Thésaurus Multilingue comprend une liste générale des termes clés ainsi que différents index et traductions. La liste générale est présentée par ordre alphabétique des termes clés en Anglais avec un numéro de référence séquentiel. Chaque entrée dans le Thésaurus se présente de la manière suivante:

- Un "terme clé" en Anglais dans la présentation initialement retenue du concept de référence. Ce terme clé est accompagné du code du champ d'application.

- Six autres entrées correspondant aux différentes versions linguistiques des thésaurus nationaux; selon la nature documentaire du terme, on peut avoir un descripteur, un non-descripteur, un mot adjectif, un mot général ou un terme non traduit. Si le terme est en lettres capitales, il s'agit d'un descripteur dans la langue considérée et il est traduit dans les autres. Si la présentation est en lettres minuscules, il s'agit d'un non-descripteur et il est alors suivi par un ou deux descripteurs (en lettres capitales) indiquant ainsi le ou les termes à employer de préférence. Les termes en relation de préférence (*UF*) avec un certain descripteur figurent sous ce descripteur dans chaque version linguistique. Ils sont précédés par le code UF et séparés par un point virgule.

- Les adjectifs (ex. Abyssal) sont écrits en majuscules pour les versions linguistiques qui les emploient en tant que descripteurs. Pour les autres langues, les adjectifs sont écrits en minuscules et suivis de la mention (adj.).

- Les termes généraux (ex. Cycle) sont suivis de la mention (gen.) pour les versions linguistiques qui ne les emploient pas dans leur système documentaire en tant que descripteurs.

- Les termes non-traduits (ex. [EN ECHELON]) sont placés entre crochets pour les versions linguistiques concernées.

Pour une information complémentaire sur la presentation, se reporter à la legende du MT.

On trouvera en plus: des index par langue, classés par ordre alphabétique dans cette langue avec référence aux numéros séquentiels de la liste générale, de référence, et un index par thèmes dont les champs d'application et l'ordre sont les suivants:

APPL	Géophysique appliquée
CHEE	Eléments (chimiques)
CHES	Composés chimiques
COMS	Minerais et substances utiles
ECON	Gîtologie
ENGI	Géologie de l'Ingénieur
ENVI	Environnement
EXTR	Géologie extraterrestre
EXTS	Météorites, planètes
GEOC	Géochimie

GEOH	Hydrologie
GEOL	Géologie générale
IGMS	Roches métamorphiques—Systématique
IGNE	Pétrologie
IGNS	Roches ignées—Systématique
INST	Instrumentation—Moyens techniques
ISOT	Géochimie isotopique
MARI	Géologie marine
MATH	Géologie mathématique
METH	Méthodologie
MINE	Minéralogie
MING	Groupes minéraux
MINI	Exploitation et économie minière
MISC	Termes généraux
PALE	Paléontologie
PALS	Paléontologie systématique
PHCH	Propriétés physiques et chimiques, processus
SEDI	Sedimentologie
SEDS	Roches sédimentaires—Systématique
SOLI	Physique du Globe
STRA	Stratigraphie
STRS	Stratigraphie—Systématique
STRU	Géologie structurale
SURF	Géomorphologie—Géologie du Quaternaire
SUSS	Sols—Systématique
TEST	Textures et structures

LES PARTICIPANTS

Dans les projets internationaux et de longue durée comme c'est le cas du Thésaurus Multilingue, le nombre des participants est généralement trés élevé et leur participation peut être variée. Le Thésaurus Multilingue ICSTI–UISG a eu la chance de compter parmi ses membres de nombreux animateurs qui ont travaillé avec enthousiasme et compétence à sa réalisation. Bien que chacun ait une spécialité (ou une langue), ils faisaient bien souvent preuve d'opinions fermes, parfois différentes, et les exprimaient avec beaucoup de conviction. Le résultat fut, finalement, plein d'enseignements pédagogiques.

Il est rappelé que cette opération a été lancée par un groupe de travail commun COGEODOC et ICSTI. Entre 1970–1976, L. Delbos fut Président du groupe de travail, suivi par H. Glashoff pour la période 1976–1981. L'actuel Président est G.N. Rassam.

La base de données du Thésaurus Multilingue a été montée sur l'ordinateur du Service géologique d'Allemagne fédérale (BGR) par H. Glashoff. Après son départ en 1981, il y eu une période de relative inactivité, suivie par la reprise de la maintenance de la base de données par le Conseil Nationale de la Recherche d'Italie (CNR). La gestion informatique opérationnelle a repris en 1983 et les logiciels spécifiques ont été crées par R. Potenza du CNR, permettant la publication de cet ouvrage.

RESPONSABLES DU GROUPE D'EDITION

République Fédérale d'Allemagne—Bundesanstalt für Geowissenschaften und Rohstoffe.

H. Glashoff, ancien Président de COGEODOC, depuis 1980 à l'IUT de Hannovre
G. Niedermeier, éditeur de la version allemande
J. Novak, membre de COGEODOC.

Espagne—Instituto Geologico y Minero de España.

J. Fernandez Tomas, membre de COGEODOC,
L. de Lamo, A. Sanchez Lencina, éditeurs de la version espagnole.

Finlande—Geological Survey of Finland.

C. Kortman, éditeur de la version finnoise, membre de COGEODOC.

France—Bureau de Recherches Géologiques et Minières, membre de l'ICSTI.

    P. Calas, éditeur.

    L. Delbos, ancien Président de COGEODOC.

    J. Gravesteijn, secrétaire de COGEODOC, coordinateur du Thésaurus, éditeur de la version française.

    —Centre National de la Recherche Scientifique.

    E. Lagarde

Italie—Consiglio Nazionale delle Ricerche.

    Milan: R. Potenza, coordinateur de la banque de données du Thésaurus Multilingue, membre de COGEODOC,

    R. Carimati, B. Testa, éditeurs de la version italienne, responsables de la banque de données du Thésaurus.

    Bologne: M. Manzoni, éditeur des versions italienne et américaine.

Etats-Unis—American Geological Institute.

    J.J. Lloyd, éditeur de la version américaine de 1970–1974

G.N. Rassam, Président de COGEODOC, représentant de l'UISG à l'ICSTI, éditeur de la version américaine.

URSS—VSEGEI (All Union Geological Institute), Léningrad.

    A.N. Oleynikov, membre de COGEODOC, éditeur de la version Russe

    Z.D. Moskalenko, éditeur de la version Russe.

    B.D. Dvorkina, éditeur de la version Russe.

    —VNIIMORGEO (All Union Institute of Marine Geology and Geophysics), Riga.

    G.I. Konshin

Tchécoslovaquie—GEOFOND-Prague.

    J. Hruska, Vice-Président de COGEODOC, éditeur de la version tchèque

    E. Kvetonova, éditeur de la version tchèque.

Participation Partielle au Thésaurus:

    Canada—Geological Survey of Canada, K. Gunn

    Hongrie—(MAFI), Hungarian Geological Survey, J. Palmai

    Suède—University of Uppsala, P. Bengtson

Einführung

Der Mehrsprachige Thesaurus der Geowissenschaften (MT) wurde in den vergangenen 14 Jahren durch die Initiative zweier internationaler Institutionen entwickelt: "International Council for Scientific and Technical Information" (ICSTI) und "Commission of Geological Documentation" (COGEODOC) der "International Union of Geological Sciences" (IUGS).

ICSTI zielt als nicht-profitgebundene Organisation wie ihre Vorgängerin ICSU AB (International Council of Scientific Unions Abstracting Board) auf die Verbesserung der Zugänglichkeit und Kenntnis von wissenschaftlicher und technischer Information. Seine Mitglieder umfassen sowohl die in der Welt führenden Datenbankproduzenten in Wissenschaft und Technik als auch die Repräsentanten aller Aktionsbereiche der Informationsvermittlung. COGEODOC fördert die Zusammenarbeit zwischen nationalen und internationalen erdwissenschaftlichen Dokumentationszentren und zielt auf die Verbesserung der Verfügbarkeit und der Nutzung von geologischen Informationen durch Geowissenschaftler und Spezialisten, die mit geowissenschaftlichen Fragen beschäftigt sind. Ergänzende und wichtige Hilfe wurde den an der Thesaurus-Entwicklung Beteiligten von ihren Organisationen gewährt. Eine Liste dieser Stellen ist im Kapitel "Herausgeber-Gruppe" am Ende dieser Einführung gegeben.

1970 wählte ICSU AB die Geologie als Thema eines Pilotprojektes, um die Methodik für den Aufbau eines mehrsprachigen Thesaurus zu entwickeln, die dann auf andere Wissenschaftsbereiche angewendet werden kann. Hierbei wurde angenommen, daß die Geologie ein enger begrenztes Vokabular als andere wissenschaftliche Disziplinen besitzt und daß ihre Terminologie und Nomenklatur weniger Mehrdeutigkeiten aufweist.

Dieses Pilotprojekt wurde von COGEODOC in Zusammenarbeit mit ICSU AB durchgeführt. Als Sprachen sollten Deutsch, Englisch, Französisch und Tschechisch berücksichtigt werden, und der Test wurde auf die Tektonik mit 300 Begriffen beschränkt. Das Ergebnis dieser Arbeit wurde auf dem Internationalen Geologen-Kongreß 1972 in Montreal (Kanada) vorgestellt. Auf diesem Kongreß sprach IUGS den Wunsch aus, daß die IUGS/ICSU AB Arbeitsgruppe (WGMT) dieses Pilotprojekt auf einen mehrsprachigen Thesaurus der gesamten Geowissenschaften ausweiten solle. Nach anfänglichen Untersuchungen und theoretischen Experimenten ging die WGMT zu einem vollkommen pragmatischen Vorgehen über, das bis zuletzt beibehalten wurde.

Als erste Arbeitsgrundlage wurde das Glossar des American Geological Institute ausgewählt. Es enthält ca 33.000 Begriffe und ihre Definitionen und dürfte die Geowissenschaften zu etwa 80% umfassen, ohne veraltete und umgangssprachliche Begriffe.

Schnell zeigte sich, daß die Annahme des begrenzten geologischen Vokabulars und weniger zweideutiger Begriffe zu optimistisch war. Tatsächlich mußte viel Zeit auf die Klärung der Begriffsinhalte in den unterschiedlichen Sprachen aufgewendet werden. Als wichtig stellte es sich heraus, eher Konzepte als einzelne Begriffe miteinander in Bezug zu setzen, in einer Wissenschaft, in der verwandte Begriffe unterschiedliche Inhalte besitzen (Falte–Fault), und in der Homonyme in einer Sprache einer stärkeren Einschränkung unterliegen als in einer anderen (Schist–Schiste).

WARUM EIN MEHRSPRACHIGER THESAURUS FÜR DIE GEOLOGIE?

Die Behandlung bibliographischer Informationen und die Schaffung manueller oder automatisierter Register erfordert die Anwendung dokumentarischer Methoden der Klassifikation und Organisation der bibliographischen Auswertung.

In einem manuellen System spielt die Klassifikation eine vorrangige Rolle. Mit der Einführung EDV-gestützter Suchmethoden wurden die Prozeduren zur Auswahl zutreffender Dokumente verfeinert, und zwar so, daß Schlagwärter an Stelle von allgemeineren Begriffen zur gezielten Recherche herangezogen werden konnten. Die dramatische Entwicklung der Möglichkeiten der Telekommunikation eröffnete eine neue Dimension für die breite Nutzung von Informationen. Sie hob die Trennung zwischen Labor und Datenbank, zwischen Nutzer und Dokumentar und damit letztlich zwischen dem Wissenschaftler als Produzenten und dem Wissenschaftler als Nutzer von Daten auf.

Es überrascht daher nicht, daß mehrere Dokumentationszentren begannen, ihre Methoden und Produkte einander anzugleichen.

1970 schlossen sich die geologischen Dienste von Frankreich, der Bundesrepublik Deutschland und der Tschechoslowakei zusammen, um eine gemeinsame Datenbank zu begründen und zu betreiben. Es wurde ein gemeinsames Auswertevokabular benutzt, und direkte Kontakte zwischen den Dokumentationszentren führten zu gemeinsamer Auswertepraxis. Andererseits begann in den siebziger Jahren GEOINFORM, das gemeinsame geowissenschaftliche Informationsprojekt der COMECON-Länder, mit der Entwicklung eines mehrsprachigen Thesaurus für die Auswertung der geowissenschaftlichen Literatur der beteiligten Länder.

Das wichtigste Ergebnis des ICSTI–IUGS Multilingual Thesaurus-Projektes (MT) war die Stärkung der internationalen Zusammenarbeit. Neun europäische Länder (Bundesrepublik Deutschland, Finnland, Frankreich, Italien, Polen, Rumänien, Spanien, Tschechoslowakei und Ungarn) und die USA (vertreten durch das American Geological Institute, AGI) schlossen sich zu einem Informationsnetz zusammen, indem sie zu den beiden internationalen Datenbasen (GeoRef und PASCAL-GEODE) gemeinsam beisteuern und eine gemeinsame Dokumentationssprache benutzen, die mit dem MT vollkommen kompatibel ist.

Auf diese Weise kann der MT in naher Zukunft der Schaltmechanismus eines weltweiten Netzwerkes für geologische Information werden. Als dynamisches und ständig fortgeführtes Projekt soll der MT Entwicklungen in der internationalen Informationsszene berücksichtigen und zur besseren und einfacheren Nutzung von geowissenschaftlichen Informationen beitragen.

WIE WIRD EIN MEHRSPRACHIGER THESAURUS AUFGEBAUT?

Die Schaffung von Thesauri ist eng gebunden an die automatisierte Verarbeitung bibliographischer Zitate und die daraus resultierenden Recherchestrategien. Die International Standards Organization (ISO) definiert für monolinguale Thesauri drei Kriterien:

1. *Kontrolliertes Auswerte-Vokabular:* Begriffe der natürlichen Sprache, um den Inhalt eines Dokumentes in einer normierten, verdichteten Form wiederzugeben (Deskriptoren).

2. *Logische Beziehungen zwischen den Begriffen:* Diese Beziehungen können hierarchisch ("Ober"- und "Unter"-Begriffe) oder assoziativ ("benutze-" oder "benutzt für-" Begriffe bzw. "siehe auch" Begriffe) sein.

3. *Formale Organisation der Beziehungen:* Funktion eines Begriffes (Deskriptor oder Nicht-Deskriptor) und seiner Beziehungen zu anderen Wörtern innerhalb eines Begriffsumfeldes.

Beim mehrsprachigen Thesaurus ist die Strukturierung und Übersetzung von Konzepten in mehrere Sprachen erlaubt. Aus einem mehrsprachigen Thesaurus können daher mehrere einsprachige strukturierte Thesauri erstellt werden. Der mehrsprachige Thesaurus ist aber mehr als nur ihre Summe.

Mehrsprachige Thesauri können *a priori* erstellt werden, wenn mehrere Dokumentationszentren beschließen, eine gemeinsame Datenbasis aufzubauen. Ein mehrsprachiger Thesaurus erlaubt jeder Sprachgruppe in der eigenen Sprache zu arbeiten, sowohl bei der Auswertung als auch bei der Datenbankrecherche. Solche Netzwerke wurden von den Vereinten Nationen, der Europäischen Gemeinschaft und den COMECON-Staaten geschaffen.

Bei der internationalen Zusammenarbeit zwischen bereits existierenden Informationssystemen basiert der methodische Ansatz auf einer *a posteriori*-Kompatibilität zwischen den Konzepten, die in den unterschiedlichen Systemen der einzelnen Sprachen benutzt werden. In diesem Fall nimmt der Vergleich der Konzepte und Begriffe und die Analyse der Auswertetechnik der beteiligten Stellen den wichtigsten Platz bei der Erstellung eines mehrsprachigen Thesaurus ein. Die a-posteriori-Methode wurde für das MT-Projekt herangezogen.

Sprachen

Die Arbeitsgruppe umfaßte folgende Sprachgruppen: Deutsch, Englisch, Finnisch, Französisch, Italienisch, Russisch, Spanisch und Tschechisch. Andere Sprachen können in Zukunft berücksichtigt werden. Ungarische und schwedische Spezialisten beteiligten sich an der Arbeit der Gruppe, ihre Sprachversionen sind jedoch noch nicht in die Datenbank aufgenommen worden. Einige Jahre lang haben auch kanadische Teilnehmer an diesem Projekt mitgearbeitet.

Prinzipien

Der MT versteht sich als Schaltmechanismus zwischen geowissenschaftlichen Dokumentationszentren. Er umfaßt den Aufbau eines gemeinsamen Auswerte-vokabulars, das mit den Arbeitsmethoden und den im Einsatz befindlichen Systemen der verschiedenen Zentren kompatibel ist. Mit anderen Worten: Es wurden Brücken gebaut zwischen Zentren mit unterschiedlichen Sprachen und mit verschiedenen Methodiken. Solche Aufgabe schien unlösbar zu sein, aber die Arbeitsgruppe begann mit einem stufenweisen Lösungsweg, wobei von Anfang an ein pragmatisches Vorgehen beschlossen wurde.

Die meisten Mitglieder der Gruppe hatten ausreichend praktische Erfahrungen, die zur Entscheidungsfindung notwendig waren. Außerdem benutzten mehrere Mitglieder der Gruppe in ihrer Kooperation bereits Methoden und Terminologie eines mehrsprachigen Thesaurus und schufen so eine "erprobte Basis" für das MT-Konzept.

Methodik

- A posteriori Ansatz: Auswahl geeigneter Begriffe aus dem AGI Glossary of Geology durch jeden Teilnehmer.

- Überprüfung der Kompatibilität der Begriffe mit jedem nationalen monolingualen Thesaurus.

- Diskussion bei jährlichen Treffen über die Definition der Begriffe und ihrer Eignung als Deskriptoren oder Nicht-Deskriptoren in jeder Sprachversion.

- Einführung oder Streichung von Begriffen nach folgenden Kriterien: Nutzen der Begriffe für die Auswertung oder Recherche, Benutzung der Begriffe als Deskriptoren durch mindestens eines der beteiligten Systeme.

- Einführung von Begriffen, die als notwendig erachtet werden, um die Kompatibilität zwischen einer individuellen Sprachversion und dem MT zu gewährleisten.

- Ordnung der Begriffe durch jede Sprachgruppe
    1. Zuweisung ihrer Rolle im eigenen Dokumentationssystem (Deskriptor, Nicht-Deskriptor).
    2. Aufstellen von Ersatzbeziehungen für Nicht-Deskriptoren, Vorzugsbegriffe können angegeben werden. Diese Beziehungen können einfach oder komplex sein, je nach der verwendeten Post- oder Präkoordination.
    3. Umkehrung der Ersatzbeziehungen. Der bevorzugte Gebrauch (UF) wird für die betreffenden Deskriptoren angegeben, sowohl für einfache als auch für multiple Vorzugsbegriffe.

- Einige Adjektive und allgemeine Begriffe wurden für die postkoordinierte Indexierung in den MT eingeführt. Diese Begriffe sind nur in Beziehung zu anderen Deskriptoren und in syntaktischer Verknüpfung, die in einigen Systemen angewandt wird, zu benutzen (z.B. Abyssal, Zyklus).

- Wenn ein Begriff nur in einer Sprache gebräuchlich ist, wird dieses Wort in den anderen Sprachen einfach wiederholt, z.B. En Echelon in Englisch.
- Zuordnung der Begriffe zu einem Fachgebiet.

Hierarchische Struktur des MT

Die hierarchische Struktur ist für jede linguistische Version eingeführt worden. Diese "nationalen" Thesauri bilden den erweiternden Anhang zum mehrsprachigen Nukleus und sind mit ihm durch die MT-Begriffe verbunden.

Beispiele von Hierarchien in "nationalen" Thesauri:

Englisch: Anticlines

BT	Folds
SA	Anticlinoria
SA	Antiform Folds
SA	Arches
SA	Diapirs
SA	Domes
SA	Geanticlines
SA	Synclines

Deutsch: Senkung

BT	Epirogenese
NT	Senkungs-Becken
RT	Hebung
RT	Syneklise
SA	Isostasie
SA	Eustatik

(BT = broader term = Oberbegriff; NT = narrower term = Unterbegriff; RT = related term = verwandter Begriff; SA = see also = siehe auch)

INHALT

Die MT-Datenbasis enthält zur Zeit ca. 5,000 Schlagwörter (Deskriptoren und Nicht-Deskriptoren) in 8 Sprachversionen. Diese Begriffe sind unterteilt in 36 Fachgebiete, von denen 20 mit den Hauptbereichen der Geowissenschaften übereinstimmen, 11 umfassen systematische Teile wie Stratigraphie, Gesteinsnamen, Elemente usw., 5 Felder beinhalten Begriffe, die für alle Fachbereiche anwendbar sind (Eigenschaften, Methoden usw.).

Bei den systematischen Unterteilungen wurde die Auswahl unter den verfügbaren Begriffen nach folgenden Kriterien getroffen:

- Stratigraphie: Epochen und Formationsnamen. Feld: STRS
- Paläontologie: Namen von Gruppen (Familien oder Ordnungen) als Funktion der dokumentarischen Wichtigkeit der Gruppe. Feld: PALS
- Böden: Begriffe aus den amerikanischen und europäischen Klassifikationen. Feld: SUSS
- Gesteine: Namen entsprechend der IUGS-Klassifikationen. Felder: IGNS, IGMS, SEDS
- Minerale: Hauptgruppen. Feld: MING
- Extraterrestrische Objekte: Planeten und Meteoriten. Feld: EXTS

Geographische und physiographische Ortsnamer sind z.Z. noch nicht aufgenommen worden.

PRÄSENTATION

Für die vorliegende gedruckte Version des Thesaurus wurden 6 Sprachen ausgewählt: Englisch, Französisch, Deutsch, Russisch, Italienisch und Spanisch. Für andere Sprachen kann aus der MT-Datenbasis eine nationale Version ausgedruckt werden. Diese kann auch eine oder mehrer weitere Sprachen enthalten.

Der MT enthält sowohl eine Hauptliste der Schlagwörter als auch verschiedene Register und Übersetzungen. Die Hauptliste führt die Schlagwörter in alphabetischer Reihenfolge mit einer Bezugsnummer auf.

Jeder Eintrag ist wie folgt aufgebaut:

- Schlagwort in Englisch in der Originalschreibweise der Quelle, der es entnommen wurde, gefolgt durch das Kürzel des Fachgebietes.

- Sechs Einträge in den verschiedenen Sprachversionen des MT, die die dokumentarische Natur der Begriffe anzeigen (Deskriptor, Nicht-Deskriptor, Adjektiv, allgemeiner oder nichtübersetzbarer Begriff). Ist der Begriff mit Großbuchstaben geschrieben, handelt es sich in dieser Sprache um einen Deskriptor und dieser wird übersetzt. Ein Begriff in Groß-Klein-Schreibweise ist ein Nicht-Deskriptor, der von ein oder zwei Deskriptoren in Großbuchstaben gefolgt sein muß, die den oder die Vorzugsbegriff(e) angeben. Werden zwei Deskriptoren aufgeführt, so werden sie durch ein Semikolon getrennt.

- Der Vorzugsgebrauch von Deskriptoren wird in jeder Sprachversion durch UF angegeben. Mehrere Vorzugsbegriffe werden durch ein Semikolon voneinander getrennt.

- Adjektive werden in den Versionen mit Post-Koordination in Großbuchstaben geschrieben. In den anderen Sprachversionen wird ein Adjektiv mit "(adj.)" gekennzeichnet, z.B. Abyssal (adj.).

- Allgemeine Begriffe befolgen die gleichen Regeln: Sie werden durch "(gen.)" in den Sprachversionen ohne Post-Koordination ergänzt, z.B. [En Echelon].

- Sprachindex: Für jede Sprache gibt ein spezielles Register in alphabetischer Ordnung den Hinweis auf die Nummer der Deskriptoren in der Hauptliste.

- Fachgebietindex: Anordnung der Schlagwörter nach Fachgebieten.

Liste der Fachgebietscodes und ihrer Bedeutung

APPL	Angewandte Geophysik
CHEE	Elemente
CHES	Chemische Verbindungen
COMS	Mineralische Rohstoffe (soweit nicht IGMS, IGNS, MING, SEDS)
ECON	Lagerstättenkunde
ENGI	Ingenieurgeologie
ENVI	Umweltgeologie
EXTR	Extraterrestrische Geologie
EXTS	Meteoriten, Planeten
GEOC	Geochemie
GEOH	Hydrogeologie
GEOL	Allgemeine Geologie
IGMS	Metamorphe Gesteine—Systematik
IGNE	Petrologie
IGNS	Magmatische Gesteine—Systematik
INST	Instrumente, Geräte
ISOT	Isotopengeochemie, physikalische Altersbestimmung, Geochronologie
MARI	Meeresgeologie
MATH	Mathematische Geologie
METH	Methodik
MINE	Mineralogie
MING	Mineralgruppen
MINI	Bergbau
MISC	Verschiedenes
PALE	Paläontologie
PALS	Paläontologie—Systematik
PHCH	Physikalische und chemische Eigenschaften und Prozesse
SEDI	Sedimentologie
SEDS	Sedimentgesteine—Systematik
SOLI	Physik der festen Erde
STRA	Stratigraphie

STRS Stratigraphie—Systematik
STRU Tektonik
SURF Geomorphologie, Quartärgeologie
SUSS Böden—Systematik
TEST Texturen, Strukturen

DIE TEILNEHMER

In langlebigen internationalen Projekten, wie dem MT-Projekt, ist normalerweise die Zahl derer, die Arbeit und Einsatz einbringen, sehr groß und wechselnd. Die ICSTI–IUGS-Arbeitsgruppe war in der glücklichen Lage, unter ihren Mitgliedern enthusiastische und energische Persönlichkeiten zu haben. Da die meisten von ihnen besondere dokumentarische Erfahrung und Sprachkenntnisse besaßen, hatten sie oft feste und klare Vorstellungen und brachten diese auch nachdrücklich zum Ausdruck. Das Ergebnis war zumindest ein Lernprozeß.

Die Arbeit begann mit einer gemeinsamen Arbeitsgruppe von COGEODOC und ICSTI (WGMT). Im Zeitraum 1970–1976 führte L. Delbos den Vorsitz, gefolgt von H. Glashoff von 1976–1981. Der jetzige Vorsitzende von WGMT ist G.N. Rassam.

Die MT-Datenbasis wurde von H. Glashoff in der Bundesanstalt für Geowissenschaften und Rohstoffe aufgebaut. Nach seinem Rücktritt 1981 folgte eine Periode relativer Inaktivität bis zur Übernahme der Datenpflege durch das National Research Council in Italien (CNR). Die maschinelle Verarbeitung begann 1983 wieder, wobei von R. Potenza (CNR) spezielle Programmierung geleistet wurde, um die Publikation dieses Werkes zu ermöglichen.

DIE HERAUSGEBER-GRUPPE

Bundesrepublik Deutschland—Bundesanstalt für Geowissenschaften und Rohstoffe
H. Glashoff, früherer Vorsitzender COGEODOC
G. Niedermeier, Herausgeberin der deutschen Version
J. Nowak, COGEODOC-Mitglied

Finnland—Geological Survey of Finland
C. Kortman, Herausgeber der finnischen Version, COGEODOC-Mitglied

Frankreich—Bureau de Recherches Géologiques et Minières, ICSTI-Mitglied
P. Calas, Herausgeber.
L. Delbos, früherer Vorsitzender COGEODOC
J. Gravesteijn, Sekretär COGEODOC, MT-Koordinator, Herausgeber der französischen Version
—Centre National de la Recherche Scientifique ICSTI-Mitglied
E. Lagarde

Italien—Consiglio Nazionale delle Ricerche
Mailand: R. Potenza, Koordinator MT-Datenbasis, COGEODOC-Mitglied
R. Carimati, B. Testa, Herausgeber der italienischen Version, Manager der MT-Datenbasis
Bologna: M. Manzoni, Herausgeber der italienischen und amerikanischen Version

Spanien—Instituto Geologico y Minero de España
J. Fernandez Tomas, COGEODOC-Mitglied
L. de Lamo, A. Sanchez Lencina, Herausgeber der spanischen Version

Tschechoslowakei—GEOFOND-Prag
J. Hruska, Vizevorsitzender COGEODOC
E. Knetonova, Herausgeberin der tschechischen Version

UdSSR—VSEGEI (All Union Geologisches Institut), Leningrad
A.N. Oleynikov, COGEODOC-Mitglied, Herausgeber der russischen Version
Z.D. Moskalenko, Herausgeberin der russischen Version
B.D. Dvorkina
—VNIIMORGEO (All Union Institut für Meeresgeologie und Geophysik), Riga
G.I. Konshin

USA—American Geological Institute
J. Lloyd, Herausgeber der amerikanischen Version 1970–1974
G.N. Rassam, Vorsitzender COGEODOC, IUGS-Repräsentant
bei ICSTI, Herausgeber der amerikanischen Version

Zeitweilige Beteiligung:
Kanada—Geological Survey of Canada, K. Gunn
Schweden—Universität Uppsala, P. Bengtson
Ungarn—MAFI (Ungarischer Geologischer Dienst), J. Palmai

Введение

Разработка Многоязычного Тезауруса по геологии (МТ) осуществлялась на протяжении последних 14 лет силами двух международных организаций: Международного Совета по научной и технической информации (МСНТИ) и Комиссии по геологической документации (КОГЕОДОК) Международного Союза геологических наук (МСГН). МСНТИ, как и его предшественник – МСНС-РБ (Реферативное бюро Международного Совета научных союзов) действует на некоммерческой основе; его задача – расширить осведомлённость о наличии научной и технической информации и сделать её более доступной. В состав этой организации входят разработчики ведущих мировых баз данных по науке и технике, а также представители всех звеньев деятельности в цепи передачи информации. КОГЕОДОК призван содействовать международному сотрудничеству национальных и интернациональных информационных центров в области наук о Земле и имеет целью усовершенствовать качество и степень использования геологической информации геологами и другими специалистами, причастными к геологической деятельности. Большую дополнительную помощь оказали организации, в которых работали создатели МТ. Перечень этих организаций приведен в конце Введения (см. раздел – Редакционная группа).

В 1970 году МСНС-РБ выбрало геологию в качестве объекта для реализации экспериментального проекта по разработке методологии создания многоязычного тезауруса, которая впоследствии могла бы быть использована в других науках. Основанием для выбора этой отрасли знания послужило предположение, что геология, по сравнению с другими научными дисциплинами, имеет более ограниченный словарь и меньшее количество неоднозначностей в области терминологии и номенклатуры.

К работам МСНС-РБ по этому проекту присоединилась КОГЕОДОК; лингвистическую базу составили английский, немецкий, французский и чешский языки; объект исследования был ограничен подмассивом «Тектоника» объемом в 300 терминов. Данная работа была завершена и представлена на Международном геологическом конгрессе в Монреале (Канада) в августе 1972 года. Во время Конгресса МСГН обратился к объединенной рабочей группе МСГН/МСНС-РБ по многоязычному тезаурусу (РГМТ) с просьбой расширить экспериментальный проект и осуществить составление полного многоязычного тезауруса по геологической терминологии. После изучения и экспериментального сравнения различных теоретических подходов РГМТ приняла общую прагматическую методологию и в дальнейшем стабильно использовала её в своей деятельности.

В качестве рабочей основы был взят Толковый словарь (Glossary of Geology) Американского геологического института, включающий около 33 000 терминов и их определений, что составляет приблизительно 80% геологической лексики, без учета устаревшей номенклатуры и сленга.

В краткие сроки выявилось, что предположения об ограниченном объеме и малом количестве неоднозначностей геологической терминологии были неоправданно оптимистичны. Больших затрат времени потребовало истолкование значений терминов, используемых в различных языках. При этом наиболее важным явилось установление соотношений не между терминами, а между научными понятиями, поскольку сходные слова могут иметь в разных языках различные значения (fault – Falte), а омонимы порой имеют неодинаковые объемы (schist – schiste).

ДЛЯ ЧЕГО СОЗДАЕТСЯ МНОГОЯЗЫЧНЫЙ ТЕЗАУРУС ПО ГЕОЛОГИИ?

Обработка библиографической информации и создание ручных и автоматизированных указателей требует использования при библиографическом анализе документалистических методов классификации и организации данных.

В системе ручного использования первостепенную роль играет классификация. Со внедрением автоматизированных методов поиска (и компьютеров) процедура отбора требуемых документов становится более «тонкой» в том смысле, что при поиске наибольшее значение обретают ключевые слова, а не широкие смысловые категории. Значительное расширение возможностей и качества телесвязи вывело интегрирование информации в новое измерение: сократились дистанция между лабораторией и базой данных, между исследователем и документалистом и, таким образом, между научным работником – **создателем** данных и научным работником – **пользователем**.

Поэтому неудивительно, что в различных информационных центрах проявилась тенденция к совершенствованию совместимости используемых методов и конечной продукции.

В 1970 году геологические службы Франции, Западной Германии и Чехословакии, предприняв совместные работы по созданию и корректировке объединенной базы данных, стали использовать при индексировании общий словарь, установили прямые контакты между информационными центрами и обеспечили единообразие индексирования. С другой стороны, в

системе ГЕОИНФОРМ (общий проект стран СЭВ по геологической информатике) в семидесятых годах была начата разработка многоязычного тезауруса для анализа геологической информации стран-участниц СЭВ.

Основным итогом проекта МСНТИ-МСГН по созданию многоязычного тезауруса (МТ) было укрепление международной кооперации в данной области. Участвуя в работе информационной сети, девять европейских стран (Венгрия, Испания, Италия, Польша, Румыния, Федеративная Республика Германии, Финляндия, Франция и Чехословакия) и Соединенные Штаты Америки (представленные Американским геологическим институтом – АГИ) внесли вклад в совершенствование двух действующих международных баз данных – ГеоРеф и ПАСКАЛЬ – ЖЕОД. и приняли к использованию общий язык индексирования, полностью совместимый с МТ.

Таким образом, в ближайшем будущем МТ может стать средством, обеспечивающим вхождение в международную сеть геологической информации.

Предполагается, что проект по МТ представит собой динамичную и непрерывную работу, которая будет учитывать международные достижения в сфере информатики и сможет улучшить и облегчить использование геологической информации.

КАК ПОСТРОИТЬ МНОГОЯЗЫЧНЫЙ ТЕЗАУРУС?

Тезаурусы.

Создание тезаурусов тесно связано с автоматизацией библиографического анализа и соответствующими процедурами документалистических исследований.

Стандарт Международной организации по стандартизации (ISO) определяет для одноязычных тезаурусов три критерия:

1.   Контролируемый язык индексирования (термины, представленные на естественном языке и используемые для передачи содержания документа в сжатой форме);

2.   Установление отношений между терминами. Эти отношения могут быть иерархическими, включающими «видовые» и «родовые» термины, или ассоциативными – с указателями «USE» или «USE FOR».

3.   Формальная организация отношений (функции термина и его связь с другими терминами, относящихся к той же группе определённых понятий).

Отличие многоязычного тезауруса состоит в том, что он позволяет осуществлять перестановку и перевод понятий на нескольких языках. Многоязычный тезаурус «слагается» из ряда тезаурусов с чётко выраженной одноязычной структурой, но он отнюдь не является простой суммой последних.

Если различные информационные центры решают создать объединённую базу данных, то многоязычный тезаурус, позволяющий каждой лингвистической группе осуществлять индексирование и поиск на родном языке, может быть создан априорно. Такие системы были созданы ООН, Комиссией Европейского сообщества и странами СЭВ.

При решении задачи международной кооперации самостоятельных информационных систем методологический подход основывается на **апостериодной** совместимости понятий, используемых в системах, действующих на разных языках.

В этом случае, в построении многоязычного тезауруса наиболее важное место занимают сравнение понятий и терминов, а также анализ применяемых участниками методов индексирования. Для проекта МТ была выбрана **апостериодная** методология.

Языки.

В рабочей группе были представлены следующие языки: английский, испанский, итальянский, немецкий, русский, финский, французский, чешский. В дальнейшем в систему могут быть включены и другие языки.

В работе группы принимали участие также венгерские и шведские специалисты, но эти лингвистические версии в настоящее время ещё не включены в базу данных. В течение ряда лет в работах по проекту участвовали канадские специалисты.

Принципы.

МТ был задуман как средство сопряжения информационных центров, специализирующихся в области наук о Земле. Он предусматривает создание общего словаря индексирования, совместимого с методами обработки и информационными системами, действующими в различных центрах. Другими словами, МТ призван воздвигнуть мосты между центрами использующими различные языки и различные методологии. Эта задача могла показаться невыполнимой, но рабочая группа, придерживаясь прагматических позиций, приступила к её поэтапному решению.

Большинство членов группы имели достаточные опыт и знания, необходимые для принятия надлежащих технических решений. Кроме того, совместное использование несколькими членами группы единой методологии и терминологии МТ, позволило создать «полигон» для испытания концепций МТ.

МЕТОДОЛОГИЯ

- Апостериодный подход: выбор каждым участником необходимых терминов из Геологического словаря АГИ.

- Сверка совместимости терминов с каждым «национальным» одноязычным тезаурусом.

- Обсуждение (на ежегодных совещаниях группы) определений терминов, разбраковка их в качестве дескрипторов и недескрипторов в каждой языковой версии.

- Введение или исключение терминов с учётом следующих критериев: пригодность терминов для решения задач индексирования и поиска; использование термина в качестве дескриптора по крайней мере в одной из представленных систем.

- Включение терминов, которые по мнению участников необходимы для обеспечения совместимости их языковых версий и МТ;

- Определение роли терминов (D – дескриптор, N – недескриптор) в документографических системах каждой линг-

вистической группы (дескрипторы записываются заглавными буквами) и установление возможностей замены отношений для недескрипторов. Принимаемые к использованию термины приводятся после обозначения UF:, если таких терминов несколько, между ними ставится точка с запятой. Такие замены могут быть простыми или сложными – в зависимости от специфики термина или его характеристик (пост- или предкоординационных).

- Введение в МТ некоторых прилагательных и общих терминов, используемых при посткоординационном индексировании. За прилагательным следует помета (adj.), за общими терминами – (gen.). Например: ABYSSAL (adj.), CYCLE (gen.). Эти термины применяются только в синтаксических связях с другими дескрипторами, используемых в конкретных национальных системах.

- Для терминов, используемых только в одном языке: повторение оригинальной формы термина в других языках с предшествующей пометой – (например, – EN ECHELON).

- Классификация терминов по тематике и полям.

Иерархическая структура МТ:

Для каждой лингвистической версии может быть установлена индивидуальная иерархическая структура. Такие «национальные» тезаурусы дополняют и расширяют многоязычное ядро и связаны с ним посредством базовых терминов МТ.

Примеры иерархий в «национальных» одноязычных те-
заурусах:

Русский:　　ДВИЖЕНИЕ ТЕКТОНИЧЕСКОЕ (относится к термину МТ *plate tectonics*)

В (ВТ)	ТЕКТОГЕНЕЗ
N (NT)	ДВИЖЕНИЕ ЭВСТАТИЧЕСКОЕ
	ДРЕЙФ КОНТИНЕНТОВ
	СКЛАДКООБРАЗОВАНИЕ
	ТАФРОГЕНЕЗ
	ЭПЕЙРОГЕНЕЗ
а! (RT!)	КОЛЕБАНИЕ
	СМЕЩЕНИЕ
	ТЕКТОНИКА ПЛИТ

Английский:	ANTICLINES
ВТ	FOLDS
SA	ANTICLINORIA
SA	ANTIFORM FOLDS
SA	ARCHES
SA	DIAPIRS
SA	DOMES
SA	GEANTICLINES
SA	SYNCLINES

[В (ВТ) = вышестоящий дескриптор; Н (NT) = нижестоящий
дескриптор; а! (RT) – односторонняя ассоциативная связь с
термином; SA= смотри также].

СОДЕРЖАНИЕ

В настоящее время МТ содержит около 5000 основных
терминов в 8 лингвистических вариантах. Основные термины
подразделены на 36 групп или полей, из которых 20 – соответ-
ствуют крупным разделам наук о Земле, 11 – представляют собой систематическую часть классификаций в таких областях
как стратиграфия, наименования горных пород, элементы и др.
Пять самостоятельных полей включают понятия, являющиеся
общими для всех отраслей (свойства, методы и т.д.).

Выбор лексики систематических полей производился из
имеющейся совокупности терминов по следующим критериям:

- Стратиграфия: названия отделов и ярусов. Поле: STRS
- Палеонтология: наименования групп организмов (до се-
  мейства или отряда) – в соответствии со значением этих
  групп для информационного поиска. Поле: PALS

- Почвы: термины, используемые в американской и евро-
  пейской классификациях. Поле: SUUS

- Горные породы: Общепринятые названия, в соответствии с классификациями МСГН. Поля: IGNS, IGMS, SEDS

- Минералы: основные группы. Поле: MINS

- Небесные тела: планеты и метеориты. Поле: EXTS

Географические названия в настоящее время в словарь не
включены.

ФОРМА МТ

В публикуемой версии тезауруса представлены шесть языков: английский, испанский, итальянский, немецкий, русский и французский. Вопрос о публикации других версий МТ, в сочетании с одним или несколькими языковыми вариантами, остается открытым.

МТ включает общий список ключевых слов, а также ряд индексов и переводов терминов. Ключевые слова располагаются в общем списке в алфавитной последовательности, с указанием их порядковых номеров. Каждая запись имеет следующую структуру:

- Ключевое слово на английском языке, представленное в форме, изначально выбранной для данного понятия. После ключевого слова указывается смысловое поле, к которому оно относится.

- Шесть записей, соответствующих различным лингвистическим версиям национальных тезаурусов и отражающих документографическую характеристику терминов (дескриптор, недескриптор, прилагательное, общий термин, непереводимый термин). Если термин записан заглавными буквами, это указывает, что данный термин является дескриптором в данной лингвистической версии и переведен. Иная форма записи означает, что термины являются недескрипторами, и за ними должна следовать отсылка к одному или двум дескрипторам, причём тот из них, которому отдаётся предпочтение, печатается заглавными буквами (см. стр. 7, отношение UF).

Кроме того, имеются также:

- Лингвистические указатели: для терминов каждого языка, расположенных в алфавитном порядке, имеется отдельный указатель, дающий отсылку к номерам ключевых слов, включённых в общий список.

- Указатель по полям: ключевые слова классифицированы по полям. Полный список кодов и наименований полей приведён ниже:

APPL	Прикладная геофизика
CHEE	Элементы
CHES	Химические соединения
COMS	Полезные ископаемые
ECON	Экономическая геология
ENGI	Инженерная геология
ENVI	Окружающая среда
EXTR	Космическая геология
EXTS	Метеориты, небесные тела
GEOC	Геохимия
GEOH	Гидрология
GEOL	Общая геология
IGMS	Метаморфические породы (систематика)
IGNE	Петрология
IGNS	Изверженные породы (систематика)
INST	Инструменты – оборудование
ISOT	Изотопная геохимия (Абсолютный возраст, геохронология)
MARI	Морская геология
MATH	Математическая геология
METH	Методы
MINE	Минералогия
MING	Группы минералов
MINI	Горное дело
MISC	Общие термины
PALE	Палеонтология
PALS	Палеонтология (систематика)

PHCH	Физические и химические свойства, процессы
SEDI	Седиментология
SEDS	Осадочные породы (систематика)
SOLI	Глубинное строение Земли
STRA	Стратиграфия
STRS	Подразделение стратиграфической шкалы
STRU	Структурная геология
SURF	Геоморфология – четвертичная геология
SUSS	Почвы (систематика)
TEST	Текстуры и структуры

УЧАСТНИКИ

Коллективы, выполняющие долгосрочные и международные проекты, подобные проекту по МТ, обычно многочисленны, и состав их весьма разнообразен. Успех проекта МСНТИ-МСГН по МТ был предопределен участием в его работе многих энергичных энтузиастов. Эти специалисты, как правило, имевшие одинаковый опыт работы и единый язык общения, нередко придерживались существенно различных мнений, которые они отстаивали со значительной твёрдостью. Полученные результаты, по крайней мере, поучительны.

Разработки были начаты объединённой рабочей группой КОГЕОДОК и МСНТИ (РГМТ). В период с 1970 по 1976 гг. председателем РГМТ был Л. Дельбос; в 1976-1981 гг. его сменил Х. Гласгофф. Ныне председателем РГМТ является Г.Н. Рассам.

База данных МТ была создана при Германской геологической службе (ФРГ) Х. Гласгоффом. После его ухода в 1981 г. наступил период относительной инертности, а затем ведение базы данных взял на себя Совет научных исследований (СНИ) Италии. В 1983 г. была возобновлена обработка данных, и Р. Потенца предпринял подготовку в СНИ программного обеспечения задач, связанных с публикацией этой книги.

РЕДАКЦИОННАЯ ГРУППА

Италия— Национальный совет научных исследований Милан:

Р. Потенца, член КОГЕОДОК, координатор базы данных МТ;

А. Каримата, Б. Теста — редакторы итальянской версии, руководители базы данных МТ; Болонья:

М. Манцони, редактор итальянской и английской версий.

Испания — Испанский институт геологии и горного дела:

Х. Фернандес Томас, член КОГЕОДОК, Л. де Ламо, А. Санчес – редакторы испанской версии;

СССР— ВСЕГЕИ (Всесоюзный научно-исследовательский геологический институт), Ленинград:

А.Н. Олейников, член КОГЕОДОК, редактор русской версии;

З.Д. Москаленко, редактор русской версии;

Б.Д. Дворкина;

СССР — ВНИИМОРГЕО (Всесоюзный научно-исследовательский институт морской геологии и геофизики), Рига:

Г.И. Коншин.

США — Американский геологический институт:

Дж. ЛЛойд, редактор английской версии 1970-1974 гг.;

Г.Н. Рассам, председатель КОГЕОДОК, представитель МСГН в МСНТИ, редактор английской версии.

Финляндия — Геологическая служба Финляндии:

К. Кортман, член КОГЕОДОК, редактор финской версии.

Франция — Бюро геологических и горнорудных исследований, член МСНТИ:

П. Калас, редактор;

Л. Дельбос, экс-председатель КОГЕОДОК;

Ж. Гравештейн, секретарь КОГЕОДОК, координатор МТ, редактор французской версии;

Франция — Национальный центр научных исследований, член МСНТИ:

Э. Лягард.

ФРГ — Федеральное управление геологических наук и минерального сырья:

Х. Гласгофф, экс-председатель КОГЕОДОК;

Г. Нидермайер, редактор немецкой версии;

Й. Новак, член КОГЕОДОК.

ЧССР — Геофонд-Прага:

И. Грушка, вице-председатель КОГЕОДОК, редактор чешской версии;

Е. Кветонова, редактор чешской версии.

Временное участие принимали:

Канада — Геологическая служба Канады, К. Ганн.

Венгрия — Венгерский геологический институт (МАФИ), Й. Палмаи.

Швеция — Университет Уппсала, П. Бенгтсон.

Introducción

El Tesauro Multilingüe de Ciencias de la Tierra (MT) ha sido desarrollado en los últimos 14 años gracias al empeño de dos organizaciones internacionales: el Consejo Internacional para Información Científica y Técnica (ICSTI) y la Comisión de Documentación Geológica (COGEODOC) de la Unión Internacional de Ciencias Geológicas (IUGS). ICSTI, que es una organización sin ánimo de lucro, como su predecesora ICSU AB (Junta de Resúmenes del Consejo Internacional de Uniones Científicas), tiene como objetivo el incrementar la accesibilidad y el conocimiento de la información científica y técnica. Entre sus miembros cuenta con los productores de bases de datos de ciencia y tecnología más importantes de mundo, así como con representantes de todos los sectores de actividad en la cadena de transferencia de la información. COGEODOC promueve la cooperación internacional entre centros documentales nacionales e internacionales sobre Ciencias de la Tierra y aspira a una mejora en la disponibilidad y utilización de la información geológica por los especialistas involucrados en actividades geológicas. Las organizaciones en que trabajaban los participantes suministraron importante ayuda adicional. Para consultar una relación de éstas organizaciones, vea el Grupo Editorial al final de ésta introducción.

En 1970 el ICSU AB escogió la Geología como tema de un proyecto piloto para desarrollar una metodología para la estructuración de un tesauro multilingüe que pudiera luego aplicarse a otras ciencias, bajo la suposición de que la Geología tiene un vocabulario más limitado que el de otras disciplinas científicas y de que había menos ambigüedad en la terminología y nomenclatura.

Al proyecto piloto de ICSU AB se unió COGEODOC, decidiéndose que la cobertura idiomática incluyera checo, inglés, francés y alemán, mientras que el campo temático se redujo a tectónica, con un subgrupo de 300 términos. Se completó el trabajo y se presentó al Congreso Geológico Internacional en Montreal (Canadá) en Agosto de 1972. En el Congreso, la IUGS solicitó que el grupo de trabajo conjunto IUGS/ICSU AB (WGMT) acometiera la extensión del proyecto piloto a un tesauro multilingüe completo del vocabulario de Geología. Tras exámenes y experimentos iniciales con enfoques teóricos, el WGMT cambió a una metodología totalmente pragmática, utilizándola de principio a fin.

Como primera herramienta de referencia se tomó el Glosario del Instituto Geológico Americano. Este Glosario contiene alrededor de 33.000 términos y sus definiciones. Está completo aproximadamente en un 80%, sin considerar la nomenclatura arcaica ni el argot.

Pronto se vió que las suposiciones de que la Geología tendría un vocabulario limitado con pocos términos ambigüos eran más que optimistas. En efecto, se invirtió mucho tiempo en explicaciones sobre el significado de los términos en diversos idiomas. Era importante relacionar conceptos mejor que términos, en una ciencia en la que palabras afines pueden tener distintos significados (Fault–Falte) y la homónimas a veces son más restringidas en un idioma que en otro (Schist–Schiste).

¿PARA QUE UN TESAURO MULTILINGUE DE GEOLOGIA?

El tratamiento de la información bibliográfica y la creación de índices manuales o automatizados requiere el uso de métodos documentales de clasificación y organización para el análisis bibliográfico.

En un sistema manual, la clasificación juega un papel primordial. Con la introducción de los ordenadores y de métodos automáticos de recuperación, los procedimientos para la selección de los documentos pertinentes se han hecho más "refinados," en el sentido de que se utilizan términos de indización para la recuperación, en lugar de amplias categorías. El subsiguiente dramático incremento en facilidad y calidad de las telecomunicaciones ha añadido una nueva dimensión a la difusión de la información: ha reducido la distancia entre el laboratorio y la base de datos y entre el investigador y el documentalista, y como resultado, entre el científico como *productor* de datos y el científico como *usuario* de datos.

No es sorprendente por tanto que comenzase un movimiento en varios centros documentales para aumentar la compatibilidad entre sus métodos y productos.

En 1970 los Servicios Geológicos de Francia, Alemánia y Checoslovaquia cooperaron para crear y actualizar una base de datos conjunta. Se utilizó un vocabulario de indización común y los contactos directos entre los centros documentales condujeron a una práctica de indización común. Por otra parte, en los setenta, GEOINFORM, el proyecto común sobre información de Ciencias de la Tierra de los paises del COMECON, comenzó desarrollando un tesauro multilingüe para el análisis de la literatura de Ciencias de la Tierra que sería tratada por los paises participantes.

El principal resultado del proyecto de tesauro multilingüe (MT) del ICSTI-IUGS fue el reforzamiento de ésta cooperación internacional. Nueve paises europeos (Checoslovaquia, Finlandia, Francia, República Federal de Alemánia, Hungría, Italia, Polonia, Rumanía y España) y los Estados Unidos (representados por el Instituto Geológico Americano, AGI) participan en una red de información contribuyendo a la actualización de las dos bases de datos internacionales en funcionamiento, GeoRef y PASCAL-GEODE y utilizando un lenguaje de indización común perfectamente compatible con el MT.

El MT podría así llegar a ser en un futuro próximo el mecanismo de intercambio de una red mundial para la información geológica.

Está previsto que el proyecto MT sea una operación dinámica y continuada que tenga en cuenta las evoluciones en el panorama de la información internacional y contribuya a un uso mejor y más sencillo de la información de Ciencias de la Tierra

¿COMO SE HACE UN TESAURO MULTILINGUE?

Tesauros

La creación de un Tesauro está estrechamente vinculada al tratamiento automático del análisis bibliográfico y al procedimiento de recuperación documental resultante.

La normativa de la Organización de Normas Internacionales (ISO) para los tesauros monolingües define tres criterios.

1. *Lenguaje controlado de indización:* Términos seleccionados y homologados (descriptores) del lenguaje natural y utilizados para representar, en forma condensada, el contenido del documento.

2. *Establecimiento de relaciones entre términos.* Estas relaciones pueden ser jerárquicas con términos "específicos" y "genéricos", o asociativas, con términos de "use" y "use for" o "see also."

3. *Organización formal de relaciones:* Función de un término (descriptor o no descriptor) y su relación con otros términos pertenecientes al mismo grupo de conceptos

Un tesauro multilingüe se diferencia en que permite la transposición y traducción de conceptos en más de un idioma. El tesauro multilingüe se puede dividir en varios tesauros estructurados monolingües distintos, aunque no se trata de una simple suma de ellos.

Los tesauros multilingües se pueden crear *a priori,* cuando diferentes centros documentales decidan crear una base de datos común. El tesauro multilingüe permite operar en su propio idioma a cada grupo linguístico, tanto para indizar como para recuperar. Tales redes han sido creadas por las Naciones Unidas, Comisión de la Comunidad Europea y paises del COMECON.

En el caso de cooperaciones internacionales entre sistemas de información existentes desunidos, el enfoque metodológico se basa en una compatibilidad *a posteriori* entre los conceptos utilizados por los distintos sistemas que operan en diferentes idiomas.

En éste caso, la comparación de conceptos y términos y el análisis de las técnicas de indización de los participantes adquieren el papel primordial en la elaboración de un tesauro multilingüe. Se eligió la metodología *a posteriori* para el proyecto MT.

Idiomas

El Grupo de Trabajo comprendía los siguientes grupos idiomáticos: checo, inglés, finlandés, francés, alemán, italiano, ruso y español. En el futuro podrán incluirse otros idiomas.

Especialistas húngaros y suecos participaron en los trabajos del grupo, pero todavía no han sido incluidas sus versiones lingüísticas en la base de datos. Durante algunos años hubo también participación canadiense en el proyecto.

Principios

El MT ha sido concebido como un mecanismo de intercambio entre centros documentales especializados en Ciencias de la Tierra. Ello requiere el crear un vocabulario de indización común compatible con los métodos de trabajo y sistemas en uso en los diversos centros. En otras palabras, deben tenderse puentes entre centros de diferentes idiomas y con distinta metodología. Tal cometido puede parecer imposible, pero el Grupo de Trabajo comenzó con un enfoque pragmático por etapas.

La mayoría de los miembros del grupo tenían la experiencia técnica necesaria para la toma de decisiones. La existencia de una cooperación entre varios miembros del grupo que habían utilizado la metodología y terminología del MT creo los "cimientos" para el concepto MT.

Metodología

- Enfoque *a posteriori:* Selección por cada participante de los términos apropiados en el Glosario de Geología de AGI.

- Verificación de la compatibilidad de términos con cada tesauro monolingüe "nacional."

- Discusión en reuniones anuales de definición de términos, en su calidad de descriptores o no descriptores para cada versión lingüística.

- Introducción o eliminación de términos en función del siguiente criterio: utilidad de los términos para la indización o recuperación; uso de los términos como descriptores por al menos uno de los sistemas representados.

- Adición de los términos juzgados necesarios por cada participante para asegurar la compatibilidad entre sus versiones lingüísticas y el MT.

- Se asigna después, por cada grupo lingüístico:
  1. El papel de los términos en su propio sistema documental (Descriptor; No Descriptor)
  2. Las relaciones de sustitución eventuales para los no descriptores. Estos sustitutos pueden ser simples (USE) o complejos (USE . . . AND . . .) en función de lo específico que sea el término y sus características, si es post- o pre-coordinado
  3. Reciprocidad de relaciones entre no descriptores y descriptores. El código *UF* genera las relaciones de preferencia para todos los descriptores implicados en la conexión con los no descriptores que dependen de ellos

- Si un término solo se utiliza en un idioma, simplemente se repite la forma original en los otros idiomas, sin traducción

- Clasificación de términos por tema o ámbito.

Estructura Jerárquica del MT

La estructura jerárquica se establece al nivel de cada versión lingüistica. Estos tesauros "nacionales" constituyen los anexos o extensiones del núcleo multilingüe y están relacionados con él por los términos básicos.

Ejemplos de jerarquías en tesauros "nacionales":

Ruso: DVIZHENIE TEKTONICHESKOE (relacionado con el término del MT TECTONICA DE PLACAS)

V(BT)	TEKTOGENEZ
N(NT)	DVIZHENIE EHSTATICHESKOE
	DREJF KONTINENTOV
	SKLADKOOBRAZOVANIE
	TAFROGENEZ
	EHPEYROGENEZ
a! (RT!)	KOLEBANIE
	SMESHCHENIE
	TEKTONIKA PLIT

Americano: ANTICLINES (equivalente al término del MT)

BT	FOLDS
SA	ANTICLINORIA
SA	ANTIFORM FOLDS
SA	ARCHES
SA	DIAPIRS
SA	DOMES
SA	GEANTICLINES
SA	SYNCLINES

(BT = término más amplio; NT = término más restringido; RT = término relacionado; SA = ver también)

CONTENIDO

La base de datos del MT contiene actualmente aproximadamente 5.000 términos clave, expresados en 8 versiones lingüisticas. Los térmi-nos básicos están clasificados en 36 grupos o campos, 20 de los cuales corresponden a las subdivisiones más importantes de las Ciencias de la Tierra; 11 están relacionadas con las divisiones sistemáticas de dominios de clasificación tales como estratigrafía, nombres de rocas, elementos, etc. 5 campos distintos describen conceptos comunes a todas las subdivisiones (propiedades, métodos, etc.).

Para las subdivisiones sistemáticas, se ha realizado una selección entre los términos disponibles. Los criterios para la selección fueron los siguientes:

- Estratigrafía: Nombres de periodo y piso. Campo : STRS
- Paleontología: Nombres de grupos (hasta familia u orden) en función de la importancia documental de los mismos. Campo : PALS
- Suelos : Términos utilizados en las clasificaciones americanas y europeas. Campo : SUSS
- Rocas : Nombres generalmente aceptados, de acuerdo con las clasificaciones de la IUGS. Campos : IGNS, IGMS, SEDS
- Minerales : Grupos principales Campo : MING
- Cuerpos extraterrestres : Planetas y meteoritos. Campo : EXTS

No se han incluído por el momento los nombres de lugares geográficos y fisiográficos

PRESENTACION

Se eligicrou seis idiomas (ingles, frances, alemania, italiano, ruso y español) para esta versión impresa del tesauro. Para otros idiomas re deja abierta la puerta para imprimir una versión nacional del MT en uno o varies de éstos idiomas.

El MT incluye una lista principal de *terminos clave* asi como varios indices y traducciones. La lista principal muestra los terminos clave por orden alfabetico junto con un número secuencial de referencia. Coda entroda cousta de lo siguiente:

Un termino clave en ingles presentado en la forma seleccionada inicialmente del concepto de referencia. El término clave va seguido por el campo de aplicación.

Seis entradas correspondientes a las diferentes versiones lingüistias del tesauro indicando el carácter documental del término (descriptor, no descriptor, adjetivo, término general o no traducible). Si un término es descriptor en un idioma, estará escrito en letras mayúsculas y traducido a los demás idiomas.

Si el término esta escrito en letras mayúsculas, es descriptor para éste idioma y está traducido.

Los términos no descriptores estarán escritos en letras minúsculas y van seguidos por uno o dos descriptores con una relación de preferencia escritos en mayúsculas. En el caso de que se utilicen dos descriptores, estos estarán separados por punto y coma.

Para cada versión lingüistica, se indica el uso prefernte mediante el código UF (Use for). Si un descriptor es un término preferente para varios no descriptores, estos estarán separados por punto y coma.

Los adjetivos están escritos en mayúsculas para aquellas versiones lingüisticas que los utilizan como descriptores. Para las versiones que no usan adjetivos como descriptores, estos están escritos con minúscula seguidos de la mención (adj), ej., (Abisal).

Los términos generales van seguidos por la mención (gen) en todas aquellas versiones que no los emplean como descriptores, ej., Paleo (gen.).

Los términos no traducidos están colocados entre corchetes, ej. [EN ECHELON].

Para mayor informacion de ésta presentación re puede ver la leyenda del MT.

Se incluyen además:

- Indices lingüisticos : Para cada idioma, un índice específico en orden alfabético proporciona referencias de los números de los términos clave utilizados en la lista principal.

- Indice por campo : Los términos clave se han clasificado por campo

temático. A continuación se proporciona una lista completa de los códigos de campo y sus equivalentes.

APPL	Geofísica aplicada
CHEE	Elementos
CHES	Compuestos químicos
COMS	Substancias útiles
ECON	Geología económica
ENGI	Geotecnia
ENVI	Medio ambiente
EXTR	Geología extraterrestre
EXTS	Meteoritos, planetas
GEOC	Geoquímica
GEOH	Hidrología
GEOL	Geología general
IGMS	Sistemática de rocas metamórficas
IGNE	Petrología
IGNS	Sistemática de rocas ígneas
INST	Instrumentos—equipos
ISOT	Geoquímica de isótopos/Edad absoluta, geocronología
MARI	Geología marina
MATH	Geología matemática
METH	Métodología
MINE	Mineralogía
MING	Grupos minerales
MINI	Minería
MISC	Varios
PALE	Paleontología
PALS	Sistemática de paleontología
PHCH	Propiedades físicas y químicas, procesos
SEDI	Sedimentología
SEDS	Sistemática de rocas sedimentarias
SOLI	Física del globo
STRA	Estratigrafía
STRS	Sistemática de estratigrafía

STRU   Geología estructural
SURF   Geomorfología—Geología del Cuaternario
SUSS   Sistemática de suelos
TEST   Texturas—estructuras

LOS PARTICIPANTES

En proyectos internacionales de larga duración, como el MT, el número de personas que contribuyen con su trabajo y esfuerzos es en general amplio y diverso. El ICSTI—IUGS MT ha tenido la fortuna de contar con la participación entusiasta y activa de muchos individuos. Aunque a menudo compartían pericia (e idiomas), a veces tenían opiniones distintas y sólidas y las expresaban con energía. El resultado fue, al menos, educativo.

El trabajo se comenzó por un grupo de trabajo conjunto de COGEODOC e ICSTI (WGMT). Durante el periodo 1970–1976 L. Delbos fue el Presidente del WGMT, seguido por H. Glashoff en el periodo 1976–1981. El actual Presidente del WGMT es G. N. Rassam.

La base de datos MT fue creada por H. Glashoff en el Servicio Geológico de Alemánia Federal (BGR). Después de su renuncia en 1981, hubo un periodo de relativa inactividad encargándose posteriormente del mantenimiento de la base el Consejo Nacional de Investigación de Italia (CNR). Se reanudó el proceso operativo en 1983 encargándose R. Potenza del CNR de la programación especial con el propósito de la publicación de éste libro.

RESPONSABLES DEL TRABAJO DE EDICION

Checoslovaquia—GEOFOND, Praga
J. Hruska, Vicepresidente de COGEODOC, editor de la versión checa
E. Kvetonova, editor de la versión checa

España—Instituto Geológico y Minero de España
J. Fernandez Tomas, miembro de COGEODOC

L. de Lamo, editor de la versión española
A. Sanchez Lencina, editor de la versión española

Estados Unidos—American Geological Institute.
J.J. Lloyd, editor de la versión Americana de 1970 a 1974
G.N. Rassam, Presidente de COGEODOC, representante del UISG en el ICSTI, editor de la versión americana

Finlandia—Geological Survey of Finland.
C. Kortman, miembro de COGEODOC, editor de la versión finlandesa.

Francia—Bureau de recherches géologiques et minières, miembro de ICSTI.
P. Calas, editor
L. Delbos, antiguo Presidente de COGEODOC
J. Gravesteijn, Secretario de COGEODOC, coordinador del tesauro, editor de la versión francesa
—Centre National de la Recherche Scientifique
E. Lagarde

Italia—Consiglio Nazionale delle Ricerche.
Milán: R. Potenza, coordinador de la base de datos del tesauro multilingüe, miembro de COGEODOC
R. Carimati, editor de la versión italiana
B. Testa, editor de la versión italiana
Bolonia: M. Manzoni, editor de las versiones italiana y americana

Republica Federal de Alemánia—Bundesanstalt für Geowissenschaften und Rohstoffe.
H. Glashoff, antiguo Presidente de COGEODOC
D. Niedermeier, editor de la versión alemana
J. Nowak, miembro de COGEODOC

URSS—VSEGEI, Leningrado.
A.N. Oleynikov, miembro de COGEODOC, editor de la versión rusa

Z.D. Moskalenko, editor de la versión rusa
B.D. Dvorkina, editor de la versión rusa
—VNIIMORGEO, Riga
G.I. Konshin

Participación parcial en el Tesauro:
Canadá—Geological Survey of Canada, K. Gunn
Hungría—MAFI, Hungarian Geological Survey, J. Palmai
Suecia—University of Uppsala, P. Bengston

Introduzione

Il Thesaurus multilingue di Scienze della Terra (MT) è stato sviluppato nel corso degli ultimi 14 anni con il contributo di due organizzazioni internazionali: il Consiglio Internazionale per la Documentazione Scientifica e Tecnica (ICSTI) e la Commissione di Documentazione Geologica (COGEODOC) dell'Unione Internazionale di Scienze Geologiche (IUGS).

L'ICSTI, che è un'organizzazione senza fini di lucro come quella che l'ha preceduta, l'ICSU AB (Consiglio Internazionale delle Unioni Scientifiche, Organizzazione per gli Abstracts), ha per scopo il progresso dell'accessibilità e della conoscenza dell'informazione scientifica e tecnica. Essa include fra i suoi membri i maggiori produttori di basi di dati scientifici e tecnologici, come pure i rappresentanti di tutti i settori di attività della catena di trasferimento delle informazioni.

COGEODOC promuove la cooperazione tra centri nazionali ed internazionali di documentazione per le Scienze della Terra ed ha per scopo il progresso dell'accessibilità e dell'uso dell'informazione geologica da parte di scienziati e specialisti.

Un importante contributo è stato inoltre fornito dalle organizzazioni in cui operavano i singoli partecipanti. Un elenco di queste organizzazioni si trova nel paragrafo relativo al Gruppo Editoriale in questa Introduzione.

Nel 1970 i servizi geologici di Francia, Germania Federale e Cecoslovacchia scelse la Geologia come argomento per un progetto pilota destinato a sviluppare le metodologie di costruzione di un thesaurus multilingue, tali da poter essere applicate alle altre scienze: questa scelta presumeva che il vocabolario della Geologia fosse più limitato di quello delle altre discipline scientifiche e che minori fossero le ambiguità di terminologia e di nomenclatura.

L'ICSU AB si unì a COGEODOC per la realizzazione del progetto pilota per le lingue Francese, Inglese e Tedesca; il campo degli argomenti era ridotto alla tettonica, per un sottoinsieme di circa 300 termini. Il lavoro fu completato e presentato al Congresso Geologico Internazionale di Montreal (Canada) nell'agosto del 1972. Al Congresso l'IUGS richiese che il gruppo di lavoro congiunto IUGS–ICSU AB intraprendesse l'espansione del progetto pilota al fine di estendere il Thesaurus Multilingue all'intero vocabolario della Geologia. Dopo alcune analisi ed esperimenti preliminari di approccio teorico, il Gruppo di Lavoro (WGMT) adottò la metodologia del tutto pragmatica seguita successivamente.

Come primo strumento di riferimento fu preso il "Glossary" dell' American Geological Institute che contiene circa 33000 termini con le rispettive definizioni; esso è approssimativamente completo così che si può calcolare che contenga circa l'80 per cento dei termini in uso, senza contare la nomenclatura arcaica o i termini gergali.

Si è rapidamente dimostrato infondato l'ottimistico assunto che il vocabolario della Geologia sia limitato, con pochi termini ambigui. Infatti si è spesa la maggior parte del tempo per chiarire il significato dei termini nelle diverse lingue, soprattutto per riuscire a mettere in relazione i concetti piuttosto che i termini in quanto, in questa scienza, termini affini possono avere differenti significati (Fault-Falte), e gli omonimi sono talvolta più restrittivi in una lingua che nell'altra (Schist-Schiste).

PERCHÉ UN THESAURUS MULTILINGUE DI GEOLOGIA?

Il trattamento dell'informazione bibliografica e la creazione di indici manuali o automatici richiede l'uso di metodi specifici di classificazione e organizzazione per l'analisi documentaria.

In un sistema manuale gioca un ruolo preminente la classificazione;

con l'introduzione dei metodi automatici di ricerca tramite calcolatori, le procedure per la selezione dei documenti pertinenti divengono più raffinate, nel senso che vengono usati nella ricerca i termini indice piuttosto che le categorie più vaste. La rapida crescita quantitativa e qualitativa delle telecomunicazioni ha aggiunto una nuova dimensione alla diffusione dell'informazione: essa infatti ha abbattuto le distanze tra il laboratorio e le basi dati e tra il ricercatore e il documentalista e, come risultato, tra lo scienziato *produttore* di dati e lo scienziato *utilizzatore* di dati.

Non c'è quindi da sorprendersi che in vari centri di documentazione sia sorto un orientamento inteso ad accrescere la compatibilità tra i rispettivi metodi e prodotti.

Nel 1970 i servizi geologici Francese, Tedesco e Cecoslovacco collaborarono per creare e aggiornare una base dati comune. Si utilizzava un unico vocabolario e, attraverso contatti diretti fra centri di documentazione, si tendeva ad una tecnica comune di indicizzazione. Negli anni settanta d'altra parte il GEOINFORM, il progetto per l'informazione nelle Scienze della Terra dei Paesi del COMECON, cominciò lo sviluppo di un thesaurus multilingue per l'analisi della letteratura.

La principale ricaduta del progetto di un Thesaurus Multilingue ICSTI-IUGS fu il consolidamento di questa collaborazione internazionale. Oggi nove Paesi Europei (Cecoslovacchia, Finlandia, Francia, Italia, Polonia, Repubblica Federale Tedesca, Romania, Spagna, Ungheria) con gli Stati Unit (rappresentati dall'American Geological Institute-AGI), partecipano ad una rete informativa contribuendo all'aggiornamento delle due basi di data internazionali, GeoRef e PASCAL-GEODE, e utilizzano un linguaggio di indicizzazione comune, perfettamente compatibile con il Thesaurus Multilingue.

Il Thesaurus Multilingue si prospetta quindi come uno strumento di conversione comune in una rete mondiale di informazione geologica. E'infatti previsto che esso abbia uno sviluppo progressivo e dinamico, per tener conto degli sviluppi dello scenario informativo internazionale, e contribuire così ad un sempre migliore e più facile uso dell'informazione nelle Scienze della Terra.

COME SI COSTRUISCE UN THESAURUS MULTILINGUE

Thesauri

La creazione di thesauri è strettamente connessa al trattamento automatico dell'analisi bibliografica e alle procedure di ricerca documentaria che ne risultano.

L'Organizzazione Internazionale per gli Standards (ISO) definisce tre criteri per i thesauri monolingui:

1. *Linguaggio d'indicizzazione controllato:* Il thesaurus include termini selezionati dal linguaggio naturale e formalizzati, usati per rappresentare in forma condensata il contenuto del documento (descrittori).

2. *Definizione delle relazioni fra i termini:* Queste relazioni possono essere gerarchiche, con termini specifici e generici, o associative, con termini "use," "used for" e "see also."

3. *Organizzazione formale delle relazioni:* Sono definiti il ruolo di ogni termine come descrittore o non descrittore ed i suoi rapporti con gli altri termini appartenenti allo stesso gruppo di concetti.

Un Thesaurus Multilingue differisce da questi, in quanto permette la trasposizione e la traduzione di concetti in più lingue. Il TM può essere "diviso" in thesauri monolingui strutturati, ma non è la semplice somma di essi.

Thesauri monolingue possono essere creati *a priori* quando diversi centri di documentazione decidono di costruire una base dati comune in cui ciascun centro, mediante il TM, può operare nella propria lingua, sia per indicizzare che per effettuare la ricerca. Reti di questo tipo sono state create dalle Nazioni Unite, dalla Commissione delle Comunità Europee e dai Paesi del COMECON.

Nel caso della collaborazione internazionale fra sistemi informativi separati già esistenti, l'approccio metodologico è invece basato sulla compatibilità definita *a posteriori* tra i concetti usati nei vari sistemi che operano nelle diverse lingue. In questo caso la comparazione di concetti e termini e l'analisi delle tecniche di indicizzazione usate dai partecipanti è la fase più importante della costruzione di un Thesaurus Multilingue.

La metodologia *a posteriori* fu dunque scelta per il progetto del TM.

Lingue

Il Gruppo di lavoro comprende i gruppi linguistici: Ceco, Finlandese, Francese, Inglese, Italiano, Russo, Spagnolo, Tedesco. L'ampliamento ad altre lingue potrà avvenire in futuro. Specialisti Ungheresi e Svedesi parteciparono al lavoro del Gruppo, ma le loro versioni linguistiche non sono ancora state incluse nella base dati; per alcuni anni ci fu pure una partecipazione Canadese al progetto.

Principi

Il TM è stato concepito come un meccanismo di conversione tra centri di documentazione specializzati in Science della Terra. La sua realizzazione quindi comporta la creazione di un vocabolario comune per l'indicizzazione, compatible con i metodi di lavoro e i sistemi usati nei vari centri: in altre parole, devono essere costituiti collegamenti fra centri di lingua e metodologia diverse. Ciò potrebbe apparire un compito impossibile, ma il Gruppo di lavoro lo ha affrontato pragmaticamente per stadi. La maggior parte dei membri aveva sufficiente esperienza tecnica per prendere le decisioni necessarie. Inoltre, la collaborazione tra alcuni dei membri del Gruppo aveva collaudato metodologie e terminologia, creando cosi un "banco di prova" per i concetti del TM.

La Metodologia

La costruzione del TM si è sviluppata per fasi, seguendo criteri comuni, di cui riassumiamo le principali:

- Scelta da parte di ogni partecipante dei termini appropriati nel "Glossary of Geology" dell'AGI.

- Discussione in riunioni annuali delle definizioni dei termini e del ruolo di descrittori o non descrittori in ciascuna versione linguistica.

- Introduzione od eliminazione di termini in funzione della loro utilità agli effetti dell'indicizzazione o della ricerca, o del loro impiego come descrittore in almeno uno dei sistemi rappresentati.

- Aggiunta dei termini chiave ritenuti necessari da parte di ogni rappresentante per assicurare la compatibilità tra il rispettivo thesaurus nazionale e il TM.

- Assegnazione a ciascun termine di:
    1. Funzione in ogni sistema documentario (descrittore o non descrittore)
    2. Eventuali rapporti di sostituzione per i non descrittori, per i quali possono essere indicati uno o più termini preferiti, in funzione della loro specificità o delle loro caratteristiche nei sistemi post- o pre-coordinati.
    3. Reciprocità dei rapporti di sostituzione: per i descrittori richiamati è indicato l'uso preferenziale ("use for")

- Introduzione nel TM di aggettivi e termini generali per l'indicizzazione post-coordinata: questi termini sono utilizzabili solo insieme ad altri descrittori nelle relazioni sintattiche usate da alcuni sistemi.

- Individuazione dei termini usati soltanto in alcune lingue: per le altre viene riportata la forma originale; ad esempio: En echelon in inglese.

- Classificazione per argomenti o per campi di tutti i termini.

Struttura del TM

Si è lasciata la definizione di strutture gerarchiche o relazionali ai singoli thesauri nazionali, ciascuno dei quali forma quindi un'estensione strutturata del nucleo multilingue cui è collegato dall'insieme dei termini base in esso contenuti.

Esempi di gerarchie nei thesauri nazionali.

Russo: DVIZHENIE TEKTONICHESKOE
V(BT)   TEKTOGENEZ
N(NT)   DVIZHENIE EHSTATICHESKOE
           DREJF KONTINENTOV
           SKLADKOOBRAZOVANIE
           TAFROGENEZ
           EHPEYROGENEZ
a!(RT!)   KOLEBANIE
           SMESHCHENIE
           TEKTONIKA PLIT

Americano: ANTICLINES
BT   FOLDS
SA   ANTICLINORIA
SA   ANTIFORM FOLDS
SA   ARCHES
SA   DIAPIRS
SA   DOMES
SA   GEANTICLINES
SA   SYNCLINES

(BT = termine generico; NT = termine specifico; RT = termine correlato; SA = vedi anche)

CONTENUTO

La base dati del TM contiene attualmente circa 5000 termini chiave, espressi come descrittori e non-descrittori in 8 versioni linguistiche. I termini sono classificati in 36 gruppi o campi, 20 dei quali corrispondono alle principali suddivisioni delle Scienze della Terra; 11 riguardano la parte sistematica di domini classificativi, come la stratigrafia, nomi delle rocce, elementi, ecc. Cinque campi distinti descrivono concetti comuni a tutte le suddivisioni (proprietà, metodi, ecc.). I termini inclusi nei campi sistematici sono stati selezionati secondo i seguenti criteri:

- Campo STRS, Stratigrafia: sono stati inclusi nel TM i nomi delle epoche e dei piani
- Campo PALS, Paleontologia: nomi dei gruppi (fino al livello dell'ordine o della famiglia) in funzione della loro importanza ai fini documentari
- Campo SUSS, Suoli: termini usati nelle classificazioni Americana ed Europea
- Campi IGNS, IGMS, SEDS, Rocce: nomi generalmente accettati dalle principali classificazioni; quando esistenti sono state seguite le classificazioni raccomandate dall'IUGS
- Campo MING, Minerali: gruppi principali
- Campo EXTS, Corpi extraterrestri: pianeti e meteoriti.

I nomi geografici e fisiografici non sono stati inclusi in questa edizione.

PRESENTAZIONE

Per questa versione a stampa del TM sono state scelte sei lingue: Francese, Inglese, Italiano, Russo, Spagnolo e Tedesco. E'stata lasciata aperta la possibilità di stampare altre versioni nazionali del TM insieme con una o più delle lingue in esso rappresentate.

Il TM include una lista principale di *termini chiave,* oltre a diversi indici e traduzioni. La lista principale è ordinata alfabeticamente e ad ogni termine è associato un numero sequenziale di riferimento. Ogni voce è composta come segue:

- Termine chiave in Inglese, presentato nella forma scelta inizialmente per rappresentare il concetto di riferimento. Per ogni termine chiave è indicato il codice del rispettivo campo di applicazione.

- Sei termini corrispondenti alle versioni contenute nel Thesaurus Multilingue. Il ruolo documentario del termine (descrittore, non descrittore, aggettivale, generale o non traducibile) è evidenziato tipograficamente: se il termine é in caratteri maiuscoli é descrittore per la rispettiva lingua ed ha la traduzione. Se é scritto in maiuscolo e minuscolo, il termine non é descrittore ed é seguito da uno o due descrittori (in lettere maiuscole) che indicano i(l) termine(i) preferito(i), eventualmente separati da un ";".

- Gli aggettivi sono in tutte maiuscole per quelle versioni che si riferiscono a sistemi post-coordinati; per le altre versioni linguistiche, gli aggettivi sono seguiti da "(adj.)" come ad es. abissale (adj.).

- I termini generali seguono le stesse regole degli aggettivi; sono seguiti da "(gen.)" nelle versioni che non usano indici post-coordinati; es. paleo (gen.).

- I termini non tradotti sono chiusi in parentesi quadre, ad es. [en echelon].

Per ulteriori informazioni sulla presentazione, vedere la legenda del MT.

All'elenco principale seguono sei indici in ordine alfabetico per ciascuna lingua, ciascuno dei quali permette di risalire al numero del rispettivo termine chiave della lista principale. I termini chiave sono infine elencati in ordine di campo: la lista completa dei campi e dei rispettivi codici è riportata qui di seguito.

APPL	Geofisica applicata
CHEE	Elementi chimici
CHES	Composti chimici
COMS	Sostanze utili
ECON	Geologia economica
ENGI	Geologia applicata
ENVI	Ambiente
EXTR	Geologia extraterrestre
EXTS	Corpi extraterrestri (meteoriti e pianeti)
GEOC	Geochimica
GEOH	Idrologia
GEOL	Geologia generale
IGMS	Sistematica delle rocce metamorfiche
IGNE	Petrologia
IGNS	Sistematica delle rocce ignee
INST	Strumenti e apparecchiature
ISOT	Geochimica isotopica, geocronologia, età assoluta
MARI	Geologia marina
MATH	Geologia matematica
METH	Metodologie
MINE	Mineralogia
MING	Gruppi minerali
MINI	Geologia e tecnica mineraria
MISC	Miscellanea
PALE	Paleontologia generale
PALS	Paleontologia sistematica
PHCH	Proprietà e processi chimico-fisici
SEDI	Sedimentologia
SEDS	Sistematica delle rocce sedimentarie
SOLI	Geofisica della Terra solida
STRA	Stratigrafia
STRS	Sistematica stratigrafica
STRU	Geologia strutturale
SURF	Geomorfologia e geologia del Quaternario
SUSS	Sistematica dei suoli
TEST	Tessiture e strutture

I PARTECIPANTI

In programmi internazionali di lunga durata come il TM il numero delle persone che portano il loro contributo di lavoro e di idee è di solito molto grande e vario. Il TM ICSTI-IUGS ha avuto la fortuna di avere la partecipazione entusiasta ed energica di molti individui. Questi spesso condividevano le medesime esperienze (e le lingue), ma altrettanto spesso avevano opinioni radicate che esprimevano con tutta energia. Il risultato è stato, se non altro, istruttivo.

Il lavoro fu avviato da un gruppo di lavoro congiunto COGEO-DOC-ICSTI (WGMT). Durante il periodo 1970–1976 L. Delbos fu presidente del WGMT, seguito da H. Glashoff nel periodo 1976–1981. Il presidente attuale è G. N. Rassam. La base dati TM fu creata presso il Servizio Geologico Tedesco (BGR) da H. Glashoff. Dopo le sue dimissioni, nel 1981, ci fu un periodo di relativa stasi seguita dall'assunzione della gestione della base dati da parte del Consiglio Nazionale delle Ricerche Italiano (CNR). Le elaborazioni operative ripartirono nel 1983 e R. Potenza del CNR provvide alla predisposizione dei programmi necessari alla preparazione per la stampa di questo libro.

GRUPPO EDITORIALE

Cecoslovacchia—GEOFOND-Praga.
J. Hruska, vicepresidente di COGEODOC;
E.Kvetonova, editori per la versione ceca.

Finlandia—Servizio Geologico Finlandese.
C. Kortman, membro di COGEODOC, editore per la versione finlandese.

Francia—Bureau de Recherches Géologiques et Minières, membro ICSTI.
P. Calas, editor. L. Delbos, ex presidente, COGEODOC.
J. Gravesteijn, segretario di COGEODOC, coordinatore del MT, editore per la versione francese.
—Consiglio Nazionale per la Ricerca Scientifica, membro ICSTI, E. Lagarde.

Germania Federale—Bundesanstalt für Geowissenschaften und Rohstoffe, Hannover.
H. Glashoff, ex presidente di COGEODOC.
G. Niedermeier, editore della versione tedesca
J. Nowak, membro di COGEODOC.

Italia—Consiglio Nazionale delle Ricerche, Milano.
R. Potenza, membro di COGEODOC, coordinatore della base dati.
R. Carimati; B. Testa: editori della versione italiana e gestori della base dati.
—Consiglio Nazionale delle Ricerche, Bologna.
M. Manzoni: editore delle edizioni italiana e americana.

Spagna—Instituto Geologico y Minero de España, Madrid.
J. Fernandez Tomas, membro di COGEODOC; L. de Lamo; A. Sanchez Lencina, editori della versione spagnola.

U.R.S.S.—VSEGEI (Istituto Geologico dell'Unione), Leningrado.
A.N. Oleynikov, membro di COGEODOC; Z.D. Moskalenko; B.D. Dvorkina, editori per la versione russa.
—VNIIMORGEO (Istituto di Geologia Marina e Geofisica dell'Unione), Riga.
G.I. Konshin.

U.S.A.—American Geological Institute, Alexandria, Va.
J.J. Lloyd, editore della versione americana (inglese) 1970–74
G.N. Rassam, presidente di COGEODOC, rappresentante dell'IUGS presso l'ICSTI; editore della versione americana (inglese).

Partecipazioni temporanee:
Canada—Servizio Geologico Canadese, Ottawa, K. Gunn
Ungheria—MAFI (Servizio Geologico Nazionale Ungherese), J. Palmai
Svezia—Università di Uppsala, P. Bengtson.

Legend and Preface to Entries

The MT is designed for transfer of information between documentation services; thus, the terms used may not always conform to current usage in geoscience. An effort has been made however to provide, whenever possible, more than one term to cover closely parallel concepts, for example, *Borderland* and *Continental borderland;* or *Hercynian Orogeny* and *Variscan Orogeny*. The bulk of the main entries on the left column were chosen from AGI's *Glossary of Geology*. Other terms were introduced by different national documentation services when the need for a particular concept was felt not met by Glossary terms.

This is *not a directory,* that is, the terms used in the different columns for different languages may not be strictly synonymous. They are merely the actual usage of terms to represent a particular concept.

MAIN ENTRIES

Main entry terms are in English. They are numbered sequentially and these numbers are used throughout, in the main body and in the various indexes. The main entry terms are always accompanied by a 4-letter abbreviation of the field to which that concept was assigned (field of study code). For a list of the fields and their abbreviations, consult the Introduction. A field index follows the various language indexes after the text.

The next six columns contain the entries for the various languages—documentation services involved. If a term is a descriptor in that particular language it is found in all *upper case* letters. An entry in upper and lower case letters indicates a non-descriptor and is followed by upper-case term(s) indicating the proper descriptor(s). A *UF* (use for) following a descriptor indicates the non-descriptors that were assigned to that descriptor.

Exceptions to this presentation may exist:

1. A term (in lower case) followed by (adj.) indicates an adjective used in some languages as a descriptor but not in others. It stands alone, that is, it is not followed by a descriptor.

2. A term (in lower case) followed by (gen.) indicates a general term used in some languages as a descriptor but not in others. It stands alone (not followed by a descriptor).

3. Very few terms, though non-descriptors for all languages, were left in to indicate an important concept.

4. Some Russian entries may have the following form: [GELIVITY] VYVETRANIE MOROZNOE indicating that a non-descriptor was not translated and therefore is given in English in brackets and is followed by the proper descriptor(s).

5. Entries with the form [ACID MINE DRAINAGE] indicate a non-translated term to be considered as a descriptor for information-transfer purposes, until such time as the particular concept is introduced in the national thesaurus.

6. Entries with the form BERMA (GEOMORFOLOGIYA) indicate a general term in a particular language, necessitating an explanation in parentheses.

7. Entries with the form DOKEMBRIJSKIJ/OROGEN/ indicate a combination of an adjective and a noun.

Note that within entries, the hyphen of a hyphenated word (e.g., Large-scale or Alpine-Type) is represented by an equals sign (Large=scale or Alpine=Type). The hyphen that occurs when a word has been artificially broken at the end of the line appears as a short hyphen.

INDEXES

There are six language indexes, followed by a field index. The numbers following all index entries refer to main entry numbers rather than page numbers. The entries in the language indexes include both descriptors and non-descriptors with no distinction made between them.

	ENGLISH	FRANCAIS	DEUTSCH	RUSSKIJ	ESPANOL	ITALIANO
1 A LAYER SOLI	A layer CRUST	Couche A CROUTE TER- RESTRE	A=Schicht ERD=KRUSTE	A sloj KORA ZEMNAYA	Capa=A CORTEZA= TERRESTRE	Strato A CROSTA TER- RESTRE
2 AA LAVA IGNE	AA LAVA	LAVE AA	SCHLACKEN=LAVA	AA=LAVA	LAVA=AA	LAVA AA
3 AALENIAN STRS	AALENIAN	AALENIEN	AALENIUM	AALEN	AALIENSE	AALENIANO
4 ABANDONED SHORELINE SURF	Abandoned shoreline BEACHES	Cote fossile PALEOGEOGRA- PHIE; LIGNE RIV- AGE	Fossil.Kuestenlinie PALAEOGEOGRA- PHIE; KUESTE	LINIYA BERE- GOVAYA DREV- NYAYA	Costa=abandonada PALEOGEOGRAFIA; LINEA=COSTA	Linea di riva abban- donata PALEOGEOGRAFIA; LINEA DI RIVA
5 ABLATION SURF	ABLATION	ABLATION	ABLATION	ABLYATSIYA	ABLACION	ABLAZIONE
6 ABRASION SURF	ABRASION *UF:* Abrasion coast; Abrasion surface; Attrition	ABRASION *UF:* Attrition; Surface abrasion	ABRASION *UF:* Abrasions=Flaeche; Attrition	ABRAZIA	ABRASION *UF:* Atricion; Superficie=de=abrasion	ABRASIONE *UF:* Sfregamento; Super- ficie d'abrasione
7 ABRASION COAST SURF	Abrasion coast ABRASION; BEACHES	Cote abrasion EROSION LITTOR- ALE	Abrasions=Kueste KUESTEN= EROSION	BEREG ABRAZION- NYJ	Costa=abrasion EROSION=LITORAL	Costa d'abrasione EROSIONE COSTI- ERA
8 ABRASION SURFACE SURF	Abrasion surface ABRASION	Surface abrasion ABRASION	Abrasions=Flaeche ABRASION	POVERKHNOST' ABRAZIONNAYA	Superficie=de=abra- sion ABRASION	Superficie d'abrasione ABRASIONE
9 ABRASIVE COMS	ABRASIVES *UF:* Tripoli	ABRASIF *UF:* Tripoli	POLIERSTOFF *UF:* Tripoli	ABRAZIV	ABRASIVO *UF:* Tripoli	ABRASIVO *UF:* Tripoli
10 ABSOLUTE AGE ISOT	ABSOLUTE AGE *UF:* Concordant age; Dating; Decay cons- tant; Discordant age; Half=life period; Over- print	Age absolu DATATION	Absolut=Alter PHYSIKAL. ALTERSBESTIM- MUNG	VOZRAST ABSO- LYUTNYJ *UF:* Datirovka; Vozrast diskordantnyj; Vozrast konkordantnyj; Voz- rast Zemli	Edad=absoluta DATACION	Eta assoluta DATAZIONE
11 ABSOLUTE GRAVITY SOLI	Absolute gravity GRAVITY FIELD	Pesanteur absolue GRAVIMETRIE	Absolut.Schwere GRAVIMETRIE	SILA TYAZHESTI ABSOLYUTNAYA	Gravedad=absoluta GRAVIMETRIA	Gravita assoluta GRAVIMETRIA
12 ABSORBENT COMS	ABSORBENT MATE- RIALS *UF:* Bleaching clay	ABSORBANT *UF:* Terre adsorbante	ABSORBER *UF:* Bleich=Erde	ABSORBENT	ABSORBENTE *UF:* Arcilla=esmectica	ASSORBENTE *UF:* Argilla sbiancante
13 ABSORPTION PHCH	ABSORPTION	ABSORPTION	ABSORPTION	ABSORBTSIYA	ABSORCION	ASSORBIMENTO
14 ABSORPTION SPECTROSCOPY METH	Absorption spectros- copy ATOMIC ABSORP- TION	SPECTROMETRIE ABSORPTION *UF:* Spectre absorption	ABSORPTIONS= SPEKTROMETRIE *UF:* Absorptions= Spektrum	SPEKTROSKOPIYA ABSORBTSION- NAYA	ESPECTROMETRIA= DE=ABSORCION *UF:* Espectro=absorcion	SPETTROMETRIA D'ASSORBIMENTO *UF:* Spettro d'assorbimento
15 ABSORPTION SPECTRUM PHCH	ATOMIC ABSORP- TION SPECTRA	Spectre absorption SPECTROMETRIE ABSORPTION	Absorptions=Spektrum ABSORPTIONS= SPEKTROMETRIE	SPEKTR POGLOSH- CHENIYA	Espectro=absorcion ESPECTROMETRIA= DE=ABSORCION	Spettro d'assorbimento SPETTROMETRIA D'ASSORBIMENTO

	ENGLISH	FRANCAIS	DEUTSCH	RUSSKIJ	ESPANOL	ITALIANO
16 ABUNDANCE MATH	ABUNDANCE	ABONDANCE	HAEUFIGKEIT	RASPROSTRANEN-NOST'	ABUNDANCIA	ABBONDANZA
17 ABYSS MARI	DEEPS	ABYSSE *UF:* Chenal mer profonde; Vallee mer profonde	TIEFSEE *UF:* Tiefsee=Rinne; Tiefsee=Tal	GLUBOKOVOD'E *UF:* Abissal'naya depressiya	SIMA=MARINA *UF:* Canal=marino=profundo; Surco=abisal	ABISSO *UF:* Canale sottomarino; Valle abissale
18 ABYSSAL GEOL	Abyssal(adj.)	Abyssal(adj.)	Abyssal(adj.)	Abissal'naya /depressiya/ GLUBOKOVOD'E	Abisal(adj.)	Abissale(adj.)
19 ABYSSAL ENVIRONMENT PALE	Abyssal environment DEEP=SEA ENVIRONMENT	Milieu abyssal MILIEU MER PROFONDE	Tiefwasser=Milieu TIEFSEE=MILIEU	ABISSAL' *UF:* Sedimentatsiya abissal'naya	Medio=abisal MEDIO=MAR=PROFUNDO	Ambiente abissale AMBIENTE DI MARE PROFONDO
20 ABYSSAL GAP MARI	Abyssal gap OCEAN FLOORS	Vallee mer profonde ABYSSE	Tiefsee=Tal TIEFSEE	USHCHEL'E ABISSAL'NOE	Surco=abisal SIMA=MARINA	Valle abissale ABISSO
21 ABYSSAL HILL MARI	Abyssal hill OCEAN FLOORS	Colline abyssale PLAINE ABYSSALE	Tiefsee=Berg TIEFSEE=BODEN	KHOLM ABISSAL'NYJ	Colina=abisal LLANURA=ABISAL	Collina abissale PIANA ABISSALE
22 ABYSSAL PLAIN MARI	ABYSSAL PLAINS	PLAINE ABYSSALE *UF:* Colline abyssale	TIEFSEE=BODEN *UF:* Tiefsee=Berg	RAVNINA ABISSAL'NAYA	LLANURA=ABISAL *UF:* Colina=abisal	PIANA ABISSALE *UF:* Collina abissale
23 ABYSSAL SEDIMENTATION SEDI	DEEP=SEA SEDIMENTATION	SEDIMENTATION MER PROFONDE	TIEFSEE=SEDIMENTATION	Sedimentatsiya abissal'naya ABISSAL'; SEDIMENTATSIYA MORSKAYA	SEDIMENTACION=AGUA=PROFUNDA	SEDIMENTAZIONE DI MARE PROFONDO
24 ACADIAN OROGENY STRU	ACADIAN PHASE	PHASE ACADIENNE	ACAD.OROGENESE	Akadskij GERTSINSKIJ	OROGENIA=ACADIENSE	FASE ACADIANA
25 ACANTHARIA PALS	ACANTHARIA	Acantharia RADIOLARIA	Acantharia RADIOLARIA	Acantharia RADIOLARIA	Acantharia RADIOLARIA	Acantharia RADIOLARIA
26 ACANTHODII PALS	ACANTHODII	ACANTHODII	ACANTHODII	ACANTHODII	ACANTHODII	ACANTHODII
27 ACCELEROGRAM APPL	ACCELEROGRAMS	ACCELEROGRAMME	BESCHLEUNIGUNGS=AUFZEICHNUNG	AKSELEROGRAMMA	ACELEROGRAMA	ACCELEROGRAMMA
28 ACCELEROMETER INST	ACCELEROMETERS	ACCELEROMETRE	BESCHLEUNIGUNGS=MESSER	MINERAL AKTSESSORNYJ	ACELEROMETRO	ACCELEROMETRO
29 ACCESSORY MINERAL IGNE	ACCESSORY MINERALS	MINERAUX ACCESSOIRES	AKZESSOR. MINERAL	AKTSESSORIJ	MINERALES=ACCESORIOS	MINERALE ACCESSORIO
30 ACCRETION GEOL	ACCRETION *UF:* Continental accretion	ACCRETION *UF:* Croissance continentale	ANLAGERUNG *UF:* Kontinent=Wachstum	AKKRETSIYA	ACRECION *UF:* Crecimiento=continental	ACCRESCIMENTO *UF:* Accrescimento continentale
31 ACCUMULATION GEOL	Accumulation RESERVOIR ROCKS	EMMAGASINEMENT	UNTERIRD. AKKUMULATION	AKKUMULYATSIYA	CAPACIDAD=ALMACENAMIENTO	IMMAGAZZINAMENTO
32 ACCUMULATIVE COAST SURF	Accumulative coast COASTAL PLAINS; BEACHES	Cote accumulation PLAINE COTIERE	Akkumulations=Kueste KUESTEN=EBENE	BEREG AKKUMULYATIVNYJ	Costa=acumulacion LLANURA=COSTERA	Costa d'accumulazione PIANA COSTIERA

	ENGLISH	FRANCAIS	DEUTSCH	RUSSKIJ	ESPANOL	ITALIANO
33 ACCUMULATIVE PLAIN SURF	Accumulative plain PLAINS	Plaine accumulation PLAINE	Aufschotterungs=Ebene FLACHLAND	RAVNINA AKKU- MULYATIVNAYA	Plano=acumulacion LLANURA	Piana d'accumulazione PIANA
34 ACCURACY MATH	ACCURACY	PRECISION	PRAEZISIONS= MESSUNG	TOCHNOST'	PRECISION=DE= MEDIDA	PRECISIONE
35 ACETOLYSIS METH	Acetolysis CHEMICAL ANAL- YSIS	Acetolyse METHODOLOGIE ANALYSE; ACIDE ORGANIQUE	Azetolyse ORGAN.SAEURE; AUFLOESUNG	ATSETOLIZ	Acetolisis DISOLUCION; ACIDO=ORGANICO	Acetolisi DISSOLUZIONE; ACIDO ORGANICO
36 ACHONDRITE EXTS	ACHONDRITES *UF:* Angrite; Aubrite; Chassignite; Diogenite; Eucrite; Howardite; Nakhlite; Ureilite	ACHONDRITE *UF:* Angrite; Aubrite; Chassignite; Diogenite; Eucrite achondrite; Howardite; Nakhlite; Ureilite	ACHONDRIT *UF:* Angrit; Aubrit; Chassignit; Diogenit; Eucrit; Howardit; Nakhlith; Ureilit	AKHONDRIT *UF:* Angrit; Chassignit; Diogenit; Ehvkrit; Govardit; Naklit; Obrit; Ureilit	ACONDRITA *UF:* Angreita; Aubrita; Casignita; Diogenita; Eucrita; Howardita; Nakhlita; Ureilita	ACONDRITE *UF:* Angrite; Aubrite; Chassignite; Diogenite; Eucrite (meteorite); Howardite; Nakhlite; Ureilite
37 ACID CHES	ACIDS	ACIDE	SAEURE	KISLOTA	ACIDO	ACIDO
38 ACID MINE DRAINAGE ENVI	ACID MINE DRAIN- AGE	EXHAURE MINE ACIDE	SAUR.GRUBEN= WASSER	[ACID MINE DRAINAGE]	DESAGUE=ACIDO= DE=MINA	DRENAGGIO ACIDO
39 ACIDIC COMPOSITION IGNE	ACIDIC COMPOSI- TION *UF:* Felsic composition	COMPOSITION ACIDE	SAUR.CHEMISMUS	KISLYJ (SOSTAV) *UF:* Fel'zicheskij	COMPOSICION= ACIDA	COMPOSIZIONE ACIDA
40 ACIDITY PHCH	Acidity PH	Acidite PH	Aziditaet PH=WERT	KISLOTNOST'	Acidez PH	Acidita PH
41 ACOUSTIC LOG APPL	ACOUSTICAL LOG- GING	DIAGRAPHIE SONIQUE	AKUSTIK=LOG	Karotazh akusticheskij KAROTAZH SEJS- MICHESKIJ	DIAGRAFIA= SONICA	DIAGRAFIA ACUS- TICA
42 ACOUSTIC PROPERTY PHCH	ACOUSTICAL PROPERTIES	PROPRIETE ACOUS- TIQUE	AKUST. EIGENSCHAFT	SVOJSTVO AKUSTI- CHESKOE	PROPIEDAD= ACUSTICA	PROPRIETA ACUS- TICA
43 ACOUSTICAL METHODS APPL	ACOUSTICAL METHODS	METHODE ACOUS- TIQUE	AKUST.METHODE	METOD AKUSTI- CHESKIJ *UF:* S'emka akusti- cheskaya	METODO= ACUSTICO	METODO ACUS- TICO
44 ACOUSTICAL SURVEY APPL	ACOUSTICAL SUR- VEYS	LEVE ACOUSTIQUE	AKUST.AUFNAHME	S'emka akusticheskaya METOD AKUSTI- CHESKIJ	CAMPAÑA PROSPECCION= ACUSTICA	RILEVAMENTO ACUSTICO
45 ACOUSTICAL WAVE SOLI	ACOUSTICAL WAVES	ONDE ACOUS- TIQUE	AKUST.WELLE	VOLNA AKUSTI- CHESKAYA	ONDA=ACUSTICA	ONDA ACUSTICA
46 ACRANIA PALS	Acrania CHORDATA	Acrania CHORDATA	Acrania CHORDATA	ACRANIA	Acrania CHORDATA	Acrania CHORDATA
47 ACRISOL SUSS	Acrisols PODZOLS	Acrisol PODZOL	Acrisol PODSOL	AKRISOL	Acrisol PODZOL	Acrisol PODSOL
48 ACRITARCHA PALS	ACRITARCHS	ACRITARCHA	ACRITARCHA	ACRITARCHA	ACRITARCHA	ACRITARCHA
49 ACROTRETIDA PALS	Acrotretida INARTICULATA	Acrotretida INARTICULATA	Acrotretida INARTICULATA	ACROTRETIDA	Acrotretida INARTICULATA	Acrotretida INARTICULATA

	ENGLISH	FRANCAIS	DEUTSCH	RUSSKIJ	ESPANOL	ITALIANO
50 ACTINIARIA PALS	ACTINIARIA	ACTINIARIA	ACTINIARIA	Actiniaria ANTHOZOA	ACTINIARIA	ACTINIARIA
51 ACTINIUM CHEE	ACTINIUM	ACTINIUM	AC	AKTINIJ	ACTINIO	ATTINIO
52 ACTINOCERATOIDEA PALS	Actinoceratoidea CEPHALOPODA	Actinoceratoidea NAUTILOIDEA	Actinoceratoidea NAUTILOIDEA	ACTINOCERAT- OIDEA	Actinoceratoidea NAUTILOIDEA	Actinoceratoidea NAUTILOIDEA
53 ACTINOLITE FACIES IGNE	ACTINOLITE FACIES	FACIES ACTINOTE	AKTINOLITH= FAZIES	FATSIYA AKTINOLITOVAYA	FACIES=ACTINOTA	FACIES ATTINOTO
54 ACTIVATION ANALYSIS METH	ACTIVATION ANALYSIS	ANALYSE ACTIVA- TION	AKTIVIERUNGS= ANALYSE	ANALIZ AKTIVAT- SIONNYJ	ANALISIS- ACTIVACION	ANALISI PER ATTIVAZIONE
55 ACTIVATION ENERGY PHCH	ACTIVATION ENERGY	ENERGIE ACTIVA- TION	AKTIVIERUNGS= ENERGIE	EHNERGIYA AKTIVATSII	ENERGIA- ACTIVACION	ENERGIA DI ATTIVAZIONE
56 ACTIVE FAULT STRU	ACTIVE FAULTS	FAILLE ACTIVE	AKTIV.STOERUNG	RAZLOM ZHIVUSH- CHIJ	FALLA-ACTIVA	FAGLIA ATTIVA
57 ACTIVE LAYER SURF	ACTIVE LAYER	COUCHE ACTIVE	AUFTAU=ZONE	SLOJ AKTIVNYJ	CAPA=ACTIVA	STRATO ATTIVO
58 ACTIVE MARGIN SOLI	ACTIVE MARGINS	MARGE CON- TINENTALE ACTIVE	AKTIV. KONTINENTAL= RAND	Zakraina aktivnaya KRAJ PLATFORMY; AKTIVIZATSIYA	MARGEN= CONTINENTAL= ACTIVO	MARGINE CON- TINENTALE ATTIVO
59 ACTIVE TECTONICS STRU	Active tectonics TECTONICS	Tectonique active TECTONIQUE	Aktuo=Tektonik TEKTONIK	AKTIVIZATSIYA UF: Reaktivizatsiya; Remagnetizatsiya; Zakraina aktivnaya	Tectonica=activa TECTONICA	Tettonica attiva TETTONICA
60 ACTIVE VOLCANO IGNE	Active volcano VOLCANOES	Volcan actif VOLCAN	Aktiv.Vulkan VULKAN	VULKAN DEJST- VUYUSHCHIJ	Volcan=activo VOLCAN	Vulcano attivo VULCANO
61 ACTIVITY PHCH	ACTIVITY	ACTIVITE	AKTIVITAET	AKTIVNOST'	ACTIVIDAD	ATTIVITA
62 ACTUALISM GEOL	Actualism UNIFORMITARIAN- ISM	ACTUALISME UF: Uniformitarianisme	AKTUALISMUS UF: Uniformitarismus	AKTUALIZM UF: Uniformizm	ACTUALISMO UF: Uniformidad	ATTUALISMO UF: Uniformitarianismo
63 ADAPTATION PALE	ADAPTATION UF: Adaptive radiation; Homeomorphy	ADAPTATION UF: Homeomorphie; Radiation adaptative	ANPASSUNG UF: Adaptiv. Strahlung; Homoeomorphie	ADAPTATSIYA	ADAPTACION UF: Homomorfo; Radiacion=adaptable	ADATTAMENTO UF: Omeomorfia; Radia- zone adattativa
64 ADAPTIVE RADIATION PALE	Adaptive radiation ADAPTATION	Radiation adaptative ADAPTATION; EVO- LUTION BIOLOGIQUE	Adaptiv. Divergenz ANPASSUNG; BIOLOG. EVOLUTION; STRAHLUNG	RADIATSIYA ADAP- TIVNAYA	Radiacion=adaptable ADAPTACION; EVOLUCION= BIOLOGICA	Radiazione adattativa ADATTAMENTO; EVOLUZIONE BIOLOGICA
65 ADIABATIC CONDITION PHCH	Adiabatic processes THERMODYNAMIC PROPERTIES	Processus adiabatique THERMODY- NAMIQUE	Adiabat.Vorgang THERMODYNAMIK	ADIABATICHESKIJ	Proceso=adiabatico TERMODINAMICA	Condizione adiabatica TERMODINAMICA
66 ADIAGNOSTIC MISC	Adiagnostic(adj.)	Adiagnostique(adj.)	Adiagnostisch(adj.)	ADIAGNOSTI- CHESKIJ	Adiagnostico(adj.)	Adiagnostico(adj.)

	ENGLISH	FRANCAIS	DEUTSCH	RUSSKIJ	ESPANOL	ITALIANO
67 ADSORBED WATER GEOH	Adsorbed water CAPILLARY WATER	Eau adsorbee EAU CAPILLAIRE	Haft-Wasser KAPILLAR=WASSER	VODA ADSOR-BIROVANNAYA	Agua=de=adsorcion AGUA=CAPILAR	Acqua di adsorbimento ACQUA CAPILLARE
68 ADSORPTION PHCH	ADSORPTION	ADSORPTION	ADSORPTION	ADSORBTSIYA	ADSORBCION	ADSORBIMENTO
69 AERATION SURF	Aeration UNSATURATED ZONES	Aeration ZONE NON SATUREE	Belueftung UNGESAETTIGT. ZONE	AEHRATSIYA	Aireacion ZONA-NO-SATURADA	Aerazione ZONA NON SATURATA
70 AERIAL MAPPING GEOL	Aerial mapping AERIAL PHOTOGRAPHY; AIRBORNE METHODS	Cartographie aerienne METHODE AEROPORTEE; PHOTOGEOLOGIE	Luftbild=Kartierung AIR=BORNE=AUFNAHME; PHOTOGEOLOGIE	Aehrokartirovanie AEHROFOTOS'EMKA	Cartografia=aerea FOTOGEOLOGIA; METODO=AEROPORTADO	Cartografia aerea FOTOGEOLOGIA; RILEVAMENTO AEROPORTATO
71 AERIAL PHOTOGRAPH GEOL	AERIAL PHOTOGRAPHY UF: Aerial mapping; Aerial survey	PHOTOGRAPHIE AERIENNE	LUFTBILD	AEHROFOTOSNI-MOK	FOTOGRAFIA=AEREA	AEROFOTOGRAFIA
72 AERIAL SURVEY GEOL	Aerial survey AERIAL PHOTOGRAPHY	Leve aeroporte METHODE AEROPORTEE; PHOTOGEOLOGIE	Luft=Vermessung AIR=BORNE=AUFNAHME	AEHROFOTOS'EMKA UF: Aehrokartirovanie	Campaña prospeccion=aeroportada METODO-AEROPORTADO	Prospezione aerea RILEVAMENTO AEROPORTATO
73 AEROBIC CONDITION PALE	AEROBIC ENVIRONMENT	MILIEU AEROBIE	AEROB.MILIEU	AEHROBNYJ	MEDIO=AEROBIO	AMBIENTE AEROBICO
74 AEROMAGNETIC SURVEY APPL	Aeromagnetic survey MAGNETIC SURVEYS; AIRBORNE METHODS	Leve aeromagnetique LEVE MAGNETIQUE; METHODE AEROPORTEE	Aeromagnet. Vermessung MAGNET. VERMESSUNG; AIR=BORNE=AUFNAHME	S'EMKA AEHRO-MAGNITNAYA	Prospeccion=aeromagnetica LEVANTAMIENTO=MAGNETICO; METODO=AEROPORTADO	Rilevamento aero-magnetico RILEVAMENTO MAGNETICO; RILEVAMENTO AEROPORTATO
75 AEROSOL SURF	AEROSOLS	AEROSOL	AEROSOL	AEHROZOL'	AEROSOL	AEROSOL
76 AFFINITY MISC	AFFINITIES	Affinite(gen.)	Affinitaet(gen.)	RODSTVO	Afininidad(gen.)	Affinita(gen.)
77 AFFLUENT SURF	Affluent streams STREAMS	Affluent RIVIERE	Nebenfluss FLUSS	PRITOCHNIJ UF: Pritok	Afluente RIO	Affluente FIUME
78 AFMAG METHOD APPL	AFMAG METHOD	METHODE AFMAG	AFMAG	METOD AFMAG	METODO=AFMAG	METODO AFMAG
79 AFTERSHOCK SOLI	AFTERSHOCKS	REPLIQUE SISMIQUE	NACHBEBEN	Aftershok ZEMLETRYASENIE	RESPUESTA-SISMICA	REPLICA SISMICA
80 AGE GEOL	AGE UF: Age of the Earth	AGE UF: Biochronologie	ALTER UF: Biochronologie	VOZRAST GEOLOGICHESKIJ	EDAD UF: Biocronologia	ETA UF: Biocronologia
81 AGE OF THE EARTH GEOL	Age of the Earth AGE; EARTH	Age de la Terre DATATION; PLANETE TERRE	Erd=Alter PHYSIKAL. ALTERSBESTIMMUNG; PLANET=ERDE	Vozrast Zemli VOZRAST ABSOLYUTNYJ; ZEMLYA	Edad=de=la=Tierra DATACION; PLANETA-TIERRA	Eta della terra DATAZIONE; PIANETA TERRA

	ENGLISH	FRANCAIS	DEUTSCH	RUSSKIJ	ESPANOL	ITALIANO
82 AGGLOMERATE SEDS	AGGLOMERATE	AGGLOMERAT	AGGLOMERAT	AGLOMERAT	AGLOMERADO	AGGLOMERATO
83 AGGLUTINATE IGNE	Agglutinates PYROCLASTICS	Pyroclastique soude PYROCLASTIQUE	Vulkan.Schweiss= Schlacke PYROKLAST. GESTEIN	AGGLYUTINAT	Piroclastico=cementado PIROCLASTICO	Agglutinato PRODOTTO PIRO- CLASTICO
84 AGGRADATION SURF	AGGRADATION	EPANDAGE	VERFUELLUNG	AGGRADATSIYA	ACRECENTA- MIENTO	[AGGRADATION]
85 AGGREGATE COMS	AGGREGATE	GRANULAT	ZUSCHLAG=STOFF	AGREGAT	AGREGADO	AGGREGATO
86 AGNATHA PALS	AGNATHA UF: Anaspida; Diplorhina; Monorhina; Osteostraci; Thelodonti	AGNATHA UF: Anaspida; Diplorhina; Monorhina; Osteostraci; Thelodonti	AGNATHA UF: Anaspida; Diplorhina; Monorhina; Osteostraci; Thelodonti	AGNATHA	AGNATHA UF: Anaspida; Diplorhina; Monorhina; Osteostraci; Thelodonti	AGNATHA UF: Anaspida; Diplorina; Monorhina; Osteostraci; Thelodonti
87 AGNOSTIDA PALS	AGNOSTIDA UF: Miomera	AGNOSTIDA UF: Miomera	AGNOSTIDA UF: Miomera	Agnostida MIOMERA	AGNOSTIDA UF: Miomera	AGNOSTIDA UF: Miomera
88 AGONIATITIDA PALS	Agoniatitida CEPHALOPODA	Agoniatitida AMMONOIDEA	Agoniatitida AMMONOIDEA	AGONIATITIDA	Agoniatitida AMMONOIDEA	Agoniatitida AMMONOIDEA
89 AGPAITIC IGNE	Agpaitic(adj.)	Agpaitique(adj.)	Agpaitisch(adj.)	AGPAITOVYJ	Agpaitico(adj.)	Agpaitico(adj.)
90 AGRICULTURE SURF	AGRICULTURE UF: Agrogeology	AGRICULTURE UF: Agrogeologie	LANDWIRTSCHAFT UF: Agrogeologie	SEL'SKOE KHO- ZYAJSTVO	AGRICULTURA UF: Agrogeologia	AGRICOLTURA UF: Agrogeologia
91 AGROGEOLOGY SURF	Agrogeology AGRICULTURE	Agrogeologie AGRICULTURE	Agrogeologie LANDWIRTSCHAFT	AGROGEOLOGIYA	Agrogeologia AGRICULTURA	Agrogeologia AGRICOLTURA
92 AHERMATYPIC TAXON PALE	AHERMATYPIC TAXA	TAXON AHERMA- TYPIQUE	AHERMATYP. TAXON	AGERMATIPNYJ	TAXON= AHERMATIPICO	TAXON AHERMA- TIPICO
93 AIR GEOL	Air ATMOSPHERE	Air ATMOSPHERE	Luft ATMOSPHAERE	Vozdukh ATMOSFERA	Aire=atmosferico ATMOSFERA	Aria ATMOSFERA
94 AIR-SEA INTERFACE MARI	AIR=SEA INTER- FACE	INTERFACE AIR MER	LUFT=MEER= GRENZFLAECHE	Svyaz vozdukh=more ATMOSFERA; MORE	INTERFASE=AIRE= MAR	INTERFACCIA ARIA=MARE
95 AIRBORNE METHOD APPL	AIRBORNE METH- ODS UF: Aerial mapping; Aeromagnetic survey	METHODE AERO- PORTEE UF: Cartographie aerienne; Leve aero- magnetique; Leve aeroporte	AIR=BORNE= AUFNAHME UF: Aeromagnet. Vermessung; Luftbild= Kartierung; Luft= Vermessung	AEHROMETOD	METODO- AEROPORTADO UF: Campaña= prospeccion= aeroportada; Cartografia=aerea; Prospeccion= aeromagnetica	RILEVAMENTO AEROPORTATO UF: Cartografia aerea; Prospezione aerea; Rilevamento aero- magnetico
96 AIRFIELD ENGI	AIRFIELDS	AERODROME	FLUGPLATZ	AEHRODROM	AERODROMO	AERODROMO
97 AIRY WAVE SOLI	AIRY WAVES	Onde Airy ONDE SURFACE	Airy=Welle OBERFLAECHEN= WELLE	Volna Ehry VOLNA; METOD SEJSMICHESKIJ	Ondas=Airy ONDA= LONGITUDINAL	Onda di Airy ONDA DI SUPERFI- CIE

	ENGLISH	FRANCAIS	DEUTSCH	RUSSKIJ	ESPANOL	ITALIANO
98 AKMOLITH IGNE	Akmolith LACCOLITHS	Acmolite LACCOLITE	Akmolith LAKKOLITH	AKMOLIT	Acmolito LACOLITO	Acmolite LACCOLITE
99 AKTCHAGYLIAN STRS	Aktchagylian PLIOCENE	Akchagylien PLIOCENE	Akchagylium PLIOZAEN	AKCHAGYL	Aktchagyliense PLIOCENO	Aktchagyliano PLIOCENE
100 ALABASTER SEDS	ALABASTER	ALBATRE	ALABASTER	ALEBASTR	ALABASTRO	ALABASTRO
101 ALBEDO PHCH	ALBEDO	ALBEDO	ALBEDO	AL'BEDO	ALBEDO	ALBEDO
102 ALBERTITE SEDS	Albertite BITUMENS	Albertite BITUME	Albertit BITUMEN	Al'bertit BITUM	Albertita BETUN	Albertite BITUME
103 ALBIAN STRS	ALBIAN	ALBIEN	ALB	AL'B	ALBIENSE	ALBIANO
104 ALBITE=EPIDOTE-AMPHIBOL.FACIES IGNE	ALBITE=EPIDOTE=AMPHIBOL.FACIES	Facies amphibolite albite epidot FACIES AMPHIBO-LITE	Albit=Epidot=Amphibolit=Fazies AMPHIBOLIT=FAZIES	Fatsiya al'bit=ehpidot=amfibolit FATSIYA EHPIDOT=AMFIBOLITOVAYA	Facies=anfibolita=albita=epidota FACIES=ANFIBOLITA	Facies anfiboliti albite epidoto FACIES DELLE ANFIBOLITI
105 ALBITE=EPIDOTE-HORNFELS FACIES IGNE	ALBITE=EPIDOTE=HORNFELS FACIES	Facies corneenne albite epidote FACIES CORN-EENNE	Albit=Epidot=Hornfels=Fazies HORNFELS=FAZIES	FATSIYA AL'BIT-EHPIDOT-ROGOVIKOVAYA	Facies=corneana=albita=epidota FACIES=CORNEANA	Facies cornubian. albite epidoto FACIES DELLE CORNUBIANITI
106 ALBITIZATION IGNE	ALBITIZATION	ALBITISATION	ALBITISIERUNG	AL'BITIZATSIYA	ALBITIZACION	ALBITIZZAZIONE
107 ALBOLL SUSS	Albolls MOLLISOLS	Alboll MOLLISOL	Alboll MOLLISOL	Alboll MOLLISOL	Alboll MOLLISOL	Alboll MOLLISOL
108 ALCYONACEA PALS	Alcyonacea OCTOCORALLIA	Alcyonacea OCTOCORALLA	Alcyonacea OCTOCORALLIA	Alcyonacea OCTOCORALLIA	Alcyonacea OCTOCORALLIA	Alcyonacea OCTOCORALLIA
109 ALDANIAN STRS	Aldanian LOWER CAMBRIAN	Aldanien CAMBRIEN INF	Aldanium U.KAMBRIUM	ALDANSKIJ	Aldaniense CAMBRICO=INF	Aldaniano CAMBRIANO INF.
110 ALFISOL SUSS	ALFISOLS *UF:* Aqualfs; Boralfs; Luvisols; Udalfs; Ustalfs; Xeralfs	ALFISOL *UF:* Aqualf; Boralf; Luvisol; Podzoluvisol; Udalf; Ustalf; Xeralf	ALFISOL *UF:* Aqualf; Boralf; Luvisol; Podzoluvisol; Udalf; Ustalf; Xeralf	AL'FISOL *UF:* Akvalf; Boralf; Kseralf; Udalf; Ustalf	ALFISOL *UF:* Aqualf; Boralf; Luvisol; Podzoluvisol; Udalf; Ustalf; Xeralf	ALFISOL *UF:* Aqualf; Boralf; Luvisol; Podzoluvisol; Udalf; Ustalf; Xeralf
111 ALGAE PALS	ALGAE *UF:* Euglenophyta; Xanthophyta	ALGAE *UF:* Euglenophycophyta; Xanthophyta	ALGAE *UF:* Euglenophyta; Xanthophyta	VODOROSLI *UF:* Algal bank; Algal biscuit; Algal mat; Bugor vodoroslevyj; Rif vodoroslevyj; Tasmanites; Vodoroslevyj izvestnyak	ALGAE *UF:* Euglenophyta; Xanthophyta	ALGAE *UF:* Euglenophytha; Xanthopyta
112 ALGAL BANK SEDI	ALGAL BANKS	BANC ALGAIRE	ALGEN=BANK	[ALGAL BANK] VODOROSLI; RIF	BANCO-DE-ALGAS	BANCO ALGALE
113 ALGAL BISCUIT SEDI	ALGAL BISCUITS	BISCUIT ALGAIRE	ALGEN=FLACHSTRUKTUR	[ALGAL BISCUIT] VODOROSLI; PORODA OSA-DOCHNAYA	BIZCOCHO=ALGAS	BISCOTTO ALGALE

	ENGLISH	FRANCAIS	DEUTSCH	RUSSKIJ	ESPANOL	ITALIANO
114 ALGAL LIMESTONE SEDS	ALGAL LIMESTONE	CALCAIRE ALGAIRE	ALGEN=KALK	Vodoroslevyj izvestnyak VODOROSLI; IZVESTNYAK	CALIZA=DE=ALGAS	CALCARE ALGALE
115 ALGAL MAT SEDI	ALGAL MATS	MATTE ALGAIRE	ALGEN=MATTE	[ALGAL MAT] VODOROSLI	CONCENTRACION=ALGAS	FELTRO ALGALE
116 ALGAL MOUND SEDI	ALGAL MOUNDS	BUTTE ALGAIRE	ALGEN=DOM	Bugor vodoroslevyj BUGOR; VODOROSLI	CERRO=DE=ALGAS	COSTRUZIONE ALGALE
117 ALGAL REEF SEDI	Algal reefs BIOHERMS; ALGAL STRUCTURES	Recif algue RECIF; STRUCTURE ALGAIRE	Algen=Riff RIFF; ALGEN=STRUKTUR	Rif vodoroslevyj VODOROSLI; BIOGERM	Arrecife=de=algas ARRECIFE; ESTRUCTURA=ALGAREA	Scogliera d'alghe SCOGLIERA; STRUTTURA ALGALE
118 ALGAL STRUCTURE SEDI	ALGAL STRUC-TURES *UF:* Algal reefs	STRUCTURE ALGAIRE *UF:* Recif algue	ALGEN=STRUKTUR *UF:* Algen=Riff	STRUKTURA VODOROSLEVAYA	ESTRUCTURA=ALGAREA *UF:* Arrecife=de=algas	STRUTTURA ALGALE *UF:* Scogliera d'alghe
119 ALGOL MATH	Algol COMPUTER PRO-GRAMS	Algol PROGRAMME ORDINATEUR	Algol RECHENPRO-GRAMM	ALGOL	Algol PROGRAMA=ORDENADOR	Algol PROGRAMMA DI CALCOLO
120 ALGOMAN OROGENY STRU	Algoman orogeny OROGENY; PRE-CAMBRIAN	Orogenie algomienne OROGENIE ANTE-CAMBRIENNE	Algoman.Orogenese PRAEKAMBR. OROGENESE	ALGOMANSKIJ	Orogenia=algomana OROGENIA=PRECAMBRICA	Orogenesi algomaniana OROGENESI PRE-CAMBRIANA
121 ALGORITHM MATH	ALGORITHMS	ALGORITHME	ALGORITHMUS	ALGORITM	ALGORITMO	ALGORITMO
122 ALIPHATIC HYDROCARBON CHES	ALIPHATIC HYDROCARBONS	HYDROCARBURE ALIPHATIQUE	ALIPHAT. KOHLENWASSER-STOFF	UGLEVODOROD ALIFATICHESKIJ	HIDROCARBURO=ALIFATICO	IDROCARBURO ALIFATICO
123 ALKALI-BASALT IGNS	ALKALI BASALTS *UF:* Basanite	BASALTE ALCALIN *UF:* Basanite	ALKALI=BASALT *UF:* Basanit	BAZAL'T SHCHELOCHNOJ	BASALTO=ALCALINO *UF:* Basanita	BASALTO ALCALINO *UF:* Basanite
124 ALKALI-CALCIC COMPOSITION IGNE	Alkali=calcic composi-tion CALC=ALKALIC COMPOSITION	Composition alcalino-calcique COMPOSITION CALCOALCALINE	Alkali=Kalk=Typ KALK=AL-KALI=TYP	IZVESTKOVO=SHCHELOCHNOJ	Composicion=alcalino-calcica COMPOSICION=CALCOALCALINA	Composizione alcalical-cica COMPOSIZIONE CALCALCALINA
125 ALKALI-GRANITE IGNS	ALKALI GRANITES	GRANITE ALCALIN	ALKALI=GRANIT	Shchelochnoj granit GRANIT; SHCHELOCHNOST'	GRANITO=ALCALINO	GRANITO ALCALINO
126 ALKALI-METALS CHES	ALKALI METALS	METAL ALCALIN	ALKALI=METALL	METALL SHCHELOCHNYJ *UF:* Metall shcheloch-nozemel'nyj	METAL=ALCALINO	METALLO ALCALINO
127 ALKALI-SYENITE IGNS	ALKALI SYENITES	SYENITE ALCA-LINE	ALKALI=SYENIT	Shchelchnoj SIENIT; SHCHELOCHNOST'	SIENITA=ALCALINA	SIENITE A FELDS-PATOIDI
128 ALKALIC COMPOSITION IGNE	ALKALIC COMPO-SITION *UF:* Alkaline metasoma-tism; Foidite; Peralka-line composition; Sub-alkalic composition; Unakite	COMPOSITION ALCALINE *UF:* Composition hyper-alcaline; Foidite; Meta-somatose alcaline; Roche subalcaline; Unakite	ALKALI=TYP *UF:* Alkali=Metasomatose; Foidit; Peralkal.Chemismus; Subalkal.Chemismus; Unakit	SHCHELOCHNOJ	COMPOSICION=ALCALINA *UF:* Composicion=peralcalina; Composicion=subalcalina; Foidita; Metasomatismo=alcalino; Unakita	COMPOSIZIONE ALCALINA *UF:* Composizione peral-calina; Composizione subalcalina; Foidite; Metasomatismo alcalino; Unakite

	ENGLISH	FRANCAIS	DEUTSCH	RUSSKIJ	ESPANOL	ITALIANO
129 ALKALINE EARTH METALS CHES	ALKALINE EARTH METALS	METAL ALCALINO TERREUX	ERDALKALI= METALL	Metall shchelochnoze- mel'nyj METALL SHCHELOCHNYJ	METAL= ALCALINOTERREO	METALLI ALCALINO TER- ROSI
130 ALKALINE METASOMATISM IGNE	Alkaline metasomatism ALKALIC COMPO- SITION; METASO- MATISM	Metasomatose alcaline METASOMATOSE; COMPOSITION ALCALINE	Alkali=Metasomatose METASOMATOSE; ALKALI=TYP	METASOMATOZ SHCHELOCHNOJ	Metasomatismo= alcalino METASOMATISMO; COMPOSICION= ALCALINA	Metasomatismo alcalino METASOMATOSI; COMPOSIZIONE ALCALINA
131 ALKALINITY PHCH	ALKALINITY	ALCALINITE	ALKALINITAET	SHCHELOCHNOST' UF: Shchelochnoj sienit; Shcheloch'; Shcheloch- noj granit	ALCALINIDAD	ALCALINITA
132 ALKANES CHES	ALKANES	Alkane COMPOSE HYDRO- CARBURE	Alkan KOHLENWASS- ERSTOFF= VERBINDUNG	Alkan UGLEVODOROD	Alcano COMPUESTO= HIDROCARBURO	Alcani GRUPPO IDROCAR- BURI
133 ALLEGHENY OROGENY STRU	ALLEGHENY OROGENY	OROGENIE ALLE- GHENY	ALLEGHENY= OROGENESE	Alleganskij GERTSINSKIJ	OROGENIA= ALLEGHENY	OROGENESI ALLEGHENIANA
134 ALLOCHEM SEDI	ALLOCHEMS	Allocheme CALCAIRE; MICRO- FACIES	Allochem KALK; MIKRO= FAZIES	ALLOKHEM	Aloquimico CALIZA; MICROFA- CIES	Allochimico CALCARE; MICRO- FACIES
135 ALLOCHTHON GEOL	ALLOCHTHONS	ALLOCHTONIE	ALLOCHTHONIE	ALLOKHTON	ALOCTONIA	ALLOCTONO
136 ALLOGENIC MINERAL MINE	Allogenic mineral AUTHIGENESIS	Mineral allothigene AUTHIGENESE	Allothigen.Mineral AUTHIGENESE	ALLOTIGENNYJ	Mineral=alotigeno AUTIGENESIS	Minerale allotigeno AUTIGENESI
137 ALLOGROMIINA PALS	ALLOGROMIINA	ALLOGROMIINA	ALLOGROMIINA	ALLOGROMIINA	ALLOGROMIINA	ALLOGROMIINA
138 ALLOMETRY PALE	ALLOMETRY	ALLOMETRIE	ALLOMETRIE	ALLOMETRIYA	ALOMETRIA	ALLOMETRIA
139 ALLOY CHES	ALLOYS	ALLIAGE	LEGIERUNG	SPLAV	ALEACION	LEGA
140 ALLUVIAL SURF	Alluvial(adj.)	Alluvial(adj.)	Alluvial(adj.)	ALLYUVIAL'NYJ UF: Razvedka allyuvi- al'naya	Aluvial(adj.)	Alluvionale(adj.)
141 ALLUVIAL FAN SURF	ALLUVIAL FANS	CONE ALLUVION	SCHWEMM- FAECHER	KONUS VYNOSA ALLYUVIAL'NYJ	ABANICOS= FLUVIALES	CONOIDE ALLU- VIONALE
142 ALLUVIAL FILL SURF	FLUVIAL FILL	REMBLAIEMENT FLUVIATILE	AUFSCHOT- TERUNG	OTLOZHENIE ALLYUVIAL'NOE	DEPOSITO= FLUVIAL	RIEMPIMENTO FLUVIALE
143 ALLUVIAL GROUNDWATER GEOH	ALLUVIAL GROUNDWATER	NAPPE ALLUVION UF: Rayon hydraulique	QUARTAER. GRUNDWASSER UF: Hydraul.Radius	VODA GRUN- TOVAYA ALLYU- VIYA	MANTO=ALUVIAL UF: Radio=hidraulico	FALDA IN ALLU- VIONI UF: Raggio idraulico
144 ALLUVIAL PLAIN SURF	ALLUVIAL PLAINS	PLAINE ALLUVI- ALE	ALLUVIAL- FLAECHE	RAVNINA ALLYU- VIAL'NAYA	LLANURA= ALUVIAL	PIANA ALLUVION- ALE

	ENGLISH	FRANCAIS	DEUTSCH	RUSSKIJ	ESPANOL	ITALIANO
145 ALLUVIAL PROSPECTING ECON	ALLUVIAL PROS-PECTING *UF:* Panning	PROSPECTION ALLUVIONNAIRE *UF:* Prospection batee	ALLUVIAL-PROSPEKTION *UF:* Goldwaesche	Razvedka allyuvi-al'naya ALLYUVIAL'NYJ; ROSSYP'	PROSPECCION=ALUVIONAR *UF:* Batea	PROSPEZIONE IN ALLUVIONI *UF:* Lavaggio a batea
146 ALLUVIUM SURF	ALLUVIUM	ALLUVION *UF:* Charge riviere; Fluvent	ALLUVION *UF:* Fluvent; Fluviatil. Fracht	ALLYUVIJ	ALUVION *UF:* Fluvent; Transporte=suspension=fluvial	ALLUVIONE *UF:* Carico fluviale; Fluvent
147 ALPHA RAY PHCH	ALPHA RAYS	RAYONNEMENT ALPHA	ALPHA=STRAHLUNG	AL'FA=LUCHI	RAYOS=ALFA	RAGGI ALFA
148 ALPHA-RAY SPECTROSCOPY METH	ALPHA=RAY SPEC-TROSCOPY	SPECTROMETRIE ALPHA	ALPHA=SPEKTROMETRIE	AL'FA SPEK-TROSKOPIYA	ESPECTROMETRIA=ALFA	SPETTROMETRIA ALFA
149 ALPINE ENVIRONMENT ENVI	ALPINE ENVIRON-MENT	MILIEU ALPIN	ALPIN.MILIEU	AL'PIJSKIJ *UF:* Trias al'pijskij	MEDIO=ALPINO	AMBIENTE ALPINO
150 ALPINE OROGENY STRU	ALPINE OROGENY *UF:* Alpine Structure; Cascadian Orogeny; Cimmerian Orogeny	OROGENIE ALPINE *UF:* Orogenie cimme-rienne; Revolution cascadienne; Tec-tonique alpine	ALPID.OROGENESE *UF:* Alpin.Tektonik; Kaskad.Umbruch; Kimmer.Orogenese	AL'PIJSKIJ /ORO-GEN/ *UF:* Kaskadnyj; Larami-jskij	OROGENIA=ALPINA *UF:* Alpino; Fase=orogenica=cascadiense; Orogenia=cimmerica	OROGENESI ALPINA *UF:* Orogenesi cim-merica; Rivoluzione cascadiana; Tettonica alpina
151 ALPINE TECTONICS STRU	Alpine structure ALPINE OROGENY	Tectonique alpine OROGENIE ALPINE	Alpin.Tektonik ALPID.OROGENESE	TEKTONIKA AL'PINOTIPNAYA	Alpino OROGENIA=ALPINA	Tettonica alpina OROGENESI ALPINA
152 ALPINE TRIASSIC STRS	Alpine triassic TRIASSIC	Trias alpin TRIAS	Alpin.Trias TRIAS	Trias al'pijskij AL'PIJSKIJ; TRIAS	Trias=alpino TRIAS	Triassico alpino TRIASSICO
153 ALPINE-TYPE STRU	ALPINE=TYPE	TYPE ALPIN	ALPINOTYP. BAUSTIL	AL'PINOTIPNYJ	TIPO=ALPINO	ALPINOTIPO
154 ALTERATION SURF	ALTERATION	ALTERATION	VERWITTERUNG	IZMENENIE	ALTERACION	ALTERAZIONE
155 ALTITUDE MISC	ALTITUDE	ALTITUDE	ERHEBUNG	VYSOTA	ALTITUD	ALTITUDINE
156 ALUMINA CHES	ALUMINA	ALUMINE	TONERDE	GLINOZEM	ALUMINA	ALLUMINA
157 ALUMINUM CHEE	ALUMINUM	ALUMINIUM *UF:* Alunitisation; Com-position hyperalu-mineuse; Composition subalumineuse	AL *UF:* Alunitisierung; Peraluminat=Gehalt; Subaluminoes. Chemismus	ALYUMINIJ	ALUMINIO *UF:* Alunitizacion; Composicion=peraluminosa; Composicion=subaluminosa	ALLUMINIO *UF:* Alunitizzazione; Composizione iperal-luminifera; Composiz-ione ipoalluminifera
158 ALUNITIZATION IGNE	ALUNITIZATION	Alunitisation ALTERATION HYDROTHERMALE; ALUMINIUM	Alunitisierung HYDROTHERMAL-UMWANDLUNG; AL	ALUNITIZATSIYA	Alunitizacion ALTERACION=HIDROTERMAL; ALUMINIO	Alunitizzazione ALTERAZIONE IDROTERMALE; ALLUMINIO
159 ALVEOLINELLIDAE PALS	ALVEOLINELLIDAE	ALVEOLINELLIDAE	ALVEOLINELLIDAE	ALVEOLINELLIDAE	ALVEOLINELLIDAE	ALVEOLINELLIDAE

	ENGLISH	FRANCAIS	DEUTSCH	RUSSKIJ	ESPANOL	ITALIANO
160 AMBER SEDS	AMBER	Ambre RESINE	Bernstein FOSSIL.HARZ	YANTAR'	Ambar RESINA	Ambra RESINA
161 AMBLYPODA PALS	AMBLYPODA	AMBLYPODA	AMBLYPODA	Amblypoda PLACENTALIA	AMBLYPODA	AMBLYPODA
162 AMERICIUM CHEE	AMERICIUM	AMERICIUM	AM	AMERITSIJ	AMERICIO	AMERICIO
163 AMINO ACID CHES	AMINO ACIDS	ACIDE AMINE	AMINO=SAEURE	AMINOKISLOTA	AMINOACIDO	AMMINOACIDO
164 AMMODISCACEA PALS	AMMODISCACEA	AMMODISCACEA	AMMODISCACEA	AMMODISCACEA	AMMODISCACEA	AMMODISCACEA
165 AMMONIA CHES	AMMONIA	COMPOSE AMMONIAQUE	AMMONIAK	AMMIAK	AMONIACO	AMMONIACA
166 AMMONITIDA PALS	Ammonitida CEPHALOPODA	Ammonitida AMMONOIDEA	Ammonitida AMMONOIDEA	AMMONITIDA	Ammonitida AMMONOIDEA	Ammonitida AMMONOIDEA
167 AMMONITINA PALS	Ammonitina CEPHALOPODA	Ammonitina AMMONOIDEA	Ammonitina AMMONOIDEA	AMMONITINA *UF:* Desmoceratida; Perisphinctida; Psiloc- eratida	Ammonitina AMMONOIDEA	Ammonitina AMMONOIDEA
168 AMMONIUM CHES	AMMONIUM	AMMONIUM	AMMONIUM	AMMONIJ	AMONIO	AMMONIO
169 AMMONOIDEA PALS	AMMONOIDEA	AMMONOIDEA *UF:* Agoniatitida; Ammonitida; Ammoni- tina; Bactritoidea	AMMONOIDEA *UF:* Agoniatitida; Ammonitida, Ammoni- tina; Bactritoidea	AMMONOIDEA	AMMONOIDEA *UF:* Agoniatitida; Am- monitida, Ammonitina, Bactritoidea	AMMONOIDEA *UF:* Agoniatitida; Ammonitida, Ammoni- tina; Bactritida
170 AMORPHOUS MINERAL TEST	AMORPHOUS MATERIALS	MATIERE AMOR- PHE	AMORPH. SUBSTANZ	AMORFNYJ	AMORFO	MATERIA AMORFA
171 AMPHIBIA PALS	AMPHIBIA *UF:* Apsidospondyli; Urodela	AMPHIBIA *UF:* Apsidospondyli; Batrachosauria	AMPHIBIA *UF:* Apsidospondyli; Batrachosauria	AMPHIBIA *UF:* Lissamphibia	AMPHIBIA *UF:* Apsidospondyli; Batrachosauria	AMPHIBIA *UF:* Apsidospondyli; Batrachosauria
172 AMPHIBOLE MING	AMPHIBOLE	AMPHIBOLE	AMPHIBOL	AMFIBOL	ANFIBOL	ANFIBOLI
173 AMPHIBOLITE IGMS	AMPHIBOLITES *UF:* Hornblendite	AMPHIBOLITE *UF:* Amphibolitisation; Hornblendite	AMPHIBOLIT *UF:* Amphibolitisierung; Hornblendit	AMFIBOLIT *UF:* Ortoamfibolit; Paraamfibolit	ANFIBOLITA *UF:* Anfibolitizacion; Hornblendita	ANFIBOLITE *UF:* Anfibolitizzazione; Orneblendite
174 AMPHIBOLITE FACIES IGNE	AMPHIBOLITE FACIES *UF:* Cordierite= amphibolite=facies	FACIES AMPHIBO- LITE *UF:* Facies amphibolite albite epidote; Facies amphibolite cordierite	AMPHIBOLIT= FAZIES *UF:* Albit=Epidot= Amphibolit=Fazies; Cordierit=Amphibolit= Fazies	FATSIYA AMFI- BOLITOVAYA *UF:* Fatsiya kordierit= amfibolitovaya	FACIES= ANFIBOLITA *UF:* Facies=anfibolita= albita=epidota; Facies= anfibolita=cordierita	FACIES DELLE ANFIBOLITI *UF:* Facies anfiboliti a cordierite; Facies anfiboliti albite epidoto
175 AMPHIBOLITIZATION IGNE	AMPHIBOLITIZA- TION	Amphibolitisation ALTERATION HYDROTHERMALE; AMPHIBOLITE	Amphibolitisierung HYDROTHERMAL= UMWANDLUNG; AMPHIBOLIT	AMFIBOLIZATSIYA	Anfibolitizacion ALTERACION= HIDROTERMAL; ANFIBOLITA	Anfibolitizzazione ALTERAZIONE IDROTERMALE; ANFIBOLITE

	ENGLISH	FRANCAIS	DEUTSCH	RUSSKIJ	ESPANOL	ITALIANO
176 AMPHINEURA PALS	Amphineura POLYPLACOPHORA	AMPHINEURA *UF:* Polyplacophora	AMPHINEURA *UF:* Polyplacophora	AMPHINEURA *UF:* Aplacophora; Poly- placophora	AMPHINEURA *UF:* Polyplacophora	AMPHINEURA *UF:* Polyplacophora
177 AMPLITUDE PHCH	AMPLITUDE	AMPLITUDE	AMPLITUDE	AMPLITUDA	AMPLITUD	AMPIEZZA
178 ANAEROBIC CONDITION PALE	ANAEROBIC ENVI- RONMENT *UF:* Euxinic environment	MILIEU ANAERO- BIE *UF:* Milieu euxinique	ANAEROB.MILIEU *UF:* Euxin.Milieu	ANAEHROBNYJ *UF:* Takson anaehrobnyj	MEDIO= ANAEROBIO *UF:* Medio=euxinico	AMBIENTE ANAEROBICO *UF:* Ambiente euxinico
179 ANAEROBIC TAXON PALE	ANAEROBIC TAXA	TAXON ANAERO- BIE	ANAEROB.TAXON	Takson anaehrobnyj ANAEHROBNYJ; TAKSONOMIYA	TAXON= ANAEROBIO	TAXON ANAERO- BICO
180 ANALOG SIMULATION MATH	ANALOG SIMULA- TION *UF:* Analog techniques	SIMULATION ANALOGIQUE *UF:* Technique analogique	ANALOG= SIMULATION *UF:* Analog=Technik	MODELIROVANIE ANALOGOVOE	SIMULACION= ANALOGICA *UF:* Tecnica=analogica	SIMULAZIONE ANALOGICA *UF:* Tecnica analogica
181 ANALOG TECHNIQUE MATH	Analog techniques ANALOG SIMULA- TION	Technique analogique SIMULATION ANALOGIQUE	Analog=Technik ANALOG= SIMULATION	TEKHNIKA ANALO- GOVAYA	Tecnica=analogica SIMULACION= ANALOGICA	Tecnica analogica SIMULAZIONE ANALOGICA
182 ANALYSIS METH	ANALYSIS	Analyse(gen.)	Analyse(gen.)	ANALIZ *UF:* Analiz rasseyannykh ehlementov; Assay value; Petrographic analysis	Analisis(gen.)	Analisi(gen.)
183 ANALYTICAL METHOD METH	CHEMICAL ANAL- YSIS *UF:* Acetolysis; Dialysis; Inert component; Iso- cons	METHODOLOGIE ANALYSE *UF:* Acetolyse; Capteur	ANALYSEN= METHODIK *UF:* Mess=Geraet	METOD ANALITI- CHESKIJ *UF:* Vacuum fusion analysis	METODOLOGIA= ANALISIS *UF:* Prueba	METODO ANALI- TICO
184 ANAPSIDA PALS	ANAPSIDA	ANAPSIDA	ANAPSIDA	Anapsida REPTILIA	ANAPSIDA	ANAPSIDA
185 ANASPIDA PALS	Anaspida AGNATHA	Anaspida AGNATHA	Anaspida AGNATHA	ANASPIDA	Anaspida AGNATHA	Anaspida AGNATHA
186 ANATEXIS IGNE	ANATEXIS *UF:* Metatexis; Syntexis	ANATEXIE *UF:* Metatexie; Syntexie	ANATEXIS *UF:* Metatexis; Syntexis	ANATEKSIS *UF:* Plavlenie chastich- noe	ANATEXIA *UF:* Metatexia; Sintexia	ANATESSI *UF:* Metatessi; Sintessi
187 ANATEXITE IGMS	ANATEXITE	ANATEXITE	ANATEXIT	ANATEKSIT	ANATEXITA	ANATESSITE
188 ANATOMY PALE	ANATOMY *UF:* Gastrozooid; Zooid	ANATOMIE *UF:* Byssus; Gastrozo- oide; Zooide	ANATOMIE *UF:* Byssus; Gastrozooid; Zooid	ANATOMIYA *UF:* Sutura	ANATOMIA *UF:* Biso; Gastrozoide; Zooide	ANATOMIA *UF:* Bisso; Gastrozooide; Zooide
189 ANCHIMETAMORPHISM IGNE	ANCHIMETAMOR- PHISM	ANCHIMETAMOR- PHISME	ANCHI= METAMORPHOSE	Ankhimetamorfizm KATAGENEZ	ANCHIMETAMOR- FISMO	ANCHIMETAMOR- FISMO
190 ANCHORING ENGI	ANCHORS	ANCRAGE	VERANKERUNG	ZAYAKORIVANIE	ANCLAJE	ANCORAGGIO
191 ANCIENT ICE AGE STRA	ANCIENT ICE AGES	AGE GLACIAIRE ANCIEN	NICHT= QUARTAER. EISZEIT	DREVNELED- NIKOVYJ	EDAD= GLACIACION	GLACIALE ANTICO

	ENGLISH	FRANCAIS	DEUTSCH	RUSSKIJ	ESPANOL	ITALIANO
192 ANDALUSITE COMS	ANDALUSITE	ANDALOUSITE	ANDALUSIT	ANDALUZIT	ANDALUCITA	ANDALUSITE
193 ANDEPT SUSS	Andepts ANDOSOLS	Andept INCEPTISOL; ANDOSOL	Andept INCEPTISOL; ANDOSOL	Andept INSEPTISOL	Andept INCEPTISOL; ANDOSOL	Andept INCEPTISOL; ANDOSOL
194 ANDESITE IGNS	ANDESITES	ANDESITE	ANDESIT	ANDEZIT	ANDESITA	ANDESITE
195 ANDESITE LINE IGNE	ANDESITE LINE	Ligne marshall ASSOCIATION MAGMATIQUE	Andesit=Linie MAGMAT. VERGESELLSCHAF- TUNG	LINIYA ANDEZI- TOVAYA	Linea=de=la=andesita ASOCIACION= MAGMATICA	Linea delle andesiti ASSOCIAZIONE MAGMATICA
196 ANDOSOL SUSS	ANDOSOLS *UF:* Andepts	ANDOSOL *UF:* Andept	ANDOSOL *UF:* Andept	ANDOSOL	ANDOSOL *UF:* Andept	ANDOSOL *UF:* Andept
197 ANELASTIC MEDIUM SOLI	ANELASTIC MATE- RIALS	MILIEU ANELAS- TIQUE	ANELAST.MEDIUM	[ANELASTIC MEDIUM]	MEDIO= INELASTICO	MEZZO ANELAS- TICO
198 ANELASTICITY PHCH	ANELASTICITY	ANELASTICITE	UNELASTIZITAET	Neuprugost' UPRUGOST'	INELASTICIDAD	ANELASTICITA
199 ANGIOSPERMAE PALS	ANGIOSPERMS	ANGIOSPERMAE	ANGIOSPERMAE	ANGIOSPERMAE *UF:* Dicotyledoneae; Monocotyledoneae	ANGIOSPERMAE	ANGIOSPERMAE
200 ANGLE OF REPOSE ENGI	ANGLE OF REPOSE	Talus equilibre VERSANT; GEOME- TRIE	Schuettungs=Winkel HANG; GEOMETRIE	UGOL ESTESTVEN- NOGO OTKOSA	Angulo=de=reposo LADERA; GEO- METRIA	Angolo di riposo VERSANTE; GEO- METRIA
201 ANGRITE EXTS	Angrite ACHONDRITES	Angrite ACHONDRITE	Angrit ACHONDRIT	Angrit AKHONDRIT	Angreita ACONDRITA	Angrite ACONDRITE
202 ANGULAR UNCONFORMITY STRU	ANGULAR UNCON- FORMITIES *UF:* Discordance	DISCORDANCE ANGULAIRE *UF:* Discordance strati- graphique; Discordance tectonique	WINKEL= DISKORDANZ *UF:* Nonconformity; Tekton.Diskordanz	NESOGLASIE UGLOVOE *UF:* Nesoglasie struk- turnoe	DISCORDANCIA= ANGULAR *UF:* Discordancia= igneo=sedimentaria; Discordancia=tectonica	DISCORDANZA ANGOLARE *UF:* Discordanza sem- plice; Discordanza tettonica
203 ANHEDRAL CRYSTAL MINE	ANHEDRAL CRYS- TALS	Mineral xenomorphe HABITUS	Anhedr.Kristall KORN=GEFUEGE	Angedral'nyj KSENOMORFNYJ	Alotriomorfo HABITO	Cristallo anedrale ABITO CRISTAL- LINO
204 ANHYDRITE COMS	ANHYDRITE	ANHYDRITE	ANHYDRIT	ANGIDRIT	ANHIDRITA	ANIDRITE
205 ANHYDROUS COMPOSITION GEOC	ANHYDROUS MATERIALS	Substance anhydre DESHYDRATATION	Anhydrid=Chemismus DEHYDRISIERUNG	BEZVODNYJ	Anhidro DESHIDRATACION	Composizione anidra DISIDRATAZIONE
206 ANHYSTERETIC REMANENT MAGNETIZATION PHCH	ANHYSTERETIC REMANENT MAG- NETIZATION	AIMANTATION REMANENTE ANHYSTERETIQUE	ANHYSTERET. REMAN. MAGNETISIERUNG	Namagnichennost' ideal'naya NAMAGNICHEN- NOST' OSTATOCH- NAYA	IMANTACION= REMANENTE= ANHISTERETICA	MAGNETIZZAZ- IONE ANISTERE- TICA RESIDUA
207 ANION PHCH	ANIONS	ANION	ANION	Anion ION	ANION	ANIONE

	ENGLISH	FRANCAIS	DEUTSCH	RUSSKIJ	ESPANOL	ITALIANO
208 ANISIAN STRS	ANISIAN	ANISIEN	ANIS	ANIZIJ	ANISIENSE	ANISICO
209 ANISOPLEURA PALS	Anisopleura GASTROPODA	Anisopleura STROPHOMENIDA	Anisopleura STROPHOMENIDA	ANISOPLEURA	Anisopleura STROPHOMENIDA	Anisopleura STROPHOMENIDA
210 ANISOTROPY PHCH	ANISOTROPY	ANISOTROPIE *UF:* Mineral biaxe	ANISOTROPIE *UF:* Zweiachsig.Kristall	ANIZOTROPIYA	ANISOTROPIA *UF:* Mineral=biaxial	ANISOTROPIA *UF:* Cristallo biassiale
211 ANKERITE SEDS	Ankerite SEDIMENTARY ROCKS	Ankerite CALCAIRE	Ankerit=Gestein KALK	ANKERIT	Roca=de=ankerita CALIZA	Ankerite CALCARE
212 ANNELIDA PALS	Annelida WORMS	Annelida VERMES	Annelida VERMES	ANNELIDA *UF:* Polychaetia	Annelida VERMES	Annelida VERMES
213 ANOMALODESMATA PALS	Anomalodesmata BIVALVIA	Anomalodesmata BIVALVIA	Anomalodesmata LAMELLI- BRANCHIATA	ANOMALODES- MATA	Anomalodesmata BIVALVIA	Anomalodesmata BIVALVIA
214 ANOMALY PHCH	ANOMALIES *UF:* Geophysical anomaly; High value anomaly	ANOMALIE *UF:* Anomalie geophysique; Depression geochimique; Maximum anomalie	ANOMALIE *UF:* Geochem. Einbruchskrater; Geophysikal.Anomalie; Hoher=Anomalie=Wert	ANOMALIYA	ANOMALIA *UF:* Anomalia=geofisica; Hundimiento=geoquimico; Maximo=anomalia	ANOMALIA *UF:* Abbattimento geochimico; Anomalia geofisica; Massimo di anomalia
215 ANOROGENIC PROCESS GEOL	ATECTONIC PRO- CESSES	PROCESSUS ATEC- TONIQUE	ATEKTON. BILDUNG	ANOROGENNYJ	PROCESO= ATECTONICO	PROCESSO ATET- TONICO
216 ANORTHOSITE IGNS	ANORTHOSITE	ANORTHOSITE *UF:* Anorthositisation	ANORTHOSIT *UF:* Anorthositisierung	ANORTOZIT	ANORTOSITA *UF:* Anortositizacion	ANORTOSITE *UF:* Anortositizzazione
217 ANORTHOSITIZATION IGNE	ANORTHOSITIZA- TION	Anorthositisation ALTERATION HYDROTHERMALE; ANORTHOSITE	Anorthositisierung HYDROTHERMAL= UMWANDLUNG; ANORTHOSIT	ANORTOZITIZAT- SIYA	Anortositizacion ALTERACION= HIDROTERMAL; ANORTOSITA	Anortositizzazione ALTERAZIONE IDROTERMALE; ANORTOSITE
218 ANTECLISE STRU	ANTECLISE	ANTECLISE	ANTEKLISE	ANTEKLIZA	ANTICLISIO	ANTECLISI
219 ANTHOCYATHEA PALS	Anthocyathea ARCHAEOCYATHA	Anthocyathea ARCHAEOCYATHA	Anthocyathea ARCHAEOCYATHA	ANTHOCYATHEA	Anthocyathea ARCHAEOCYATHA	Anthocyathea ARCHAEOCYATA
220 ANTHOZOA PALS	ANTHOZOA *UF:* Corallite	ANTHOZOA *UF:* Corallite	ANTHOZOA *UF:* Corallit	ANTHOZOA *UF:* Actiniaria; Antipatharia; Ceriantharia; Ceriantipatharia; Corallimorpharia; Heterocorallia; Hexactiniaria; Zoanthiniaria	ANTHOZOA *UF:* Coralico	ANTHOZOA *UF:* Corallite
221 ANTHRACITE COMS	ANTHRACITE	ANTHRACITE	ANTHRAZIT	ANTRATSIT	ANTRACITA	ANTRACITE
222 ANTHRAXOLITE SEDS	ANTHRAXOLITE	Anthraxolite BITUME	Anthraxolith BITUMEN	Antraksolit ASFAL'TIT	Antraxolita BETUN	Antraxolite BITUME
223 ANTIARCHI PALS	Antiarchi PLACODERMI	Antiarchi PLACODERMI	Antiarchi PLACODERMI	ANTIARCHI	Antiarchi PLACODERMI	Antiarchi PLACODERMI

	ENGLISH	FRANCAIS	DEUTSCH	RUSSKIJ	ESPANOL	ITALIANO
224 ANTICLINE STRU	ANTICLINES *UF:* Nose (fold); Periclines; Placanticlines	ANTICLINAL *UF:* Nose; Periclinal; Placanticlinal	ANTIKLINALE *UF:* Falten=Stirn; Perikline; Placanticline	ANTIKLINAL'	ANTICLINAL *UF:* Periclinal; Placa=anticlinal; Promontorio=anticlinal	ANTICLINALE *UF:* Nose; Periclinale; Placanticline
225 ANTICLINORIUM STRU	ANTICLINORIA	ANTICLINORIUM	ANTIKLINORIUM	ANTIKLINORIJ	ANTICLINORIO	ANTICLINORIO
226 ANTIDUNE SURF	ANTIDUNES	ANTIDUNE	GEGENRIPPEL	ANTIDYUNA	ANTIDUNA	ANTIDUNA
227 ANTIFORM STRU	ANTIFORM FOLDS	PLI ANTIFORME	ANTIFORM	Antiforma STRUKTURA POLOZHITEL'NAYA	PLIEGUE=ANTIFORME	ANTIFORME
228 ANTIMONATES MING	ANTIMONATES	ANTIMONIATE	ANTIMONIAT	ANTIMONAT	ANTIMONIATO	ANTIMONIATO
229 ANTIMONY CHEE	ANTIMONY	ANTIMOINE	SB	SUR'MA	ANTIMONIO	ANTIMONIO
230 ANTIPATHARIA PALS	Antipatharia CERIANTIPA-THARIA	Antipatharia CERIANTIPA-THARIA	Antipatharia CERIANTIPA-THARIA	Antipatharia ANTHOZOA	Antipatharia CERIANTIPA-THARIA	Antipatharia CERIANTIPA-THARIA
231 ANTIPERTHITE TEST	ANTIPERTHITE	ANTIPERTHITE	ANTIPERTHIT	Antipertit PERTIT	ANTIPERTITA	ANTIPERTITE
232 ANTISTRESS MINERAL MINE	ANTISTRESS MINERALS	Mineral antistress PROPRIETE MECANIQUE	Antistress=Mineral MECHAN. EIGENSCHAFT	[ANTISTRESS=MINERAL]	Mineral=antitensional PROPIEDAD=MECANICA	Minerale antistress PROPRIETA MECCANICA
233 ANTITHETIC FAULT STRU	Antithetic faults FAULTS	Faille contraire FAILLE	Antithet. Verwerfung STOERUNG	ANTITETICHESKIJ	Falla=antitetica FALLA	Faglia antitetica FAGLIA
234 ANUROMORPHA PALS	ANUROMORPHA	ANUROMORPHA	ANUROMORPHA	ANUROMORPHA	ANUROMORPHA	ANUROMORPHA
235 APERTURE PALE	APERTURES	Ouverture ANATOMIE SQUE-LETTE	Apertur SKELETT	APERTURA *UF:* Ust'e golostomnoe	Apertura ANATOMIA=ESQUELETO	Apertura ANATOMIA DELLO SCHELETRO
236 APHANITIC TEXTURE TEST	APHANITIC TEXTURE	TEXTURE APHANI-TIQUE	APHANIT.GEFUEGE	AFANITOVYJ *UF:* Skrytokristallicheskij	TEXTURA=AFANITICA	TESSITURA AFANI-TICA
237 APICAL MISC	Apical(adj.)	Apical(adj.)	Apikal(adj.)	APIKAL'NYJ	Apical(adj.)	Apicale(adj.)
238 APLACOPHORA PALS	APLACOPHORA	APLACOPHORA	APLACOPHORA	Aplacophora AMPHINEURA	APLACOPHORA	APLACOPHORA
239 APLITE IGNS	APLITE *UF:* Aplitic texture	APLITE *UF:* Texture aplitique	APLIT *UF:* Aplit=Gefuege	APLIT	APLITA *UF:* Textura=aplitica	APLITE *UF:* Tessitura aplitica
240 APLITIC TEXTURE TEST	Aplitic texture APLITE	Texture aplitique APLITE	Aplit=Gefuege APLIT	APLITOVYJ	Textura=aplitica APLITA	Tessitura aplitica APLITE
241 APOGEE EXTR	APOGEE	Apogee GEOLOGIE EXTRA TERRESTRE	Apogaeum EXTRATERRESTR. GEOLOGIE	APOGEJ	Apogeo GEOLOGIA=EXTRATERRESTRE	Apogeo GEOLOGIA EXTRATERRESTRE

	ENGLISH	FRANCAIS	DEUTSCH	RUSSKIJ	ESPANOL	ITALIANO
242 APOGRANITE IGNE	APOGRANITE	APOGRANITE	APOMAGMAT. GESTEIN	APOGRANIT	APOGRANITO	APOGRANITO
243 APOLLO PROGRAM EXTR	APOLLO PROGRAM	PROGRAMME APOLLO	APOLLO= PROGRAMM	APOLLON PRO- GRAMMA	PROGRAMA= APOLO	PROGRAMMA APOLLO
244 APOMAGMATIC DEPOSIT ECON	Apomagmatic deposit HYDROTHERMAL PROCESSES	Gite apomagmatique GITE HYDROTHER- MAL	Apomagmat. Vorkommen HYDROTHERMAL. VORKOMMEN	APOMAGMATI- CHESKIJ	Deposito= apomagmatico YACIMIENTO= HIDROTERMAL	Giacimento apomagma- tico GIACIMENTO IDROTERMALE
245 APOPHYSIS IGNE	Apophysis INTRUSIONS	Apophyse INTRUSION	Apophyse INTRUSIONS= KOERPER	APOFIZA	Apofisis INTRUSION	Apofisi INTRUSIONE
246 APPLIED GEOLOGY GEOL	Applied geology GEOLOGY	Geologie appliquee GEOLOGIE	Angewandt.Geologie GEOLOGIE	Geologiya prikladnaya GEOLOGIYA	Geologia=aplicada GEOLOGIA	Geologia applicata GEOLOGIA
247 APPLIED GEOPHYSICS APPL	APPLIED GEOPHYS- ICS	GEOPHYSIQUE APPLIQUEE	ANGEWANDT. GEOPHYSIK	GEOFIZIKA RAZVE- DOCHNAYA	GEOFISICA= APLICADA	GEOFISICA APPLI- CATA
248 APPLIED HYDROGEOLOGY GEOH	APPLIED HYDROGEOLOGY	HYDROGEOLOGIE APPLIQUEE	ANGEWANDT. HYDROGEOLOGIE	GIDROGEOLOGIYA (PRIKLADNAYA)	OBJETIVO= HIDROGEOLOGICO	IDROGEOLOGIA APPLICATA
249 APPROXIMATION MATH	APPROXIMATION	Approximation METHODE STATIS- TIQUE	Naeherung MATHEMAT. METHODE	APPROKSIMATSIYA	Aproximacion METODO= ESTADISTICO	Approssimazione METODO STATIS- TICO
250 APRON SURF	Apron OUTWASH PLAINS	Bordure detritique CONE DEJECTION	Sander SCHUTT=KEGEL	SHLEJF	Abanico=fluvioglaciar CONO=DE= DEYECCION	Fascia detritica CONO DI DETRITO
251 APSHERONIAN STRS	Apsheronian VILLAFRANCHIAN	Apsheronien VILLAFRANCHIEN	Apsheronium VILLAFRANCIUM	APSHERON	Apsheroniense VILLAFRAN- QUIENSE	Apsheroniano VILLAFRANCHI- ANO
252 APSIDOSPONDYLI PALS	Apsidospondyli AMPHIBIA	Apsidospondyli AMPHIBIA	Apsidospondyli AMPHIBIA	APSIDOSPONDYLI	Apsidospondyli AMPHIBIA	Apsidospondyli AMPHIBIA
253 APTIAN STRS	APTIAN	APTIEN	APT	APT	APTIENSE	APTIANO
254 AQUALF SUSS	Aqualfs ALFISOLS	Aqualf ALFISOL; PSEU- DOGLEY	Aqualf ALFISOL; PSEUDO- GLEY	Akvalf AL'FISOL	Aqualf ALFISOL; PSEU- DOGLEY	Aqualf ALFISOL; PSEU- DOGLEY
255 AQUENT SUSS	Aquents ENTISOLS	Aquent ENTISOL; MILIEU HUMIDE	Aquent ENTISOL; FEUCHT. MILIEU	Akvent EHNTISOL	Aquent ENTISOL; MEDIO- HUMEDO	Aquent ENTISOL; AMBI- ENTE UMIDO
256 AQUEOUS SOLUTION PHCH	AQUEOUS SOLU- TIONS	SOLUTION AQUEUSE	WAESSRIG. LOESUNG	RASTVORENIE	SOLUCION= ACUOSA	SOLUZIONE ACQU- OSA
257 AQUEPT SUSS	Aquepts INCEPTISOLS	Aquept INCEPTISOL; PSEU- DOGLEY	Aquept INCEPTISOL; PSEUDO=GLEY	Akvept INSEPTISOL	Aquept INCEPTISOL; PSEU- DOGLEY	Aquept INCEPTISOL
258 AQUERT SUSS	Aquerts VERTISOLS	Aquert VERTISOL; MILIEU HUMIDE	Aquert VERTISOL; FEUCHT.MILIEU	Akvert EHNTISOL	Aquert VERTISOL; MEDIO- HUMEDO	Aquert VERTISUOLO; AMBIENTE UMIDO

	ENGLISH	FRANCAIS	DEUTSCH	RUSSKIJ	ESPANOL	ITALIANO
259 AQUICLUDE GEOH	Aquicludes AQUIFERS; PERME-ABILITY	Aquiclude NAPPE EAU; PER-MEABILITE	Aquiclud GRUNDWASSER=LEITER; PERMEA-BILITAET	VODOUPOR *UF:* Vodoupor pronitsae-myj	Capa=confinante ACUIFERO; PER-MEABILIDAD	Strato impermeabile ACQUIFERO; PER-MEABILITA
260 AQUIFER GEOH	AQUIFERS *UF:* Aquicludes; Aqui-fuges; Aquitards; Arte-sian aquifer; Hydraulic radius	NAPPE EAU *UF:* Aquiclude; Couche impermeable; Couche semipermeable	GRUNDWASSER=LEITER *UF:* Aquiclud; Aquitard; Grundwasser=Stauer	GORIZONT VODONOSNYJ *UF:* Multilayer system	ACUIFERO *UF:* Acuifugo; Capa=confinante; Capa=semipermeable	ACQUIFERO *UF:* Aquitardo; Livello impermeabile; Strato impermeabile
261 AQUIFER TEST GEOH	Aquifer test PUMP TESTS	ESSAI DEBIT *UF:* Pompage essai	PUMP=VERSUCH *UF:* Pump=Test	ISPYTANIE VODONOSNOGO GORIZONTA	ENSAYO=BOMBEO *UF:* Bombeo=de=ensayo	PROVA DI PORTATA *UF:* Prova di pompaggio
262 AQUIFUGE GEOH	Aquifuges AQUIFERS; PERME-ABILITY	Couche impermeable NAPPE EAU; PER-MEABILITE	Grundwasser=Stauer GRUNDWASSER=LEITER; PERMEA-BILITAET	VODOUPOR NEPRONITSAEMYJ	Acuifugo ACUIFERO; PER-MEABILIDAD	Livello impermeabile ACQUIFERO; PER-MEABILITA
263 AQUITANIAN STRS	AQUITANIAN	AQUITANIEN	AQUITAN	AKVITAN	AQUITANIENSE	AQUITANIANO
264 AQUITARD GEOH	Aquitards AQUIFERS; PERME-ABILITY	Couche semipermeable NAPPE EAU; PER-MEABILITE	Aquitard GRUNDWASSER=LEITER; PERMEA-BILITAET	Vodoupor pronitsaemyj VODOUPOR	Capa=semipermeable ACUIFERO; PER-MEABILIDAD	Aquitardo ACQUIFERO; PER-MEABILITA
265 AQUOD SUSS	Aquods SPODOSOLS	Aquod SPODOSOL; GLEY	Aquod SPODOSOL; GLEY	Akvod SPODOSOL	Aquod SPODOSOL; GLEY	Aquod SPODOSOL; GLEY
266 AQUOLL SUSS	Aquolls MOLLISOLS	Aquoll MOLLISOL; GLEY	Aquoll MOLLISOL; GLEY	Akvoll MOLLISOL	Aquoll MOLLISOL; GLEY	Aquoll MOLLISOL; GLEY
267 AQUOX SUSS	Aquox OXISOLS	Aquox OXISOL; MILIEU HUMIDE	Aquox OXISOL; FEUCHT. MILIEU	Akvoks OKSISOL	Aquox OXISOL; MEDIO=HUMEDO	Aquox OXISOL; AMBIENTE UMIDO
268 AQUULT SUSS	Aquults ULTISOLS	Aquult ULTISOL; MILIEU HUMIDE	Aquult ULTISOL; FEUCHT. MILIEU	Akvult UL'TISOL	Aquult ULTISOL; MEDIO=HUMEDO	Aquult ULTISOL; AMBI-ENTE UMIDO
269 ARACHNIDA PALS	ARACHNIDA	ARACHNIDA	ARACHNIDA	ARACHNIDA	ARACHNIDA	ARACHNIDA
270 ARCH STRU	ARCHES	STRUCTURE ARQUEE	GIRLANDEN=FORM	SVOD	ESTRUCTURA=ARQUEADA	STRUTTURA AD ARCO
271 ARCH DAM ENGI	ARCH DAMS	BARRAGE VOUTE	BOGEN=STAUDAMM	Plotina arochnaya PLOTINA	PRESA=ARQUEADA	DIGA AD ARCO
272 ARCHAEOCOPIDA PALS	ARCHAEOCOPIDA	ARCHAEOCOPIDA	ARCHAEOCOPIDA	Archaeocopida OSTRACODA	ARCHAEOCOPIDA	ARCHEOCOPIDA
273 ARCHAEOCYATHA PALS	ARCHAEOCYATHA *UF:* Anthocyathea; Archaeocyathea; Mono-cyathea	ARCHAEOCYATHA *UF:* Anthocyathea; Archaeocyathea; Mono-cyathea	ARCHAEOCYATHA *UF:* Anthocyathea; Archaeocyathea; Mono-cyathea	ARCHAEOCYATHA	ARCHAEOCYATHA *UF:* Anthocyathea; Archaeocyathea; Mono-cyathea	ARCHAEOCYATA *UF:* Anthocyathea; Archaeocyathea; Mono-cyathea
274 ARCHAEOCYATHEA PALS	Archaeocyathea ARCHAEOCYATHA	Archaeocyathea ARCHAEOCYATHA	Archaeocyathea ARCHAEOCYATHA	ARCHAEOCYA-THEA	Archaeocyathea ARCHAEOCYATHA	Archaeocyathea ARCHAEOCYATA

	ENGLISH	FRANCAIS	DEUTSCH	RUSSKIJ	ESPANOL	ITALIANO
275 ARCHAEOGASTROPODA PALS	ARCHAEO-GASTROPODA	ARCHAEO-GASTROPODA	ARCHAEO-GASTROPODA	Archaeogastropoda GASTROPODA	ARCHAEO-GASTROPODA	ARCHAEO-GASTROPODA
276 ARCHAEOPTERYGES PALS	Archaeopteryges ARCHAEORNITHES	Archaeopteryges ARCHAEORNITHES	Archaeopteryges ARCHAEORNITHES	ARCHAEOPTERY-GES	Archaeopteryges ARCHAEORNITHES	Archaeopteryges ARCHAEORNITHES
277 ARCHAEORNITHES PALS	ARCHAEORNITHES *UF:* Archaeopteryges; Saurornithes	ARCHAEORNITHES *UF:* Archaeopteryges; Saurornithes	ARCHAEORNITHES *UF:* Archaeopteryges; Saurornithes	Archaeornithes AVES	ARCHAEORNITHES *UF:* Archaeopteryges; Saurornithes	ARCHAEORNITHES *UF:* Archaeopteryges; Saurornithes
278 ARCHEAN STRS	ARCHEAN	ARCHEEN	ARCHAIKUM	ARKHEJ	ARCAICO	ARCHEANO
279 ARCHEIDES STRU	Archeides OROGENY; PRE-CAMBRIAN	Archeides OROGENIE ANTE-CAMBRIENNE	Archaeiden PRAEKAMBR. OROGENESE	ARKHEIDY	Archeides OROGENIA=PRECAMBRICA	Archeides OROGENESI PRE-CAMBRIANA
280 ARCHEOLOGICAL SITE MISC	ARCHAEOLOGICAL SITES	SITE ARCHEOLOGIQUE	ARCHAEOLOG. STAETTE	STOYANKA ARKHEOLOGI-CHESKAYA	LUGAR=ARQUEOLOGICO	SITO ARCHEOLOGICO
281 ARCHEOLOGY MISC	ARCHAEOLOGY *UF:* Archaeomagnetism	ARCHEOLOGIE *UF:* Archeomagnetisme	ARCHAEOLOGIE *UF:* Archaeomagnetismus	ARKHEOLOGIYA	ARQUEOLOGIA *UF:* Arqueomagnetismo	ARCHEOLOGIA *UF:* Archeomagnetismo
282 ARCHEOMAGNETISM SOLI	Archaeomagnetism MAGNETIC METH-ODS; ARCHAEOL-OGY	Archeomagnetisme ARCHEOLOGIE; PALEOMAGNE-TISME	Archaeomagnetismus ARCHAEOLOGIE; PALAEOMAGNE-TISMUS	Arkheomagnetizm PALEOMAGNE-TIZM	Arqueomagnetismo ARQUEOLOGIA; PALEOMAGNE-TISMO	Archeomagnetismo ARCHEOLOGIA; PALEOMAGNE-TISMO
283 ARCHIPOLYPODA PALS	Archipolypoda MYRIAPODA	Archipolypoda MYRIAPODA	Archipolypoda MYRIAPODA	ARCHIPOLYPODA	Archipolypoda MYRIAPODA	Archipolypoda MYRIAPODA
284 ARCHOSAURIA PALS	ARCHOSAURIA	ARCHOSAURIA	ARCHOSAURIA	ARCHOSAURIA	ARCHOSAURIA	ARCHOSAURIA
285 ARCOIDA PALS	Arcoida BIVALVIA	Arcoida BIVALVIA	Arcoida LAMELLI-BRANCHIATA	ARCOIDA	Arcoida BIVALVIA	Arcoida BIVALVIA
286 ARCTIC SURF	Arctic(adj.)	Arctique(adj.)	Arktisch(adj.)	ARKTICHESKIJ *UF:* Arkticheskaya seriya	Artico(adj.)	Artico(adj.)
287 ARCTIC SUITE IGNE	Arctic suite VOLCANOLOGY	Serie arctique VOLCANOLOGIE; ASSOCIATION MAGMATIQUE	Arkt.Abfolge VULKANOLOGIE; MAGMAT. VERGESELLSCHAF-TUNG	Arkticheskaya seriya ARKTICHESKIJ; VULKANOLOGIYA	Serie=artica VULCANOLOGIA; ASOCIACION=MAGMATICA	Serie artica VULCANOLOGIA; ASSOCIAZIONE MAGMATICA
288 ARCUATE MISC	Arcuate(adj.)	Arque(adj.)	Bogig(adj.)	DUGOOBRAZNYJ	Arqueado(adj.)	Arcuato(adj.)
289 ARCUATE FAULT STRU	ARCUATE FAULTS	FAILLE ARQUEE	GEBOGEN. VERWERFUNG	RAZLOM DUGOO-BRAZNYJ	FALLA=ARQUEADA	FAGLIA ARCUATA
290 ARENACEOUS TEST	ARENACEOUS TEX-TURE	TEXTURE ARE-NACEE	SANDIG.GEFUEGE	PESCHANISTYJ	TEXTURA=ARENOSA	TESSITURA ARE-NACEA
291 ARENIGIAN STRS	ARENIGIAN	ARENIG	ARENIG	ARENIG	ARENIGIENSE	ARENIGIANO

	ENGLISH	FRANCAIS	DEUTSCH	RUSSKIJ	ESPANOL	ITALIANO
292 ARENITE SEDS	Arenite CLASTIC ROCKS	Arenite ROCHE CLASTIQUE	Arenit KLAST.GESTEIN	Arenit PSAMMITOVYJ	Arenita ROCA=CLASTICA	Arenite ROCCIA CLASTICA
293 ARENOSOL SUSS	Arenosols SOILS	Arenosol SOL PEU EVOLUE; SABLE	Arenosol AZONAL.BODEN; SAND	POCHVA PES- CHANAYA	Arenosol SUELO= INMADURO; ARENA	Arenosol SUOLO IMMA- TURO; SABBIA
294 ARGID SUSS	Argids ARIDISOLS	Argid ARIDISOL	Argid ARIDISOL	ARGID	Argid ARIDISOL	Argid ARIDISOL
295 ARGILLACEOUS COMPOSITION SEDI	ARGILLACEOUS TEXTURE	Roche argileuse ARGILE	Ton=Gehalt TON	GLINISTYJ	Roca=arcillosa ARCILLA	Composizione argillosa ARGILLA
296 ARGON CHEE	ARGON UF: Excess argon	ARGON UF: Argon atmospherique; Exces argon	AR UF: Atmosphaer.Argon; Ueberschuss=Argon	ARGON	ARGON UF: Argon=atmosferico; Exceso=argon	ARGO UF: Argon in eccesso; Argon atmosferico
297 ARGON=ARGON DATING ISOT	AR/AR	AR=AR	AR=AR= DATIERUNG	METOD ARGONOVYJ AKTIVATSIONNYJ	DATACION=AR	AR/AR
298 ARHEISM GEOH	Arheism ARIDITY	Areisme HYDROLOGIE SUR- FACE; ARIDITE	Abflusslos.Gebiet HYDROLOGIE; ARIDITAET	Areizm ARIDNOST'	Endorreismo HIDROLOGIA= SUPERFICIE; ARIDEZ	Areismo IDROLOGIA; ARIDITA
299 ARID ENVIRONMENT SURF	ARID ENVIRON- MENT	MILIEU ARIDE UF: Plante xerophile; Ustox; Xeralf; Xero- phyte; Xerult	ARID.MILIEU UF: Ustox; Xeralf; Xerophil.Pflanze; Xero- phyt; Xerult	ARIDNYJ UF: Semi=arid environ- ment; Zasukha	MEDIO=ARIDO UF: Planta=xerofila; Ustox; Xeralf; Xerofita; Xerult	AMBIENTE ARIDO UF: Pianta xerofila; Xerofita
300 ARIDISOL SUSS	ARIDISOLS UF: Argids; Orthids	ARIDISOL UF: Argid; Orthid	ARIDISOL UF: Argid; Orthid	ARIDISOL UF: Ortid	ARIDISOL UF: Argid; Orthid	ARIDISOL UF: Argid; Orthid
301 ARIDITY SURF	ARIDITY UF: Arheism; Xerophile plant; Xerophyte	ARIDITE UF: Areisme	ARIDITAET UF: Abflusslos.Gebiet	ARIDNOST' UF: Areizm	ARIDEZ UF: Endorreismo	ARIDITA UF: Areismo; Ustox; Xeralf; Xerult
302 ARKOSE SEDS	ARKOSE	ARKOSE	ARKOSE	ARKOZ	ARCOSA	ARCOSE
303 ARMORED MUD BALL SEDI	ARMORED MUD BALLS	GALET ARGILE INDURE	VERFESTIGT. SCHLICKGEROELL	Katun glinyanyj PSEVDOKONGLOM- ERAT; KONKRET- SIYA	CANTO=ARCILLA= ENDURECIDA	CIOTTOLO D'ARGILLA INDURITO
304 ARMORICAN OROGENY STRU	Armorican orogeny HERCYNIAN OROGENY	Orogenie armoricaine OROGENIE HERCY- NIENNE	Armorikan.Orogenese HERZYN. OROGENESE	Armorikanskij GERTSINSKIJ	Orogenia=armoricana OROGENIA= HERCINICA	Orogenesi armoricana OROGENESI ERCINICA
305 AROMATIC HYDROCARBON CHES	AROMATIC HYDROCARBONS	HYDROCARBURE AROMATIQUE	AROMAT. KOHLENWASSER- STOFF	UGLEVODOROD AROMATICHESKIJ	HIDROCARBURO= AROMATICO	IDROCARBURO AROMATICO
306 ARRAY METH	ARRAYS	RESEAU SISMIQUE	SEISM.NETZ	RASSTANOVKA	RED=SISMICA	RETE SISMICA

	ENGLISH	FRANCAIS	DEUTSCH	RUSSKIJ	ESPANOL	ITALIANO
307 ARRIVAL TIME APPL	ARRIVAL TIME	TEMPS ARRIVEE	ANKUNFTS=ZEIT	VREMYA VSTU-PLENIYA	TIEMPO=DE-LLEGADA	TEMPO D'ARRIVO
308 ARROYO SURF	ARROYOS	ARROYO	ARROYO	Arroio REKA	ARROYO	ARROYO
309 ARSENATES MING	ARSENATES	ARSENIATE	ARSENAT	ARSENAT	ARSENIATO	ARSENIATO
310 ARSENIC CHEE	ARSENIC	ARSENIC	AS	MYSH'YAK	ARSENICO	ARSENICO
311 ARTERITE IGMS	Arterite METAMORPHIC ROCKS	Arterite MIGMATITE	Arterit MIGMATIT	Arterit MIGMATIT	Arterita MIGMATITA	Arterite MIGMATITE
312 ARTESIAN GEOH	Artesian(adj.)	Artesien(adj.)	Artesisch(adj.)	ARTEZIANSKIJ	Artesiano(adj.)	Artesiano(adj.)
313 ARTESIAN AQUIFER GEOH	Artesian aquifer CONFINED AQUI-FERS; AQUIFERS	Nappe artesienne NAPPE CAPTIVE	Artes.Grundwasser GESPANNT. GRUNDWASSER	GORIZONT ARTEZ-IANSKIJ UF: Gorizont napornyj	Manto=artesiano MANTO=CAUTIVO	Falda artesiana FALDA CIR-COSCRITTA
314 ARTESIAN WELL GEOH	ARTESIAN WELLS	PUITS ARTESIEN	ARTES.BRUNNEN	SKVAZHINA ARTEZIANSKAYA	POZO=ARTESIANO	POZZO ARTESIANO
315 ARTHRODIRA PALS	Arthrodira PLACODERMI	Arthrodira PLACODERMI	Arthrodira PLACODERMI	ARTHRODIRA	Arthrodira PLACODERMI	Arthrodira PLACODERMI
316 ARTHROPLEURIDA PALS	Arthropleurida MYRIAPODA	Arthropleurida MYRIAPODA	Arthropleurida MYRIAPODA	Arthropleurida DIPLOPODA	Arthropleurida MYRIAPODA	Arthropleurida MYRIAPODA
317 ARTHROPODA PALS	ARTHROPODA UF: Hemicrustacea; Merostomoidea; Ony-chophora; Pentas-tomida; Pycnogonida; Tardigrada	ARTHROPODA UF: Hemicrustacea; Merostomoidea; Ony-chophora; Pentas-tomida; Pycnogonida; Tardigrada	ARTHROPODA UF: Hemicrustacea; Merostomoidea; Ony-chophora; Pentas-tomida; Pycnogonida; Tardigrada	ARTHROPODA	ARTHROPODA UF: Hemicrustacea; Merostomoidea; Ony-chophora; Pentas-tomida; Pycnogonida; Tardigrada	ARTHROPODA UF: Hemicrustacea; Merostomoidea; Ony-chophora; Pentas-tomida; Pycnogonida; Tardigrada
318 ARTICULATA PALS	ARTICULATA	ARTICULATA UF: Kutorginida	ARTICULATA UF: Kutorginida	ARTICULATA	ARTICULATA UF: Kutorginida	ARTICULATA UF: Kutorginida
319 ARTIFACT PALE	ARTIFACTS	ARTEFACT	ARTEFAKT	ARTEFAKT	UTIL-PREHISTORICO	ARTEFATTO
320 ARTIFICIAL ASH COMS	ASH	Cendre industrielle SUBSTANCE UTILE	Industrie=Asche NUTZBAR. GESTEINE	Tekhnologicheskij pepel MATERIAL	Ceniza=artificial SUSTANCIA=UTIL	Cenere artificiale SOSTANZA UTILE
321 ARTIFICIAL RECHARGE GEOH	ARTIFICIAL RECHARGE	ALIMENTATION ARTIFICIELLE	KUENSTL. GRUNDWASSER-ANREICHERUNG	PITANIE PODZEM-NYKH VOD ISKUSTVENNOE	RECARGA=ARTIFICIAL	ALIMENTAZIONE ARTIFICIALE
322 ARTIFICIAL SATELLITE INST	SATELLITE METH-ODS UF: Orbit	SATELLITE ARTI-FICIEL UF: Orbite	KUENSTL. SATELLIT UF: Orbit	SPUTNIK	SATELITE=ARTIFICIAL UF: Orbita	SATELLITE ARTIFI-CIALE UF: Orbita
323 ARTINSKIAN STRS	ARTINSKIAN	ARTINSKIEN	ARTINSK	ARTINSKIJ	ARTINSKIENSE	ARTINSKIANO

	ENGLISH	FRANCAIS	DEUTSCH	RUSSKIJ	ESPANOL	ITALIANO
324 ARTIODACTYLA PALS	ARTIODACTYLA	ARTIODACTYLA	ARTIODACTYLA	ARTIODACTYLA *UF:* Suiformes; Tylopoda	ARTIODACTYLA	ARTIODACTYLA
325 ASBESTOS COMS	ASBESTOS	AMIANTE	ASBEST	ASBEST	AMIANTO	AMIANTO
326 ASCENSION THEORY ECON	Ascension theory THEORETICAL STUDIES; ECO- NOMIC GEOLOGY	Theorie mouvement par ascension GENESE	Aszendenz=Theorie GENESE	TEORIYA VOSKHO- DYASHCHIKH RASTVOROV	Teoria=movimiento= ascendente GENESIS	Teoria ascensionale GENESI
327 ASEISMIC DESIGN ENGI	ASEISMIC DESIGN	OUVRAGE ASISMIQUE	ERDBEBENSICHER. BAU	Asejsmichnoe sooruzhe- nie ASEJSMICHNYJ; SOORUZHENIE	CONSTRUCCION- ANTISISMICA	COSTRUZIONE ASISMICA
328 ASEISMIC MARGINS SOLI	ASEISMIC MAR- GINS	MARGE CON- TINENTALE ASISMIQUE	ASEISM. KONTINENTAL= RAND	Okraina asejsmi- cheskaya ASEJSMICHNYJ; OKRAINA MATERIKOVAYA	MARGEN= CONTINENTAL= ASISMICA	MARGINE CON- TINENTALE ASISMICO
329 ASEISMIC REGION SOLI	ASEISMIC REGIONS	Zone asismique SISMICITE	Aseism.Region SEISMIZITAET	ASEJSMICHNYJ *UF:* Asejsmichnoe sooruzhenie; Okraina asejsmicheskaya	Asismico SISMICIDAD	Regione asismica SISMICITA
330 ASH CONTENT SEDI	Ash content COAL	Teneur cendre ANALYSE CHIMIQUE; ROCHE CARBONEE	Aschegehalt CHEM.ANALYSE; KOHLENSTOFFHALTIG. GESTEIN	ZOL'NOST'	Contenido=ceniza ANALISIS= QUIMICO; ROCA= CARBONOSA	Contenuto in cenere COMPOSIZIONE CHIMICA; ROCCIA CARBONIOSA
331 ASH FALL IGNE	ASH FALLS	PLUIE CENDRE	ASCHEN=REGEN	Peplopad PIROKLAST	LLUVIA=CENIZA	PIOGGIA DI CENERE
332 ASH FLOW IGNE	ASH FLOWS	COULEE CENDRE	ASCHEN=STROM	POTOK PEPLOVYJ	COLADA=CENIZA	COLATA DI CENERE
333 ASHGILLIAN STRS	ASHGILLIAN	ASHGILL	ASHGILL	ASHGILL	ASGILLIENSE	ASHGILLIANO
334 ASPHALT COMS	ASPHALT	ASPHALTE *UF:* Asphaltite	ASPHALT *UF:* Asphaltit	ASFAL'T	ASFALTO *UF:* Asfaltita	ASFALTO *UF:* Asfaltite
335 ASPHALTITE COMS	Asphaltite BITUMENS	Asphaltite ASPHALTE	Asphaltit ASPHALT	ASFAL'TIT *UF:* Antraksolit	Asfaltita ASFALTO	Asfaltite ASFALTO
336 ASSAY ECON	ASSAYS *UF:* Assay value	Essai minerai QUALITE MINERAI	Erz=Analyse ERZ=QUALITAET	Analiz kolichestvennyj rudy ANALIZ KOLI- CHESTVENNYJ	Ensayo CALIDAD= MINERAL	Saggio QUALITA DEL MINERALE
337 ASSAY VALUE ECON	Assay value ASSAYS	Teneur metal QUALITE MINERAI	Metall=Gehalt ERZ=QUALITAET	[ASSAY VALUE] ANALIZ; RUDA	Analisis=contenido CALIDAD= MINERAL	Tenore in metallo QUALITA DEL MINERALE
338 ASSEMBLAGE PALE	ASSEMBLAGES *UF:* Assemblage zones; Associations	Association FAUNE; FLORE	Vergesellschaftung FAUNA; FLORA	Sovokupnost' BIOTSENOZ	Asociacion FAUNA; FLORA	Associazione di fossili FAUNA; FLORA

	ENGLISH	FRANCAIS	DEUTSCH	RUSSKIJ	ESPANOL	ITALIANO
339 ASSEMBLAGE ZONE STRA	Assemblage zones ASSEMBLAGES	Assise biostrati- graphique FAUNE SPECI- FIQUE; FLORE SPE- CIFIQUE	Faunen=Zone SPEZIF.FAUNA; SPEZIF.FLORA	ZONA KOMPLEKS- NAYA *UF:* Tsenozona	Zona=bioestratigrafica FAUNA= ESPECIFICA; FLORA= ESPECIFICA	Zona d'associazione FAUNA SPECIFICA; FLORA SPECIFICA
340 ASSIMILATION IGNE	ASSIMILATION	ASSIMILATION MAGMATIQUE *UF:* Basification	MAGMAT. ASSIMILATION *UF:* Basifizierung	ASSIMILYATSIYA	ASIMILACION= MAGMATICA *UF:* Basificacion	ASSIMILAZIONE MAGMATICA *UF:* Basificazione
341 ASSOCIATION PAL	Associations ASSEMBLAGES	ASSOCIATION FOS- SILE	FOSSIL=VER- GESELLSCHAFTUNG	ASSOTSIATSIYA	ASOCIACIONES= FOSILES	ASSOCIAZIONE
342 ASSYNTIAN OROGENY STRU	ASSYNTIC OROG- ENY	OROGENIE ASSYN- TIENNE	ASSYNT. OROGENESE	ASSINTSKIJ /ORO- GEN/	OROGENIA= ASINTICA	OROGENESI ASSIN- TICA
343 ASTATINE CHEE	ASTATINE	ASTATE	AT	ASTATIN	ASTATO	ASTATO
344 ASTERISM MINE	ASTERISM	ASTERISME	ASTERISMUS	ASTERIZM	DIAGRAMA=LAUE	ASTERISMO
345 ASTEROID EXTR	ASTEROIDS	ASTEROIDE	ASTEROID	ASTEROID	ASTEROIDE	ASTEROIDE
346 ASTEROIDEA PALS	ASTEROIDEA *UF:* Asterozoa	ASTEROIDEA *UF:* Asterozoa	ASTEROIDEA *UF:* Asterozoa	ASTEROIDEA	ASTEROIDEA *UF:* Asterozoa	ASTEROIDEA *UF:* Asterozoa
347 ASTEROZOA PALS	Asterozoa ASTEROIDEA	Asterozoa ASTEROIDEA	Asterozoa ASTEROIDEA	ASTEROZOA	Asterozoa ASTEROIDEA	Asterozoa ASTEROIDEA
348 ASTHENOSPHERE SOLI	ASTHENOSPHERE	ASTHENOSPHERE	ASTENOSPHAERE	ASTENOSFERA *UF:* B sloj	ASTENOSFERA	ASTENOSFERA
349 ASTRAPOTHERIA PALS	ASTRAPOTHERIA	ASTRAPOTHERIA	ASTRAPOTHERIA	ASTRAPOTHERIA	ASTRAPOTHERIA	ASTRAPOTHERIA
350 ASTROBLEME EXTR	ASTROBLEMES	ASTROBLEME	ASTROBLEM	Astroblema KRATER UDARA	ASTROBLEMA	ASTROBLEMA
351 ASTROGEOLOGY EXTR	Astrogeology PLANETOLOGY	Astrogeologie GEOLOGIE EXTRA TERRESTRE	Astrogeologie EXTRATERRESTR. GEOLOGIE	ASTROGEOLOGIYA	Astrogeologia GEOLOGIA= EXTRATERRESTRE	Astrogeologia GEOLOGIA EXTRATERRESTRE
352 ASTRONOMY EXTR	ASTRONOMY	ASTRONOMIE	ASTRONOMIE	ASTRONOMIYA	ASTRONOMIA	ASTRONOMIA
353 ASYMMETRIC FOLD STRU	ASYMMETRIC FOLDS	PLI ASYMETRIQUE	ASYMMETR.FALTE	SKLADKA ASIM- METRICHNAYA	PLIEGUE= ASIMETRICO	PIEGA ASIM- METRICA
354 ATAXITE EXTS	Ataxite METEORITES	Ataxite METEORITE MET- ALLIQUE	Ataxit METALL= METEORIT	Ataksit METEORIT ZHEL- EZNYJ	Ataxita METEORITO= METALICO	Ataxite METEORITE MET- ALLICA
355 ATLANTIC SUITE IGNE	Atlantic suite VOLCANOLOGY	Serie atlantique VOLCANOLOGIE; ASSOCIATION MAGMATIQUE	Atlant.Abfolge VULKANOLOGIE; MAGMAT. VERGESELLSCHAF- TUNG	ATLANTICHESKIJ	Serie=atlantica VULCANOLOGIA; ASOCIACION= MAGMATICA	Serie atlantica VULCANOLOGIA; ASSOCIAZIONE MAGMATICA

	ENGLISH	FRANCAIS	DEUTSCH	RUSSKIJ	ESPANOL	ITALIANO
356 ATLAS MISC	ATLAS	ATLAS	ATLAS	ATLAS	ATLAS	ATLANTE
357 ATMOGENIC ORIGIN SEDI	Atmogenic EOLIAN SEDIMEN- TATION	Origine atmogenique SEDIMENTATION EOLIENNE	Atmogen.Herkunft AEOL. SEDIMENTATION	ATMOGENNYJ	Origen=atmogenico SEDIMENTACION= EOLICA	Origine atmogena SEDIMENTAZIONE EOLICA
358 ATMOSPHERE ENVI	ATMOSPHERE UF: Air; Greenhouse effect; Subaerial envi- ronment	ATMOSPHERE UF: Air; Argon atmospherique; Effet serre; Milieu atmospherique	ATMOSPHAERE UF: Atmosphaer.Argon; Atmosphaer.Milieu; Gewaechshaus=Effekt; Luft	ATMOSFERA UF: Paleoatmosfera; Sreda nazemnaya; Svyaz vozdukh=more; Vozdukh	ATMOSFERA UF: Aire=atmosferico; Argon=atmosferico; Efecto=invernadero; Medio=subaereo	ATMOSFERA UF: Ambiente subaereo; Argon atmosferico; Aria; Effetto di serra
359 ATMOSPHERIC ARGON ISOT	Atmospheric Argon K/AR	Argon atmospherique ARGON; ATMO- SPHERE	Atmosphaer.Argon AR; ATMOSPHAERE	ARGON ATMOS- FERNYJ	Argon=atmosferico ARGON; ATMOSF- ERA	Argon atmosferico ARGON; ATMOSF- ERA
360 ATMOSPHERIC PRECIPITATION GEOH	ATMOSPHERIC PRECIPITATION UF: Hail	PRECIPITATION ATMOSPHERIQUE UF: Grele	ATMOSPHAER. NIEDERSCHLAG UF: Hagel	OSADKI ATMOS- FERNYE UF: Grad; Rosa	PRECIPITACION= ATMOSFERICA UF: Granizo	PRECIPITAZIONE ATMOSFERICA UF: Grandine
361 ATMOSPHERIC PRESSURE SURF	ATMOSPHERIC PRESSURE	PRESSION BARO- METRIQUE	LUFT=DRUCK	DAVLENIE ATMOS- FERNOE	PRESION= ATMOSFERICA	PRESSIONE BARO- METRICA
362 ATOLL MARI	ATOLLS	ATOLL	ATOLL	ATOLL	ATOLON	ATOLLO
363 ATOMIC ABSORPTION METH	ATOMIC ABSORP- TION UF: Absorption spectros- copy	ABSORPTION ATOMIQUE	ATOM= ABSORPTION	ABSORBTSIYA ATOMNAYA	ABSORCION= ATOMICA	ASSORBIMENTO ATOMICO
364 ATOMIC BOND PHCH	ATOMIC BOND	LIAISON INTERA- TOMIQUE	BINDUNGSART	SVYAZ' ATOM- NAYA	ENLACE= INTERATOMICO	LEGAME INTERA- TOMICO
365 ATOMIC PACKING MINE	ATOMIC PACKING UF: Close packing	EMPILEMENT ATOME UF: Empilement dense	ATOM=PACKUNG UF: Dichtest.Packung	[ATOMIC PACKING] STRUKTURA KRISTALLICHESKAYA	ESTRUCTURA= ATOMO UF: Empaquetado= atomico=cerrado	IMPILAMENTO ATOMICO UF: Impacchettamento fitto
366 ATTENUATION PHCH	ATTENUATION	ATTENUATION	DAEMPFUNG	UMEN'SHENIE	ATENUACION= ONDA	ATTENUAZIONE
367 ATTERBERG LIMIT ENGI	ATTERBERG LIM- ITS	LIMITE ATTER- BERG	ATTERBERG= GRENZEN	ATTERBERGA PRE- DEL	LIMITE= ATTERBERG	LIMITI DI ATTER- BERG
368 ATTRITION SURF	Attrition ABRASION	Attrition ABRASION	Attrition ABRASION	ISTIRANIE	Atricion ABRASION	Sfregamento ABRASIONE
369 AUBRITE EXTS	Aubrite ACHONDRITES	Aubrite ACHONDRITE	Aubrit ACHONDRIT	Obrit AKHONDRIT	Aubrita ACONDRITA	Aubrite ACONDRITE
370 AUGEN STRUCTURE TEST	Augen structure PORPHYRITIC TEX- TURE	Structure oeillee TEXTURE POR- PHYRIQUE	Augen=Gefuege PORPHYR= GEFUEGE	OCHKOVYJ	Estructura=glandular TEXTURA= PORFIDICA	Tessitura occhiadina TESSITURA POR- FIRICA
371 AUGER DRILLING MINI	Auger drilling BOREHOLES	Sondage tariere SONDAGE	Hand=Bohrung BOHRUNG	BURENIE SHNEKOVOE	Sondeo=auger POZO=SONDEO	Perforazione auger SONDAGGIO

	ENGLISH	FRANCAIS	DEUTSCH	RUSSKIJ	ESPANOL	ITALIANO
372 AULACOGEN STRU	AULACOGENS *UF:* Border facies; Meta- morphic zone	AULACOGENE	AULAKOGEN	AVLAKOGEN	AULACOGENO	AULACOGENO
373 AUREOLE GEOC	HALOES	AUREOLE GEOCHIMIQUE	GEOCHEM. AUREOLE	OREOL *UF:* Oreol rasseyaniya pervichnyj	AUREOLA= GEOQUIMICA	AUREOLA GEOCHIMICA
374 AUSTRALITE EXTS	Australite TEKTITES	Australite TECTITE	Australit TEKTIT	Avstralit TEKTIT	Australita TECTITA	Australite TECTITE
375 AUTHIGENESIS MINE	AUTHIGENESIS *UF:* Allogenic mineral; Authigenic minerals	AUTHIGENESE *UF:* Mineral allothigene; Mineral authigene	AUTHIGENESE *UF:* Allothigen.Mineral; Authigen.Mineral	Autigenez AUTIGENNYJ	AUTIGENESIS *UF:* Mineral=alotigeno; Mineral=autigeno	AUTIGENESI *UF:* Minerale allotigeno; Minerale autigeno
376 AUTHIGENIC MINERAL MINE	Authigenic minerals AUTHIGENESIS	Mineral authigene AUTHIGENESE	Authigen.Mineral AUTHIGENESE	AUTIGENNYJ *UF:* Autigenez	Mineral=autigeno AUTIGENESIS	Minerale autigeno AUTIGENESI
377 AUTOCHTHON GEOL	AUTOCHTHONS	AUTOCHTONIE	AUTOCHTHONIE	AVTOKHTON	AUTOCTONIA	AUTOCTONIA
378 AUTOCORRELATION MATH	AUTOCORRELA- TION	AUTOCORRELA- TION	AUTOKORRELA- TION	AVTOKORRELYAT- SIYA	AUTOCORRELA- CION	AUTOCORRELAZ- IONE
379 AUTOLITH IGNE	Autoliths XENOLITHS	Autolite ENCLAVE ROCHE	Magmat.Einschluss XENOLITH	AVTOLIT	Autolito ENCLAVE=ROCA	Autolite XENOLITE
380 AUTOLYSIS IGNE	Autolysis AUTOMETAMOR- PHISM	Autolyse METAMORPHISME AUTO	Autolyse AUTO= METAMORPHOSE	Avtoliz AVTOMETASOMA- TOZ	Autolisis AUTOMETAMOR- FISMO	Autolisi AUTOMETAMOR- FISMO
381 AUTOMATED CARTOGRAPHY METH	AUTOMATIC CAR- TOGRAPHY	CARTOGRAPHIE AUTOMATIQUE	AUTOMAT. KARTOGRAPHIE	KARTOGRAFIYA AVTOMATIZ- IROVANNAYA	CARTOGRAFIA= AUTOMATICA	CARTOGRAFIA AUTOMATIZZATA
382 AUTOMATION METH	Automation DATA PROCESSING	Informatisation ORDINATEUR	Automation RECHNER	AVTOMATIZAT- SIYA	Automatizacion ORDENADOR	Automazione CALCOLATORE
383 AUTOMETAMORPHISM IGNE	AUTOMETAMOR- PHISM *UF:* Autolysis; Auto- metasomatism; Deuteric composition; Lateral secretion	METAMORPHISME AUTO *UF:* Autolyse; Auto- metasomatose; Effet deuterique; Secretion lateral	AUTO= METAMORPHOSE *UF:* Auto= Metasomatose; Auto- lyse; Epimagmat. Effekt; Lateral= Sekretion	AVTOMETAMOR- FIZM *UF:* Dejtericheskij ehf- fekt	AUTOMETAMOR- FISMO *UF:* Autolisis; Autometa- somatismo; Epimagma- tico; Secrecion=lateral	AUTOMETAMOR- FISMO *UF:* Autolisi; Autometa- somatismo; Effetto deuterico; Secrezione laterale
384 AUTOMETASOMATISM IGNE	Autometasomatism AUTOMETAMOR- PHISM	Autometasomatose METAMORPHISME AUTO	Auto=Metasomatose AUTO= METAMORPHOSE	AVTOMETASOMA- TOZ *UF:* Avtoliz	Autometasomatismo AUTOMETAMOR- FISMO	Autometasomatismo AUTOMETAMOR- FISMO
385 AUTORADIOGRAPHY APPL	AUTORADIOGRA- PHY	AUTORADIOGRA- PHIE	AUTORADIOGRA- PHIE	AVTORADIOGRA- FIYA	AUTORADIOGRA- FIA	AUTORADIOGRA- FIA
386 AUTOTROPHIC ORGANISM PALE	Autotrophic organism NUTRITION	Organisme autotrophe NUTRITION	Autotroph.Organismus ERNAEHRUNG	AVTOTROFY	Organismo=autotrofico NUTRICION	Organismo autotrofico ANATOMIA DELLA NUTRIZIONE
387 AUTUNIAN STRS	AUTUNIAN	AUTUNIEN	AUTUNIUM	OTEHN	AUTUNIENSE	AUTUNIANO

	ENGLISH	FRANCAIS	DEUTSCH	RUSSKIJ	ESPANOL	ITALIANO
388 AVALANCHE SURF	AVALANCHES	AVALANCHE	LAWINE	LAVINA	AVALANCHA	VALANGA
389 AVERAGE MATH	Average STATISTICAL DIS- TRIBUTION	Moyenne DISTRIBUTION STATISTIQUE	Durchschnittswert STATIST. VERTEILUNG	ZNACHENIE SRED- NEE	Media DISTRIBUCION= ESTADISTICA	Media DISTRIBUZIONE STATISTICA
390 AVES PALS	AVES *UF:* Odontornithes	AVES *UF:* Odontornithes	AVES *UF:* Odontornithes	AVES *UF:* Archaeornithes	AVES *UF:* Odontornithes	AVES *UF:* Odontornithes
391 AXIAL PLANE STRU	AXIAL=PLANE STRUCTURES *UF:* Axial=plane cleav- age	PLAN AXIAL *UF:* Clivage plan axial	ACHSENEBENE *UF:* Tekton.Schieferung	PLOSKOST' OSEVAYA	PLANO=AXIAL *UF:* Exfoliation=segun= plan=axial	PIANO ASSIALE *UF:* Clivaggio di piano assiale
392 AXIAL=PLANE CLEAVAGE STRU	Axial=plane cleavage CLEAVAGE; AXIAL=PLANE STRUCTURES	Clivage plan axial SCHISTOSITE; PLAN AXIAL	Tekton.Schieferung SCHIEFERUNG; ACHSENBENE	KLIVAZH OSEVOJ POVERKHNOSTI	Exfoliacion=segun= plano=axial ESQUISTOSIDAD; PLANO=AXIAL	Clivaggio di piano assi- ale SCISTOSITA; PIANO ASSIALE
393 AXIS MATH	Axis GEOMETRY	Axe GEOMETRIE	Achse GEOMETRIE	OS'	Eje GEOMETRIA	Asse GEOMETRIA
394 AZONAL SOIL SUSS	AZONAL SOILS	SOL BRUT *UF:* Lithosol; Orthent; Psamment; Regosol; Sol desertique	AZONAL.BODEN *UF:* Arenosol; Orthent; Psamment; Regosol; Skelett=Boden; Xero- sol; Yermosol	POCHVA AZON- AL'NAYA *UF:* Litosol; Pochva nepolnorazvitaya	SUELO=BRUTO *UF:* Litosol; Orthent; Psamment; Regosol; Suelo=desertico	SUOLO AZONALE *UF:* Orthent; Psamment; Regosuolo; Suolo sche- letrico; Yermosol
395 B LAYER SOLI	B layer LOW=VELOCITY ZONES	Couche B MANTEAU SUP	B=Schicht OBER.ERD= MANTEL	B sloj ASTENOSFERA	Capa=B MANTO=GLOBO= SUP	Strato B MANTELLO SUPERIORE
396 BACK ARC BASIN STRU	BACK ARC BASINS	BASSIN MARGINAL INTERNE	INNER.RAND= BECKEN	[BACK ARC BASIN]	CUENCA= MARGINAL= INTERNA	BACINO DI RETR- OARCO
397 BACK REEF SEDI	Back reef REEFS	Recif interne RECIF	Rueck=Riff RIFF	Rif tylovoj RIF BAR'ERNYJ	Arrecife=interno ARRECIFE	Retroscogliera SCOGLIERA
398 BACKGROUND GEOC	BACKGROUND	TENEUR FOND	SPIEGELWERT	FON	FONDO=REGIONAL	TENORE DI FONDO
399 BACKSET BED SEDI	BACKSET BEDS	Couche contrepentee STRATIFICATION ENTRECROISEE; SEDIMENTATION DELTAIQUE	[BACKSET=BED] KREUZSCHICH- TUNG; DELTA= SEDIMENTATION	Sloj obratnyj SLOISTOST'	[BACKSET=BED] ESTRATIFIC- ACION=CRUZADA; SEDIMENTACION= DELTAICA	[BACKSET BED] STRATIFICAZIONE INCROCIATA; SEDI- MENTAZIONE DEL- TIZIA
400 BACTERIA PALS	BACTERIA *UF:* Bacteriogenic mate- rials	BACTERIA *UF:* Origine bacteriogene	BACTERIA *UF:* Bakteriogen. Entstehung	BACTERIA	BACTERIA *UF:* Bacteriogenico	BACTERIA *UF:* Origine batteriogena
401 BACTERIOGENIC ORIGIN GEOL	Bacteriogenic materials BIOGENIC EFFECTS; BACTE- RIA	Origine bacteriogene ORIGINE BIOGENE; BACTERIA	Bakteriogen.Entstehung BIOGEN.BILDUNG; BACTERIA	BAKTERIOGENNYJ	Bacteriogenico ORIGEN= BIOGENICO; BAC- TERIA	Origine batteriogena ORIGINE ORGANO- GENA; BACTERIA
402 BACTRITOIDEA PALS	Bactritoidea CEPHALOPODA	Bactritoidea AMMONOIDEA	Bactritoidea AMMONOIDEA	BACTRITOIDEA	Bactritoidea AMMONOIDEA	Bactritida AMMONOIDEA

	ENGLISH	FRANCAIS	DEUTSCH	RUSSKIJ	ESPANOL	ITALIANO
403 BADLANDS SURF	Badlands SOIL EROSION	Mauvaises terres EROSION SOL	[BADLANDS] BODEN=EROSION	Bedlend EHROZIYA; POCHVA	Malpais EROSION=SUELO	[BADLANDS] EROSIONE DEL SUOLO
404 BAIKALIAN OROGENY STRU	BAIKALIAN PHASE	PHASE BAIKA- LIENNE	BAIKAL= OROGENESE	BAJKAL'SKIJ	FASE=BAIKALIANA	OROGENESI BAI- KALIANA
405 BAJADA SURF	BAJADAS	Bajada CONE DEJECTION	Bajada SCHUTT=KEGEL	RAVNINA PRED- GORNAYA	Bajada CONO=DE= DEYECCION	Bajada CONO DI DETRITO
406 BAJOCIAN STRS	BAJOCIAN	BAJOCIEN	BAJOCIUM	BAJOS	BAJOCIENSE	BAJOCIANO
407 BALANCE MISC	BALANCE	BILAN	BILANZ	BALANS *UF:* Mass balance	BALANCE	BILANCIO
408 BALLAST ENGI	BALLAST	Ballast MATERIAU VIABI- LITE	Schuett=Gut STRASSENBAU= MATERIAL	BALLAST	Balasto MATERIAL= CARRETERA	Ballast MATERIALE PER MASSICCIATA
409 BANDING TEST	BANDED STRUC- TURES	STRUCTURE RUBANEE	BAENDER= STRUKTUR	POLOSCHATOST'	ESTRUCTURA=EN= BANDAS	STRUTTURA A BANDE
410 BAR SURF	BARS *UF:* Barrier beaches; Barriers; Rock bars	BARRE *UF:* Barriere	BARRE *UF:* Barriere	BAR *UF:* Bar'ernyj; Val bar'ernyj	BARRA *UF:* Barrera	BARRA *UF:* Barriera
411 BARCHAN SURF	BARCHANS	Barkhane DUNE CONTINEN- TALE	Barchan KONTINENTAL= DUENE	BARKHAN *UF:* Dyuna pustynnaya	Medano=semilunar DUNA= CONTINENTAL	Barcana DUNA CONTINEN- TALE
412 BARITE COMS	BARITE	BARITE	BARYT	BARIT	BARITA	BARITE
413 BARIUM CHEE	BARIUM	BARYUM	BA	BARIJ	BARIO	BARIO
414 BARREMIAN STRS	BARREMIAN	BARREMIEN	BARREME	BARREM	BARREMIENSE	BARREMIANO
415 BARREN DEPOSIT ECON	BARREN DEPOSITS	Sterile MORTS TERRAINS	Taub.Lagerstaette ABRAUM	[BARREN DEPOSIT] BEDNYJ; RUDA	Deposito=esteril RECUBRIMIENTO= ESTERIL	Roccia sterile TERRENO STERILE
416 BARRIER SURF	Barriers BARS	Barriere BARRE	Barriere BARRE	Bar'ernyj BAR	Barrera BARRA	Barriera BARRA
417 BARRIER BAR MARI	BARRIER BARS	Barre littorale CORDON LITTO- RAL	Strand=Barriere STRAND=WALL	BAR BAR'ERNYJ	Barra=barrera CORDON=LITORAL	Barra littorale CORDONE LITOR- ALE
418 BARRIER BEACH SURF	Barrier beaches BARS; BEACHES	Plage barriere CORDON LITTO- RAL	Barriere=Strand STRAND=WALL	Val bar'ernyj VAL BEREGOVOJ; BAR	Barrera=playa CORDON=LITORAL	Lido CORDONE LITOR- ALE
419 BARRIER LAGOON SURF	Barrier lagoons LAGOONS	Lagon barriere LAGUNE	Barriere=Lagune LAGUNE	Laguna bar'ernaya LAGUNA	Albufera=de=atolon ALBUFERA	Laguna chiusa LAGUNA

	ENGLISH	FRANCAIS	DEUTSCH	RUSSKIJ	ESPANOL	ITALIANO
420 BARRIER REEF SEDI	BARRIER REEFS	RECIF BARRIERE	BARRIERE=RIFF	RIF BAR'ERNYJ *UF:* Rif tylovoj	BARRERA= ARRECIFAL	BARRIERA CORAL- LINA
421 BARTONIAN STRS	Bartonian UPPER EOCENE	Bartonien EOCENE SUP	Bartonium O.EOZAEN	BARTON	Bartoniense EOCENO=SUP	Bartoniano EOCENE SUP.
422 BASAL GEOL	Basal(adj.)	Basal(adj.)	Basal(adj.)	BAZAL'NYJ	Basal(adj.)	Basale(adj.)
423 BASAL CLEAVAGE MINE	Basal cleavage MINERAL CLEAV- AGE	Clivage isometrique CLIVAGE MINERAL	Basis=Spaltbarkeit SPALTBARKEIT	Klivazh bazal'nyj KLIVAZH POSLOJ- NYJ; SPAJNOST'	Exfoliacion=basal EXFOLIACION= MINERAL	Sfaldatura basale SFALDATURA
424 BASAL CONGLOMERATE SEDI	Basal conglomerate CONGLOMERATE	Conglomerat base CONGLOMERAT	Basal=Konglomerat KONGLOMERAT	KONGLOMERAT BAZAL'NYJ	Conglomerado=base CONGLOMERADO	Conglomerato basale CONGLOMERATO
425 BASALT IGNS	BASALTS *UF:* Deccan traps; Picrite; Plateau basalt	BASALTE *UF:* Basalte plateau; Kreep; Picrite; Trapp Deccan	BASALT *UF:* Dekkan=Trapp; Kreep; Pikrit; Plateau= Basalt	BAZAL'T *UF:* Metabazal't	BASALTO *UF:* Basalto=de=Dec- can; Basalto=de=mese- ta; Kreep; Picrita	BASALTO *UF:* Basalto del Deccan; Basalto di plateau; Kreep; Picrite
426 BASALTIC LAYER SOLI	BASALTIC LAYER *UF:* Sima	COUCHE BASAL- TIQUE	BASALT=SCHICHT	SLOJ BAZAL'TOVYJ *UF:* Kora nizhnyaya	CAPA=BASALTICA	LIVELLO BASAL- TICO
427 BASANITE IGNS	Basanite ALKALI BASALTS	Basanite BASALTE ALCALIN	Basanit ALKALI=BASALT	BAZANIT	Basanita BASALTO= ALCALINO	Basanite BASALTO ALCALINO
428 BASE METAL ECON	BASE METALS	METAL BASE	BASIS=METALL	[BASE METAL] METALL TSVETNOJ	METAL=BASE	METALLO BASE
429 BASEMENT STRU	BASEMENT *UF:* Infrastructure	SOCLE	BASEMENT	FUNDAMENT	ZOCALO	BASAMENTO
430 BASEMENT TECTONICS STRU	BASEMENT TEC- TONICS	TECTONIQUE SOCLE *UF:* Infrastructure	GRUNDGEBIRGS- TEKTONIK *UF:* Infra=Struktur	TEKTONIKA FUN- DAMENTA	TECTONICA=DE= ZOCALO *UF:* Infraestructura	TETTONICA DEL BASAMENTO *UF:* Infrastruttura
431 BASHKIRIAN STRS	BASHKIRIAN	BASHKIRIEN	BASCHKIR	BASHKIRSKIJ	BASHKIRIENSE	BASHKIRIANO
432 BASIFICATION IGNE	Basification OCEANIZATION	Basification ASSIMILATION MAGMATIQUE	Basifizierung MAGMAT. ASSIMILATION	BAZIFIKATSIYA	Basificacion ASIMILACION= MAGMATICA	Basificazione ASSIMILAZIONE
433 BASIN PLANNING GEOH	BASIN=PLANNING	AMENAGEMENT BASSIN	WASSERWIRT- SCHAFTS=PLAN	Vodokhozyastvennoe planirovanie PLANIROVANIE; VODA	PLANIFICACION= CUENCA= HIDROGEOLOG	GESTIONE DI BACINO
434 BATHOLITH IGNE	BATHOLITHS *UF:* Roof pendant	BATHOLITE *UF:* Pluton roche ignee; Roof pendant	BATHOLITH *UF:* Magmakammer= Dach; Pluton	BATOLIT *UF:* Ostanets krovli; Pluton (magm.)	BATOLITO *UF:* Pluton; Techo= rebajado (magm.)	BATOLITE *UF:* Pendente; Plutone
435 BATHONIAN STRS	BATHONIAN	BATHONIEN	BATHONIUM	BAT	BATONIENSE	BATONIANO
436 BATHYAL ENVIRONMENT PALE	Bathyal environment DEEP=SEA ENVI- RONMENT	Milieu bathyal MILIEU MER PRO- FONDE	Bathyal=Milieu TIEFSEE=MILIEU	Batial'nyj GLUBOKOVODNYJ	Medio=batial MEDIO=MAR= PROFUNDO	Ambiente batiale AMBIENTE DI MARE PROFONDO

	ENGLISH	FRANCAIS	DEUTSCH	RUSSKIJ	ESPANOL	ITALIANO
437 BATHYMETRIC CHART MARI	BATHYMETRIC MAPS	CARTE BATHY-METRIQUE	BATHYMETR. KARTE	KARTA BATI-METRICHESKAYA	MAPA=BATIMETRICO	CARTA BATI-METRICA
438 BATHYMETRY MARI	BATHYMETRY	BATHYMETRIE	BATHYMETRIE	BATIMETRIYA *UF:* Paleobatimetriya	BATIMETRIA	BATIMETRIA
439 BATRACHOSAURIA PALS	Batrachosauria LABYRINTHODON-TIA	Batrachosauria AMPHIBIA	Batrachosauria AMPHIBIA	BATRACHOSAURIA	Batrachosauria AMPHIBIA	Batrachosauria AMPHIBIA
440 BAUXITE COMS	BAUXITE	BAUXITE	BAUXIT	BOKSIT	BAUXITA	BAUXITE
441 BAUXITIZATION ECON	BAUXITIZATION	BAUXITISATION	BAUXITISIERUNG	BOKSITIZATSIYA	BAUXITIZACION	BAUXITIZZAZIONE
442 BAVENO TWIN LAW MINE	Baveno twin law TWINNING	Macle Baveno MACLE	Baveno=Gesetz ZWILLING	Dvojnik bavenskij /zakon/ DVOJNIK	Macla=de=Baveno MACLA	Legge di Baveno GEMINAZIONE
443 BAY SURF	Bays GULFS	Baie GOLFE	Bay GOLF	Bukhta ZALIV	Bahia GOLFO	Baia GOLFO
444 BEACH SURF	BEACHES *UF:* Abandoned shore-line; Abrasion coast; Accumulative coast; Barrier beaches; Beach scarps; Cheniers; Coast; Cusps; Foreshore	PLAGE *UF:* Avant=plage; Crois-sant; Microfalaise plage	STRAND *UF:* Cusp; Strand=Abbruch; Vorstrand	PLYAZH *UF:* Plyazh nizhnij; Ustup plyazhevyj; Val plyazhovoj	PLAYA *UF:* Anteplaya; Cuspide; Escarpe=de=playa	SPIAGGIA *UF:* Antispiaggia; Cus-pide; Scarpata litorale
445 BEACH PLACER ECON	BEACH PLACERS	MINERALISATION LITTORALE	STRAND=SEIFE	ROSSYP' BERE-GOVAYA	MINERALIZACION=LITORAL	GIACIMENTO LIT-TORALE
446 BEACH RIDGE SURF	BEACH RIDGES	CRETE PLAGE	UFERWALL	VAL BEREGOVOJ *UF:* Val bar'ernyj; Val plyazhovoj	CRESTA=PLAYA	RUGA DI SPIAGGIA
447 BEACH SCARP SURF	Beach scarps SCARPS; BEACHES	Microfalaise plage FALAISE; PLAGE	Strand=Abbruch KLIFF; STRAND	Ustup plyazhevyj PLYAZH	Escarpe=de=playa ACANTILADO; PLAYA	Scarpata litorale FALESIA; SPIAGGIA
448 BEACHROCK SEDI	BEACHROCK	BEACH ROCK	BEACH=ROCK	BICH=ROK	BEACH=ROCK	BEACHROCK
449 BEAN ORE COMS	Bean ore PISOLITIC TEX-TURE; IRON	Minerai pisolitique TEXTURE PISOLI-TIQUE	Bohnerz PISOLITH=GEFUEGE	RUDA BOBOVAYA	Mineral=pisolitico TEXTURA=PISOLITICA	Minerale pisolitica TESSITURA PISOLI-TICA
450 BEARING CAPACITY ENGI	BEARING CAPAC-ITY	PORTANCE	TRAGFAEHIGKEIT	SPOSOBNOST' NESUSHCHAYA	CAPACIDAD=DE=CARGA	PORTANZA
451 BECKE LINE MINE	Becke line REFRACTIVE INDEX	Lisere Becke INDICE REFRAC-TION	Becke=Linie BRECHUNGS=INDEX	Poloska Bekke KRISTALLOOPTIKA	Linea=de=Becke INDICE=DE=REFRACCION	Linea di Becke INDICE DI RIFRAZ-IONE
452 BED LOAD SURF	BEDLOAD	CHARGE FOND	BODENFRACHT	STOK TVERDYJ *UF:* Stok vzveshennyj	CARGA=FONDO	CARICO DI FONDO
453 BEDDING STRU	BEDDING *UF:* Parallel bedding	LITAGE *UF:* Litage parallele	FEIN=SCHICHTUNG *UF:* Parallel=Schichtung	SLOISTOST' *UF:* Layered medium; Massive bedding; Sloj golovnoj; Sloj obratnyj; Sloj peredovoj; Sloj pridonnyj	ESTRATIFIC-ACION=FINA *UF:* Estratificacion=concordante	STRATIFICAZIONE SOTTILE *UF:* Stratificazione parallela

	ENGLISH	FRANCAIS	DEUTSCH	RUSSKIJ	ESPANOL	ITALIANO
454 BEDDING PLANE STRU	Bedding planes STRATIFICATION	Plan stratification STRATIFICATION	Schicht=Flaeche SCHICHTUNG	POVERKHNOST' NAPLASTOVANIYA	Plano=estratificacion ESTRATIFICACION	Piano di stratificazione STRATIFICAZIONE
455 BEDDING-PLANE CLEAVAGE STRU	Bedding=plane cleav- age BEDDING; CLEAV- AGE	Clivage parallele STRATIFICATION	Schichtungs= Schieferung SCHICHTUNG	KLIVAZH POSLOJ- NYJ UF: Klivazh bazal'nyj	Exfoliacion= concordante ESTRATIFICACION	Clivaggio parallelo STRATIFICAZIONE
456 BEDDING-PLANE FAULT STRU	BEDDING FAULTS UF: Bedding=plane cleavage	FAILLE STRATIFI- CATION	SCHICHTPAR- ALLEL.STOERUNG	RAZRYV SOGLAS- NYJ	FALLA- ESTRATIFICADA	FAGLIA DI STRATO
457 BEDROCK GEOL	BEDROCK	[BEDROCK]	ANSTEHENDES	PORODA KOREN- NAYA	ROCA=FIRME	[BEDROCK]
458 BEHEADED STREAM SURF	Beheaded streams STREAM CAPTURE	Riviere decapitee CAPTURE COURS EAU	Gekoepft.Tal FLUSS= ANZAPFUNG	Potok obezglavlennyj REKA	Corriente=decapitada CAPTURA	Decapitazione fluviale CATTURA DI CORSO D'ACQUA
459 BELEMNOIDEA PALS	BELEMNOIDEA	BELEMNOIDEA	BELEMNOIDEA	BELEMNOIDEA	BELEMNOIDEA	BELEMNOIDEA
460 BELTIAN OROGENY STRU	Beltian orogeny OROGENY; PRE- CAMBRIAN	Orogenie beltienne OROGENIE ANTE- CAMBRIENNE	Belt.Orogenese PRAEKAMBR. OROGENESE	BELTSKIJ	Orogenia=beltica OROGENIA= PRECAMBRICA	Orogenesi beltiana OROGENESI PRE- CAMBRIANA
461 BENCH SURF	BENCHES	BANQUETTE	GEHWEG	BENCH	BANCO=TERRAZA	BANCHINA
462 BENIOFF ZONE SOLI	BENIOFF ZONE	ZONE BENIOFF	BENIOFF=ZONE	BEN'OFFA ZONA	ZONA=BENIOFF	ZONA DI BENIOFF
463 BENNETTITALES PALS	BENNETTITALES	BENNETTITALES	BENNETTITALES	BENNETTITALES	BENNETTITALES	BENNETTITALES
464 BENTHONIC ENVIRONMENT PALE	BENTHONIC ENVI- RONMENT UF: Benthos	MILIEU BEN- THIQUE UF: Benthos	BENTHON.MILIEU UF: Benthos	Bentonnyj BENTOS	MEDIO= BENTONICO UF: Bentos	AMBIENTE BEN- TONICO UF: Benthos
465 BENTHOS PALE	Benthos BENTHONIC ENVI- RONMENT; FAUNA	Benthos MILIEU BEN- THIQUE; FAUNE	Benthos BENTHON.MILIEU; FAUNA	BENTOS UF: Bentonnyj	Bentos MEDIO= BENTONICO; FAUNA	Benthos AMBIENTE BEN- TONICO; FAUNA
466 BENTONITE COMS	BENTONITE	BENTONITE	BENTONIT	BENTONIT	BENTONITA	BENTONITE
467 BERGSCHRUND SURF	Bergschrund GLACIAL FEA- TURES	Rimaye GLACIER; RUP- TURE	Bergschrund GLETSCHER; RUP- TUR	Bergshrund TRESHCHINA LED- NIKOVAYA	Rimaya GLACIAR; RUP- TURA	Crepaccio terminale GHIACCIAIO; ROT- TURA
468 BERKELIUM CHEE	BERKELIUM	BERKELIUM	BK	BERKELIJ	BERKELIO	BERKELIO
469 BERM SURF	Berms MARINE TER- RACES	Banquette plage TERRASSE MARINE	Berme STRAND= TERRASSE	BERMA (GEOMOR- FOLOGIYA)	Terraza=costera TERRAZA= MARINA	Berma TERRAZZO MARINO
470 BERRIASIAN STRS	BERRIASIAN	BERRIASIEN	BERRIAS	BERRIAS	BERRIASIENSE	BERRIASIANO

	ENGLISH	FRANCAIS	DEUTSCH	RUSSKIJ	ESPANOL	ITALIANO
471 BERYLLIUM CHEE	BERYLLIUM	BERYLLIUM	BE	BERILLYJ	BERILIO	BERILLIO
472 BERYLLIUM DATING ISOT	BE=10/BE=9	BE 10=BE 9	BE 10=BE 9	METOD BERIL-LIEVYJ	BE=10=BE=9	BE=10/BE=9
473 BETA RAY PHCH	BETA RAYS	RAYONNEMENT BETA	BETA=STRAHLUNG	BETA=LUCHI	RAYOS=BETA	RAGGI BETA
474 BEYRICHICOPINA PALS	BEYRICHICOPINA	BEYRICHICOPINA	BEYRICHICOPINA	Beyrichicopina PALEOCOPIDA	BEYRICHICOPINA	BEYRICHICOPINA
475 BIAXIAL MINE	Biaxial(adj.)	Biaxial(adj.)	Biaxial(adj.)	DVUOSNYJ	Biaxial(adj.)	Biassiale(adj.)
476 BIAXIAL CRYSTAL MINE	Biaxial crystals OPTICAL PROPER-TIES	Mineral biaxe PROPRIETE OPTIQUE; ANI-SOTROPIE	Zweiachsig.Kristall OPT.EIGENSCHAFT; ANISOTROPIE	KRISTALL DVUOS-NYJ	Mineral=biaxial PROPIEDAD=OPTICA; ANI-SOTROPIA	Cristallo biassiale PROPRIETA OTTICA; ANI-SOTROPIA
477 BIBLIOGRAPHY MISC	BIBLIOGRAPHY	BIBLIOGRAPHIE	BIBLIOGRAPHIE	BIBLIOGRAFIYA	BIBLIOGRAFIA	BIBLIOGRAFIA
478 BIGHT SURF	Bights GULFS	Anse GOLFE	Bucht GOLF	Bukhta ZALIV	Comba GOLFO	Ansa GOLFO
479 BINARY SYSTEM PHCH	Binary system PHASE EQUILIBRIA	Systeme binaire DIAGRAMME EQUILIBRE	Binaer.System PHASEN=GLEICHGEWICHT	SISTEMA BINARNAYA	Sistema=binario DIAGRAMA=EQUILIBRIO	Sistema binario DIAGRAMMA D'EQUILIBRIO
480 BINOMIAL DISTRIBUTION MATH	Binomial distribution STATISTICAL DIS-TRIBUTION	Distribution binomiale DISTRIBUTION STATISTIQUE	Binomial=Verteilung STATIST. VERTEILUNG	RASPREDELENIE BINOMIAL'NOE	Distribucion=binomial DISTRIBUCION=ESTADISTICA	Distribuzione binomiale DISTRIBUZIONE STATISTICA
481 BIOCENOSIS PALE	BIOCENOSES	BIOCENOSE	BIOCOENOSE	BIOTSENOZ *UF:* Biota; Soobsh-chestvo; Sovokupnost'	BIOCENOSIS	BIOCENOSI
482 BIOCHEMICAL SEDIMENTATION SEDI	BIOCHEMICAL SEDIMENTATION	SEDIMENTATION BIOCHIMIQUE	BIOCHEM. SEDIMENTATION	Sedimentatsiya biokhimicheskaya BIOGEOKHIMIYA; SEDIMENTATSIYA	SEDIMENTACION=BIOQUIMICA	SEDIMENTAZIONE BIOCHIMICA
483 BIOCHORE PALE	Biochore BIOTOPES	Biochore BIOTOPE	Biochore BIOTOP	Biokhor BIOLOGICHESKIJ	Biocoro BIOTOPO	Biocora BIOTOPO
484 BIOCHRON STRA	Biochron CHRONOSTRATI-GRAPHY	Duree vie FAUNE SPECI-FIQUE	Biochrone SPEZIF.FAUNA	BIOKHRON	Biocrona FAUNA=ESPECIFICA	Biocrona FAUNA SPECIFICA
485 BIOCHRONOLOGY STRA	Biochronology CHRONOSTRATI-GRAPHY	Biochronologie AGE	Biochronologie ALTER	BIOKHRONOL-OGIYA	Biocronologia EDAD	Biocronologia ETA
486 BIOCLAST SEDI	Bioclasts BIOCLASTIC SEDI-MENTATION	Roche bioclastique SEDIMENTATION BIOCLASTIQUE	Bioklast BIOKLAST. SEDIMENTATION	BIOKLAST *UF:* Sedimentatsiya bioklasticheskaya	Bioclasto SEDIMENTACION=BIOCLASTICA	Bioclasto SEDIMENTAZIONE BIOCLASTICA
487 BIOCLASTIC LIMESTONE SEDS	BIOCALCARENITE	CALCAIRE BIO-CLASTIQUE *UF:* Biosparite; Calcaire entroques	BIOKLAST.KALK *UF:* Biosparit; Encrinit	IZVESTNYAK BIOK-LASTICHESKIJ	CALIZA=BIOCLASTICA *UF:* Biosparita; Encrinita	BIOCALCARENITE *UF:* Biosparite; Encrinite

	ENGLISH	FRANCAIS	DEUTSCH	RUSSKIJ	ESPANOL	ITALIANO
488 BIOCLASTIC SEDIMENTATION SEDI	BIOCLASTIC SEDI-MENTATION *UF:* Bioclasts	SEDIMENTATION BIOCLASTIQUE *UF:* Roche bioclastique	BIOKLAST. SEDIMENTATION *UF:* Bioklast	Sedimentatsiya bioklas-ticheskaya SEDIMENTATSIYA; BIOKLAST	SEDIMENTACION= BIOCLASTICA *UF:* Bioclasto	SEDIMENTAZIONE BIOCLASTICA *UF:* Bioclasto
489 BIODEGRADATION SURF	BIODEGRADATION	BIODEGRADATION	BIODEGRADATION	BIODEGRADAT-SIYA	BIODEGRADACION	BIODEGRADAZ-IONE
490 BIOFACIES SEDI	BIOFACIES *UF:* Biofacies map; Biosomes	BIOFACIES *UF:* Biosome; Carte biofacies	BIO=FAZIES *UF:* Biofazies=Karte; Biosom	BIOFATSIYA *UF:* Biosom; Karta biofatsial'naya	BIOFACIES *UF:* Biosoma; Mapa=biofacies	BIOFACIES *UF:* Biosoma; Carta delle biofacies
491 BIOFACIES MAP SEDI	Biofacies map BIOFACIES; MAPS	Carte biofacies BIOFACIES	Biofazies=Karte BIO=FAZIES	Karta biofatsial'naya KARTA; BIOFAT-SIYA	Mapa=biofacies BIOFACIES	Carta delle biofacies BIOFACIES
492 BIOGENESIS GEOL	Biogenesis BIOGENIC EFFECTS	Biogenese ORIGINE BIOGENE	Biogenese BIOGEN.BILDUNG	BIOGENEZ *UF:* Struktura biogen-naya	Biogenesis ORIGEN=BIOGENICO	Biogenesi ORIGINE ORGANO-GENA
493 BIOGENIC EFFECT GEOL	BIOGENIC EFFECTS *UF:* Bacteriogenic mate-rials; Biogenesis; Biolith; Biomicrite; Biosparite	ACTION BIOGENE	BIOGEN.WIRKUNG	BIOGENNYJ PROT-SESS	ACCION=BIOGENICA	AZIONE BIO-GENICA
494 BIOGENIC STRUCTURE TEST	BIOGENIC STRUC-TURES *UF:* Organic structure	STRUCTURE BIO-GENE	BIOGEN.STRUKTUR	Struktura biogennaya STRUKTURA; BIO-GENEZ	ESTRUCTURA=BIOGENA	STRUTTURA BIO-GENICA
495 BIOGEOCHEMICAL EXPLORATION ECON	BIOGEOCHEMICAL METHODS	PROSPECTION BIOGEOCHIMIQUE	BIO=GEOCHEM. PROSPEKTION	POISK BIOGEOKHIMI-CHESKIJ *UF:* Poisk geobotani-cheskij	PROSPECCION=BIOGEOQUIMICA	PROSPEZIONE BIOGEOCHIMICA
496 BIOGEOCHEMISTRY GEOC	BIOCHEMISTRY	BIOCHIMIE	BIO=GEOCHEMIE	BIOGEOKHIMIYA *UF:* Sedimentatsiya biokhimicheskaya	BIOGEOQUIMICA	BIOGEOCHIMICA
497 BIOGEOGRAPHY PALE	BIOGEOGRAPHY *UF:* Geobios; Paleobo-tanic province; Provin-ciality	BIOGEOGRAPHIE *UF:* Geobios; Province paleobotanique; Provin-cialite	BIOGEOGRAPHIE *UF:* Geobios; Palaeobotan.Provinz; Provinzialitaet	BIOGEOGRAFIYA *UF:* Provintsii i zony	BIOGEOGRAFIA *UF:* Geobios; Provincia=paleobotanica; Provin-cialidad	BIOGEOGRAFIA *UF:* Geobios; Provincia paleobotanica; Provin-cialita
498 BIOGRAPHY MISC	BIOGRAPHY *UF:* Personal bibliogra-phy	Biographie BIBLIOGRAPHIE PERSONNELLE	Biographie PERSONAL=BIBLIOGRAPHIE	BIOGRAFIYA	Biografia BIBLIOGRAFIA=PERSONAL	Biografia BIBLIOGRAFIA PERSONALE
499 BIOHERM SEDI	BIOHERMS *UF:* Algal reefs; Causto-biolith	BIOHERME *UF:* Boundstone	BIOHERM *UF:* Organ.Kalk	BIOGERM *UF:* Rif vodoroslevyj	BIOHERMES *UF:* Boundstone	BIOHERMA *UF:* Boundstone
500 BIOLITH SEDI	Biolith BIOGENIC EFFECTS	Biolite ORIGINE BIOGENE	Biolith BIOGEN.BILDUNG	BIOLIT *UF:* Biomikrit; Biosparit; Ehnkrinit	Biolito ORIGEN=BIOGENICO	Biolite ORIGINE ORGANO-GENA
501 BIOLOGICAL CYCLE PALE	BIOLOGICAL CYCLE	CYCLE BIOLOGIQUE	BIOLOG.ZYKLUS	TSIKL BIOLOGI-CHESKIJ *UF:* Biokhor	CICLO=BIOLOGICO	CICLO BIOLOGICO

	ENGLISH	FRANCAIS	DEUTSCH	RUSSKIJ	ESPANOL	ITALIANO
502 BIOLOGY MISC	BIOLOGY *UF:* Heredity; Heterotrophic; Internal structure (pale); Organisms	BIOLOGIE *UF:* Heredite; Organisme; Organisme heterotrophe; Structure interne	BIOLOGIE *UF:* Heterotroph. Organismus; Internstruktur; Organismus; Vererbung	BIOLOGIYA *UF:* Gistologiya	BIOLOGIA *UF:* Afininidad; Estructura=interna; Herencia; Organismo; Organismo heterotrofico	BIOLOGIA *UF:* Ereditarieta; Organismo; Organismo eterotrofo; Struttura interna
503 BIOMASS PALE	BIOMASS	BIOMASSE	BIOMASSE	BIOMASSA	BIOMASA	BIOMASSA
504 BIOMETRY PALE	BIOMETRY	BIOMETRIE	BIOMETRIE	BIOMETRIYA	BIOMETRIA	BIOMETRIA
505 BIOMICRITE SEDS	Biomicrite LIMESTONE; BIOGENIC EFFECTS	Biomicrite CALCAIRE MICROCRISTALLIN; ORIGINE BIOGENE	Biomikrit MIKROKRISTALLIN.KALK; BIOGEN.BILDUNG	Biomikrit IZVESTNYAK; BIOLIT	Biomicrita CALIZA= MICROCRISTALINA; ORIGEN= BIOGENICO	Biomicrite CALCARE MICROCRISTALLINO; ORIGINE ORGANOGENA
506 BIORHEXISTASY SURF	BIORHEXISTASY	BIORHEXISTASIE	BIORHEXISTASIE	BIOREKSISTAZIYA	BIORRESISTASIA	BIORESISTASI
507 BIOSOME STRA	Biosomes BIOFACIES	Biosome BIOFACIES	Biosom BIO=FAZIES	Biosom BIOFATSIYA	Biosoma BIOFACIES	Biosoma BIOFACIES
508 BIOSPARITE SEDS	Biosparite LIMESTONE; BIOGENIC EFFECTS	Biosparite CALCAIRE BIOCLASTIQUE	Biosparit BIOKLAST.KALK	Biosparit IZVESTNYAK; BIOLIT	Biosparita CALIZA= BIOCLASTICA	Biosparite BIOCALCARENITE
509 BIOSPHERE GEOL	BIOSPHERE	BIOSPHERE	BIOSPHAERE	BIOSFERA	BIOSFERA	BIOSFERA
510 BIOSTRATIGRAPHIC UNIT STRA	Biostratigraphic unit BIOSTRATIGRAPHY; STRATIGRAPHIC UNITS	Unite biostratigraphique UNITE STRATIGRAPHIQUE	Biostratigraph.Einheit STRATIGRAPH. EINHEIT	PODRAZDELENIE BIOSTRATIGRAF.	Unidad= bioestratigrafica UNIDAD= ESTRATIGRAFICA	Unita biostratigrafica UNITA STRATIGRAFICA
511 BIOSTRATIGRAPHY STRA	BIOSTRATIGRAPHY *UF:* Biostratigraphic unit; Biozones	BIOSTRATIGRAPHIE	BIOSTRATIGRAPHIE	BIOSTRATIGRAFIYA	BIOESTRATIGRAFIA	BIOSTRATIGRAFIA
512 BIOSTROME SEDI	BIOSTROMES	BIOSTROME	BIOSTROM	BIOSTROM	BIOSTROMA	BIOSTROMA
513 BIOTA PALE	BIOTA	BIOS	BIOTA	Biota BIOTSENOZ	BIOTA	BIOTA
514 BIOTOPE PALE	BIOTOPES *UF:* Biochore	BIOTOPE *UF:* Biochore	BIOTOP *UF:* Biochore	BIOTOP *UF:* Ehkosistema	BIOTOPO *UF:* Biocoro	BIOTOPO *UF:* Biocora
515 BIOTURBATION SEDI	BIOTURBATION	BIOTURBATION	BIOTURBATION	Bioturbatsiya POSTROJKA ORGANOGENNAYA	BIOTURBACION	BIOTURBAZIONE
516 BIOTYPE PALE	BIOTYPES *UF:* Genotypes	BIOTYPE *UF:* Genotype	BIOTYPUS *UF:* Genotyp	BIOTIP	BIOTIPO *UF:* Genotipo	BIOTIPO *UF:* Genotipo
517 BIOZONE STRA	Biozones BIOSTRATIGRAPHY; ZONING	Biozone FAUNE SPECIFIQUE	Biozone SPEZIF.FAUNA	BIOZONA *UF:* Zona ehkologicheskya; Zona faunisticheskaya	Biozona FAUNA= ESPECIFICA	Biozona FAUNA SPECIFICA

	ENGLISH	FRANCAIS	DEUTSCH	RUSSKIJ	ESPANOL	ITALIANO
518 BIRCH DISCONTINUITY SOLI	Birch discontinuity DISCONTINUITIES; MANTLE	Discontinuite Birch DISCONTINUITE; MANTEAU	Birch=Diskontinuitaet DISKONTINUIT- AET; ERD=MANTEL	Bercha poverkhnost' MANTIYA; GRANITSA SEJSMI- CHESKAYA	Discontinuidad=de= Birch DISCONTINUIDAD; MANTO=GLOBO	Discontinuita di Birch DISCONTINUITA; MANTELLO
519 BIREFRINGENCE MINE	BIREFRINGENCE	BIREFRINGENCE	DOPPELBRECHUNG	DVUPRELOMLENIE	BIRREFRINGENCIA	BIRIFRANGENZA
520 BISMUTH CHEE	BISMUTH	BISMUTH	BI	VISMUT	BISMUTO	BISMUTO
521 BITUMEN COMS	BITUMENS UF: Albertite; Asphaltite	BITUME UF: Albertite; Anthraxo- lite; Bituminisation	BITUMEN UF: Albertit; Anthrax- olith; Bituminisierung	BITUM UF: Al'bertit	BETUN UF: Albertita; Antrax- olita; Bituminizacion	BITUME UF: Albertite; Antraxo- lite; Bituminizzazione
522 BITUMINIZATION SEDI	BITUMINIZATION	Bituminisation BITUME	Bituminisierung BITUMEN	BITUMINIZATSIYA	Bituminizacion BETUN	Bituminizzazione BITUME
523 BITUMINOUS COAL COMS	BITUMINOUS COAL	CHARBON BITU- MINEUX	BITUMINOES. KOHLE	UGOL' BITUMINOZ- NYJ	CARBON= BITUMINOSO	CARBONE BITUMINOSO
524 BITUMINOUS SHALE COMS	Bituminous shale OIL SHALE	SCHISTE BITU- MINEUX UF: Schiste a huile	BITUMINOES. SCHIEFER UF: Oel=Schiefer	Slanets bituminoznyj SLANETS GORYUCHIJ	ESQUISTO= BITUMINOSO UF: Esquisto=petrolifero	SCISTO BITUMINOSO UF: Scisto a olio
525 BLASTIC TEXTURE TEST	Blastic texture TEXTURES	Texture blastique TEXTURE	Blast.Gefuege KORN=GEFUEGE	BLASTICHESKIJ	Textura=blastica TEXTURA	Tessitura blastica TESSITURA
526 BLASTOIDEA PALS	BLASTOIDEA	BLASTOIDEA	BLASTOIDEA	BLASTOIDEA	BLASTOIDEA	BLASTOIDEA
527 BLATTOIDEA PALS	BLATTODEA	BLATTOPTEROIDA	BLATTOPTEROIDA	BLATTODEA	BLATTOPTEROIDA	BLATTOPTEROIDA
528 BLEACHING CLAY COMS	Bleaching clay ABSORBENT MATE- RIALS; CLAYS	Terre adsorbante ARGILE; ABSOR- BANT	Bleich=Erde TON; ABSORBER	GLINA OTBELIVAYUSH- CHAYA UF: Fullerova zemlya	Arcilla=esmectica ARCILLA; ABSOR- BENTE	Argilla sbiancante ARGILLA; ASSOR- BENTE
529 BLIND DEPOSIT ECON	BLIND DEPOSITS	GITE CACHE	VERDECKT. LAGERSTAETTE	SLEPOJ	YACIMIENTO=NO= AFLORANTE	GIACIMENTO CHI- USO
530 BLOCK GEOL	BLOCKS	BLOC UF: Bloc isole	BLOCK UF: Geschiebe	BLOK	BLOQUE UF: Bloque=aislado	BLOCCO UF: Blocco isolato
531 BLOCK DIAGRAM GEOL	BLOCK DIAGRAMS	Bloc diagramme MAQUETTE	Block=Diagramm MODELL	BLOK= DIAGRAMMA	Bloque=diagrama MAQUETA	Block=diagramma MODELLO IN SCALA
532 BLOCK FAULTING STRU	BLOCK STRUC- TURES	TECTONIQUE CAS- SANTE	BLOCK=BILDUNG	TEKTONIKA BLOKOVAYA UF: Mikroplita	TECTONICA=DE= FRACTURA	TETTONICA DISGI- UNTIVA
533 BLOWHOLE SURF	BLOWHOLES	Trou souffleur EROSION LITTOR- ALE	[BLOWHOLE] KUESTEN= EROSION	KANAL PRODU- VANIYA	Bufadero EROSION=LITORAL	Sfiatatoio EROSIONE COSTI- ERA
534 BLUESCHIST IGMS	BLUESCHIST UF: Glaucophane schist facies	SCHISTE BLEU	GLAUKOPHAN= SCHIEFER	Slanets goluboj SLANETS GLAUKO- FANOVYJ	ESQUISTO=AZUL	SCISTI BLU

	ENGLISH	FRANCAIS	DEUTSCH	RUSSKIJ	ESPANOL	ITALIANO
535 BODY WAVE SOLI	BODY WAVES	ONDE VOLUME	RAUM=WELLE	Volna ob'emnaya VOLNA SEJSMI- CHESKAYA	ONDA=TRANSVER- SAL	ONDA DI VOLUME
536 BOG SURF	BOGS	MARECAGE *UF:* Palse	MORAST *UF:* Palsa	BOLOTO *UF:* Swamp; Swamp sedimentation; Wetland	PANTAÑO *UF:* Palsa	ACQUITRINO *UF:* Palsa
537 BOG IRON ORE COMS	Bog iron ore IRON; LIMONITE	Fer marais LIMONITE	Rasen=Eisenerz LIMONIT	RUDA BOLOTNAYA	Hierro=de=los= pantanos LIMONITA	. Ferro delle paludi LIMONITE
538 BOGHEAD COAL COMS	Boghead coal PEAT	Charbon boghead CHARBON SAPRO- PELIQUE	Boghead=Kohle SAPROPEL=KOHLE	BOGKHED	Carbon=boghead · CARBON= SAPROPELICO	Carbone boghead CARBONE SAPRO- PELITICO
539 BOMB IGNE	Bombs PYROCLASTICS	Bombe volcanique PYROCLASTIQUE	Bombe PYROKLAST. GESTEIN	BOMBA VULKANI- CHESKAYA	Bomba=volcanica PIROCLASTICO	Bomba PRODOTTO PIRO- CLASTICO
540 BONE BED SEDI	BONE BEDS	BONE BED	BONE=BED	BREKCHIYA KOS- TYANAYA	BONE=BED	BONE BED
541 BOOK REVIEW MISC	BOOK REVIEWS	ANALYSE LIVRE	BUCH= BESPRECHUNG	REFERAT/KNIGI/	ANALISIS=LIBRO	RECENSIONE
542 BOOLEAN ALGEBRA MATH	Boolean algebra MATHEMATICAL METHODS	Algebre booleenne METHODE MATHE- MATIQUE	Boole'sche=Algebra MATHEMAT. METHODE	BULEVA ALGEBRA	Algebra=booleana METODO= MATEMATICO	Algebra booleana METODO MATEMA- TICO
543 BORALF SUSS	Boralfs ALFISOLS	Boralf ALFISOL; ZONE FROIDE	Boralf ALFISOL; KALT. ZONE	Boralf AL'FISOL	Boralf ALFISOL; ZONA= FRIA	Boralf ALFISOL; ZONA FREDDA
544 BORATES MING	BORATES *UF:* Borax	BORATE *UF:* Borax	BORAT *UF:* Borax	BORAT	BORATO *UF:* Borax	BORATO *UF:* Borace
545 BORAX COMS	Borax BORATES; SODIUM	Borax BORATE; SODIUM	Borax BORAT; NA	BURA	Borax BORATO; SODIO	Borace BORATO; SODIO
546 BORDER FACIES IGNE	Border facies METAMORPHISM	Facies bordure INTRUSION	Rand=Fazies INTRU- SIONS=KOERPER	EHNDOKONTAKT *UF:* Brekchiya kontak- tovaya	Facies=borde INTRUSION	Facies di contatto INTRUSIONE
547 BORDERLAND STRU	Borderland CONTINENTAL SHELF	Bordure continentale PLATEFORME CON- TINENTALE	Borderland KONTINENTAL= TAFEL	Borderland SHEL'F KONTINEN- TAL'NYJ	Borde=continental PLATAFORMA= CONTINENTAL	Area marginale PIATTAFORMA CONTINENTALE
548 BOREAL SURF	BOREAL *UF:* Subboreal	ZONE FROIDE *UF:* Boralf; Milieu froid; Subboreal	KALT.ZONE *UF:* Boralf; Kalt.Milieu; Subboreal	BOREAL'NYJ *UF:* Subboreal'nyj	ZONA=FRIA *UF:* Boralf; Medio=frio; Subboreal	ZONA FREDDA *UF:* Ambiente freddo; Boralf; Subboreale

549 BOREHOLE
METH
	BOREHOLES
	UF: Auger drilling; Bore-
	hole section; Develop-
	ment drilling; Direc-
	tional drilling; Drill
	sludge; Drilling; Drill-
	ing equipment; Drilling
	vessel; Exploratory well;
	Full hole drilling; Mo-
	hole project; Offshore
	drilling; Rotary drilling;
	Rotary=percussion
	drilling; Turbodrilling;
	Vibratory drilling; Well
	head
		SONDAGE
		UF: Boue forage; Fora-
		bilite; Forage percussion
		rotary; Forage rotary;
		Materiel sondage; Nav-
		ire sondage; Projet Mo-
		hole; Sondage carottage
		complet; Sondage car-
		otte; Sondage develop-
		pement; Sondage off-
		shore; Sondage oriente;
		Sondage reconnais-
		sance; Sondage tariere;
		Sondage turbo; Sondage
		vibration; Tete forage
			BOHRUNG
			UF: Aufschluss=Boh-
			rung; Bohr-Geraet;
			Bohr=Kopf;
			Bohr=Schiff;
			Bohr=Schlamm; Bohr-
			barkeit; Entwick-
			lungs=Bohrung;
			Hand=Bohrung;
			Kern=Bohrung; Mo-
			hole=Projekt; Off-
			shore=Bohrung; Rich-
			tungs=Bohren; Rota-
			ry=Bohrmethode; Rota-
			ry=Bohrung; Tur-
			bo=Bohrung; Vibra-
			tions=Bohrung; Voll-
			kern=Bohrung
				SKVAZHINA
				UF: Well screen
					POZO=SONDEO
					UF: Barco=sondeo;
					Cruz=produccion;
					Lodo=de=sondeo; Ma-
					quinaria=sondeo; Perfo-
					rabilidad; Pozo=explo-
					racion; Proyecto=Mo-
					hole; Sondeo=auger;
					Sondeo=direccion; Son-
					deo=explotacion; Son-
					deo=off=shore; Son-
					deo=rotacion=percu-
					sion; Sondeo=rotary;
					Sondeo=testigo=con-
					tinuo; Sondeo=vibra-
					cion; Testigo=sondeo;
					Turbo-sondeo
						SONDAGGIO
						UF: Apparecchi per per-
						forazione; Carotaggio;
						Fango di perforazione;
						Nave per trivellazione;
						Perforabilita; Perfora-
						zione; Peforazione aug-
						er; Perforazione dire-
						zionata; Perforazione
						offshore; Perforazione
						rotary; Perforazione ro-
						topercussione; Perfora-
						zione vibratoria; Pro-
						getto Mohole; Sondag-
						gio esplorativo; Sviluppo
						di pozzo; Testa di pozzo;
						Turboperforazione

	ENGLISH	FRANCAIS	DEUTSCH	RUSSKIJ	ESPANOL	ITALIANO
550 BOREHOLE SECTION GEOL	Borehole section BOREHOLES	COUPE SONDAGE *UF:* Log	BOHR=PROFIL *UF:* Bohr=Log	RAZREZ GEOLOGI- CHESKIJ SKVAZHINY	CORTE=GEOL= SONDEO *UF:* Testificacion	SEZIONE DI SON- DAGGIO *UF:* Log
551 BOROLL SUSS	Borolls MOLLISOLS	Boroll MOLLISOL; CHER- NOZEM	Boroll MOLLISOL; TSCH- ERNOZEM	Boroll MOLLISOL	Boroll MOLLISOL; CHER- NOZEM	Boroll MOLLISOL; CERNOZEM
552 BORON CHEE	BORON	BORE	B	BOR	BORO	BORO
553 BOTRYOIDAL TEST	Botryoidal texture TEXTURES; COL- LOIDAL MATERI- ALS	Forme botryoidale TEXTURE; COLL- OIDE	Glaskopf=Gefuege KORN=GEFUEGE; KOLLOID	GROZD'EVIDNYJ	Forma=botrioidal TEXTURA; COL- OIDE	Botrioidale TESSITURA; COLL- OIDE
554 BOTTOM SEDIMENT SEDI	Bottom sediments SEDIMENTS	Sediment fond SEDIMENT	Boden=Sediment SEDIMENT	OSADOK DONNYJ	Sedimento=base SEDIMENTO	Sedimento di fondo SEDIMENTO
555 BOTTOMSET BED SEDI	BOTTOMSET BEDS	Couche basale STRATIFICATION ENTRECROISEE; SEDIMENTATION DELTAIQUE	[BOTTOMSET= BED] KREUZSCHICH- TUNG; DELTA= SEDIMENTATION	Sloj pridonnyj SLOISTOST'	Capa=basal ESTRATIFIC- ACION=CRUZADA; SEDIMENTACION= DELTAICA	Strato di letto STRATIFICAZIONE INCROCIATA; SEDI- MENTAZIONE DEL- TIZIA
556 BOUDINAGE STRU	BOUDINAGE	BOUDINAGE	BOUDINAGE	BUDINAZH	BOUDINAGE	BOUDINAGE
557 BOUGUER ANOMALY APPL	BOUGUER ANOMA- LIES	ANOMALIE BOUGUER	BOUGUER= ANOMALIE	Buge anomaliya ANOMALIYA GRAVITATSION- NAYA	ANOMALIA= BOUGUER	ANOMALIA DI BOUGUER
558 BOULDER SEDI	BOULDERS	Bloc isole BLOC	Geschiebe BLOCK	VALUN	Bloque=aislado BLOQUE	Blocco isolato BLOCCO
559 BOULDER CLAY SURF	BOULDER CLAY	ARGILE A BLO- CAUX	GESCHIEBE= MERGEL	GLINA VALUN- NAYA	ARCILLA=CON= BLOQUES	ARGILLA A BLOC- CHI
560 BOULDER TRAIN SURF	BOULDER TRAINS	ALIGNEMENT BLOC GLACIAIRE	BLOCK=STROM	Gryada valunnaya MORENA	ALINEACION= BLOQUE=GLACIAR	ALLINEAMENTO DI ERRATICI
561 BOUNDARY LAYER GEOC	BOUNDARY LAYER	COUCHE LIMITE	GRENZSCHICHT	SLOJ POGRANICH- NYJ	CAPA=LIMITE	STRATO LIMITE
562 BOUNDSTONE SEDS	Boundstone LIMESTONE	[BOUNDSTONE] CALCAIRE; BIO- HERME	Organ.Kalk KALK; BIOHERM	[BOUNDSTONE] PORODA KAR- BONATNAYA	[BOUNDSTONE] CALIZA; BIOHER- MES	[BOUNDSTONE] CALCARE; BIO- HERMA
563 BOWEN'S REACTION SERIES IGNE	Bowen's reaction series FRACTIONAL CRYSTALLIZATION	Suite reactionnelle Bowen CRISTALLISATION FRACTIONNEE; FUSION	Bowen's=Reaktions= Serie FRAKTIONIERT. KRISTALLISATION; SCHMELZE	Ryad reaktsionnyj Bouehna RYAD REAKTSION- NYJ	Series=de=reaccion= de=Bowen CRISTALIZACION= FRACCIONADA; FUSION	Serie reazionale di Bowen CRISTALLIZZAZ- IONE FRA- ZIONATA; FUSIONE
564 BOXWORK TEST	Boxwork texture TEXTURES; EXO- GENE PROCESSES	Texture cloisonnee TEXTURE; PROCES- SUS EXOGENE	Zellen=Gefuege KORN=GEFUEGE; EXOGEN.VORGANG	YACHEISTYJ	Textura=tabicada TEXTURA; PROCESO= EXOGENO	Tessitura reticolare TESSITURA; EVOLUZIONE SUPERGENICA

	ENGLISH	FRANCAIS	DEUTSCH	RUSSKIJ	ESPANOL	ITALIANO
565 BRACHIOPODA PALS	BRACHIOPODA *UF:* Kutorginida	BRACHIOPODA	BRACHIOPODA	BRACHIOPODA	BRACHIOPODA	BRACHIOPODA
566 BRACHIOPTERYGII PALS	BRACHIOPTERYGII	BRACHIOPTERYGII	BRACHIOPTERYGII	Brachiopterygii OSTEICHTHYES	BRACHIOPTERYGII	BRACHIOPTERYGII
567 BRACHYFOLD STRU	Brachyfold FOLDS	Brachypli PLI	Brachy=Falte FALTE	BRAKHISKLADKA	Braquipliegue PLIEGUE	Brachipiega PIEGA
568 BRACKISH WATER GEOH	BRACKISH WATER	EAU SAUMATRE	BRACK=WASSER	VODA SOLONOVATAYA	AGUA=SALOBRE	ACQUA SAL- MASTRA
569 BRACKISH-WATER ENVIRONMENT PALE	BRACKISH=WATER ENVIRONMENT	MILIEU SAU- MATRE	BRACK=WASSER- MILIEU	SOLONOVATOVOD- NYJ	MEDIO=SALOBRE	AMBIENTE SAL- MASTRO
570 BRAGG ANGLE MINE	Bragg angle X=RAY DIFFRAC- TION ANALYSIS	Reflection Bragg DIFFRACTION RX	Bragg=Winkel ROENTGEN- BEUGUNG	Ugol otrazheniya DIFRAKTSIYA	Angulo=de=Bragg DIFRACCION=RX	Angolo di Bragg DIFFRAZIONE RAGGI X
571 BRAIDED CHANNEL SURF	Braided channels CHANNELS	Reseau anastomose CHENAL	Wild=Wasser RINNE	RUSLO VETVYASH- CHEESYA	Red=de=canales CAUCE	Drenaggio anastomiz- zato CANALE
572 BRANCHIOPODA PALS	BRANCHIOPODA *UF:* Cladocera; Con- chostraca; Notostraca; Phyllopoda	Branchiopoda CONCHOSTRACA	Branchiopoda CONCHOSTRACA	BRANCHIOPODA *UF:* Phyllopoda	Branchiopoda CONCHOSTRACA	Branchiopoda CONCHOSTRACA
573 BRANCHIURA PALS	Branchiura CRUSTACEA	Branchiura CRUSTACEA	Branchiura CRUSTACEA	Branchiura CRUSTACEA	Branchiura CRUSTACEA	Branchiura CRUSTACEA
574 BRAVAIS LATTICE MINE	Bravais lattice CRYSTAL STRUC- TURE	Reseau Bravais CONSTANTE RETICULAIRE	Bravais=Gitter GITTERKON- STANTE	Reshetka Brave RESHETKA KRIS- TALLICHESKAYA	Red=de=Bravais CONSTANTE= RETICULAR	Reticolo di Bravais COSTANTE RETI- COLARE
575 BRECCIA GEOL	BRECCIA *UF:* Collapse breccia; Contact breccia; Shock breccia	BRECHE *UF:* Breche choc; Breche contact; Breche effon- drement; Minerai disloque	BREKZIE *UF:* Einsturz=Brekzie; Geschleppt.Erz; Kontakt=Brekzie; Schock=Brekzie	BREKCHIYA *UF:* Brekchiya kontak- tovaya; Collapse brec- cia	BRECHA *UF:* Brecha=de= contacto; Brecha=de= desplome; Brecha=de= impacto; Escape	BRECCIA *UF:* Breccia d'impatto; Breccia di collasso; Breccia di contatto; Drag ore
576 BRICK CLAY COMS	Brick clay CLAYS; CON- STRUCTION MATE- RIALS	Argile a brique ARGILE; MATERIAU CON- STRUCTION	Ziegel=Ton TON; BAUSTOFF	GLINA KIRPICH- NAYA	Arcilla=de=ladrillo ARCILLA; MATERIAL= CONSTRUCCION	Argilla per laterizi ARGILLA; MATERI- ALE DA COSTRUZ- IONE
577 BRIDGE ENGI	BRIDGES	PONT	BRUECKE	MOST	PUENTE	PONTE
578 BRINE SEDS	BRINES	SAUMURE	SALZ=LAUGE	RAPA	SALMUERA	BRINA
579 BRITTLE PHCH	Brittle(adj.)	Friable(adj.)	Sproede(adj.)	KHRUPKOST'	Fragil(adj.)	Fragile(adj.)
580 BROMIDES MING	BROMIDES	BROMURE	BROMID	BROMID	BROMURO	BROMURO

	ENGLISH	FRANCAIS	DEUTSCH	RUSSKIJ	ESPANOL	ITALIANO
581 BROMINE CHEE	BROMINE	BROME	BR	BROM	BROMO	BROMO
582 BROWN COAL COMS	Brown coal LIGNITE	[BROWN COAL] LIGNITE	BRAUNKOHLE *UF:* Lignit	UGOL' BURYJ *UF:* Lignit	Carbon=pardo LIGNITO	[BROWN COAL] LIGNITE
583 BROWN SOIL SUSS	BROWN SOILS *UF:* Cambisols; Phaeozem	SOL BRUN *UF:* Cambisol; Ochrept; Phaeozem; Udoll	BRAUN=ERDE *UF:* Cambisol; Ochrept; Phaeozem; Udoll	BUROZEM	SUELO=PARDO *UF:* Cambisol; Faeozem; Ochrept; Udoll	SUOLO BRUNO *UF:* Cambisol; Ochrept; Phaeozem; Udoll
584 BRYOPHYTA PALS	BRYOPHYTES	BRYOPHYTA	BRYOPHYTA	BRYOPSIDA	BRYOPHYTA	BRYOPHYTA
585 BRYOZOA PALS	BRYOZOA *UF:* Gymnolaemata; Phylactolaemata; Zooe- cium	BRYOZOA *UF:* Gymnolaemata; Phylactolaemata; Zoe- cium	BRYOZOA *UF:* Gymnolaemata; Phylactolaemata; Zoez- ium	BRYOZOA	BRYOZOA *UF:* Gymolaemata; Phylactolaemata; Zoe- cia	BRYOZOA *UF:* Gymnolaemata; Phylactolaemata; Zooe- cium
586 BUILDING ENGI	BUILDINGS	BATIMENT	BAUWESEN	STROENIE	EDIFICIO	EDIFICIO
587 BUILDING CEMENT COMS	CEMENT MATERI- ALS	CIMENT INDUSTR- IEL	INDUSTRIE= ZEMENT	TSEMENT STROI- TEL'NYJ *UF:* Syr'e tsementnoe	CEMENTO= INDUSTRIAL	CEMENTO INDUSTRIALE
588 BUILDING STONE COMS	BUILDING STONE	PIERRE TAILLE	BAUSTEIN	STROJMATERIALY	PIEDRA=DE= CONSTRUCCION	PIETRA DA TAGLIO
589 BULIMINACEA PALS	BULIMINACEA	BULIMINACEA	BULIMINACEA	BULIMINACEA	BULIMINACEA	BULIMINACEA
590 BULK DENSITY PHCH	Bulk density DENSITY	Densite brute DENSITE	Roh=Dichte DICHTE	Ves ob'emnyj VES UDEL'NYJ	Densidad=aparente DENSIDAD	Densita apparente DENSITA
591 BULK MODULUS PHCH	BULK MODULUS	MODULE ELASTI- CITE *UF:* Module rigidite	ELASTIZITAETS= MODUL *UF:* Schubmodul	MODUL' OB'EMNYJ	MODULO= ELASTICIDAD *UF:* Modulo=rigidez	MODULO D'ELASTICITA *UF:* Modulo di rigidita
592 BULLARD DISCONTINUITY SOLI	Bullard discontinuity DISCONTINUITIES; CORE	Discontinuite Bullard DISCONTINUITE; NOYAU TER- RESTRE	Bullard= Diskontinuitaet DISKONTINUIT- AET; ERD=KERN	BULLARDA POVERKHNOST'	Discontinuidad=de= Bullard DISCONTINUIDAD; NUCLEO=GLOBO	Discontinuita di Bul- lard DISCONTINUITA; NUCLEO TER- RESTRE
593 BUNTSANDSTEIN STRS	BUNTER	BUNTSANDSTEIN	BUNTSANDSTEIN	Buntzandshtejn TRIAS NIZHNIJ	BUNTSANDSTEIN	BUNTSANDSTEIN
594 BURDIGALIAN STRS	BURDIGALIAN	BURDIGALIEN	BURDIGAL	BURDIGAL	BURDIGALIENSE	BURDIGALIANO
595 BUREAU OF MINES MISC	BUREAU OF MINES	SERVICE DES MINES	BERGBAU= BEHOERDE	UCHREZHDENIE GEOLOGICHESKOE	SERVICIO=DE= MINAS	CORPO DELLE MINIERE
596 BURIAL GEOL	Burial(gen.)	Enfouissement(gen.)	Versenkung(gen.)	POGREBENNYJ	Entierro(gen.)	Seppellimento(gen.)
597 BURIAL METAMORPHISM IGNE	BURIAL META- MORPHISM	METAMORPHISME ENFOUISSEMENT	VERSENKUNGS= METAMORPHOSE	METAMORFIZM NAGRUZKI	METAMORFISMO= DE=SUBSIDENCIA	METAMORFISMO DI AFFOSSAMENTO

	ENGLISH	FRANCAIS	DEUTSCH	RUSSKIJ	ESPANOL	ITALIANO
598 BURIED CHANNEL SURF	BURIED CHAN- NELS	CHENAL ENFOUI	BEGRABEN.RINNE	Ruslo pogrebennoe DOLINA POGRE- BENNAYA	CANAL=FOSIL	CANALE SEPOLTO
599 BURIED VALLEY SURF	BURIED VALLEYS	VALLEE ENFOUIE	BEGRABEN.TAL	DOLINA POGRE- BENNAYA *UF:* Ruslo pogrebennoe	VALLE= FOSILIZADO	VALLE SEPOLTA
600 BURROW PALE	BURROWS	TERRIER	WOHN=SPUR	KHODY CHERVEJ	MADRIGUERA	[BURROW]
601 BYSSUS PALE	Byssus MORPHOLOGY	Byssus ANATOMIE	Byssus ANATOMIE	BISSUS	Biso ANATOMIA	Bisso ANATOMIA
602 C LAYER SOLI	C layer UPPER MANTLE	Couche C MANTEAU SUP	C=Schicht OBER.ERD= MANTEL	C sloj MANTIYA VERKH- NYAYA	Capa=C MANTO=GLOBO= SUP	Strato C MANTELLO SUPERIORE
603 C.I.P.W. CLASSIFICATION IGNE	C.I.P.W. Classification PETROGRAPHY	Classification CIPW CALCUL PETRO- GRAPHIQUE	CIPW=Norm PETROGRAPH. ANALYSE	KLASSIFIKATSIYA C.I.P.W. *UF:* Sostav normativnyj	Clasificacion=C.I.P.W. PETROGRAFIA	Classificazione C.I.P.W. CALCOLO PETROGRA- FICO
604 CADMIUM CHEE	CADMIUM	CADMIUM	CD	KADMIJ	CADMIO	CADMIO
605 CAFEMIC COMPOSITION IGNE	Cafemic composition MAFIC COMPOSI- TION	Composition cafemique COMPOSITION CALCOALCALINE	Cafem. Zusammensetzung KALK=ALKALI= TYP	KAFEMICHESKIJ	Composicion=cafemica COMPOSICION= CALCOALCALINA	Composizione cafemica COMPOSIZIONE CALCALCALINA
606 CALABRIAN STRS	CALABRIAN	CALABRIEN	CALABRIUM	Kalabrij CHETVERTICHNYJ	CALABRIENSE	CALABRIANO
607 CALC=ALKALIC COMPOSITION IGNE	CALC=ALKALIC COMPOSITION *UF:* Alkali=calcic com- position	COMPOSITION CALCOALCALINE *UF:* Composition alcalino=calcique; Composition cafemique	KALK=ALKALI= TYP *UF:* Alkali=Kalk=Type; Cafem. Zusammensetzung	KAL'TSEVO= SHCHELOCHNOJ	COMPOSICION= CALCOALCALINA *UF:* Composicion alcalino=calcica; Composicion=cafemica	COMPOSIZIONE CALCALCALINA *UF:* Composizione alcali- calcica; Composizione cafemica
608 CALCAREOUS ALGAE PALS	CALCAREOUS ALGAE	ALGUE CALCAIRE	KALK=ALGEN	VODOROSLI IZVESTKOVYE	ALGA=CALCAREA	ALGHE CALCAREE
609 CALCAREOUS COMPOSITION SEDI	CALCAREOUS COMPOSITION	COMPOSITION CARBONATEE	KALK=GEHALT	IZVESTKOVISTYJ	COMPOSICION= CARBONATADA	COMPOSIZIONE CALCAREA
610 CALCIC COMPOSITION IGNE	CALCIC COMPOSI- TION	COMPOSITION CALCIQUE	KALK= CHEMISMUS	KAL'TSIEVYJ	COMPOSICION= CALCICA	COMPOSIZIONE CALCICA
611 CALCIPHYRE IGMS	Calciphyre MARBLES	Calciphyre MARBRE	Kalziphyr MARMOR	KAL'TSIFIR	Calcifido MARMOL	Calcefiro MARMO
612 CALCISPONGEA PALS	Calcispongea PORIFERA	Calcispongea PORIFERA	Calcispongea PORIFERA	CALCISPONGEA	Calcispongea PORIFERA	Calcispongea PORIFERA
613 CALCITE COMS	CALCITE	CALCITE	KALZIT	KAL'TSIT	CALCITA	CALCITE

	ENGLISH	FRANCAIS	DEUTSCH	RUSSKIJ	ESPANOL	ITALIANO
614 CALCITIZATION GEOL	CALCITIZATION	CALCITISATION	KALZITISIERUNG	KAL'TSITIZATSIYA	CALCITIZACION	CALCITIZZAZIONE
615 CALCIUM CHEE	CALCIUM	CALCIUM	CA	KAL'TSIJ	CALCIO	CALCIO
616 CALCRETE SURF	CALCRETE	CROUTE CALCAIRE	KALK=KRUSTE	Kal'kret KORA VYVETRIVANIYA	COSTRA= CALCAREA	CROSTA CAL- CAREA
617 CALDERA SURF	CALDERAS UF: Cauldrons; Collapse caldera; Volcano= tectonic depression	CALDEIRA UF: Caldeira effondre- ment; Depression volca- notectonique; Subsi- dence en chaudron	CALDERA UF: Einsturz=Caldera; Vulkan.Einbruch; Vulkano=Tekton.Senke	KAL'DERA UF: Kal'dera obrus- heniya; Kotlovina vulkanicheskaya	CALDERA UF: Caldera=de= hundimiento; Depresion= vulcanotectonica; Subsidencia=volcanica	CALDERA UF: Caldera di sprofon- damento; Depressione vulcanotettonica; Spro- fondamento vulcanico
618 CALEDONIAN OROGENY STRU	CALEDONIAN OROGENY	OROGENIE CALE- DONIENNE	KALEDON. OROGENESE	KALEDONSKIJ UF: Takonskij orogen	OROGENIA= CALEDONIANA	OROGENESI CALE- DONIANA
619 CALICHE SURF	Caliche WEATHERING CRUST	Caliche INCRUSTATION; CALCAIRE	Caliche INKRUSTATION; KALK	Kaliche KORA VYVETRIVANIYA	Caliche INCRUSTACION; CALIZA	Caliche INCROSTAZIONE; CALCARE
620 CALIFORNIUM CHEE	CALIFORNIUM	CALIFORNIUM	CF	KALIFORNYJ	CALIFORNIO	CALIFORNIO
621 CALIPER LOGGING APPL	CALIPER LOGGING	DIAMETRAGE	KALIBER= BOHRLOCH= MESSUNG	KAVERNO- METRIYA	DIAMETRO- SONDEO	DIAMETRAGGIO
622 CALLOVIAN STRS	CALLOVIAN	CALLOVIEN	CALLOVIUM	KELLOVEJ	CALLOVIENSE	CALLOVIANO
623 CAMAROIDEA PALS	CAMAROIDEA	CAMAROIDEA	CAMAROIDEA	CAMAROIDEA	CAMAROIDEA	CAMAROIDEA
624 CAMBISOL SUSS	Cambisols BROWN SOILS	Cambisol SOL BRUN	Cambisol BRAUN=ERDE	KAMBISOL	Cambisol SUELO=CASTAÑO; SUELO=PARDO	Cambisol SUOLO BRUNO
625 CAMBRIAN STRS	CAMBRIAN	CAMBRIEN	KAMBRIUM	KEMBRIJ	CAMBRICO	CAMBRIANO
626 CAMPANIAN STRS	CAMPANIAN	CAMPANIEN	CAMPAN	KAMPAN	CAMPANIENSE	CAMPANIANO
627 CAMPTOSTROMATOIDEA PALS	CAMPTOSTROMAT- OIDEA	CAMPTOSTROMAT- OIDEA	CAMPTOSTROMAT- OIDEA	Camptostromatoidea ECHINOZOA	CAMPTOSTROMAT- OIDEA	CAMPTOSTROMAT- OIDEA
628 CANAL ENGI	CANALS	CANAL	KANAL	KANAL	CANAL	CANALE ARTIFI- CIALE
629 CANNEL COAL COMS	Cannel coal COAL	[CANNEL COAL] CHARBON SAPRO- PELIQUE	Kennel=Kohle SAPROPEL=KOHLE	KENNEL'	Carbon=cannel CARBON= SAPROPELICO	Carbone a fiamma lunga CARBONE SAPRO- PELITICO
630 CANYON SURF	CANYONS	CANYON UF: Cours eau encaisse	CANYON UF: Entrenched stream	KAN'ON	CAÑON UF: Curso=encajado	CANYON UF: Corso d'acqua incassato

	ENGLISH	FRANCAIS	DEUTSCH	RUSSKIJ	ESPANOL	ITALIANO
631 CAP ROCK ECON	CAP ROCKS	[CAP ROCK]	[CAP=ROCK]	KEHPROK	[CAP=ROCK]	[CAP ROCK]
632 CAPE SURF	CAPES	CAP	KAP	MYS	CABO	CAPO
633 CAPILLARITY PHCH	CAPILLARITY	CAPILLARITE	KAPILLARITAET	KAPILLYARNOST'	CAPILARIDAD	CAPILLARITA
634 CAPILLARY PERCOLATION PHCH	SUCTION	SUCCION	KAPILLAR=SOG	Vpityvanie INFIL'TRATSIYA	SUCCION	SUZIONE
635 CAPILLARY PRESSURE PHCH	CAPILLARY PRES- SURE	PRESSION CAPIL- LAIRE	KAPILLAR=DRUCK	DAVLENIE KAPILL- YARNOE	PRESION=CAPILAR	PRESSIONE CAPIL- LARE
636 CAPILLARY WATER GEOH	CAPILLARY WATER UF: Adsorbed water	EAU CAPILLAIRE UF: Eau adsorbee	KAPILLAR= WASSER UF: Haft=Wasser	VODA KAPILLA- YARNAYA	AGUA=CAPILAR UF: Agua=de=adsorcion	ACQUA CAPILLARE UF: Acqua di adsorbi- mento
637 CAPTURED STREAM SURF	STREAM CAPTURE UF: Beheaded streams	CAPTURE COURS EAU UF: Riviere decapitee	FLUSS- ANZAPFUNG UF: Gekoepft.Tal	Potok perekhvachennyj REKA	CAPTURA UF: Corriente= decapitada	CATTURA DI CORSO D'ACQUA UF: Decapitazione fluvi- ale
638 CARADOCIAN STRS	CARADOCIAN	CARADOC	CARADOC	KARADOK	CARODICIENSE	CARADOCIANO
639 CARAPACE PALE	Carapace MORPHOLOGY	Carapace ANATOMIE SQUE- LETTE	Panzer SKELETT	Karapaks PANTSIR'	Caparazon ANATOMIA= ESQUELETO	Carapace ANATOMIA DELLO SCHELETRO
640 CARAT ECON	Carat ORE QUALITY	Carat QUALITE MINERAI	Karat ERZ=QUALITAET	KARAT	Quilate CALIDAD= MINERAL	Carato QUALITA DEL MINERALE
641 CARBOHYDRATE CHES	CARBOHYDRATES	SUCRE	KOHLEHYDRAT	UGLEVOD	HIDRATO=DE= CARBONO	CARBOIDRATO
642 CARBON CHEE	CARBON	CARBONE	C	UGLEROD UF: Carbonaceous rock	CARBONO	CARBONIO
643 CARBON DIOXIDE CHES	CARBON DIOXIDE	GAZ CARBONIQUE	KOHLENDIOXID	UGLEKISLOTA	GAS=CARBONICO	ANIDRIDE CAR- BONICA
644 CARBON MONOXIDE CHES	CARBON MONOX- IDE	OXYDE CARBONE	KOHLENMONOXID	OKIS' UGLERODA	OXIDO=DE= CARBONO	MONOSSIDO DI CARBONIO
645 CARBON RATIO ISOT	C=13/C=12	C 13=C 12	C13=C12	C13/C12	RELACION= CARBONO	C13/C12
646 CARBON-14 DATING ISOT	C=14 UF: Radiocarbon	C 14 UF: Radiocarbone	C=14=DATIERUNG UF: Radio=Kohlenstoff	METOD RADIOUGLEROD- NYJ	C=14 UF: Carbono= radioactivo	C=14 UF: Radiocarbonio
647 CARBONACEOUS CHONDRITE EXTS	CARBONACEOUS CHONDRITES	Chondrite carbonee CHONDRITE	Kohlenstoffhaltig. Chondrit CHONDRIT	Khondrit uglerodistyj KHONDRIT	Condrita=carbonosa CONDRITA	Condrite carboniosa CONDRITE

	ENGLISH	FRANCAIS	DEUTSCH	RUSSKIJ	ESPANOL	ITALIANO
648 CARBONACEOUS COMPOSITION CHES	CARBONACEOUS COMPOSITION	COMPOSITION CARBONEE *UF:* Caustobiolite	KOHLEN=GEHALT *UF:* Kaustobiolith	UGLISTYJ	COMPOSICIÓN=CARBONOSA *UF:* Caustobiolita	COMPOSIZIONE CARBONIOSA *UF:* Caustobiolite
649 CARBONACEOUS ROCK SEDS	Carbonaceous rock ORGANIC RESI-DUES	ROCHE CARBONEE *UF:* Residu organique; Sediment organique; Teneur cendre	KOHLENSTOF-FHALTIG.GESTEIN *UF:* Aschegehalt; Organ. Loesungsrueckstand; Organogen.Sediment	[CARBONACEOUS ROCK] PORODA KARBON-ATNAYA; UGLE-ROD	ROCA=CARBONOSA *UF:* Contenido=ceniza; Residuo=organico; Sedimento=organico	ROCCIA CAR-BONIOSA *UF:* Contenuto in cenere; Residuo organico; Sedimento organico
650 CARBONATE ROCKS SEDS	CARBONATE ROCKS	ROCHE CARBON-ATEE	KARBONAT=GESTEIN *UF:* Karbonatisierung	PORODA KAR-BONATNAYA *UF:* Boundstone; Carbo-naceous rock	ROCA=CARBONATADA	ROCCIA CARBONA-TICA
651 CARBONATES MING	CARBONATES	CARBONATE *UF:* Carbonatation	KARBONAT	KARBONAT	CARBONATO *UF:* Carbonatizacion	CARBONATO *UF:* Carbonatazione
652 CARBONATITE IGMS	CARBONATITES	CARBONATITE	KARBONATIT	KARBONATIT	CARBONATITA	CARBONATITE
653 CARBONATIZATION SEDI	CARBONATIZA-TION	Carbonatation CARBONATE	Karbonatisierung KARBONAT=GESTEIN	KARBONATIZAT-SIYA	Carbonatizacion CARBONATO	Carbonatazione CARBONATO
654 CARBONIFEROUS STRS	CARBONIFEROUS	CARBONIFERE	KARBON	KARBON	CARBONIFERO	CARBONIFERO
655 CARNIAN STRS	CARNIAN	CARNIEN	KARN	KARNIJ	CARNIENSE	CARNICO
656 CARNIEULE SEDS	Cellular dolomite DOLOSTONE	Carnieule DOLOMIE	Kiesel=Kalk DOLOMIT=GESTEIN	[CARNIEULE]	Carniola DOLOMITA	Carniola DOLOMIA
657 CARNIVORA PALS	CARNIVORA	CARNIVORA	CARNIVORA	CARNIVORA *UF:* Creodonta; Fis-sipeda	CARNIVORA	CARNIVORA
658 CARTERINACEA PALS	CARTERINACEA	CARTERINACEA	CARTERINACEA	CARTERINACEA	CARTERINACEA	CARTERINACEA
659 CARTOGRAM MISC	CARTOGRAMS	CARTE OMBREE	RELIEF=KARTE	KARTOGRAMMA	MAPA=RELIEVE=SOMBREADO	CARTA A OMBREG-GIO
660 CARTOGRAPHY METH	CARTOGRAPHY *UF:* Cartographic scales; Large=scale; Legend; Mapping; Surveys	CARTOGRAPHIE *UF:* A grande echelle; A petite echelle; Echelle cartographique; Leve carte; Lever; Travaux terrain	KARTOGRAPHIE *UF:* Gelaende=Aufnahme; Gelaende=Untersuchung; Gross. Massstab; Karten=Massstab; Kartierung; Klein.Massstab	KARTOGRAFIYA	CARTOGRAFIA *UF:* Escala; Estudio-de=campo; Gran-escala; Investigacion; Levantamiento carto-grafico; Pequeña=es-cala	CARTOGRAFIA *UF:* A grande scala; A piccola scala; Rileva-mento cartografico; Rilevamento geologico; Scala; Studio di cam-pagna
661 CASCADE FOLD STRU	Cascade folds FOLDS	Pli en cascade PLI	Kaskaden=Falte FALTE	Skladka kaskadnaya FLEKSURA	Pliegue=en=cascada PLIEGUE	Piega a cascata PIEGA
662 CASCADIAN REVOLUTION STRU	Cascadian orogeny ALPINE OROGENY	Revolution cascadienne OROGENIE ALPINE	Kaskad.Umbruch ALPID.OROGENESE	Kaskadnyj AL'PIJSKIJ /ORO-GEN/	Fase=orogenica=cascadiense OROGENIA=ALPINA	Rivoluzione cascadiana OROGENESI ALPINA

	ENGLISH	FRANCAIS	DEUTSCH	RUSSKIJ	ESPANOL	ITALIANO
663 CASE STUDIES MISC	CASE STUDIES	ETUDE CAS	FALL=STUDIE	[CASE STUDIES]	ESTUDIO=DE=UN= CASO	STUDIO DEL CASO
664 CASSIDULINACEA PALS	CASSIDULINACEA	CASSIDULINACEA	CASSIDULINACEA	CASSIDULINACEA	CASSIDULINACEA	CASSIDULINACEA
665 CAST PALE	CASTS	MOULAGE	ABGUSS	SLEPOK	VACIADO	IMPRONTA
666 CASTANOZEM SUSS	CHESTNUT SOILS	SOL CHATAIN	KASTANNOZEM	POCHVA KASH- TANOVAYA	SUELO=CASTAÑO *UF:* Cambisol	SUOLO CASTANO
667 CATACLASIS STRU	Cataclasis TECTONIC BREC- CIA; MYLONITES	Cataclase BRECHE TEC- TONIQUE; MYLO- NITE	Kataklase TEKTON.BREKZIE; MYLONIT	KATAKLAZ	Cataclasis BRECHA= TECTONICA; MILONITA	Cataclasite BRECCIA TET- TONICA; MILONITE
668 CATAGENESIS SEDI	CATAGENESIS	CATAGENESE	KATAGENESE	KATAGENEZ *UF:* Ankhimetamorfizm; Early diagenesis; Katater- mal'noe mestorozhdenie	CATAGENESIS	CATAGENESI
669 CATAGRAPHIA PALS	Catagraphia PROBLEMATIC MICROFOSSILS	Catagraphia PROBLEMATICA MICRO	Catagraphia MIKRO= PROBLEMATICA	CATAGRAPHIA	Catagrafia PROBLEMATICA= MICRO	Catagraphia MICROFOSSILE PROBLEMATICO
670 CATALYSIS PHCH	CATALYSIS	CATALYSE	KATALYSE	KATALIZ	CATALISIS	CATALISI
671 CATASTROPHISM GEOL	CATASTROPHISM	CATASTROPHISME	KATASTROPHEN= THEORIE	KATASTROFIZM	CATASTROFISMO	CATASTROFISMO
672 CATCHMENT HYDRODYNAMICS GEOH	CATCHMENT HYDRODYNAMICS	HYDRODY- NAMIQUE CAPT- AGE	FASSUNGS= HYDRODYNAMIK	[CATCHMENT HYDRODYNAM- ICS] GIDRODI- NAMIKA; BASSEYN VODOSBORNYJ	HIDRODINAMICA= CAPTACION	IDRODINAMICA DI CAPTAZIONE
673 CATENA SURF	CATENAS	ZONEOGRAPHIE SOL	CATENA	Katena POCHVA	ZONOGRAFIA= SUELO	ZONEOGRAFIA DEL SUOLO
674 CATION PHCH	CATIONS	CATION	KATION	Kation ION	CATION	CATIONE
675 CAULDRON SURF	Cauldrons CALDERAS	Subsidence en chau- dron CALDEIRA	Vulkan.Einbruch CALDERA	Kotlovina vulkani- cheskaya KAL'DERA	Subsidencia=volcanica CALDERA	Sprofondamento vul- canico CALDERA
676 CAUSTOBIOLITH SEDS	Caustobiolith BIOHERMS	Caustobiolite COMPOSITION CARBONEE	Kaustobiolith KOHLEN=GEHALT	KAUSTOBIOLIT	Caustobiolita COMPOSICION= CARBONOSA	Caustobiolite COMPOSIZIONE CARBONIOSA
677 CAVE SURF	CAVES *UF:* Caverns	CAVERNE *UF:* Grotte	HOEHLE *UF:* Kaverne	PESHCHERA *UF:* Kaverna; Replenish- ment (speleo); Speleo- them	CAVERNA *UF:* Cueva	CAVERNA *UF:* Grotta
678 CAVERN SURF	Caverns CAVES	Grotte CAVERNE	Kaverne HOEHLE	Kaverna PESHCHERA	Cueva CAVERNA	Grotta CAVERNA

	ENGLISH	FRANCAIS	DEUTSCH	RUSSKIJ	ESPANOL	ITALIANO
679 CAVERNOUS TEXTURE TEST	Cavernous texture TEXTURES	Texture caverneuse TEXTURE	Kavernoes.Gefuege KORN=GEFUEGE	KAVERNOZNOST'	Textura=cavernosa TEXTURA	Tessitura cavernosa TESSITURA
680 CAVINGS MINI	CAVINGS	Exploitation par eboulement EXPLOITATION	[CAVINGS] GEWINNUNG	Razrabotka obrushe- niem RAZRABOTKA RUDNYKH MESTOROZHDENIJ	Explotacion=por= hundimiento EXPLOTACION	Sbancamento SFRUTTAMENTO
681 CAYTONIALES PALS	CAYTONIALES	CAYTONIALES	CAYTONIALES	CAYTONIALES	CAYTONIALES	CAYTONIALES
682 CELLULAR POROSITY PHCH	Cellular porosity POROSITY	Porosite cellulaire POROSITE	Zellular=Porositaet POROSITAET	Poristost' yacheistaya PORISTOST'	Porosidad=celular POROSIDAD	Porosita cellulare POROSITA
683 CELLULOSE CHES	CELLULOSE	CELLULOSE	ZELLULOSE	Tsellyuloza VESHCHESTVO ORGANICHESKOE; FLORA	CELULOSA	CELLULOSA
684 CEMENT SEDI	CEMENT	CIMENT ROCHE	BINDEMITTEL	TSEMENT *UF:* Matriks	CEMENTO=ROCA	CEMENTO SEDI- MENTARIO
685 CEMENTATION SEDI	CEMENTATION	CEMENTATION	ZEMENTATION	TSEMENTATSIYA *UF:* Tamponazh	CEMENTACION	CEMENTAZIONE
686 CENOMANIAN STRS	CENOMANIAN	CENOMANIEN	CENOMAN	SENOMAN	CENOMANIENSE	CENOMANIANO
687 CENOTYPAL IGNE	Cenotypal(adj.)	Neovolcanique(adj.)	Neovulkanisch(adj.)	[CENOTYPAL]	Cenotipico(adj.)	Cenotipico(adj.)
688 CENOZOIC STRS	CENOZOIC	CENOZOIQUE	KAENOZOIKUM	KAJNOZOJ *UF:* Tretichnyj	CENOZOICO	CENOZOICO
689 CENOZONE STRA	Cenozones FAUNA; ZONING	Cenozone FAUNE SPECI- FIQUE	Cenozone SPEZIF.FAUNA	Tsenozona ZONA KOMPLEKS- NAYA; FAUNA	Cenozona FAUNA= ESPECIFICA	Cenozona FAUNA SPECIFICA
690 CENTROCLINE STRU	Centroclines SYNCLINES	Cuvette synclinale SYNCLINAL	Zentroklinale SYNKLINALE	TSENTROKLINAL'	Centroclinal SINCLINAL	Struttura centroclinale SINCLINALE
691 CEPHALOCARIDA PALS	Cephalocarida CRUSTACEA	Cephalocarida CRUSTACEA	Cephalocarida CRUSTACEA	CEPHALOCARIDA	Cephalocarida CRUSTACEA	Cephalocarida CRUSTACEA
692 CEPHALODISCIDA PALS	Cephalodiscida HEMICHORDATA	Cephalodiscida STOMOCHORDATA	Cephalodiscida STOMOCHORDATA	CEPHALODISCIDA	Cephalodiscida STOMOCHORDATA	Cephalodiscida HEMICHORDATA
693 CEPHALOPODA PALS	CEPHALOPODA *UF:* Actinoceratoidea; Agoniatitida; Ammoni- tida; Ammonitina; Bactritoidea	CEPHALOPODA	CEPHALOPODA	CEPHALOPODA	CEPHALOPODA	CEPHALOPODA
694 CERAMICS COMS	CERAMIC MATERI- ALS *UF:* Potter's clay	CERAMIQUE *UF:* Argile poterie	KERAM.ROHSTOFF *UF:* Potters=Ton	SYR'E KERAMI- CHESKOE	CERAMICA *UF:* Arcilla=de= ceramica	CERAMICA *UF:* Argilla da ceramica
695 CERATITIDA PALS	CERATITIDA	CERATITIDA	CERATITIDA	CERATITIDA	CERATITIDA	CERATITIDA

	ENGLISH	FRANCAIS	DEUTSCH	RUSSKIJ	ESPANOL	ITALIANO
696 CERATOMORPHA PALS	CERATOMORPHA	CERATOMORPHA	CERATOMORPHA	CERATOMORPHA	CERATOMORPHA	CERATOMORPHA
697 CERIANTHARIA PALS	Ceriantharia CERIANTIPA- THARIA	Ceriantharia CERIANTIPA- THARIA	Ceriantharia CERIANTIPA- THARIA	Ceriantharia ANTHOZOA	Ceriantharia CERIANTIPA- THARIA	Ceriantharia CERIANTIPA- THARIA
698 CERIANTIPATHARIA PALS	CERIANTIPA- THARIA *UF:* Antipatharia; Ceri- antharia	CERIANTIPA- THARIA *UF:* Antipatharia; Ceri- antharia	CERIANTIPA- THARIA *UF:* Antipatharia; Ceri- antharia	Ceriantipatharia ANTHOZOA	CERIANTIPA- THARIA *UF:* Antipatharia; Ceri- antharia	CERIANTIPA- THARIA *UF:* Antipatharia; Ceri- antharia
699 CERIUM CHEE	CERIUM	CERIUM	CE	TSERIJ	CERIO	CERIO
700 CESIUM CHEE	CESIUM	CESIUM	CS	TSEZIJ	CESIO	CESIO
701 CETACEA PALS	CETACEA	CETACEA	CETACEA	CETACEA	CETACEA	CETACEA
702 CHAETOGNATHA PALS	Chaetognatha WORMS	Chaetognatha VERMES	Chaetognatha VERMES	CHAETOGNATHA	Chaetognatha VERMES	Chaetognatha VERMES
703 CHALCOPHILE ELEMENT CHES	CHALCOPHILE ELEMENTS	ELEMENT CHAL- COPHILE	CHALKOPHIL. ELEMENT	KHAL'KOFIL'NYJ	ELEMENTO- CALCOFILO	ELEMENTO CAL- COFILO
704 CHALCOSPHERE SOLI	Chalcosphere CRUST; UPPER MANTLE	Chalcosphere CROUTE TER- RESTRE; MANTEAU SUP	Chalkosphaere ERD=KRUSTE; OBER.ERD= MANTEL	KHAL'KOSFERA	Calcosfera CORTEZA= TERRESTRE; MANTO=GLOBO= SUP	Calcosfera CROSTA TER- RESTRE; MAN- TELLO SUPERIORE
705 CHALK SEDS	CHALK	CRAIE	SCHREIBKREIDE	MEL (PORODA)	CRETA	GESSO
706 CHANDLER WOBBLE SOLI	CHANDLER WOB- BLE	NUTATION CHAN- DLER	CHANDLER= SCHWANKUNG	Dvizhenie chandlerovskoe NUTATSIYA	NUTACION= CHANDLER	NUTAZIONE CHANDLER
707 CHANNEL SURF	CHANNELS *UF:* Braided channels	CHENAL *UF:* Reseau anastomose	RINNE *UF:* Wild=Wasser	RUSLO *UF:* Stream sediment	CAUCE *UF:* Red=de=canales	CANALE *UF:* Drenaggio anas- tomizzato
708 CHANNEL GEOMETRY SURF	CHANNEL GEOME- TRY	GEOMETRIE CHE- NAL	RINNEN= GEOMETRIE	GEOMETRIYA RUSLA	GEOMETRIA- CANAL	GEOMETRIA DEL CANALE
709 CHAOTIC STRUCTURE TEST	CHAOTIC MATERI- ALS	Structure chaotique TEXTURE; PYRO- CLASTIQUE	Chaot.Gefuege KORN=GEFUEGE; PYROKLAST. GESTEIN	KHAOTICHESKIJ	Estructura=caotica TEXTURA; PIRO- CLASTICO	Caotico TESSITURA; PRO- DOTTO PIROCLAS- TICO
710 CHARACTERISTICS MISC	Characteristics(gen.)	Caracteristique(gen.)	Charakteristik(gen.)	KHARAKTERIS- TIKA	Caracteristica(gen.)	Carattere(gen.)
711 CHARNIAN OROGENY STRU	CHARNIAN OROG- ENY	OROGENIE CHAR- NIENNE	CHARN. OROGENESE	CHARNIJSKIJ	OROGENIA= CHARNIA	OROGENESI CHARNIANA
712 CHARNOCKITE IGMS	CHARNOCKITE	CHARNOCKITE	CHARNOCKIT	CHARNOKIT	CHARNOCKITA	CHARNOCKITE

	ENGLISH	FRANCAIS	DEUTSCH	RUSSKIJ	ESPANOL	ITALIANO
713 CHAROPHYTA PALS	CHAROPHYTA	CHAROPHYCEAE	CHAROPHYTA	CHAROPHYTA	CHAROPHYTA	CHAROPHYTA
714 CHASSIGNITE EXTS	Chassignite ACHONDRITES	Chassignite ACHONDRITE	Chassignit ACHONDRIT	Shassin'it AKHONDRIT	Casignita ACONDRITA	Chassignite ACONDRITE
715 CHATTERMARK TEST	TOOL MARKS	TRACE MECANIQUE	MECHAN.SPUR	BOROZDA= TSARAPINA	HUELLA= MECANICA	IMPRONTA MEC- CANICA
716 CHATTIAN STRS	Chattian UPPER OLIGOCENE	Chattien OLIGOCENE SUP	Chattium O.OLIGOZAEN	KHATT *UF:* Oligotsen verkhnij	Chattiense OLIGOCENO=SUP	Cattiano OLIGOCENE SUP.
717 CHEILOSTOMATA PALS	CHEILOSTOMATA	CHEILOSTOMATA	CHEILOSTOMATA	CHEILOSTOMATA	CHEILOSTOMATA	CHEILOSTOMATA
718 CHELATION GEOC	Chelation COMPLEXOMETRY	Chelation COMPLEXOMETRIE	Chelat=Bildung KOMPLEXOMETRIE	KHELATOOBRA- ZOVANIE	Quelacion COMPLEXOMETRIA	Chelazione COMPLESSO- METRIA
719 CHELICERATA PALS	CHELICERATA	CHELICERATA	CHELICERATA	CHELICERATA	CHELICERATA	CHELICERATA
720 CHELONIA PALS	CHELONIA *UF:* Testudinata	CHELONIA *UF:* Testudinata	CHELONIA *UF:* Testudinata	CHELONIA	CHELONIA *UF:* Testudinata	CHELONIA *UF:* Testudinata
721 CHEMICAL COMPOSITION GEOC	CHEMICAL COM- POSITION	ANALYSE CHIMIQUE *UF:* Compose inerte; Composition; Ligne isoconcentration; Teneur cendre	CHEM.ANALYSE *UF:* Aschegehalt; Inaktiv.Komponente; Isocone; Zusammen- setzung	KHIMICHESKIJ SOSTAV	ANALISIS= QUIMICO *UF:* Componente=inerte; Composicion; Contenido=ceniza; Isosalinidad	COMPOSIZIONE CHIMICA *UF:* Componente inerte; Composizione; Con- tenuto in cenere
722 CHEMICAL CONTROLS ECON	CHEMICAL CON- TROLS	CONTROLE GEOCHIMIQUE GITE	CHEM.EINFLUSS	KONTROL' KHIMI- CHESKIJ	YACIMIENTO= QUIMICO	CONTROLLO GEOCHIMICO GIA- CIMENTO
723 CHEMICAL DEPOSITION SEDI	CHEMICAL SEDI- MENTATION *UF:* Chemogenic origin	SEDIMENTATION CHIMIQUE *UF:* Origine chimique	CHEM. SEDIMENTATION *UF:* Chem.Entstehung	OSAZHDENIE KHIMICHESKOE	SEDIMENTACION= QUIMICA *UF:* Origen=quimico	SEDIMENTAZIONE CHIMICA *UF:* Origine chimica
724 CHEMICAL EQUILIBRIUM GEOC	Chemical equilibrium CHEMICAL PROP- ERTIES	Equilibre chimique DIAGRAMME EQUILIBRE	Chem.Gleichgewicht PHASEN= GLEICHGEWICHT	RAVNOVESIE KHIMICHESKOE	Equilibrio=quimico DIAGRAMA= EQUILIBRIO	Equilibrio chimico DIAGRAMMA D'EQUILIBRIO
725 CHEMICAL EVOLUTION GEOL	Chemical evolution CHEMICAL PROP- ERTIES	Evolution chimique PROPRIETE CHIMIQUE	Chemismus CHEM. EIGENSCHAFT	[CHEMICAL EVO- LUTION] EHVOLYUTSIYA; SVOJSTVO KHIMI- CHESKOE	Quimismo PROPIEDAD= QUIMICA	Chimismo PROPRIETA CHIMICA
726 CHEMICAL EXPLOSION ENGI	CHEMICAL EXPLO- SIONS	EXPLOSION CHIMIQUE	CHEM.EXPLOSION	[CHEMICAL EXPLOSION] VZRYV; SVOJSTVO KHIMICHESKOE	EXPLOSION= QUIMICA	ESPLOSIONE CHIMICA
727 CHEMICAL FOSSILS PALE	CHEMICAL FOS- SILS	FOSSILE CHIMIQUE	CHEM.FOSSIL	Khemofossilii KHEMOGENNYJ; FOSSILIZATSIYA	FOSIL=QUIMICO	FOSSILE CHIMICO
728 CHEMICAL INDUSTRY ECON	CHEMICAL INDUS- TRY	INDUSTRIE CHIMIQUE	CHEM.INDUSTRIE	PROMYSHLEN- NOST' KHIMI- CHESKAYA	INDUSTRIA= QUIMICA	INDUSTRIA CHIMICA

	ENGLISH	FRANCAIS	DEUTSCH	RUSSKIJ	ESPANOL	ITALIANO
729 CHEMICAL PROPERTY PHCH	CHEMICAL PROP- ERTIES *UF:* Chemical equilib- rium; Chemical evolu- tion; Chemical system	PROPRIETE CHIMIQUE *UF:* Evolution chimique	CHEM. EIGENSCHAFT *UF:* Chemismus	SVOJSTVO KHIMI- CHESKOE *UF:* Chemical evolution; Chemical explosion	PROPIEDAD= QUIMICA *UF:* Quimismo	PROPRIETA CHIMICA *UF:* Chimismo
730 CHEMICAL RATIO GEOC	RATIOS *UF:* Isotope ratio	RAPPORT CHIMIQUE *UF:* Rapport isotopique	CHEM. VERHAELTNIS *UF:* Isotopen= Verhaeltnis	SOOTNOSHENIE (KHIM)	RELACION= QUIMICA *UF:* Relacion=isotopica	RAPPORTO CHIMICO *UF:* Rapporto isotopico
731 CHEMICAL REDUCTION PHCH	REDUCTION	REDUCTION CHIMIQUE	CHEM.REDUKTION	VOSSTANOVLENIE (KHIM)	REDUCCION= QUIMICA	RIDUZIONE CHIMICA
732 CHEMICAL REMANENT MAGNETIZATION PHCH	CHEMICAL REMA- NENT MAGNE- TIZATION	AIMANTATION REMANENTE CHIMIQUE	CHEMOREMANENT. MAGNETISIERUNG	Namagnichennost' khimicheskaya NAMAGNICHEN- NOST' OSTATOCH- NAYA	IMANTACION= REMANENTE= QUIMICA	MAGNETIZZAZ- IONE CHIMICA RESIDUA
733 CHEMICAL SYSTEM PHCH	Chemical system CHEMICAL PROP- ERTIES	Systeme chimique DIAGRAMME EQUILIBRE	Chem.System PHASEN= GLEICHGEWICHT	SISTEMA KHIMI- CHESKAYA	Sistema=quimico DIAGRAMA= EQUILIBRIO	Sistema chimico DIAGRAMMA D'EQUILIBRIO
734 CHEMICAL VALENCY PHCH	VALENCY	VALENCE CHIMIQUE	WERTIGKEIT	VALENTNOST'	VALENCIA= QUIMICA	VALENZA CHIMICA
735 CHEMICAL WEATHERING SURF	CHEMICAL WEATHERING	ALTERATION CHIMIQUE	CHEM. VERWITTERUNG	VYVETRIVANIE KHIMICHESKOE	ALTERACION= QUIMICA	ALTERAZIONE CHIMICA
736 CHEMOGENIC ORIGIN SEDI	Chemogenic origin CHEMICAL SEDI- MENTATION	Origine chimique SEDIMENTATION CHIMIQUE	Chem.Entstehung CHEM. SEDIMENTATION	KHEMOGENNYJ *UF:* Khemofossilii	Origen=quimico SEDIMENTACION= QUIMICA	Origine chimica SEDIMENTAZIONE CHIMICA
737 CHENIER SURF	Cheniers BEACHES	Chenier CORDON LITTO- RAL	Chenier STRAND=WALL	Val plyazhovoj VAL BEREGOVOJ; PLYAZH	Restinga CORDON=LITORAL	Chenier CORDONE LITOR- ALE
738 CHERNOZEM SUSS	CHERNOZEMS	CHERNOZEM *UF:* Boroll	TSCHERNOZEM *UF:* Boroll	CHERNOZEM	CHERNOZEM *UF:* Boroll	CERNOZEM *UF:* Boroll
739 CHERT SEDS	CHERT *UF:* Novaculite; Opoka	SILEXITE *UF:* Novaculite; Opoka; Porcelanite	FLINT *UF:* Novaculith; Opoka; Porcellanit	Slanets kremnistyj FLINT	SILEXITA *UF:* Novaculita; Opoka; Porcelanita	SILEXITE *UF:* Novaculite; Opoka; Porcellanite
740 CHERTIFICATION SEDI	CHERTIFICATION	CHERTIFICATION	HORNSTEIN= BILDUNG	OKREMNENIE (CHERT)	CHERTIFICACION	SELCIFICAZIONE
741 CHEVRON FOLD STRU	KINK FOLDS	PLI EN CHEVRON *UF:* Pli accordeon	HARMONIKA= FALTE	Skladka strel'chataya SKLADKA ZIGZA- GOOBRAZNAYA	PLIEGUE=EN=ZIG= ZAG	PIEGA A KINK *UF:* Piega a zigzag
742 CHILOPODA PALS	Chilopoda MYRIAPODA	Chilopoda MYRIAPODA	Chilopoda MYRIAPODA	CHILOPODA	Chilopoda MYRIAPODA	Chilopoda MYRIAPODA
743 CHINA CLAY COMS	China clay KAOLIN	Terre a porcelaine KAOLIN	Porzellan=Erde KAOLIN	GLINA FAR- FOROVAYA	Caolin=china CAOLIN	Argilla per porcellana CAOLINO
744 CHIROPTERA PALS	CHIROPTERA	CHIROPTERA	CHIROPTERA	CHIROPTERA	CHIROPTERA	CHIROPTERA

	ENGLISH	FRANCAIS	DEUTSCH	RUSSKIJ	ESPANOL	ITALIANO
745 CHITIN CHES	Chitin ORGANIC MATERI- ALS	Chitine MATIERE ORGANIQUE	Chitin ORGAN.SUBSTANZ	KHITIN	Quitina MATERIA= ORGANICA	Chitina SOSTANZA ORGANICA
746 CHITINOZOA PALS	CHITINOZOA	CHITINOZOA	CHITINOZOA	Chitinozoa PROTOZOA	CHITINOZOA	CHITINOZOA
747 CHLORIDES MING	CHLORIDES	CHLORURE	CHLORID	KHLORID	CLORURO	CLORURO
748 CHLORINATION GEOC	Chlorination CHLORINE	Chloration ATTAQUE CHIMIQUE; CHLORE	Chlorinierung AETZUNG; CL	KHLORIROVANIE	Cloracion ATAQUE=QUIMICO; CLORO	Clorurazione ATTACCO CHIMICO; CLORO
749 CHLORINE CHEE	CHLORINE *UF:* Chlorination; Chlorinity	CHLORE *UF:* Chlorinite; Chloruration	CL *UF:* Chlorinierung; Chlorinitaet	KHLOR	CLORO *UF:* Cloracion; Clorini- dad	CLORO *UF:* Clorinita; Cloruraz- ione
750 CHLORINITY PHCH	Chlorinity CHLORINE	Chlorinité CHLORE	Chlorinitaet CL	Khlornost' SOLENOST'	Clorinidad CLORO	Clorinita CLORO
751 CHLORITE SCHIST IGMS	Chlorite schist GREENSCHIST	Schiste chlorite SCHISTE VERT	Chlorit=Schiefer GRUEN=SCHIEFER	Slanets khloritovyj SLANETS ZELENYJ	Cloritoesquisto ESQUISTO=VERDE	Scisto a clorite SCISTO VERDE
752 CHLORITIZATION IGNE	CHLORITIZATION	CHLORITISATION	CHLORITISIERUNG	KHLORITIZATSIYA	CLORITIZACION	CLORITIZZAZIONE
753 CHLOROPHYLL CHES	CHLOROPHYLL	CHLOROPHYLLE	CHLOROPHYLL	KHLOROFILL	CLOROFILA	CLOROFILLA
754 CHLOROPHYTA PALS	CHLOROPHYTA	CHLOROPHYCO- PHYTA	CHLOROPHYTA	CHLOROPHYTA *UF:* Dasycladaceae; Desmids	CHLOROPHYTA	CHLOROPHYTA
755 CHONDRICHTHYES PALS	CHONDRICHTHYES	CHONDRICHTHYES	CHONDRICHTHYES	CHONDRICHTHYES	CHONDRICHTHYES	CHONDRICHTHYES
756 CHONDRITE EXTS	CHONDRITES *UF:* Enstatite chondrites	CHONDRITE *UF:* Chondrite a ensta- tite; Chondrite car- bonee	CHONDRIT *UF:* Enstatithaltig. Chondrit; Kohlenstoffhaltig. Chondrit	KHONDRIT *UF:* Enstatite chondrite; Khondrit uglerodistyj	CONDRITA *UF:* Condrita= carbonosa; Condrita= enstatita	CONDRITE *UF:* Condrite carboniosa; Condrite enstatitica
757 CHONDRULE EXTR	CHONDRULES	CHONDRULE	CHONDRE	KHONDRA KHONDRIT	CONDRULO	CONDRULA
758 CHORDATA PALS	CHORDATA *UF:* Acrania; Prochor- data	CHORDATA *UF:* Acrania; Prochor- data	CHORDATA *UF:* Acrania; Prochor- data	CHORDATA	CHORDATA *UF:* Acrania; Prochor- data	CHORDATA *UF:* Acrania; Prochor- data
759 CHROMATES MING	CHROMATES	CHROMATE	CHROMAT	KHROMAT	CROMATO	CROMATO
760 CHROMATOGRAPHY METH	CHROMATOGRA- PHY	CHROMATOGRA- PHIE	CHROMATOGRA- PHIE	KHROMATOGRA- FIYA	CROMATOGRAFIA	CROMATOGRAFIA
761 CHROMIUM CHEE	CHROMIUM	CHROME	CR	KHROM	CROMO	CROMO

	ENGLISH	FRANCAIS	DEUTSCH	RUSSKIJ	ESPANOL	ITALIANO
762 CHRONOSTRATIGRAPHIC UNIT STRA	Chronostratigraphic unit CHRONOSTRATI-GRAPHY; STRATI-GRAPHIC UNITS	Unite chronostrati-graphique UNITE STRATI-GRAPHIQUE	Chronostratigraph. Einheit STRATIGRAPH. EINHEIT	Khronostratigrafi-cheskoe podrazdelenie KHRONOSTRATI-GRAFIYA	Unidad=cronoestratigrafica UNIDAD=ESTRATIGRAFICA	Unita cronostratigra-fica UNITA STRATI-GRAFICA
763 CHRONOSTRATIGRAPHY STRA	CHRONOSTRATI-GRAPHY *UF:* Biochron; Biochronology; Chronostratigraphic unit; Chronozones	CHRONOSTRATI-GRAPHIE	CHRONOSTRATI-GRAPHIE	KHRONOSTRATI-GRAFIYA *UF:* Khronostratigraficheskoe podrazdelenie	CRONOESTRATI-GRAFIA	CRONOSTRATI-GRAFIA
764 CHRONOZONE STRA	Chronozones CHRONOSTRATI-GRAPHY; ZONING	Chronozone ECHELLE STRATI-GRAPHIQUE	Chronozone STRATIGRAPH. SKALA	KHRONOZONA	Cronozona ESCALA=ESTRATIGRAFICA	Cronozona SCALA STRATI-GRAFICA
765 CHRYSOPHYTA PALS	CHRYSOPHYTA	CHRYSOPHYCO-PHYTA	CHRYSOPHYTA	CHRYSOPHYTA *UF:* Coccolithophoraceae; Diskoastery	CHRYSOPHYCO-PHYTA	CHRYSOPHYTA
766 CILIATA PALS	Ciliata PROTISTA	Ciliata PROTOZOA	Ciliata PROTOZOA	CILIATA	Ciliata PROTOZOA	Ciliata PROTOZOA
767 CILIOPHORA PALS	Ciliophora PROTISTA	Ciliophora PROTOZOA	Ciliophora PROTOZOA	CILIOPHORA	Ciliophora PROTOZOA	Ciliophora PROTOZOA
768 CIMMERIAN OROGENY STRU	Cimmerian orogeny ALPINE OROGENY	Orogenie cimmerienne OROGENIE ALPINE	Kimmer.Orogenese ALPID.OROGENESE	KIMMERIJSKIJ	Orogenia=cimmerica OROGENIA=ALPINA	Orogenesi cimmerica OROGENESI ALPINA
769 CINDER IGNE	Cinder SCORIA	Scorie volcanique SCORIE	Vulkan. Schlacke SCHLACKE	Shlak SHLAK VULKANI-CHESKIJ	Escoria vulcanica ESCORIA	Scoria SCORIE
770 CINDER CONE SURF	CINDER CONES	CONE CENDRE	ASCHEN=KEGEL	Konus shlakovyj KONUS VULKANI-CHESKIJ; SHLAK VULKANICHESKIJ	CONO=DE=CENIZAS	CONO DI CENERE
771 CIPOLIN IGMS	Cipolin MARBLES	Cipolin MARBRE	Cipolin MARMOR	TSIPOLIN	Cipolino MARMOL	Cipollino MARMO
772 CIRCULAR POLARIZATION MINE	CIRCULAR POLAR-IZATION	POUVOIR ROTA-TOIRE	DREHUNGS=VERMOEGEN	POLYARIZATSIYA KRUGOVAYA	POLARIZACION=CIRCULAR	POTERE ROTA-TORIO
773 CIRCULATION MARI	Circulation OCEAN CIRCULA-TION	Circulation COURANT	Zirkulation STROEMUNG	TSIRKULYATSIYA *UF:* Okeanicheskaya tsirkyulatsiya; Paleot-sirkulyatsiya	Circulacion CORRIENTE	Circolazione CORRENTE
774 CIRCUM PACIFIC BELT GEOL	Circum=pacific region VOLCANOLOGY	Ceinture circum=pacifique CHAINE GEOSYN-CLINALE	Zirkumpazif.Guertel GEOSYNKLINAL=KETTE	TIKHOOKEANSKIJ PODVIZHNYJ POYAS	Cinturon=circumpacifico CORDILLERA=GEOSINCLINAL	Cintura circumpacifica CATENA GEOSIN-CLINALE
775 CIRQUE SURF	CIRQUES *UF:* Hanging valleys	CIRQUE GLA-CIAIRE *UF:* Vallee suspendue	KAR	TSIRK LED-NIKOVYJ	CIRCO=GLACIAR	CIRCO *UF:* Valle sospesa

	ENGLISH	FRANCAIS	DEUTSCH	RUSSKIJ	ESPANOL	ITALIANO
776 CIRRIPEDIA PALS	CIRRIPEDIA	CIRRIPEDIA	CIRRIPEDIA	CIRRIPEDIA	CIRRIPEDIA	CIRRIPEDIA
777 CIVIL ENGINEERING ENGI	CIVIL ENGINEER-ING *UF:* Public works	Genie civil TRAVAUX PUBLICS	Tiefbau INGENIEUR=BAU	STROITEL'STVO GRAZHDANSKOE	Ingenieria=civil OBRA=PUBLICA	Ingegneria civile LAVORI PUBBLICI
778 CLADOCERA PALS	Cladocera BRANCHIOPODA	Cladocera CONCHOSTRACA	Cladocera CONCHOSTRACA	CLADOCERA	Cladocera CONCHOSTRACA	Cladocera CONCHOSTRACA
779 CLAIM ECON	Claims MINING GEOLOGY	Permis exploration TITRE MINIER	Schuerf=Recht KONZESSION	ZAYAVKA	Permiso=investigacion=mina CONCESION=MINERA	Permesso di esplora-zone TITOLO MINE-RARIO
780 CLARAIN SEDS	Clarain MACERALS	Clarain MACERAL	Clarain MAZERAL	KLAREN	Clareno MACERAL	Clarain MACERALE
781 CLARKE GEOC	Clarke GEOCHEMICAL METHODS	Clarke PROSPECTION GEOCHIMIQUE	Clarke GEOCHEM. PROSPEKTION	KLARK	Clarke PROSPECCION=GEOQUIMICA	Clarke PROSPEZIONE GEOCHIMICA
782 CLASSIFICATION MISC	CLASSIFICATION *UF:* Holmes' classifica-tion; Phylum	CLASSIFICATION *UF:* Classification Holmes; Phylum	KLASSIFIKATION *UF:* Holmes' Klassifikation; Phylum	KLASSIFIKATSIYA	CLASIFICACION *UF:* Clasificacion=de=Holmes; Filum	CLASSIFICAZIONE *UF:* Classificazione di Holmes; Phylum
783 CLAST SEDI	CLASTS	CLASTE	BRUCHSTUECK	OBLOMOK	CLASTO	CLASTO
784 CLASTIC SEDI	Clastic(adj.)	Clastique(adj.)	Klastisch(adj.)	OBLOMOCHNYJ	Clastico(adj.)	Clastico(adj.)
785 CLASTIC DIKE SEDI	CLASTIC DIKES	FILON CLASTIQUE	SEDIMENT=GANG	DAIKA KLASTI-CHESKAYA *UF:* Daika pes-chanikovaya	FILON=CLASTICO	FILONE CLASTICO
786 CLASTICS SEDS	CLASTIC ROCKS *UF:* Arenite; Detrital materials; Rudite; Sparagmite	ROCHE CLASTIQUE *UF:* Arenite; Composi-tion detritique; Diamic-tite; Polymicte; Rudite; Sparagmite	KLAST.GESTEIN *UF:* Arenit; Detrit. Zusammensetzung; Diamiktit; Polymikt. Gefuege; Rudit; Sparagmit	KLASTIT *UF:* Diamiktit	ROCA=CLASTICA *UF:* Arenita; Composicion=detritica; Diamictita; Esparag-mita; Polimicto; Rudita	ROCCIA CLASTICA *UF:* Arenite; Composiz-ione detritica; Diamic-tite; Polimittico; Rudite; Sparagmite
787 CLAY SEDS	CLAYS *UF:* Bleaching clay; Brick clay	ARGILE *UF:* Argile a brique; Galet argile; Roche argileuse; Terre adsor-bante	TON *UF:* Bleich=Erde; Ton=Gehalt; Ton=Geroell; Ziegel=Ton	GLINA	ARCILLA *UF:* Arcilla=de=ladrillo; Arcilla=esmectica; Bola=de=arcilla; Roca=arcillosa	ARGILLA *UF:* Argilla per laterizi; Argilla sbiancante; Composizione argillosa; Palla d'argilla
788 CLAY BALL SEDI	CLAY BALLS	Galet argile GALET; ARGILE	Ton=Geroell GEROELL; TON	Sharik glinistyj PSEVDOKONGLOM-ERAT; KONKRET-SIYA	Bola=de=arcilla CANTO=RODADO; ARCILLA	Palla d'argilla CIOTTOLO; ARGILLA
789 CLAY MINERALOGY MINE	CLAY MINERAL-OGY	MINERALOGIE ARGILE	TONMINERALOGIE	MINERALOGIYA GLIN	MINERALOGIA=ARCILLAS	MINERALOGIA DELLE ARGILLE

	ENGLISH	FRANCAIS	DEUTSCH	RUSSKIJ	ESPANOL	ITALIANO
790 CLAY MINERALS MING	CLAY MINERALS	ARGILE MINERAL	TONMINERAL	MINERAL GLINIS-TYJ	ARCILLA=MINERAL	MINERALI DELLE ARGILLE
791 CLAYSTONE SEDS	CLAYSTONE	ARGILITE	TONIG.GESTEIN	ARGILLIT	ARCILLITA	ARGILLITE
792 CLEAVAGE STRU	CLEAVAGE *UF:* Axial=plane cleav-age; Bedding=plane cleavage	CLIVAGE TEC-TONIQUE	GESTEINS=SPALTBARKEIT	KLIVAZH	CLIVAGE=TECTONICO	CLIVAGGIO TET-TONICO
793 CLIFF SURF	CLIFFS	FALAISE *UF:* Microfalaise plage	KLIFF *UF:* Strand=Abbruch	KLIF	ACANTILADO *UF:* Escarpe=de=playa	FALESIA *UF:* Scarpata litorale
794 CLIMATE SURF	CLIMATE *UF:* Cyclones; Green-house effect; Humid climate	CLIMAT	KLIMA	KLIMAT *UF:* Ehffekt klimati-cheskij; Izmenenie sezonnoe; Nastuplenie pustyn'; Zonal'nost' klimaticheskaya	CLIMA	CLIMA
795 CLIMATIC EFFECT GEOL	CLIMATE EFFECTS *UF:* Climate trace; Climate zonation	ACTION CLIMA-TIQUE *UF:* Retrecissement	KLIMA=WIRKUNG *UF:* Schrumpfung	Ehffekt klimaticheskij KLIMAT	ACCION=CLIMATICA *UF:* Retraccion	AZIONE CLIMA-TICA *UF:* Ritiro
796 CLIMATIC TRACE GEOL	Climate trace CLIMATE EFFECTS	TRACE CLIMA-TIQUE	KLIMA=SPUR	Sled paleoklimata INDIKATOR; PALEOKLIMAT	HUELLA=CLIMATICA	TRACCIA CLIMA-TICA
797 CLIMATIC ZONATION SURF	Climate zonation CLIMATE EFFECTS	ZONE CLIMATIQUE	KLIMA=ZONE	Zonal'nost' klimati-cheskaya ZONAL'NOST; KLI-MAT	ZONA=CLIMATICA	ZONALITA CLIMA-TICA
798 CLOSE PACKING MINE	Close packing ATOMIC PACKING	Empilement dense EMPILEMENT ATOME	Dichtest.Packung ATOM=PACKUNG	UPAKOVKA PLOT-NEJSHAYA	Empaquetado=atomico=cerrado ESTRUCTURA=ATOMO	Impacchettamento fitto IMPILAMENTO ATOMICO
799 CLUSTER ANALYSIS MATH	CLUSTER ANALY-SIS	ANALYSE GROUPE	CLUSTER=ANALYSE	ANALIZ KLAS-TERNYJ	ANALISIS=GRUPO	CLUSTER ANALY-SIS
800 CLYMENIIDA PALS	CLYMENIIDA	CLYMENIIDA	CLYMENIIDA	CLYMENIIDA	CLYMENIDA	CLYMENIIDA
801 COAL COMS	COAL *UF:* Ash content; Cannel coal; Coal gas; Coal measures; Gas coal	CHARBON *UF:* Charbon gaz; Houil-ler productif	KOHLE *UF:* Gas=Kohle; Kohle=Horizont	UGOL' *UF:* Coal ball; Karta ugol'nykh mestorozh-denij; Ugol' saprope-lievyj	CARBON *UF:* Capas=de=carbon; Hulla=grasa	CARBONE *UF:* Carbone da gas; Livelli carboniferi
802 COAL BALL SEDI	COAL BALLS	COAL BALL	TORF=KOHLE=KONKRETION	[COAL BALL] KONKRETSIYA; UGOL'	BOLA=DE=CARBON	COAL BALL
803 COAL FIELD ECON	COAL FIELDS	BASSIN HOUILLER	KOHLE=BECKEN	BASSEJN UGOL'NYJ	CUENCA=HULLERA	BACINO CAR-BONIFERO
804 COAL GAS COMS	Coal gas COAL; GASES	Gaz charbon GAZ	Kohlen=Gas GAS	GAZ UGOL'NYJ	Gas=carbon GAS	Gas di carbone GAS

	ENGLISH	FRANCAIS	DEUTSCH	RUSSKIJ	ESPANOL	ITALIANO
805 COAL MEASURES ECON	Coal measures COAL	Houiller productif CHARBON	Kohle=Horizont KOHLE	FORMATSIYA UGLENOSNAYA	Capas=de=carbon CARBON	Livelli carboniferi CARBONE
806 COAL SEAM ECON	COAL SEAMS *UF:* Seams	COUCHE CHARBON *UF:* Veine charbon	KOHLE=FLOEZ *UF:* Floez	PLAST UGOL'NYJ	CAPA=CARBON *UF:* Capa=productiva	LIVELLO DI CAR- BONE *UF:* Letto produttivo
807 COAL=DEPOSIT MAP ECON	COAL=DEPOSIT MAPS	CARTE HOUILLERE	KOHLEN= LAGERSTAETTEN= KARTE	Karta ugol'nykh mestorozhdenij KARTA MESTOROZHDENIJ; UGOL'	MAPA=HULLERO	CARTA GIACI- MENTI DI CAR- BONE
808 COALIFICATION SEDI	COALIFICATION	HOUILLIFICATION	INKOHLUNG	UGLEFIKATSIYA	CARBONIFICACION	CARBONIZZAZ- IONE
809 COARSE GRAVEL SEDS	Coarse gravel GRAVEL	Gravier grossier GRAVIER	Grob.Geroell KIES	[COARSE GRAVEL] GRAVIJ; KRUPNOZ- ERNISTYJ	Grava=gruesa GRAVA	Ghiaia grossolana GHIAIA
810 COARSE=GRAINED FRACTION SEDI	COARSE=GRAINED MATERIALS	FRACTION GROS- SIERE	GROBFRAKTION	KRUPNOZERNIS- TYJ *UF:* Coarse gravel	FRACCION= GRUESA	FRAZIONE GROS- SOLANA
811 COAST SURF	Coast BEACHES	Cote LIGNE RIVAGE	Meeres=Kueste KUESTE	POBEREZH'E	Costa LINEA=COSTA	Costa LINEA DI RIVA
812 COASTAL DUNE SURF	COASTAL DUNES	DUNE COTIERE	KUESTEN=DUENE	DYUNA BERE- GOVAYA	DUNA=COSTERA	DUNA COSTIERA
813 COASTAL EROSION SURF	LITTORAL ERO- SION	EROSION LITTOR- ALE *UF:* Cote abrasion; Trou souffleur	KUESTEN= EROSION *UF:* Abrasions=Kueste; Blowhole	RAZMYV PLYAZHA	EROSION=LITORAL *UF:* Bufadero; Costa= abrasion	EROSIONE COSTI- ERA *UF:* Costa d'abrasione; Sfiatatoio
814 COASTAL GEOMORPHOLOGY SURF	Coastal geomorphology SHORE FEATURES	Morphologie cotiere MORPHOLOGIE COTE	Kuesten=Morphologie KUESTEN= MERKMAL	GEOMOR- FOLOGIYA BERE- GOV	Morfologia=costera MORFOLOGIA= COSTA	Morfologia di costa MORFOLOGIA COS- TIERA
815 COASTAL PLAIN SURF	COASTAL PLAINS *UF:* Accumulative coast; Marine plains	PLAINE COTIERE *UF:* Cote accumulation; Plaine marine	KUESTEN=EBENE *UF:* Akkumulations= Kueste; Marin.Ebene; Salz=Marsch	RAVNINA PRI- BREZHNAYA	LLANURA= COSTERA *UF:* Costa=acumulacion; Llanura=de=abrasion	PIANA COSTIERA *UF:* Costa d'accumulazione; Piana d'origine marina
816 COASTLINE SURF	COASTLINES	Ligne cotiere LIGNE RIVAGE	Kuesten=Linie KUESTE	Granitsa poberezh'ya LINIYA BERE- GOVAYA	Linea=costera LINEA=COSTA	Linea di costa LINEA DI RIVA
817 COATED GRAIN TEST	Coated grain GRAINS; TEX- TURES	Grain enduit TEXTURE	[COATED=GRAIN] KORN=GEFUEGE	ZERNO OKUTAN- NOE	Grano=revestido TEXTURA	Granulo rivestito TESSITURA
818 COBALT CHEE	COBALT	COBALT	CO	KOBAL'T	COBALTO	COBALTO
819 COBOL MATH	Cobol COMPUTER PRO- GRAMS	Cobol PROGRAMME ORDINATEUR	Cobol RECHENPRO- GRAMM	Cobol PROGRAMMIROVA- NIE	Cobol PROGRAMA= ORDENADOR	Cobol PROGRAMMA DI CALCOLO

	ENGLISH	FRANCAIS	DEUTSCH	RUSSKIJ	ESPANOL	ITALIANO
820 COCCOLITHOPHO-RACEAE PALS	COCCOLITHOPHO-RACEAE *UF:* Discoasters	COCCOLITHO-PHORALES *UF:* Discoaster	COCCOLITHO-PHORALES *UF:* Discoasteroidea	Coccolithophoraceae CHRYSOPHYTA	COCCOLITHO-PHORALES *UF:* Discoaster	COCCOLITHOPHO-RACEAE *UF:* Discoaster
821 CODA WAVE SOLI	CODA WAVES	CAUDA	CODA=WELLE	KODA VOLNA	ONDA=TERMINAL	ONDA DI CODA
822 COEFFICIENT PHCH	Coefficient PARAMETERS	Coefficient PARAMETRE	Koeffizient PARAMETER	KOEHFFITSIENT *UF:* Transmissibility coefficient	Coeficiente PARAMETRO	Coefficiente PARAMETRO
823 COELENTERATA PALS	COELENTERATA *UF:* Dipleurozoa; Pro-tomedusae	COELENTERATA *UF:* Dipleurozoa; Pro-tomedusae	COELENTERATA *UF:* Dipleurozoa; Pro-tomedusae	COELENTERATA	COELENTERATA *UF:* Dipleurozoa; Pro-tomedusae	COELENTERATA *UF:* Dipleurozoa; Pro-tomedusae
824 COENOTHECALIA PALS	Coenothecalia OCTOCORALLIA	Coenothecalia OCTOCORALLA	Coenothecalia OCTOCORALLA	Coenothecalia OCTOCORALLIA	Coenothecalia OCTOCORALLA	Coenothecalia OCTOCORALLIA
825 COERCIVITY SOLI	COERCIVITY	COERCIVITE	KOERZITIV=KRAFT	Sila koehrtsitivnaya RAZMAGNICHIVA-NIE	COERCITIVIDAD	COERCITIVITA
826 COEXISTING MINERALS MINE	COEXISTING MIN-ERALS	MINERAL COEXI-STANT	KOEXISTIEREND. MINERAL	Mineraly sosushchest-vuyushchie PARAGENEZIS	MINERALES=COEXISTENTES	MINERALE COESIS-TENTE
827 COHESION PHCH	COHESION *UF:* Cohesionless soil; Tenacity	COHESION *UF:* Sol coherent; Sol non coherent; Tenacite	KOHAESION *UF:* Bindig.Boden; Nichtbindig.Boden; Zaehigkeit	STSEPLENIE	COHESION *UF:* Suelo=cohesivo; Suelo=sin=cohesion; Tenacidad	COESIONE *UF:* Suolo coerente; Suolo incoerente; Tena-cita
828 COHESIONLESS SOIL ENGI	Cohesionless soils COHESION; SOILS	Sol non coherent COHESION; SOL	Nichtbindig.Boden KOHAESION; BODEN	GRUNT NESVYAS-NYJ	Suelo=sin=cohesion COHESION; SUELO	Suolo incoerente COESIONE; SUOLO
829 COHESIVE MATERIAL ENGI	COHESIVE MATE-RIALS *UF:* Cohesive soils	MATERIAU COHER-ENT	KOHAESIV.STOFF	[COHESIVE MATE-RIAL] GRUNT SVYAZNYJ	MATERIAL=COHERENTE	MATERIALE COER-ENTE
830 COHESIVE SOIL ENGI	Cohesive soils COHESIVE MATE-RIALS; SOILS	Sol coherent COHESION; SOL	Bindig.Boden KOHAESION; BODEN	GRUNT SVYAZNYJ *UF:* Cohesive material	Suelo=cohesivo COHESION; SUELO	Suolo coerente COESIONE; SUOLO
831 COKING COAL COMS	COKE COAL	CHARBON COKE	KOKS=KOHLE	UGOL' KOKSOVYJ	CARBON=COQUE	CARBONE COKE
832 COLD ENVIRONMENT SURF	COLD ENVIRON-MENT	Milieu froid ZONE FROIDE	Kalt.Milieu KALT.ZONE	KHOLODNYJ	Medio=frio ZONA=FRIA	Ambiente freddo ZONA FREDDA
833 COLEOPTEROIDA PALS	COLEOPTEROIDA	COLEOPTEROIDA	COLEOPTEROIDA	COLEOPTEROIDA	COLEOPTEROIDA	COLEOPTEROIDA
834 COLLAPSE SURF	Collapse COLLAPSE STRUC-TURES	EFFONDREMENT *UF:* Breche effondre-ment; Ecroulement toit; Structure effondre-ment	EINSTURZ *UF:* Einsturz=Brekzie; Einsturz=Struktur; Magmakammer=Einbruch	OBRUSHENIE *UF:* Collapse breccia	HUNDIMIENTO *UF:* Brecha=de-desplome; Estructura=de=desplome; Hundimiento=techo	SPROFONDA-MENTO *UF:* Breccia di collasso; Crollo di tetto; Strut-tura di collasso
835 COLLAPSE BRECCIA GEOL	Collapse breccia BRECCIA	Breche effondrement BRECHE; EFFON-DREMENT	Einsturz=Brekzie BREKZIE; EINS-TURZ	[COLLAPSE BREC-CIA] BREKCHIYA; OBRU-SHENIE	Brecha=de=desplome BRECHA; HUNDI-MIENTO	Breccia di collasso BRECCIA; SPRO-FONDAMENTO

	ENGLISH	FRANCAIS	DEUTSCH	RUSSKIJ	ESPANOL	ITALIANO
836 COLLAPSE CALDERA SURF	Collapse caldera CALDERAS	Caldeira effondrement CALDEIRA	Einsturz=Caldera CALDERA	Kal'dera obrusheniya KAL'DERA	Caldera=de- hundimiento CALDERA	Caldera di sprofonda- mento CALDERA
837 COLLAPSE STRUCTURE STRU	COLLAPSE STRUC- TURES *UF:* Collapse	Structure effondrement EFFONDREMENT	Einsturz=Struktur EINSTURZ	STRUKTURA OBRUSHENIYA	Estructura=de= desplome HUNDIMIENTO	Struttura di collasso SPROFONDA- MENTO
838 COLLECTING GEOL	COLLECTING	RECOLTE	AUFSAMMLUNG	SBOR *UF:* Mineralogical locality	RECOGIDA	RACCOLTA
839 COLLECTION GEOL	COLLECTIONS	COLLECTION	SAMMLUNG	KOLLEKTSIYA	COLECCION	COLLEZIONE
840 COLLEMBOLA PALS	Collembola INSECTA	Collembola INSECTA	Collembola INSECTA	COLLEMBOLA	Collembola INSECTA	Collembola INSECTA
841 COLLOFORM TEXTURE TEST	Colloform textures TEXTURES; COL- LOIDAL MATERI- ALS	Texture colloforme TEXTURE; COLL- OIDE	Kolloid=Gefuege KORN=GEFUEGE; KOLLOID	KOLLOFORMNYJ	Textura=coloforme TEXTURA; COL- OIDE	Tessitura colloforme TESSITURA; COLL- OIDE
842 COLLOID PHCH	COLLOIDAL MATE- RIALS *UF:* Botryoidal texture; Colloform textures; Mineral gel	COLLOIDE *UF:* Forme botryoidale; Mineral gel; Texture colloforme	KOLLOID *UF:* Gel=Mineral; Glaskopf=Gefuege; Kolloid=Gefuege	KOLLOID *UF:* Flokkulyatsiya; Mineraloid	COLOIDE *UF:* Forma=botrioidal; Mineral=gel; Textura= coloforme	COLLOIDE *UF:* Botrioidale; Gel minerale; Tessitura colloforme
843 COLLUVIUM SURF	COLLUVIUM	COLLUVION	COLLUVIUM	KOLLYUVIJ	COLUVION	COLLUVIUM
844 COLONIAL TAXON PALE	COLONIAL TAXA	TAXON COLONIAL	KOLONIEBILDEND. TAXON	KOLONIAL'NYJ	TAXON=COLONIAL	TAXON COLONI- ALE
845 COLOR PHCH	COLOR *UF:* Color index	COULEUR *UF:* Indice couleur; Leucosome	FARBE *UF:* Farb=Index; Leuco- some	TSVET	COLOR *UF:* Indice=color; Leu- cosoma	COLORE *UF:* Indice di colore; Leucosoma
846 COLOR CENTER MINE	COLOR CENTERS	CENTRE COLORE	FARB=ZENTRUM	Tsentr okraski DEFEKT KRISTAL- LICHESKOJ STRUK- TURY	CENTRO= COLOREADO	CENTRO CROMA- TICO
847 COLOR INDEX IGNE	Color index COLOR	Indice couleur COULEUR	Farb=Index FARBE	INDEKS TSVE- TOVOJ	Indice=color COLOR	Indice di colore COLORE
848 COLOR TRACER METH	DYE TRACERS	TRACEUR COLO- RANT	FARB- MARKIERUNG	TRASSER TSVET- NOY	TRAZADOR= COLORANTE	TRACCIANTE COLORANTE
849 COLORIMETRY METH	COLORIMETRY	COLORIMETRIE	KOLORIMETRIE	KOLORIMETRIYA	COLORIMETRIA	COLORIMETRIA
850 COLUMNAR JOINTING TEST	COLUMNAR JOINTS	FISSURATION PRISMATIQUE	PRISMAT. SAEULENBILDUNG	OTDEL'NOST' STOLBCHATAYA	FISURACION= PRISMATICA	FESSURAZIONE PRISMATICA
851 COLUMNARIINA PALS	Columnariina RUGOSA	Columnariina RUGOSA	Columnariina RUGOSA	Columnariina RUGOSA	Columnariina RUGOSA	Columnariina RUGOSA
852 COMAGMATIC GENESIS IGNE	Comagmatic materials MAGMAS	Genese comagmatique MAGMA	Comagmat.Genese MAGMA	KOMAGMATICH- NYJ *UF:* Rodstvo geneti- cheskoe	Comagmatico MAGMA	Comagmatico MAGMA

	ENGLISH	FRANCAIS	DEUTSCH	RUSSKIJ	ESPANOL	ITALIANO
853 COMMENSALISM PALE	Commensalism EPIBIOTISM	Commensalisme EPIBIOTISME	Kommensalismus EPIBIOTISMUS	Kommensalizm SIMBIOZ	Comensalismo EPIBIOTISMO	Commensalismo EPIBIOSI
854 COMMON-DEPTH-POINT METHOD APPL	COMMON-DEPTH-POINT METHOD	COUVERTURE MULTIPLE	MEHRFACH-UEBERTRAGUNG	METOD OBSHCHEJ GLUBINNOJ TOCHKI	COVERTURA=MULTIPLE	COPERTURA MULTIPLA
855 COMMUNITY PALE	COMMUNITIES	POPULATION UF: Phytocenose	POPULATION UF: Phytocoenose	Soobshchestvo BIOTSENOZ	POBLACION UF: Fitocenosis	COMUNITA UF: Fitocenosi
856 COMPACTIBILITY PHCH	COMPACTIBILITY	COMPACITÉ	KOMPAKTIONS=VERMOEGEN	SZHIMAEMOST' GORNYKH POROD	COMPACIDAD	COMPATTABILITA
857 COMPACTION SEDI	COMPACTION UF: Differential compaction	COMPACTAGE	KOMPAKTION	UPLOTNENIE (OSADKOV)	COMPACTACION	COMPATTAZIONE
858 COMPANY ECON	COMPANIES	SOCIETE	FIRMA	PREDPRIYATIE	SOCIEDAD	SOCIETA
859 COMPETENT MATERIAL ENGI	COMPETENT MATERIALS	MATERIAU COMPETENT	KOMPETENZ	KOMPETENTNYJ	MATERIAL=COMPETENTE	MATERIALE COMPETENTE
860 COMPLEX GEOL	COMPLEXES	Complexe(gen.)	Komplex(gen.)	KOMPLEKS UF: Magmatic association	Complejo(gen.)	Complesso(gen.)
861 COMPLEXING PHCH	COMPLEXING	COMPLEXATION	KOMPLEX=BILDUNG	KOMPLEKSOOBRA-ZOVANIE	COMPLEXACION	COMPLESSAZIONE
862 COMPLEXOMETRY METH	COMPLEXOMETRY UF: Chelation	COMPLEXOMETRIE UF: Chelation	KOMPLEXOMETRIE UF: Chelat=Bildung	KOMPLEKSO-METRIYA	COMPLEXOMETRIA UF: Quelacion	COMPLESSO-METRIA UF: Chelazione
863 COMPOSITION GEOC	COMPOSITION	Composition ANALYSE CHIMIQUE	Zusammensetzung CHEM.ANALYSE	SOSTAV UF: Lunar constitution	Composicion ANALISIS=QUIMICO	Composizione COMPOSIZIONE CHIMICA
864 COMPRESSIBILITY ENGI	COMPRESSIBILITY	COMPRESSIBILITE	KOMPRESSIBILIT-AET	SZHIMAEMOST'	COMPRESIBILIDAD	COMPRESSIBILITA
865 COMPRESSION PHCH	COMPRESSION UF: Compression test; Compressive stress; Horizontal compression	COMPRESSION UF: Contrainte compression; Essai compression	KOMPRESSION UF: Kompressions=Spannung; Kompressions=Test	SZHATIE UF: Kontraktsiya; Napryazhenie szhatiya; Szhatie gorizontal'noe; Tektonika szatiyan /tang/	COMPRESION UF: Ensayo=comprension; Tension=de=compresion	COMPRESSIONE UF: Forza di compressione; Prova di compressione
866 COMPRESSION STRENGTH ENGI	COMPRESSIVE STRENGTH	RESISTANCE COMPRESSION	DRUCK=WIDERSTAND	PROCHNOST' NA SZHATIE	RESISTENCIA=A-LA=COMPRESION	RESISTENZA ALLA COMPRESSIONE
867 COMPRESSION TECTONICS STRU	COMPRESSION TECTONICS	TECTONIQUE TANGENTIELLE UF: Compression horizontale; Raccourcissement	TANGENTIAL=TEKTONIK UF: Einengung; Horizontal=Kompression	Tektonika szhatiya TEKTONIKA; SZHATIE	TECTONICA=TANGENCIAL UF: Acortamiento; Compresion=horizontal	TETTONICA COMPRESSIVA UF: Compressione orizzontale; Raccorciamento crostale
868 COMPRESSION TEST ENGI	Compression test COMPRESSION	Essai compression COMPRESSION	Kompressions=Test KOMPRESSION	ISPYTANIE KOMPRESSIONNOE	Ensayo=comprension COMPRESION	Prova di compressione COMPRESSIONE

	ENGLISH	FRANCAIS	DEUTSCH	RUSSKIJ	ESPANOL	ITALIANO
869 COMPRESSIVE STRESS PHCH	Compressive stress COMPRESSION; STRESS	Contrainte compression COMPRESSION	Kompressions= Spannung KOMPRESSION	Napryazhenie szhatiya SZHATIE	Tension=de= compresion COMPRESION	Forza di compressione COMPRESSIONE
870 COMPUTER MATH	COMPUTERS	ORDINATEUR *UF:* Informatisation	RECHNER *UF:* Automation	EHVM *UF:* Tekhnika vychisli- tel'naya	ORDENADOR *UF:* Automatizacion	CALCOLATORE *UF:* Automazione
871 COMPUTER LANGUAGE MATH	Language (math) COMPUTER PRO- GRAMS	Langage informatique PROGRAMME ORDINATEUR	Rechensprache RECHENPRO- GRAMM	YAZYK EHVM	Lenguaje=informatico PROGRAMA= ORDENADOR	Linguaggio informatico PROGRAMMA DI CALCOLO
872 COMPUTER PROGRAM MATH	COMPUTER PRO- GRAMS *UF:* Algol; Cobol; Lan- guage (math)	PROGRAMME ORDINATEUR *UF:* Algol; Cobol; Lang- age informatique	RECHENPRO- GRAMM *UF:* Algol; Cobol; Rec- hensprache	PROGRAMMIROVA- NIE *UF:* Cobol	PROGRAMA= ORDENADOR *UF:* Algol; Cobol; Lenguaje=informatico	PROGRAMMA DI CALCOLO *UF:* Algol; Cobol; Linguaggio informatico
873 CONCENTRATION PHCH	CONCENTRATION	CONCENTRATION	KONZENTRATION	KONTSENTRAT- SIYA *UF:* Kontsentratsiya gravimetricheskaya	CONCENTRACION	CONCENTRAZIONE
874 CONCENTRIC FOLD STRU	CONCENTRIC FOLDS	PLI CON- CENTRIQUE	BIEGE=FALTE	SKLADKA KONT- SENTRICHESKAYA	PLIEGUE= CONCENTRICO	PIEGA CON- CENTRICA
875 CONCHOIDAL FRACTURE TEST	Conchoidal materials MINERAL CLEAV- AGE	Fracture conchoidale CLIVAGE MINERAL	Muschelig.Bruch SPALTBARKEIT	IZLOM RAKOVIS- TYJ	Fractura=concoidea EXFOLIACION= MINERAL	Frattura concoide SFALDATURA
876 CONCHOSTRACA PALS	Conchostraca BRANCHIOPODA	CONCHOSTRACA *UF:* Branchiopoda; Cladocera; Lipostraca; Phyllopoda	CONCHOSTRACA *UF:* Branchiopoda; Cladocera; Lipostraca; Phyllopoda	CONCHOSTRACA	CONCHOSTRACA *UF:* Branchiopoda; Cladocera; Lipostraca; Phyllopoda	CONCHOSTRACA *UF:* Branchiopoda; Cladocera; Lipostraca; Phyllopoda
877 CONCORDANCE STRA	Concordance STRATIFICATION	Couche concordante STRATIFICATION	Konkordanz SCHICHTUNG	SOGLASNYJ	Capa=concordante ESTRATIFICACION	Concordanza di stratifi- cazione STRATIFICAZIONE
878 CONCORDANT AGE ISOT	Concordant age CORRELATION; ABSOLUTE AGE	Age concordant CORRELATION; DATATION	Konkordant.Alter KORRELATION; PHYSIKAL. ALTERSBESTIM- MUNG	Vozrast konkordantnyj VOZRAST ABSO- LYUTNYJ	Edad=acorde CORRELACION; DATACION	Eta concordante CORRELAZIONE; DATAZIONE
879 CONCRETE ENGI	CONCRETE	BETON	BETON	BETON	CEMENTO	CALCESTRUZZO
880 CONCRETE DAM ENGI	CONCRETE DAMS	BARRAGE BETON	BETON=DAMM	PLOTINA BETON- NAYA	PRESA=HORMIGON	DIGA IN CALC- ESTRUZZO
881 CONCRETION SEDI	CONCRETIONS	CONCRETION	KONKRETION	KONKRETSIYA *UF:* Coal ball; Katun glinyanyj; Sharik glinis- tyj	CONCRECION	CONCREZIONE
882 CONDENSATION PHCH	CONDENSATION	CONDENSATION	KONDENSATION	KONDENSATSIYA	CONDENSACION	CONDENSAZIONE
883 CONDUCTIVITY PHCH	CONDUCTIVI- TY	Conductivite(gen.)	Leitfaehigkeit(gen.)	PROVODIMOST'	Conductividad(gen.)	Conducibilita(gen.)

	ENGLISH	FRANCAIS	DEUTSCH	RUSSKIJ	ESPANOL	ITALIANO
884 CONDYLARTHRA PALS	CONDYLARTHRA	CONDYLARTHRA	CONDYLARTHRA	CONDYLARTHRA	CONDYLARTHRA	CONDYLARTHRA
885 CONE OF DEPRESSION GEOH	CONE OF DEPRES- SION	Cone depression ABAISSEMENT NIVEAU EAU	Absenkungs=Trichter GRUNDWASSER- SPIEGEL- ABSENKUNG	Voronka depression- naya PONIZHENIE UROVNYA	Cono=de=depresion DESCENSO= NIVEL=DE=AGUA	Cono di depressione ABBASSAMENTO DELLA FALDA
886 CONE-IN-CONE TEST	CONE-IN=CONE	[CONE IN CONE]	[CONE-IN=CONE]	KONUS V KONUSE (STRUKTURA)	[CONE-IN=CONE]	[CONE=IN=CONE]
887 CONFINED AQUIFER GEOH	CONFINED AQUI- FERS UF: Artesian aquifer	NAPPE CAPTIVE UF: Nappe artesienne	GESPANNT. GRUNDWASSER UF: Artes.Grundwasser	Gorizont napornyj GORIZONT ARTEZ- IANSKIJ	MANTO=CAUTIVO UF: Manto=artesiano	FALDA CIR- COSCRITTA UF: Falda artesiana
888 CONFLUENCE SURF	Confluent streams STREAMS	Confluent RIVIERE	Zusammenfluss FLUSS	Pritok PRITOCHNYJ	Confluente RIO	Confluenza FIUME
889 CONFORMITY STRA	Conformity STRATIFICATION	Concordance strati- graphique STRATIFICATION	Konkordant.Lagerung SCHICHTUNG	ZALEGANIE SOGLASNOE	Concordancia= estratigrafica ESTRATIFICACION	Concordanza STRATIFICAZIONE
890 CONGELATION ENGI	Congelation FROST ACTION	Congelation ACTION FROID	Gefrieren KAELTE= WIRKUNG	ZATVERDENIE	Congelacion ACCION=FRIO	Congelamento AZIONE DEL GELO
891 CONGELIFRACTION SURF	CONGELIFRAC- TION	CONGELIFRAC- TION	SPALTENFROST= VERWITTERUNG	VYVETRIVANIE MOROZNOE UF: Gelivity; Vozdejstvie moroznoe	METEORIZACION= POR=HELADA	CONGELIFRAZ- IONE
892 CONGLOMERATE SEDS	CONGLOMERATE UF: Basal conglomerate; Fanglomerate; Pseudo- tillite	CONGLOMERAT UF: Conglomerat base; Fanglomerat; Pseudotil- lite	KONGLOMERAT UF: Basal=Konglomerat; Fanglomerat; Pseudotil- lit	KONGLOMERAT	CONGLOMERADO UF: Conglomerado= base; Fanglomerado; Pseudotillita	CONGLOMERATO UF: Conglomerato bas- ale; Fanglomerato; Pseudotillite
893 CONGRUENT MELTING IGNE	Congruent melting MELTING	Fusion congruente FUSION	Kongruent.Schmelze SCHMELZE	PLAVLENIE KONG- RUEHNTNOE	Mezcla=congruente FUSION	Fusione congruente FUSIONE
894 CONIACIAN STRS	CONIACIAN	CONIACIEN	CONIAC	KON'YAK	CONIACIENSE	CONIACIANO
895 CONIFERALES PALS	CONIFERALES	CONIFERALES	CONIFERALES	CONIFERALES	CONIFERALES	CONIFERALES
896 CONJUGATE GEOL	Conjugate(adj.)	Conjugue(adj.)	Konjugiert(adj.)	SOPRYAZHENNYJ	Conjugado(adj.)	Coniugato(adj.)
897 CONJUGATE FOLD STRU	CONJUGATE FOLDS	PLI CONJUGUE	[CONJUGATE= FOLD]	SKLADKA SOPRYAZHEN- NAYA	PLIEGUE= CONJUGADO	PIEGA CONIUGATA
898 CONNATE WATER SEDI	CONNATE WATERS UF: Edge water	EAU CONNEE UF: Eau gisement	[CONNATE= WATER] UF: Lagerstaetten= Wasser	VODA POGREBEN- NAYA	AGUA= CONGENITA UF: Agua=marginal	ACQUA CONNATA UF: Acqua di giacimento
899 CONOCARDIOIDA PALS	Conocardioida BIVALVIA	Conocardioida BIVALVIA	Conocardioida LAMELLI- BRANCHIATA	CONOCARDIOIDA	Conocardioida BIVALVIA	Conocardioida BIVALVIA

	ENGLISH	FRANCAIS	DEUTSCH	RUSSKIJ	ESPANOL	ITALIANO
900 CONODONT APPARATUS PALE	Conodont apparatus CONODONTA	Appareil conodonte CONODONTA	Conodonten=Apparat CONODONTA	APPARAT KONO- DONTOVYJ	Aparato=masticador= conodontos CONODONTA	Apparato conodonte CONODONTA
901 CONODONTA PALS	CONODONTA *UF:* Conodont apparatus	CONODONTA *UF:* Appareil conodonte	CONODONTA *UF:* Conodonten= Apparat	CONODONTA	CONODONTA *UF:* Aparato= masticador=conodontos	CONODONTA *UF:* Apparato conodonte
902 CONRAD DISCONTINUITY SOLI	CONRAD DISCON- TINUITY	DISCONTINUITE CONRAD	CONRAD= DISKONTINUITAET	KONRADA POVERKHNOST'	DISCONTINUIDAD- DE=CONRAD	DISCONTINUITA DI CONRAD
903 CONSANGUINITY IGNE	Consanguinity MAGMATIC ASSO- CIATIONS	Consanguinite ASSOCIATION MAGMATIQUE	Konsanguinitaet MAGMAT. VERGESELLSCHAF- TUNG	Rodstvo geneticheskoe KOMAGMATICH- NYJ	Consanguineidad ASOCIACION= MAGMATICA	Consanguineita ASSOCIAZIONE MAGMATICA
904 CONSEQUENT STREAM SURF	Consequent streams STREAMS; RELIEF	Vallee consequente RIVIERE; RELIEF STRUCTURAL	Konsequent.Strom FLUSS; STRUKTUR= RELIEF	Reka konsekventnaya REKA; REL'EF STRUKTURNYJ	Rio=consecuente RIO; RELIEVE= ESTRUCTURAL	Corso d'acqua consegu- ente FIUME; MOR- FOSTRUTTURA
905 CONSERVATION ENVI	CONSERVATION	PROTECTION ENVIRONNEMENT *UF:* Protection rivage	UMWELT=SCHUTZ	OKHRANA *UF:* Human ecology; Protected zone; Water authority	PROTECCION= MEDIO=AMBIENTE *UF:* Proteccion=costera	PROTEZIONE AMBIENTALE *UF:* Protezione costiera
906 CONSISTENCY PHCH	CONSISTENCY	CONSISTANCE	KONSISTENZ	KONSISTENTSIYA	CONSISTENCIA	CONSISTENZA
907 CONSOLIDATION GEOL	CONSOLIDATION	CONSOLIDATION	KONSOLIDIERUNG	KONSOLIDATSIYA	CONSOLIDACION	CONSOLIDAMENTO
908 CONSOLIDOMETER TEST METH	CONSOLIDOMETER TESTS	COMPRESSION OEDOMETRIQUE	OEDOMETRIE	KONSOLIDO- METRIYA	COMPRESION= EDOMETRICA	PROVA EDO= METRICA
909 CONSTANT MISC	Constant PARAMETERS	Constante PARAMETRE	Konstante PARAMETER	KONSTANTA	Constante PARAMETRO	Costante PARAMETRO
910 CONSTRUCTION ENGI	Construction STRUCTURES	Construction OUVRAGE	Konstruktion INGENIEUR=BAU	SOORUZHENIE INZHENERNOE	Construccion OBRAS	Costruzione OPERA INGEGNERISTICA
911 CONSTRUCTION MATERIALS COMS	CONSTRUCTION MATERIALS *UF:* Brick clay; Road material	MATERIAU CON- STRUCTION *UF:* Argile a brique	BAUSTOFF *UF:* Ziegel=Ton	STROITEL'NOE SYR'E *UF:* Kamen' dorozhnyj	MATERIAL= CONSTRUCCION *UF:* Arcilla=de=ladrillo	MATERIALE DA COSTRUZIONE *UF:* Argilla per laterizi
912 CONSUMPTION ECON	CONSUMPTION *UF:* Water consumption	CONSOMMATION *UF:* Consommation eau	VERBRAUCH *UF:* Wasser=Verbrauch	POTREBLENIE	CONSUMO *UF:* Consumo=agua	CONSUMO *UF:* Consumo idrico
913 CONTACT GEOL	Boundary(gen.)	Contact(gen.)	Kontakt(gen.)	KONTAKT	Contacto(gen.)	Contatto(gen.)
914 CONTACT BRECCIA GEOL	Contact breccia BRECCIA	Breche contact BRECHE	Kontakt=Brekzie BREKZIE	Brekchiya kontak- tovaya BREKCHIYA; EHN- DOKONTAKT	Brecha=de=contacto BRECHA	Breccia di contatto BRECCIA
915 CONTACT METAMORPHISM IGNE	CONTACT META- MORPHISM *UF:* Pyrometamorphism; Sanidinite facies	METAMORPHISME CONTACT *UF:* Facies a sanidine; Pyrometamorphisme	KONTAKT= METAMORPHOSE *UF:* Pyro= Metamorphose; Sanidinit=Fazies	METAMORFIZM KONTAKTOVYJ	METAMORFISMO= CONTACTO *UF:* Facies=sanidina; Pirometamorfismo	METAMORFISMO DI CONTATTO *UF:* Facies delle sanidiniti; Metasoma- tismo di contatto; Pirometamorfismo

	ENGLISH	FRANCAIS	DEUTSCH	RUSSKIJ	ESPANOL	ITALIANO
916 CONTACT METASOMATISM IGNE	Contact metasomatism METASOMATISM	Metasomatose contact PYROMETASOMA-TOSE	Kontakt=Metasomatose PYRO-METASOMATOSE	METASOMATOZ KONTAKTOVYJ	Metasomatismo=de=contacto PIROMETASOMA-TISMO	Metasomatismo di contatto PIROMETASOMA-TOSI; METAMOR-FISMO DI CON-TATTO
917 CONTAMINATION IGNE	CONTAMINATION	CONTAMINATION MAGMATIQUE	MAGMAT. KONTAMINATION	KONTAMINATSIYA	CONTAMINACION=MAGMATICA	CONTAMINAZIONE
918 CONTINENT GEOL	CONTINENTS	Continent RELIEF CONTI-NENT	Kontinent KONTINENTAL=RELIEF	KONTINENT *UF:* Nazemnyj	Continente RELIEVE=CONTINENTAL	Continente FORMA DEL RILIEVO
919 CONTINENTAL ACCRETION SOLI	Continental accretion ACCRETION	Croissance continentale ACCRETION	Kontinent=Wachstum ANLAGERUNG	AKKRETSIYA KON-TINENTOV	Crecimiento=continental ACRECION	Accrescimento continentale ACCRESCIMENTO
920 CONTINENTAL BORDERLAND STRU	CONTINENTAL BORDERLAND	MARGE CON-TINENTALE COM-PLEXE	KONTINENTAL-RAND	KRAJ PLATFORMY *UF:* Zakraina aktivnaya	MARGEN=CONTINENTAL=COMPLEJO	MARGINE CON-TINENTALE COM-PLESSO
921 CONTINENTAL CRUST SOLI	CONTINENTAL CRUST	CROUTE CON-TINENTALE	KONTINENTAL. KRUSTE	KORA KONTINEN-TAL'NAYA	CORTEZA=CONTINENTAL	CROSTA CON-TINENTALE
922 CONTINENTAL DRIFT SOLI	CONTINENTAL DRIFT *UF:* Displacement theory; Fixism	DERIVE CON-TINENTALE *UF:* Mobilisme	KONTINENTAL-DRIFT *UF:* Verschiebungs=Theorie	DRIFT KONTINEN-TOV	DERIVA=CONTINENTAL *UF:* Teoria=del=desplazamiento	DERIVA CON-TINENTALE *UF:* Teoria deriva continentale
923 CONTINENTAL ENVIRONMENT PALE	CONTINENTAL ENVIRONMENT	MILIEU CONTI-NENTAL *UF:* Milieu terrestre	KONTINENTAL. MILIEU *UF:* Terrestr.Milieu	Usloviya kontinen-tal'nye FATSIYA KON-TINENTAL'NAYA	MEDIO=CONTINENTAL *UF:* Medio=terrestre	AMBIENTE CON-TINENTALE *UF:* Ambiente terrestre
924 CONTINENTAL GLACIATION SURF	Continental glaciation GLACIATION	Glaciation continentale GLACIATION	Kontinental=Vereisung VEREISUNG	OLEDENENIE KON-TINENTAL'NOE	Glaciacion=continental GLACIACION	Glaciazione continen-tale GLACIAZIONE
925 CONTINENTAL MARGIN MARI	CONTINENTAL MARGIN	MARGE CON-TINENTALE	SCHELF	OKRAINA MATERIKOVAYA *UF:* Okraina asejsmi-cheskaya	MARGEN=CONTINENTAL	MARGINE CON-TINENTALE
926 CONTINENTAL NUCLEUS STRU	Continental nucleus CRATONS	Noyau continental CRATON	Kontinental=Kern KRATON	YADRO KON-TINENTAL'NOE	Nucleo=continental CRATON	Nucleo continentale CRATONE
927 CONTINENTAL RISE MARI	CONTINENTAL RISE	GLACIS CONTI-NENTAL	KONTINENTAL=SOCKEL	PODNOZH'E MATERIKOVOE	GLACIS=LLANURA=ABISAL	SCARPATA CON-TINENTALE
928 CONTINENTAL SEDIMENTATION SEDI	CONTINENTAL SEDIMENTATION	SEDIMENTATION CONTINENTALE	KONTINENTAL-SEDIMENTATION	FATSIYA KON-TINENTAL'NAYA *UF:* Usloviya kontinen-tal'nye	SEDIMENTACION=CONTINENTAL	SEDIMENTAZIONE CONTINENTALE
929 CONTINENTAL SHELF MARI	CONTINENTAL SHELF *UF:* Borderland; Shelf; Shelf sea	PLATEFORME CON-TINENTALE *UF:* Bordure continen-tale; Plateau continental	KONTINENTAL-TAFEL *UF:* Borderland; Kontinental=Schelf	SHEL'F KONTINEN-TAL'NYJ *UF:* Borderland; Inner shelf; Outer shelf; Shel'f	PLATAFORMA=CONTINENTAL *UF:* Borde=continental; Llanura=continental	PIATTAFORMA CONTINENTALE *UF:* Area marginale; Piattaforma

	ENGLISH	FRANCAIS	DEUTSCH	RUSSKIJ	ESPANOL	ITALIANO
930 CONTINENTAL SLOPE MARI	CONTINENTAL SLOPE	TALUS MARIN	SCHELF=HANG	SKLON KONTINEN-TAL'NYJ *UF:* Inner slope	TALUD=MARINO	TALUS MARINO
931 CONTINENTAL ZONE SURF	CONTINENTAL ZONE	ZONE CONTINEN-TALE	KONTINENTAL=ZONE	[CONTINENTAL ZONE] SUSHA	ZONA=CONTINENTAL	ZONA CONTINEN-TALE
932 CONTOUR MAP MISC	CONTOUR MAPS	CARTE COURBE NIVEAU	UMRISS=KARTE	Karta konturnaya KARTA TOPOGRA-FICHESKAYA	MAPA=DE=CURVAS=DE=NIVEL	CARTA A ISOLINEE
933 CONTRACTING EARTH SOLI	Contracting earth CONTRACTION; EARTH	Contraction terre CONTRACTION; PLANETE TERRE	Erd=Schrumpfung KONTRAKTION; PLANET=ERDE	GIPOTEZA KON-TRAKTSIONNAYA	Contraccion=tierra CONTRACCION; PLANETA=TIERRA	Contrazione terrestre CONTRAZIONE; PIANETA TERRA
934 CONTRACTION PHCH	CONTRACTION *UF:* Contracting earth	CONTRACTION *UF:* Contraction terre	KONTRAKTION *UF:* Erd=Schrumpfung	Kontraktsiya SZHATIE	CONTRACCION *UF:* Contraccion=tierra	CONTRAZIONE *UF:* Contrazione ter-restre
935 CONTRAST PHCH	Contrast OPTICAL PROPER-TIES	Contraste PROPRIETE OPTIQUE	Kontrast OPT.EIGENSCHAFT	CONTRASTNOST'	Contraste PROPIEDAD=OPTICA	Contrasto PROPRIETA OTTICA
936 CONTROL MISC	Control(gen.)	Controle(gen.)	Kontrolle(gen.)	KONTROL' *UF:* Erosion control; Kontrol' mekhani-cheskij; Kontrol' paleogeograficheskij; Magmatic controls; Production control	Control(gen.)	Controllo(gen.)
937 CONULARIDA PALS	CONULARIDA	CONULARIDA	CONULARIDA	CONULATA	CONULARIDA	CONULARIDA
938 CONVECTION PHCH	CONVECTION	CONVECTION	KONVEKTION	KONVEKTSIYA	CONVECCION=TERMICA	CONVEZIONE
939 CONVECTION CURRENT SOLI	CONVECTION CUR-RENTS	COURANT CON-VECTION	KONVEKTIONS-STROEMUNG	KONVEKTSIYA TEPLOVAYA	CORRIENTE=CONVECCION	CORRENTE DI CONVEZIONE
940 CONVERGENCE GEOL	CONVERGENCE	CONVERGENCE *UF:* Organisme homeo-morphe	KONVERGENZ *UF:* Isomorph. Organismus	KONVERGENTSIYA	CONVERGENCIA	CONVERGENZA *UF:* Organismo isomorfo
941 CONVOLUTE BEDDING TEST	CONVOLUTE BEDS	STRATE CONVO-LUTEE	WICKEL=SCHICHT	SLOJCHATOST' KONVOLYUTNAYA	ESTRATO=CONVOLUTO	STRATI CON-VOLUTI
942 COOLING IGNE	COOLING	REFROIDISSEMENT	ABKUEHLUNG	OKHLAZHDENIE	ENFRIAMIENTO	RAFFREDDA-MENTO
943 COORDINATION PHCH	COORDINATION	COORDINENCE	CHEM. KOORDINATION	KOORDINATSIYA	COORDINACION	COORDINAZIONE
944 COPEPODA PALS	COPEPODA	COPEPODA	COPEPODA	COPEPODA	COPEPODA	COPEPODA
945 COPERNICAN EXTR	COPERNICAN	Copernicien ECHELLE STRATI-GRAPHIQUE; LUNE	Kopernikan.Alter STRATIGRAPH. SKALA; MOND	KOPERNIKANSKIJ	Coperniciense ESCALA=ESTRATIGRAFICA; LUNA	Copernicano SCALA STRATI-GRAFICA; LUNA

	ENGLISH	FRANCAIS	DEUTSCH	RUSSKIJ	ESPANOL	ITALIANO
946 COPPER CHEE	COPPER	CUIVRE	CU	MED'	COBRE	RAME
947 COPROLITE PALE	COPROLITES *UF:* Fecal pellets	COPROLITE *UF:* Pelote fecale	KOPROLITH *UF:* Kot=Pille	KOPROLITY *UF:* Komochki fekal'nye	COPROLITO *UF:* Pelet=fecal	COPROLITE *UF:* Grumo fecale
948 COQUINA SEDS	COQUINA	LUMACHELLE	SCHILL=KALK	RAKUSHNYAK	LUMAQUELA	LUMACHELLA
949 CORAL REEF MARI	Coral reefs REEFS	Recif coralliaire RECIF	Korallen=Riff RIFF	RIF KORALLOVYJ *UF:* Rif okajmlyayush- chij	Arrecife=coral ARRECIFE	Scogliera corallina SCOGLIERA
950 CORALLIMORPHARIA PALS	CORALLIMOR- PHARIA	CORALLIMOR- PHARIA	CORALLIMOR- PHARIA	Corallimorpharia ANTHOZOA	CORALLIMOR- PHARIA	CORALLIMOR- PHARIA
951 CORALLINACEAE PALS	CORALLINACEAE	CORALLINACEAE	CORALLINACEAE	Corallinaceae RHODOPHYTA	CORALLINACEAE	CORALLINACEAE
952 CORALLITE PALE	Corallite ANTHOZOA	Corallite ANTHOZOA	Corallit ANTHOZOA	KORALLIT	Coralico ANTHOZOA	Corallite ANTHOZOA
953 CORDAITALES PALS	CORDAITALES	CORDAITALES	CORDAITALES	CORDAITALES	CORDAITALES	CORDAITALES
954 CORDIERITE- **AMPHIBOLITE FACIES** IGNE	Cordierite= amphibolite=facies AMPHIBOLITE FACIES	Facies amphibolite cor- dierite FACIES AMPHIBO- LITE	Cordierit=Amphibolit= Fazies AMPHIBOLIT= FAZIES	Fatsiya kordierit= amfibolitovaya FATSIYA AMFI- BOLITOVAYA	Facies=anfibolita= cordierita FACIES= ANFIBOLITA	Facies anfiboliti a cor- dierite FACIES DELLE ANFIBOLITI
955 CORDILLERA STRU	CORDILLERA	CORDILLERE	CORDILLERE	CORDIL'ERA	CORDILLERA	CORDIGLIERA
956 CORE DRILLING MINI	Core drilling CORES	Sondage carotte SONDAGE	Kern=Bohrung BOHRUNG	BURENIE KOLONKOVOE	Testigo=sondeo POZO=SONDEO	Carotaggio SONDAGGIO
957 CORE OF THE EARTH SOLI	CORE *UF:* Bullard discontinu- ity; Gutenberg disconti- nuity	NOYAU TER- RESTRE *UF:* Discontinuite Bul- lard; Discontinuite Gutenberg	ERD=KERN *UF:* Bullard= Diskontinuitaet; Gutenberg= Diskontinuitaet	YADRO ZEMLI *UF:* E sloj; F sloj; G sloj	NUCLEO=GLOBO *UF:* Discontinuidad= de=Bullard; Discontinuidad=de= Gutenberg	NUCLEO TER- RESTRE *UF:* Discontinuita di Bullard; Discontinuita di Gutenberg
958 CORE SAMPLE GEOL	CORES *UF:* Core drilling	CAROTTE	BOHR=KERN	KERN	TESTIGO	CAROTA
959 CORIOLIS EFFECT SOLI	CORIOLIS FORCE	ACCELERATION CORIOLIS	CORIOLIS- BESCHLEUNIGUNG	ZAKON KORIOLISA	ACELERACION= CORIOLIS	EFFETTO DI CORI- OLIS
960 CORRECTION MATH	CORRECTIONS *UF:* Overprint	CORRECTION *UF:* Superposition age	KORREKTUR *UF:* Ueberpraegung	KORREKTSIYA	CORRECCION *UF:* Superposicion=edad	CORREZIONE *UF:* Sovrapposizione di eta
961 CORRELATION STRA	CORRELATION *UF:* Concordant age; Discordant age; Homotaxy; Law of correlation of facies	CORRELATION *UF:* Age concordant; Age discordant; Equiv- alence; Loi correlation facies	KORRELATION *UF:* Diskordant.Alter; Fazies=Abfolge= Gesetz; Homotaxie; Konkordant.Alter	KORRELYATSIYA *UF:* Interplanetary comparison	CORRELACION *UF:* Edad=acorde; Edad=no=acorde; Homotaxia; Ley=de= correlacion=de=facies	CORRELAZIONE *UF:* Eta concordante; Eta discordante; Legge della correlaz.di facies; Omotassia
962 CORRELATION **COEFFICIENT** MATH	CORRELATION COEFFICIENT	COEFFICIENT COR- RELATION	KORRELATIONS- KOEFFIZIENT	KOEHFFITSIENT KORRELYATSII	COEFICIENTE= CORRELACION	COEFFICIENTE DI CORRELAZIONE

	ENGLISH	FRANCAIS	DEUTSCH	RUSSKIJ	ESPANOL	ITALIANO
963 CORRESPONDENCE ANALYSIS MATH	CORRESPONDENCE ANALYSIS	ANALYSE COR-RESPONDANCES	KORRESPONDENZ= ANALYSE	Analiz sootvestvij ANALIZ FAK-TORNYJ	ANALISIS= CORRESPONDENCIA	ANALISI DI COR-RISPONDENZA
964 CORROSION SURF	CORROSION	CORROSION	KORROSION	KORROZIYA	CORROSION	CORROSIONE
965 CORUNDUM COMS	CORUNDUM	CORINDON	KORUND	KORUND	CORINDON	CORINDONE
966 CORYNEXOCHIDA PALS	CORYNEXOCHIDA	CORYNEXOCHIDA	CORYNEXOCHIDA	Corynexochida POLYMERA	CORYNEXOCHIDA	CORYNEXOCHIDA
967 COSMIC DUST EXTR	COSMIC DUST *UF:* Cosmic particle; Magnetic spherules	POUSSIERE COSMIQUE *UF:* Particule cosmique; Spherule magnetique	KOSM.STAUB *UF:* Kosm.Partikel; Magnet.Kugel	PYL' KOSMI-CHESKAYA	POLVO=COSMICO *UF:* Particula=cosmica; Polvo=magnetico	POLVERE COSMICA *UF:* Particella cosmica; Sferula magnetica
968 COSMIC NEBULA EXTR	Cosmic nebula EXTRATERRES-TRIAL GEOLOGY	Nebuleuse GEOLOGIE EXTRA TERRESTRE	Kosm.Nebel EXTRATERRESTR. GEOLOGIE	TUMANNOST' KOSMICHESKAYA	Nebulosa GEOLOGIA= EXTRATERRESTRE	Nebulosa GEOLOGIA EXTRATERRESTRE
969 COSMIC PARTICLE EXTR	Cosmic particle COSMIC DUST	Particule cosmique POUSSIERE COSMIQUE	Kosm.Partikel KOSM.STAUB	Chastitsy kosmicheskie KOSMOS; CHAS-TITSA	Particula=cosmica POLVO=COSMICO	Particella cosmica POLVERE COSMICA
970 COSMIC RADIATION EXTR	COSMIC RAYS	RAYONNEMENT COSMIQUE	KOSM.STRAHLUNG	RADIATSIYA KOSMICHESKAYA	RAYOS=COSMICOS	RAGGI COSMICI
971 COSMIC SYSTEM EXTR	Cosmic system EXTRATERRES-TRIAL GEOLOGY	Systeme cosmique GEOLOGIE EXTRA TERRESTRE	Kosm.System EXTRATERRESTR. GEOLOGIE	KOSMOS *UF:* Chastitsy kosmi-cheskie	Sistema=cosmico GEOLOGIA= EXTRATERRESTRE	Sistema cosmico GEOLOGIA EXTRATERRESTRE
972 COSMOCHEMISTRY EXTR	COSMOCHEM-ISTRY	COSMOCHIMIE	KOSMOCHEMIE	KOSMOKHIMIYA	COSMOQUIMICA	COSMOCHIMICA
973 COSMOGENIC ELEMENT CHES	COSMOGENIC ELE-MENTS	ELEMENT COSMO-GENIQUE	KOSMOGEN. ELEMENT	KOSMOGENNYJ	ELEMENTO= COSMOGENICO	ELEMENTO COS-MOGENICO
974 COSMOLOGICAL PRINCIPLE EXTR	Cosmological principle EXTRATERRES-TRIAL GEOLOGY	Principe cosmologique GEOLOGIE EXTRA TERRESTRE	Kosmolog.Prinzip EXTRATERRESTR. GEOLOGIE	POSTULAT KOS-MOLOGICHESKIJ	Principio=cosmologico GEOLOGIA= EXTRATERRESTRE	Principio cosmologico GEOLOGIA EXTRATERRESTRE
975 COSMOLOGY EXTR	Cosmology EXTRATERRES-TRIAL GEOLOGY	Cosmologie GEOLOGIE EXTRA TERRESTRE	Kosmologie EXTRATERRESTR. GEOLOGIE	KOSMOLOGIYA	Cosmologia GEOLOGIA= EXTRATERRESTRE	Cosmologia GEOLOGIA EXTRATERRESTRE
976 COSTS MISC	COSTS	COUT	KOSTEN	Stoimost' TSENA	COSTE	COSTO
977 COTYLOSAURIA PALS	COTYLOSAURIA	COTYLOSAURIA	COTYLOSAURIA	COTYLOSAURIA	COTYLOSAURIA	COTYLOSAURIA
978 COUNTRY ROCK GEOL	COUNTRY ROCKS	ROCHE ENCAIS-SANTE	RAHMEN=GESTEIN	Poroda vmeshchayush-chaya obramlyeniya PORODA VMESH-CHAYUSHCHAYA	ROCA=ENCAJANTE	ROCCIA INCAS-SANTE
979 COVER STRU	COVER BEDS	FORMATION RECOUVREMENT	DECKGEBIRGE	CHEKHOL PLAT-FORMENNYJ	FORMACION= RECUBRIMIENTO	LIVELLO DI COPERTURA

	ENGLISH	FRANCAIS	DEUTSCH	RUSSKIJ	ESPANOL	ITALIANO
980 CRANIUM PALE	CRANIUM	CRANE	SCHAEDEL	CHEREP	CRANEO	CRANIO
981 CRATER SURF	CRATERS *UF:* Crater lakes; Cratering; Impact crater; Impact structures	CRATERE *UF:* Cratere impact; Formation cratere; Lac cratere; Structure impact	KRATER *UF:* Einschlag=Krater; Einschlag=Struktur; Krater=Bildung; Krater=See	KRATER *UF:* Krateroobrazovanie; Maar; Ozero kraternoe	CRATER *UF:* Crater=de=impacto; Crateres=impacto; Estructura=impacto; Lago=de=crater	CRATERE *UF:* Cratere d'impatto; Formazione di cratere; Lago craterico; Struttura d' impatto
982 CRATER LAKE SURF	Crater lakes LAKES; CRATERS	Lac cratere LAC; CRATERE	Krater=See SEE; KRATER	Ozero kraternoe OZERO; KRATER	Lago=de=crater LAGO; CRATER	Lago craterico LAGO; CRATERE
983 CRATERING SURF	Cratering CRATERS	Formation cratere CRATERE	Krater=Bildung KRATER	Krateroobrazovanie KRATER; GENEZIS	Crateres=impacto CRATER	Formazione di cratere CRATERE
984 CRATON STRU	CRATONS *UF:* Continental nucleus; Forelands; Platforms; Thalassocraton	CRATON *UF:* Avant=pays; Noyau continental; Plateforme; Thalassocraton	KRATON *UF:* Kontinental=Kern; Plattform; Thalassokraton; Vorland	KRATON	CRATON *UF:* Antepais; Nucleocontinental; Plataforma; Thalassocraton	CRATONE *UF:* Avampaese; Nucleo continentale; Piattaforma intracratonica; Talassocratone
985 CREEK SURF	Creeks STREAMS	Ruisseau RIVIERE	Bach FLUSS	Ruchej REKA	Cala RIO	Ruscello FIUME
986 CREEP SURF	CREEP	FLUAGE	KRIECH=VORGANG	OPOLZANIE *UF:* Opolzanie gravitatsionnoe	REPTACION	CREEP
987 CREODONTA PALS	CREODONTA	CREODONTA	CREODONTA	Creodonta CARNIVORA	CREODONTA	CREODONTA
988 CREST SURF	Crests MOUNTAINS	Crete MASSIF MONTAGNEUX	Berg=Kamm BERGMASSIV	Greben' MASSIV	Cresta MACIZO=MONTAÑOSO	Cresta MASSICCIO MONTUOSO
989 CRETACEOUS STRS	CRETACEOUS	CRETACE	KREIDE	MEL	CRETACICO	CRETACEO
990 CREVASSE SURF	Crevasse GLACIERS; RUPTURE	Crevasse GLACIER; RUPTURE	Gletscher=Spalte GLETSCHER; RUPTUR	TRESHCHINA LEDNIKOVAYA *UF:* Bergshrund	Grieta=de=hielo GLACIAR; RUPTURA	Crepaccio GHIACCIAIO; ROTTURA
991 CRINOIDEA PALS	CRINOIDEA	CRINOIDEA	CRINOIDEA	CRINOIDEA	CRINOIDEA	CRINOIDEA
992 CRINOZOA PALS	Crinozoa ECHINODERMATA	Crinozoa ECHINODERMATA	Crinozoa ECHINODERMATA	CRINOZOA *UF:* Edrioblastoidea; Lepidocystoidea; Parablastoidea	Crinozoa ECHINODERMATA	Crinozoa ECHINODERMATA
993 CRITICAL ANGLE ENGI	Critical angle SLOPES; GEOMETRY	Angle critique VERSANT; GEOMETRIE	Krit.Winkel HANG; GEOMETRIE	Ugol kriticheskij SKLON; GEOMETRIYA	Angulo=critico LADERA; GEOMETRIA	Angolo critico VERSANTE; GEOMETRIA
994 CRITICAL FLOW GEOH	CRITICAL FLOW	REGIME CRITIQUE	KRIT.STROEMUNG	POTOK KRITICHESKIJ	REGIMEN=CRITICO	REGIME CRITICO
995 CRITICAL REVIEW MISC	CRITICAL REVIEW	ETUDE CRITIQUE	KRIT. DARSTELLUNG	RETSENZIYA	ESTUDIO=CRITICO	STUDIO CRITICO

	ENGLISH	FRANCAIS	DEUTSCH	RUSSKIJ	ESPANOL	ITALIANO
996 CROCODILIA PALS	CROCODILIA	CROCODILIA	CROCODILIA	CROCODILIA	CROCODILIA	CROCODILIA
997 CROSS=BEDDING SEDI	CROSS=BEDDING	STRATIFICATION OBLIQUE *UF:* Couche frontale	SCHRAEGSCHICH-TUNG *UF:* Foreset=bed	SLOJCHATOST' KOSAYA	ESTRATIFIC-ACION-OBLICUA *UF:* Capa-frontal	STRATIFICAZIONE OBLIQUA *UF:* Strato frontale
998 CROSS=STRATIFICATION TEST	CROSS=STRATIFICATION	STRATIFICATION ENTRECROISEE *UF:* Couche basale; Couche contrepentee; Couche sommitale	KREUZSCHICH-TUNG *UF:* Backset=bed; Bottomset=bed; Topset=bed	SLOISTOST' KOSAYA	ESTRATIFIC-ACION=CRUZADA *UF:* Backset=bed; Capa=basal; Capa=de=techo	STRATIFICAZIONE INCROCIATA *UF:* Backset bed; Strato di letto; Strato di tetto
999 CROSSOPTERYGII PALS	Crossopterygii OSTEICHTHYES	Crossopterygii OSTEICHTHYES	Crossopterygii OSTEICHTHYES	CROSSOPTERYGII *UF:* Rhipidistii	Crossopterygii OSTEICHTHYES	Crossopterygii OSTEICHTHYES
1000 CRUSH BRECCIA GEOL	Crush breccia TECTONIC BREC-CIA	Breche broyage BRECHE TEC-TONIQUE	Reibungs=Brekzie TEKTON.BREKZIE	Brekchiya treniya BREKCHIYA TEK-TONICHESKAYA	Brecha=cataclastica BRECHA=TECTONICA	Breccia di frizione BRECCIA TET-TONICA
1001 CRUST SOLI	CRUST *UF:* A layer; Chal-cosphere; Pyrosphere	CROUTE TER-RESTRE *UF:* Chalcosphere; Couche A; Pyrosphere	ERD=KRUSTE *UF:* A=Schicht; Chalkosphaere; Pyrosphaere	KORA ZEMNAYA *UF:* A sloj	CORTEZA=TERRESTRE *UF:* Calcosfera; Capa=A; Pirosfera	CROSTA TER-RESTRE *UF:* Calcosfera; Pirosf-era; Strato A
1002 CRUSTACEA PALS	CRUSTACEA *UF:* Branchiura; Cephalocarida; Gnathostraca; Lipostraca; Maxil-lopoda; Metanauplius; Mystacocarida	CRUSTACEA *UF:* Branchiura; Cephalocarida; Gnathostraca; Maxil-lopoda; Metanauplius; Mystacocarida; Notostraca	CRUSTACEA *UF:* Branchiura; Cephalocarida; Gnathostraca; Maxil-lopoda; Metanauplius; Mystacocarida; Notostraca	CRUSTACEA *UF:* Branchiura	CRUSTACEA *UF:* Branchiura; Cephalocarida; Gnathostraca; Maxil-lopoda; Metanauplius; Mystacocarida; Notostraca	CRUSTACEA *UF:* Branchiura; Cephalocarida; Gnathostraca; Maxil-lopoda; Metanauplius; Mystacocarida; Notostraca
1003 CRYOPEDOLOGY SURF	Cryopedology FROZEN GROUND	Cryopedologie SOL GELE	Kryopedologie FROST=BODEN	MERZLOTOVEDE-NIE	Criopedologia SUELO=HELADO	Criopedologia SUOLO GELATO
1004 CRYOTECTONICS SURF	Cryotectonics GLACIOTECTONICS	Cryotectonique GLACIOTEC-TONIQUE	Kryotektonik EIS=TEKTONIK	GLYATSIOTEK-TONICHESKIJ	Criotectonico GLACIOTEC-TONICO	Criotettonica GLACIOTET-TONICA
1005 CRYOTURBATION TEST	CRYOTURBATION	CRYOTURBATION	KRYOTURBATION	KRIOTURBATSIYA	CRIOTURBACION	CRIOTURBAZIONE
1006 CRYPTOCRYSTALLINE TEXTURE TEST	Cryptocrystalline tex-ture TEXTURES; CRYS-TALLINITY	Texture cryptocristal-line TEXTURE; CRIS-TALLINITE	Kryptokristallin. Gefuege KORN=GEFUEGE; KRISTALLINITAET	Skrytokristallicheskij AFANITOVYJ	Textura=criptocristalina TEXTURA; CRIS-TALINIDAD	Criptocristallino TESSITURA; CRIS-TALLINITA
1007 CRYPTODONTA PALS	Cryptodonta BIVALVIA	Cryptodonta BIVALVIA	Cryptodonta LAMELLI-BRANCHIATA	CRYPTODONTA	Cryptodonta BIVALVIA	Cryptodonta BIVALVIA
1008 CRYPTOGENIC GEOL	Cryptogenic(adj.)	Cryptogene(adj.)	Kryptogen(adj.)	KRIPTOGENNYJ	Criptogeno(adj.)	Criptogenico(adj.)
1009 CRYPTOSTOMATA PALS	CRYPTOSTOMATA	CRYPTOSTOMATA	CRYPTOSTOMATA	CRYPTOSTOMATA	CRYPTOSTOMATA	CRYPTOSTOMATA
1010 CRYPTOVOLCANIC STRUCTURE SURF	Cryptovolcanic struc-ture IMPACT FEATURES	Structure cryptovol-canique STRUCTURE ANNULAIRE	Kryptovulkan.Struktur RING=STRUKTUR	Struktura kriptovul-kanicheskaya STRUKTURA KRIP-TOEHKSPLOZIV-NAYA	Estructura=criptovolcanica ESTRUCTURA=ANULAR	Struttura criptovul-canica STRUTTURA ANU-LARE

	ENGLISH	FRANCAIS	DEUTSCH	RUSSKIJ	ESPANOL	ITALIANO
1011 CRYSTAL MINE	CRYSTALS	CRISTAL	KRISTALL	KRISTALL	CRISTAL	CRISTALLO
1012 CRYSTAL CHEMISTRY MINE	CRYSTAL CHEMIS- TRY *UF:* Isostructural miner- als	CRISTALLOCHIMIE *UF:* Isostructure	KRISTALLCHEMIE *UF:* Struktur=Gleich. Mineral	KRISTAL- LOKHIMIYA	CRISTALOQUIMICA *UF:* Isoestructural	CRISTAL- LOCHIMICA *UF:* Minerale isostrut- turale
1013 CRYSTAL DEFECT MINE	DEFECTS	DEFAUT CRISTAL- LIN	KRISTALL=FEHLER	DEFEKT KRISTAL- LICHESKOJ STRUK- TURY *UF:* Surface defect; Tsentr okraski	IMPERFECCION= CRISTALINA	DIFETTO CRISTAL- LINO
1014 CRYSTAL DISLOCATION MINE	CRYSTAL DISLO- CATIONS	DISLOCATION CRISTALLINE	KRISTALL= VERSETZUNG	Dislokatsiya kristalli- cheskoj reshetki RESHETKA KRIS- TALLICHESKAYA	DISLOCACION= CRISTALINA	DISLOCAZIONE CRISTALLINA
1015 CRYSTAL FIELD MINE	CRYSTAL FIELD	CHAMP CRISTAL- LIN	KRISTALL=FELD	Pole kristallicheskoe STRUKTURA KRIS- TALLICHESKAYA	CAMPO= CRISTALINO	CAMPO CRISTAL- LINO
1016 CRYSTAL FORM MINE	CRYSTAL FORM *UF:* Miller indices	FORME CRISTAL- LINE *UF:* Indice Miller	TRACHT *UF:* Miller=Indizes	Forma kristallografi- cheskaya GABITUS	FORMA= CRISTALINA *UF:* Indice=de=Miller	FORMA CRISTAL- LINA *UF:* Indici di Miller
1017 CRYSTAL GROWTH MINE	CRYSTAL GROWTH *UF:* Nucleus	CROISSANCE CRIS- TALLINE *UF:* Noyau cristallisa- tion; Striation mineral	KRISTALL= WACHSTUM *UF:* Kern; Riefung	ROST KRISTALLA *UF:* Nucleation	CRECIMIENTO= CRISTALINO *UF:* Estriacion=mineral; Nucleo	ACCRESCIMENTO CRISTALLINO *UF:* Germe cristallino; Striatura di cristallo
1018 CRYSTAL LATTICE MINE	LATTICE	MAILLE CRISTAL- LINE	KRISTALL=GITTER	RESHETKA KRIS- TALLICHESKAYA *UF:* Dislokatsiya kritalli- cheskoj reshetki; Re- shetka Brave	MALLA= CRISTALINA	CELLA CRISTAL- LINA
1019 CRYSTAL OPTICS MINE	Crystal optics CRYSTALLOGRA- PHY; OPTICAL PROPERTIES	Optique cristal CRISTALLOGRA- PHIE; PROPRIETE OPTIQUE	Kristall=Optik KRISTALLOGRA- PHIE; OPT. EIGENSCHAFT	KRISTALLOOPTIKA *UF:* Poloska Bekke	Optica=cristalina CRISTALOGRAFIA; PROPIEDAD= OPTICA	Ottica mineralogica CRISTALLOGRA- FIA; PROPRIETA OTTICA
1020 CRYSTAL STRUCTURE MINE	CRYSTAL STRUC- TURE *UF:* Bravais lattice; Poly- hedra; Structural crys- tallography	STRUCTURE CRIS- TALLINE *UF:* Polyedre	KRISTALL= STRUKTUR *UF:* Polyeder	STRUKTURA KRIS- TALLICHESKAYA *UF:* Atomic packing; Pole kristallicheskoe; Poliehdr; Surface defect	ESTRUCTURA= CRISTALINA *UF:* Indice=de=Miller; Poliedro	STRUTTURA CRIS- TALLINA *UF:* Poliedro
1021 CRYSTAL SYSTEM MINE	CRYSTAL SYSTEMS *UF:* Cubic system; Hex- agonal system; Isomet- ric system; Orthorhom- bic system; Pyramidal system; Rhombic sys- tem; Triclinic system; Trigonal system	SYSTEME CRISTAL- LIN *UF:* Systeme cubique; Systeme hexagonal; Systeme isometrique; Systeme orthorhom- bique; Systeme pyrami- dal; Systeme rhom- bique; Systeme triclinique; Systeme trigonal	KRISTALL= SYSTEM *UF:* Hexagonal.System; Isometr.System; Kub. System; Orthorhomb. System; Pyramidal. System; Rhomb. System; Trigonal. System; Triklin.System	SINGONIYA	SISTEMA= CRISTALINO *UF:* Sistema=cubico; Sistema=hexagonal; Sistema=isometrico; Sistema=ortorombico; Sistema=piramidal; Sistema=regular; Sistema=rombico; Sistema=triclinico; Sistema=trigonal	SISTEMA CRISTAL- LINO *UF:* Sistema esagonale; Sistema isometrico; Sistema monometrico; Sistema ortorombico; Sistema piramidale; Sistema rombico; Sistema triclino; Sistema trigonale
1022 CRYSTALLINE ROCK IGNE	CRYSTALLINE ROCKS	ROCHE CRISTAL- LINE *UF:* Phyllonite	KRISTALLIN. GESTEIN *UF:* Phyllonit	Kristallicheskij ZERNISTYJ	ROCA= CRISTALINA *UF:* Filonita	ROCCIA CRISTAL- LINA *UF:* Fillonite

	ENGLISH	FRANCAIS	DEUTSCH	RUSSKIJ	ESPANOL	ITALIANO
1023 CRYSTALLINITY MINE	CRYSTALLINITY *UF:* Cryptocrystalline texture; Crystallization index	CRISTALLINITE *UF:* Indice cristallisation; Texture cryptocristalline; Texture hyalocristalline; Texture hypocristalline	KRISTALLINITAET *UF:* Hyalokristallin. Gefuege; Hypokristallin.Gefuege; Kristallisierungs=Index; Kryptokristallin. Gefuege	KRISTALLICH-NOST'	CRISTALINIDAD *UF:* Indice=cristalizacion; Textura=criptocristalina; Textura=hialocristalina; Textura=hipocristalina	CRISTALLINITA *UF:* Criptocristallino; Indice di cristallizzazione; Tessitura ialocristallina; Tessitura ipocristallina
1024 CRYSTALLITE MINE	CRYSTALLITES	CRISTALLITE	KRISTALLIT	KRISTALLIT	CRISTALITICO	CRISTALLITE
1025 CRYSTALLIZATION IGNE	CRYSTALLIZATION *UF:* Heteromorphism; Hydatogenesis; Intratelluric processes; Liquation; Piezocrystallization	CRISTALLISATION *UF:* Anneaux Liesegang; Genese intratellurique; Heteromorphisme; Hydatogenese; Piezocristallisation; Separation phase	KRISTALLISATION *UF:* Heteromorphismus; Hydatogenese; Intratellur.Genese; Liesegang=Ringe; Liquation; Piezo=Kristallisation	KRISTALLIZAT-SIYA *UF:* Kumulat	CRISTALIZACION *UF:* Anillos=Liesegang; Heteromorfismo; Hidatogenesis; Intratelurico; Licuacion; Piezocristalizacion	CRISTALLIZZAZ-IONE *UF:* Anelli di Liesegang; Eteromorfismo; Genesi intratellurica; Idatogenesi; Piezocristallizzazione; Separazione del liquido
1026 CRYSTALLIZATION INDEX IGNE	Crystallization index CRYSTALLINITY	Indice cristallisation CRISTALLINITE	Kristallisierungs=Index KRISTALLINITAET	INDEKS KRISTALL-IZATSII	Indice=cristalizacion CRISTALINIDAD	Indice di cristallizzazione CRISTALLINITA
1027 CRYSTALLOBLAST IGNE	Crystalloblast PORPHYRITIC TEXTURE	Cristalloblaste TEXTURE PORPHYRIQUE	Kristalloblast PORPHYR=GEFUEGE	KRISTALLOBLAST	Cristaloblasto TEXTURA=PORFIDICA	Cristalloblasto TESSITURA PORFIRICA
1028 CRYSTALLOGRAPHY MINE	CRYSTALLOGRA-PHY *UF:* Crystal optics; Hauy's law	CRISTALLOGRA-PHIE *UF:* Cristallographie structure; Loi Hauy; Optique cristal	KRISTALLOGRA-PHIE *UF:* Hauy=Gesetz; Kristall=Optik; Struktur. Kristallographie	KRISTALLOGRA-FIYA *UF:* Gayui zakon	CRISTALOGRAFIA *UF:* Cristalografia=estructural; Ley=de=Hauy; Optica=cristalina	CRISTALLOGRAFIA *UF:* Cristallografia strutturale; Legge di Hauy; Ottica mineralogica
1029 CTENOSTOMATA PALS	CTENOSTOMATA	CTENOSTOMATA	CTENOSTOMATA	CTENOSTOMATA	CTENOSTOMATA	CTENOSTOMATA
1030 CUBIC SYSTEM MINE	Cubic system CRYSTAL SYSTEMS	Systeme cubique SYSTEME CRISTAL-LIN	Kub.System KRISTALL=SYSTEM	SINGONIYA KUBI-CHESKAYA *UF:* Didodekaehdr; Sistema izometricheskaya	Sistema=cubico SISTEMA=CRISTALINO	Sistema monometrico SISTEMA CRISTAL-LINO
1031 CUESTA SURF	CUESTAS	CUESTA	SCHICHTSTUFE	KUESTA	CUESTA	CUESTA
1032 CUMULATE IGNE	CUMULATES	CUMULAT	KUMULAT=GESTEIN	Kumulat KRISTALLIZAT-SIYA; FRAK-TSIONIROVANIE	CUMULO	ROCCIA DI CUMULO
1033 CURIE POINT PHCH	CURIE POINT	POINT CURIE	CURIE=PUNKT	KYURI TOCHKA	PUNTO-DE=CURIE	PUNTO DI CURIE
1034 CURIUM CHEE	CURIUM	CURIUM	CM	KYURIJ	CURIO	CURIO
1035 CURRENT GEOL	CURRENTS	COURANT *UF:* Circulation	STROEMUNG *UF:* Zirkulation	TECHENIE	CORRIENTE *UF:* Circulacion	CORRENTE *UF:* Circolazione

	ENGLISH	FRANCAIS	DEUTSCH	RUSSKIJ	ESPANOL	ITALIANO
1036 CURRENT MARK SEDI	CURRENT MARK- INGS	TRACE COURANT	STROEMUNGS= SPUR	ZNAKI TECHENIYA	HUELLA= CORRIENTE	TRACCIA DI COR- RENTE
1037 CUSP SURF	Cusps BEACHES	Croissant PLAGE	Cusp STRAND	FESTON	Cuspide PLAYA	Cuspide SPIAGGIA
1038 CUTOFF GRADE ECON	CUTOFF GRADE	TENEUR LIMITE	ARMERZ	SODERZHANIE BORTOVOE	CONTENIDO= LIMITE	TENORE LIMITE
1039 CUTTINGS ECON	CUTTINGS	DEBLAI FORAGE	BOHR=GUT	MUKA BUROVAYA	ESCOMBRERA=DE= TESTIGOS	[CUTTINGS]
1040 CYANOPHYTA PALS	CYANOPHYTA	CYANOPHYTA	CYANOPHYTA	CYANOPHYTA	CIANOFITA	CYANOPHYTA
1041 CYBERNETICS MATH	Cybernetics MATHEMATICAL METHODS	Cybernetique METHODE MATHE- MATIQUE	Kybernetik MATHEMAT. METHODE	KIBERNETIKA	Cibernetica METODO= MATEMATICO	Cibernetica METODO MATEMA- TICO
1042 CYCADALES PALS	CYCADALES	CYCADALES	CYCADALES	CYCADALES	CYCADALES	CYCADALES
1043 CYCADOFILICALES PALS	CYCADOFILICALES	CYCADOFILICALES	CYCADOFILICALES	CYCADOFILICALES	CYCADOFILICALES	CYCADOFILICALES
1044 CYCLE GEOL	Cycles(gen.)	Cycle(gen.)	Zyklus(gen.)	TSIKL *UF:* Tsikl gidrologi- cheskij; Tsiklichnost'	Ciclo(gen.)	Ciclo(gen.)
1045 CYCLE OF SEDIMENTATION SEDI	Cycle of sedimentation CYCLIC PRO- CESSES	CYCLE SEDI- MENTAIRE *UF:* Cyclotheme	SEDIMENTATIONS= ZYKLUS *UF:* Zyklothem	TSIKL SEDI- MENTATSII *UF:* Tsiklotema	CICLO= SEDIMENTARIO *UF:* Ciclotema	CICLO SEDIMEN- TARIO *UF:* Ciclotema
1046 CYCLIC LOADING ENGI	CYCLIC LOADING	MISE EN CHARGE CYCLIQUE	ZYKL.BELASTUNG	NAGRUZKA TSIK- LICHESKAYA	CARGA=CICLICA	CARICAMENTO CICLICO
1047 CYCLIC PROCESS GEOL	CYCLIC PRO- CESSES *UF:* Cycle of sedimenta- tion	Processus cyclique PERIODICITE	Zykl.Prozess PERIODIZITAET	Tsiklichnost' TSIKL	Procesos=ciclicos PERIODICIDAD	Processo ciclico PERIODICITA
1048 CYCLOCYSTOIDEA PALS	CYCLOCYSTOIDEA	CYCLOCYSTOIDEA	CYCLOCYSTOIDEA	Cyclocystoidea ECHINOZOA	CYCLOCYSTOIDEA	CYCLOCYSTOIDEA
1049 CYCLONE SURF	Cyclones CLIMATE	Cyclone VENT	Zyklone WIND	Tsiklon VETER	Ciclon VIENTO	Ciclone VENTO
1050 CYCLOSTOMATA PALS	CYCLOSTOMATA	CYCLOSTOMATA	CYCLOSTOMATA	CYCLOSTOMATA	CYCLOSTOMATA	CYCLOSTOMATA
1051 CYCLOTHEM SEDI	CYCLOTHEMS	Cyclotheme CYCLE SEDI- MENTAIRE	Zyklothem SEDIMENTATIONS= ZYKLUS	Tsiklotema TSIKL SEDI- MENTATSII	Ciclotema CICLO= SEDIMENTARIO	Ciclotema CICLO SEDIMEN- TARIO
1052 CYLINDRICAL FOLD STRU	CYLINDRICAL FOLDS	PLI CYCLIN- DRIQUE	ZYLINDR.FALTE	Skladka tsilindri- cheskaya SKLADKA; TSILIN- DRICHESKIJ	PLIEGUE= CILINDRICO	PIEGA CILINDRICA

	ENGLISH	FRANCAIS	DEUTSCH	RUSSKIJ	ESPANOL	ITALIANO
1053 CYLINDRICAL STRUCTURE TEST	CYLINDRICAL STRUCTURES	Structure cylindrique GEOMETRIE	Zylindr.Gefuege GEFUEGE	TSILINDRICHESKIJ *UF:* Skladka tsilindri-cheskaya	Estructura=cilindrica GEOMETRIA	Struttura cilindrica GEOMETRIA
1054 CYNOMORPHA PALS	CYNOMORPHA	CYNOMORPHA	CYNOMORPHA	Cynomorpha SIMIAN	CYNOMORPHA	CYNOMORPHA
1055 CYSTIPHYLLINA PALS	Cystiphyllina RUGOSA	Cystiphyllina RUGOSA	Cystiphyllina RUGOSA	Cystiphyllina RUGOSA	Cystiphyllina RUGOSA	Cystiphyllina RUGOSA
1056 CYSTOIDEA PALS	CYSTOIDEA	CYSTOIDEA	CYSTOIDEA	CYSTOIDEA	CYSTOIDEA	CYSTOIDEA
1057 D LAYER SOLI	D layer LOWER MANTLE	Couche D MANTEAU INF	D=Schicht UNTER.ERD=MANTEL	D sloj MANTIYA NIZH-NYAYA	Capa=D MANTO−GLOBO=INF	Strato D MANTELLO INFERIORE
1058 D/H RATIO ISOT	D/H	D=H	D/H=VERHAELTNIS	D/H	D/H	D/H
1059 DAM ENGI	DAMS	BARRAGE	STAU=DAMM	PLOTINA *UF:* Groundwater dam; Plotina arochnaya; Plotina s kamennoj nasypkoj	PRESA	DIGA
1060 DANIAN STRS	DANIAN	DANIEN	DAN	DAT	DANIENSE	DANIANO
1061 DARCY'S LAW GEOH	DARCY'S LAW	LOI DARCY	DARCY=GESETZ	Darsi zakon FIL'TRATSIYA	LEY=DE=DARCY	LEGGE DI DARCY
1062 DARWINISM PALE	Darwinism THEORETICAL STUDIES; BIO-LOGIC EVOLUTION	Darwinisme THEORIE; EVOLU-TION BIOLOGIQUE	Darwinismus THEORIE; BIOLOG. EVOLUTION	Darvinizm EHVOLYUTSIYA; TEORIYA	Darwinismo TEORIA; EVOLUCION=BIOLOGICA	Darwinismo TEORIA; EVOLUZ-IONE BIOLOGICA
1063 DASYCLADACEAE PALS	DASYCLADACEAE	DASYCLADACEAE	DASYCLADACEAE	Dasycladaceae CHLOROPHYTA	DASYCLADACEAE	DASYCLADACEAE
1064 DATA MISC	DATA	DONNEE	DATEN	DANNYE *UF:* Analiz dannykh; Electron microscopy data; Electron probe data; SEM data; TEM data	DATOS	DATO
1065 DATA ACQUISITION MATH	DATA ACQUISI-TION	SAISIE DONNEE	DATEN=ERFASSUNG	SBOR DANNYKH	TOMA=DE=DATOS	ACQUISIZIONE DEI DATI
1066 DATA ANALYSIS MATH	Data analysis DATA PROCESSING	Analyse donnee TRAITEMENT DON-NEE	Daten=Analyse DATEN=AUFBEREITUNG	Analiz dannykh ANALIZ MATEMA-TICHESKIJ; DAN-NYE	Analisis=de=datos TRATAMIENTO=DATOS	Analisi dei dati TRATTAMENTO DEI DATI
1067 DATA BANK MATH	DATA BASES	BASE DONNEE	DATENBANK	BANK DANNYKH	BANCO=DE=DATOS	BANCA DI DATI
1068 DATA HANDLING MATH	DATA HANDLING	MANIPULATION DONNEE	DATEN=AUFBEREITUNG *UF:* Daten=Analyse	OBRABOTKA DAN-NYKH (INFORM)	MANIPULACION=DATOS	MANIPOLAZIONE DI DATI

	ENGLISH	FRANCAIS	DEUTSCH	RUSSKIJ	ESPANOL	ITALIANO
1069 DATA PROCESSING MATH	DATA PROCESSING *UF:* Automation; Data analysis	TRAITEMENT DON-NEE *UF:* Analyse donnee; Lissage	DATEN=VERARBEITUNG *UF:* Glaettung	OBRABOTKA DAN-NYKH	TRATAMIENTO=DATOS *UF:* Analisis=de=datos; Suavizado	TRATTAMENTO DEI DATI *UF:* Analisi dei dati; Lisciamento
1070 DATA RETRIEVAL MATH	DATA RETRIEVAL	RESTITUTION DONNEE	DATEN=WIEDERGEWINNUNG	POISK DANNYKH	BUSQUEDA=DE=DATOS	REPERIMENTO DATI
1071 DATA STORAGE MATH	DATA STORAGE	STOCKAGE DON-NEE	DATEN=SPEICHERUNG	KHRANENIE DAN-NYKH	ALMACENAMIENTO=DATOS	ARCHIVIAZIONE DATI
1072 DATA TABLE MATH	TABLES	TABLE DONNEES	TABELLE	Tablitsa dannykh MATRITSA	TABLA=DE=DATOS	TABELLA
1073 DATING ISOT	Dating ABSOLUTE AGE	DATATION *UF:* Age absolu; Age concordant; Age de la terre; Age discordant; Constante desintegration; Datation trace fission; Datation trace particule; Superposition age	PHYSIKAL. ALTERSBESTIM-MUNG *UF:* Absolut=Alter; Diskordant.Alter; Erd=Alter; Konkordant.Alter; Spaltspuren=Datierung; Teilchen=Spur=Datierung; Ueberpraegung; Zerfalls=Konstante	Datirovka VOZRAST ABSO-LYUTNYJ	DATACION *UF:* Constante=de=desintegracion; Datacion=traza=fision; Datacion=traza=particula; Edad=absoluta; Edad=acorde; Edad=de=la=tierra; Edad=no=acorde; Superposicion=edad	DATAZIONE *UF:* Costante di decadimento; Eta assoluta; Eta concordante; Eta della terra; Eta discordante; Eta tracce di fissione; Sovrapposizione di eta; Traccia di particella
1074 DAUGHTER PRODUCT ISOT	Daughter product RADIOACTIVITY; ISOTOPES	Produit intermediaire RADIOACTIVITE	Tochter=Element RADIOAKTIVITAET	PRODUKT DOCH-ERNIJ	Producto=hijo RADIOACTIVIDAD	Prodotto di decadimento RADIOATTIVITA
1075 DEAD ICE SURF	Dead ice ICE	Glace morte GLACE	Toteis EIS	LED MERTVYJ	Glaciar=colgado HIELO	Ghiaccio morto GHIACCIO
1076 DEBRIS CONE SURF	DEBRIS CONES	CONE DEJECTION *UF:* Bajada; Bordure detritique	SCHUTT=KEGEL *UF:* Bajada; Sander	[DEBRIS CONE] KONUS VYNOSA	CONO=DE=DEYECCION *UF:* Abanico=fluvioglaciar; Bajada	CONO DI DETRITO *UF:* Bajada; Fascia detritica
1077 DEBRIS FLOW SURF	DEBRIS FLOWS	COULEE CLAS-TIQUE	SCHUTT=STROM	SEL'	FLUJO=MATERIAL=DETRITICO	COLATA DI DETRITO
1078 DECAPODA PALS	Decapoda MALACOSTRACA	DECAPODA *UF:* Malacostraca	DECAPODA *UF:* Malacostraca	Decapoda MALACOSTRACA	DECAPODA *UF:* Malacostraca	DECAPODA *UF:* Malacostraca
1079 DECAY CONSTANT PHCH	Decay constant RADIOACTIVITY; ABSOLUTE AGE	Constante desintegration RADIOACTIVITE; DATATION	Zerfalls=Konstante RADIOAKTIVITAET; PHYSIKAL. ALTERSBESTIM-MUNG	POSTOYANNAYA RASPADA	Constante=de=desintegracion RADIOACTIVIDAD; DATACION	Costante di decadimento RADIOATTIVITA; DATAZIONE
1080 DECCAN BASALT IGNE	Deccan traps BASALTS	Trapp Deccan BASALTE; COMPOSITION THOLEITIQUE	Dekkan=Trapp BASALT; THOLEIIT. CHEMISMUS	DEKANA TRAPPY	Basalto=de=Deccan BASALTO; COMPOSICION=TOLEITICA	Basalto del Deccan BASALTO; COMPOSIZIONE THOLEITICA
1081 DECLINATION SOLI	DECLINATION	DECLINAISON	DEKLINATION	SKLONENIE	DECLINACION	DECLINAZIONE

	ENGLISH	FRANCAIS	DEUTSCH	RUSSKIJ	ESPANOL	ITALIANO
1082 DECOLLEMENT STRU	DECOLLEMENT	DECOLLEMENT	ABSCHERUNG	SKLADCHATOST' SRYVA	DESPEGUE	SCOLLAMENTO
1083 DECONTAMINATION ENVI	DECONTAMINA- TION	DECONTAMINA- TION	ENTGIFTUNG	Dekontaminatsiya VODA; OCHISTKA	DESCONTAMINA- CION	DECONTAMINAZ- IONE
1084 DECONVOLUTION MATH	DECONVOLUTION	DECONVOLUTION	DEKONVOLUTION	DEKONVOLUTSIYA	DECONVOLUCION	DECONVOLUZIONE
1085 DECORATIVE STONE COMS	Decorative stones ORNAMENTAL MATERIALS	Pierre ornementale MATERIAU ORNE- MENTATION	Verblend=Stein WERKSTEIN	KAMEN' DEKORA- TIVNYJ	Piedra=ornamentacion MATERIAL= ORNAMENTACION	Pietra decorativa MATERIALE DA ORNAMENTAZ- IONE
1086 DECREPITATION ISOT	DECREPITATION	DECREPITOMETRIE	DEKREPITATIONS= METHODE	RASTRESKIVANIE	DECREPITACION	DECREPITO- METRIA
1087 DEDOLOMITIZATION SEDI	DEDOLOMITIZA- TION	DEDOLOMITISA- TION	ENTDOLOMITI- SIERUNG	DEDOLOMITIZAT- SIYA	DEDOLOMITIZA- CION	DEDOLOMITIZZAZ- IONE
1088 DEEP-FOCUS EARTHQUAKE SOLI	DEEP=FOCUS EARTHQUAKES	SEISME PROFOND	TIEFEN=BEBEN	ZEMLETRYASENIE GLUBOKOFOKUS- NOE	SISMO=PROFUNDO	SISMA PROFONDO
1089 DEEP-SEA CHANNEL MARI	Deep=sea channel DEEP=SEA ENVI- RONMENT	Chenal mer profonde ABYSSE	Tiefsee=Rinne TIEFSEE	KANAL GLUBOKO- VODNYJ	Canal=marino= profundo SIMA=MARINA	Canale sottomarino ABISSO
1090 DEEP-SEA ENVIRONMENT PALE	DEEP=SEA ENVI- RONMENT *UF:* Abyssal environ- ment; Bathyal environ- ment; Deep=sea chan- nel; Deep=water zone	MILIEU MER PRO- FONDE *UF:* Milieu abyssal; Milieu bathyal; Milieu eau profonde	TIEFSEE=MILIEU *UF:* Bathyal=Milieu; Tiefwasser=Milieu; Tiefwasser=Zone	GLUBOKOVODNYJ *UF:* Batial'nyj	MEDIO=MAR= PROFUNDO *UF:* Medio=abisal; Medio=batial; Zona= mar=profundo	AMBIENTE DI MARE PROFONDO *UF:* Ambiente abissale; Ambiente batiale; Mare profondo
1091 DEEP-SEISMIC SOUNDING APPL	DEEP SEISMIC SOUNDING	SONDAGE SISMIQUE PRO- FOND	SEISM. TIEFENSON- DIERUNG	ZONDIROVANIE SEJSMICHESKOE GLUBINNOE	SONDEO= SISMICO= PROFUNDO	SONDAGGIO SISMICO PRO- FONDO
1092 DEEP-TOW METHOD APPL	DEEP=TOW METH- ODS	METHODE TRAC- TION FOND	TIEF=SCHLEPP= METHODE	[DEEP=TOW METHOD] ISSLEDOVANIE	METODO= TRACCION=FONDO	METODO TRAZ- IONE DI FONDO
1093 DEEP-WATER ZONE MARI	Deep=water zone DEEP=SEA ENVI- RONMENT	Milieu eau profonde MILIEU MER PRO- FONDE	Tiefwasser=Zone TIEFSEE=MILIEU	GLUBOKOVOD'E *UF:* Abbisal'naya /de- pressiya/; Abyss	Zona=mar=profundo MEDIO=MAR= PROFUNDO	Mare profondo AMBIENTE DI MARE PROFONDO
1094 DEEPSEATED TECTONICS STRU	DEEP=SEATED STRUCTURES	TECTONIQUE FOND	TIEFEN=TEKTONIK	GLUBINNYJ	TECTONICA=DE= FONDO	TETTONICA PRO- FONDA
1095 DEFLATION SURF	Deflation WIND EROSION	Deflation EROSION EOLIENNE	Deflation WIND=EROSION	DEFLYATSIYA	Deflacion EROSION=EOLICA	Deflazione EROSIONE EOLICA
1096 DEFORMATION STRU	DEFORMATION *UF:* Heave; Yield strength	DEFORMATION *UF:* Deformation sedi- ment; Deformation sol; Deformation ver- sant; Point deforma- tion; Soulevement	VERFORMUNG *UF:* Boden=Hebung; Boden=Verformung; Fruehdiagenet. Verformung; Hang= Verformung; Streck= Grenze	DEFORMATSIYA	DEFORMACION *UF:* Abombamiento; Deformacion= sedimento=blando; Deformacion=suelo; Deformacion=talud; Limite=elastico	DEFORMAZIONE *UF:* Deformazione del suolo; Deformazione di sedimenti molli; Deformazione di ver- sante; Soglia di defor- mazione; Sollevamento del terreno

	ENGLISH	FRANCAIS	DEUTSCH	RUSSKIJ	ESPANOL	ITALIANO
1097 DEGASSING PHCH	DEGASSING	DEPART GAZ	ENTGASUNG	DEGAZATSIYA	SALIDA=GAS	DEGASSAZIONE
1098 DEGLACIATION SURF	DEGLACIATION	DEGLACIATION	VEREISUNGS= RUECKGANG	DEGLYATSIATSIYA	DESGLACIACION	DEGLACIAZIONE
1099 DEGRADATION SURF	DEGRADATION	DESAGREGATION	DEGRADATION	DEGRADATSIYA	DEGRADACION	DISAGGREGAZ- IONE
1100 DEGREE OF FREEDOM PHCH	Degree of freedom THERMODYNAMIC PROPERTIES	Degre liberte THERMODY- NAMIQUE	Freiheitsgrad THERMODYNAMIK	STEPEN' SVOBODY	Grado=de=libertad TERMODINAMICA	Grado di liberta TERMODINAMICA
1101 DEHYDRATION PHCH	DEHYDRATION	DESHYDRATATION *UF:* Substance anhydre	DEHYDRATISIERUNG *UF:* Anhydrid= Chemismus	DEGIDRATATSIYA	DESHIDRATACION *UF:* Anhidro	DISIDRATAZIONE *UF:* Composizione ani- dra
1102 DEINOTHERIOIDEA PALS	DEINOTHERIOIDEA	DEINOTHERIOIDEA	DEINOTHERIOIDEA	Deinotherioidea PROBOSCIDEA	DEINOTHERIOIDEA	DEINOTHERIOIDEA
1103 DELTA SURF	DELTAS	DELTA	DELTA	DEL'TA	DELTA	DELTA
1104 DELTAIC ENVIRONMENT PALE	DELTAIC ENVI- RONMENT	MILIEU DELTAIQUE	DELTA=MILIEU	DEL'TOVYJ *UF:* Mud lump	MEDIO=DELTAICO	AMBIENTE DEL- TIZIO
1105 DELTAIC SEDIMENTATION SEDI	DELTAIC SEDI- MENTATION	SEDIMENTATION DELTAIQUE *UF:* Couche basale; Couche contrepentee; Couche frontale; Cou- che sommitale	DELTA= SEDIMENTATION *UF:* Backset=bed; Bottomset=bed; Foreset=bed; Topset= bed	SEDIMENTATSIYA DEL'TOVAYA	SEDIMENTACION= DELTAICA *UF:* Backset=bed; Capa=basal; Capa= de=techo; Capa= frontal	SEDIMENTAZIONE DELTIZIA *UF:* Backset bed; Strato di letto; Strato di tetto; Strato frontale
1106 DEMAGNETIZATION PHCH	DEMAGNETIZA- TION	DESAIMANTATION	ENTMAGNETI- SIERUNG	RAZMAGNICHIVA- NIE *UF:* Sila koehrtsitivnaya	DESIMANTACION	SMAGNETIZZAZ- IONE
1107 DEMOSPONGEA PALS	Demospongea PORIFERA	Demospongea PORIFERA	Demospongea PORIFERA	DEMOSPONGEA	Demospongea PORIFERA	Demospongea PORIFERA
1108 DENDRITE SEDI	Dendrite DIAGENESIS; MAN- GANESE	Dendrite STRUCTURE DIA- GENETIQUE; MAN- GANESE	Dendrit DIAGENET. GEFUEGE; MN	DENDRIT	Dendrita ESTRUCTURA= DIAGENETICA; MANGANESO	Dendrite STRUTTURA DIA- GENETICA; MAN- GANESE
1109 DENDRITIC DRAINAGE PATTERN GEOH	Dendritic drainage pat- tern DRAINAGE PAT- TERNS	Drainage rayonnant RESEAU HYDRO- GRAPHIQUE	Dendrit.Entwaesserung HYDROGRAPH. NETZ	Set' rechnaya drevovid- naya SET' RECHNAYA	Avenamiento= dendritico RED= HIDROGRAFICA	Rete idrografica den- dritica RETICOLATO IDRO- GRAFICO
1110 DENDROCHRONOLOGY STRA	Dendrochronology TREE RINGS	DEN- DROCHRONOLO- GIE *UF:* Anneau croissance	DENDRO= CHRONOLOGIE *UF:* Jahresring	DEN- DROKHRO- NOLOGIYA	DEN- DROCRONOLOGIA *UF:* Anillo=crecimiento	DEN- DROCRONOLOGIA *UF:* Anello di accresci- mento
1111 DENDROGRAM MATH	DENDROGRAMS	DENDROGRAMME	DENDROGRAMM	DENDROGRAMMA	DENDROGRAMA	DENDROGRAMMA

	ENGLISH	FRANCAIS	DEUTSCH	RUSSKIJ	ESPANOL	ITALIANO
1112 DENDROIDEA PALS	DENDROIDEA	DENDROIDEA	DENDROIDEA	DENDROIDEA	DENDROIDEA	DENDROIDEA
1113 DENSITY PHCH	DENSITY *UF:* Bulk density; Pyc- nometer; Pycnometry	DENSITE *UF:* Densite brute; Differenciation par gravite; Pycnocline; Pycnometre; Pycnome- trie	DICHTE *UF:* Dichte=Gradient; Pyknometer; Pyknome- trie; Roh=Dichte; Schwere= Differenzierung	PLOTNOST'	DENSIDAD *UF:* Densidad=aparente; Diferenciacion= gravitacional; Picno- clina; Picnometria; Picnometro	DENSITA *UF:* Densita apparente; Differenziazione gravitativa; Picnoclino; Picnometria; Picno- metro
1114 DENSITY CONCENTRATION ECON	GRAVITY CONCEN- TRATION	CONCENTRATION GRAVITE	SCHWERE= TRENNUNG	Kontsentratsiya gravi- metricheskaya KONTSENTRAT- SIYA; GRAVI- METRIYA	CONCENTRACION= POR=GRAVEDAD	CONCENTRAZIONE PER GRAVITA
1115 DENTITION PALE	Dentition TEETH	Denture MACHOIRE; DENT	Dentition KIEFER; ZAHN	APPARAT ZUBNOJ	Denticion MANDIBULA; DIENTE	Dentizione MASCELLA; DENTE
1116 DENUDATION SURF	DENUDATION	DENUDATION	DENUDATION	DENUDATSIYA	DENUDACION	DENUDAMENTO
1117 DEPOSIT EVALUATION ECON	ECONOMIC EVALU- ATION	EVALUATION GITE	LAGERSTAETTEN= BEWERTUNG	PODSCHET ZAPA- SOV	EVALUACION= YACIMIENTO	VALUTAZIONE DI GIACIMENTO
1118 DEPOSIT GENESIS ECON	MINERAL DEPOS- ITS GENESIS *UF:* Postmagmatic materials	GENESE GITE *UF:* Genese postmagma- tique	LAGERSTAETTEN= GENESE *UF:* Postmagmat. Bildung	GENEZIS MESTOROZHDENIJ	GENESIS= YACIMIENTO *UF:* Postmagmatico	GENESI DI GIACI- MENTO *UF:* Origine postmagma- tica
1119 DEPOSIT MORPHOLOGY ECON	Ore morphology ORE BODIES	MORPHOLOGIE GITE *UF:* Colonne; Corps mineralise	LAGERSTAETTEN= FORM *UF:* Erzfall; Erzkoerper	MORFOLOGIYA MESTOROZHDENIJ	MORFOLOGIA= YACIMIENTO *UF:* Columna; Cuerpo= mineralizado	MORFOLOGIA DI GIACIMENTO *UF:* Colonna; Corpo mineralizzato
1120 DEPOSITION SEDI	DEPOSITION	DEPOT	ABLAGERUNG	Otlozhenie SEDIMENTATSIYA	RELLENO	DEPOSITO
1121 DEPRESSION GEOL	DEPRESSIONS	DEPRESSION	DEPRESSION	DEPRESSIYA (GEOMORF)	DEPRESION	DEPRESSIONE
1122 DEPTH MISC	DEPTH	PROFONDEUR	TIEFE	GLUBINA *UF:* Depth indicator	PROFUNDIDAD	PROFONDITA
1123 DEPTH INDICATOR MARI	DEPTH INDICA- TORS	INDICATEUR BATHYMETRIQUE	TIEFEN= INDIKATOR	[DEPTH INDICA- TOR] GLUBINA; INDIKATOR	INDICADOR= BATIMETRICO	INDICATORE BATI- METRICO
1124 DEPTH OF COMPENSATION MARI	CARBONATE COM- PENSATION DEPTH	PROFONDEUR COMPENSATION CARBONATE	KARBONAT= SAETTIGUNGS= ZONE	GLUBINA KOMPEN- SATSII KARBONAT- NOJ	PROFUNDIDAD= COMPENSACION= CARBONA	PROFONDITA DI COMPENSAZIONE
1125 DERIVATIVE MATH	Derivative MATHEMATICAL MODELS	Derivation METHODE MATHE- MATIQUE	Mathemat. Differenzierung MATHEMAT. METHODE	PROIZVODNAYA	Derivacion METODO= MATEMATICO	Derivata METODO MATEMA- TICO
1126 DERMAL STRUCTURE PALE	DERMAL STRUC- TURES	TEGUMENT VERTE- BRE	HAUT=BILDUNGEN	Dermal'nyj SKELET	TEGUMENTO= VERTEBRADO	TEGUMENTO

	ENGLISH	FRANCAIS	DEUTSCH	RUSSKIJ	ESPANOL	ITALIANO
1127 DERMAPTERA PALS	DERMAPTEROIDA *UF:* Protelytroptera	DERMAPTEROIDA *UF:* Protelytroptera	DERMAPTEROIDA *UF:* Protelytroptera	DERMAPTEROIDA	DERMAPTEROIDA *UF:* Protelytroptera	DERMAPTEROIDA *UF:* Protelytroptera
1128 DERMOPTERA PALS	DERMOPTERA	DERMOPTERA	DERMOPTERA	DERMOPTERA	DERMOPTERA	DERMOPTERA
1129 DESALINATION METH	DESALINIZATION	DESSALEMENT	ENTSALZUNG	RASSOLONENIE	DESALINIZACION	DISSALAZIONE
1130 DESCENSION THEORY ECON	Descension theory THEORETICAL STUDIES; ECO- NOMIC GEOLOGY	Theorie mouvement par descente GENESE	Deszendenz=Theorie GENESE	[DESCENSION THE- ORY] GIPERGENEZ; TEO- RIYA	Teoria=movimiento= descendente GENESIS	Teoria discendente GENESI
1131 DESERT SURF	DESERTS *UF:* Desert dune; Ham- madas; Oasis; Wadi	DESERT *UF:* Hammada; Oasis; Oued	WUESTE *UF:* Hammada; Oase; Wadi	PUSTYNYA *UF:* Ehrg; Eolian fea- ture; Nastuplenie pustyn'; Vadi	DESIERTO *UF:* Hammada; Oasis; Rambla	DESERTO *UF:* Erg; Hammada; Oasi; Pavimentazione desertica; Uadi
1132 DESERT DUNE SURF	Desert dune DUNES; DESERTS	DUNE CONTINEN- TALE *UF:* Barkhane; Erg	KONTINENTAL= DUENE *UF:* Barchan; Erg	Dyuna pustynnaya BARKHAN	DUNA= CONTINENTAL *UF:* Erg; Medano= semilunar	DUNA CONTINEN- TALE *UF:* Barcana; Erg
1133 DESERT PAVEMENT SURF	DESERT PAVE- MENTS	Cuticule deflation EBOULIS; EROSION EOLIENNE	Wuesten=Pflaster SCHUTT; WIND- EROSION	MOSTOVAYA PUSTYNNAYA	Pavimento=desertico DERRUBIO; EROSION=EOLICA	Pavimentazione deser- tica DESERTO; ERO- SIONE EOLICA
1134 DESERTIFICATION SURF	DESERTIFICATION	DESERTIFICATION	WUESTEN= BILDUNG	Nastuplenie pustyn' KLIMAT; PUS- TYNYA	DESERTIZACION	DESERTIFICAZ- IONE
1135 DESICCATION PHCH	DESICCATION *UF:* Shrinkage; Shrink- age cracks	DESSICATION	AUSTROCKNUNG	OBEZVOZHIVANIE	DESECACION	ESSICCAZIONE
1136 DESILICATION GEOL	Desilication LEACHING; SILICA	Desilicification LESSIVAGE; SILICE	Entkieselung AUFLOESUNG; KIESELSAEURE	DESILIKATSIYA	Desilizacion LAVADO; SILICE	Desilicizzazione LISCIVIAZIONE; SILICE
1137 DESMIDS PALS	DESMIDS	FLORE DESMIDIEE	DESMIDEEN= FLORA	Desmidy CHLOROPHYTA	[DESMIDS]	FLORA A DESMIDI
1138 DESMOCERATIDA PALS	DESMOCERATIDA	DESMOCERATIDA	DESMOCERATIDA	Desmoceratida AMMONITINA	DESMOCERATIDA	DESMOCERATIDA
1139 DESMOSTYLIA PALS	DESMOSTYLIA	DESMOSTYLIA	DESMOSTYLIA	DESMOSTYLIA	DESMOSTYLIA	DESMOSTYLIA
1140 DETERMINISM GEOL	Determinism THEORETICAL STUDIES	Determinisme THEORIE	Determinismus THEORIE	DETERMINIZM	Determinismo TEORIA	Determinismo TEORIA
1141 DETERMINISTIC MODEL MATH	Deterministic model MODELS	Modele deterministe MODELE	Determinist.Modell MODELL	MODEL' DETER- MINIROVANNAYA	Modelo=determinativo MODELO	Modello deterministico MODELLO
1142 DETRITAL COMPOSITION SEDI	Detrital materials CLASTIC ROCKS	Composition detritique ROCHE CLASTIQUE	Detrit. Zusammensetzung KLAST.GESTEIN	DETRITOVYJ	Composicion=detritica ROCA=CLASTICA	Composizione detritica ROCCIA CLASTICA

	ENGLISH	FRANCAIS	DEUTSCH	RUSSKIJ	ESPANOL	ITALIANO
1143 DETRITAL REMANENT MAGNETIZATION PHCH	DETRITAL REMANENT MAGNETIZATION	AIMANTATION REMANENTE DETRITIQUE	DETRIT. REMANENT. MAGNETISIERUNG	Namagnichennost' detritovaya NAMAGNICHENNOST' OSTATOCHNAYA	IMANTACION= REMANENTE= DETRITICA	MAGNETIZZAZIONE DETRITICA RESIDUA
1144 DETRITAL SEDIMENTATION SEDI	DETRITAL SEDIMENTATION	SEDIMENTATION DETRITIQUE *UF:* Origine mecanique	DETRIT. SEDIMENTATION *UF:* Mechan.Ursprung	SEDIMENTATSIYA DETRITOVAYA	SEDIMENTACION= DETRITICA *UF:* Origen=mecanico	SEDIMENTAZIONE DETRITICA *UF:* Origine meccanica
1145 DEUTERIC EFFECT IGNE	Deuteric composition AUTOMETAMORPHISM	Effet deuterique METAMORPHISME AUTO	Epimagmat.Effekt AUTO- METAMORPHOSE	Dejtericheskij ehffekt AVTOMETAMORFIZM	Epimagmatico AUTOMETAMORFISMO	Effetto deuterico AUTOMETAMORFISMO
1146 DEUTERIUM ISOT	DEUTERIUM	DEUTERIUM	DEUTERIUM	DEJTERIJ	DEUTERIO	DEUTERIO
1147 DEVELOPMENT MISC	DEVELOPMENT *UF:* Development drilling	DEVELOPPEMENT	ENTWICKLUNG	RAZVITIE	DESARROLLO	SVILUPPO
1148 DEVELOPMENT DRILLING MINI	Development drilling BOREHOLES; DEVELOPMENT	Sondage developpement SONDAGE; EXPLOITATION	Entwicklungs=Bohrung BOHRUNG; GEWINNUNG	BURENIE EHKSPLUATATSIONNOE	Sondeo=explotacion POZO=SONDEO; EXPLOTACION	Sviluppo di pozzo SONDAGGIO; SFRUTTAMENTO
1149 DEVIATION MATH	Deviation STATISTICAL DISTRIBUTION	Ecart DISTRIBUTION STATISTIQUE	Abweichung STATIST. VERTEILUNG	OTKLONENIE	Desviacion DISTRIBUCION= ESTADISTICA	Deviazione DISTRIBUZIONE STATISTICA
1150 DEVITRIFICATION PHCH	DEVITRIFICATION	DEVITRIFICATION	ENTGLASUNG	DEVITRIFIKATSIYA	DESVITRIFICACION	DEVETRIFICAZIONE
1151 DEVONIAN STRS	DEVONIAN	DEVONIEN	DEVON	DEVON	DEVONICO	DEVONIANO
1152 DEW SURF	Dew MOISTURE	Rosee HUMIDITE	Tau FEUCHTIGKEIT	Rosa OSADKI ATMOSFERNYE	Rocio HUMEDAD	Rugiada UMIDITA
1153 DIACHRONISM STRA	DIACHRONISM	DIACHRONISME	DIACHRONISMUS	DIAKHRONIZM	DIACRONISMO	DIACRONISMO
1154 DIADOCHY MINE	Diadochy SUBSTITUTION	Diadochie ISOMORPHISME	Diadochie ISOMORPHIE	Diadokhiya IZOMORFIZM	Diadoquia ISOMORFISMO	Diadochia ISOMORFISMO
1155 DIAGENESIS SEDI	DIAGENESIS *UF:* Dendrite; Hypergenesis; Induration; Metagenesis; Organic structure	DIAGENESE	DIAGENESE	DIAGENES	DIAGENESIS	DIAGENESI
1156 DIAGNOSIS MISC	Diagnosis MODELS	Diagnose MODELE	Diagnose MODELL	DIAGNOZ	Diagnosis MODELO	Diagnosi MODELLO
1157 DIAGRAM MISC	Diagrams GRAPHIC METHODS	Diagramme REPRESENTATION GRAPHIQUE	Diagramm GRAPH. DARSTELLUNG	DIAGRAMMA *UF:* Diagramma pyl'tsevaya; Strukturnaya diagramma	Diagrama REPRESENTACION= GRAFICA	Diagramma RAPPRESENTAZIONE GRAFICA

	ENGLISH	FRANCAIS	DEUTSCH	RUSSKIJ	ESPANOL	ITALIANO
1158 DIALYSIS PHCH	Dialysis CHEMICAL ANAL- YSIS	Dialyse SEPARATION; DIF- FUSION	Dialyse TRENNUNG; DIF- FUSION	Dializ DIFFUZIYA	Dialisis SEPARACION; DIFUSION	Dialisi SEPARAZIONE; DIFFUSIONE
1159 DIAMAGNETIC PROPERTY PHCH	Diamagnetic property MAGNETIC PROP- ERTIES	Propriete diamagne- tique PROPRIETE MAGNETIQUE	Diamagnet.Eigenschaft MAGNET. EIGENSCHAFT	DIAMAGNETIK	Propiedad= diamagnetica PROPIEDAD= MAGNETICA	Proprieta diamagnetica PROPRIETA MAGNETICA
1160 DIAMICTITE SEDS	DIAMICTITE	Diamictite ROCHE CLAS- TIQUE; HETERO- GENEITE	Diamiktit KLAST.GESTEIN; HETEROGENITAET	Diamiktit KLASTIT; GETERO- GENNYJ	Diamictita ROCA=CLASTICA; HETEROGENEIDAD	Diamictite ROCCIA CLASTICA; ETEROGENEITA
1161 DIAMOND COMS	DIAMONDS	DIAMANT	DIAMANT	ALMAZ	DIAMANTE	DIAMANTE
1162 DIAPHTHORESIS IGNE	Diaphthoresis RETROGRADE METAMORPHISM	Diaphthorese METAMORPHISME RETRO	Diaphthorese RETRO= METAMORPHOSE	DIAFTOREZ UF: Metamorfizm retro- gradnyj	Diaftoresis RETROMETAMOR- FISMO	Diaftoresi RETROMETAMOR- FISMO
1163 DIAPHTHORITE IGMS	Diaphtorite METAMORPHIC ROCKS	Diaphthorite ROCHE METAMOR- PHIQUE; METAM- ORPHISME RETRO	Diaphthorit METAMORPHIT; RETRO= METAMORPHOSE	DIAFTORIT	Diaftorita ROCA= METAMORFICA; RETROMETAMOR- FISMO	Diaftorite ROCCIA METAM- ORFICA; RETRO- METAMORFISMO
1164 DIAPIR STRU	DIAPIRS	DIAPIR	DIAPIR	DIAPIR	DIAPIRO	DIAPIRO
1165 DIAPIRISM STRU	DIAPIRISM	DIAPIRISME	DIAPIRISMUS	DIAPIRIZM	DIAPIRISMO	DIAPIRISMO
1166 DIASTEM STRA	Diastems STRATIGRAPHIC GAPS	Diasteme LACUNE STRATI- GRAPHIQUE	Diastem LUECKE	Diastema PERERYV	Diastema LAGUNA	Diastema LACUNA
1167 DIASTROPHISM STRU	Diastrophism TECTONICS	Diastrophisme TECTOGENESE	Diastrophismus TEKTOGENESE	DVIZHENIE TEK- TONICHESKOE	Diastrofismo TECTOGENESIS	Diastrofismo TETTOGENESI
1168 DIATOMACEOUS EARTH COMS	DIATOMACEOUS EARTH	TERRE DIATOMITE	DIATOMEEN=ERDE	ZEMLYA DIATO- MOVAYA UF: Tripoli	TIERRA=DE= DIATOMEAS	TERRA A DIA- TOMEE
1169 DIATOMEAE PALS	DIATOMS	DIATOMEAE	DIATOMEAE	DIATOMEAE	DIATOMEAE	DIATOMEAE
1170 DIATOMITE SEDS	DIATOMITE	DIATOMITE	DIATOMIT	DIATOMIT	DIATOMITA	DIATOMITE UF: Tripoli
1171 DIATREME IGNE	DIATREMES	DIATREME	DIATREMA	TRUBKA VZRYVA	DIATREMA	DIATREMA
1172 DICOTYLEDONEAE PALS	DICOTYLEDONEAE	DICOTYLEDONES	DICOTYLEDONEAE	Dicotyledoneae ANGIOSPERMAE	DICOTYLEDONEAE	DICOTILEDONI
1173 DICTYONELLIDINA PALS	DICTYONELLIDINA	DICTYONELLIDINA	DICTYONELLIDINA	DICTYONELLIDINA	DICTYONELLIDINA	DICTYONELLIDINA
1174 DIELECTRIC CONSTANT PHCH	DIELECTRIC CONS- TANT	CONSTANTE DIE- LECTRIQUE	DIELEKTR. KONSTANTE	POSTOYANNAYA DIEHLEKTRI- CHESKAYA	CONSTANTE= DIELECTRICA	COSTANTE DIELET- TRICA

	ENGLISH	FRANCAIS	DEUTSCH	RUSSKIJ	ESPANOL	ITALIANO
1175 DIFFERENTIAL COMPACTION SEDI	Differential compaction COMPACTION	Tassement differentiel TASSEMENT	Differential= Kompaktion SETZUNG	UPLOTNENIE DIF- FERENTSIROVAN- NOE	Compactacion= diferencial ASIENTO	Costipamento differen- ziale ASSESTAMENTO
1176 DIFFERENTIAL THERMAL ANALYSIS METH	DIFFERENTIAL THERMAL ANALY- SIS	ANALYSE THER- MIQUE DIFFER- ENTIELLE	DTA	ANALIZ TERMI- CHESKIJ DIFFER- ENTS.	TERMO= DIFERENCIAL	ANALISI TER- MODIFFERENZI- ALE
1177 DIFFERENTIAL WEATHERING SURF	DIFFERENTIAL WEATHERING	ALTERATION DIF- FERENTIELLE	SELEKTIV. VERWITTERUNG	Vyvetrivanie selektiv- noe VYVETRIVANIE	ALTERACION= DIFERENCIAL	ALTERAZIONE DIFFERENZIALE
1178 DIFFERENTIATION IGNE	DIFFERENTIATION _UF:_ Differentiation index; Gravitational differentiation; Meta- morphic differentiation	DIFFERENCIATION MAGMATIQUE _UF:_ Differenciation par gravite; Indice differenciation; Segre- gation	MAGMAT. DIFFERENTIATION _UF:_ Differentiations= Index; Schwere= Differenzierung; Segre- gation	DIFFERENTSIAT- SIYA	DIFERENCIACION= MAGMATICA _UF:_ Diferenciacion= gravitacional; Indice= diferenciacion; Segre- gacion	DIFFERENZIAZ- IONE MAGMATICA _UF:_ Differenziazione gravitativa; Indice di differenziazione; Segregazione
1179 DIFFERENTIATION INDEX IGNE	Differentiation index DIFFERENTIATION	Indice differenciation DIFFERENCIATION MAGMATIQUE	Differentiations=Index MAGMAT. DIFFERENTIATION	INDEKS DIFFER- ENTSIATSII	Indice=diferenciacion DIFERENCIACION= MAGMATICA	Indice di differenziaz- ione DIFFERENZIAZ- IONE MAGMATICA
1180 DIFFRACTION PHCH	DIFFRACTION	DIFFRACTION	BEUGUNG	DIFRAKTSIYA _UF:_ Ugol otrazheniya	DIFRACCION	DIFFRAZIONE
1181 DIFFUSION PHCH	DIFFUSION _UF:_ Diffusivity	DIFFUSION _UF:_ Dialyse; Diffusivite	DIFFUSION _UF:_ Dialyse; Diffusions= Vermoegen	DIFFUZIYA _UF:_ Dializ	DIFUSION _UF:_ Dialisis; Difusividad	DIFFUSIONE _UF:_ Dialisi; Diffusivita
1182 DIFFUSIVITY PHCH	Diffusivity DIFFUSION	Diffusivite DIFFUSION	Diffusions=Vermoegen DIFFUSION	TEMPERATURO- PROVODNOST'	Difusividad DIFUSION	Diffusivita DIFFUSIONE
1183 DIGITAL TECHNIQUE MATH	DIGITAL SIMULA- TION	SIMULATION NUMERIQUE	DIGITAL=TECHNIK	Tekhnika vychisli- tel'naya EHVM	SIMULACION= NUMERICA	TECNICA DIGITALE
1184 DIKE IGNE	DIKES _UF:_ Sole injection	DYKE _UF:_ Injection faille	DYKE _UF:_ Sohlflaechen= Injektion	DAIKA	DIQUE _UF:_ Intrusion=plano= falla	DICCO _UF:_ Iniezione basale
1185 DIKE SWARM IGNE	DIKE SWARMS	ESSAIM DYKE	DYKE=SCHWARM	SISTEMA DAEK	ASOCIACION= DIQUES	SCIAME DI FILONI
1186 DILATANCY PHCH	DILATANCY	DILATANCE	DILATANZ	DILATANTSIYA	DILATANCIA	DILATANZA
1187 DILATION PHCH	DILATION	DILATATION	DEHNUNG	DILATATSIYA	DILATACION	DILATAZIONE
1188 DILUTION PHCH	DILUTION _UF:_ Isotope dilution	DILUTION _UF:_ Dilution isotopique	VERDUENNUNG _UF:_ Isotopen= Verduennung	RAZBAVLENIE	DILUCION _UF:_ Medida=isotopica	DILUIZIONE _UF:_ Diluizione isotopica
1189 DIMORPHISM PALE	Dimorphism POLYMORPHISM	Dimorphisme POLYMORPHISME	Dimorphismus POLYMORPHIE	DIMORFIZM _UF:_ Polovoj dimorfizm	Dimorfismo POLIMORFISMO	Dimorfismo POLIMORFISMO
1190 DINOCERATA PALS	DINOCERATA	DINOCERATA	DINOCERATA	DINOCERATA	DINOCERATA	DINOCERATA

	ENGLISH	FRANCAIS	DEUTSCH	RUSSKIJ	ESPANOL	ITALIANO
1191 DINOFLAGELLATA PALS	DINOFLAGELLATA	DINOFLAGELLATA	DINOFLAGELLATA	Dinoflagellata PYRROPHYTA	DINOFLAGELATA	DINOFLAGELLATA
1192 DINOSAURIA PALS	DINOSAURS	DINOSAURIA	DINOSAURIA	DINOSAURIA	DINOSAURIA	DINOSAURIA
1193 DIOGENITE EXTS	Diogenite ACHONDRITES	Diogenite ACHONDRITE	Diogenit ACHONDRIT	Diogenit AKHONDRIT	Diogenita ACONDRITA	Diogenite ACONDRITE
1194 DIORITE IGNS	DIORITES UF: Essexite; Intermedi- ate composition	DIORITE UF: Essexite; Roche intermediaire	DIORIT UF: Intermediaer. Chemismus	DIORIT	DIORITA UF: Composicion= intermedia; Essexita	DIORITE UF: Composizione inter- media; Essexite
1195 DIP STRU	DIP UF: Dip slope	PENDAGE UF: Angle avec la verti- cale	EINFALLEN UF: Flaechen=Normale	PADENIE UF: Ugol naklona	BUZAMIENTO UF: Angulo=con=la= vertical	IMMERSIONE UF: Angolo con la verti- cale
1196 DIP FAULT STRU	DIP FAULTS	FAILLE TRANS- VERSALE	SCHRAEG- VERWERFUNG	SBROS POPERECH- NYJ	FALLA= TRANSVERSAL	FAGLIA TRASVERS- ALE
1197 DIP METER INST	Dip meter DIPMETER LOG- GING	Pendagemetre CLINOMETRE	Neigungs=Messer KLINOMETER	IZMERITEL' PADENIYA	Pendientometro CLINOMETRO	[DIP METER]; INCLINOMETRO
1198 DIP SLOPE SURF	Dip slope SLOPES; DIP	Revers RELIEF STRUC- TURAL	Einfalls=Hang STRUKTUR- RELIEF	Sklon konsekventnyj SKLON	Ladera=estructural RELIEVE= ESTRUCTURAL	Pendio strutturale MORFOSTRUT- TURA
1199 DIP-SLIP FAULT STRU	DIP-SLIP FAULTS	FAILLE EFFONDRE- MENT	STOERUNGS- EINFALLEN	[DIP-SLIP FAULT] SBROSO-SDVIG	FALLA- HUNDIMIENTO	FAGLIA D'IMMERSIONE
1200 DIPLEUROZOA PALS	Dipleurozoa COELENTERATA	Dipleurozoa COELENTERATA	Dipleurozoa COELENTERATA	DIPLEUROZOA	Dipleurozoa COELENTERATA	Dipleurozoa COELENTERATA
1201 DIPLOID MINE	Diploid minerals SYMMETRY	Mineral diploide SYMETRIE	Diploid.Mineral SYMMETRIE	Didodekaehdr SINGONIYA KUBI- CHESKAYA; FORMA KRISTAL- LOGRAFI- CHESKAYA	Diploide SIMETRIA	Diploide SIMMETRIA
1202 DIPLOPODA PALS	Diplopoda MYRIAPODA	Diplopoda MYRIAPODA	Diplopoda MYRIAPODA	DIPLOPODA UF: Arthropleurida	Diplopoda MYRIAPODA	Diplopoda MYRIAPODA
1203 DIPLORHINA PALS	Diplorhina AGNATHA	Diplorhina AGNATHA	Diplorhina AGNATHA	DIPLORHINA	Diplorhina AGNATHA	Diplorina AGNATHA
1204 DIPMETER LOGGING APPL	DIPMETER LOG- GING UF: Dip meter	PENDAGEMETRIE	GEFAELLE= MESSUNG	NAKLONO- METRIYA	PENDIENTO- METRIA	MISURAZIONE DI PENDENZA
1205 DIPNOI PALS	DIPNOI	DIPNOI	DIPNOI	DIPNOI	DIPNOI	DIPNOI
1206 DIPOLE MOMENT APPL	DIPOLE MOMENT	MOMENT DIPOLAIRE	DIPOL=MOMENT	MOMENT DIPOL'NYJ	MOMENTO= DIPOLAR	MOMENTO DIPO- LARE
1207 DIPTERA PALS	DIPTERA	DIPTERA	DIPTERA	Diptera INSECTA	DIPTERA	DIPTERA

	ENGLISH	FRANCAIS	DEUTSCH	RUSSKIJ	ESPANOL	ITALIANO
1208 DIRECTIONAL DRILLING MINI	Directional drilling BOREHOLES	Sondage oriente SONDAGE	Richtungs=Bohren BOHRUNG	BURENIE NAPRAV- LENNOE	Sondeo=direccion POZO=SONDEO	Perforazione dire- zionata SONDAGGIO
1209 DISCHARGE GEOH	DISCHARGE UF: Groundwater dis- charge	DEBIT UF: Debit eau souter- raine	SCHUETTUNG UF: Grundwasser= Ergiebigkeit	RASKHOD	CAUDAL UF: Descarga=agua= subterranea	PORTATA UF: Portata d'acqua sotterranea
1210 DISCOASTERS PALS	Discoasters COCCOLITHOPHO- RACEAE	Discoaster COCCOLITHO- PHORALES	Discoasteroidea COCCOLITHO- PHORALES	Diskoastery CHRYSOPHYTA	Discoaster COCCOLITHO- PHORALES	Discoaster COCCOLITHOPHO- RACEAE
1211 DISCONFORMITY STRA	DISCONFORMITIES	DISCORDANCE EROSION	EROSIONS- DISKORDANZ	NESOGLASIE STRA- TIGRAFICHESKOE	DISCORDANCIA= EROSION	DISCORDANZA EROSIVA
1212 DISCONTINUITY SOLI	DISCONTINUITIES UF: Birch discontinuity; Bullard discontinuity; Gutenberg discontinu- ity	DISCONTINUITE UF: Discontinuite Birch; Discontinuite Bullard; Discontinuite Guten- berg	DISKONTINUITAET UF: Birch= Diskontinuitaet; Bullard= Diskontinuitaet; Gutenberg= Diskontinuitaet	GRANITSA SEJSMI- CHESKAYA UF: Bercha poverkhnost'	DISCONTINUIDAD UF: Discontinuidad= de=Birch; Discontinuidad=de= Bullard; Discontinuidad=de= Gutenberg	DISCONTINUITA UF: Discontinuita di Birch; Discontinuita di Bullard; Discon- tinuita di Gutenberg
1213 DISCORBACEA PALS	DISCORBACEA	DISCORBACEA	DISCORBACEA	DISCORBACEA	DISCORBACEA	DISCORBACEA
1214 DISCORDANCE STRA	Discordance ANGULAR UNCON- FORMITIES	Discordance tectonique DISCORDANCE ANGULAIRE	Tekton.Diskordanz WINKEL= DISKORDANZ	DISKORDANTNYJ	Discordancia=tectonica DISCORDANCIA= ANGULAR	Discordanza tettonica DISCORDANZA ANGOLARE
1215 DISCORDANT GEOL	Discordant(adj.)	Discordant(adj.)	Diskordant(adj.)	NESOGLASNYJ	Discordante(adj.)	Discordante(adj.)
1216 DISCORDANT AGE ISOT	Discordant age CORRELATION; ABSOLUTE AGE	Age discordant CORRELATION; DATATION	Diskordant.Alter KORRELATION; PHYSIKAL. ALTERSBESTIM- MUNG	Vozrast diskordantnyj VOZRAST ABSO- LYUTNYJ	Edad=no=acorde CORRELACION; DATACION	Eta discordante CORRELAZIONE; DATAZIONE
1217 DISCOVERY MISC	DISCOVERIES	DECOUVERTE	ENTDECKUNG	OTKRYTIE	DESCUBRIMIENTO	SCOPERTA
1218 DISCRIMINANT ANALYSIS MATH	DISCRIMINANT ANALYSIS	ANALYSE DIS- CRIMINANTE	DISKRIMINANZ= ANALYSE	ANALIZ DISKRIMI- NANTNYJ	ANALISIS= DISCRIMINANTE	ANALISI DISCRIMI- NANTE
1219 DISHARMONIC FOLD STRU	DISHARMONIC FOLDS	PLI DISHAR- MONIQUE	DISHARMON. FALTE	SKLADKA DISGAR- MONICHNAYA	PLIEGUE= DISARMONICO	PIEGA DISAR- MONICA
1220 DISJUNCTIVE FOLD STRU	DISJUNCTIVE FOLDS	PLI FAILLE	BRUCH=FALTE	SKLADKA PRIRA- ZLOMNAYA	PLIEGUE=FALLA	PIEGA=FAGLIA
1221 DISLOCATION BRECCIA GEOL	Dislocation breccia TECTONIC BREC- CIA	Breche dislocation BRECHE TEC- TONIQUE	Dislokations=Brekzie TEKTON.BREKZIE	Brekchiya dislokatsion- naya BREKCHIYA TEK- TONICHESKAYA	Brecha=de=dislocacion BRECHA= TECTONICA	Breccia di dislocazione BRECCIA TET- TONICA
1222 DISPERSION PHCH	OPTICAL DISPER- SION UF: Rotatory dispersion	DISPERSION OPTIQUE UF: Dispersion rotatoire	OPT.DISPERSION UF: Rotierend. Dispersion	DISPERSIYA UF: Dispersiya svetla rotuyushchaya; Wave dispersal	DISPERSION= OPTICA UF: Dispersion=rotatoria	DISPERSIONE OTTICA UF: Dispersione rota- toria

	ENGLISH	FRANCAIS	DEUTSCH	RUSSKIJ	ESPANOL	ITALIANO
1223 DISPERSION PATTERN GEOC	DISPERSION PAT-TERNS	AUREOLE DISPER-SION	DISPERSIONS=AUREOLE	[DISPERSION PAT-TERN]	AUREOLA=DISPERSION	AUREOLA DI DIS-PERSIONE
1224 DISPLACEMENT STRU	DISPLACEMENTS	DEPLACEMENT FAILLE	VERWERFUNGS=VERSCHIEBUNG	SMESHCHENIE UF: Smeshchenie gorizontal'noe	DESPLAZAMIENTO	RIGETTO
1225 DISPLACEMENT THEORY STRU	Displacement theory CONTINENTAL DRIFT	Mobilisme DERIVE CON-TINENTALE	Verschiebungs=Theorie KONTINENTAL=DRIFT	GIPOTEZA PER-EMESHCHENIYA KONTINENTOV	Teoria=del=desplazamiento DERIVA=CONTINENTAL	Teoria deriva continentale DERIVA CON-TINENTALE
1226 DISSEMINATION ECON	DISSEMINATED DEPOSITS	GITE DISSEMINA-TION	DISSEMINATION	VKRAPLENIE	YACIMIENTO-DISEMINACION	GIACIMENTO DIS-SEMINATO
1227 DISSOCIATION PHCH	DISSOCIATION	DISSOCIATION IONIQUE	IONEN=DISSOZIIERUNG	DISSOTSIATSIYA	DISOCIACION-IONICA	DISSOCIAZIONE IONICA
1228 DISSOLUTION PHCH	SOLUTION	DISSOLUTION	AUFLOESUNG UF: Azetolyse; Entkie-selung	RASTVORENIE	DISOLUCION UF: Acetolisis	DISSOLUZIONE UF: Acetolisi
1229 DISSOLVED LOAD SEDI	DISSOLVED MATE-RIALS	MATIERE DISS-OUTE	GELOEST.FRACHT	VESHCHESTVO RASTVORENNOE	MATERIA=DISUELTA	CARICO IN SOLUZ-IONE
1230 DISTRIBUTION MATH	Distribution STATISTICAL DIS-TRIBUTION	Distribution DISTRIBUTION STATISTIQUE	Verteilung STATIST. VERTEILUNG	RASPREDELENIE UF: Partitioning	Distribucion DISTRIBUCION=ESTADISTICA	Distribuzione DISTRIBUZIONE STATISTICA
1231 DIURNAL VARIATION SOLI	DIURNAL VARIA-TIONS	VARIATION DIURNE	TAGES=SCHWANKUNG	VARIATSII MAGNITNYE SUTOCHNYE	VARIACION=DIURNA	VARIAZIONE DIURNA
1232 DIVERGENCE GEOL	Divergence(gen.)	Divergence(gen.)	Divergenz(gen.)	DIVERGENTSIYA	Divergencia(gen.)	Divergenza(gen.)
1233 DOCODONTA PALS	DOCODONTA	DOCODONTA	DOCODONTA	Docodonta MAMMALIA	DOCODONTA	DOCODONTA
1234 DOLERITE IGNS	DIABASE	DOLERITE	DOLERIT	DOLERIT	DOLERITA	DIABASE
1235 DOLINE SURF	DOLINES	DOLINE	DOLINE	VORONKA KARS-TOVAYA UF: Truba karstovaya	DOLINA	DOLINA
1236 DOLOMITIC LIMESTONE SEDS	DOLOMITIC LIME-STONE	CALCAIRE DOLOMITIQUE	DOLOMIT=KALK	IZVESTNYAK DOLOMITOVYJ	CALIZA=DOLOMITICA	CALCARE DOLOMI-TICO
1237 DOLOMITIZATION SEDI	DOLOMITIZATION	DOLOMITISATION	DOLOMITI-SIERUNG	DOLOMITIZATSIYA	DOLOMITIZACION	DOLOMITIZZAZ-IONE
1238 DOLOSTONE SEDS	DOLOSTONE UF: Cellular dolomite	DOLOMIE UF: Carnieule	DOLOMIT=GESTEIN UF: Kiesel=Kalk	DOLOMIT	DOLOMIA UF: Carniola	DOLOMIA UF: Carniola
1239 DOMAIN STRUCTURE MINE	Domain structure MAGNETIC DOMAINS	Structure domaine DOMAINE MAGNE-TIQUE	Domaenen=Struktur MAGNET.BEREICH	Struktura dommennaya FERROMAGNETIK	Dominio=estructural DOMINIO-MAGNETICO	Struttura a dominii DOMINIO MAGNE-TICO

	ENGLISH	FRANCAIS	DEUTSCH	RUSSKIJ	ESPANOL	ITALIANO
1240 DOME GEOL	DOMES UF: Gneissic dome	DOME UF: Dome gneissique; Dome volcanique	DOM UF: Gneiskuppel; Vulkan=Kuppe	KUPOL	DOMO UF: Cupula=volcanica; Domo=gneisico	DUOMO UF: Cupola di gneiss; Duomo vulcanico
1241 DOMERIAN STRS	DOMERIAN	DOMERIEN	DOMERIUM	Domer PLINSBAKH	DOMERIENSE	DOMERIANO
1242 DOWNWARPING STRU	Downwarping SUBSIDENCE	Affaissement regional SUBSIDENCE	Absenkung SENKUNG	DVIZHENIE NISKHODYASCH- CHEE	Subsidencia=regional SUBSIDENCIA	Abbassamento region- ale SUBSIDENZA
1243 DRAG FOLD STRU	DRAG FOLDS	PLI ETIREMENT	SCHLEPP=FALTE	SKLADKA VOLOC- HENIYA	PLIEGUE=DE= ARRASTRE	PIEGA DI TRAS- CINAMENTO
1244 DRAG ORE ECON	Drag ore ORE BODIES	Minerai disloque BRECHE; MINER- ALISATION	Geschleppt.Erz BREKZIE; MINER- ALISATION	BREKCHIYA RUD- NAYA	Escape BRECHA; MINER- ALIZACION	[Drag Ore] BRECCIA; MINER- ALIZZAZIONE
1245 DRAINAGE BASIN GEOH	DRAINAGE BASINS UF: Divides	BASSIN VERSANT UF: Ligne partage eaux	EINZUGSGEBIET UF: Wasser=Scheide	BASSEJN VODOS- BORNYJ UF: Stream order; Catchment hydrody- namics	DIVISORIA=DE= AGUAS UF: Linea=de= separacion=de=aguas	BACINO IDROGRA- FICO UF: Spartiacque
1246 DRAINAGE SYSTEM SURF	DRAINAGE PAT- TERNS UF: Dendritic drainage pattern; Endorheism; Interfluves	RESEAU HYDRO- GRAPHIQUE UF: Drainage rayonnant; Interfluve; Rang reseau hydrographique	HYDROGRAPH. NETZ UF: Dendrit. Entwaesserung; Gewaesser=Anordnung; Zwischenstromland	SET' RECHNAYA UF: Set' rechnaya drevovidnaya	RED= HIDROGRAFICA UF: Avenamiento= dendritico; Categoria= cursos=agua; Interflu- vio	RETICOLATO IDRO- GRAFICO UF: Gerarchia fluviale; Interfluvio; Rete idro- grafica dendritica
1247 DRAINING ENGI	DRAINING	DRAINAGE TER- RAIN	DRAINUNG	DRENAZH ISKUST- VENNYJ UF: Reclamation; Sifon (gidrogeol)	DRENAJE= TERRENO	DRENAGGIO
1248 DRAPE FOLD STRU	DRAPE FOLDS	PLI CHAPE	DRAPIER=FALTE	SKLADKA CHEKHLA	PLIEGUE=CAPA	PIEGA DI RIVESTI- MENTO
1249 DRAWDOWN GEOH	DRAWDOWN	ABAISSEMENT NIVEAU EAU UF: Cone depression	GRUNDWASSER= SPIEGEL= ABSENKUNG UF: Absenkungs= Trichter	PONIZHENIE UROVNYA UF: Voronka depression- naya	DESCENSO= NIVEL=DE=AGUA UF: Cono=de=depresion	ABBASSAMENTO DELLA FALDA UF: Cono di depressione
1250 DRAWINGS MISC	Drawings(gen.)	Dessin(gen.)	Zeichnung(gen.)	GRAFIKA UF: Grafik	Dibujo(gen.)	Disegno(gen.)
1251 DREDGING METH	DREDGING	DRAGAGE	BAGGERN	DRAGA	DRAGADO	DRAGAGGIO
1252 DRESSING MINI	DRESSING UF: Ore dressing plant	VALORISATION UF: Usine traitement	AUFBEREITUNG UF: Erz= Aufbereitungs=Anlage	OBOGASHCHENIE (ZAVOD)	BENEFICIACION UF: Planta= preparacion=mena	PREPARAZIONE DEI MINERALI UF: Impianto di tratta- mento
1253 DRILLABILITY MINI	DRILLABILITY	Forabilite RESISTANCE MECANIQUE; SONDAGE	Bohrbarkeit FESTIGKEIT; BOHRUNG	BURIMOST'	Perforabilidad RESISTENCIA= MECANICA; POZO= SONDEO	Perforabilita RESISTENZA MEC- CANICA; SONDAG- GIO

	ENGLISH	FRANCAIS	DEUTSCH	RUSSKIJ	ESPANOL	ITALIANO
1254 DRILLING METH	Drilling BOREHOLES	FORAGE	WASSER= BOHRUNG	BURENIE	SONDEO= MECANICO	TRIVELLAZIONE
1255 DRILLING EQUIPMENT INST	Drilling equipment BOREHOLES; INSTRUMENTS	Materiel sondage INSTRUMENTA- TION; SONDAGE	Bohr=Geraet INSTRUMEN- TIERUNG; BOHRUNG	OBORUDOVANIE BUROVOE	Maquinaria=sondeo INSTRUMENTA- CION; POZO= SONDEO	Apparecchi per per- forazione STRUMENTAZ- IONE; SONDAGGIO
1256 DRILLING SLUDGE ENGI	Drill sludge BOREHOLES	Boue forage SONDAGE	Bohr=Schlamm BOHRUNG	SHLAM BUROVOJ	Lodo=de=sondeo POZO=SONDEO	Fango di perforazione SONDAGGIO
1257 DRILLING VESSEL INST	Drilling vessel BOREHOLES; OFF- SHORE	Navire sondage SONDAGE; OFF- SHORE	Bohr=Schiff BOHRUNG; OFF= SHORE	SUDNO BUROVOE	Barco=sondeo POZO=SONDEO; OFF=SHORE	Nave per trivellazione SONDAGGIO; OFF- SHORE
1258 DROUGHT SURF	DROUGHT	SECHERESSE	DUERRE	Zasukha ARIDNYJ	SEQUIA	SICCITA
1259 DRUMLIN SURF	DRUMLINS	DRUMLIN	DRUMLIN	DRUMLIN	DRUMLING	DRUMLIN
1260 DRUSE MINE	Druse TEXTURES	Druse GEODE	Druse GEODE	DRUZA	Drusa GEODA	Drusa GEODE
1261 DUNE SURF	DUNES UF: Desert dune; Ergs	DUNE	DUENE	DYUNA UF: Ehrg	DUNA	DUNA
1262 DURAIN SEDS	Durain MACERALS	Durain MACERAL	Durain MAZERAL	DYUREN	Dureno MACERAL	Durite MACERALE
1263 DUST SEDI	DUST	POUSSIERE	STAUB	PYL'	POLVO	POLVERE
1264 DYNAMIC EQUILIBRIUM PHCH	Dynamic equilibrium EQUILIBRIUM	Equilibre dynamique EQUILIBRE	Dynam.Gleichgewicht PHASEN- GLEICHGEWICHT	RAVNOVESIE DINAMICHESKOE	Equilibrio=dinamico DIAGRAMA= EQUILIBRIO	Equilibrio dinamico EQUILIBRIO
1265 DYNAMIC GEOLOGY GEOL	Dynamic geology GEODYNAMICS	Geologie dynamique GEODYNAMIQUE	Dynam.Geologie GEODYNAMIK	GEOLOGIYA DINAMICHESKAYA	Geologia=dinamica GEODINAMICA	Geologia dinamica GEODINAMICA
1266 DYNAMIC METAMORPHISM IGNE	DYNAMIC META- MORPHISM UF: Dynamothermal metamorphism	METAMORPHISME DYNAMO UF: Metamorphisme dynamothermal	DYNAMO- METAMORPHOSE UF: Dynamo=Therm. Metamorphose	DINAMOMETAM- ORFIZM	DINAMOMETAM- ORFISMO UF: Metamorfismo= dinamotermal	METAMORFISMO DINAMICO UF: Termodinamometamorfismo
1267 DYNAMICS PHCH	DYNAMICS	DYNAMIQUE	DYNAMIK	DINAMIKA UF: Ice movement	DINAMICA	DINAMICA
1268 DYNAMO THEORY SOLI	Dynamo theory MAGNETOHYDRO- DYNAMICS	Theorie dynamo THEORIE; CHAMP MAGNETIQUE	Dynamo=Theorie THEORIE; MAG- NETFELD	Dinamo teoriya GEOMAGNETIZM	Teoria=de=la=dinamo TEORIA; CAMPO MAGNETICO	Teoria della dinamo TEORIA; CAMPO MAGNETICO
1269 DYNAMOTHERMAL METAMORPHISM IGNE	Dynamothermal meta- morphism DYNAMIC META- MORPHISM	Metamorphisme dyna- mothermal METAMORPHISME DYNAMO	Dynamo=Therm. Metamorphose DYNAMO- METAMORPHOSE	METAMORFIZM DINAMOTER- MAL'NYJ	Metamorfismo= dinamotermal DINAMOMETAM- ORFISMO	Termodinamometamor- fismo METAMORFISMO DINAMICO
1270 DYSPROSIUM CHEE	DYSPROSIUM	DYSPROSIUM	DY	DISPROSIJ	DISPROSIO	DISPROSIO

	ENGLISH	FRANCAIS	DEUTSCH	RUSSKIJ	ESPANOL	ITALIANO
1271 E LAYER SOLI	E layer OUTER CORE	Couche E NOYAU TER- RESTRE EXTERNE	E=Schicht AEUSSER.ERD= KERN	E sloj YADRO ZEMLI	Capa=E NUCLEO= EXTERNO	Strato E NUCLEO ESTERNO
1272 EARLY DIAGENESIS SEDI	EARLY DIAGENE- SIS	DIAGENESE PRE- COCE	FRUEH= DIAGENESE	[EARLY DIAGENE- SIS] KATAGENEZ	DIAGENESIS= PRECOZ	DIAGENESI PRE- COCE
1273 EARLY **PROTEROZOIDES** STRU	Early proterozoides OROGENY; PRO- TEROZOIC	Proterozoides inf OROGENIE ANTE- CAMBRIENNE	Frueh=Proterozoiden PRAEKAMBR. OROGENESE	PROTEROZOIDY RANNIE	Proterozoides=inf OROGENIA= PRECAMBRICA	Proterozoidi inf. OROGENESI PRE- CAMBRIANA
1274 EARTH CURRENT SOLI	EARTH=CURRENT METHODS UF: Telluric methods	METHODE COU- RANT TELLURIQUE	TELLUR.STROM	TOKI TELLURI- CHESKIE	CORRIENTE= TELURICA	CORRENTE TEL- LURICA
1275 EARTH DAM ENGI	EARTH DAMS	BARRAGE TERRE	ERD=DAMM	PLOTINA ZEMLY- ANAYA	PRESA=TIERRA	DIGA IN TERRA
1276 EARTH INTERIOR SOLI	INTERIOR	CONSTITUTION INTERNE	ERD=AUFBAU	STROENIE GLUBIN- NOE (ZEMLI)	CONSTITUCION= GLOBO	INTERNO DELLA TERRA
1277 EARTH PRESSURE ENGI	EARTH PRESSURE	PRESSION TER- RAIN	ERDDRUCK	DAVLENIE GRUNTA UF: Uplift pressure	PRESION= TERRENO	PRESSIONE DEL TERRENO
1278 EARTH ROTATION SOLI	Earth rotation EARTH; ROTATION	Rotation terre ROTATION	Erd=Rotation ROTATION	VRASHCHENIE ZEMLI	Rotacion=tierra ROTACION	Rotazione terrestre ROTAZIONE
1279 EARTH SCIENCE GEOL	Earth science GEOLOGY	Science de la Terre GEOLOGIE	Geowissenschaften GEOLOGIE	Nauki o Zemle GEOLOGIYA	Ciencia=de=la=Tierra GEOLOGIA	Scienze della Terra GEOLOGIA
1280 EARTH TIDE SOLI	EARTH TIDES	MAREE TER- RESTRE	ERD=GEZEITEN	PRILIVY ZEMLI	MAREA= TERRESTRE	MAREA TER- RESTRE
1281 EARTH WORK ENGI	EARTHWORKS	TERRASSEMENT	ERDARBEITEN	RABOTY ZEMLY- ANYE	MOVIMIENTO=DE= TIERRAS	TERRAZZAMENTO ARTIFICIALE
1282 EARTH-MOON COUPLE EXTR	EARTH=MOON COUPLE	COUPLE TERRE= LUNE	ERDE=MOND= KOPPLUNG	[EARTH=MOON COUPLE]	PAR=TIERRA= LUNA	COPPIA TERRA= LUNA
1283 EARTHFLOW ENGI	Earthflows LANDSLIDES	Glissement terre GLISSEMENT TER- RAIN	Erdfliessen ERDRUTSCH	[EARTHFLOW] OPOLZEN'	Flujo=de=tierra DESLIZAMIENTO= TERRENO	Scivolamento di terra SCIVOLAMENTO DI TERRENO
1284 EARTHQUAKE SOLI	EARTHQUAKES UF: Earthquake engi- neering	SEISME UF: Essaim seisme; Ingenierie sismique; Isoseiste	ERDBEBEN UF: Erdbeben= Schwarm; Erdbeben= Technik; Isoseiste	ZEMLETRYASENIE UF: Aftershock; For- shok; Roj zemletryase- nij; Teleseismic signal	TERREMOTO UF: Ingenieria=sismica; Linea=isosismica; Serie=sismos	TERREMOTO UF: Ingegneria sismica; Isosista; Sciame di terremoti
1285 EARTHQUAKE **ENGINEERING** ENGI	Earthquake engineering ENGINEERING GEOLOGY; EARTH- QUAKES	Ingenierie sismique SEISME; ACTION PREVENTIVE	Erdbeben=Technik ERDBEBEN; VORBEUGEND. MASSNAHME	SEJSMOLOGIYA INZHENERNAYA	Ingenieria=sismica TERREMOTO; ACCION= PREVENTIVA	Ingegneria sismica TERREMOTO; AZIONE PREVEN- TIVA
1286 EARTHQUAKE **MAGNITUDE** SOLI	MAGNITUDE	INTENSITE SISMIQUE	SEISM. INTENSITAET	SILA ZEMLETR- YASENIYA	SISMODINAMO- METRIA	MAGNITUDO SISMICA

	ENGLISH	FRANCAIS	DEUTSCH	RUSSKIJ	ESPANOL	ITALIANO
1287 EARTHQUAKE SWARM SOLI	SWARMS SEISME	Essaim seisme SEISME	Erdbeben=Schwarm ERDBEBEN	Roj zemletryasenij ZEMLETRYASENIE	Serie=sismos TERREMOTO	Sciame di terremoti TERREMOTO
1288 EAST MISC	East GEOGRAPHY	Est GEOGRAPHIE	Osten GEOGRAPHIE	VOSTOK	Este GEOGRAFIA	Est GEOGRAFIA
1289 EBURNEAN OROGENY STRU	Eburnean orogeny OROGENY; PRE-CAMBRIAN	Orogenie eburneenne OROGENIE ANTE-CAMBRIENNE	Eburn.Orogenese PRAEKAMBR. OROGENESE	EHBURNEJSKIJ	Orogenia=eburnea OROGENIA=PRECAMBRICA	Orogenesi eburnea OROGENESI PRE-CAMBRIANA
1290 ECCENTRICITY MISC	ECCENTRICITY	EXCENTRICITE	EXZENTRIZITAET	EHKSTSENTRISI-TET	EXCENTRICIDAD	ECCENTRICITA
1291 ECHINODERMATA PALS	ECHINODERMATA *UF:* Crinozoa; Echino-zoa; Homalozoa	ECHINODERMATA *UF:* Crinozoa; Echino-zoa; Homalozoa	ECHINODERMATA *UF:* Crinozoa; Echino-zoa; Homalozoa	ECHINODERMATA	ECHINODERMATA *UF:* Crinozoa; Echino-zoa; Homalozoa	ECHINODERMATA *UF:* Crinozoa; Echino-zoa; Homalozoa
1292 ECHINOIDEA PALS	ECHINOIDEA	ECHINOIDEA	ECHINOIDEA	ECHINOIDEA	ECHINOIDEA	ECHINOIDEA
1293 ECHINOZOA PALS	Echinozoa ECHINODERMATA	Echinozoa ECHINODERMATA	Echinozoa ECHINODERMATA	ECHINOZOA *UF:* Camptostromatoidea; Cyclocystoidea; Edrioasteroidea; Heli-coplacoidea	Echinozoa ECHINODERMATA	Echinozoa ECHINODERMATA
1294 ECHO SOUNDING APPL	ECHO SOUNDING	ECHOSONDAGE	ECHO=LOTUNG	EHKHO-ZONDIROVANIE	ECOSONDEO	ECOSONDAGGIO
1295 ECLOGITE IGMS	ECLOGITE	ECLOGITE	EKLOGIT	EHKLOGIT	ECLOGITA	ECLOGITE
1296 ECLOGITE FACIES IGNE	ECLOGITE FACIES	FACIES ECLOGITE	EKLOGIT=FAZIES	FATSIYA EHKLOGI-TOVAYA	FACIES=ECLOGITA	FACIES DELLE ECLOGITI
1297 ECOLOGY PALE	ECOLOGY *UF:* Euryhaline taxa; Eurythermal taxa; Halophilic taxa; Hekis-totherm plant; Holo-phyte; Lithophyte; Marine ecology; Meso-phyle organism; Meta-phyte; Oxylophyte; Phreatophyte; Psammo-fauna; Psychrophyte; Substrates	ECOLOGIE *UF:* Ecologie marine; Holophyte; Lithophyte; Metaphyte; Organisme euryhalin; Organisme eurytherme; Organisme halophile; Organisme mesophyle; Oxylophyte; Phreatophyte; Plante hekistotherme; Psam-mofaune; Psychrophyte; Substrat	OEKOLOGIE *UF:* Euryhalin. Organismus; Eurythermal. Organismus; Halophil. Organismus; Hekistotherm.Pflanze; Holophyt; Lithophyt; Marin. Oekologie; Mesophyl.Organismus; Metaphyt; Oxylophyt; Phreatophyt; Psammo-fauna; Psychrophyt; Substrat	EHKOLOGIYA *UF:* Ehkologiya mor-skaya; Epibiotism; Paleoehkologiya; Sreda estestvennaya	ECOLOGIA *UF:* Arena=fosilifera; Ecologia=marina; Eurihalino; Euritermo; Freatofita; Halofilo; Holofita; Litofita; Metafita; Organismo=mesofilo; Oxilofita; Planta=hekistotermica; Psicrofita; Substratum	ECOLOGIA *UF:* Ecologia marina; Holophyte; Lithophyte; Metaphyte; Organismo alofilo; Organismo eurialino; Organismo euritermico; Organismo mesofilo; Oxylophyte; Phreatophyte; Pianta echistoterma; Psammo-fauna; Psychrophyte; Substrato
1298 ECONOMIC GEOLOGY ECON	ECONOMIC GEOL-OGY *UF:* Ascension theory; Descension theory	GEOLOGIE ECO-NOMIQUE	WIRTSCHAFTS=GEOLOGIE	GEOLOGIYA POLEZNYKH ISKO-PAEMYKH	GEOLOGIA=ECONOMICA	GEOLOGIA ECO-NOMICA
1299 ECONOMIC GEOLOGY MAP ECON	ECONOMIC GEOL-OGY MAPS	CARTE GITES MIN-ERAUX	MONTANGEOLOG. KARTE	KARTA MESTOROZHDENIJ *UF:* Karta nerudnykh poleznykh iskop; Karta ugol'nykh mestorozh-denij	MAPA-GEOLOGICO=ECONOMICO	CARTA MATERIE PRIME

	ENGLISH	FRANCAIS	DEUTSCH	RUSSKIJ	ESPANOL	ITALIANO
1300 ECONOMIC MINERAL ECON	Economic mineral MINERAL RESOURCES	Mineral economique SUBSTANCE UTILE	Mineral.Rohstoff ROHSTOFF	MINERAL RUDNYJ	Mineral=industrial SUSTANCIA=UTIL	Minerale sfruttabile SOSTANZA UTILE
1301 ECOSYSTEM PALE	ECOSYSTEMS	ECOSYSTEME	OEKOSYSTEM	Ehkosistema BIOTOP	ECOSISTEMA	ECOSISTEMA
1302 ECOZONE STRA	Ecozones PALEOECOLOGY; ZONING	Unite ecostrati- graphique FAUNE SPECI- FIQUE	Oekozone SPEZIF.FAUNA	Zona ehkologicheskya BIOZONA	Ecozona FAUNA= ESPECIFICA	Ecozona FAUNA SPECIFICA
1303 ECTOTROPHA PALS	ECTOTROPHA *UF:* Thysanura	ECTOTROPHA *UF:* Thysanura	ECTOTROPHA *UF:* Thysanura	Ectotropha INSECTA	ECTOTROPHA *UF:* Thysanura	ECTOTROPHA *UF:* Thysanura
1304 EDDY FLUX PHCH	EDDY FLOW	REGIME TURBU- LENT	TURBULENT. STROEMUNG	POTOK VIKHREVOJ	REGIMEN= TURBULENTO	REGIME TUR- BOLENTO
1305 EDENTATA PALS	Edentata MAMMALIA	Edentata MAMMALIA	Edentata MAMMALIA	EDENTATA	Edentata MAMMALIA	Edentata MAMMALIA
1306 EDGE WATER ECON	Edge water CONNATE WATERS; PETRO- LEUM	Eau gisement EAU CONNEE; HYDROCARBURE	Lagerstaetten=Wasser CONNATE=WATER; KOHLENWASSER- STOFF	VODY NEFTYANYE	Agua=marginal AGUA= CONGENITA; HIDROCARBURO	Acqua di giacimento ACQUA CONNATA; RISERVE COMBUS- TIBILI
1307 EDRIOASTEROIDEA PALS	EDRIOASTEROIDEA	EDRIOASTEROIDEA	EDRIOASTEROIDEA	Edrioasteroidea ECHINOZOA	EDRIOASTEROIDEA	EDRIOASTEROIDEA
1308 EDRIOBLASTOIDEA PALS	EDRIOBLASTOIDEA	EDRIOBLASTOIDEA	EDRIOBLASTOIDEA	Edrioblastoidea CRINOZOA	EDRIOBLASTOIDEA	EDRIOBLASTOIDEA
1309 EDUCATION MISC	EDUCATION	ENSEIGNEMENT	BERUFS- AUSBILDUNG	OBRAZOVANIE	ENSEÑANZA	INSEGNAMENTO
1310 EFFLUENT GEOH	Effluents RIVERS	Cours eau derive RIVIERE	See=Ausfluss FLUSS	Potok vytekayushchij REKA	Emisario RIO	Emissario FIUME
1311 EGG PALE	EGGS	OEUF	EI	YAJTSO	HUEVO	UOVO
1312 EH PHCH	EH	EH	REDOX= POTENTIAL	EH	EH	EH
1313 EIFELIAN STRS	EIFELIAN	EIFELIEN	EIFEL	EHJFEL'	EIFELIENSE	EIFELIANO
1314 EINSTEINIUM CHEE	EINSTEINIUM	EINSTEINIUM	ES	EHJNSHTEJNIJ	EINSTEINIO	EINSTEINIO
1315 EJECTA IGNE	Ejecta PYROCLASTICS	Projection volcanique PYROCLASTIQUE	Auswuerfling PYROKLAST. GESTEIN	EHZHEKTIT	Proyeccion=volcanica PIROCLASTICO	Proiezione vulcanica PRODOTTO PIRO- CLASTICO
1316 ELASMOBRANCHII PALS	ELASMOBRANCHII	ELASMOBRANCHII	ELASMOBRANCHII	ELASMOBRANCHII *UF:* Euselachii	ELASMOBRANCHII	ELASMOBRANCHII
1317 ELASTIC CONSTANT PHCH	ELASTIC CON- STANTS	CONSTANTE ELAS- TICITE *UF:* Constante Lame	ELASTIZITAETS= KONSTANTE *UF:* Lame=Konstante	Konstanta uprugosti MODUL' UPRU- GOSTI	CONSTANTE= ELASTICIDAD *UF:* Constante=Lame	COSTANTE DI ELASTICITA *UF:* Costante di Lame

	ENGLISH	FRANCAIS	DEUTSCH	RUSSKIJ	ESPANOL	ITALIANO
1318 ELASTIC LIMIT STRU	ELASTIC LIMIT	LIMITE ELASTI-CITE	ELAST.GRENZE	PREDEL UPRU-GOSTI	LIMITE=ELASTICIDAD	LIMITE DI ELASTI-CITA
1319 ELASTIC MATERIAL ENGI	ELASTIC MATERI-ALS	MILIEU ELASTIQUE	ELAST.MEDIUM	[ELASTIC MATE-RIAL] UPRUGOST'	MEDIO=ELASTICO	MEZZO ELASTICO
1320 ELASTIC PROPERTY PHCH	ELASTIC PROPER-TIES	PROPRIETE ELAS-TIQUE	ELAST. EIGENSCHAFT	Svojstvo ehlasticheskoe UPRUGOST'; SVOJS-TVO	PROPIEDAD=ELASTICA	PROPRIETA ELAS-TICA
1321 ELASTIC STRAIN PHCH	ELASTIC STRAIN	DEFORMATION ELASTIQUE	ELAST. VERFORMUNG	DEFORMATSIYA UPRUGAYA	DEFORMACION=ELASTICA	DEFORMAZIONE ELASTICA
1322 ELASTIC WAVE SOLI	ELASTIC WAVES *UF:* Seismic velocity; Seismic waves	Onde elastique ONDE SISMIQUE	Elast.Welle SEISM.WELLE	Volna uprugaya VOLNA SEJSMI-CHESKAYA	Onda=elastica ONDA=SISMICA	Onda elastica ONDA SISMICA
1323 ELASTICITY PHCH	ELASTICITY *UF:* Lame constant	ELASTICITE	ELASTIZITAET	UPRUGOST' *UF:* Elastic material; Neuprugost'; Svojstvo ehlasticheskoe	ELASTICIDAD	ELASTICITA
1324 ELECTRIC LOG APPL	ELECTRICAL LOG-GING *UF:* Lateral log; Micro-laterolog; Microlog	DIAGRAPHIE ELEC-TRIQUE *UF:* Diagraphie resisti-vite; Microlaterolog; Microlog	ELEKTRO=LOG *UF:* Lateral=Log; Mikrolaterolog; Mikrolog	KAROTAZH EHLEKTRICHESKIJ	DIAGRAFIA=ELECTRICA *UF:* Microlog; Testificacion=de=microrresistividad; Testificacion=de=resistividad	DIAGRAFIA ELET-TRICA *UF:* Log laterale; Micro-laterolog; Microlog
1325 ELECTRIC POTENTIAL PHCH	Electric potential ELECTRICAL PROP-ERTIES	Potentiel electrique PROPRIETE ELEC-TRIQUE	Elektr.Potential ELEKTR. EIGENSCHAFT	POTENTSIAL EHLEKTRICHESKIJ	Potencial=electrico PROPIEDAD=ELECTRICA	Potenziale elettrico PROPRIETA ELET-TRICA
1326 ELECTRICAL CONDUCTIVITY PHCH	ELECTRICAL CON-DUCTIVITY	CONDUCTIVITE ELECTRIQUE	ELEKTR. LEITFAEHIGKEIT	EHLEKTROPRO-VODNOST'	CONDUCTIVIDAD=ELECTRICA	CONDUTTIVITA ELETTRICA
1327 ELECTRICAL EXPLORATION APPL	Electrical exploration ELECTRICAL METHODS; MIN-ERAL EXPLORA-TION	Prospection electrique METHODE ELEC-TRIQUE	Elektr.Prospektion ELEKTR.METHODE	EHLEKTRORAZ-VEDKA *UF:* Ehlektrometriya; Ground method; S'emka ehlektricheskaya	Prospeccion=electrica METODO=ELECTRICO	Prospezione geoelettrica METODO ELET-TRICO
1328 ELECTRICAL FIELD PHCH	ELECTRICAL FIELD	CHAMP ELEC-TRIQUE	ELEKTR.FELD	POLE EHLEKTRI-CHESKOE	CAMPO=ELECTRICO	CAMPO ELETTRICO
1329 ELECTRICAL METHOD APPL	ELECTRICAL METHODS *UF:* Electrical explora-tion; Resistivity method	METHODE ELEC-TRIQUE *UF:* Methode resistivite; Methode tellurique; Prospection electrique	ELEKTR.METHODE *UF:* Elektr.Prospektion; Tellurik; Widerstands=Messung	Ehlektrometriya EHLEKTRORAZ-VEDKA	METODO=ELECTRICO *UF:* Metodo=resistividad; Metodo=telurico; Prospeccion=electrica	METODO ELET-TRICO *UF:* Metodo della resis-tivita; Metodo tellurico; Prospezione geoelettrica
1330 ELECTRICAL POLARIZATION PHCH	Electric polarization ELECTRICAL PROP-ERTIES	Polarisation electrique PROPRIETE ELEC-TRIQUE	Elektr.Polarisation ELEKTR. EIGENSCHAFT	POLYARIZATSIYA EHLEKTRI-CHESKAYA	Polarizacion=electrica PROPIEDAD=ELECTRICA	Polarizzazione elettrica PROPRIETA ELET-TRICA
1331 ELECTRICAL PROPERTY PHCH	ELECTRICAL PROP-ERTIES *UF:* Electric polariza-tion; Electric potential	PROPRIETE ELEC-TRIQUE *UF:* Polarisation elec-trique; Potentiel elec-trique	ELEKTR. EIGENSCHAFT *UF:* Elektr.Polarisation; Elektr.Potential	SVOJSTVO EHLEK-TRICHESKOE	PROPIEDAD=ELECTRICA *UF:* Polarizacion=electrica; Potencial=electrico	PROPRIETA ELET-TRICA *UF:* Polarizzazione elettrica; Potenziale elettrico

	ENGLISH	FRANCAIS	DEUTSCH	RUSSKIJ	ESPANOL	ITALIANO
1332 ELECTRICAL SOUNDING APPL	ELECTRICAL SOUNDING	SONDAGE ELEC-TRIQUE	GEOELEKTR. SONDIERUNG	EHLEKTRO-ZONDIROVANIE	SONDEO= ELECTRICO	SONDAGGIO ELET-TRICO
1333 ELECTRICAL SURVEYS APPL	ELECTRICAL SUR-VEYS	LEVE ELECTRIQUE	ELEKTR. VERMESSUNG	S'emka ehlektri-cheskaya EHLEKTRORAZ-VEDKA	LEVANTAMIENTO=ELECTRICO	RILEVAMENTO ELETTRICO
1334 ELECTRICITY PHCH	ELECTRICITY	ELECTRICITE	ELEKTRIZITAET	EHLEKTRI-CHESTVO	ELECTRICIDAD	ELETTRICITA
1335 ELECTRO-OSMOSIS PHCH	ELECTRO-OSMOSIS	ELECTRO OSMOSE	ELEKTRO=OSMOSE	EHLEKTROOSMOS	ELECTROOSMOSIS	ELETTROOSMOSI
1336 ELECTROCHEMICAL PROPERTY PHCH	ELECTROCHEMI-CAL PROPERTIES	PROPRIETE ELEC-TROCHIMIQUE	ELEKTROCHEM. EIGENSCHAFT	SVOJSTVO EHLEK-TROKHIMI-CHESKOE	PROPIEDAD=ELECTROQUIMICA	PROPRIETA ELET-TROCHIMICA
1337 ELECTRODE APPL	ELECTRODES	ELECTRODE	ELEKTRODE	EHLEKTROD	ELECTRODO	ELETTRODO
1338 ELECTROLYSIS PHCH	ELECTROLYSIS	ELECTROLYSE	ELEKTROLYSE	EHLEKTROLIZ	ELECTROLISIS	ELETTROLISI
1339 ELECTROLYTE PHCH	ELECTROLYTES	ELECTROLYTE	ELEKTROLYT	EHLEKTROLIT	ELECTROLITO	ELETTROLITA
1340 ELECTROMAGNETIC FIELD SOLI	ELECTROMAG-NETIC FIELD *UF:* Electromagnetism	CHAMP ELECTRO-MAGNETIQUE *UF:* Electromagnetisme	ELEKTRO=MAGNET.FELD *UF:* Elektro=Magnetik	POLE EHLEKTRO-MAGNITNOE	CAMPO=ELECTROMAGNETICO *UF:* Electromagnetismo	CAMPO ELETTRO-MAGNETICO *UF:* Elettromagnetismo
1341 ELECTROMAGNETIC METHOD APPL	ELECTROMAG-NETIC METHODS	METHODE ELEC-TROMAGNETIQUE	ELEKTRO-MAGNET. METHODE	METOD EHLEK-TROMAGNITNYJ *UF:* S'emka ehlektro-magnitnaya	METODO=ELECTROMAGNETICO	METODO ELETTRO-MAGNETICO
1342 ELECTROMAGNETIC SURVEY APPL	ELECTROMAG-NETIC SURVEYS	LEVE ELECTRO-MAGNETIQUE	ELEKTRO-MAGNET. VERMESSUNG	S'emka ehlektromagnit-naya METOD EHLEK-TROMAGNITNYJ	LEVANTAMIENTO=ELECTROMAGNETICO	RILEVAMENTO ELETTROMAGNE-TICO
1343 ELECTROMAGNETICAL LOGGING APPL	ELECTROMAG-NETIC LOGGING	DIAGRAPHIE ELEC-TROMAGNETIQUE	ELEKTRO-MAGNET. BOHRLOCH=MESSUNG	EHLEKTRO-MAGNITNYJ KAROTAZH	DIAGRAFIA=ELECTROMAGNETICA	DIAGRAFIA ELET-TROMAGNETICA
1344 ELECTROMAGNETISM PHCH	Electromagnetism ELECTROMAG-NETIC FIELD	Electromagnetisme CHAMP ELECTRO-MAGNETIQUE	Elektro=Magnetik ELEKTRO-MAGNET.FELD	EHLEKTROMAGNE-TIZM	Electromagnetismo CAMPO=ELECTROMAGNETICO	Elettromagnetismo CAMPO ELETTRO-MAGNETICO
1345 ELECTRON PHCH	ELECTRONS	ELECTRON	ELEKTRON	EHLEKTRON	ELECTRON	ELETTRONE
1346 ELECTRON DIFFRACTION ANALYSIS METH	ELECTRON DIF-FRACTION ANALY-SIS	DIFFRACTION ELECTRON	ELEKTRONEN=BEUGUNGS=ANALYSE	EHLEKTRONOGRA-FIYA	ANALISIS=DIFRACCION=ELECTRON	ANALISI DIFFRAZ-IONE ELET-TRONICA

	ENGLISH	FRANCAIS	DEUTSCH	RUSSKIJ	ESPANOL	ITALIANO
1347 ELECTRON MICROPROBE METH	ELECTRON PROBE	MICROSONDE ELECTRONIQUE	MIKRO=SONDE	MIKROZOND EHLEKTRONNYJ *UF:* Electron probe data	MICROSONDA= ELECTRONICA	MICROSONDA ELETTRONICA
1348 ELECTRON MICROSCOPE INST	Electron microscope ELECTRON MICROSCOPY; INSTRUMENTS	Microscope electronique INSTRUMENTA- TION; MICROSCO- PIE ELEC- TRONIQUE	Elektronen=Mikroskop INSTRUMEN- TIERUNG; ELEKTRONEN= MIKROSKOPIE	MIKROSKOP EHLEKTRONNYJ	Microscopio= electronico INSTRUMENTA- CION; MICROSCOPIA= ELECTRONICA	Microscopio elettronico STRUMENTAZ- IONE; MICRO- SCOPIA ELET- TRONICA
1349 ELECTRON MICROSCOPY METH	ELECTRON MICROSCOPY *UF:* Electron microscope	MICROSCOPIE ELECTRONIQUE *UF:* Microscope elec- tronique	ELEKTRONEN= MIKROSKOPIE *UF:* Elektronen= Mikroskop	MIKROSKOPIYA EHLEKTRONNAYA *UF:* Electron microscopy data	MICROSCOPIA= ELECTRONICA *UF:* Microscopio= electronico	MICROSCOPIA ELETTRONICA *UF:* Microscopio elet- tronico
1350 ELECTRON MICROSCOPY DATA METH	ELECTRON MICROSCOPY DATA	DONNEE MICROS- COPIE ELEC- TRONIQUE	ELEKTRONEN= MIKROSKOP. DATEN	[ELECTRON MICROSCOPY DATA] DANNYE; MIKROSKOPIYA EHLEKTRONNAYA	DATO= MICROSCOPIA= ELECTRONICA	DATO MICROS- COPIO ELET- TRONICO
1351 ELECTRON PARAMAGNETIC RESONANCE PHCH	ELECTRON PARA- MAGNETIC RESO- NANCE	RPE	PARAMAGNET. ELEKTRONEN= RESONANZ	REZONANS PARA- MAGNITNYJ EHLEKTRONNYJ	RPE	RPE
1352 ELECTRON=PROBE DATA METH	ELECTRON PROBE DATA	DONNEE MICRO- SONDE ELEC- TRONIQUE	ELEKTRONEN= MIKRO=SONDEN- DATEN	[ELECTRON PROBE DATA] DANNYE; MIKROZOND EHLEKTRONNYJ	DATO= MICROSONDA= ELECTRONICA	DATO MICRO- SONDA ELET- TRONICA
1353 ELEMENTARY PARTICLE PHCH	ELEMENTARY PARTICLES	Particule elementaire MATIERE PAR- TICULAIRE	Elementar=Teilchen PARTIKEL	CHASTITSA EHLE- MENTARNAYA	Materia=elemental PARTICULA= ELEMENTAL	Particella elementare PARTICELLA
1354 ELEPHANTOIDEA PALS	ELEPHANTOIDEA	ELEPHANTOIDEA	ELEPHANTOIDEA	Elephantoidea PROBOSCIDEA	ELEPHANTOIDEA	ELEPHANTOIDEA
1355 ELONGATE MINERAL TEST	ELONGATE MINER- ALS	MINERAL ALLONGE	NADELIG.MINERAL	[ELONGATE MIN- ERAL] GABITUS; PRIZMATICHESKIJ	MINERAL= ALARGADO	MINERALE ALLUN- GATO
1356 ELUTRIATION SEDI	Elutriation PREPARATION	Elutriation PREPARATION	Ausschlaemmung PRAEPARATION	Otmuchivanie GRANULOMETRIYA	Elutriacion PREPARACION	Elutriazione PREPARAZIONE
1357 ELUVIUM SURF	ELUVIUM	ELUVION	ELUVIUM	EHLYUVIJ *UF:* Stone line	ELUVION	ELUVIUM
1358 EMBANKMENT ENGI	EMBANKMENTS	REMBLAI	VERFUELL- MATERIAL	NASYP'	RELLENO- TERRAPLEN	ARGINE
1359 EMBRECHITE IGMS	EMBRECHITE	EMBRECHITE	EMBRECHIT	Ehmbrekhit MIGMATIT	EMBRECHITA	EMBRECHITE
1360 EMBRITHOPODA PALS	EMBRITHOPODA	EMBRITHOPODA	EMBRITHOPODA	EMBRITHOPODA	EMBRITHOPODA	EMBRITHOPODA

	ENGLISH	FRANCAIS	DEUTSCH	RUSSKIJ	ESPANOL	ITALIANO
1361 EMERGENCE SURF	EMERGENCE	EMERSION	EMERGENZ	NASTUPLENIE SUSHI	EMERSION	EMERSIONE
1362 EMISSION SPECTROSCOPY METH	EMISSION SPEC-TROSCOPY	SPECTROMETRIE EMISSION	EMISSIONS-SPEKTROMETRIE	ANALIZ SPEK-TRAL'NYJ EHMIS-SIONNYJ	ESPECTROMETRIA=DE=EMISION	SPETTROMETRIA D'EMISSIONE
1363 EMPLACEMENT IGNE	EMPLACEMENT	MISE EN PLACE	PLATZNAHME	Vnedrenie INTRUZIYA	EMPLAZAMIENTO	MESSA IN POSTO
1364 EMSIAN STRS	EMSIAN	EMSIEN	EMS	EHMS	EMSIENSE	EMSIANO
1365 EN ECHELON STRU	EN ECHELON FAULTS	FAILLE EN ECHE-LON	STAFFEL=BRUCH	STUPENCHATYJ	FALLA=EN=ESCALON	FAGLIE A GRADINATA
1366 ENCRINITE SEDS	Encrinite LIMESTONE	Calcaire entroques CALCAIRE BIO-CLASTIQUE	Encrinit BIOKLAST.KALK	Ehnkrinit IZVESTNYAK; BIOLIT	Encrinita CALIZA=BIOCLASTICA	Encrinite BIOCALCARENITE
1367 END PRODUCT ISOT	End product RADIOACTIVITY; ISOTOPES	Produit final RADIOACTIVITE	End=Produkt RADIOAKTIVITAET	PRODUKT KONECHNYJ	Producto=final RADIOACTIVIDAD	Prodotto finale RADIOATTIVITA
1368 ENDEMIC POPULATION PALE	ENDEMIC TAXA	POPULATION ENDEMIQUE	ENDEM. POPULATION	EHNDEMICHNYJ	POBLACION=ENDEMICA	POPOLAZIONE ENDEMICA
1369 ENDOCERATOIDEA PALS	Endoceratoidea NAUTILOIDEA	Endoceratoidea NAUTILOIDEA	Endoceratoidea NAUTILOIDEA	ENDOCERATOIDEA	Endoceratoidea NAUTILOIDEA	Endoceratoidea NAUTILOIDEA
1370 ENDOGENE PROCESS ECON	ENDOGENE PRO-CESSES	PROCESSUS ENDO-GENE	ENDOGEN. VORGANG	EHNDOGENNYJ	PROCESO=ENDOGENO	EVOLUZIONE IPO-GENICA
1371 ENDORHEISM GEOH	Endorheism DRAINAGE PAT-TERNS	Endoreisme HYDROLOGIE SUR-FACE	Abflusslosigkeit HYDROLOGIE	DRENAZH (VNUTRENNIJ)	Cuenca=endorreica HIDROLOGIA=SUPERFICIE	Endoreismo IDROLOGIA
1372 ENDOTHERMIC REACTION PHCH	ENDOTHERMIC REACTIONS	REACTION ENDO-THERMIQUE	ENDOTHERM. REAKTION	Reaktsiya ehndotermi-cheskaya TERMODINAMIKA	REACCION=ENDOTERMICA	REAZIONE ENDOTERMICA
1373 ENERGY SOURCE ECON	ENERGY SOURCES	ENERGIE *UF:* Combustible fossile	ENERGIE *UF:* Fossil.Brennstoff	EHNERGIYA	ENERGIA *UF:* Fuel=fosil	ENERGIA *UF:* Combustibile fossile
1374 ENGINEERING GEOLOGY ENGI	ENGINEERING GEOLOGY *UF:* Earthquake engi-neering; Geotechnics; Mining engineering	Geologie ingenieur GEOTECHNIQUE	Ingenieur=Geologie GEOTECHNIK	GEOLOGIYA INZ-HENERNAYA *UF:* Geotekhnika; Karta inzhenerno=geologicheskaya	Geologia=del=ingeniero GEOTECNIA	Geologia tecnica GEOTECNICA
1375 ENGINEERING WORKS ENGI	STRUCTURES *UF:* Construction	OUVRAGE *UF:* Construction	INGENIEUR=BAU *UF:* Konstruktion; Tief-bau	SOORUZHENIE *UF:* Asejsmichnoe sooruzhenie; Fluid injection; Marine installation; Public works; Submarine installation; Under-ground canalization	OBRAS *UF:* Construccion	OPERA INGEGNERISTICA *UF:* Costruzione

	ENGLISH	FRANCAIS	DEUTSCH	RUSSKIJ	ESPANOL	ITALIANO
1376 ENRICHMENT ECON	ENRICHMENT	ENRICHISSEMENT	ANREICHERUNG	OBOGASHCHENIE	ENRIQUECI- MIENTO	ARRICCHIMENTO
1377 ENSIALIC GEOSYNCLINE STRU	Ensialic geosynclines GEOSYNCLINES	Geosynclinal ensialique GEOSYNCLINAL	Ensial.Geosynklinale GEOSYNKLINALE	GEOSINKLINAL' EHNSIALI- CHESKAYA	Geosinclinal=sialico GEOSINCLINAL	Geosinclinale intracra- tonico GEOSINCLINALE
1378 ENSIMATIC GEOSYNCLINE STRU	Ensimatic geosynclines GEOSYNCLINES	Geosynclinal ensima- tique GEOSYNCLINAL	Ensimat.Geosynklinale GEOSYNKLINALE	GEOSINKLINAL' EHNSIMATI- CHESKAYA	Geosinclinal=simatico GEOSINCLINAL	Geosinclinale intercra- tonico GEOSINCLINALE
1379 ENSTATITE CHONDRITE EXTS	Enstatite chondrites CHONDRITES	Chondrite a enstatite CHONDRITE	Enstatithaltig.Chondrit CHONDRIT	[ENSTATITE CHON- DRITE] KHONDRIT	Condrita=enstatita CONDRITA	Condrite enstatitica CONDRITE
1380 ENTEROLITHIC FOLD STRU	ENTEROLITHIC FOLDS	PLI ENTEROLITHIQUE	ENTEROLITH. FALTE	Skladka ehnteroliti- cheskaya SKLADKA; TEKS- TURA	PLIEGUE= ENTEROLITICO	PIEGA ENTEROLI- TICA
1381 ENTEROPNEUSTA PALS	ENTEROPNEUSTA	ENTEROPNEUSTA	ENTEROPNEUSTA	ENTEROPNEUSTA	ENTEROPNEUSTA	ENTEROPNEUSTA
1382 ENTHALPY PHCH	ENTHALPY	ENTHALPIE	ENTHALPIE	EHNTAL'PIYA	ENTALPIA	ENTALPIA
1383 ENTISOL SUSS	ENTISOLS *UF:* Aquents; Fluvents; Orthents; Psamments	ENTISOL *UF:* Aquent; Fluvent; Orthent; Psamment	ENTISOL *UF:* Aquent; Fluvent; Orthent; Psamment	EHNTISOL *UF:* Akvent; Flyuvent; Ortent; Psamment	ENTISOL *UF:* Aquent; Fluvent; Orthent; Psamment	ENTISOL *UF:* Aquent; Fluvent; Orthent; Psamment
1384 ENTRENCHED STREAM SURF	Entrenched streams STREAMS; VAL- LEYS	Cours eau encaisse CANYON	[ENTRENCHED STREAM] CANYON	Reka vrezannaya REKA	Curso=encajado CAÑON	Corso d'acqua incassato CANYON
1385 ENTROPY PHCH	ENTROPY	ENTROPIE	ENTROPIE	EHNTROPIYA	ENTROPIA	ENTROPIA
1386 ENVIRONMENT ENVI	ENVIRONMENT	MILIEU	MILIEU	SREDA OKRUZHAYUSH- CHAYA *UF:* Human ecology; Paleosreda; Sreda nazemnaya	MEDIO=AMBIENTE	AMBIENTE
1387 ENVIRONMENTAL ANALYSIS SEDI	ENVIRONMENTAL ANALYSIS	ETUDE MILIEU	MILIEU=STUDIE	ANALIZ FATSI- AL'NYJ	ESTUDIO=MEDIO	STUDIO AMBIEN- TALE
1388 ENVIRONMENTAL GEOLOGY ENVI	ENVIRONMENTAL GEOLOGY	GEOLOGIE ENVIRONNEMENT	UMWELT= GEOLOGIE	GEOLOGIYA OKRUZHAYUSH- CHEJ SREDY *UF:* Geologiya meditsin- skaya	GEOLOGIA=DEL= MEDIO=AMBIENTE	GEOLOGIA AMBIENTALE
1389 EOCENE STRS	EOCENE	EOCENE	EOZAEN	EHOTSEN *UF:* Ehotsen nizhnij; Ehotsen srednij; Ehotsen verkhnij	EOCENO	EOCENE
1390 EOCRINOIDEA PALS	EOCRINOIDEA	EOCRINOIDEA	EOCRINOIDEA	EOCRINOIDEA	EOCRINOIDEA	EOCRINOIDEA

	ENGLISH	FRANCAIS	DEUTSCH	RUSSKIJ	ESPANOL	ITALIANO
1391 EOLIAN ENVIRONMENT PALE	Eolian environment EOLIAN SEDIMEN- TATION	Milieu eolien SEDIMENTATION EOLIENNE	Aeol.Milieu AEOL. SEDIMENTATION	Sreda eholovaya SEDIMENTATSIYA EHOLOVAYA; VETER	Medio=eolico SEDIMENTACION= EOLICA	Ambiente eolico SEDIMENTAZIONE EOLICA
1392 EOLIAN FEATURE SURF	EOLIAN FEATURES	MORPHOLOGIE EOLIENNE	AEOL. MORPHOLOGIE	[EOLIAN FEA- TURE] FORMA REL'EFA; PUSTYNYA	MORFOLOGIA= EOLICA	MORFOLOGIA EOLICA
1393 EOLIAN SEDIMENTATION SEDI	EOLIAN SEDIMEN- TATION UF: Atmogenic; Eolian environment	SEDIMENTATION EOLIENNE UF: Milieu eolien; Ori- gine atmogenique; Transport eolien	AEOL. SEDIMENTATION UF: Aeol.Milieu; Atmogen.Herkunft	SEDIMENTATSIYA EHOLOVAYA UF: Sreda eholovaya; Wind transport	SEDIMENTACION= EOLICA UF: Medio=eolico; Origen=atmogenico; Transporte=eolico	SEDIMENTAZIONE EOLICA UF: Ambiente eolico; Origine atmogena; Trasporto eolico
1394 EON STRA	Eons STRATIGRAPHIC UNITS	Eon ECHELLE STRATI- GRAPHIQUE	Aeon STRATIGRAPH. SKALA	EHON	Eon ESCALA= ESTRATIGRAFICA	Eone SCALA STRATI- GRAFICA
1395 EOSUCHIA PALS	EOSUCHIA	EOSUCHIA	EOSUCHIA	EOSUCHIA	EOSUCHIA	EOSUCHIA
1396 EPEIROGENY STRU	EPEIROGENY	EPIROGENESE	EPIROGENESE	EHPEJROGENEZ	EPIROGENESIS	EPIROGENESI
1397 EPHEMERAL STREAM SURF	EPHEMERAL STREAMS	RUISSEAU TEM- PORAIRE	KURZDAUERND. FLUSS	POTOK VREMEN- NYJ	CORRIENTE= TEMPORAL	RUSCELLO TEMPO- RANEO
1398 EPIBIOTISM PALE	EPIBIOTISM UF: Commensalism	EPIBIOTISME UF: Commensalisme	EPIBIOTISMUS UF: Kommensalismus	[EPIBIOTISM] EHKOLOGIYA	EPIBIOTISMO UF: Comensalismo	EPIBIOSI UF: Commensalismo
1399 EPIBOLITE IGMS	Epibolite METAMORPHIC ROCKS	Epibolite MIGMATITE	Epibolit MIGMATIT	Ehpibolit MIGMATIT	Epibolita MIGMATITA	Epibolite MIGMATITE
1400 EPICENTER SOLI	EPICENTERS	EPICENTRE	EPIZENTRUM	EHPITSENTR	EPICENTRO	EPICENTRO
1401 EPIDOTE=AMPHIBOLITE FACIES IGNE	EPIDOTE= AMPHIBOLITE FACIES	FACIES AMPHIBO- LITE EPIDOTE	EPIDOT- AMPHIBOLIT= FAZIES	FATSIYA EHPIDOT= AMFIBOLITOVAYA UF: Fatsiya al'bit= ehpidot=amfibolitovaya	FACIES= ANFIBOLITA= EPIDOTA	FACIES ANFIBO- LITE A EPIDOTO
1402 EPIDOTIZATION IGNE	EPIDOTIZATION	Epidotisation ALTERATION HYDROTHERMALE	Epidotisierung HYDROTHERMAL= UMWANDLUNG	EHPIDOTIZATSIYA	Epidotizacion ALTERACION= HIDROTERMAL	Epidotizzazione ALTERAZIONE IDROTERMALE
1403 EPIEUGEOSYNCLINE STRU	Epieugeosynclines GEOSYNCLINES	Epieugeosynclinal GEOSYNCLINAL	Epieugeosynklinale GEOSYNKLINALE	EHPIEHVGEOSINK- LINAL'	Epieugeosinclinal GEOSINCLINAL	Epieugeosinclinale GEOSINCLINALE
1404 EPIGENESIS GEOL	EPIGENE PRO- CESSES	EPIGENESE UF: Metagenese	EPIGENESE UF: Metagenese	EHPIGENEZ UF: Pozdnij diagenez	EPIGENESIS UF: Metagenesis	EPIGENESI UF: Metagenesi
1405 EPIMAGMATIC ORIGIN IGNE	Epimagatic origin MAGMAS; GENESIS	Origine epimagmatique GENESE	Epimagmat.Genese GENESE	EHPIMAGMATI- CHESKIJ	Origen=epimagmatico GENESIS	Origine epimagmatica GENESI
1406 EPITAXY MINE	EPITAXY	EPITAXIE	EPITAXIE	EHPITAKSIYA	EPITAXIA	EPITASSIA

	ENGLISH	FRANCAIS	DEUTSCH	RUSSKIJ	ESPANOL	ITALIANO
1407 EPITHERMAL DEPOSIT ECON	EPITHERMAL PROCESSES	GITE EPITHERMAL	EPITHERMAL. VORKOMMEN	Ehpitermal'noe mestorozhdenie MESTOROZHDENIE MAGMATICHESKOE	YACIMIENTO= EPITERMAL	GIACIMENTO EPITERMALE
1408 EPIZONE IGNE	EPIZONAL METAMORPHISM	METAMORPHISME EPIZONAL	EPIZONE	EHPIZONA	METAMORFISMO= EPIZONAL	EPIZONA
1409 EPOCH STRA	Epochs STRATIGRAPHIC UNITS	Epoque ECHELLE STRATIGRAPHIQUE	Epoche STRATIGRAPH. SKALA	EHPOKHA	Epoca ESCALA= ESTRATIGRAFICA	Epoca SCALA STRATIGRAFICA
1410 EPR SPECTRUM PHCH	EPR SPECTRA	SPECTRE RPE	PARAMAGN. ELEKTR.RESON. SPEKTRUM	SPEKTR EHPR	ESPECTRO-RPE	SPETTRO RPE
1411 EQUATION MATH	EQUATIONS	EQUATION MATHEMATIQUE *UF:* Fonction	MATHEMAT. GLEICHUNG *UF:* Funktion	URAVNENIE	ECUACION= MATEMATICA *UF:* Funcion	EQUAZIONE *UF:* Funzione
1412 EQUATION OF STATE PHCH	EQUATIONS OF STATE	EQUATION ETAT	ZUSTANDS= GLEICHUNG	Uravnenie sostoyaniya RAVNOVESIE FAZOVOE; SOSTOYANIE (FIZICHESKOE)	RAZON=DE= ESTADO	EQUAZIONE DI STATO
1413 EQUATOR MISC	Equator EQUATORIAL REGION	Ligne equatoriale ZONE EQUATORIALE	Aequator AEQUATORIAL= ZONE	EHKVATOR	Linea=ecuatorial ZONA= ECUATORIAL	Equatore AMBIENTE EQUATORIALE
1414 EQUATORIAL ZONE SURF	EQUATORIAL REGION *UF:* Equator	ZONE EQUATORIALE *UF:* Ligne equatoriale	AEQUATORIAL= ZONE *UF:* Aequator	ZONA EHKVATORIAL'NAYA	ZONA= ECUATORIAL *UF:* Linea=ecuatorial	AMBIENTE EQUATORIALE *UF:* Equatore
1415 EQUIGRANULAR TEXTURE TEST	Equigranular texture TEXTURES	Texture equigranulaire TEXTURE	Gleichkoernig.Gefuege KORN=GEFUEGE	STRUKTURA RAVNOMERNOZERNISTYJ	Textura=isogranular TEXTURA	Tessitura equigranulare TESSITURA
1416 EQUILIBRIUM PHCH	EQUILIBRIUM *UF:* Dynamic equilibrium	EQUILIBRE *UF:* Equilibre dynamique	GLEICHGEWICHT	RAVNOVESIE	EQUILIBRIO	EQUILIBRIO *UF:* Equilibrio dinamico
1417 EQUISETALES PALS	Equisetales PTERIDOPHYTES	Equisetales PTERIDOPHYTA	Equisetales PTERIDOPHYTA	ARTHROPSIDA	Equisetales PTERIDOPHYTA	Equisetales PTERIDOFITE
1418 ERA STRA	Eras STRATIGRAPHIC UNITS	Ere ECHELLE STRATIGRAPHIQUE	Aera STRATIGRAPH. SKALA	EHRA	Era ESCALA= ESTRATIGRAFICA	Era SCALA STRATIGRAFICA
1419 ERATHEM STRA	Erathems STRATIGRAPHIC UNITS	Eratheme ECHELLE STRATIGRAPHIQUE	Aerathem STRATIGRAPH. SKALA	EHRATEMA	Eratema ESCALA= ESTRATIGRAFICA	Eratema SCALA STRATIGRAFICA
1420 ERATOSTHENIAN EXTR	ERATOSTHENIAN	Eratosthenien ECHELLE STRATIGRAPHIQUE; LUNE	Eratosthen.Alter STRATIGRAPH. SKALA; MOND	EHRATOSFENSKIJ	Eratostensiense ESCALA= ESTRATIGRAFICA; LUNA	Eratosteniano SCALA STRATIGRAFICA; LUNA
1421 ERBIUM CHEE	ERBIUM	ERBIUM	ER	EHRBIJ	ERBIO	ERBIO

	ENGLISH	FRANCAIS	DEUTSCH	RUSSKIJ	ESPANOL	ITALIANO
1422 ERG SURF	Ergs DUNES	Erg DUNE CONTINEN- TALE	Erg KONTINENTAL= DUENE	Ehrg DYUNA; PUS- TYNYA	Erg DUNA= CONTINENTAL	Erg DUNA CONTINEN- TALE; DESERTO
1423 ERODIBILITY SURF	ERODIBILITY	ERODABILITE	ERODIERBARKEIT	[ERODIBILITY] EHROZIYA; SVOJS- TVO	RESISTENCIA=A= LA=EROSION	ERODIBILITA
1424 EROSION SURF	EROSION *UF:* Marine erosion; Scouring; Sheet erosion	EROSION *UF:* Barriere rocheuse; Decapage; Erosion en nappe; Erosion marine; Monolithe; Sillon erosion	EROSION *UF:* Abschwemmung; Furche; Gesteins= riegel; Marin.Erosion; Monolith; Scouring	EHROZIYA *UF:* Bedlend; Ehroziya pochv; Erodibility; Erosion control; Ero- sion feature; Inversiya rel'efa; Planatsiya	EROSION *UF:* Burilado; Erosion= marina; Erosion=por= arroyada; Estria; Monolito; Umbral= rocoso	EROSIONE *UF:* Barriera rocciosa; Dilavamento; Erosione marina; Monolito; Solco d'erosione; Stria- mento
1425 EROSION CONTROL SURF	EROSION CON- TROL	CONTROLE ERO- SION	EROSIONS= EINFLUSS	[EROSION CON- TROL] EHROZIYA; KON- TROL'	CONTROL= EROSION	CONTROLLO DELL'EROSIONE
1426 EROSION CYCLE SURF	EROSION CYCLE	CYCLE EROSION	EROSIONS= ZYKLUS	TSIKL EHROZION- NYJ	CICLO=DE= EROSION	CICLO D'EROSIONE
1427 EROSION FEATURE SURF	EROSION FEA- TURES *UF:* Erosion remnant; Monolith	MORPHOLOGIE EROSION *UF:* Temoin erosion	EROSIONS= MORPHOLOGIE	[EROSION FEA- TURE] EHROZIYA; FORMA REL'EFA	MORFOLOGIA= EROSION *UF:* Relicto=erosion	MORFOLOGIA D'EROSIONE *UF:* Relitto d'erosione
1428 EROSION REMNANT SURF	Erosion remnant EROSION FEA- TURES	Temoin erosion MORPHOLOGIE EROSION	Erosions=Rest EROSIONS= OBERFLAECHE	OSTANETS EHRO- ZIONNYJ *UF:* Gora ostrovnaya	Relicto=erosion MORFOLOGIA= EROSION	Relitto d'erosione MORFOLOGIA D'EROSIONE
1429 EROSION SURFACE SURF	EROSION SUR- FACES	SURFACE EROSION	EROSIONS= OBERFLAECHE *UF:* Erosions=Rest	POVERKHNOST' EHROZIONNAYA	SUPERFICIE=DE= EROSION	SUPERFICIE D'EROSIONE
1430 ERRATIC BLOCK SEDI	ERRATICS *UF:* Indicative boulder	BLOC ERRATIQUE *UF:* Bloc indicateur	ERRAT.BLOCK *UF:* Leit=Geschiebe	VALUN EHRRATI- CHESKIJ	BLOQUE= ERRATICO *UF:* Bloque=indicador	MASSO ERRATICO *UF:* Erratico indicatore
1431 ERROR MATH	ERRORS	ERREUR	FEHLER	OSHIBKA	ERROR	ERRORE
1432 ERUPTION IGNE	ERUPTIONS *UF:* Hydroexplosion	ERUPTION *UF:* Explosion phrea- tique	ERUPTION *UF:* Hydro=Explosion	IZVERZHENIE	ERUPCION *UF:* Hidroexplosion	ERUZIONE *UF:* Idroesplosione
1433 ESCARPMENT SURF	SCARPS *UF:* Beach scarps	ESCARPEMENT	[ESCARPMENT]	USTUP *UF:* Ustup sbrosovyj	ESCARPE	SCARPATA D'EROSIONE
1434 ESKER SURF	ESKERS	ESKER	ESKER	OZ	ESKER	ESKER
1435 ESSEXITE IGNS	Essexite DIORITES; FELDS- PATHOID COMPO- SITION	Essexite DIORITE; ROCHE FELDS- PATHOIDIQUE	Essexit FOID=GESTEIN	EHSSEKSIT	Essexita DIORITA; ROCA= FELDESPATOIDICA	Essexite DIORITE; ROCCIA A FELDSPATOIDI
1436 ESTUARINE ENVIRONMENT PALE	ESTUARINE ENVI- RONMENT	MILIEU ESTUAIRE	AESTUAR=MILIEU	FATSIYA EHSTUA- RIEVAYA	MEDIO=ESTUARIO	AMBIENTE D'ESTUARIO

	ENGLISH	FRANCAIS	DEUTSCH	RUSSKIJ	ESPANOL	ITALIANO
1437 ESTUARINE SEDIMENTATION SEDI	ESTUARINE SEDI-MENTATION	SEDIMENTATION ESTUAIRE	AESTUAR=SEDIMENTATION	SEDIMENTATSIYA EHSTUARIEVAYA	SEDIMENTACION=ESTUARIO	SEDIMENTAZIONE D'ESTUARIO
1438 ESTUARY SURF	ESTUARIES	ESTUAIRE	AESTUAR	EHSTUARIJ	ESTUARIO	ESTUARIO
1439 ETCHING METH	ETCHING	ATTAQUE CHIMIQUE *UF:* Chloruration; Fluoruration	AETZUNG *UF:* Chlorinierung; Fluorinierung	TRAVLENIE	ATAQUE=QUIMICO *UF:* Cloracion; Fluoruracion	ATTACCO CHIMICO *UF:* Clorurazione; Fluorurazione
1440 ETHANE CHES	ETHANE	ETHANE	AETHAN	EHTAN	ETANO	ETANO
1441 ETHERS CHES	ETHERS	Ether MATIERE ORGANIQUE	Aether ORGAN.SUBSTANZ	Ehfir VESHCHESTVO ORGANICHESKOE	Eter MATERIA=ORGANICA	Etere MATERIA ORGANICA
1442 EUCILIATA PALS	Euciliata PROTISTA	Euciliata PROTOZOA	Euciliata PROTOZOA	EUCILIATA	Euciliata PROTOZOA	Euciliata PROTOZOA
1443 EUCRITE EXTS	Eucrite ACHONDRITES	Eucrite achondrite ACHONDRITE	Eucrit ACHONDRIT	Ehvkrit AKHONDRIT	Eucrita ACONDRITA	Eucrite (meteorite) ACONDRITE
1444 EUCRYSTALLINE TEXTURE TEST	Eucrystalline texture TEXTURES	Texture grenue ROCHE GRENUE	Eukristallin.Gefuege PLUTONIT	Yavnokristallicheskij POLNOKRISTALLI-CHESKIJ	Textura=granuda ROCA=GRANUDA	Tessitura eucristallina ROCCIA PLU-TONICA
1445 EUGEOSYNCLINE STRU	EUGEOSYNCLINES	EUGEOSYNCLINAL	EUGEOSYNKLI-NALE	EHVGEOSINKLI-NAL'	EUGEOSINCLINAL	EUGEOSINCLI-NALE
1446 EUGLENOPHYTA PALS	Euglenophyta ALGAE	Euglenophycophyta ALGAE	Euglenophyta ALGAE	EUGLENOPHYTA	Euglenophyta ALGAE	Euglenophytha ALGAE
1447 EUHEDRAL CRYSTAL MINE	Euhedral crystals TEXTURES	Cristal idiomorphe TEXTURE	Idiomorph.Kristall KORN=GEFUEGE	Ehvgedral'nyj IDIOMORFNYJ	Cristal=idiomorfo TEXTURA	Cristallo euedrale TESSITURA
1448 EUROPIUM CHEE	EUROPIUM	EUROPIUM	EU	EVROPIJ	EUROPIO	EUROPIO
1449 EURYHALINE ORGANISM PALE	Euryhaline taxa ECOLOGY; SALIN-ITY	Organisme euryhalin ECOLOGIE; SALI-NITE	Euryhalin.Organismus OEKOLOGIE; SALINITAET	EHVRIGALINNYJ	Eurihalino ECOLOGIA; SALINI-DAD	Organismo eurialino ECOLOGIA; SALINITA
1450 EURYTHERMAL ORGANISM PALE	Eurythermal taxa ECOLOGY; TEM-PERATURE	Organisme eurytherme ECOLOGIE; TEM-PERATURE	Eurythermal.Organismus OEKOLOGIE; TEMP-ERATUR	EHVRITERMNYJ	Euritermo ECOLOGIA; TEMP-ERATURA	Organismo euritermico ECOLOGIA; TEMP-ERATURA
1451 EUSELACHII PALS	EUSELACHII	EUSELACHII	EUSELACHII	Euselachii ELASMOBRANCHII	EUSELACHII	EUSELACHII
1452 EUSTATISM STRU	EUSTACY	EUSTATISME	EUSTATIK	EHVSTAZIYA	EUSTATISMO	EUSTATISMO
1453 EUTAXITIC TEXTURE TEST	Eutaxitic texture TEXTURES	Texture eutaxique TEXTURE	Taxit.Gefuege KORN=GEFUEGE	EHVTAKSITOVYJ	Textura=eutaxitica TEXTURA	Tessitura eutassitica TESSITURA

	ENGLISH	FRANCAIS	DEUTSCH	RUSSKIJ	ESPANOL	ITALIANO
1454 EUTECTIC CONDITION PHCH	Eutectic conditions PHASE EQUILIBRIA	Systeme eutectique DIAGRAMME EQUILIBRE	Eutektikum PHASEN= GLEICHGEWICHT	EHVTEKTIKA	Eutectica DIAGRAMA= EQUILIBRIO	Eutettico DIAGRAMMA D'EQUILIBRIO
1455 EUTROPHICATION ENVI	EUTROPHICATION	EUTROPHISATION	EUTROPHIERUNG	EHVTROFIKATSIYA	EUTROFIZACION	EUTROFIZZAZ- IONE
1456 EUXINIC ENVIRONMENT PALE	Euxinic environment ANAEROBIC ENVI- RONMENT	Milieu euxinique MILIEU ANAERO- BIE	Euxin.Milieu ANAEROB.MILIEU	EHVKSINNYJ	Medio=euxinico MEDIO= ANAEROBIO	Ambiente euxinico AMBIENTE ANAEROBICO
1457 EVALUATION MISC	EVALUATION	Evaluation QUALITE	Bewertung QUALITAET	OTSENKA	Evaluacion CALIDAD	Stima QUALITA
1458 EVAPORATION GEOH	EVAPORATION	EVAPORATION	EVAPORATION	ISPARENIE	EVAPORACION	EVAPORAZIONE
1459 EVAPORITE SEDS	EVAPORITES UF: Hydatogenesis	EVAPORITE UF: Hydatogenese; Sylvinite	EVAPORIT UF: Hydatogenese; Sylvinit	EHVAPORIT	EVAPORITA UF: Hidatogenesis; Silvinita	EVAPORITE UF: Idatogenesi; Silvi- nite
1460 EVAPOTRANSPIRATION GEOH	EVAPOTRANSPIRA- TION	EVAPOTRANSPIRA- TION	EVAPOTRANSPIRA- TION	ISPARENIE SUM- MARNOE	EVAPOTRANSPIRA- CION	EVAPOTRASPIRAZ- IONE
1461 EVOLUTION PALE	BIOLOGIC EVOLU- TION UF: Darwinism; Geno- morphs; Homology; Lamarckism; Mutation; Nomogenesis; Parallel- ism; Prototype	EVOLUTION BIOLOGIQUE UF: Archetype; Darwin- isme; Fixisme; Geno- morphe; Homologie; Lamarckisme; Muta- tion; Nomogenese; Parallelisme; Rayonne- ment adaptative	BIOLOG. EVOLUTION UF: Adaptiv.Divergenz; Darwinismus; Fixismus; Genomorph; Homolo- gie; Lamarckismus; Mutation; Nomogenese; Parallelitaet; Prototyp	EHVOLYUTSIYA UF: Chemical evolution; Darvinizm; Genomorfa; Nasledstvennost'	EVOLUCION= BIOLOGICA UF: Darwinismo; Fijismo; Genomorfo; Homologia; Mutacion; Nomogenesis; Para- lelismo; Prototipo; Radiacion=adaptable; Teoria=de=lamarck	EVOLUZIONE BIOLOGICA UF: Darwinismo; Fis- sismo; Genomorfo; Lamarckismo; Mutaz- ione; Nomogenesi; Omologia; Parallelismo; Prototipo; Radiazione adattativa
1462 EXCAVATION ENGI	EXCAVATIONS	FOUILLE GENIE CIVIL	AUSSCHACHTUNG	EHKSKAVATSIYA	EXCAVACION	SCAVO
1463 EXCESS ARGON ISOT	Excess argon ARGON; ISOTOPES	Exces argon ARGON	Ueberschuss=Argon AR	ARGON IZBYTOCH- NYJ	Exceso=argon ARGON	Argon in eccesso ARGO
1464 EXCHANGE CAPACITY PHCH	Exchange capacity ION EXCHANGE	Capacite echange ECHANGE ION	Austausch=Kapazitaet IONEN= AUSTAUSCH	Sposobnost' obmennaya OBMEN IONNYJ	Capacidad=de=cambio CAMBIO=IONICO	Capacita di scambio SCAMBIO IONICO
1465 EXCURSION GEOL	FIELD TRIPS	EXCURSION	EXKURSION	EHKSKURSIYA	EXCURSION	ESCURSIONE
1466 EXFOLIATION SURF	EXFOLIATION	DESQUAMATION	EXFOLIATION	DESKVAMATSIYA	EXFOLIACION	DESQUAMAZIONE
1467 EXHALATIVE PROCESS ECON	EXHALATIVE PRO- CESSES	GITE EXHALATIF	EXHALATIONS= LAGERSTAETTE	EHKSGALYATSIYA	YACIMIENTO= EXHALATIVO	GIACIMENTO DI ESALAZIONE
1468 EXINE PALE	EXINE	EXINE	EXINE	Ehkzina SPORA	EXINA	EXINE
1469 EXINITE SEDS	Exinite MACERALS	Exinite MACERAL	Exinit MAZERAL	EHKZINIT	Exinita MACERAL	Exinite MACERALE
1470 EXOBIOLOGY EXTR	EXOBIOLOGY	EXOBIOLOGIE	EXOBIOLOGIE	EHKZOBIOLOGIYA	EXOBIOLOGIA	ESOBIOLOGIA

	ENGLISH	FRANCAIS	DEUTSCH	RUSSKIJ	ESPANOL	ITALIANO
1471 EXOGENE PROCESS ECON	EXOGENE PROCESSES *UF:* Boxwork texture; Kanga	PROCESSUS EXOGENE *UF:* Kanga; Texture cloisonnee	EXOGEN.VORGANG *UF:* Kanga; Zellen=Gefuege	EHKZOGENNYJ	PROCESO=EXOGENO *UF:* Kanga; Textura=tabicada	EVOLUZIONE SUPERGENICA *UF:* Kanga; Tessitura reticolare
1472 EXORHEISM GEOH	Exorheism HYDROLOGY	Exoreisme HYDROLOGIE SURFACE	Extern.Entwaesserung HYDROLOGIE	Vneshnij drenazh GIDROLOGIYA; STOK	Exorreismo HIDROLOGIA=SUPERFICIE	Esoreismo IDROLOGIA
1473 EXOTHERMIC REACTION PHCH	Exothermic reactions THERMODYNAMIC PROPERTIES	Reaction exothermique THERMODYNAMIQUE	Exotherm.Reaktion THERMODYNAMIK	Reaktsiya ehkzotermi-cheskaya TERMODINAMIKA	Reaccion=exotermica TERMODINAMICA	Reazione esotermica TERMODINAMICA
1474 EXPANDABLE MATERIAL COMS	EXPANSIVE MATERIALS	MATERIAU EXPANSE	EXPANDIEREND. MATERIAL	MATERIAL VSPUCHIVAYUSH-CHIJSYA	MATERIAL=DILATABLE	MATERIALE ESPANDIBILE
1475 EXPANDING EARTH SOLI	EXPANDING EARTH	EXPANSION	EXPANSION	GIPOTEZA RASSHIRYAYUSH-CHAYASYA ZEMLI	EXPANSION	ESPANSIONE TERRESTRE
1476 EXPEDITION-CRUISE GEOL	EXPEDITIONS	EXPEDITION CROISIERE	EXPEDITION	EHKSPEDITSIYA	EXPEDICION=CRUCERO	SPEDIZIONE=CROCIERA
1477 EXPERIMENTAL PETROLOGY IGNE	Experimental petrology EXPERIMENTAL STUDIES; PETROLOGY	Petrologie experimentale EXPERIENCE	Experimentell. Petrologie EXPERIMENT	PETROLOGIYA EHKSPERIMEN-TAL'NAYA	Petrologia=experimental EXPERIMENTO	Petrologia sperimentale ESPERIMENTO
1478 EXPERIMENTAL SEISMOLOGY SOLI	Experimental seismology EXPERIMENTAL STUDIES; SEISMOLOGY	Seismologie experimentale METHODE SISMIQUE	Experimental=Seismik SEISM.METHODE	Ehksperimental'naya sejsmologiya SEJSMOLOGIYA; EHKSPERIMENT	Sismologia=experimental METODO=SISMICO	Sismologia sperimentale METODO SISMICO
1479 EXPERIMENTAL STUDIES GEOL	EXPERIMENTAL STUDIES *UF:* Experimental petrology; Experimental seismology	EXPERIENCE *UF:* Petrologie experimentale	EXPERIMENT *UF:* Experimentell. Petrologie	EHKSPERIMENT *UF:* Ehksperimental'naya sejsmologiya	EXPERIMENTO *UF:* Petrologia=experimental	ESPERIMENTO *UF:* Petrologia sperimentale
1480 EXPLANATORY NOTE MISC	EXPLANATORY TEXT	NOTICE EXPLICATIVE	KARTEN=ERLAEUTERUNG	ZAPISKA OB'YASNITEL'NAYA	NOTA=EXPLICATIVA	NOTA ILLUSTRATIVA
1481 EXPLOITABILITY ECON	EXPLOITABILITY	DECISION EXPLOITABILITE	BAUWUERDIGKEIT	RENTABEL'NOST'	DECISION=EXPLOTABILIDAD	SFRUTTABILITA
1482 EXPLOITATION MINI	EXPLOITATION *UF:* Mining; Plugging; Producing horizon; Producing well	EXPLOITATION *UF:* Art des mines; Bouchage; Completion puits; Exploitation miniere; Exploitation par eboulement; Horizon productif; Puits a gaz; Puits petrole; Puits production; Sondage developpement	GEWINNUNG *UF:* Bergbau; Bergbau=Technik; Bohr=Vollendung; Cavings; Entwicklungs=Bohrung; Erdoel=Bohrung; Gas=Bohrung; Pfropfen=Bildung; Produktiv. Horizont; Produktiv. Schacht	EHKSPLUATAT-SIYA *UF:* Gornoe delo; Mining licence; Petroleum engineering	EXPLOTACION *UF:* Explotacion=minera; Explotacion-por=hundimiento; Horizonte=productivo; Ingenieria=minera; Pozo=de=gas; Pozo=de=petroleo; Pozo=productivo; Sondeo=explotacion; Taponado; Terminacion=pozo	SFRUTTAMENTO *UF:* Coltivazione mineraria; Completamento di pozzo; Ingegneria mineraria; Orizzonte produttivo; Pozzo di metano; Pozzo petrolifero; Pozzo produttivo; Sbancamento; Sviluppo di pozzo; Tamponatura
1483 EXPLORATION ECON	EXPLORATION *UF:* Groundwater exploration; Prospecting	PROSPECTION *UF:* Exploration; Prospection eau souterraine; Puits	PROSPEKTION *UF:* Exploration; Grundwasser=Erkundung; Schacht	RAZVEDKA	PROSPECCION *UF:* Exploracion; Exploracion=agua=subterranea; Pozo=minero	PROSPEZIONE *UF:* Esplorazione; Esplorazione di falde idriche; Pozzo di miniera

	ENGLISH	FRANCAIS	DEUTSCH	RUSSKIJ	ESPANOL	ITALIANO
1484 EXPLORATORY WELL ECON	Exploratory well BOREHOLES	Sondage reconnaissance SONDAGE	Aufschluss=Bohrung BOHRUNG	SKVAZHINA POISKOVAYA	Pozo=de=exploracion POZO=SONDEO	Sondaggio esplorativo SONDAGGIO
1485 EXPLOSION APPL	EXPLOSIONS	EXPLOSION	EXPLOSION	VZRYV *UF:* Chemical explosion	EXPLOSION	ESPLOSIONE
1486 EXPLOSION BRECCIA GEOL	Explosion breccia VOLCANIC BREC- CIA	Breche explosion BRECHE VOL- CANIQUE	Explosions=Brekzie VULKAN.BREKZIE	BREKCHIYA EHKS- PLOZIVNAYA	Brecha=de=explosion BRECHA= VOLCANICA	Breccia d'esplosione BRECCIA VUL- CANICA
1487 EXPLOSIVE SOURCE APPL	Explosive source SEISMIC METHODS	Source explosion METHODE SISMIQUE	Explosions=Quelle SEISM.METHODE	ISTOCHNIK VZRY- VNOJ	Foco=explosiones METODO=SISMICO	Sorgente d'esplosione METODO SISMICO
1488 EXPORT ECON	EXPORT	EXPORTATION	EXPORT	EHKSPORT	EXPORTACION	ESPORTAZIONE
1489 EXPOSURE AGE ISOT	EXPOSURE AGE	AGE EXPOSITION	AUFSCHLUSS= ALTER	VOZRAST KOSMI- CHESKIJ	EDAD=EXPOSICION	ETA D'ESPOSIZIONE
1490 EXSOLUTION PHCH	EXSOLUTION	EXSOLUTION	ENTMISCHUNG	RASPAD TVER- DOGO RASTVORA	EXSOLUCION	ESSOLUZIONE
1491 EXSURGENCE GEOH	EXSURGENCE	EXSURGENCE	EXSURGENZ	[EXSURGENCE] VODA PODZEM- NAYA; ISTOCHNIK	SURGENCIA	EMERGENZA
1492 EXTENSOMETER INST	EXTENSOMETERS	EXTENSOMETRE	EXTENSOMETER	EHKSTENZOMETR	EXTENSOMETRO	ESTENSOMETRO
1493 EXTERNIDES STRU	Externides GEOSYNCLINES	Externides CHAINE GEOSYN CLINALE	Externiden GEOSYNKLINAL- KETTE	EHKSTERNIDY	Externides CORDILLERA= GEOSINCLINAL	Esternidi CATENA GEOSIN- CLINALE
1494 EXTINCT LAKE SURF	EXTINCT LAKES	LAC MORT	VERLANDET.SEE	OZERO OTMER- SHEE	LAGO=CERRADO	LAGO ESTINTO
1495 EXTINCT TAXON PALE	EXTINCT TAXA	TAXON ETEINT	AUSGESTORBEN. TAXON	VYMERSHIJ	TAXON= EXTINGUIDO	TAXON ESTINTO
1496 EXTINCTION PALE	EXTINCTION	EXTINCTION	AUSSTERBEN	Vymiranie TAFONOMIYA	EXTINCION	ESTINZIONE
1497 EXTRAPOLATION METH	Extrapolation MATHEMATICAL METHODS	Extrapolation INTERPRETATION	Extrapolation INTERPRETATION	EHKSTRAPOLYAT- SIYA	Extrapolacion INTERPRETACION	Extrapolazione INTERPRETAZIONE
1498 EXTRATERRESTRIAL GEOLOGY EXTR	EXTRATERRES- TRIAL GEOLOGY *UF:* Cosmic nebula; Cosmic system; Cosmo- logical principle; Cos- mology; Galactic year; Interstellar space; Perigee	GEOLOGIE EXTRA TERRESTRE *UF:* Apogee; Astrogeolo- gie; Cosmologie; Espace interstellaire; Nebuleuse; Perigee; Principe cosmologique; Systeme cosmique; Univers	EXTRATERRESTR. GEOLOGIE *UF:* Apogaeum; Astrogeologie; Kosm. Nebel; Kosm.System; Kosmolog.Prinzip; Kosmologie; Peri- gaeum; Universum; Weltraum	GEOLOGIYA KOSMICHESKAYA	GEOLOGIA= EXTRATERRESTRE *UF:* Apogeo; Astrogeologia; Cos- mologia; Espacio= interestelar; Nebulosa; Perigeo; Principio= cosmologico; Sistema= cosmico; Universo	GEOLOGIA EXTRATERRESTRE *UF:* Apogeo; Astrogeologia; Cos- mologia; Nebulosa; Perigeo; Principio cosmologico; Sistema cosmico; Spazio inter- stellare; Universo
1499 F LAYER SOLI	F layer INNER CORE	Couche F NOYAU TER- RESTRE INTERNE	F=Schicht INNER.ERD=KERN	F sloj YADRO ZEMLI	Capa=F NUCLEO=INTERNO	Strato F NUCLEO INTERNO

	ENGLISH	FRANCAIS	DEUTSCH	RUSSKIJ	ESPANOL	ITALIANO
1500 FABRIC TEST	FABRIC	FABRIQUE	GEFUEGE *UF:* Zylindr.Gefuege	[FABRIC] STRUKTURA	FABRICA	STRUTTURA
1501 FACIES GEOL	FACIES	Facies(gen.)	Fazies(gen.)	FATSIYA *UF:* Reef environment; Sebkha environment	Facies(gen.)	Facies(gen.)
1502 FACTOR ANALYSIS MATH	FACTOR ANALYSIS	ANALYSE FACTOR- IELLE	FAKTOREN= ANALYSE	ANALIZ FAK- TORNYJ *UF:* Analiz sootvestvij	ANALISIS= FACTORIAL	ANALISI FATTORI- ALE
1503 FAMENNIAN STRS	FAMENNIAN	FAMENNIEN	FAMENNE	FAMEN	FAMENIENSE	FAMENNIANO
1504 FAN SURF	Fans(gen.)	Cone(gen.)	Faecher(gen.)	KONUS VYNOSA	Cono(gen.)	Conoide(gen.)
1505 FANGLOMERATE SEDI	Fanglomerate CONGLOMERATE	Fanglomerat CONGLOMERAT	Fanglomerat KONGLOMERAT	FANGLOMERAT	Fanglomerado CONGLOMERADO	Fanglomerato CONGLOMERATO
1506 FATTY ACID CHES	FATTY ACIDS *UF:* Lipids	ACIDE GRAS *UF:* Lipide	FETT=SAEURE *UF:* Lipid	KISLOTA ZHIRNAYA	ACIDO=GRASO *UF:* Lipido	ACIDO GRASSO *UF:* Lipide
1507 FAULT STRU	FAULTS *UF:* Antithetic faults; Faulting; Hanging wall; Horizontal fault; Lateral faults; Oblique fault; Upthrow	FAILLE *UF:* Faille contraire; Faille oblique; Levre relevee	STOERUNG *UF:* Antithet. Verwerfung; Gehoben. Scholle; Schraeg. Stoerung	RAZLOM *UF:* Brekchiya razlom- naya; Razlom gorizon- tal'nyj; Zona razloma	FALLA *UF:* Falla=antitetica; Falla=oblicua; Labio= levantado	FAGLIA *UF:* Faglia antitetica; Faglia obliqua; Lembo rialzato
1508 FAULT BRECCIA GEOL	Fault breccia TECTONIC BREC- CIA	Breche faille BRECHE TEC- TONIQUE	Verwerfungs=Brekzie TEKTON.BREKZIE	Brekchiya razlomnaya BREKCHIYA TEK- TONICHESKAYA; RAZLOM	Brecha=de=falla BRECHA= TECTONICA	Breccia di faglia BRECCIA TET- TONICA
1509 FAULT GOUGE STRU	GOUGE	ARGILE FAILLE	GANG=LETTE	GLINKA TRENIYA	PULVERIZADO= FALLA	ARGILLA DI FAGLIA
1510 FAULT PLANE STRU	FAULT PLANES *UF:* Shear planes	PLAN FAILLE *UF:* Plan cisaillement	VERWERFUNGS= FLAECHE *UF:* Scher=Flaeche	SBRASYVATEL' *UF:* Smestitel'	PLANO=FALLA *UF:* Plano=de= cizallamiento	PIANO DI FAGLIA *UF:* Piano di taglio
1511 FAULT SCARP SURF	FAULT SCARPS	ESCARPEMENT FAILLE	BRUCH=STUFE	Ustup sbrosovyj USTUP; SBROS	ESCARPE=DE= FALLA	SCARPATA DI FAGLIA
1512 FAULT SYSTEM STRU	FAULT SYSTEMS	RESEAU FRAC- TURE *UF:* Zone fracturee	STOERUNGS= SYSTEM *UF:* Bruch=System	ZONA NARUS- HENIYA *UF:* Zona razloma	RED=FRACTURA *UF:* Zona=fracturada	RETICOLO DI FRATTURE *UF:* Zona fratturata
1513 FAULT ZONE STRU	FAULT ZONES	ZONE FAILLEE	BRUCH=ZONE	Zona razloma ZONA NARUS- HENIYA; RAZLOM	ZONA=FALLA	ZONA FAGLIATA
1514 FAULT-PLANE STRIATION STRU	Fault=plane striations SLICKENSIDES	STRIATION *UF:* Strie glaciaire	STRIEMUNG *UF:* Glazial=Striemung	BOROZDA SKOL'ZHENIYA	ESTRIACION *UF:* Estria=glaciar	STRIATURA *UF:* Striatura glaciale
1515 FAULTING STRU	Faulting FAULTS	TECTONIQUE FRACTURATION	BRUCH=TEKTONIK	DVIZHENIE DIZ'YUNKTIVNOE	FALLADO	FAGLIAMENTO

	ENGLISH	FRANCAIS	DEUTSCH	RUSSKIJ	ESPANOL	ITALIANO
1516 FAUNA PALE	FAUNA *UF:* Benthos; Cenozones; Faunal list; Faunizones; Geobios; Nekton; Planctivorous animal; Psammofauna	FAUNE *UF:* Association; Benthos; Faune planctivore; Geobios; Liste faunistique; Nekton; Psammofaune	FAUNA *UF:* Benthos; Faunen= Liste; Geobios; Nekton; Planktonfresser; Psammofauna; Vergesellschaftung	FAUNA *UF:* Tsenozona; Zona faunisticheskaya	FAUNA *UF:* Animal=planctivoro; Arena=fosilifera; Asociacion; Bentos; Geobios; Lista=faunistica; Necton	FAUNA *UF:* Animale plantivoro; Associazione di fossili; Benthos; Geobios; Lista faunistica; Necton; Psammofauna
1517 FAUNAL LIST PALE	Faunal list FAUNA	Liste faunistique FAUNE	Faunen=Liste FAUNA	SPISOK FAUNY	Lista=faunistica FAUNA	Lista faunistica FAUNA
1518 FAUNAL PROVINCE PALE	FAUNAL PROVINCES	PROVINCE FAUNISTIQUE	FAUNEN=PROVINZ	PROVINTSIYA FAUNISTICHESKAYA	PROVINCIA= FAUNISTICA	PROVINCIA FAUNISTICA
1519 FAUNIZONE STRA	Faunizones FAUNA; ZONING	Zone faunistique FAUNE SPECIFIQUE	Faunizone SPEZIF.FAUNA	Zona faunisticheskaya BIOZONA; FAUNA	Zona=faunistica FAUNA= ESPECIFICA	Faunizona FAUNA SPECIFICA
1520 FEASIBILITY STUDY MISC	FEASIBILITY STUDIES	ETUDE FAISABILITE	DURCHFUEHRBARKEITS=STUDIE	OBOSNOVANIE TEKHNIKO= EHKONOMICHESKOE	ESTUDIO= FACTIBILIDAD	STUDIO DI FATTIBILITA
1521 FECAL PELLET PALE	Fecal pellets COPROLITES	Pelote fecale COPROLITE	Kot=Pille KOPROLITH	Komochki fekal'nye KOPROLITY	Pelet=fecal COPROLITO	Grumo fecale COPROLITE
1522 FEEDER IGNE	Feeder VENTS	Fissure magmatique NECK	Foerder=Schlot SCHLOT	Kanal podvodyashchij NEKK	Conducto=alimentador CHIMENEA	Condotto alimentatore NECK
1523 FEEDING TRACK PALE	NUTRITION TRACES	TRACE NUTRITION	FRASS=SPUR	SLED PITANYA	HUELLA= NUTRICION	PISTE DI NUTRIZIONE
1524 FELDSPAR MING	FELDSPAR GROUP	FELDSPATH	FELDSPAT	SHPAT POLEVOJ	FELDESPATO	FELDSPATO
1525 FELDSPATHIZATION IGNE	FELDSPATHIZATION	FELDSPATHISATION	FELDSPATISIERUNG	FEL'DSHPATIZATSIYA	FELDESPATIZACION	FELDSPATIZZAZIONE
1526 FELDSPATHOIDIC ROCK IGNS	FELDSPATHOID COMPOSITION *UF:* Essexite	ROCHE FELDSPATHOIDIQUE *UF:* Essexite; Theralite	FOID=GESTEIN *UF:* Essexit; Theralith	PORODA SHCHELOCHNAYA *UF:* Foidit	ROCA= FELDESPATOIDICA *UF:* Essexita; Teralita	ROCCIA A FELDSPATOIDI *UF:* Essexite; Teralite
1527 FELSIC COMPOSITION IGNE	Felsic composition ACIDIC COMPOSITION	Roche felsique GRANITE	Felsisch.Chemismus GRANIT	Fel'zicheskij KISLYJ (SOSTAV)	Composicion=felsica GRANITO	Composizione felsica GRANITO
1528 FEMIC COMPOSITION IGNE	Femic MAFIC COMPOSITION	Roche mafique COMPOSITION MAFIQUE	Femisch.Chemismus MAFISCH. CHEMISMUS	FEMICHESKIJ	Composicion=femica COMPOSICION= MAFICA	Composizione femica COMPOSIZIONE MAFICA
1529 FENITE IGMS	Fenite SYENITES	Fenite SYENITE	Fenit SYENIT	FENIT	Fenita SIENITA	Fenite SIENITE
1530 FENITIZATION IGNE	FENITIZATION	FENITISATION	FENITISIERUNG	FENITIZATSIYA	FENITIZACION	FENITIZZAZIONE
1531 FERMIUM CHEE	FERMIUM	FERMIUM	FM	FERMIJ	FERMIO	FERMIO
1532 FERRALSOL SUSS	FERRALSOLS	SOL FERRALLITIQUE	FERRALSOL	FERRALIT	SUELO= FERRALITICO	FERRALSOL

	ENGLISH	FRANCAIS	DEUTSCH	RUSSKIJ	ESPANOL	ITALIANO
1533 FERRIC IRON CHES	FERRIC IRON	FER FERRIQUE	FERRI=EISEN	ZHELEZO ZAKIS-NOE	HIERRO=FERRICO	FERRO FERRICO
1534 FERRICRETE SURF	FERRICRETE	CROUTE FERRU-GINEUSE	EISEN=KRUSTE	Ferrikret KORA VYVETRIVANIYA	COSTRA=FERRUGINOSA	CROSTA FERRUG-GINOSA
1535 FERROD SUSS	Ferrods SPODOSOLS	Ferrod SPODOSOL; POD-ZOL	Ferrod SPODOSOL; POD-SOL	Ferrod SPODOSOL	Ferrod SPODOSOL; POD-ZOL	Ferrod SPODOSOL; POD-SOL
1536 FERROMAGNETIC PROPERTY PHCH	Ferromagnetic property MAGNETIC PROP-ERTIES	Propriete ferromagne-tique PROPRIETE MAGNETIQUE	Ferromagnet. Eigenschaft MAGNET. EIGENSCHAFT	FERROMAGNETIK *UF:* Struktura domen-naya	Propiedad=ferromagnetica PROPIEDAD=MAGNETICA	Proprieta ferromagne-tica PROPRIETA MAGNETICA
1537 FERROMANGANESE COMPOSITION SEDI	FERROMAN-GANESE COMPOSI-TION	COMPOSITION FERROMAN-GANIFERE	EISEN=MANGAN=GEHALT	[FERROMAN-GANESE COMPOSI-TION] ZHELEZO; MARGA-NETS	COMPOSICION=FERROMANGANIFERA	COMPOSIZIONE FERROMANGANE-SIFERA
1538 FERROUS IRON CHES	FERROUS IRON	FER FERREUX	FERRO=EISEN	ZHELEZO OKISNOE	HIERRO=FERROSO	FERRO FERROSO
1539 FERRUGINOUS COMPOSITION GEOC	FERRUGINOUS COMPOSITION	COMPOSITION FERRUGINEUSE	EISEN=GEHALT	ZHELEZISTYJ	COMPOSICION=FERRUGINOSA	COMPOSIZIONE FERRUGGINOSA
1540 FERRUGINOUS SOIL SUSS	FERRUGINOUS SOILS	SOL FERRUGINEUX	EISENHALTIG. BODEN	POCHVA ZHELEZ-ISTAYA	SUELO=FERRUGINOSO	SUOLO FER-RUGINOSO
1541 FERTILIZER COMS	FERTILIZERS	ENGRAIS	DUENGEMITTEL	UDOBRENIE	ABONO	FERTILIZZANTE
1542 FIELD STUDY GEOL	FIELD STUDIES	Travaux terrain CARTOGRAPHIE	Gelaende=Untersuchung KARTOGRAPHIE	GEOLOGIYA POLEVAYA	Estudio=de=campo CARTOGRAFIA	Studio di campagna CARTOGRAFIA
1543 FILICALES PALS	FILICOPSIDA	FILICOPSIDA	FILICOPSIDA	PTEROPSIDA	FILICOPSIDA	FILICALES
1544 FILLING MATERIAL COMS	FILLING MATERI-ALS	PRODUIT CHARGE	FUELLSTOFF	NAPOLNITEL'	PRODUCTO=CARGA	MATERIALE DI RIEMPIMENTO
1545 FILTER INST	Well filter WELLS; FILTERS	Filtre INSTRUMENTA-TION; FILTRATION	Filter INSTRUMEN-TIERUNG; FILTRA-TION	FIL'TR *UF:* Well screen	Filtro=tamiz INSTRUMENTA-CION; FILTRACION	Filtro di pozzo STRUMENTAZ-IONE; FILTRAZ-IONE
1546 FILTERING MATH	FILTERS	FILTRAGE SIGNAL	FILTERUNG	FIL'TROVANIE	FILTRADO	FILTRAGGIO DI SEGNALE
1547 FILTRATION GEOH	FILTRATION	FILTRATION *UF:* Filtre	FILTRATION *UF:* Filter	FIL'TRATSIYA *UF:* Darsi zakon	FILTRACION *UF:* Filtro=tamiz	FILTRAZIONE *UF:* Filtro di pozzo
1548 FINANCING MISC	FINANCING	FINANCEMENT	FINANZIERUNG	FINANSIROVANIE	FINANCIACION	FINANZIAMENTO
1549 FINE=GRAINED FRACTION SEDI	FINE=GRAINED MATERIALS	FRACTION FINE	FEINFRAKTION	FRAKTSIYA TONKAYA	FRACCION=FINA	FRAZIONE FINE

	ENGLISH	FRANCAIS	DEUTSCH	RUSSKIJ	ESPANOL	ITALIANO
1550 FINITE DIFFERENCE ANALYSIS MATH	FINITE DIFFERENCE ANALYSIS	METHODE DIFFERENCE FINIE	FINITE= DIFFERENZ= ANALYSE	METOD KONECH-NYKH RAZNOSTEJ	METODO-DIFERENCIA-FINITA	METODO DIFFERENZE FINITE
1551 FINITE ELEMENTS MATH	FINITE ELEMENT ANALYSIS	METHODE ELEMENT FINI	FINIT.ELEMENTE=ANALYSE	EHLEMENT KONECHNYJ	ELEMENTO=FINITO	ELEMENTO FINITO
1552 FIRECLAY COMS	FIRECLAY	ARGILE REFRACTAIRE	FEUERFEST.TON	GLINA OGNEU-PORNAYA *UF:* Underclay	ARCILLA-REFRACTARIA	ARGILLA REFRATTARIA
1553 FIRN SURF	FIRN	NEVE	FIRN	FIRN	CAMPO=DE=NIEVE	NEVATO
1554 FISSION PHCH	FISSION	FISSION NUCLEAIRE	KERNSPALTUNG	DELENIE	FISION=NUCLEAR	FISSIONE
1555 FISSION TRACKS ISOT	FISSION TRACKS	TRACE FISSION *UF:* Datation trace fission; Datation trace particule	SPALTSPUREN *UF:* Spaltspuren=Datierung; Teilchen=Spur=Datierung	TREK DELENIYA	TRAZA=FISION *UF:* Datacion=traza=fision; Datacion=traza=particula	TRACCE DI FISSIONE *UF:* Eta tracce di fissione; Traccia di particella
1556 FISSION-TRACK DATING ISOT	FISSION=TRACK DATING	Datation trace fission DATATION; TRACE FISSION	Spaltspuren=Datierung PHYSIKAL. ALTERSBESTIM-MUNG; SPALTS-PUREN	METOD TREKOV DELENIYA *UF:* Metod trekov	Datacion=traza=fision DATACION; TRAZA=FISION	Eta tracce di fissione DATAZIONE; TRACCE DI FISSIONE
1557 FISSIPEDA PALS	FISSIPEDA	FISSIPEDIA	FISSIPEDA	Fissipeda CARNIVORA	FISSIPEDIA	FISSIPEDA
1558 FIXATION PHCH	FIXATION	FIXATION ION	IONEN=BINDUNG	[FIXATION]	FIJACION=ION	FISSAZIONE IONICA
1559 FIXISM GEOL	Fixism THEORETICAL STUDIES; CONTINENTAL DRIFT	Fixisme THEORIE; EVOLUTION BIOLOGIQUE	Fixismus THEORIE; BIOLOG. EVOLUTION	FIKSIZM	Fijismo TEORIA; EVOLUCION=BIOLOGICA	Fissismo TEORIA; EVOLUZIONE BIOLOGICA
1560 FJORD SURF	FJORDS	FJORD	FJORD	FIORD	FIORDO	FIORDO
1561 FLAGGY TEXTURE TEST	Flaggy texture TEXTURES	Texture dalle TEXTURE	Wolken=Gefuege KORN=GEFUEGE	PLITCHATYJ *UF:* Otdel'nost' plitchataya	Textura=laminar TEXTURA	Tessitura listata TESSITURA
1562 FLAME PHOTOMETRY METH	FLAME PHOTOMETRY	PHOTOMETRIE FLAMME	FLAMMEN=PHOTOMETRIE	FOTOMETRIYA PLAMENNAYA	FOTOMETRIA-LLAMA	FOTOMETRIA DI FIAMMA
1563 FLAME STRUCTURE TEST	FLAME STRUCTURES	STRUCTURE FLAMME	FLAMMEN=STRUKTUR	PLAMENNYJ	ESTRUCTURA-EN-LLAMA	STRUTTURA A FIAMME
1564 FLANDRIAN STRS	FLANDRIAN	FLANDRIEN	FLANDRIUM	Flandrij CHETVERTICHNYJ	FLANDRIENSE	FLANDRIANO
1565 FLASER STRUCTURE TEST	FLASER STRUCTURES	Texture filamenteuse TEXTURE	Flaser=Gefuege KORN=GEFUEGE	POLOSCHATYJ	Estructura=lenticular TEXTURA	Stratificazione flaser TESSITURA

	ENGLISH	FRANCAIS	DEUTSCH	RUSSKIJ	ESPANOL	ITALIANO
1566 FLEXURE STRU	FLEXURE	FLEXURE	FLEXUR	FLEKSURA *UF:* Skladka kaskad- naya; Skladka stu- penchataya	FLEXURA	FLESSURA
1567 FLINT SEDS	FLINT	SILEX	HORNSTEIN	FLINT *UF:* Slanets kremnistyj	SILEX	SELCE
1568 FLOCCULATION PHCH	FLOCCULATION	FLOCULATION	AUSFLOCKUNG	Flokkulyatsiya KOLLOID	FLOCULACION	FLOCCULAZIONE
1569 FLOOD GEOH	FLOODS	CRUE	HOCH=WASSER	NAVODNENIE	CRECIDA=RIO	PIENA
1570 FLOOD PLAIN SURF	FLOODPLAINS	PLAINE INOND- ABLE	UEBERSCHWE- MMUNGS= FLAECHE	POJMA	LECHO=MAYOR	PIANA INONDA- BILE
1571 FLORA PALE	FLORA *UF:* Florizones; Halo- phyte; Hekistotherm plant; Heterosporous taxa; Holophyte; Litho- phyte; Metaphyte; Oxylophyte; Phreato- phyte; Phytocoenoses; Psychrophyte; Xero- phile plant; Xerophyte	FLORE *UF:* Association; Halo- phyte; Holophyte; Lithophyte; Metaphyte; Oxylophyte; Phreato- phyte; Phytocenose; Plante hekistotherme; Plante heterosporee; Plante xerophile; Prov- ince paleobotanique; Psychrophyte; Xero- phyte	FLORA *UF:* Halophyt; Hekistotherm.Pflanze; Heterospor.Pflanze; Holophyt; Lithophyt; Metaphyt; Oxylophyt; Palaeobotan.Provinz; Phreatophyt; Phyto- coenose; Psychrophyt; Vergesellschaftung; Xerophil.Pflanze; Xero- phyt	FLORA *UF:* Tsellyuloza	FLORA *UF:* Asociacion; Fito- cenosis; Freatofita; Halofita; Holofita; Litofita; Metafita; Oxilofita; Planta= hekistotermica; Planta= heterospora; Planta= xerofila; Provincia= paleobotanica; Psicro- fita; Xerofita	FLORA *UF:* Alofita; Associaz- ione di fossili; Fito- cenosi; Holophyte; Lithophyte; Metaphyte; Oxylophyte; Phreato- phyte; Pianta echis- toterma; Pianta etero- sporica; Pianta xerofila; Provincia paleobo- tanica; Psychrophyte; Xerofita
1572 FLORAL PROVINCE PALE	FLORAL PROV- INCES *UF:* Paleobotanic prov- ince	PROVINCE FLORIS- TIQUE	FLOREN=PROVINZ	Provintsiya floristi- cheskaya PROVINTSIYA PALEOBOTANI- CHESKAYA	PROVINCIA= FLORISTICA	PROVINCIA FLORISTICA
1573 FLORIZONE STRA	Florizones FLORA; ZONING	Florizone FLORE SPECIFIQUE	Florizone SPEZIF.FLORA	ZONA FLORISTI- CHESKAYA	Florizona FLORA= ESPECIFICA	Florizona FLORA SPECIFICA
1574 FLOTATION METH	FLOTATION	FLOTTATION	FLOTATION	FLOTATSIYA	FLOTACION	FLOTTAZIONE
1575 FLOW PHCH	FLOW	COULEE	FLIESSEN	POTOK	COLADA	COLATA
1576 FLOW CLEAVAGE STRU	FLOW CLEAVAGE	SCHISTOSITE FLUX	FLIESS- SCHIEFERUNG	KLIVAZH TEC- HENIYA *UF:* Klivazh istecheniya	EXFOLIACION= FLUIDAL	SCISTOSITA FLUID- ALE
1577 FLOW FOLD STRU	FLOW FOLDS	PLI FLUAGE	FLIESS=FALTE	SKLADKA TEC- HENIYA	PLIEGUE=DE- FLUENCIA	PIEGA DI FLUSSO
1578 FLUCTUATION GEOH	FLUCTUATIONS	FLUCTUATION	FLUKTUATION	KOLEBANIE UROV- NYA (GIDROGEOL)	FLUCTUACION	FLUTTUAZIONE
1579 FLUID INCLUSION MINE	FLUID INCLUSIONS *UF:* Gas inclusion	INCLUSION FLUIDE	FLUESSIGKEITS= EINSCHLUSS	VKLYUCHENIE GAZOVOZHIDKOE	INCLUSION= FLUIDA	INCLUSIONE FLUIDA

	ENGLISH	FRANCAIS	DEUTSCH	RUSSKIJ	ESPANOL	ITALIANO
1580 FLUID INJECTION ENGI	FLUID INJECTION	INJECTION FLUIDE	FLUESSIGKEITS= INJEKTION	[FLUID INJEC- TION] SOORUZHENIE; STABILIZATSIYA	INYECCION= FLUIDA	INIEZIONE DI FLUIDO
1581 FLUID MECHANICS PHCH	FLUID MECHANICS	MECANISME ECOULEMENT	ABFLUSS= VORGANG	GIDROMEK- HANIKA	MECANISMO= ESCORRENTIA	MECCANICA DEI FLUIDI
1582 FLUID PHASE PHCH	FLUID PHASE	PHASE FLUIDE	FLUESSIG.PHASE	FAZA ZHIDKAYA (RASPLAV)	FASE=FLUIDA	FASE FLUIDA
1583 FLUID PRESSURE ENGI	FLUID PRESSURE	PRESSION FLUIDE	FLUESSIGKEITS= DRUCK	[FLUID PRESSURE] DAVLENIE GIDROSTATI- CHESKOE	PRESION=FLUIDO	PRESSIONE DI FLUIDO
1584 FLUIDAL TEXTURE TEST	Fluidal texture TEXTURES	FLUIDALITE	FLUIDAL= GEFUEGE	FLYUIDAL'NYJ	TEXTURA= FLUIDAL	TESSITURA FLUID- ALE
1585 FLUORESCENCE PHCH	FLUORESCENCE	FLUORESCENCE	FLUORESZENZ	FLYUORESTSENT- SIYA	FLUORESCENCIA	FLUORESCENZA
1586 FLUORIDES MING	FLUORIDES	FLUORURE	FLUORID	FTORIDIT	FLUORATO	FLUORURO
1587 FLUORIMETRY METH	FLUORIMETRY	SPECTROMETRIE FLUORESCENCE	FLUORIMETRIE	FLUORIMETRIYA	FLUORIMETRIA	SPETTROMETRIA DI FLUORESCENZA
1588 FLUORINATION GEOC	FLUORINATION	Fluoruration ATTAQUE CHIMIQUE; FLUOR	Fluorinierung AETZUNG; F	FTORIROVANIE	Fluoruracion ATAQUE=QUIMICO; FLUOR	Fluorurazione ATTACCO CHIMICO; FLUORO
1589 FLUORINE CHEE	FLUORINE	FLUOR UF: Fluoruration	F UF: Fluorinierung	FTOR	FLUOR UF: Fluoruracion	FLUORO UF: Fluorurazione
1590 FLUORSPAR COMS	FLUORSPAR	FLUORITE	FLUORIT	FLYUORIT	ESPATO=FLUOR	FLUORITE
1591 FLUTE CAST TEST	FLUTE CASTS	STRUCTURE CAN- NELEE	RIFFEL= STRUKTUR	Turboglif GIEROGLIF	ESTRUCTURA= ACANALADA	CONTROIM- PRONTA LOBATA
1592 FLUVENT SUSS	Fluvents ENTISOLS	Fluvent ENTISOL; ALLU- VION	Fluvent ENTISOL; ALLU- VION	Flyuvent EHNTISOL	Fluvent ENTISOL; ALUVION	Fluvent ENTISOL; ALLUV- IONE
1593 FLUVIAL ENVIRONMENT PALE	FLUVIAL ENVI- RONMENT	MILIEU FLUVIA- TILE	FLUVIATIL.MILIEU	FATSIYA RECH- NAYA	MEDIO=FLUVIAL	AMBIENTE FLUVI- ALE
1594 FLUVIAL EROSION SURF	Fluvial erosion WATER EROSION	EROSION FLUVIA- TILE	FLUSS=EROSION	EHROZIYA RECH- NAYA	EROSION=FLUVIAL	EROSIONE FLUVI- ALE
1595 FLUVIAL SEDIMENTATION SEDI	FLUVIAL SEDI- MENTATION	SEDIMENTATION= FLUVIATILE	FLUVIATIL. SEDIMENTATION	SEDIMENTATSIYA RECHNAYA	SEDIMENTACION= FLUVIAL	SEDIMENTAZIONE FLUVIALE
1596 FLUVIOGLACIAL SURF	GLACIOFLUVIAL FEATURES	FLUVIOGLACIAIRE	FLUVIO=GLAZIAL	FLYUVIOGLYAT- SIAL'NYJ	FLUVIO=GLACIAR	FLUVIOGLACIALE
1597 FLUVIOGLACIAL SEDIMENTATION SEDI	GLACIOFLUVIAL SEDIMENTATION	SEDIMENTATION FLUVIOGLACIAIRE	FLUVIO=GLAZIAL- SEDIMENTATION	SEDIMENTATSIYA FLYUVIOGLATSI- AL'NAYA	SEDIMENTACION= FLUVIOGLACIAR	SEDIMENTAZIONE FLUVIOGLACIALE

	ENGLISH	FRANCAIS	DEUTSCH	RUSSKIJ	ESPANOL	ITALIANO
1598 FLUVIOSOL SUSS	FLUVIOSOLS	FLUVIOSOL	FLUVIOSOL	POCHVA POJMEN-NAYA	FLUVIOSOL	FLUVIOSOL
1599 FLYSCH SEDI	FLYSCH *UF:* Wildflysch	FLYSCH *UF:* Wildflysch	FLYSCH *UF:* Wildflysch	FLISH	FLYSCH *UF:* Wildflysch	FLYSCH *UF:* Wildflysch
1600 FOCAL MECHANISM SOLI	FOCAL MECHA-NISM	MECANISME FOCAL	HERD-MECHANISMUS	Mekhanizm zemletr-yaseniya OCHAG ZEM-LETRYASENIYA	MECANISMO-FOCAL	MECCANISMO FOCALE
1601 FOIDITE IGNS	Foidite GABBROS; ALKALIC COMPO-SITION	Foidite GABBRO; COMPOSI-TION ALCALINE	Foidit GABBRO; ALKALI-TYP	Foidit NEFELIN; PORODA SHCHELOCHNAYA	Foidita GABRO; COMPOSICION-ALCALINA	Foidite GABBRO; COMPO-SIZIONE ALCALINA
1602 FOLD STRU	FOLDS *UF:* Brachyfold; Cascade folds; Fold system; Fold tectonics; Folding; Gravity folds; Intra-folial folds; Limb; Plunge; Step folds; Virgation; Zigzag fold	PLI *UF:* Brachypli; Flanc pli; Pli en cascade; Pli escalier; Pli gravitaire; Pli intra-foliaire; Vergence; Virgation	FALTE *UF:* Brachy=Falte; Faeltelung; Gravitations=Falte; Kaskaden=Falte; Schenkel; Treppen=Falte; Vergenz; Virga-tion; Zick=Zack=Falte	SKLADKA *UF:* Os' skladki; Skladka ehnteroliticheskaya; Skladka stupenchataya; Skladka tsilindri-cheskaya	PLIEGUE *UF:* Braquipliegue; Crenulacion; Flanco; Pliegue=de=gravedad; Pliegue=en=acordeon; Pliegue=en=cascada; Pliegue=monoclinal; Trampa=litologica; Vergencia; Virgacion	PIEGA *UF:* Brachipiega; Fianco; Piega a cascata; Piega a gradino; Piega gravitativa; Piega intrafoliare; Vergenza; Virgazione
1603 FOLD AXIS STRU	FOLD AXES	AXE PLI *UF:* Plongement	FALTENACHSE *UF:* Abtauchen	Os' skladki SKLADKA; GEO-METRIYA	EJE=DE=PLEGAMIENTO *UF:* Pinchamiento	ASSE DI PIEGA *UF:* Immersione assiale di piega
1604 FOLD BELT STRU	FOLD BELTS *UF:* Orogenic belt	Chaine plissee CHAINE GEOSYN-CLINALE	Falten=Guertel GEOSYNKLINAL-KETTE	POYAS SKLADCHA-TYJ *UF:* Poyas podvizhnyj; Sistema skladchataya	Cinturon=plegado CORDILLERA=GEOSINCLINAL	Catena a pieghe CATENA GEOSIN-CLINALE
1605 FOLD BRECCIA GEOL	Fold breccia TECTONIC BREC-CIA	Breche plissement BRECHE TEC-TONIQUE	Falten=Brekzie TEKTON.BREKZIE	Brekchiya vnutrisklad-chataya BREKCHIYA TEK-TONICHESKAYA	Brecha=de=pliegue BRECHA=TECTONICA	Breccia di piega BRECCIA TET-TONICA
1606 FOLD SYSTEM STRU	Fold system FOLDS	Systeme plisse TECTONIQUE SOU-PLE	Falten=System FALTEN=TEKTONIK	Sistema skladchataya GEOSINKLINAL'; POYAS SKLADCHA-TYJ	Sistema=de=pliegues TECTONICA=DE=PLEGAMIENTO	Sistema di pieghe TETTONICA A PIEGHE
1607 FOLD TECTONICS STRU	Fold tectonics FOLDS	TECTONIQUE SOU-PLE *UF:* Plissement; Systeme plisse	FALTEN=TEKTONIK *UF:* Falten=System; Faltung	Tektonika plikativnaya DVIZHENIE SKLAD-CHATOE	TECTONICA=DE=PLEGAMIENTO *UF:* Plegamiento; Sistema=de=pliegues	TETTONICA A PIEGHE *UF:* Piegamento; Sistema di pieghe
1608 FOLDING STRU	Folding FOLDS	Plissement TECTONIQUE SOU-PLE	Faltung FALTEN=TEKTONIK	DVIZHENIE SKLAD-CHATOE *UF:* Tektonika plikativ-naya	Plegamiento TECTONICA=DE=PLEGAMIENTO	Piegamento TETTONICA A PIEGHE
1609 FOLIATION TEST	FOLIATION	FOLIATION *UF:* Texture lepidoblas-tique	KRISTALLIS-ATIONS=SCHIEFERUNG *UF:* Lepidoblast.Gefuege	LISTOVATOST'	FOLIACION *UF:* Lepidoblastica	FOLIAZIONE *UF:* Tessitura lepidoblas-tica
1610 FORAMINIFERA PALS	FORAMINIFERA	FORAMINIFERA	FORAMINIFERA	FORAMINIFERA	FORAMINIFERA	FORAMINIFERA

	ENGLISH	FRANCAIS	DEUTSCH	RUSSKIJ	ESPANOL	ITALIANO
1611 FORECASTING MAP MISC	PREDICTION MAPS	CARTE PREVISION-NELLE	HOEFFIGKEITS=KARTE	KARTA PROGNOZ-NAYA	MAPA=PREVISORIO	CARTA PREVISION-ALE
1612 FOREDEEP STRU	Foredeeps GEOSYNCLINES	Avant=fosse GEOSYNCLINAL	Vortiefe GEOSYNKLINALE	PROGIB KRAEVOJ	Antefosa GEOSINCLINAL	Avanfossa GEOSINCLINALE
1613 FORELAND STRU	Forelands CRATONS	Avant=pays CRATON	Vorland KRATON	FORLAND	Antepais CRATON	Avampaese CRATONE
1614 FORESET BED SEDI	FORESET BEDS	Couche frontale STRATIFICATION OBLIQUE; SEDI-MENTATION DELTAIQUE	[FORESET=BED] SCHRAEGSCHICH-TUNG; DELTA=SEDIMENTATION	Sloj peredovoj SLOISTOST'	Capa=frontal ESTRATIFIC-ACION=OBLICUA; SEDIMENTACION=DELTAICA	Strato frontale STRATIFICAZIONE OBLIQUA; SEDI-MENTAZIONE DEL-TIZIA
1615 FORESHOCK SOLI	FORESHOCKS	ONDE PRECUR-SEUR	VORBEBEN	Forshok ZEMLETRYASENIE	ONDA=PRECURSORA	ONDA PREMONI-TRICE
1616 FORESHORE MARI	Foreshore BEACHES	Avant=plage PLAGE	Vorstrand STRAND	Plyazh nizhnij PLYAZH	Anteplaya PLAYA	Antispiaggia SPIAGGIA
1617 FORMATION STRA	Formations STRATIGRAPHIC UNITS	Formation ECHELLE STRATI-GRAPHIQUE	Formation STRATIGRAPH. SKALA	PODSVITA	Formacion ESCALA=ESTRATIGRAFICA	Formazione SCALA STRATI-GRAFICA
1618 FORMULA PHCH	FORMULA	FORMULE	FORMEL	FORMULA	FORMULA	FORMULA
1619 FOSSIL PALE	FOSSILS UF: Internal cast	FOSSILE	FOSSIL	OKAMENELOST'	FOSIL	FOSSILE
1620 FOSSIL FUEL COMS	Fossil fuels FUEL RESOURCES	Combustible fossile ENERGIE	Fossil.Brennstoff ENERGIE	POLEZNYE ISKO-PAEMYE GORYU-CHIE	Fuel=fosil ENERGIA	Combustibile fossile ENERGIA
1621 FOSSIL ICE WEDGE SURF	FOSSIL ICE WEDGES	COIN GLACE FOS-SILE	FOSSIL. FROSTSPALTE	KLIN LEDYANOJ OKAMENELYJ	CUÑA=DE=HIELO=FOSIL	CUNEO DI GHIAC-CIO FOSSILE
1622 FOSSIL MAN PALE	FOSSIL MAN	HOMME FOSSILE	FOSSIL.MENSCH	CHELOVEK ISKO-PAEMYJ	HOMBRE=FOSIL	UOMO FOSSILE
1623 FOSSIL WATER GEOH	FOSSIL WATERS	EAU FOSSILE	FOSSIL.WASSER	VODA ISKOPAE-MAYA	AGUA=FOSIL	ACQUA FOSSILE
1624 FOSSIL WOOD PALE	FOSSIL WOOD UF: Silicified wood	BOIS FOSSILE UF: Bois silicifie	FOSSIL.HOLZ UF: Verkieselt.Holz	DEREVO OKAME-NELOE UF: Derevo okremneloe	MADERA=FOSIL UF: Madera=silificada	LEGNO FOSSILE UF: Legno silicizzato
1625 FOSSILIFEROUS DEPOSIT PALE	FOSSIL LOCALI-TIES	GISEMENT FOS-SILIFERE	FOSSIL=FUNDPUNKT	PALEONTOLOGI-CHESKI OKHARAK-TERIZOVANNOE MESTO	YACIMIENTO=FOSILIFERO	LIVELLO FOS-SILIFERO
1626 FOSSILIZATION PALE	FOSSILIZATION	FOSSILISATION UF: Moule interne	FOSSILISATION UF: Inner.Abdruck	FOSSILIZATSIYA UF: Khemofossilii	FOSILIZACION UF: Molde=interno	FOSSILIZZAZIONE UF: Modello interno
1627 FOUNDATION ENGI	FOUNDATIONS	FONDATION	BAUGRUND	OSNOVANIE SOORUZHENIJ	CIMENTACION	FONDAZIONE

103 • MULTILINGUAL THESAURUS OF GEOSCIENCES

	ENGLISH	FRANCAIS	DEUTSCH	RUSSKIJ	ESPANOL	ITALIANO
1628 FOURIER ANALYSIS MATH	FOURIER ANALY- SIS	ANALYSE HAR- MONIQUE	FOURIER= ANALYSE	Fur'e ryad ANALIZ GARMONI- CHESKIJ	ANALISIS= ARMONICO	ANALISI DI FOU- RIER
1629 FOYAITE IGNS	FOYAITE	FOYAITE	FOYAIT	FOJYAIT	FOYAITA	FOYAITE
1630 FRACTIONATION PHCH	CHEMICAL FRAC- TIONATION	FRACTIONNEMENT	FRAKTIONIERUNG	FRAKTSIONIROVA- NIE *UF:* Kumulat	FRACCIONA- MIENTO	FRAZIONAMENTO
1631 FRACTIONNAL **CRYSTALLIZATION** IGNE	FRACTIONAL CRYSTALLIZATION *UF:* Bowen's reaction series	CRISTALLISATION FRACTIONNEE *UF:* Suite reactionnelle Bowen	FRAKTIONIERT. KRISTALLISATION *UF:* Bowen's=Reak- tions=Serie	KRISTALLIZAT- SIYA FRAKTSION- NAYA	CRISTALIZACION= FRACCIONADA *UF:* Series=de= reaccion=de=Bowen	CRISTALLIZZAZ- IONE FRAZIONATA *UF:* Serie reazionale di Bowen
1632 FRACTURE STRU	FRACTURES	FRACTURE *UF:* Geofracture	BRUCH *UF:* Geofraktur	TRESHCHINA *UF:* Diaklaza; Mik- rotreshchina	FRACTURA *UF:* Geofractura	FRATTURA *UF:* Parafora
1633 FRACTURE STRENGTH ENGI	FRACTURE STRENGTH	RESISTANCE RUP- TURE	BRUCH= FESTIGKEIT	[FRACTURE STRENGTH] PROCHNOST'	RESISTENCIA=A= LA=RUPTURA	RESISTENZA A ROTTURA
1634 FRACTURE ZONE STRU	FRACTURE ZONES *UF:* Geofracture	Zone fracturee RESEAU FRAC- TURE	Bruch=System STOERUNGS= SYSTEM	ZONA TRESH- CHINOVATOSTI	Zona=fracturada RED=FRACTURA	Zona fratturata RETICOLO DI FRATTURE
1635 FRACTURING STRU	FRACTURING	FRACTURATION	BRUCH=BILDUNG *UF:* Granulierung	TRESHCHINOVA- TOST'	FRACTURACION	FRATTURAZIONE
1636 FRAGMENT SEDI	FRAGMENTS	FRAGMENT	FRAGMENT	GRANULA	FRAGMENTO	FRAMMENTO
1637 FRAGMENTATION METH	FRAGMENTATION *UF:* Granulation	FRAGMENTATION *UF:* Granulation	PROBEN= ZERKLEINERUNG	DROBLENIE	FRAGMENTACION *UF:* Granulacion	FRAMMENTAZ- IONE *UF:* Granulazione
1638 FRAMBOIDAL TEXTURE TEST	FRAMBOIDAL TEX- TURE	TEXTURE FRAM- BOIDALE	FRAMBOIDAL= GEFUEGE	FRAMBOID	TEXTURA= FRAMBOIDAL	TESSITURA FRAM- BOIDALE
1639 FRANCIUM CHEE	FRANCIUM	FRANCIUM	FR	FRANTSIJ	FRANCIO	FRANCIO
1640 FRASNIAN STRS	FRASNIAN	FRASNIEN	FRASNE	FRAN	FRASNIENSE	FRASNIANO
1641 FREE ENERGY PHCH	FREE ENERGY	ENERGIE LIBRE	FREI.ENERGIE	EHNERGIYA SVO- BODNAYA	ENERGIA=LIBRE	ENERGIA LIBERA
1642 FREE OSCILLATION SOLI	FREE OSCILLA- TIONS	OSCILLATION PRO- PRE	EIGEN= SCHWINGUNG	Kolebanie svobodnoe ZEMLYA; KOLEBA- NIE	OSCILACION= PROPIA	OSCILLAZIONE PROPRIA
1643 FREE-AIR ANOMALY APPL	FREE=AIR ANOMA- LIES	ANOMALIE AIR LIBRE	FREILUFT= ANOMALIE	Anomaliya v svobod- nom vozdukhe ANOMALIYA GRAVITATSION- NAYA	ANOMALIA=AIRE= LIBRE	ANOMALIA IN ARIA LIBERA

	ENGLISH	FRANCAIS	DEUTSCH	RUSSKIJ	ESPANOL	ITALIANO
1644 FREQUENCY MATH	FREQUENCY	FREQUENCE	FREQUENZ	CHASTOTA	FRECUENCIA	FREQUENZA
1645 FREQUENCY DISTRIBUTION MATH	Frequency distribution STATISTICAL DIS- TRIBUTION	Distribution frequence DISTRIBUTION STATISTIQUE	Haeufigkeits= Verteilung STATIST. VERTEILUNG	RASPREDELENIE VEROYATNOSTEJ	Distribucion=de= frecuencia DISTRIBUCION= ESTADISTICA	Distribuzione di fre- quenza DISTRIBUZIONE STATISTICA
1646 FREQUENCY SOUNDING APPL	FREQUENCY SOUNDING	SONDAGE FRE- QUENCE	FREQUENZ= SONDIERUNG	ZONDIROVANIE CHASTOTNOE EHLEKTRICHESKOE	SONDEO= FRECUENCIA= VARIABLE	SONDAGGIO DI FREQUENZA
1647 FRESH WATER GEOH	FRESH WATER	EAU DOUCE	SUESS=WASSER	VODA PRESNAYA *UF:* Voda pit'evaya	AGUA=DULCE	ACQUA DOLCE
1648 FRESHWATER ENVIRONMENT PALE	FRESH=WATER ENVIRONMENT	MILIEU EAU DOUCE	SUESS=WASSER= MILIEU	PRESNOVODNYJ *UF:* Freshwater sedimen- tation	MEDIO=AGUA= DULCE	AMBIENTE D'ACQUA DOLCE
1649 FRESHWATER SEDIMENTATION SEDI	FRESH=WATER SEDIMENTATION	SEDIMENTATION EAU DOUCE	SUESS=WASSER= SEDIMENTATION	[FRESHWATER SEDIMENTATION] PRESNOVODNYJ; SEDIMENTATSIYA	SEDIMENTACION= AGUA=DULCE	SEDIMENTAZIONE IN ACQUA DOLCE
1650 FRICTION PHCH	FRICTION	FROTTEMENT	REIBUNG	TRENIE	FRICCION	FRIZIONE
1651 FRINGING REEF SEDI	FRINGING REEFS	RECIF FRANGEANT	SAUMRIFF	Rif okajmlyayushchij RIF KORALLOVYJ	ARRECIFE=DE= BARRERA	SCOGLIERA FRAN- GENTE
1652 FRONT IGNE	Fronts METASOMATISM	Front METASOMATOSE	Front METASOMATOSE	FRONT	Frente METASOMATISMO	Fronte METASOMATOSI
1653 FROST ACTION SURF	FROST ACTION *UF:* Congelation; Frost features	ACTION FROID *UF:* Congelation	KAELTE= WIRKUNG *UF:* Gefrieren	Vozdejstvie moroznoe VYVETRIVANIE MOROZNOE	ACCION=FRIO *UF:* Congelacion	AZIONE DEL GELO *UF:* Congelamento
1654 FROST FEATURE SURF	Frost features FROST ACTION	Morphologie perigla- ciaire PERIGLACIAIRE	Frost=Erscheinung PERIGLAZIAER	REL'EF MERZLOT- NYJ	Morfologia=periglaciar PERIGLACIAR	Struttura da gelo PERIGLACIALE
1655 FROST HEAVING SURF	FROST HEAVING	SOULEVEMENT GEL	FROST=HEBUNG	VSPUCHIVANIE MOROZNOE	ABOMBAMIENTO= POR=HIELO	SOLLEVAMENTO DA GELO
1656 FROZEN GROUND SURF	FROZEN GROUND *UF:* Cryopedology; Palsa	SOL GELE *UF:* Cryopedologie; Palse; Pergelisol; Sol polygonal	FROST=BODEN *UF:* Kryopedologie; Palsa; Permafrost; Polygon=Boden	GRUNT MERZLYJ	SUELO=HELADO *UF:* Criopedologia; Palsa; Permafrost; Suelo=poligonal	SUOLO GELATO *UF:* Criopedologia; Palsa; Permafrost; Suolo poligonale
1657 FRUCTIFICATION PALE	FRUITS	FRUCTIFICATION	FRUKTIFIKATION *UF:* Heterospor.Pflanze	ORGAN GENERA- TIVNYJ	FRUCTIFICACION	FRUTTIFICAZIONE
1658 FUEL RESOURCES COMS	FUEL RESOURCES *UF:* Fossil fuels	HYDROCARBURE *UF:* Eau gisement	KOHLENWASSER- STOFF *UF:* Lagerstaetten= Wasser	SYR'E TOPLIVOE	HIDROCARBURO *UF:* Agua=marginal	RISERVE COMBUS- TIBILI *UF:* Acqua di giacimento
1659 FUGACITY PHCH	FUGACITY	FUGACITE	FUGAZITAET	FUGITIVNOST'	FUGACIDAD	FUGACITA

	ENGLISH	FRANCAIS	DEUTSCH	RUSSKIJ	ESPANOL	ITALIANO
1660 FULGURITE TEST	FULGURITE	FULGURITE	FULGURIT	FUL'GURIT	FULGURITA	FULGURITE
1661 FULL HOLE DRILLING MINI	Full hole drilling BOREHOLES	Sondage carottage complet SONDAGE	Vollkern=Bohrung BOHRUNG	BURENIE BESK- ERNOVOE	Sondeo=testigo= continuo POZO=SONDEO	Perforazione SONDAGGIO
1662 FULLER'S EARTH COMS	FULLER'S EARTH	TERRE A FOULON	WALK=ERDE	Fullerova zemlya GLINA OTBELIVAYUSH- CHAYA	TIERRA=DE= BATAN	CRETA
1663 FULVIC ACID CHES	FULVIC ACIDS	ACIDE FULVIQUE	FULVO=SAEURE	FUL'VOKISLOTA	ACIDO=FULVICO	ACIDO FULVICO
1664 FUMAROLE IGNE	FUMAROLES	FUMEROLLE	FUMAROLE	FUMAROLA	FUMAROLA	FUMAROLA
1665 FUNCTION MATH	FUNCTIONS	Fonction EQUATION MATHEMATIQUE	Funktion MATHEMAT. GLEICHUNG	FUNKTSIYA *UF:* Funktsiya statisti- cheskaya; Spherical harmonic analysis	Funcion ECUACION= MATEMATICA	Funzione EQUAZIONE
1666 FUNCTIONAL MORPHOLOGY PALE	FUNCTIONAL MORPHOLOGY	MORPHOLOGIE FONCTIONNELLE	FUNKTIONELL. ANPASSUNGS= FORM	MORFOLOGIYA FUNKTSIAL'NAYE	MORFOLOGIA= FUNCIONAL	MORFOLOGIA FUN- ZIONALE
1667 FUNGI PALS	FUNGI	FUNGI	FUNGI	FUNGI	FUNGI	FUNGI
1668 FURROW SURF	FURROWS	Sillon erosion EROSION	Furche EROSION	BOROZDA *UF:* Zhelob lednikovyj	Estria EROSION	Solco d'erosione EROSIONE
1669 FUSINITE SEDS	Fusinite MACERALS	Fusinite MACERAL	Fusinit MAZERAL	FYUZINIT	Fusinita MACERAL	Fusinite MACERALE
1670 FUSULINIDAE PALS	FUSULINIDAE	FUSULINIDAE	FUSULINIDAE	FUSULINIDAE	FUSULINIDAE	FUSULINIDAE
1671 G LAYER SOLI	G layer INNER CORE	Couche G NOYAU TER- RESTRE INTERNE	G=Schicht INNER.ERD=KERN	G sloj YADRO ZEMLI	Capa=G NUCLEO=INTERNO	Strato G NUCLEO INTERNO
1672 GABBRO IGNS	GABBROS *UF:* Foidite; Granogab- bro	GABBRO *UF:* Foidite; Granogab- bro; Theralite	GABBRO *UF:* Foidit; Theralith	GABBRO *UF:* Granogabbro; Meta- gabbro	GABRO *UF:* Foidita; Granoga- bro; Teralita	GABBRO *UF:* Foidite; Granogab- bro; Teralite
1673 GADOLINIUM CHEE	GADOLINIUM	GADOLINIUM	GD	GADOLINIJ	GADOLINIO	GADOLINIO
1674 GALACTIC YEAR MISC	Galactic year EXTRATERRES- TRIAL GEOLOGY	Annee galactique FACTEUR TEMPS	Galakt.Jahr ZEITFAKTOR	GOD GALAKTI- CHESKIJ	Año=galactico FACTOR=TIEMPO	Anno galattico FATTORE TEMPO
1675 GALLIUM CHEE	GALLIUM	GALLIUM	GA	GALLIJ	GALIO	GALLIO
1676 GAMMA RAY PHCH	GAMMA RAYS	RAYONNEMENT GAMMA	GAMMA= STRAHLUNG	GAMMA-LUCHI	RAYOS=GAMMA	RAGGI GAMMA

	ENGLISH	FRANCAIS	DEUTSCH	RUSSKIJ	ESPANOL	ITALIANO
1677 GAMMA RAY SPECTROSCOPY METH	GAMMA=RAY SPECTROSCOPY	SPECTROMETRIE GAMMA	GAMMA=SPEKTROMETRIE	GAMMA=SPEKTROSKOPIYA	ESPECTROMETRIA=GAMMA	SPETTROMETRIA GAMMA
1678 GAMMA-GAMMA LOG APPL	GAMMA=GAMMA METHODS	DIAGRAPHIE GAMMA GAMMA	GAMMA=GAMMA=LOG	Karotazh gamma=gamma KAROTAZH; RAZ-VEDKA RADIO-METRICHESKAYA	TESTIFICACION=GAMMA=GAMMA	DIAGRAFIA GAMMA=GAMMA
1679 GAMMA-RAY LOG APPL	GAMMA=RAY METHODS	DIAGRAPHIE GAMMA	GAMMA=LOG	Karotazh gamma KAROTAZH; RAZ-VEDKA RADIO-METRICHESKAYA	TESTIFICACION=GAMMA	DIAGRAFIA GAMMA
1680 GANGUE IGNE	GANGUE	GANGUE	GANGART	PORODA PUSTAYA *UF:* Otval	GANGA	GANGA
1681 GANISTER COMS	Ganister REFRACTORY MATERIALS	Ganister REFRACTAIRE	Ganister FEUERFEST. ROHSTOFF	GANISTER	Ganister REFRACTARIO	Galestro REFRATTARIO
1682 GAP STRA	STRATIGRAPHIC GAPS *UF:* Diastems	LACUNE STRATI-GRAPHIQUE *UF:* Diasteme; Hiatus	LUECKE *UF:* Diastem; Hiatus	PERERYV (STRATIGR)	LAGUNA *UF:* Diastema; Hiato	LACUNA *UF:* Diastema; Hiatus
1683 GARNET MING	GARNET GROUP	GRENAT	GRANAT	GRANAT	GRANATE	GRANATO
1684 GAS CHES	GASES *UF:* Coal gas; Hydrogen gas; Oxygen gas; Phreatic gas	GAZ *UF:* Champ gaz; Charbon gaz; Gaz charbon; Gaz hydrogene; Gaz oxygene; Gaz phrea-tique; Inclusion gazeuse	GAS *UF:* Gas=Einschluss; Gas-Feld; Gas=Kohle; Kohlen=Gas; Phreat. Gas; Sauerstoff; Wasserstoff	GAZ *UF:* Karta neftegazonos-nosti	GAS *UF:* Campo=gas=hidrocarburo; Gas=carbon; Gas=freatico; Gas=hidrogeno; Gas=oxigeno; Hulla=grasa; Inclusion=gas	GAS *UF:* Campo di metano; Carbone da gas; Gas di carbone; Gas frea-tico; Gas idrogeno; Gas ossigeno; Inclu-sione gassosa
1685 GAS CAP ECON	Gas cap NATURAL GAS; TRAPS	Accumulation gaz GAZ NATUREL; PIEGE	Gaskappe ERDGAS; FALLE	SHAPKA GAZOVAYA	Acumulacion=gas GAS=NATURAL; TRAMPA	Cappello di gas GAS NATURALE; TRAPPOLA
1686 GAS CHROMATOGRAPHY METH	GAS CHROMATOG-RAPHY	CHROMATOGRA-PHIE PHASE GAZEUSE	GAS=CHROMATOGRAPHIE	KHROMATOGRA-FIYA GAZOVAYA	CROMATOGRAFIA=FASE=GASEOSA	GASCROMATOGRA-FIA
1687 GAS COAL COMS	Gas coal NATURAL GAS; COAL	Charbon gaz GAZ; CHARBON	Gas=Kohle GAS; KOHLE	UGOL' GAZOVYJ	Hulla=grasa GAS; CARBON	Carbone da gas GAS; CARBONE
1688 GAS CONDENSATE COMS	CONDENSATES	Condense gaz GAZ NATUREL	Gas=Kondensat ERDGAS	GAZOKONDENSAT	Condensado GAS=NATURAL	Condensato di gas GAS NATURALE
1689 GAS FIELD ECON	Gas fields OIL AND GAS FIELDS	Champ gaz CHAMP HYDRO-CARBURE; GAZ	Gas=Feld KW=FELD; GAS	BASSEJN GAZONOSNYJ	Campo=gas-hidrocarburo CAMPO=HIDROCARBURO; GAS	Campo di metano CAMPO D'IDROCARBURI; GAS
1690 GAS INCLUSION MINE	Gas inclusion FLUID INCLUSIONS	Inclusion gazeuse INCLUSION; GAZ	Gas=Einschluss EINSCHLUSS; GAS	VKLYUCHENIE GAZOVOE	Inclusion=gas INCLUSION; GAS	Inclusione gassosa INCLUSIONE; GAS

	ENGLISH	FRANCAIS	DEUTSCH	RUSSKIJ	ESPANOL	ITALIANO
1691 GAS STORAGE ENGI	GAS STORAGE	STOCKAGE GAZ	GAS= SPEICHERUNG	GAZOKHRANILISH- CHE	ALMACENAMIENTO= GAS	STOCCAGGIO DEL GAS
1692 GAS WELL ECON	Gas well NATURAL GAS; WELLS	Puits a gaz EXPLOITATION; GAZ NATUREL	Gas=Bohrung GEWINNUNG; ERD- GAS	SKVAZHINA GAZOVAYA	Pozo=de=gas EXPLOTACION; GAS=NATURAL	Pozzo di metano SFRUTTAMENTO; GAS NATURALE
1693 GAS=OIL RATIO ECON	GAS=OIL RATIO	Rapport gaz=petrole PETROLE; GAZ NATUREL	Gas=Oel=Verhaeltnis ERDOEL; ERDGAS	FAKTOR GAZOVYJ	Relacion=gas=petroleo PETROLEO; GAS= NATURAL	Rapporto gas=petrolio PETROLIO; GAS NATURALE
1694 GASEOUS PHASE PHCH	GASEOUS PHASE	PHASE GAZEUSE	GAS=PHASE	FAZA GAZOVAYA	FASE=GASEOSA	FASE GASSOSA
1695 GASIFICATION METH	GASIFICATION	GAZEIFICATION	VERGASUNG	GAZIFIKATSIYA	GASIFICACION	GASSIFICAZIONE
1696 GASTROPODA PALS	GASTROPODA *UF:* Anisopleura; Iso- pleura; Opistho- branchia; Pulmonata	GASTROPODA *UF:* Isopleura; Opistho- branchia; Pulmonata	GASTROPODA *UF:* Isopleura; Opistho- branchia; Pulmonata	GASTROPODA *UF:* Archaeogastropoda; Mesogastropoda; Neo- gastropoda; Opistho- branchia	GASTROPODA *UF:* Isopleura; Opistho- branchia; Pulmonata	GASTROPODA *UF:* Isopleura; Opistho- branchia; Pulmonata
1697 GASTROZOOID PALE	Gastrozooid ANATOMY	Gastrozooide ANATOMIE	Gastrozooid ANATOMIE	GASTROZOOID	Gastrozoide ANATOMIA	Gastrozooide ANATOMIA
1698 GAUGING GEOH	GAUGING	JAUGEAGE	FLUESSIGKEITS= MESSUNG	IZMERENIE	AFORO	CALIBRAZIONE
1699 GAUSS EPOCH STRA	GAUSS EPOCH	EPOQUE GAUSS	GAUSS=EPOCHE	GAUSSA EHPOKHA	EPOCA=GAUSS	EPOCA GAUSS
1700 GEANTICLINE STRU	GEANTICLINES	GEANTICLINAL	GEANTIKLINE	GEOANTIKLINAL'	GEANTICLINAL	GEOANTICLINALE
1701 GEDINNIAN STRS	GEDINNIAN	GEDINNIEN	GEDINNE	ZHEDIN	GEDINIENSE	GEDINNIANO
1702 GEL PHCH	GELS	GEL COLLOIDAL	GEL	GEL'	GEL=COLOIDAL	GEL
1703 GEL MINERAL MINE	Mineral gel COLLOIDAL MATE- RIALS	Mineral gel COLLOIDE	Gel=Mineral KOLLOID	Mineraloid KOLLOID; MIN- ERAL	Mineral=gel COLOIDE	Gel minerale COLLOIDE
1704 GELIVITY SURF	GELIVITY	GELIVITE	FROST= BESTAENDIGKEIT	[GELIVITY] VYVETRIVANIE MOROZNOE	GELIVACION	GELIVITA
1705 GEMMATION PALE	Gemmation ONTOGENY	Bourgeonnement ONTOGENIE	Knospung ONTOGENESE	POCHKOVANIE	Gemacion ONTOGENIA	Gemmazione ONTOGENESI
1706 GEMMOLOGY MINE	Gemmology GEMS	Gemmologie GEMME	Gemmologie SCHMUCKSTEIN	Gemmologiya KAMEN' DRA- GOTSENNYJ	Gemologia GEMA	Gemmologia GEMMA
1707 GEMSTONE COMS	GEMS *UF:* Gemmology; Pre- cious stone; Semipre- cious stone	GEMME *UF:* Gemmologie; Pierre precieuse; Pierre semi- precieuse	SCHMUCKSTEIN *UF:* Edelstein; Gem- mologie; Halbedelstein	KAMEN' DRA- GOTSENNYJ *UF:* Gemmologiya; Kamen poludragotsen- nyj; Kamen' yuvelirnyj	GEMA *UF:* Gemologia; Piedra= preciosa; Piedra= semipreciosa	GEMMA *UF:* Gemmologia; Pietra preziosa; Pietra semi- preziosa

	ENGLISH	FRANCAIS	DEUTSCH	RUSSKIJ	ESPANOL	ITALIANO
1708 GENERATION MISC	Generation(gen.)	Generation(gen.)	Generation(gen.)	GENERATSIYA	Generacion(gen.)	Generazione(gen.)
1709 GENESIS GEOL	GENESIS *UF:* Epimagmatic origin; Igneous rock genesis; Lithogenesis; Petro- genesis	GENESE *UF:* Genese sedi- mentaire; Lithogenese; Origine epimagma- tique; Theorie mouve- ment par ascension; Theorie mouvement par descente	GENESE *UF:* Aszendenz=Theorie; Deszendenz=Theorie; Epimagmat.Genese; Lithogenese; Sediment=Genese	GENEZIS *UF:* Genezis izverzhen- nykh porod; Krateroo- brazovanie; Nucleation; Ore source; Pedo- genezis; Roll=type deposit; Tip mestorozh- denij pnevmatolitovyj	GENESIS *UF:* Litogenesis; Origen=epimagmatico; Sedimentogenesis; Teoria=movimiento= ascendente; Teoria= movimiento= descendente	GENESI *UF:* Genesi di sedimento; - Litogenesi; Origine epimagmatica; Teoria ascensionale; Teoria discendente
1710 GENOMORPH PALE	Genomorphs BIOLOGIC EVOLU- TION	Genomorphe EVOLUTION BIOLOGIQUE	Genomorph BIOLOG. EVOLUTION	Genomorfa EHVOLYUTSIYA	Genomorfo EVOLUCION= BIOLOGICA	Genomorfo EVOLUZIONE BIOLOGICA
1711 GENOTYPE PALE	Genotypes BIOTYPES	Genotype BIOTYPE	Genotyp BIOTYPUS	GENOTIP *UF:* Paratip	Genotipo BIOTIPO	Genotipo BIOTIPO
1712 GEOBAROMETRY GEOL	GEOLOGIC BAROM- ETRY	GEOBAROMETRIE	GEOBAROMETRIE	GEOBAROMETRIYA	GEOBAROMETRIA	GEOBAROMETRIA
1713 GEOBIOS PALE	Geobios BIOGEOGRAPHY; FAUNA	Geobios BIOGEOGRAPHIE; FAUNE	Geobios BIOGEOGRAPHIE; FAUNA	GEOBIOS	Geobios BIOGEOGRAFIA; FAUNA	Geobios BIOGEOGRAFIA; FAUNA
1714 GEOBOTANICAL **EXPLORATION** ECON	GEOBOTANICAL METHODS	PROSPECTION GEOBOTANIQUE	GEOBOTAN. PROSPEKTION	Poisk geobotanicheskij POISK BIOGEOKHIMI- CHESKIJ	PROSPECCION= GEOBOTANICA	PROSPEZIONE GEOBOTANICA
1715 GEOCHEMICAL CYCLE GEOC	GEOCHEMICAL CYCLE	CYCLE GEOCHIMIQUE	GEOCHEM.ZYKLUS	TSIKL GEOKHIMI- CHESKIJ	CICLO= GEOQUIMICO	CICLO GEOCHIMICO
1716 GEOCHEMICAL **EXPLORATION** ECON	GEOCHEMICAL METHODS *UF:* Clarke; Geochemi- cal facies; Geochemical threshold	PROSPECTION GEOCHIMIQUE *UF:* Clarke; Depression geochimique; Facies geochimique; Seuil anormal	GEOCHEM. PROSPEKTION *UF:* Clarke; Geochem. Einbruchskrater; Geochem.Fazies; Sch- wellenwert	POISK GEOKHIMI- CHESKIJ	PROSPECCION= GEOQUIMICA *UF:* Clarke; Facies= geoquimica; Hundimiento= geoquimico; Umbral= anomalia	PROSPEZIONE GEOCHIMICA *UF:* Abbattimento geochimico; Clarke; Facies geochimica; Soglia geochimica
1717 GEOCHEMICAL FACIES GEOC	Geochemical facies GEOCHEMICAL METHODS	Facies geochimique PROSPECTION GEOCHIMIQUE	Geochem.Fazies GEOCHEM. PROSPEKTION	FATSIYA GEOKHIMI- CHESKAYA	Facies=geoquimica PROSPECCION= GEOQUIMICA	Facies geochimica PROSPEZIONE GEOCHIMICA
1718 GEOCHEMICAL **INDICATOR** ECON	GEOCHEMICAL INDICATORS	INDICATEUR GEOCHIMIQUE	GEOCHEM. INDIKATOR	INDIKATOR GEOKHIMICHESKIJ	INDICADOR= GEOQUIMICO	INDICATORE GEOCHIMICO
1719 GEOCHEMICAL MAP GEOC	GEOCHEMICAL MAPS	CARTE GEOCHIMIQUE	GEOCHEM.KARTE	KARTA GEOKHIMI- CHESKAYA	MAPA= GEOQUIMICO	CARTA GEOCHIMICA
1720 GEOCHEMICAL **PROFILE** GEOC	GEOCHEMICAL PROFILES	PROFIL GEOCHIMIQUE	GEOCHEM.PROFIL	PROFIL' GEOKHIMICHESKIJ	PERFIL= GEOQUIMICO	PROFILO GEOCHIMICO
1721 GEOCHEMICAL SINK ECON	SINKS	Depression geochimique PROSPECTION GEOCHIMIQUE; ANOMALIE	Geochem. Einbruchskrater GEOCHEM. PROSPEKTION; ANOMALIE	[GEOCHEMICAL SINK] GEOKHIMIYA	Hundimiento= geoquimico PROSPECCION= GEOQUIMICA; ANOMALIA	Abbattimento geochimico PROSPEZIONE GEOCHIMICA; ANOMALIA

	ENGLISH	FRANCAIS	DEUTSCH	RUSSKIJ	ESPANOL	ITALIANO
1722 GEOCHEMISTRY GEOC	GEOCHEMISTRY *UF:* Isotope geochemis-try	GEOCHIMIE	GEOCHEMIE	GEOKHIMIYA *UF:* Geochemical sink	GEOQUIMICA	GEOCHIMICA
1723 GEOCHRONOLOGIC OVERPRINT ISOT	Overprint ABSOLUTE AGE; CORRECTIONS	Superposition age DATATION; COR-RECTION	Ueberpraegung PHYSIKAL. ALTERSBESTIM-MUNG; KORREK-TUR	NALOZHENNOST'	Superposicion=edad DATACION; COR-RECCION	Sovrapposizione di eta DATAZIONE; COR-REZIONE
1724 GEOCHRONOLOGIC UNIT STRA	Geochronologic unit GEOCHRONOLOGY	Unite geochronologique UNITE STRATI-GRAPHIQUE	Geochronolog.Einheit STRATIGRAPH. EINHEIT	PODRAZDELENIE GEOKHRONOLOGI-CHESKOE	Unidad=geocronologica UNIDAD=ESTRATIGRAFICA	Unita geocronologica UNITA STRATI-GRAFICA
1725 GEOCHRONOLOGY ISOT	GEOCHRONOLOGY *UF:* Geochronologic unit; Selenochronology; Varve chronology	GEOCHRONOLOGIE *UF:* Chronologie varve; Selenochronologie	GEOCHRONOLOGIE *UF:* Selenochronologie; Warwen=Chronologie	GEOKHRONOL-OGIYA	GEOCRONOLOGIA *UF:* Selenocronologia; Varva=cronologia	GEOCRONOLOGIA *UF:* Selenocronologia; Varvecronologia
1726 GEODE SEDI	GEODES	GEODE *UF:* Druse	GEODE *UF:* Druse	ZHEODA	GEODA *UF:* Drusa	GEODE *UF:* Drusa
1727 GEODESY SOLI	GEODESY *UF:* Gnomonic projec-tion; Hypsometry; Radio interferometry	GEODESIE *UF:* Hypsometrie; Inter-ferometrie onde radio; Projection gnomonique	GEODAESIE *UF:* Gnomon.Projektion; Hoehen=Messung; Radio=Interferometrie	GEODEZIYA	GEODESIA *UF:* Hipsometria; Proyeccion=gnomonica; Radio=interferometria	GEODESIA *UF:* Ipsometria; Proiez-ione gnomonica; Radiointerferometria
1728 GEODETIC COORDINATE MISC	GEODETIC COOR-DINATES *UF:* Latitude; Longitude	COORDONNEE GEODESIQUE *UF:* Latitude; Longitude	GEODAET. KOORDINATE *UF:* Geograph.Breite; Geograph.Laenge	KOORDINATA (GEODET)	COORDENADAS=GEODESICAS *UF:* Latitud; Longitud=geografica	COORDINATE GEODETICHE *UF:* Latitudine; Longitu-dine
1729 GEODYNAMICS SOLI	GEODYNAMICS *UF:* Dynamic geology	GEODYNAMIQUE *UF:* Geologie dynamique; Geoondula-tion	GEODYNAMIK *UF:* Dynam.Geologie; Geoundation	GEODINAMIKA	GEODINAMICA *UF:* Geologia=dinamica; Geoondulacion	GEODINAMICA *UF:* Geologia dinamica; Geoundazione
1730 GEOFRACTURE STRU	Geofracture FRACTURE ZONES	Geofracture FRACTURE	Geofraktur BRUCH	ZONA SOCH-LENENIYA	Geofractura FRACTURA	Parafora FRATTURA
1731 GEOGRAPHY GEOL	GEOGRAPHY *UF:* East; North; South; West	GEOGRAPHIE *UF:* Est; Nord; Ouest; Sud	GEOGRAPHIE *UF:* Norden; Osten; Sueden; Westen	GEOGRAFIYA	GEOGRAFIA *UF:* Este; Norte; Oeste; Sur	GEOGRAFIA *UF:* Est; Nord; Ovest; Sud
1732 GEOID SOLI	GEOID	GEOIDE	GEOID	GEOID	GEOIDE	GEOIDE
1733 GEOISOTHERM SOLI	Geoisotherm GEOTHERMAL GRADIENT	Geoisotherme GRADIENT GEO-THERMIQUE	Geoisotherme GEOTHERM. GRADIENT	GEOIZOTERMA	Geoisoterma GRADIENTE=GEOTERMICO	Geoisoterma GRADIENTE GEOTERMICO
1734 GEOLOGIC COLUMN STRA	STRATIGRAPHIC COLUMNS	ECHELLE STRATI-GRAPHIQUE *UF:* Chronozone; Cop-ernicien; Eon; Epoque; Eratheme; Eratosthe-nien; Ere; Etage; For-mation; Imbrien; Mem-bre; Periode; Pre-imbrien; Procellarien; Serie	STRATIGRAPH. SKALA *UF:* Aeon; Aera; Aera-them; Chronozone; Epoche; Eratosthen. Alter; Formation; Imbrium; Kopernikan. Alter; Oceanus=Procellarius; Periode; Prae=Imbrium; Sch-ichtglied; Serie; Stufe	KOLONKA STRATI-GRAFICHESKAYA	ESCALA=ESTRATIGRAFICA *UF:* Coperniciense; Cronozona; Eon; Epoca; Era; Eratema; Eratostensiense; For-macion; Imbriano; Miembro; Periodo; Piso=estratigrafico; Preimbriense; Procella-riense; Serie	SCALA STRATI-GRAFICA *UF:* Copernicano; Crono-zona; Eone; Epoca; Era; Eratema; Eratos-teniano; Formazione; Imbriano; Membro; Periodo; Piano strati-grafico; Preimbriano; Procellariano; Serie

	ENGLISH	FRANCAIS	DEUTSCH	RUSSKIJ	ESPANOL	ITALIANO
1735 GEOLOGIC MAP GEOL	GEOLOGIC MAPS *UF:* Lithologic maps	CARTE GEOLOGIQUE *UF:* Carte petro- graphique	GEOLOG.KARTE *UF:* Petrograph.Karte	KARTA GEOLOGI- CHESKAYA *UF:* Karta petrografi- cheskaya	MAPA=GEOLOGICO *UF:* Mapa=petrografico	CARTA GEOLOGICA *UF:* Carta litologica
1736 GEOLOGIC SECTION GEOL	SECTIONS	COUPE GEOLOGIQUE	PROFIL	RAZREZ GEOLOGI- CHESKIJ	CORTE= GEOLOGICO	PROFILO GEOLOGICO
1737 GEOLOGIC STRUCTURE STRU	Geologic structure TECTONICS	Structure geologique TECTONIQUE	Tekton.Struktur TEKTONIK	STRUKTURA TEK- TONICHESKAYA *UF:* Ehlement struk- turnyj; Stroenie tek- tonicheskoe; Struk- turnaya diagramma	Estructura=geologica TECTONICA	Struttura tettonica TETTONICA
1738 GEOLOGICAL HAZARDS GEOL	GEOLOGIC HAZ- ARDS	CATASTROPHE NATURELLE	NATURKATAS- TROPHE	OPASNOST' GEOLOGI- CHESKAYA	CATASTROFE= NATURAL	RISCHIO GEOLOGICO
1739 GEOLOGICAL SURVEY MISC	SURVEY ORGANI- ZATIONS	SERVICE GEOLOGIQUE	GEOLOG.DIENST	SLUZHBA GEOLOGI- CHESKAYA	SERVICIO= GEOLOGICO	SERVIZIO GEOLOGICO
1740 GEOLOGIST MISC	GEOLOGISTS	ROLE GEOLOGUE	GEOLOG. MITWIRKUNG	GEOLOG	PAPEL=DEL= GEOLOGO	RUOLO DEL GEOLOGO
1741 GEOLOGY GEOL	GEOLOGY *UF:* Applied geology; Earth science; Geonomy; History of geology; Subsurface geology	GEOLOGIE *UF:* Geologie appliquee; Geologie subsurface; Geonomie; Science de la terre	GEOLOGIE *UF:* Angewandt. Geologie; Geonomie; Geowissenschaften; Subsurface=geology	GEOLOGIYA *UF:* Geologija priklad- naya; Nauki o zemle	GEOLOGIA *UF:* Ciencia=de=la= tierra; Geologia= aplicada; Geologia= del=subsuelo; Geo- nomia	GEOLOGIA *UF:* Geologia applicata; Geologia del substrato; Geonomia; Scienze della terra
1742 GEOMAGNETIC POLE SOLI	POLE POSITIONS *UF:* Paleomagnetic pole	POLE GEOMAGNE- TIQUE *UF:* Derive pole; Pole paleomagnetique	GEOMAGNET.POL *UF:* Palaeomagnet.Pol; Pol=Wanderung	POLYUS MAGNIT- NYJ	POLO= GEOMAGNETICO *UF:* Deriva=polar; Polo= paleomagnetico	POLO GEOMAGNE- TICO *UF:* Deriva del polo; Polo paleomagnetico
1743 GEOMAGNETIC REVERSAL SOLI	Geomagnetic reversal REVERSALS	INVERSION CHAMP *UF:* Inversion	FELDINVERSION *UF:* Umkehrung	INVERSIYA MAGNITNOGO POLYA *UF:* Izmeneniya magnit- nogo polya; Reversal	INVERSION=DE= CAMPO *UF:* Inversion	INVERSIONE DEL CAMPO *UF:* Inversione
1744 GEOMAGNETIC SECULAR VARIATION SOLI	SECULAR VARIA- TIONS	VARIATION SECULAIRE	MAGNET. SAEKULARVARIA- TION	Izmeneniya magnitnogo polya INVERSIYA MAGNITNOGO POLYA	VARIACION= SECULAR	VARIAZIONE SECO- LARE
1745 GEOMAGNETISM SOLI	Geomagnetism MAGNETIC FIELD	Geomagnetisme CHAMP MAGNE- TIQUE	Geomagnetismus MAGNETFELD	GEOMAGNETIZM *UF:* Dinamo teoriya; Magnetizm	Geomagnetismo CAMPO= MAGNETICO	Geomagnetismo CAMPO MAGNE- TICO
1746 GEOMETRY MISC	GEOMETRY *UF:* Axis; Critical angle; Length; Planes; Radius; Shape; Upthrow	GEOMETRIE *UF:* Angle critique; Axe; Forme; Levre relevee; Longueur; Plan; Radius; Structure cylindrique; Talus equilibre	GEOMETRIE *UF:* Achse; Flaeche; Form; Gehoben.Scholle; Krit.Winkel; Laenge; Radius; Schuettungs= Winkel	GEOMETRIYA *UF:* Os' skladki; Ugol kriticheskij	GEOMETRIA *UF:* Angulo=critico; Angulo=de=reposo; Eje; Estructura= cilindrica; Forma; Labio=levantado; Longitud; Plano; Radius	GEOMETRIA *UF:* Angolo critico; Angolo di riposo; Asse; Forma; Lembo rialzato; Lunghezza; Piano; Radius; Struttura cilindrica

	ENGLISH	FRANCAIS	DEUTSCH	RUSSKIJ	ESPANOL	ITALIANO
1747 GEOMORPHOLOGICAL MAP SURF	GEOMORPHO- LOGIC MAPS	CARTE GEOMOR- PHOLOGIQUE	GEOMORPHOLOG. KARTE	Karta geomorfologi- cheskaya KARTA; GEOMOR- FOLOGIYA	MAPA= GEOMORFOLOGICO	CARTA GEOMOR- FOLOGICA
1748 GEOMORPHOLOGY SURF	GEOMORPHOLOGY	GEOMORPHOLOGIE	GEOMORPHOLOGIE	GEOMOR- FOLOGIYA *UF:* Geomorfologiya podvodnaya; Karta geomorfologicheskya	GEOMORFOLOGIA	GEOMORFOLOGIA
1749 GEONOMY GEOL	Geonomy GEOLOGY	Geonomie GEOLOGIE	Geonomie GEOLOGIE	GEONOMIYA	Geonomia GEOLOGIA	Geonomia GEOLOGIA
1750 GEOPHONE INST	GEOPHONES	GEOPHONE	GEOPHON	Geofon SEJSMOPRIEMNIK	GEOFONO	GEOFONO
1751 GEOPHYSICAL ANOMALY SOLI	Geophysical anomaly ANOMALIES	Anomalie geophysique ANOMALIE	Geophysikal.Anomalie ANOMALIE	ANOMALIYA GEO- FIZICHESKAYA	Anomalia=geofisica ANOMALIA	Anomalia geofisica ANOMALIA
1752 GEOPHYSICAL EXPLORATION APPL	Geophysical exploration GEOPHYSICAL METHODS; MIN- ERAL EXPLORA- TION	Prospection geophy- sique METHODE GEO- PHYSIQUE	Geophysikal. Prospektion GEOPHYSIKAL. METHODE	RAZVEDKA GEO- FIZICHESKAYA	Prospeccion=geofisica METODO= GEOFISICO	Prospezione geofisica METODO GEOFI- SICO
1753 GEOPHYSICAL INSTRUMENTS INST	Geophysical instru- ments INSTRUMENTS; GEOPHYSICAL METHODS	Instrument geophysique INSTRUMENTA- TION; METHODE GEOPHYSIQUE	Geophysikal.Instrument INSTRUMEN- TIERUNG; GEOPHYSIKAL. METHODE	TEKHNIKA GEO- FIZICHESKAYA	Instrumentos=geofisicos INSTRUMENTA- CION; METODO= GEOFISICO	Strumento geofisico STRUMENTAZ- IONE; METODO GEOFISICO
1754 GEOPHYSICAL MAP SOLI	GEOPHYSICAL MAPS	CARTE GEOPHY- SIQUE	GEOPHYSIKAL. KARTE	KARTA GEOFIZI- CHESKAYA	MAPA=GEOFISICO	CARTA GEOFISICA
1755 GEOPHYSICAL METHOD APPL	GEOPHYSICAL METHODS *UF:* Geophysical explo- ration; Geophysical instruments; Geother- mal method; Reception equipment	METHODE GEO- PHYSIQUE *UF:* Equipement recep- tion; Instrument geo- physique; Prospection geophysique	GEOPHYSIKAL. METHODE *UF:* Empfangs=Geraet; Geophysikal. Instrument; Geophysikal. Prospektion	METOD GEOFIZI- CHESKIJ *UF:* Marine methods	METODO= GEOFISICO *UF:* Instrumentos= geofisicos; Prospeccion= geofisica; Recepcion= señal	METODO GEOFI- SICO *UF:* Prospezione geofi- sica; Ricezione; Stru- mento geofisico
1756 GEOPHYSICAL PROFILE SOLI	GEOPHYSICAL PROFILES	PROFIL GEOPHY- SIQUE	GEOPHYSIKAL. PROFIL	PROFIL' GEOFIZI- CHESKIJ	PERFIL=GEOFISICO	PROFILO GEOFI- SICO
1757 GEOPHYSICAL SURVEY APPL	GEOPHYSICAL SURVEYS	LEVE GEOPHY- SIQUE	GEOPHYSIKAL. VERMESSUNG	S'EMKA GEOFIZI- CHESKAYA	LEVANTAMIENTO= GEOFISICO	RILEVAMENTO GEOFISICO
1758 GEOPHYSICIST SOLI	Geophysicists GEOPHYSICS; MANPOWER	Geophysicien GEOPHYSIQUE; MAIN D'OEUVRE	Geophysiker GEOPHYSIK; ARBEITSWESEN	GEOFIZIK	Geofisico GEOFISICA; MANO=DE=OBRA	Geofisico GEOFISICA; MANO D'OPERA
1759 GEOPHYSICS SOLI	GEOPHYSICS *UF:* Geophysicists	GEOPHYSIQUE *UF:* Geophysicien	GEOPHYSIK *UF:* Geophysiker	GEOFIZIKA	GEOFISICA *UF:* Geofisico	GEOFISICA *UF:* Geofisico
1760 GEOPRESSURE ENGI	GEOPRESSURE	GEOPRESSION	PORENWASSER- DRUCK	[GEOPRESSURE] DAVLENIE GEOSTATI- CHESKOE	GEOPRESION	GEOPRESSIONE

	ENGLISH	FRANCAIS	DEUTSCH	RUSSKIJ	ESPANOL	ITALIANO
1761 GEOSECS MARI	GEOSECS	GEOSECS	GEOSECS	[GEOSECS]	GEOSECS	GEOSECS
1762 GEOSTATISTICS MATH	GEOSTATISTICS	GEOSTATISTIQUE	GEOSTATISTIK	GEOSTATISTIKA	GEOESTADISTICA	GEOSTATISTICA
1763 GEOSYNCLINAL RIDGE STRU	Geosynclinal ridge GEOSYNCLINES	Ride geosynclinale GEOSYNCLINAL	Geosynklinal=Schwelle GEOSYNKLINALE	[GEOSYNCLINAL RIDGE] GEOSINKLINAL'	Cadena=geosinclinal GEOSINCLINAL	Cordigliera intrageosinclinale GEOSINCLINALE
1764 GEOSYNCLINAL SEA MARI	Geosynclinal sea GEOSYNCLINES	Mer geosynclinale GEOSYNCLINAL	Geosynklinal=Meer GEOSYNKLINALE	MORE GEOSINKLI-NAL'NOE	Mar=geosinclinal GEOSINCLINAL	Mare geosinclinale GEOSINCLINALE
1765 GEOSYNCLINAL SEDIMENTATION SEDI	GEOSYNCLINAL SEDIMENTATION	SEDIMENTATION GEOSYNCLINALE	GEOSYNKLINAL= SEDIMENTATION	[GEOSYNCLINAL SEDIMENTATION] GEOSINKLINAL'; SEDIMENTATSIYA	SEDIMENTACION= GEOSINCLINAL	SEDIMENTAZIONE GEOSINCLINALE
1766 GEOSYNCLINAL TRENCH STRU	Geosynclinal trench GEOSYNCLINES	Sillon GEOSYNCLINAL	Geosynklinal=Furche GEOSYNKLINALE	[GEOSYNCLINAL TRENCH] GEOSINKLINAL'	Surco GEOSINCLINAL	Fossa intrageosinclinale GEOSINCLINALE
1767 GEOSYNCLINE STRU	GEOSYNCLINES *UF:* Ensialic geosynclines; Ensimatic geosynclines; Epieugeosynclines; Externides; Foredeeps; Geosynclinal ridge; Geosynclinal sea; Geosynclinal trench; Great=circle belt; Idiogeosynclines; Leptogeosyncline; Mesogeosyncline; Parageosynclines; Paraliageosynclines; Taphrogeosynclines	GEOSYNCLINAL *UF:* Avant=fosse; Epieugeosynclinal; Geosynclinal ensialique; Geosynclinal ensimatique; Idiogeosynclinal; Leptogeosynclinal; Mer geosynclinale; Mesogeosynclinal; Parageosynclinal; Paraliageosynclinal; Ride geosynclinale; Sillon; Taphrogeosynclinal	GEOSYNKLINALE *UF:* Ensial. Geosynklinale; Ensimat.Geosynklinale; Epieugeosynklinale; Geosynklinal=Furche; Geosynklinal=Meer; Geosynklinal=Schwelle; Idiogeosynklinale; Lepto=Geosynklinale; Mesogeosynklinale; Parageosynklinale; Paral.Geosynklinale; Taphro=Geosynklinale; Vortiefe	GEOSINKLINAL' *UF:* Geosynclinal ridge; Geosynclinal sedimentation; Geosynclinal trench; Poyas orogenicheskij; Poyas podvizhnyj; Sistema skladchataya	GEOSINCLINAL *UF:* Antefosa; Cadena=geosinclinal; Epieugeosinclinal; Geosinclinal-simatico; Geosinclinal=sialico; Idiogeosinclinal; Leptogeosinclinal; Mar=geosinclinal; Mesogeosinclinal; Parageosinclinal; Paraliageosinclinal; Surco; Tafrogeosinclinal	GEOSINCLINALE *UF:* Avanfossa; Cordigliera intrageosinclinale; Epieugeosinclinale; Fossa intrageosinclinale; Geosinclinale intercratonico; Geosinclinale intracratonico; Idiogeosinclinale; Leptogeosinclinale; Mare geosinclinale; Mesogeosinclinale; Parageosinclinale; Paraliageosinclinale; Tafrogeosinclinale
1768 GEOTECHNICAL MAP ENGI	GEOTECHNICAL MAPS *UF:* Geotechnical survey	CARTE GEOTECHNIQUE *UF:* Leve geotechnique	GEOTECHN.KARTE *UF:* Geotechn.Aufnahme	Karta inzhenerno=geologicheskaya KARTA; GEOLOGIYA INZHENERNAYA	MAPA=GEOTECNICO *UF:* Estudio=geotecnico	CARTA GEOTECNICA *UF:* Rilevamento geotecnico
1769 GEOTECHNICAL PROPERTY ENGI	ENGINEERING PROPERTIES	PROPRIETE GEOTECHNIQUE	GEOTECHN. EIGENSCHAFT	SVOJSTVO GEOTEKHNICHESKOE	PROPIEDAD=GEOTECNICA	PROPRIETA GEOTECNICA
1770 GEOTECHNICAL SURVEY ENGI	Geotechnical survey GEOTECHNICAL MAPS	Leve geotechnique CARTE GEOTECHNIQUE	Geotechn.Aufnahme GEOTECHN.KARTE	S'EMKA INZHENERNO= GEOLOGICHESKAYA	Estudio=geotecnico MAPA=GEOTECNICO	Rilevamento geotecnico CARTA GEOTECNICA
1771 GEOTECHNICS ENGI	Geotechnics ENGINEERING GEOLOGY	GEOTECHNIQUE *UF:* Geologie ingenieur	GEOTECHNIK *UF:* Ingenieur=Geologie	Geotekhnika GEOLOGIYA INZHENERNAYA	GEOTECNIA *UF:* Geologia=del=ingeniero	GEOTECNICA *UF:* Geologia tecnica
1772 GEOTECTONICS STRU	Geotectonics TECTONICS	Geotectonique TECTONIQUE	Geotektonik TEKTONIK	GEOTEKTONIKA	Geotectonica TECTONICA	Geotettonica TETTONICA

113 • MULTILINGUAL THESAURUS OF GEOSCIENCES

	ENGLISH	FRANCAIS	DEUTSCH	RUSSKIJ	ESPANOL	ITALIANO
1773 GEOTHERMAL ENERGY ECON	GEOTHERMAL ENERGY *UF:* Geothermics	ENERGIE GEO-THERMIQUE	GEOTHERM. ENERGIE	EHNERGIYA GEOTERMAL'NAYA	ENERGIA=GEOTERMICA	ENERGIA GEOTER-MICA
1774 GEOTHERMAL FIELD ECON	GEOTHERMAL FIELDS	CHAMP GEOTHER-MIQUE	GEOTHERM.FELD	POLE TEPLOVOE	CAMPO=GEOTERMICO	CAMPO GEOTER-MICO
1775 GEOTHERMAL GRADIENT SOLI	GEOTHERMAL GRADIENT *UF:* Geoisotherm	GRADIENT GEO-THERMIQUE *UF:* Geoisotherme	GEOTHERM. GRADIENT *UF:* Geoisotherme	GRADIENT GEOTERMICHESKIJ	GRADIENTE=GEOTERMICO *UF:* Geoisoterma	GRADIENTE GEOTERMICO *UF:* Geoisoterma
1776 GEOTHERMAL METAMORPHISM IGNE	Geothermal metamorphism THERMAL META-MORPHISM	Metamorphisme geothermique METAMORPHISME	Geotherm. Metamorphose METAMORPHOSE	METAMORFIZM GEOTERMAL'NYJ	Metamorfismo-geotermico METAMORFISMO	Metamorfismo geoter-mico TIPO DI METAMOR-FISMO
1777 GEOTHERMAL METHOD APPL	Geothermal method GEOPHYSICAL METHODS	Methode geothermique GEOTHERMIE	Geotherm.Methode GEOTHERMIK	METOD GEOTERMI-CHESKIJ	Metodo=geotermico GEOTERMIA	Metodo geotermico GEOTERMIA
1778 GEOTHERMAL SYSTEM ECON	GEOTHERMAL SYSTEMS *UF:* Geothermal method	SYSTEME GEO-THERMIQUE	GEOTHERM. SYSTEM	[GEOTHERMAL SYSTEM]	SISTEMA=GEOTERMICO	SISTEMA GEOTER-MICO
1779 GEOTHERMICS APPL	Geothermics GEOTHERMAL ENERGY	GEOTHERMIE *UF:* Methode geother-mique; Prospection thermique; Source chaleur	GEOTHERMIK *UF:* Geotherm.Methode; Waerme=Quelle	GEOTERMIKA	GEOTERMIA *UF:* Fuentes=de=calor; Metodo=geotermico; Prospeccion=termica	GEOTERMIA *UF:* Metodo geotermico; Prospezione termica; Sorgente di calore
1780 GEOTHERMOMETRY GEOL	GEOLOGIC THER-MOMETRY *UF:* Paleothermometry	GEOTHERMOME-TRIE	GEOTHERMOME-TRIE	GEOTERMO-METRIYA	GEOTERMO-METRIA	GEOTERMO-METRIA
1781 GEOUNDATION STRU	UNDATION	Geoondulation THEORIE; GEODY-NAMIQUE	Geoundation THEORIE; GEODY-NAMIK	GEOUNDATSIYA	Geoondulacion TEORIA; GEODI-NAMICA	Geoundazione TEORIA; GEODI-NAMICA
1782 GERMANIUM CHEE	GERMANIUM	GERMANIUM	GE	GERMANIJ	GERMANIO	GERMANIO
1783 GERMANOTYPE TECTONICS STRU	Germanotype tectonics TECTONICS	Tectonique germano-type TECTOGENESE	Germanotyp.Tektonik TEKTONIK	TEKTONIKA GER-MANOTIPNAYA	Tectonica=germanica TECTONICA	Tettonica germano-tipica TETTOGENESI
1784 GEYSER IGNE	GEYSERS	GEYSER	GEYSER	GEJZER	GEYSER	GEYSER
1785 GEYSERITE SEDS	Geyserite SILICEOUS SINTER	GEYSERITE	GEYSERIT	GEJZERIT	GEYSERITA	GEYSERITE
1786 GIANT FIELD ECON	GIANT FIELDS	CHAMP PETROLIER GEANT	KW=GROSS-LAGERSTAETTE	MESTOROZHDENIE-GIGANT	CAMPO=PETROLIFERO=GIGANTE	CAMPO PETROLIFERO GIGANTE
1787 GIGANTISM PALE	Gigantism ONTOGENY	Gigantisme ONTOGENIE	Gigantismus ONTOGENESE	GIGANTIZM	Gigantismo ONTOGENIA	Gigantismo ONTOGENESI

	ENGLISH	FRANCAIS	DEUTSCH	RUSSKIJ	ESPANOL	ITALIANO
1788 GIGANTOSTRACA PALS	Gigantostraca MEROSTOMATA	Gigantostraca MEROSTOMATA	Gigantostraca MEROSTOMATA	Gigantostraca MEROSTOMATA	Gigantostraca MEROSTOMATA	Gigantostraca MEROSTOMATA
1789 GINKGOALES PALS	GINKGOALES	GINKGOALES	GINKGOALES	GINKGOALES	GINKGOALES	GINKGOALES
1790 GIVETIAN STRS	GIVETIAN	GIVETIEN	GIVET	ZHIVET	GIVETIENSE	GIVETIANO
1791 GLACIAL SURF	Glacial(adj.)	Glaciaire(adj.)	Glazial(adj.)	LEDNIKOVYJ	Glaciar(adj.)	Glaciale(adj.)
1792 GLACIAL DRIFT SURF	DRIFT	DEPOT GLACIAIRE	GLAZIAL= ABLAGERUNG	OTLOZHENIYA LEDNIKOVYE	DEPOSITO= GLACIAR	DETRITO GLA- CIALE
1793 GLACIAL ENVIRONMENT SURF	GLACIAL ENVI- RONMENT	MILIEU GLACIAIRE	GLAZIAL=MILIEU	FATSIYA LED- NIKOVAYA	MEDIO=GLACIAR	AMBIENTE GLA- CIALE
1794 GLACIAL EROSION SURF	GLACIAL EROSION *UF:* Riegel	EROSION GLA- CIAIRE *UF:* Verrou	GLAZIAL= EROSION *UF:* Riegel	EHROZIYA LED- NIKOVAYA	EROSION= GLACIAR *UF:* Umbral=glaciar	EROSIONE GLA- CIALE *UF:* Soglia
1795 GLACIAL FEATURE SURF	GLACIAL FEA- TURES *UF:* Bergschrund	MORPHOLOGIE GLACIAIRE	GLAZIAL= MORPHOLOGIE	FORMA REL'EFA LEDNIKOVAYA	MORFOLOGIA= GLACIAR	MORFOLOGIA GLA- CIALE
1796 GLACIAL GEOLOGY SURF	GLACIAL GEOL- OGY *UF:* Glaciology	GEOLOGIE GLA- CIAIRE *UF:* Glaciologie	GLAZIAL= GEOLOGIE *UF:* Glaziologie	GEOLOGIYA LED- NIKOVAYA	GEOLOGIA= GLACIAR *UF:* Glaciologia	GEOLOGIA GLA- CIALE *UF:* Glaciologia
1797 GLACIAL LAKE SURF	GLACIAL LAKES	LAC GLACIAIRE	GLETSCHER=SEE	OZERO LED- NIKOVOE	LAGO=GLACIAR	LAGO GLACIALE
1798 GLACIAL SEDIMENTATION SEDI	GLACIAL SEDI- MENTATION	SEDIMENTATION GLACIAIRE	GLAZIAL= SEDIMENTATION	[GLACIAL SEDI- MENTATION] LEDNIK; SEDIMEN- TATSIYA	SEDIMENTACION= GLACIAR	SEDIMENTAZIONE GLACIALE
1799 GLACIAL STAGE STRA	Glacial stages GLACIATION	Stade glaciaire GLACIATION	Gletscher=Stadium VEREISUNG	STADIYA LED- NIKOVAYA	Etapa=glaciar GLACIACION	Stadio glaciale GLACIAZIONE
1800 GLACIAL STRIATION SURF	STRIATIONS	Strie glaciaire STRIATION; GLACI- ATION	Glazial=Striemung STRIEMUNG; VEREISUNG	SHTRIKHOVKA LEDNIKOVAYA	Estria=glaciar ESTRIACION; GLA- CIACION	Striatura glaciale STRIATURA; GLA- CIAZIONE
1801 GLACIAL TRANSPORT SURF	GLACIAL TRANS- PORT	TRANSPORT GLA- CIAIRE	GLAZIAL= TRANSPORT	[GLACIAL TRANS- PORT] LEDNIK; TRANS- PORTIROVKA (GEOL)	TRANSPORTE= GLACIAR	TRASPORTO GLA- CIALE
1802 GLACIAL VALLEY SURF	GLACIAL VALLEYS	VALLEE GLA- CIAIRE	GLETSCHER=TAL *UF:* Hangend.Tal	DOLINA LED- NIKOVAYA	VALLE=GLACIAR *UF:* Valle=colgado	VALLE GLACIALE
1803 GLACIATION SURF	GLACIATION *UF:* Continental glacia- tion; Glacial stages; Interpluvial stages; Pluvial processes	GLACIATION *UF:* Glaciation continen- tale; Interpluvial; Plu- vial; Stade glaciaire; Strie glaciaire	VEREISUNG *UF:* Glazial=Striemung; Gletscher=Stadium; Interpluvial; Kontinental=Vereisung; Pluvial	OLEDENENIE	GLACIACION *UF:* Estria=glaciar; Etapa=glaciar; Glaciacion=continental; Interpluvial; Pluvial	GLACIAZIONE *UF:* Glaciazione con- tinentale; Interpluviale; Pluviale; Stadio gla- ciale; Striatura glaciale

	ENGLISH	FRANCAIS	DEUTSCH	RUSSKIJ	ESPANOL	ITALIANO
1804 GLACIER SURF	GLACIERS *UF:* Crevasse; Ice fields	GLACIER *UF:* Champ glace; Crevasse; Rimaye	GLETSCHER *UF:* Bergschrund; Eis=Feld; Gletscher=Spalte	LEDNIK *UF:* Glacial sedimentation; Glacial transport; Glacier surge; Ice movement; Polar ice-cap; Surge	GLACIAR *UF:* Campo=de=hielo; Grieta=de=hielo; Rimaya	GHIACCIAIO *UF:* Crepaccio; Crepaccio terminale; Vedretta
1805 GLACIER SURGE SURF	GLACIER SURGES	CRUE GLACIAIRE	GLETSCHER-ABBRUCH	[GLACIER SURGE] LEDNIK	BATIENTE=GLACIAR	AVANZAMENTO GLACIALE
1806 GLACIOLACUSTRINE SEDIMENTATION SEDI	GLACIOLACUS-TRINE SEDIMENTATION	SEDIMENTATION GLACIOLACUSTRE	GLAZIO-LIMN. SEDIMENTATION	LEDNIKOVO-OZERNYJ	SEDIMENTACION=GLACIOLACUSTRE	SEDIMENTAZIONE GLACIOLACUSTRE
1807 GLACIOLOGY SURF	Glaciology GLACIAL GEOLOGY	Glaciologie GEOLOGIE GLACIAIRE	Glaziologie GLAZIAL-GEOLOGIE	GLYATSIOLOGIYA	Glaciologia GEOLOGIA=GLACIAR	Glaciologia GEOLOGIA GLACIALE
1808 GLACIOMARINE SEDIMENTATION SEDI	GLACIOMARINE SEDIMENTATION	SEDIMENTATION GLACIOMARINE	GLAZIOMARIN. SEDIMENTATION	LEDNIKOVO-MORSKOJ	SEDIMENTACION=GLACIOMARINA	SEDIMENTAZIONE GLACIOMARINA
1809 GLACIOTECTONICS STRU	GLACIOTECTONICS *UF:* Cryotectonics	GLACIOTEC-TONIQUE *UF:* Cryotectonique	EIS-TEKTONIK *UF:* Kryotektonik	GLYATSIODIS-LOKATSIYA	GLACIOTEC-TONICO *UF:* Criotectonico	GLACIOTET-TONICA *UF:* Criotettonica
1810 GLACIS SURF	GLACIS	GLACIS	GLACIS	GLASIS	GLACIS	PENDIO
1811 GLASS MISC	GLASSES *UF:* Glassy texture; Natural glass; Vitreous materials; Vitrification; Vitrophyric texture; Volcanic glass	Verre VERRE NATUREL	Glas NATUERL.GLAS	STEKLO *UF:* Hydration of glass	Vidrio VIDRIO=NATURAL	Vetro VETRO NATURALE
1812 GLASS-SAND COMS	Glass=sand SAND; GLASS MATERIALS	Sable verrerie SABLE; VERRERIE	Glassand GLAS=ROHSTOFF; SAND	PESOK STEKOL'NYJ	Arena=de=vidrio ARENA; MATERIAL=VIDRIO	Sabbia da vetro SABBIA; VETRERIA
1813 GLASS-MAKING MATERIAL COMS	GLASS MATERIALS *UF:* Glass=sand	VERRERIE *UF:* Sable verrerie	GLAS=ROHSTOFF *UF:* Glassand	SYR'E STEKOL'NOE	MATERIAL=VIDRIO *UF:* Arena=de=vidrio	VETRERIA *UF:* Sabbia da vetro
1814 GLASSY TEXTURE TEST	Glassy texture TEXTURES; GLASSES	Texture hyaline TEXTURE; VERRE NATUREL	Glasig.Gefuege KORN=GEFUEGE; NATUERL.GLAS	STRUKTURA STEK-LOVATAYA	Textura=vitrea TEXTURA; VIDRIO=NATURAL	Tessitura vetrosa TESSITURA; VETRO NATURALE
1815 GLAUCONITE COMS	GLAUCONITE	GLAUCONITE	GLAUKONIT	GLAUKONIT	GLAUCONITA	GLAUCONITE
1816 GLAUCONITIC COMPOSITION SEDI	GLAUCONITIC COMPOSITION	COMPOSITION GLAUCONITIQUE	GLAUKONIT. CHEMISMUS	GLAUKONITOVYJ	COMPOSICION=GLAUCONITICA	COMPOSIZIONE GLAUCONITICA
1817 GLAUCONITIZATION SEDI	GLAUCONITIZA-TION	GLAUCONITISA-TION	GLAUKONITI-SIERUNG	GLAUKONITOO-BRAZOVANIE	GLAUCONITIZA-CION	GLAUCONITIZZAZ-IONE
1818 GLAUCOPHANE SCHIST FACIES IGNE	Glaucophane schist facies BLUESCHIST	FACIES GLAUCO-PHANITE	GLAUKOPHAN-FAZIES	FATSIYA SLANTSEV GLAUKO-FANOVYKH	FACIES=GLAUCOFANA	FACIES SCISTI BLU

	ENGLISH	FRANCAIS	DEUTSCH	RUSSKIJ	ESPANOL	ITALIANO
1819 GLEYSOL SUSS	GLEYS	GLEY *UF:* Aquod; Aquoll	GLEY *UF:* Aquod; Aquoll	POCHVA GLEEVAYA	GLEY *UF:* Aquod; Aquoll	GLEY *UF:* Aquod; Aquoll
1820 GLIDE TWIN MINE	Glide twin TWINNING	Macle glissement MACLE	Gleitungs=Zwilling ZWILLING	Dvonik skol'zheniya DVOJNIK	Macla=deformada MACLA	Geminato per scorri- mento GEMINAZIONE
1821 GLOBAL TECTONICS STRU	Global tectonics PLATE TECTONICS	Tectonique globale TECTONIQUE PLAQUE	Global=Tektonik PLATTEN= TEKTONIK	Tektonika global'naya TEKTONIKA PLIT	Megatectonica TECTONICA=DE= PLACAS	Tettonica globale TETTONICA A PLACCHE
1822 GLOBIGERINACEA PALS	GLOBIGERINACEA	GLOBIGERINACEA	GLOBIGERINACEA	GLOBIGERINACEA	GLOBIGERINACEA	GLOBIGERINACEA
1823 GLOMERO- TEST	Glomero=(adj.)	Glomero(adj.)	Glomero=(adj.)	GLOMERO=	Glomero(adj.)	Glomero=(adj.)
1824 GLOSSARY MISC	GLOSSARIES	GLOSSAIRE	NOMENKLATUR= WOERTERBUCH	SLOVAR'	GLOSARIO	GLOSSARIO
1825 GLOSSOPTERIDALES PALS	GLOSSOPTERID- ALES	GLOSSOPTERID- ALES	GLOSSOPTERID- ALES	GLOSSOPTERID- ALES	GLOSSOPTERID- ALES	GLOSSOPTERID- ALES
1826 GNATHOSTOMI PALS	Gnathostomi PISCES	Gnathostomi PISCES	Gnathostomi PISCES	GNATHOSTOMI	Gnathostomi PISCES	Gnatostomi PISCES
1827 GNATHOSTRACA PALS	Gnathostraca CRUSTACEA	Gnathostraca CRUSTACEA	Gnathostraca CRUSTACEA	GNATHOSTRACA	Gnathostraca CRUSTACEA	Gnathostraca CRUSTACEA
1828 GNEISS IGMS	GNEISSES *UF:* Gneissic dome; Granite gneiss; Schlie- ren	GNEISS *UF:* Dome gneissique; Granitogneiss; Schlie- ren	GNEIS *UF:* Gneiskuppel; Granit=Gneis; Schlieren=Gefuege	GNEJS *UF:* Ortognejs	GNEIS *UF:* Domo=gneisico; Franja=de= segregacion; Gneis= granitico	GNEISS *UF:* Cupola di gneiss; Granito gneissico; Schlieren
1829 GNEISSIC DOME STRU	Gneissic dome DOMES; GNEISSES	Dome gneissique DOME; GNEISS	Gneiskuppel DOM; GNEIS	KUPOL GNEJSOVYJ	Domo=gneisico DOMO; GNEIS	Cupola di gneiss DUOMO; GNEISS
1830 GNEISSIC TEXTURE TEST	GNEISSIC TEX- TURE	STRUCTURE GNEISSIQUE	GNEIS=GEFUEGE	GNEJSOVYJ	TEXTURA= GNEISICA	TESSITURA GNEIS- SICA
1831 GNETALES PALS	GNETALES	GNETALES	GNETALES	Gnetales GYMNOSPERMAE	GNETALES	GNETALES
1832 GNOMONIC PROJECTION SOLI	Gnomonic projection GEODESY	Projection gnomonique GEODESIE	Gnomon.Projektion GEODAESIE	GNOMOGRAMMA	Proyeccion=gnomonica GEODESIA	Proiezione gnomonica GEODESIA
1833 GOLD CHEE	GOLD *UF:* Nugget	OR *UF:* Pepite	AU *UF:* Goldklumpen	ZOLOTO	ORO *UF:* Pepita	ORO *UF:* Pepita
1834 GONDWANA STRA	GONDWANA	GONDWANA	GONDWANA	GONDVANA	GONDWANA	GONDWANA
1835 GONIATITIDA PALS	GONIATITIDA	GONIATITIDA	GONIATITIDA	GONIATITIDA	GONIATITIDA	GONIATITIDA
1836 GONIOMETRY METH	GONIOMETRY	GONIOMETRIE	GONIOMETRIE	GONIOMETRIYA	GONIOMETRIA	GONIOMETRIA

	ENGLISH	FRANCAIS	DEUTSCH	RUSSKIJ	ESPANOL	ITALIANO
1837 GORGE SURF	Gorges VALLEYS	Gorge VALLEE	Tobel TAL	USHCHEL'E	Garganta VALLE	Gola VALLE
1838 GOSSAN ECON	GOSSAN	CHAPEAU FER	EISERN.HUT	SHLYAPA ZHELEZ- NAYA	MONTERA=DE- HIERRO	CAPPELLO DI FERRO
1839 GRAB SAMPLING METH	Grab sampling SAMPLING	Echantillonnage par grappin ECHANTILLON- NAGE	Greifer=Probe PROBENNAHME	OTBOR PROB GRE- JFERNYJ	Muestreo=aleatorio MUESTREO	Bennata CAMPIONATURA
1840 GRABEN STRU	GRABENS	GRABEN UF: Taphrogeosynclinal	GRABEN UF: Taphro= Geosynklinale	GRABEN	FOSA=TECTONICA UF: Tafrogeosinclinal	GRABEN UF: Tafrogeosinclinale
1841 GRADED BEDDING TEST	GRADED BEDDING	GRANOCLASSE- MENT	GRADIERT. SCHICHTUNG	SLOISTOST' RIT- MICHNAYA	GRANOSELECCION	GRANOCLASSAZ- IONE
1842 GRADIENT MISC	GRADIENT UF: Gradiometers; Hade; Hydraulic gradi- ent	GRADIENT UF: Gradient hydraulique; Gradio- metre	GRADIENT UF: Gradiometer; Hydraul.Gradient	GRADIENT UF: Gradient gidravli- cheskij	GRADIENTE UF: Gradiente= hidraulico; Gradio- metro	GRADIENTE UF: Gradiente idraulico; Gradimetro
1843 GRADIOMETER INST	Gradiometers INSTRUMENTS; GRADIENT	Gradiometre INSTRUMENTA- TION; GRADIENT	Gradiometer INSTRUMEN- TIERUNG; GRADI- ENT	GRADIOMETR	Gradiometro INSTRUMENTA- CION; GRADIENTE	Gradimetro STRUMENTAZ- IONE; GRADIENTE
1844 GRAIN TEST	GRAINS UF: Coated grain	GRAIN	KORN	ZERNO	GRANO	GRANA
1845 GRANITE IGNS	GRANITES	GRANITE UF: Roche felsique; Unakite	GRANIT UF: Felsisch.Chemismus; Unakit	GRANIT UF: Mikrogranit; Shchelochnoj granit	GRANITO UF: Composicion= felsica; Unakita	GRANITO UF: Composizione fel- sica; Unakite
1846 GRANITE GNEISS IGMS	Granite gneiss GNEISSES; GRA- NITIC COMPOSI- TION	Granitogneiss GNEISS	Granit=Gneis GNEIS	GRANITO=GNEJS	Gneis=granitico GNEIS	Granito gneissico GNEISS
1847 GRANITIC IGNE	Granitic(adj.)	Granitique(adj.)	Granitisch(adj.)	GRANITOVYJ	Granitico(adj.)	Granitico(adj.)
1848 GRANITIC COMPOSITION IGNE	GRANITIC COMPO- SITION UF: Granite gneiss; Granogabbro	ROCHE GRANI- TIQUE	GRANIT. CHEMISMUS	GRANITNYJ	COMPOSICION= GRANITICA	COMPOSIZIONE GRANITICA
1849 GRANITIC LAYER SOLI	GRANITIC LAYER	COUCHE GRANI- TIQUE	GRANIT=SCHICHT	SLOJ GRANITNYJ UF: Kora verkhnyaya	CAPA=GRANITICA	LIVELLO GRANI- TICO
1850 GRANITIZATION IGNE	GRANITIZATION UF: Transformism	GRANITISATION UF: Transformisme	GRANITISIERUNG UF: Transformismus	GRANITIZATSIYA	GRANITIZACION UF: Transformismo	GRANITIZZAZIONE UF: Trasformismo
1851 GRANITOID IGNS	GRANITOID COM- POSITION	GRANITOIDE UF: Granogabbro	GRANITOID UF: Granogabbro	GRANITOID	GRANITOIDE UF: Granogabro	GRANITOIDE UF: Granogabbro
1852 GRANOBLASTIC TEXTURE TEST	Granoblastic texture TEXTURES	Texture granoblastique TEXTURE	Granoblast.Gefuege KORN=GEFUEGE	GRANOBLASTOVYJ	Textura=granoblastica TEXTURA	Tessitura granoblastica TESSITURA

	ENGLISH	FRANCAIS	DEUTSCH	RUSSKIJ	ESPANOL	ITALIANO
1853 GRANODIORITE IGNS	GRANODIORITES	GRANODIORITE	GRANODIORIT	GRANODIORIT *UF:* Granogabbro	GRANODIORITA	GRANODIORITE
1854 GRANOGABBRO IGNS	Granogabbro GRANITIC COMPO- SITION; GABBROS	Granogabbro GRANITOIDE; GAB- BRO	Granogabbro GRANITOID	Granogabbro GABBRO; GRA- NODIORIT	Granogabro GRANITOIDE; GABRO	Granogabbro GRANITOIDE; GAB- BRO
1855 GRANOPHYRIC TEXTURE TEST	Granophyric texture TEXTURES	Texture granophyrique TEXTURE	Granophyr=Gefuege KORN=GEFUEGE	GRANOFIROVYJ	Textura=granofidica TEXTURA	Tessitura granofirica TESSITURA
1856 GRANULAR TEST	Granular texture TEXTURES	Texture granulaire TEXTURE	Koernig KORN=GEFUEGE	ZERNISTYJ *UF:* Kristallicheskij	Textura=granular TEXTURA	Tessitura granulare TESSITURA
1857 GRANULATION TEST	Granulation FRAGMENTATION	Granulation FRAGMENTATION	Granulierung BRUCH=BILDUNG	GRANULYATSIYA	Granulacion FRAGMENTACION	Granulazione FRAMMENTAZ- IONE
1858 GRANULITE IGMS	GRANULITES	GRANULITE	GRANULIT	GRANULIT	GRANULITA	GRANULITE
1859 GRANULITE FACIES IGNE	GRANULITE FACIES	FACIES GRANU- LITE	GRANULIT=FAZIES	FATSIYA GRANULITOVAYA	FACIES= GRANULITA	FACIES DELLE GRANULITI
1860 GRANULITIC TEST	Granulitic(adj.)	Granulitique(adj.)	Granulitisch(adj.)	GRANULITOVYJ	Granulitico(adj.)	Granulitico(adj.)
1861 GRANULOMETRY SEDI	GRANULOMETRY *UF:* Sieves	GRANULOMETRIE *UF:* Rapport sable= argile; Tamis	GRANULOMETRIE	GRANULO- METRIYA *UF:* Analiz granulo- metricheskij; Otmuchivanie	GRANULOMETRIA *UF:* Relacion=arenita= lutita; Tamiz	GRANULOMETRIA *UF:* Rapporto sabbia= argilla; Setaccio
1862 GRAPESTONE SEDS	Grapestone LIMESTONE	[GRAPESTONE] CALCAIRE	Karbonat.Sandstein KALK	KHONDROLIT	Bahamita CALIZA	[GRAPESTONE] CALCARE
1863 GRAPHIC MATH	Graphic display GRAPHIC METH- ODS	Graphique REPRESENTATION GRAPHIQUE	Graph.Ausgabe GRAPH. DARSTELLUNG	Grafik GRAFIKA	Grafico REPRESENTACION= GRAFICA	Grafico RAPPRESENTAZ- IONE GRAFICA
1864 GRAPHIC METHOD MISC	GRAPHIC METH- ODS *UF:* Diagrams; Graphic display	REPRESENTATION GRAPHIQUE *UF:* Diagramme; Graphique; Diagramme structural	GRAPH. DARSTELLUNG *UF:* Diagramm; Gefuege=Diagramm; Graph.Ausgabe	[GRAPHIC METHOD] GRAFICHESKIJ; METODOLOGIYA	REPRESENTACION= GRAFICA *UF:* Diagrama; Diagrama=estructural; Grafico	RAPPRESENTAZ- IONE GRAFICA *UF:* Diagramma; Dia- gramma strutturale; Grafico
1865 GRAPHIC TEXTURE TEST	GRAPHIC TEXTURE	TEXTURE GRAPHIQUE	GRAPH.GEFUEGE	GRAFICHESKIJ *UF:* Graphic method	TEXTURA= GRAFICA	TESSITURA GRA- FICA
1866 GRAPHITE MING	GRAPHITE	GRAPHITE *UF:* Graphitisation	GRAPHIT *UF:* Graphitisierung	GRAFIT	GRAFITO *UF:* Grafitizacion	GRAFITE *UF:* Grafitizzazione
1867 GRAPHITIZATION IGNE	GRAPHITIZATION	Graphitisation GRAPHITE	Graphitisierung GRAPHIT	GRAFITIZATSIYA	Grafitizacion GRAFITO	Grafitizzazione GRAFITE
1868 GRAPTOLITHINA PALS	GRAPTOLITHINA	GRAPTOLITHINA	GRAPTOLITHINA	GRAPTOLITHINA	GRAPTOLITHINA	GRAPTOLITHINA
1869 GRAPTOLOIDEA PALS	GRAPTOLOIDEA	GRAPTOLOIDEA	GRAPTOLOIDEA	GRAPTOLOIDEA	GRAPTOLOIDEA	GRAPTOLOIDEA

	ENGLISH	FRANCAIS	DEUTSCH	RUSSKIJ	ESPANOL	ITALIANO
1870 GRASSLAND ENVI	GRASSLANDS	PRAIRIE	GRASLAND	[GRASSLAND]	PRADERA	PRATERIA
1871 GRAVEL SEDS	GRAVEL *UF:* Coarse gravel; Rubble	GRAVIER *UF:* Gravier grossier	KIES *UF:* Grob.Geroell	GRAVIJ *UF:* Coarse gravel; Peschanik grubozernis- tyj	GRAVA *UF:* Grava=gruesa	GHIAIA *UF:* Ghiaia grossolana
1872 GRAVEL FILTER GEOH	GRAVEL FILTERS	MASSIF FILTRANT	KIES=FILTER	FIL'TR GRAVIJNYJ	EMPAQUETADURA= GRAVA	MASSA FILTRANTE
1873 GRAVIMETER INST	GRAVIMETERS	GRAVIMETRE	GRAVIMETER	GRAVIMETR	GRAVIMETRO	GRAVIMETRO
1874 GRAVIMETRY SOLI	GRAVIMETRY	GRAVIMETRIE *UF:* Constante gravita- tion; Gravitation; Pesanteur absolue	GRAVIMETRIE *UF:* Absolut.Schwere; Gravitation; Gravitations= Konstante	GRAVIMETRIYA *UF:* Kontsentratsiya gravimetricheskaya	GRAVIMETRIA *UF:* Constante= gravitacional; Gravedad=absoluta; Gravitacion	GRAVIMETRIA *UF:* Costante gravita- zionale; Gravitazione; Gravita assoluta
1875 GRAVITATION SOLI	Gravitation GRAVITY FIELD	Gravitation GRAVIMETRIE	Gravitation GRAVIMETRIE	GRAVITATSIYA	Gravitacion GRAVIMETRIA	Gravitazione GRAVIMETRIA
1876 GRAVITATIONAL CONSTANT PHCH	Gravitational constant GRAVITY FIELD	Constante gravitation GRAVIMETRIE; PARAMETRE	Gravitations= Konstante GRAVIMETRIE; PARAMETER	POSTOYANNAYA GRAVITATSION- NAYA	Constante= gravitacional GRAVIMETRIA; PARAMETRO	Costante gravitazionale GRAVIMETRIA; PARAMETRO
1877 GRAVITATIONAL DIFFERENTIATION IGNE	Gravitational differenti- ation DIFFERENTIATION	Differenciation par gra- vite DIFFERENCIATION MAGMATIQUE; DENSITE	Schwere= Differenzierung MAGMAT. DIFFERENTIATION; DICHTE	DIFFERENTSIAT- SIYA GRAVITAT- SIONNAYA	Diferenciacion= gravitacional DIFERENCIACION= MAGMATICA; DEN- SIDAD	Differenziazione gravitativa DIFFERENZIAZ- IONE MAGMATICA; DENSITA
1878 GRAVITY ANOMALY APPL	GRAVITY ANOMA- LIES	ANOMALIE GRAVI- METRIQUE	SCHWERE= ANOMALIE	ANOMALIYA GRAVITATSION- NAYA *UF:* Anomaliya v svo- bodnom vozdukhe; Buge anomaliya	ANOMALIA= GRAVIMETRICA	ANOMALIA GRAVI- METRICA
1879 GRAVITY EXPLORATION APPL	Gravity exploration GRAVITY METH- ODS; MINERAL EXPLORATION	Prospection gravi- metrique METHODE GRAVI- METRIQUE	Gravimetr.Prospektion GRAVIMETR. METHODE	GRAVIRAZVEDKA *UF:* Metod gravimetri- cheskij	Prospeccion= gravimetrica METODO= GRAVIMETRICO	Prospezione gravi- metrica METODO GRAVI- METRICO
1880 GRAVITY FIELD SOLI	GRAVITY FIELD *UF:* Absolute gravity; Gravitation; Gravita- tional constant	CHAMP GRAVI- METRIQUE	GRAVIMETR.FELD	POLE GRAVITAT- SIONNOE	CAMPO= GRAVITATORIO	CAMPO GRAVI- METRICO
1881 GRAVITY FOLD STRU	Gravity folds FOLDS; GRAVITY SLIDING	Pli gravitaire PLI; ECOULEMENT GRAVITE	Gravitations=Falte FALTE; SCHWERE= GLEITUNG	SKLADKA GRAVITATSION- NAYA	Pliegue=de=gravedad PLIEGUE; CORRIMIENTO= GRAVEDAD	Piega gravitativa PIEGA; COLATA PLASTICA
1882 GRAVITY MAP SOLI	GRAVITY MAPS	CARTE GRAVI- METRIQUE	GRAVIMETR. KARTE	KARTA GRAVITAT- SIONNAYA	MAPA= GRAVIMETRICO	CARTA GRAVI- METRICA
1883 GRAVITY METHOD APPL	GRAVITY METH- ODS *UF:* Gravity exploration	METHODE GRAVI- METRIQUE *UF:* Prospection gravi- metrique	GRAVIMETR. METHODE *UF:* Gravimetr. Prospektion	Metod gravimetri- cheskij GRAVIRAZVEDKA	METODO= GRAVIMETRICO *UF:* Prospeccion= gravimetrica	METODO GRAVI- METRICO *UF:* Prospezione gravi- metrica

	ENGLISH	FRANCAIS	DEUTSCH	RUSSKIJ	ESPANOL	ITALIANO
1884 GRAVITY PLATFORM ENGI	GRAVITY PLAT- FORMS	PLATEFORME GRAVITAIRE	SCHWERE= PLATTFORM	PLATFORMA MOR- SKAYA STATSIO- NARNAYA	PLATAFORMA= GRAVIMETRICA	PIATTAFORMA GRAVIMETRICA
1885 GRAVITY SLIDING STRU	GRAVITY SLIDING *UF:* Gravity folds; Grav- ity tectonics	ECOULEMENT GRAVITE *UF:* Pli gravitaire; Tec- tonique gravite	SCHWERE= GLEITUNG *UF:* Gravitations=Falte; Gravitations=Tektonik	Opolzanie gravitatsion- noe TEKTONIKA GRAVITATSION- NAYA; OPOLZANIE	CORRIMIENTO= GRAVEDAD *UF:* Pliegue=de= gravedad; Tectonica= gravedad	SCIVOLAMENTO GRAVITATIVO *UF:* Tettonica gravita- tiva
1886 GRAVITY SURVEY APPL	GRAVITY SURVEYS	LEVE GRAVI- METRIQUE	GRAVIMETR. VERMESSUNG	S'EMKA GRAVI- METRICHESKAYA	LEVANTAMIENTO= GRAVIMETRICO	RILEVAMENTO GRAVIMETRICO
1887 GRAVITY TECTONICS STRU	Gravity tectonics GRAVITY SLIDING	Tectonique gravite ECOULEMENT GRAVITE	Gravitations=Tektonik SCHWERE= GLEITUNG	TEKTONIKA GRAVITATSION- NAYA *UF:* Opolzanie gravitat- sionnoe	Tectonica=gravedad CORRIMIENTO= GRAVEDAD	Tettonica gravitativa SCIVOLAMENTO GRAVITATIVO
1888 GRAYWACKE SEDS	GRAYWACKE	GRAUWACKE	GRAUWACKE	GRAUVAKKA	GRAUWACA	GROVACCA
1889 GREASY LUSTER PHCH	Greasy luster OPTICAL PROPER- TIES	Eclat gras PROPRIETE OPTIQUE	Fett=Glanz OPT.EIGENSCHAFT	ZHIRNYJ BLESK	Brillo=graso PROPIEDAD= OPTICA	Lucentezza grassa PROPRIETA OTTICA
1890 GREAT-CIRCLE BELT GEOL	Great=circle belt GEOSYNCLINES	Ceinture grand cercle CHAINE GEOSYN- CLINALE	Grosskreis=Guertel GEOSYNKLINAL= KETTE	POYAS KOL'TSEVOJ OKEANICHESKIJ	Cinturon=tectonico CORDILLERA= GEOSINCLINAL	Cintura massima CATENA GEOSIN- CLINALE
1891 GREENHOUSE EFFECT SURF	Greenhouse effect ATMOSPHERE; CLIMATE	Effet serre ATMOSPHERE	Gewaechshaus=Effekt ATMOSPHAERE	EHFFEKT PARNIKOVYJ	Efecto=invernadero ATMOSFERA	Effetto di serra ATMOSFERA
1892 GREENSCHIST IGMS	GREENSCHIST *UF:* Chlorite schist	SCHISTE VERT *UF:* Schiste chlorite; Sericitoschiste; Talc- schiste	GRUEN=SCHIEFER *UF:* Chlorit=Schiefer; Serizit=Schiefer; Talk=Schiefer	SLANETS ZELENYJ *UF:* Slanets khloritovyj; Talc schist	ESQUISTO=VERDE *UF:* Cloritoesquisto; Esquisto=sericitico; Talcoesquisto	SCISTO VERDE *UF:* Scisto a clorite; Scisto sericitico; Tal- coscisto
1893 GREENSCHIST FACIES IGNE	GREENSCHIST FACIES	FACIES SCHISTE VERT *UF:* Roche verte	GRUEN= SCHIEFER=FAZIES *UF:* Gruenstein	FATSIYA SLANTSEV ZELENYKH	FACIES= ESQUISTO=VERDE *UF:* Roca=verde	FACIES DEGLI SCISTI VERDI *UF:* Rocce verdi
1894 GREENSTONE IGMS	Greenstone GREENSTONE BELTS	Roche verte FACIES SCHISTE VERT	Gruenstein GRUEN= SCHIEFER=FAZIES	PORODA ZELENOKAMEN- NAYA	Roca=verde FACIES= ESQUISTO=VERDE	Rocce verdi FACIES DEGLI SCISTI VERDI
1895 GREENSTONE BELT STRU	GREENSTONE BELTS *UF:* Greenstone	CEINTURE ROCHE VERTE	GRUENSTEIN= GUERTEL	POYAS ZELENOKA- MENNYJ	CINTURON= OFIOLITICO	CINTURA DI ROCCE VERDI
1896 GREISEN IGNE	GREISEN	GREISEN *UF:* Greisenisation	GREISEN *UF:* Vergreisung	GREJZEN	GREISEN *UF:* Greisenizacion	GREISEN *UF:* Greisenizzazione
1897 GREISENIZATION IGNE	Greisenization	Greisenisation GREISEN	Vergreisung GREISEN	GREJZENIZATSIYA	Greisenizacion GREISEN	Greisenizzazione GREISEN
1898 GRENVILLE OROGENY STRU	GRENVILLIAN OROGENY	OROGENIE GREN- VILLE	GRENVILLE= OROGENESE	GRENVIL'SKIJ	OROGENIA= GRENVILLE	OROGENESI GREN- VILLIANA

	ENGLISH	FRANCAIS	DEUTSCH	RUSSKIJ	ESPANOL	ITALIANO
1899 GRINDING METH	GRINDING	BROYAGE	ABREIBUNG	IZMEL'CHENIE	MOLIENDA	FRANTUMAZIONE
1900 GRIT SEDS	Grit SEDIMENTARY ROCKS	Gres grossier GRES	Grit SANDSTEIN	Peschanik grubozernis- tyj GRAVIJ	Arenisca=grosera ARENISCA	Arenaria grossolana ARENARIA
1901 GROIN ENGI	GROINS	ESTACADE	BUHNE	[GROIN] PORT	ESTACADA	PENNELLO
1902 GROOVE SURF	GROOVES	CANNELURE	SCHLEIF=MARKE	Zhelob lednikovyj BOROZDA	SUPERFICIE= ACANALADURA	SCANALATURA
1903 GROOVE CAST SEDI	GROOVE CASTS	EMPREINTE CAN- NELURE	SCHLEIF= MARKEN- AUSGUSS	Znak volocheniya GIEROGLIF	IMPRESION=FOSIL	SOLCO DA TRAS- CINAMENTO
1904 GROUND SURF	Ground RELIEF	Surface terre RELIEF CONTI- NENT	Erdoberflaeche KONTINENTAL= RELIEF	GRUNT UF: Grouting; Strong motion; Zakreplenie gruntov	Terreno RELIEVE= CONTINENTAL	Terreno FORMA DEL RILIEVO
1905 GROUND ICE SURF	GROUND ICE	GLACE SOL	GRUNDEIS	LED POGREBEN- NYJ	HIELO=SUELO	GHIACCIO DEL SUOLO
1906 GROUND METHOD APPL	GROUND METH- ODS	MESURE AU SOL	GELAENDE- VERMESSUNG	[GROUND METHOD] EHLEKTRORAZ- VEDKA	MEDIDA=EN=EL= SUELO	MISURA AL SUOLO
1907 GROUND MORAINE SURF	Ground moraines MORAINES	Moraine fond MORAINE	Grund=Moraene MORAENE	MORENA DON- NAYA	Morrena=de=fondo MORRENA	Morena di fondo MORENA
1908 GROUND MOTION SOLI	GROUND MOTION	MOUVEMENT SOL	BODEN= BEWEGUNG	DVIZHENIE GRUNTA	MOVIMIENTO= SUELO	MOVIMENTO DEL SUOLO
1909 GROUND PRESSURE SOLI	Ground pressure PRESSURE	Pression geostatique PRESSION; SOL	Geostat.Druck DRUCK; BODEN	DAVLENIE GEOSTATI- CHESKOE UF: Geopressure	Presion=geostatica PRESION; SUELO	Pressione geostatica PRESSIONE; SUOLO
1910 GROUND TRUTH APPL	GROUND TRUTH	VERITE TERRAIN	GELAENDE= KONTROLLE	[GROUND TRUTH] ZONDIROVANIE DISTANTSIONNOE	TERRENO= VERDADERO	EVIDENZA DI TER- RENO
1911 GROUND WATER GEOH	GROUND WATER UF: Groundwater dis- charge; Groundwater exploration; Groundwa- ter level; Groundwater runoff; Hydrogeological survey; Phreatic water; Water balance	EAU SOUTER- RAINE UF: Debit eau souter- raine; Prospection eau souterraine	GRUNDWASSER UF: Grundwasser= Ergiebigkeit; Grundwasser= Erkundung	VODA PODZEM- NAYA UF: Exsurgence; Pro- tected zone	AGUA= SUBTERRANEA UF: Descarga=agua= subterranea; Exploracion=agua= subterranea	ACQUA SOTTER- RANEA UF: Esplorazione di falde idriche; Portata d'acqua sotterranea
1912 GROUND-WATER BUDGET GEOH	Water balance GROUND WATER	Bilan eau souterraine BILAN EAU	Grundwasser=Bilanz WASSER=BILANZ	BALANS PODZEM- NIKH VOD	Balance=de=agua= subterranea BALANCE=DE= AGUA	Bilancio idrico sotter- raneo BILANCIO IDROGEOLOGICO
1913 GROUND-WATER DAM GEOH	GROUNDWATER DAMS	BARRAGE SOUTER- RAIN	GRUNDWASSER- SPERRE	[GROUNDWATER DAM] VODA GRUNTOV- AYA; PLOTINA	EMBALSE= SUBTERRANEO	SBARRAMENTO IDROGEOLOGICO

	ENGLISH	FRANCAIS	DEUTSCH	RUSSKIJ	ESPANOL	ITALIANO
1914 GROUND=WATER DISCHARGE GEOH	Groundwater discharge DISCHARGE; GROUND WATER	Debit eau souterraine DEBIT; EAU SOUTERRAINE	Grundwasser= Ergiebigkeit SCHUETTUNG; GRUNDWASSER	RAZGRUZKA PODZEMNYKH VOD	Descarga=agua= subterranea CAUDAL; AGUA= SUBTERRANEA	Portata d'acqua sotter- ranea PORTATA; ACQUA SOTTERRANEA
1915 GROUND=WATER EXPLORATION GEOH	Groundwater explora- tion EXPLORATION; GROUND WATER	Prospection eau souter- raine PROSPECTION; EAU SOUTER- RAINE	Grundwasser= Erkundung PROSPEKTION; GRUNDWASSER	RAZVEDKA PODZEMNYKH VOD	Exploracion=agua= subterranea PROSPECCION; AGUA= SUBTERRANEA	Esplorazione di falde idriche PROSPEZIONE; ACQUA SOTTER- RANEA
1916 GROUND=WATER LEVEL GEOH	Groundwater level GROUND WATER LEVELS	Niveau eau souterraine SURFACE PIEZO- METRIQUE	Grundwasser=Spiegel HYDROSTAT. OBERFLAECHE	UROVEN' PODZEM- NYKH VOD	Nivel=agua= subterranea SUPERFICIE= PIEZOMETRICA	Livello della falda idrica SUPERFICIE PIEZO- METRICA
1917 GROUND=WATER RECHARGE GEOH	NATURAL RECHARGE	ALIMENTATION NATURELLE	NATUERL. GRUNDWASSER= SPEISUNG	PITANIE PODZEM- NYKH VOD	ALIMENTACION= NATURAL	ALIMENTAZIONE NATURALE
1918 GROUND=WATER RUNOFF GEOH	Groundwater runoff GROUND WATER; RUNOFF	Ecoulement eau souter- raine HYDRODY- NAMIQUE	Grundwasser=Abfluss HYDRODYNAMIK	STOK PODZEMNYJ	Escorrentia=agua= subterranea HIDRODINAMICA	Deflusso di falda idrica IDRODINAMICA
1919 GROUNDMASS TEST	Groundmass TEXTURES	Mesostase TEXTURE	Grundmasse KORN=GEFUEGE	MASSA OSNOV- NAYA	Mesostasis TEXTURA	Massa di fondo TESSITURA
1920 GROUTING ENGI	GROUTING	FONCAGE CIMENTATION	ZEMENT= EINPRESSUNG	[GROUTING] GRUNT; STABIL- IZATSIYA	INYECCION= LECHADA	CEMENTAZIONE PENETRANTE
1921 GROWTH FAULT STRU	GROWTH FAULTS	FAILLE SYNGENE- TIQUE	SYNSEDIMENTAER. STOERUNG	RAZRYV KONSEDI- MENTATSIONNYJ	FALLA= SINGENETICA	FAGLIA SINGENE- TICA
1922 GROWTH LINE PALE	Growth lines ONTOGENY	Ligne croissance ONTOGENIE	Anwachs=Streifen ONTOGENESE	LINIYA NARA- STANIYA	Linea=de=crecimiento ONTOGENIA	Stria di accrescimento ONTOGENESI
1923 GRUS SEDS	GRUS	ARENE	GRUS	DRESVA	ELUVION= GRANITICO	SABBIONE
1924 GUANO SEDS	GUANO	GUANO	GUANO	GUANO	GUANO	GUANO
1925 GUIDE FOSSIL STRA	Guide fossil INDEX FOSSILS	Fossile specifique FAUNE SPECI- FIQUE	Leit=Fossil SPEZIF.FAUNA	ISKOPAEMOE RUKOVODYASH- CHEE *UF:* Vid=indeks	Fosil=especifico FAUNA= ESPECIFICA	Fossile specifico FAUNA SPECIFICA
1926 GUIDEBOOK GEOL	GUIDEBOOK	LIVRET GUIDE	FUEHRER	PUTEVODITEL'	LIBRO=GUIA	LIBRETTO GUIDA
1927 GULF SURF	GULFS *UF:* Bays; Bights	GOLFE *UF:* Anse; Baie	GOLF *UF:* Bay; Bucht	ZALIV *UF:* Bukhta; Inlet	GOLFO *UF:* Bahia; Comba	GOLFO *UF:* Ansa; Baia
1928 GULLY SURF	GULLIES	RAVIN	SCHLUCHT	OVRAG	BARRANCO	FORRA
1929 GUTENBERG DISCONTINUITY SOLI	Gutenberg discontinu- ity CORE; DISCONTI- NUITIES	Discontinuite Gutenberg DISCONTINUITE; NOYAU TER- RESTRE	Gutenberg= Diskontinuitaet DISKONTINUIT- AET; ERD=KERN	GUTENBERGA POVERKHNOST' *UF:* Sloj ponizhennykh skorostej	Discontinuidad=de= Gutenberg DISCONTINUIDAD; NUCLEO=GLOBO	Discontinuita di Guten- berg DISCONTINUITA; NUCLEO TER- RESTRE

	ENGLISH	FRANCAIS	DEUTSCH	RUSSKIJ	ESPANOL	ITALIANO
1930 GYMNOLAEMATA PALS	Gymnolaemata BRYOZOA	Gymnolaemata BRYOZOA	Gymnolaemata BRYOZOA	GYMNOLAEMATA	Gymolaemata BRYOZOA	Gymnolaemata BRYOZOA
1931 GYMNOSPERMAE PALS	GYMNOSPERMS *UF:* Pteridospermae	GYMNOSPERMAE *UF:* Pteridospermae	GYMNOSPERMAE *UF:* Pteridospermae	GYMNOSPERMAE *UF:* Gnetales	GYMNOSPERMAE *UF:* Pteridospermae	GIMNOSPERME *UF:* Pteridospermae
1932 GYPSIFICATION SEDI	Gypsification GYPSUM	Gypsification GYPSE	Gips=Bildung GIPS	OGIPSOVANIE	Yesificacion YESO	Gessificazione GESSO
1933 GYPSUM COMS	GYPSUM *UF:* Gypsification	GYPSE *UF:* Gypsification	GIPS *UF:* Gips=Bildung	GIPS	YESO *UF:* Yesificacion	GESSO *UF:* Gessificazione
1934 GYTTJA SEDS	GYTTJA	GYTTJA	GYTTJA	Gitt'ya SAPROPEL'	GYTJA	GYTTJA
1935 GZHELIAN STRS	GZHELIAN	GZHELIEN	GSHEL	GZHEL'SKIJ	GZHELIENSE	GZHELIANO
1936 HABIT MINE	HABIT *UF:* Homeomorphism	HABITUS *UF:* Homeomorphisme; Mineral xenomorphe	HABITUS *UF:* Homoeomorphismus	GABITUS *UF:* Elongate mineral; Forma kristallografi-cheskaya	HABITO *UF:* Alotriomorfo; Homomorfismo	ABITO CRISTAL-LINO *UF:* Cristallo anedrale; Omeomorfismo
1937 HABITAT PALE	HABITAT	HABITAT	HABITAT	Sreda estestvennaya EHKOLOGIYA	HABITAT	HABITAT
1938 HADE STRU	Hade GRADIENT	Angle avec la verticale PENDAGE	Flaechen=Normale EINFALLEN	Ugol naklona PADENIE	Angulo=con=la= vertical BUZAMIENTO	Angolo con la verticale IMMERSIONE
1939 HAECKEL'S LAW PALE	Haeckel's law ONTOGENY	Loi Haeckel ONTOGENIE	Haeckel's=Gesetz ONTOGENESE	GEKKELYA ZAKON	Ley=de=Haeckel ONTOGENIA	Legge di Haeckel ONTOGENESI
1940 HAFNIUM CHEE	HAFNIUM	HAFNIUM	HF	GAFNIJ	HAFNIO	AFNIO
1941 HAIL SURF	Hail ATMOSPHERIC PRECIPITATION	Grele PRECIPITATION ATMOSPHERIQUE	Hagel ATMOSPHAER. NIEDERSCHLAG	Grad OSADKI ATMOS-FERNYE	Granizo PRECIPITACION= ATMOSFERICA	Grandine PRECIPITAZIONE ATMOSFERICA
1942 HALF=LIFE PERIOD ISOT	Half=life period ABSOLUTE AGE	Demi periode RADIOACTIVITE	Halbwerts=Zeit RADIOAKTIVITAET	PERIOD POLURAS-PADA	Vida=media= radioactiva RADIOACTIVIDAD	Periodo di dimezza-mento RADIOATTIVITA
1943 HALF=SPACE SOLI	HALF=SPACE	DEMI ESPACE	HALBRAUM	POLUPROSTRAN-STVO	SEMIESPACIO	SEMISPAZIO
1944 HALIDES MING	HALIDES	HALOGENURE	HALID	GALOID	HALURO	ALOGENURO
1945 HALITIC ENVIRONMENT SEDI	Halitic environment SALINITY	Milieu salin SALINITE	Halit.Milieu SALINITAET	GALITOVYJ	Halitico SALINIDAD	Ambiente salato SALINITA
1946 HALMYROLYSIS MARI	HALMYROLYSIS	ALTERATION SOUS MARINE	HALMYROLYSE	GAL'MIROLIZ	ALTERACION= SUBMARINA	ALTERAZIONE SOTTOMARINA
1947 HALOGEN CHES	HALOGENS	HALOGENE	HALOGEN	GALOGEN	HALOGENO	ALOGENO

	ENGLISH	FRANCAIS	DEUTSCH	RUSSKIJ	ESPANOL	ITALIANO
1948 HALOPHILIC ORGANISM PALE	Halophilic taxa ECOLOGY; SALIN- ITY	Organisme halophile ECOLOGIE; SALI- NITE	Halophil.Organismus OEKOLOGIE; SALINITAET	GALOFIL'NYJ	Halofilo ECOLOGIA; SALINI- DAD	Organismo alofilo ECOLOGIA; SALINITA
1949 HALOPHYTE PALE	Halophyte SALINITY; FLORA	Halophyte SALINITE; FLORE	Halophyt SALINITAET; FLORA	GALOFIT	Halofita SALINIDAD; FLORA	Alofita SALINITA; FLORA
1950 HAMMADA SURF	Hammadas DESERTS	Hammada DESERT; PLATEAU	Hammada WUESTE; PLATEAU	KHAMADA	Hammada DESIERTO; MESETA	Hammada DESERTO; ALTOPI- ANO
1951 HAMMER SEISMICS APPL	Hammer seismics SEISMIC METHODS	Sismique marteau METHODE SISMIQUE	Hammer=Seismik SEISM.METHODE	[HAMMER SEISMICS] METOD SEJSMICHES- KIJ	Sismica=martillo METODO=SISMICO	Sismica a percussione METODO SISMICO
1952 HANGING VALLEY SURF	Hanging valleys CIRQUES	Vallee suspendue CIRQUE GLA- CIAIRE	Hangend.Tal GLETSCHER=TAL	DOLINA VISYACHAYA	Valle=colgado VALLE=GLACIAR	Valle sospesa CIRCO
1953 HANGING WALL ECON	Hanging wall VEINS; FAULTS	Toit FILON	Hangend.Gestein GANG	BOK VISYACHIJ	Techo FILON	Tetto FILONE
1954 HARBOR ENGI	HARBORS *UF:* Piers	PORT *UF:* Quai	HAFEN *UF:* Hafenmauer	PORT *UF:* Groin; Mol	PUERTO *UF:* Muelle	PORTO *UF:* Palo-pila
1955 HARDGROUND SEDI	HARDGROUND	[HARDGROUND]	[HARDGROUND]	PODOSHVA USTO- JCHIVAYA	[HARDGROUND]	[HARDGROUND]
1956 HARDNESS PHCH	HARDNESS	DURETE *UF:* Materiau incompe- tent	HAERTE *UF:* Inkompetent.Ges- tein	TVERDOST' *UF:* Mikrotverdost'	DUREZA *UF:* Roca=incompetente	DUREZZA *UF:* Roccia incompetente
1957 HARMONICS MATH	HARMONICS	HARMONIQUE	HARMON. ANALYSE	ANALIZ GARMONI- CHESKIJ *UF:* Fur'e ryad	ARMONICO	ARMONICA
1958 HARPOLITH IGNE	Harpoliths LACCOLITHS	Harpolite LACCOLITE	Harpolith LAKKOLITH	GARPOLIT	Harpolito LACOLITO	Arpolite LACCOLITE
1959 HAUTERIVIAN STRS	HAUTERIVIAN	HAUTERIVIEN	HAUTERIVE	GOTERIV	HAUTERIVIENSE	HAUTERIVIANO
1960 HAUY'S LAW MINE	Hauy's law CRYSTALLOGRA- PHY	Loi Hauy CRISTALLOGRA- PHIE	Hauy=Gesetz KRISTALLOGRA- PHIE	Gayui zakon KRISTALLOGRA- FIYA; ZAKON	Ley-de=Hauy CRISTALOGRAFIA	Legge di Hauy CRISTALLOGRAFIA
1961 HAWAIIAN-TYPE **ERUPTION** IGNE	HAWAIAN=TYPE ERUPTIONS	HAWAIEN	HAWAII=TYP	IZVERZHENIE GAVAJSKOGO TIPA *UF:* Vulkan shchitovoj	HAWAIANO	ERUZIONE HAWAI- ANA
1962 HEAT PHCH	Heat THERMODYNAMIC PROPERTIES	Chaleur THERMODY- NAMIQUE	Waerme THERMODYNAMIK	TEPLOTA *UF:* Heat source	Calor TERMODINAMICA	Calore TERMODINAMICA
1963 HEAT CAPACITY PHCH	HEAT CAPACITY	CAPACITE CALORI- FIQUE	WAERME= KAPAZITAET	TEPLOEMKOST'	CAPACIDAD= CALORIFICO	CAPACITA TER- MICA
1964 HEAT CONDUCTION PHCH	Heat conduction THERMAL CON- DUCTIVITY	Conduction chaleur CONDUCTIVITE THERMIQUE	Waerme=Leitung WAERME= LEITFAEHIGKEIT	TEPLOPROVOD- NOST'	Conduccion=calor CONDUCTIVIDAD= TERMICA	Conducibilita di calore CONDUTTIVITA TERMICA

	ENGLISH	FRANCAIS	DEUTSCH	RUSSKIJ	ESPANOL	ITALIANO
1965 HEAT CRACK TEST	Heat crack RUPTURE	Fissure contraction ACTION CHALEUR; CRAQUELURE BOUE	Hitze=Sprung WAERME= WIRKUNG; SCHLAMM= SCHWUND=RISS	TRESHCHINA KON- TRAKTSIONNAYA	Rotura=termica ACCION=CALOR; GRIETA= DESECACION	Fessura di dissecca- mento AZIONE TERMICA; FESSURAZIONE DA DISSECCA- MENTO
1966 HEAT EQUIVALENT OF FUSION PHCH	Heat equivalent of fusion THERMODYNAMIC PROPERTIES	Equivalent calorifique fusion FUSION; THERMO- DYNAMIQUE	Schmelz=Waerme SCHMELZE; THER- MODYNAMIK	TEPLOTA PLAV- LENIYA UF: Teplota plavleniya udel'naya	Equivalente=calor=de= fusion FUSION; TERMODI- NAMICA	Equivalente calorico di fusione FUSIONE; TER- MODINAMICA
1967 HEAT FLOW SOLI	HEAT FLOW UF: Heat sources; Ther- mal prospecting	FLUX GEOTHER- MIQUE	WAERMESTROM	POTOK TEPLOVOJ	FLUJO= GEOTERMICO	FLUSSO DI CALORE
1968 HEAT SOURCE SOLI	Heat sources HEAT FLOW	Source chaleur GEOTHERMIE	Waerme=Quelle GEOTHERMIK	[HEAT SOURCE] ISTOCHNIK; TEPLOTA	Fuentes=de=calor GEOTERMIA	Sorgente di calore GEOTERMIA
1969 HEAT TRANSFER SOLI	HEAT TRANSFER	TRANSFERT CHA- LEUR	WAERME= TRANSFER	TEPLOPERENOS	TRASMISION	TRASFERIMENTO DI CALORE
1970 HEAVE SURF	Heave DEFORMATION	Soulevement DEFORMATION; SOL	Boden=Hebung VERFORMUNG; BODEN	VSPUCHIVANIE	Abombamiento DEFORMACION; SUELO	Sollevamento del ter- reno DEFORMAZIONE; SUOLO
1971 HEAVY ISOTOPE ISOT	Heavy isotopes ISOTOPES	Isotope lourd ISOTOPE	Schwer=Isotop ISOTOP	IZOTOP TYAZHE- LYJ	Isotopo=pesado ISOTOPO	Isotopo pesante ISOTOPO
1972 HEAVY METALS CHES	HEAVY METALS	METAL LOURD	SCHWER=METALL	METALL TYAZHE- LYJ	METALES= PESADOS	METALLO PESANTE
1973 HEAVY MINERALS SEDI	HEAVY MINERALS	MINERAL LOURD	SCHWER= MINERAL	FRAKTSIYA TYAZ- HELAYA	MINERALES= PESADOS	MINERALI PESANTI
1974 HEAVY OIL COMS	HEAVY OIL	PETROLE LOURD	SCHWER=OEL	NEFT' TYAZ- HELAYA	ACEITE=PESADO	PETROLIO PESANTE
1975 HEKISTOTHERM PLANT PALE	Hekistotherm plant ECOLOGY; FLORA	Plante hekistotherme ECOLOGIE; FLORE	Hekistotherm.Pflanze OEKOLOGIE; FLORA	GEKISTOTERM	Planta=hekistotermica ECOLOGIA; FLORA	Pianta echistoterma ECOLOGIA; FLORA
1976 HELICOPLACOIDEA PALS	HELICOPLACOIDEA	HELICOPLACOIDEA	HELICOPLACOIDEA	Helicoplacoidea ECHINOZOA	HELICOPLACOIDEA	HELICOPLACOIDEA
1977 HELIOZOA PALS	Heliozoa PROTISTA	Heliozoa PROTOZOA	Heliozoa PROTOZOA	HELIOZOA	Heliozoa PROTOZOA	Heliozoa PROTOZOA
1978 HELIUM CHEE	HELIUM	HELIUM	HE	GELIJ	HELIO	ELIO
1979 HELIUM AGE METHOD ISOT	HE=4/HE=3	HE=HE	HELIUM= DATIERUNG	METOD GELIEVYJ	HE=HE	HE/HE
1980 HELVETIAN STRS	HELVETIAN	HELVETIEN	HELVET	GEL'VET	HELVETIENSE	ELVEZIANO
1981 HEMATITE COMS	HEMATITE	HEMATITE	HAEMATIT	GEMATIT	HEMATITES	EMATITE

	ENGLISH	FRANCAIS	DEUTSCH	RUSSKIJ	ESPANOL	ITALIANO
1982 HEMICRUSTACEA PALS	Hemicrustacea ARTHROPODA	Hemicrustacea ARTHROPODA	Hemicrustacea ARTHROPODA	HEMICRUSTACEA	Hemicrustacea ARTHROPODA	Hemicrustacea ARTHROPODA
1983 HEMIPTEROIDA PALS	HEMIPTEROIDA	HEMIPTEROIDA	HEMIPTEROIDA	Hemipteroida INSECTA	HEMIPTEROIDA	HEMIPTEROIDEA
1984 HERCYNIAN OROGENY STRU	HERCYNIAN OROGENY *UF:* Armorican orogeny; Hercynides; Sudetic orogeny; Variscan orogeny	OROGENIE HERCY- NIENNE *UF:* Hercynides; Oroge- nie armoricaine; Oroge- nie sudete; Orogenie varisque	HERZYN. OROGENESE *UF:* Armorikan. Orogenese; Sudet. Orogenese; Varist. Orogenese; Varistiden	GERTSINSKIJ *UF:* Akadskij; Allegan- skij; Armorikanskij; Sudetskij; Varijskij	OROGENIA= HERCINICA *UF:* Hercinides; Orogenia=armoricana; Orogenia=sudetica; Orogenia=varisca	OROGENESI ERCINICA *UF:* Orogenesi armori- cana; Orogenesi sude- tica; Orogenesi varisica; Variscidi
1985 HERCYNIDES STRU	Hercynides HERCYNIAN OROGENY	Hercynides OROGENIE HERCY- NIENNE	Varistiden HERZYN. OROGENESE	GERTSINIDY	Hercinides OROGENIA= HERCINICA	Variscidi OROGENESI ERCINICA
1986 HEREDITY PALE	Heredity BIOLOGY	Heredite BIOLOGIE	Vererbung BIOLOGIE	Nasledstvennost' EVOLUTSIYA	Herencia BIOLOGIA	Ereditarieta BIOLOGIA
1987 HERMATYPIC TAXON PALE	HERMATYPIC TAXA	TAXON HERMA- TYPIQUE	HERMATYP.TAXON	GERMATIPNYJ	TAXON= HERMATIPICO	TAXON HERMA- TIPICO
1988 HETEROBLASTIC TEXTURE TEST	Heteroblastic texture TEXTURES	Texture heteroblastique TEXTURE	Heteroblast.Gefuege KORN=GEFUEGE	GETEROBLAS- TOVYJ	Textura=heteroblastica TEXTURA	Tessitura eteroblastica TESSITURA
1989 HETEROCHRONISM STRA	Heterochronism LITHOFACIES	Heterochronisme LITHOFACIES	Heterochronie LITHO=FAZIES	GETEROKHRONIZM	Heterocronismo LITOFACIES	Eterocronismo LITOFACIES
1990 HETEROCHRONOUS STRA	Heterochronous(adj.)	Heterochrone(adj.)	Heterochron(adj.)	RAZNOVREMEN- NYJ	Heterocrono(adj.)	Eterocrono(adj.)
1991 HETEROCORALLIA PALS	HETEROCORALLIA	HETEROCORALLIA	HETEROCORALLIA	Heterocorallia ANTHOZOA	HETEROCORALLIA	HETEROCORALLIA
1992 HETERODONTA PALS	Heterodonta BIVALVIA	Heterodonta BIVALVIA	Heterodonta LAMELLI- BRANCHIATA	HETERODONTA	Heterodonta BIVALVIA	Heterodonta BIVALVIA
1993 HETEROGENEITY MISC	HETEROGENEITY *UF:* Inhomogeneity	HETEROGENEITE *UF:* Diamictite; Inhomo- geneite	HETEROGENITAET *UF:* Diamiktit; Inhomo- genitaet	GETEROGENNOST'	HETEROGENEIDAD *UF:* Diamictita; Inhomo- geneidad	ETEROGENEITA *UF:* Diamictite; Inomo- geneita
1994 HETEROGENEOUS MISC	Heterogeneous(adj.)	Heterogene(adj.)	Heterogen(adj.)	GETEROGENNYJ *UF:* Diamiktit	Heterogeneo	Eterogeneo(adj.)
1995 HETEROMORPH ORGANISM PALE	HETEROMORPHIC TAXA	TAXON HETERO- MORPHE	HETEROMORPH. ORGANISMUS	GETEROMORFNYJ	HETEROMORFO	TAXON ETERO- MORFO
1996 HETEROMORPHISM IGNE	Heteromorphism CRYSTALLIZATION	Heteromorphisme CRISTALLISATION	Heteromorphismus KRISTALLISATION	GETEROMORFIZM	Heteromorfismo CRISTALIZACION	Eteromorfismo CRISTALLIZZAZ- IONE
1997 HETEROSPOROUS PLANT PALE	Heterosporous taxa REPRODUCTION; FLORA	Plante heterosporee REPRODUCTION; FLORE	Heterospor.Pflanze FRUKTIFIKATION; FLORA	GETEROSPOROVYJ	Planta=heterospora REPRODUCCION; FLORA	Pianta eterosporica RIPRODUZIONE; FLORA

	ENGLISH	FRANCAIS	DEUTSCH	RUSSKIJ	ESPANOL	ITALIANO
1998 HETEROTROPHIC ORGANISM PALE	Heterotrophic organism BIOLOGY	Organisme hete- rotrophe BIOLOGIE	Heterotroph. Organismus BIOLOGIE	GETEROTROFNYJ	Organismo= heterotrofico BIOLOGIA	Organismo eterotrofo BIOLOGIA
1999 HETTANGIAN STRS	HETTANGIAN	HETTANGIEN	HETTANGIUM	GEITANG	HETTANGIENSE	HETTANGIANO
2000 HEXACORALLA PALS	SCLERACTINIA	HEXACORALLA	HEXACORALLA	HEXACORALLA	HEXACORALLA	HEXACORALLA
2001 HEXACTINIARIA PALS	HEXACTINIARIA	HEXACTINIARIA	HEXACTINIARIA	Hexactiniaria ANTHOZOA	HEXACTINIARIA	HEXACTINIARIA
2002 HEXAGONAL SYSTEM MINE	Hexagonal system CRYSTAL SYSTEMS	Systeme hexagonal SYSTEME CRISTAL- LIN	Hexagonal.System KRISTALL= SYSTEM	SINGONIYA GEK- SAGONAL'NAYA	Sistema=hexagonal SISTEMA= CRISTALINO	Sistema esagonale SISTEMA CRISTAL- LINO
2003 HEXAHEDRITE EXTS	Hexahedrite METEORITES	Hexahedrite METEORITE MET- ALLIQUE	Hexahedrit METALL= METEORIT	Geksaehdrit METEORIT ZHELEZNYJ	Exaedrita METEORITO- METALICO	Esaedrite METEORITE ME- TALLICA
2004 HIATUS STRA	Hiatus UNCONFORMITIES	Hiatus LACUNE STRATI- GRAPHIQUE	Hiatus LUECKE	PERERYV *UF:* Diastema; Miscibil- ity gap	Hiato LAGUNA	Hiatus LACUNA
2005 HIEROGLYPH TEST	Hieroglyphs SEDIMENTARY STRUCTURES	Hieroglyphe STRUCTURE SEDI- MENTAIRE	Hieroglyphe SEDIMENT= GEFUEGE	GIEROGLIF *UF:* Sole mark; Tur- boglif; Znak voloc- heniya	Jeroglifico ESTRUCTURA= SEDIMENTARIA	Geroglifico STRUTTURA SEDI- MENTARIA
2006 HIGH ANOMALY VALUE SOLI	High value anomaly ANOMALIES	Maximum anomalie ANOMALIE	Hoher=Anomalie= Wert ANOMALIE	ANOMALIYA POLOZHITEL'NAYA	Maximo=anomalia ANOMALIA	Massimo di anomalia ANOMALIA
2007 HIGH PRESSURE PHCH	HIGH PRESSURE	HAUTE PRESSION	HOCH=DRUCK	DAVLENIE VYSOKOE	PRESION=ALTA	ALTA PRESSIONE
2008 HIGH TEMPERATURE PHCH	HIGH TEMPERA- TURE	HAUTE TEMPERA- TURE	HOCH= TEMPERATUR	VYSOKOTEMPERA- TURNYJ	TEMPERATURA- ALTA	ALTA TEMPERA- TURA
2009 HIGH-ENERGY ENVIRONMENT SEDI	HIGH=ENERGY ENVIRONMENT	MILIEU HAUTE ENERGIE	HOCH=ENERGET. MILIEU	Sreda volnennaya SEDIMENTATSIYA	MEDIO=ALTA= ENERGIA	AMBIENTE AD ALTA ENERGIA
2010 HIGH-GRADE ECON	High=grade(adj.)	Riche(adj.)	Hochgradig(adj.)	BOGATYJ	Alto=grado(adj.)	Di alto grado(adj.)
2011 HIGH-GRADE METAMORPHISM IGNE	HIGH=GRADE METAMORPHISM	METAMORPHISME FORT	HOCHGRADIG. METAMORPHOSE	[HIGH=GRADE METAMORPHISM] STEPEN' METAM- ORFIZMA	METAMORFISMO- FUERTE	METAMORFISMO FORTE
2012 HIGHLAND SURF	Highlands MOUNTAINS	Hautes=terres MASSIF MON- TAGNEUX	Hochland BERGMASSIV	NAGOR'E	Tierras=altas MACIZO= MONTAÑOSO	Alteterre MASSICCIO MON- TUOSO
2013 HILL SURF	HILLS	COLLINE	HUEGEL	Kholm MASSIV GORNYJ	COLINA	COLLINA

	ENGLISH	FRANCAIS	DEUTSCH	RUSSKIJ	ESPANOL	ITALIANO
2014 HINGE PALE	Hinge (pale) MORPHOLOGY	Charniere test TEST	Schloss SCHALE	ZAMOK	Charnela TEST	Cerniera (pale) GUSCIO
2015 HINGE FAULT STRU	HINGE FAULTS	FAILLE CHAR- NIERE	SCHARNIER= VERWERFUNG	RAZRYV SHARNIRNYJ	FALLA=DE= CHARNELA	FAGLIA A CERNI- ERA
2016 HIPPOMORPHA PALS	HIPPOMORPHA	HIPPOMORPHA	HIPPOMORPHA	HIPPOMORPHA	HIPPOMORPHA	HIPPOMORPHA
2017 HISTOGRAM MATH	HISTOGRAMS	HISTOGRAMME	HISTOGRAMM	GISTOGRAMMA	HISTOGRAMA	ISTOGRAMMA
2018 HISTOLOGY PALE	HISTOLOGY	HISTOLOGIE	HISTOLOGIE	Gistologiya BIOLOGIYA	HISTOLOGIA	ISTOLOGIA
2019 HISTORICAL GEOLOGY STRA	HISTORICAL GEOL- OGY	GEOLOGIE HISTORIQUE	HISTOR.GEOLOGIE	GEOLOGIYA ISTORICHESKAYA	GEOLOGIA= HISTORICA	GEOLOGIA STORICA
2020 HISTORY MISC	HISTORY *UF:* History of geology	HISTORIQUE *UF:* Histoire de la geolo- gie	GESCHICHTE *UF:* Geschichte=der= Geologie	ISTORIYA	HISTORICA *UF:* Historia=de=la= geologia	STORIA *UF:* Storia della geologia
2021 HISTORY OF GEOLOGY GEOL	History of geology GEOLOGY; HIS- TORY	Histoire de la geologie HISTORIQUE	Geschichte=der= Geologie GESCHICHTE	ISTORIYA GEOLOGII	Historia=de=la= geologia HISTORICA	Storia della geologia STORIA
2022 HISTOSOL SUSS	HISTOSOLS	HISTOSOL	HISTOSOL	KHISTOSOL	HISTOSOL	HISTOSOL
2023 HODOGRAPH SOLI	Hodographs INSTRUMENTS; TRAVELTIME CURVES	Dromochronique INSTRUMENTA- TION; COURBE TEMPS PARCOURS	Hodograph INSTRUMEN- TIERUNG; LAUFZEIT=KURVE	GODOGRAF	Hodografo INSTRUMENTA- CION; CURVA- TIEMPO= RECORRIDO	Odografo STRUMENTAZ- IONE; DRO- MOCRONA
2024 HOLMES' **CLASSIFICATION** IGNE	Holmes' classification CLASSIFICATION; PETROLOGY	Classification Holmes CLASSIFICATION; PETROLOGIE	Holmes Klassifikation KLASSIFIKATION; PETROGRAPHIE	KLASSIFIKATSIYA KHOLMSA	Clasificacion=de= Holmes CLASIFICACION; PETROGRAFIA	Classificazione di Holmes CLASSIFICAZIONE; PETROGRAFIA
2025 HOLMIUM CHEE	HOLMIUM	HOLMIUM	HO	GOL'MIJ	HOLMIO	OLMIO
2026 HOLOCENE STRS	HOLOCENE *UF:* Recent sedimenta- tion	HOLOCENE	HOLOZAEN	GOLOTSEN	HOLOCENO	OLOCENE
2027 HOLOCEPHALI PALS	HOLOCEPHALI	HOLOCEPHALI	HOLOCEPHALI	HOLOCEPHALI	HOLOCEPHALI	HOLOCEPHALI
2028 HOLOCRYSTALLINE **TEXTURE** TEST	Holocrystalline texture TEXTURES	Texture holocristalline TEXTURE	Holokristallin.Gefuege KORN=GEFUEGE	POLNOKRISTALLI- CHESKIJ *UF:* Yavnokristallicheskij	Textura=holocristalina TEXTURA	Tessitura olocristallina TESSITURA
2029 HOLOGRAPHY METH	HOLOGRAPHY	HOLOGRAPHIE	HOLOGRAPHIE	GOLOGRAFIYA	HOLOGRAFIA	OLOGRAFIA
2030 HOLOPHYTE PALE	Holophyte ECOLOGY; FLORA	Holophyte ECOLOGIE; FLORE	Holophyt OEKOLOGIE; FLORA	GOLOFIT	Holofita ECOLOGIA; FLORA	Holophyte ECOLOGIA; FLORA

	ENGLISH	FRANCAIS	DEUTSCH	RUSSKIJ	ESPANOL	ITALIANO
2031 HOLOSTEI PALS	HOLOSTEI	HOLOSTEI	HOLOSTEI	HOLOSTEI	HOLOSTEI	HOLOSTEI
2032 HOLOSTOMATOUS PALE	Holostomatous taxa MORPHOLOGY	Holostome ANATOMIE SQUE- LETTE	Holostom SKELETT	Ust'e golostomnoe APERTURA	Holostomado ANATOMIA= ESQUELETO	Olostomato ANATOMIA DELLO SCHELETRO
2033 HOLOTHURIOIDEA PALS	HOLOTHUROIDEA	HOLOTHUROIDEA	HOLOTHUROIDEA	HOLOTHUROIDEA	HOLOTHUROIDEA	HOLOTHUROIDEA
2034 HOLOTYPE PALE	Holotypes NOMENCLATURE; PALEONTOLOGY	Holotype NOMENCLATURE	Holotyp NOMENKLATUR	GOLOTIP *UF:* Ehkzemplyar tipovoj; Neotip	Holotipo NOMENCLATURA	Olotipo NOMENCLATURA
2035 HOMALOZOA PALS	Homalozoa ECHINODERMATA	Homalozoa ECHINODERMATA	Homalozoa ECHINODERMATA	HOMALOZOA	Homalozoa ECHINODERMATA	Homalozoa ECHINODERMATA
2036 HOMEOMORPHISM MINE	Homeomorphism HABIT	Homeomorphisme HABITUS	Homoeomorphismus HABITUS	GOMEOMORFISM	Homomorfismo HABITO	Omeomorfismo ABITO CRISTAL- LINO
2037 HOMEOMORPHY PALE	Homeomorphy ADAPTATION	Homeomorphie ADAPTATION	Homoeomorphie ANPASSUNG	GOMEOMORFIYA	Homomorfo ADAPTACION	Omeomorfia ADATTAMENTO
2038 HOMINIDAE PALS	HOMINIDAE	HOMINIDAE	HOMINIDAE	Hominidae SIMIAN	HOMINIDAE	HOMINIDAE
2039 HOMO SAPIENS PALS	HOMO SAPIENS	HOMO SAPIENS	HOMO–SAPIENS	HOMO SAPIENS	HOMO–SAPIENS	HOMO SAPIENS
2040 HOMOCLINE STRU	Homoclines MONOCLINES	Homoclinal MONOCLINAL	Homoklinale MONOKLINALE	Gomoklinal' MONOKLINAL'	Homoclinal MONOCLINAL	Omoclinale MONOCLINALE
2041 HOMOGENEITY MISC	HOMOGENEITY *UF:* Homogenization	HOMOGENEITE *UF:* Homogeneisation	HOMOGENITAET *UF:* Homogenisierung	GOMOGENNOST'	HOMOGENEIDAD *UF:* Homogeneizacion	OMOGENEITA *UF:* Omogeneizzazione
2042 HOMOGENIZATION MISC	Homogenization HOMOGENEITY	Homogeneisation HOMOGENEITE	Homogenisierung HOMOGENITAET	GOMOGENIZAT- SIYA	Homogeneizacion HOMOGENEIDAD	Omogeneizzazione OMOGENEITA
2043 HOMOIOSTELEA PALS	HOMOIOSTELEA	HOMOIOSTELEA	HOMOIOSTELEA	HOMOIOSTELEA	HOMOIOSTELEA	HOMOIOSTELEA
2044 HOMOLOGY PALE	Homology BIOLOGIC EVOLU- TION	Homologie EVOLUTION BIOLOGIQUE	Homologie BIOLOG. EVOLUTION	GOMOLOGIYA	Homologia EVOLUCION= BIOLOGICA	Omologia EVOLUZIONE BIOLOGICA
2045 HOMONYMY PALE	HOMONYMY	HOMONYMIE	HOMONYMIE	Gomonimiya TAKSONOMIYA	HOMONIMO	OMONIMIA
2046 HOMOSTELEA PALS	HOMOSTELEA	HOMOSTELEA	HOMOSTELEA	HOMOSTELEA	HOMOSTELEA	HOMOSTELEA
2047 HOMOTAXY STRA	Homotaxy CORRELATION	Equivalence CORRELATION	Homotaxie KORRELATION	GOMOTAKSI- AL'NOST'	Homotaxia CORRELACION	Omotassia CORRELAZIONE
2048 HOOKE'S LAW PHCH	HOOKE'S LAW	LOI HOOKE	HOOKES=GESETZ	GUKA ZAKON *UF:* Differentsiatsiya gorizontal'naya	LEY–DE–HOOKE	LEGGE DI HOOKE

	ENGLISH	FRANCAIS	DEUTSCH	RUSSKIJ	ESPANOL	ITALIANO
2049 HORIZON DIFFERENTIATION SURF	HORIZON DIFFERENTIATION	DIFFERENCIATION HORIZON	HORIZONT= DIFFERENZIERUNG	Differentsiatsiya gorizontal'naya GUKA ZAKON; GORIZONTAL'NYJ	DIFERENCIACION= HORIZONTE	ORIZZONTE DIFFERENZIATO
2050 HORIZONTAL MISC	Horizontal(adj.)	Horizontal(adj.)	Horizontal(adj.)	GORIZONTAL'NYJ *UF:* Differentsiatsiya gorizontal'naya; Napravlenie gorizontal'noe; Razlom gorizontal'nyj; Smeshchenie gorizontal'noe; Szhatie gorizontal'noe	Horizontal(adj.)	Orizzontale(adj.)
2051 HORIZONTAL COMPRESSION STRU	Horizontal compression HORIZONTAL ORIENTATION; COMPRESSION	Compression horizontale TECTONIQUE TANGENTIELLE	Horizontal= Kompression TANGENTIAL= TEKTONIK	Szhatie gorizontal'noe GORIZONTAL'NYJ; SZHATIE	Compresion=horizontal TECTONICA= TANGENCIAL	Compressione orizzontale TETTONICA COMPRESSIVA
2052 HORIZONTAL DIRECTION GEOL	HORIZONTAL ORIENTATION *UF:* Horizontal compression; Horizontal fault	Direction horizontale ORIENTATION	Horizontal=Richtung ORIENTIERUNG	Napravlenie gorizontal'noe ORIENTIROVKA; GORIZONTAL'NYJ	Direccion=horizontal ORIENTACION	Direzione orizzontale ORIENTAZIONE
2053 HORIZONTAL DISPLACEMENT STRU	HORIZONTAL DISPLACEMENTS	MOUVEMENT HORIZONTAL	HORIZONTAL= VERSCHIEBUNG	Smeshchenie gorizontal'noe SMESHCHENIE; GORIZONTAL'NYJ	DESPLAZAMIENTO= HORIZONTAL	MOVIMENTO ORIZZONTALE
2054 HORIZONTAL FAULT STRU	Horizontal fault FAULTS; HORIZONTAL ORIENTATION	Faille horizontale FAILLE CHEVAUCHEMENT	Horizontal= Verwerfung AUFSCHIEBUNG	Razlom gorizontal'nyj RAZLOM; GORIZONTAL'NYJ	Falla=desplazamiento= horizontal CABALGAMIENTO	Faglia a piano orizzontale ACCAVALLAMENTO
2055 HORNBLENDE- HORNFELS FACIES IGNE	HORNBLENDE- HORNFELS FACIES	Facies corneenne hornblende FACIES CORNEENNE	Hornblende=Hornfels= Fazies HORNFELS=FAZIES	Fatsiya rogovo= obmanko=rogovikovaya FATSIYA ROGOVIKOVAYA	Facies=corneana= horneblenda FACIES= CORNEANA	Facies cornubianiti a orneblenda FACIES DELLE CORNUBIANITI
2056 HORNBLENDITE IGNS	Hornblendite AMPHIBOLITES	Hornblendite AMPHIBOLITE	Hornblendit AMPHIBOLIT	GORNBLENDIT	Hornblendita ANFIBOLITA	Orneblendite ANFIBOLITE
2057 HORNFELS IGMS	HORNFELS	CORNEENNE	HORNFELS	ROGOVIK	CORNEANA	CORNUBIANITE
2058 HORNFELS FACIES IGNE	HORNFELS FACIES *UF:* Pyroxene=hornfels facies	FACIES CORNEENNE *UF:* Facies corneenne albite epidote; Facies corneenne a hornblende; Facies corneenne pyroxene	HORNFELS=FAZIES *UF:* Albit=Epidot= Hornfels=Fazies; Hornblende=Hornfels= Fazies; Pyroxen= Hornfels=Fazies	FATSIYA ROGOVIKOVAYA *UF:* Fatsiya rogovo= obmanko=rogovikovaya	FACIES= CORNEANA *UF:* Facies=corneana= albita=epidota; Facies= corneana=horneblenda; Facies=corneana= piroxenica	FACIES DELLE CORNUBIANITI *UF:* Facies cornubian. albite epidoto; Facies cornubianiti a orneblenda; Facies cornubianiti pirosseniche
2059 HORSETAIL ECON	Horsetail ORE BODIES; VEINS	Filon en queue de cheval FILON	[HORSETAIL] GANG	ZHILA TIPA KONSKOGO KHVOSTA	Cola=de=caballo FILON	Filone a coda di cavallo FILONE
2060 HORST STRU	HORSTS *UF:* Median masses	HORST *UF:* Massif median	HORST *UF:* Zentral.Massiv	GORST	HORST *UF:* Bloque=central	HORST *UF:* Massiccio centrale
2061 HOST MATERIAL IGNE	Host materials HOST ROCKS	Materiau hote ROCHE HOTE	Wirts=Material WIRTS=GESTEIN	VMESHCHAYUSHCHIJ	Material=huesped ROCA=HUESPED	Materiale incassante ROCCIA OSPITE

	ENGLISH	FRANCAIS	DEUTSCH	RUSSKIJ	ESPANOL	ITALIANO
2062 HOST ROCK IGNE	HOST ROCKS *UF:* Host materials	ROCHE HOTE *UF:* Materiau hote	WIRTS=GESTEIN *UF:* Wirts=Material	PORODA VMESH- CHAYUSHCHAYA *UF:* Poroda vmeshchay- ushchaya obramlyeniya	ROCA=HUESPED *UF:* Materiales=huesped	ROCCIA OSPITE *UF:* Materiale incassante
2063 HOT PHCH	Hot(adj.)	Chaud(adj.)	Heiss(adj.)	GORYACHIJ	Caliente(adj.)	Caldo(adj.)
2064 HOT BRINE ECON	HOT BRINES	SAUMURE CHAUDE	ERZSCHLAMM	RASSOL GORYACHIJ	SALMUERA= CALIENTE	BRINA CALDA
2065 HOT SPOT SOLI	HOT SPOTS	POINT CHAUD	[HOT=SPOT]	TOCHKA PERE- GREVA	PUNTO=CALIENTE	PUNTO CALDO
2066 HOT SPRING GEOH	HOT SPRINGS	SOURCE THER- MALE	HEISS.QUELLE	ISTOCHNIK GORYACHIJ	FUENTE=TERMAL	SORGENTE TER- MALE
2067 HOWARDITE EXTS	Howardite ACHONDRITES	Howardite ACHONDRITE	Howardit ACHONDRIT	Govardit AKHONDRIT	Howardita ACONDRITA	Howardite ACONDRITE
2068 HUDSONIAN OROGENY STRU	HUDSONIAN OROGENY	OROGENIE HUDSO- NIENNE	HUDSON= OROGENESE	GUDZONSKIJ	OROGENIA= HUDSONIANA	OROGENESI HUD- SONIANA
2069 HUMAN ECOLOGY ENVI	HUMAN ECOLOGY	ECOLOGIE HUMAINE	SOZIAL= OEKOLOGIE	[HUMAN ECOL- OGY] OKHRANA; SREDA OKRUZHAYUSH- CHAYA	ECOLOGIA= HUMANA	ECOLOGIA UMANA
2070 HUMAN WASTE ENVI	HUMAN WASTE	DECHET DOMES- TIQUE	SIEDLUNGS= MUELL	[HUMAN WASTE] OTKHODY	RESIDUO=URBANO	RIFIUTI DOMES- TICI
2071 HUMIC ACID CHES	HUMIC ACIDS	ACIDE HUMIQUE	HUMIN=SAEURE	GUMINOKISLOTA	ACIDO=HUMICO	ACIDO UMICO
2072 HUMIC SOIL SUSS	HUMIC SOILS	SOL RICHE EN HUMUS *UF:* Humod	HUMUSREICH. BODEN *UF:* Humod	POCHVA GUMU- SOVAYA	SUELO=RICO=EN= HUMUS *UF:* Humod	SUOLO UMICO *UF:* Humod
2073 HUMID CLIMATE SURF	Humid climate CLIMATE; HUMID ENVIRONMENT	Climat humide MILIEU HUMIDE	Feucht.Zone FEUCHT.MILIEU	KLIMAT GUMID- NYJ	Clima=humedo MEDIO=HUMEDO	Clima umido AMBIENTE UMIDO
2074 HUMID ENVIRONMENT SURF	HUMID ENVIRON- MENT *UF:* Humid climate	MILIEU HUMIDE *UF:* Aquent; Aquert; Aquox; Aquult; Climat humide; Orthox; Plano- sol; Udalf; Udult; Ustert	FEUCHT.MILIEU *UF:* Aquent; Aquert; Aquox; Aquult; Feucht. Zone; Orthox; Planosol; Udalf; Udult; Ustert	GUMIDNYJ	MEDIO=HUMEDO *UF:* Aquent; Aquert; Aquox; Aquult; Clima=humedo; Orthox; Planosol; Udalf; Udult; Ustert	AMBIENTE UMIDO *UF:* Aquent; Aquert; Aquox; Aquult; Clima umido; Orthox; Plano- sol; Udalf; Udult; Ustert
2075 HUMIDITY PHCH	HUMIDITY *UF:* Hygroscopic water	HUMIDITE *UF:* Eau hygroscopique; Rosee	FEUCHTIGKEIT *UF:* Hygroskop.Wasser; Tau	VLAZHNOST'	HUMEDAD *UF:* Agua=higroscopica; Rocio	UMIDITA *UF:* Acqua igroscopica; Rugiada
2076 HUMIFICATION SURF	Humification HUMUS	Humification HUMUS	Humifizierung HUMUS	Gumifikatsiya GUMUS	Humificacion HUMUS	Umificazione HUMUS
2077 HUMIN CHES	Humin HUMUS	Humine MATIERE ORGANIQUE	Humin ORGAN.SUBSTANZ	GUMIN	Humin MATERIA= ORGANICA	Umina SOSTANZA ORGANICA

	ENGLISH	FRANCAIS	DEUTSCH	RUSSKIJ	ESPANOL	ITALIANO
2078 HUMITE COAL SEDS	HUMITE	CHARBON HUMIQUE	HUMUS=KOHLE	UGOL' GUMUSOVYJ	HUMITA	HUMITE
2079 HUMMOCK SURF	HUMMOCKS	HUMMOCK	HUMMOCK	BUGOR *UF:* Bugor torfyannoj; Bugor vodoroslevyj	HUMMOCK	HUMMOCK
2080 HUMOD SUSS	Humods SPODOSOLS	Humod SPODOSOL; SOL RICHE EN HUMUS	Humod SPODOSOL; HUMUSREICH. BODEN	Gumod SPODOSOL	Humod SPODOSOL; SUELO=RICO=EN= HUMUS	Humod SPODOSOL; SUOLO UMICO
2081 HUMOX SUSS	Humox OXISOLS	Humox OXISOL	Humox OXISOL	Gumoks OKSISOL	Humox OXISOL	Humox OXISOL
2082 HUMULT SUSS	Humults ULTISOLS	Humult ULTISOL	Humult ULTISOL	Gumult UL'TISOL	Humult ULTISOL	Humult ULTISOL
2083 HUMUS SURF	HUMUS *UF:* Humification; Humin	HUMUS *UF:* Humification	HUMUS *UF:* Humifizierung	GUMUS *UF:* Gumifikatsiya	HUMUS *UF:* Humificacion	HUMUS *UF:* Umificazione
2084 HYALOCRYSTALLINE TEXTURE TEST	Hyalocrystalline texture TEXTURES	Texture hyalocrystal-line TEXTURE; CRIS-TALLINITE	Hyalokristallin.Gefuege KORN=GEFUEGE; KRISTALLINITAET	GIALOKRISTALLI-CHESKIJ	Textura=hialocristalina TEXTURA; CRIS-TALINIDAD	Tessitura ialocristallina TESSITURA; CRIS-TALLINITA
2085 HYALOSPONGEA PALS	Hyalospongea PORIFERA	Hyalospongea PORIFERA	Hyalospongea PORIFERA	HYALOSPONGEA	Hyalospongea PORIFERA	Hyalospongea PORIFERA
2086 HYBRIDISM IGNE	HYBRIDIZATION	HYBRIDATION MAGMATIQUE	MAGMAT. HYBRIDISIERUNG	GIBRIDIZATSIYA	HIBRIDO	IBRIDAZIONE MAGMATICA
2087 HYDATOGENESIS GEOL	Hydatogenesis CRYSTALLIZA-TION; EVAPORITES	Hydatogenese CRISTALLISATION; EVAPORITE	Hydatogenese KRISTALLISATION; EVAPORIT	GALOGENEZ	Hidatogenesis CRISTALIZACION; EVAPORITA	Idatogenesi CRISTALLIZZAZ-IONE; EVAPORITE
2088 HYDRARCH PALE	Hydrarch(adj.)	Hydrarch(adj.)	Hydrarch(adj.)	GIDRICHESKIJ	Hydrarch(adj.)	Hydrarch(adj.)
2089 HYDRATION PHCH	HYDRATION	HYDRATATION	HYDRATISIERUNG	GIDRATATSIYA *UF:* Hydration of glass	HIDRATACION	IDRATAZIONE
2090 HYDRATION OF GLASS ISOT	HYDRATION OF GLASS	HYDRATATION VERRE	GLAS=HYDRATISIERUNG	[HYDRATION OF GLASS] GIDRATATSIYA; STEKLO	HIDRATACION=VIDRIO	IDRATAZIONE DEL VETRO
2091 HYDRAULIC CONDUCTIVITY GEOH	HYDRAULIC CON-DUCTIVITY	CONDUCTIVITE HYDRAULIQUE	HYDRAUL. LEITFAEHIGKEIT	VODOPRONITSAE-MOST' *UF:* Transmissibility coefficient	CONDUCTIVIDAD=HIDRAULICA	CONDUTTIVITA IDRAULICA
2092 HYDRAULIC FRACTURING ENGI	HYDRAULIC FRAC-TURING	FRACTURATION HYDRAULIQUE	HYDRAUL.RISS=ERZEUGUNG	[HYDRAULIC FRACTURING]	FRACTURACION=HIDRAULICA	FRATTURAZIONE IDRAULICA
2093 HYDRAULIC GRADIENT GEOH	Hydraulic gradient GRADIENT; STREAMS	Gradient hydraulique GRADIENT; HYDRAULIQUE	Hydraul.Gradient GRADIENT; HYDRAULIK	Gradient gidravlicheskij GIDRAVLIKA; GRA-DIENT	Gradiente=hidraulico GRADIENTE; HIDRAULICA	Gradiente idraulico GRADIENTE; IDRAULICA

	ENGLISH	FRANCAIS	DEUTSCH	RUSSKIJ	ESPANOL	ITALIANO
2094 HYDRAULIC HEAD GEOH	Head PRESSURE; WATER	Charge hydraulique PRESSION; EAU	Hydraul.Belastung DRUCK; WASSER	Napor gidrostaticheskij DAVLENIE GIDROSTATI- CHESKOE	Carga=hidraulica PRESION; AGUA	Carico idraulico PRESSIONE; ACQUA
2095 HYDRAULIC MAP GEOH	HYDRAULIC MAPS	CARTE HYDRAULIQUE	HYDRAUL.KARTE	[HYDRAULIC MAP] GIDRAVLIKA; KARTA	MAPA= HIDRAULICO	CARTA IDRAULICA
2096 HYDRAULIC PRESSURE GEOH	HYDRAULIC PRES- SURE	PRESSION HYDRAULIQUE	HYDRAUL.DRUCK	Davlenie gidrodinami- cheskoe GIDRODINAMIKA; DAVLENIE	PRESION= HIDRAULICA	PRESSIONE IDRAULICA
2097 HYDRAULIC RADIUS GEOH	Hydraulic radius AQUIFERS	Rayon hydraulique INFILTRATION; NAPPE ALLUVION	Hydraul.Radius INFILTRATION; QUARTAER. GRUNDWASSER	RADIUS GIDRAVLI- CHESKIJ	Radio=hidraulico INFILTRACION; MANTO=ALUVIAL	Raggio idraulico INFILTRAZIONE; FALDA IN ALLU- VIONI
2098 HYDRAULICS GEOH	HYDRAULICS	HYDRAULIQUE UF: Gradient hydraulique	HYDRAULIK UF: Hydraul.Gradient	GIDRAVLIKA UF: Gradient gidravli- cheskij; Hydraulic map	HIDRAULICA UF: Gradiente= hidraulico	IDRAULICA UF: Gradiente idraulico
2099 HYDROCARBON CHES	HYDROCARBONS	COMPOSE HYDRO- CARBURE UF: Alkane	KOHLENWASS- ERSTOFF= VERBINDUNG UF: Alkan	UGLEVODOROD UF: Fitan; Isoprenoid	COMPUESTO= HIDROCARBURO UF: Alcano	GRUPPO IDROCAR- BURI UF: Alcani
2100 HYDROCHEMICAL FACIES GEOC	Hydrochemical facies HYDROCHEM- ISTRY	Facies hydrochimique HYDROCHIMIE	Hydrochem.Fazies HYDROCHEMIE	FATSIYA GIDROCHIMI- CHESKAYA	Facies=hidroquimica HIDROQUIMICA	Facies idrochimica IDROCHIMICA
2101 HYDROCHEMICAL MAP GEOH	HYDROCHEMICAL MAPS	CARTE HYDROCHIMIQUE	HYDROCHEM. KARTE	KARTA GIDROKHIMI- CHESKAYA	MAPA= HIDROQUIMICO	CARTA IDROCHIMICA
2102 HYDROCHEMISTRY GEOH	HYDROCHEM- ISTRY UF: Hydrochemical facies	HYDROCHIMIE UF: Facies hydrochimique	HYDROCHEMIE UF: Hydrochem.Fazies	GIDROGEOKH- IMIYA	HIDROQUIMICA UF: Facies= hidroquimica	IDROCHIMICA UF: Facies idrochimica
2103 HYDRODYNAMIC MODEL GEOH	Hydrodynamic models HYDRODYNAMICS; MODELS	Modele hydrody- namique MODELE; HYDRO- DYNAMIQUE	Hydrodynam.Modell MODELL; HYDRO- DYNAMIK	MODEL' GIDRODI- NAMICHESKAYA	Modelo=hidrodinamico MODELO; HIDRODI- NAMICA	Modello idrodinamico MODELLO; IDRODI- NAMICA
2104 HYDRODYNAMICS GEOH	HYDRODYNAMICS UF: Hydrodynamic models	HYDRODY- NAMIQUE UF: Ecoulement eau souterraine; Hydrologie karstique; Modele hydrodynamique	HYDRODYNAMIK UF: Grundwasser= Abfluss; Hydrodynam. Modell; Karst= Hydrologie	GIDRODINAMIKA UF: Catchment hydrody- namics; Davlenie gi- drodinamicheskoe	HIDRODINAMICA UF: Escorrentia=agua= subterranea; Hidrogeologia=karst; Modelo=hidrodinamico	IDRODINAMICA UF: Deflusso di falda idrica; Idrologia car- sica; Modello idrodi- namico
2105 HYDROEXPLOSION IGNE	Hydroexplosion ERUPTIONS; WATER	Explosion phreatique ERUPTION; EAU	Hydro=Explosion ERUPTION; WASSER	GIDROEHKSPLOZ- IYA	Hidroexplosion ERUPCION; AGUA	Idroesplosione ERUZIONE; ACQUA
2106 HYDROGEN CHEE	HYDROGEN UF: Hydrogen gas	HYDROGENE UF: Gaz hydrogene	H UF: Wasserstoff	VODOROD	HIDROGENO UF: Gas=hidrogeno	IDROGENO UF: Gas idrogeno

	ENGLISH	FRANCAIS	DEUTSCH	RUSSKIJ	ESPANOL	ITALIANO
2107 HYDROGEN GAS CHES	Hydrogen gas HYDROGEN; GASES	Gaz hydrogene HYDROGENE; GAZ	Wasserstoff H; GAS	VODOROD GAZ	Gas=hidrogeno HIDROGENO; GAS	Gas idrogeno IDROGENO; GAS
2108 HYDROGEN SULFIDE CHES	HYDROGEN SUL- FIDE	GAZ HYDROGENE SULFURE	SCHWEFELW- ASSERSTOFF	SEROVODOROD	GAS=SULFUROSO	SOLFURO DI IDRO- GENO
2109 HYDROGEOCHEMI- CAL PROSPECTING ECON	HYDROGEOCHEMI- CAL METHODS	PROSPECTION HYDROGEOCHIMIQUE	HYDROGEOCHEM. PROSPEKTION	POISK GIDROGEOKHIMI- CHESKIJ	PROSPECCION= HIDROGEOQUIMICA	PROSPEZIONE IDROGEOCHIMICA
2110 HYDROGEOLOGICAL CONTROL ECON	HYDROGEOLOGI- CAL CONTROLS	CONTROLE HYDROGEOLOGIQUE GITE	HYDROGEOLOG. EINFLUSS	KONTROL' GIDROGEOLOGI- CHESKIJ	CONTROL= HIDROGEOLOGICO	CONTROLLO IDROGEOLOGICO GIACIMEN
2111 HYDROGEOLOGICAL MAP GEOH	HYDROGEOLOGIC MAPS	CARTE HYDROGEOLOGIQUE	HYDROGEOLOG. KARTE	KARTA GIDROGEOLOGI- CHESKAYA	MAPA= HIDROGEOLOGICO	CARTA IDROGEOLOGICA
2112 HYDROGEOLOGICAL PROVINCE GEOH	HYDROGEOLOGIC PROVINCES	PROVINCE HYDROGEOLOGIQUE	HYDROGEOLOG. PROVINZ	PROVINTSIYA GIDROGEOLOGI- CHESKAYA	PROVINCIA= HIDROGEOLOGICA	PROVINCIA IDROGEOLOGICA
2113 HYDROGEOLOGICAL SURVEY GEOH	Hydrogeological survey GROUND WATER	Leve hydrogeologique HYDROGEOLOGIE REGIONALE	Hydrogeolog. Erkundung REGIONAL= HYDROGEOLOGIE	S'EMKA GIDROGEOLOGI- CHESKAYA	Estudio=hidrogeologico HIDROGEOLOGIA= REGIONAL	Rilevamento idrogeologico IDROGEOLOGIA REGIONALE
2114 HYDROGEOLOGICAL TECHNOLOGY GEOH	Hydrogeological tech- nology HYDROGEOLOGY; TECHNOLOGY	Technologie hydrogeologique TECHNOLOGIE	Wasser=Technik TECHNOLOGIE	TEKHNIKA GIDROGEOLOGI- CHESKAYA	Tecnologia= hidrogeologica HIDROGEOLOGIA; TECNOLOGIA	Tecnologia idrogeologica TECNOLOGIA
2115 HYDROGEOLOGY GEOH	HYDROGEOLOGY *UF:* Hydrogeological technology; Regional hydrogeology	HYDROGEOLOGIE	HYDROGEOLOGIE	GIDROGEOLOGIYA	HIDROGEOLOGIA *UF:* Tecnologia= hidrogeologica	IDROGEOLOGIA
2116 HYDROGRAPH GEOH	HYDROGRAPHS	HYDROGRAMME	PEGEL=KURVE	GIDROGRAF	HIDROGRAMA	IDROGRAMMA
2117 HYDROGRAPHY GEOH	Hydrography HYDROSPHERE	Hydrographie HYDROSPHERE	Hydrographie HYDROSPHAERE	GIDROGRAFIYA	Hidrografia HIDROSFERA	Idrografia IDROSFERA
2118 HYDROIDA PALS	Hydroida HYDROZOA	Hydroida HYDROZOA	Hydroida HYDROZOA	HYDROIDA	Hydroida HYDROZOA	Hydroida HYDROZOA
2119 HYDROLOGIC CYCLE GEOH	HYDROLOGIC CYCLE	CYCLE EAU	WASSER= KREISLAUF	Tsikl gidrologicheskij GIDROLOGIYA; TSIKL	CICLO=AGUA	CICLO DELL'ACQUA
2120 HYDROLOGICAL MAP GEOH	HYDROLOGIC MAPS	CARTE HYDROLOGIQUE	HYDROLOG.KARTE	Karta gidrologi- cheskaya KARTA; GIDROLOGIYA	MAPA= HIDROLOGICO	CARTA IDROLOGICA
2121 HYDROLOGY GEOH	HYDROLOGY *UF:* Exorheism; Karst hydrology	HYDROLOGIE SUR- FACE *UF:* Areisme; Endore- isme; Exoreisme	HYDROLOGIE *UF:* Abflusslos.Gebiet; Abflusslosigkeit; Extern.Entwaesserung	GIDROLOGIYA *UF:* Karta gidrologi- cheskaya; Tsikl gidrologicheskij; Vnesh- nij drenazh	HIDROLOGIA= SUPERFICIE *UF:* Cuenca=endorreica; Endorreismo; Exor- reismo	IDROLOGIA *UF:* Areismo; Endoreismo; Esoreismo

#	Term	ENGLISH	FRANCAIS	DEUTSCH	RUSSKIJ	ESPANOL	ITALIANO
2122	**HYDROMETALLURGY** METH	HYDROMETAL-LURGY	HYDROMETALLUR-GIE	HYDRO=METALLURGIE	GIDROMETALLUR-GIYA	HIDROMETALLUR-GIA	IDROMETALLUR-GIA
2123	**HYDROMETER** INST	Hydrometers INSTRUMENTS; HYDROMETRY	Hydrometre INSTRUMENTA-TION; HYDROME-TRIE	Hydrometer INSTRUMEN-TIERUNG; HYDROMETRIE	GIDROMETR	Hidrometro INSTRUMENTA-CION; HIDRO-METRIA	Idrometro STRUMENTAZ-IONE; IDROMETRIA
2124	**HYDROMETRY** GEOH	HYDROMETRY *UF:* Hydrometers	HYDROMETRIE *UF:* Hydrometre	HYDROMETRIE *UF:* Hydrometer	GIDROMETRIYA	HIDROMETRIA *UF:* Hidrometro	IDROMETRIA *UF:* Idrometro
2125	**HYDROPHONE** INST	Hydrophone INSTRUMENTS; SEISMIC METHODS	Hydrophone INSTRUMENTA-TION; METHODE SISMIQUE	Hydrophon INSTRUMEN-TIERUNG; SEISM. METHODE	GIDROFON	Hidrofono INSTRUMENTA-CION; METODO-SISMICO	Idrofono STRUMENTAZ-IONE; METODO SISMICO
2126	**HYDROSPHERE** SOLI	HYDROSPHERE *UF:* Hydrography	HYDROSPHERE *UF:* Hydrographie	HYDROSPHAERE *UF:* Hydrographie	GIDROSFERA	HIDROSFERA *UF:* Hidrografia	IDROSFERA *UF:* Idrografia
2127	**HYDROSPIRE** PALE	Hydrospires RESPIRATION; MORPHOLOGY	Hydrospire RESPIRATION	Hydrospire ATMUNG	GIDROSPIRA	Hidrospira RESPIRACION	Idrospira ANATOMIA APPARATO RESPIRATORIO
2128	**HYDROSTATIC LEVEL** GEOH	Potentiometric surface PIEZOMETRY	Niveau hydrostatique SURFACE PIEZO-METRIQUE	Hydrostat.Niveau HYDROSTAT. OBERFLAECHE	UROVEN' GIDROSTATI-CHESKIJ *UF:* Poverkhnost' p'ezometricheskaya	Nivel=hidrostatico SUPERFICIE=PIEZOMETRICA	Livello idrostatico SUPERFICIE PIEZO-METRICA
2129	**HYDROSTATIC PRESSURE** GEOH	HYDROSTATIC PRESSURE	PRESSION HYDROSTATIQUE	HYDROSTAT. DRUCK *UF:* Grundwasser=Spiegel	DAVLENIE GIDROSTATI-CHESKOE *UF:* Fluid pressure; Napor gidrostaticheskij; Uplift pressure	PRESION=HIDROSTATICA	PRESSIONE IDROSTATICA
2130	**HYDROTHERMAL** IGNE	Hydrothermal(adj.)	Hydrothermal(adj.)	Hydrothermal(adj.)	GIDROTER-MAL'NYJ	Hidrotermal(adj.)	Idrotermale(adj.)
2131	**HYDROTHERMAL ALTERATION** IGNE	HYDROTHERMAL ALTERATION	ALTERATION HYDROTHERMALE *UF:* Alunitisation; Amphibolitisation; Anorthositisation; Epidotisation; Palagoni-tisation; Saussuritisa-tion; Scapolitisation; Steatitisation	HYDROTHERMAL-UMWANDLUNG *UF:* Alunitisierung; Amphibolitisierung; Anorthositisierung; Epidotisierung; Pala-gonitisierung; Saussuri-tisierung; Skapoliti-sierung; Steatitisierung	METAMORFIZM GIDROTER-MAL'NYJ	ALTERACION=HIDROTERMAL *UF:* Alunitizacion; Anfi-bolitizacion; Anortosit-izacion; Epidotizacion; Escapolitizacion; Esteatizacion; Pala-gonitizacion; Sausurit-izacion	ALTERAZIONE IDROTERMALE *UF:* Alunitizzazione; Anfibolitizzazione; Anortositizzazione; Epidotizzazione; Pala-gonitizzazione; Saus-suritizzazione; Scapoli-tizzazione; Steatitizzazione
2132	**HYDROTHERMAL DEPOSIT** ECON	HYDROTHERMAL PROCESSES *UF:* Apomagmatic deposit	GITE HYDROTHER-MAL *UF:* Gite apomagma-tique; Gite catathermal; Gite hypothermal	HYDROTHERMAL. VORKOMMEN *UF:* Apomagmat. Vorkommen; Hypothermal. Vorkommen; Katathermal. Vorkommen	GIDROTER-MAL'NOE MESTOROZHDENIE	YACIMIENTO=HIDROTERMAL *UF:* Deposito=apomagmatico; Deposito=catatermal; Yacimiento=hipotermal	GIACIMENTO IDROTERMALE *UF:* Catatermale; Giaci-mento apomagmatico; Giacimento catatermale
2133	**HYDROTHERMAL STAGE** IGNE	HYDROTHERMAL CONDITIONS	CONDITION HYDROTHERMALE	HYDROTHERMAL-BEDINGUNG	PROTSESS GIDROTER-MAL'NYJ	ESTADO=HIDROTERMAL	FASE IDROTER-MALE

	ENGLISH	FRANCAIS	DEUTSCH	RUSSKIJ	ESPANOL	ITALIANO
2134 HYDROTHERMAL WATER GEOH	Hydrothermal waters THERMAL WATERS	Eau hydrothermale EAU THERMO-MINERALE	Hydrothermal=Wasser THERMAL-WASSER	RASTVOR GIDROTER-MAL'NYJ	Agua=hidrotermal AGUA=TERMOMINERAL	Acqua termominerale ACQUA CURATIVA
2135 HYDROUS GEOL	Hydrous(adj.)	Hydrate(adj.)	Waesserig(adj.)	VODO-SODERZHASHCHIJ	Acuoso(adj.)	Idratato(adj.)
2136 HYDROXIDES MING	HYDROXIDES	HYDROXYDE	HYDROXID	GIDROOKIS'	HIDROXIDO	IDROSSIDO
2137 HYDROZOA PALS	HYDROZOA UF: Hydroida	HYDROZOA UF: Hydroida	HYDROZOA UF: Hydroida	HYDROZOA	HYDROZOA UF: Hydroida	HYDROZOA UF: Hydroida
2138 HYGROPHYLE ORGANISM PALE	Hygrophyle(adj.)	Hygrophyle(adj.)	Hygrophil(adj.)	GIGROFIL'NYJ	Higrofilo(adj.)	Igrofilo(adj.)
2139 HYGROSCOPIC WATER SURF	Hygroscopic water HUMIDITY	Eau hygroscopique HUMIDITE	Hygroskop.Wasser FEUCHTIGKEIT	VODA GIGROSKOPI-CHESKAYA	Agua=higroscopica HUMEDAD	Acqua igroscopica UMIDITA
2140 HYMENOPTEROIDA PALS	HYMENOPTERA	HYMENOP-TEROIDA	HYMENOP-TEROIDA	Hymenoptera INSECTA	HYMENOPTERA	HYMENOPTERA
2141 HYOLITHES PALS	HYOLITHES	HYOLITHES	HYOLITHES	HYOLITHES	HYOLITHES	HYOLITHES
2142 HYPABYSSAL ORIGIN IGNE	HYPABYSSAL ROCKS	Roche hypabyssale ROCHE MICRO-GRENUE	Hypabyssal=Genese SUBVULKAN. GESTEIN	GIPABISSAL'NYJ	Roca=hipoabisal ROCA=MICROGRANUDA	Origine ipoabissale TESSITURA MICRO-GRANULARE
2143 HYPERGENESIS SURF	Hypergenesis DIAGENESIS	Formation supergene FORMATION SUPERFICIELLE	Hypergenese OBERFLAECHEN=BILDUNGEN	GIPERGENEZ UF: Descension theory	Alteracion=supergenica FORMACION=SUPERFICIAL	Rielaborazione esogena FORMAZIONE SUPERFICIALE
2144 HYPIDIOBLASTIC TEXTURE TEST	Hypidioblastic texture TEXTURES	Texture hypidioblas-tique TEXTURE	Hypidioblast.Gefuege KORN=GEFUEGE	GIPIDIOBLASTOVYJ	Textura=hipidioblastica TEXTURA	Tessitura ipidioblastica TESSITURA
2145 HYPIDIOMORPHIC TEXTURE TEST	Hypidiomorphic texture TEXTURES	Texture hypidiomor-phique TEXTURE	Hypidiomorph.Gefuege KORN=GEFUEGE	GIPIDIOMORFNYJ	Textura=hipidiomorfica TEXTURA	Tessitura ipidiomorfa TESSITURA
2146 HYPOCENTER SOLI	FOCUS	HYPOCENTRE	HYPOZENTRUM	OCHAG ZEMLETR-YASENIYA UF: Mekhanizm zem-letryaseniya	HIPOCENTRO	IPOCENTRO
2147 HYPOCRYSTALLINE TEXTURE TEST	Hypocrystalline texture TEXTURES	Texture hypocristalline TEXTURE; CRIS-TALLINITE	Hypokristallin.Gefuege KRISTALLINITAET; KORN=GEFUEGE	GIPOKRISTALLI-CHESKIJ	Textura=hipocristalina TEXTURA; CRIS-TALINIDAD	Tessitura ipocristallina TESSITURA; CRIS-TALLINITA
2148 HYPODIGM PALE	Hypodigms NOMENCLATURE	Hypodigme NOMENCLATURE	Typ=Material NOMENKLATUR	GIPODIGM	Hipodigma NOMENCLATURA	Ipodigma NOMENCLATURA
2149 HYPOTHERMAL DEPOSIT ECON	HYPOTHERMAL PROCESSES UF: Katathermal depos-its	Gite hypothermal GITE HYDROTHER-MAL	Hypothermal. Vorkommen HYDROTHERMAL. VORKOMMEN	GIPOTERMAL'NYJ	Yacimiento=hipotermal YACIMIENTO=HIDROTERMAL	Catatermale GIACIMENTO IDROTERMALE

	ENGLISH	FRANCAIS	DEUTSCH	RUSSKIJ	ESPANOL	ITALIANO
2150 HYPOTHESIS MISC	Hypothesis THEORETICAL STUDIES	Hypothese THEORIE	Hypothese THEORIE	GIPOTEZA	Hipotesis TEORIA	Ipotesi TEORIA
2151 HYPSOMETRY SURF	Hypsometry GEODESY	Hypsometrie GEODESIE	Hoehen=Messung GEODAESIE	Gipsometriya TOPOGRAFIYA	Hipsometria GEODESIA	Ipsometria GEODESIA
2152 HYRACOIDEA PALS	HYRACOIDEA	HYRACOIDEA	HYRACOIDEA	HYRACOIDEA	HYRACOIDEA	HYRACOIDEA
2153 HYSTRICHOMORPHA PALS	HYSTRICHOMOR- PHA	HYSTRICOMORPHA	HYSTRICOMORPHA	HYSTRICHOMOR- PHA	HYSTRICHOMOR- PHA	HYSTRICHOMOR- PHA
2154 IAPETUS SOLI	IAPETUS	PROTOATLAN- TIQUE	IAPETUS=OZEAN	[IAPETUS]	PROTOATLANTICO	PROTOATLANTICO
2155 ICE SURF	ICE *UF:* Dead ice	GLACE *UF:* Glace morte	EIS *UF:* Toteis	LED *UF:* Mass balance	HIELO *UF:* Glaciar=colgado	GHIACCIO *UF:* Ghiaccio morto
2156 ICE CAP SURF	ICE CAPS *UF:* Polar ice cap	CALOTTE GLA- CIAIRE *UF:* Calotte glaciaire polaire	POL=KALOTTE *UF:* Polar.Eis=Kalotte	SHAPKA LEDY- ANAYA *UF:* Polyarnyj lednik pokrovnyj	CASQUETE= GLACIAR *UF:* Casquete=polar	CALOTTA GLA- CIALE *UF:* Calotta polare
2157 ICE FIELD SURF	Ice fields GLACIERS	Champ glace GLACIER	Eis=Feld GLETSCHER	POLE LEDOVOE	Campo=de=hielo GLACIAR	Vedretta GHIACCIAIO
2158 ICE MOVEMENT SURF	ICE MOVEMENT	MOUVEMENT GLACE	EIS=BEWEGUNG	[ICE MOVEMENT] LEDNIK; DINAMIKA	MOVIMIENTO- HIELO	MOVIMENTO DI GHIACCIO
2159 ICE RAFTING SURF	ICE RAFTING	TRANSPORT GLACE	EIS=TRANSPORT	[ICE RAFTING] TRANSPOR- TIROVKA (GEOL)	TRANSPORTE= HIELO	TRASPORTO DI GHIACCIO
2160 ICE SHEET SURF	ICE SHEETS	INLANDSIS	INLAND=EIS	POKROV LED- NIKOVYJ	GLACIAR= CONTINENTAL	INLANDSIS
2161 ICE SHELF SURF	ICE SHELVES	PLATEAU LITTO- RAL GLACE	EIS=BARRIERE	LEDNIK SHEL'FOVYJ	MESETA= LITORAL=HELADA	PIANA LITORALE GLACIALE
2162 ICE WEDGE SURF	ICE WEDGES	COIN GLACE	EIS=KEIL	Klin ledyanoj MERZLOTA VECH- NAYA	CUÑA=DE=HIELO	CUNEO DI GHIAC- CIO
2163 ICEBERG MARI	ICEBERGS	[ICEBERG]	EISBERG	AJSBERG	[ICEBERG]	[ICEBERG]
2164 ICHNOFOSSIL PALE	ICHNOFOSSILS *UF:* Ichnology; Rock borer; Scolite	ICHNITES	ICHNITES	Ikhnofossilii SLED ZHIZ- NEDEYATEL'NOSTI	ICHNITES	ICNITE *UF:* Organismo litofago; Scolite; Struttura vermiforme
2165 ICHNOLOGY PALE	Ichnology ICHNOFOSSILS	Ichnologie TRACE ORGANIQUE	Ichnologie LEBENS=SPUR	IKHNOLOGIYA	Icnologia HUELLA= ORGANICA	Icnologia TRACCIA FOSSILE
2166 ICHTHYOPTERYGIA PALS	ICHTHYOPTERY- GIA	ICHTHYOPTERY- GIA	ICHTHYOPTERY- GIA	ICHTHYOPTERY- GIA *UF:* Ichthyosauria	ICHTHYOPTERY- GIA	ICHTHYOPTERY- GIA

	ENGLISH	FRANCAIS	DEUTSCH	RUSSKIJ	ESPANOL	ITALIANO
2167 ICHTHYOSAURIA PALS	ICHTHYOSAURIA	ICHTHYOSAURIA	ICHTHYOSAURIA	Ichthyosauria ICHTHYOPTERY- GIA	ICHTHYOSAURIA	ICHTYOSAURIA
2168 IDIOBLAST TEST	Idioblastic texture TEXTURES	Idioblaste TEXTURE	Idioblast KORN=GEFUEGE	IDIOBLAST	Idioblasto TEXTURA	Idioblasto TESSITURA
2169 IDIOGEOSYNCLINE STRU	Idiogeosynclines GEOSYNCLINES	Idiogeosynclinal GEOSYNCLINAL	Idiogeosynklinale GEOSYNKLINALE	IDIOGEOSINKLI- NAL'	Idiogeosinclinal GEOSINCLINAL	Idiogeosinclinale GEOSINCLINALE
2170 IDIOMORPHIC TEXTURE TEST	Idiomorphic texture TEXTURES	Texture idiomorphe TEXTURE	Idiomorph.Gefuege KORN=GEFUEGE	IDIOMORFNYJ *UF:* Ehvgedral'nyj	Textura=idiomorfica TEXTURA	Tessitura idiomorfa TESSITURA
2171 IDIOMORPHISM MINE	Idiomorphism ISOMORPHISM	Idiomorphisme ISOMORPHISME	Idiomorphie ISOMORPHIE	IDIOMORFIZM	Idiomorfismo ISOMORFISMO	Idiomorfismo ISOMORFISMO
2172 IGNEOUS ACTIVITY IGNE	IGNEOUS ACTIV- ITY *UF:* Magmatism	Activite ignee MAGMATISME	Magmat.Aktivitaet MAGMATISMUS	Izverzhenie (aktiv.) MAGMATIZM	Actividad=ignea MAGMATISMO	Attivita magmatica MAGMATISMO
2173 IGNEOUS COMPLEX IGNE	Igneous complex INTRUSIONS	Complexe igne INTRUSION	Magmatit=Komplex INTRUSIONS= KOERPER	KOMPLEKS MAG- MATICHESKIJ	Complejo=igneo INTRUSION	Complesso eruttivo INTRUSIONE
2174 IGNEOUS FACIES IGNE	Igneous facies IGNEOUS ROCKS; FACIES	Facies igne ROCHE IGNEE	Magmat.Fazies MAGMATIT	Fatsiya magmati- cheskikh porod MAGMATICHESKIJ	Facies=ignea ROCA=IGNEA	Facies magmatica ROCCIA IGNEA
2175 IGNEOUS ROCK GENESIS IGNE	Igneous rock genesis IGNEOUS ROCKS; GENESIS	Genese roche ignee MAGMATISME	Magmatit=Genese MAGMATISMUS	Genezis izverzhennykh porod GENEZIS	Genesis=roca=ignea MAGMATISMO	Genesi di roccia intru- siva MAGMATISMO
2176 IGNEOUS ROCKS IGNS	IGNEOUS ROCKS *UF:* Igneous facies; Igneous rock genesis; Leucocratic composi- tion; Melanocratic composition; Mesocra- tic composition; Thera- lite	ROCHE IGNEE *UF:* Composition melanocrate; Composi- tion mesocrate; Facies igne; Monolithe; Roche leucocrate	MAGMATIT *UF:* Leukokrat. Chemismus; Magmat. Fazies; Melanokrat. Zusammensetzung; Mesokrat. Zusammensetzung; Monolith	PORODA MAGMATI= CHESKAYA	ROCA=IGNEA *UF:* Composicion= leucocratica; Composicion= melanocratica; Composicion= mesocratica; Facies= ignea; Monolito	ROCCIA IGNEA *UF:* Composizione leu- cocratica; Composiz- ione melanocratica; Composizione mesocra- tica; Facies magmatica; Monolito
2177 IGNIMBRITE IGNS	IGNIMBRITE	IGNIMBRITE	IGNIMBRIT	IGNIMBRIT	IGNIMBRITA	IGNIMBRITE
2178 IMAGERY APPL	IMAGERY	IMAGERIE	ABBILD	[IMAGERY]	IMAGEN	IMMAGINE
2179 IMBRIAN EXTR	IMBRIAN	Imbrien ECHELLE STRATI- GRAPHIQUE; LUNE	Imbrium STRATIGRAPH. SKALA; MOND	IMBRIJ	Imbriense ESCALA= ESTRATIGRAFICA; PLANETA=LUNA	Imbriano SCALA STRATI- GRAFICA; LUNA
2180 IMBRICATE STRUCTURE STRU	IMBRICATE STRUCTURES	TECTONIQUE IMBRIQUEE	SCHUPPEN= TEKTONIK	CHESHUJCHATYJ	TECTONICA= IMBRICADA	TETTONICA IMBRI- CATA
2181 IMBRICATION SEDI	IMBRICATION	IMBRICATION	VERSCHUPPUNG	CHESHUJCHATOST'	IMBRICACION	IMBRICAZIONE
2182 IMMATURE GEOL	Immature(adj.)	Immature(adj.)	Unreif(adj.)	NEZRELYJ	Inmaduro(adj.)	Immaturo(adj.)

	ENGLISH	FRANCAIS	DEUTSCH	RUSSKIJ	ESPANOL	ITALIANO
2183 IMMATURE SOIL SUSS	IMMATURE SOILS	SOL PEU EVOLUE *UF:* Arenosol; Xerosol	UNREIF.BODEN	Pochva nepolnoraz-vitaya POCHVA AZON-AL'NAYA	SUELO=INMADURO *UF:* Arenosol; Xerosol	SUOLO IMMATURO *UF:* Arenosol; Xerosol
2184 IMMERSION METHOD METH	Immersion method OPTICAL PROPER-TIES	Methode par immersion METHODOLOGIE; INDICE REFRAC-TION	Immersions=Mikroskopie METHODIK; BRECHUNGS=INDEX	METOD IMMER-SIONNYJ	Metodo=de=inmersion METODOLOGIA; INDICE=DE=REFRACCION	Metodo a immersione METODOLOGIA; INDICE DI RIFRAZ-IONE
2185 IMMISCIBILITY PHCH	IMMISCIBILITY *UF:* Immiscible materi-als	IMMISCIBILITE *UF:* Substance non=miscible	NICHT=MISCHBARKEIT *UF:* Nicht=mischbar. Material	NESMESIMOST'	INMISCIBILIDAD *UF:* Inmiscible	IMMISCIBILITA *UF:* Materiale immisci-bile
2186 IMMISCIBLE MATERIAL PHCH	Immiscible materials IMMISCIBILITY	Substance non=miscible IMMISCIBILITE	Nicht=mischbar. Material NICHTMISCH-BARKEIT	NESMESHIVA-YUSHCHIJSYA	Inmiscible INMISCIBILIDAD	Materiale immiscibile IMMISCIBILITA
2187 IMPACT CRATER EXTR	Impact crater CRATERS; IMPACT FEATURES	Cratere impact CRATERE; EFFET CHOC	Einschlag=Krater KRATER; STOSSWELLEN=EFFEKT	KRATER UDARA *UF:* Astroblema; Krater meteoritnyj; Struktura impaktnaya; Struktura udara	Crater=de=impacto CRATER; EFECTO=DE=CHOQUE	Cratere d'impatto CRATERE; EFFETTO DI SHOCK
2188 IMPACT FEATURE EXTR	IMPACT FEATURES *UF:* Cryptovolcanic structure; Impact crater	EFFET CHOC *UF:* Breche choc; Choc; Cratere impact; Metamorphisme impact; Structure impact	STOSSWELLEN=EFFEKT *UF:* Einschlag=Krater; Einschlag=Struktur; Schock; Schock=Brekzie; Stoss=Metamorphose	Struktura udara KRATER UDARA	EFECTO=DE=CHOQUE *UF:* Brecha=de=impacto; Crater=de=impacto; Estructura=impacto; Impacto; Metamorfismo=impacto	EFFETTO DI SHOCK *UF:* Breccia d'impatto; Cratere d'impatto; Impatto; Metamorfismo d'impatto; Struttura d' impatto
2189 IMPACT METAMORPHISM IGNE	Impact metamorphism SHOCK METAMOR-PHISM	Metamorphisme impact METAMORPHISME; EFFET CHOC	Stoss=Metamorphose METAMORPHOSE; STOSSWELLEN=EFFEKT	Metamorfizm impakt-nyj METAMORFIZM UDARNYJ	Metamorfismo=impacto METAMORFISMO; EFECTO=DE=CHOQUE	Metamorfismo d'impatto TIPO DI METAMOR-FISMO; EFFETTO DI SHOCK
2190 IMPACT SEISMICS APPL	Impact seismics SEISMIC METHODS	Sismique poussee METHODE SISMIQUE	Stoss=Seismik SEISM.METHODE	[IMPACT SEISMICS] METOD SEJSMICHES-KIJ	Sismica=impacto METODO=SISMICO	Sismica avanzata METODO SISMICO
2191 IMPACT STATEMENT ENVI	IMPACT STATE-MENTS	ETUDE IMPACT MILIEU	UMFELD=STUDIE	[IMPACT STATE-MENT]	ESTUDIO=IMPACTO=MEDIO	IMPATTO AMBIEN-TALE
2192 IMPACT STRUCTURE TEST	Impact structures CRATERS	Structure impact CRATERE; EFFET CHOC	Einschlag=Struktur KRATER; STOSSWELLEN=EFFEKT	Struktura impaktnaya KRATER UDARA	Estructura=impacto CRATER; EFECTO=DE=CHOQUE	Struttura d' impatto CRATERE; EFFETTO DI SHOCK
2193 IMPACTITE EXTR	IMPACTITE	IMPACTITE	IMPAKTIT	IMPAKTIT	IMPACTITA	IMPACTITE
2194 IMPORT ECON	IMPORT	IMPORTATION	IMPORT	IMPORT	IMPORTACION	IMPORTAZIONE
2195 IMPREGNATION ECON	IMPREGNATED DEPOSITS	GITE IMPREGNA-TION	IMPRAEGNATIONS=LAGERSTAETTE	IMPREGNATSIYA	YACIMIENTO=IMPREGNACION	GIACIMENTO D'IMPREGNAZIONE

	ENGLISH	FRANCAIS	DEUTSCH	RUSSKIJ	ESPANOL	ITALIANO
2196 IMPURITY PHCH	IMPURITIES	IMPURETE	UNREINHEIT	PRIMES'	IMPUREZAS=AGUA	IMPUREZZA
2197 IN SITU MISC	IN SITU	IN SITU	IN=SITU= UNTERSUCHUNG	IN SITU	IN=SITU	PROVA IN SITO
2198 INARTICULATA PALS	INARTICULATA *UF:* Acrotretida; Lingulida; Obolellida	INARTICULATA *UF:* Acrotretida; Lingulida; Obolellida	INARTICULATA *UF:* Acrotretida; Lingulida; Obolellida	INARTICULATA	INARTICULATA *UF:* Acrotretida; Lingulida; Obolellida	INARTICULATA *UF:* Acrotretida; Lingulida; Obolellida
2199 INCEPTISOL SUSS	INCEPTISOLS *UF:* Aquepts; Ochrepts; Umbrepts	INCEPTISOL *UF:* Andept; Aquept; Ochrept; Umbrept	INCEPTISOL *UF:* Andept; Aquept; Ochrept; Umbrept	INSEPTISOL *UF:* Akvept; Andept; Okhrept; Umbrept	INCEPTISOL *UF:* Andept; Aquept; Ochrept; Umbrept	INCEPTISOL *UF:* Andept; Aquept; Ochrept; Umbrept
2200 INCLUSION MINE	INCLUSIONS	INCLUSION *UF:* Inclusion gazeuse	EINSCHLUSS *UF:* Gas=Einschluss	VKLYUCHENIE *UF:* Mineral inclusion	INCLUSION *UF:* Inclusion=gas	INCLUSIONE *UF:* Inclusione gassosa
2201 INCOHERENT MISC	Incoherent(adj.)	Incoherent(adj.)	Inkohaerent(adj.)	RYKHLYJ *UF:* Material rykhlyj	Incoherente(adj.)	Incoerente(adj.)
2202 INCOHERENT MATERIAL COMS	COHESIONLESS MATERIALS *UF:* Cohesionless soils	MATERIAU NON COHERENT *UF:* Materiau non conso- lide	LOCKER=GESTEIN *UF:* Unverfestigt.Gestein	Material rykhlyj RYKHLYJ; MATE- RIAL OSADOCHNYJ	MATERIAL=NO- COHERENTE *UF:* Material=no= consolidado	MATERIALE INCOERENTE *UF:* Materiale sciolto
2203 INCOMPETENT ROCK ENGI	INCOMPETENT MATERIALS	Materiau incompetent DURETE	Inkompetent.Gestein HAERTE	NEKOMPETENT- NYJ	Roca=incompetente DUREZA	Roccia incompetente DUREZZA
2204 INCONGRUENT MELTING IGNE	Incongruent melting PARTIAL MELTING	Fusion incomplete FUSION PARTIELLE	Inkongruent.Schmelzen PARTIELL. SCH- MELZEN	PLAVLENIE INKONGRUEHNT- NOE	Fusion=incongruente FUSION=PARCIAL	Fusione incongruente FUSIONE PARZI- ALE
2205 INCRUSTATION SURF	DURICRUST	INCRUSTATION *UF:* Caliche; Depot concretionne; Indura- tion	INKRUSTATION *UF:* Caliche; Sinter; Verhaertung	INKRUSTATSIYA *UF:* Otverdenie	INCRUSTACION *UF:* Caliche; Endureci- miento; Sinter	INCROSTAZIONE *UF:* Caliche; Deposito concrezionato; Induri- mento
2206 INDEX FOSSIL STRA	INDEX FOSSILS *UF:* Guide fossil	FOSSILE CARAC- TERISTIQUE	LEITFOSSIL	Vid=indeks ISKOPAEMOE RUKOVODYASH- CHEE	FOSIL= CARACTERISTICO	FOSSILE GUIDA
2207 INDEX MAPS MISC	INDEX MAPS	CARTE ASSEM- BLAGE	INDEX=KARTE	Karta registratsionnaya KARTA; OBZOR	MAPA=INDICE	CARTA D'INSIEME
2208 INDEX MINERAL IGNE	Index minerals FACIES	Mineral repere FACIES METAMOR- PHISME	Leit=Mineral METAMORPHOSE= FAZIES	INDEKS=MINERAL	Mineral=indice FACIES= METAMORFISMO	Minerale indice FACIES METAMOR- FICA
2209 INDEX OF REFRACTION PHCH	REFRACTIVE INDEX *UF:* Becke line; Refrac- tivity; Refractometers; Refractometry	INDICE REFRAC- TION *UF:* Lisere Becke; Met- hode par immersion; Pouvoir refracteur; Refractometre; Refrac- tometrie	BRECHUNGS= INDEX *UF:* Becke=Linie; Immersions= Mikroskopie; Refrak- tometrie	POKAZATEL' PRE- LOMLENIYA	INDICE=DE= REFRACCION *UF:* Linea=de=Becke; Metodo=de=inmersion; Refractividad; Refrac- tometria; Refracto- metro	INDICE DI RIFRAZ- IONE *UF:* Linea di Becke; Metodo a immersione; Potere rifrangente; Rifrattometria; Rifrat- tometro
2210 INDIAN STRS	Olenekian LOWER TRIASSIC	Indien TRIAS INF	Indien=Stufe U.TRIAS	IND	Indiense TRIAS=INF	Indiano TRIASSICO INF.

	ENGLISH	FRANCAIS	DEUTSCH	RUSSKIJ	ESPANOL	ITALIANO
2211 INDICATIVE BOULDER SURF	Indicative boulder ERRATICS	Bloc indicateur BLOC ERRATIQUE; ORIGINE	Leit=Geschiebe ERRAT.BLOCK; HERKUNFT	VALUN RUKOVO- DYASHCHIJ	Bloque=indicador BLOQUE= ERRATICO; ORI- GEN	Erratico indicatore MASSO ERRATICO; ORIGINE
2212 INDICATOR MISC	INDICATORS	Indicateur GUIDE	Indikator INDIKATOR= MINERAL	INDIKATOR *UF:* Depth indicator; Sled paleoklimata	Indicador GUIA	Indicatore GUIDA
2213 INDIUM CHEE	INDIUM	INDIUM	IN	INDIJ	INDIO	INDIO
2214 INDOCHINITE EXTS	Indochinite TEKTITES	Indochinite TECTITE	Indochinit TEKTIT	Indoshinit TEKTIT	Indochinita TECTITA	Indochinite TECTITE
2215 INDUCED MAGNETIZATION PHCH	INDUCED MAGNE- TIZATION	AIMANTATION INDUITE	INDUZIERT. MAGNETISIERUNG	NAMAGNICHEN- NOST' INDUTSIROVANNAYA	IMANTACION= INDUCIDA	MAGNETIZZAZ- IONE INDOTTA
2216 INDUCED POLARIZATION APPL	INDUCED POLAR- IZATION	POLARISATION PROVOQUEE	INDUZIERT. POLARISATION	METOD VYZVAN- NOJ POLYARIZAT- SII	POLARIZACION= PROVOCADA	POLARIZZAZIONE INDOTTA
2217 INDUCTIVE METHOD APPL	INDUCTION	METHODE INDUC- TION	INDUKTIONS= METHODE	METOD INDUKTIV- NYJ *UF:* Metod vyzvannogo potentsiala; Vozrast ionevouranov'j	METODO=DE= INDUCCION	METODO D'INDUZIONE
2218 INDURATION SEDI	Induration DIAGENESIS	Induration INCRUSTATION	Verhaertung INKRUSTATION	Otverdenie INKRUSTATSIYA	Endurecimiento INCRUSTACION	Indurimento INCROSTAZIONE
2219 INDUSTRIAL MINERAL COMS	INDUSTRIAL MIN- ERALS	SUBSTANCE UTILE *UF:* Cendre industrielle; Mineral economique	NUTZBAR. GESTEINE *UF:* Industrie=Asche	NEMETALLI- CHESKOE SYR'E	SUSTANCIA=UTIL *UF:* Ceniza=artificial; Mineral=industrial	SOSTANZA UTILE *UF:* Cenere artificiale; Minerale sfruttabile
2220 INDUSTRIAL USE ECON	Industrial use UTILIZATION	Utilisation industrielle UTILISATION SUB- STANCE	Industriell.Nutzung VERWENDUNG	ISPOL'ZOVANIE PROMYSHLENNOE	Utilizacion=industrial UTILIZACION= SUSTANCIA	Utilizzazione industri- ale UTILIZZAZIONE DI SOSTANZA
2221 INDUSTRIAL WASTE ENVI	INDUSTRIAL WASTE *UF:* Spoils	DECHET INDUSTR- IEL *UF:* Deblai sterile	INDUSTRIE= MUELL *UF:* Taub.Gestein	OTKHODY PRO- MYSHLENNYE	RESIDUO= INDUSTRIAL *UF:* Escombrera	RIFIUTI INDUSTRI- ALI *UF:* Rifiuto inerte
2222 INDUSTRY ECON	INDUSTRY	INDUSTRIE	INDUSTRIE	PROMYSHLEN- NOST'	INDUSTRIA	INDUSTRIA
2223 INERT COMPONENT PHCH	Inert component CHEMICAL ANAL- YSIS	Compose inerte ANALYSE CHIMIQUE	Inaktiv.Komponente CHEM.ANALYSE	KOMPONENT INERTNYJ	Componente=inerte ANALISIS= QUIMICO	Componente inerte COMPOSIZIONE CHIMICA
2224 INERT GAS CHES	INERT GASES	GAZ RARE	EDELGAS	GAZ INERTNYJ	GAS=RARO	GAS INERTE
2225 INFERRED ORE ECON	Inferred ore RESERVES	Minerai probable RESERVE	Hergeleitet.Erz RESERVE	ZAPASY PROGNOZ- NYE *UF:* Subeconomic deposit	Mineral=probable RESERVA= MINERAL	Minerale presunto RISERVA

	ENGLISH	FRANCAIS	DEUTSCH	RUSSKIJ	ESPANOL	ITALIANO
2226 INFILTRATION GEOH	INFILTRATION	INFILTRATION *UF:* Rayon hydraulique	INFILTRATION *UF:* Hydraul.Radius	INFIL'TRATSIYA *UF:* Leakage; Pro- sachivanie; Seepage; Vpityvanie	INFILTRACION *UF:* Radio=hidraulico	INFILTRAZIONE *UF:* Raggio idraulico
2227 INFILTRATION GALLERY GEOH	INFILTRATION GALLERY	GALERIE CAP- TANTE	STOLLEN= FASSUNG	GALEREYA VODOS- BORNAYA	GALERIA=DE= CAPTACION	GALLERIA DI CAP- TAZIONE
2228 INFORMATICS MATH	INFORMATICS	INFORMATIQUE	INFORMATIK	INFORMATIKA	INFORMATICA= GEOLOGICA	INFORMATICA
2229 INFORMATION SYSTEM MISC	INFORMATION SYSTEMS	DOCUMENTATION *UF:* Publication	DOKUMENTATION *UF:* Publikation	SISTEMA INFOR- MATSIONNAYA	DOCUMENTACION *UF:* Publicacion	DOCUMENTAZ- IONE *UF:* Pubblicazione
2230 INFRACAMBRIAN STRS	INFRACAMBRIAN	INFRACAMBRIEN	INFRAKAMBRIUM	INFRAKEMBRIJ	INFRACAMBRICO	INFRACAMBRIANO
2231 INFRARED PHCH	Infrared(adj.)	Infra=rouge(adj.)	Infrarot(adj.)	INFRAKRASNYJ	Infrarrojo(adj.)	Infrarosso(adj.)
2232 INFRARED PHOTOGRAPHY METH	INFRARED PHO- TOGRAPHY	PHOTO IR	IR= PHOTOGRAPHIE	FOTOGRAFIROVA- NIE INFRAKRAS- NOE	FOTO=IR	FOTOGRAFIA ALL'INFRAROSSO
2233 INFRARED RADIATION PHCH	INFRARED SPEC- TRA	RAYON IR	IR=STRAHLUNG	IZLUCHENIE INFRAKRASNOE	RAYOS=IR	RAGGI INFRAROSSI
2234 INFRARED SPECTROSCOPY METH	INFRARED SPEC- TROSCOPY	SPECTROMETRIE IR	IR= SPEKTROMETRIE	SPEKTROSKOPIYA INFRAKRASNAYA	ESPECTROMETRIA= IR	SPETTROMETRIA ALL'INFRAROSSO
2235 INFRASTRUCTURE STRU	Infrastructure BASEMENT	Infrastructure TECTONIQUE SOCLE	Infra=Struktur GRUNDGEBIRGS= TEKTONIK	INFRASTRUKTURA	Infraestructura TECTONICA=DE= ZOCALO	Infrastruttura TETTONICA DEL BASAMENTO
2236 INGRESSION SURF	Ingression TRANSGRESSION	Ingression TRANSGRESSION	Ingression TRANSGRESSION	INGRESSIYA	Ingresion TRANSGRESION	Ingressione TRASGRESSIONE
2237 INHERITED FEATURES SURF	Inherited features REGENERATION	Heritage REGENERATION	[INHERITED] REGENERATION	UNASLEDOVAN- NYJ	Heredado REGENERACION	Morfologia ereditata RIGENERAZIONE
2238 INHOMOGENEITY PHCH	Inhomogeneity HETEROGENEITY	Inhomogeneite HETEROGENEITE	Inhomogenitaet HETEROGENITAET	NEODNORODNOST'	Inhomogeneidad HETEROGENEIDAD	Inomogeneita ETEROGENEITA
2239 INJECTION ENGI	INJECTION	INJECTION	INJEKTION	IN'EKTSIYA	INYECCION	INIEZIONE
2240 INJECTION BRECCIA GEOL	Injection breccia TECTONIC BREC- CIA	Breche injection BRECHE TEC- TONIQUE	Injektions=Brekzie TEKTON.BREKZIE	Brekchiya in'ektsionnaya BREKCHIYA TEK- TONICHESKAYA	Brecha=de=inyeccion BRECHA= TECTONICA	Breccia d'iniezione BRECCIA TET- TONICA
2241 INLET SURF	INLETS	BRAS MER	MEERESARM	[INLET] ZALIV	BRAZO=DE=MAR	BRACCIO DI MARE
2242 INLIER STRU	Inliers WINDOWS	Boutonniere FENETRE	[INLIER] FENSTER	OKNO EHROZION- NOE	Ojal VENTANA= TECTONICA	Inlier FINESTRA TET- TONICA

	ENGLISH	FRANCAIS	DEUTSCH	RUSSKIJ	ESPANOL	ITALIANO
2243 INNER CORE SOLI	INNER CORE *UF:* F layer; G layer	NOYAU TER- RESTRE INTERNE *UF:* Couche F; Couche G	INNER.ERD=KERN *UF:* F-Schicht; G=Schicht	YADRO VNUTREN- NEE	NUCLEO=INTERNO *UF:* Capa=F; Capa=G	NUCLEO INTERNO *UF:* Strato F; Strato G
2244 INNER SHELF MARI	INNER SHELF	PLATEFORME CON- TINENTALE IN- TERNE	INNER.SCHELF	[INNER SHELF] SHEL'F KONTINEN- TAL'NYJ	PLATAFORMA= CONTINENTAL= INTERNA	PIATTAFORMA CONTINENTALE INTERNA
2245 INNER SLOPE MARI	INNER SLOPE	TALUS MARIN INTERNE	INNER. KONTINENTAL= ABHANG	[INNER SLOPE] SKLON KONTINEN- TAL'NYJ	TALUD=MARINO= INTERNO	TALUS MARINO INTERNO
2246 INOCERAMI PALS	INOCERAMI	INOCERAMI	INOCERAMI	INOCERAMI	INOCERAMI	INOCERAMI
2247 INORGANIC MISC	Inorganic(adj.)	Inorganique(adj.)	Anorganisch(adj.)	NEORGANICHES- KIJ	Inorganico(adj.)	Inorganico(adj.)
2248 INORGANIC ACID CHES	INORGANIC ACIDS	ACIDE MINERAL	ANORGAN.SAEURE	KISLOTA NEOGAN- ICHESKAYA	ACIDO=MINERAL	ACIDO INORGAN- ICO
2249 INORGANIC MATERIAL CHES	INORGANIC MATE- RIALS *UF:* Inorganic origin	MATIERE MINER- ALE *UF:* Origine inorganique	ANORGAN. SUBSTANZ *UF:* Anorgan.Entstehung	MATERIAL NEOR- GANICHESKIJ	MATERIAL=INOR- GANICO *UF:* Origen=inorganico	MATERIA INOR- GANICA *UF:* Origine inorganica
2250 INORGANIC ORIGIN GEOL	Inorganic origin INORGANIC MATE- RIALS	Origine inorganique MATIERE MINER- ALE	Anorgan.Entstehung ANORGAN. SUBSTANZ	NEORGANICHES- KIJ (GENEZIS)	Origen=inorganico MATERIAL=INOR- GANICO	Origine inorganica MATERIA INOR- GANICA
2251 INSECTA PALS	INSECTA *UF:* Collembola; Man- todea; Protura	INSECTA *UF:* Collembola; Man- todea; Protura	INSECTA *UF:* Collembola; Man- todea; Protura	INSECTA *UF:* Diptera; Ectotropha; Hemipteroida; Hyme- noptera; Protelytrop- tera; Siphonaptera	INSECTA *UF:* Collembola; Man- todea; Protura	INSECTA *UF:* Collembola; Man- todea; Protura
2252 INSECTIVORA PALS	INSECTIVORA	INSECTIVORA	INSECTIVORA	INSECTIVORA	INSECTIVORA	INSECTIVORA
2253 INSELBERG SURF	INSELBERGS	[INSELBERG]	INSELBERG	Gora ostrovnaya OSTANETS EHRO- ZIONNYJ	MONTE=ISLA	[INSELBERG]
2254 INSEQUENT VALLEY SURF	Insequent valleys VALLEYS	Vallee insequente VALLEE	Insequent.Tal TAL	INSEKVENTNYJ	Insecuente VALLE	Valle insequente VALLE
2255 INSOLATION SURF	INSOLATION	ENSOLEILLEMENT	SONNEN= EINSTRAHLUNG	INSOLYATSIYA	INSOLACION	INSOLAZIONE
2256 INSOLUBLE RESIDUE CHES	INSOLUBLE RESI- DUES	RESIDU INSOLU- BLE	UNLOESL. RUECKSTAND	OSTATOK NERAST- VORIMYJ	RESIDUO= INSOLUBLE	RESIDUO INSOLU- BILE
2257 INSTITUTION MISC	INSTITUTIONS	INSTITUTION	INSTITUTION	UCHREZHDENIE *UF:* Learned society	INSTITUCION	ISTITUZIONE
2258 INSTRUMENTATION INST	INSTRUMENTS *UF:* Drilling equipment; Electron microscope; Geophysical instru- ments; Gradiometers; Hodographs; Hydrom- eters; Hydrophone; Magnetic compass; Mi- croscopes; Penetrometers; Probe; Pycnometer; Radiometer; Reception equipment; Refractom- eters; Sampler; Seismic equipment; Sieves; Sparker; Spectrometers	INSTRUMENTA- TION *UF:* Boussole; Capteur; Dromochronique; Echantillonneur; Equipement reception; Equipement scissome- trie; Equipement sismi- que; Filtre; Gradiom- etre; Hydrometre; Hy- drophone; Instrument geophysique; Materiel sondage; Microscope; Microscope electroni- que; Penetrometre; Pyc- nometre; Radiometre; Refractometre; Spark- er; Spectrometre; Tamis	INSTRUMENTIE- RUNG *UF:* Bohr=Geraet; Elek- tronen=Mikroskop; Empfangs=Geraet; Fil- ter; Fluegelsonde; Geo- physikal. Instrument; Gradiome- ter; Hodograph; Hy- drometer; Hydrophon; Magnet=Kompass; Mess=Geraet; Mikros- kop; Penetrometer; Pyk- nometer; Radiometer; Refraktometer; Sam- pler; Seism.Instrument; Sieb; Sparker; Spek- trometer	OBORUDOVANIE	INSTRUMENTA- CION *UF:* Brujula; Equi- po=sismico; Espectrom- etro; Filtro=tamiz; Generador=chispa; Gradiometro; Hidro- fono; Hidrometro; Ho- dografo; Instrumen- tos=geofisicos; Maqui- naria=sondeo; Micro- scopio; Microscop- io=electronico; Pene- trometro; Picnometro; Prueba; Radiometro; Recepcion=señal; Re- fractometro; Saca=muestras; Tamiz; Tensiometro=aspa	STRUMENTAZIONE *UF:* Apparecchi per per- forazione; Apparecchia- tura sismica; Bussola magnetica; Campiona- tore; Filtro di pozzo; Gradimetro; Idrofono; Idrometro; Microscopio; Microscopio elettronico; Odografo; Penetrome- tro; Picnometro; Ra- diometro; Ricezione; Rifrattometro; Scissom- etro; Setaccio; Sonda; Sparker; Spettrometro; Strumento geofisico

	ENGLISH	FRANCAIS	DEUTSCH	RUSSKIJ	ESPANOL	ITALIANO
2259 INSULANT COMS	INSULATION MATERIALS	ISOLANT	ISOLIERSTOFF	MATERIAL IZO-LYATSIONNYJ	AISLANTE	ISOLANTE
2260 INTERCALATION SEDI	Intercalation STRATIFICATION	Intercalation STRATIFICATION	Wechsellagerung SCHICHTUNG	PERESLAIVANIE	Intercalacion ESTRATIFICACION	Intercalazione STRATIFICAZIONE
2261 INTERFACE PHCH	INTERFACES	INTERFACE UF: Interface huile eau	GRENZFLAECHE UF: Oel=Wasser=Grenze	POVERKHNOST' RAZDELA	INTERFASE UF: Interfase=agua=petroleo	INTERFACCIA UF: Interfaccia olio=acqua
2262 INTERFERENCE PHCH	Interference OPTICAL PROPER-TIES	Interference PROPRIETE OPTIQUE	Interferenz OPT.EIGENSCHAFT	INTERFERENTSIYA	Interferencia PROPIEDAD= OPTICA	Interferenza PROPRIETA OTTICA
2263 INTERFEROMETRY METH	INTERFEROMETRY	INTERFEROME-TRIE	INTERFEROME-TRIE	INTERFERO-METRIYA	INTERFERO-METRIA	INTERFERO-METRIA
2264 INTERFLUVE SURF	Interfluves DRAINAGE PAT-TERNS; RELIEF	Interfluve RESEAU HYDRO-GRAPHIQUE; RELIEF CONTI-NENT	Zwischenstromland HYDROGRAPH. NETZ; KONTINENTAL=RELIEF	MEZHDURECH'E	Interfluvio RED=HIDROGRAFICA; RELIEVE= CONTINENTAL	Interfluvio RETICOLATO IDRO-GRAFICO; FORMA DEL RILIEVO
2265 INTERFORMATIONAL SEDI	Interformational(adj.)	Interformationnel(adj.)	Interformationell(adj.)	MEZHFORMAT-SIONNYJ	Interformacional(adj.)	Interformazionale(adj.)
2266 INTERGLACIAL STRA	INTERGLACIAL STAGES	MILIEU INTERGLA-CIAIRE	INTERGLAZIAL	MEZHLEDNIKOVYJ	MEDIO=INTERGLACIAR	INTERGLACIALE
2267 INTERGROWTH TEST	INTERGROWTHS	CROISSANCE ENCHEVETREE	VERWACHSUNG	PRORASTANIE	INTERCRECI-MIENTO	CONCRESCIMENTO
2268 INTERLAYER MINE	MINERAL INTER-LAYERS	INTERCOUCHE	ZWISCHENSCH-ICHT	PROSLOJ	ESPACIO-INTERPLANAR	INTERSTRATO
2269 INTERMEDIATE COMPOSITION IGNE	Intermediate composition DIORITES	Roche intermediaire DIORITE	Intermediaer. Chemismus DIORIT	SREDNIJ (SOSTAV)	Composicion=intermedia DIORITA	Composizione inter-media DIORITE
2270 INTERMEDIATE-FOCUS EARTHQUAKE SOLI	INTERMEDIATE=FOCUS EARTH-QUAKES	SEISME INTER-MEDIAIRE	MITTELTIEF. ERDBEBEN	ZEMLETRYASENIE SREDNEFOKUSNOE	SISMO=INTERMEDIO	SISMA A FUOCO INTERMEDIO
2271 INTERMITTENT SPRING GEOH	INTERMITTENT SPRINGS	SOURCE INTER-MITTENTE	INTERMITTI-EREND.QUELLE	ISTOCHNIK VRE-MENNYJ	FUENTE=INTERMITENTE	SORGENTE INTER-MITTENTE
2272 INTERMITTENT STREAM GEOH	Intermittent stream STREAMS	Ecoulement intermit-tent RIVIERE; VARIA-TION SAISON-NIERE	Intermittierend.Strom FLUSS; JAHRESZEITL. SCHWANKUNG	REKA PER-ESYKHAYUSH-CHAYA UF: Vadi	Corriente=intermitente=agua RIO; VARIACION=ESTACIONARIA	Corso d'acqua intermit-tente FIUME; VARIAZ-IONE STAGIONALE
2273 INTERMONTANE UNIT SURF	INTERMONTANE BASINS	ENTREMONT	INTRAMONTAN. BECKEN	MEZHGORNYJ	CUENCA=ENTRE-MONTANAS	BACINO INTER-MONTANO
2274 INTERNAL CAST PALE	Internal cast FOSSILS	Moule interne FOSSILISATION	Inner.Abdruck FOSSILISATION	YADRO (PALE)	Molde=interno FOSILIZACION	Modello interno FOSSILIZZAZIONE
2275 INTERNAL STRUCTURE PALE	Internal structure (pale) BIOLOGY	Structure interne BIOLOGIE	Internstruktur BIOLOGIE	STROENIE VNU-TRENNEE	Estructura=interna BIOLOGIA	Struttura interna BIOLOGIA

	ENGLISH	FRANCAIS	DEUTSCH	RUSSKIJ	ESPANOL	ITALIANO
2276 INTERNAT. AGREEMENT MISC	INTERNATIONAL COOPERATION *UF:* Technical cooperation	COOPERATION INTERNATIONALE *UF:* Cooperation technique	INTERNATIONAL. ZUSAMMENAR-BEIT *UF:* Techn.Hilfe	SOGLASHENIE MEZHDUNAROD-NOE	COOPERACION=INTERNACIONAL *UF:* Cooperacion=tecnica	COOPERAZIONE INTERNAZIONALE *UF:* Cooperazione tecnica
2277 INTERNAT. GEOL. CONGRESS MISC	INTERNATIONAL GEOLOGICAL CONGRESS	CONGRES GEOLOGIQUE INTERNATIONAL	INTERNATIONAL. GEOLOG. KONGRESS	MEZHDUNAR. GEOL. KONGRESS	CONGRESO=GEOL=INTER	CONGRESSO GEOL. INTERNAZIONALE
2278 INTERNAT. GEOL. CORREL. PROGR. MISC	IGCP	PROGRAMME CORRELATION PICG	INTERNAT.GEOL. KORRELAT. PROGRAMM	MEZHDUNAR. GEOL. KORRELATS. PROGR.	PROGRAMA=CORRELACION=PICG	IGCP
2279 INTERNIDES STRU	Internides OROGENY	Internides CHAINE GEOSYN-CLINALE	Interniden GEOSYNKLINAL=KETTE	INTERNIDY	Internidas CORDILLERA=GEOSINCLINAL	Internidi CATENA GEOSIN-CLINALE
2280 INTERPLANETARY COMPARISON EXTR	INTERPLANETARY COMPARISON	COMPARAISON PLANETE	INTERPLANETAR. VERGLEICH	[INTERPLANE-TARY COMPARI-SON] PLANETA; KORRE-LYATSIYA	COMPARACION=PLANETA	COMPARAZIONE INTERPLANE-TARIA
2281 INTERPLUVIAL STRA	Interpluvial stages GLACIATION	Interpluvial GLACIATION	Interpluvial VEREISUNG	MEZHPLYUVI-AL'NYJ	Interpluvial GLACIACION	Interpluviale GLACIAZIONE
2282 INTERPOLATION MATH	Interpolation STATISTICAL DIS-TRIBUTION	Interpolation DISTRIBUTION STATISTIQUE	Interpolation STATIST. VERTEILUNG	INTERPOLYATSIYA	Interpolacion DISTRIBUCION=ESTADISTICA	Interpolazione DISTRIBUZIONE STATISTICA
2283 INTERPRETATION MISC	INTERPRETATION	INTERPRETATION *UF:* Extrapolation	INTERPRETATION *UF:* Extrapolation	INTERPRETATSIYA	INTERPRETACION *UF:* Extrapolacion	INTERPRETAZIONE GEOCHIMICA *UF:* Extrapolazione
2284 INTERSERTAL TEXTURE TEST	Intersertal texture TEXTURES	Texture intersertale TEXTURE POR-PHYRIQUE	Intersertal.Gefuege PORPHYR=GEFUEGE	INTERSERTAL'NYJ	Intersertal TEXTURA=PORFIDICA	Tessitura intersertale TESSITURA POR-FIRICA
2285 INTERSTELLAR SPACE EXTR	Interstellar space EXTRATERRES-TRIAL GEOLOGY	Espace interstellaire GEOLOGIE EXTRA TERRESTRE	Weltraum EXTRATERRESTR. GEOLOGIE	PROSTRANSTVO MEZHZVEZDNOE	Espacio=interestelar GEOLOGIA=EXTRATERRESTRE	Spazio interstellare GEOLOGIA EXTRATERRESTRE
2286 INTERTIDAL ENVIRONMENT PALE	INTERTIDAL ENVI-RONMENT *UF:* Tidal environment	MILIEU INTER-TIDAL	GEZEITEN=MILIEU *UF:* Tidal=Milieu	PRILIVNO=OTLIVNYJ *UF:* Intertidal sedimentation	MEDIO=MAREA	AMBIENTE INTER-TIDALE
2287 INTERTIDAL SEDIMENTATION SEDI	INTERTIDAL SEDI-MENTATION	SEDIMENTATION INTERTIDALE	GEZEITEN=SEDIMENTATION	[INTERTIDAL SEDI-MENTATION] PRILIVNO=OTLIVNYJ; SEDI-MENTATSIYA	SEDIMENTACION=MAREA	SEDIMENTAZIONE INTERTIDALE
2288 INTRACLAST SEDI	INTRACLASTS	INTRACLASTE	INTRAKLAST	INTRAKLAST	INTRACLASTO	INTRACLASTO
2289 INTRACONTINENTAL BELT STRU	INTRACONTINEN-TAL BELTS	CHAINE INTRA-CONTINENTALE	KONTINENTAL=KETTE	GEOSINKLINAL' VNUTRIKON-TINENTAL'NAYA	CORDILLERA-INTRACONTINENTAL	CATENA INTRA-CONTINENTALE

	ENGLISH	FRANCAIS	DEUTSCH	RUSSKIJ	ESPANOL	ITALIANO
2290 INTRACRATONIC BASIN STRU	INTRACRATONIC BASINS	BASSIN INTRACRA-TONIQUE	EPIKONTINENTAL. BECKEN	GEOSINKLINAL' VNUTRIKRA-TONOVAYA	CUENCA= INTRACRATONICA	BACINO INTRACRATONICO
2291 INTRAFORMATIONAL DEPOSITION SEDI	Intraformational deposition SEDIMENTATION	Sedimentation intrafor-mationnelle SEDIMENTATION	Synsedimentaer. Bildung SEDIMENTATION	VNUTRIFORMAT-SIONNYJ	Sedimentacion= intraformacional SEDIMENTACION	Intraformazionale SEDIMENTAZIONE
2292 INTRATELLURIC GENESIS IGNE	Intratelluric processes CRYSTALLIZATION	Genese intratellurique CRISTALLISATION	Intratellur.Genese KRISTALLISATION	INTRATELLURI-CHESKIJ	Intratelurico CRISTALIZACION	Genesi intratellurica CRISTALLIZZAZ-IONE
2293 INTRUSION IGNE	INTRUSIONS *UF:* Apophysis; Igneous complex; Phacolith; Sphenolith	INTRUSION *UF:* Apophyse; Complexe igne; Facies bordure; Phacolite; Sphenolite	INTRUSIONS-KOERPER *UF:* Apophyse; Magmatit=Komplex; Phacolith; Rand= Fazies; Sphenulithus	INTRUZIYA *UF:* Plutonicheskij; Vnedrenie	INTRUSION *UF:* Apofisis; Complejo-igneo; Esfenolito; Facies=borde; Facolito	INTRUSIONE *UF:* Apofisi; Complesso eruttivo; Facies di contatto; Facolite; Sfenolite
2294 INTRUSIVE IGNE	Intrusive(adj.)	Intrusif(adj.)	Intrusiv(adj.)	INTRUZIYA	Intrusivo(adj.)	Intrusivo(adj.)
2295 INVENTORY MISC	INVENTORY	INVENTAIRE	INVENTAR	IZOBRETENIE	INVENTARIO	INVENTARIO
2296 INVERSE PROBLEM APPL	INVERSE PROBLEM	PROBLEME INVERSE	INVERS.PROBLEM	ZADACHA OBRAT-NAYA	PROBLEMA-INVERSO	PROBLEMA INVERSO
2297 INVERTEBRATA PALS	INVERTEBRATA	INVERTEBRATA	INVERTEBRATA	INVERTEBRATA	INVERTEBRATA	INVERTEBRATA
2298 IODINE CHEE	IODINE	IODE	J	IOD	YODO	IODIO
2299 ION PHCH	IONS *UF:* Ionization	ION *UF:* Ionisation	ION *UF:* Ionisierung	ION *UF:* Anion; Kation	ION *UF:* Ionizacion	IONE *UF:* Ionizzazione
2300 ION EXCHANGE PHCH	ION EXCHANGE *UF:* Exchange capacity	ECHANGE ION *UF:* Capacite echange	IONEN-AUSTAUSCH *UF:* Austausch= Kapazitaet	OBMEN IONNYJ *UF:* Sposobnost' obmen-naya	CAMBIO=IONICO *UF:* Capacidad=de= cambio	SCAMBIO IONICO *UF:* Capacita di scambio
2301 ION PROBE METH	ION PROBE	MICROSONDE IONIQUE	IONEN=SONDE	ZOND IONNYJ	MICROSONDA= IONICA	MICROSONDA A IONI
2302 IONIUM-THORIUM DATING ISOT	IO/TH	IO=TH	IO=TH= DATIERUNG	METOD IONIEVYJ *UF:* Metod ionievo-uranovyj	IO=TH	IO/TH
2303 IONIUM-URANIUM DATING ISOT	IO/U	IO=U	IO=U=DATIERUNG	Metod ionievo= uranovyj METOD IONIEVYJ	IO=U	IO/U
2304 IONIZATION PHCH	Ionization IONS	Ionisation ION	Ionisierung ION	IONIZATSIYA	Ionizacion ION	Ionizzazione IONE
2305 IONOSPHERE SOLI	IONOSPHERE	IONOSPHERE	IONOSPHAERE	IONOSFERA	IONOSFERA	IONOSFERA

	ENGLISH	FRANCAIS	DEUTSCH	RUSSKIJ	ESPANOL	ITALIANO
2306 IRIDESCENCE PHCH	Iridescence OPTICAL PROPER- TIES	Irisation PROPRIETE OPTIQUE	Irideszenz OPT.EIGENSCHAFT	IRIZATSIYA	Iridiscencia PROPIEDAD= OPTICA	Iridescenza PROPRIETA OTTICA
2307 IRIDIUM CHEE	IRIDIUM	IRIDIUM	IR	IRIDIJ	IRIDIO	IRIDIO
2308 IRON CHEE	IRON *UF:* Bean ore; Bog iron ore	FER *UF:* Roche ferrugineuse	FE *UF:* Eisenstein	ZHELEZO *UF:* Ferromanganese composition	HIERRO *UF:* Roca=ferruginosa	FERRO *UF:* Roccia ferrifera
2309 IRON AND STEEL INDUSTRY ECON	STEEL INDUSTRY	SIDERURGIE	EISEN=INDUSTRIE	PROMYSHLEN- NOST' STALELITEJ- NAYA	SIDERURGIA	SIDERURGIA
2310 IRON BACTERIA GEOL	IRON BACTERIA	BACTERIE FER	EISEN=BAKTERIEN	ZHELEZOBAK- TERIYA	BACTERIA- HIERRO	FERROBATTERI
2311 IRON FORMATION SEDI	IRON FORMA- TIONS *UF:* Ironstone	FORMATION FER- RIFERE *UF:* Roche ferrugineuse	EISEN= FORMATION *UF:* Eisenstein	FORMATSIYA ZHELEZORUD- NAYA	FORMACION= FERRIFERA *UF:* Roca=ferruginosa	FORMAZIONE FER- RIFERA *UF:* Roccia ferrifera
2312 IRONSTONE SEDS	Ironstone IRON FORMA- TIONS	Roche ferrugineuse FORMATION FER- RIFERE	Eisenstein EISEN= FORMATION	PORODA ZHELEZ- ISTAYA	Roca=ferruginosa FORMACION= FERRIFERA	Roccia ferrifera FORMAZIONE FER- RIFERA
2313 IRREVERSIBILITY MISC	Irreversibility TRANSFORMA- TIONS	Irreversibilite TRANSFORMATION	Irreversibilitaet TRANSFORMATION	NEOBRATIMOST'	Irreversibilidad TRANSFORMA- CION	Irreversibilita TRASFORMAZIONE
2314 IRRIGATION GEOH	IRRIGATION	IRRIGATION	BEWAESSERUNG	IRRIGATSIYA	IRRIGACION	IRRIGAZIONE
2315 ISLAND SURF	ISLANDS *UF:* Island volcano	ILE *UF:* Volcan insulaire	INSEL *UF:* Insel=Vulkan	OSTROV	ISLA *UF:* Volcan=insular	ISOLA *UF:* Vulcano insulare
2316 ISLAND ARC STRU	ISLAND ARCS	ARC INSULAIRE	INSELBOGEN	DUGA OSTROV- NAYA	ARCO=INSULAR	ARCO INSULARE
2317 ISLAND VOLCANO IGNE	Island volcano VOLCANOES; ISLANDS	Volcan insulaire VOLCAN; ILE	Insel=Vulkan VULKAN; INSEL	OSTROV VULKANI- CHESKIJ	Volcan=insular VOLCAN; ISLA	Vulcano insulare VULCANO; ISOLA
2318 ISOANOMALY PHCH	Isoanomaly ISOPLETH MAPS	Isanomale ISOPLETHE	Isanomale ISOLINIE	IZOANOMALA	Isoanomalia ISOPLETA	Isoanomala CARTA DELLE ISOPLETE
2319 ISOBAR PHCH	Isobars ISOPLETH MAPS; PRESSURE	Isobare ISOPLETHE; PRES- SION	Isobare ISOLINIE; DRUCK	IZOBARA	Isobara ISOPLETA; PRE- SION	Isobara CARTA DELLE ISOPLETE; PRES- SIONE
2320 ISOBASE STRU	Isobase ISOPLETH MAPS	Isobase ISOPLETHE; EXHAUSSEMENT	Isobase ISOLINIE; HEBUNG	IZOBAZA	Isobase ISOPLETA; LEVANTAMIENTO	Isobase CARTA DELLE ISOPLETE; SOL- LEVAMENTO
2321 ISOBATH GEOL	ISOBATH	ISOBATHE	ISOBATHE	IZOBATA	ISOBATA	ISOBATA

	ENGLISH	FRANCAIS	DEUTSCH	RUSSKIJ	ESPANOL	ITALIANO
2322 ISOCHEMICAL METAMORPHISM IGNE	Isochemical metamorphism METAMORPHISM	Metamorphisme isochimique METAMORPHISME	Isochem.Metamorphose METAMORPHOSE	METAMORFIZM IZOKHIMICHESKIJ	Metamorfismo=isoquimico METAMORFISMO	Metamorfismo isochimico TIPO DI METAMORFISMO
2323 ISOCHRON ISOT	ISOCHRONS	ISOCHRONE	ISOCHRONE	IZOKHRONA	ISOCRONA	ISOCRONA
2324 ISOCLINE STRU	ISOCLINES _UF:_ Isoclinic line	PLI ISOCLINAL	ISOKLINALE	IZOKLINAL'	ISOCLINAL	ISOCLINALE
2325 ISOCLINIC LINE SOLI	Isoclinic line ISOCLINES	Isocline ISOPLETHE; CHAMP MAGNETIQUE	Isokline ISOLINIE; MAGNETFELD	IZOKLINA	Linea=isoclina ISOPLETA; CAMPO=MAGNETICO	Linea isoclina CARTA DELLE ISOPLETE; CAMPO MAGNETICO
2326 ISOCON PHCH	Isocons ISOPLETH MAPS; CHEMICAL ANALYSIS	Ligne isoconcentration ISOPLETHE; ANALYSE CHIMIQUE	Isocone ISOLINIE; CHEM. ANALYSE	IZOKONT-SENTRATY	Isosalinidad ISOPLETA; ANALISIS=QUIMICO	Linea d'isosalinita CARTA DELLE ISOPLETE; COMPOSIZIONE CHIMICA
2327 ISOGON SOLI	Isogon ISOPLETH MAPS; MAGNETIC FIELD	Isogone ISOPLETHE; CHAMP MAGNETIQUE	Isogone ISOLINIE; MAGNETFELD	IZOGONA	Isogona ISOPLETA; CAMPO=MAGNETICO	Isogona CARTA DELLE ISOPLETE; CAMPO MAGNETICO
2328 ISOGRAD IGNE	ISOGRADS	ISOGRADE	ISOGRADE	IZOGRADA	ISOGRADA	ISOGRADA
2329 ISOHYPSE SURF	Isohypse ISOPLETH MAPS; TOPOGRAPHIC MAPS	Isohypse ISOPLETHE; CARTE TOPOGRAPHIQUE	Isohypse ISOLINIE; TOPOGRAPH. KARTE	Izogipsy TOPOGRAFIYA	Curva=de=nivel ISOPLETA; MAPA=TOPOGRAFICO	Isoipsa CARTA DELLE ISOPLETE; CARTA TOPOGRAFICA
2330 ISOLATION PALE	ISOLATION	ISOLEMENT	ISOLIERUNG	IZOLIROVANNOST'	AISLAMIENTO	ISOLAMENTO
2331 ISOLITH SEDI	Isolith ISOPLETH MAPS; LITHOFACIES	Isolithe ISOPLETHE; LITHOFACIES	Isolith ISOLINIE; LITHO=FAZIES	IZOLITA	Isolito ISOPLETA; LITOFACIES	Isolito CARTA DELLE ISOPLETE; LITOFACIES
2332 ISOMETRIC SYSTEM MINE	Isometric system CRYSTAL SYSTEMS	Systeme isometrique SYSTEME CRISTALLIN	Isometr.System KRISTALL=SYSTEM	Sistema izometricheskaya SINGONIYA KUBICHESKAYA	Sistema=regular SISTEMA=CRISTALINO	Sistema isometrico SISTEMA CRISTALLINO
2333 ISOMORPH ORGANISM PALE	Isomorph organism ISOMORPHISM	Organisme homeomorphe CONVERGENCE	Isomorph.Organismus KONVERGENZ	IZOMORFIYA	Isomorfo ISOMORFISMO	Organismo isomorfo CONVERGENZA
2334 ISOMORPHISM GEOL	ISOMORPHISM _UF:_ Idiomorphism; Isomorph organism	ISOMORPHISME _UF:_ Diadochie; Idiomorphisme	ISOMORPHIE _UF:_ Diadochie; Idiomorphie	IZOMORFIZM _UF:_ Diadokhiya	ISOMORFISMO _UF:_ Diadoquia; Idiomorfismo; Isomorfo	ISOMORFISMO _UF:_ Diadochia; Idiomorfismo
2335 ISOPACH SEDI	ISOPACH MAPS	ISOPAQUE	ISOPACHE	IZOPAKHITA	ISOPACA	CARTA DELLE ISOPACHE
2336 ISOPIESTIC LINE PHCH	Isopiestic line ISOPLETH MAPS; PIEZOMETRY	Ligne isopieze ISOPLETHE; PIEZOMETRIE	Aequipotential=Linie ISOLINIE; PIEZOMETRIE	GIDROIZOP'EZA	Isopieza ISOPLETA; PIEZOMETRIA	Linea isopiestica CARTA DELLE ISOPLETE; PIEZOMETRIA

	ENGLISH	FRANCAIS	DEUTSCH	RUSSKIJ	ESPANOL	ITALIANO
2337 ISOPLETH MISC	ISOPLETH MAPS *UF:* Isoanomaly; Isobars; Isobase; Isochore; Isochores; Isocons; Isogon; Isohypse; Isolith; Isopiestic line; Isorad; Isorat; Isotherms	ISOPLETHE *UF:* Isanomale; Isobare; Isobase; Isochore; Isocline; Isogone; Isohypse; Isolithe; Isorad; Isorat; Isoseiste; Isotherme; Ligne isoconcentration; Ligne isopieze; Phase isochore	ISOLINIE *UF:* Aequipotential=Linie; Isanomale; Isobare; Isobase; Isochore; Isocone; Isogone; Isohypse; Isokline; Isolith; Isorad; Isorat; Isoseiste; Isotherme; Stratigraph.Isochore	IZOLINIYA *UF:* Structure contour map	ISOPLETA *UF:* Curva=de=nivel; Isoanomalia; Isobara; Isobase; Isocora; Isocora=estratigrafica; Isogona; Isolito; Isopieza; Isorad; Isorat; Isosalinidad; Isoterma; Linea=isoclina; Linea=isosismica	CARTA DELLE ISOPLETE *UF:* Isoanomala; Isobara; Isobase; Isocora; Isocora di fase; Isogona; Isoipsa; Isolito; Isorad; Isorat; Isosista; Isoterma; Linea d'isosalinita; Linea isoclina; Linea isopiestica
2338 ISOPLEURA PALS	Isopleura GASTROPODA	Isopleura GASTROPODA	Isopleura GASTROPODA	ISOPLEURA	Isopleura GASTROPODA	Isopleura GASTROPODA
2339 ISOPRENOID CHES	ISOPRENOIDS	ISOPRENOIDE	ISOPRENOID	[ISOPRENOID] UGLEVODOROD	ISOPRENOIDE	ISOPRENOIDE
2340 ISORAD ISOT	Isorad ISOPLETH MAPS; RADIOACTIVITY	Isorad ISOPLETHE; RADIOACTIVITE	Isorad ISOLINIE; RADIOAKTIVITAET	IZORADA	Isorad ISOPLETA; RADIOACTIVIDAD	Isorad CARTA DELLE ISOPLETE; RADIOATTIVITA
2341 ISORAT ISOT	Isorat ISOPLETH MAPS; ISOTOPES	Isorat ISOPLETHE; ISOTOPE	Isorat ISOLINIE; ISOTOP	IZORATA	Isorat ISOPLETA; ISOTOPO	Isorat CARTA DELLE ISOPLETE; ISOTOPO
2342 ISOSEISM SOLI	Isoseismic line ISOSEISMIC MAPS	Isoseiste ISOPLETHE; SEISME	Isoseiste ISOLINIE; ERDBEBEN	IZOSEJSMA *UF:* Karta izosejsm	Linea=isosismica ISOPLETA; TERREMOTO	Isosista CARTA DELLE ISOPLETE; TERREMOTO
2343 ISOSEISMIC MAP SOLI	ISOSEISMIC MAPS *UF:* Isoseismic line	CARTE ISOSISMIQUE	ISOSEISM.KARTE	Karta izosejsm KARTA SEJSMOLOGICHESKAYA; IZOSEJSMA	MAPA=ISOSISMICO	CARTA ISOSISMICA
2344 ISOSTASY SOLI	ISOSTASY	ISOSTASIE	ISOSTASIE	IZOSTAZIYA	ISOSTASIA	ISOSTASIA
2345 ISOSTRUCTURAL MINERAL MINE	Isostructural minerals CRYSTAL CHEMISTRY	Isostructure CRISTALLOCHIMIE	Struktur=Gleich. Mineral KRISTALLCHEMIE	IZOSTRUKTURNYJ	Isoestructural CRISTALOQUIMICA	Minerale isostrutturale CRISTALLOCHIMICA
2346 ISOTHERM PHCH	Isotherms ISOPLETH MAPS; TEMPERATURE	Isotherme ISOPLETHE; TEMPERATURE	Isotherme ISOLINIE; TEMPERATUR	IZOTERMA *UF:* Namagnichennost' izotermicheskaya	Isoterma ISOPLETA; TEMPERATURA	Isoterma CARTA DELLE ISOPLETE; TEMPERATURA
2347 ISOTHERMAL REMANENT MAGNETIZATION PHCH	ISOTHERMAL REMANENT MAGNETIZATIO	AIMANTATION ISOTHERMOREMANENTE	ISOTHERMOREMANENT. MAGNETISIERUNG	Namagnichennost' izotermicheskaya NAMAGNICHENNOST' OSTATOCHNAYA; IZOTERMA	IMANTACION=ISOTERMORREMANENTE	MAGNETIZZAZIONE ISOTERMICA RESIDUA
2348 ISOTOPE ISOT	ISOTOPES *UF:* Daughter product; End product; Excess argon; Heavy isotopes; Isorat; Isotope dilution; Isotope effect; Isotope geochemistry; Isotope ratio; Light isotope; Parent product	ISOTOPE *UF:* Dilution isotopique; Effet isotopique; Isorat; Isotope leger; Isotope lourd; Rapport isotopique	ISOTOP *UF:* Isorat; Isotopen=Effekt; Isotopen=Verduennung; Isotopen=Verhaeltnis; Leicht=Isotop; Schwer=Isotop	IZOTOP *UF:* Oxygen=isotope ratio; Sulphur=isotope ratio	ISOTOPO *UF:* Efecto=isotopico; Isorat; Isotopo=ligero; Isotopo=pesado; Medida=isotopica; Relacion=isotopica	ISOTOPO *UF:* Diluizione isotopica; Effetto isotopico; Isorat; Isotopo leggero; Isotopo pesante; Rapporto isotopico

	ENGLISH	FRANCAIS	DEUTSCH	RUSSKIJ	ESPANOL	ITALIANO
2349 ISOTOPE DILUTION ISOT	Isotope dilution ISOTOPES; DILU-TION	Dilution isotopique DILUTION; ISO-TOPE	Isotopen=Verduennung VERDUENNUNG; ISOTOP	METOD IZOTOP-NOGO RAZBAV-LENIYA	Medida=isotopica DILUCION; ISO-TOPO	Diluizione isotopica DILUIZIONE; ISO-TOPO
2350 ISOTOPE EFFECT ISOT	Isotope effect ISOTOPES	Effet isotopique ISOTOPE	Isotopen=Effekt ISOTOP	EHFFEKT IZOTOP-NYJ	Efecto=isotopico ISOTOPO	Effetto isotopico ISOTOPO
2351 ISOTOPE GEOLOGY ISOT	Isotope geochemistry ISOTOPES	GEOCHIMIE ISO-TOPIQUE	ISOTOPEN=GEOCHEMIE	GEOLOGIYA IZO-TOPNAYA	GEOQUIMICA=ISOTOPICA	GEOCHIMICA ISO-TOPICA
2352 ISOTOPE RATIO ISOT	Isotope ratio ISOTOPES; RATIOS	Rapport isotopique ISOTOPE; RAPPORT CHIMIQUE	Isotopen=Verhaeltnis ISOTOP; CHEM. VERHAELTNIS	OTNOSHENIE IZO-TOPNOE	Relacion=isotopica ISOTOPO; RELACION=QUIMICA	Rapporto isotopico ISOTOPO; RAP-PORTO CHIMICO
2353 ISOTOPIC FRACTIONATION ISOT	FRACTIONATION	FRACTIONNEMENT ISOTOPIQUE	ISOTOPEN=FRAKTIONIERUNG	FRAKTSIONIROVA-NIE IZOTOPOV	FRACCIONAMIENTO=ISOTOPICO	FRAZIONAMENTO ISOTOPICO
2354 ISOTROPY PHCH	ISOTROPY	ISOTROPIE	ISOTROPIE	IZOTROPIYA	ISOTROPIA	ISOTROPIA
2355 ITABIRITE IGMS	ITABIRITE *UF:* Taconite	ITABIRITE *UF:* Taconite	ITABIRIT *UF:* Taconit	Itabirit KVARTSIT	ITABIRITA *UF:* Taconita	ITABIRITE *UF:* Taconite
2356 ITACOLUMITE IGMS	ITACOLUMITE	Itacolumite QUARTZITE	Itacolumit QUARZIT	ITAKOLUMIT	Itacolumita CUARCITA	Itacolumite QUARZITE
2357 JADEITITE IGMS	Jadeitite METAMORPHIC ROCKS	Jadeitite PYROXENITE	Jadeitit PYROXENIT	ZHADEITIT	Jadeitita PIROXENITA	Giadeitite PIROSSENITE
2358 JARAMILLO EVENT STRA	JARAMILLO EVENT	EPISODE JARAMILLO	JARAMILLO=EPISODE	YARAMILLO EHPOKHA	EPISODIO=JARAMILLO	EVENTO DI JARAMILLO
2359 JASPER SEDS	JASPER	Jaspe QUARTZ	Jaspis QUARZ	YASHMA	Jaspe CUARZO	Diaspro QUARZO
2360 JAW PALE	JAWS	MACHOIRE *UF:* Denture	KIEFER *UF:* Dentition	CHELYUST'	MANDIBULA *UF:* Denticion	MASCELLA *UF:* Dentizione
2361 JOINT STRU	JOINTS *UF:* Jointing; Shear joints	DIACLASE *UF:* Diaclase cisaille-ment; Reseau diaclase; Structure prismatique	KLUFT *UF:* Klueftung; Prismat. Gefuege; Scher=Kluft	Diaklaza TRESHCHINA	DIACLASA *UF:* Diaclasamiento; Fractura=de=cizallamiento; Prisma-tica	DIACLASI *UF:* Diaclasi di taglio; Fratturazione a giunti; Struttura prismatica
2362 JOINTING STRU	Jointing JOINTS	Reseau diaclase DIACLASE	Klueftung KLUFT	TRESHCHINOVA-TOST' OTDEL'NOST'	Diaclasamiento DIACLASA	Fratturazione a giunti DIACLASI
2363 JURASSIC STRS	JURASSIC	JURASSIQUE	JURA	YURA	JURASICO	GIURASSICO
2364 JUVENILE IGNE	Juvenile(adj.)	Juvenile(adj.)	Juvenil(adj.)	YUVENIL'NYJ	Juvenil(adj.)	Giovanile(adj.)
2365 JUVENILE WATER GEOH	JUVENILE WATERS	EAU JUVENILE	JUVENIL.WASSER	VODA YUVENIL'NAYA	AGUA=JUVENIL	ACQUA GIOVANILE

	ENGLISH	FRANCAIS	DEUTSCH	RUSSKIJ	ESPANOL	ITALIANO
2366 KAME SURF	KAMES	KAME	KAMES	KAM	MONTICULO= MATERIALES= GLACIARES	KAME
2367 KANGA ECON	Kanga MANGANESE; EXO- GENE PROCESSES	Kanga MANGANESE; PROCESSUS EXO- GENE	Kanga MN; EXOGEN. VORGANG	KANGA	Kanga MANGANESO; PROCESO= EXOGENO	Kanga MANGANESE; EVOLUZIONE SUPERGENICA
2368 KAOLIN COMS	KAOLIN *UF:* White clay	KAOLIN *UF:* Argile blanche; Terre a porcelaine	KAOLIN *UF:* Porzellan=Erde; Porzellan=Ton	KAOLIN	CAOLIN *UF:* Arcilla=blanca; Caolin=china	CAOLINO *UF:* Argilla bianca; Argilla per porcellana
2369 KAOLINIZATION SEDI	KAOLINIZATION	KAOLINISATION	KAOLINISIERUNG	KAOLINIZATSIYA	CAOLINIZACION	CAOLINIZZAZIONE
2370 KARELIAN OROGENY STRU	KARELIAN OROG- ENY	OROGENIE CARE- LIENNE	KAREL. OROGENESE	KAREL'SKIJ	OROGENIA= CARELIANA	OROGENESI CAR- ELIANA
2371 KARREN SURF	KARREN	LAPIEZ	KARREN	Karry KARST	LAPIAZ	CAMPO SOLCATO
2372 KARROO STRA	KARROO SYSTEM	KARROO	KARROO	KARRU	KARROO	KARROO
2373 KARST SURF	KARST *UF:* Karst hydrology; Karst spring	KARST *UF:* Hydrologie kars- tique; Source karstique	KARST *UF:* Karst=Hydrologie; Karst=Quelle	KARST *UF:* Karry; Landshaft karstovyj; Solution cavity	KARST *UF:* Fuente=karstica; Hidrogeologia=karst	CARSISMO *UF:* Idrologia carsica; Sorgente carsica
2374 KARST FILLING SURF	KARST FILLING	REMPLISSAGE KARSTIQUE	KARST=FUELLUNG	[REPLENISHMENT (SPELEO)] PESHCHERA	RELLENO= KARSTICO	RIEMPIMENTO CARSICO
2375 KARST HYDROLOGY GEOH	Karst hydrology HYDROLOGY; KARST	Hydrologie karstique HYDRODY- NAMIQUE; KARST	Karst=Hydrologie HYDRODYNAMIK; KARST	GIDROLOGIYA KARSTOVAYA	Hidrogeologia=karst HIDRODINAMICA; KARST	Idrologia carsica IDRODINAMICA; CARSISMO
2376 KARST SPRING GEOH	Karst spring SPRINGS; KARST	Source karstique SOURCE; KARST	Karst=Quelle QUELLE; KARST	ISTOCHNIK KARS- TOVYJ	Fuente=karstica MANANTIAL; KARST	Sorgente carsica SORGENTE; CAR- SISMO
2377 KATAMORPHISM IGNE	Katamorphism CATAZONAL METAMORPHISM	Catamorphisme METAMORPHISME CATAZONAL	Kata=Metamorphose KATAZONE	KATAMORFIZM	Catamorfismo METAMORFISMO= CATAZONAL	Catametamorfismo CATAZONA
2378 KATANGA OROGENY STRU	KATANGAN OROG- ENY	OROGENIE KATAN- GIENNE	KATANGA= OROGENESE	KATANGSKIJ	OROGENIA= KATANGIANA	OROGENESI KATANGHIANA
2379 KATATHERMAL DEPOSIT ECON	Katathermal deposits HYPOTHERMAL PROCESSES	Gite catathermal GITE HYDROTHER- MAL	Katathermal. Vorkommen HYDROTHERMAL. VORKOMMEN	Katatermal'noe mestorozhdenie MESTOROZHDENIE MAGMATI- CHESKOE; KATA- GENEZ	Deposito=catatermal YACIMIENTO= HIDROTERMAL	Giacimento catatermale GIACIMENTO IDROTERMALE
2380 KATAZONE IGNE	CATAZONAL METAMORPHISM *UF:* Katamorphism	METAMORPHISME CATAZONAL *UF:* Catamorphisme	KATAZONE *UF:* Kata= Metamorphose	KATAZONA	METAMORFISMO= CATAZONAL *UF:* Catamorfismo	CATAZONA *UF:* Catametamorfismo

	ENGLISH	FRANCAIS	DEUTSCH	RUSSKIJ	ESPANOL	ITALIANO
2381 KAZANIAN STRS	KAZANIAN *UF:* Ufimian	KAZANIEN *UF:* Ufien	KASAN *UF:* Ufimium	KAZANSKIJ	KAZANIENSE *UF:* Ufimiense	KAZANIANO *UF:* Ufimiano
2382 KAZIMOVIAN STRS	KASIMOVIAN	KASIMOVIEN	KASIMOV	KASIMOVSKIJ	KASIMOVIENSE	KASIMOVIANO
2383 KELYPHITIC RIM IGNE	Kelyphytic rim REACTION RIMS	Frange kelyphitique FRANGE REAC- TIONNELLE	Kelyphit=Rinde REAKTIONSSAUM	KAJMA KELIFI- TOVAYA	Corona=quelifitica FRANJA=DE= REACCION	Anello chelifitico ANELLO DI REAZ- IONE
2384 KENORAN OROGENY STRU	KENORAN OROG- ENY	OROGENIE KENO- RIENNE	KENORA= OROGENESE	KENORANSKIJ	OROGENIA= KENORAN	OROGENESI KEN- ORIANA
2385 KERATOPHYRE IGMS	KERATOPHYRE	KERATOPHYRE	KERATOPHYR	KERATOFIR	QUERATOFIDO	CHERATOFIRO
2386 KEROGEN CHES	KEROGEN	KEROGENE	KEROGEN	KEROGEN	KEROGENO	KEROGENE
2387 KETTLE SURF	KETTLES	Marmite MARMITE GEANT	Kessel STRUDELTOPF	KOTLOVINA TER- MOKARSTOVAYA	Marmita=glaciar MARMITA=DE= GIGANTE	Marmitta MARMITTA DEI GIGANTI
2388 KEUPER STRS	KEUPER	KEUPER	KEUPER	Kejper TRIAS VERKHNIJ	KEUPER	KEUPER
2389 KEY BED STRA	MARKER BEDS	NIVEAU REPERE	LEITHORIZONT	GORIZONT MARKIRUYUSH- CHIJ	NIVEL=GUIA	REPERE SISMICO
2390 KIBARA OROGENY STRU	Kibara orogeny OROGENY; PRE- CAMBRIAN	Orogenie kibarienne OROGENIE ANTE- CAMBRIENNE	Kibara=Orogenese PRAEKAMBR. OROGENESE	KIBARSKIJ	Orogenia=kibara OROGENIA= PRECAMBRICA	Orogenesi kibariana OROGENESI PRE- CAMBRIANA
2391 KIMBERLITE IGNS	KIMBERLITE	KIMBERLITE	KIMBERLIT	KIMBERLIT	KIMBERLITA	KIMBERLITE
2392 KIMMERIDGIAN STRS	KIMMERIDGIAN	KIMMERIDGIEN	KIMMERIDGE	KIMERIDZH	KIMMERIDGIENSE	KIMMERIDGIANO
2393 KINEMATICS PHCH	KINEMATICS	CINEMATIQUE	KINEMATIK	KINEMATIKA	CINEMATICA	CINEMATICA
2394 KINK BAND STRU	KINK=BAND STRUCTURES	KINK BAND	[KINK=BAND]	KINKBAND	KINK=BAND	KINK BAND
2395 KINZIGITE IGMS	Kinzigite LEPTYNITE	Kinzigite LEPTYNITE	Kinzigit LEPTIT	KINTSIGIT	Kinzigita LEPTINITA	Kinzigite LEPTINITE
2396 KLIPPE STRU	KLIPPEN	LAMBEAU	KLIPPE	KLIPP	KLIPPE	KLIPPE
2397 KREEP EXTR	KREEP	Kreep BASALTE; LUNE	Kreep BASALT; MOND	KREEP	Kreep BASALTO; LUNA	Kreep BASALTO; LUNA
2398 KRYPTON CHEE	KRYPTON	KRYPTON	KR	KRIPTON	CRIPTON	KRIPTON
2399 KUJALNIKIAN STRS	Kujalnikian UPPER PLIOCENE	Kuyalnik PLIOCENE SUP	Kujalnikium O.PLIOZAEN	KUYAL'NITSKIJ *UF:* Pliotsen srednij	Kujalnikiense PLIOCENO=SUP	Kujalnikiano PLIOCENE SUP.

	ENGLISH	FRANCAIS	DEUTSCH	RUSSKIJ	ESPANOL	ITALIANO
2400 KUNGURIAN STRS	KUNGURIAN *UF*: Ufimian	KOUNGOURIEN *UF*: Ufien	KUNGUR *UF*: Ufimium	KUNGUR	KUNGURIENSE *UF*: Ufimiense	KUNGURIANO *UF*: Ufimiano
2401 KURTOSIS MATH	KURTOSIS	KURTOSIS	KURTOSIS	Ehkstsess STATISTIKA	CURTOSIS	KURTOSI
2402 KUTORGINIDA PALS	Kutorginida BRACHIOPODA	Kutorginida ARTICULATA	Kutorginida ARTICULATA	KUTORGINIDA	Kutorginida ARTICULATA	Kutorginida ARTICULATA
2403 LABORATORY TEST ENGI	LABORATORY STUDIES	ESSAI LABORA- TOIRE	LABOR= UNTERSUCHUNG	ISPYTANIE LABORATORNOE *UF*: Uniaxial compres- sion test	ENSAYO= LABORATORIO	PROVA DI LABORA- TORIO
2404 LABYRINTHODONTIA PALS	LABYRINTHODON- TIA *UF*: Batrachosauria	LABYRINTHODON- TIA	LABYRINTHODON- TIA	LABYRINTHODON- TIA	LABYRINTHODON- TIA	LABYRINTHODON- TIA
2405 LACCOLITH IGNE	LACCOLITHS *UF*: Akmolith; Har- poliths; Sheets	LACCOLITE *UF*: Acmolite; Harpolite; Intrusion couche	LAKKOLITH *UF*: Akmolith; Decken= Intrusion; Harpolith	LAKKOLIT	LACOLITO *UF*: Acmolito; Harpolito; Lamina=magmatica	LACCOLITE *UF*: Acmolite; Arpolite; Intrusione stratificata
2406 LACERTILIA PALS	Lacertilia LEPIDOSAURIA	Lacertilia SQUAMATA	Lacertilia SQUAMATA	LACERTILIA	Lacertilia SQUAMATA	Lacertilia SQUAMATA
2407 LACUSTRINE ENVIRONMENT PALE	LACUSTRINE ENVI- RONMENT	MILIEU LACUSTRE	LIMN.MILIEU	OZERNYJ	MEDIO=LACUSTRE	AMBIENTE LACUSTRE
2408 LACUSTRINE SEDIMENT SEDI	LAKE SEDIMENTS	SEDIMENT LACUSTRE	LIMN.SEDIMENT	OSADOK OZERNYJ	SEDIMENTO= LACUSTRE	SEDIMENTO LACUSTRE
2409 LACUSTRINE SEDIMENTATION SEDI	LACUSTRINE SEDI- MENTATION	SEDIMENTATION LACUSTRE	LIMN. SEDIMENTATION	SEDIMENTATSIYA OZERNAYA	SEDIMENTACION= LACUSTRE	SEDIMENTAZIONE LACUSTRE
2410 LADINIAN STRS	LADINIAN	LADINIEN	LADIN	LADIN	LADINIENSE	LADINICO
2411 LAGOON SURF	LAGOONS *UF*: Barrier lagoons	LAGUNE *UF*: Lagon barriere	LAGUNE *UF*: Barriere=Lagune	LAGUNA *UF*: Laguna bar'ernaya; Sreda lagunnaya	ALBUFERA *UF*: Albufera=de=atolon	LAGUNA *UF*: Laguna chiusa
2412 LAGOONAL ENVIRONMENT PALE	LAGOONAL ENVI- RONMENT	MILIEU LAGUNAIRE	LAGUNAER.MILIEU	Sreda lagunnaya LAGUNA	MEDIO=ALBUFERA	AMBIENTE LAGU- NARE
2413 LAGOONAL SEDIMENTATION SEDI	LAGOONAL SEDI- MENTATION	SEDIMENTATION LAGUNAIRE	LAGUNAER. SEDIMENTATION	SEDIMENTATSIYA LAGUNNAYA	SEDIMENTACION= ALBUFERA	SEDIMENTAZIONE DI LAGUNA
2414 LAHAR IGNE	LAHARS	Lahar COULEE BOUE	Lahar SCHLAMM=STROM	LAKHAR	Corriente=de=barro= volcanico COLADA=DE= BARRO	Lahar COLATA DI FANGO
2415 LAKE SURF	LAKES *UF*: Crater lakes; Lake water level	LAC *UF*: Lac cratere; Niveau lac	SEE *UF*: Krater=See; See= Wasserspiegel	OZERO *UF*: Ozero kraternoe; Ozero lavovoe; Uroven' vody ozera	LAGO *UF*: Lago=de=crater; Nivel=lago	LAGO *UF*: Lago craterico; Livello lacustre

	ENGLISH	FRANCAIS	DEUTSCH	RUSSKIJ	ESPANOL	ITALIANO
2416 LAKE WATER LEVEL GEOH	Lake water level LAKES	Niveau lac VARIATION; LAC	See=Wasserspiegel VARIATION; SEE	Uroven' vody ozera OZERO; UROVEN'	Nivel=lago VARIACION; LAGO	Livello lacustre VARIAZIONE; LAGO
2417 LAMARCKISM PALE	Lamarckism BIOLOGIC EVOLU- TION; THEORETI- CAL STUDIES	Lamarckisme THEORIE; EVOLU- TION BIOLOGIQUE	Lamarckismus THEORIE; BIOLOG. EVOLUTION	LAMARKIZM	Teoria=de=Lamarck TEORIA; EVOLUCION= BIOLOGICA	Lamarckismo TEORIA; EVOLUZ- IONE BIOLOGICA
2418 LAME CONSTANT PHCH	Lame constant ELASTICITY	Constante Lame CONSTANTE ELAS- TICITE	Lame=Konstante ELASTIZITAETS= KONSTANTE	LAMEH KON- STANTA	Constante=Lame CONSTANTE= ELASTICIDAD	Costante di Lame COSTANTE DI ELASTICITA
2419 LAMELLAE TEST	Lamellae LAMINATIONS	Structure lamellaire LAMINATION	Lamellar.Gefuege LAMINIERUNG	Plastina PLASTINCHATYJ	Laminar LAMINACION	Lamina LAMINAZIONE
2420 LAMINAR FLOW GEOH	LAMINAR FLOW	REGIME LAMINAIRE	LAMINAR. STROEMUNG	POTOK LAMI- NARNYJ	REGIMEN= LAMINAR	REGIME LAMI- NARE
2421 LAMINATED STRUCTURE TEST	Laminated structure LAMINATIONS	Structure laminee LAMINATION	Laminiert.Gefuege LAMINIERUNG	PLASTINCHATYJ UF: Plastina	Estructura=laminar LAMINACION	Struttura laminata LAMINAZIONE
2422 LAMINATION TEST	LAMINATIONS UF: Lamellae; Lami- nated structure	LAMINATION UF: Structure lamellaire; Structure laminee	LAMINIERUNG UF: Lamellar.Gefuege; Laminiert.Gefuege	Otdel'nost' plitchataya PLITCHATYJ	LAMINACION UF: Estructura=laminar; Laminar	LAMINAZIONE UF: Lamina; Struttura laminata
2423 LAMPROPHYRE IGNS	LAMPROPHYRES UF: Lamprophyric texture	LAMPROPHYRE UF: Texture lampro- phyrique	LAMPROPHYR UF: Lamprophyr= Gefuege	LAMPROFIR	LAMPROFIDO UF: Textura= lamprofidica	LAMPROFIRO UF: Tessitura lampro- firica
2424 LAMPROPHYRIC TEXTURE TEST	Lamprophyric texture TEXTURES; LAM- PROPHYRES	Texture lampro- phyrique LAMPROPHYRE	Lamprophyr=Gefuege LAMPROPHYR	LAMPROFIROVYJ	Textura=lamprofidica LAMPROFIDO	Tessitura lamprofirica LAMPROFIRO
2425 LAND SURF	Land RELIEF	Terre emergee RELIEF CONTI- NENT	Festland KONTINENTAL= RELIEF	SUSHA UF: Continental zone	Area=continental RELIEVE= CONTINENTAL	Terra emersa FORMA DEL RILIEVO
2426 LAND SUBSIDENCE ENGI	LAND SUBSIDENCE	AFFAISSEMENT	BODENSENKUNG	PROSADKA	DESLIZAMIENTO= LADERA	AFFOSSAMENTO
2427 LAND USE ENVI	LAND USE	UTILISATION TER- RAIN	NATURRAEUML. NUTZUNG	[LAND USE]	UTILIZACION= TERRENO	UTILIZZAZIONE DEL TERRITORIO
2428 LANDFILL ENGI	LANDFILLS	REMBLAIEMENT	AUFSCHUETTUNG	[LANDFILL] ZAKHORONENIE OTKHODOV	TERRAPLENA- MIENTO	COLMATAZIONE
2429 LANDFORM SURF	LANDFORMS UF: Morphogenesis	RELIEF CONTI- NENT UF: Continent; Inter- fluve; Morphogenese; Morphometrie; Surface terre; Terre emergee	KONTINENTAL= RELIEF UF: Erdoberflaeche; Festland; Kontinent; Morphogenese; Mor- phometrie; Zwischens- tromland	FORMA REL'EFA UF: Eolian feature; Erosion feature	RELIEVE= CONTINENTAL UF: Area=continental; Continente; Interfluvio; Morfogenesis; Morfo- metria; Terreno	FORMA DEL RILIEVO UF: Continente; Interflu- vio; Morfogenesi; Mor- fometria; Terra emersa; Terreno
2430 LANDSAT GEOL	LANDSAT	LANDSAT	LANDSAT	LANDSAT	LANDSAT	LANDSAT

	ENGLISH	FRANCAIS	DEUTSCH	RUSSKIJ	ESPANOL	ITALIANO
2431 LANDSLIDE ENGI	LANDSLIDES *UF:* Earthflows	GLISSEMENT TER- RAIN *UF:* Ecroulement; Glisse- ment terre	ERDRUTSCH *UF:* Erdfliessen; Fels- rutsch	OPOLZEN' *UF:* Earthflow; Opolzen' gryazevoj; Osov bloch- nyj	DESLIZAMIENTO= TERRENO *UF:* Deslizamiento de masas rocosas; Flujo=de=tierra	SCIVOLAMENTO DI TERRENO *UF:* Frana di scivola- mento in roccia; Scivolamento di terra
2432 LANTHANUM CHEE	LANTHANUM	LANTHANE	LA	LANTAN	LANTANO	LANTANIO
2433 LAPILLI IGNE	Lapilli PYROCLASTICS	Lapilli PYROCLASTIQUE	Lapilli PYROKLAST. GESTEIN	LAPILLI	Lapilli PIROCLASTICO	Lapilli PRODOTTO PIRO- CLASTICO
2434 LAPLACE **TRANSFORMATION** MATH	LAPLACE TRANS- FORMATIONS	TRANSFORMATION LAPLACE	LAPLACE= TRANSFORMATION	LAPLASA PREO- BRAZOVANIE	TRANSFORMACION= LAPLACE	TRASFORMATA DI LAPLACE
2435 LARAMIDE OROGENY STRU	LARAMIDE OROG- ENY	OROGENIE LARA- MIENNE	LARAM. OROGENESE	Laramijskij AL'PIJSKIJ /ORO- GEN/	OROGENIA= LARAMICA	OROGENESI LARAMICA
2436 LARGE=SCALE MISC	Large=scale CARTOGRAPHY	A grande echelle CARTOGRAPHIE	Gross.Massstab KARTOGRAPHIE	MASSHTAB KRUP- NYJ	Gran=escala CARTOGRAFIA	A grande scala CARTOGRAFIA
2437 LASER PHCH	LASER METHODS	METHODE LASER	LASER	LAZER	LASER	LASER
2438 LATE DIAGENESIS SEDI	LATE DIAGENESIS	DIAGENESE TAR- DIVE	SPAET=DIAGENESE	Pozdnij diagenez EHPIGENEZ	DIAGENESIS= TARDIA	DIAGENESI TARDIVA
2439 LATE PROTEROZOIDES STRU	Late proterozoides PROTEROZOIC; OROGENY	Proterozoides sup OROGENIE ANTE- CAMBRIENNE	Spaet=Proterozoiden PRAEKAMBR. OROGENESE	PROTEROZOIDY POZDNIE	Proterozoides=sup OROGENIA=PRE- CAMBRICA	Proterozoidi sup. OROGENESI PRE- CAMBRIANA
2440 LATENT HEAT OF **FUSION** PHCH	Latent heat of fusion THERMODYNAMIC PROPERTIES	Chaleur equivalente fusion THERMODY- NAMIQUE; FUSION	Latent.Schmelz= Waerme THERMODYNAMIK; SCHMELZE	Teplota plavleniya udel'naya TEPLOTA PLAV- LENIYA; TEPLOTA UDEL'NAYA	Calor=latente=de= fusion TERMODINAMICA; FUSION	Calore latente di fusione TERMODINAMICA; FUSIONE
2441 LATERAL GEOL	Lateral(adj.)	Lateral(adj.)	Lateral(adj.)	LATERAL'NYJ	Lateral(adj.)	Laterale(adj.)
2442 LATERAL FAULT STRU	Lateral faults FAULTS	Faille laterale FAILLE DECRO- CHEMENT	Lateral=Stoerung BLATT= VERSCHIEBUNG	SDVIG *UF:* Wrench fault	Falla=lateral FALLA= HORIZONTAL	Faglia a separaz.sec. direzione FAGLIA A SCORRI- MENTO ORIZZON- TALE
2443 LATERAL LOG APPL	Lateral log ELECTRICAL LOG- GING	Diagraphie resistivite DIAGRAPHIE ELEC- TRIQUE	Lateral=Log ELEKTRO=LOG	Karotazh bokovoj KAROTAZH	Testificacion=de= resistividad DIAGRAFIA= ELECTRICA	Log laterale DIAGRAFIA ELET- TRICA
2444 LATERAL MORAINE SURF	Lateral moraines MORAINES	Moraine laterale MORAINE	Seiten=Moraene MORAENE	Morena bokovaya MORENA	Morrena=lateral MORRENA	Morena laterale MORENA
2445 LATERAL SECRETION ECON	Lateral secretion AUTOMETAMOR- PHISM	Secretion lateral METAMORPHISME AUTO	Lateral=Sekretion AUTO= METAMORPHOSE	SEKRETSIYA LAT- ERAL'NAYA	Secrecion=lateral AUTOMETAMOR- FISMO	Secrezione laterale AUTOMETAMOR- FISMO

	ENGLISH	FRANCAIS	DEUTSCH	RUSSKIJ	ESPANOL	ITALIANO
2446 LATERITE SEDS	LATERITES *UF:* Lateritic soil	LATERITE *UF:* Sol lateritique	LATERIT *UF:* Laterit=Boden	LATERIT	LATERITA *UF:* Suelo=lateritico	LATERITE *UF:* Suolo lateritico
2447 LATERITIC SOIL SUSS	Lateritic soil LATERITES	Sol lateritique LATERITE	Laterit=Boden LATERIT	POCHVA LATERIT- NAYA	Suelo=lateritico LATERITA	Suolo lateritico LATERITE
2448 LATERIZATION SURF	LATERIZATION	LATERITISATION	LATERITISIERUNG	LATERITIZATSIYA	LATERITIZACION	LATERITIZZAZ- IONE
2449 LATITE IGNS	Latite TRACHYANDESITE	Latite TRACHYANDESITE	Latit TRACHYANDESIT	LATIT	Latita TRAQUIANDESITA	Latite TRACHIANDESITE
2450 LATITUDE MISC	Latitude GEODETIC COOR- DINATES	Latitude COORDONNEE GEODESIQUE	Geograph.Breite GEODAET. KOORDINATE	SHIROTA *UF:* Paleolatitude	Latitud COORDENADAS= GEODESICAS	Latitudine COORDINATE GEODETICHE
2451 LATTICE CONSTANT MINE	CELL DIMENSIONS	CONSTANTE RETICULAIRE *UF:* Reseau Bravais	GITTERKON- STANTE *UF:* Bravais=Gitter	PARAMETRY RESHETKI	CONSTANTE= RETICULAR *UF:* Red=de=Bravais	COSTANTE RETI- COLARE *UF:* Reticolo di Bravais
2452 LATTICE TEXTURE TEST	Lattice texture TEXTURES	Texture maillee TEXTURE	Gitter=Gefuege KORN=GEFUEGE	RESHETCHATYJ	Textura=reticulada TEXTURA	Struttura a maglie TESSITURA
2453 LAURASIA STRA	LAURASIA	LAURASIE	LAURASIA	LAVRAZIYA	LAURASIA	LAURASIA
2454 LAVA IGNE	LAVA *UF:* Tuff lava	LAVE *UF:* Tufo=lave	LAVA *UF:* Tuffo=Lava	LAVA *UF:* Kanal lavovoyj podvodyashchij; Ozero lavovoe	LAVA *UF:* Toba=volcanica	LAVA *UF:* Agglomerato di lava
2455 LAVA CHANNEL IGNE	LAVA CHANNELS	CHENAL LAVE	LAVA=KANAL	Kanal vulkana NEKK	CANAL=DE=LAVA	CANALE DI LAVA
2456 LAVA FIELD IGNE	LAVA FIELDS	CHAMP LAVE	LAVA=FELD	Pole lavovoe POTOK LAVOVYJ	CAMPO=DE=LAVA	CAMPO DI LAVA
2457 LAVA FLOW IGNE	LAVA FLOWS	COULEE LAVE	LAVA=STROM	POTOK LAVOVYJ *UF:* Pole lavovoe	COLADA=LAVA	COLATA LAVICA
2458 LAVA LAKE IGNE	LAVA LAKES	LAC LAVE	LAVA=SEE	Ozero lavovoe OZERO; LAVA	LAGO=LAVA	LAGO DI LAVA
2459 LAVA TUBE IGNE	LAVA TUBES	TUNNEL LAVE	LAVA=TUNNEL	Kanal lavovoyj podvo- dyashchij LAVA	TUNEL=DE=LAVA	GALLERIA LAVICA
2460 LAW MISC	Laws LEGISLATION	Loi LEGISLATION	Gesetz RECHT	ZAKON *UF:* Gayui zakon; Zakon matematicheskij	Ley LEGISLACION	Legge LEGISLAZIONE
2461 LAW OF CORRELATION **OF FACIES** STRA	Law of correlation of facies CORRELATION; FACIES	Loi correlation facies CORRELATION	Fazies=Abfolge= Gesetz KORRELATION	ZAKON KORRE- LYATSII FATSIJ	Ley=de=correlacion= de=facies CORRELACION	Legge della correlaz.di facies CORRELAZIONE
2462 LAYER TEST	Layers STRATIFICATION	Couche STRATIFICATION	Schicht SCHICHTUNG	SLOJ *UF:* Mud bank	Estrato ESTRATIFICACION	Livello STRATIFICAZIONE

	ENGLISH	FRANCAIS	DEUTSCH	RUSSKIJ	ESPANOL	ITALIANO
2463 LAYERED INTRUSION IGNE	LAYERED INTRU-SIONS	ROCHE IGNEE STRATOIDE	[LAYERED=INTRUSION]	INTRUZIYA RASS-LOENNAYA	ROCA=IGNEA=ESTRATOIDE	ROCCIA IGNEA STRATOIDE
2464 LAYERED MEDIUM SOLI	LAYERED MEDIA	MILIEU STRATIFIE	GESCHICHTET. KOMPLEX	[LAYERED MEDIUM] SLOISTOST'	MEDIO=MATERIAL=DEPOSITADO	MEZZO STRATIFI-CATO
2465 LEACHING PHCH	LEACHING UF: Desilication; Residual deposits; Soil leaching	LESSIVAGE UF: Desilicification; Gite residuel; Lessivage sol; Phaeozem	AUSWASCHUNG UF: Boden=Auswaschung; Phaeozem; Residual=Lagerstaette	VYSH-CHELACHIVANIE UF: Lixiviation	LAVADO UF: Deposito=residual; Desilizacion; Faeozem; Lavado=suelo	[LEACHING]
2466 LEAD CHEE	LEAD	PLOMB	PB	SVINETS	PLOMO	PIOMBO
2467 LEAD-ALPHA AGE METHOD ISOT	Lead=alpha age method U/PB	Datation plomb=alpha U=PB	Alpha=Blei=Datierung U=PB=DATIERUNG	METOD AL'FA=SVINTSOVYJ	Datacion=pb=alfa U=PB	Pb=alpha U/PB
2468 LEAD-LEAD DATING ISOT	PB/PB	PB=PB	PB=PB=DATIERUNG	Metod svintsovo-svintsovyj METOD SVINTSOVO=URANOVIJ	PB=PB	PB/PB
2469 LEAF PALE	LEAVES	FEUILLE	BLATT	LIST	HOJA	FOGLIA
2470 LEAKAGE GEOH	LEAKAGE	DRAINANCE	DURCHSICKERUNG	[LEAKAGE] INFIL'TRATSIYA	FUGA	PERDITA
2471 LEAN ECON	Lean(adj.)	Pauvre(adj.)	Arm(adj.)	BEDNYJ UF: Barren deposit	Bajo=grado(adj.)	Magro(adj.)
2472 LEARNED SOCIETY MISC	LEARNED SOCIE-TIES	SOCIETE SAVANTE	WISSENSCHAFTL. GESELLSCHAFT	[LEARNED SOCI-ETY] UCHREZHDENIE	SOCIEDAD=CIENTIFICA	ASSOCIAZIONE CULTURALE
2473 LEAST SQUARES MATH	LEAST=SQUARES ANALYSIS	MOINDRES CAR-RES	KLEINST. QUADRATE=METHODE	METOD NAI-MEN'SHIKH KVA-DRATOV	MINIMOS=CUADRADOS	MINIMI QUADRATI
2474 LECTOTYPE PALE	Lectotypes NOMENCLATURE	Lectotype NOMENCLATURE	Lectotyp NOMENKLATUR	LEKTOTIP	Lectotipo NOMENCLATURA	Lectotipo NOMENCLATURA
2475 LEGEND MISC	Legend MAPS; CARTOGRA-PHY	Legende CARTE; NOMEN-CLATURE	Legende KARTE; NOMENK-LATUR	LEGENDA	Leyenda MAPA; NOMEN-CLATURA	Legenda CARTA; NOMEN-CLATURA
2476 LEGISLATION MISC	LEGISLATION UF: Laws	LEGISLATION UF: Loi	RECHT UF: Gesetz	ZAKONODA-TEL'STVO	LEGISLACION UF: Ley	LEGISLAZIONE UF: Legge
2477 LENGTH MISC	Length GEOMETRY	Longueur GEOMETRIE	Laenge GEOMETRIE	DLINA	Longitud GEOMETRIA	Lunghezza GEOMETRIA
2478 LENGTH OF DAY SOLI	LENGTH OF DAY	DUREE JOUR	TAGESDAUER	[LENGTH OF DAY]	DURACION=DEL=DIA	DURATA DEL GIORNO

	ENGLISH	FRANCAIS	DEUTSCH	RUSSKIJ	ESPANOL	ITALIANO
2479 LENIAN STRS	Lenian LOWER CAMBRIAN	Lenien CAMBRIEN INF	Lenium U.KAMBRIUM	LENSKIJ	Leniense CAMBRICO=INF	Leniano CAMBRIANO INF.
2480 LENS GEOL	LENSES	LENTILLE	LINSE	LINZA	LENTEJON	LENTE
2481 LEPERDITOCOPIDA PALS	LEPERDITOCOPIDA	LEPERDITOCOPIDA	LEPERDITOCOPIDA	Leperditocopida OSTRACODA	LEPERDITOCOPIDA	LEPERDITOCOPIDA
2482 LEPIDOBLASTIC **TEXTURE** TEST	Lepidoblastic texture TEXTURES	Texture lepidoblastique FOLIATION	Lepidoblast.Gefuege KRISTALLIS- ATIONS= SCHIEFERUNG	LEPIDOBLASTOVYJ	Lepidoblastica FOLIACION	Tessitura lepidoblastica FOLIAZIONE
2483 LEPIDOCYSTOIDEA PALS	LEPIDOCYSTOIDEA	LEPIDOCYSTOIDEA	LEPIDOCYSTOIDEA	Lepidocystoidea CRINOZOA	LEPIDOCYSTOIDEA	LEPIDOCYSTOIDEA
2484 LEPIDOSAURIA PALS	LEPIDOSAURIA UF: Lacertilia; Ophidia	LEPIDOSAURIA	LEPIDOSAURIA	LEPIDOSAURIA UF: Squamata	LEPIDOSAURIA	LEPIDOSAURIA
2485 LEPOSPONDYLI PALS	LEPOSPONDYLI	LEPOSPONDYLI UF: Urodela	LEPOSPONDYLI UF: Urodela	LEPOSPONDYLI	LEPOSPONDYLI UF: Urodela	LEPOSPONDYLI UF: Urodela
2486 LEPTITE IGMS	Leptite LEPTYNITE	Leptite LEPTYNITE	Leptit	Leptit LEPTINIT	Leptita LEPTINITA	Leptite LEPTINITE
2487 LEPTOGEOSYNCLINE STRU	Leptogeosyncline GEOSYNCLINES	Leptogeosynclinal GEOSYNCLINAL	Lepto=Geosynklinale GEOSYNKLINALE	LEPTOGEOSINKLI- NAL'	Leptogeosinclinal GEOSINCLINAL	Leptogeosinclinale GEOSINCLINALE
2488 LEPTYNITE IGMS	LEPTYNITE UF: Kinzigite, Leptite	LEPTYNITE UF: Kinzigite, Leptite	LEPTYNIT	LEPTINIT UF: Leptit	LEPTINITA UF: Kinzigita; Leptita	LEPTINITE UF: Kinzigite; Leptite
2489 LEUCOCRATIC **COMPOSITION** IGNE	Leucocratic composi- tion IGNEOUS ROCKS	Roche leucocrate ROCHE IGNEE	Leukokrat.Chemismus MAGMATIT	LEJKOKRATOVYJ	Composicion= leucocratica ROCA=IGNEA	Composizione leucocra- tica ROCCIA IGNEA
2490 LEUCOSOMES IGNE	Leucosome MIGMATITES	Leucosome COULEUR; MIGMA- TITE	Leucosome FARBE; MIGMATIT	LEJKOSOM	Leucosoma COLOR; MIGMA- TITA	Leucosoma COLORE; MIGMA- TITE
2491 LEVEE ENGI	LEVEES	LEVEE	FLUSS=DEICH	[LEVEE] DAMBA	DIQUE=DE= CONTENCION	RILEVAMENTO
2492 LEVELING METH	LEVELING	NIVELLEMENT	NIVELLEMENT	NIVELIROVKA	NIVELACION	LIVELLAMENTO
2493 LEVELS GEOL	Level(gen.) UF: Groundwater level	Niveau(gen.)	Niveau(gen.)	UROVEN' UF: Uroven' vody ozera	Nivel(gen.)	Livello(gen.)
2494 LEXICON MISC	LEXICONS	LEXIQUE	LEXIKON	LEKSIKA	LEXICO	LESSICO
2495 LICHENES PALS	LICHENS	LICHENES	LICHENES	LICHENES	LIQUENES	LICHENES
2496 LICHENOMETRY STRA	LICHENOMETRY	LICHENOMETRIE	LICHENOMETRIE	LIKHENOMETRIYA	LIQUENOMETRIA	LICHENOMETRIA

	ENGLISH	FRANCAIS	DEUTSCH	RUSSKIJ	ESPANOL	ITALIANO
2497 LICHIDA PALS	LICHIDA	LICHIDA	LICHIDA	Lichida POLYMERA	LICHIDA	LICHIDA
2498 LIESEGANG RINGS TEST	Liesegang rings TEXTURES	Anneaux Liesegang CRISTALLISATION	Liesegang=Ringe KRISTALLISATION	KOL'TSA LIZE- GANGA	Anillos=Liesegang CRISTALIZACION	Anelli di Liesegang CRISTALLIZZAZ- IONE
2499 LIFE PALE	LIFE *UF:* Organic life	VIE *UF:* Vie organique	LEBEN *UF:* Organ.Leben	ZHIZN'	VIDA *UF:* Vida=organica	VITA *UF:* Vita organica
2500 LIGHT ISOTOPE ISOT	Light isotope ISOTOPES	Isotope leger ISOTOPE	Leicht=Isotop ISOTOP	IZOTOP LEGKIJ	Isotopo=ligero ISOTOPO	Isotopo leggero ISOTOPO
2501 LIGHT MINERALS SEDI	LIGHT MINERALS	MINERAL LEGER	LEICHT=MINERAL	FRAKTSIYA LEG- KAYA	MINERALES= LIGEROS	MINERALI LEG- GERI
2502 LIGHT OIL COMS	Light oil PETROLEUM	Huile legere PETROLE	Leicht=Oel ERDOEL	NEFT' LEGKAYA	Petroleo=ligero PETROLEO	Petrolio leggero PETROLIO
2503 LIGNIN CHES	Lignin ORGANIC MATERI- ALS	Lignine MATIERE ORGANIQUE	Lignin ORGAN.SUBSTANZ	LIGNIN	Lignina MATERIA= ORGANICA	Lignina SOSTANZA ORGANICA
2504 LIGNITE COMS	LIGNITE *UF:* Brown coal	LIGNITE *UF:* Brown coal	Lignit BRAUNKOHLE	Lignit UGOL' BURYJ	LIGNITO *UF:* Carbon=pardo	LIGNITE *UF:* Brown coal
2505 LIMB STRU	Limb FOLDS	Flanc pli PLI	Schenkel FALTE	KRYLO	Flanco PLIEGUE	Fianco PIEGA
2506 LIME COMS	LIME	CHAUX	KALK=ERDE	IZVEST'	CAL	CALCE
2507 LIMESTONE SEDS	LIMESTONE *UF:* Biomicrite; Biospar- ite; Boundstone; Encri- nite; Grapestone	CALCAIRE *UF:* Allocheme; Anker- ite; Boundstone; Cali- che; Grapestone	KALK *UF:* Allochem; Ankerit= Gestein; Caliche; Karbonat.Sandstein; Organ.Kalk	IZVESTNYAK *UF:* Biomikrit; Biosparit; Ehnkrinit; Microcrys- talline limestone; Vodoroslevyj izvestnyak	CALIZA *UF:* Aloquimico; Bahamita; Boundstone; Caliche; Roca=de= ankerita	CALCARE *UF:* Allochimico; Anker- ite; Boundstone; Cali- che; Grapestone
2508 LIMNOLOGY SURF	LIMNOLOGY	LIMNOLOGIE	LIMNOLOGIE	LIMNOLOGIYA	LIMNOLOGIA	LIMNOLOGIA
2509 LIMON SEDS	Loam SILT	Limon SILT	Lehm SILT	SUGLINOK	Limon LIMO	Suolo limoso SILT
2510 LIMONITE COMS	LIMONITE *UF:* Bog iron ore	LIMONITE *UF:* Fer marais	LIMONIT *UF:* Rasen=Eisenerz	LIMONIT	LIMONITA *UF:* Hierro=de=los= pantanos	LIMONITE *UF:* Ferro delle paludi
2511 LINEAMENT STRU	LINEAMENTS	LINEAMENT	LINEAMENT	RAZLOM GLUBIN- NYJ *UF:* Sbroso=sdvig regional'nyj	LINEAMIENTO	LINEAMENTO
2512 LINEATION STRU	LINEATION	LINEATION	LINEAR=GEFUEGE	LINEJNOST'	LINEACION	LINEAZIONE
2513 LINGULIDA PALS	Lingulida INARTICULATA	Lingulida INARTICULATA	Lingulida INARTICULATA	LINGULIDA	Lingulida INARTICULATA	Lingulida INARTICULATA

	ENGLISH	FRANCAIS	DEUTSCH	RUSSKIJ	ESPANOL	ITALIANO
2514 LINKED VEINS ECON	Linked veins VEINS	Filon anastomose FILON	Gang=System GANG	ZHILA STU- PENCHATAYA	Filones=ramificados FILON	Sistema di vene FILONE
2515 LINOPHYRIC TEXTURE TEST	Linophyric texture TEXTURES	Texture linophyrique TEXTURE POR- PHYRIQUE	Linophyr.Gefuege PORPHYR- GEFUEGE	LINOFIROVYJ	Linofidica TEXTURA= PORFIDICA	Tessitura linofirica TESSITURA POR- FIRICA
2516 LIPID CHES	Lipids FATTY ACIDS	Lipide ACIDE GRAS	Lipid FETT=SAEURE	LIPIDY	Lipido ACIDO=GRASO	Lipide ACIDO GRASSO
2517 LIPOSTRACA PALS	Lipostraca CRUSTACEA	Lipostraca CONCHOSTRACA	Lipostraca CONCHOSTRACA	LIPOSTRACA	Lipostraca CONCHOSTRACA	Lipostraca CONCHOSTRACA
2518 LIQUATION IGNE	Liquation CRYSTALLIZATION	Separation phase CRISTALLISATION	Liquation KRISTALLISATION	LIKVATSIYA	Licuacion CRISTALIZACION	Separazione del liquido CRISTALLIZZAZ- IONE
2519 LIQUEFACTION PHCH	LIQUEFACTION	LIQUEFACTION	VERFLUESSIGUNG	Razzhizhenie TIKSOTROPIYA	LICUEFACCION	LIQUEFAZIONE
2520 LIQUID PHASE PHCH	LIQUID PHASE	PHASE LIQUIDE	SCHMELZ=PHASE	FAZA ZHIDKAYA	FASE=LIQUIDA	FASE LIQUIDA
2521 LIQUID WASTE ENVI	LIQUID WASTE	DECHET LIQUIDE	FLUESSIG.ABFALL	OTKHODY ZHIDKIE	RESIDUO=LIQUIDO	RIFIUTI LIQUIDI
2522 LIQUIDUS IGNE	Liquidus PHASE EQUILIBRIA	Liquidus DIAGRAMME EQUILIBRE	Liquidus PHASEN- GLEICHGEWICHT	LIKVIDUS	Liquidus DIAGRAMA- EQUILIBRIO	Liquidus DIAGRAMMA D'EQUILIBRIO
2523 LISSAMPHIBIA PALS	LISSAMPHIBIA	LISSAMPHIBIA	LISSAMPHIBIA	Lissamphibia AMPHIBIA	LISSAMPHIBIA	LISSAMPHIBIA
2524 LITHIFICATION SEDI	LITHIFICATION	LITHIFICATION	VERSTEINERUNG	LITIFIKATSIYA *UF:* Otverdenie	LITIFICACION	LITIFICAZIONE
2525 LITHIUM CHEE	LITHIUM	LITHIUM	LI	LITIJ	LITIO	LITIO
2526 LITHOFACIES SEDI	LITHOFACIES *UF:* Heterochronism; Isolith; Lithology; Lithostromes; Litho- topes; Lithotype; Magnafacies; Megafa- cies; Origofacies	LITHOFACIES *UF:* Facies primaire; Heterochronisme; Isolithe; Lithologie; Lithostrome; Lithotope; Lithotype; Magnafa- cies; Megafacies	LITHO=FAZIES *UF:* Heterochronie; Isolith; Lithologie; Lithostrom; Lithotop; Lithotyp; Magna= Fazies; Mega=Fazies; Origo=Fazies	LITOFATSIYA	LITOFACIES *UF:* Heterocronismo; Isolito; Litologia; Litostroma; Litotipo; Litotopo; Magnafacies; Megafacies; Origofa- cies	LITOFACIES *UF:* Eterocronismo; Isolito; Litologia; Litostroma; Litotipo; Litotopo; Magnafacies; Megafacies; Origofa- cies
2527 LITHOFACIES MAP SEDI	LITHOFACIES MAPS	CARTE LITHOLOGIQUE	LITHOFAZIES- KARTE	KARTA LITOLOGO= FATSIAL'NAYA	MAPA=LITOFACIES	CARTA DELLE LITOFACIES
2528 LITHOGENESIS SEDI	Lithogenesis GENESIS; SEDI- MENTARY ROCKS	Lithogenese GENESE	Lithogenese GENESE	LITOGENEZ	Litogenesis GENESIS	Litogenesi GENESI
2529 LITHOLOGIC GEOL	Lithologic(adj.)	Lithologique(adj.)	Lithologisch(adj.)	LITOLOGICHESKIJ	Litologico(adj.)	Litologico(adj.)
2530 LITHOLOGIC CONTROLS ECON	LITHOLOGIC CON- TROLS	CONTROLE LITHOLOGIQUE	LITHOLOG. EINFLUSS	KONTROL' LITOLOGICHESKIJ	CONTROL= LITOLOGICO	CONTROLLO LITOLOGICO

	ENGLISH	FRANCAIS	DEUTSCH	RUSSKIJ	ESPANOL	ITALIANO
2531 LITHOLOGIC TRAP ECON	Lithologic traps TRAPS	Piege lithologique PIEGE	Litholog.Falle FALLE	LOVUSHKA LITOLOGI- CHESKAYA	Trampa=litologica PLIEGUE	Trappola litologica TRAPPOLA
2532 LITHOLOGY GEOL	Lithology LITHOFACIES	Lithologie LITHOFACIES	Lithologie LITHO=FAZIES	LITOLOGIYA UF: Petrologiya osadoch- naya	Litologia LITOFACIES	Litologia LITOFACIES
2533 LITHOPHILE ELEMENT CHES	LITHOPHILE ELE- MENTS	ELEMENT LITHOPHILE	LITHOPHIL. ELEMENT	LITOFIL'NYJ	ELEMENTO= LITOFILO	ELEMENTO LITO- FILO
2534 LITHOPHYTE PALE	Lithophyte ECOLOGY; FLORA	Lithophyte ECOLOGIE; FLORE	Lithophyt OEKOLOGIE; FLORA	LITOFIT	Litofita ECOLOGIA; FLORA	Lithophyte ECOLOGIA; FLORA
2535 LITHOSOL SUSS	Lithosols SOILS	Lithosol SOL BRUT	Skelett=Boden AZONAL.BODEN	Litosol POCHVA AZON- AL'NAYA	Litosol SUELO=BRUTO	Suolo scheletrico SUOLO AZONALE
2536 LITHOSPHERE SOLI	LITHOSPHERE	LITHOSPHERE UF: Origine supracrus- tale	LITHOSPHAERE UF: Suprakrustal.Genese	LITOSFERA UF: A sloj	LITOSFERA UF: Origen= supracortical	LITOSFERA UF: Origine sopracros- tale
2537 LITHOSTRATIGRAPHIC UNIT STRA	Lithostratigraphic unit STRATIGRAPHIC UNITS	Unite lithostrati- graphique UNITE STRATI- GRAPHIQUE	Lithostratigraph. Einheit STRATIGRAPH. EINHEIT	PODRAZDELENIE LITOSTRATIGRAF.	Unidad= litoestratigrafica UNIDAD= ESTRATIGRAFICA	Unita litostratigrafica UNITA STRATI- GRAFICA
2538 LITHOSTRATIGRAPHY STRA	LITHOSTRATIGRA- PHY	Lithostratigraphie STRATIGRAPHIE	Litho=Stratigraphie STRATIGRAPHIE	LITOSTRATIGRA- FIYA	Litoestratigrafia ESTRATIGRAFIA	Litostratigrafia STRATIGRAFIA
2539 LITHOSTROME SEDI	Lithostromes LITHOFACIES	Lithostrome LITHOFACIES	Lithostrom LITHO=FAZIES	Litostrom LITOTOP	Litostroma LITOFACIES	Litostroma LITOFACIES
2540 LITHOTOPE SEDI	Lithotopes LITHOFACIES	Lithotope LITHOFACIES	Lithotop LITHO=FAZIES	LITOTOP UF: Litostrom	Litotopo LITOFACIES	Litotopo LITOFACIES
2541 LITHOTYPE GEOL	Lithotype LITHOFACIES	Lithotype LITHOFACIES	Lithotyp LITHO=FAZIES	LITOTIP	Litotipo LITOFACIES	Litotipo LITOFACIES
2542 LITOPTERNA PALS	LITOPTERNA	LITOPTERNA	LITOPTERNA	LITOPTERNA	LITOPTERNA	LITOPTERNA
2543 LITTORAL SURF	Littoral(adj.)	Littoral(adj.)	Litoral(adj.)	LITORAL' UF: Otmel'ilistaya; Sreda sublitoral'naya; Subtidal environment	Litoral(adj.)	Littorale(adj.)
2544 LITTORAL DRIFT SURF	LITTORAL DRIFT	DERIVE LITTOR- ALE	KUESTEN= ABDRIFT	POTOK NANOSOV	DERIVA=LITORAL	DERIVA LITORALE
2545 LITTORAL ENVIRONMENT PALE	LITTORAL ENVI- RONMENT UF: Paralic environment	MILIEU LITTORAL UF: Milieu paralique; Milieu tidal	LITORAL=MILIEU UF: Paral.Milieu	FATSIYA PRI- BREZHNAYA	MEDIO=LITORAL UF: Medio=mareal; Medio=paralico	AMBIENTE LITOR- ALE UF: Ambiente paralico; Ambiente tidale
2546 LITTORAL SEDIMENTATION SEDI	NEARSHORE SEDI- MENTATION	SEDIMENTATION LITTORALE	LITORAL= SEDIMENTATION	SEDIMENTATSIYA BEREGOVAYA	SEDIMENTACION= LITORAL	SEDIMENTAZIONE LITORALE

	ENGLISH	FRANCAIS	DEUTSCH	RUSSKIJ	ESPANOL	ITALIANO
2547 LITUOLACEA PALS	LITUOLACEA	LITUOLACEA	LITUOLACEA	LITUOLACEA	LITUOLACEA	LITUOLACEA
2548 LIVING PALE	Living(adj.)	Vivant(adj.)	Lebend(adj.)	ZHIVUSHCHIJ	Viviente(adj.)	Vivente(adj.)
2549 LIVING FOSSIL PALE	LIVING FOSSILS	FOSSILE VIVANT	LEBEND.FOSSIL	ISKOPAEMOE ZHIVUSHCHEE	FOSIL=VIVIENTE	FOSSILE VIVENTE
2550 LIXIVIATION PHCH	LIXIVIATION	LIXIVIATION	AUSLAUGUNG	[LIXIVIATION] VYSHCHE LACHIVANIE	LIXIVIACION	LISCIVIAZIONE *UF:* Desilicizzazione; Giacimento residuale; Leaching; Lisciviazione del suolo; Phaeozem
2551 LLANDEILIAN STRS	LLANDEILIAN	LLANDEILO	LLANDEILO	LLANDEJLO	LLANDEILOIENSE	LLANDEILIANO
2552 LLANDOVERIAN STRS	LLANDOVERIAN	LLANDOVERY	LLANDOVERY	LLANDOVERI	LLANDOVERIENSE	LLANDOVERIANO
2553 LLANVIRNIAN STRS	LLANVIRNIAN	LLANVIRN	LLANVIRN	LLANVIRN	LLANVIRNIENSE	LLANVIRNIANO
2554 LOAD ENGI	LOADING	CHARGEMENT	BELASTUNG	MATERIAL OSA-DOCHNYJ *UF:* Material rykhlyj; Rechnoj stok tverdyj	CARGA	CARICAMENTO
2555 LOAD CAST SEDI	LOAD CASTS	EMPREINTE CHARGE	BELASTUNGS=MARKE	ZNAKI VNE-DRENIYA	MOLDE=DE=CARGA	IMPRONTA DI CAR-ICO
2556 LOAD TEST ENGI	LOAD TESTS	PRESSIOMETRIE	DRUCK=MESSUNG	NAGRUZKA OPYT-NAYA	PRESIOMETRIA	PRESSOMETRIA
2557 LOCAL MISC	Local(adj.)	Local(adj.)	Lokal(adj.)	LOKAL'NYJ	Local(adj.)	Locale(adj.)
2558 LOCOMOTION PALE	LOCOMOTION	LOCOMOTION	FORTBEWEGUNG	PEREDVIZHENIE	ANATOMIA=LOCOMOCION	LOCOMOZIONE
2559 LODE ECON	Lode VEINS	Gite filonien FILON	Erz=Gang GANG	Rudnye zhily TELO RUDNOE	Yacimiento=filoniano FILON	Giacimento filoniano FILONE
2560 LODRANITE EXTS	Lodranite METEORITES	Lodranite SIDEROLITE	Lodranit STEIN=EISEN=METEORIT	Lodranit SIDEROLIT	Lodranito SIDEROLITO	Lodranite SIDEROLITE
2561 LOESS SEDS	LOESS	LOESS	LOESS	LESS	LOES	LOESS
2562 LOG GEOL	Logs WELL=LOGGING	Log COUPE SONDAGE	Bohr=Log BOHR=PROFIL	KAROTAZH *UF:* Karotazh bokovoj; Karotazh gamma; Karotazh gamma=gamma; Karotazh nejtronnyj; Karotazh skvazhinnyj; Mikroka-rotazh; Mikrokarotazh bokovoj	Testificacion CORTE=GEOL=SONDEO	Log SEZIONE DI SON-DAGGIO

	ENGLISH	FRANCAIS	DEUTSCH	RUSSKIJ	ESPANOL	ITALIANO
2563 LOGNORMAL DISTRIBUTION MATH	Lognormal distribution STATISTICAL DISTRIBUTION	Distribution lognormale DISTRIBUTION STATISTIQUE	Normallog=Verteilung STATIST. VERTEILUNG	RASPREDELENIE LOGARIFMICH.-NORM.	Distribucion=lognormal DISTRIBUCION=ESTADISTICA	Distribuzione lognormale DISTRIBUZIONE STATISTICA
2564 LONG-PERIOD WAVE SOLI	LONG=PERIOD WAVES	ONDE LONGUE PERIODE	LANGPERIOD. WELLE	Volna dlinnoperiodnaya VOLNA SEJSMICHESKAYA	ONDA=LARGO=PERIODO	ONDA A LUNGO PERIODO
2565 LONGITUDE MISC	Longitude GEODETIC COORDINATES	Longitude COORDONNEE GEODESIQUE	Geograph.Laenge GEODAET. KOORDINATE	DOLGOTA	Longitud=geografica COORDENADAS=GEODESICAS	Longitudine COORDINATE GEODETICHE
2566 LONGITUDINAL PROFILE SURF	STREAM GRADIENT *UF:* Subsequent valleys; Thalweg	PROFIL EN LONG *UF:* Profil equilibre; Thalweg; Vallee obsequente; Vallee structurale; Vallee subsequente	LAENGS=PROFIL *UF:* Gefaelle=Kurve; Gleichgewichts=Profil; Obsequent.Tal; Subsequent.Tal; Tekton.Tal	PROFIL' PRODOL'NYJ	PERFIL=LONGITUDINAL *UF:* Perfil=de=equilibrio; Vaguada; Valle=estructural; Valle=obsecuente; Valle=subsecuente	PROFILO LONGITUDINALE *UF:* Fondovalle; Profilo d'equilibrio; Valle obsequente; Valle susseguente; Valle tettonica
2567 LONGSHORE BAR SURF	LONGSHORE BARS	CORDON LITTORAL *UF:* Barre littorale; Chenier; Plage barriere; Tombolo	STRAND=WALL *UF:* Barriere=Strand; Chenier; Strand=Barriere; Tombolo	BAR BEREGOVOJ	CORDON=LITORAL *UF:* Barra=barrera; Barrera=playa; Restinga; Tombolo	CORDONE LITTORALE *UF:* Barra litorale; Chenier; Lido; Tombolo
2568 LONGSHORE CURRENT MARI	LONGSHORE CURRENTS	COURANT LITTORAL	KUESTEN=STROEMUNG	TECHENIE BEREGOVOE	CORRIENTE=LITORAL	CORRENTE LITTORALE
2569 LOPOLITH IGNE	LOPOLITHS	LOPOLITE	LOPOLITH	LOPOLIT	LOPOLITO	LOPOLITE
2570 LOVE WAVE SOLI	LOVE WAVES	ONDE LOVE	LOVE=WELLE	VOLNA LOVA	ONDA=LOVE	ONDA LOVE
2571 LOW PRESSURE PHCH	LOW PRESSURE	BASSE PRESSION	NIEDER=DRUCK	Davlenie nizkoe DAVLENIE; MINIMUM	PRESION=BAJA	BASSA PRESSIONE
2572 LOW TEMPERATURE PHCH	LOW TEMPERATURE	BASSE TEMPERATURE	TIEF=TEMPERATUR	Temperatura nizkaya TEMPERATURA; MINIMUM	TEMPERATURA=BAJA	BASSA TEMPERATURA
2573 LOW WATER GEOH	LOW=WATER LEVELS	ETIAGE	NIEDRIG.WASSER	MEZHEN'	ESTIAJE	LIVELLO DI MAGRA
2574 LOW-ENERGY ENVIRONMENT SEDI	LOW=ENERGY ENVIRONMENT	MILIEU BASSE ENERGIE	NIEDER=ENERGET. MILIEU	Sreda spokojnaya SEDIMENTATSIYA	MEDIO=BAJA=ENERGIA	AMBIENTE A BASSA ENERGIA
2575 LOW-GRADE METAMORPHISM IGNE	LOW=GRADE METAMORPHISM	METAMORPHISME FAIBLE	NIEDRIGGRADIG. METAMORPHOSE	[LOW=GRADE METAMORPHISM] STEPEN' METAMORFIZMA	METAMORFISMO=BAJO=GRADO	METAMORFISMO DEBOLE
2576 LOW-VELOCITY LAYER SOLI	LOW=VELOCITY ZONES *UF:* B layer	ZONE FAIBLE VITESSE	ZONE=NIEDRIG. GESCHWINDIGKEIT	Sloj ponizhennykh skorostej GUTENBERGA POVERKHNOST'	ZONA=DE=BAJA=VELOCIDAD	ZONA A BASSA VELOCITA

	ENGLISH	FRANCAIS	DEUTSCH	RUSSKIJ	ESPANOL	ITALIANO
2577 LOWER CAMBRIAN STRS	LOWER CAMBRIAN *UF:* Aldanian; Lenian	CAMBRIEN INF *UF:* Aldanien; Lenien	U.KAMBRIUM *UF:* Aldanium; Lenium	KEMBRIJ NIZHNIJ	CAMBRICO–INF *UF:* Aldaniense; Leniense	CAMBRIANO INF. *UF:* Aldaniano; Leniano
2578 LOWER CARBONIFEROUS STRS	LOWER CARBONIF-EROUS	CARBONIFERE INF	U.KARBON	KARBON NIZHNIJ *UF:* Missisipij	CARBONIFERO-INF	CARBONIFERO INF.
2579 LOWER CRETACEOUS STRS	LOWER CRETA-CEOUS	CRETACE INF	U.KREIDE	MEL NIZHNIJ	CRETACICO–INF	CRETACEO INF.
2580 LOWER CRUST SOLI	LOWER CRUST	CROUTE TER-RESTRE INF	UNTER.ERD=KRUSTE	Kora nizhnyaya SLOJ BAZAL'TOVYJ	CORTEZA=TERRESTRE=INF	CROSTA PRO-FONDA
2581 LOWER DEVONIAN STRS	LOWER DEVONIAN	DEVONIEN INF	U.DEVON	DEVON NIZHNIJ	DEVONICO–INF	DEVONIANO INF.
2582 LOWER EOCENE STRS	LOWER EOCENE *UF:* Ypresian	EOCENE INF *UF:* Ypresien	U.EOZAEN *UF:* Ypresium	Ehotsen nizhnij EHOTSEN	EOCENO–INF *UF:* Ipresiense	EOCENE INF. *UF:* Ypresiano
2583 LOWER JURASSIC STRS	LOWER JURASSIC	JURASSIQUE INF	U.JURA	YURA NIZHNYAYA *UF:* Lejas nizhnij	JURASICO–INF	GIURASSICO INF.
2584 LOWER LIASSIC STRS	LOWER LIASSIC	LIAS INF	U.LIAS	Lejas nizhnij YURA NIZHNYAYA	LIAS–INF	LIASSICO INF.
2585 LOWER MANTLE SOLI	LOWER MANTLE *UF:* D layer	MANTEAU INF *UF:* Couche D	UNTER.ERD=MANTEL *UF:* D=Schicht	MANTIYA NIZH-NYAYA *UF:* D sloj	MANTO=GLOBO=INF *UF:* Capa–D	MANTELLO INFERIORE *UF:* Strato D
2586 LOWER MIOCENE STRS	LOWER MIOCENE	MIOCENE INF	U.MIOZAEN	Miotsen nizhnij MIOTSEN	MIOCENO–INF	MIOCENE INF.
2587 LOWER OLIGOCENE STRS	LOWER OLIGO-CENE	OLIGOCENE INF	U.OLIGOZAEN	Oligotsen nizhnij RYUPEL'	OLIGOCENO–INF	OLIGOCENE INF.
2588 LOWER ORDOVICIAN STRS	LOWER ORDOVI-CIAN	ORDOVICIEN INF	U.ORDOVIZIUM	ORDOVIK NIZHNIJ	ORDOVICICO–INF	ORDOVICIANO INF.
2589 LOWER PERMIAN STRS	LOWER PERMIAN	PERMIEN INF	U.PERM	PERM' NIZH-NYAYA *UF:* Lezhen' krasnyj	PERMICO=INF	PERMIANO INF.
2590 LOWER PLEISTOCENE STRS	LOWER PLEISTO-CENE	PLEISTOCENE INF	ALT=PLEISTOZAEN	Plejstotsen nizhnij PLEJSTOTSEN	PLEISTOCENO–INF	PLEISTOCENE INF.
2591 LOWER PLIOCENE STRS	LOWER PLIOCENE *UF:* Pontian	PLIOCENE INF *UF:* Pontien	U.PLIOZAEN *UF:* Pontium	Pliotsen nizhnij PLIOTSEN	PLIOCENO–INF *UF:* Pontiense	PLIOCENE INF. *UF:* Pontiano
2592 LOWER PROTEROZOIC STRS	LOWER PROTERO-ZOIC	PROTEROZOIQUE INF	U.PROTEROZOIKUM	PROTEROZOJ NIZHNIJ	PROTEROZOICO–INF	PROTEROZOICO INF.
2593 LOWER SILURIAN STRS	LOWER SILURIAN	SILURIEN INF	U.SILUR	SILUR NIZHNIJ *UF:* Tarannon	SILURICO–INF	SILURIANO INF.
2594 LOWER TRIASSIC STRS	LOWER TRIASSIC *UF:* Olenekian	TRIAS INF *UF:* Indien	U.TRIAS *UF:* Indien=Stufe	TRIAS NIZHNIJ *UF:* Buntzandshtejn	TRIAS=INF *UF:* Indiense	TRIASSICO INF. *UF:* Indiano
2595 LOWLAND SURF	Lowlands PLAINS	Basses=terres PLAINE	Ebene FLACHLAND	NIZMENNOST'	Tierras=bajas LLANURA	Pianura PIANA

	ENGLISH	FRANCAIS	DEUTSCH	RUSSKIJ	ESPANOL	ITALIANO
2596 LUDLOVIAN STRS	LUDLOVIAN	LUDLOW	LUDLOW	LUDLOV	LUDLOWIENSE	LUDLOWIANO
2597 LUMINESCENCE PHCH	LUMINESCENCE *UF:* Phosphorescence	LUMINESCENCE *UF:* Phosphorescence	LUMINESZENZ *UF:* Phosphoreszenz	LYUMINESTSENT- SIYA	LUMINISCENCIA *UF:* Fosforescencia	LUMINESCENZA *UF:* Fosforescenza
2598 LUNA EXTR	LUNA PROGRAM	PROGRAMME LUNA	LUNA=PROGRAMM	LUNA (INST)	PROGRAMA=LUNA	PROGRAMMA LUNA
2599 LUNAR EXTR	Lunar(adj.)	Lunaire(adj.)	Lunar(adj.)	LUNNYJ	Lunar(adj.)	Lunare(adj.)
2600 LUNAR CONSTITUTION EXTR	LUNAR CONSTITU- TION	CONSTITUTION LUNE	MOND=AUFBAU	[LUNAR CONSTI- TUTION] LUNA; SOSTAV	CONSTITUCION= LUNA	COSTITUZIONE DELLA LUNA
2601 LUNAR CRATER EXTR	LUNAR CRATERS	CRATERE LUNAIRE	MOND=KRATER	KRATER LUNNYJ	CRATER=LUNAR	CRATERE LUNARE
2602 LUNAR GEOLOGY EXTR	Lunar geology MOON	Geologie lunaire LUNE	Mond=Geologie MOND	GEOLOGIYA LUNY	Geologia=lunar LUNA	Geologia lunare LUNA
2603 LUNAR SOIL EXTR	LUNAR SOILS	Sol lunaire REGOLITHE	Mond=Boden REGOLITH	POCHVA LUNNAYA	Suelo=lunar REGOLITO	Suolo lunare REGOLITE
2604 LUNAR STATION EXTR	LUNAR STATIONS	STATION LUNAIRE	MOND=STATION	STANTSIYA LUN- NAYA	ESTACION=LUNAR	STAZIONE LUNARE
2605 LUNAR TERRA EXTR	LUNAR MASSIFS	MASSIF LUNAIRE	MOND=TERRA	KONTINENT (LUNNYJ)	MACIZO=LUNAR	MASSICCIO LUNARE
2606 LUSTER PHCH	Luster OPTICAL PROPER- TIES	Eclat PROPRIETE OPTIQUE	Glanz OPT.EIGENSCHAFT	BLESK	Brillo PROPIEDAD= OPTICA	Lucentezza PROPRIETA OTTICA
2607 LUTETIUM CHEE	LUTETIUM	LUTETIUM	LU	LYUTETSIJ	LUTECIO	LUTEZIO
2608 LUTETIUM-HAFNIUM DATING ISOT	LU/HF	LU=HF	LU=HF= DATIERUNG	METOD LYUTETSIEVO- GAFNIEVYJ	LU=HF	LU/HF
2609 LUVISOL SUSS	Luvisols ALFISOLS	Luvisol ALFISOL	Luvisol ALFISOL	LUVISOL	Luvisol ALFISOL	Luvisol ALFISOL
2610 LYCOPODIALES PALS	LYCOPSIDA	LYCOPSIDA	LYCOPODIALES	LYCOPSIDA *UF:* Noeggerathiales	LYCOPSIDA	LYCOPODIALES
2611 LYTOCERATIDA PALS	LYTOCERATIDA	LYTOCERATIDA	LYTOCERATIDA	LYTOCERATINA	LYTOCERATIDA	LYTOCERATIDA
2612 MAAR SURF	MAARS	MAAR	MAAR	Maar KRATER	MAAR	MAAR
2613 MACERAL SEDI	MACERALS *UF:* Clarain; Durain; Exinite; Fusinite; Vitrain	MACERAL *UF:* Clarain; Durain; Exinite; Fusinite; Vitrain	MAZERAL *UF:* Clarain; Durain; Exinit; Fusinit; Glanz= Kohle	MATSERAL	MACERAL *UF:* Clareno; Dureno; Exinita; Fusinita; Vitreno	MACERALE *UF:* Clarain; Durite; Exinite; Fusinite; Vitrina

	ENGLISH	FRANCAIS	DEUTSCH	RUSSKIJ	ESPANOL	ITALIANO
2614 MACHAERIDIA PALS	MACHAERIDIA	MACHAERIDIA	MACHAERIDIA	MACHAERIDIA	MACHAERIDIA	MACHAERIDIA
2615 MAESTRICHTIAN STRS	MAESTRICHTIAN	MAESTRICHTIEN	MAASTRICHT	MAASTRIKHT	MAESTRICHTIENSE	MAASTRICHTIANO
2616 MAFIC COMPOSITION IGNE	MAFIC COMPOSI- TION *UF:* Cafemic composi- tion; Femic	COMPOSITION MAFIQUE *UF:* Roche mafique	MAFISCH. CHEMISMUS *UF:* Femisch.Chemismus	OSNOVNOJ	COMPOSICION= MAFICA *UF:* Composicion= femica	COMPOSIZIONE MAFICA *UF:* Composizione femica
2617 MAGMA IGNE	MAGMAS *UF:* Comagmatic materi- als; Epimagmatic origin; Magmatic origin; Plutonism; Pyrosphere; Residual magma; Sial	MAGMA *UF:* Genese comagma- tique; Magma residuel; Origine magmatique; Pyrosphere; Sial; Sima	MAGMA *UF:* Comagmat.Genese; Magmat.Entstehung; Pyrosphaere; Rest= Magma; Sial; Sima	MAGMA	MAGMA *UF:* Comagmatico; Magma=residual; Origen=magmatico; Pirosfera; Sial; Sima	MAGMA *UF:* Comagmatico; Magma residuo; Ori- gine magmatica; Pirosf- era; Sial; Sima
2618 MAGMA CHAMBER IGNE	MAGMA CHAM- BERS *UF:* Roof collapse	CHAMBRE MAG- MATIQUE *UF:* Ecroulement toit	MAGMAKAMMER *UF:* Magmakammer= Einbruch	KAMERA	CAMARA= MAGMATICA *UF:* Hundimiento=techo	CAMERA MAGMA- TICA *UF:* Crollo di tetto
2619 MAGMATIC IGNE	Magmatic(adj.)	Magmatique(adj.)	Magmatisch(adj.)	MAGMATICHESKIJ *UF:* Fatziya magmati- cheskikh porod; Mag- matic association; Magmatic controls; Magmatic origin	Magmatico(adj.)	Magmatico(adj.)
2620 MAGMATIC ASSOCIATION IGNE	MAGMATIC ASSO- CIATIONS *UF:* Consanguinity; Pacific suite	ASSOCIATION MAGMATIQUE *UF:* Consanguinite; Ligne marshall; Serie arctique; Serie atlan- tique; Serie pacifique	MAGMAT. VERGESELLSCHAF- TUNG *UF:* Andesit=Linie; Arkt.Abfolge; Atlant. Abfolge; Konsanguinit- aet; Pazif.Abfolge	[MAGMATIC ASSO- CIATION] KOMPLEKS, MAG- MATICHESKIJ	ASOCIACION= MAGMATICA *UF:* Consanguineidad; Linea=de=la=andesita; Serie=artica; Serie= atlantica; Serie= pacifica	ASSOCIAZIONE MAGMATICA *UF:* Consanguineita; Linea delle andesiti; Serie artica; Serie atlantica; Serie pacifica
2621 MAGMATIC CONTROLS ECON	IGNEOUS PRO- CESSES	CONTROLE IGNE GITE	MAGMAT. EINFLUSS	[MAGMATIC CON- TROLS] KONTROL'; MAG- MATICHESKIJ	YACIMIENTO= IGNEO	CONTROLLO IGNEO GIACI- MENTO
2622 MAGMATIC ORE DEPOSIT ECON	INTRAMAGMATIC DEPOSITS	GITE INTRAMAG- MATIQUE	LIQUID=MAGMAT. VORKOMMEN	MESTOROZHDENIE MAGMATI- CHESKOE *UF:* Ehpitermal'noe mestor- ozhdenie; Katathermal'noe mestorozhdenie	YACIMIENTO= INTRAMAGMATICO	GIACIMENTO INTRAMAGMA- TICO
2623 MAGMATIC ORIGIN IGNE	Magmatic origin MAGMAS	Origine magmatique MAGMA	Magmat.Entstehung MAGMA	[MAGMATIC ORI- GIN] MAGMATICHESKIJ	Origen=magmatico MAGMA	Origine magmatica MAGMA
2624 MAGMATISM IGNE	Magmatism IGNEOUS ACTIV- ITY	MAGMATISME *UF:* Activite ignee; Genese roche ignee; Plutonisme	MAGMATISMUS *UF:* Magmat.Aktivitaet; Magmatit=Genese; Plutonismus	MAGMATIZM *UF:* Izverzhenie (aktiv.)	MAGMATISMO *UF:* Actividad=ignea; Genesis=roca=ignea; Plutonismo	MAGMATISMO *UF:* Attivita magmatica; Genesi di roccia intru- siva; Plutonismo
2625 MAGNAFACIES SEDI	Magnafacies LITHOFACIES	Magnafacies LITHOFACIES	Magna=Fazies LITHO=FAZIES	Magnafatsiya MAKROFATSIYA	Magnafacies LITOFACIES	Magnafacies LITOFACIES

	ENGLISH	FRANCAIS	DEUTSCH	RUSSKIJ	ESPANOL	ITALIANO
2626 MAGNESITE COMS	MAGNESITE	MAGNESITE	MAGNESIT	MAGNEZIT	MAGNESITA	MAGNESITE
2627 MAGNESIUM CHEE	MAGNESIUM	MAGNESIUM	MG	MAGNIJ	MAGNESIO	MAGNESIO
2628 MAGNETIC ANOMALY APPL	MAGNETIC ANO- MALIES	ANOMALIE MAGNETIQUE	MAGNET. ANOMALIE	ANOMALIYA MAGNITNAYA	ANOMALIA= MAGNETICA	ANOMALIA MAGNETICA
2629 MAGNETIC AZIMUTH SOLI	Magnetic azimuth MAGNETIC FIELD	Azimut magnetique CHAMP MAGNE- TIQUE	Magnet.Azimut MAGNETFELD	AZIMUT MAGNIT- NYJ	Azimut=magnetico CAMPO- MAGNETICO	Azimut magnetico CAMPO MAGNE- TICO
2630 MAGNETIC COMPASS INST	Magnetic compass MAGNETIC FIELD; INSTRUMENTS	Boussole INSTRUMENTA- TION; ORIENTA- TION	Magnet=Kompass INSTRUMEN- TIERUNG; ORIEN- TIERUNG	KOMPAS (MAGNITNYJ)	Brujula INSTRUMENTA- CION; ORIENTA- CION	Bussola magnetica STRUMENTAZ- IONE; ORIENTA- MENTO
2631 MAGNETIC DOMAIN MINE	MAGNETIC DOMAINS *UF:* Domain structure	DOMAINE MAGNE- TIQUE *UF:* Structure domaine	MAGNET.BEREICH *UF:* Domaenen= Struktur	DOMEN	DOMINIO= MAGNETICO *UF:* Dominio= estructural	DOMINIO MAGNE- TICO *UF:* Struttura a dominii
2632 MAGNETIC ELEMENT PHCH	Magnetic element MAGNETIC PROP- ERTIES	Element magnetique PROPRIETE MAGNETIQUE	Magnet.Element MAGNET. EIGENSCHAFT	EHLEMENT MAGNETIZMA	Elemento=magnetico PROPIEDAD= MAGNETICA	Elemento magnetico PROPRIETA MAGNETICA

2633 MAGNETIC FIELD
SOLI
	MAGNETIC FIELD
	UF: Geomagnetism;
	Isogon; Magnetic
	azimuth; Magnetic
	compass; Magnetic
	field variation; Mag-
	netic storms
		CHAMP MAGNE-
		TIQUE
		UF: Azimut magnetique;
		Geomagnetisme; Iso-
		cline; Isogone; Magne-
		tisme; Theorie dynamo;
		Variation champ
		magnetique
			MAGNETFELD
			UF: Dynamo=Theorie;
			Geomagnetismus;
			Isogone; Isokline;
			Magnet.Azimut;
			Magnet.Feldvariation;
			Magnetik
				POLE MAGNITNOE
				UF: Reversal
					CAMPO=
					MAGNETICO
					UF: Azimut=magnetico;
					Geomagnetismo; Iso-
					gona; Linea=isoclina;
					Magnetismo; Teoria=
					de=la=dinamo;
					Variacion=campo=
					magnetico
						CAMPO MAGNE-
						TICO
						UF: Azimut magnetico;
						Geomagnetismo; Iso-
						gona; Linea isoclina;
						Magnetismo; Teoria
						della dinamo; Variaz-
						ione di campo magne-
						tico

2634 MAGNETIC FIELD VARIATION SOLI	Magnetic field variation MAGNETIC FIELD; VARIATIONS	Variation champ magnetique CHAMP MAGNE- TIQUE	Magnet.Feldvariation MAGNETFELD	VARIATSII MAGNITNYE	Variacion=campo= magnetico CAMPO= MAGNETICO	Variazione di campo magnetico CAMPO MAGNE- TICO
2635 MAGNETIC INCLINATION SOLI	INCLINATION	INCLINAISON MAGNETIQUE	MAGNET. INKLINATION	NAKLONENIE MAGNITNOE	INCLINACION= MAGNETICA	INCLINAZIONE MAGNETICA
2636 MAGNETIC MAP SOLI	Magnetic map MAGNETIC SUR- VEY MAPS	Carte magnetique CARTE GEO- MAGNETIQUE	Magnet.Karte GEOMAGNET. KARTE	KARTA MAGNIT- NAYA *UF:* Magnetic survey map	Mapa=magnetico MAPA= GEOMAGNETICO	Carta magnetica CARTA GEO- MAGNETICA
2637 MAGNETIC METHOD APPL	MAGNETIC METH- ODS *UF:* Archaeomagnetism	METHODE MAGNE- TIQUE	MAGNET. METHODE	MAGNITORAZ- VEDKA	METODO= MAGNETICO	METODO MAGNE- TICO
2638 MAGNETIC MINERALS MINE	MAGNETIC MINER- ALS	MINERAL MAGNE- TIQUE	MAGNET.MINERAL	MINERAL FERRO- MAGNITNYJ	MINERAL= MAGNETICO	MINERALE MAGNETICO
2639 MAGNETIC PERMEABILITY PHCH	Magnetic permeability MAGNETIC PROP- ERTIES	Permeabilite magne- tique PROPRIETE MAGNETIQUE	Magnet.Permeabilitaet MAGNET. EIGENSCHAFT	PRONITSAEMOST' MAGNITNAYA	Permeabilidad= magnetica PROPIEDAD= MAGNETICA	Permeabilita magnetica PROPRIETA MAGNETICA

	ENGLISH	FRANCAIS	DEUTSCH	RUSSKIJ	ESPANOL	ITALIANO
2640 MAGNETIC PROPERTY PHCH	MAGNETIC PROP-ERTIES UF: Diamagnetic property; Ferromagnetic property; Magnetic element; Magnetic permeability; Magnetostriction; Paramagnetic property	PROPRIETE MAGNETIQUE UF: Element magnetique; Magnetostriction; Permeabilite magnetique; Propriete diamagnetique; Propriete ferromagnetique; Propriete paramagnetique	MAGNET. EIGENSCHAFT UF: Diamagnet. Eigenschaft; Ferromagnet. Eigenschaft; Magnet. Element; Magnet. Permeabilitaet; Magnetostriktion; Paramagnet. Eigenschaft	SVOJSTVO MAGNITNOE	PROPIEDAD= MAGNETICA UF: Elemento= magnetico; Magnetoestriccion; Permeabilidad= magnetica; Propiedad= diamagnetica; Propiedad= ferromagnetica; Propiedad= paramagnetica	PROPRIETA MAGNETICA UF: Elemento magnetico; Magnetostrizione; Permeabilita magnetica; Proprieta diamagnetica; Proprieta ferromagnetica; Proprieta paramagnetica
2641 MAGNETIC SPHERULE EXTR	Magnetic spherules COSMIC DUST	Spherule magnetique POUSSIERE COSMIQUE	Magnet.Kugel KOSM.STAUB	SHARIKI MAGNIT-NYE	Polvo=magnetico POLVO=COSMICO	Sferula magnetica POLVERE COSMICA
2642 MAGNETIC STORM SOLI	Magnetic storms MAGNETIC FIELD	Orage magnetique VARIATION SAI-SONNIERE	Magnet.Sturm JAHRESZEITL. SCHWANKUNG	BURYA MAGNIT-NAYA	Tormenta=magnetica VARIACION= ESTACIONARIA	Tempesta magnetica VARIAZIONE STA-GIONALE
2643 MAGNETIC SURVEY APPL	MAGNETIC SUR-VEYS UF: Aeromagnetic survey	LEVE MAGNE-TIQUE UF: Leve aeromagnetique	MAGNET. VERMESSUNG UF: Aeromagnet. Vermessung	S'EMKA MAGNIT-NAYA	LEVANTAMIENTO= MAGNETICO UF: Prospeccion= aeromagnetica	RILEVAMENTO MAGNETICO UF: Rilevamento aero-magnetico
2644 MAGNETIC SURVEY MAP SOLI	MAGNETIC SUR-VEY MAPS UF: Magnetic map	CARTE GEO-MAGNETIQUE UF: Carte magnetique	GEOMAGNET. KARTE UF: Magnet.Karte	KARTA GEO-MAGNITNAYA	MAPA= GEOMAGNETICO UF: Mapa=magnetico	CARTA GEO-MAGNETICA UF: Carta magnetica
2645 MAGNETIC SUSCEPTIBILITY PHCH	MAGNETIC SUS-CEPTIBILITY	SUSCEPTIBILITE MAGNETIQUE	MAGNET. SUSZEPTIBILITAET	VOSPRIIMCHIVOST' MAGNITNAYA	SUSCEPTIBILIDAD= MAGNETICA	SUSCETTIVITA MAGNETICA
2646 MAGNETISM SOLI	Magnetism MAGNETIZATION	Magnetisme CHAMP MAGNE-TIQUE	Magnetik MAGNETFELD	Magnetizm GEOMAGNETIZM	Magnetismo CAMPO= MAGNETICO	Magnetismo CAMPO MAGNE-TICO
2647 MAGNETITE COMS	MAGNETITE	MAGNETITE	MAGNETIT	MAGNETIT	MAGNETITA	MAGNETITE
2648 MAGNETIZATION PHCH	MAGNETIZATION UF: Magnetism	AIMANTATION	MAGNETISIERUNG	NAMAGNICHEN-NOST'	IMANTACION	MAGNETIZZAZ-IONE
2649 MAGNETOHYDRODY-NAMICS SOLI	MAGNETOHYDRO-DYNAMICS UF: Dynamo theory	MAGNETOHYDRO-DYNAMIQUE	MAGNETO= HYDRODYNAMIK	MAGNITOGIDRODI-NAMIKA	MAGNETOHID-RODINAMICA	MAGNETOIDRODI-NAMICA
2650 MAGNETOMETER INST	MAGNETOMETERS	MAGNETOMETRE	MAGNETOMETER	MAGNITOMETR UF: Remagnetizatsiya	MAGNETOMETRO	MAGNETOMETRO
2651 MAGNETOSPHERE SOLI	MAGNETOSPHERE	MAGNETOSPHERE	MAGNE-TOSPHAERE	MAGNITOSFERA	MAGNETOSFERA	MAGNETOSFERA
2652 MAGNETOSTRATIGRA-PHY STRA	MAGNETOSTRATI-GRAPHY	MAGNETOSTRATI-GRAPHIE	MAGNETOSTRATI-GRAPHIE	STRATIGRAFIYA PALEOMAGNIT-NAYA	MAGNETOESTRA-TIGRAFIA	STRATIGRAFIA MAGNETICA
2653 MAGNETOSTRICTION PHCH	Magnetostriction MAGNETIC PROP-ERTIES	Magnetostriction PROPRIETE MAGNETIQUE	Magnetostriktion MAGNET. EIGENSCHAFT	MAGNITOSTRIKT-SIYA	Magnetoestriccion PROPIEDAD= MAGNETICA	Magnetostrizione PROPRIETA MAGNETICA

	ENGLISH	FRANCAIS	DEUTSCH	RUSSKIJ	ESPANOL	ITALIANO
2654 MAGNETOTELLURIC METHOD APPL	MAGNETOTELLU-RIC METHODS	METHODE MAGNE-TOTELLURIQUE	MAGNETO=TELLUR.METHODE	METOD MAGNITO-TELLURICHESKIJ *UF:* S'emka magnitotel-luricheskaya	METODO=MAGNETOTELURICO	METODO MAGNE-TOTELLURICO
2655 MAGNETOTELLURIC SURVEY APPL	MAGNETOTELLU-RIC SURVEYS	LEVE MAGNETO-TELLURIQUE	MAGNETO=TELLUR. VERMESSUNG	S'emka magnitotelluri-cheskaya METOD MAGNITO-TELLURICHESKIJ	PROSPECCION=MAGNETOTELURICA	RILEVAMENTO MAGNETOTEL-LURICO
2656 MAJIAN STRS	Majian MIDDLE CAM-BRIAN	Mayien CAMBRIEN MOYEN	Majium M.KAMBRIUM	MAJSKIJ	Majiense CAMBRICO=MEDIO	Majiano CAMBRIANO MED.
2657 MAJOR ELEMENT CHES	MAJOR ELEMENTS	ANALYSE MAJEURS	HAUPTELEMENT=ANALYSE	EHLEMENT GLAV-NYJ	ANALISIS=MAYORES	ELEMENTI MAGGI-ORI
2658 MAJOR-ELEMENT ANALYSIS METH	MAJOR=ELEMENT ANALYSES	METHODE ANA-LYSE MAJEURS	HAUPTELEMENT=ANALYSEN=METHODIK	ANALIZ GLAV-NYKH EHLEMEN-TOV	METODO=ANALISIS=MAYORES	ANALISI ELE-MENTI MAGGIORI
2659 MALACOSTRACA PALS	MALACOSTRACA *UF:* Decapoda	Malacostraca DECAPODA	Malacostraca DECAPODA	MALACOSTRACA *UF:* Decapoda	Malacostraca DECAPODA	Malacostraca DECAPODA
2660 MALLEABLE CONSISTENCE PHCH	Malleable materials PLASTICITY	Substance malleable PLASTICITE	Duktil.Substanz PLASTIZITAET	KOVKIJ	Maleable PLASTICIDAD	Malleabilita PLASTICITA
2661 MAMMALIA PALS	MAMMALIA *UF:* Edentata	MAMMALIA *UF:* Edentata	MAMMALIA *UF:* Edentata	MAMMALIA *UF:* Docodonta; Mars-upialia; Pantotheria; Symmetrodonta; Tri-conodonta	MAMMALIA *UF:* Edentata	MAMMALIA *UF:* Edentata
2662 MANGANESE CHEE	MANGANESE *UF:* Dendrite; Kanga	MANGANESE *UF:* Dendrite; Kanga	MN *UF:* Dendrit; Kanga	MARGANETS *UF:* Ferromanganese composition	MANGANESO *UF:* Dendrita; Kanga	MANGANESE *UF:* Dendrite; Kanga
2663 MANGROVE SWAMP SURF	MANGROVE SWAMPS	MARAIS MAN-GROVE	MANGROVE=SUMPF	MANGR	MANGLAR	PALUDE A MAN-GROVIA
2664 MANPOWER MISC	MANPOWER *UF:* Geophysicists	MAIN D'OEUVRE *UF:* Geophysicien	ARBEITSWESEN *UF:* Geophysiker	SILA RABOCHAYA	MANO=DE=OBRA *UF:* Geofisico	MANO D'OPERA *UF:* Geofisico
2665 MANTLE SOLI	MANTLE *UF:* Birch discontinuity	MANTEAU *UF:* Discontinuite Birch	ERD=MANTEL *UF:* Birch=Diskontinuitaet	MANTIYA *UF:* Bercha poverkhnost'	MANTO=GLOBO *UF:* Discontinuidad=de=Birch	MANTELLO *UF:* Discontinuita di Birch
2666 MANTODEA PALS	Mantodea INSECTA	Mantodea INSECTA	Mantodea INSECTA	MANTODEA	Mantodea INSECTA	Mantodea INSECTA
2667 MANUAL MISC	MANUALS	MANUEL	HANDBUCH	SPRAVOCHNIK	MANUAL	MANUALE
2668 MAP GEOL	MAPS *UF:* Biofacies map; Legend; Small scale	CARTE *UF:* Legende	KARTE *UF:* Legende	KARTA *UF:* Hydraulic map; Karta biofatsial'naya; Karta geomorfologi-cheskaya; Karta gidrologicheskaya; Karta inzhenerno=geologicheskaya; Karta pochvennaya; Karta registratsionnaya; Water map	MAPA *UF:* Leyenda	CARTA *UF:* Legenda

	ENGLISH	FRANCAIS	DEUTSCH	RUSSKIJ	ESPANOL	ITALIANO
2669 MAP SCALE MISC	Cartographic scales CARTOGRAPHY	Echelle cartographique CARTOGRAPHIE	Karten=Massstab KARTOGRAPHIE	MASSHTAB	Escala CARTOGRAFIA	Scala CARTOGRAFIA
2670 MAPPING METH	Mapping CARTOGRAPHY	Leve carte CARTOGRAPHIE	Kartierung KARTOGRAPHIE	KARTIROVANIE	Levantamiento=cartografico CARTOGRAFIA	Rilevamento cartografico CARTOGRAFIA
2671 MARBLE IGMS	MARBLES *UF:* Calciphyre; Cipolin	MARBRE *UF:* Calciphyre; Cipolin	MARMOR *UF:* Cipolin; Kalziphyr	MRAMOR	MARMOL *UF:* Calcifido; Cipolino	MARMO *UF:* Calcefiro; Cipollino
2672 MARE EXTR	MARIA *UF:* Thalassoid	MER LUNAIRE *UF:* Bassin lunaire	MOND=MARE *UF:* Thalassoid	MORE LUNNOE	MAR=LUNAR *UF:* Thalassoide	MARE LUNARE *UF:* Thalassoide
2673 MARGINAL PLATEAU MARI	Marginal plateau MARGINAL SEAS	Plateforme marginale MER BORDIERE	Rand=Plateau RAND=MEER	PLATO KRAEVOE	Plataforma=marginal MAR=MARGINAL	Piattaforma marginale MARE MARGINALE
2674 MARGINAL SEA MARI	MARGINAL SEAS *UF:* Marginal plateau	MER BORDIERE *UF:* Mer plateau continental; Plateforme marginal	RAND=MEER *UF:* Rand=Plateau; Schelf=Meer	MORE OKRAINNOE	MAR=MARGINAL *UF:* Mar=plataforma=continental; Plataforma=marginal	MARE MARGINALE *UF:* Mare di piattaforma; Piattaforma marginale
2675 MARGINATION TEXTURE TEST	Margination texture TEXTURES	Texture marge cristaux TEXTURE	Rand=Gefuege KORN=GEFUEGE	Tekstura marginatsionnyj TEKSTURA	Textura=marginal TEXTURA	Tessitura marginale TESSITURA
2676 MARINE ECOLOGY PALE	Marine ecology MARINE ENVIRONMENT; ECOLOGY	Ecologie marine MILIEU MARIN; ECOLOGIE	Marin.Oekologie MARIN.MILIEU; OEKOLOGIE	Ehkologiya morskaya MORSKOJ; EHKOLOGIYA	Ecologia=marina MEDIO=MARINO; ECOLOGIA	Ecologia marina AMBIENTE MARINO; ECOLOGIA
2677 MARINE ENVIRONMENT PALE	MARINE ENVIRONMENT *UF:* Marine ecology; Marine erosion; Seas; Submarine volcano	MILIEU MARIN *UF:* Ecologie marine; Erosion marine; Mer; Vallee sous marine; Volcan sous marin	MARIN.MILIEU *UF:* Marin.Erosion; Marin.Oekologie; Meer; Submarin.Tal; Submarin.Vulkan	MORSKOJ *UF:* Ehkologiya morskaya; Geomorfologiya podvodnaya; Marine installation; Marine methods; Marine transport; Submarine installation	MEDIO=MARINO *UF:* Ecologia=marina; Erosion=marina; Mar; Valle=submarino; Volcan=submarino	AMBIENTE MARINO *UF:* Ecologia marina; Erosione marina; Mare; Valle sottomarina; Vulcano sottomarino
2678 MARINE EROSION SURF	Marine erosion EROSION; MARINE ENVIRONMENT	Erosion marine EROSION; MILIEU MARIN	Marin.Erosion EROSION; MARIN. MILIEU	EHROZIYA MORSKAYA	Erosion=marina EROSION; MEDIO=MARINO	Erosione marina EROSIONE; AMBIENTE MARINO
2679 MARINE GEOLOGY MARI	MARINE GEOLOGY	GEOLOGIE MARINE	MEERES=GEOLOGIE	GEOLOGIYA MORSKAYA	GEOLOGIA=MARINA	GEOLOGIA MARINA
2680 MARINE INSTALLATION ENGI	MARINE INSTALLATIONS	INSTALLATION MARINE	MEERES=ANLAGE	[MARINE INSTALLATION] MORSKOJ; SOORUZHENIE	INSTALACION=MARINA	INSTALLAZIONE MARINA
2681 MARINE MARSH SURF	Marine marsh SALT MARSHES	Marais marin MARAIS LITTORAL	Salz=Marsch KUESTEN=SALZ=MARSCH; KUESTEN=EBENE	Boloto morskoe MARSH	Pantaño=marino LAGO=LITORAL	Palude marina PALUDE LITORALE
2682 MARINE METHODS APPL	MARINE METHODS	ETUDE EN MER	OZEANOGRAPH. METHODE	[MARINE METHODS] MORSKOJ; METOD GEOFIZICHESKIJ	ESTUDIO=EN=MAR	STUDIO IN MARE

	ENGLISH	FRANCAIS	DEUTSCH	RUSSKIJ	ESPANOL	ITALIANO
2683 MARINE PLAIN SURF	Marine plains COASTAL PLAINS	Plaine marine PLAINE COTIERE	Marin.Ebene KUESTEN=EBENE	RAVNINA BERE- GOVAYA	Llanura=de=abrasion LLANURA= COSTERA	Piana d'origine marina PIANA COSTIERA
2684 MARINE PLATFORM ENGI	MARINE PLAT- FORMS	PLATEFORME MARINE *UF:* Puits offshore	MARIN. PLATTFORM *UF:* Offshore=Bohrloch	PLATFORMA BUROVAYA	PLATAFORMA= MARINA *UF:* Perforacion=off= shore	PIATTAFORMA MARINA *UF:* Pozzo offshore
2685 MARINE SEDIMENT SEDI	MARINE SEDI- MENTS	SEDIMENT MARIN	MARIN.SEDIMENT	OSADOK MORSKOJ	SEDIMENTO= MARINO	SEDIMENTO MARINO
2686 MARINE SEDIMENTATION SEDI	MARINE SEDIMEN- TATION	SEDIMENTATION MARINE	MARIN. SEDIMENTATION	SEDIMENTATSIYA MORSKAYA *UF:* Sedimentatsiya abissal'naya; Sreda sublitoral'naya	SEDIMENTACION= MARINA	SEDIMENTAZIONE MARINA
2687 MARINE SILL MARI	Marine sills OCEAN FLOORS	Seuil marin RELIEF SOUS MARIN	Meeres=Furche SUBMARIN.RELIEF	POROG PODVOD- NYJ	Umbral RELIEVE= SUBMARINO	Solco sottomarino RILIEVO SOTTO- MARINO
2688 MARINE TERRACE SURF	MARINE TER- RACES *UF:* Berms	TERRASSE MARINE *UF:* Banquette plage	STRAND- TERRASSE *UF:* Berme	PRIBREZHNO MORSKOJ	TERRAZA= MARINA *UF:* Terraza=costera	TERRAZZO MARINO *UF:* Berma
2689 MARINE TRANSPORT SEDI	MARINE TRANS- PORT	TRANSPORT MARIN	MARIN. TRANSPORT	[MARINE TRANS- PORT] MORSKOJ; TRANS- PORTIROVKA (GEOL)	TRANSPORTE= MARINO	TRASPORTO MARINO
2690 MARINER EXTR	MARINER PRO- GRAM	PROGRAMME MARINER	MARINER= PROGRAMM	MARINER (SPUTNIK)	MARINER	MARINER
2691 MARKOV PROCESS MATH	MARKOV CHAIN ANALYSIS	CHAINE MARKOV	MARKOV=KETTE	MARKOVSKIJ PROTSESS	CADENA=DE= MARKOV	CATENA DI MARKOV
2692 MARL SEDS	MARL	MARNE	MERGEL	MERGEL' *UF:* Opoka	MARGA	MARNA
2693 MARSH SURF	MARSHES	MARAIS LITTORAL *UF:* Marais marin	MARSCH	MARSH *UF:* Boloto morskoe; Salt marsh	LAGO=LITORAL *UF:* Pantaño=marino	PALUDE LITORALE *UF:* Palude marina
2694 MARSH GAS CHES	Marsh gas METHANE	Gaz marais METHANE	Sumpfgas METHAN	GAZ BOLOTNYJ	Gas=pantanos METANO	Gas di palude METANO
2695 MARSUPIALIA PALS	MARSUPIALIA	MARSUPIALIA	MARSUPIALIA	Marsupialia MAMMALIA	MARSUPIALIA	MARSUPIALIA
2696 MASCON EXTR	MASCONS	MASCON	MASCON	MASKON	MASCON	MASCON
2697 MASS PHCH	MASS	MASSE	MASSE	MASSA	MASA	MASSA
2698 MASS BALANCE SURF	MASS BALANCE	BILAN MASSE	MASSEN=BILANZ	[MASS BALANCE] BALANS; LED	BALANCE=MASA	BILANCIO DI MASSA
2699 MASS MOVEMENT SURF	MASS MOVEMENTS *UF:* Mass wasting	MOUVEMENT MASSE *UF:* Mouvement matiere	MASSEN= BEWEGUNG *UF:* Massen=Abfall	PEREMESHCHENIE GRAVITATSION- NOE	MOVIMIENTO= MASA *UF:* Transporte=en= masas	MOVIMENTO DI MASSA *UF:* Disgregazione in massa

	ENGLISH	FRANCAIS	DEUTSCH	RUSSKIJ	ESPANOL	ITALIANO
2700 MASS SPECTROSCOPY METH	MASS SPECTROS-COPY	SPECTROMETRIE MASSE	MASSEN=SPEKTROMETRIE	MASS=SPEKTROSKOPIYA	ESPECTROME-TRIA=DE=MASA	SPETTROMETRIA DI MASSA
2701 MASS TRANSFER GEOL	MASS TRANSFER	TRANSFERT MASSE	MASSEN=TRANSFER	[MASS TRANSFER] MATERIAL; TRAN-SPORTIROVKA (GEOL)	TRANSFERENCIA=DE=MASA	TRASFERIMENTO DI MASSA
2702 MASS WASTING SURF	Mass wasting MASS MOVEMENTS	Mouvement matiere MOUVEMENT MASSE	Massen=Abfall MASSEN=BEWEGUNG	OSYP'	Transporte=en=masas MOVIMIENTO=MASA	Disgregazione in massa MOVIMENTO DI MASSA
2703 MASSIVE TEST	Massive(adj.)	Massif(adj.)	Massiv(adj.)	MASSIVNYJ *US:* Massive bedding	Masivo(adj.)	Massivo(adj.)
2704 MASSIVE BEDDING SEDI	MASSIVE BEDDING	STRATIFICATION MASSIVE	BANKUNG	[MASSIVE BED-DING] MASSIVNYJ; SLOISTOST'	ESTRATIFIC-ACION=MASIVA	STRATIFICAZIONE MASSICCIA
2705 MASSIVE DEPOSIT ECON	MASSIVE DEPOSITS	AMAS MINERALISE	MASSIG. ERZKOERPER	MESTOROZHDENIE MASSIVNOE	MASA=MINERALIZADA	AMMASSO MINER-ALIZZATO
2706 MASTIGOPHORA PALS	Mastigophora PROTISTA	Mastigophora PROTOZOA	Mastigophora PROTOZOA	MASTIGOPHORA	Mastigophora PROTOZOA	Mastigophora PROTOZOA
2707 MATERIAL GEOL	Materials(gen.)	Materiau(gen.)	Material(gen.)	MATERIAL *UF:* Mass transfer; Tekhnologicheskij pepel	Material(gen.)	Materiali(gen.)
2708 MATHEMATICAL ANALYSIS MATH	Mathematical analysis MATHEMATICAL METHODS	Analyse mathematique METHODE MATHE-MATIQUE	Mathemat.Analyse MATHEMAT. METHODE	ANALIZ MATEMA-TICHESKIJ *UF:* Analiz dannykh	Analisis=matematico METODO=MATEMATICO	Analisi matematica METODO MATEMA-TICO
2709 MATHEMATICAL EXPECTATION MATH	Expectation MATHEMATICAL METHODS	Prevision probabilite METHODE MATHE-MATIQUE	Erwartung MATHEMAT. METHODE	OZHIDANIE MATE-MATICHESKOE	Prevision=probabilidad METODO=MATEMATICO	Speranza matematica METODO MATEMA-TICO
2710 MATHEMATICAL GEOGRAPHY MATH	Mathematical geography MATHEMATICAL METHODS	Geographie mathema-tique METHODE MATHE-MATIQUE	Mathemat.Geographie MATHEMAT. METHODE	GEOGRAFIYA MATEMATI-CHESKAYA	Geografia matematica METODO=MATEMATICO	Geografia matematica METODO MATEMA-TICO
2711 MATHEMATICAL GEOLOGY MATH	MATHEMATICAL GEOLOGY *UF:* Mathematical law	GEOLOGIE MATHE-MATIQUE	MATHEMAT. GEOLOGIE	GEOLOGIYA MATE-MATICHESKAYA	GEOLOGIA=MATEMATICA	GEOLOGIA MATE-MATICA
2712 MATHEMATICAL LAW MATH	Mathematical law MATHEMATICAL GEOLOGY	Loi mathematique METHODE MATHE-MATIQUE	Mathemat.Gesetz MATHEMAT. METHODE	Zakon matematicheskij ZAKON; MATEMA-TIKA	Ley=matematica METODO=MATEMATICO	Legge matematica METODO MATEMA-TICO
2713 MATHEMATICAL MATRIX MATH	Mathematical matrix MATHEMATICAL METHODS	Matrice mathematique METHODE STATIS-TIQUE	Mathemat.Matrix STATIST.METHODE	MATRITSA *UF:* Tablitsa dannykh	Matriz=matematica METODO=ESTADISTICO	Matrice matematica METODO STATIS-TICO
2714 MATHEMATICAL METHODS MATH	MATHEMATICAL METHODS *UF:* Boolean algebra; Cybernetics; Expecta-tion; Extrapolation; Mathematical analysis; Mathematical geogra-phy; Mathematical matrix; Mathematics; Matrix algebra; System analysis; Vector; Weighting	METHODE MATHE-MATIQUE *UF:* Algebre booleenne; Analyse mathematique; Analyse systeme; Cyb-ernetique; Derivation; Geographie mathema-tique; Loi mathema-tique; Mathematique; Ponderation; Prevision probabilite; Vecteur	MATHEMAT. METHODE *UF:* Boole'sche=Algebra; Erwartung; Kybernetik; Mathemat.Analyse; Mathemat. Differenzierung; Mathemat.Geographie; Mathemat.Gesetz; Mathematik; Nae-herung; System=Analyse; Vektor; Wich-tung	Metod matematicheskij MATEMATIKA	METODO=MATEMATICO *UF:* Algebra=booleana; Analisis=matematico; Analisis=ponderado; Analisis=sistema; Cibernetica; Deriva-cion; Geografia=matematica; Ley=matematica; Matemati-cas; Prevision=probabilidad; Vector	METODO MATEMA-TICO *UF:* Algebra booleana; Analisi dei sistemi; Analisi matematica; Cibernetica; Derivata; Geografia matematica; Legge matematica; Matematica; Peso statistico; Speranza matematica; Vettore

	ENGLISH	FRANCAIS	DEUTSCH	RUSSKIJ	ESPANOL	ITALIANO
2715 MATHEMATICAL MODEL MATH	MATHEMATICAL MODELS *UF:* Derivative	MODELE MATHE-MATIQUE	MATHEMAT. MODELL	MODEL' MATEMA-TICHESKAYA *UF:* Geostatistika	MODELO=MATEMATICO	MODELLO MATE-MATICO
2716 MATHEMATICS MATH	Mathematics MATHEMATICAL METHODS	Mathematique METHODE MATHE-MATIQUE	Mathematik MATHEMAT. METHODE	MATEMATIKA *UF:* Metod matematicheskij; Systems analog; Zakon matematicheskij	Matematicas METODO=MATEMATICO	Matematica METODO MATEMA-TICO
2717 MATRIX ALGEBRA MATH	Matrix algebra MATHEMATICAL METHODS	Algebre matricielle METHODE STATIS-TIQUE	Matrix=Algebra STATIST.METHODE	ALGEBRA MATRICHNAYA	Algebra=matricial METODO=ESTADISTICO	Algebra matriciale METODO STATIS-TICO
2718 MATURITY GEOL	MATURITY *UF:* Maturity index	MATURATION *UF:* Indice maturite	REIFUNG *UF:* Reifegrad	ZRELOST'	MADURACION *UF:* Indice=de=madurez	MATURITA *UF:* Indice di maturita
2719 MATURITY INDEX SEDI	Maturity index MATURITY	Indice maturite MATURATION	Reifegrad REIFUNG	INDEKS ZRELOSTI	Indice=de=madurez MADURACION	Indice di maturita MATURITA
2720 MATUYAMA EPOCH STRA	MATUYAMA EPOCH	EPOQUE MATUYAMA	MATUYAMA=EPOCHE	MATUYAMA EHPOKHA	EPOCA=MATUYAMA	EPOCA MATUYAMA
2721 MAXILLOPODA PALS	Maxillopoda CRUSTACEA	Maxillopoda CRUSTACEA	Maxillopoda CRUSTACEA	MAXILLOPODA	Maxillopoda CRUSTACEA	Maxillopoda CRUSTACEA
2722 MAXIMUM MATH	Maximum STATISTICAL DIS-TRIBUTION	Maximum DISTRIBUTION STATISTIQUE	Maximum STATIST. VERTEILUNG	MAKSIMUM	Maximo DISTRIBUCION=ESTADISTICA	Massimo DISTRIBUZIONE STATISTICA
2723 MAZATSAL REVOLUTION STRU	Mazatsal revolution OROGENY; PRE-CAMBRIAN	Revolution Mazatsal OROGENIE ANTE-CAMBRIENNE	Mazatsal=Revolution PRAEKAMBR. OROGENESE	MAZATTSAL'SKIJ	Revolucion=Mazatsal OROGENIA=PRECAMBRICA	Rivoluzione Mazatsali-ana OROGENESI PRE-CAMBRIANA
2724 MEANDER SURF	MEANDERS	MEANDRE	MAEANDER	MEANDR	MEANDRO	MEANDRO
2725 MECHANICAL CONTROL ECON	MECHANICAL CONTROLS	GITE DETRITIQUE	DETRIT. LAGERSTAETTE	Kontrol' mekhanicheskij KONTROL'	YACIMIENTO=DETRITICO	GIACIMENTO DETRITICO
2726 MECHANICAL ORIGIN SEDI	Mechanical origin SEDIMENTATION	Origine mecanique SEDIMENTATION DETRITIQUE	Mechan.Ursprung DETRIT. SEDIMENTATION	MEKHANOGENEZ	Origen=mecanico SEDIMENTACION=DETRITICA	Origine meccanica SEDIMENTAZIONE DETRITICA
2727 MECHANICAL PROPERTY PHCH	MECHANICAL PROPERTIES	PROPRIETE MECANIQUE *UF:* Mineral antistress	MECHAN. EIGENSCHAFT *UF:* Antistress=Mineral	SVOJSTVO MEK-HANICHESKOE	PROPIEDAD=MECANICA *UF:* Mineral=antitensional	PROPRIETA MEC-CANICA *UF:* Minerale antistress
2728 MECHANICAL WEATHERING SURF	MECHANICAL WEATHERING	ALTERATION MECANIQUE	MECHAN. VERWITTERUNG	Vyvetrivanie mekhanicheskoe VYVETRIVANIE	ALTERACION=MECANICA	ALTERAZIONE MECCANICA
2729 MECHANICS PHCH	MECHANICS	MECANIQUE	MECHANIK	MEKHANIKA	MECANICA	MECCANICA
2730 MECHANISM GEOL	MECHANISM	Mecanisme(gen.)	Mechanismus(gen.)	MEKHANIZM	Mecanismo(gen.)	Meccanismo(gen.)

	ENGLISH	FRANCAIS	DEUTSCH	RUSSKIJ	ESPANOL	ITALIANO
2731 MECOPTEROIDA PALS	MECOPTEROIDA	MECOPTEROIDA	MECOPTEROIDA	MECOPTEROIDA	MECOPTEROIDA	MECOPTEROIDA
2732 MEDIAN MASS STRU	Median masses HORSTS	Massif median HORST	Zentral.Massiv HORST	MASSIV SREDIN- NYJ	Bloque=central HORST	Massiccio centrale HORST
2733 MEDICAL GEOLOGY GEOL	MEDICAL GEOL- OGY	MEDECINE	MEDIZIN	Geologiya meditsin- skaya GEOLOGIYA OKRUZHAYUSH- CHEJ SREDY	MEDICINA	GEOLOGIA SANI- TARIA
2734 MEDICINAL WATER GEOH	MEDICINAL WATERS	EAU THERMO- MINERALE UF: Eau hydrothermale	MINERALISIERT. THERMAL= WASSER	Voda lechebnaya VODA MINER- AL'NAYA	AGUA= TERMOMINERAL UF: Agua=hidrotermal	ACQUA CURATIVA UF: Acqua termominer- ale
2735 MEDITERRANEAN SOIL SUSS	MEDITERRANEAN SOILS	SOL MEDITERRAN- EEN UF: Terra rossa	MEDITERRAN. BODEN UF: Terra=rossa	POCHVA SRE- DIZEMNOMOR- SKAYA	SUELO= MEDITERRANEO UF: Terra=rossa	SUOLO MEDITER- RANEO UF: Terra rossa
2736 MEDIUM-GRAINED TEST	Medium=grained(adj.)	Grain moyen(adj.)	Mittelkoernig(adj.)	SREDNEZERNISTYJ	Grano=medio(adj.)	A grana media(adj.)
2737 MEETING MISC	Meeting SYMPOSIA	REUNION UF: Symposium	TAGUNG UF: Symposium	SOVESHCHANIE UF: Simpozium	REUNION UF: Simposio	RIUNIONE UF: Simposio
2738 MEGAFACIES SEDI	Megafacies LITHOFACIES	Megafacies LITHOFACIES	Mega=Fazies LITHO=FAZIES	MAKROFATSIYA UF: Magnafatsiya	Megafacies LITOFACIES	Megafacies LITOFACIES
2739 MEGASPORES PALE	MEGASPORES	MEGASPORE	MEGASPORE	MEGASPORA	MEGAESPORA	MEGASPORA
2740 MELANGE GEOL	MELANGE	MELANGE	MELANGE	MELANZH	ROCA=DE= MEZCLA	MELANGE
2741 MELANOCRATIC COMPOSITION IGNE	Melanocratic composi- tion IGNEOUS ROCKS	Composition melano- crate ROCHE IGNEE	Melanokrat. Zusammensetzung MAGMATIT	MELANOKRA- TOVYJ	Composicion= melanocratica ROCA=IGNEA	Composizione melanocratica ROCCIA IGNEA
2742 MELIORATION SURF	AMELIORATION	AMENDEMENT UF: Amendement sol	MELIORATION	MELIORATSIYA	MEJORA=SUELO	AMMENDAMENTO
2743 MELT IGNE	MELTS	MATIERE FONDUE	SCHMELZE UF: Bowen's= Reaktions=Serie; Kongruent.Schmelze; Latent.Schmelz= Waerme; Schmelz= Waerme	RASPLAV	MEZCLA=FUNDIDA	MATERIA FUSA
2744 MELTING PHCH	MELTING UF: Congruent melting	FUSION UF: Chaleur equivalente fusion; Equivalent calorifique fusion; Fusion congruente; Suite reactionnelle Bowen	SCHMELZEN	PLAVLENIE	FUSION UF: Calor=latente=de= fusion; Equivalente= calor=de=fusion; Mezcla=congruente; Series=de=reaccion= de-Bowen	FUSIONE UF: Calore latente di fusione; Equivalente calorico di fusione; Fusione congruente; Serie reazionale di Bowen
2745 MELTWATER GEOH	MELTWATER	EAU FONTE	SCHMELZ= WASSER	VODA TALAYA	AGUA=DE= FUNDICION	ACQUA DI FUSIONE

	ENGLISH	FRANCAIS	DEUTSCH	RUSSKIJ	ESPANOL	ITALIANO
2746 MEMBER STRA	Members STRATIGRAPHIC UNITS	Membre ECHELLE STRATI- GRAPHIQUE	Schichtglied STRATIGRAPH. SKALA	PACHKA	Miembro ESCALA= ESTRATIGRAFICA	Membro SCALA STRATI- GRAFICA
2747 MENDELEVIUM CHEE	MENDELEVIUM	MENDELEVIUM	MD	MENDELEVIJ	MENDELEVIO	MENDELEVIO
2748 MEOTIAN STRS	MEOTIAN	MEOTIEN	MAEOT	MEOTICHESKIJ	MEOTIENSE	MEOZIANO
2749 MERCURY CHEE	MERCURY	MERCURE	HG	RTUT'	MERCURIO	MERCURIO
2750 MEROSTOMATA PALS	MEROSTOMATA *UF:* Gigantostraca	MEROSTOMATA *UF:* Gigantostraca	MEROSTOMATA *UF:* Gigantostraca	MEROSTOMATA *UF:* Gigantostraca	MEROSTOMATA *UF:* Gigantostraca	MEROSTOMATA *UF:* Gigantostraca
2751 MEROSTOMOIDEA PALS	Merostomoidea ARTHROPODA	Merostomoidea ARTHROPODA	Merostomoidea ARTHROPODA	MEROSTOMOIDEA	Merostomoidea ARTHROPODA	Merostomoidea ARTHROPODA
2752 MESOCRATIC COMPOSITION IGNE	Mesocratic composition IGNEOUS ROCKS	Composition mesocrate ROCHE IGNEE	Mesokrat. Zusammensetzung MAGMATIT	MEZOKRATOVYJ	Composicion= mesocratica ROCA=IGNEA	Composizione mesocra- tica ROCCIA IGNEA
2753 MESOGAEA STRA	MESOGAEA	MESOGEE	MESOGAEA	MEZOGEA	MESOGEA	MESOGEA
2754 MESOGASTROPODA PALS	MESO- GASTROPODA	MESO- GASTROPODA	MESO- GASTROPODA	Mesogastropoda GASTROPODA	MESO- GASTROPODA	MESO- GASTROPODA
2755 MESOGEOSYNCLINE STRU	Mesogeosyncline GEOSYNCLINES	Mesogeosynclinal GEOSYNCLINAL	Mesogeosynklinale GEOSYNKLINALE	MEZOGEOSINKLI- NAL'	Mesogeosinclinal GEOSINCLINAL	Mesogeosinclinale GEOSINCLINALE
2756 MESOPHYLE ORGANISM PALE	Mesophyle organism ECOLOGY	Organisme mesophyle ECOLOGIE	Mesophyl.Organismus OEKOLOGIE	MEZOFILY	Organismo=mesofilo ECOLOGIA	Organismo mesofilo ECOLOGIA
2757 MESOSAURIA PALS	MESOSAURIA	MESOSAURIA	MESOSAURIA	MESOSAURIA	MESOSAURIA	MESOSAURIA
2758 MESOSIDERITE EXTS	Mesosiderites METEORITES	Mesosiderite SIDEROLITE	Mesosiderit STEIN=EISEN= METEORIT	Mezosiderit SIDEROLIT	Mesosiderita SIDEROLITO	Mesosiderite SIDEROLITE
2759 MESOTHERMAL DEPOSIT ECON	MESOTHERMAL PROCESSES	GITE MESOTHER- MAL	MESOTHERMAL. VORKOMMEN	MEZOTERMAL'NYJ	YACIMIENTO= MESOTERMAL	GIACIMENTO MESOTERMALE
2760 MESOZOIC STRS	MESOZOIC	SECONDAIRE	MESOZOIKUM	MEZOZOJ	MESOZOICO	SECONDARIO
2761 MESOZONE IGNE	MESOZONAL METAMORPHISM	METAMORPHISME MESOZONAL	MESOZONE	MEZOZONA	METAMORFISMO= MESOZONAL	METAMORFISMO MESOZONALE
2762 METABASALT IGMS	METABASALT	METABASALTE	META=BASALT	Metabazalt' BAZAL'T; METAM- ORFIZM	METABASALTO	METABASALTO
2763 METACOPINA PALS	Metacopina OSTRACODA	Metacopina OSTRACODA	Metacopina OSTRACODA	Metacopina PODOCOPIDA	Metacopina OSTRACODA	Metacopina OSTRACODA

	ENGLISH	FRANCAIS	DEUTSCH	RUSSKIJ	ESPANOL	ITALIANO
2764 METAGABBRO IGMS	METAGABBRO	METAGABBRO	META=GABBRO	Metagabbro GABBRO; METAM- ORFIZM	METAGABRO	METAGABBRO
2765 METAGENESIS SEDI	Metagenesis DIAGENESIS	Metagenese EPIGENESE	Metagenese EPIGENESE	METAGENEZ	Metagenesis EPIGENESIS	Metagenesi EPIGENESI
2766 METAL CHES	METALS	ELEMENT METAL- LIQUE	METALL	METALL *UF:* Rare metal	METAL	ELEMENTO MET- ALLICO
2767 METALLIC METEORITE EXTS	IRON METEORITES	METEORITE MET- ALLIQUE *UF:* Ataxite; Hexadrite; Octahedrite	METALL= METEORIT *UF:* Ataxit; Hexahedrit; Oktaehdrit	METEORIT ZHELEZNYJ *UF:* Ataksit; Geksaeh- drit; Oktaehdrit	METEORITO- METALICO *UF:* Ataxita; Exaedrita; Octaedrita	METEORITE MET- ALLICA *UF:* Ataxite; Esaedrite; Ottaedrite
2768 METALLIC PROPERTY ECON	Metallic properties OPTICAL PROPER- TIES	Propriete metallique PROPRIETE OPTIQUE	Metallisch.Eigenschaft OPT.EIGENSCHAFT	METALLICHESKIJ	Metalico PROPIEDAD= OPTICA	Proprieta metallica PROPRIETA OTTICA
2769 METALLIFEROUS ECON	Metalliferous(adj.)	Metallifere(adj.)	Erzhaltig(adj.)	METALLO- SODERZHASHCHIJ	Metalifero(adj.)	Metallifero(adj.)
2770 METALLOGENIC DISTRICT ECON	Metallogenic district METALLOGENIC PROVINCES	District metallogenique PROVINCE METAL- LOGENIQUE	Distrikt METALL=PROVINZ	[METALLOGENIC DISTRICT] PROVINTSIYA METALLOGENI- CHESKAYA	Distrito=metalogenico PROVINCIA= METALOGENICA	Distretto metallogenico PROVINCIA MET- ALLOGENICA
2771 METALLOGENIC EPOCH ECON	METALLOGENIC EPOCHS	EPOQUE METALLO- GENIQUE	METALLOGENET. EPOCHE	EHPOKHA METAL- LOGENICHESKAYA	EPOCA= METALOGENICA	EPOCA METALLO- GENICA
2772 METALLOGENIC MAP ECON	METALLOGENIC MAPS	CARTE METALLO- GENIQUE	METALLOGENET. KARTE	KARTA METALLO- GENICHESKAYA	MAPA= METALOGENETICO	CARTA METALLO- GENICA
2773 METALLOGENIC PROVINCE ECON	METALLOGENIC PROVINCES *UF:* Metallogenic district	PROVINCE METAL- LOGENIQUE *UF:* District metallo- genique	METALL=PROVINZ *UF:* Distrikt	PROVINTSIYA METALLOGENI- CHESKAYA *UF:* Metallogenic district	PROVINCIA= METALOGENICA *UF:* Distrito= metalogenico	PROVINCIA MET- ALLOGENICA *UF:* Distretto metallo- genico
2774 METALLOGENY ECON	METALLOGENY	METALLOGENIE	METALLOGENESE	METALLOGENIYA *UF:* Prognoz metallo- geneticheskij	METALOGENIA	METALLOGENESI
2775 METALLOID CHES	Metalloid NONMETALS	Metalloide ELEMENT NON METALLIQUE	Metalloid NICHT=METALL	METALLOID	Metaloide ELEMENTO=NO= METALICO	Metalloide ELEMENTO NON METALLICO
2776 METALLOTECT ECON	Metallotect ORE BODIES	Metallotecte CONTROLE GITE	Metallotekt LAGERSTAETTEN= MILIEU	METALLOTEKT	Metalotecto CONTROL= YACIMIENTO	Metallotecto CONTROLLO DI MINERALIZZAZ- IONE
2777 METALLURGY METH	METALLURGY	METALLURGIE	METALLURGIE	METALLURGIYA	METALURGIA	METALLURGIA
2778 METAMICT MINE	METAMICT MINER- ALS	METAMICTE	METAMIKT	METAMIKTNYJ	METAMICTO	MINERALE METAMITTICO
2779 METAMICTIZATION IGNE	METAMICTIZA- TION	METAMICTISA- TION	METAMIKTI- SIERUNG	RASPAD METAMIK- TNYJ	METAMICTIZA- CION	METAMITTIZZAZ- IONE

	ENGLISH	FRANCAIS	DEUTSCH	RUSSKIJ	ESPANOL	ITALIANO
2780 METAMORPHIC BELT IGNE	METAMORPHIC BELTS	CEINTURE METAMORPHIQUE	METAMORPH. GUERTEL	POYAS METAMOR-FICHESKIJ	CINTURON= METAMORFICO	CINTURA METAM-ORFICA
2781 METAMORPHIC COMPLEX IGNE	Metamorphic complex METAMORPHIC ROCKS	Complexe metamor-phique ROCHE METAMOR-PHIQUE	Metamorph.Komplex METAMORPHIT	KOMPLEKS METAMORFI-CHESKIJ	Complejo=metamorfico ROCA= METAMORFICA	Complesso metamorfico ROCCIA METAM-ORFICA
2782 METAMORPHIC DIFFERENTIATION IGNE	Metamorphic differen-tiation METAMORPHISM; DIFFERENTIATION	Differenciation metam-orphique FACIES METAMOR-PHISME	Metamorph. Differentiation METAMORPHOSE= FAZIES	DIFFERENTSIAT-SIYA METAMORFICHES-KAYA	Diferenciacion= metamorfica FACIES= METAMORFISMO	Differenziazione metamorfica FACIES METAMOR-FICA
2783 METAMORPHIC FACIES IGNE	FACIES	FACIES METAMOR-PHISME UF: Differenciation metamorphique; Facies mineral; Grade metam-orphique; Mineral repere; Zone metamor-phique	METAMORPHOSE= FAZIES UF: Leit=Mineral; Metamorph. Differentiation; Metamorph.Zone; Metamorphose=Grad; Mineral=Fazies	FATSIYA METAM-ORFICHESKAYA UF: Stepen' metamor-fizma	FACIES= METAMORFISMO UF: Diferenciacion= metamorfica; Facies= mineral; Grado= metamorfico; Mineral= indice; Zona= metamorfica	FACIES METAMOR-FICA UF: Differenziazione metamorfica; Facies minerale; Grado metamorfico; Minerale indice; Zona metamor-fica
2784 METAMORPHIC GRADE IGNE	Grade FACIES	Grade metamorphique FACIES METAMOR-PHISME	Metamorphose=Grad METAMORPHOSE= FAZIES	Stepen' metamorfizma FATSIYA METAM-ORFICHESKAYA	Grado=metamorfico FACIES= METAMORFISMO	Grado metamorfico FACIES METAMOR-FICA
2785 METAMORPHIC ROCK IGMS	METAMORPHIC ROCKS UF: Arterite; Diaphto-rite; Epibolite; Jadei-tite; Metamorphic complex; Unakite; Venite	ROCHE METAMOR-PHIQUE UF: Complexe metamor-phique; Diaphthorite	METAMORPHIT UF: Diaphthorit; Metamorph.Komplex	PORODA METAM-ORPHICHESKAYA	ROCA= METAMORFICA UF: Complejo= metamorfico; Diaftorita	ROCCIA METAM-ORFICA UF: Complesso metam-orfico; Diaftorite
2786 METAMORPHIC ZONE IGNE	Metamorphic zone METAMORPHISM	Zone metamorphique FACIES METAMOR-PHISME	Metamorph.Zone METAMORPHOSE= FAZIES	ZONA GLUBIN-NAYA	Zona=metamorfica FACIES= METAMORFISMO	Zona metamorfica FACIES METAMOR-FICA
2787 METAMORPHISM IGNE	METAMORPHISM UF: Border facies; Iso-chemical metamor-phism; Metamorphic differentiation; Meta-morphic zone; Plutonic metamorphism	METAMORPHISME UF: Metamorphisme geothermique; Metam-orphisme impact; Metamorphisme isochimique; Metamor-phisme plutonique	METAMORPHOSE UF: Geotherm. Metamorphose; Isochem. Metamorphose; Pluton. Metamorphose; Stoss= Metamorphose	METAMORFIZM UF: Metabazal't; Meta-gabbro	METAMORFISMO UF: Metamorfismo= impacto; Metamorfismo= geotermico; Metamorfismo= isoquimico; Metamorfismo= plutonico	TIPO DI METAMOR-FISMO UF: Metamorfismo d'impatto; Metamor-fismo geotermico; Metamorfismo isochimico; Metamor-fismo plutonico
2788 METAMORPHOGENIC DEPOSIT ECON	METAMORPHIC PROCESSES	GITE METAMOR-PHOGENE	METAMORPH. LAGERSTAETTE	METAMORFOGEN-NYJ	YACIMIENTO= METAMORFOGENO	GIACIMENTO METAMORFOGENO
2789 METAMORPHOSIS PALE	Metamorphosis ONTOGENY	Metamorphose ONTOGENIE	Biolog.Metamorphose ONTOGENESE	METAMORFOZ	Metamorfosis ONTOGENIA	Metamorfosi ONTOGENESI
2790 METANAUPLIUS PALE	Metanauplius ONTOGENY; CRUS-TACEA	Metanauplius ONTOGENIE; CRUS-TACEA	Metanauplius ONTOGENESE; CRUSTACEA	METANAUPLIUS	Metanauplius ONTOGENIA; CRUSTACEA	Metanauplius ONTOGENESI; CRUSTACEA
2791 METAPHYTE PALE	Metaphyte ECOLOGY; FLORA	Metaphyte ECOLOGIE; FLORE	Metaphyt OEKOLOGIE; FLORA	METAFIT	Metafita ECOLOGIA; FLORA	Metaphyte ECOLOGIA; FLORA

	ENGLISH	FRANCAIS	DEUTSCH	RUSSKIJ	ESPANOL	ITALIANO
2792 METASEDIMENTARY ROCK IGMS	METASEDIMEN-TARY ROCKS	ROCHE METASEDI-MENTAIRE	META=SEDIMENT	METAOSADOCH-NYJ	SECUENCIA=METASEDIMENTARIA	ROCCIA METASEDIMEN-TARIA
2793 METASOMA IGNE	Metasoma METASOMATISM	Metasome METASOMATOSE	Metasom METASOMATOSE	METASOMA	Metasoma METASOMATISMO	Metasoma METASOMATOSI
2794 METASOMATISM IGNE	METASOMATISM UF: Alkaline metasomatism; Contact metasomatism; Fronts; Metasoma	METASOMATOSE UF: Front; Metasomatose alcaline; Metasome	METASOMATOSE UF: Alkali=Metasomatose; Front; Metasom	METASOMATOZ	METASOMATISMO UF: Frente; Metasoma; Metasomatismo=alcalino	METASOMATOSI UF: Fronte; Metasoma; Metasomatismo alcalino
2795 METASOMATITE IGMS	METASOMATIC ROCKS	ROCHE METASO-MATIQUE	METASOMAT. GESTEIN	METASOMATIT	METASOMATITA	ROCCIA METASO-MATICA
2796 METATEXIS IGNE	Metatexis ANATEXIS	Metatexie ANATEXIE	Metatexis ANATEXIS	METATEKSIS	Metatexia ANATEXIA	Metatessi ANATESSI
2797 METEOR CRATER EXTR	METEOR CRATERS	CRATERE METE-ORIQUE	METEORITEN=KRATER	Krater meteoritnyj KRATER UDARA	CRATER=METEORICO	CRATERE METE-ORITICO
2798 METEORITE EXTS	METEORITES UF: Ataxite; Aubrite; Hexahedrite; Lodranite; Mesosiderites; Octahedrite; Pallasites; Regmaglypt; Siderophyres	METEORITE UF: Regmaglypte	METEORIT UF: Regmaglypt	METEORIT	METEORITO UF: Regmaglypto	METEORITE UF: Regmaglipto
2799 METEOROIDS EXTS	METEOROIDS	METEOROIDE	METEOROID	METEOROID	METEOROIDE	METEOROIDE
2800 METEOROLOGY SURF	METEOROLOGY	METEOROLOGIE	METEOROLOGIE	METEOROLOGIYA	METEOROLOGIA	METEOROLOGIA
2801 METHANE CHES	METHANE UF: Marsh gas	METHANE UF: Gaz marais	METHAN UF: Sumpfgas	METAN	METANO UF: Gas=pantanos	METANO UF: Gas di palude
2802 METHODOLOGY METH	METHODS UF: Morphometry; Operations; Pycnometry; Standard method	METHODOLOGIE UF: Methode par immersion; Methode standard; Morphometrie; Procedure; Pycnometrie; Technique; Trou explosion sismique	METHODIK UF: Immersions=Mikroskopie; Morphometrie; Pyknometrie; Schusspunkt; Standard=Methode; Technik; Verfahren	METODOLOGIYA UF: Graphic method	METODOLOGIA UF: Metodo=de=inmersion; Metodo=standard; Morfometria; Picnometria; Pozo=de=disparo; Procedimiento; Tecnicas	METODOLOGIA UF: Metodo a immersione; Metodo standard; Morfometria; Operazioni; Picnometria; Punto di scoppio; Tecnica
2803 MIAROLITIC TEXTURE TEST	Miarolitic texture TEXTURES	Texture miarolitique TEXTURE	Miarolith=Gefuege KORN=GEFUEGE	MIAROLITOVYJ	Textura=miarolitica TEXTURA	Tessitura miarolitica TESSITURA
2804 MICA MING	MICA	MICA	GLIMMER	SLYUDA	MICA	MICA
2805 MICASCHIST IGMS	MICA SCHIST	MICASCHISTE UF: Schiste metamorphique; Schiste tachete	GLIMMER=SCHIEFER UF: Flecken=Schiefer; Schiefer	SLYUDYANOJ SLA-NETS	MICAESQUISTO UF: Esquisto; Esquisto=moteado	MICASCISTO UF: Scisto; Scisto maculato
2806 MICRITE SEDS	MICRITE	MICRITE	MIKRIT	MIKRIT	MICRITA	MICRITE

	ENGLISH	FRANCAIS	DEUTSCH	RUSSKIJ	ESPANOL	ITALIANO
2807 MICROCOMPUTER MATH	MICROCOMPUTERS	MICROORDI- NATEUR	MIKRO= PROZESSOR	MIKRO=EHVM	MICROORDENA- DOR	MICROCALCOLA- TORE
2808 MICROCRATER TEST	MICROCRATERS	MICROCRATERE	MIKRO=KRATER	MIKROKRATER	MICROCRATER	MICROCRATERE
2809 MICROCRYSTALLINE **LIMESTONE** SEDS	MICROCRYSTAL- LINE LIMESTONE	CALCAIRE MICROCRISTALLIN *UF:* Biomicrite	MIKROKRIST- ALLIN.KALK *UF:* Biomikrit	[MICROCRYSTAL- LINE LIMESTONE] IZVESTNYAK; MIKROZERNISTYJ	CALIZA= MICROCRISTALINA *UF:* Biomicrita	CALCARE MICROCRISTAL- LINO *UF:* Biomicrite
2810 MICROEARTHQUAKE SOLI	MICROEARTH- QUAKES	MICROTREMBLE- MENT DE TERRE	MIKRO=BEBEN	MIKROZEMLE- TRYASENIE	MICROTERRE- MOTO	MICROSCOSSA
2811 MICROFACIES SEDI	MICROFACIES	MICROFACIES *UF:* Allocheme	MIKRO=FAZIES *UF:* Allochem	MIKROFATSIYA	MICROFACIES *UF:* Aloquimico	MICROFACIES *UF:* Allochimico
2812 MICROFAUNA PALE	Microfauna MICROFOSSILS	Microfaune MICROFOSSILE	Mikro=Fauna MIKRO=FOSSIL	MIKROFAUNA	Microfauna MICROFOSIL	Microfauna MICROFOSSILE
2813 MICROFISSURE TEST	MICROCRACKS	MICROFISSURE	MIKRO=FISSUR	Mikrotreshchina TRESHCHINA	MICROFISURA	MICROFESSURA
2814 MICROFLORA PALE	Microflora MICROFOSSILS	Microflore MICROFOSSILE	Mikro=Flora MIKRO=FOSSIL	MIKROFLORA	Microflora MICROFOSIL	Microflora MICROFOSSILE
2815 MICROFOLD STRU	MICROFOLDS	MICROPLI	MIKRO=FALTE	MIKROSKLADKA	MICROPLIEGUE	MICROPIEGA
2816 MICROFOSSIL PALE	MICROFOSSILS *UF:* Microfauna; Micro- flora	MICROFOSSILE *UF:* Microfaune; Micro- flore	MIKRO=FOSSIL *UF:* Mikro=Fauna; Mikro=Flora	MIKROFOSSILII	MICROFOSIL *UF:* Microfauna; Micro- flora	MICROFOSSILE *UF:* Microfauna; Micro- flora
2817 MICROGRAINED **TEXTURE** TEST	MICROGRAINED TEXTURE	ROCHE MICRO- GRENUE *UF:* Roche hypabyssale	SUBVULKAN. GESTEIN *UF:* Hypabyssal=Genese	MIKROZERNISTYJ *UF:* Microcrystalline limestone	ROCA= MICROGRANUDA *UF:* Roca=hipoabisal	TESSITURA MICRO- GRANULARE *UF:* Origine ipoabissale
2818 MICROGRANITE IGNS	MICROGRANITE	MICROGRANITE	MIKRO=GRANIT	Mikrogranit GRANIT	MICROGRANITO	MICROGRANITO
2819 MICROGRANITIC **TEXTURE** TEST	Microgranitic texture TEXTURES	Texture micrograni- tique TEXTURE POR- PHYRIQUE	Mikrogranit.Gefuege PORPHYR= GEFUEGE	MIKROGRANI- TOVYJ	Textura= microgranitica TEXTURA= PORFIDICA	Tessitura micrograni- tica TESSITURA POR- FIRICA
2820 MICROHARDNESS PHCH	MICROHARDNESS	MICRODURETE	MIKRO=HAERTE	Mikrotverdost' TVERDOST'	MICRODUREZA	MICRODUREZZA
2821 MICROLATEROLOG APPL	Microlaterolog ELECTRICAL LOG- GING	Microlaterolog DIAGRAPHIE ELEC- TRIQUE	Mikrolaterolog ELEKTRO=LOG	Mikrokarotazh bokovoj KAROTAZH	Testificacion=de= microrresistividad DIAGRAFIA= ELECTRICA	Microlaterolog DIAGRAFIA ELET- TRICA
2822 MICROLITIC TEXTURE TEST	MICROLITE	MICROLITE	MIKROLITH. GEFUEGE	MIKROLITOVYJ	TEXTURA= MICROLITICA	TESSITURA MICROLITICA
2823 MICROLOG APPL	Microlog ELECTRICAL LOG- GING	Microlog DIAGRAPHIE ELEC- TRIQUE	Mikrolog ELEKTRO=LOG	Mikrokarotazh KAROTAZH	Microlog DIAGRAFIA= ELECTRICA	Microlog DIAGRAFIA ELET- TRICA

	ENGLISH	FRANCAIS	DEUTSCH	RUSSKIJ	ESPANOL	ITALIANO
2824 MICROMETEORITE EXTS	MICROMETEOR-ITES	MICROMETEORITE	MIKRO=METEORIT	MIKROMETEORIT *UF:* Mikrotektit	MICROMETEORITO	MICROMETEORITE
2825 MICROMORPHOLOGY SURF	MICROMOR-PHOLOGY	MICROMORPHOLO-GIE SOL	MIKRO-MORPHOLOGIE	MIKROREL'EF	MICROMORFO-LOGIA=SUELO	MICROMOR-FOLOGIA DEL SUOLO
2826 MICROORGANISM PALE	MICROORGANISMS	MICROORGANISME	MIKRO=ORGANISMEN	MIKROORGANIZM	MICROOR-GANISMO	MICROOR-GANISMO
2827 MICROPALEONTOLOGY PALE	MICROPALEON-TOLOGY	MICROPALEON-TOLOGIE	MIKRO-PALAEONTOLOGIE	MIKROPALEON-TOLOGIYA	MICROPALEON-TOLOGIA	MICROPALEON-TOLOGIA
2828 MICROPHYTOLITH PALS	Microphytoliths PROBLEMATIC MICROFOSSILS	Microphytolithi PROBLEMATICA MICRO	[Microphytoliths] MIKRO=PROBLEMATICA	Mikrofitolity MIKROPROBLEMA-TIKA (ISKOP)	Microphytolithi PROBLEMATICA=MICRO	Microphytolithi MICROFOSSILE PROBLEMATICO
2829 MICROPLATE SOLI	MICROPLATES	MICROPLAQUE	MIKRO=PLATTE	Mikroplita PLITA LITOS-FERNAYA; TEK-TONIKA BLOKOVAYA	MICROPLACA	MICROPLACCA
2830 MICROPROBLEMATICA PALS	PROBLEMATIC MICROFOSSILS *UF:* Catagraphia; Micro-phytoliths	PROBLEMATICA MICRO *UF:* Catagraphia; Micro-phytolithi	MIKRO=PROBLEMATICA *UF:* Catagraphia; Micro-phytolithi	MIKROPROBLEMA-TIKA (ISKOP) *UF:* Mikrofitolity	PROBLEMATICA=MICRO *UF:* Catagrafia; Micro-phytolithi	MICROFOSSILE PROBLEMATICO *UF:* Catagraphia; Micro-phytolithi
2831 MICROSCOPE INST	Microscopes INSTRUMENTS; MICROSCOPE METHODS	Microscope INSTRUMENTA-TION; MICROSCO-PIE	Mikroskop INSTRUMEN-TIERUNG; MIKROSKOPIE	MIKROSKOP	Microscopio INSTRUMENTA-CION; MICROS-COPIA	Microscopio STRUMENTAZ-IONE; MICROS-COPIA
2832 MICROSCOPIC SIZE TEST	Microscopic size TEXTURES	Texture microscopique TEXTURE	Mikroskop.Gefuege KORN=GEFUEGE	MIKROSKOPI-CHESKIJ	Textura=microscopica TEXTURA	Scala microscopica TESSITURA
2833 MICROSCOPY METH	MICROSCOPE METHODS *UF:* Microscopes	MICROSCOPIE *UF:* Microscope	MIKROSKOPIE *UF:* Mikroskop	ANALIZ MIKROSKOPI-CHESKIJ	MICROSCOPIA *UF:* Microscopio	MICROSCOPIA *UF:* Microscopio
2834 MICROSEISM SOLI	MICROSEISMS	MICROSEISME	MIKRO=SEISMIK	MIKROSEJSMY	MICROSISMO	MICROSISMA
2835 MICROSTYLOLITE TEST	MICROSTYLOLITES	MICROSTYLOLITE	MIKRO=STYLOLITH	Mikrostilolit STILOLIT	MICRO=ESTILOLITO	MICROSTILOLITE
2836 MICROTECTONICS STRU	Microtectonics STRUCTURAL ANALYSIS	MICROTEC-TONIQUE	MIKRO=TEKTONIK	MIKROTEKTONIKA	MICROTECTONICA	MICROTETTONICA
2837 MICROTEKTITE EXTS	Microtektites TEKTITES	Microtectite TECTITE	Mikro=Tektit TEKTIT	Mikrotektit TEKTIT; MIKRO-METEORIT	Microtectita TECTITA	Microtectite TECTITE
2838 MICROWAVE METHOD APPL	MICROWAVE METHODS	METHODE MICROONDE	MIKROWELLEN=METHODE	Metod mikrovolnovoj METOD RADIOVOL-NOVYJ	METODO=MICROONDAS	METODO MICROONDE
2839 MICROWAVE SPECTROSCOPY METH	MICROWAVE SPEC-TROSCOPY	SPECTROMETRIE MICROONDE	MIKROWELLEN=SPEKTROMETRIE	SPEKTROMETRIYA MIKROVOL-NOVAYA	ESPECTROMETRIA=MICROONDA	SPETTROMETRIA MICROONDE

	ENGLISH	FRANCAIS	DEUTSCH	RUSSKIJ	ESPANOL	ITALIANO
2840 MID-OCEANIC RIDGE MARI	MID=OCEAN RIDGES *UF:* Submarine ridge	DORSALE OCEANIQUE *UF:* Vulcanorium	SUBMARIN. RUECKEN	KHREBET SREDINNO=OKEANICHESKIJ	DORSAL=OCEANICA *UF:* Vulcanorio	DORSALE OCEAN-ICA *UF:* Vulcanorium
2841 MIDDLE CAMBRIAN STRS	MIDDLE CAM-BRIAN *UF:* Majian	CAMBRIEN MOYEN *UF:* Mayien	M.KAMBRIUM *UF:* Majium	KEMBRIJ SREDNIJ	CAMBRICO=MEDIO *UF:* Majiense	CAMBRIANO MED. *UF:* Majiano
2842 MIDDLE DEVONIAN STRS	MIDDLE DEVO-NIAN	DEVONIEN MOYEN	M.DEVON	DEVON SREDNIJ	DEVONICO=MEDIO	DEVONIANO MED.
2843 MIDDLE EOCENE STRS	MIDDLE EOCENE	EOCENE MOYEN	M.EOZAEN	Ehotsen srednij EHOTSEN	EOCENO=MEDIO	EOCENE MED.
2844 MIDDLE JURASSIC STRS	MIDDLE JURASSIC	JURASSIQUE MOYEN	M.JURA	YURA SREDNYAYA	JURASICO=MEDIO	GIURASSICO MED.
2845 MIDDLE LIASSIC STRS	MIDDLE LIASSIC	LIAS MOYEN	M.LIAS	LEJAS SREDNIJ	LIAS=MEDIO	LIASSICO MED.
2846 MIDDLE MIOCENE STRS	MIDDLE MIOCENE	MIOCENE MOYEN	M.MIOZAEN	Miotsen srednij MIOTSEN	MIOCENO=MEDIO	MIOCENE MED.
2847 MIDDLE OLIGOCENE STRS	MIDDLE OLIGO-CENE	OLIGOCENE MOYEN	M.OLIGOZAEN	Oligotsen srednij RYUPEL'	OLIGOCENO=MEDIO	OLIGOCENE MED.
2848 MIDDLE ORDOVICIAN STRS	MIDDLE ORDOVI-CIAN	ORDOVICIEN MOYEN	M.ORDOVIZIUM	ORDOVIK SREDNIJ	ORDOVICICO=MEDIO	ORDOVICIANO MED.
2849 MIDDLE PLEISTOCENE STRS	MIDDLE PLEISTO-CENE	PLEISTOCENE MOYEN	M.PLEISTOZAEN	Plejstotsen srednij PLEJSTOTSEN	PLEISTOCENO=MEDIO	PLEISTOCENE MED.
2850 MIDDLE PLIOCENE STRS	MIDDLE PLIOCENE	PLIOCENE MOYEN	M.PLIOZAEN	Pliotsen srednij KUYAL'NITSKIJ	PLIOCENO=MEDIO	PLIOCENE MED.
2851 MIDDLE PROTEROZOIC STRS	MIDDLE PROTERO-ZOIC	PROTEROZOIQUE MOYEN	M. PROTEROZOIKUM	PROTEROZOJ SREDNIJ	PROTEROZOICO=MEDIO	PROTEROZOICO MED.
2852 MIDDLE TRIASSIC STRS	MIDDLE TRIASSIC	TRIAS MOYEN	M.TRIAS	TRIAS SREDNIJ	TRIAS=MEDIO	TRIASSICO MED.
2853 MIGMATITE IGMS	MIGMATITES *UF:* Leucosome	MIGMATITE *UF:* Arterite; Epibolite; Leucosome; Venite	MIGMATIT *UF:* Arterit; Epibolit; Leucosome; Venit	MIGMATIT *UF:* Arterit; Ehmbrekhit; Ehpibolit	MIGMATITA *UF:* Arterita; Epibolita; Leucosoma; Venita	MIGMATITE *UF:* Arterite; Epibolite; Leucosoma; Venite
2854 MIGMATIZATION IGNE	MIGMATIZATION	MIGMATISATION	MIGMATISIERUNG	MIGMATIZATSIYA	MIGMATIZACION	MIGMATIZZAZ-IONE
2855 MIGRATION GEOL	MIGRATION	MIGRATION	MIGRATION	MIGRATSIYA *UF:* Migratsiya ehlemen-tov	MIGRACION	MIGRAZIONE
2856 MIGRATION OF ELEMENT GEOC	MIGRATION OF ELEMENTS	MIGRATION ELE-MENT	ELEMENTEN=MIGRATION	Migratsiya ehlementov MIGRATSIYA	MIGRACION=DE=ELEMENTOS	MIGRAZIONE DI ELEMENTI
2857 MILIOLACEA PALS	MILIOLACEA	MILIOLACEA	MILIOLACEA	MILIOLACEA	MILIOLACEA	MILIOLACEA

	ENGLISH	FRANCAIS	DEUTSCH	RUSSKIJ	ESPANOL	ITALIANO
2858 MILIOLINA PALS	MILIOLINA	MILIOLINA	MILIOLINA	MILIOLINA	MILIOLINA	MILIOLINA
2859 MILITARY GEOLOGY GEOL	MILITARY GEOL-OGY	GUERRE	MILITAER=GEOLOGIE	GEOLOGIYA VOEN-NAYA	GUERRA	GEOLOGIA MILI-TARE
2860 MILLER INDICES MINE	Miller indices CRYSTAL FORM	Indice Miller FORME CRISTAL-LINE	Miller=Indizes TRACHT	SIMVOL KRISTAL-LICHESKOJ GRANI	Indice=de=Miller FORMA=CRISTALINA; ESTRUCTURA=CRISTALINA	Indici di Miller FORMA CRISTAL-LINA
2861 MINE MINI	MINES	MINE *UF:* Mine souterraine	BERGWERK *UF:* Unterird.Gewinnung	VYRABOTKA GORNAYA	MINA *UF:* Mineria=subterranea	MINIERA *UF:* Coltivazione sotter-ranea
2862 MINERAL MINE	MINERALS *UF:* Stress mineral	MINERAL *UF:* Mineral stress	MINERAL *UF:* Stress=Mineral	MINERAL *UF:* Mineral inclusion; Mineralogical locality; Mineraloid; New min-eral data; Speleothem	MINERALES *UF:* Mineral=tensional	MINERALE *UF:* Minerale stress
2863 MINERAL ASSEMBLAGE MINE	MINERAL ASSEM-BLAGES	ASSOCIATION MINERALE *UF:* Mineral essentiel	MINERAL=VER-GESELLSCHAFTUNG *UF:* Gesteinsbildend. Mineral	Assotsiatsiya miner-al'naya PARAGENEZIS	ASOCIACION=MINERAL *UF:* Minerales=constituyentes	ASSOCIAZIONE MINERALOGICA *UF:* Minerale essenziale
2864 MINERAL CLEAVAGE MINE	MINERAL CLEAV-AGE *UF:* Basal cleavage; Conchoidal materials; Mineral striations	CLIVAGE MINERAL *UF:* Clivage isometrique; Fracture conchoidal; Striation mineral	SPALTBARKEIT *UF:* Basis=Spaltbarkeit; Muschelig.Bruch; Riefung	SPAJNOST' *UF:* Klivazh bazal'nyj	EXFOLIACION=MINERAL *UF:* Estriacion=mineral; Exfoliacion=basal; Fractura=concoidea	SFALDATURA *UF:* Frattura concoide; Sfaldatura basale; Striatura di cristallo
2865 MINERAL COMPOSITION MINE	MINERAL COMPO-SITION	COMPOSITION MINERALOGIQUE	MINERALOG. ANALYSE	ANALIZ MINER-ALOGICHESKIJ	DETERMINACION=MINERAL	COMPOSIZIONE MINERALOGICA
2866 MINERAL DEPOSIT ECON	Mineral deposits MINERAL RESOURCES	Gisement RESSOURCE MINERALE	Lagerstaette MINERAL. ROHSTOFFE	MESTOROZHDENIE	Deposito=mineral RECURSO=MINERO	Giacimento RISORSE MINERA-RIE
2867 MINERAL ECONOMICS ECON	MINERAL ECO-NOMICS	ECONOMIE MINIERE	BERGWIRTSCHAFT	EHKONOMIKA MINERAL'NOGO SYR'YA	ECONOMIA=MINERA	ECONOMIA MINE-RARIA
2868 MINERAL EXPLORATION ECON	MINERAL EXPLO-RATION *UF:* Electrical exploration; Geophysical exploration; Gravity exploration	PROSPECTION MINIERE	ERZ=PROSPEKTION	RAZVEDKA MESTOROZHDENIJ	PROSPECCION=MINERA	PROSPEZIONE MINERARIA
2869 MINERAL FACIES IGNE	Mineral facies P=T CONDITIONS; FACIES	Facies mineral FACIES METAMOR-PHISME	Mineral=Fazies METAMORPHOSE=FAZIES	FATSIYA MINER-AL'NAYA	Facies=mineral FACIES=METAMORFISMO	Facies minerale FACIES METAMOR-FICA
2870 MINERAL INCLUSION IGNE	MINERAL INCLU-SIONS	INCLUSION MINERALE	MINERAL=EINSCHLUSS	[MINERAL INCLU-SION] VKLYUCHENIE; MINERAL	INCLUSION=MINERAL	INCLUSIONE MINERALE
2871 MINERAL PHASE PHCH	Mineral phase PHASE EQUILIBRIA	Phase minerale DIAGRAMME EQUILIBRE	Mineral=Phase PHASEN=GLEICHGEWICHT	FAZA MINER-AL'NAYA	Fase=mineral DIAGRAMA=EQUILIBRIO	Fase mineralogica DIAGRAMMA D'EQUILIBRIO

	ENGLISH	FRANCAIS	DEUTSCH	RUSSKIJ	ESPANOL	ITALIANO
2872 MINERAL RESOURCE ECON	MINERAL RESOURCES *UF:* Economic mineral; Mineral deposits	RESSOURCE MINERALE *UF:* Gisement; Gite mineral	MINERAL. ROHSTOFFE *UF:* Erz=Lagerstaette; Lagerstaette	POLEZNYE ISKO-PAEMYE	RECURSO=MINERO *UF:* Deposito=mineral; Yacimiento=mineral	RISORSE MINERA-RIE *UF:* Giacimento; Giacimento metallifero
2873 MINERAL SPRING GEOH	Mineral spring SPRINGS	Source eau minerale EAU MINERALE	Mineral=Wasser=Quelle MINERAL=WASSER	ISTOCHNIK MIN-ERAL'NYJ	Fuente=agua=mineral AGUA=MINERAL	Sorgente minerale ACQUA MINERALE
2874 MINERAL STRIATION MINE	Mineral striations MINERAL CLEAV-AGE	Striation mineral CLIVAGE MIN-ERAL; CROIS-SANCE CRISTAL-LINE	Riefung SPALTBARKEIT; KRISTALL-WACHSTUM	SHTRIKHOVKA MINERALA	Estriacion=mineral EXFOLIACION=MINERAL; CRECIMIENTO=CRISTALINO	Striatura di cristallo SFALDATURA; ACCRESCIMENTO CRISTALLINO
2875 MINERAL WATER GEOH	MINERAL WATERS	EAU MINERALE *UF:* Source eau minerale	MINERAL=WASSER *UF:* Mineral=Wasser=Quelle	VODA MINER-AL'NAYA *UF:* Voda lechebnaya	AGUA=MINERAL *UF:* Fuente=agua=mineral	ACQUA MINERALE *UF:* Sorgente minerale
2876 MINERALIZATION ECON	MINERALIZATION	MINERALISATION *UF:* Minerai; Minerai disloque	MINERALISATION *UF:* Erz; Geschleppt.Erz	MINERALIZATSIYA	MINERALIZACION *UF:* Escape; Mena	MINERALIZZAZ-IONE *UF:* Drag ore; Giacimento minerario
2877 MINERALIZER ECON	ORE=FORMING FLUIDS	FLUIDE MINERALI-SATEUR	MINERALISI-EREND.LOESUNG	MINERALIZATOR	FLUIDO=MINERALIZADOR	FLUIDO MINER-ALIZZANTE
2878 MINERALOGICAL LOCALITY MINE	MINERALOGICAL LOCALITY	GITE MINER-ALOGIQUE	MINERAL=FUNDPUNKT	Mineralogical locality SBOR; MINERAL	YACIMIENTO=MINERALOGICO	LOCALITA MINER-ALOGICA
2879 MINERALOGICAL PHASE RULE PHCH	Mineralogical phase rule PHASE RULE	Loi phases miner-alogiques DIAGRAMME EQUILIBRE	Mineralog.Phasen=Regel PHASEN-GLEICHGEWICHT	PRAVILO FAZ MINERALOGI-CHESKIKH	Regla=mineralogica=de=las=fases DIAGRAMA=EQUILIBRIO	Regola delle fasi miner-alogiche DIAGRAMMA D'EQUILIBRIO
2880 MINERALOGY MINE	MINERALOGY	MINERALOGIE	MINERALOGIE	MINERALOGIYA	MINERALOGIA	MINERALOGIA
2881 MINICOMPUTER MATH	MINICOMPUTERS	MINIORDINATEUR	KLEINRECHNER	MINI=EHVM	MINIORDENADOR	MINICALCOLA-TORE
2882 MINIMUM MATH	Minimum STATISTICAL DIS-TRIBUTION	Minimum DISTRIBUTION STATISTIQUE	Minimum STATIST. VERTEILUNG	MINIMUM *UF:* Davlenie nizkoe; Temperatura nizkaya	Minimo DISTRIBUCION=ESTADISTICA	Minimo DISTRIBUZIONE STATISTICA
2883 MINING MINI	Mining EXPLOITATION	Exploitation miniere EXPLOITATION	Bergbau GEWINNUNG	RAZRABOTKA RUDNYKH MESTOROZHDENI *UF:* Razrabotka obrusheniem	Explotacion=minera EXPLOTACION	Coltivazione mineraria SFRUTTAMENTO
2884 MINING ENGINEERING MINI	Mining engineering MINING GEOLOGY; ENGINEERING GEOLOGY	Art des mines EXPLOITATION	Bergbau=Technik GEWINNUNG	Gornoe delo EHKSPLUATAT-SIYA	Ingenieria=minera EXPLOTACION	Ingegneria mineraria SFRUTTAMENTO
2885 MINING GEOLOGY MINI	MINING GEOLOGY *UF:* Claims; Mining engineering; Underground mining	GEOLOGIE MINIERE	MONTAN=GEOLOGIE	GEOLOGIYA RUD-NYKH MESTOROZHDENIJ	GEOLOGIA=MINERA	GEOLOGIA MINE-RARIA

	ENGLISH	FRANCAIS	DEUTSCH	RUSSKIJ	ESPANOL	ITALIANO
2886 MINING LICENSE ECON	MINING LICENSE	TITRE MINIER *UF:* Permis exploration	KONZESSION *UF:* Schuerf=Recht	[MINING LICENSE] EHKSPLUATAT- SIYA	CONCESION= MINERA *UF:* Permiso= investigacion=mina	TITOLO MINE- RARIO *UF:* Permesso di esplora- zone
2887 MINOR ELEMENT CHES	MINOR ELEMENTS	ANALYSE MINEURS	NEBENELEMENT= ANALYSE	EHLEMENT MALYJ	ELEMENTOS= MENORES	ELEMENTI MINORI
2888 MINOR-ELEMENT ANALYSIS METH	MINOR=ELEMENT ANALYSES	METHODE ANA- LYSE MINEURS	NEBENELEMENT= ANALYSEN= METHODIK	ANALIZ NA MALYE EHLE- MENTY	METODO= ANALISIS= MENORES	ANALISI ELE- MENTI MINORI
2889 MIOCENE STRS	MIOCENE	MIOCENE	MIOZAEN	MIOTSEN *UF:* Miotsen nizhnij; Miotsen srednij; Miotsen verkhnij	MIOCENO	MIOCENE
2890 MIOGEOSYNCLINE STRU	MIOGEOSYN- CLINES	MIOGEOSYNCLI- NAL	MIOGEOSYNKLI- NALE	MIOGEOSINKLI- NAL'	MIOGEOSINCLI- NAL	MIOGEOSINCLI- NALE
2891 MIOMERA PALS	Miomera AGNOSTIDA	Miomera AGNOSTIDA	Miomera AGNOSTIDA	MIOMERA *UF:* Agnostida	Miomera AGNOSTIDA	Miomera AGNOSTIDA
2892 MIOSPORE PALE	MIOSPORES	MIOSPORE	MIOSPORE	Miospora SPORA	MIOSPORA	MIOSPORA
2893 MISCIBILITY PHCH	MISCIBILITY	MISCIBILITE	MISCHBARKEIT	SMESHIVAEMOST' *UF:* Miscibility gap	MISCIBILIDAD	MISCIBILITA
2894 MISCIBILITY GAP MINE	MISCIBILITY GAP	LACUNE MISCIBI- LITE	MISCHUNGS= LUECKE	[MISCIBILITY GAP] PERERYV; SMESHIVAEMOST'	LAGUNA=DE= MISCIBILIDAD	LACUNA DI MISCI- BILITA
2895 MISE A LA MASSE APPL	MISE=A=LA= MASSE	MISE A LA MASSE	[MISE=A=LA= MASSE]	Metod vyzvannogo potentsiala METOD INDUKTIV- NYJ	PUESTA=A=MASA	MESSA A TERRA
2896 MISSISSIPPI VALLEY-TYPE ECON	MISSISSIPPI VALLEY=TYPE	GITE TYPE MISSIS- SIPPI VALLEY	MISSISSIPPI=TAL= LAGERSTAETTENTYP	MESTOROZHDENIE TIPA MISSISSIPPI	YACIMIENTO= TIPO=MISSISSIPPI= VALLEY	GIACIMENTO TIPO VALLE DEL MISSIS- SIPPI
2897 MISSISSIPPIAN STRS	MISSISSIPPIAN	MISSISSIPPIEN	MISSISSIPPIAN	Missisipij KARBON NIZHNIJ	MISSISSIPPIENSE	MISSISSIPPIANO
2898 MIXED-LAYER MINERAL MINE	MIXED=LAYER MINERALS	MINERAL INTER- STRATIFIE	MISCHGITTER= MINERAL	MINERAL SMESHANNO= SLOJNYJ	ARCILLA= INTERESTRATIFI- CADA	MINERALE A STRATI MISTI
2899 MOBILE BELT STRU	MOBILE BELTS	ZONE MOBILE	MOBIL.GUERTEL	Poyas podvizhnyj GEOSINKLINAL'; POYAS SKLADCHA- TYJ	ZONA=MOVIL	ZONA MOBILE
2900 MOBILITY PHCH	MOBILITY	MOBILITE	MOBILITAET	MOBIL'NOST'	MOVILIDAD	MOBILITA
2901 MOBILIZATION PHCH	MOBILIZATION	MOBILISATION GEOCHIMIQUE	MOBILISIERUNG	MOBILIZATSIYA	MOVILIZACION	MOBILIZZAZIONE GEOCHIMICA

	ENGLISH	FRANCAIS	DEUTSCH	RUSSKIJ	ESPANOL	ITALIANO
2902 MODAL ANALYSIS METH	MODAL ANALYSIS	ANALYSE MODALE	MODAL=ANALYSE	ANALIZ MODAL'NYJ	ANALISIS=MODAL	ANALISI MODALE
2903 MODE IGNE	Mode PETROGRAPHY	Mode petrographique CALCUL PETRO- GRAPHIQUE	[MODE] PETROGRAPH. ANALYSE	MINERALOGI- CHESKIJ (SOSTAV)	Moda PETROGRAFIA	Moda CALCOLO PETRO- GRAFICO
2904 MODEL MISC	MODELS *UF:* Deterministic model; Diagnosis; Hydrodynamic models	MODELE *UF:* Diagnose; Modele deterministe; Modele hydrodynamique	MODELL *UF:* Block=Diagramm; Determinist.Modell; Diagnose; Hydrodynam.Modell	MODEL'	MODELO *UF:* Diagnosis; Modelo- determinativo; Modelo=hidrodinamico	MODELLO *UF:* Diagnosi; Modello deterministico; Modello idrodinamico
2905 MODIFIED MERCALLI SCALE SOLI	MODIFIED MER- CALLI SCALE	ECHELLE MER- CALLI MODIFIEE	MODIFIZIERT. MERCALLI=SKALA	SHKALA ZEMLETR- YASENIJ *UF:* Shkala Richtera	ESCALA= MERCALLI= MODIFICADA	SCALA MERCALLI MODIFICATA
2906 MODIOMORPHOIDA PALS	Modiomorphoida BIVALVIA	Modiomorphoida BIVALVIA	Modiomorphoida LAMELLI- BRANCHIATA	MODIOMOR- PHOIDA	Modiomorphoida BIVALVIA	Modiomorphoida BIVALVIA
2907 MODULUS PHCH	Modulus PARAMETERS	Module PARAMETRE	Modul PARAMETER	MODUL'	Modulo PARAMETRO	Modulo PARAMETRO
2908 MODULUS OF RIGIDITY PHCH	Modulus of rigidity SHEAR MODULUS	Module rigidite MODULE ELASTI- CITE	Schubmodul ELASTIZITAETS= MODUL	Modul' sdviga MODUL' UPRU- GOSTI	Modulo=rigidez MODULO= ELASTICIDAD	Modulo di rigidita MODULO D'ELASTICITA
2909 MOHOLE PROJECT SOLI	Mohole project MOHOROVICIC DISCONTINUITY; BOREHOLES	Projet Mohole MOHOROVICIC; SONDAGE	Mohole=Projekt MOHO	MOKHO PROEKT	Proyecto=Mohole MOHOROVICIC	Progetto Mohole DISCONTINUITA DI MOHOROVICIC; SONDAGGIO
2910 MOHOROVICIC DISCONTINUITY SOLI	MOHOROVICIC DISCONTINUITY *UF:* Mohole project	MOHOROVICIC *UF:* Projet Mohole	MOHO *UF:* Mohole=Projekt	MOKHOROVICH- ICHA POVERKH- NOST'	MOHOROVICIC *UF:* Proyecto=Mohole	DISCONTINUITA DI MOHOROVICIC *UF:* Progetto Mohole
2911 MOISTURE GEOH	MOISTURE *UF:* Dew; Zone of aera- tion	HUMIDITE SOL	BODEN=FEUCHTE	VLAGA POCHVEN- NAYA	HUMEDAD=SUELO	UMIDITA DEL SUOLO
2912 MOLASSE SEDI	MOLASSE	MOLASSE	MOLASSE	MOLASSA	MOLASA	MOLASSA
2913 MOLDAVITE EXTS	Moldavite TEKTITES	Moldavite TECTITE	Moldavit TEKTIT	Moldavit TEKTIT	Moldavita TECTITA	Moldavite TECTITE
2914 MOLECULAR SPECTROSCOPY METH	Molecular spectroscopy SPECTROSCOPY	Spectrometrie moleculaire SPECTROMETRIE	Molekular= Spektroskopie SPEKTROMETRIE	SPEKTROSKOPIYA MOLEKUL- YARNAYA	Espectrometria= molecular ESPECTROMETRIA	Spettrometria moleco- lare SPETTROMETRIA
2915 MOLLISOL SUSS	MOLLISOLS *UF:* Albolls; Aquolls; Borolls; Rendolls; Udolls; Ustolls; Xerolls	MOLLISOL *UF:* Alboll; Aquoll; Boroll; Rendoll; Udoll; Ustoll; Xeroll	MOLLISOL *UF:* Alboll; Aquoll; Boroll; Rendoll; Udoll; Ustoll; Xeroll	MOLLISOL *UF:* Akvoll; Alboll; Boroll; Kseroll; Ren- doll; Udoll; Ustoll	MOLLISOL *UF:* Alboll; Aquoll; Boroll; Rendoll; Udoll; Ustoll; Xeroll	MOLLISOL *UF:* Alboll; Aquoll; Boroll; Rendoll; Udoll; Ustoll; Xeroll
2916 MOLLUSCA PALS	MOLLUSCA	MOLLUSCA	MOLLUSCA	MOLLUSCA *UF:* Rostroconchia	MOLLUSCA	MOLLUSCA

	ENGLISH	FRANCAIS	DEUTSCH	RUSSKIJ	ESPANOL	ITALIANO
2917 MOLYBDATES MING	MOLYBDATES	MOLYBDATE	MOLYBDAT	MOLIBDAT	MOLIBDATO	MOLIBDATO
2918 MOLYBDENUM CHEE	MOLYBDENUM	MOLYBDENE	MO	MOLIBDEN	MOLIBDENO	MOLIBDENO
2929 MONAZITE COMS	MONAZITE	MONAZITE	MONAZIT	MONATSIT	MONACITA	MONAZITE
2920 MONOCLINE STRU	MONOCLINES *UF:* Homoclines	MONOCLINAL *UF:* Homoclinal	MONOKLINALE *UF:* Homoklinale	MONOKLINAL' *UF:* Gomoklinal'	MONOCLINAL *UF:* Homoclinal	MONOCLINALE *UF:* Omoclinale
2921 MONOCLINIC SYSTEM MINE	MONOCLINIC SYS- TEM	SYSTEME MONO- CLINIQUE	MONOKLIN. SYSTEM	SINGONIYA MONOKLINNAYA	SISTEMA= MONOCLINICO	SISTEMA MONO- CLINO
2922 MONOCOTYLEDONEAE PALS	MONOCOTYLE- DONEAE	MONOCOTYLE- DONES	MONOCOTYLE- DONEAE	Monocotyledoneae ANGIOSPERMAE	MONOCOTYLE- DONES	MONOCOTILEDONI
2923 MONOCYATHEA PALS	Monocyathea ARCHAEOCYATHA	Monocyathea ARCHAEOCYATHA	Monocyathea ARCHAEOCYATHA	MONOCYATHEA	Monocyathea ARCHAEOCYATHA	Monocyathea ARCHAEOCYATA
2924 MONOGRAPH MISC	MONOGRAPHS	MONOGRAPHIE	MONOGRAPHIE	MONOGRAFIYA *UF:* Treatise	MONOGRAFIA	MONOGRAFIA
2925 MONOLETE TAXON PALE	MONOLETE TAXA	TAXON MONOLETE	MONOLET.TAXON	[MONOLETE TAXON]	TAXON= MONOLETE	TAXON MONOLETO
2926 MONOLITH SURF	Monolith EROSION FEA- TURES	Monolithe EROSION; ROCHE IGNEE	Monolith EROSION; MAGMA- TIT	MONOLIT	Monolito EROSION; ROCA= IGNEA	Monolito EROSIONE; ROCCIA IGNEA
2927 MONOMORPHIC PALE	Monomorphic(adj.)	Monomorphe(adj.)	Monomorph(adj.)	MONOMORFNYJ	Monomorfico(adj.)	Monomorfo(adj.)
2928 MONOPLACOPHORA PALS	MONOPLACO- PHORA	MONOPLACO- PHORA	MONOPLACO- PHORA	MONOPLACO- PHORA	MONOPLACO- PHORA	MONOPLACO- PHORA
2929 MONORHINA PALS	Monorhina AGNATHA	Monorhina AGNATHA	Monorhina AGNATHA	MONORHINA	Monorhina AGNATHA	Monorhina AGNATHA
2930 MONOTREMATA PALS	MONOTREMATA	MONOTREMATA	MONOTREMATA	MONOTREMATA	MONOTREMATA	MONOTREMATA
2931 MONSOON SURF	MONSOONS	MOUSSON	MONSUN	MUSSON	VIENTO=MONZON	MONSONE
2932 MONTE CARLO METHOD MATH	MONTE CARLO ANALYSIS	METHODE MONTE CARLO	MONTE=CARLO= METHODE	METOD MONTE= KARLO	METODO=MONTE= CARLO	METODO DI MONTE CARLO
2933 MONZONITE IGNS	MONZONITES *UF:* Monzonitic texture	MONZONITE *UF:* Texture monzoni- tique	MONZONIT *UF:* Monzonit=Gefuege	MONTSONIT	MONZONITA *UF:* Textura= monzonitica	MONZONITE *UF:* Tessitura monzoni- tica
2934 MONZONITIC TEXTURE TEST	Monzonitic texture MONZONITE; TEX- TURES	Texture monzonitique MONZONITE	Monzonit=Gefuege MONZONIT	MONTSONITOVYJ	Textura=monzonitica MONZONITA	Tessitura monzonitica MONZONITE
2935 MOON EXTS	MOON *UF:* Lunar geology; Preimbrian; Selenochronology; Selenology	LUNE *UF:* Copernicien; Eratos- thenien; Geologie lunaire; Imbrien; Kreep; Pre=imbrien; Procellarien; Selenochronologie; Selenologie	MOND *UF:* Eratosthen.Alter; Imbrium; Kopernikan. Alter; Kreep; Mond= Geologie; Oceanus= Procellarius; Prae= Imbrium; Selenochronologie; Selenologie	LUNA *UF:* Lunar constitution	LUNA *UF:* Coperniciense; Eratostensiense; Geologie=lunar; Imbriense; Kreep; Preimbriense; Procella- riense; Selenocronologia; Selenologia	LUNA *UF:* Copernicano; Era- tosteniano; Geologia lunare; Imbriano; Kreep; Preimbriano; Procellariano; Selenocronologia; Selenologia

	ENGLISH	FRANCAIS	DEUTSCH	RUSSKIJ	ESPANOL	ITALIANO
2936 MOONQUAKE EXTR	MOONQUAKES	SEISME LUNAIRE	MOND=BEBEN	LUNOTRYASENIE	SISMO=LUNAR	SISMA LUNARE
2937 MOONSCAPE EXTR	LUNAR RELIEF	RELIEF LUNAIRE	MOND=RELIEF	REL'EF LUNNYJ	RELIEVE=LUNAR	RILIEVO LUNARE
2938 MORAINE SURF	MORAINES *UF:* Ground moraines; Lateral moraines	MORAINE *UF:* Moraine fond; Moraine laterale	MORAENE *UF:* Grund=Moraene; Seiten=Moraene	MORENA *UF:* Gryada valunnaya; Morena bokovaya	MORRENA *UF:* Morrena=de=fondo; Morrena=lateral	MORENA *UF:* Morena di fondo; Morena laterale
2939 MORPHOGENESIS SURF	Morphogenesis LANDFORMS	Morphogenese RELIEF CONTI- NENT	Morphogenese KONTINENTAL= RELIEF	MORFOGENEZ	Morfogenesis RELIEVE= CONTINENTAL	Morfogenesi FORMA DEL RILIEVO
2940 MORPHOLOGY GEOL	MORPHOLOGY *UF:* Byssus; Carapace; Hinge (pale); Holosto- matous taxa; Hydrospires	MORPHOLOGIE	MORPHOLOGIE	MORFOLOGIYA *UF:* Forma	MORFOLOGIA	MORFOLOGIA
2941 MORPHOMETRY METH	Morphometry METHODS; RELIEF	Morphometrie METHODOLOGIE; RELIEF CONTI- NENT	Morphometrie METHODIK; KONTINENTAL= RELIEF	Morfometriya TOPOGRAFIYA; REL'EF	Morfometria METODOLOGIA; RELIEVE= CONTINENTAL	Morfometria METODOLOGIA; FORMA DEL RILIEVO
2942 MORPHOSCOPY METH	MORPHOSCOPY	MORPHOSCOPIE *UF:* Forme grain	MORPHOSKOPIE *UF:* Korn=Gestalt	MORFOSKOPIYA *UF:* Forma chastits	MORFOSCOPIA *UF:* Forma=grano	MORFOSCOPIA *UF:* Forma del granulo
2943 MORPHOSTRUCTURE SURF	MORPHOSTRUC- TURES *UF:* Structural relief	Relief morphostructural RELIEF STRUC- TURAL	Morphostruktur STRUKTUR= RELIEF	MORFOSTRUK- TURA	Morfoestructura RELIEVE= ESTRUCTURAL	Rilievo morfostruttur- ale MORFOSTRUT- TURA
2944 MORTAR MATERIAL COMS	BINDERS	LIANT	MOERTEL= MATERIAL	Syr'e tsementnoe TSEMENT STROI- TEL'NYJ	AGLOMERANTE	LEGANTE
2945 MOSAIC STRUCTURE TEST	Mosaic structure TEXTURES	Texture en mosaique TEXTURE	Mosaik=Gefuege KORN=GEFUEGE	MOZAICHNYJ	Textura=en=mosaico TEXTURA	Tessitura a mosaico TESSITURA
2946 MOSCOVIAN STRS	MOSCOVIAN	MOSCOVIEN	MOSKOV	MOSKOVSKIJ	MOSCOVIENSE	MOSCOVIANO
2947 MOSSBAUER **SPECTROSCOPY** METH	MOSSBAUER SPEC- TROSCOPY	SPECTROMETRIE MOESSBAUER	MOESSBAUER= SPEKTROMETRIE	SPEKTROMETRIYA MESSBAUE- HROVSKAYA	ESPECTROMETRIA= MOSSBAUER	SPETTROMETRIA MOESSBAUER
2948 MOSSBAUER **SPECTRUM** PHCH	MOSSBAUER SPEC- TRA	SPECTRE MOESS- BAUER	MOESSBAUER= SPEKTRUM	SPEKTR MESS- BAUEHROVSKIJ	ESPECTRO= MOSSBAUER	SPETTRO MOESS- BAUER
2949 MOTTLING TEST	Mottling TEXTURES	Tacheture TEXTURE	Flecken=Gefuege KORN=GEFUEGE	PYATNISTOST'	Roca=moteada TEXTURA	Tessitura a chiazze TESSITURA
2950 MOUNTAINS SURF	MOUNTAINS *UF:* Crests; Highlands	MASSIF MON- TAGNEUX *UF:* Crete; Hautes= terres	BERGMASSIV *UF:* Berg=Kamm; Hoch- land	MASSIV GORNYJ *UF:* Gora=igla; Greben'; Kholm	MACIZO= MONTAÑOSO *UF:* Cresta; Tierras= altas	MASSICCIO MON- TUOSO *UF:* Alteterre; Cresta

	ENGLISH	FRANCAIS	DEUTSCH	RUSSKIJ	ESPANOL	ITALIANO
2951 MOVEMENT MISC	Movement(gen.)	Mouvement(gen.)	Bewegung(gen.)	DVIZHENIE	Movimientos(gen.)	Movimento(gen.)
2952 MUD SEDS	MUD	VASE	SCHLAMM	IL UF: Mud bank	FANGO	FANGO
2953 MUD BANK SEDI	MUD BANKS	BANC VASE	SCHLAMM=BANK	[MUD BANK] IL; SLOJ	BANDAS=DE= BARRO	BANCO DI FANGO
2954 MUD CRACK SURF	MUDCRACKS	CRAQUELURE BOUE UF: Craquelure retre- cissement; Fissure contraction	SCHLAMM= SCHWUND=RISS UF: Hitze=Sprung; Schwundriss	TRESHCHINA USYKHANIYA	GRIETA= DESECACION UF: Grieta=retraccion; Rotura=termica	FESSURAZIONE DA DISSECCA- MENTO UF: Fessura di contraz- ione; Fessura di dissec- camento
2955 MUD FLAT SURF	MUD FLATS	VASIERE	SCHLICK=KUESTE	Otmel' ilistaya LITORAL'	MARISMA	PIANA DI FANGO
2956 MUD FLOW SEDI	MUDFLOWS	COULEE BOUE UF: Glissement masse boueuse; Lahar	SCHLAMM=STROM UF: Lahar; Schlamm= Rutschung	POTOK GRYAZEVOJ	COLADA=DE= BARRO UF: Corriente=de= barro=volcanico; Deslizamiento=de= barro	COLATA DI FANGO UF: Lahar; Smottamento
2957 MUD LUMP SEDI	MUD LUMPS	LOUPE BOUE	SCHLAMM=DIAPIR	[MUD LUMP] DEL'TOVYJ; OSA- DOK	ISLA=DE=DELTA	GRUMO DI FANGO
2958 MUD VOLCANO IGNE	MUD VOLCANOES	VOLCAN BOUE	SCHLAMM= VULKAN	VULKAN GRYAZEVOJ	VOLCAN=DE= BARRO	VULCANO DI FANGO
2959 MUDSLIDE SURF	MUDSLIDES	Glissement masse boueuse COULEE BOUE	Schlamm=Rutschung SCHLAMM=STROM	Opolzen' gryazevoj OPOLZEN'	Deslizamiento=de= barro COLADA=DE= BARRO	Smottamento COLATA DI FANGO
2960 MULL SOIL SUSS	Mull SOILS	SOL A MULL	MULL=BODEN	MULL'	MULL=SOIL	SUOLO A MULL
2961 MULLION STRUCTURE TEST	MULLIONS	STRUCTURE EN MENEAU	MULLION= STRUKTUR	STRUKTURA MUL- LION	ESTRUCTURA= ALMOHADILLADA	STRUTTURA A MULLION
2962 MULTILAYER SYSTEM GEOH	MULTILAYER SYS- TEMS	SYSTEME MULTI- COUCHE	STOCKWERKS= SYSTEM	[MULTILAYER SYSTEM] GORIZONT VODON- OSNYJ	SISTEMA= MULTICAPA	SISTEMA MULT- ISTRATO
2963 MULTIPHASE GEOL	Multiphase(gen.)	Multiphase(gen.)	Mehrphasen(gen.)	MNOGOFAZNYJ	Polifasica(gen.)	Polifasico(gen.)
2964 MULTIPHASE FLOW GEOH	MULTIPHASE FLOW	ECOULEMENT POLYPHASIQUE	MEHRPHASEN= STROEMUNG	POTOK MNOGO- FAZNYJ	ESCORRENTIA= POLIFASICA	COLATA POLIFA- SICA
2965 MULTIPHASE TECTONICS STRU	MULTIPHASE TEC- TONICS	TECTONIQUE SUPERPOSEE UF: Superposition	UEBERPRAEGUNGS= TEKTONIK UF: Ueberlagerung	TEKTONIKA MNO- GOFAZNAYA	TECTONICA= SUPERPUESTA UF: Superposicion	TETTONICA SOVRAIMPOSTA UF: Sovrapposizione

	ENGLISH	FRANCAIS	DEUTSCH	RUSSKIJ	ESPANOL	ITALIANO
2966 MULTISPECTRAL ANALYSIS METH	MULTISPECTRAL ANALYSIS	TELEDETECTION MULTISPECTRALE	MULTISPEKTRAL-ANALYSE	Analiz multispek-tral'nyj ZONDIROVANIE DISTANTSIONNOE	TELEDETECCION=MULTIESPECTRAL	TELEOSSERVAZ-IONE MULTISPET-TRALE
2967 MULTITUBERCULATA PALS	MULTITUBER-CULATA	MULTITUBER-CULATA	MULTITUBER-CULATA	ALLOTHERIA *UF:* Taeniolaboidea	MULTITUBER-CULATA	MULTITUBER-CULATA
2968 MULTIVARIATE STATISTICS MATH	MULTIVARIATE ANALYSIS	ANALYSE MULTI-VARIABLE	MULTIVARIAT. ANALYSE	STATISTIKA MNO-GOMERNAYA	ANALISIS=MULTIVARIABLE	ANALISI MULTI-VARIATA
2969 MUSCHELKALK STRS	MUSCHELKALK	MUSCHELKALK	MUSCHELKALK	MUSHEL'KAL'K	MUSCHELKALK	MUSCHELKALK
2970 MUSEUM MISC	MUSEUMS	MUSEE	MUSEUM	MUZEJ	MUSEO	MUSEO
2971 MUTATION PALE	Mutation BIOLOGIC EVOLU-TION	Mutation EVOLUTION BIOLOGIQUE	Mutation BIOLOG. EVOLUTION	MUTATSIYA	Mutacion EVOLUCION=BIOLOGICA	Mutazione EVOLUZIONE BIOLOGICA
2972 MYLONITE GEOL	MYLONITES *UF:* Cataclasis; Myloni-tic structure; Phyllo-nites; Ultramylonite	MYLONITE *UF:* Cataclase; Phyllo-nite; Structure myloni-tique; Ultramylonite	MYLONIT *UF:* Kataklase; Mylonit=Gefuege; Phyllonit; Ultra=Mylonit	MILONIT	MILONITA *UF:* Cataclasis; Estructura=milonitica; Filonita; Ultramilonita	MILONITE *UF:* Cataclasite; Fillo-nite; Struttura miloni-tica; Ultramilonite
2973 MYLONITIC STRUCTURE TEST	Mylonitic structure MYLONITES	Structure mylonitique MYLONITE	Mylonit=Gefuege MYLONIT	MILONITOVYJ	Estructura=milonitica MILONITA	Struttura milonitica MILONITE
2974 MYLONITIZATION STRU	MYLONITIZATION	MYLONITISATION	MYLONITI-SIERUNG	MILONITIZATSIYA	MILONITIZACION	MILONITIZZAZ-IONE
2975 MYODOCOPIDA PALS	Myodocopida OSTRACODA	Myodocopina OSTRACODA	Myodocopida OSTRACODA	Myodocopida OSTRACODA	Myodocopida OSTRACODA	Myodocopida OSTRACODA
2976 MYOIDA PALS	Myoida BIVALVIA	Myoida BIVALVIA	Myoida LAMELLI-BRANCHIATA	MYOIDA	Myoida BIVALVIA	Myoida BIVALVIA
2977 MYOMORPHA PALS	MYOMORPHA	MYOMORPHA	MYOMORPHA	MYOMORPHA	MYOMORPHA	MYOMORPHA
2978 MYRIAPODA PALS	MYRIAPODA *UF:* Archipolypoda; Chilopoda; Diplopoda; Pauropoda; Symphyla	MYRIAPODA *UF:* Archipolypoda; Arthropleurida; Chilopoda; Diplopoda; Pauropoda; Symphyla	MYRIAPODA *UF:* Archipolypoda; Arthropleurida; Chilopoda; Diplopoda; Pauropoda; Symphyla	MYRIAPODA	MYRIAPODA *UF:* Archipolypoda; Arthropleurida; Chilopoda; Diplopoda; Pauropoda; Symphyla	MYRIAPODA *UF:* Archipolypoda; Arthropleurida; Chilopoda; Diplopoda; Pauropoda; Symphyla
2979 MYRMEKITE TEST	MYRMEKITE	MYRMEKITE	MYRMEKIT	MIRMEKIT	MIRMEKITA	MIRMECHITE
2980 MYSTACOCARIDA PALS	Mystacocarida CRUSTACEA	Mystacocarida CRUSTACEA	Mystacocarida CRUSTACEA	MYSTACOCARIDA	Mystacocarida CRUSTACEA	Mystacocarida CRUSTACEA
2981 MYTILOIDA PALS	Mytiloida BIVALVIA	Mytiloida BIVALVIA	Mytiloida LAMELLI-BRANCHIATA	MYTILOIDA	Mytiloida BIVALVIA	Mytiloida BIVALVIA

	ENGLISH	FRANCAIS	DEUTSCH	RUSSKIJ	ESPANOL	ITALIANO
2982 MYXOMYCETES PALS	MYXOMYCETES	MYCETOZOA	MYXOMYCETES	MYXOMYCETES	MYXOMYCETES	MYXOMYCETES
2983 NAKHLITE EXTS	Nakhlite ACHONDRITES	Nakhlite ACHONDRITE	Nakhlith ACHONDRIT	Naklit AKHONDRIT	Nakhlita ACONDRITA	Nakhlite ACONDRITE
2984 NAMURIAN STRS	NAMURIAN	NAMURIEN	NAMUR	NAMYUR	NAMURIENSE	NAMURIANO
2985 NANNOFOSSIL PALE	NANNOFOSSILS	NANOFOSSILE	NANNO=FOSSIL	NANOFOSSILII	NANOFOSIL	NANNOFOSSILE
2986 NAPPE STRU	NAPPES *UF:* Outliers; Roots	NAPPE *UF:* Avant=butte; Racine	DECKE *UF:* Outlier; Wurzel	POKROV TEKTONI- CHESKIJ	MANTO *UF:* Raiz=de=manto; Testigo=de=erosion	FALDA *UF:* Outlier; Radice
2987 NATANTES PALS	Natantes NEORNITHES	Natantes NEORNITHES	Natantes NEORNITHES	NATANTES	Natantes NEORNITHES	Natantes NEORNITHES
2988 NATIVE ELEMENTS MING	NATIVE ELEMENTS	ELEMENT NATIF	GEDIEGEN. ELEMENT	EHLEMENT SAMO- RODNYJ	ELEMENTO= NATIVO	ELEMENTO NATIVO
2989 NATURAL GAS COMS	NATURAL GAS *UF:* Gas cap; Gas coal; Gas well	GAZ NATUREL *UF:* Accumulation gaz; Condense gaz; Puits a gaz; Rapport gaz= petrole	ERDGAS *UF:* Gas=Bohrung; Gas=Kondensat; Gas= Oel=Verhaeltnis; Gaskappe	GAZ PRIRODNYJ	GAS=NATURAL *UF:* Acumulacion=gas; Condensado; Pozo=de= gas; Relacion=gas= petroleo	GAS NATURALE *UF:* Cappello di gas; Condensato di gas; Pozzo di metano; Rap- porto gas=petrolio
2990 NATURAL GLASS IGNE	Natural glass GLASSES	VERRE NATUREL *UF:* Obsidienne; Texture hyaline; Texture vitreuse; Verre; Verre volcanique; Vitrifica- tion	NATUERL.GLAS *UF:* Glas; Glas= Gefuege; Glasig. Gefuege; Obsidian; Verglasung; Vulkan. Glas	Steklo prirodnoe STEKLO VULKANI- CHESKOE	VIDRIO=NATURAL *UF:* Obsidiana; Textura=vitrea; Vidrio; Vidrio=volcanico; Vitrea; Vitrificacion	VETRO NATURALE *UF:* Ossidiana; Tessitura vetrosa; Vetrificazione; Vetro; Vetro vulcanico; Vetroso
2991 NATURAL REMANENT MAGNETIZATION PHCH	NATURAL REMA- NENT MAGNE- TIZATION	AIMANTATION REMANENTE NATURELLE	NATUERL. REMANENT. MAGNETISIERUNG	NAMAGNICHEN- NOST' OSTATOCH- NAYA ESTESTVENNAYA	IMANTACION= REMANENTE= NATURAL	MAGNETIZZAZ- IONE NATURALE RESIDUA
2992 NATURAL RESOURCE COMS	NATURAL RESOURCES	RESSOURCE NATU- RELLE	BODENSCHAETZE *UF:* Vorkommen	RESURSY PRIROD- NYE	RECURSO= NATURAL	RISORSE NATUR- ALI
2993 NAUTILOIDEA PALS	NAUTILOIDEA *UF:* Endoceratoidea	NAUTILOIDEA *UF:* Actinoceratoidea; Endoceratoidea	NAUTILOIDEA *UF:* Actinoceratoidea; Endoceratoidea	NAUTILOIDEA	NAUTILOIDEA *UF:* Actinoceratoidea; Endoceratoidea	NAUTILOIDEA *UF:* Actinoceratoidea; Endoceratoidea
2994 NEANDERTHALIAN PALS	NEANDERTHAL	NEANDERTHALIEN	NEANDERTHALER	[NEANDER- THALIAN]	NEANDERTHA- LIENSE	NEANDERTHALI- ANO
2995 NECK IGNE	VOLCANIC NECKS	NECK *UF:* Fissure magmatique	SCHLOT *UF:* Foerder=Schlot	NEKK *UF:* Kanal podvodyash- chij; Kanal vulkana; Zherlo	CHIMENEA *UF:* Conducto= alimentador	NECK *UF:* Condotto alimenta- tore
2996 NEEDLE SURF	Needles VOLCANIC DOMES; RELIEF	Aiguille VOLCAN; RELICTE	Nadel VULKAN; RELIKT	Gora=igla MASSIV GORNYJ	Aguja VOLCAN; RELICTA	Guglia vulcanica VULCANO; RELITTO
2997 NEKTON PALE	Nekton PELAGIC ENVI- RONMENT; FAUNA	Necton MILIEU PELAGIQUE; FAUNE	Nekton PELAG.MILIEU; FAUNA	NEKTON	Necton MEDIO=PELAGICO; FAUNA	Necton AMBIENTE PELAGICO; FAUNA

	ENGLISH	FRANCAIS	DEUTSCH	RUSSKIJ	ESPANOL	ITALIANO
2998 NEMERTA PALS	Nemerta WORMS	Nemerta VERMES	Nemerta VERMES	NEMERTA	Nemerta VERMES	Nemerta VERMES
2999 NEOCOMIAN STRS	NEOCOMIAN	NEOCOMIEN	NEOKOM	NEOKOM	NEOCOMIENSE	NEOCOMIANO
3000 NEODYMIUM CHEE	NEODYMIUM	NEODYME	ND	NEODIM	NEODIMIO	NEODIMIO
3001 NEOGASTROPODA PALS	NEOGASTROPODA	NEOGASTROPODA	NEOGASTROPODA	Neogastropoda GASTROPODA	NEOGASTROPODA	NEOGASTROPODA
3002 NEOGENE STRS	NEOGENE	NEOGENE	NEOGEN	NEOGEN	NEOGENO	NEOGENE
3003 NEON CHEE	NEON	NEON	NE	NEON	NEON	NEON
3004 NEORNITHES PALS	NEORNITHES *UF:* Natantes; Ratites; Volantes	NEORNITHES *UF:* Natantes; Ratites; Volantes	NEORNITHES *UF:* Natantes; Ratites; Volantes	NEORNITHES	NEORNITHES *UF:* Natantes; Ratites; Volantes	NEORNITHES *UF:* Natantes; Ratites; Volantes
3005 NEOTECTONICS STRU	NEOTECTONICS	NEOTECTONIQUE	NEO–TEKTONIK	NEOTEKTONIKA	NEOTECTONICA	NEOTETTONICA
3006 NEOTENY PALE	Neoteny ONTOGENY	Neotenie ONTOGENIE	Neotenie ONTOGENESE	NEOTENIYA	Neotenia ONTOGENIA	Neotenia ONTOGENESI
3007 NEOTYPE PALE	NEOTYPES	NEOTYPE	NEOTYP	Neotip GOLOTIP	NEOTIPO	NEOTIPO
3008 NEPHELINE MING	NEPHELINE	NEPHELINE	NEPHELIN	NEFELIN *UF:* Foidit	NEFELINA	NEFELINA
3009 NEPHELOID LAYER MARI	NEPHELOID LAYER	COUCHE NEPHEL- OIDE	NEPHELOID– HORIZONT	SLOJ NEFELOID- NYJ	CAPA–NEFELOIDE	LIVELLO NEFEL- OIDE
3010 NEPTUNISM GEOL	Neptunism THEORETICAL STUDIES; SEDI- MENTATION	Neptunisme THEORIE; SEDI- MENTATION	Neptunismus THEORIE; SEDI- MENTATION	NEPTUNIZM	Neptunismo TEORIA; SEDI- MENTACION	Nettunismo TEORIA; SEDI- MENTAZIONE
3011 NEPTUNIUM CHEE	NEPTUNIUM	NEPTUNIUM	NP	NEPTUNIJ	NEPTUNIO	NETTUNIO
3012 NERVOUS SYSTEM PALE	NERVOUS SYSTEM	ANATOMIE APPAREIL NER- VEUX	NERVENSYSTEM	SISTEMA NERV- NAYA	ANATOMIA– APARATO- NERVIOSO	ANATOMIA SISTEMA NERVOSO
3013 NETWORK METH	GEOMETRIC NET- WORKS	MAILLE GEO- METRIQUE	ANALYSEN–NETZ	SET'	MALLA– GEOMETRICA	MAGLIA GEO- METRICA
3014 NEUTRON PHCH	NEUTRONS	NEUTRON	NEUTRON	NEJTRON	NEUTRON	NEUTRONE
3015 NEUTRON ACTIVATION METH	NEUTRON ACTIVA- TION ANALYSIS	ANALYSE ACTIVA- TION NEUTRONIQUE	NEUTRONEN- AKTIVIERUNGS- ANALYSE	ANALIZ NEJTRONNO- AKTIVATSIONNYJ	ACTIVACION- NEUTRONICA	ANALISI ATTIVAZ- IONE NEUTRONICA

	ENGLISH	FRANCAIS	DEUTSCH	RUSSKIJ	ESPANOL	ITALIANO
3016 NEUTRON DIFFRACTION ANALYSIS METH	NEUTRON DIF-FRACTION ANALY-SIS	DIFFRACTION NEUTRON	NEUTRONEN=BEUGUNGS=ANALYSE	NEJTRONOGRA-FIYA	ANALISIS=DIFRACCION=NEUTRON	ANALISI DIFFRAZ-IONE NEUTRONICA
3017 NEUTRON LOG APPL	NEUTRON LOG-GING	DIAGRAPHIE NEU-TRON	NEUTRON=LOG	Karotazh nejtronnyj KAROTAZH; RAZ-VEDKA RADIO-METRICHESKAYA	DIAGRAFIA=NEUTRON	DIAGRAFIA NEUTRONICA
3018 NEUTRON-SOIL-MOISTURE METER INST	NEUTRON PROBE	SONDE NEUTRON	NEUTRONEN=SONDE	VLAGOMER POCH-VENNYJ NEJTRON-NYJ	SONDA=NEUTRON	SONDA A NEUTRONI
3019 NEW MISC	New(gen.)	Nouveau(gen.)	Neu(gen.)	NOVYJ *UF:* New description; New mineral data	Nuevo(gen.)	Nuovo(gen.)
3020 NEW DATA MISC	NEW DATA *UF:* New description	NOUVELLE DON-NEE *UF:* Nouvelle descrip-tion; Nouvelle donnee minerale	NEU.DATEN *UF:* Neu.Mineraldaten; Neubeschreibung	NOVYE DANNYE	NUEVO=DATO *UF:* Novedad; Nuevo-dato=mineral	DATO NUOVO *UF:* Minerale nuovo; Novita
3021 NEW DESCRIPTION MISC	New description NEW DATA	Nouvelle description NOUVELLE DON-NEE	Neubeschreibung NEU.DATEN	[NEW DESCRIP-TION] NOVYJ	Novedad NUEVO=DATO	Novita DATO NUOVO
3022 NEW METHODS MISC	NEW METHODS	METHODE NOU-VELLE	NEU.METHODE	METOD NOVYJ	METODO=NUEVO	METODO NUOVO
3023 NEW MINERAL DATA MINE	NEW MINERALS	Nouvelle donnee miner-ale NOUVELLE DON-NEE	Neu.Mineraldaten NEU.DATEN	[NEW MINERAL DATA] MINERAL; NOVYJ	Nuevo=dato=mineral NUEVO=DATO	Minerale nuovo DATO NUOVO
3024 NEW NAME MISC	NEW NAMES	NOM NOUVEAU	NEU.NAME	NAZVANIE NOVOE *UF:* Takson novyj	NOMBRE=NUEVO	NOME NUOVO
3025 NEW TAXON PALE	NEW TAXA	TAXON NOUVEAU	NEU.TAXON	Takson novyj NAZVANIE NOVOE	NUEVO=TAXON	TAXON NUOVO
3026 NICKEL CHEE	NICKEL	NICKEL	NI	NIKEL'	NIQUEL	NICKEL
3027 NIGGLI'S CLASSIFICATION IGNE	Niggli's classification PETROGRAPHY	Classification Niggli CALCUL PETRO-GRAPHIQUE	Niggli=Klassifikation PETROGRAPH. ANALYSE	KLASSIFIKATSIYA NIGGLI	Clasificacion=de=Niggli PETROGRAFIA	Classificazione di Niggli CALCOLO PETRO-GRAFICO
3028 NILSSONIALES PALS	NILSSONIALES	NILSSONIALES	NILSSONIALES	NILSSONIALES	NILSSONIALES	NILSSONIALES
3029 NIOBATES MING	NIOBATES	NIOBATE	NIOBAT	Niobat TANTALO=NIOBAT	NIOBATO	NIOBATO
3030 NIOBIUM CHEE	NIOBIUM	NIOBIUM	NB	NIOBIJ	NIOBIO	NIOBIO

	ENGLISH	FRANCAIS	DEUTSCH	RUSSKIJ	ESPANOL	ITALIANO
3031 NIOBOTANTALATES MING	NIOBOTANTA- LATES	NIOBOTANTALATE	NIOB=TANTALAT	TANTALO=NIOBAT *UF:* Niobat; Tantalat	NIOBOTANTALATO	NIOBATO- TANTALATO
3032 NITRATES MING	NITRATES *UF:* Nitrification	NITRATE *UF:* Nitrification	NITRAT *UF:* Nitrifizierung	NITRAT	NITRATO *UF:* Nitrificacion	NITRATO *UF:* Nitrificazione
3033 NITRIFICATION SEDI	Nitrification NITRATES	Nitrification NITRATE	Nitrifizierung NITRAT	NITRIFIKATSIYA	Nitrificacion NITRATO	Nitrificazione NITRATO
3034 NITROGEN CHEE	NITROGEN *UF:* Nitrogen gas	AZOTE *UF:* Azote organique; Gaz azote	N *UF:* Nitrogen=Gas; Organ.Stickstoff	AZOT	NITROGENO *UF:* Gas=nitrogeno; Nitrogeno=organico	AZOTO *UF:* Azoto organico; Gas azoto
3035 NITROGEN GAS CHES	Nitrogen gas NITROGEN	Gaz azote AZOTE	Nitrogen=Gas N	AZOT (GAZ)	Gas=nitrogeno NITROGENO	Gas azoto AZOTO
3036 NIVATION SURF	NIVATION	NIVATION	SCHNEE=EROSION	NIVATSIYA	NIVACION	NIVAZIONE
3037 NOBLE METAL CHES	PRECIOUS METALS	METAL PRECIEUX	EDEL=METALL	METALL BLAGO- RODNYJ	METAL=PRECIOSO	METALLO PREZ- IOSO
3038 NODOSARIACEA PALS	NODOSARIACEA	NODOSARIACEA	NODOSARIACEA	NODOSARIACEA	NODOSARIACEA	NODOSARIACEA
3039 NODULE SEDI	NODULES	NODULE	KNOLLE	NODUL'	NODULO	NODULO
3040 NOEGGERATHIALES PALS	NOEGGERATHI- ALES	NOEGGERATHI- ALES	NOEGGERATHI- ALES	Noeggerathiales LYCOPSIDA	NOEGGERATHI- ALES	NOEGGERATHI- ALES
3041 NOISE APPL	NOISE	BRUIT FOND	RAUSCHEN	SHUM	RUIDO=DE=FONDO	RUMORE DI FONDO
3042 NOMENCLATURE MISC	NOMENCLATURE *UF:* Holotypes; Hypodigms; Lectotypes; Nominal taxon; Para- types; Terminology	NOMENCLATURE *UF:* Holotype; Hypodigme; Lectotype; Legende; Paratype; Taxon nominale; Ter- minologie	NOMENKLATUR *UF:* Holotyp; Lectotyp; Legende; Nominal= Taxon; Paratyp; Ter- minologie; Typ= Material	NOMENKLATURA	NOMENCLATURA *UF:* Hipodigma; Holo- tipo; Lectotipo; Leyenda; Paratipo; Taxon=nominal; Ter- minologia	NOMENCLATURA *UF:* Ipodigma; Lectotipo; Legenda; Olotipo; Paratipo; Taxon nomi- nale; Terminologia
3043 NOMINAL TAXON PALE	Nominal taxon NOMENCLATURE	Taxon nominale NOMENCLATURE	Nominal=Taxon NOMENKLATUR	TAKSON NOMI- NAL'NYJ	Taxon=nominal NOMENCLATURA	Taxon nominale NOMENCLATURA
3044 NOMOGENESIS PALE	Nomogenesis BIOLOGIC EVOLU- TION	Nomogenese EVOLUTION BIOLOGIQUE	Nomogenese BIOLOG. EVOLUTION	NOMOGENEZ	Nomogenesis EVOLUCION= BIOLOGICA	Nomogenesi EVOLUZIONE BIOLOGICA
3045 NOMOGRAM MATH	NOMOGRAMS	ABAQUE	VERGLEICHS= DIAGRAMM	NOMOGRAMMA	ABACO	ABACO
3046 NON-EXPLOSIVE SOURCE APPL	Non=explosive sources SEISMIC METHODS	Source nonexplosive METHODE SISMIQUE	Nicht=Explosiv.Quelle SEISM.METHODE	ISTOCHNIK NEVZ- RYVNOJ	Fuente=no=explosiva METODO=SISMICO	Fonte non d'esplosione METODO SISMICO
3047 NON-FERROUS METALS CHES	NON=FERROUS METALS	METAL NON FER- REUX	NE=METALL	METALL TSVETNOJ *UF:* Base metal	METALES=NO= FERREOS	METALLO NON FERROSO
3048 NON-METALS CHES	NONMETALS *UF:* Metalloid	ELEMENT NON METALLIQUE *UF:* Metalloide	NICHT=METALL *UF:* Metalloid	NEMETALL *UF:* Karta nerudnykh poleznykh iskop	ELEMENTO=NO= METALICO *UF:* Metaloide	ELEMENTO NON METALLICO *UF:* Metalloide

	ENGLISH	FRANCAIS	DEUTSCH	RUSSKIJ	ESPANOL	ITALIANO
3049 NON-SATURATED ZONE GEOH	UNSATURATED ZONES *UF:* Aeration	ZONE NON SATUREE *UF:* Aeration; Zone aeration	UNGESAETTIGT. ZONE *UF:* Belueftung; Nichtgesaettigt.Zone	[NON-SATURATED ZONE] ZONA AEHRATSII	ZONA-NO-SATURADA *UF:* Aireacion; Zona=de=aireacion	ZONA NON SATURATA *UF:* Aerazione; Zona d'areazione
3050 NONCONFORMITY STRA	Nonconformities UNCONFORMITIES	Discordance stratigraphique DISCORDANCE ANGULAIRE	[NONCONFORMITY] WINKEL=DISKORDANZ	Nesoglasie strukturnoe NESOGLASIE UGLOVOE	Discordancia=igneo-sedimentaria DISCORDANCIA=ANGULAR	Discordanza semplice DISCORDANZA ANGOLARE
3051 NONMETAL DEPOSIT ECON	NONMETAL DEPOSITS	SUBSTANCE NON METALLIQUE	NICHT=METALL. ROHSTOFF	MESTOROZHDENIE NEMETALLI-CHESKOE	NO-METAL	SOSTANZA NON METALLICA *UF:* Chitina; Lignina; Umina
3052 NONMETALLIC-DEPOSIT MAP ECON	NONMETALLIC=DEPOSIT MAPS	CARTE SUB-STANCE UTILE	NICHT=METALL=LAGERSTAETTEN=KARTE	Karta nerudnykh poleznykh iskop KARTA MESTOROZHDENIJ; NEMETALL	MAPA=SUSTANCIAS=UTILES	CARTA MINERALI NON METALLIFERI
3053 NORIAN STRS	NORIAN	NORIEN	NOR	NORIJ	NORIENSE	NORICO
3054 NORM IGNE	Norm PETROGRAPHY	Norme petrographique CALCUL PETRO-GRAPHIQUE	Norm PETROGRAPH. ANALYSE	Sostav normativnyj KLASSIFIKATSIYA C.I.P.W.	Norma CALCULO-PETROGRAFICO	Norma petrografica CALCOLO PETRO-GRAFICO
3055 NORMAL DISTRIBUTION MATH	Normal distribution STATISTICAL DIS-TRIBUTION	Distribution normale DISTRIBUTION STATISTIQUE	Normal=Verteilung STATIST. VERTEILUNG	RASPREDELENIE NORMAL'NOE	Distribucion=normal DISTRIBUCION=ESTADISTICA	Distribuzione normale DISTRIBUZIONE STATISTICA
3056 NORMAL FAULT STRU	NORMAL FAULTS	FAILLE NORMALE	ABSCHIEBUNG	SBROS *UF:* Ustup sbrosovyj	FALLA=NORMAL	FAGLIA NORMALE
3057 NORTH MISC	North GEOGRAPHY	Nord GEOGRAPHIE	Norden GEOGRAPHIE	SEVER	Norte GEOGRAFIA	Nord GEOGRAFIA
3058 NOSE STRU	Nose (fold) ANTICLINES	[NOSE] ANTICLINAL	Falten=Stirn ANTIKLINALE	NOS STRUK-TURNYJ	Promontorio=anticlinal ANTICLINAL	[NOSE] ANTICLINALE
3059 NOTOSTRACA PALS	Notostraca BRANCHIOPODA	Notostraca CRUSTACEA	Notostraca CRUSTACEA	NOTOSTRACA	Notostraca CRUSTACEA	Notostraca CRUSTACEA
3060 NOTOUNGULATA PALS	NOTOUNGULATA	NOTOUNGULATA	NOTOUNGULATA	NOTOUNGULATA	NOTOUNGULATA	NOTOUNGULATA
3061 NOVACULITE SEDS	Novaculite CHERT	Novaculite SILEXITE	Novaculith FLINT	NOVAKULIT	Novaculita SILEXITA	Novaculite SILEXITE
3062 NUCLEAR ENERGY ECON	NUCLEAR ENERGY	ENERGIE NUCLEAIRE	KERN=ENERGIE	EHNERGIYA YAD-ERNAYA	ENERGIA=NUCLEAR	ENERGIA NUCL-EARE
3063 NUCLEAR EXPLOSION ENVI	NUCLEAR EXPLO-SIONS	EXPLOSION NUCLEAIRE	KERN=EXPLOSION	VZRYV YADERNYJ	EXPLOSION=NUCLEAR	ESPLOSIONE NUCLEARE
3064 NUCLEAR MAGNETIC RESONANCE METH	NUCLEAR MAG-NETIC RESO-NANCE	RESONANCE MAGNETIQUE NUCLEAIRE	KERN=MAGNET. RESONANZ	REZONANS MAGNITNYJ YAD-ERNYJ	RESONANCIA=MAGNETICA=NUCLEAR	RISONANZA MAGNETICO=NUCLEARE

	ENGLISH	FRANCAIS	DEUTSCH	RUSSKIJ	ESPANOL	ITALIANO
3065 NUCLEAR POWER PLANT ENGI	NUCLEAR FACILITIES	CENTRALE NUCLEAIRE	KERNKRAFTWERK	EHLEKTROSTANTSIYA ATOMNAYA	CENTRAL=NUCLEAR	CENTRALE NUCLEARE
3066 NUCLEATION PHCH	NUCLEATION	NUCLEATION	KEIMBILDUNG	[NUCLEATION] GENEZIS; ROST KRISTALLA	NUCLEACION	NUCLEAZIONE
3067 NUCLEUS MINE	Nucleus CRYSTAL GROWTH	Noyau cristallisation CROISSANCE CRISTALLINE	Kern KRISTALL=WACHSTUM	YADRO	Nucleo CRECIMIENTO=CRISTALINO	Germe cristallino ACCRESCIMENTO CRISTALLINO
3068 NUCULOIDEA PALS	Nuculoidea BIVALVIA	Nuculoidea BIVALVIA	Nuculoidea LAMELLIBRANCHIATA	NUCULOIDEA	Nuculoidea BIVALVIA	Nuculoidea BIVALVIA
3069 NUEE ARDENTE IGNE	NUEE ARDENTE	NUEE ARDENTE	GLUTWOLKE	Tucha palyashchaya IZVERZHENIE PELEJSKOGO TIPA	NUBE=ARDIENTE	NUBE ARDENTE
3070 NUGGET COMS	Nugget GOLD	Pepite OR	Goldklumpen AU	SAMORODOK	Pepita ORO	Pepita ORO
3071 NUMMULITIDAE PALS	NUMMULITIDAE	NUMMULITIDAE	NUMMULITIDAE	NUMMULITIDAE	NUMMULITIDAE	NUMMULITIDAE
3072 NUTATION SOLI	Nutation ROTATION	Nutation ROTATION	Nutation ROTATION	NUTATSIYA *UF:* Dvizhenie chandlerovskoe	Nutacion ROTACION	Nutazione ROTAZIONE
3073 NUTRITION PALE	NUTRITION *UF:* Autotrophic organism; Phagotrophic organism; Phytophagous organism; Planctivorous animal	NUTRITION *UF:* Faune planctivore; Organisme autotrophe; Organisme phagotrophe; Organisme phytophague	ERNAEHRUNG *UF:* Autotroph. Organismus; Planktonfresser; Phagotroph. Organismus; Phytophag.Organismus	PITANIE (PALE)	NUTRICION *UF:* Animal=planctivoro; Organismo=autotrofico; Organismo=fagotrofico; Organismo=fitofago	ANATOMIA DELLA NUTRIZIONE *UF:* Animale plantivoro; Organismo autotrofico; Organismo fagotropico; Organismo fitofago
3074 OASIS SURF	Oasis DESERTS; SPRINGS	Oasis DESERT; SOURCE	Oase WUESTE; QUELLE	OAZIS	Oasis DESIERTO; MANANTIAL	Oasi DESERTO; SORGENTE
3075 OBDUCTION SOLI	OBDUCTION	OBDUCTION	OBDUKTION	Obduktsiya SUBDUKTSIYA	OBDUCCION	OBDUZIONE
3076 OBJECTIVE MISC	Objectives PROJECTS	Objectif PROJET	Zweck PROJEKT	OB'EKTIV	Objetivos PROYECTO	Obbiettivo PROGETTO
3077 OBLIQUE FAULT STRU	Oblique fault OBLIQUE ORIENTATION; FAULTS	Faille oblique FAILLE	Schraeg.Stoerung STOERUNG	SBROS KOSOJ	Falla=oblicua FALLA	Faglia obliqua FAGLIA
3078 OBLIQUE ORIENTATION STRU	OBLIQUE ORIENTATION *UF:* Oblique fault	ORIENTATION OBLIQUE	SCHRAEG. ORIENTIERUNG	[OBLIQUE ORIENTATION] ORIENTIROVKA; POPERECHNYJ	ORIENTACION=OBLIGUA	ORIENTAZIONE OBLIQUA
3079 OBOLELLIDA PALS	Obolellida INARTICULATA	Obolellida INARTICULATA	Obolellida INARTICULATA	OBOLELLIDA	Obolellida INARTICULATA	Obolellida INARTICULATA

	ENGLISH	FRANCAIS	DEUTSCH	RUSSKIJ	ESPANOL	ITALIANO
3080 OBSEQUENT VALLEY SURF	Obsequent valleys VALLEYS	Vallee obsequente PROFIL EN LONG	Obsequent.Tal LAENGS=PROFIL	Dolina obsekventnaya DOLINA	Valle=obsecuente PERFIL= LONGITUDINAL	Valle obsequente PROFILO LONGI- TUDINALE
3081 OBSERVATION MISC	OBSERVATIONS	OBSERVATION GEOPHYSIQUE	BEOBACHTUNG	NABLYUDENIE	OBSERVACION- GEOFISICA	OSSERVAZIONE GEOFISICA
3082 OBSERVATORY MISC	OBSERVATORIES	OBSERVATOIRE	OBSERVATORIUM	OBSERVATORIYA	OBSERVATORIO	OSSERVATORIO
3083 OBSIDIAN IGNS	OBSIDIAN	Obsidienne VERRE NATUREL; RHYOLITE	Obsidian NATUERL.GLAS; RHYOLITH	OBSIDIAN	Obsidiana VIDRIO=NATURAL; RIOLITA	Ossidiana VETRO NATURALE; RIOLITE
3084 OCEAN MARI	WORLD OCEAN	OCEAN *UF:* Niveau mer	OZEAN *UF:* Meeresspiegel	OKEAN *UF:* Okeanicheskaya tsirkyulatsiya	OCEANO *UF:* Nivel=del=mar	OCEANO *UF:* Livello marino
3085 OCEAN BASIN MARI	OCEAN BASINS	BASSIN OCEANIQUE	OZEAN=BECKEN	KOTLOVINA OKEANICHESKAYA	CUENCA- OCEANICA	BACINO OCEANICO
3086 OCEAN CIRCULATION MARI	OCEAN CIRCULA- TION *UF:* Circulation	CIRCULATION OCEANIQUE	MEERES= ZIRKULATION	Okeanicheskaya tsirkyulatsiya OKEAN; TSIRKU- LYATSIYA	CIRCULACION= OCEANICA	CIRCOLAZIONE OCEANICA
3087 OCEAN DRILLING PROGRAM MARI	OCEAN DRILLING PROGRAM	[OCEAN DRILLING PROGRAM	[OCEAN DRILLING PROGRAM]	PROEKT GLUBOKO- VODNOGO BURENIYA	[OCEAN= DRILLING= PROGRAM]	[OCEAN DRILLING PROGRAM]
3088 OCEAN FLOOR MARI	OCEAN FLOORS *UF:* Abyssal gap; Abys- sal hill; Marine sills; Sea bottom; Straits; Submarine morphology	FOND MARIN	MEERES=BODEN	LOZHE OKEANA	FONDO=MARINO	FONDO MARINO
3089 OCEAN=BOTTOM SEISMOGRAPH INST	OCEAN=BOTTOM SEISMOGRAPHS	SISMOGRAPHE FOND SOUS MARIN	SUBMARIN. SEISMOGRAPH	Sejsmograf okeani- cheskij SEJSMOPRIEMNIK	SISMOGRAFO- FONDO= SUBMARINO	SISMOGRAFO SOT- TOMARINO
3090 OCEANIC MARI	Oceanic(adj.)	Oceanique(adj.)	Ozeanisch(adj.)	OKEANICHESKIJ	Oceanico(adj.)	Oceanico(adj.)
3091 OCEANIC CRUST SOLI	OCEANIC CRUST	CROUTE OCEANIQUE	OZEAN.KRUSTE	KORA OKEANI- CHESKAYA	CORTEZA= OCEANICA	CROSTA OCEAN- ICA
3092 OCEANIZATION SOLI	OCEANIZATION *UF:* Basification	OCEANISATION	OZEANISATION	OKEANIZATSIYA	OCEANIZACION	OCEANIZZAZIONE
3093 OCEANOGRAPHY MARI	OCEANOGRAPHY *UF:* Oceanology	OCEANOGRAPHIE *UF:* Oceanologie	OZEANOGRAPHIE *UF:* Ozeanologie	OKEANOGRAFIYA *UF:* Okeanologiya; Paleookeanografiya	OCEANOGRAFIA *UF:* Oceanologia	OCEANOGRAFIA *UF:* Oceanologia
3094 OCEANOLOGY MARI	Oceanology OCEANOGRAPHY	Oceanologie OCEANOGRAPHIE	Ozeanologie OZEANOGRAPHIE	Okeanologiya OKEANOGRAFIYA	Oceanologia OCEANOGRAFIA	Oceanologia OCEANOGRAFIA
3095 OCHER COMS	OCHER	OCRE	OCKER *UF:* Coenothecalia	OKHRA	OCRE	OCRA

	ENGLISH	FRANCAIS	DEUTSCH	RUSSKIJ	ESPANOL	ITALIANO
3096 OCHREPT SUSS	Ochrepts INCEPTISOLS	Ochrept INCEPTISOL; SOL BRUN	Ochrept INCEPTISOL; BRAUN=ERDE	Okhrept INSEPTISOL	Ochrept INCEPTISOL; SUELO=PARDO	Ochrept INCEPTISOL; SUOLO BRUNO
3097 OCTAHEDRITE EXTS	Octahedrite METEORITES	Octahedrite METEORITE MET- ALLIQUE	Oktahedrit METALL= METEORIT	Oktaehdrit METEORIT ZHELEZNYJ	Octaedrita METEORITO- METALICO	Ottaedrite METEORITE MET- ALLICA
3098 OCTOCORALLA PALS	OCTOCORALLIA UF: Alcyonacea; Coeno- thecalia; Trachypsam- miacea	OCTOCORALLIA UF: Alcyonacea; Coeno- thecalia; Trachypsam- miacea	OCTOCORALLIA UF: Alcyonacea; Coeno- thecalia; Trachypsam- miacea	OCTOCORALLIA UF: Alcyonacea; Coeno- thecalia; Trachypsam- miacea	OCTOCORALLA UF: Alcyonacea; Coeno- thecalia; Trachypsam- miacea	OCTOCORALLIA UF: Alcyonacea; Coeno- thecalia; Trachypsam- miacea
3099 ODONTOLOGY PALE	Odontology TEETH	Odontologie DENT	Odontologie ZAHN	ODONTOLOGIYA	Odontologia DIENTE	Odontologia DENTE
3100 ODONTOPLEURIDA PALS	ODONTOPLEURIDA	ODONTOPLEURIDA	ODONTOPLEURIDA	Odontopleurida POLYMERA	ODONTOPLEURIDA	ODONTOPLEURIDA
3101 ODONTORNITHES PALS	Odontornithes AVES	Odontornithes AVES	Odontornithes AVES	ODONTORNITHES	Odontornithes AVES	Odontornithes AVES
3102 OFFSHORE MARI	OFFSHORE UF: Drilling vessel; Offshore drilling; Off- shore well	OFFSHORE UF: Navire sondage; Sondage offshore	OFF=SHORE UF: Bohr=Schiff; Offshore=Bohrung	PRIBREZH'E	OFF=SHORE UF: Barco=sondeo; Sondeo=off=shore	OFFSHORE UF: Nave per trivellaz- ione; Perforazione offshore
3103 OFFSHORE DRILLING MINI	Offshore drilling BOREHOLES; OFF- SHORE	Sondage offshore SONDAGE; OFF- SHORE	Offshore=Bohrung BOHRUNG; OFF= SHORE	BURENIE MOR- SKOE	Sondeo=off=shore POZO=SONDEO; OFF=SHORE	Perforazione offshore SONDAGGIO; OFF- SHORE
3104 OFFSHORE WELL MINI	Offshore well WELLS; OFFSHORE	Puits offshore PLATEFORME MARINE	Offshore=Bohrloch MARIN. PLATTFORM	SKVAZHINA MOR- SKAYA	Perforacion=off=shore PLATAFORMA= MARINA	Pozzo offshore PIATTAFORMA MARINA
3105 OIL AND GAS FIELD ECON	OIL AND GAS FIELDS UF: Gas fields; Oil fields	CHAMP HYDRO- CARBURE UF: Champ gaz; Champ petrole	KW=FELD UF: Gas=Feld; Oelfeld	RAJON NEFTEGA- ZONOSNYJ	CAMPO= HIDROCARBURO UF: Campo=gas= hidrocarburo; Campo= petrolifero	CAMPO D'IDROCARBURI UF: Campo di metano; Campo petrolifero
3106 OIL AND GAS MAP ECON	OIL=AND=GAS MAPS	CARTE HYDRO- CARBURE	KW= LAGERSTAETTEN= KARTE	Karta neftegazonos- nosti GAZ; NEFT'	MAPA= HIDROCARBURO	CARTA GIACI- MENTI D'IDROCARBURI
3107 OIL FIELD ECON	Oil fields OIL AND GAS FIELDS	Champ petrole CHAMP HYDRO- CARBURE; PETROLE	Oelfeld KW=FELD; ERDOEL	BASSEJN NEF- TENOSNYJ	Campo=petrolifero CAMPO= HIDROCARBURO; PETROLEO	Campo petrolifero CAMPO D'IDROCARBURI; PETROLIO
3108 OIL SAND COMS	OIL SANDS	SABLE BITU- MINEUX	OEL=SAND	PESOK NEFTENOS- NYJ	ARENA= BITUMINOSA	SABBIA BITUMINOSA
3109 OIL SEEP ECON	OIL SEEPS	SUINTEMENT PETROLE	ERDOEL= SICKERUNG	ISTOCHNIK NEF- TYANOJ	EXUDACION= PETROLEO	STILLICIDIO DI OLIO
3110 OIL SHALE COMS	OIL SHALE UF: Bituminous shale	Schiste a huile SCHISTE BITU- MINEUX	Oel=Schiefer BITUMINOES. SCHIEFER	SLANETS GORYUCHIJ UF: Slanets bituminoz- nyj	Esquisto=petrolifero ESQUISTO= BITUMINOSO	Scisto a olio SCISTO BITUMINOSO

	ENGLISH	FRANCAIS	DEUTSCH	RUSSKIJ	ESPANOL	ITALIANO
3111 OIL TRAP ECON	Oil trap TRAPS; PETRO- LEUM	Piege a petrole PIEGE; PETROLE	Erdoel=Falle FALLE; ERDOEL	LOVUSHKA NEFTI	Trampa=petrolifera TRAMPA; PETROLEO	Trappola petrolifera TRAPPOLA; PETROLIO
3112 OIL WELL MINI	Oil well PETROLEUM; WELLS	Puits petrole EXPLOITATION; PETROLE	Erdoel=Bohrung GEWINNUNG; ERD- OEL	SKVAZHINA NEF- TYANAYA	Pozo=de=petroleo EXPLOTACION; PETROLEO	Pozzo petrolifero SFRUTTAMENTO; PETROLIO
3113 OIL=GAS INTERFACE ECON	OIL=GAS INTER- FACE	INTERFACE HUILE GAZ	OEL=GAS= GRENZFLACHE	KONTAKT GAZONEFTYANOJ	INTERFASE= ACEITE=GAS	INTERFACCIA OLIO=GAS
3114 OIL=WATER INTERFACE ECON	OIL=WATER INTERFACE	Interface huile eau INTERFACE	Oel=Wasser=Grenze GRENZFLAECHE	RAZDEL VODO- NEFTYANOJ	Interfase=agua= petroleo INTERFASE	Interfaccia olio=acqua INTERFACCIA
3115 OLIGOCENE STRS	OLIGOCENE	OLIGOCENE	OLIGOZAEN	OLIGOTSEN	OLIGOCENO	OLIGOCENE
3116 OLISTOLITH SEDI	OLISTOLITHS	OLISTOLITHE	OLISTHOLITH	OLISTOLIT	OLISTOLITO	OLISTOLITE
3117 OLISTOSTROME SEDI	OLISTOSTROMES	OLISTOSTROME	OLISTHOSTROM	OLISTOSTROM	OLISTOSTROMA	OLISTOSTROMA
3118 OLIVINE MING	OLIVINE	OLIVINE	OLIVIN	OLIVIN	OLIVINO	OLIVINA
3119 ONCOLITE SEDI	ONCOLITES	ONCOLITE	ONKOLITH	ONKOLIT	ONCOLITO	ONCOLITE
3120 ONE=DIMENSIONAL MODEL MATH	ONE= DIMENSIONAL MODELS	MODELE 1 DIMEN- SION	EINDIMENSIONAL. MODELL	MODEL' ODNOM- ERNAYA	MODELO= UNIDIMENSIONAL	MODELLO MONODIMENSION- ALE
3121 ONTOGENY PALE	ONTOGENY *UF:* Gemmation; Gigan- tism; Growth lines; Haeckel's law; Meta- morphosis; Metanau- plius; Neoteny; Ortho- genesis	ONTOGENIE *UF:* Bourgeonnement; Gigantisme; Ligne croissance; Loi Haeckel; Metamor- phose; Metanauplius; Neotenie; Orthogenese	ONTOGENESE *UF:* Anwachs=Streifen; Biolog.Metamorphose; Gigantismus; Haeckel's=Gesetz; Knospung; Metanau- plius; Neotenie; Ortho- genese	ONTOGENIYA	ONTOGENIA *UF:* Gemacion; Gigan- tismo; Ley=de=Haeck- el; Linea=de=creci- miento; Metamorfosis; Metanauplius; Neote- nia; Ortogenesis	ONTOGENESI *UF:* Gemmazione; Gigantismo; Legge di Haeckel; Metamor- fosi; Metanauplius; Neotenia; Ortogenesi; Stria di accrescimento
3122 ONYCHOPHORA PALS	Onychophora ARTHROPODA	Onychophora ARTHROPODA	Onychophora ARTHROPODA	ONYCHOPHORA	Onychophora ARTHROPODA	Onychophora ARTHROPODA
3123 OOID TEST	Ooid SPHERULES	Ooide SPHERULE	Ooid SPHAERULIT	Ooid OOLIT	Ooide ESFERULA	Ooide SFERULA
3124 OOLITE SEDI	OOLITE	OOLITE	OOLITH	OOLIT *UF:* Ooid	OOLITO	OOLITE
3125 OOLITIC LIMESTONE SEDS	OOLITIC LIME- STONE	CALCAIRE OOLI- TIQUE	OOLITH=KALK	IZVESTNYAK OOLI- TOVYJ	CALIZA=OOLITICA	CALCARE OOLI- TICO
3126 OOLITIC TEXTURE TEST	OOLITIC TEXTURE	TEXTURE OOLI- TIQUE	OOLITH=GEFUEGE	OOLITOVYJ *UF:* Pizolitovyj	TEXTURA= OOLITICA	TESSITURA OOLI- TICA

	ENGLISH	FRANCAIS	DEUTSCH	RUSSKIJ	ESPANOL	ITALIANO
3127 OOZE SEDS	OOZE	BOUE ORGANIQUE	ORGAN.SCHLAMM	IL GLUBOKOVOD-NYJ	BARRO=ORGANICO	FANGO ORGANICO
3128 OPAQUE PHCH	Opaque(adj.)	Opaque(adj.)	Opak(adj.)	NEPROZRACHNYJ	Opaco(adj.)	Opaco(adj.)
3129 OPEN-PIT MINING MINI	SURFACE MINING *UF:* Strip mining	CIEL OUVERT *UF:* Exploitation en decouverte	TAGEBAU *UF:* Strip-mining	RAZRABOTKA OTKRYTAYA	CIELO=ABIERTO *UF:* Mina=cielo=abierto	MINIERA A CIELO APERTO *UF:* Sfruttamento in superficie
3130 OPERATIONS MISC	Operations METHODS	Procedure METHODOLOGIE	Verfahren METHODIK	MEROPRIYATIYA	Procedimiento METODOLOGIA	Operazioni METODOLOGIA
3131 OPHIDIA PALS	Ophidia LEPIDOSAURIA	Ophidia SQUAMATA	Ophidia SQUAMATA	OPHIDIA	Ophidia SQUAMATA	Ophidia SQUAMATA
3132 OPHIOCISTIOIDEA PALS	OPHIOCISTOIDEA	OPHIOCISTIOIDEA	OPHIOCISTIOIDEA	OPHIOCISTIOIDEA	OPHIOCISTIOIDEA	OPHIOCISTIOIDEA
3133 OPHIOLITE IGNE	OPHIOLITE	OPHIOLITE	OPHIOLITH	FORMATSIYA OFIOLITOVAYA	OFIOLITA	OFIOLITE
3134 OPHITIC TEXTURE TEST	OPHITIC TEXTURE	TEXTURE OPHI-TIQUE	OPHIT=GEFUEGE	OFITOVYJ	TEXTURA=OFITICA	TESSITURA OFI-TICA
3135 OPHIUROIDEA PALS	OPHIUROIDEA	OPHIUROIDEA	OPHIUROIDEA	OPHIUROIDEA	OPHIUROIDEA	OPHIUROIDEA
3136 OPISTHOBRANCHIA PALS	Opisthobranchia GASTROPODA	Opisthobranchia GASTROPODA	Opisthobranchia GASTROPODA	Opisthobranchia GASTROPODA	Opisthobranchia GASTROPODA	Opisthobranchia GASTROPODA
3137 OPOKA SEDS	Opoka CHERT	Opoka SILEXITE	Opoka FLINT	Opoka KREMNEZEM; MERGEL'	Opoka SILEXITA	Opoka SILEXITE
3138 OPTICAL EXTINCTION PHCH	Optical extinction OPTICAL PROPER-TIES	Extinction optique PROPRIETE OPTIQUE	Ausloeschung OPT.EIGENSCHAFT	POGASANIE	Extincion=optica PROPIEDAD=OPTICA	Estinzione ottica PROPRIETA OTTICA
3139 OPTICAL POLARIZATION PHCH	POLARIZATION	POLARISATION	POLARISATION	POLYARIZATSIYA	POLARIZACION	POLARIZZAZIONE
3140 OPTICAL PROPERTY PHCH	OPTICAL PROPER-TIES *UF:* Biaxial crystals; Contrast; Crystal optics; Greasy luster; Immersion method; Interference; Irides-cence; Luster; Metallic properties; Optical extinction; Scintilla-tions	PROPRIETE OPTIQUE *UF:* Contraste; Eclat; Eclat gras; Extinction optique; Interference; Irisation; Mineral biaxe; Optique cristal; Propriete metallique; Scintillation	OPT.EIGENSCHAFT *UF:* Ausloeschung; Brechungs-Vermoegen; Fett=Glanz; Glanz; Interferenz; Irideszenz; Kontrast; Kristall=Optik; Metallisch. Eigenschaft; Refrakto-meter; Szintillation; Zweiachsig.Kristall	SVOJSTVO OPTI-CHESKOE	PROPIEDAD=OPTICA *UF:* Brillo; Brillo=graso; Centelleo; Contraste; Extincion=optica; Interferencia; Iridiscen-cia; Metalico; Mineral=biaxial; Optica=cristalina	PROPRIETA OTTICA *UF:* Contrasto; Cristallo biassiale; Estinzione ottica; Interferenza; Iridescenza; Lucentezza; Lucentezza grassa; Ottica mineralogica; Proprieta metallica; Scintillazione
3141 OPTICAL ROTATION PHCH	OPTICAL ROTA-TION	ROTATION OPTIQUE	OPT.DREHUNG	VRASHCHENIE OPTICHESKOE	PODER-ROTATORIO	ROTAZIONE OTTICA

	ENGLISH	FRANCAIS	DEUTSCH	RUSSKIJ	ESPANOL	ITALIANO
3142 OPTICAL SPECTROSCOPY METH	OPTICAL SPEC-TROSCOPY	SPECTROMETRIE OPTIQUE	OPT. SPEKTROMETRIE	SPEKTROMETRIYA OPTICHESKAYA	ESPECTROMETRIA=OPTICA	SPETTROMETRIA OTTICA
3143 OPTICAL SPECTRUM PHCH	OPTICAL SPECTRA	SPECTRE OPTIQUE	OPT.SPEKTRUM	SPEKTR OPTI-CHESKIJ	ESPECTRO=OPTICO	SPETTRO OTTICO
3144 ORANGE MATERIAL EXTR	ORANGE MATE-RIAL	MATIERE ORANGE LUNE	ORANGES=MATERIAL	[ORANGE MATE-RIAL]	MATERIA=LUNAR=NARANJA	MATERIALE ARANCIONE
3145 ORBICULAR TEXTURE TEST	ORBICULAR TEX-TURE	TEXTURE ORBICULAIRE	SPHAEROIDAL=GEFUEGE	ORBIKULYARNYJ	TEXTURA=ORBICULAR	TESSITURA ORBI-COLARE
3146 ORBIT EXTR	Orbit SATELLITE METH-ODS	Orbite SATELLITE ARTI-FICIEL	Orbit KUENSTL. SATELLIT	ORBITA	Orbita SATELITE=ARTIFICIAL	Orbita SATELLITE ARTIFI-CIALE
3147 ORBITOIDACEA PALS	ORBITOIDACEA	ORBITOIDACEA	ORBITOIDACEA	ORBITOIDACEA	ORBITOIDACEA	ORBITOIDACEA
3148 ORBITOIDIDAE PALS	ORBITOIDIDAE	ORBITOIDIDAE	ORBITOIDIDAE	ORBITOIDIDAE	ORBITOIDIDAE	ORBITOIDIDAE
3149 ORDER–DISORDER MINE	ORDER=DISORDER	ORDRE DESORDRE	ORDER=DISORDER	UPORYADOCHEN-NOST'	ORDEN=DESORDEN	ORDINE=DISORDINE
3150 ORDOVICIAN STRS	ORDOVICIAN	ORDOVICIEN	ORDOVIZIUM	ORDOVIK	ORDOVICICO	ORDOVICIANO
3151 ORE ECON	Ore ORE BODIES	Minerai MINERALISATION	Erz MINERALISATION	RUDA *UF:* Assay value; Barren deposit; Kachestvo rudy; Ore source; Ore transport; Roll=type deposit	Mena MINERALIZACION	Giacimento minerario MINERALIZZAZ-IONE
3152 ORE BODY ECON	ORE BODIES *UF:* Drag ore; Horsetail; Metallotect; Ore; Ore deposit; Ore shoots; Ore sources; Sedimen-tary ore	Corps mineralise MORPHOLOGIE GITE	Erzkoerper LAGERSTAETTEN=FORM	TELO RUDNOE *UF:* Rudnye zhily	Cuerpo=mineralizado MORFOLOGIA=YACIMIENTO	Corpo mineralizzato MORFOLOGIA DI GIACIMENTO
3153 ORE CONTROL ECON	ORE CONTROLS	CONTROLE GITE *UF:* Metallotecte	LAGERSTAETTEN=MILIEU *UF:* Metallotekt	KONTROL' ORU-DENENIYA	CONTROL=YACIMIENTO *UF:* Metalotecto	CONTROLLO DI MINERALIZZAZ-IONE *UF:* Metallotecto
3154 ORE DEPOSIT ECON	Ore deposit ORE BODIES	Gite mineral RESSOURCE MINERALE	Erz=Lagerstaette MINERAL. ROHSTOFFE	ZALEZH' RUD-NAYA	Yacimiento=mineral RECURSO=MINERO	Giacimento metallifero RISORSE MINERA-RIE
3155 ORE DRESSING PLANT MINI	Ore dressing plant DRESSING	Usine traitement VALORISATION	Erz=Aufbereitungs=Anlage AUFBEREITUNG	FABRIKA OBOGATI-TEL'NAYA	Planta=preparacion=mena BENEFICIACION	Impianto di tratta-mento PREPARAZIONE DEI MINERALI
3156 ORE GRADE ECON	Grade ORE QUALITY	Teneur QUALITE MINERAI	Erz=Gehalt ERZ=QUALITAET	SODERZHANIE	Ley=mineral CALIDAD=MINERAL	Tenore QUALITA DEL MINERALE

	ENGLISH	FRANCAIS	DEUTSCH	RUSSKIJ	ESPANOL	ITALIANO
3157 ORE GUIDE ECON	ORE GUIDES	GUIDE *UF:* Guide tectonique; Indicateur	INDIKATOR= MINERAL *UF:* Indikator; Tekton. Indikator	KRITERIJ POISKOVYJ	GUIA *UF:* Guia=estructural; Indicador	GUIDA *UF:* Guida strutturale; Indicatore
3158 ORE MICROSCOPY METH	ORE MICROSCOPY	MICROSCOPIE MINERAI	ERZ= MIKROSKOPIE	MINERAGRAFIYA	MICROSCOPIA= MINERAL	MICROSCOPIA MINERARIA
3159 ORE MINERAL ECON	ORE MINERALS	CONSTITUANT MINERAI	ERZ=BESTANDTEIL	RUDNYJ	CONSTITUYENTE= MINERAL	MINERALE UTILE
3160 ORE QUALITY ECON	ORE QUALITY *UF:* Carat; Grade	QUALITE MINERAI *UF:* Carat; Essai min- erai; Teneur; Teneur metal	ERZ=QUALITAET *UF:* Erz=Analyse; Erz= Gehalt; Karat; Metall= Gehalt	Kachestvo rudy RUDA	CALIDAD= MINERAL *UF:* Analisis=contenido; Ensayo; Ley=mineral; Quilate	QUALITA DEL MINERALE *UF:* Carato; Saggio; Tenore; Tenore in metallo
3161 ORE SHOOT ECON	Ore shoots ORE BODIES	Colonne MORPHOLOGIE GITE	Erzfall LAGERSTAETTEN= FORM	STOLB RUDNYJ	Columna MORFOLOGIA= YACIMIENTO	Colonna MORFOLOGIA DI GIACIMENTO
3162 ORE SOURCE ECON	Ore sources ORE BODIES	Source minerai ORIGINE	Erz=Quelle HERKUNFT	[ORE SOURCE] GENEZIS; RUDA	Yacimiento ORIGEN	Fonte mineraria ORIGINE
3163 ORE TRANSPORT ECON	ORE TRANSPORT	REMANIEMENT MINERAI	ERZ=FRACHT	[ORE TRANSPORT] RUDA; TRANSPOR- TIROVKA (GEOL)	TRANSPORTE=DE= MINERAL	RIMANEGGIA- MENTO DI MINER- ALE
3164 ORGANIC MISC	Organic(adj.)	Organique(adj.)	Organisch(adj.)	ORGANICHESKIJ *UF:* Organic residue; Organic sediment	Organico(adj.)	Organico(adj.)
3165 ORGANIC ACID CHES	ORGANIC ACIDS	ACIDE ORGANIQUE *UF:* Acetolyse	ORGAN.SAEURE *UF:* Azetolyse	KISLOTA ORGANI- CHESKAYA	ACIDO=ORGANICO *UF:* Acetolisis	ACIDO ORGANICO *UF:* Acetolisi
3166 ORGANIC CARBON CHES	ORGANIC CARBON	CARBONE ORGANIQUE	ORGAN. KOHLENSTOFF	UGLEROD ORGANI- CHESKIJ	CARBONO= ORGANICO	CARBONIO ORGANICO
3167 ORGANIC COMPOUND CHES	ORGANIC COM- POUNDS	MINERAL ORGANIQUE	ORGAN. VERBINDUNG	SOEDINENIE ORGANICHESKOE	MINERAL= ORGANICO	COMPOSTO ORGANICO
3168 ORGANIC GEOCHEMISTRY GEOC	Organic geochemistry ORGANIC MATERI- ALS	Geochimie organique MATIERE ORGANIQUE	Organ.Geochemie ORGAN.SUBSTANZ	GEOKHIMIYA ORGANICHESKAYA	Geoquimica=organica MATERIA= ORGANICA	Geochimica organica MATERIA ORGANICA
3169 ORGANIC LIFE PALE	Organic life LIFE	Vie organique VIE	Organ.Leben LEBEN	ZHIZN' ORGANI- CHESKAYA	Vida=organica VIDA	Vita organica VITA
3170 ORGANIC MATERIAL CHES	ORGANIC MATERI- ALS *UF:* Chitin; Lignin; Organic geochemistry; Phytane; Sterols	MATIERE ORGANIQUE *UF:* Chitine; Ether; Geochimie organique; Humine; Lignine; Phytane; Sterol	ORGAN.SUBSTANZ *UF:* Aether; Chitin; Humin; Lignin; Organ. Geochemie; Phytan; Sterol	VESHCHESTVO ORGANICHESKOE *UF:* Ehfir; Sterol; Tsel- lyuloza	MATERIA= ORGANICA *UF:* Esteroles; Eter; Fitan; Geoquimica= organica; Humin; Lignina; Quitina	MATERIA ORGANICA *UF:* Etere; Fitano; Geochimica organica; Sterolo
3171 ORGANIC NITROGEN CHES	ORGANIC NITRO- GEN	Azote organique AZOTE; ORIGINE BIOGENE	Organ.Stickstoff N; BIOGEN. BILDUNG	AZOT ORGANI- CHESKIJ	Nitrogeno=organico NITROGENO; ORIGEN= BIOGENICO	Azoto organico AZOTO; ORIGINE ORGANOGENA

	ENGLISH	FRANCAIS	DEUTSCH	RUSSKIJ	ESPANOL	ITALIANO
3172 ORGANIC ORIGIN GEOL	BIOGENIC ORIGIN	ORIGINE BIOGENE *UF:* Azote organique; Biogenese; Biolite; Biomicrite; Origine bacteriogene; Structure organique	BIOGEN.BILDUNG *UF:* Bakteriogen. Entstehung; Biogenese; Biolith; Biomikrit; Organ.Gefuege; Organ. Stickstoff	ORGANOGENNYJ	ORIGEN= BIOGENICO *UF:* Bacteriogenico; Biogenesis; Biolito; Biomicrita; Estructura=organica; Nitrogeno=organico	ORIGINE ORGANO- GENA *UF:* Azoto organico; Biogenesi; Biolite; Biomicrite; Origine batteriogena; Struttura organica
3173 ORGANIC RESIDUE SEDI	ORGANIC RESI- DUES *UF:* Carbonaceous rock; Organic sediments	Residu organique ROCHE CARBONEE	Organ. Loesungsrueckstand KOHLENSTOF- FHALTIG.GESTEIN	[ORGANIC RESI- DUE] ORGANICHESKIJ; OSADOK	Residuo=organico ROCA= CARBONOSA	Residuo organico ROCCIA CAR- BONIOSA
3174 ORGANIC SEDIMENT SEDS	Organic sediments ORGANIC RESI- DUES	Sediment organique ROCHE CARBONEE	Organogen.Sediment KOHLENSTOF- FHALTIG.GESTEIN	[ORGANIC SEDI- MENT] ORGANICHESKIJ; OSADOK	Sedimento=organico ROCA= CARBONOSA	Sedimento organico ROCCIA CAR- BONIOSA
3175 ORGANIC STRUCTURE TEST	Organic structure BIOGENIC STRUC- TURES	Structure organique ORIGINE BIOGENE	Organ.Gefuege BIOGEN.BILDUNG	POSTROJKA ORGANOGENNAYA *UF:* Bioturbatsiya	Estructura=organica ORIGEN= BIOGENICO	Struttura organica ORIGINE ORGANO- GENA
3176 ORGANISM PALE	Organisms BIOLOGY	Organisme BIOLOGIE	Organismus BIOLOGIE	ORGANIZM	Organismo BIOLOGIA	Organismo BIOLOGIA
3177 ORIENTATION GEOL	ORIENTATION	ORIENTATION *UF:* Boussole; Direction horizontale; Vergence	ORIENTIERUNG *UF:* Horizontal= Richtung; Magnet= Kompass; Vergenz	ORIENTIROVKA *UF:* Napravlenie gori- zontal'noe; Oblique orientation; Preferred orientation	ORIENTACION *UF:* Brujula; Direccion= horizontal; Vergencia	ORIENTAZIONE *UF:* Direzione orizzon- tale
3178 ORIENTATION SURVEY ECON	PRELIMINARY STUDIES	ENQUETE PRE- LIMINAIRE	VORLAEUFIG. UNTERSUCHUNG	Predvaritel'nyj poisk POISKI	ESTUDIO= PRELIMINAR	STUDIO PRELIMI- NARE
3179 ORIGIN GEOL	Origin PROVENANCE	ORIGINE *UF:* Bloc indicateur; Origine supracrustale; Provenance; Source minerai	HERKUNFT *UF:* Erz=Quelle; Leit= Geschiebe; Lieferge- biet; Suprakrustal. Genese	PROISKHOZHDE- NIE	ORIGEN *UF:* Bloque=indicador; Origen=supracortical; Procedencia; Yaci- miento	ORIGINE *UF:* Erratico indicatore; Fonte mineraria; Ori- gine sopracrostale; Provenienza
3180 ORIGOFACIES SEDI	Origofacies LITHOFACIES	Facies primaire LITHOFACIES	Origo=Fazies LITHO=FAZIES	ORIGOFATSIYA	Origofacies LITOFACIES	Origofacies LITOFACIES
3181 ORNAMENTAL STONE COMS	ORNAMENTAL MATERIALS *UF:* Decorative stones	MATERIAU ORNE- MENTATION *UF:* Pierre ornementale	WERKSTEIN *UF:* Verblend=Stein	KAMEN' PODELOCHNYJ	MATERIAL= ORNAMENTACION *UF:* Piedra= ornamentacion	MATERIALE DA ORNAMENTAZ- IONE *UF:* Pietra decorativa
3182 ORNAMENTATION PALE	ORNAMENTATION	ORNEMENTATION TEST	SKULPTUR	SKUL'PTURA	ORNAMENTACION= EXTERIOR	ORNAMENTAZ- IONE
3183 ORNITHISCHIA PALS	ORNITHISCHIA	ORNITHISCHIA	ORNITHISCHIA	ORNITHISCHIA	ORNITHISCHIA	ORNITHISCHIA
3184 OROGENIC BELT STRU	Orogenic belt FOLD BELTS	CHAINE GEOSYN- CLINALE *UF:* Ceinture circum= pacifique; Ceinture grand cercle; Chaine plissee; Externides; Internides	GEOSYNKLINAL= KETTE *UF:* Externiden; Falten= Guertel; Grosskreis= Guertel; Interniden; Zirkumpazif.Guertel	Poyas orogenicheskij GEOSINKLINAL'	CORDILLERA= GEOSINCLINAL *UF:* Cinturon= circumpacifico; Cinturon=plegado; Cinturon=tectonico; Externides; Internidas	CATENA GEOSIN- CLINALE *UF:* Catena a pieghe; Cintura circumpacifica; Cintura massima; Esternidi; Internidi

	ENGLISH	FRANCAIS	DEUTSCH	RUSSKIJ	ESPANOL	ITALIANO
3185 OROGENY STRU	OROGENY *UF:* Algoman orogeny; Archeides; Beltian orogeny; Early proterozoides; Eburnean orogeny; Internides; Kibara orogeny; Late proterozoides; Mazatsal revolution; Proterozoides; Svecofennian orogeny; Tectonic cycle	OROGENESE *UF:* Cycle tectonique; Superstructure tectonique	OROGENESE *UF:* Tekton. Ueberstruktur; Tekton. Zyklus	OROGENEZ *UF:* Takonskij orogen	OROGENESIS *UF:* Ciclo=tectonico; Superestructura=tectonica	OROGENESI *UF:* Ciclo tettonico; Sovrastruttura
3186 ORTHENT SUSS	Orthents ENTISOLS	Orthent ENTISOL; SOL BRUT	Orthent ENTISOL; AZONAL. BODEN	Ortent EHNTISOL	Orthent ENTISOL; SUELO=BRUTO	Orthent ENTISOL; SUOLO AZONALE
3187 ORTHID SUSS	Orthids ARIDISOLS	Orthid ARIDISOL	Orthid ARIDISOL	Ortid ARIDISOL	Orthid ARIDISOL	Orthid ARIDISOL
3188 ORTHIDA PALS	ORTHIDA	ORTHIDA	ORTHIDA	ORTHIDA	ORTHIDA	ORTHIDA
3189 ORTHOAMPHIBOLITE IGMS	ORTHOAMPHIBO-LITE	ORTHO AMPHIBO-LITE	ORTHO=AMPHIBOLIT	Ortoamfibolit AMFIBOLIT	ORTOANFIBOLITA	ORTOANFIBOLITE
3190 ORTHOD SUSS	Orthods SPODOSOLS	Orthod SPODOSOL; PODZOL	Orthod SPODOSOL; PODSOL	Ortod SPODOSOL	Orthod SPODOSOL; PODZOL	Orthod SPODOSOL; PODSOL
3191 ORTHOGENESIS PALE	Orthogenesis ONTOGENY	Orthogenese ONTOGENIE	Orthogenese ONTOGENESE	ORTOGENEZ	Ortogenesis ONTOGENIA	Ortogenesi ONTOGENESI
3192 ORTHOGNEISS IGMS	ORTHOGNEISS	ORTHO GNEISS	ORTHO=GNEIS	Ortognejs GNEJS	ORTOGNEIS	ORTOGNEISS
3193 ORTHOPHYRIC TEXTURE TEST	Orthophyric texture TEXTURES	Texture orthophyrique TEXTURE	Orthophyr=Gefuege KORN=GEFUEGE	ORTOFIROVYJ	Textura=ortofidica TEXTURA	Tessitura ortofirica TESSITURA
3194 ORTHORHOMBIC SYSTEM MINE	Orthorhombic system CRYSTAL SYSTEMS	Systeme orthorhombique SYSTEME CRISTALLIN	Orthorhomb.System KRISTALL=SYSTEM	Singoniya ortorombicheskaya SINGONIYA ROMBICHESKAYA	Sistema=ortorombico SISTEMA=CRISTALINO	Sistema ortorombico SISTEMA CRISTALLINO
3195 ORTHOX SUSS	Orthox OXISOLS	Orthox OXISOL; MILIEU HUMIDE	Orthox OXISOL; FEUCHT. MILIEU	Ortoks OKSISOL	Orthox OXISOL; MEDIO=HUMEDO	Orthox OXISOL; AMBIENTE UMIDO
3196 ORYCTOCOENOSIS PALE	Oryctocoenosis THANATOCENOSES	Oryctocoenose THANATO-COENOSE	Oryktocoenose THANATO-COENOSE	ORIKTOTSENOZ	Orictocenosis TANATOCENOSIS	Oryctocoenosis TANATOCENOSI
3197 OSCILLATION SOLI	OSCILLATIONS	OSCILLATION	OSZILLATION	KOLEBANIE *UF:* Kolebanie svobodnoe; Strong motion	OSCILACION	OSCILLAZIONE
3198 OSCILLATION THEORY STRU	Oscillation theory THEORETICAL STUDIES; TECTONICS	Theorie oscillatoire THEORIE; TECTOGENESE	Oszillations=Theorie THEORIE; TEKTOGENESE	GIPOTEZA OSTSILYATSIONNAYA	Teoria=oscilatoria TEORIA; TECTOGENESIS	Teoria delle oscillazioni TEORIA; TETTOGENESI

	ENGLISH	FRANCAIS	DEUTSCH	RUSSKIJ	ESPANOL	ITALIANO
3199 OSMIUM CHEE	OSMIUM	OSMIUM	OS	OSMIJ	OSMIO	OSMIO
3200 OSTEICHTHYES PALS	OSTEICHTHYES *UF:* Crossopterygii; Sarcopterygii	OSTEICHTHYES *UF:* Crossopterygii; Sarcopterygii	OSTEICHTHYES *UF:* Crossopterygii; Sarcopterygii	OSTEICHTHYES *UF:* Brachiopterygii	OSTEICHTHYES *UF:* Crossopterygii; Sarcopterygii	OSTEICHTHYES *UF:* Crossopterygii; Sarcopterygii
3201 OSTEOLOGY PALE	OSTEOLOGY	OSTEOLOGIE	OSTEOLOGIE	OSTEOLOGIYA	OSTEOLOGIA	OSTEOLOGIA
3202 OSTEOSTRACI PALS	Osteostraci AGNATHA	Osteostraci AGNATHA	Osteostraci AGNATHA	OSTEOSTRACI	Osteostraci AGNATHA	Osteostraci AGNATHA
3203 OSTRACODA PALS	OSTRACODA *UF:* Metacopina; Myo- docopida; Paleocopida; Podocopida	OSTRACODA *UF:* Metacopina; Myo- docopida; Paleocopida; Podocopida	OSTRACODA *UF:* Metacopina; Myo- docopida; Paleocopida; Podocopida	OSTRACODA *UF:* Archaeocopida; Leperditocopida; Myo- docopida	OSTRACODA *UF:* Metacopina; Myo- docopida; Paleocopida; Podocopida	OSTRACODA *UF:* Metacopina; Myo- docopida; Paleocopida; Podocopida
3204 OSTREACEA PALS	OSTREACEA	OSTREACEA	OSTREACEA	Ostreacea PTERIOIDA	OSTREACEA	OSTRACEA
3205 OTOLITH PALE	OTOLITHS	OTOLITE	OTOLITH	OTOLIT	OTOLITO	OTOLITE
3206 OUTCROP GEOL	OUTCROPS	AFFLEUREMENT	AUFSCHLUSS	OBNAZHENIE *UF:* Relief exposure	AFLORAMIENTO	AFFIORAMENTO
3207 OUTER CORE SOLI	OUTER CORE *UF:* E layer	NOYAU TER- RESTRE EXTERNE *UF:* Couche E	AEUSSER.ERD- KERN *UF:* E=Schicht	YADRO VNESHNEE ZEMLI	NUCLEO- EXTERNO *UF:* Capa=E	NUCLEO ESTERNO *UF:* Strato E
3208 OUTER PLANET EXTR	OUTER PLANETS	PLANETE EXTERI- EURE	AEUSSER.PLANET	Planeta yupiterskoj gruppy PLANETA	PLANETA= EXTERIOR	PIANETA ESTERNO
3209 OUTER SHELF MARI	OUTER SHELF	PLATEFORME CON- TINENTALE EXTERNE	AEUSSER.SCHELF	[OUTER SHELF] SHEL'F KONTINEN- TAL'NYJ	PLATAFORMA= CONTINENTAL= EXTERNA	PIATTAFORMA CONTINENTALE ESTERNA
3210 OUTLIER STRU	Outliers NAPPES	Avant=butte RELICTE; NAPPE	[Outlier] RELIKT; DECKE	OSTANETS	Testigo=de=erosion RELICTA; MANTO	[OUTLIER] RELITTO; FALDA
3211 OUTWASH SURF	OUTWASH	EPANDAGE FLU- VIOGLACIAIRE	SCHMELZ- WASSER= ABLAGERUNG	OTLOZHENIYA FLYUVIOGLYATSI- AL'NYE	EXPANSION= FLUVIOGLACIAR	ESPANDIMENTO FLUVIOGLACIALE
3212 OUTWASH PLAIN SURF	OUTWASH PLAINS *UF:* Apron	PLAINE FLU- VIOGLACIAIRE	GLAZIAL= AUFSCHUETTUNGS= EBENE	ZANDR	LLANURA= FLUVIOGLACIAR	PIANA FLU- VIOGLACIALE
3213 OVERBURDEN MINI	OVERBURDEN	MORTS TERRAINS *UF:* Sterile	ABRAUM *UF:* Taub.Lagerstaette	VSKRYSHA	RECUBRIMIENTO= ESTERIL *UF:* Deposito=esteril	TERRENO STERILE *UF:* Roccia sterile
3214 OVERCONSOLIDATED MATERIAL ENGI	OVERCONSOLI- IDATED MATERI- ALS	MATERIAU SUR- CONSOLIDE	STARK= VERFESTIGT. MATERIAL	PEREUPLOTNEN- NYJ	MATERIAL= SOBRECONSOLIDADO	MATERIALE SOVRACONSOLI- DATO
3215 OVERGROWTH MINE	OVERGROWTHS	SURCROISSANCE	UEBERWACHSUNG	NAROST	SUPERCRECI- MIENTO	SOVRACCRESCITA

	ENGLISH	FRANCAIS	DEUTSCH	RUSSKIJ	ESPANOL	ITALIANO
3216 OVERSATURATED PHCH	Oversaturated(adj.)	Sursature(adj.)	Uebersaettigt(adj.)	PERESYSHCHEN- NYJ *UF:* Rastvor peresyshc- hennyj	Sobresaturado(adj.)	Soprasaturo(adj.)
3217 OVERSATURATED **SOLUTION** PHCH	Oversaturated solution SATURATION	Sursaturation SATURATION	Uebersaettigung SAETTIGUNG	Rastvor peresyshc- hennyj RASTVOR; PER- ESYSHCHENNYJ	Sobresaturacion SATURACION	Sovrasaturazione SATURAZIONE
3218 OVERTHRUST STRU	OVERTHRUST FAULTS	Charriage FAILLE CHEVAU- CHEMENT	Ueberschiebung AUFSCHIEBUNG	SHAR'YAZH	Corrimiento CABALGAMIENTO	Sovrascorrimento ACCAVALLA- MENTO
3219 OVERTURNED FOLD STRU	OVERTURNED FOLDS *UF:* Vergence	PLI DEVERSE	UEBERKIPPT. FALTE	SKLADKA OPROKINUTNAYA	PLIEGUE= INVERTIDO	PIEGA ROVES- CIATA
3220 OXFORDIAN STRS	OXFORDIAN	OXFORDIEN	OXFORD	OKSFORD	OXFORDIENSE	OXFORDIANO
3221 OXIDATION PHCH	OXIDATION	OXYDATION	OXIDATION	OKISLENIE	OXIDACION	OSSIDAZIONE
3222 OXIDES MING	OXIDES	OXYDE	OXID	OKISEL	OXIDO	OSSIDO
3223 OXIDIZED ZONE ECON	OXIDATION ZONES	ZONE OXYDATION	OXIDATIONS= ZONE	ZONA OKISLENIYA	ZONA=OXIDACION	ZONA DI OSSIDAZ- IONE
3224 OXISOL SUSS	OXISOLS *UF:* Aquox; Humox; Orthox; Ustox	OXISOL *UF:* Aquox; Humox; Orthox; Ustox	OXISOL *UF:* Aquox; Humox; Orthox; Ustox	OKSISOL *UF:* Akvoks; Gumoks; Ortoks; Ustoks	OXISOL *UF:* Aquox; Humox; Orthox; Ustox	OXISOL *UF:* Aquox; Humox; Orthox; Ustox
3225 OXYGEN CHEE	OXYGEN *UF:* Oxygen gas	OXYGENE *UF:* Gaz oxygene	O *UF:* Sauerstoff	KISLOROD *UF:* Oxygen ratio	OXIGENO *UF:* Gas=oxigeno	OSSIGENO *UF:* Gas ossigeno
3226 OXYGEN GAS CHES	Oxygen gas OXYGEN; GASES	Gaz oxygene OXYGENE; GAZ	Sauerstoff O; GAS	KISLOROD GAZ	Gas=oxigeno OXIGENO; GAS	Gas ossigeno OSSIGENO; GAS
3227 OXYGEN RATIO ISOT	O=18/O=16	O 18=O 16	O18=O16	[OXYGEN RATIO] IZOTOP; KISLOROD	O=18=O=16	O=18/O=16
3228 OXYLOPHYTE PALE	Oxylophyte ECOLOGY; FLORA	Oxylophyte ECOLOGIE; FLORE	Oxylophyt OEKOLOGIE; FLORA	OKSILOFIT	Oxilofita ECOLOGIA; FLORA	Oxylophyte ECOLOGIA; FLORA
3229 P-T CONDITIONS PHCH	P=T CONDITIONS *UF:* Mineral facies	CONDITION PRES- SION TEMPERA- TURE	DRUCK= TEMPERATUR= BEDINGUNGEN	P=T USLOVIYA	CONDICION= PRESION= TEMPERATURA	CONDIZIONI P=T
3230 P-WAVES SOLI	P=WAVES	ONDE P	P=WELLE	Volna prodol'naya VOLNA SEJSMI- CHESKAYA	ONDA=P	ONDA P
3231 PACIFIC SUITE IGNE	Pacific suite VOLCANOLOGY; MAGMATIC ASSO- CIATIONS	Serie pacifique VOLCANOLOGIE; ASSOCIATION MAGMATIQUE	Pazif.Abfolge VULKANOLOGIE; MAGMAT. VERGESELLSCHAF- TUNG	TIKHOOKEANSKIJ	Serie=pacifica VULCANOLOGIA; ASOCIACION= MAGMATICA	Serie pacifica VULCANOLOGIA; ASSOCIAZIONE MAGMATICA

	ENGLISH	FRANCAIS	DEUTSCH	RUSSKIJ	ESPANOL	ITALIANO
3232 PACKING TEST	PACKING	ARRANGEMENT GRAIN	PACKUNG	UPAKOVKA	ORDENACION=GRANOS	COSTIPAZIONE
3233 PAHOEHOE IGNE	PAHOEHOE	PAHOEHOE	PAHOEHOE	PAKHOEKHOE	PAHOEHOE	PAHOEHOE
3234 PAINT MATERIAL COMS	PAINT MINERALS	COLORANT	FARBSTOFF	PIGMENT MINERAL'NYJ	COLORANTE	COLORANTE
3235 PALAEOHETERODONTA PALS	Palaeoheterodonta BIVALVIA	Palaeoheterodonta BIVALVIA	Palaeoheterodonta LAMELLI-BRANCHIATA	PALAEOHETERODONTA	Palaeoheterodonta BIVALVIA	Palaeoheterodonta BIVALVIA
3236 PALAEOPHYTICUM STRS	Palaeophyticum PALEOZOIC	Paleophytique PRIMAIRE	Palaeophytikum PALAEOZOIKUM	PALEOFIT	Paleofitico PRIMARIO	Paleofitico PALEOZOICO
3237 PALAEOTAXODONTA PALS	Palaeotaxodonta BIVALVIA	Palaeotaxodonta BIVALVIA	Palaeotaxodonta LAMELLI-BRANCHIATA	PALAEOTAXODONTA	Palaeotaxodonta BIVALVIA	Palaeotaxodonta BIVALVIA
3238 PALAGONITIZATION IGNE	PALAGONITIZATION	Palagonitisation ALTERATION HYDROTHERMALE	Palagonitisierung HYDROTHERMAL=UMWANDLUNG	PALAGONITIZATSIYA	Palagonitizacion ALTERACION=HIDROTERMAL	Palagonitizzazione ALTERAZIONE IDROTERMALE
3239 PALEO GEOL	Paleo(gen.)	Paleo(gen.)	Palaeo(gen.)	PALEO *UF:* Paleoatmosfera; Paleobatimetriya; Paleoehkologiya; Paleolatitude; Paleookeanografiya; Paleotsirkulyatsiya	Paleo(gen.)	Paleo(gen.)
3240 PALEO-ATMOSPHERE STRA	PALEOATMOSPHERE	PALEOATMOSPHERE	PALAEOATMOSPHAERE	Paleoatmosfera ATMOSFERA; PALEO	PALEOATMOSFERA	PALEOATMOSFERA
3241 PALEO-OCEANOGRAPHY STRA	PALEO=OCEANOGRAPHY	PALEOOCEANOGRAPHIE	PALAEOOZEANOGRAPHIE	Paleookeanografiya PALEO; OKEANOGRAFIYA	PALEO=OCEANOGRAFIA	PALEOOCEANOGRAFIA
3242 PALEOBATHYMETRY STRA	PALEOBATHYMETRY	PALEOBATHYMETRIE	PALAEOBATHYMETRIE	Paleobatimetriya BATIMETRIYA; PALEO	PALEOBATIMETRIA	PALEOBATIMETRIA
3243 PALEOBIOLOGY PALE	PALEOBIOLOGY	PALEOBIOLOGIE	PALAEOBIOLOGIE	Paleobiologiya PALEONTOLOGIYA	PALEOBIOLOGIA	PALEOBIOLOGIA
3244 PALEOBOTANIC PROVINCE PALE	Paleobotanic province FLORAL PROVINCES; BIOGEOGRAPHY	Province paleobotanique BIOGEOGRAPHIE; FLORE	Palaeobotan.Provinz BIOGEOGRAPHIE; FLORA	PROVINTSIYA PALEOBOTANICHESKAYA *UF:* Provintsiya floristicheskaya	Provincia=paleobotanica BIOGEOGRAFIA; FLORA	Provincia paleobotanica BIOGEOGRAFIA; FLORA
3245 PALEOBOTANY PALE	PALEOBOTANY	PALEOBOTANIQUE	PALAEOBOTANIK	PALEOBOTANIKA	PALEOBOTANICA	PALEOBOTANICA
3246 PALEOCENE STRS	PALEOCENE	PALEOCENE	PALEOZAEN	PALEOTSEN	PALEOCENO	PALEOCENE

	ENGLISH	FRANCAIS	DEUTSCH	RUSSKIJ	ESPANOL	ITALIANO
3247 PALEOCIRCULATION STRA	PALEOCIRCULA-TION	PALEOCIRCULA-TION	PALAEOZIRKULA-TION	Paleotsirkulyatsiya PALEO; TSIRKU-LYATSIYA	PALEOCIRCULA-CION	PALEOCIRCOLAZ-IONE
3248 PALEOCLIMATE STRA	Paleoclimate PALEOCLIMA-TOLOGY	PALEOCLIMAT *UF:* Paleoclimatologie	PALAEOKLIMA *UF:* Palaeoklimatologie	PALEOKLIMAT *UF:* Sled paleoklimata	PALEOCLIMA *UF:* Paleoclimatologia	PALEOCLIMA
3249 PALEOCLIMATOLOGY STRA	PALEOCLIMA-TOLOGY *UF:* Paleoclimate	Paleoclimatologie PALEOCLIMAT	Palaeoklimatologie PALAEOKLIMA	PALEOKLIMA-TOLOGIYA	Paleoclimatologia PALEOCLIMA	Paleoclimatologia PALEORILIEVO
3250 PALEOCOPIDA PALS	Paleocopida OSTRACODA	Paleocopida OSTRACODA	Paleocopida OSTRACODA	PALEOCOPIDA *UF:* Beyrichicopina	Paleocopida OSTRACODA	Paleocopida OSTRACODA
3251 PALEOCURRENT SEDI	PALEOCURRENTS	PALEOCOURANT	PALAEOSTROE-MUNG	PALEOTECHENIE	PALEOCORRIENTE	PALEOCORRENTE
3252 PALEOECOLOGY PALE	PALEOECOLOGY *UF:* Ecozones	PALEOECOLOGIE	PALOEKOLOGIE	Paleoehkologiya EHKOLOGIYA; PALEO	PALEOECOLOGIA	PALEOECOLOGIA
3253 PALEOENVIRONMENT PALE	PALEOENVIRON-MENT	PALEOENVIRON-NEMENT	PALAEOMILIEU	Paleosreda PALEOGEOGRA-FIYA; SREDA OKRUZHAYUSH-CHAYA	PALEOAMBIENTE	PALEOAMBIENTE
3254 PALEOGENE STRS	PALEOGENE	PALEOGENE	PALAEOGEN	PALEOGEN	PALEOGENO	PALEOGENE
3255 PALEOGEOGRAPHIC CONTROLS ECON	PALEOGEO-GRAPHIC CON-TROLS	CONTROLE PALEOGEO-GRAPHIQUE	PALAEOGEOGRAPH. EINFLUSS	Kontrol' paleogeografi-cheskij KONTROL'; PALEOGEOGRA-FIYA	CONTROL= PALEOGEOGRAFICO	CONTROLLO PALEOGEOGRA-FICO
3256 PALEOGEOGRAPHIC MAP STRA	PALEOGEO-GRAPHIC MAPS	CARTE PALEOGEO-GRAPHIQUE	PALAEOGEOGRAPH. KARTE	KARTA PALEOGEO-GRAFICHESKAYA	MAPA= PALEOGEOGRAFICO	CARTA PALEOGEO-GRAFICA
3257 PALEOGEOGRAPHY STRA	PALEOGEOGRAPHY	PALEOGEOGRA-PHIE *UF:* Cote fossile	PALAEOGEOGRA-PHIE *UF:* Fossil.Kuestenlinie	PALEOGEOGRA-FIYA *UF:* Kontrol' paleogeo-graficheskij; Paleosreda	PALEOGEOGRAFIA *UF:* Costa=abandonada	PALEOGEOGRAFIA *UF:* Linea di riva abban-donata
3258 PALEOGEOMOR-PHOLOGY STRA	Paleogeomorphology PALEORELIEF	Paleomorphologie PALEORELIEF	Palaeomorphologie PALAEORELIEF	PALEOGEOMOR-FOLOGIYA	Paleomorfologia PALEORELIEVE	Paleomorfologia PALEORILIEVO
3259 PALEOKARST STRA	PALEOKARST	PALEOKARST	PALAEOKARST	PALEOKARST	PALEOKARST	PALEOCARSISMO
3260 PALEOLATITUDE STRA	PALEOLATITUDE	PALEOLATITUDE	PALAEOLATITUDE	PALEOSHIROTA	PALEOLATITUD	PALEOLATITUDINE
3261 PALEOLIMNOLOGY STRA	PALEOLIMNOLOGY	PALEOLIMNOLO-GIE	PALAEOLIMNOLO-GIE	PALEOLIM-NOLOGIYA	PALEOLIM-NOLOGIA	PALEOLIM-NOLOGIA

	ENGLISH	FRANCAIS	DEUTSCH	RUSSKIJ	ESPANOL	ITALIANO
3262 PALEOMAGNETIC POLE SOLI	Paleomagnetic pole PALEOMAGNE- TISM; POLE POSI- TIONS	Pole paleomagnetique POLE GEOMAGNE- TIQUE; PALEO- MAGNETISME	Palaeomagnet.Pol GEOMAGNET.POL; PALAEOMAGNE- TISMUS	POLYUS PALEO- MAGNITNYJ	Polo=paleomagnetico POLO= GEOMAGNETICO; PALEOMAGNE- TISMO	Polo paleomagnetico POLO GEOMAGNE- TICO; PALEO- MAGNETISMO
3263 PALEOMAGNETISM SOLI	PALEOMAGNETISM UF: Paleomagnetic pole	PALEOMAGNE- TISME UF: Archeomagnetisme; Pole paleomagnetique	PALAEOMAGNE- TISMUS UF: Archaeomagnetismus; Palaeomagnet.Pol	PALEOMAGNE- TIZM UF: Arkheomagnetizm	PALEOMAGNE- TISMO UF: Arqueomagnetismo; Polo=paleomagnetico	PALEOMAGNE- TISMO UF: Archeomagnetismo; Polo paleomagnetico
3264 PALEONTOLOGICAL RECONSTRUCTION PALE	PALEONTOLOGI- CAL RECONSTRUC- TION	RECONSTITUTION PALEON- TOLOGIQUE	PALAEONTOLOG. REKONSTRUKTION	Rekonstruktsiya paleontologicheskaya PALEONTOLOGIYA; REKONSTRUKT- SIYA	RECONSTRUCCION= PALEONTOLOGICA	RICOSTRUZIONE PALEONTOLOGICA
3265 PALEONTOLOGY PALE	PALEONTOLOGY UF: Holotypes; Phylum	PALEONTOLOGIE	PALAEONTOLOGIE	PALEONTOLOGIYA UF: Paleobiologiya; Rekonstruktsiya paleontologiya	PALEONTOLOGIA	PALEONTOLOGIA
3266 PALEORELIEF STRA	PALEORELIEF UF: Paleogeomorphology	PALEORELIEF UF: Paleomorphologie	PALAEORELIEF UF: Palaeomorphologie	PALEOREL'EF	PALEORELIEVE UF: Paleomorfologia	PALEORILIEVO UF: Paleoclimatologia; Paleomorfologia
3267 PALEOSALINITY SEDI	PALEOSALINITY	PALEOSALINITE	PALAEOSALINIT- AET	Paleosolenost' SOLENOST'	PALEOSALINIDAD	PALEOSALINITA
3268 PALEOSOL STRA	PALEOSOLS UF: Planosols	PALEOSOL UF: Planosol	PALAEOBODEN UF: Planosol	PALEOPOCHVA	PALEOSUELO UF: Planosol	PALEOSUOLO UF: Planosol
3269 PALEOTEMPERATURE STRA	PALEOTEMPERA- TURE	PALEOTEMPERA- TURE UF: Paleothermometrie	PALAEOTEMPERA- TUR UF: Palaeothermometrie	PALEOTEMPERA- TURA	PALEOTEMPERA- TURA UF: Paleotermometria	PALEOTEMPERA- TURA UF: Paleotermometria
3270 PALEOTHERMOMETRY METH	Paleothermometry GEOLOGIC THER- MOMETRY	Paleothermometrie PALEOTEMPERA- TURE	Palaeothermometrie PALAEOTEMPERA- TUR	PALEOTERMO- METRIYA	Paleotermometria PALEOTEMPERA- TURA	Paleotermometria PALEOTEMPERA- TURA
3271 PALEOZOIC STRS	PALEOZOIC UF: Palaeophyticum	PRIMAIRE UF: Paleophytique	PALAEOZOIKUM UF: Palaeophytikum	PALEOZOJ	PRIMARIO UF: Paleofitico	PALEOZOICO UF: Paleofitico
3272 PALEOZOOLOGY PALE	PALEOZOOLOGY	PALEOZOOLOGIE	PALAEOZOOLOGIE	PALEOZOOLOGIYA	PALEOZOOLOGIA	PALEOZOOLOGIA
3273 PALIMPSEST IGNE	Palimpsest RELICT MATERI- ALS	Structure fantome RELICTE	Palimpsest RELIKT	PALIMPSESTOVYJ	Palimpsestica RELICTA	Palinsesto RELITTO
3274 PALINGENESIS IGNE	PALINGENESIS	PALINGENESE	PALINGENESE	PALINGENEZ	PALINGENESIS	PALINGENESI
3275 PALLADIUM CHEE	PALLADIUM	PALLADIUM	PD	PALLADIJ	PALADIO	PALLADIO
3276 PALLASITE EXTS	Pallasites METEORITES	Pallasite SIDEROLITE	Pallasit STEIN=EISEN= METEORIT	Pallasit SIDEROLIT	Pallasita SIDEROLITO	Pallasite SIDEROLITE

	ENGLISH	FRANCAIS	DEUTSCH	RUSSKIJ	ESPANOL	ITALIANO
3277 PALSA SURF	Palsa FROZEN GROUND	Palse SOL GELE; MAREC- AGE	Palsa FROST=BODEN; MORAST	Bugor torfyannoj BUGOR; TORF	Palsa SUELO=HELADO; PANTANO	Palsa SUOLO GELATO; ACQUITRINO
3278 PALYNOLOGY PALE	PALYNOLOGY	PALYNOLOGIE	PALYNOLOGIE	PALINOLOGIYA	PALINOLOGIA	PALINOLOGIA
3279 PALYNOMORPH PALS	PALYNOMORPHS	PALYNOMORPHE	PALYNOMORPHA	PALYNOMORPHA	PALINOMORFO	PALINOMORFO
3280 PAMPA SURF	Pampas STEPPES	Pampa ZONE STEPPIQUE	Pampa STEPPEN=ZONE	Pampa STEP'	Pampa ZONA=ESTEPARIA	Pampa AMBIENTE DI STEPPA
3281 PANGEA STRA	PANGAEA	PANGEE	PANGAEA	PANGEA	PANGEA	PANGEA
3282 PANIDIOMORPHIC **TEXTURE** TEST	Panidiomorphic texture TEXTURES	Texture panidiomorphe TEXTURE	Panidiomorph.Gefuege KORN=GEFUEGE	PANIDIOMORFNYJ	Textura=panidiomorfa TEXTURA	Tessitura panidiomor- fica TESSITURA
3283 PANNING METH	Panning ALLUVIAL PROS- PECTING	Prospection batee PROSPECTION ALLUVIONNAIRE	Goldwaesche ALLUVIAL= PROSPEKTION	SHLIKHOVANIE	Batea PROSPECCION= ALUVIONAR	Lavaggio a batea PROSPEZIONE IN ALLUVIONI
3284 PANTODONTA PALS	PANTODONTA	PANTODONTA	PANTODONTA	PANTODONTA	PANTODONTA	PANTODONTA
3285 PANTOTHERIA PALS	PANTOTHERIA	EUPANTOTHERIA	PANTOTHERIA	Pantotheria MAMMALIA	PANTOTHERIA	PANTOTHERIA
3286 PARA- IGNE	Para=(adj.)	Para=(adj.)	Para=(adj.)	PARA (PORODA)	Para=(adj.)	Para=(adj.)
3287 PARAAMPHIBOLITE IGMS	PARAAMPHIBO- LITE	PARA AMPHIBO- LITE	PARA= AMPHIBOLIT	Paraamfibolit AMFIBOLIT	PARA=ANFIBOLITA	PARAANFIBOLITE
3288 PARABLASTOIDEA PALS	PARABLASTOIDEA	PARABLASTOIDEA	PARABLASTOIDEA	Parablastoidea CRINOZOA	PARABLASTOIDEA	PARABLASTOIDEA
3289 PARACRINOIDEA PALS	PARACRINOIDEA	PARACRINOIDEA	PARACRINOIDEA	PARACRINOIDEA	PARACRINOIDEA	PARACRINOIDEA
3290 PARAGENESIS MINE	PARAGENESIS	PARAGENESE	PARAGENESE	PARAGENEZIS UF: Assotsiatsiya miner- al'naya; Mineraly sosushchestvuyushchie	PARAGENESIS	PARAGENESI
3291 PARAGEOSYNCLINE STRU	Parageosynclines GEOSYNCLINES	Parageosynclinal GEOSYNCLINAL	Parageosynklinale GEOSYNKLINALE	PARAGEOSINKLI- NAL'	Parageosinclinal GEOSINCLINAL	Parageosinclinale GEOSINCLINALE
3292 PARAGNEISS IGMS	PARAGNEISS	PARA GNEISS	PARA=GNEIS	PARAGNEJS	PARAGNEIS	PARAGNEISS
3293 PARALIAGEOSYNCLINE STRU	Paraliageosynclines GEOSYNCLINES	Paraliageosynclinal GEOSYNCLINAL	Paral.Geosynklinale GEOSYNKLINALE	PARALIAGEOSINK- LINAL'	Paraliageosinclinal GEOSINCLINAL	Paraliageosinclinale GEOSINCLINALE
3294 PARALIC ENVIRONMENT PALE	Paralic environment LITTORAL ENVI- RONMENT	Milieu paralique MILIEU LITTORAL	Paral.Milieu LITORAL=MILIEU	PARALICHESKIJ	Medio=paralico MEDIO=LITORAL	Ambiente paralico AMBIENTE LITOR- ALE

	ENGLISH	FRANCAIS	DEUTSCH	RUSSKIJ	ESPANOL	ITALIANO
3295 PARALLEL BEDDING TEST	Parallel bedding BEDDING	Litage parallele LITAGE	Parallel=Schichtung FEIN= SCHICHTUNG	SLOISTOST' PARAL- LEL'NAYA	Estratificacion= concordante ESTRATIFIC- ACION=FINA	Stratificazione parallela STRATIFICAZIONE SOTTILE
3296 PARALLEL TEXTURE TEST	Parallel texture TEXTURES	Texture parallele TEXTURE	Parallel=Gefuege KORN=GEFUEGE	PARALLEL'NYJ	Textura=paralela TEXTURA	Tessitura parallela TESSITURA
3297 PARALLELISM PALE	Parallelism THEORETICAL STUDIES; BIO- LOGIC EVOLUTION	Parallelisme THEORIE; EVOLU- TION BIOLOGIQUE	Parallelitaet THEORIE; BIOLOG. EVOLUTION	PARALLELIZM	Paralelismo TEORIA; EVOLUCION= BIOLOGICA	Parallelismo TEORIA; EVOLUZ- IONE BIOLOGICA
3298 PARAMAGNETIC **PROPERTY** PHCH	Paramagnetic property MAGNETIC PROP- ERTIES	Propriete paramagne- tique PROPRIETE MAGNETIQUE	Paramagnet. Eigenschaft MAGNET. EIGENSCHAFT	PARAMAGNETIK	Propiedad= paramagnetica PROPIEDAD= MAGNETICA	Proprieta paramagne- tica PROPRIETA MAGNETICA
3299 PARAMETER MISC	PARAMETERS UF: Coefficient; Cons- tant; Modulus	PARAMETRE UF: Coefficient; Con- stante; Constante gravitation; Module	PARAMETER UF: Gravitations= Konstante; Koeffizient; Konstante; Modul	PARAMETER	PARAMETRO UF: Coeficiente; Con- stante; Constante= gravitacional; Modulo	PARAMETRO UF: Coefficiente; Costante; Costante gravitazionale; Modulo
3300 PARASITISM PALE	PARASITES	PARASITISME	PARASITISMUS	PARAZITIZM	PARASITISMO	PARASSITISMO
3301 PARATYPE PALE	Paratypes NOMENCLATURE	Paratype NOMENCLATURE	Paratyp NOMENKLATUR	Paratip GENOTIP	Paratipo NOMENCLATURA	Paratipo NOMENCLATURA
3302 PARENT MATERIAL **SOILS** SURF	PARENT MATERI- ALS	ROCHE MERE SOL	BODEN= AUSGANGSGESTEIN	Pochvoobrazuyush- chaya poroda mat PORODA MATERIN- SKAYA	ROCA=MADRE= SUELO	ROCCIA MADRE SUOLO
3303 PARENT PRODUCT ISOT	Parent product ISOTOPES	Produit depart RADIOACTIVITE	Ausgangs=Produkt RADIOAKTIVITAET	PRODUKT RODI- TEL'SKIJ	Producto=padre RADIOACTIVIDAD	Radionuclide iniziale RADIOATTIVITA
3304 PARENT ROCK GEOL	Parent rock PROTOLITHS	ROCHE MERE	MUTTER=GESTEIN	Poroda materinskaya PORODA NEF- TENOSNAYA	ROCA=MADRE	ROCCIA MADRE
3305 PARTIAL MELTING IGNE	PARTIAL MELTING UF: Incongruent melting	FUSION PARTIELLE UF: Fusion incomplete	PARTIELL. SCHMELZEN UF: Inkongruent. Schmelzen	Plavlenie chastichnoe ANATEKSIS	FUSION=PARCIAL UF: Fusion= incongruente	FUSIONE PARZI- ALE UF: Fuzione incongru- ente
3306 PARTIAL PRESSURE PHCH	PARTIAL PRES- SURE	PRESSION PART- IELLE	PARTIAL=DRUCK	DAVLENIE PARTSI- ALNOE	PRESION=PARCIAL	PRESSIONE PARZI- ALE
3307 PARTICLE SEDI	PARTICLES	MATIERE PAR- TICULAIRE UF: Particule ele- mentaire	PARTIKEL UF: Elementar=Teilchen	CHASTITSA UF: Chastitsy kosmi- cheskie	PARTICULA= ELEMENTAL UF: Materia=elemental	PARTICELLA UF: Particella elemen- tare
3308 PARTICLE SHAPE SEDI	SHAPE ANALYSIS	Forme grain MORPHOSCOPIE	Korn=Gestalt MORPHOSKOPIE	Forma chastits MORFOSKOPIYA	Forma=grano MORFOSCOPIA	Forma del granulo MORFOSCOPIA
3309 PARTICLE-TRACK **DATING** ISOT	PARTICLE=TRACK DATING	Datation trace particule DATATION; TRACE FISSION	Teilchen=Spur= Datierung PHYSIKAL. ALTERSBESTIM- MUNG; SPALTS- PUREN	Metod trekov METOD TREKOV DELENIYA	Datacion=traza= particula DATACION; TRAZA=FISION	Traccia di particella DATAZIONE; TRACCE DI FIS- SIONE

	ENGLISH	FRANCAIS	DEUTSCH	RUSSKIJ	ESPANOL	ITALIANO
3310 PARTING TEST	PARTING LINEA-TION	LINEATION COU-RANT	ABSONDERUNG	OTDEL'NOST'	LINEACION=CORRIENTE	SUDDIVISIBILITA
3311 PARTITION COEFFICIENT GEOC	PARTITION COEF-FICIENTS	COEFFICIENT PARTAGE	VERTEILUNGS=KOEFFIZIENT	KOEHFFITSIENT RASPREDELENIYA	COEFICIENTE=REPARTICION	COEFFICIENTE DI PARTIZIONE
3312 PARTITIONING PHCH	PARTITIONING	REPARTITION ELE-MENT	ELEMENT=VERTEILUNG	[PARTITIONING] RASPREDELENIE	REPARTICION=FASE	RIPARTIZIONE DI ELEMENTI
3313 PATCH REEF SEDI	PATCH REEFS	BANC RECIF	RIFF=BANK	[PATCH REEF]	BANCO-ARRECIFAL	[PATCH REEF]
3314 PATENT MISC	PATENTS	BREVET	PATENT	PATENT	PATENTE	BREVETTO
3315 PATHOLOGY PALE	PATHOLOGY	PATHOLOGIE	PATHOLOGIE	PATOLOGIYA	PATOLOGIA	PATOLOGIA
3316 PATTERN MISC	PATTERNS(gen.)	Pattern(gen.)	Muster(gen.)	[PATTERN]	Patron(gen.)	Pattern(gen.)
3317 PATTERNED GROUND SURF	PATTERNED GROUND UF: Polygonal ground	SOL FIGURE	STRUKTUR=BODEN	GRUNT STRUK-TURNYJ	SUELO=CON=FIGURAS	SUOLO FIGURATO
3318 PAUROPODA PALS	Pauropoda MYRIAPODA	Pauropoda MYRIAPODA	Pauropoda MYRIAPODA	PAUROPODA MYRIAPODA	Pauropoda MYRIAPODA	Pauropoda MYRIAPODA
3319 PEARL SEDI	PEARLS	PERLE	PERLE	ZHEMCHUG	PERLA	PERLA
3320 PEAT COMS	PEAT UF: Boghead coal	TOURBE	TORF	TORF UF: Bugor torfyannoj	TURBA	TORBA
3321 PEAT BOG SURF	PEAT BOGS	TOURBIERE	MOOR	BOLOTO TORFY-ANOE	TURBERA	TORBIERA
3322 PEBBLE SEDI	PEBBLES	GALET UF: Galet argile	GEROELL UF: Ton=Geroell	GAL'KA	CANTO=RODADO UF: Bola=de=arcilla	CIOTTOLO UF: Palla d'argilla
3323 PECTINACEA PALS	PECTINACEA	PECTINACEA UF: Pterioida	PECTINACEA UF: Pterioida	Pectinacea PTERIOIDA	PECTINACEA	PECTINACEA UF: Pterioida
3324 PEDIMENT SURF	PEDIMENTS	PEDIMENT	PEDIMENT	Pediment PENEPLEN	PEDIMENTO	PEDIMENTO
3325 PEDIPLAIN SURF	PEDIPLAINS	PEDIPLAINE	PEDIPLAIN	Pediplen PENEPLEN	PEDIPLANO	PEDEPIANO
3326 PEDOGENESIS SURF	PEDOGENESIS	PEDOGENESE	BODEN=BILDUNG	Pedogenezis POCHVA; GENEZIS	EDAFOGENESIS	PEDOGENESI
3327 PEGMATITE IGNS	PEGMATITE UF: Pegmatitic composi-tion; Pegmatoid texture	PEGMATITE UF: Pegmatoide; Struc-ture pegmatitique	PEGMATIT UF: Pegmatit=Gefuege; Pegmatoid	PEGMATIT UF: Pegmatitovyj	PEGMATITA UF: Estructura=pegmatitica; Pegmat-oide	PEGMATITE UF: Pegmatoide; Tessi-tura pegmatitica
3328 PEGMATITIC TEST	Pegmatitic (adj.) PEGMATITE	Pegmatitique(adj.) PEGMATITE	Pegmatitisch(adj.) PEGMATIT	Pegmatitovyj(adj.) PEGMATIT	Pegmatitica(adj.) PEGMATITA	Pegmatiticao(adj.) PEGMATITE

	ENGLISH	FRANCAIS	DEUTSCH	RUSSKIJ	ESPANOL	ITALIANO
3329 PEGMATOID IGNE	Pegmatoid texture PEGMATITE	Pegmatoide PEGMATITE	Pegmatoid PEGMATIT	PEGMATOID	Pegmatoide PEGMATITA	Pegmatoide PEGMATITE
3330 PELAGIC ENVIRONMENT PALE	PELAGIC ENVI-RONMENT *UF:* Nekton	MILIEU PELAGIQUE *UF:* Necton	PELAG.MILIEU *UF:* Nekton	PELAGICHESKIJ *UF:* Pelagic sedimenta-tion	MEDIO=PELAGICO *UF:* Necton	AMBIENTE PELAGICO *UF:* Necton
3331 PELAGIC SEDIMENTATION SEDI	PELAGIC SEDIMEN-TATION	SEDIMENTATION PELAGIQUE	PELAG. SEDIMENTATION	[PELAGIC SEDI-MENTATION] PELAGICHESKIJ; SEDIMENTATSIYA	SEDIMENTACION=PELAGICA	SEDIMENTAZIONE PELAGICA
3332 PELECYPODA PALS	BIVALVIA *UF:* Anomalodesmata; Arcoida; Conocar-dioida; Cryptodonta; Heterodonta; Modio-morphoida; Myoida; Mytiloida; Nuculoidea; Palaeoheterodonta; Palaeotaxodonta; Pterioida; Pteromor-phia; Trigoniidae	BIVALVIA *UF:* Anomalodesmata; Arcoida; Conocar-dioida; Cryptodonta; Heterodonta; Modio-morphoida; Myoida; Mytiloida; Nuculoidea; Palaeoheterodonta; Palaeotaxodonta; Ptero-morphia; Trigoniidae	LAMELLI-BRANCHIATA *UF:* Anomalodesmata; Arcoida; Conocar-dioida; Cryptodonta; Heterodonta; Modio-morphoida; Myoida; Mytiloida; Nuculoidea; Palaeoheterodonta; Palaeotaxodonta; Ptero-morphia; Trigoniidae	BIVALVIA	BIVALVIA *UF:* Anomalodesmata; Arcoida; Conocar-dioida; Cryptodonta; Heterodonta; Modio-morphoida; Myoida; Mytiloida; Nuculoidea; Palaeoheterodonta; Palaeotaxodonta; Pterioida; Pteromor-phia; Trigoniidae	BIVALVIA *UF:* Anomalodesmata; Arcoida; Conocar-dioida; Cryptodonta; Heterodonta; Myoida; Mytiloida; Nuculoidea; Palaeoheterodonta; Palaeotaxodonta; Ptero-morphia; Trigoniidae; Modiomorphoida
3333 PELEE-TYPE ERUPTION IGNE	PELEAN-TYPE ERUPTIONS	PELEEN	PELEE=TYP	IZVERZHENIE PELEJSKOGO TIPA *UF:* Tucha palyashchaya	PELEANO	PELEEANO
3334 PELITE SEDS	Pelite SHALE	Pelite SILTSTONE	Pelit SILTSTEIN	PELIT	Pelita LIMOLITA	Pelite SILTITE
3335 PELITIC TEXTURE TEST	PELITIC TEXTURE	TEXTURE PELI-TIQUE	PELIT=GEFUEGE	PELITOVYJ	TEXTURA=PELITICA	TESSITURA PELI-TICA
3336 PELITIC-METAMORPHIC SEQUENCE IGMS	METAPELITE	METAPELITE	META=PELIT	[PELITIC-METAMORPHIC SEQUENCE]	SECUENCIA=PELITICO=METAMORFICA	METAPELITE
3337 PELLET SEDI	PELLETS	PELOTE	PELLET	PELLETY	BOLA	PELLETS
3338 PELLETIZING METH	PELLETIZATION	BOULETAGE	PELLETIERUNG	BRIKETIROVANIE	PELETIZACION	PELLETIZZAZIONE
3339 PENEPLAIN SURF	PENEPLAINS	PENEPLAINE	PENEPLAIN	PENEPLEN *UF:* Pediment; Pediplen; Planatsiya	PENILLANURA	PENEPIANO
3340 PENETRATION TEST ENGI	PENETRATION TESTS *UF:* Penetrometers	PENETROMETRIE *UF:* Penetrometre	DRUCK=SONDIERUNG *UF:* Penetrometer	PENETRATSIYA	PENETROMETRIA *UF:* Penetrometro	PENETROMETRIA *UF:* Penetrometro
3341 PENETROMETER INST	Penetrometers INSTRUMENTS; PENETRATION TESTS	Penetrometre INSTRUMENTA-TION; PENE-TROMETRIE	Penetrometer INSTRUMEN-TIERUNG; DRUCK=SONDIERUNG	PENETROMETR	Penetrometro INSTRUMENTA-CION; PENETRO-METRIA	Penetrometro STRUMENTAZ-IONE; PENETRO-METRIA
3342 PENNSYLVANIAN STRS	PENNSYLVANIAN	PENNSYLVANIEN	PENNSYLVANIAN	Pensil'vanij KARBON VERKHNIJ	PENSYLVANIENSE	PENNSILVANIANO

	ENGLISH	FRANCAIS	DEUTSCH	RUSSKIJ	ESPANOL	ITALIANO
3343 PENOKEAN OROGENY STRU	PENOKEAN OROG- ENY	OROGENIE PENOK- EENNE	PENOKE= OROGENESE	PENOKENSKIJ	OROGENIA= PENOKEENSE	OROGENESI PENOKEANA
3344 PENTAMERIDA PALS	PENTAMERIDA	PENTAMERIDA	PENTAMERIDA	PENTAMERIDA	PENTAMERIDA	PENTAMERIDA
3345 PENTASTOMIDA PALS	Pentastomida ARTHROPODA	Pentastomida ARTHROPODA	Pentastomida ARTHROPODA	PENTASTOMIDA	Pentastomida ARTHROPODA	Pentastomida ARTHROPODA
3346 PENTOXYLALES PALS	PENTOXYLALES	PENTOXYLALES	PENTOXYLALES	PENTOXYLALES	PENTOXYLALES	PENTOXYLALES
3347 PERALKALINE **COMPOSITION** IGNE	Peralkaline composition ALKALIC COMPO- SITION	Composition hyperalca- line COMPOSITION ALCALINE	Peralkal.Chemismus ALKALI=TYP	UL'TRASHCH- ELOCHNOJ (SOSTAV)	Composicion= peralcalina COMPOSICION= ALCALINA	Composizione peral- calina COMPOSIZIONE ALCALINA
3348 PERALUMINOUS **COMPOSITION** IGNE	PERALUMINOUS COMPOSITION	Composition hyperalu- mineuse ALUMINIUM	Peraluminat=Gehalt AL	Perglinozemistyj GLINOZEMISTYJ	Composicion= peraluminosa ALUMINIO	Composizione iperal- luminifera ALLUMINIO
3349 PERCHED AQUIFER GEOH	PERCHED AQUI- FERS	NAPPE PERCHEE	SCHWEBEND. GRUNDWASSER	VERKHOVODKA	MANTO=COLGADO	FALDA SOSPESA
3350 PERCOLATION GEOH	PERCOLATION	PERCOLATION	PERKOLATION	Prosachivanie INFIL'TRATSIYA	PERCOLACION	PERCOLAZIONE
3351 PERICLINE STRU	Periclines ANTICLINES	Periclinal ANTICLINAL	Perikline ANTIKLINALE	PERIKLINAL'	Periclinal ANTICLINAL	Periclinale ANTICLINALE
3352 PERIDOTITE IGNS	PERIDOTITES	PERIDOTITE	PERIDOTIT	PERIDOTIT	PERIDOTITA	PERIDOTITE
3353 PERIGEE EXTR	Perigee EXTRATERRES- TRIAL GEOLOGY	Perigee GEOLOGIE EXTRA TERRESTRE	Perigaeum EXTRATERRESTR. GEOLOGIE	PERIGEJ	Perigeo GEOLOGIA= EXTRATERRESTRE	Perigeo GEOLOGIA EXTRATERRESTRE
3354 PERIGLACIAL **FEATURES** SURF	PERIGLACIAL FEA- TURES	PERIGLACIAIRE *UF:* Morphologie per- iglaciaire	PERIGLAZIAER *UF:* Frost=Erscheinung	PERIGLYATSI- AL'NYJ	PERIGLACIAR *UF:* Morfologia= periglaciar	PERIGLACIALE *UF:* Struttura da gelo
3355 PERIGLACIAL ZONE SURF	PERIGLACIAL ENVIRONMENT	MILIEU PERIGLA- CIAIRE	PERIGLAZIAL= ZONE	ZONA OLE- DENENIYA KRAEVAYA	MEDIO= PERIGLACIAR	ZONA PERIGLA- CIALE
3356 PERIOD STRA	Periods STRATIGRAPHIC UNITS	Periode ECHELLE STRATI- GRAPHIQUE	Periode STRATIGRAPH. SKALA	PERIOD	Periodo ESCALA= ESTRATIGRAFICA	Periodo SCALA STRATI- GRAFICA
3357 PERIODICITY GEOL	PERIODICITY	PERIODICITE *UF:* Processus cyclique	PERIODIZITAET *UF:* Zykl.Prozess	PERIODICHNOST'	PERIODICIDAD *UF:* Procesos=ciclicos	PERIODICITA *UF:* Processo ciclico
3358 PERISPHINCTIDA PALS	PERISPHINCTIDA	PERISPHINCTIDA	PERISPHINCTIDA	Perisphinctida AMMONITINA	PERISPHINCTIDA	PERISPHINCTIDA
3359 PERISSODACTYLA PALS	PERISSODACTYLA	PERISSODACTYLA	PERISSODACTYLA	PERISSODACTYLA	PERISSODACTYLA	PERISSODACTYLA
3360 PERLITE IGNE	PERLITE	PERLITE *UF:* Texture perlitique	PERLIT *UF:* Perlit=Gefuege	PERLIT	PERLITA *UF:* Textura=perlitica	PERLITE *UF:* Tessitura perlitica

	ENGLISH	FRANCAIS	DEUTSCH	RUSSKIJ	ESPANOL	ITALIANO
3361 PERLITIC TEXTURE TEST	Perlitic texture TEXTURES	Texture perlitique PERLITE	Perlit=Gefuege PERLIT	PERLITOVYJ	Textura=perlitica PERLITA	Tessitura perlitica PERLITE
3362 PERMAFROST SURF	PERMAFROST	Pergelisol SOL GELE	Permafrost FROST=BODEN	MERZLOTA VECH- NAYA UF: Klin ledyanoj	Permafrost SUELO=HELADO	Permafrost SUOLO GELATO
3363 PERMEABILITY PHCH	PERMEABILITY UF: Aquicludes; Aqui- fuges; Aquitards	PERMEABILITE UF: Aquiclude; Couche impermeable; Couche semipermeable	PERMEABILITAET UF: Aquiclud; Aquitard; Grundwasser=Stauer	PRONITSAEMOST'	PERMEABILIDAD UF: Acuifugo; Capa= confinante; Capa= semipermeable	PERMEABILITA UF: Aquitardo; Livello impermeable; Strato impermeabile
3364 PERMIAN STRS	PERMIAN	PERMIEN	PERM	PERM'	PERMICO	PERMIANO
3365 PERSONAL BIBLIOGRAPHY MISC	Personal bibliography BIOGRAPHY	BIBLIOGRAPHIE PERSONNELLE UF: Biographie	PERSONAL= BIBLIOGRAPHIE UF: Biographie	BIBLIOGRAFIYA PERSONAL'NAYA	BIBLIOGRAFIA= PERSONAL UF: Biografia	BIBLIOGRAFIA PERSONALE UF: Biografia
3366 PERSPECTIVE ECON	Perspective POSSIBILITIES	PERSPECTIVE UF: Perspective exploita- tion	PERSPEKTIVE UF: Moeglichkeit	PERSPEKTIVNOST'	PERSPECTIVA UF: Perspectiva= explotacion	PROSPETTIVA UF: Possibilita
3367 PERTHITE TEST	PERTHITE UF: Perthitic texture	PERTHITE UF: Texture perthitique	PERTHIT UF: Perthit=Gefuege	PERTIT UF: Antipertit	PERTITA UF: Textura=pertitica	PERTITE UF: Tessitura pertitica
3368 PERTHITIC TEXTURE TEST	Perthitic texture TEXTURES; PER- THITE	Texture perthitique PERTHITE	Perthit=Gefuege PERTHIT	PERTITOVYJ	Textura=pertitica PERTITA	Tessitura pertitica PERTITE
3369 PESTICIDE ENVI	PESTICIDES	PESTICIDE	PESTIZID	PESTITSID UF: Yad	PESTICIDA	PESTICIDA
3370 PETROCHEMICAL COEFFICIENT IGNE	Petrochemical coeffi- cient PETROGRAPHY	Coefficient petrochimique CALCUL PETRO- GRAPHIQUE	Petrochem.Koeffizient PETROGRAPH. ANALYSE	KOEHFFITSIENT PETROKHIMI- CHESKIJ	Coeficiente= petroquimico CALCULO= PETROGRAFICO	Coefficiente petrochimico CALCOLO PETRO- GRAFICO
3371 PETROCHEMICAL DIAGRAM IGNE	Petrochemical diagram PHASE EQUILIBRIA	Diagramme petrochimique DIAGRAMME EQUILIBRE	Petrochem.Diagramm PHASEN- GLEICHGEWICHT	DIAGRAMMA PETROKHIMI- CHESKAYA	Diagrama= petroquimico DIAGRAMA= EQUILIBRIO	Diagramma petrochimico DIAGRAMMA D'EQUILIBRIO
3372 PETROCHEMISTRY IGNE	Petrochemistry PETROLOGY	Petrochimie PETROLOGIE	Petrochemie PETROLOGIE	PETROKHIMIYA	Petroquimica PETROLOGIA	Petrochimica PETROLOGIA
3373 PETROFABRIC TEST	PETROFABRICS UF: Structural petrology	PETROFABRIQUE UF: Petrologie structur- ale	GESTEINS- GEFUEGE UF: Gefuege=Petrologie	Petrostruktura STRUKTURA	PETROFABRICA UF: Petrologia= estructural	STRUTTURA DELLA ROCCIA UF: Petrologia struttur- ale
3374 PETROGENESIS GEOL	Petrogenesis GENESIS; ROCKS	Petrogenese PETROLOGIE	Petrogenese PETROLOGIE	PETROGENEZIS	Petrogenesis PETROLOGIA	Petrogenesi PETROLOGIA
3375 PETROGRAPHIC ANALYSIS IGNE	Petrographic analysis PETROGRAPHY	CALCUL PETRO- GRAPHIQUE UF: Classification C.I.P. W.; Classification Niggli; Coefficient petrochimique; Mode petrographique; Norme petrographique	PETROGRAPH. ANALYSE UF: C.I.P.W.=Norm; Mode; Niggli= Klassifikation; Norm; Petrochem.Koeffizient	[PETROGRAPHIC ANALYSIS] ANALIZ; PETRO- GRAFIYA	CALCULO= PETROGRAFICO UF: Coeficiente= petroquimico; Norma	CALCOLO PETRO- GRAFICO UF: Classificazione C.I.P.W.; Classificaz- ione di Niggli; Coeffi- ciente petrochimico; Moda; Norma petro- grafica

	ENGLISH	FRANCAIS	DEUTSCH	RUSSKIJ	ESPANOL	ITALIANO
3376 PETROGRAPHIC MAP IGNE	Lithologic maps GEOLOGIC MAPS	Carte petrographique CARTE GEOLOGIQUE	Petrograph.Karte GEOLOG.KARTE	Karta petrografi-cheskaya KARTA GEOLOGI-CHESKAYA; PETROGRAFIYA	Mapa=petrografico MAPA=GEOLOGICO	Carta litologica CARTA GEOLOGICA
3377 PETROGRAPHIC PROVINCE IGNE	Petrographic province PETROGRAPHY	Province petro-graphique PETROGRAPHIE	Petrograph.Provinz PETROGRAPHIE	PROVINTSIYA PETROGRAFI-CHESKAYA	Provincia=petrografia PETROGRAFIA	Provincia petrografica PETROGRAFIA
3378 PETROGRAPHY GEOL	PETROGRAPHY *UF:* C.I.P.W. classifica-tion; Mode; Niggli's classification; Norm; Petrochemical coeffi-cient; Petrographic analysis; Petrographic province	PETROGRAPHIE *UF:* Province petro-graphique	PETROGRAPHIE *UF:* Holmes' Klassifika-tion; Petrograph. Provinz	PETROGRAFIYA *UF:* Karta petrografi-cheskaya; Petrographic analysis	PETROGRAFIA *UF:* Clasificacion=C.I.P. W.; Clasificacion=de= Niggli; Clasificacion= de=Holmes; Moda; Provincia=petrografia	PETROGRAFIA *UF:* Classificazione di Holmes; Provincia petrografica
3379 PETROLEUM COMS	PETROLEUM *UF:* Edge water; Light oil; Oil trap; Oil well; Petroleum geology	PETROLE *UF:* Champ petrole; Geologie petroliere; Huile legere; Piege a petrole; Puits petrole; Rapport gaz=petrole	ERDOEL *UF:* Erdoel=Bohrung; Erdoel=Falle; Erdoel= Geologie; Gas=Oel= Verhaeltnis; Leicht= Oel; Oelfeld	NEFT' *UF:* Karta neftegazonos-nosti; Petroleum engi-neering	PETROLEO *UF:* Campo=petrolifero; Geologia=del=petroleo; Petroleo=ligero; Pozo= de=petroleo; Relacion= gas=petroleo; Trampa= petrolifera	PETROLIO *UF:* Campo petrolifero; Geologia del petrolio; Petrolio leggero; Pozzo petrolifero; Rapporto gas=petrolio; Trappola petrolifera
3380 PETROLEUM ENGINEERING ENGI	PETROLEUM ENGI-NEERING	GENIE PETROLIER	ERDOEL=TECHNIK	[PETROLEUM ENGINEERING] NEFT'; EHKS-PLUATATSIYA	OBRAS=DEL= PETROLEO	INGEGNERIA DEL PETROLIO
3381 PETROLEUM GEOLOGY ECON	Petroleum geology PETROLEUM	Geologie petroliere PETROLE	Erdoel=Geologie ERDOEL	GEOLOGIYA NEF-TYANAYA	Geologia=del=petroleo PETROLEO	Geologia del petrolio PETROLIO
3382 PETROLOGY GEOL	PETROLOGY *UF:* Experimental petrol-ogy; Holmes' classifica-tion; Petrochemistry	PETROLOGIE *UF:* Classification Holmes; Petrochimie; Petrogenese	PETROLOGIE *UF:* Petrochemie; Petro-genese	PETROLOGIYA	PETROLOGIA *UF:* Petrogenesis; Petro-quimica	PETROLOGIA *UF:* Petrochimica; Petro-genesi
3383 PETROPHYSICS ENGI	Petrophysics PHYSICAL PROPER-TIES; ROCKS	Petrophysique PROPRIETE PHY-SIQUE; ROCHE	Petrophysik PHYSIKAL. EIGENSCHAFT; GESTEIN	PETROFIZIKA	Petrofisica PROPIEDAD= FISICA; ROCA	Petrofisica PROPRIETA FISICA; ROCCIA
3384 PH PHCH	PH *UF:* Acidity	PH *UF:* Acidite	PH=WERT *UF:* Aziditaet	PH	PH *UF:* Acidez	PH *UF:* Acidita
3385 PHACOLITH IGNE	Phacolith INTRUSIONS	Phacolite INTRUSION	Phacolith INTRUSIONS= KOERPER	FAKOLIT	Facolito INTRUSION	Facolite INTRUSIONE
3386 PHACOPIDA PALS	PHACOPIDA	PHACOPIDA	PHACOPIDA	Phacopida POLYMERA	PHACOPIDA	PHACOPIDA
3387 PHAEOPHYTA PALS	PHAEOPHYTA	PHAEOPHYCO-PHYTA	PHAEOPHYTA	PHAEOPHYTA	PHAEOPHYCO-PHYTA	PHAEOPHYTA
3388 PHAEOZEM SUSS	Phaeozem BROWN SOILS	Phaeozem SOL BRUN; LESSIV-AGE	Phaeozem BRAUN=ERDE; AUSWASCHUNG	FEOZEM	Faeozem SUELO=PARDO; LAVADO	Phaeozem SUOLO BRUNO; LISCIVIAZIONE

	ENGLISH	FRANCAIS	DEUTSCH	RUSSKIJ	ESPANOL	ITALIANO
3389 PHAGOTROPHIC ORGANISM PALE	Phagotrophic organism NUTRITION	Organisme phagotrophe NUTRITION	Phagotroph.Organismus ERNAEHRUNG	FAGOTROFNYJ	Organismo=fagotrofico NUTRICION	Organismo fagotropico ANATOMIA DELLA NUTRIZIONE
3390 PHANEROZOIC STRS	PHANEROZOIC	PHANEROZOIQUE	PHANEROZOIKUM	FANEROZOJ	FANEROZOICO	FANEROZOICO
3391 PHASE GEOL	Phases(gen.)	Phase(gen.)	Phase(gen.)	FAZA	Fase(gen.)	Fase(gen.)
3392 PHASE EQUILIBRIA PHCH	PHASE EQUILIBRIA *UF:* Binary system; Eutectic conditions; Isochores; Liquidus; Mineral phase; Petrochemical diagram; Quaternary system; Reaction series; Reactions; Solidus; Ternary system	DIAGRAMME EQUILIBRE *UF:* Diagramme petrochimique; Equilibre chimique; Liquidus; Loi phases mineralogiques; Phase isochore; Phase minerale; Solidus; Systeme binaire; Systeme chimique; Systeme eutectique; Systeme quaternaire; Systeme ternaire	PHASEN=GLEICHGEWICHT *UF:* Binaer.System; Chem.Gleichgewicht; Chem.System; Dynam. Gleichgewicht; Eutektikum; Isochore; Liquidus; Mineral= Phase; Mineralog. Phasen=Regel; Petrochem.Diagramm; Quartaer.System; Solidus; Ternaer. System	RAVNOVESIE FAZOVOE *UF:* Uravnenie sostoyaniya	DIAGRAMA=EQUILIBRIO *UF:* Diagrama= petroquimico; Equilibrio=dinamico; Equilibrio=quimico; Eutectica; Fase= mineral; Isocora; Liquidus; Regla= mineralogica=de=las= fases; Sistema=binario; Sistema=cuaternario; Sistema=quimico; Sistema=ternario; Solidus	DIAGRAMMA D'EQUILIBRIO *UF:* Diagramma petrochimico; Equilibrio chimico; Eutettico; Fase mineralogica; Isocora di fase; Liquidus; Regola delle fasi mineralogiche; Sistema binario; Sistema chimico; Sistema quaternario; Sistema ternario; Solidus
3393 PHASE ISOCHORE PHCH	Isochores ISOPLETH MAPS; PHASE EQUILIBRIA	Phase isochore ISOPLETHE; DIAGRAMME EQUILIBRE	Isochore ISOLINIE; PHASEN=GLEICHGEWICHT	IZOCHORNYJ	Isocora ISOPLETA; DIAGRAMA=EQUILIBRIO	Isocora di fase CARTA DELLE ISOPLETE; DIAGRAMMA D'EQUILIBRIO
3394 PHASE RULE PHCH	PHASE RULE *UF:* Mineralogical phase rule	REGLE PHASE	PHASEN=REGEL	PRAVILO FAZ	REGLA=FASES	REGOLA DELLE FASI
3395 PHENOCRYST TEST	PHENOCRYSTS	PHENOCRISTAL	EINSPRENGLING	FENOKRISTALL	FENOCRISTAL	FENOCRISTALLO
3396 PHENOLS CHES	PHENOLS	PHENOL	PHENOL	FENOL	FENOL	FENOLO
3397 PHI SCALE METH	PHI SCALE	ECHELLE PHI	PHI=SKALA	SHKALA FI	ESCALA=FI	SCALA PHI
3398 PHILOSOPHY MISC	PHILOSOPHY	PHILOSOPHIE	PHILOSOPHIE	FILOSOFIYA	FILOSOFIA	FILOSOFIA
3399 PHOLADOMYIDA PALS	PHOLADOMYIDA	PHOLADOMYIDA	PHOLADOMYIDA	PHOLADOMYIDA	PHOLADOMYIDA	PHOLADOMYIDA
3400 PHOLIDOTA PALS	PHOLIDOTA	PHOLIDOTA	PHOLIDOTA	PHOLIDOTA	PHOLIDOTA	PHOLIDOTA
3401 PHONOLITE IGNS	PHONOLITES	PHONOLITE	PHONOLITH	FONOLIT	FONOLITA	FONOLITE
3402 PHOSPHATE ROCK COMS	PHOSPHATE ROCKS *UF:* Phosphorite	ROCHE PHOSPHATEE *UF:* Phosphorite	PHOSPHAT=GESTEIN *UF:* Phosphorit	PORODA FOSFAT-NAYA	ROCA=FOSFATADA *UF:* Fosforita	ROCCIA FOSFATICA *UF:* Fosforite

	ENGLISH	FRANCAIS	DEUTSCH	RUSSKIJ	ESPANOL	ITALIANO
3403 PHOSPHATES MING	PHOSPHATES	PHOSPHATE	PHOSPHAT	FOSFAT	FOSFATO	FOSFATO
3404 PHOSPHATIZATION SEDI	PHOSPHATIZATION	PHOSPHATISATION	PHOSPHATI- SIERUNG	FOSFATIZATSIYA	FOSFATIZACION	FOSFATIZZAZIONE
3405 PHOSPHORESCENCE PHCH	Phosphorescence LUMINESCENCE	Phosphorescence LUMINESCENCE	Phosphoreszenz LUMINESZENZ	FOSFORESTSENT- SIYA	Fosforescencia LUMINISCENCIA	Fosforescenza LUMINESCENZA
3406 PHOSPHORITE SEDS	Phosphorite PHOSPHATE ROCKS	Phosphorite ROCHE PHOS- PHATEE	Phosphorit PHOSPHAT= GESTEIN	FOSFORITY	Fosforita ROCA=FOSFATADA	Fosforite ROCCIA FOSFA- TICA
3407 PHOSPHOROUS CHEE	PHOSPHOROUS	PHOSPHORE	P	FOSFOR	FOSFORO	FOSFORO
3408 PHOTOCHEMICAL REACTION PHCH	Photochemical reaction THERMODYNAMIC PROPERTIES	Reaction pho- tochimique THERMODY- NAMIQUE	Photochem.Reaktion THERMODYNAMIK	Reaktsiya fotokhimi- cheskaya FOTOSINTEZ	Reaccion=fotoquimica TERMODINAMICA	Reazione fotochimica TERMODINAMICA
3409 PHOTOGEOLOGIC MAP GEOL	PHOTOGEOLOGIC MAPS	CARTE PHO- TOGEOLOGIQUE	PHOTOGEOLOG. KARTE	KARTA FOTOGEOLOGI- CHESKAYA	MAPA= FOTOGEOLOGICO	CARTA FOTOGEOLOGICA
3410 PHOTOGEOLOGY GEOL	PHOTOGEOLOGY	PHOTOGEOLOGIE *UF:* Cartographie aerienne	PHOTOGEOLOGIE *UF:* Luftbild= Kartierung	FOTOGEOLOGIYA	FOTOGEOLOGIA *UF:* Cartografia=aerea	FOTOGEOLOGIA *UF:* Cartografia aerea
3411 PHOTOGRAMMETRY METH	PHOTOGRAMME- TRY	PHOTOGRAMME- TRIE	PHOTOGRAMME- TRIE	FOTOGRAM- METRIYA	FOTOGRAMETRIA	FOTOGRAM- METRIA
3412 PHOTOGRAPHY METH	PHOTOGRAPHY	PHOTOGRAPHIE	PHOTOGRAPHIE	FOTOGRAFIROVA- NIE	FOTOGRAFIA	FOTOGRAFIA
3413 PHOTOMETRY METH	PHOTOMETRY	PHOTOMETRIE	PHOTOMETRIE	FOTOMETRIYA	FOTOMETRIA	FOTOMETRIA
3414 PHOTOSYNTHESIS PHCH	PHOTOSYNTHESIS	PHOTOSYNTHESE	PHOTOSYNTHESE	FOTOSINTEZ *UF:* Reaktsiya fotokhimicheskaya	FOTOSINTESIS	FOTOSINTESI
3415 PHREATIC GAS GEOL	Phreatic gas GASES	Gaz phreatique GAZ	Phreat.Gas GAS	GAZ FREATI- CHESKIJ	Gas=freatico GAS	Gas freatico GAS
3416 PHREATIC WATER GEOH	Phreatic water GROUND WATER	Nappe phreatique NAPPE LIBRE	Phreat.Grundwasser UNGESPANNT. GRUNDWASSER	Voda gruntovaya VODA PODZEM- NAYA BEZNA- PORNAYA	Agua=freatica MANTO=LIBRE	Falda freatica ACQUIFERO NON CONFINATO
3417 PHREATOPHYTE PALE	Phreatophyte ECOLOGY; FLORA	Phreatophyte ECOLOGIE; FLORE	Phreatophyt OEKOLOGIE; FLORA	FREATOFIT	Freatofita ECOLOGIA; FLORA	Phreatophyte ECOLOGIA; FLORA
3418 PHYLACTOLAEMATA PALS	Phylactolaemata BRYOZOA	Phylactolaemata BRYOZOA	Phylactolaemata BRYOZOA	PHYLACTOLAE- MATA	Phylactolaemata BRYOZOA	Phylactolaemata BRYOZOA
3419 PHYLLITE IGMS	PHYLLITES	PHYLLITE	PHYLLIT	FILLIT	FILITA	FILLITE

	ENGLISH	FRANCAIS	DEUTSCH	RUSSKIJ	ESPANOL	ITALIANO
3420 PHYLLOCERATIDA PALS	PHYLLOCERATIDA	PHYLLOCERATIDA	PHYLLOCERATIDA	PHYLLOCERATINA	PHYLLOCERATIDA	PHYLLOCERATIDA
3421 PHYLLONITE IGMS	Phyllonites MYLONITES	Phyllonite MYLONITE; ROCHE CRISTALLINE	Phyllonit MYLONIT; KRISTALLIN. GESTEIN	FILLONIT	Filonita MILONITA; ROCA= CRISTALINA	Fillonite MILONITE; ROCCIA CRISTALLINA
3422 PHYLLOPODA PALS	Phyllopoda BRANCHIOPODA	Phyllopoda CONCHOSTRACA	Phyllopoda CONCHOSTRACA	Phyllopoda BRANCHIOPODA	Phyllopoda CONCHOSTRACA	Phyllopoda CONCHOSTRACA
3423 PHYLOGENY PALE	PHYLOGENY	PHYLOGENIE	PHYLOGENESE	FILOGENIYA	FILOGENIA	FILOGENESI
3424 PHYLUM PALE	Phylum CLASSIFICATION; PALEONTOLOGY	Phylum CLASSIFICATION	Phylum KLASSIFIKATION	FILUM	Filum CLASIFICACION	Phylum CLASSIFICAZIONE
3425 PHYSICAL CHEMISTRY PHCH	Physical chemistry PHYSICOCHEMI-CAL PROPERTIES	Chimie physique PROPRIETE PHYSI-COCHIMIQUE	Physikochemie PHYSIKOCHEM. EIGENSCHAFT	FIZKHIMIYA	Fisicoquimica PROPIEDAD=FISICO=QUIMICA	Chimica fisica PROPRIETA FISICO=CHIMICA
3426 PHYSICAL MODEL GEOL	PHYSICAL MODELS	MODELE PHY-SIQUE	PHYSIKAL.MODELL	MODEL' FIZI-CHESKAYA	MODELO=FISICO	MODELLO FISICO
3427 PHYSICAL PROPERTY PHCH	PHYSICAL PROPER-TIES UF: Petrophysics	PROPRIETE PHY-SIQUE UF: Petrophysique	PHYSIKAL. EIGENSCHAFT UF: Petrophysik	SVOJSTVO FIZI-CHESKOE	PROPIEDAD=FISICA UF: Petrofisica	PROPRIETA FISICA UF: Petrofisica
3428 PHYSICAL STATE PHCH	Physical state THERMODYNAMIC PROPERTIES	Etat physique THERMODY-NAMIQUE	Physikal.Status THERMODYNAMIK	SOSTOYANIE (FIZICHESKOE) UF: Uravnenie sostoy-aniya	Estado=fisico TERMODINAMICA	Stato fisico TERMODINAMICA
3429 PHYSICAL WEATHERING SURF	PHYSICAL WEATH-ERING	ALTERATION PHY-SIQUE	PHYSIKAL. VERWITTERUNG	VYVETRIVANIE FIZICHESKOE	ALTERACION=FISICA	ALTERAZIONE FISICA
3430 PHYSICOCHEMICAL PROPERTY PHCH	PHYSICOCHEMI-CAL PROPERTIES UF: Physical chemistry	PROPRIETE PHYSI-COCHIMIQUE UF: Chimie physique	PHYSIKOCHEM. EIGENSCHAFT UF: Physikochemie	SVOJSTVO FIZ-KHIMICHESKOE	PROPIEDAD=FISICO=QUIMICA UF: Fisicoquimica	PROPRIETA FISICO=CHIMICA UF: Chimica fisica
3431 PHYSIOGRAPHIC PROVINCE SURF	PHYSIOGRAPHIC PROVINCES	PROVINCE PHYSIO-GRAPHIQUE	PHYSIOGRAPH. PROVINZ	PROVINTSIYA FIZIKO-GEOGRAFICHESKAYA	PROVINCIA=FISOGRAFICA	PROVINCIA FISIO-GRAFICA
3432 PHYSIOLOGY PALE	PHYSIOLOGY	PHYSIOLOGIE	PHYSIOLOGIE	FIZIOLOGIYA	FISIOLOGIA	FISIOLOGIA
3433 PHYTANE CHES	Phytane ORGANIC MATERI-ALS	Phytane MATIERE ORGANIQUE	Phytan ORGAN.SUBSTANZ	Fitan UGLEVODOROD	Fitan MATERIA=ORGANICA	Fitano MATERIA ORGANICA
3434 PHYTOCOENOSIS PALE	Phytocoenoses FLORA	Phytocenose POPULATION; FLORE	Phytocoenose POPULATION; FLORA	FITOTSENOZ	Fitocenosis POBLACION; FLORA	Fitocenosi COMUNITA; FLORA
3435 PHYTOLITHES PALS	Phytoliths PROBLEMATIC FOSSILS	Phytolithes PROBLEMATICA	Phytolithes PROBLEMATICA	PHYTOLITHES	Phytolithes PROBLEMATICA	Phytolithes PROBLEMATICA

	ENGLISH	FRANCAIS	DEUTSCH	RUSSKIJ	ESPANOL	ITALIANO
3436 PHYTOPHAGOUS ORGANISM PALE	Phytophagous organism NUTRITION	Organisme phytophague NUTRITION	Phytophag.Organismus ERNAEHRUNG	TRAVOYADNYJ	Organismo=fitofago NUTRICION	Organismo fitofago ANATOMIA DELLA NUTRIZIONE
3437 PHYTOPLANKTON PALE	PHYTOPLANKTON	PHYTOPLANCTON	PHYTOPLANKTON	Fitoplankton PLANKTON	FITOPLANCTON	FITOPLANCTON
3438 PICRITE IGNS	Picrite BASALTS	Picrite BASALTE	Pikrit BASALT	PIKRIT	Picrita BASALTO	Picrite BASALTO
3439 PIEDMONT SURF	PIEDMONT	PIEDMONT	PIEDMONT	PREDGOR'E	PIE=DE=MONTE	PEDEMONTE
3440 PIER ENGI	Piers HARBORS	Quai PORT	Hafenmauer HAFEN	Mol PORT	Muelle PUERTO	Palo=pila PORTO
3441 PIEZOCRYSTALLIZA-TION IGNE	Piezocrystallization CRYSTALLIZATION	Piezocristallisation CRISTALLISATION	Piezo=Kristallisation KRISTALLISATION	P'EZOKRIST-ALLIZATSIYA	Piezocristalizacion CRISTALIZACION	Piezocristallizzazione CRISTALLIZZAZ-IONE
3442 PIEZOELECTRIC EFFECT PHCH	PIEZOELECTRIC EFFECTS	PROPRIETE PIE-ZOELECTRIQUE	PIEZO=ELEKTR. EIGENSCHAFT	P'EZOEHLEK-TRICHESTVO	PROPIEDAD=PIEZOELECTRICA	PROPRIETA PIE-ZOELETTRICA
3443 PIEZOMETRY GEOH	PIEZOMETRY *UF:* Isopiestic line	PIEZOMETRIE *UF:* Ligne isopieze	PIEZOMETRIE *UF:* Aequipotential=Linie	P'EZOMETRIYA	PIEZOMETRIA *UF:* Isopieza	PIEZOMETRIA *UF:* Linea isopiestica
3444 PIGMENTS CHES	PIGMENTS	PIGMENT	PIGMENT	PIGMENT	PIGMENTO	PIGMENTO
3445 PILE ENGI	PILES	PIEU	PFAHL	SVAYA *UF:* Kolonna	PILOTE=DE=CIMENTACION	PALO
3446 PILLAR ENGI	PILLARS	PILIER	PFEILER	Kolonna SVAYA	PILASTRA	PILASTRO
3447 PILLOW LAVA IGNE	PILLOW LAVA *UF:* Pillow structure	PILLOW LAVA *UF:* Structure en coussin	PILLOW=LAVA *UF:* Kissen=Gefuege	PILLOU=LAVA	PILLOW=LAVA *UF:* Estructura=pillow	PILLOW LAVA *UF:* Struttura a cuscini
3448 PILLOW STRUCTURE TEST	Pillow structure PILLOW LAVA	Structure en coussin PILLOW LAVA	Kissen=Gefuege PILLOW=LAVA	PILLOU	Estructura=pillow PILLOW=LAVA	Struttura a cuscini PILLOW LAVA
3449 PINCH OUT SEDI	PINCH OUTS	Amincissement BISEAU	Ausduennen AUSKEILEN	VYKLINIVANIE *UF:* Klin	Acuñado BORDE=CUENCA	Terminazione laterale CUNEO STRATI-GRAFICO
3450 PINGO SURF	PINGOS	PINGO	PINGO	GIDROLAKKOLIT	HIDROLACOLITO	PINGO
3451 PINNACLE REEF SEDI	PINNACLE REEFS	RECIF PINACLE	KRUSTENRIFF	Rifovyj pik RIF	PINACULO=DE=ARRECIFES	SCOGLIERA A PIN-NACOLO
3452 PINNIPEDIA PALS	PINNIPEDIA	PINNIPEDIA	PINNIPEDIA	PINNIPEDIA	PINNIPEDIA	PINNIPEDIA
3453 PIPE IGNE	PIPES	PIPE	PIPE	TRUBA	PIPA	TUBO

	ENGLISH	FRANCAIS	DEUTSCH	RUSSKIJ	ESPANOL	ITALIANO
3454 PIPELINE ENGI	PIPELINES	PIPELINE	PIPELINE	TRUBOPROVOD	OLEODUCTO	CONDOTTA
3455 PIPING ENGI	PIPING	RENARD	GRUNDBRUCH	EHROZIYA TUN- NEL'NAYA	SOCAVON	PIPING
3456 PISCES PALS	PISCES UF: Gnathostomi	PISCES UF: Gnathostomi	PISCES UF: Gnathostomi	PISCES	PISCES UF: Gnathostomi	PISCES UF: Gnatostomi
3457 PISOLITE SEDI	PISOLITHS	PISOLITE	PISOLITH	PIZOLIT	PISOLITO	PISOLITE
3458 PISOLITIC TEXTURE TEST	PISOLITIC TEX- TURE UF: Bean ore	TEXTURE PISOLI- TIQUE UF: Minerai pisolitique	PISOLITH= GEFUEGE UF: Bohnerz	Pizolitovyj OOLITOVYJ	TEXTURA= PISOLITICA UF: Mineral=pisolitico	TESSITURA PISOLI- TICA UF: Minerale pisolitica
3459 PIT MINI	Pit PIT SECTIONS	Puits PROSPECTION	Schacht PROSPEKTION	SHURF	Pozo=minero PROSPECCION	Pozzo di miniera PROSPEZIONE
3460 PIT SECTION GEOL	PIT SECTIONS UF: Pit	COUPE PUITS	AUFSCHLUSS= PROFIL	RAZREZ GORNOJ VYRABOTKI	CORTE=GEOL= POZO	SEZIONE DI POZZO
3461 PLACANTICLINE STRU	Placanticlines ANTICLINES	Placanticlinal ANTICLINAL	Placanticline ANTIKLINALE	PLAKANTIKLINAL'	Placa=anticlinal ANTICLINAL	Placanticline ANTICLINALE
3462 PLACENTALIA PALS	PLACENTALIA	PLACENTALIA	PLACENTALIA	PLACENTALIA UF: Amblypoda	PLACENTALIA	PLACENTALIA
3463 PLACER ECON	PLACERS	PLACER	SEIFE	ROSSYP' UF: Razvedka allyuvi- al'naya	PLACER	PLACER
3464 PLACODERMI PALS	PLACODERMI UF: Antiarchi; Arth- rodira	PLACODERMI UF: Antiarchi; Arth- rodira	PLACODERMI UF: Antiarchi; Arth- rodira	PLACODERMI	PLACODERMI UF: Antiarchi; Arth- rodira	PLACODERMI UF: Antiarchi; Arth- rodira
3465 PLAIN SURF	PLAINS UF: Accumulative plain; Lowlands	PLAINE UF: Basses=terres; Plaine accumulation	FLACHLAND UF: Aufschotterungs= Ebene; Ebene	RAVNINA	LLANURA UF: Plano=acumulacion; Tierras=bajas	PIANA UF: Piana d'accumulazione; Pianura
3466 PLANAR BEDDING STRUCTURE TEST	PLANAR BEDDING STRUCTURES UF: Planar features	STRATIFICATION PLANE UF: Structure plane	HORIZONTAL= SCHICHTUNG UF: Horizontal= Merkmal	[PLANAR BEDDING STRUCTURE] STRUKTURA KON- SEDIMENTATSION- NAYA	ESTRATIFIC- ACION=PLANA UF: Microplanos= impacto	STRATIFICAZIONE PIANA UF: Struttura piana
3467 PLANAR FEATURES TEST	Planar features PLANAR BEDDING STRUCTURES	Structure plane STRATIFICATION PLANE	Horizontal=Merkmal HORIZONTAL= SCHICHTUNG	PLOSKO= PARALLEL'NYJ	Microplanos=impacto ESTRATIFIC- ACION=PLANA	Struttura piana STRATIFICAZIONE PIANA
3468 PLANATION SURF	PLANATION	PENEPLANATION	EINEBNUNG	Planatsiya PENEPLEN; EHROZ- IYA	PENEPLANACION	PENEPLANAZIONE
3469 PLANCTOSPHAEROIDEA PALS	Planctosphaeroidea HEMICHORDATA	Planctosphaeroidea STOMOCHORDATA	Planctosphaeroidea STOMOCHORDATA	PLANCTOSPHAER- OIDEA	Planctosphaeroidea STOMOCHORDATA	Planctosphaeroidea HEMICHORDATA
3470 PLANE MISC	Planes GEOMETRY	Plan GEOMETRIE	Flaeche GEOMETRIE	PLOSKOST'	Plano GEOMETRIA	Piano GEOMETRIA

	ENGLISH	FRANCAIS	DEUTSCH	RUSSKIJ	ESPANOL	ITALIANO
3471 PLANET EXTS	PLANETS	PLANETE	PLANET	PLANETA *UF:* Interplanetary comparison; Planeta yupiterskoj gruppy; Planeta zemnoj gruppy	PLANETA	PIANETA
3472 PLANET EARTH SOLI	EARTH *UF:* Age of the earth; Contracting earth; Earth rotation	PLANETE TERRE *UF:* Age de la terre; Contraction terre; Forme globe	PLANET=ERDE *UF:* Erd=Alter; Erd=Gestalt; Erd=Schrumpfung	ZEMLYA *UF:* Forma Zemli; Kolebanie svobodnoe; Planeta zemnoj gruppy; Vozrast Zemli	PLANETA=TIERRA *UF:* Contraccion=tierra; Edad=de=la=tierra; Forma=globo	PIANETA TERRA *UF:* Contrazione terrestre; Eta della terra; Forma della terra
3473 PLANET JUPITER EXTS	JUPITER	PLANETE JUPITER	JUPITER	YUPITER	PLANETA=JUPITER	PIANETA GIOVE
3474 PLANET MARS EXTS	MARS	PLANETE MARS	MARS	MARS	PLANETA=MARTE	PIANETA MARTE
3475 PLANET MERCURY EXTS	MERCURY PLANET	PLANETE MER-CURE	MERKUR	MERKURIJ	PLANETA=MERCURIO	PIANETA MER-CURIO
3476 PLANET NEPTUNE EXTS	NEPTUNE	PLANETE NEP-TUNE	NEPTUN	NEPTUN	PLANETA=NEPTUNO	PIANETA NET-TUNO
3477 PLANET PLUTO EXTS	PLUTO	PLANETE PLUTON	PLUTO	PLUTON (PLANETA)	PLANETA=PLUTON	PIANETA PLUTONE
3478 PLANET SATURN EXTS	SATURN	PLANETE SATURNE	SATURN	SATURN	PLANETA=SATURNO	PIANETA SATURNO
3479 PLANET URANUS EXTS	URANUS	PLANETE URANUS	URANUS	URAN (PLANETA)	PLANETA=URANO	PIANETA URANO
3480 PLANET VENUS EXTS	VENUS	PLANETE VENUS	VENUS	VENERA	PLANETA=VENUS	PIANETA VENERE
3481 PLANET-EARTH COMPARISON EXTR	TERRESTRIAL COMPARISON	COMPARAISON TERRE	VERGLEICH=ERDE	[PLANET=EARTH COMPARISON]	COMPARACION=TIERRA	COMPARAZIONE TERRESTRE
3482 PLANETARY EXTR	Planetary(adj.)	Planetaire(adj.)	Planetar(adj.)	PLANETARNYJ	Planetario(adj.)	Planetario(adj.)
3483 PLANETARY CONSTITUTION EXTR	PLANETARY INTE-RIORS	CONSTITUTION PLANETE	PLANETEN=AUFBAU	PLANETARNYJ SOSTAV	CONSTITUCION=PLANETA	COSTITUZIONE DI PIANETA
3484 PLANETARY SATELLITE EXTR	SATELLITES	SATELLITE PLA-NETE	PLANETEN=SATELLIT	SPUTNIK (ESTESTV)	SATELITE	SATELLITE
3485 PLANETOLOGY EXTR	PLANETOLOGY *UF:* Astrogeology	PLANETOLOGIE	PLANETOLOGIE	PLANETOLOGIYA	PLANETOLOGIA	PLANETOLOGIA
3486 PLANKTIVOROUS ANIMAL PALE	Planctivorous animal NUTRITION; FAUNA	Faune planctivore NUTRITION; FAUNE	Planktonfresser ERNAEHRUNG; FAUNA	PITAYUSHCHIJSYA PLANKTONOM	Animal=planctivoro NUTRICION; FAUNA	Animale plantivoro ANATOMIA DELLA NUTRIZIONE; FAUNA
3487 PLANKTON PALE	PLANKTON	PLANCTON	PLANKTON	PLANKTON *UF:* Fitoplankton	PLANCTON	PLANCTON

	ENGLISH	FRANCAIS	DEUTSCH	RUSSKIJ	ESPANOL	ITALIANO
3488 PLANKTONIC TAXON PALE	PLANKTONIC TAXA	TAXON PLANC-TONIQUE	PLANKTON.TAXON	TAKSONOMIYA PLANKTONA	TAXON=PLANCTONICO	TAXON PLANC-TONICO
3489 PLANNING MISC	PLANNING	PLANIFICATION	PLANUNG	PLANIROVANIE *UF:* Vodokhozyastvennoe planirovanie	PLANIFICACION	PIANIFICAZIONE
3490 PLANOSOL SUSS	Planosols PALEOSOLS	Planosol PALEOSOL; MILIEU HUMIDE	Planosol PALAEOBODEN; FEUCHT.MILIEU	PLANOSOL	Planosol PALEOSUELO; MEDIO=HUMEDO	Planosol PALEOSUOLO; AMBIENTE UMIDO
3491 PLANTAE PALE	PLANTAE	PLANTE	PFLANZE	RASTENIE	PLANTA=FOSIL	PIANTA
3492 PLASTIC FLOW ENGI	PLASTIC FLOW	ECOULEMENT PLASTIQUE	PLAST.FLIESSEN	DEFORMATSIYA PLASTICHESKAYA	CORRIMIENTO=PLASTICO	COLATA PLASTICA *UF:* Piega gravitativa
3493 PLASTIC MATERIAL ENGI	PLASTIC MATERI-ALS	MATERIAU PLAS-TIQUE	PLAST.MATERIAL	PLASTICHNYJ	MATERIAL=PLASTICO	MATERIALE PLAS-TICO
3494 PLASTICITY PHCH	PLASTICITY *UF:* Malleable materials; Zone of flow	PLASTICITE *UF:* Substance mallea-ble; Zone plastique	PLASTIZITAET *UF:* Duktil.Substanz; Fliess=Zone	PLASTICHNOST'	PLASTICIDAD *UF:* Maleable; Zona=de=flujo	PLASTICITA *UF:* Malleabilita; Zona di scorrimento
3495 PLATE SOLI	PLATES	PLAQUE	PLATTE	PLITA LITOS-FERNAYA *UF:* Mikroplita	PLACA	PLACCA
3496 PLATE COLLISION SOLI	PLATE COLLISION	COLLISION PLAQUE	PLATTEN-KOLLISION	STOLKNOVENIE PLIT	COLISION=PLACA	COLLISIONE DI PLACCHE
3497 PLATE LIMIT SOLI	PLATE BOUNDA-RIES	LIMITE PLAQUE	PLATTEN=GRENZE	PREDEL PLITY	LIMITE=PLACA	LIMITE DI PLACCA
3498 PLATE ROTATION STRU	PLATE ROTATION	ROTATION PLAQUE	PLATTEN-ROTATION	Vrashchenie plit TEKTONIKA PLIT	ROTACION=PLACA	ROTAZIONE DI PLACCA
3499 PLATE TECTONICS STRU	PLATE TECTONICS *UF:* Global tectonics	TECTONIQUE PLAQUE *UF:* Tectonique globale	PLATTEN=TEKTONIK *UF:* Global=Tektonik	TEKTONIKA PLIT *UF:* Tektonika glob-al'naya; Vrashchenie plit	TECTONICA=DE=PLACAS *UF:* Megatectonica	TETTONICA A PLACCHE *UF:* Tettonica globale
3500 PLATE=BEARING TEST ENGI	PLATE=BEARING TESTS	ESSAI PLAQUE	PLATTENDRUCK=VERSUCH	ISPYTANIE SHTAM-POM	ENSAYO=DE=PLACA	PROVA ALLA PIASTRA
3501 PLATEAU SURF	PLATEAUS *UF:* Plateau basalt	PLATEAU *UF:* Basalte plateau; Hammada	PLATEAU *UF:* Hammada; Plateau=Basalt	PLATO	MESETA *UF:* Basalto=de=meseta; Hammada	ALTOPIANO *UF:* Basalto di plateau; Hammada
3502 PLATEAU BASALT IGNE	Plateau basalt PLATEAUS; BASALTS	Basalte plateau PLATEAU; BAS-ALTE	Plateau=Basalt PLATEAU; BASALT	PLATOBAZAL'T	Basalto=de=meseta MESETA; BASALTO	Basalto di plateau ALTOPIANO; BAS-ALTO
3503 PLATFORM STRU	Platforms CRATONS	Plateforme CRATON	Plattform KRATON	PLATFORMA	Plataforma CRATON	Piattaforma intracra-tonica CRATONE
3504 PLATINUM CHEE	PLATINUM	PLATINE	PT	PLATINA	PLATINO	PLATINO

	ENGLISH	FRANCAIS	DEUTSCH	RUSSKIJ	ESPANOL	ITALIANO
3505 PLATINUM GROUP MING	PLATINUM MINER-ALS	MINE DU PLATINE	PLATIN=GRUPPE	PLATINOID	GRUPO=DEL=PLATINO	PLATINOIDI
3506 PLATYRRHINA PALS	PLATYRRHINA	PLATYRRHINA	PLATYRRHINA	PLATYRRHINA	PLATYRRHINA	PLATYRRHINA
3507 PLAYA SURF	PLAYAS	BASSIN SALE	PLAYA	PLAJYA	SALINA	BACINO SALINO
3508 PLEISTOCENE STRS	PLEISTOCENE	PLEISTOCENE	PLEISTOZAEN	PLEJSTOTSEN *UF:* Plejstotsen nizhnij; Plejstotsen srednij; Plejstotsen verkhnij; Villafrank	PLEISTOCENO	PLEISTOCENE
3509 PLEOCHROISM PHCH	PLEOCHROISM	PLEOCHROISME	PLEOCHROISMUS	PLEOKHROIZM	PLEOCROISMO	PLEOCROISMO
3510 PLICATION STRU	Intrafolial folds FOLDS	Pli intrafoliaire PLI	Faeltelung FALTE	PLOJCHATOST'	Crenulacion PLIEGUE	Piega intrafoliare PIEGA
3511 PLIENSBACHIAN STRS	PLIENSBACHIAN	PLIENSBACHIEN	PLIENSBACHIUM	PLINSBAKH *UF:* Domer	PLIENSBACHIENSE	PLIENSBACHIANO
3512 PLIOCENE STRS	PLIOCENE *UF:* Aktchagylian	PLIOCENE *UF:* Akchagylien	PLIOZAEN *UF:* Akchagylium	PLIOTSEN *UF:* Pliotsen nizhnij; Pliotsen verkhnij	PLIOCENO *UF:* Aktchagyliense	PLIOCENE *UF:* Aktchagyliano
3513 PLUG IGNE	Plugs VOLCANOES	Bouchon VOLCAN	Pfropfen VULKAN	[PLUG] KUPOL VULKANI-CHESKIJ	Tapon VOLCAN	Tappo VULCANO
3514 PLUGGING MINI	Plugging EXPLOITATION	Bouchage EXPLOITATION	Pfropfen=Bildung GEWINNUNG	TAMPONIROVANIE SKVAZHINY	Taponado EXPLOTACION	Tamponatura SFRUTTAMENTO
3515 PLUME SOLI	PLUMES	PANACHE	AUFWOELBUNG	[PLUME]	PLUMA	PENNACCHIO
3516 PLUNGE STRU	Plunge FOLDS	Plongement AXE PLI	Abtauchen FALTENACHSE	POGRUZHENIE (SKLADKI)	Pinchamiento EJE=DE=PLEGAMIENTO	Immersione assiale di piega ASSE DI PIEGA
3517 PLUTON IGNE	PLUTONS	Pluton roche ignee BATHOLITE	Pluton BATHOLITH	Pluton (magm.) BATOLIT	Pluton BATOLITO	Plutone BATOLITE
3518 PLUTONIC METAMORPHISM IGNE	Plutonic metamorphism METAMORPHISM	Metamorphisme plutonique METAMORPHISME	Pluton.Metamorphose METAMORPHOSE	METAMORFIZM PLUTONICHESKIJ	Metamorfismo=plutonico METAMORFISMO	Metamorfismo plutonico TIPO DI METAMORFISMO
3519 PLUTONIC ROCK IGNS	PLUTONIC ROCKS	ROCHE GRENUE *UF:* Texture grenue	PLUTONIT *UF:* Eukristallin.Gefuege	Plutonicheskij INTRUZIYA	ROCA=GRANUDA *UF:* Textura=granuda	ROCCIA PLU-TONICA *UF:* Tessitura eucristallina
3520 PLUTONISM GEOL	Plutonism MAGMAS	Plutonisme MAGMATISME	Plutonismus MAGMATISMUS	PLUTONIZM	Plutonismo MAGMATISMO	Plutonismo MAGMATISMO
3521 PLUTONIUM CHEE	PLUTONIUM	PLUTONIUM	PU	PLUTONIJ	PLUTONIO	PLUTONIO

	ENGLISH	FRANCAIS	DEUTSCH	RUSSKIJ	ESPANOL	ITALIANO
3522 PLUVIAL STRA	Pluvial processes GLACIATION	Pluvial GLACIATION	Pluvial VEREISUNG	PLUVIAL'NYJ	Pluvial GLACIACION	Pluviale GLACIAZIONE
3523 PNEUMATOLYSIS IGNE	PNEUMATOLYSIS *UF:* Pneumatolytic deposit	CONDITION PNEU- MATOLYTIQUE *UF:* Gite pneumatoly- tique	PNEUMATOLYT. BEDINGUNG *UF:* Pneumatolyt. Lagerstaette	PNEVMATOLIZ *UF:* Tip mestorozhdenij pnevmatolitov	CONDICION= NEUMATOLITICA *UF:* Yacimiento= neumatolitico	CONDIZIONI PNEU- MATOLITICHE *UF:* Giacimento pneu- matolitico
3524 PNEUMATOLYTIC **DEPOSIT** ECON	Pneumatolytic deposit PNEUMATOLYSIS	Gite pneumatolytique CONDITION PNEU- MATOLYTIQUE	Pneumatolyt. Lagerstaette PNEUMATOLYT. BEDINGUNG	Tip mestorozhdenij pnevmatolitovyj GENEZIS; PNEVMA- TOLIZ	Yacimiento= neumatolitico CONDICION= NEUMATOLITICA	Giacimento pneuma- tolitico CONDIZIONI PNEU- MATOLITICHE
3525 PODOCOPIDA PALS	Podocopida OSTRACODA	Podocopida OSTRACODA	Podocopida OSTRACODA	PODOCOPIDA *UF:* Metacopina	Podocopida OSTRACODA	Podocopida OSTRACODA
3526 PODZOL SUSS	PODZOLS *UF:* Acrisols	PODZOL *UF:* Acrisol; Ferrod; Orthod	PODSOL *UF:* Acrisol; Ferrod; Orthod	PODZOL *UF:* Podzoluvisol	PODZOL *UF:* Acrisol; Ferrod; Orthod	PODSOL *UF:* Acrisol; Ferrod; Orthod
3527 PODZOLUVISOL SUSS	Podzoluvisols SPODOSOLS	Podzoluvisol SPODOSOL; ALFI- SOL	Podzoluvisol SPODOSOL; ALFI- SOL	Podzoluvisol PODZOL	Podzoluvisol SPODOSOL; ALFI- SOL	Podzoluvisol SPODOSOL; ALFI- SOL
3528 POGONOPHORA PALS	Pogonophora WORMS	Pogonophora VERMES	Pogonophora VERMES	POGONOPHORA	Pogonophora VERMES	Pogonophora VERMES
3529 POIKILITIC TEXTURE TEST	Poikilitic texture TEXTURES	Texture poecilitique TEXTURE	Poikilit.Gefuege KORN=GEFUEGE	POJKILITOVYJ	Textura=poikilitica TEXTURA	Tessitura pecilitica TESSITURA
3530 POIKILOBLASTIC **TEXTURE** TEST	Poikiloblastic texture TEXTURES	Texture poeciloblas- tique TEXTURE	Poikiloblast.Gefuege KORN=GEFUEGE	POJKILOBLAS- TOVYJ	Textura=poikiloblastica TEXTURA	Tessitura peciloblastica TESSITURA
3531 POISSON DISTRIBUTION MATH	Poisson distribution STATISTICAL DIS- TRIBUTION	Distribution Poisson DISTRIBUTION STATISTIQUE	Poisson=Verteilung STATIST. VERTEILUNG	PUASSONA RAS- PREDELENIE	Distribucion=Poisson DISTRIBUCION= ESTADISTICA	Distribuzione di Pois- son DISTRIBUZIONE STATISTICA
3532 POISSON'S RATIO STRU	POISSON'S RATIO	COEFFICIENT POIS- SON	POISSON= KOEFFIZIENT	PUASSONA KOEHF- FITSIENT	COEFICIENTE= POISSON	COEFFICIENTE DI POISSON
3533 POLAR SURF	Polar(adj.)	Polaire(adj.)	Polar(adj.)	POLYARNYJ	Polar(adj.)	Polare(adj.)
3534 POLAR ICECAP SURF	Polar ice cap ICE CAPS	Calotte glaciaire polaire CALOTTE GLA- CIAIRE	Polar.Eis=Kalotte POL=KALOTTE	Polyarnyj lednik pokrovnyj SKAPHA LEDYNAYA	Casquete=polar CASQUETE= GLACIAR	Calotta polare CALOTTA GLA- CIALE
3535 POLAR WANDERING SOLI	POLAR WANDER- ING	Derive pole POLE GEOMAGNE- TIQUE	Pol=Wanderung GEOMAGNET.POL	PEREMESHCHENIE POLYUSA	Deriva=polar POLO= GEOMAGNETICO	Deriva del polo POLO GEOMAGNE- TICO
3536 POLAROGRAPHY METH	POLAROGRAPHY	POLAROGRAPHIE	POLAROGRAPHIE	POLYAROGRAFIYA	POLAROGRAFIA	POLAROGRAFIA
3537 POLICY MISC	POLICY	POLITIQUE	POLITIK	POLITIKA	POLITICA	POLITICA

	ENGLISH	FRANCAIS	DEUTSCH	RUSSKIJ	ESPANOL	ITALIANO
3538 POLISHED SECTION METH	POLISHED SEC- TIONS	SECTION POLIE	ANSCHLIFF	ANSHLIF	SECCION=PULIDA	SEZIONE LUCIDA
3539 POLLEN PALE	POLLEN	POLLEN	POLLEN	PYL'TSA *UF:* Diagramma pyl'tsevaya	POLEN	POLLINE
3540 POLLEN DIAGRAM STRA	POLLEN DIA- GRAMS	PALYNODIA- GRAMME	POLLEN= DIAGRAMM	Diagramma pyl'tsevaya PYL'TSA; DIA- GRAMMA	PALINODIAGRAMA	DIAGRAMMA POL- LINICO
3541 POLLUTANT ENVI	POLLUTANTS	POLLUANT	SCHADSTOFF	ZAGRYAZNITEL'	CONTAMINANTE	INQUINANTE
3542 POLLUTED WATER ENVI	POLLUTED WATER	EAU POLLUEE	VERUNREINIGT. WASSER	Voda zagryaznennaya VODA; ZAGRYAZ- NENIE	AGUA= CONTAMINADA	ACQUA INQUINATA
3543 POLLUTION ENVI	POLLUTION	POLLUTION	VERUNREINIGUNG	ZAGRYAZNENIE *UF:* Voda zagryaznen- naya	CONTAMINACION	INQUINAMENTO
3544 POLONIUM CHEE	POLONIUM	POLONIUM	PO	POLONIJ	POLONIO	POLONIO
3545 POLYCHAETA PALS	POLYCHAETIA	POLYCHAETA	POLYCHAETIA	Polychaetia ANNELIDA	POLYCHAETA	POLYCHAETIA
3546 POLYGENETIC GEOL	Polygenetic(adj.)	Polygenique(adj.)	Polygenetisch(adj.)	POLIGENETI- CHESKIJ	Poligenetico(adj.)	Poligenico(adj.)
3547 POLYGONAL GROUND SURF	Polygonal ground PATTERNED GROUND	Sol polygonal SOL GELE	Polygon=Boden FROST=BODEN	POCHVA POLIGON- AL'NAYA	Suelo=poligonal SUELO=HELADO	Suolo poligonale SUOLO GELATO
3548 POLYHEDRA MINE	Polyhedra CRYSTAL STRUC- TURE	Polyedre STRUCTURE CRIS- TALLINE	Polyeder KRISTALL= STRUKTUR	Poliehdr STRUKTURA KRIS- TALLICHESKAYA	Poliedro ESTRUCTURA= CRISTALINA	Poliedro STRUTTURA CRIS- TALLINA
3549 POLYMERA PALS	Polymera PTYCHOPARIIDA	Polymera PTYCHOPARIIDA	Polymera PTYCHOPARIIDA	POLYMERA *UF:* Corynexochida; Lichida; Odonto- pleurida; Phacopida; Ptychopariida; Redlichiida	Polymera PTYCHOPARIIDA	Polymera PTYCHOPARIIDA
3550 POLYMERIZATION PHCH	POLYMERIZATION	POLYMERISATION	POLYMERI- SIERUNG	POLIMERIZATSIYA	POLIMERIZACION	POLIMERIZZAZ- IONE
3551 POLYMETALLIC DEPOSIT ECON	POLYMETALLIC ORES	GITE POLYMETAL- LIQUE	POLYMETALL. LAGERSTAETTE	POLIMETALLI- CHESKIJ	YACIMIENTO= POLIMETALICO	GIACIMENTO POLI- METALLICO
3552 POLYMETAMORPHISM IGNE	POLYMETAMOR- PHISM	METAMORPHISME POLY	POLY= METAMORPHOSE	POLIMETAMOR- FIZM *UF:* Prograde metamor- phism	POLIMETAMOR- FISMO	POLIMETAMOR- FISMO
3553 POLYMICTIC TEXTURE SEDI	Polymictic texture SEDIMENTARY STRUCTURES	Polymicte STRUCTURE SEDI- MENTAIRE; ROCHE CLASTIQUE	Polymikt.Gefuege SEDIMENT= GEFUEGE; KLAST. GESTEIN	POLIMIKTOVYJ	Polimicto ESTRUCTURA= SEDIMENTARIA; ROCA=CLASTICA	Polimittico STRUTTURA SEDI- MENTARIA; ROC- CIA CLASTICA

	ENGLISH	FRANCAIS	DEUTSCH	RUSSKIJ	ESPANOL	ITALIANO
3554 POLYMINERALIC IGNE	Polymineralic(adj.)	Polymineral(adj.)	Polymineralisch(adj.)	POLIMINERAL'NYJ	Polimineral(adj.)	Poliminerale(adj.)
3555 POLYMORPHISM MINE	POLYMORPHISM *UF:* Dimorphism	POLYMORPHISME *UF:* Dimorphisme	POLYMORPHIE *UF:* Dimorphismus	POLIMORFIZM	POLIMORFISMO *UF:* Dimorfismo	POLIMORFISMO *UF:* Dimorfismo
3556 POLYPHASE PROCESS GEOL	POLYPHASE PRO- CESSES	DEFORMATION POLYPHASEE	STOCKWERKS= TEKTONIK	PROTSES MNOGO- FASOVYJ	DEFORMACION= POLIFASICA	DEFORMAZIONE POLIFASICA
3557 POLYPLACOPHORA PALS	POLYPLACOPHORA *UF:* Amphineura	Polyplacophora AMPHINEURA	Polyplacophora AMPHINEURA	Polyplacophora AMPHINEURA	Polyplacophora AMPHINEURA	Polyplacophora AMPHINEURA
3558 POLYTYPISM MINE	POLYTYPISM	POLYTYPISME	POLYTYPIE	POLITIPIYA	POLITIPISMO	POLITIPISMO
3559 PONTIAN STRS	Pontian LOWER PLIOCENE	Pontien PLIOCENE INF	Pontium U.PLIOZAEN	PONT	Pontiense MIOCENO=SUP; PLIOCENO=INF	Pontiano PLIOCENE INF.
3560 POPULAR GEOLOGY GEOL	POPULAR GEOL- OGY	GEOLOGIE VUL- GARISATION	POPULAERWI- SSENSCHAFTL. ARTIKEL	[POPULAR GEOL- OGY]	GEOLOGIA= DIVULGACION	DIVULGAZIONE GEOLOGICA
3561 PORCELLANITE SEDS	PORCELLANITE	Porcelanite SILEXITE	Porcellanit FLINT	PORTSELLANIT	Porcelanita SILEXITA	Porcellanite SILEXITE
3562 PORE PRESSURE GEOH	PORE PRESSURE	PRESSION PORES	POREN=DRUCK	DAVLENIE POROVOE	PRESION=POROS	PRESSIONE NEI PORI
3563 PORE WATER SEDI	PORE WATER	EAU INTERSTIT- IELLE	POREN=WASSER	VODA POROVAYA	AGUA= INTERSTICIAL	ACQUA INTERSTIZ- IALE
3564 PORIFERA PALS	PORIFERA *UF:* Calcispongea; Demospongea; Hyalospongea	PORIFERA *UF:* Calcispongea; Demospongea; Hyalospongea	PORIFERA *UF:* Calcispongea; Demospongea; Hyalospongea	PORIFERA	PORIFERA *UF:* Calcispongea; Demospongea; Hyalospongea	PORIFERA *UF:* Calcispongea; Demospongea; Hyalospongea
3565 POROSITY PHCH	POROSITY *UF:* Cellular porosity	POROSITE *UF:* Porosite cellulaire	POROSITAET *UF:* Zellular=Porositaet	PORISTOST' *UF:* Poristost' yac- heistaya	POROSIDAD *UF:* Porosidad=celular	POROSITA *UF:* Porosita cellulare
3566 POROUS MEDIUM ENGI	POROUS MATERI- ALS	MILIEU POREUX	POROES.MEDIUM	PORISTYJ	MEDIO=POROSO	MEZZO POROSO
3567 PORPHYRITIC TEXTURE TEST	PORPHYRITIC TEX- TURE *UF:* Augen structure; Crystalloblast	TEXTURE POR- PHYRIQUE *UF:* Cristalloblaste; Structure oeillee; Tex- ture intersertale; Tex- ture linophyrique; Texture micrograni- tique; Texture vitro- phyrique	PORPHYR= GEFUEGE *UF:* Augen=Gefuege; Intersertal.Gefuege; Kristalloblast; Linophyr.Gefuege; Mikrogranit.Gefuege; Vitrophyr.Gefuege	PORFIROVYJ	TEXTURA= PORFIDICA *UF:* Cristaloblasto; Estructura=glandular; Intersertal; Linofidica; Textura= microgranitica; Textura=vitrofidica	TESSITURA POR- FIRICA *UF:* Cristaloblasto; Tessitura intersertale; Tessitura linofirica; Tessitura micrograni- tica; Tessitura occhiadina; Tessitura vitrofirica
3568 PORPHYROBLASTIC TEXTURE TEST	PORPHYROBLAS- TIC TEXTURE	TEXTURE POR- PHYROBLASTIQUE	PORPHYROBLAST. GEFUEGE	PORFIROBLAS- TOVYJ	TEXTURA= PORFIDOBLASTICA	TESSITURA POR- FIROBLASTICA
3569 PORPHYRY COPPER ECON	PORPHYRY COP- PER	PORPHYRY COP- PER	[PORPHYRY= COPPER]	MEDNOPOR- FIROVYJ	COBRE=PORFIDICO	PORPHYRY COP- PER

	ENGLISH	FRANCAIS	DEUTSCH	RUSSKIJ	ESPANOL	ITALIANO
3570 PORTLANDIAN STRS	PORTLANDIAN	PORTLANDIEN	PORTLAND	PORTLAND	PORTLANDIENSE	PORTLANDIANO
3571 POSITIVE ORE ECON	Positive ore RESERVES	Minerai prouve RESERVE	Positiv.Erz RESERVE	ZAPASY PRO- MYSHLENNYE	Yacimiento=positivo RESERVA= MINERAL	Giacimento accertato RISERVA
3572 POSSIBILITIES ECON	POSSIBILITIES UF: Perspective	Perspective exploitation PERSPECTIVE	Moeglichkeit PERSPEKTIVE	Vozmozhnost' PROGNOZ	Perspectiva= explotacion PERSPECTIVA	Possibilita PROSPETTIVA
3573 POSTGLACIAL SURF	Postglacial(adj.)	Postglaciaire(adj.)	Postglazial(adj.)	POSLELEDNIKOVYJ	Postglaciar(adj.)	Postglaciale(adj.)
3574 POSTMAGMATIC **GENESIS** IGNE	Postmagmatic materials MINERAL DEPOSITS GENESIS	Genese postmagmatique GENESE GITE	Postmagmat.Bildung LAGERSTAETTEN= GENESE	POSTMAGMATI- CHESKIJ	Postmagmatico GENESIS= YACIMIENTO	Origine postmagmatica GENESI DI GIACIMENTO
3575 POSTTECTONIC **PROCESS** STRU	POSTTECTONIC PROCESSES	PROCESSUS POST TECTONIQUE	POSTTEKTON. BILDUNG	POSTTEKTONI- CHESKIJ	POSTECTONICO	POSTTETTONICO
3576 POTABLE WATER GEOH	POTABILITY	POTABILITE	TRINKBARKEIT	Voda pit'evaya VODA PRESNAYA	POTABILIDAD	POTABILITA
3577 POTASH COMS	POTASH	POTASSE UF: Sylvinite	KALI UF: Sylvinit	POTASH	SALES=POTASICAS UF: Silvinita	POTASSA UF: Silvinite
3578 POTASSIC **COMPOSITION** IGNE	POTASSIC COMPOSITION	COMPOSITION POTASSIQUE	KALIUM=GEHALT	KALIEVYJ	COMPOSICION= POTASICA	COMPOSIZIONE POTASSICA
3579 POTASSIUM CHEE	POTASSIUM	POTASSIUM	K	KALIJ	POTASIO	POTASSIO
3580 POTASSIUM-ARGON **DATING** ISOT	K/AR UF: Atmospheric argon	K=AR UF: K=Ca	K=AR=DATIERUNG UF: K=Ca=Datierung	METOD KALIJ- ARGONOVYJ	K=AR UF: K=Ca	K/AR UF: K/Ca
3581 POTASSIUM-CALCIUM **DATING** ISOT	K/CA	K=Ca K=AR	K=Ca=Datierung K=AR=DATIERUNG	METOD KALIJ- KAL'TSIEVYJ	K=Ca K=AR	K/Ca K/AR
3582 POTENTIAL ORE ECON	POTENTIAL DEPOSITS	GISEMENT POTENTIEL	POTENTIELL. LAGERSTAETTE	[SUBECONOMIC DEPOSIT] ZAPASY PROGNOZ- NYE	YACIMIENTO= POTENCIAL	GIACIMENTO POTENZIALE
3583 POTENTIOMETRIC **SURFACE** GEOH	POTENTIOMETRIC SURFACE	SURFACE PIEZO- METRIQUE UF: Niveau eau souterraine; Niveau hydrostatique	HYDROSTAT. OBERFLAECHE UF: Grundwasser= Spiegel; Hydrostat. niveau	Poverkhnost' p'ezometricheskaya UROVEN' GIDROSTATI- CHESKIJ	SUPERFICIE= PIEZOMETRICA UF: Nivel=agua= subterranea; Nivel= hidrostatico	SUPERFICIE PIEZO- METRICA UF: Livello della falda idrica; Livello idrostatico
3584 POTENTIOMETRY METH	POTENTIOMETERS	POTENTIOMETRIE	POTENTIOMETRIE	POTENTSIOMETR	POTENCIOMETRIA	POTENZIOMETRO
3585 POTHOLE SURF	POTHOLES	MARMITE GEANT UF: Marmite	STRUDELTOPF UF: Kessel	KOTEL	MARMITA=DE= GIGANTE UF: Marmita=glaciar	MARMITTA DEI GIGANTI UF: Marmitta

	ENGLISH	FRANCAIS	DEUTSCH	RUSSKIJ	ESPANOL	ITALIANO
3586 POTTER'S CLAY COMS	Potter's clay CERAMIC MATERI-ALS	Argile poterie CERAMIQUE	Potters=Ton KERAM.ROHSTOFF	GLINA GON-CHARNAYA	Arcilla=de=ceramica CERAMICA	Argilla da ceramica CERAMICA
3587 POWDER DIFFRACTION METH	X-RAY POWDER DIFFRACTION	DIAGRAMME POU-DRE	PULVER=AUFNAHME	Metod poroshka ANALIZ RENTGENO-SPEKTRAL'NYJ	DIAGRAMA=DE=POLVO	DIFFRAZIONE DI POLVERI
3588 POWER PLANT ENGI	POWER PLANTS	CENTRALE ELEC-TRIQUE	KRAFTWERK	EHLEKTROSTANT-SIYA	CENTRAL=ELECTRICA	CENTRALE ELET-TRICA
3589 POZZOLAN COMS	POZZOLAN	POUZZOLANE	PUZZOLAN	PUTSTSOLAN	PUZOLANA	POZZOLANA
3590 PRAECARDIOIDA PALS	PRAECARDIIDA	PRAECARDIIDA	PRAECARDIIDA	PRAECARDIOIDA	PRAECARDIIDA	PRAECARDIIDA
3591 PRASEODYMIUM CHEE	PRASEODYMIUM	PRASEODYME	PR	PRAZEODIM	PRASEODIMIO	PRASEODIMIO
3592 PRE-IMBRIAN EXTR	Preimbrian STRATIGRAPHIC UNITS; MOON	Pre=imbrien ECHELLE STRATI-GRAPHIQUE; LUNE	Prae=Imbrium STRATIGRAPH. SKALA; MOND	DOIMBRIJ	Preimbriense ESCALA-ESTRATIGRAFICA; LUNA	Preimbriano SCALA STRATI-GRAFICA; LUNA
3593 PRECAMBRIAN STRS	PRECAMBRIAN *UF:* Algoman orogeny; Archeides; Beltian orogeny; Eburnean orogeny; Kibara orog-eny; Mazatsal revolu-tion; Proterozoides; Svecofennian orogeny	ANTECAMBRIEN	PRAEKAMBRIUM	DOKEMBRIJ	PRECAMBRICO	PRECAMBRIANO
3594 PRECAMBRIAN OROGENY STRU	PRECAMBRIAN OROGENY	OROGENIE ANTE-CAMBRIENNE *UF:* Archeides; Orogenie algomienne; Orogenie beltienne; Orogenie eburneenne; Orogenie kibarienne; Orogenie svecofennienne; Proterozoides; Proteroz-oides inf; Proterozoides sup; Revolution Maza-tsal	PRAEKAMBR. OROGENESE *UF:* Algoman.Orogenese; Archaeiden; Belt. Orogenese; Eburn. Orogenese; Frueh-Proterozoiden; Kibara=Orogenese; Mazatsal=Revolution; Proterozoi-den; Spaet=Proterozoiden; Svekofenniden=Orogenese	DOKEMBRIJSKIJ /OROGEN/	OROGENIA=PRECAMBRICA *UF:* Archeides; Orogenia=algomana; Orogenia=beltica; Orogenia=eburnea; Orogenia=kibara; Orogenia=svecofenniana; Proterozoides; Proterozoides=inf; Proterozoides=sup; Revolucion=Mazatsal	OROGENESI PRE-CAMBRIANA *UF:* Archeides; Oro-genesi algomaniana; Orogenesi beltiana; Orogenesi eburnea; Orogenesi kibariana; Orogenesi svecofenni-ana; Proterozoidi; Proterozoidi inf.; Proterozoidi sup.; Rivoluzione Mazatsali-ana
3595 PRECIOUS STONE COMS	Precious stone GEMS	Pierre precieuse GEMME	Edelstein SCHMUCKSTEIN	Kamen' yuvelirnyj KAMEN' DRA-GOTSENNYJ	Piedra=preciosa GEMA	Pietra preziosa GEMMA
3596 PRECIPITATION PHCH	PRECIPITATION	PRECIPITATION CHIMIQUE	CHEM. AUSFAELLUNG	OSAZHDENIE	PRECIPITACION=QUIMICA	PRECIPITAZIONE CHIMICA
3597 PREDATION PALE	PREDATION	PREDATION	RAUBTIERSPUR	KHISHCHNYJ	DEPREDACION	PREDAZIONE
3598 PREDICTION MISC	PREDICTION *UF:* Prognosis	PREVISION *UF:* Prognostic	Vorhersage PROGNOSE	Predskazanie PROGNOZ	PREVISION *UF:* Pronostico	PREVISIONE *UF:* Prognosi

	ENGLISH	FRANCAIS	DEUTSCH	RUSSKIJ	ESPANOL	ITALIANO
3599 PREDICTIVE METALLOGENY ECON	PREDICTIVE MET-ALLOGENY	METALLOGENIE PREVISIONNELLE	ERZ=HOEFFIGKEIT	Prognoz metallogeneti-cheskij METALLOGENIYA; PROGNOZ	METALOGENIA=PREVISORIA	METALLOGENESI PREVISIONALE
3600 PREFERRED ORIENTATION TEST	PREFERRED ORI-ENTATION	ORIENTATION PREFERENTIELLE	VORZUGS=ORIENTIERUNG	[PREFERRED ORIENTATION] ORIENTIROVKA	ORIENTACION=PREFERENTE	ORIENTAZIONE PREFERENZIALE
3601 PREGLACIAL SURF	Preglacial(adj.)	Preglaciaire(adj.)	Praeglazial(adj.)	DOLEDNIKOVYJ	Preglaciar(adj.)	Preglaciale(adj.)
3602 PREHISTORIC AGE STRA	PREHISTORY	PREHISTOIRE	VORGESCHICHTE	DOISTORICHESKIJ	PREHISTORIA	PREISTORIA
3603 PREHNITE-PUMPELLYITE FACIES IGNE	PREHNITE=PUMPELLYITE FACIES	FACIES PREHNITE PUMPELLYITE	PREHNIT=PUMPELLYIT=FAZIES	Fatsiya prenit=pumpelliitovaya FATSIYA TSEOLI-TOVAYA	FACIES=PREHNITA=PUMPELLYITA	FACIES PREHNITE=PUMPELLYITE
3604 PRENEANDERTHALIAN PALS	PRE=NEANDERTHAL	PRENEANDERTHA-LIEN	PRAE=NEANDERTHALER	[PRENEANDER-THALIAN]	PRENEANDER-THALIENENSE	PRENEANDER-THALIANO
3605 PREPARATION METH	PREPARATION *UF:* Elutriation	PREPARATION *UF:* Elutriation	PRAEPARATION *UF:* Ausschlaemmung	PODGOTOVKA PROB	PREPARACION *UF:* Elutriacion	PREPARAZIONE *UF:* Elutriazione
3606 PRESERVATION PALE	PRESERVATION	CONSERVATION	ERHALTUNG	SOKHRANNOST'	CONSERVACION	CONSERVAZIONE DI FOSSILI
3607 PRESSURE PHCH	PRESSURE *UF:* Ground pressure; Head; Isobars	PRESSION *UF:* Charge hydraulique; Isobare; Pression geostatique	DRUCK *UF:* Auftriebs=Druck; Geostat.Druck; Hydraul.Belastung; Isobare	DAVLENIE *UF:* Davlenie gidrodi-namicheskoe; Davlenie nizkoe; Pressure solu-tion; Water pressure	PRESION *UF:* Carga=hidraulica; Isobara; Presion=geostatica	PRESSIONE *UF:* Carico idraulico; Isobara; Pressione geostatica
3608 PRESSURE SOLUTION SEDI	PRESSURE SOLU-TION	DISSOLUTION SOUS PRESSION	DRUCK=LOESUNG	[PRESSURE SOLU-TION] DAVLENIE; RAST-VOR	DISOLUCION=BAJO=PRESION	DISSOLUZIONE SOTTO PRESSIONE
3609 PREVENTION GEOL	PREVENTION	ACTION PREVEN-TIVE *UF:* Ingenierie sismique	VORBEUGEND. MASSNAHME *UF:* Erdbeben=Technik; Kuesten=Schutz	ZASHCHITA	ACCION=PREVENTIVA *UF:* Ingenieria=sismica	AZIONE PREVEN-TIVA *UF:* Ingegneria sismica
3610 PRICE ECON	PRICE	PRIX	PREIS	TSENA *UF:* Stoimost'	PRECIO	PREZZO
3611 PRIMARY DISPERSION PATTERN GEOC	PRIMARY DISPER-SION	AUREOLE PRI-MAIRE	PRIMAER=AUREOLE	OREOL RASSEY-ANIYA PERVICH-NYJ	AUREOLA=PRIMARIA	AUREOLA PRI-MARIA
3612 PRIMARY STRUCTURE TEST	PRIMARY STRUC-TURES	STRUCTURE PRI-MAIRE	PRIMAER=GEFUEGE	[PRIMARY STRUC-TURE] STRUKTURA OSAD-OCHNAYA	ESTRUCTURA=PRIMARIA	STRUTTURA PRI-MARIA

	ENGLISH	FRANCAIS	DEUTSCH	RUSSKIJ	ESPANOL	ITALIANO
3613 PRIMATES PALS	PRIMATES	PRIMATES	PRIMATES	PRIMATES	PRIMATES	PRIMATES
3614 PRIMITIVE MISC	Primitive(adj.)	Primitif(adj.)	Primitiv(adj.)	PRIMITIVNYJ	Primitivo(adj.)	Primitivo(adj.)
3615 PRIORITY MISC	Priority(gen.)	Priorite(gen.)	Prioritaet(gen.)	PRIORITET	Prioridad(gen.)	Priorita(gen.)
3616 PRISMATIC STRUCTURE TEST	Prismatic structure TEXTURES	Structure prismatique DIACLASE	Prismat.Gefuege KLUFT	PRIZMATICHESKIJ *UF:* Elongate mineral	Prismatica DIACLASA	Struttura prismatica DIACLASI
3617 PROBABILITY MATH	PROBABILITY	PROBABILITE	WAHRSCHEIN- LICHKEITS= RECHNUNG	TEORIYA VEROYATNOSTEJ	CALCULO= PROBABILIDAD	PROBABILITA
3618 PROBE METH	Probe INSTRUMENTS	Capteur INSTRUMENTA- TION; METHO- DOLOGIE ANALYSE	Mess=Geraet INSTRUMEN- TIERUNG; ANALYSEN= METHODIK	ZOND	Prueba INSTRUMENTA- CION; METODOLOGIA= ANALISIS	Sonda STRUMENTAZIONE
3619 PROBLEMATICA PALS	PROBLEMATIC FOSSILS *UF:* Phytoliths	PROBLEMATICA *UF:* Phytolithes	PROBLEMATICA *UF:* Phytolithes	PROBLEMATIKI	PROBLEMATICA *UF:* Phytolithes	PROBLEMATICA *UF:* Phytolithes
3620 PROBOSCIDEA PALS	PROBOSCIDEA	PROBOSCIDEA	PROBOSCIDEA	PROBOSCIDEA *UF:* Deinotherioidea; Elephantoidea	PROBOSCIDEA	PROBOSCIDEA
3621 PROCELLARIAN EXTR	PROCELLARIAN	Procellarien ECHELLE STRATI- GRAPHIQUE; LUNE	Oceanus=Procellarius STRATIGRAPH. SKALA; MOND	PROTSELLYARIJ	Procellariense ESCALA= ESTRATIGRAFICA; LUNA	Procellariano SCALA STRATI- GRAFICA; LUNA
3622 PROCESS GEOL	Processes(gen.)	Processus(gen.)	Prozess(gen.)	PROTSESS	Proceso(gen.)	Processo(gen.)
3623 PROCHORDATA PALS	Prochordata CHORDATA	Prochordata CHORDATA	Prochordata CHORDATA	PROCHORDATA	Prochordata CHORDATA	Prochordata CHORDATA
3624 PRODUCTION ECON	PRODUCTION	PRODUCTION	PRODUKTION	DOBYCHA *UF:* Production control	PRODUCCION	PRODUZIONE
3625 PRODUCTION CONTROL ECON	PRODUCTION CON- TROL	GESTION PRODUC- TION	PRODUKTIONS= KONTROLLE	[PRODUCTION CONTROL] DOBYCHA; KON- TROL'	CONTROL= PRODUCCION	GESTIONE DELLA PRODUZIONE
3626 PRODUCTION HORIZON ECON	Producing horizon EXPLOITATION	Horizon productif EXPLOITATION	Produktiv.Horizont GEWINNUNG	GORIZONT PRODUKTIVNYJ	Horizonte=productivo EXPLOTACION	Orizzonte produttivo SFRUTTAMENTO
3627 PRODUCTION WELL MINI	Producing well EXPLOITATION; WELLS	Puits production EXPLOITATION	Produktiv.Schacht GEWINNUNG	SKVAZHINA EHKS- PLUATATSION- NAYA	Pozo=productivo EXPLOTACION	Pozzo produttivo SFRUTTAMENTO
3628 PROFILE OF EQUILIBRIUM SURF	Profile of equilibrium STREAMS	Profil equilibre PROFIL EN LONG	Gleichgewichts=Profil LAENGS=PROFIL	PROFIL' RAVNOVE- SIYA *UF:* Tal'veg	Perfil=de=equilibrio PERFIL= LONGITUDINAL	Profilo d'equilibrio PROFILO LONGI- TUDINALE

	ENGLISH	FRANCAIS	DEUTSCH	RUSSKIJ	ESPANOL	ITALIANO
3629 PROFIT ECON	PROFITABILITY	RENTABILITE	PROFIT	PRIBYL'	RENTABILIDAD	REDDITIVITA
3630 PROGANOSAURIA PALS	PROGANOSAURIA	PROGANOSAURIA	PROGANOSAURIA	PROGANOSAURIA	PROGANOSAURIA	PROGANOSAURIA
3631 PROGNOSIS MISC	Prognosis PREDICTION	Prognostic PREVISION	PROGNOSE UF: Vorhersage	PROGNOZ UF: Predskazanie; Prognoz metallogeneticheskij; Vozmozhnost'	Pronostico PREVISION	Prognosi PREVISIONE
3632 PROGRADATION SURF	PROGRADATION	Progression rivage LIGNE RIVAGE	Seewaertig. Kuestenverlagerung KUESTE	NASTUPANIE	Progresion=costa LINEA=COSTA	Progradazione di riva LINEA DI RIVA
3633 PROGRADE METAMORPHISM IGNE	PROGRADE META-MORPHISM	METAMORPHISME PROGRESSIF	PROGRAD. METAMORPHOSE	[PROGRADE META-MORPHISM] POLIMETAMORFIZM	METAMORFISMO=PROGRESIVO	METAMORFISMO PROGRESSIVO
3634 PROGRAM MISC	PROGRAMS	PROGRAMME	PROGRAMM	PROGRAMMA	PROGRAMA	PROGRAMMA
3635 PROGRESS REPORT MISC	PROGRESS REPORT	RAPPORT ACTI-VITE	TAETIGKEITS=BERICHT	OTCHET INFOR-MATSIONNYJ	INFORME=ACTIVIDAD	RAPPORTO D'ATTIVITA
3636 PROJECT MISC	PROJECTS UF: Objectives	PROJET UF: Objectif	PROJEKT UF: Zweck	PROEKT	PROYECTO UF: Objetivos	PROGETTO UF: Obbiettivo
3637 PROMETHIUM CHEE	PROMETHIUM	PROMETHIUM	PM	PROMETIJ	PROMETIO	PROMEZIO
3638 PROPERTY PHCH	Properties(gen.)	Propriete(gen.)	Eigenschaft(gen.)	SVOJSTVO UF: Erodibility; Svojstvo ehlasticheskoe	Propiedad(gen.)	Proprieta(gen.)
3639 PROPYLITIZATION IGNE	PROPYLITIZATION	PROPYLITISATION	PROPYLITI-SIERUNG	PROPILITIZATSIYA	PROPILITIZACION	PROPILITIZZAZ-IONE
3640 PROSIMII PALS	PROSIMII	PROSIMII	PROSIMII	PROSIMII	PROSIMII	PROSIMII
3641 PROSPECTING ECON	Prospecting EXPLORATION	Exploration PROSPECTION	Exploration PROSPEKTION	POISKI UF: Predvaritel'nyj poisk	Exploracion PROSPECCION	Esplorazione PROSPEZIONE
3642 PROTACTINIUM CHEE	PROTACTINIUM	PROTOACTINIUM	PA	PROTAKTINIJ	PROTACTINIO	PROTOATTINIO
3643 PROTACTINIUM-IONIUM DATING ISOT	PA/IO	PA=IO	PA=IO=DATIERUNG	METOD PROTOAKTINIEVO-IONIEVYJ	PA=IO	PA/IO
3644 PROTECTED ZONE GEOH	PROTECTED WELL ZONES	PERIMETRE PRO-TECTION	SCHUTZBEZIRK	[PROTECTED ZONE] VODA PODZEM-NAYA; OKHRANA	PERIMETRO=PROTECCION	PERIMETRO DI PROTEZIONE
3645 PROTEINS CHES	PROTEINS	PROTEINE	PROTEIN	PROTEIN	PROTEINA	PROTEINA
3646 PROTELYTROPTERA PALS	Protelytroptera DERMAPTEROIDA	Protelytroptera DERMAPTEROIDA	Protelytroptera DERMAPTEROIDA	Protelytroptera INSECTA	Protelytroptera DERMAPTEROIDA	Protelytroptera DERMAPTEROIDA

	ENGLISH	FRANCAIS	DEUTSCH	RUSSKIJ	ESPANOL	ITALIANO
3647 PROTEROZOIC STRS	PROTEROZOIC *UF:* Early proterozoides; Late proterozoides	PROTEROZOIQUE	PROTEROZOIKUM	PROTEROZOJ	PROTEROZOICO *UF:* Proterozoides=sup	PROTEROZOICO
3648 PROTEROZOIDES STRU	Proterozoides OROGENY; PRE- CAMBRIAN	Proterozoides OROGENIE ANTE- CAMBRIENNE	Proterozoiden PRAEKAMBR. OROGENESE	PROTEROZOIDY	Proterozoides OROGENIA= PRECAMBRICA	Proterozoidi OROGENESI PRE- CAMBRIANA
3649 PROTOCILIATA PALS	Protociliata PROTISTA	Protociliata PROTOZOA	Protociliata PROTOZOA	PROTOCILIATA	Protociliata PROTOZOA	Protociliata PROTOZOA
3650 PROTOLITH IGNE	PROTOLITHS *UF:* Parent rock	PROTOLITHE	PROTOLITH	[PROTOLITH] PORODA MATERIN- SKAYA	PROTOLITO	PROTOLITO
3651 PROTOMEDUSAE PALS	Protomedusae COELENTERATA	Protomedusae COELENTERATA	Protomedusae COELENTERATA	PROTOMEDUSAE	Protomedusae COELENTERATA	Protomedusae COELENTERATA
3652 PROTON PHCH	PROTONS	PROTON	PROTON	PROTON	PROTON	PROTONE
3653 PROTOTYPE PALE	Prototype BIOLOGIC EVOLU- TION	Archetype EVOLUTION BIOLOGIQUE	Prototyp BIOLOG. EVOLUTION	PROTOTIP	Prototipo EVOLUCION= BIOLOGICA	Prototipo EVOLUZIONE BIOLOGICA
3654 PROTOZOA PALS	PROTISTA *UF:* Ciliata; Ciliophora; Euciliata; Heliozoa; Mastigophora; Proto- ciliata; Reticularea; Rhizopodea; Sarcodina; Sporozoa; Suctoria	PROTOZOA *UF:* Ciliata; Ciliophora; Euciliata; Heliozoa; Mastigophora; Proto- ciliata; Reticularea; Rhizopodea; Sarcodina; Sporozoa; Suctoria	PROTOZOA *UF:* Ciliata; Ciliophora; Euciliata; Heliozoa; Mastigophora; Proto- ciliata; Reticularea; Rhizopodea; Sarcodina; Sporozoa; Suctoria	PROTOZOA *UF:* Chitinozoa	PROTOZOA *UF:* Ciliata; Ciliophora; Euciliata; Heliozoa; Mastigophora; Proto- ciliata; Reticularea; Rhizopodea; Sarcodina; Sporozoa; Suctoria	PROTOZOA *UF:* Ciliata; Ciliophora; Euciliata; Heliozoa; Mastigophora; Proto- ciliata; Reticularea; Rhizopodea; Sarcodina; Sporozoa; Suctoria
3655 PROTURA PALS	Protura INSECTA	Protura INSECTA	Protura INSECTA	PROTURA	Protura INSECTA	Protura INSECTA
3656 PROVENANCE SEDI	PROVENANCE *UF:* Origin	Provenance ORIGINE	Liefergebiet HERKUNFT	OBLAST' SNOSA	Procedencia ORIGEN	Provenienza ORIGINE
3657 PROVINCIALITY PALE	Provinciality BIOGEOGRAPHY	Provincialite BIOGEOGRAPHIE	Provinzialitaet BIOGEOGRAPHIE	Provintsii i zony BIOGEOGRAFIYA	Provincialidad BIOGEOGRAFIA	Provincialita BIOGEOGRAFIA
3658 PSAMMENT SUSS	Psamments ENTISOLS	Psamment ENTISOL; SOL BRUT	Psamment ENTISOL; AZONAL. BODEN	Psamment EHNTISOL	Psamment ENTISOL; SUELO- BRUTO	Psamment ENTISOL; SUOLO AZONALE
3659 PSAMMITE SEDS	Psammite SANDSTONE	Psammite GRES	Psammit SANDSTEIN	Psammit PESCHANIK	Psamita ARENISCA	Psammite ARENARIA
3660 PSAMMITIC TEST	Psammitic(adj.)	Psammitique(adj.)	Psammitisch(adj.)	PSAMMITOVYJ *UF:* Arenit	Psamitico(adj.)	Psammitico(adj.)
3661 PSAMMOFAUNA PALE	Psammofauna ECOLOGY; FAUNA	Psammofaune ECOLOGIE; FAUNE	Psammofauna OEKOLOGIE; FAUNA	PSAMMOFAUNA	Arena=fosilifera ECOLOGIA; FAUNA	Psammofauna ECOLOGIA; FAUNA
3662 PSEPHITIC TEST	Psephitic(adj.)	Psephitique(adj.)	Psephitisch(adj.)	PSEFITOVYJ	Psefitico(adj.)	Psefitico(adj.)
3663 PSEUDOCONGLOMER- ATE STRU	Pseudoconglomerate TECTONIC BREC- CIA	Pseudoconglomerat BRECHE TEC- TONIQUE	Pseudo=Konglomerat TEKTON.BREKZIE	PSEVDOKONGLOM- ERAT *UF:* Katun glinyanyj; Sharik glinistyj	Pseudoconglomerado BRECHA= TECTONICA	Pseudoconglomerato BRECCIA TET- TONICA

	ENGLISH	FRANCAIS	DEUTSCH	RUSSKIJ	ESPANOL	ITALIANO
3664 PSEUDOGLEY SUSS	PSEUDOGLEYS	PSEUDOGLEY *UF:* Aqualf; Aquept	PSEUDO=GLEY *UF:* Aqualf; Aquept	PSEVDOGLEJ	PSEUDOGLEY *UF:* Aqualf; Aquept	PSEUDOGLEY *UF:* Aqualf
3665 PSEUDOMORPHISM MINE	PSEUDOMORPHISM	PSEUDOMORPHOSE	PSEUDOMORPHOSE	PSEVDOMORFOZA	PSEUDOMORFOSIS	PSEUDOMORFOSI
3666 PSEUDOTILLITE SEDS	Pseudotillite CONGLOMERATE	Pseudotillite CONGLOMERAT	Pseudotillit KONGLOMERAT	PSEVDOTILLIT	Pseudotillita CONGLOMERADO	Pseudotillite CONGLOMERATO
3667 PSILOCERATIDA PALS	PSILOCERATIDA	PSILOCERATIDA	PSILOCERATIDA	Psiloceratida AMMONITINA	PSILOCERATIDA	PSILOCERATIDA
3668 PSILOPSIDA PALS	PSILOPSIDA	PSILOPSIDA	PSILOPSIDA	PSILOPSIDA	PSILOPSIDA	PSILOPSIDA
3669 PSYCHROPHYTE PALE	Psychrophyte ECOLOGY; FLORA	Psychrophyte ECOLOGIE; FLORE	Psychrophyt OEKOLOGIE; FLORA	PSIKHROFIT	Psicrofita ECOLOGIA; FLORA	Psychrophyte ECOLOGIA; FLORA
3670 PTERIDOPHYLLA PALS	PTERIDOPHYLLEN	PTERIDOPHYLLA	PTERIDOPHYLLA	PTERIDOPHYLLA	PTERIDOPHYLLA	PTERIDOPHYLLA
3671 PTERIDOPHYTA PALS	PTERIDOPHYTES *UF:* Equisetales	PTERIDOPHYTA *UF:* Equisetales	PTERIDOPHYTA *UF:* Equisetales	PTERIDOPHYTA	PTERIDOPHYTA *UF:* Equisetales	PTERIDOFITE *UF:* Equisetales
3672 PTERIDOSPERMAE PALS	Pteridospermae GYMNOSPERMS	Pteridospermae GYMNOSPERMAE	Pteridospermae GYMNOSPERMAE	PTERIDOSPERMI- DAE	Pteridospermae GYMNOSPERMAE	Pteridospermae GIMNOSPERME
3673 PTERIOIDA PALS	Pterioida BIVALVIA	Pterioida PECTINACEA	Pterioida PECTINACEA	PTERIOIDA *UF:* Ostreacea; Pecti- nacea	Pterioida PECTINACEA	Pterioida PECTINACEA
3674 PTEROBRANCHIA PALS	PTEROBRANCHIA *UF:* Rhabdopleurida	PTEROBRANCHIA *UF:* Rhabdopleurida	PTEROBRANCHIA *UF:* Rhabdopleurida	PTEROBRANCHIA	PTEROBRANCHIA *UF:* Rhabdopleurida	PTEROBRANCHIA *UF:* Rhabdophleurida
3675 PTEROMORPHIA PALS	Pteromorphia BIVALVIA	Pteromorphia BIVALVIA	Pteromorphia LAMELLI- BRANCHIATA	PTEROMORPHIA	Pteromorphia BIVALVIA	Pteromorphia BIVALVIA
3676 PTEROPODA PALS	PTEROPODA	PTEROPODA	PTEROPODA	PTEROPODA	PTEROPODA	PTEROPODA
3677 PTEROSAURIA PALS	PTEROSAURIA	PTEROSAURIA	PTEROSAURIA	PTEROSAURIA	PTEROSAURIA	PTEROSAURIA
3678 PTYCHOPARIIDA PALS	PTYCHOPARIIDA *UF:* Polymera	PTYCHOPARIIDA *UF:* Polymera	PTYCHOPARIIDA *UF:* Polymera	Ptychopariida POLYMERA	PTYCHOPARIIDA *UF:* Polymera	PTYCHOPARIIDA *UF:* Polymera
3679 PTYGMATIC FOLD IGNE	PTYGMATIC FOLDS	PLI PTYGMATIQUE	PTYGMAT.FALTE	SKLADKA PTIGMA- TITOVAYA	PLIEGUE= PTIGMATICO	PIEGA PTIGMA- TICA
3680 PUBLIC WORKS ENGI	Public works CIVIL ENGINEER- ING	TRAVAUX PUBLICS *UF:* Genie civil	ERSCHLIESSUNGS= ARBEITEN	[PUBLIC WORKS] SOORUZHENIE	OBRA=PUBLICA *UF:* Ingeniere=civil	LAVORI PUBBLICI *UF:* Ingegneria civile
3681 PUBLICATION MISC	PUBLICATIONS	Publication DOCUMENTATION	Publikation DOKUMENTATION	PUBLIKATSIYA	Publicacion DOCUMENTACION	Pubblicazione DOCUMENTAZ- IONE

	ENGLISH	FRANCAIS	DEUTSCH	RUSSKIJ	ESPANOL	ITALIANO
3682 PULMONATA PALS	Pulmonata GASTROPODA	Pulmonata GASTROPODA	Pulmonata GASTROPODA	PULMONATA	Pulmonata GASTROPODA	Pulmonata GASTROPODA
3683 PUMICE IGNE	PUMICE	PONCE	BIMS	PEMZA	PIEDRA=POMEZ	POMICE
3684 PUMPAGE GEOH	Pumpage PUMP TESTS	POMPAGE	PUMPEN	OTKACHKA	BOMBEO	POMPAGGIO
3685 PUMPING TEST METH	PUMP TESTS *UF:* Aquifer test; Pump- age	Pompage essai ESSAI DEBIT	Pump=Test PUMP=VERSUCH	OTKACHKA OPYT- NAYA	Bombeo=de=ensayo ENSAYO=BOMBEO	Prova di pompaggio PROVA DI PORTATA
3686 PUNCH CARD MATH	PUNCH CARDS	CARTE PERFOREE	LOCHKARTE	PERFOKARTA	TARJETA= PERFORADA	SCHEDA PER- FORATA
3687 PURIFICATION ENVI	PURIFICATION	EPURATION	REINIGUNG	OCHISTKA *UF:* Dekontaminatsiya; Refining	DEPURACION	DEPURAZIONE
3688 PURINE CHES	PURINE	PURINE	PURIN	PURIN	PURINA	PURINA
3689 PYCNOCLINE MARI	PYCNOCLINES	Pycnocline DENSITE; EAU MER	Dichte=Gradient DICHTE; MEER= WASSER	PIKNOKLIN	Picnoclina DENSIDAD; AGUA= DE=MAR	Picnoclino DENSITA; ACQUA DI MARE
3690 PYCNOGONIDA PALS	Pycnogonida ARTHROPODA	Pycnogonida ARTHROPODA	Pycnogonida ARTHROPODA	PYCNOGONIDA	Pycnogonida ARTHROPODA	Pycnogonida ARTHROPODA
3691 PYCNOMETER INST	Pycnometer INSTRUMENTS; DENSITY	Pycnometre INSTRUMENTA- TION; DENSITE	Pyknometer INSTRUMEN- TIERUNG; DICHTE	PIKNOMETR	Picnometro INSTRUMENTA- CION; DENSIDAD	Picnometro STRUMENTAZ- IONE; DENSITA
3692 PYCNOMETRY METH	Pycnometry METHODS; DEN- SITY	Pycnometrie METHODOLOGIE; DENSITE	Pyknometrie METHODIK; DICHTE	PIKNOMETRIYA	Picnometria METODOLOGIA; DENSIDAD	Picnometria METODOLOGIA; DENSITA
3693 PYRAMIDAL SYSTEM MINE	Pyramidal system CRYSTAL SYSTEMS	Systeme pyramidal SYSTEME CRISTAL- LIN	Pyramidal.System KRISTALL= SYSTEM	SINGONIYA TETRAGO- NAL'NAYA	Sistema=piramidal SISTEMA= CRISTALINO	Sistema piramidale SISTEMA CRISTAL- LINO
3694 PYRITIZATION SEDI	PYRITIZATION	PYRITISATION	PYRITISIERUNG	PIRITIZATSIYA	PIRITIZACION	PIRITIZZAZIONE
3695 PYROCLASTIC ROCK IGNE	PYROCLASTICS *UF:* Agglutinates; Bombs; Cinder; Ejecta; Lapilli; Spatter cones; Vitroclastic texture	PYROCLASTIQUE *UF:* Bombe volcanique; Cendre volcanique; Cone scories; Lapilli; Projection volcanique; Pyroclastique soude; Scorie volcanique; Structure chaotique; Texture vitroclastique	PYROKLAST. GESTEIN *UF:* Asche; Auswuer- fling; Bombe; Chaot. Gefuege; Lapilli; Schlacken=Kegel; Vitroklast.Gefuege; Vulkan.Asche; Vulkan. Schweiss=Schlacke	PIROKLAST *UF:* Peplopad; Vul- kanoklasticheskij	PIROCLASTICO *UF:* Bomba=volcanica; Ceniza; Ceniza= volcanica; Cono=de= escorias; Estructura= caotica; Lapilli; Piroclastico= cementado; Proyeccion=volcanica; Textura=vitroclastica	PRODOTTO PIRO- CLASTICO *UF:* Agglutinato; Bomba; Caotico; Cenere vulcanica; Cono di eiezione; Lapilli; Proiezione vulcanica; Scoria; Tessitura vitroclastica
3696 PYROLYSIS METH	PYROLYSIS	PYROLYSE	PYROLYSE	PIROLIZ	PIROLISIS	PIROLISI
3697 PYROMETALLURGY METH	PYROMETAL- LURGY	PYROMETALLUR- GIE	PYRO= METALLURGIE	PIROMETALLUR- GIYA	PIROMETALURGIA	PIROMETALLUR- GIA

	ENGLISH	FRANCAIS	DEUTSCH	RUSSKIJ	ESPANOL	ITALIANO
3698 PYROMETAMORPHISM IGNE	Pyrometamorphism CONTACT META-MORPHISM	Pyrometamorphisme METAMORPHISME CONTACT	Pyro=Metamorphose KONTAKT=METAMORPHOSE	PIROMETAMOR-FIZM	Pirometamorfismo METAMORFISMO-CONTACTO	Pirometamorfismo METAMORFISMO DI CONTATTO
3699 PYROMETASOMATISM IGNE	PYROMETASOMA-TISM	PYROMETASOMA-TOSE UF: Metasomatose contact	PYRO-METASOMATOSE UF: Kontakt=Metasomatose	PIROMETASOMA-TIZM	PIROMETASOMA-TISMO UF: Metasomatismo=de=contacto	PIROMETASOMA-TOSI UF: Metasomatismo di contatto
3700 PYROSPHERE SOLI	Pyrosphere CRUST; MAGMAS	Pyrosphere CROUTE TER-RESTRE; MAGMA	Pyrosphaere ERD=KRUSTE; MAGMA	PIROSFERA	Pirosfera CORTEZA=TERRESTRE; MAGMA	Pirosfera CROSTA TER-RESTRE; MAGMA
3701 PYROTHERIA PALS	PYROTHERIA	PYROTHERIA	PYROTHERIA	PYROTHERIA	PYROTHERIA	PYROTHERIA
3702 PYROXENE MING	PYROXENE	PYROXENE	PYROXEN	PIROKSEN	PIROXENO	PIROSSENO
3703 PYROXENE-HORNFELS FACIES IGNE	Pyroxene=hornfels facies HORNFELS FACIES	Facies corneenne pyroxene FACIES CORN-EENNE	Pyroxen=Hornfels=Fazies HORNFELS=FAZIES	FATSIYA PIROKSEN-ROGOVIKOVAYA	Facies=corneana=piroxenica FACIES=CORNEANA	Facies cornubianiti pirosseniche FACIES DELLE CORNUBIANITI
3704 PYROXENITE IGNS	PYROXENITE	PYROXENITE UF: Jadeitite	PYROXENIT UF: Jadeitit	PIROKSENIT	PIROXENITA UF: Jadeitita	PIROSSENITE UF: Giadeitite
3705 PYRROPHYTA PALS	PYRROPHYTA	PYRROPHYCO-PHYTA	PYRROPHYTA	PYRROPHYTA UF: Dinoflagellata	PYRROPHYCO-PHYTA	PYRROPHYTA
3706 QUALITATIVE METHOD METH	QUALITATIVE ANALYSIS	ANALYSE QUALI-TATIVE	QUALITATIV. METHODE	ANALIZ KACHEST-VENNYJ	ANALISIS=CUALITATIVO	ANALISI QUALITA-TIVA
3707 QUALITY ECON	QUALITY	QUALITE UF: Evaluation	QUALITAET UF: Bewertung	KACHESTVO	CALIDAD UF: Evaluacion	QUALITA UF: Stima
3708 QUANTITATIVE METHOD METH	QUANTITATIVE ANALYSIS	ANALYSE QUANTI-TATIVE	QUANTITATIV. METHODE	ANALIZ KOLI-CHESTVENNYJ UF: Analiz kolichestven-nyj rudy	ANALISIS=CUANTITATIVO	ANALISI QUAN-TITATIVA
3709 QUARRY MINI	QUARRIES	CARRIERE	STEINBRUCH	KAR'ER	CANTERA	CAVA
3710 QUARTZ MING	QUARTZ	QUARTZ UF: Jaspe	QUARZ UF: Jaspis	KVARTS	CUARZO UF: Jaspe	QUARZO UF: Diaspro
3711 QUARTZ DIORITE IGNS	QUARTZ DIORITES	DIORITE QUARTZ-IQUE	QUARZ=DIORIT	DIORIT KVARTSEVYJ	CUARZO=DIORITA	DIORITE QUARZIF-ERA
3712 QUARTZITE IGMS	QUARTZITES	QUARTZITE UF: Itacolumite	QUARZIT UF: Itacolumit	KVARTSIT UF: Itabirit; Takonit	CUARCITA UF: Itacolumita	QUARZITE UF: Itacolumite
3713 QUATERNARY STRS	QUATERNARY	QUATERNAIRE	QUARTAER	CHETVERTICHNYJ UF: Flandrij; Kalabrij; Sitsilijskij	CUATERNARIO	QUATERNARIO
3714 QUATERNARY SYSTEM PHCH	Quaternary system PHASE EQUILIBRIA	Systeme quaternaire DIAGRAMME EQUILIBRE	Quartaer.System PHASEN-GLEICHGEWICHT	SISTEMA CHE-TYREKHKOMPO-NENTNAYA	Sistema=cuaternario DIAGRAMA=EQUILIBRIO	Sistema quaternario DIAGRAMMA D'EQUILIBRIO

	ENGLISH	FRANCAIS	DEUTSCH	RUSSKIJ	ESPANOL	ITALIANO
3715 QUICK CLAY ENGI	QUICK CLAY	Argile fluide ARGILE SENSIBLE	Quick=Ton FLIESS=TON	GLINA VYSOKOCHUVSTVI-TEL'NAYA	Arcilla=fluida ARCILLA=SENSIBLE	Argilla fluida ARGILLA SENSIBILE
3716 QUICKSAND ENGI	QUICKSAND	BOULANCE	FLIESS=SAND	PLYVUN	ARENA=MOVEDIZA	SABBIA MOBILE
3717 RACEMIZATION ISOT	RACEMIZATION	RACEMISATION	RACEMISIERUNG	RATSEMIZATSIYA	RACEMISACION	RACEMIZZAZIONE
3718 RADAR METH	RADAR METHODS	METHODE RADAR	RADAR	RADAR	RADAR	RADAR
3719 RADIAL TEST	Radial(adj.)	Radial(adj.)	Radial(adj.)	RADIAL'NYJ	Radial(adj.)	Radiale(adj.)
3720 RADIATION PHCH	RADIATION	RAYONNEMENT	STRAHLUNG	RADIATSIYA	RADIACION	RADIAZIONE
3721 RADIATION DAMAGE ISOT	RADIATION DAMAGE	DESTRUCTION RADIATION	STRAHLUNGS=SCHADEN	NARUSHENIE RADIATSIONNOE	DETERIORO=POR=RADIACION	DANNO DA RADIAZIONE
3722 RADIO INTERFEROMETRY APPL	Radio interferometry GEODESY	Interferometrie onde radio GEODESIE	Radio=Interferometrie GEODAESIE	RADIOINTERFERO-METRIYA	Radio=interferometria GEODESIA	Radiointerferometria GEODESIA
3723 RADIO SPECTROSCOPY METH	RADIO SPECTROSCOPY	SPECTROMETRIE HERTZIENNE	RADIOFREQUENZ=SPEKTROMETRIE	RADIOSPEKTRO-METRIYA	ESPECTROMETRIA=HERTZIANA	SPETTROMETRIA RADIOFREQUENZE
3724 RADIO WAVE APPL	RADIO=WAVE METHODS	METHODE RADIO-FREQUENCE	HOCHFREQUENZ	METOD RADIOVOL-NOVOJ *UF:* Metod mikrovolnovoj	METODO=RADIOFRECUENCIA	RADIOFREQUENZA
3725 RADIOACTIVE DECAY ISOT	RADIOACTIVE DECAY	DESINTEGRATION RADIOACTIVE *UF:* Chaine radioactive	RADIOAKTIV. ZERFALL *UF:* Radioaktiv.Kette	RASPAD RADIOAK-TIVNYJ	DESINTEGRACION=RADIOACTIVA *UF:* Series=radioactivas	DECADIMENTO RADIOATTIVO *UF:* Catena radioattiva
3726 RADIOACTIVE DUST EXTR	Radioactive dust RADIOACTIVITY	Poussiere radioactive RADIOACTIVITE	Radioaktiv.Staub RADIOAKTIVITAET	PYL' RADIOAKTIV-NAYA	Polvo=radioactivo RADIOACTIVIDAD	Polvere radioattiva RADIOATTIVITA
3727 RADIOACTIVE ISOTOPE ISOT	RADIOACTIVE ISOTOPES	ISOTOPE RADIOAC-TIF	RADIOAKTIV. ISOTOP	IZOTOP RADIOAK-TIVNYJ	ISOTOPO=RADIOACTIVO	RADIOISOTOPO
3728 RADIOACTIVE SERIES ISOT	Radioactive series RADIOACTIVITY	Chaine radioactive DESINTEGRATION RADIOACTIVE	Radioaktiv.Kette RADIOAKTIV. ZERFALL	TSEPOCHKA RADIOAKTIVNAYA	Series=radioactivas DESINTEGRACION=RADIOACTIVA	Catena radioattiva DECADIMENTO RADIOATTIVO
3729 RADIOACTIVE TRACER METH	RADIOACTIVE TRACERS	TRACEUR RADIOACTIF	RADIOAKTIV. MARKIERUNG	INDIKATOR RADIOAKTIVNYJ	TRAZADOR=RADIOACTIVO	TRACCIANTE RADIOATTIVO
3730 RADIOACTIVE WASTE ENVI	RADIOACTIVE WASTE	DECHET RADIOAC-TIF	RADIOAKTIV. ABFALL	OTKHODY RADIOAKTIVNYE	RESIDUO=RADIOACTIVO	RIFIUTI RADIOATTIVI
3731 RADIOACTIVITY PHCH	RADIOACTIVITY *UF:* Daughter product; Decay constant; End product; Isorad; Radioactive dust; Radioactive series; Radiometer	RADIOACTIVITE *UF:* Constante desintegration; Demi periode; Isorad; Poussiere radioactive; Produit depart; Produit final; Produit intermediaire; Radiometre	RADIOAKTIVITAET *UF:* Ausgangs=Produkt; End=Produkt; Halbwerts=Zeit; Isorad; Radioaktiv. Staub; Radiometer; Tochter=Element; Zerfalls=Konstante	RADIOAKTIVNOST'	RADIOACTIVIDAD *UF:* Constante=de=desintegracion; Isorad; Polvo=radioactivo; Producto=final; Producto=hijo; Producto=padre; Radiometro; Vida=media=radioactiva	RADIOATTIVITA *UF:* Costante di decadimento; Isorad; Periodo di dimezzamento; Polvere radioattiva; Prodotto di decadimento; Prodotto finale; Radiometro; Radionuclide iniziale

	ENGLISH	FRANCAIS	DEUTSCH	RUSSKIJ	ESPANOL	ITALIANO
3732 RADIOACTIVITY METHOD APPL	RADIOACTIVITY METHODS	METHODE RADIO-METRIQUE	RADIOAKTIV. METHODE	Metod radiometri-cheskij RAZVEDKA RADIO-METRICHESKAYA	METODO=RADIOMETRICO	METODO RADIO-METRICO
3733 RADIOCARBON ISOT	Radiocarbon C=14	Radiocarbone C 14	Radio=Kohlenstoff C=14=DATIERUNG	RADIOUGLEROD	Carbono=radioactivo C=14	Radiocarbonio C=14
3734 RADIOGRAPHY METH	X=RAY RADIOGRA-PHY	RADIOGRAPHIE RX	ROENTGEN-AUFNAHME	SNIMOK RENT-GENOVSKIJ	RADIOGRAFIA=RX	RADIOGRAFIA R=X
3735 RADIOLARIA PALS	RADIOLARIA	RADIOLARIA *UF:* Acantharia	RADIOLARIA *UF:* Acantharia	RADIOLARIA *UF:* Acantharia	RADIOLARIA *UF:* Acantharia	RADIOLARIA *UF:* Acantharia
3736 RADIOLARITE SEDS	RADIOLARITE	RADIOLARITE	RADIOLARIT	RADIOLYARIT	RADIOLARITA	RADIOLARITE
3737 RADIOMETER INST	Radiometer INSTRUMENTS; RADIOACTIVITY	Radiometre INSTRUMENTA-TION; RADIOACTI-VITE	Radiometer INSTRUMEN-TIERUNG; RADIOAKTIVITAET	RADIOMETR	Radiometro INSTRUMENTA-CION; RADIOAC-TIVIDAD	Radiometro STRUMENTAZ-IONE; RADIOAT-TIVITA
3738 RADIOMETRIC PROSPECTING APPL	RADIOACTIVITY SURVEYS	LEVE RADIO-METRIQUE	RADIOMETR. PROSPEKTION	RAZVEDKA RADIO-METRICHESKAYA *UF:* Karotazh gamma; Karotazh gamma=gamma; Karotazh nejtronnyj; Metod radiometricheskij	PROSPECCION=RADIOMETRICA	PROSPEZIONE RADIOMETRICA
3739 RADIUM CHEE	RADIUM	RADIUM	RA	RADIJ	RADIO	RADIO
3740 RADIUS MATH	Radius GEOMETRY	Radius GEOMETRIE	Radius GEOMETRIE	RADIUS	Radius GEOMETRIA	Raggio GEOMETRIA
3741 RADON CHEE	RADON	RADON	RN	RADON	RADON	RADON
3742 RAIN GEOH	RAIN	PLUIE *UF:* Trace pluie	REGEN *UF:* Fossil.Regentropfen	DOZHD'	LLUVIA *UF:* Huella=de=lluvia	PIOGGIA *UF:* Impronta di pioggia
3743 RAIN PRINT SEDI	RAIN PRINTS	Trace pluie PLUIE	Fossil.Regentropfen REGEN	OTPECHATKI KAPEL' DOZHDYA	Huella=de=lluvia LLUVIA	Impronta di pioggia PIOGGIA
3744 RAIN WASH GEOH	Rain wash RUNOFF	Ruissellement pluie RUISSELLEMENT	Regen=Auswaschung ABFLUSS	Smyv ploskostnoj EHROZIYA PLOSKOSTNAYA	Escorrentia=lluvia ESCORRENTIA	Ruscellamento pluviale RUSCELLAMENTO
3745 RAMAN SPECTROSCOPY METH	RAMAN SPECTROS-COPY	SPECTROMETRIE RAMAN	RAMAN=SPEKTROMETRIE	SPEKTROMETRIYA RAMANOVSKAYA	ESPECTROMETRIA=RAMAN	SPETTROMETRIA RAMAN
3746 RANGE STRA	RANGE *UF:* Range zone	EXTENSION STRA-TIGRAPHIQUE *UF:* Zone extension	STRATIGRAPH. VERBREITUNG *UF:* Verbreitungs=Zone	RASPROSTRANE-NIE	EXTENSION=ESTRATIGRAFICA *UF:* Acrozona	ESTENSIONE *UF:* Range zone
3747 RANGE ZONE STRA	Range zone RANGE; ZONING	Zone extension EXTENSION STRA-TIGRAPHIQUE	Verbreitungs=Zone STRATIGRAPH. VERBREITUNG	ZONA RAN-GOVAYA	Acrozona EXTENSION=ESTRATIGRAFICA	Range zone ESTENSIONE

	ENGLISH	FRANCAIS	DEUTSCH	RUSSKIJ	ESPANOL	ITALIANO
3748 RANK OF COAL ECON	RANK	DEGRE HOUILLIFI-CATION	INKOHLUNGS=GRAD	KLASSIFIKATSIYA UGLEY PRO-MYSHLENNAYA	GRADO=CARBONIFICACION	GRADO DI CAR-BONIZZAZIONE
3749 RANKER SUSS	RANKER	RANKER *UF:* Umbrept	RANKER *UF:* Umbrept	RANKER	RANKER *UF:* Umbrept	RANKER *UF:* Umbrept
3750 RAPAKIVI TEST	RAPAKIVI	RAPAKIVI	RAPAKIWI	RAPAKIVI STRUK-TURA	RAPAKIVI	RAPAKIVI
3751 RARE EARTHS CHES	RARE EARTHS	TERRE RARE	SELTEN.ERDEN	ZEMLI REDKIE	TIERRAS=RARAS	TERRE RARE
3752 RARE METAL CHES	RARE METALS	METAL RARE	SELTEN.METALL	METALL REDKIJ	METALES=RAROS	METALLO RARO
3753 RATE MISC	RATES(gen.)	Taux(gen.)	Rate(gen.)	STEPEN' *UF:* High=grade meta-morphism	Proporcion(gen.)	Tasso(gen.)
3754 RATITES PALS	Ratites NEORNITHES	Ratites NEORNITHES	Ratites NEORNITHES	RATITES	Ratites NEORNITHES	Ratites NEORNITHES
3755 RAW MATERIAL COMS	RAW MATERIALS *UF:* Resources	MATIERE PRE-MIERE *UF:* Ressource	ROHSTOFF *UF:* Mineral.Rohstoff	SYR'E	MATERIA=PRIMA *UF:* Recursos	MATERIA PRIMA *UF:* Risorse
3756 RAYLEIGH WAVE SOLI	RAYLEIGH WAVES	ONDE RAYLEIGH	RAYLEIGH=WELLE	Rehleya volna VOLNA POVERKH-NOSTNAYA	ONDA=RAYLEIGH	ONDA DI RAY-LEIGH
3757 REACTION PHCH	Reactions PHASE EQUILIBRIA	Reaction THERMODY-NAMIQUE	Reaktion THERMODYNAMIK	REAKTSIYA	Reaccion TERMODINAMICA	Reazione TERMODINAMICA
3758 REACTION RIM MINE	REACTION RIM *UF:* Kelyphytic rim	FRANGE REAC-TIONNELLE *UF:* Frange kelyphitique	REAKTIONSSAUM *UF:* Kelyphit=Rinde	KAEMKA REAK-TSIONNAYA	FRANJA=DE=REACCION *UF:* Corona=quelifitica	ANELLO DI REAZ-IONE *UF:* Anello chelifitico
3759 REACTION SERIES IGNE	Reaction series PHASE EQUILIBRIA	Serie reaction THERMODY-NAMIQUE	Reaktions=Serie THERMODYNAMIK	RYAD REAKTSION-NYJ *UF:* Ryad reaktsionnyj Bouehna	Series=de=reaccion TERMODINAMICA	Serie di reazione TERMODINAMICA
3760 REACTIVATION STRU	REACTIVATION	REACTIVATION	REAKTIVIERUNG	Reaktivizatsiya AKTIVIZATSIYA	REACTIVACION	RIATTIVAZIONE
3761 REAGENT PHCH	REAGENTS	REACTIF	REAGENZ	REAGENT	REACTIVO	REAGENTE
3762 RECENT SEDIMENTATION SEDI	Recent sedimentation SEDIMENTATION; HOLOCENE	Sedimentation actuelle SEDIMENTATION	Aktuo=Sedimentation SEDIMENTATION	Osadkoobrazovenie sovremennoe SEDIMENTATSIYA	Sedimentacion=actual SEDIMENTACION	Sedimentazione attuale SEDIMENTAZIONE
3763 RECEPTACULITIDA PALS	RECEPTACULI-TACEAE	RECEPTACULI-TACEAE	RECEPTACULITIDA	RECEPTACULITIDA	RECEPTACULITIDA	RECEPTACULITIDA
3764 RECEPTION EQUIPMENT INST	Reception equipment INSTRUMENTS; GEOPHYSICAL METHODS	Equipement reception INSTRUMENTA-TION; METHODE GEOPHYSIQUE	Empfangs=Geraet INSTRUMEN-TIERUNG; GEOPHYSIKAL. METHODE	USTROJSTVO PRIEMNOE	Recepcion=señal INSTRUMENTA-CION; METODO-GEOFISICO	Ricezione STRUMENTAZ-IONE; METODO GEOFISICO

	ENGLISH	FRANCAIS	DEUTSCH	RUSSKIJ	ESPANOL	ITALIANO
3765 RECHARGE GEOH	RECHARGE	RECHARGE NAPPE	GRUNDWASSER= ERNEUERUNG	PITANIE (GIDRO)	RECARGA	RICARICA DI FALDA
3766 RECLAMATION ENVI	RECLAMATION	REAMENAGEMENT	REKULTIVIERUNG	[RECLAMATION] DRENAZH ISKUST- VENNYJ	REACONDICIONA- MIENTO	RECUPERO AMBIENTALE
3767 RECONSTRUCTION GEOL	RECONSTRUCTION	RECONSTITUTION PALEOGEO- GRAPHIQUE	PALAEOGEOGRAPH. REKONSTRUKTION	REKONSTRUKT- SIYA UF: Rekonstruktsiya paleontologicheskaya	RECONSTRUCCION= PALEOGEOGRAFICA	RICOSTRUZIONE PALEOGEOGRA- FICA
3768 RECOVERY MINI	RECOVERY	RECUPERATION	GEWINNUNGS- GRAD	IZVLECHENIE	RECUPERACION	RECUPERO
3769 RECREATION ENVI	RECREATION	LOISIR	ERHOLUNG	DOSUG	SUELO= RECREATIVO	TEMPO LIBERO
3770 RECRYSTALLIZATION MINE	RECRYSTALLIZA- TION	RECRISTALLISA- TION	REKRISTALLISA- TION	PEREKRISTALL- IZATSIYA	RECRISTALIZA- CION	RICRISTALLIZZAZ- IONE
3771 RECUMBENT FOLD STRU	RECUMBENT FOLDS	PLI COUCHE	LIEGEND.FALTE	SKLADKA LEZHACHAYA	PLIEGUE- TUMBADO	PIEGA CORICATA
3772 RECYCLING ECON	RECYCLING	RECYCLAGE	[RECYCLING]	RETSIRKULYAT- SIYA	RECICLAJE	RICICLAGGIO
3773 RED BEDS SEDI	RED BEDS	COUCHE ROUGE	[RED=BEDS]	KRASNOTSVETY	CAPA=ROJA	[RED BEDS]
3774 REDEPOSITION SEDI	REDEPOSITION	REMANIEMENT ROCHE	AUFGEARBEITET. GESTEIN	PEREOTLOZHENIE UF: Pererabotannyj; Perotlozhennyj	REMOCION= SEDIMENTARIA	RIDEPOSIZIONE
3775 REDLICHIIDA PALS	REDLICHIIDA	REDLICHIIDA	REDLICHIIDA	Redlichiida POLYMERA	REDLICHIIDA	REDLICHIIDA
3776 REEF SEDI	REEFS UF: Back reef; Coral reefs; Reef environ- ment; Reef sedimenta- tion	RECIF UF: Recif algue; Recif coralliaire; Recif interne	RIFF UF: Algen=Riff; Korallen=Riff; Rueck= Riff	RIF UF: Algal bank; Pinna- cle reef; Reef environ- ment; Reef sedimenta- tion	ARRECIFE UF: Arrecife=coral; Arrecife=de=algas; Arrecife=interno	SCOGLIERA UF: Retroscogliera; Scogliera corallina; Scogliera d'alghe
3777 REEF ENVIRONMENT PALE	Reef environment REEFS	MILIEU RECIFAL	RIFF=MILIEU	[REEF ENVIRON- MENT] FATSIYA; RIF	MEDIO= ARRECIFAL	AMBIENTE DI SCOGLIERA
3778 REEF SEDIMENTATION SEDI	Reef sedimentation REEFS	SEDIMENTATION RECIFALE	RIFF= SEDIMENTATION	[REEF SEDIMENTA- TION] RIF; SEDIMENTAT- SIYA	SEDIMENTACION= ARRECIFAL	SEDIMENTAZIONE DI SCOGLIERA
3779 REEFBUILDER PALE	REEF BUILDERS	CONSTRUCTEUR RECIF	RIFFBILDNER	[REEFBUILDER]	CONSTRUCTOR= DE=ARRECIFES	COSTRUTTORE DI SCOGLIERA
3780 REFINEMENT MINE	REFINEMENT	AFFINEMENT	VERFEINERUNG	[REFINEMENT]	AFINO	AFFINAMENTO
3781 REFINING METH	REFINING	RAFFINAGE	RAFFINIERUNG	[REFINING] OCHISTKA	REFINO	RAFFINAZIONE

	ENGLISH	FRANCAIS	DEUTSCH	RUSSKIJ	ESPANOL	ITALIANO
3782 REFLECTANCE PHCH	REFLECTANCE *UF:* Reflectivity	POUVOIR REFLEC-TEUR *UF:* Reflectivite	REFLEXIONS=VERMOEGEN *UF:* Reflektivitaet	KOEHFFITSIENT OTRAZHENIYA	PODER=REFLECTOR *UF:* Reflectividad	POTERE RIFLET-TENTE *UF:* Riflettivita
3783 REFLECTION PHCH	REFLECTION	REFLECTION ONDE	REFLEXION	OTRAZHENIE	REFLEXION=ONDA	RIFLESSIONE D'ONDA
3784 REFLECTION SEISMICS APPL	REFLECTION METHODS	SISMIQUE REFLEX-ION *UF:* Sparker	REFLEXIONS=SEISMIK *UF:* Sparker	METOD OTRAZ-HENNYKH VOLN	SISMICA=REFLEXION *UF:* Generador=chispa	SISMICA A RIFLES-SIONE *UF:* Sparker
3785 REFLECTIVITY PHCH	Reflectivity REFLECTANCE	Reflectivite POUVOIR REFLEC-TEUR	Reflektivitaet REFLEXIONS=VERMOEGEN	SPOSOBNOST' OTRAZHA-TEL'NAYA	Reflectividad PODER=REFLECTOR	Riflettivita POTERE RIFLET-TENTE
3786 REFRACTION PHCH	REFRACTION	REFRACTION ONDE	WELLEN=BRECHUNG	REFRAKTSIYA	REFRACCION=ONDA	RIFRAZIONE
3787 REFRACTION SEISMICS APPL	REFRACTION METHODS	SISMIQUE REFRAC-TION	REFRAKTIONS=SEISMIK	METOD PRELOM-LENNYKH VOLN	SISMICA=REFRACCION	SISMICA A RIFRAZIONE
3788 REFRACTIVITY PHCH	Refractivity REFRACTIVE INDEX	Pouvoir refracteur INDICE REFRAC-TION	Brechungs=Vermoegen OPT.EIGENSCHAFT	PRELOMLENIE	Refractividad INDICE=DE=REFRACCION	Potere rifrangente INDICE DI RIFRAZ-IONE
3789 REFRACTOMETER INST	Refractometers INSTRUMENTS; REFRACTIVE INDEX	Refractometre INSTRUMENTA-TION; INDICE REFRACTION	Refraktometer INSTRUMEN-TIERUNG; OPT. EIGENSCHAFT	REFRAKTOMETR	Refractometro INSTRUMENTA-CION; INDICE=DE=REFRACCION	Rifrattometro STRUMENTAZ-IONE; INDICE DI RIFRAZIONE
3790 REFRACTOMETRY METH	Refractometry REFRACTIVE INDEX	Refractometrie INDICE REFRAC-TION	Refraktometrie BRECHUNGS=INDEX	REFRAKTO-METRIYA	Refractometria INDICE=DE=REFRACCION	Rifrattometria INDICE DI RIFRAZ-IONE
3791 REFRACTORY COMS	REFRACTORY MATERIALS *UF:* Ganister	REFRACTAIRE *UF:* Ganister	FEUERFEST. ROHSTOFF *UF:* Ganister	TUGOPLAVKIJ	REFRACTARIO *UF:* Ganister	REFRATTARIO *UF:* Galestro
3792 REGENERATION GEOL	REGENERATION *UF:* Inherited features; Rejuvenation	REGENERATION *UF:* Heritage; Rajeunissement	REGENERATION *UF:* Inherited; Rejuvena-tion	REGENERATSIYA	REGENERACION *UF:* Heredado; Reju-venecimiento	RIGENERAZIONE *UF:* Morfologia ereditata; Ringiovani-mento
3793 REGIONAL GEOLOGY GEOL	AREAL GEOLOGY	GEOLOGIE REGIONALE	REGIONAL. GEOLOGIE	GEOLOGIYA REGIONAL'NAYA	GEOLOGIA=REGIONAL	GEOLOGIA REGIONALE
3794 REGIONAL HYDROGEOLOGY GEOH	Regional hydrogeology HYDROGEOLOCY	HYDROGEOLOGIE REGIONALE *UF:* Leve hydrogeologique	REGIONAL=HYDROGEOLOGIE *UF:* Hydrogeolog. Erkundung	GIDROGEOLOGIYA REGIONAL'NAYA	HIDROGEOLOGIA=REGIONAL *UF:* Estudio=hidrogeologico	IDROGEOLOGIA REGIONALE *UF:* Rilevamento idrogeologico
3795 REGIONAL METAMORPHISM IGNE	REGIONAL META-MORPHISM	METAMORPHISME REGIONAL	REGIONAL=METAMORPHOSE	METAMORFIZM REGIONAL'NYJ	METAMORFISMO=REGIONAL	METAMORFISMO REGIONALE
3796 REGIONAL UNCONFORMITY STRA	Regional unconformity UNCONFORMITIES	Discordance regionale DISCORDANCE	Regional=Diskordanz DISKORDANZ	NESOGLASIE REGIONAL'NOE	Discordancia=regional DISCORDANCIA	Discordanza regionale LACUNA STRATI-GRAFICA

	ENGLISH	FRANCAIS	DEUTSCH	RUSSKIJ	ESPANOL	ITALIANO
3797 REGMAGLYPT EXTR	Regmaglypt METEORITES	Regmaglypte METEORITE	Regmaglypt METEORIT	P'EZOGLIPT	Regmaglypto METEORITO	Regmaglipto METEORITE
3798 REGOLITH SURF	REGOLITH	REGOLITHE *UF:* Sol lunaire	REGOLITH *UF:* Mond=Boden	Regolit POCHVA	REGOLITO *UF:* Suelo=lunar	REGOLITE *UF:* Suolo lunare
3799 REGOSOL SUSS	Regosols SOILS	Regosol SOL BRUT	Regosol AZONAL.BODEN	REGOSOL	Regosol SUELO=BRUTO	Regosuolo SUOLO AZONALE
3800 REGRESSION STRA	REGRESSION	REGRESSION	REGRESSION	REGRESSIYA	REGRESION	REGRESSIONE
3801 REGRESSION STATISTICS MATH	REGRESSION ANALYSIS	REGRESSION STA- TISTIQUE	STATIST. REGRESSION	REGRESSIYA (MAT)	REGRESION- ESTADISTICA	REGRESSIONE STATISTICA
3802 REJUVENATION GEOL	Rejuvenation REGENERATION	Rajeunissement REGENERATION	Rejuvenation REGENERATION	OMOLOZHENIE	Rejuvenecimiento REGENERACION	Ringiovanimento RIGENERAZIONE
3803 RELATIONSHIP MISC	Relationship(gen.)	Relation(gen.)	Verhaeltnis(gen.)	SVYAZ'	Relacion(gen.)	Relazione(gen.)
3804 RELATIVE AGE STRA	RELATIVE AGE	AGE RELATIF	RELATIV.ALTER	VOZRAST OTNOSI- TEL'NYJ	EDAD=RELATIVA	ETA RELATIVA
3805 RELAXATION ENERGY PHCH	RELAXATION	RELAXATION ENERGIE	RELAXATIONS= ENERGIE	RELAKSATSIYA	ENERGIA=RELAJA- CION	RILASSAMENTO
3806 RELICT GEOL	RELICT MATERI- ALS *UF:* Palimpsest; Relict texture	RELICTE *UF:* Aiguille; Avant= butte; Structure fan- tome; Texture relicte	RELIKT *UF:* Nadel; Outlier; Palimpsest; Relikt= Gefuege	RELIKT	RELICTA *UF:* Aguja; Palimpses- tica; Testigo=de= erosion; Textura= relicta	RELITTO *UF:* Guglia vulcanica; Outlier; Palinsesto; Tessitura relitta
3807 RELICT TEXTURE TEST	Relict texture TEXTURES; RELICT MATERIALS	Texture relicte RELICTE	Relikt=Gefuege RELIKT	RELIKTOVYJ	Textura=relicta RELICTA	Tessitura relitta RELITTO
3808 RELIEF SURF	RELIEF *UF:* Consequent streams; Ground; Interfluves; Land; Morphometry; Needles	RELIEF	RELIEF	REL'EF *UF:* Morfometriya	RELIEVE	RILIEVO
3809 RELIEF EXPOSURE SURF	RELIEF EXPOSURE	EXPOSITION RELIEF	EXPOSITIONS= RELIEF	[RELIEF EXPO- SURE] OBNAZHENIE	EXPOSICION= RELIEVE	ESPOSIZIONE
3810 RELIEF INVERSION SURF	RELIEF INVERSION	INVERSION RELIEF	RELIEF=INVER- SION	Inversiya rel'efa REL'EF STRUK- TURNYJ; EHROZ- IYA	INVERSION= RELIEVE	INVERSIONE DEL RILIEVO
3811 REMAGNETIZATION PHCH	REMAGNETIZA- TION	REAIMANTATION	REMAGNETI- SIERUNG	Remagnetizatsiya MAGNITOMETR; AKTIVIZATSIYA	REIMANTACION	RIMAGNETIZZAZ- IONE
3812 REMANENT MAGNETIZATION PHCH	REMANENT MAG- NETIZATION	AIMANTATION REMANENTE	REMANENT. MAGNETISIERUNG	NAMAGNICHEN- NOST' OSTATOCH- NAYA *UF:* Namagnichennost' detritovaya; Namagnic- hennost' ideal'naya; Namagnichennost' izotermicheskaya; Namagnichennost' khimicheskaya; Namagnichennost' vyazkaya; Raz- magnichivanie termi- cheskoe	IMANTACION= REMANENTE	MAGNETIZZAZ- IONE RESIDUA

	ENGLISH	FRANCAIS	DEUTSCH	RUSSKIJ	ESPANOL	ITALIANO
3813 REMOTE SENSING METH	REMOTE SENSING	TELEDETECTION *UF:* Balayage lateral	[REMOTE= SENSING]	ZONDIROVANIE DISTANTSIONNOE *UF:* Analiz multispek- tral'nyj; Ground truth	TELEDETECCION *UF:* Metodo=de= barrido=lateral	TELEOSSERVAZ- IONE *UF:* Metodo a scansione laterale
3814 REMOTENESS MISC	Remoteness(gen.)	Eloignement(gen.)	Ferne(gen.)	UDALENNOST'	Distancia(gen.)	Distanza(gen.)
3815 RENDOLL SUSS	Rendolls MOLLISOLS	Rendoll MOLLISOL; REND- ZINE	Rendoll MOLLISOL; REN- DZINA	Rendoll MOLLISOL	Rendoll MOLLISOL; REN- DZINA	Rendoll MOLLISOL; REN- DZINA
3816 RENDZINA SUSS	RENDZINAS	RENDZINE *UF:* Rendoll	RENDZINA *UF:* Rendoll	RENDZINA	RENDZINA *UF:* Rendoll	RENDZINA *UF:* Rendoll
3817 REPLACEMENT MINE	Replacement SUBSTITUTION	Remplacement SUBSTITUTION	Ersatz SUBSTITUTION	ZAMESHCHENIE *UF:* Substitution	Reemplazamiento SUSTITUCION	Rimpiazzo SOSTITUZIONE
3818 REPORT MISC	Report(gen.)	Rapport(gen.)	Bericht(gen.)	OTCHET	Informe(gen.)	Rapporto(gen.)
3819 REPRESENTATIVE BASIN GEOH	REPRESENTATIVE BASINS	BASSIN REPRE- SENTATIF	REPRAESENTATIV= BECKEN	[REPRESENTATIVE BASIN]	CUENCA= REPRESENTATIVA	BACINO RAPPRE- SENTATIVO
3820 REPRODUCTION PALE	REPRODUCTION *UF:* Heterosporous taxa	REPRODUCTION *UF:* Plante heterosporee	FORTPFLANZUNG	REPRODUKTSIYA	REPRODUCCION *UF:* Planta=heterospora	RIPRODUZIONE *UF:* Pianta eterosporica
3821 REPTILIA PALS	REPTILIA	REPTILIA	REPTILIA	REPTILIA *UF:* Anapsida	REPTILIA	REPTILIA
3822 RESEARCH MISC	RESEARCH	RECHERCHE SCIENTIFIQUE	FORSCHUNG	ISSLEDOVANIE *UF:* Deep=tow method	INVESTIGACION= CIENTIFICA	RICERCA SCIENTI- FICA
3823 RESEQUENT VALLEY SURF	Resequent valleys VALLEYS	Vallee resequente VALLEE	Rueckschreitend.Tal TAL	RESEKVENTNYJ	Valle=resecuente VALLE	Valle resequente VALLE
3824 RESERVE ECON	RESERVES *UF:* Inferred ore; Posi- tive ore; Possibilities	RESERVE *UF:* Minerai probable; Minerai prouve	RESERVE *UF:* Hergeleitet.Erz; Positiv.Erz	Rezerv ZAPASY	RESERVA= MINERAL *UF:* Mineral=probable; Yacimiento=positivo	RISERVA *UF:* Giacimento accertato; Minerale presunto
3825 RESERVOIR GEOH	RESERVOIRS	LAC ARTIFICIEL	KUENSTL.SEE	VODOKHRANILISH- CHE *UF:* Water storage	LAGO=ARTIFICAL	LAGO ARTIFICIALE
3826 RESERVOIR ROCK ECON	RESERVOIR ROCKS *UF:* Accumulation	ROCHE MAGASIN	SPEICHER= GESTEIN	PORODA NEF- TENOSNAYA	ROCA=ALMACEN	ROCCIA SERBA- TOIO
3827 RESIDUAL SURF	Residual(adj.)	Residuel(adj.)	Residual(adj.)	OSTATOCHNYJ *UF:* Pererabotannyj	Residual(adj.)	Residuale(adj.)
3828 RESIDUAL CLAY SEDS	RESIDUAL CLAYS	ARGILE RESID- UELLE	RESIDUAL=TON	GLINA OSTATOCH- NAYA	ARCILLA= RESIDUAL	ARGILLA RESIDU- ALE
3829 RESIDUAL DEPOSIT ECON	Residual deposits LEACHING	Gite residuel LESSIVAGE	Residual=Lagerstaette AUSWASCHUNG	MESTOROZHDENIE OSTATOCHNOE	Deposito=residual LAVADO	Giacimento residuale LISCIVIAZIONE

	ENGLISH	FRANCAIS	DEUTSCH	RUSSKIJ	ESPANOL	ITALIANO
3830 RESIDUAL MAGMA IGNE	Residual magma MAGMAS	Magma residuel MAGMA	Rest=Magma MAGMA	IKHOR	Magma=residual MAGMA	Magma residuo MAGMA
3831 RESIN SEDS	RESINS	RESINE *UF:* Ambre	FOSSIL.HARZ *UF:* Bernstein	SMOLA	RESINA *UF:* Ambar	RESINA *UF:* Ambra
3832 RESISTIVITY PHCH	RESISTIVITY *UF:* Resistivity method	RESISTIVITE ELEC- TRIQUE	ELEKTR. WIDERSTAND	SOPROTIVLENIE	RESISTIVIDAD= ELECTRICA	RESISTIVITA ELET- TRICA
3833 RESISTIVITY METHOD APPL	Resistivity method ELECTRICAL METHODS; RESIS- TIVITY	Methode resistivite METHODE ELEC- TRIQUE	Widerstands=Messung ELEKTR.METHODE	METOD SOPROTIV- LENIJ	Metodo=resistividad METODO= ELECTRICO	Metodo della resistivita METODO ELET- TRICO
3834 RESOURCE COMS	Resources RAW MATERIALS	Ressource MATIERE PRE- MIERE	Vorkommen BODENSCHAETZE	ZAPASY *UF:* Rezerv	Recursos MATERIA=PRIMA	Risorse MATERIA PRIMA
3835 RESPIRATION PALE	RESPIRATION *UF:* Hydrospires	RESPIRATION *UF:* Hydrospire	ATMUNG *UF:* Hydrospire	DYKHANIE	RESPIRACION *UF:* Hidrospira	ANATOMIA APPARATO RESPIRATORIO *UF:* Idrospira
3836 RESURGENCE GEOH	RESURGENCE *UF:* Resurgent water	RESURGENCE *UF:* Eau resurgence	RESURGENZ *UF:* Karst=Wasser	RESURGENTSIYA	RESURGENCIA *UF:* Agua=resurgente	RISORGIVA *UF:* Acqua di risorgiva
3837 RESURGENT WATER GEOH	Resurgent water RESURGENCE	Eau resurgence RESURGENCE	Karst=Wasser RESURGENZ	VODA OSVOBOZH- DENNAYA	Agua=resurgente RESURGENCIA	Acqua di risorgiva RISORGIVA
3838 RETENTION GEOH	RETENTION	RETENTION	ZURUECKHAL- TUNG	[RETENTION]	RETENCION	RITENZIONE
3839 RETICULAREA PALS	Reticularea PROTISTA	Reticularea PROTOZOA	Reticularea PROTOZOA	RETICULAREA	Reticularea PROTOZOA	Reticularea PROTOZOA
3840 RETICULATE STRUCTURE TEST	Reticulate structure TEXTURES	Structure reticulee TEXTURE	Retikular=Gefuege KORN=GEFUEGE	SETCHATYJ	Estructura=reticulada TEXTURA	Struttura reticolare TESSITURA
3841 RETROGRADE METAMORPHISM IGNE	RETROGRADE METAMORPHISM *UF:* Diaphthoresis	METAMORPHISME RETRO *UF:* Diaphthorese; Diaphthorite	RETRO= METAMORPHOSE *UF:* Diaphthorese; Diaphthorit	Metamorfizm retro- gradnyj DIAFTOREZ	RETROMETAMOR- FISMO *UF:* Diaftoresis; Diaf- torita	RETROMETAMOR- FISMO *UF:* Diaftoresi; Diafto- rite
3842 REVERSAL PHCH	REVERSALS *UF:* Geomagnetic rever- sal	Inversion INVERSION CHAMP	Umkehrung FELDINVERSION	[REVERSAL] INVERSIYA MAGNITOGO POLYA; POLE MAGNITNOE	Inversion INVERSION=DE= CAMPO	Inversione INVERSIONE DEL CAMPO
3843 REVIEW ARTICLE MISC	BIBLIOGRAPHIC REVIEW	SYNTHESE BIBLIO- GRAPHIQUE	LITERATUR= BERICHT	OBZOR *UF:* Karta registratsion- naya	SINTESIS= BIBLIOGRAFICA	SINTESI BIBLIO- GRAFICA
3844 REVISION MISC	REVISION	REVISION	REVISION	REVIZIYA	REVISION	REVISIONE
3845 REWORKED GEOL	Reworked(adj.)	Remanie(adj.)	Aufgearbeitet(adj.)	Pererabotannyj PEREOTLOZHENIE; OSTATOCHNYJ	Retrabajado(adj.)	Rimaneggiato(adj.)

	ENGLISH	FRANCAIS	DEUTSCH	RUSSKIJ	ESPANOL	ITALIANO
3846 REWORKED FOSSIL PALE	REWORKING	REMANIEMENT FOSSILE	AUFGEARBEITET. FOSSIL	Perotlozhennyj PEREOTLOZHENIE	REMOCION=FOSIL	RIMANEGGIA-MENTO DI FOSSILE
3847 RHABDOPLEURIDA PALS	Rhabdopleurida PTEROBRANCHIA	Rhabdopleurida PTEROBRANCHIA	Rhabdopleurida PTEROBRANCHIA	RHAB-DOPHLEURIDA	Rhabdopleurida PTEROBRANCHIA	Rhabdophleurida PTEROBRANCHIA
3848 RHAETIAN STRS	RHAETIAN	RHETIEN	RAET	REHT	RHETIENSE	RETICO
3849 RHENIUM CHEE	RHENIUM	RHENIUM	RE	RENIJ	RENIO	RENIO
3850 RHENIUM-OSMIUM DATING ISOT	RE/OS	RE=OS	RE=OS=DATIERUNG	METOD RENIEVO-OSMIEVYJ	RE=OS	RE/OS
3851 RHEOLOGY PHCH	RHEOLOGY	RHEOLOGIE	RHEOLOGIE	REOLOGIYA	REOLOGIA	REOLOGIA
3852 RHEOMORPHISM IGNE	RHEOMORPHISM	RHEOMORPHISME	RHEOMORPHOSE	REOMORFIZM	REOMORFISMO	REOMORFISMO
3853 RHIPIDISTII PALS	RHIPIDISTIA	RHIPIDISTII	RHIPIDISTII	Rhipidistii CROSSOPTERYGII	RHIPIDISTII	RHIPIDISTII
3854 RHIZOPODEA PALS	Rhizopodea PROTISTA	Rhizopodea PROTOZOA	Rhizopodea PROTOZOA	RHIZOPODEA *UF:* Thecamoeba	Rhizopodea PROTOZOA	Rhizopodea PROTOZOA
3855 RHODIUM CHEE	RHODIUM	RHODIUM	RH	RODIJ	RODIO	RODIO
3856 RHODOPHYTA PALS	RHODOPHYTA	RHODOPHYCO-PHYTA	RHODOPHYTA	RHODOPHYTA *UF:* Corallinaceae	RHODOPHYCO-PHYTA	RHODOPHYTA
3857 RHOMBIC SYSTEM MINE	Rhombic system CRYSTAL SYSTEMS	Systeme rhombique SYSTEME CRISTAL-LIN	Rhomb.System KRISTALL=SYSTEM	SINGONIYA ROM-BICHESKAYA *UF:* Singoniya ortorom-bicheskaya	Sistema=rombico SISTEMA=CRISTALINO	Sistema rombico SISTEMA CRISTAL-LINO
3858 RHYNCHOCEPHALIA PALS	RHYNCHOCE-PHALIA	RHYNCHOCE-PHALIA	RHYNCHOCE-PHALIA	RHYNCHOCE-PHALIA	RHYNCHOCE-PHALIA	RHYNCHOCE-PHALIA
3859 RHYNCHONELLIDA PALS	RHYNCHONEL-LIDA	RHYNCHONEL-LIDA	RHYNCHONEL-LIDA	RHYNCHONEL-LIDA	RHYNCHONEL-LIDA	RHYNCHONEL-LIDA
3860 RHYODACITE IGNS	RHYODACITES	RHYODACITE	RHYODACIT	RIODATSIT	RIODACITA	RIODACITE
3861 RHYOLITE IGNS	RHYOLITES	RHYOLITE *UF:* Obsidienne	RHYOLITH *UF:* Obsidian	RIOLIT	RIOLITA *UF:* Obsidiana	RIOLITE *UF:* Ossidiana
3862 RICHTER SCALE SOLI	RICHTER SCALE	ECHELLE RICHTER	RICHTER=SKALA	Shkala Richtera SHKALA ZEMLETR-YASENIJ	ESCALA=RICHTER	SCALA RICHTER
3863 RIEGEL SURF	Riegel GLACIAL EROSION	Verrou EROSION GLA-CIAIRE	Riegel GLAZIAL=EROSION	RIGEL' *UF:* Rock bar	Umbral=glaciar EROSION=GLACIAR	Soglia EROSIONE GLA-CIALE

	ENGLISH	FRANCAIS	DEUTSCH	RUSSKIJ	ESPANOL	ITALIANO
3864 RIFT STRU	RIFT ZONES *UF:* Rift valleys	RIFT *UF:* Vallee rift	RIFT *UF:* Rift=valley	RIFT	RIFT *UF:* Fosa=de= hundimiento	RIFT *UF:* Valle di rift
3865 RIFT VALLEY STRU	Rift valleys RIFT ZONES	Vallee rift RIFT	[Rift=valley] RIFT	DOLINA RIF- TOVAYA	Fosa=de=hundimiento RIFT	Valle di rift RIFT
3866 RIGIDITY PHCH	RIGIDITY	RIGIDITE	RIGIDITAET	ZHESTKOST'	RIGIDEZ	RIGIDITA
3867 RILLES EXTR	RILLES	SILLON LUNAIRE	MOND=FURCHE	RILL	VALLE=LUNAR	VALLE LUNARE
3868 RIM GEOL	Rims(gen.)	Frange(gen.)	Saum(gen.)	KAJMA	Orla(gen.)	Orlo(gen.)
3869 RING DIKE IGNE	RING DIKES	[RING DYKE]	[RING=DYKE]	DAIKA KOL'TSEVAYA	DIQUE=ANULAR	FILONE AD ANELLO
3870 RING FRACTURE STRU	Ring fracture RING STRUCTURES	Fracture annulaire STRUCTURE ANNULAIRE	Ringfoermig.Bruch RING=STRUKTUR	TRESHCHINA KOL'TSEVAYA	Fractura=anular ESTRUCTURA= ANULAR	Frattura ad anello STRUTTURA ANU- LARE
3871 RING STRUCTURE GEOL	RING STRUCTURES *UF:* Ring fracture	STRUCTURE ANNULAIRE *UF:* Fracture annulaire; Structure cryptovol- canique	RING=STRUKTUR *UF:* Kryptovulkan. Struktur; Ringfoermig. Bruch	STRUKTURA KOL'TSEVAYA	ESTRUCTURA= ANULAR *UF:* Estructura= criptovolcanica; Fractura=anular	STRUTTURA ANU- LARE *UF:* Frattura ad anello; Struttura criptovul- canica
3872 RIPPLE MARK TEST	RIPPLE MARKS *UF:* Wind ripple	RIPPLE MARK *UF:* Wind ripple	RIPPELMARKE *UF:* Wind=Rippel	ZNAKI RYABI	RIPPLE=MARK *UF:* Ripple=eolico	RIPPLE MARK *UF:* Increspatura eolica
3873 RIVER SURF	RIVERS *UF:* Effluents; Torrents	RIVIERE *UF:* Affluent; Confluent; Cours eau derive; Ecoulement intermit- tent; Oued; Ruisseau; Vallee consequente	FLUSS *UF:* Bach; Intermittierend.Strom; Konsequent.Strom; Nebenfluss; See= Ausfluss; Wadi; Zusammenfluss	REKA *UF:* Arroiko; Potok obezglavlennyj; Potok perekhvachennyj; Potok rechnoj; Potok vytekayushchij; Rech- noj stok tverdyj; Reka konsekventnaya; Reka vrezannaya; Ruchej; Sediment yield; Stream transport	RIO *UF:* Afluente; Cala; Confluente; Corriente= intermitente=agua; Emisario; Rambla; Rio=consecuente	FIUME *UF:* Affluente; Con- fluenza; Corso d'acqua conseguente; Corso d'acqua intermittente; Emissario; Ruscello; Uadi
3874 RIVER LOAD SURF	River load TRANSPORT	Charge riviere TRANSPORT; ALLUVION	Fluviatil.Fracht TRANSPORT; ALLUVION	Rechnoj stok tverdyj MATERIAL OSA- DOCHNYJ; REKA	Transporte= suspension=fluvial TRANSPORTE; ALUVION	Carico fluviale TRASPORTO; ALLU- VIONE
3875 RIVER TERRACE SURF	River terraces TERRACES	Terrasse fleuve TERRASSE	Fluss=Terrasse TERRASSE	TERRASA RECH- NAYA	Terraza=de=rio TERRAZA	Terrazzo TERRAZZO ALLU- VIONALE
3876 ROAD LOG METH	ROAD LOG	LOG ROUTIER	STRASSEN= VERMESSUNG	OPISANIE MAR- SHRUTNOE	ITINERARIO= EXCURSION	PROFILO STRA- DALE
3877 ROAD MATERIAL COMS	Road material CONSTRUCTION MATERIALS	MATERIAU VIABI- LITE *UF:* Ballast	STRASSENBAU= MATERIAL *UF:* Schuett=Gut	Kamen' dorozhnyj STROITEL'NOE SYR'E	MATERIAL= CARRETERA *UF:* Balasto	MATERIALE PER MASSICCIATA *UF:* Ballast

	ENGLISH	FRANCAIS	DEUTSCH	RUSSKIJ	ESPANOL	ITALIANO
3878 ROAD TEST ENGI	ROAD TESTS	ESSAI ROUTIER	STRASSEN=TEST	IZYSKANIE DOROZHNOE	ENSAYO= CARRETERA	COLLAUDO STRA- DALE
3879 ROADWAY ENGI	ROADWAY	ROUTE	STRASSE	SHOSSE	AUTOPISTA	TRACCIATO
3880 ROBERTINACEA PALS	ROBERTINACEA	ROBERTINACEA	ROBERTINACEA	ROBERTINACEA	ROBERTINACEA	ROBERTINACEA
3881 ROCK GEOL	ROCKS *UF:* Petrogenesis; Petro- physics	ROCHE *UF:* Petrophysique	GESTEIN *UF:* Petrophysik	PORODA GORNAYA	ROCA *UF:* Petrofisica	ROCCIA *UF:* Petrofisica
3882 ROCK BAR SURF	Rock bars BARS	Barriere rocheuse EROSION	Gesteins=Riegel EROSION	[ROCK BAR] RIGEL'	Umbral=rocoso EROSION	Barriera rocciosa EROSIONE
3883 ROCK BURST ENGI	ROCK BURSTS	COUP CHARGE	NATUERL. GESTEINS= SPRENGUNG	VZRYV GORNYJ	ROTURA=ROCA	COLPO DI TEN- SIONE
3884 ROCK FALL SURF	ROCKFALLS	EBOULEMENT	FELSSTURZ	OBVAL	DESPRENDI- MIENTO	FRANA DI CROLLO
3885 ROCK FILL DAM ENGI	ROCKFILL DAMS	BARRAGE ENRO- CHEMENT	STEINSCHUE- TTUNGS=DAMM	Plotina s kamennoj nasypkoj PLOTINA	PRESA= ESCOLLERA	DIGA ROCK FILL
3886 ROCK GLACIER SURF	ROCK GLACIERS	GLACIER ROCHE	BLOCK= GLETSCHER	GLETCHER KAMENNYJ	GLACIAR=ROCAS	ROCK GLACIER
3887 ROCK MATRIX TEST	ROCK MATRIX	MATRICE	MATRIX	Matriks TSEMENT	MATRIZ=ROCA	MATRICE
3888 ROCK MECHANICS ENGI	ROCK MECHANICS	MECANIQUE ROCHE	FELS=MECHANIK	Mekhanika porod MEKHANIKA GRUNTOV	MECANICA=ROCAS	MECCANICA DELLE ROCCE
3889 ROCK-BORING ORGANISM PALE	Rock borer ICHNOFOSSILS	Organisme perforant TRACE ORGANIQUE	Bohr=Organismen LEBENS=SPUR	KAMNETOCHETS	Litofago HUELLA= ORGANICA	Organismo litofago ICNITE
3890 ROCK-FORMING MINERALS IGNE	ROCK=FORMING MINERALS	Mineral essentiel ASSOCIATION MINERALE	Gesteinsbildend. Mineral MINERAL= VERGESELLSCHAFTUNG	MINERAL PORO- DOOBRAZUYUSH- CHIJ	Minerales= constituyentes ASOCIACION= MINERAL	Minerale essenziale ASSOCIAZIONE MINERALOGICA
3891 ROCKSLIDE SURF	ROCKSLIDES	Ecroulement GLISSEMENT TER- RAIN	Felsrutsch ERDRUTSCH	OSOV BLOCHNYJ	Deslisamiento de masas rocosas DESLISAMIENTO= TERRENO	Frana di scivolamento in roccia SCIVOLAMENTO DI TERRENO
3892 RODENTIA PALS	RODENTIA	RODENTIA	RODENTIA	RODENTIA	RODENTIA	RODENTIA
3893 ROLL-TYPE DEPOSIT ECON	ROLL=TYPE DEPOSITS	GITE TYPE ROU- LEAU	ROLL=TYP- LAGERSTAETTE	[ROLL=TYPE DEPOSIT] GENEZIS; RUDA	YACIMIENTO= TIPO=ROLL	GIACIMENTO ROLL=TYPE
3894 ROOF COLLAPSE IGNE	Roof collapse MAGMA CHAM- BERS	Ecroulement toit EFFONDREMENT; CHAMBRE MAG- MATIQUE	Magmakammer= Einbruch EINSTURZ; MAG- MAKAMMER	OBRUSHENIE KROVLI	Hundimiento=techo HUNDIMIENTO; CAMARA= MAGMATICA	Crollo di tetto SPROFONDA- MENTO; CAMERA MAGMATICA

	ENGLISH	FRANCAIS	DEUTSCH	RUSSKIJ	ESPANOL	ITALIANO
3895 ROOF PENDANT IGNE	Roof pendant BATHOLITHS	[ROOF PENDANT] BATHOLITE; ENCLAVE ROCHE	Magmakammer=Dach BATHOLITH; XENO- LITH	Ostanets krovli BATOLIT	Techo=rebajado BATOLITO; ENCLAVE=ROCA	Pendente BATOLITE; XENO- LITE
3896 ROOT STRU	Roots NAPPES	Racine NAPPE	Wurzel DECKE	KORNI GOR	Raiz=de=manto MANTO	Radice FALDA
3897 ROSTROCONCHIA PALS	ROSTROCONCHIA	ROSTROCONCHIA	ROSTROCONCHIA	Rostroconchia MOLLUSCA	ROSTROCONCHIA	ROSTROCONCHIA
3898 ROTALIACEA PALS	ROTALIACEA	ROTALIACEA	ROTALIACEA	ROTALIACEA	ROTALIACEA	ROTALIACEA
3899 ROTALIINA PALS	ROTALIINA	ROTALIINA	ROTALIINA	ROTALIINA	ROTALIINA	ROTALIINA
3900 ROTARY DRILLING METH	Rotary drilling BOREHOLES	Forage rotary SONDAGE	Rotary=Bohrung BOHRUNG	BURENIE VRASH- CHATEL'NOE	Sondeo=rotary POZO=SONDEO	Perforazione rotary SONDAGGIO
3901 ROTARY-PERCUSSION DRILLING MINI	Rotary=percussion drilling BOREHOLES	Forage percussion rotary SONDAGE	Rotary=Bohrmethode BOHRUNG	BURENIE UDARNOVRASH- CHATEL'NOE	Sondeo=rotacion= percusion POZO=SONDEO	Perforazione rotoper- cussione SONDAGGIO
3902 ROTATION GEOL	ROTATION *UF:* Earth rotation; Nutation	ROTATION *UF:* Nutation; Rotation terre	ROTATION *UF:* Erd=Rotation; Nutation	VRASHCHENIE	ROTACION *UF:* Nutacion; Rotacion=tierra	ROTAZIONE *UF:* Nutazione; Rotaz- ione terrestre
3903 ROTATORY DISPERSION PHCH	Rotatory dispersion OPTICAL DISPER- SION	Dispersion rotatoire DISPERSION OPTIQUE	Rotierend.Dispersion OPT.DISPERSION	Dispersiya svetla rotuyushchaya DISPERSIYA	Dispersion=rotatoria DISPERSION= OPTICA	Dispersione rotatoria DISPERSIONE OTTICA
3904 ROTLIEGENDES STRS	ROTLIEGENDES	ROTLIEGENDE	ROTLIEGENDES	Lezhen' krasnyj PERM' NIZH- NYAYA	ROTLIEGENDE	ROTLIEGENDES
3905 ROUGHNESS PHCH	ROUGHNESS	RUGOSITE	RAUHIGKEIT	SHEROKHOVA- TOST'	RUGOSIDAD	RUGOSITA
3906 ROUNDNESS SEDI	ROUNDNESS	DEGRE ARRONDI	RUNDUNG	OKATANNOST'	GRADO= REDONDEZ	ARROTONDA- MENTO
3907 RUBBLE SURF	Rubble GRAVEL	Cailloutis EBOULIS	Schotter SCHUTT	SHCHEBEN'	Guijarro DERRUBIO	Pietrisco DETRITO DI FALDA
3908 RUBEFACTION SURF	RUBEFACTION	RUBEFACTION	RUBEFIZIERUNG	[RUBEFACTION]	RUBEFACCION	RUBEFAZIONE
3909 RUBIDIUM CHEE	RUBIDIUM	RUBIDIUM	RB	RUBIDIJ	RUBIDIO	RUBIDIO
3910 RUBIDIUM-STRONTIUM DATING ISOT	SR/RB	SR=RB	RB=SR= DATIERUNG	METOD RUBIDIEVO- STRONTSIEVYJ	RB=SR	SR/RB
3911 RUDISTAE PALS	RUDISTAE	RUDISTAE	RUDISTAE	HIPPURITOIDA	RUDISTAE	RUDISTAE
3912 RUDITE SEDS	Rudite CLASTIC ROCKS	Rudite ROCHE CLASTIQUE	Rudit KLAST.GESTEIN	RUDIT	Rudita ROCA=CLASTICA	Rudite ROCCIA CLASTICA

	ENGLISH	FRANCAIS	DEUTSCH	RUSSKIJ	ESPANOL	ITALIANO
3913 RUMINANTIA PALS	RUMINANTIA	RUMINANTIA	RUMINANTIA	RUMINANTIA	RUMINANTIA	RUMINANTIA
3914 RUN ECON	RUN	[RUN]	[RUN]	ZALEZH'	CORRIDA	[RUN]
3915 RUNOFF GEOH	RUNOFF *UF:* Groundwater runoff; Rain wash; Surface runoff	RUISSELLEMENT *UF:* Ruissellement pluie; Ruissellement surface	ABFLUSS *UF:* Oberflaechen= Abfluss; Regen= Auswaschung	STOK *UF:* Vneshnij drenazh	ESCORRENTIA *UF:* Escorrentia=lluvia; Escorrentia=superficial	RUSCELLAMENTO *UF:* Ruscellamento pluviale; Scorrimento superficiale
3916 RUPELIAN STRS	Rupelian STAMPIAN	Rupelien STAMPIEN	Rupel=Stufe STAMPIEN	RYUPEL' *UF:* Oligotsen nizhnij; Oligotsen srednij; Stamp	Rupeliense STAMPIENSE	Rupeliano STAMPIANO
3917 RUPTURE STRU	RUPTURE *UF:* Crevasse; Heat crack; Rupture strength	RUPTURE *UF:* Crevasse; Rimaye	RUPTUR *UF:* Bergschrund; Gletscher=Spalte	RAZRUSHENIE	RUPTURA *UF:* Grieta=de=hielo; Rimaya	ROTTURA *UF:* Crepaccio; Crepac- cio terminale
3918 RUPTURE STRENGTH PHCH	Rupture strength RUPTURE	Resistance a la rupture RESISTANCE MECANIQUE	Ruptur=Festigkeit FESTIGKEIT	Napryazhenie razrus- cayushchee PROCHNOST'	Tension=de=ruptura RESISTENCIA= MECANICA	Forza di rottura RESISTENZA MEC- CANICA
3919 RUTHENIUM CHEE	RUTHENIUM	RUTHENIUM	RU	RUTENIJ	RUTENIO	RUTENIO
3920 S WAVE SOLI	S=WAVES	ONDE S	S=WELLE	Volna poperechnaya VOLNA SEJSMI- CHESKAYA	ONDA=S	ONDA S
3921 SAKMARIAN STRS	SAKMARIAN	SAKMARIEN	SAKMAR	SAKMAR	SAKMARIENSE	SAKMARIANO
3922 SALIFEROUS GEOL	Saliferous(adj.)	Salifere(adj.)	Salzhaltig(adj.)	SOLENOSNYJ	Salino(adj.)	Salifero(adj.)
3923 SALINITY PHCH	SALINITY *UF:* Euryhaline taxa; Halitic environment; Halophilic taxa; Halo- phyte	SALINITE *UF:* Halophyte; Milieu salin; Organ- isme euryhalin; Organ- isme halophile	SALINITAET *UF:* Euryhalin. Organismus; Halit. Milieu; Halophil. Organismus; Halophyt	SOLENOST' *UF:* Khlornost'; Paleo- solenost'	SALINIDAD *UF:* Eurihalino; Halitico; Halofilo; Halofita	SALINITA *UF:* Alofita; Ambiente salato; Organismo alofi- lo; Organismo eurialino
3924 SALT COMS	SALT *UF:* Salt flats	SEL *UF:* Saline	STEINSALZ *UF:* Salzebene	SOL'	SAL=GEMA *UF:* Llanura=salina	SALGEMMA *UF:* Piana salina
3925 SALT DOME STRU	SALT DOMES	DOME SEL	SALZSTOCK	KUPOL SOLYANOJ	DOMO=DE=SAL	DUOMO SALINO
3926 SALT FLAT SURF	Salt flats SALT	Saline SEL	Salzebene STEINSALZ	OTMEL' SOLY- ANAYA	Llanura=salina SAL=GEMA	Piana salina SALGEMMA
3927 SALT LAKE SURF	SALT LAKES	LAC SALE	SALZ=SEE	OZERO SOLYANOE	LAGO=SALADO	LAGO SALATO
3928 SALT MARSH SURF	SALT MARSHES *UF:* Marine marsh	MARAIS SALE	KUESTEN=SALZ= MARSCH	[SALT MARSH] MARSH; SOLON- CHAK *UF:* Salz=Marsch	MARISMA= SALADA	PALUDE SAL- MASTRA
3929 SALT TECTONICS STRU	SALT TECTONICS	TECTONIQUE SALIFERE	SALZ=TEKTONIK	TEKTONIKA SOLY- ANAYA	TECTONICA= SALIFERA	TETTONICA DEL SALE

	ENGLISH	FRANCAIS	DEUTSCH	RUSSKIJ	ESPANOL	ITALIANO
3930 SALT-WATER INTRUSION GEOH	SALT=WATER INTRUSION	INTRUSION EAU SALEE	SALZWASSER= INTRUSION	VTORZHENIE SOLENYKH VOD	INTRUSION= AGUA=SALADA	INTRUSIONE DI ACQUA SALATA
3931 SALTATION SEDI	SALTATION	SALTATION	SALTATION	SAL'TATSIYA	SALTACION	SALTAZIONE
3932 SAMARIUM CHEE	SAMARIUM	SAMARIUM	SM	SAMARIJ	SAMARIO	SAMARIO
3933 SAMPLE METH	Samples SAMPLING	Echantillon ECHANTILLON-NAGE	Probe PROBENNAHME	OBRAZETS	Muestra MUESTREO	Campione CAMPIONATURA
3934 SAMPLER INST	Sampler INSTRUMENTS; SAMPLING	Echantillonneur INSTRUMENTA-TION; ECHANTIL-LONNAGE	[SAMPLER] INSTRUMEN-TIERUNG; PROBEN-NAHME	PROBOOTBORNIK	Saca=muestras INSTRUMENTA-CION; MUESTREO	Campionatore STRUMENTAZ-IONE; CAMPIONA-TURA
3935 SAMPLING METH	SAMPLING UF: Grab sampling; Sampler; Samples	ECHANTILLON-NAGE UF: Echantillon; Echantillonnage par grappin; Echantillonneur	PROBENNAHME UF: Greifer=Probe; Probe; Sampler	OPROBOVANIE UF: Soil sampling	MUESTREO UF: Muestra; Muestreo=aleatorio; Saca=muestras	CAMPIONATURA UF: Bennata; Campiona-tore; Campione
3936 SAND SEDS	SAND UF: Glass=sand	SABLE UF: Arenosol; Corps sable; Filet sable; Sable verrerie	SAND UF: Arenosol; Glassand; Sand=Koerper; Shoestring=Sand	PESOK UF: Sand body	ARENA UF: Arena=de=vidrio; Arenosol; Cuerpo= arena; Lentejon=de= arena	SABBIA UF: Arenosol; Corpo sabbioso; Laccio da scarpa; Sabbia da vetro
3937 SAND BODY SEDI	Sand bodies SANDSTONE	Corps sable SABLE	Sand=Koerper SAND	[SAND BODY] PESOK; PLAST	Cuerpo=arena ARENA	Corpo sabbioso SABBIA
3938 SAND-SHALE RATIO SEDI	SAND=SHALE RATIO	Rapport sable=argile GRANULOMETRIE	Sand=Schiefer= Verhaeltnis KORN= VERTEILUNG	KOEHFFITSIENT PESCHANISTOSTI	Relacion=arenita= lutita GRANULOMETRIA	Rapporto sabbia= argilla GRANULOMETRIA
3939 SANDSTONE SEDS	SANDSTONE UF: Psammite; Sand bodies; Shoestring sands	GRES UF: Gres grossier; Psam-mite	SANDSTEIN UF: Grit; Psammit	PESCHANIK UF: Psammit	ARENISCA UF: Arenisca=grosera; Psamita	ARENARIA UF: Arenaria grossolana; Psammite
3940 SANDSTONE DIKE SEDI	SANDSTONE DIKES	DYKE GRESEUX	SANDSTEIN=GANG	Daika peschanikovaya DAIKA KLASTI-CHESKAYA	DIQUE=DE= ARENISCAS	DICCO DI ARE-NARIA
3941 SANDWAVE MARI	SAND WAVES	ONDE SABLE	GROSSRIPPEL	VOLNA PES-CHANAYA	ONDA=ARENA	ONDA DI SABBIA
3942 SANIDINITE FACIES IGNE	Sanidinite facies CONTACT META-MORPHISM; FACIES	Facies a sanidine METAMORPHISME CONTACT	Sanidinit=Fazies KONTAKT= METAMORPHOSE	FATSIYA SANIDINITOVAYA	Facies=sanidina METAMORFISMO= CONTACTO	Facies delle sanidiniti METAMORFISMO DI CONTATTO
3943 SANTONIAN STRS	SANTONIAN	SANTONIEN	SANTON	SANTON	SANTONIENSE	SANTONIANO

	ENGLISH	FRANCAIS	DEUTSCH	RUSSKIJ	ESPANOL	ITALIANO
3944 SAPROPEL SEDS	SAPROPEL	SAPROPELE	SAPROPEL	SAPROPEL' *UF:* Gitt'ya; Ugol' sapro- pelevyj	SAPROPEL	SAPROPEL
3945 SAPROPELITIC COAL COMS	SAPROPELITE	CHARBON SAPRO- PELIQUE *UF:* Cannel coal; Char- bon boghead	SAPROPEL=KOHLE *UF:* Boghead=Kohle; Kennel=Kohle	Ugol' sapropelevyj SAPROPEL'; UGOL'	CARBON= SAPROPELICO *UF:* Carbon=boghead; Carbon=cannel	CARBONE SAPRO- PELITICO *UF:* Carbone a fiamma lunga; Carbone bog- head
3946 SARCODINA PALS	Sarcodina PROTISTA	Sarcodina PROTOZOA	Sarcodina PROTOZOA	SARCODINA	Sarcodina PROTOZOA	Sarcodina PROTOZOA
3947 SARCOPTERYGII PALS	Sarcopterygii OSTEICHTHYES	Sarcopterygii OSTEICHTHYES	Sarcopterygii OSTEICHTHYES	SARCOPTERYGII	Sarcopterygii OSTEICHTHYES	Sarcopterygii OSTEICHTHYES
3948 SARMATIAN STRS	SARMATIAN	SARMATIEN	SARMAT	SARMAT	SARMATIENSE	SARMATIANO
3949 SATURATED ZONE GEOH	SATURATED ZONE	ZONE SATUREE	SAETTIGUNGS= ZONE	ZONA NASYSHC- HENIYA	ZONA=SATURADA	ZONA DI SATURAZIONE
3950 SATURATION PHCH	SATURATION *UF:* Oversaturated solution; Undersatura- tion	SATURATION *UF:* Soussaturation; Sursaturation	SAETTIGUNG *UF:* Uebersaettigung; Untersaettigt.Loesung	NASYSHCHENIE	SATURACION *UF:* No=saturado; Sobresaturacion	SATURAZIONE *UF:* Sottosaturazione; Sovrasaturazione
3951 SAURISCHIA PALS	SAURISCHIA	SAURISCHIA	SAURISCHIA	SAURISCHIA	SAURISCHIA	SAURISCHIA
3952 SAURORNITHES PALS	Saurornithes ARCHAEORNITHES	Saurornithes ARCHAEORNITHES	Saurornithes ARCHAEORNITHES	SAURORNITHES	Saurornithes ARCHAEORNITHES	Saurornithes ARCHAEORNITHES
3953 SAUSSURITIZATION IGNE	SAUSSURITIZA- TION	Saussuritisation ALTERATION HYDROTHERMALE	Saussuritisierung HYDROTHERMAL= UMWANDLUNG	SOSSYURITIZAT-. SIYA	Sausuritizacion ALTERACION= HIDROTERMAL	Saussuritizzazione ALTERAZIONE IDROTERMALE
3954 SAXONIAN STRS	SAXONIAN	SAXONIEN	SAXONIUM	SAKSONIJ	SAXONIENSE	SASSONIANO
3955 SCALE MODEL GEOL	SCALE MODELS	MAQUETTE *UF:* Bloc diagramme	MASSSTABS= MODELL	MODEL' MASSHTABNAYA	MAQUETA *UF:* Bloque=diagrama	MODELLO IN SCALA *UF:* Block=diagramma
3956 SCANDIUM CHEE	SCANDIUM	SCANDIUM	SC	SKANDIJ	ESCANDIO	SCANDIO
3957 SCANNING ELECTRON MICROSCOPY METH	SCANNING METHOD	METHODE MEB	RASTER=ELEKTR. MIKROSKOPIE	MIKROSKOPIYA EHLEKTRONNAYA SKANIRUYU= SHCHAYA *UF:* SEM data	METODO=MEB	MICROSCOPIA ELETTR. SCANSIONE
3958 SCAPHOPODA PALS	SCAPHOPODA	SCAPHOPODA	SCAPHOPODA	SCAPHOPODA	SCAPHOPODA	SCAPHOPODA
3959 SCAPOLITIZATION IGNE	SCAPOLITIZATION	Scapolitisation ALTERATION HYDROTHERMALE	Skapolitisierung HYDROTHERMAL= UMWANDLUNG	SKAPOLITIZAT- SIYA	Escapolitizacion ALTERACION= HIDROTERMAL	Scapolitizzazione ALTERAZIONE IDROTERMALE

	ENGLISH	FRANCAIS	DEUTSCH	RUSSKIJ	ESPANOL	ITALIANO
3960 SCHIST IGMS	SCHISTS *UF:* Sericite schist; Talc scist	Schiste metamorphique MICASCHISTE	Schiefer GLIMMER= SCHIEFER	SLANETS KRISTAL- LICHESKIJ *UF:* Slanets seritsitovyj	Esquisto MICAESQUISTO	Scisto MICASCISTO
3961 SCHISTOSITY STRU	SCHISTOSITY	SCHISTOSITE *UF:* Clivage plan axial	SCHIEFERUNG *UF:* Tekton.Schieferung	SLANTSEVATOST'	ESQUISTOSIDAD *UF:* Exfoliacion=segun= plano=axial	SCISTOSITA *UF:* Clivaggio di piano assiale
3962 SCHLIEREN TEST	Schlieren GNEISSES	[Schlieren] GNEISS	Schlieren=Gefuege GNEIS	SHLIR	Franja=de=segregacion GNEIS	[Schlieren] GNEISS
3963 SCHUPPE STRU	TECTONIC WEDGES	ECAILLE	SCHUPPE	CHESHUYA (TECTONI- CHESKAYA)	ESCAMA	SCAGLIA
3964 SCINTILLATION PHCH	Scintillations OPTICAL PROPER- TIES	Scintillation PROPRIETE OPTIQUE	Szintillation OPT.EIGENSCHAFT	STSINTILLYAT- SIYA	Centelleo PROPIEDAD= OPTICA	Scintillazione PROPRIETA OTTICA
3965 SCIUROMORPHA PALS	SCIUROMORPHA	SCIUROMORPHA	SCIUROMORPHA	SCIUROMORPHA	SCIUROMORPHA	SCIUROMORPHA
3966 SCLERITE PALE	SCLERITES	SCLERITE	SKLERIT	Sklerit SKELET	ESCLERITES	SCLERITE
3967 SCOLECODONT PALE	SCOLECODONTS	SCOLECODONTE	SCOLECODONTA	SKOLEKODONTY	ESCOLECODONTO	SCOLECODONTI
3968 SCOLITE PALE	Scolite ICHNOFOSSILS; WORMS	Scolite TRACE ORGANIQUE; VER- MES	Skolithus LEBENS=SPUR; VERMES	SKOLIT (PALEONT)	Escolito HUELLA= ORGANICA; VER- MES	Scolite ICNITE; VERMES
3969 SCORIA IGNE	SCORIA	SCORIE	SCHLACKE	SHLAK VULKANI- CHESKIJ *UF:* Konus shlakovyj; Shlak	ESCORIA	SCORIE
3970 SCOUR MARK TEST	SCOUR MARKS	TRACE AFFOUILLE- MENT	EROSIONS=MARKE	SHTRIKHOVKA	HUELLA= DERRUBIO	TRACCIA DI ESCAVAZIONE
3971 SCOURING SURF	Scouring EROSION	Decapage EROSION	[SCOURING] EROSION	EHKZARATSIYA	Burilado EROSION	Striamento EROSIONE
3972 SCYPHOMEDUSAE PALS	Scyphomedusae SCYPHOZOA	Scyphomedusae SCYPHOZOA	Scyphomedusae SCYPHOZOA	SCYPHOMEDUSAE	Scyphomedusae SCYPHOZOA	Scyphomedusae SCYPHOZOA
3973 SCYPHOZOA PALS	SCYPHOZOA *UF:* Scyphomedusae	SCYPHOZOA *UF:* Scyphomedusae	SCYPHOZOA *UF:* Scyphomedusae	SCYPHOZOA	SCYPHOZOA *UF:* Scyphomedusae	SCYPHOZOA *UF:* Scyphomedusae
3974 SEA MARI	Seas MARINE ENVIRON- MENT	Mer MILIEU MARIN	Meer MARIN.MILIEU	MORE *UF:* Svyaz vozdukh= more	Mar MEDIO=MARINO	Mare AMBIENTE MARINO
3975 SEA BOTTOM MARI	Sea bottom OCEAN FLOORS	RELIEF SOUS MARIN *UF:* Detroit; Morpholo- gie sous marine; Seuil marin	SUBMARIN.RELIEF *UF:* Meerenge; Meeres= Furche; Submarin. Morphologie	DNO MORSKOE	RELIEVE= SUBMARINO *UF:* Estrecho; Morfologia=submarina; Umbral	RILIEVO SOTTO- MARINO *UF:* Morfologia sotto- marina; Solco sotto- marino; Stretto

	ENGLISH	FRANCAIS	DEUTSCH	RUSSKIJ	ESPANOL	ITALIANO
3976 SEA ICE MARI	SEA ICE	GLACE MARINE	MEERES=EIS	LED MORSKOJ	HIELO=MARINO	GHIACCIO MARINO
3977 SEA LEVEL MARI	Sea level CHANGES OF LEVEL	Niveau mer VARIATION; OCEAN	Meeresspiegel VARIATION; OZEAN	UROVEN' MORYA	Nivel=del=mar VARIACION; OCEANO	Livello marino VARIAZIONE; OCEANO
3978 SEA WATER MARI	SEA WATER	EAU MER *UF:* Pycnocline; Thermocline	MEER=WASSER *UF:* Dichte=Gradient; Thermokline	VODA MORSKAYA	AGUA=DE=MAR *UF:* Picnoclina; Termoclina	ACQUA DI MARE *UF:* Picnoclino; Termoclina
3979 SEA WAVE MARI	OCEAN WAVES *UF:* Surges	HOULE *UF:* Deferlante	DUENUNG *UF:* Woge	VOLNA MORSKAYA	OLEAJE *UF:* Batiente	ONDA MARINA *UF:* Flutto
3980 SEA-FLOOR SPREADING STRU	SEA=FLOOR SPREADING	EXPANSION FOND OCEANIQUE	[SEAFLOOR=SPREADING]	RASSHIRENIE DNA OKEANA *UF:* Spreading center	EXPANSION=FONDO=OCEANICO	ESPANSIONE DI FONDO OCEANICO
3981 SEA-LEVEL CHANGE MARI	CHANGES OF LEVEL *UF:* Sea level	Variation niveau mer VARIATION	Meeresspiegel=Schwankung VARIATION	IZMENENIE UROVNYA MORYA	Variacion=nivel=mar VARIACION	Livello del mare VARIAZIONE
3982 SEALING ENGI	SEALING	ETANCHEMENT	ABDICHTUNG	Tamponazh TSEMENTATSIYA	COLMATACION	SIGILLATURA
3983 SEAM ECON	Seams COAL SEAMS	Veine charbon COUCHE CHARBON	Floez KOHLE=FLOEZ	PLAST *UF:* Sand body	Capa=productiva CAPA=CARBON	Letto produttivo LIVELLO DI CARBONE
3984 SEAMOUNT MARI	SEAMOUNTS	GUYOT	[SEA=MOUNT]	GORA PODVODNAYA	GUYOT	GUYOT
3985 SEASONAL VARIATION GEOL	SEASONAL VARIATIONS	VARIATION SAISONNIERE *UF:* Ecoulement intermittent; Orage magnetique	JAHRESZEITL. SCHWANKUNG *UF:* Intermittierend. Strom; Magnet.Sturm	Izmenenie sezonnoe KLIMAT	VARIACION=ESTACIONARIA *UF:* Corriente=intermitente=agua; Tormenta=magnetica	VARIAZIONE STAGIONALE *UF:* Corso d'acqua intermittente; Tempesta magnetica
3986 SEAWALL ENGI	SEAWALLS	DIGUE	DEICH	DAMBA *UF:* Levee	MALECON	SBARRAMENTO
3987 SEBKHA ENVIRONMENT SEDI	SEBKHA ENVIRONMENT	SEBKHA	SEBKHA=MILIEU	[SEBKHA ENVIRONMENT] SOLONCHAK; FATSIYA	MEDIO=SEBKHA	SEBKHA
3988 SECONDARY AUREOLE GEOC	SECONDARY DISPERSION	AUREOLE SECONDAIRE	SEKUNDAER=AUREOLE	OREOL VTORICHNYJ	AUREOLA=SECUNDARIA	AUREOLA SECONDARIA
3989 SECONDARY MINERAL MINE	SECONDARY MINERALS	MINERAL SECONDAIRE	SEKUNDAER=MINERAL	MINERAL VTORICHNYJ	MINERAL=SECUNDARIO	MINERALE SECONDARIO
3990 SECONDARY SEDIMENTARY STRUCTURE TEST	SECONDARY STRUCTURES	STRUCTURE DIAGENETIQUE *UF:* Dendrite	DIAGENET. GEFUEGE *UF:* Dendrit	[SECONDARY SEDIMENTARY STRUCTURE] STRUKTURA OZADOCHNAYA	ESTRUCTURA=DIAGENETICA *UF:* Dendrita	STRUTTURA DIAGENETICA *UF:* Dendrite
3991 SEDIMENT SEDI	SEDIMENTS *UF:* Bottom sediments; Unconsolidated materials	SEDIMENT *UF:* Deformation sediment; Sediment fond	SEDIMENT *UF:* Boden=Sediment; Fruehdiagenet. Verformung	OSADOK *UF:* Mud lump; Organic residue; Organic sediment; Stream sediment	SEDIMENTO *UF:* Deformacion=sedimento=blando; Sedimento=base	SEDIMENTO *UF:* Deformazione di sedimenti molli; Sedimento di fondo

	ENGLISH	FRANCAIS	DEUTSCH	RUSSKIJ	ESPANOL	ITALIANO
3992 SEDIMENT TRACTION SEDI	Sediment traction SEDIMENTATION	Traction sedimentation TRACTION	Sediment=Auslaen- gung AUSLAENGUNG	Volochenie TRANSPORTI- ROVKA(GEOL)	Traccion=sedimenta- cion TRACCION	Trazione di sedimenti TENSIONE
3993 SEDIMENT YIELD SEDI	SEDIMENT YIELD	VOLUME SEDI- MENT	SEDIMENT=VOLU- MEN	[SEDIMENT YIELD] REKA; STOK POV- ERKHNOSTNYJ	VOLUMEN=DE= SEDIMENTOS	VOLUME DI SEDI- MENTI
3994 SEDIMENT=WATER INTERFACE SEDI	SEDIMENT=WATER INTERFACE	INTERFACE SEDI- MENT EAU	SEDIMENT= WASSER= GRENZFLAECHE	[SEDIMENT= WATER INTER- FACE]	INTERFASE= SEDIMENTO=AGUA	INTERFACCIA ACQUA= SEDIMENTO
3995 SEDIMENTARY BASIN SEDI	SEDIMENTARY BASINS	BASSIN SEDI- MENTAIRE	SEDIMENT= BECKEN	BASSEJN SEDI- MENTATSII	CUENCA= SEDIMENTARIA= FOSIL	BACINO SEDIMEN- TARIO
3996 SEDIMENTARY COVER STRU	SEDIMENTARY COVER	TECTONIQUE COU- VERTURE	DECKGEBIRGS= TEKTONIK	POKROV OSA- DOCHNYJ	TECTONICA=DE= COBERTERA	TETTONICA DELLA COPERTURA
3997 SEDIMENTARY ORE ECON	Sedimentary ore ORE BODIES; SEDI- MENTARY PRO- CESSES	Gite sedimentaire CONTROLE SEDI- MENTAIRE GITE	Sedimentaer. Lagerstaette SEDIMENTAER. EINFLUSS	MESTOROZHDENIE OSADOCHNOE	Deposito=sedimentario YACIMIENTO= SEDIMENTARIO	Giacimento sedimento- geno CONTROLLO SEDI- MENTARIO GIAGI- MENTO
3998 SEDIMENTARY PETROLOGY SEDI	SEDIMENTARY PETROLOGY	PETROLOGIE ROCHE SEDI- MENTAIRE	SEDIMENT= PETROLOGIE	Petrologiya osadoch- naya LITOLOGIYA	PETROLOGIA= ROCA= SEDIMENTARIA	PETROLOGIA DEL SEDIMENTARIO
3999 SEDIMENTARY PROCESS ECON	SEDIMENTARY PROCESSES UF: Sedimentary ore	CONTROLE SEDI- MENTAIRE GITE UF: Gite sedimentaire	SEDIMENTAER. EINFLUSS UF: Sedimentaer. Lagerstaette	OSADOCHNYJ	YACIMIENTO= SEDIMENTARIO UF: Deposito= sedimentario	CONTROLLO SEDI- MENTARIO GIACI- MENTO UF: Giacimento sedi- mentogeno
4000 SEDIMENTARY ROCK SEDS	SEDIMENTARY ROCKS UF: Ankerite; Grit; Lithogenesis	ROCHE SEDI- MENTAIRE	SEDIMENT= GESTEIN	PORODA OSA- DOCHNAYA UF: Algal biscuit	ROCA= SEDIMENTARIA	ROCCIA SEDIMEN- TARIA
4001 SEDIMENTARY STRUCTURE TEST	SEDIMENTARY STRUCTURES UF: Hieroglyphs; Polymictic texture; Synsedimentary struc- ture	STRUCTURE SEDI- MENTAIRE UF: Hieroglyphe; Polymicte; Structure synsedimentaire	SEDIMENT= GEFUEGE UF: Hieroglyphe; Polymikt.Gefuege; Synsedimentaer. Gefuege	STRUKTURA OSA- DOCHNAYA UF: Primary structure; Secondary sedimentary structure	ESTRUCTURA= SEDIMENTARIA UF: Estructura=sinsedi- mentaria; Jerogflifico; Polimicto	STRUTTURA SEDI- MENTARIA UF: Geroglifico; Polimit- tico; Struttura sinsedi- mentaria
4002 SEDIMENTARY TRAP SEDI	Sedimentary trap STRATIGRAPHIC TRAPS	Piege sedimentaire PIEGE	Sedimentaer.Falle FALLE	OBLAST'NAKO- PLENIYA	Trampa=sedimentaria TRAMPA	Trappola sedimentaria TRAPPOLA
4003 SEDIMENTATION SEDI	SEDIMENTATION UF: Intraformational deposition; Mechanical origin; Neptunism; Recent sedimentation; Sediment traction; Sedimentogenesis; Zone of sedimentation	SEDIMENTATION UF: Neptunisme; Sedi- mentation acutelle; Sedimentation intrafor- mationnelle; Zone sedimentation	SEDIMENTATION UF: Aktuo=Sedimenta- tion; Nep- tunismus; Sedimentations=Zone; Synsedimentaer. Bildung	SEDIMENTATSIYA UF: Freshwater sedimen- tation; Geosynclinal sedimentation; Glacial sedimentation; High= energy environment; Intertidal sedimenta- tion; Otlozhenie; Pelagic sedimentation; Recent sedimentation; Reef sedimentation; Sedimentatsiya biokhimicheskaya; Sentimentatsiya biok- lasticheskaya; Soft sediment deformation; Sreda spokojnaya; Sreda volnennaya; Swamp sedimentation	SEDIMENTACION UF: Neptunismo; Sedimentacion=actual; Sedimentacion= intraformacional; Zona=de= sedimentacion	SEDIMENTAZIONE UF: Intraformazionale; Nettunismo; Sedi- mentazione attuale; Zona di sedimentazione

	ENGLISH	FRANCAIS	DEUTSCH	RUSSKIJ	ESPANOL	ITALIANO
4004 SEDIMENTATION RATE SEDI	SEDIMENTATION RATES	TAUX SEDIMENTA-TION	SEDIMENTATIONS-RATE	SKOROST' OSAD-KONAKOPLENIYA	PROPORCION=MATERIAL=SEDIMENTADO	TASSO DI SEDI-MENTAZIONE
4005 SEDIMENTOGENESIS SEDI	Sedimentogenesis SEDIMENTATION	Genese sedimentaire GENESE	Sediment=Genese GENESE	SEDIMENTOGENEZ	Sedimentogenesis GENESIS	Genesi di sedimento GENESI
4006 SEDIMENTOLOGY SEDI	SEDIMENTOLOGY	SEDIMENTOLOGIE	SEDIMENTOLOGIE	SEDIMEN-TOLOGIYA	SEDIMENTOLOGIA	SEDIMENTOLOGIA
4007 SEED PALE	SEEDS	GRAINE	SAMEN	SEMYA	SEMILLA	SEME
4008 SEEPAGE ENGI	SEEPAGE	SUINTEMENT	AUSSICKERUNG	[SEEPAGE] INFIL'TRATSIYA	EXUDACION	STILLICIDIO
4009 SEGREGATED VEIN ECON	Segregated vein VEINS	Filon segregation FILON	Segregations=Gang GANG	ZHILA SEGREGAT-SIONNAYA	Filon=segregado FILON	Vena di segregazione FILONE
4010 SEGREGATION IGNE	SEGREGATION	Segregation DIFFERENCIATION MAGMATIQUE	Segregation MAGMAT. DIFFERENTIATION	SEGREGATSIYA	Segregacion DIFERENCIACION=MAGMATICA	Segregazione DIFFERENZIAZ-IONE MAGMATICA
4011 SEICHE GEOH	SEICHES	SEICHE	SEICHE	SEJSHI	SEICHE	SESSA
4012 SEISLOG APPL	SEISLOG	SEISLOG	SEISM.LOG	KAROTAZH SEJS-MICHESKIJ *UF:* Karotazh akusti-cheskij	REGISTRO=SISMICO	SEISLOG
4013 SEISMIC EQUIPMENT INST	Seismic equipment SEISMIC METHODS; INSTRUMENTS	Equipement sismique INSTRUMENTA-TION; METHODE SISMIQUE	Seism.Instrument INSTRUMEN-TIERUNG; SEISM. METHODE	TEKHNIKA SEJSMI-CHESKAYA	Equipo=sismico INSTRUMENTA-CION; METODO=SISMICO	Apparecchiatura sismica STRUMENTAZ-IONE; METODO SISMICO
4014 SEISMIC METHOD APPL	SEISMIC METHODS *UF:* Explosive source; Hammer seismics; Hydrophone; Impact seismics; Non=explosive sources; Seismic equipment; Seismic stations; Shot holes; Sparker	METHODE SISMIQUE *UF:* Equipement sismique; Hydrophone; Seismologie experimen-tale; Sismique marteau; Sismique poussee; Source explosion; Source nonexplosive; Trou explosion sismique	SEISM.METHODE *UF:* Experimental=Seismik; Explosions=Quelle; Hammer=Seismik; Hydrophon; Nicht=Explosiv.Quelle; Schusspunkt; Seism. Instrument; Stoss=Seismik	METOD SEJSMI-CHESKIJ *UF:* Hammer seismics; Impact seismics; Seis-mic spectral analysis; Volna Ehry	METODO=SISMICO *UF:* Equipo=sismico; Foco=explosiones; Fuente=no=explosiva; Hidrofono; Pozo=de=disparo; Sismica=impacto; Sismica=martillo; Sismologia=experimental	METODO SISMICO *UF:* Apparecchiatura sismica; Fonte non d'esplosione; Idrofono; Punto di scoppio; Sismica a percussione; Sismica avanzata; Sismologia sperimen-tale; Sorgente d'esplosione
4015 SEISMIC RISK ENGI	SEISMIC RISK	RISQUE SISMIQUE	SEISM.RISIKO	OPASNOST' SEJS-MICHESKAYA	RIESGO=SISMICO	RISCHIO SISMICO
4016 SEISMIC SOURCE SOLI	SEISMIC SOURCES	SOURCE SISMIQUE	SEISM.QUELLE	ISTOCHNIK SEJS-MICHESKIJ	FUENTE=SISMICA	SORGENTE SISMICA
4017 SEISMIC SPECTRAL=ANALYSIS METH	SPECTRAL ANALY-SIS	ANALYSE SPECTRE SISMIQUE	SEISM.SPEKTRAL=ANALYSE	[SEISMIC SPEC-TRAL ANALYSIS] METOD SEJSMI-CHESKIJ	ANALISIS=ESPECTRO=SISMICO	ANALISI SISMICA SPETTRALE

	ENGLISH	FRANCAIS	DEUTSCH	RUSSKIJ	ESPANOL	ITALIANO
4018 SEISMIC STATION APPL	Seismic stations SEISMIC METHODS	Station sismique LEVE SISMIQUE	Seism.Observatorium SEISM. VERMESSUNG	SEJSMOSTANTSIYA	Estacion=sismica LEVANTAMIENTO= SISMICO	Stazione sismica RILEVAMENTO SISMICO
4019 SEISMIC SURVEY APPL	SEISMIC SURVEYS	LEVE SISMIQUE *UF:* Station sismique	SEISM. VERMESSUNG *UF:* Seism. Observatorium	S'EMKA SEJSMI-CHESKAYA	LEVANTAMIENTO= SISMICO *UF:* Estacion=sismica	RILEVAMENTO SISMICO *UF:* Stazione sismica
4020 SEISMIC VELOCITY SOLI	Seismic velocity ELASTIC WAVES; VELOCITY	Velocite sismique VITESSE; ONDE SISMIQUE	Seism.Geschwindigkeit GESCHWINDIG-KEIT; SEISM. WELLE	SKOROST' SEJSMI-CHESKAYA	Velocidad=sismica VELOCIDAD; ONDA=SISMICA	Velocita d'onda sismica VELOCITA; ONDA SISMICA
4021 SEISMIC WAVE SOLI	Seismic waves ELASTIC WAVES	ONDE SISMIQUE *UF:* Onde elastique; Velocite sismique	SEISM.WELLE *UF:* Elast.Welle; Seism. Geschwindigkeit	VOLNA SEISMI-CHESKAYA *UF:* Sh=volna; Volna dlinnoperiodnaya; Volna ob'emnaya; Volna poperechnaya; Volna prodol'naya; Volna uprugaya	ONDA=SISMICA *UF:* Onda=elastica; Velocidad=sismica	ONDA SISMICA *UF:* Onda elastica; Velocita d'onda sismica
4022 SEISMICITY SOLI	SEISMICITY	SISMICITE *UF:* Zone asismique	SEISMIZITAET *UF:* Aseism.Region	SEJSMICHNOST'	SISMICIDAD *UF:* Asismico	SISMICITA *UF:* Regione asismica
4023 SEISMOGRAM SOLI	SEISMOGRAMS	SISMOGRAMME	SEISMOGRAMM	SEJSMOGRAMMA	SISMOGRAMA	SISMOGRAMMA
4024 SEISMOGRAPH INST	SEISMOGRAPHS	SISMOGRAPHE	SEISMOGRAPH	SEJSMOGRAF	SISMOGRAFO	SISMOGRAFIA
4025 SEISMOLOGICAL MAP SOLI	SEISMOLOGICAL MAPS	CARTE SISMICITE	SEISMOLOG.KARTE	KARTA SEJS-MOLOGI-CHESKAYA *UF:* Karta izosejsm	MAPA=SISMICO	CARTA DELLA SISMICITA
4026 SEISMOLOGY SOLI	SEISMOLOGY *UF:* Experimental seis-mology	SEISMOLOGIE	SEISMOLOGIE	SEJSMOLOGIYA *UF:* Ehksperimental'naya sejsmologiya	SISMOLOGIA	SISMOLOGIA
4027 SEISMOMETER INST	SEISMOMETERS	SISMOMETRE	SEISMOMETER	SEJSMOPRIEMNIK *UF:* Geofon; Sejsmograf okeanicheskij	SISMOMETRO	SISMOMETRO
4028 SEISMOTECTONICS SOLI	SEISMOTECTONICS	SISMOTEC-TONIQUE	SEISMOTEKTONIK	SEJSMOTEK-TONIKA	SISMOTECTONICA	SISMOTETTONICA
4029 SELENATES MING	SELENATES	SELENIATE	SELENAT	Selenat SELEN	SELENIATO	SELENIATO
4030 SELENIUM CHEE	SELENIUM	SELENIUM	SE	SELEN *UF:* Selenat	SELENIO	SELENIO
4031 SELENOCHRONOLOGY EXTR	Selenochronology GEOCHRONOLOGY; MOON	Selenochronologie GEOCHRONOLO-GIE; LUNE	Selenochronologie GEOCHRONOLO-GIE; MOND	SELENOKHRO-NOLOGIYA	Selenocronologia GEOCRONOLOGIA; LUNA	Selenocronologia GEOCRONOLOGIA; LUNA
4032 SELENOLOGY EXTR	Selenology MOON	Selenologie LUNE	Selenologie MOND	SELENOLOGIYA	Selenologia LUNA	Selenologia LUNA

	ENGLISH	FRANCAIS	DEUTSCH	RUSSKIJ	ESPANOL	ITALIANO
4033 SELF=POTENTIAL METHOD APPL	SELF=POTENTIAL METHODS	POLARISATION SPONTANEE	EIGENPOTENTIAL	METOD ESTEST-VENNOGO EHLEK-TRICHESKOGO POLYA	POLARIZACION=ESPONTANEA	POLARIZZAZIONE SPONTANEA
4034 SEM DATA METH	SEM DATA	DONNEE MEB	RASTER=ELEKTR. MIKROSKOPIE=DATEN	[SEM DATA] DANNYE; MIKROS-KOPIYA EHLEK-TRONNAYA SKANI-RUYUSHCHAYA	DATO=MEB	DATO MES
4035 SEMI=ARID ENVIRONMENT SURF	SEMI=ARID ENVI-RONMENT	MILIEU SEMI ARIDE	SEMIARID.MILIEU	[SEMI=ARID ENVI-RONMENT] ARIDNYJ	MEDIO=SEMIARIDO	AMBIENTE SEMI-ARIDO
4036 SEMIPRECIOUS STONE COMS	Semiprecious stone GEMS	Pierre semiprecieuse GEMME	Halbedelstein SCHMUCKSTEIN	Kamen' poludragotsen-nyj KAMEN' DRA-GOTSENNYJ	Piedra=semipreciosa GEMA	Pietra semipreziosa GEMMA
4037 SENONIAN STRS	SENONIAN	SENONIEN	SENON	SENON	SENONIENSE	SENONIANO
4038 SENSITIVE CLAY ENGI	SENSITIVE CLAYS	ARGILE SENSIBLE *UF:* Argile fluide	FLIESS=TON *UF:* Quick=Ton	GLINA NEUSTO-JCHIVAYA	ARCILLA=SENSIBLE *UF:* Arcilla=fluida	ARGILLA SENSI-BILE *UF:* Argilla fluida
4039 SEPARATION METH	SEPARATION	SEPARATION *UF:* Dialyse	TRENNUNG *UF:* Dialyse	SEPARATSIYA	SEPARACION *UF:* Dialisis	SEPARAZIONE *UF:* Dialisi
4040 SEPTARIUM SEDI	SEPTARIA	SEPTARIUM	SEPTARIE	SEPTARIYA	SEPTARIA	SEPTARIA
4041 SEPTUM PALE	SEPTA	SEPTUM	SEPTUM	SEPTA	SEPTA	SETTO
4042 SERICITE SCHIST IGMS	Sericite schist SCHISTS	Sericitoschiste SCHISTE VERT	Serizit=Schiefer GRUEN=SCHIEFER	Slanets seritsitovyj SLANETS KRISTAL-LICHESKIJ	Esquisto=sericitico ESQUISTO=VERDE	Scisto sericitico SCISTO VERDE
4043 SERICITIZATION IGNE	SERICITIZATION	SERICITISATION	SERIZITISIERUNG	SERITSITIZATSIYA	SERICITIZACION	SERICITIZZAZIONE
4044 SERIES STRA	Series STRATIGRAPHIC UNITS	Serie ECHELLE STRATI-GRAPHIQUE	Serie STRATIGRAPH. SKALA	SERIYA	Serie ESCALA=ESTRATIGRAFICA	Serie SCALA STRATI-GRAFICA
4045 SERPENTINE MING	SERPENTINE GROUP	SERPENTINE	SERPENTIN	SERPENTIN	SERPENTINA	SERPENTINO
4046 SERPENTINITE IGMS	SERPENTINITE	SERPENTINITE	SERPENTINIT	SERPENTINIT	SERPENTINITA	SERPENTINITE
4047 SERPENTINIZATION IGNE	SERPENTINIZA-TION	SERPENTINISA-TION	SERPENTINI-SIERUNG	SERPENTINIZAT-SIYA	SERPENTINIZA-CION	SERPENTINIZZAZ-IONE
4048 SETTLEMENT ENGI	SETTLEMENT	TASSEMENT *UF:* Tassement different-iel	SETZUNG *UF:* Differential=Kompaktion	OSADKA SOORUZ-HENIJ	ASIENTO *UF:* Compactacion=diferencial	ASSESTAMENTO *UF:* Costipamento dif-ferenziale

	ENGLISH	FRANCAIS	DEUTSCH	RUSSKIJ	ESPANOL	ITALIANO
4049 SEXUAL DIMORPHISM PALE	SEXUAL DIMOR- PHISM	DIMORPHISME SEXUEL	SEXUAL= DIMORPHISMUS	Polovoj dimorfizm DIMORFIZM	DIMORFISMO= SEXUAL	DIMORFISMO SES- SUALE
4050 SH-WAVE SOLI	SH=WAVES	ONDE SH	SH=WELLE	Sh=volna VOLNA SEJSMI- CHESKAYA	ONDA=SH	ONDA SH
4051 SHALE SEDS	SHALE *UF:* Pelite	SHALE	SCHIEFER=TON	SLANETS GLINIS- TYJ	PIZARRA=NO= METAMORFICA	ARGILLITE LAMINATA
4052 SHALLOW FOCUS EARTHQUAKE SOLI	SHALLOW=FOCUS EARTHQUAKES	SEISME SUPERFIC- IEL	OBERFLAECH- ENNAH.BEBEN	ZEMLETRYASENIE MELKOFOKUSNOE *UF:* Stick=slip	SISMO= SUPERFICIAL	SISMA SUPERFI- CIALE
4053 SHALLOW-WATER ENVIRONMENT PALE	SHALLOW=WATER ENVIRONMENT	MILIEU EAU PEU PROFONDE	FLACHWASSER- MILIEU	MELKOVODNYJ	MEDIO=AGUA= POCO=PROFUNDA	AMBIENTE DI ACQUA POCO PRO- FONDA
4054 SHAPE MISC	Shape GEOMETRY	Forme GEOMETRIE	Form GEOMETRIE	Forma MORFOLOGIYA	Forma GEOMETRIA	Forma GEOMETRIA
4055 SHAPE OF THE EARTH SOLI	FIGURE OF EARTH	Forme globe PLANETE TERRE	Erd=Gestalt PLANET=ERDE	Forma Zemli ZEMLYA	Forma=globo PLANETA=TIERRA	Forma della terra PIANETA TERRA
4056 SHATTER BRECCIA GEOL	Shatter breccia TECTONIC BREC- CIA	Breche friction BRECHE TEC- TONIQUE	Schlag=Brekzie TEKTON.BREKZIE	Brekchiya drobleniya BREKCHIYA TEK- TONICHESKAYA	Brecha=de= desmenuzamiento BRECHA= TECTONICA	Breccia spigolosa BRECCIA TET- TONICA
4057 SHATTER CONE TEST	SHATTER CONES	SHATTER CONE	[SHATTER=CONE]	Konus drobleniya METAMORFIZM UDARNYJ	ROCA=DE= IMPACTO	CONO D'ESPLOSIONE
4058 SHEAR PHCH	SHEAR *UF:* Shear joints; Shear planes	CISAILLEMENT *UF:* Diaclase cisaille- ment; Plan cisaillement	SCHERUNG *UF:* Scher=Flaeche; Scher=Kluft	SDVIG (MEK- HANIKA)	CIZALLAMIENTO *UF:* Fractura=de= cizallamiento; Plano= de=cizallamiento	FENOMENO DI TAGLIO *UF:* Diaclasi di taglio; Piano di taglio
4059 SHEAR JOINT STRU	Shear joints SHEAR; JOINTS	Diaclase cisaillement CISAILLEMENT; DIACLASE	Scher=Kluft SCHERUNG; KLUFT	TRESHCHINA SKA- LYVANIYA	Fractura=de= cizallamiento CIZALLAMIENTO; DIACLASA	Diaclasi di taglio FENOMENO DI TAGLIO; DIACLASI
4060 SHEAR MODULUS ENGI	SHEAR MODULUS *UF:* Modulus of rigidity	MODULE CISAILLE- MENT	SCHER=MODUL	MODUL'UPRU- GOSTI	MODULO= CIZALLAMIENTO	MODULO DI TAGLIO
4061 SHEAR PLANE STRU	Shear planes FAULT PLANES; SHEAR	Plan cisaillement PLAN FAILLE; CISAILLEMENT	Scher=Flaeche VERWERFUNGS- FLAECHE; SCHERUNG	Smestitel' SBRASYVATEL'	Plano=de= cizallamiento PLANO=FALLA; CIZALLAMIENTO	Piano di taglio PIANO DI FAGLIA; FENOMENO DI TAGLIO
4062 SHEAR STRENGTH PHCH	SHEAR STRENGTH	RESISTANCE CISAILLEMENT	SCHER= WIDERSTAND	SOPROTIVLENIE SDVIGU	RESISTENCIA= CIZALLAMIENTO	RESISTENZA AL TAGLIO
4063 SHEAR STRESS PHCH	SHEAR STRESS	CONTRAINTE CISAILLEMENT	SCHER= SPANNUNG	NAPRYAZHENIE SKALYVAYUSH- CHEE	TENSION= CIZALLAMIENTO	SFORZO DI TAGLIO
4064 SHEAR ZONE STRU	SHEAR ZONES	ZONE CISAILLE- MENT	SCHER=ZONE	ZONA DRO- BLENIYA	ZONA=CIZALLA	ZONA DI TAGLIO

	ENGLISH	FRANCAIS	DEUTSCH	RUSSKIJ	ESPANOL	ITALIANO
4065 SHEET EROSION SURF	Sheet erosion EROSION	Erosion en nappe EROSION	Abschwemmung EROSION	EHROZIYA PLOSKOSTNAYA *UF:* Smyv ploskostnoj	Erosion=por=arroyada EROSION	Dilavamento EROSIONE
4066 SHEET INTRUSION IGNE	Sheets LACCOLITHS	Intrusion couche LACCOLITE	Decken=Intrusion LAKKOLITH	INTRUZIYA PLAS- TOVAYA *UF:* Sole injection	Lamina=magmatica LACOLITO	Intrusione stratificata LACCOLITE
4067 SHELF MARI	Shelf CONTINENTAL SHELF	Plateau continental PLATEFORME CON- TINENTALE	Kontinental=Schelf KONTINENTAL= TAFEL	Shel'f SHEL'F KONTINEN- TAL'NYJ	Llanura=continental PLATAFORMA= CONTINENTAL	Piattaforma PIATTAFORMA CONTINENTALE
4068 SHELF ENVIRONMENT PALE	SHELF ENVIRON- MENT	MILIEU MARGE CONTINENTALE	SCHELF=MILIEU	FATSIYA SHEL'FOVAYA	MEDIO=MARGEN= CONTINENTAL	AMBIENTE MAR- GINE CONTINEN- TALE
4069 SHELF SEA MARI	Shelf sea CONTINENTAL SHELF	Mer plateau continen- tal MER BORDIERE	Schelf=Meer RAND=MEER	MORE SHEL'FOVOE	Mar=plataforma= continental MAR=MARGINAL	Mare di piattaforma MARE MARGINALE
4070 SHELF SEDIMENTATION SEDI	SHELF SEDIMEN- TATION	SEDIMENTATION MARGE CON- TINENTALE	SCHELF= SEDIMENTATION	SEDIMENTATSIYA SHEL'FOVAYA	SEDIMENTACION= MARGEN= CONTINENTAL	SEDIMENTAZIONE MARGINE CON- TINENTALE
4071 SHELL PALE	SHELLS	COQUILLE	SCHALE *UF:* Schloss	RAKOVINA	CONCHA	CONCHIGLIA
4072 SHIELD STRU	SHIELDS	BOUCLIER	SCHILD	SHCHIT	ESCUDO	SCUDO
4073 SHIELD VOLCANO IGNE	SHIELD VOLCA- NOES	VOLCAN EN BOU- CLIER	SCHILD=VULKAN	Vulkan shchitovoj IZVERZHENIE GAVAJSKOGO TIPA	VOLCAN=EN= ESCUDO	VULCANO A SCUDO
4074 SHOAL MARI	SHOALS	HAUT FOND	UNTIEFE	MEL'	FONDO=SOMERO	ALTOFONDO
4075 SHOCK PHCH	Shock SHOCK WAVES	Choc EFFET CHOC	Schock STOSSWELLEN= EFFEKT	UDAR	Impacto EFECTO=DE= CHOQUE	Impatto EFFETTO DI SHOCK
4076 SHOCK BRECCIA GEOL	Shock breccia BRECCIA	Breche choc BRECHE; EFFET CHOC	Schock=Brekzie BREKZIE; STOSSWELLEN= EFFEKT	BREKCHIYA UDARA	Brecha=de=impacto BRECHA; EFECTO= DE=CHOQUE	Breccia d'impatto BRECCIA; EFFETTO DI SHOCK
4077 SHOCK METAMORPHISM IGNE	SHOCK METAMOR- PHISM *UF:* Impact metamor- phism	METAMORPHISME CHOC	SCHOCK= METAMORPHOSE	METAMORFIZM UDARNYJ *UF:* Konus drobleniya; Metamorfizm impakt- nyj	METAMORFISMO= CHOQUE	METAMORFISMO DA IMPATTO
4078 SHOCK WAVE PHCH	SHOCK WAVES *UF:* Shock	ONDE CHOC	SCHOCK=WELLE	VOLNA UDARNAYA	ONDA=CHOQUE	ONDA D'URTO
4079 SHOESTRING SAND SEDI	Shoestring sands SANDSTONE	Filet sable SABLE	[SHOESTRING= SAND] SAND	OSADKONAKOPLE- NIE RUSLOVOE	Lentejon=de=arena ARENA	Laccio da scarpa SABBIA
4080 SHONKINITE IGNS	SHONKINITE	Shonkinite SYENITE	Shonkinit SYENIT	SHONKINIT	Shonkinita SIENITA	Shonkinite SIENITE

	ENGLISH	FRANCAIS	DEUTSCH	RUSSKIJ	ESPANOL	ITALIANO
4081 SHORE FEATURE SURF	SHORE FEATURES *UF:* Coastal geomorphology; Tombolos	MORPHOLOGIE COTE *UF:* Morphologie cotiere	KUESTEN= MERKMAL *UF:* Kuesten= Morphologie	FORMA REL'EFA BEREGOVAYA	MORFOLOGIA= COSTA *UF:* Morfologia=costera	MORFOLOGIA COS- TIERA *UF:* Morfologia di costa
4082 SHORE PROTECTION ENGI	Shore protection SHORELINES	Protection rivage PROTECTION ENVIRONNEMENT; LIGNE RIVAGE	Kuesten=Schutz VORBEUGEND. MASSNAHME; KUESTE	UKREPLENIE BEREGOV	Proteccion=costera PROTECCION= MEDIO=AMBIENTE; LINEA=COSTA	Protezione costiera PROTEZIONE AMBIENTALE; LINEA DI RIVA
4083 SHORELINE SURF	SHORELINES *UF:* Shore protection	LIGNE RIVAGE *UF:* Cote; Cote fossile; Ligne cotiere; Progression rivage; Protection rivage	KUESTE *UF:* Fossil.Kuestenlinie; Kuesten=Linie; Kuesten=Schutz; Meeres=Kueste; Seewaertig. Kuestenverlagerung	LINIYA BERE- GOVAYA *UF:* Granitsa poberezh'ya	LINEA=COSTA *UF:* Costa; Costa= abandonada; Linea= costera; Progresion= costa; Proteccion= costera	LINEA DI RIVA *UF:* Costa; Linea di costa; Linea di riva abbandonata; Progradazione di riva; Protezione costiera
4084 SHORTENING STRU	Shortening TECTONICS	Raccourcissement TECTONIQUE TAN- GENTIELLE	Einengung TANGENTIAL= TEKTONIK	SOKRASHCHENIE	Acortamiento TECTONICA= TANGENCIAL	Raccorciamento crostale TETTONICA COM- PRESSIVA
4085 SHOT HOLE APPL	Shot holes SEISMIC METHODS	Trou explosion sismique METHODOLOGIE; METHODE SISMIQUE	Schusspunkt METHODIK; SEISM. METHODE	SKVAZHINA VZRY- VNAYA	Pozo=de=disparo METODOLOGIA; METODO=SISMICO	Punto di scoppio METODOLOGIA; METODO SISMICO
4086 SHOWING ECON	MINERAL SHOWS	INDICE MINERAL	LAGERSTAETTEN= INDIZ	PRIZNAK POISKOVYJ PRYA- MOJ	INDICIO=MINERAL	INDIZIO MINE- RARIO
4087 SHRINKAGE SURF	Shrinkage DESICCATION	Retrecissement ACTION CLIMA- TIQUE	Schrumpfung KLIMA=WIRKUNG	USADKA	Retraccion ACCION= CLIMATICA	Ritiro AZIONE CLIMA- TICA
4088 SHRINKAGE CRACK TEST	Shrinkage cracks DESICCATION	Craquelure retrecissement CRAQUELURE BOUE	Schwundriss SCHLAMM= SCHWUND=RISS	TRESHCHINA USA- DOCHNAYA	Grieta=retraccion GRIETA= DESECACION	Fessura di contrazione FESSURAZIONE DA DISSECCA- MENTO
4089 SIAL SOLI	Sial MAGMAS	Sial MAGMA	Sial MAGMA	SIAL'	Sial MAGMA	Sial MAGMA
4090 SICILIAN STRS	SICILIAN	SICILIEN	SICILIUM	Sitsilijskij CHETVERTICHNYJ	SICILIENSE	SICILIANO
4091 SIDE-SCANNING METHOD APPL	SIDE=SCANNING METHODS	Balayage lateral TELEDETECTION	Side=scan=methode REMOTE=SENSING	LOKATSIYA BOKOVOGO OBZORA	Metodo=de=barrido= lateral TELEDETECCION	Metodo a scansione laterale TELEOSSERVAZ- IONE
4092 SIDEROPHILE ELEMENT CHES	SIDEROPHILE ELE- MENTS	ELEMENT SIDEROPHILE	SIDEROPHIL. ELEMENT	SIDEROFIL'NYJ	ELEMENTO= SIDEROFILO	ELEMENTO SIDEROFILO
4093 SIDEROPHYRE EXTS	Siderophyres METEORITES	Siderophyre SIDEROLITE	Siderophyr STEIN=EISEN= METEORIT	Siderofir SIDEROLIT	Siderofiro SIDEROLITO	Siderofiro SIDEROLITE

	ENGLISH	FRANCAIS	DEUTSCH	RUSSKIJ	ESPANOL	ITALIANO
4094 SIEGENIAN STRS	SIEGENIAN	SIEGENIEN	SIEGEN	ZIGEN	SIEGENINSE	SIEGENIANO
4095 SIEROZEM SUSS	SIEROZEMS	SIEROZEM	SIEROZEM	SEROZEM	SIEROZEM	SIEROZEM
4096 SIEVE INST	Sieves INSTRUMENTS; GRANULOMETRY	Tamis INSTRUMENTA- TION; GRANU- LOMETRIE	Sieb INSTRUMEN- TIERUNG; KORN= VERTEILUNG	SITO	Tamiz INSTRUMENTA- CION; GRANULO- METRIA	Setaccio STRUMENTAZ- IONE; GRANULO- METRIA
4097 SILICA MING	SILICA *UF:* Desilication	SILICE *UF:* Desilicification	KIESELSAEURE *UF:* Entkieselung	KREMNEZEM *UF:* Opoka	SILICE *UF:* Desilizacion	SILICE *UF:* Desilicizzazione
4098 SILICATES MING	SILICATES	SILICATE	SILIKAT	SILIKAT	SILICATÒ	SILICATO
4099 SILICEOUS COMPOSITION GEOC	SILICEOUS COMPO- SITION	COMPOSITION SILICEUSE	KIESEL=GEHALT	KREMNISTYJ	COMPOSICION= SILICEA	COMPOSIZIONE SILICEA
4100 SILICIFICATION GEOL	SILICIFICATION	SILICIFICATION *UF:* Bois silicifie	VERKIESELUNG *UF:* Verkieselt.Holz	OKREMNENIE	SILICIFICACION *UF:* Madera=silificada	SILICIZZAZIONE *UF:* Legno silicizzato
4101 SILICIFIED WOOD PALE	Silicified wood FOSSIL WOOD	Bois silicifie BOIS FOSSILE; SILICIFICATION	Verkieselt.Holz FOSSIL.HOLZ; VER- KIESELUNG	Derevo okremneloe DEREVO OKAME- NELOE	Madera=silificada MADERA=FOSIL; SILICIFICACION	Legno silicizzato LEGNO FOSSILE; SILICIZZAZIONE
4102 SILICOFLAGELLATA PALS	SILICOFLAGEL- LATA	SILICOFLAGEL- LATA	SILICOFLAGEL- LATA	SILICOFLAGEL- LATA	SILICOFLAGEL- LATA	SILICOFLAGEL- LATA
4103 SILICON CHEE	SILICON	SILICIUM	SI	KREMNIJ	SILICIO	SILICIO
4104 SILL IGNE	SILLS	SILL	LAGER=GANG	SILL	SILL	SILL
4105 SILLIMANITE COMS	SILLIMANITE DEPOSITS	SILLIMANITE	SILLIMANIT	SILLIMANIT	SILLIMANITA	SILLIMANITE
4106 SILT SEDS	SILT *UF:* Loam; Silt load	SILT *UF:* Charge silt; Limon	SILT *UF:* Lehm; Silt=Fracht	ALEVRIT	LIMO *UF:* Limon; Transporte= limo=suspendido	SILT *UF:* Carico solido; Suolo limoso
4107 SILT LOAD SURF	Silt load TRANSPORT; SILT	Charge silt TRANSPORT; SILT	Silt=Fracht TRANSPORT; SILT	VZVES' *UF:* Stok vzveshennyj	Transporte=limo= suspendido TRANSPORTE; LIMO	Carico solido TRASPORTO; SILT
4108 SILTING ENGI	SILTATION	ENVASEMENT	SCHLUFF= SEDIMENTATION	ZAILENIE	SEDIMENTACION= DE=LIMOS	INFANGAMENTO
4109 SILTSTONE SEDS	SILTSTONE	SILTSTONE *UF:* Pelite	SILTSTEIN *UF:* Pelit	ALEVROLIT	LIMOLITA *UF:* Pelita	SILTITE *UF:* Pelite
4110 SILURIAN STRS	SILURIAN	SILURIEN	SILUR	SILUR	SILURICO	SILURIANO
4111 SILVER CHEE	SILVER	ARGENT	AG	SEREBRO	PLATA	ARGENTO

	ENGLISH	FRANCAIS	DEUTSCH	RUSSKIJ	ESPANOL	ITALIANO
4112 SIMA SOLI	Sima BASALTIC LAYER	Sima MAGMA	Sima MAGMA	SIMA	Sima MAGMA	Sima MAGMA
4113 SIMIAN PALS	SIMIANS	SIMIEN	AFFENARTIGE	SIMIAN *UF:* Cynomorpha; Hominidae	SIMIEN	SIMIANI
4114 SIMULATION MATH	SIMULATION	SIMULATION	SIMULATION	SIMULATSIYA	SIMULACION	SIMULAZIONE
4115 SINEMURIAN STRS	SINEMURIAN	SINEMURIEN	SINEMURIUM	SINEMYUR	SINEMURIENSE	SINEMURIANO
4116 SINGLE=CRYSTAL **ANALYSIS** METH	SINGLE=CRYSTAL METHOD	DIAGRAMME MONOCRISTAL	EINKRISTALL- AUFNAHME	ANALIZ MONOKRISTALLA	DIAGRAMA- MONOCRISTAL	DIFFRAZIONE CRISTALLO SIN- GOLO
4117 SINKHOLE SURF	SINKHOLES	DOLINE EFFON- DREMENT	KARST=TRICHTER	Truba karstovaya VORONKA KARS- TOVAYA	DOLINA=DE= HUNDIMIENTO	DOLINA DI SPRO- FONDAMENTO
4118 SINTER SEDI	SILICEOUS SINTER *UF:* Geyserite	Depot concretionne INCRUSTATION	Sinter INKRUSTATION	TUF OSADOCHNYJ	[Sinter] INCRUSTACION	Deposito concrezionato INCROSTAZIONE
4119 SINUOSITY SURF	SINUOSITY	SINUOSITE RIVI- ERE	MAEANDRIEREN	[SINUOSITY]	SINUOSIDAD=RIO	SINUOSITA
4120 SIPHON GEOH	SIPHON	SIPHON	SIPHON	Sifon (gidrogeol) DRENAZH ISKUST- VENNYJ	SIFON	SIFONE
4121 SIPHONAPTERA PALS	SIPHONAPTEROIDA	SIPHONAPTEROIDA	SIPHONAPTERA	Siphonaptera INSECTA	SIPHONAPTEROIDA	SIPHONAPTERA
4122 SIPUNCULOIDA PALS	Sipunculoida WORMS	Sipunculoida VERMES	Sipunculoida VERMES	SIPUNCULOIDA	Sipunculoida VERMES	Sipunculoida VERMES
4123 SIRENIA PALS	SIRENIA	SIRENIA	SIRENIA	SIRENIA	SIRENIA	SIRENIA
4124 SITE SELECTION ENGI	SITE EXPLORA- TION	CHOIX SITE	STANDORTWAHL	VYBOR MESTA ZALOZHENIYA	ELECCION=DE= LUGAR	SCELTA DEL SITO
4125 SIZE DISTRIBUTION TEST	SIZE DISTRIBU- TION	REPARTITION GRAIN	KORN= VERTEILUNG *UF:* Sand=Schiefer= Verhaeltnis; Sieb	Analiz granulometri- cheskij GRANULO- METRIYA	REPARTICION- GRANOS	RIPARTIZIONE DI GRANA
4126 SKARN IGMS	SKARN *UF:* Tactite	SKARN *UF:* Tactite	SKARN *UF:* Tactit	SKARN *UF:* Taktit	SKARN *UF:* Tactita	SKARN *UF:* Tactite
4127 SKELETON PALE	SKELETONS	ANATOMIE SQUE- LETTE *UF:* Carapace; Holos- tome; Ouverture	SKELETT *UF:* Apertur; Holostom; Panzer	SKELET *UF:* Dermal'nyj; Ship; Sklerit	ANATOMIA= ESQUELETO *UF:* Apertura; Capara- zon; Holostomado	ANATOMIA DELLO SCHELETRO *UF:* Apertura; Carapace; Olostomato
4128 SKEWNESS MATH	SKEWNESS	OBLIQUITE	SCHIEFE	ASIMMETRIYA	OBLICUIDAD	DISSIMMETRIA

	ENGLISH	FRANCAIS	DEUTSCH	RUSSKIJ	ESPANOL	ITALIANO
4129 SLAB SOLI	SLABS	DALLE	STEINPLATTE	PORODA PLAS-TINCHATAYA	LAJA=TECTONICA	LASTRA
4130 SLATE IGMS	SLATES UF: Spotted slate	ARDOISE	TON=SCHIEFER	SLANETS	PIZARRA	ARDESIA
4131 SLATY CLEAVAGE STRU	SLATY CLEAVAGE	CLIVAGE ARDOI-SIER	SCHIEFER=SPALTBARKEIT	Klivazh istecheniya KLIVAZH TEC-HENIYA	EXFOLIACION=PIZARROSA	FISSILITA
4132 SLICKENSIDE TEST	SLICKENSIDES UF: Fault=plane stria-tions	MIROIR FAILLE	HARNISCH	ZERKALO SKOL'ZHENIYA	ESPEJO=FALLA	SPECCHIO DI FAGLIA
4133 SLIP CLEAVAGE STRU	SLIP CLEAVAGE	FAUX CLIVAGE	PSEUDO=SCHIEFERUNG	KLIVAZH SKOL'ZHENIYA	FALSO=CRUCERO	FALSO CLIVAGGIO
4134 SLOPE SURF	SLOPES UF: Critical angle; Dip slope	VERSANT UF: Angle critique; Deformation versant; Talus equilibre	HANG UF: Hang=Verformung; Krit.Winkel; Schuettungs=Winkel	SKLON UF: Sklon konsekvent-nyj; Talus slope; Ugol kriticheskij; Usto-jchivost' sklona	LADERA UF: Angulo=critico; Angulo=de=reposo; Deformacion=talud	VERSANTE UF: Angolo critico; Angolo di riposo; Deformazione di ver-sante
4135 SLOPE DEFORMATION SURF	Slope deformation SLOPE STABILITY	Deformation versant DEFORMATION; VERSANT	Hang=Verformung VERFORMUNG; HANG	DEFORMATSIYA SKLONOV	Deformacion=talud DEFORMACION; LADERA	Deformazione di ver-sante DEFORMAZIONE; VERSANTE
4136 SLOPE ENVIRONMENT PALE	SLOPE ENVIRON-MENT	MILIEU TALUS MARIN	SCHELF=HANG=MILIEU	SREDA SKLONA KONTINEN-TAL'NOGO	MEDIO=TALUD=MARINO	AMBIENTE DI TALUS MARINO
4137 SLOPE STABILITY ENGI	SLOPE STABILITY UF: Slope deformation	STABILITE VER-SANT	HANG=STABILITAET	Ustojchivost' sklona SKLON; STA-BIL'NOST'	ESTABILIDAD=TALUD	STABILITA DI VER-SANTE
4138 SLUMP STRUCTURE TEST	SLUMP STRUC-TURES	STRUCTURE SLUMPING	RUTSCH=GEFUEGE	TEKSTURA POD-VODNOGO OPOL-ZANIYA	ESTRUCTURA=SLUMPING	STRUTTURA A SLUMPING
4139 SLUMPING SEDI	SLUMPING	[SLUMPING]	RUTSCHUNG	OBRUSHENIE (OPOLZ)	[SLUMPING]	[SLUMPING]
4140 SMALL=SCALE MISC	Small scale MAPS	A petite echelle CARTOGRAPHIE	Klein.Massstab KARTOGRAPHIE	MASSHTAB MELKIJ	Pequeña=escala CARTOGRAFIA	A piccola scala CARTOGRAFIA
4141 SMOOTHING MATH	Statistical smoothing STATISTICAL ANALYSIS	Lissage TRAITEMENT DON-NEE	Glaettung DATEN=VERARBEITUNG	SGLAZHIVANIE	Suavizado TRATAMIENTO=DATOS	Lisciamento TRATTAMENTO DEI DATI
4142 SNOW SURF	SNOW	NEIGE	SCHNEE	SNEG	NIEVE	NEVE
4143 SODIUM CHEE	SODIUM UF: Borax	SODIUM UF: Borax	NA UF: Borax	NATRIJ	SODIO UF: Borax	SODIO UF: Borace
4144 SOFT CLAY ENGI	SOFT CLAYS	ARGILE MOLLE	WEICH.TON	GLINA PLASTI-CHESKAYA	ARCILLA=FRIABLE	ARGILLA MOLLE
4145 SOFT SEDIMENT DEFORMATION SEDI	SOFT SEDIMENT DEFORMATION	Deformation sediment DEFORMATION; SEDIMENT	Fruehdiagenet. Verformung VERFORMUNG; SEDIMENT	[SOFT SEDIMENT DEFORMATION] SEDIMENTATSIYA; SINGENETI-CHESKIJ	Deformacion=sedimento=blando DEFORMACION; SEDIMENTO	Deformazione di sedi-menti molli DEFORMAZIONE; SEDIMENTO

	ENGLISH	FRANCAIS	DEUTSCH	RUSSKIJ	ESPANOL	ITALIANO
4146 SOIL SURF	SOILS *UF:* Arenosols; Cohesionless soils; Cohesive soils; Lithosols; Mull; Regosols; Soil horizons; Soil leaching; Soil stabilization; Xerosols; Yermosols; Zone of aeration	SOL *UF:* Deformation sol; Lessivage sol; Pression geostatique; Sol coherent; Sol non coherent; Soulevement; Stabilisation sol	BODEN *UF:* Bindig.Boden; Boden=Auswaschung; Boden=Hebung; Boden=Stabilisierung; Boden=Verformung; Geostat.Druck; Nichtbindig.Boden	POCHVA *UF:* Bedlend; Ehroziya pochv; Karta pochvennaya; Katena; Pedogenezis; Regolit; Soil sampling	SUELO *UF:* Abombamiento; Deformacion=suelo; Estabilizacion=terreno; Lavado=suelo; Presion=geostatica; Suelo=cohesivo; Suelo=sin=cohesion	SUOLO *UF:* Deformazione del suolo; Lisciviazione del suolo; Pressione geostatica; Sollevamento del terreno; Stabilizzazione del suolo; Suolo coerente; Suolo incoerente
4147 SOIL DEFORMATION ENGI	Soil deformation SOIL MECHANICS	Deformation sol DEFORMATION; SOL	Boden=Verformung VERFORMUNG; BODEN	DEFORMATSIYA GRUNTOV	Deformacion=suelo DEFORMACION; SUELO	Deformazione del suolo DEFORMAZIONE; SUOLO
4148 SOIL EROSION SURF	SOIL EROSION *UF:* Badlands	EROSION SOL *UF:* Mauvaises terres	BODEN=EROSION *UF:* Badlands	Ehroziya pochv POCHVA; EHROZIYA	EROSION=SUELO *UF:* Malpais	EROSIONE DEL SUOLO *UF:* Badlands
4149 SOIL HORIZON SURF	Soil horizons SOILS	Horizon sol PROFIL SOL	Boden=Horizont BODEN=PROFIL	GORIZONT POCHVENNYJ	Horizonte=suelo PERFIL=SUELO	Orizzonte del suolo PROFILO DEL SUOLO
4150 SOIL IMPROVEMENT SURF	SOIL IMPROVEMENT	Amendement sol AMENDEMENT	Boden=Verbesserung MELIORATION	MELIORATSIYA POCHVY	ENMIENDA=SUELO	Ammendamento SUOLO AMMENDAMENTO
4151 SOIL LEACHING SURF	Soil leaching LEACHING; SOILS	Lessivage sol LESSIVAGE; SOL	Boden=Auswaschung AUSWASCHUNG; BODEN	Vyshchelachivanie pochvy VYSHCHELACHIVANIE; POCHVA	Lavado=suelo LAVADO; SUELO	Lisciviazione del suolo LISCIVIAZIONE; SUOLO
4152 SOIL MANAGEMENT ENVI	SOIL MANAGEMENT	AMENAGEMENT SOL	BODEN=BEWIRTSCHAFTUNG	OBRABOTKA POCHV	PLANIFICACION=SUELO	TRATTAMENTO DEL SUOLO
4153 SOIL MAP SURF	SOIL MAPS	CARTE PEDOLOGIQUE	BODEN=KARTE	Karta pochvennaya KARTA; POCHVA	MAPA=EDAFOLOGICO	CARTA PEDOLOGICA
4154 SOIL MECHANICS ENGI	SOIL MECHANICS *UF:* Soil deformation	MECANIQUE SOL	BODEN=MECHANIK	MEKHANIKA GRUNTOV *UF:* Mekhanika porod	MECANICA=SUELO	MECCANICA DEL SUOLO
4155 SOIL PROFILES SURF	SOIL PROFILES	PROFIL SOL *UF:* Horizon sol	BODEN=PROFIL *UF:* Boden=Horizont	PROFIL' POCHVY	PERFIL=SUELO *UF:* Horizonte=suelo	PROFILO DEL SUOLO *UF:* Orizzonte del suolo
4156 SOIL SAMPLING METH	SOIL SAMPLING	ECHANTILLONNAGE SOL	BODEN=PROBENNAHME	OPROBOVANIE POCHVY	MUESTREO=SUELO	CAMPIONATURA DEL SUOLO
4157 SOIL SCIENCE SURF	SOIL SCIENCES	PEDOLOGIE	BODENKUNDE	POCHVOVEDENIE	EDAFOLOGIA	SCIENZA DEI SUOLI
4158 SOIL STABILIZATION ENGI	Soil stabilization STABILIZATION; SOILS	Stabilisation sol STABILISATION; SOL	Boden=Stabilisierung STABILISATION; BODEN	Zakreplenie gruntov STABILIZATSIYA; GRUNT	Estabilizacion=terreno ESTABILIZACION; SUELO	Stabilizzazione del suolo STABILIZZAZIONE; SUOLO
4159 SOLAR RADIATION PHCH	Solar radiation SUN	Rayonnement solaire SOLEIL	Solar=Strahlung SONNE	RADIATSIYA SOLNECHNAYA	Radiacion=solar SOL=ASTRO	Radiazione solare SOLE

	ENGLISH	FRANCAIS	DEUTSCH	RUSSKIJ	ESPANOL	ITALIANO
4160 SOLAR SYSTEM EXTR	SOLAR SYSTEM	SYSTEME SOLAIRE	SONNEN=SYSTEM	SISTEMA SOL-NECHNAYA	SISTEMA=SOLAR	SISTEMA SOLARE
4161 SOLAR WIND EXTR	SOLAR WIND	VENT SOLAIRE	SONNEN=WIND	VETER SOLNECH-NYJ	VIENTO=SOLAR	VENTO SOLARE
4162 SOLE INJECTION IGNE	Sole injection DIKES	Injection faille DYKE	Sohlflaechen=Injektion DYKE	[SOLE INJECTION] INTRUZIYA PLAS-TOVAYA	Intrusion=plano=falla DIQUE	Iniezione basale DICCO
4163 SOLE MARK TEST	SOLE MARKS	FIGURE BASALE	SOHLMARKE	[SOLE MARK] GIEROGLIF	FIGURA=BASAL	FIGURA BASALE
4164 SOLEMYIDA PALS	SOLEMYIDA	SOLEMYIDA	SOLEMYIDA	SOLEMYOIDA	SOLEMYIDA	SOLEMYIDA
4165 SOLFATARA IGNE	SOLFATARAS	SOLFATARE	SOLFATARA	SOL'FATARA	SOLFATARA	SOLFATARA
4166 SOLID PHASE PHCH	SOLID PHASE	PHASE SOLIDE	FEST.PHASE	FAZA TVERDAYA	FASE=SOLIDA	FASE SOLIDA
4167 SOLID SOLUTION PHCH	SOLID SOLUTION	SOLUTION SOLIDE	FEST.LOESUNG	RASTVOR TVERDYJ	SOLUCION=SOLIDA	SOLUZIONE SOLIDA
4168 SOLID WASTE ENVI	SOLID WASTE	DECHET SOLIDE	FEST.ABFALL	OTKHODY TVER-DYE	RESIDUO=SOLIDO	RIFIUTI SOLIDI
4169 SOLID-EARTH GEOPHYSICS SOLI	SOLID=EARTH GEOPHYSICS	PHYSIQUE GLOBE	PHYSIK=DES-ERDKOERPERS	STROENIE GLUBIN-NOE	FISICA=DEL=GLOBO	GEOFISICA DELLA TERRA SOLIDA
4170 SOLIDUS IGNE	Solidus PHASE EQUILIBRIA	Solidus DIAGRAMME EQUILIBRE	Solidus PHASEN=GLEICHGEWICHT	SOLIDUS	Solidus DIAGRAMA=EQUILIBRIO	Solidus DIAGRAMMA D'EQUILIBRIO
4171 SOLIFLUCTION SURF	SOLIFLUCTION	SOLIFLUXION	SOLIFLUKTION	SOLIFLYUKTSIYA	SOLIFLUXION	SOLIFLUSSO
4172 SOLONCHAK SUSS	SOLONCHAK SOILS	SOLONCHAK	SOLONTSCHAK	SOLONCHAK UF: Salt marsh; Sebkha environment	SOLONCHAK	SUOLO SOLON-CHAK
4173 SOLONETZ SUSS	SOLONETZ SOILS	SOLONETZ	SOLONETZ	SOLONETS	SOLONEZ	SUOLO SOLONETZ
4174 SOLUBILITY PHCH	SOLUBILITY	SOLUBILITE	LOESLICHKEIT	RASTVORIMOST'	SOLUBILIDAD	SOLUBILITA
4175 SOLUTE PHCH	SOLUTES	SOLUTE	AUFGELOEST. STOFF	[SOLUTE] RASTVOR	SOLUTO	SOLUTO
4176 SOLUTION PHCH	SOLUTIONS	MATIERE SOLU-TION	LOESUNG	RASTVOR UF: Pressure solution; Rastvor peresyshchen-nyj; Solute	SOLUCION	SOLUZIONE
4177 SOLUTION CAVITY SURF	SOLUTION CAVI-TIES	CAVITE DISSOLU-TION	LOESUNGS=HOHLRAUM	[SOLUTION CAV-ITY] KARST	CAVIDAD=DISOLUCION	CAVITA DI DIS-SOLUZIONE

	ENGLISH	FRANCAIS	DEUTSCH	RUSSKIJ	ESPANOL	ITALIANO
4178 SOLUTION FEATURE SURF	SOLUTION FEATURES	MORPHOLOGIE DISSOLUTION	AUFLOESUNGS=MERKMAL	Landshaft karstovyj KARST	MORFOLOGIA=DE=DISOLUCION	MORFOLOGIA DI DISSOLUZIONE
4179 SOLUTION MINING MINI	SOLUTION MINING	EXPLOITATION PAR DISSOLUTION	LOESUNGS=BERGBAU	DOBYCHA VYSH-CHELACHIVANIEM	EXPLOTACION=POR=DISOLUCION	SFRUTTAMENTO PER DISSOLUZ-IONE
4180 SOMASTEROIDEA PALS	SOMASTEROIDEA	SOMASTEROIDEA	SOMASTEROIDEA	SOMASTEROIDEA	SOMASTEROIDEA	SOMASTEROIDEA
4181 SONAR APPL	SONAR METHODS	SONAR	SONAR	SONAR	SONAR	METODO SONAR
4182 SONOBUOY INST	SONOBUOYS	SONOBUOY	AKUST.BOJE	SONOBUJ	BOYA=ACUSTICA	BOA SONORA
4183 SORPTION PHCH	SORPTION	SORPTION	SORPTION	SORBTSIYA	SORBCION	SORPTION
4184 SORTING TEST	SORTING	CLASSEMENT GRANULO-METRIQUE	SORTIERUNG	OTSORTIROVAN-NOST'	CLASIFICACION=GRANULOMETRICA	CLASSAZIONE GRANULO-METRICA
4185 SOUNDING APPL	SOUNDING	SONDAGE GEO-PHYSIQUE	GEOPHYSIKAL. SONDIERUNG	ZONDIROVANIE	SONDEO=GEOFISICO	SONDAGGIO GEO-FISICO
4186 SOURCE ROCK SEDI	SOURCE ROCKS	ROCHE MERE HYDROCARBURE	KW=MUTTER-GESTEIN	PORODA MATERIN-SKAYA *UF:* Pochvoobrazuyushchaya poroda materinskaya; Protolith	ROCA=MADRE=HIDROCARBURO	ROCCIA MADRE IDROCARBURI
4187 SOUTH MISC	South GEOGRAPHY	Sud GEOGRAPHIE	Sueden GEOGRAPHIE	YUG	Sur GEOGRAFIA	Sud GEOGRAFIA
4188 SPACE GROUP MINE	SPACE GROUPS	GROUPE SPATIAL	RAUMGRUPPE	GRUPPA SIM-METRII	GRUPO=ESPACIAL	GRUPPO SPAZIALE
4189 SPALLATION PHCH	SPALLATION	SPALLATION	SPALLATION	RASSHCHEPLENIE	ESPALACION	SPALLAZIONE
4190 SPARAGMITE SEDS	Sparagmite CLASTIC ROCKS	Sparagmite ROCHE CLASTIQUE	Sparagmit KLAST.GESTEIN	SPARAGMIT	Esparagmita ROCA=CLASTICA	Sparagmite ROCCIA CLASTICA
4191 SPARKER INST	Sparker INSTRUMENTS; SEISMIC METHODS	Sparker INSTRUMENTA-TION; SISMIQUE REFLEXION	Sparker INSTRUMEN-TIERUNG; REFLEXIONS=SEISMIK	SPARKER	Generador=chispa INSTRUMENTA-CION; SISMICA=REFLEXION	Sparker STRUMENTAZ-IONE; SISMICA A RIFLESSIONE
4192 SPATTER CONE IGNE	Spatter cones PYROCLASTICS	Cone scories VOLCAN; PYRO-CLASTIQUE	Schlacken=Kegel VULKAN; PYROKLAST. GESTEIN	Konus peplovyj KONUS VULKANI-CHESKIJ	Cono=de=escorias VOLCAN; PIRO-CLASTICO	Cono di eiezione VULCANO; PRO-DOTTO PIROCLAS-TICO
4193 SPECIES DIVERSITY PALE	SPECIES DIVER-SITY	DIVERSITE ESPECE	ART=VERSCHIEDENHEIT	IZMENCHIVOST' VNUTRIVIDOVAYA	DIVERSIDAD=ESPECIES	DIVERSITA DI SPE-CIE
4194 SPECIFIC FAUNA PALE	SPECIFIC FAUNA	FAUNE SPECI-FIQUE *UF:* Assise biostrati-graphique; Biozone; Cenozone; Duree vie; Fossile specifique; Unite ecostrati-graphique; Zone faunis-tique	SPEZIF.FAUNA *UF:* Biochrone; Biozone; Cenozone; Faunen-Zone; Faunizone; Leit=Fossil; Oekozone	EHNDEMIKI	FAUNA=ESPECIFICA *UF:* Biocrona; Biozona; Cenozona; Ecozona; Fosil=especifico; Zona=bioestratigrafica; Zona=faunistica	FAUNA SPECIFICA *UF:* Biocrona; Biozona; Cenozona; Ecozona; Faunizona; Fossile specifico; Zona d'associazione

	ENGLISH	FRANCAIS	DEUTSCH	RUSSKIJ	ESPANOL	ITALIANO
4195 SPECIFIC FLORA PALE	SPECIFIC FLORA	FLORE SPECIFIQUE *UF:* Assise biostrati- graphique; Florizone	SPEZIF.FLORA *UF:* Faunen=Zone; Florizone	[SPECIFIC FLORA] EHNDEMIKI	FLORA= ESPECIFICA *UF:* Florizona; Zona= bioestratigrafica	FLORA SPECIFICA *UF:* Florizona; Zona d'associazione
4196 SPECIFIC GRAVITY PHCH	SPECIFIC GRAVITY	POIDS SPECIFIQUE	SPEZIF.GEWICHT	VES UDEL'NYJ *UF:* Ves ob'emnyj	PESO=ESPECIFICO	PESO SPECIFICO
4197 SPECIFIC HEAT PHCH	SPECIFIC HEAT	CHALEUR SPECI- FIQUE	SPEZIF.WAERME	TEPLOTA UDEL'NAYA *UF:* Teplota plavleniya udel'naya	CALOR= ESPECIFICO	CALORE SPECIFICO
4198 SPECIFIC SURFACE PHCH	SPECIFIC SURFACE	SURFACE SPECI- FIQUE	SPEZIF. OBERFLAECHE	UDEL'NAYA POVERKHNOST'	SUPERFICIE= ESPECIFICA	SUPERFICIE SPECI- FICA
4199 SPECTROMETER INST	Spectrometers INSTRUMENTS; SPECTROSCOPY	Spectrometre INSTRUMENTA- TION; SPECTROME- TRIE	Spektrometer INSTRUMEN- TIERUNG; SPEK- TROMETRIE	SPEKTROMETR	Espectrometro INSTRUMENTA- CION; ESPECTRO- METRIA	Spettrometro STRUMENTAZ- IONE; SPETTROS- COPIA
4200 SPECTROPHOTOMETRY METH	Spectrophotometry SPECTROSCOPY	Spectrophotometrie SPECTROMETRIE	Spektrophotometrie SPEKTROMETRIE	SPEKTROFOTO- METRIYA	Espectrofotometria ESPECTROMETRIA	Spettrofotometria SPETTROSCOPIA
4201 SPECTROSCOPY METH	SPECTROSCOPY *UF:* Molecular spectros- copy; Spectrometers; Spectrophotometry	SPECTROMETRIE *UF:* Spectrometre; Spec- trometrie moleculaire; Spectrophotometrie	SPEKTROMETRIE *UF:* Molekular= Spektroskopie; Spektro- meter; Spektropho- tometrie	SPEKTROSKOPIYA	ESPECTROMETRIA *UF:* Espectrofotometria; Espectrometria= molecular; Espectro- metro	SPETTROMETRIA *UF:* Spettrofotometria; Spettrometro; Spettro- metria molecolare
4202 SPECTRUM PHCH	SPECTRA	SPECTRE	SPEKTRUM	SPEKTR	ESPECTRO	SPETTRO
4203 SPELEOLOGY SURF	SPELEOLOGY	SPELEOLOGIE	HOEHLENKUNDE	SPELEOLOGIYA	ESPELEOLOGIA	SPELEOLOGIA
4204 SPELEOTHEM SURF	SPELEOTHEMS	SPELEOTHEME	SPELAEOTHEM	[SPELEOTHEM] MINERAL; PESH- CHERA	ESPELEOTEMA	SPELEOTEMA
4205 SPHENOLITH IGNE	Sphenolith INTRUSIONS	Sphenolite INTRUSION	Sphenulithus INTRUSIONS= KOERPER	SFENOLIT	Esfenolito INTRUSION	Sfenolite INTRUSIONE
4206 SPHERICAL HARMONIC **ANALYSIS** MATH	SPHERICAL HAR- MONIC ANALYSIS	ANALYSE HAR- MONIQUE SPHERIQUE	SPHAER.HARMON. ANALYSE	[SPHERICAL HAR- MONIC ANALYSIS] STATISTIKA; FUNKTSIYA	ANALISIS= ARMONICO= ESFERICO	ANALISI ARMONICA SFERICA
4207 SPHERICITY TEST	SPHERICITY	SPHERICITE	RUNDUNGSGRAD	SFERICHNOST'	ESFERICIDAD	SFERICITA
4208 SPHERULE TEST	SPHERULES *UF:* Ooid	SPHERULE *UF:* Ooide	SPHAERULIT *UF:* Ooid	SFERULA	ESFERULA *UF:* Ooide	SFERULA *UF:* Ooide
4209 SPHERULITE TEST	SPHERULITES	SPHEROLITE	SPHAEROLITH	SFEROLIT	ESFERULITO	SFERULITE
4210 SPILITIZATION IGNE	SPILITIZATION	SPILITISATION	SPILITISIERUNG	SPILITIZATSIYA	ESPILITIZACION	SPILITIZZAZIONE

	ENGLISH	FRANCAIS	DEUTSCH	RUSSKIJ	ESPANOL	ITALIANO
4211 SPINE PALE	SPINAL COLUMN	ANATOMIE COLONNE VERTE- BRALE	WIRBELSAEULE	Ship SKELET	ANATOMIA- COLUMNA- VERTEBRAL	COLONNA VERTE- BRALE
4212 SPINIFEX TEST	SPINIFEX TEX- TURE	SPINIFEX	SPINIFEX- GEFUEGE	SPINIFEKS	SPINIFEX	SPINIFEX
4213 SPIRIFERIDA PALS	SPIRIFERIDA	SPIRIFERIDA	SPIRIFERIDA	SPIRIFERIDA	SPIRIFERIDA	SPIRIFERIDA
4214 SPIRILLINACEA PALS	SPIRILLINACEA	SPIRILLINACEA	SPIRILLINACEA	SPIRILLINACEA	SPIRILLINACEA	SPIRILLINACEA
4215 SPIT SURF	SPITS	FLECHE LITTOR- ALE	NEHRUNGS- BOGEN	KOSA	FLECHA=LITORAL	LINGUA LITORALE
4216 SPODOSOL SUSS	SPODOSOLS *UF:* Aquods; Ferrods; Humods; Orthods; Podzoluvisols	SPODOSOL *UF:* Aquod; Ferrod; Humod; Orthod; Pod- zoluvisol	SPODOSOL *UF:* Aquod; Ferrod; Humod; Orthod; Pod- zoluvisol	SPODOSOL *UF:* Akvod; Ferrod; Gumod; Ortod	SPODOSOL *UF:* Aquod; Ferrod; Humod; Orthod; Pod- zoluvisol	SPODOSOL *UF:* Aquod; Ferrod; Humod; Orthod; Pod- zoluvisol
4217 SPOIL MINI	Spoils INDUSTRIAL WASTE	Deblai sterile DECHET INDUSTR- IEL	Taub.Gestein INDUSTRIE- MUELL	Otval PORODA PUSTAYA	Escombrera RESIDUO- INDUSTRIAL	Rifiuto inerte RIFIUTI INDUSTRI- ALI
4218 SPONGOLITE SEDS	SPONGOLITE	SPONGOLITE	SPONGIOLITH	SPONGOLIT	ESPONJOLITA	SPONGOLITE
4219 SPORE PALE	SPORES *UF:* Sporomorph	SPORE *UF:* Sporomorphe	SPORE *UF:* Sporomorph	SPORA *UF:* Ehkzina; Miospora; Sporomorfa	ESPORA *UF:* Esporomorfo	SPORA *UF:* Sporomorfo
4220 SPOROMORPH PALE	Sporomorph SPORES	Sporomorphe SPORE	Sporomorph SPORE	Sporomorfa SPORA	Esporomorfo ESPORA	Sporomorfo SPORA
4221 SPOROZOA PALS	Sporozoa PROTISTA	Sporozoa PROTOZOA	Sporozoa PROTOZOA	SPOROZOA	Sporozoa PROTOZOA	Sporozoa PROTOZOA
4222 SPOTTED SLATE IGMS	Spotted slate SLATES	Schiste tachete MICASCHISTE	Flecken=Schiefer GLIMMER- SCHIEFER	SLANETS PYATNIS- TYJ	Esquisto=moteado MICAESQUISTO	Scisto maculato MICASCISTO
4223 SPREADING CENTER SOLI	SPREADING CEN- TERS	CENTRE EXPAN- SION	AUSBREITUNGS- ZENTRUM	[SPREADING CEN- TER] RASSHIRENIE DNA OKEANA	CENTRO- EXPANSION	CENTRO DI ESPAN- SIONE
4224 SPRING GEOH	SPRINGS *UF:* Karst spring; Min- eral spring; Oasis	SOURCE *UF:* Oasis; Source kars- tique	QUELLE *UF:* Karst=Quelle; Oase	ISTOCHNIK *UF:* Exsurgence; Heat source	MANANTIAL *UF:* Fuente=karstica; Oasis	SORGENTE *UF:* Oasi; Sorgente carsica
4225 SQUAMATA PALS	SQUAMATA	SQUAMATA *UF:* Lacertilia; Ophidia	SQUAMATA *UF:* Lacertilia; Ophidia	Squamata LEPIDOSAURIA	SQUAMATA *UF:* Lacertilia; Ophidia	SQUAMATA *UF:* Lacertilia; Ophidia
4226 STABILITY GEOL	STABILITY	STABILITE	STABILITAET	STABIL'NOST' *UF:* Ustojchivost' sklona	ESTABILIDAD	STABILITA
4227 STABILIZATION ENGI	STABILIZATION *UF:* Soil stabilization	STABILISATION *UF:* Stabilisation sol	STABILISATION *UF:* Boden- Stabilisierung	STABILIZATSIYA *UF:* Fluid injection; Grouting; Zakreplenie gruntov	ESTABILIZACION *UF:* Estabilizacion= terreno	STABILIZZAZIONE *UF:* Stabilizzazione del suolo

	ENGLISH	FRANCAIS	DEUTSCH	RUSSKIJ	ESPANOL	ITALIANO
4228 STABLE ISOTOPE ISOT	STABLE ISOTOPES	ISOTOPE STABLE	STABIL.ISOTOP	IZOTOP STA-BIL'NYJ	ISOTOPO=ESTABLE	ISOTOPO STABILE
4229 STAGE STRA	Stratigraphic stage STRATIGRAPHIC UNITS	Etage ECHELLE STRATI-GRAPHIQUE	Stufe STRATIGRAPH. SKALA	YARUS	Piso=estratigrafico ESCALA=ESTRATIGRAFICA	Piano stratigrafico SCALA STRATI-GRAFICA
4230 STAINING METH	STAINING	COLORATION ECHANTILLON	FAERBUNG	METOD OKRASHIVANIYA	COLORACION-MUESTRA	COLORAZIONE
4231 STALACTITE SURF	STALACTITES *UF:* Stalagmite	STALACTITE *UF:* Stalagmite	STALAKTIT *UF:* Stalagmit	STALAKTIT *UF:* Stalagmit	ESTALACTITA *UF:* Estalagmita	STALATTITE *UF:* Stalagmite
4232 STALAGMITE SURF	Stalagmite STALACTITES	Stalagmite STALACTITE	Stalagmit STALAKTIT	Stalagmit STALAKTIT	Estalagmita ESTALACTITA	Stalagmite STALATTITE
4233 STAMPIAN STRS	STAMPIAN *UF:* Rupelian	STAMPIEN *UF:* Rupelien	STAMPIEN *UF:* Rupel=Stufe	Stamp. RYUPEL'	STAMPIENSE *UF:* Rupeliense	STAMPIANO *UF:* Rupeliano
4234 STANDARD DEVIATION MATH	STANDARD DEVIA-TION	ECART TYPE	STANDARD=ABWEICHUNG	OTKLONENIE STANDARTNOE	DESVIACION=TIPICA	DEVIAZIONE STANDARD
4235 STANDARD MATERIALS MISC	STANDARD MATE-RIALS	STANDARD CHIMIQUE	STANDARD	STANDART *UF:* Standard rock	PATRON=QUIMICO	STANDARD CHIMICO
4236 STANDARD METHOD METH	Standard method METHODS	Methode standard METHODOLOGIE	Standard=Methode METHODIK	STANDARTA METOD	Metodo=standard METODOLOGIA	Metodo standard METODOLOGIA
4237 STANDARD ROCK GEOL	STANDARD ROCKS	ROCHE STANDARD	STANDARD=GESTEIN	[STANDARD ROCK] STANDART; POR-ODA	ROCA=PATRON	ROCCIA STAN-DARD
4238 STANDARDIZATION MISC	STANDARDIZA-TION	NORMALISATION	NORMUNG	STANDARTIZAT-SIYA	NORMALIZACION	NORMALIZZAZ-IONE
4239 STATISTICAL ANALYSIS MATH	STATISTICAL ANALYSIS *UF:* Statistical functions; Statistical model; Statistical smoothing; Trend (math); Variable	METHODE STATIS-TIQUE *UF:* Algebre matricielle; Approximation; Loi statistique; Matrice mathematique; Modele statistique; Variable; Variogramme	STATIST.METHODE *UF:* Mathemat.Matrix; Matrix=Algebra; Statist.Gesetz; Statist. Modell; Variable; Variogramm	Analiz statisticheskij STATISTIKA	METODO=ESTADISTICO *UF:* Algebra=matricial; Aproximacion; Ley=estadistica; Matriz=matematica; Modelo=estadistico; Variable; Variograma	METODO STATIS-TICO *UF:* Algebra matriciale; Approssimazione; Funzione statistica; Matrice matematica; Modello statistico; Variabile; Vario-gramma
4240 STATISTICAL DISTRIBUTION MATH	STATISTICAL DIS-TRIBUTION *UF:* Average; Binomial distribution; Deviation; Distribution; Frequency distribution; Interpolation; Lognormal distribution; Maximum; Minimum; Normal distribution; Poisson distribution; Populations	DISTRIBUTION STATISTIQUE *UF:* Distribution; Distribution binomiale; Distribution frequence; Distribution lognor-male; Distribution normale; Distribution Poisson; Ecart; Interpolation; Maximum; Minimum; Moyenne; Population statistique	STATIST. VERTEILUNG *UF:* Abweichung; Binomial=Verteilung; Durchschnittswert; Haeufigkeits-Verteilung; Interpolation; Maximum; Mini-mum; Normal=Verteilung; Normallog=Verteilung; Poisson=Verteilung; Statist.Haeufigkeit; Verteilung	RASPREDELENIE STATISTICHESKOE	DISTRIBUCION=ESTADISTICA *UF:* Desviacion; Distribucion; Distribucion=binomial; Distribucion=de=frecuencia; Distribucion=lognormal; Distribucion=normal; Distribucion=Poisson; Interpolacion; Maximo; Media; Minimo; Poblacion=estadistica	DISTRIBUZIONE STATISTICA *UF:* Deviazione; Distri-buzione; Distribuzione binomiale; Distribuz-ione di frequenza; Distribuzione di Pois-son; Distribuzione lognormale; Distribuz-ione normale; Inter-polazione; Massimo; Media; Minimo; Popolazione statistica

	ENGLISH	FRANCAIS	DEUTSCH	RUSSKIJ	ESPANOL	ITALIANO
4241 STATISTICAL FUNCTION MATH	Statistical functions STATISTICAL ANALYSIS	Loi statistique METHODE STATIS-TIQUE	Statist.Gesetz STATIST.METHODE	Funktsiya statisti-cheskaya STATISTIKA; FUNKTSIYA	Ley=estadistica METODO=ESTADISTICO	Funzione statistica METODO STATIS-TICO
4242 STATISTICAL MODEL MATH	Statistical model STATISTICAL ANALYSIS	Modele statistique METHODE STATIS-TIQUE	Statist.Modell STATIST.METHODE	MODEL' STATISTI-CHESKAYA	Modelo=estadistico METODO=ESTADISTICO	Modello statistico METODO STATIS-TICO
4243 STATISTICAL POPULATION MATH	Populations STATISTICAL DIS-TRIBUTION	Population statistique DISTRIBUTION STATISTIQUE	Statist.Haeufigkeit STATIST. VERTEILUNG	SOVOKUPNOST' GENERAL'NAYA	Poblacion=estadistica DISTRIBUCION=ESTADISTICA	Popolazione statistica DISTRIBUZIONE STATISTICA
4244 STATISTICAL TREND MATH	Trend (math) STATISTICAL ANALYSIS	Tendance ANALYSE TEN-DANCE	Trend TREND=ANALYSE	TREND (STATISTICHESKIJ)	Tendencia ANALISIS=TENDENCIA	Tendenza statistica ANALISI DELLE TENDENZE
4245 STATISTICS MATH	STATISTICS	STATISTIQUE	STATISTIK	STATISTIKA *UF:* Analiz statisti-cheskij; Ehkstsess; Funktsiya statisti-cheskaya; Spherical harmonic analysis	ESTADISTICA	STATISTICA
4246 STEADY FLOW GEOH	STEADY FLOW	REGIME PERMA-NENT	STATIONAER. STROEMUNG	FIL'TRATSIYA STATSIONARNAYA	REGIMEN=PERMANENTE	REGIME STAZIO-NARIO
4247 STEATITIZATION IGNE	STEATITIZATION	Steatitisation ALTERATION HYDROTHERMALE; TALC	Steatitisierung HYDROTHERMAL=UMWANDLUNG; TALK	OTAL'KOVANIE	Esteatizacion ALTERACION=HIDROTERMAL; TALCO	Steatitizzazione ALTERAZIONE IDROTERMALE; TALCO SOSTANZA
4248 STEP FAULT STRU	STEP FAULTS	FAILLE ESCALIER	STAFFEL=STOERUNG	RAZRYV STU-PENCHATYJ	FALLA=EN=ESCALERA	FAGLIA A GRADINI
4249 STEP FOLD STRU	Step folds FOLDS	Pli escalier PLI	Treppen=Falte FALTE	Skladka stupenchataya FLEKSURA; SKLADKA	Pliegue=monoclinal PLIEGUE	Piega a gradino PIEGA
4250 STEPHANIAN STRS	STEPHANIAN	STEPHANIEN	STEFAN	STEFAN	ESTEFANIENSE	STEFANIANO
4251 STEPPE ZONE SURF	STEPPES *UF:* Pampas	ZONE STEPPIQUE *UF:* Pampa; Ustoll; Xeroll	STEPPEN=ZONE *UF:* Pampa; Ustoll; Xeroll	STEP' *UF:* Pampa	ZONA=ESTEPARIA *UF:* Pampa; Ustoll; Xeroll	AMBIENTE DI STEPPA *UF:* Pampa; Ustoll; Xeroll
4252 STEREOCHEMISTRY PHCH	STEREOCHEMIS-TRY	STEREOCHIMIE	STEREOCHEMIE	STEREOKHIMIYA	ESTEREOQUIMICA	STEREOCHIMICA
4253 STEROLS CHES	Sterols ORGANIC MATERI-ALS	Sterol MATIERE ORGANIQUE	Sterol ORGAN.SUBSTANZ	Sterol VESHCHESTVO ORGANICHESKOE	Esteroles MATERIA=ORGANICA	Sterolo MATERIA ORGANICA
4254 STICK SLIP SOLI	STICK=SLIP	GLISSEMENT SAC-CADE	RUTSCH=AUSLOESEND. ANSTOSS	[STICK=SLIP] ZEMLETRYASENIE MELKOFOKUSNOE	MOVIMIENTO=A=TIRONES	SCIVOLAMENTO DA SCOSSA
4255 STIFF CLAY ENGI	STIFF CLAYS	ARGILE RAIDE	STEIF.TON	GLINA PROCH-NAYA	ARCILLA=TENAZ	ARGILLA DURA

	ENGLISH	FRANCAIS	DEUTSCH	RUSSKIJ	ESPANOL	ITALIANO
4256 STOCHASTIC PROCESS MATH	STOCHASTIC PROCESSES	PROCESSUS STOCHASTIQUE	STOCHAST. PROZESS	SLUCHAJNYJ PROTSESS	PROCESO= ESTOCASTICO	PROCESSO STO-CASTICO
4257 STOCK IGNE	STOCKS	STOCK	STOCK	SHTOK	STOCK	STOCK
4258 STOCKWORK IGNE	STOCKWORK DEPOSITS	STOCKWERK	STOCKWERK	SHTOKVERK	STOCKWERK	STOCKWORK
4259 STOICHIOMETRY PHCH	STOICHIOMETRY	STOECHIOMETRIE	STOECHIOMETRIE	STEKHIOMETRIYA	ESTEQUIOMETRIA	STECHIOMETRIA
4260 STOLONOIDEA PALS	STOLONOIDEA	STOLONOIDEA	STOLONOIDEA	STOLONOIDEA	STOLONOIDEA	STOLONOIDEA
4261 STOMOCHORDATA PALS	HEMICHORDATA *UF:* Cephalodiscida; Planctosphaeroidea	STOMOCHORDATA *UF:* Cephalodiscida; Planctosphaeroidea	STOMOCHORDATA *UF:* Cephalodiscida; Planctosphaeroidea	HEMICHORDATA	STOMOCHORDATA *UF:* Cephalodiscida; Planctosphaeroidea	HEMICHORDATA *UF:* Cephalodiscida; Planctosphaeroidea
4262 STONE LINE SURF	STONE LINES	STONELINE	[STONE=LINE]	[STONE LINE] EHLYUVIJ	STONELINE	STONELINE
4263 STONY IRON METEORITE EXTS	STONY IRONS	SIDEROLITE *UF:* Lodranite; Meso-siderite; Pallasite; Siderophyre	STEIN=EISEN= METEORIT *UF:* Lodranit; Meso-siderit; Pallasit; Sidero-phyr	SIDEROLIT *UF:* Lodranit; Mezo-siderit; Pallasit; Siderofir	SIDEROLITO *UF:* Lodranite; Meso-siderita; Pallasita; Siderofiro	SIDEROLITE *UF:* Lodranite; Meso-siderite; Pallasite; Siderofiro
4264 STONY METEORITE EXTS	STONY METEOR-ITES	METEORITE PIER-REUSE	STEIN=METEORIT	METEORIT KAMENNYJ	METEORITO= LITICO	METEORITE ROC-CIOSA
4265 STORAGE ECON	STORAGE	STOCKAGE	SPEICHERUNG	KHRANENIE	ALMACENA-MIENTO	STOCCAGGIO
4266 STORM SURF	STORMS	TEMPETE	STURM	SHTORM	TEMPESTAD	TEMPESTA
4267 STRAIN PHCH	STRAIN	DEFORMATION SOUS CONTRAINTE	ZUG	DEFORMATSIYA (TEKTONI-CHESKAYA)	DEFORMACION= BAJO=TENSION	DEFORMAZIONE SOTTO SFORZO
4268 STRAINMETER INST	STRAINMETERS	JAUGE DEFORMA-TION	VERFORMUNGS= MESSGERAET	TENZOMETR	DEFORMACION= MEDIDA	MISURA DI DEFOR-MAZIONE
4269 STRAIT MARI	Straits OCEAN FLOORS	Detroit RELIEF SOUS MARIN	Meerenge SUBMARIN.RELIEF	PROLIV	Estrecho RELIEVE= SUBMARINO	Stretto RILIEVO SOTTO-MARINO
4270 STRATABOUND DEPOSIT ECON	STRATABOUND DEPOSITS	GITE STRATOIDE	SCHICHTGEB-UNDEN. LAGERSTAETTE	MESTOROZH-DENIE STRATIFIT-SIROVANNOE	YACIMIENTO= ESTRATOIDE	GIACIMENTO INTERSTRATO
4271 STRATEGY ECON	STRATEGY	STRATEGIE	STRATEGIE	STRATEGIYA	ESTRATEGIA	STRATEGIA
4272 STRATIFICATION SEDI	STRATIFICATION *UF:* Bedding planes; Concordance; Conformity; Intercalation; Layers; Stratified structure	STRATIFICATION *UF:* Clivage parallele; Concordance strati-graphique; Couche; Couche concordante; Intercalation; Plan stratification; Structure stratifiee	SCHICHTUNG *UF:* Geschichtet. Gefuege; Konkordant. Lagerung; Konkordanz; Schicht; Schicht= Flaeche; Schichtung= Schieferung; Wechsel-lagerung	STRATIFIKATSIYA	ESTRATIFICACION *UF:* Capa=concordante; Concordancia= estratigrafica; Estrato; Estructura= estratificada; Exfoliacion= concordante; Intercala-cion; Plano= estratificacion	STRATIFICAZIONE *UF:* Clivaggio parallelo; Concordanza; Concor-danza di stratificazione; Intercalazione; Livello; Piano di stratificazione; Struttura stratificata

	ENGLISH	FRANCAIS	DEUTSCH	RUSSKIJ	ESPANOL	ITALIANO
4273 STRATIFIED STRUCTURE TEST	Stratified structure STRATIFICATION	Structure stratifiee STRATIFICATION	Geschichtet.Gefuege SCHICHTUNG	STRATIFIT-SIROVANNYJ	Estructura=estratificada ESTRATIFICACION	Struttura stratificata STRATIFICAZIONE
4274 STRATIFORM DEPOSIT ECON	STRATIFORM DEPOSITS	GITE STRATI-FORME	LAGER	MESTOROZHDENIE STRATIFORMNOE	YACIMIENTO=ESTRATIFORME	GIACIMENTO STRATIFORME
4275 STRATIGRAPHIC BOUNDARY STRA	BOUNDARY	LIMITE STRATI-GRAPHIQUE	STRATIGRAPH. GRENZE	GRANITSA STRATI-GRAFICHESKAYA	LIMITE=ESTRATIGRAFICO	LIMITE STRATI-GRAFICO
4276 STRATIGRAPHIC CONTROL ECON	STRATIGRAPHIC CONTROLS	CONTROLE STRA-TIGRAPHIQUE	STRATIGRAPH. BEZUG	KONTROL' STRATI-GRAFICHESKIJ	CONTROL=ESTRATIGRAFICO	CONTROLLO STRA-TIGRAFICO
4277 STRATIGRAPHIC ISOCHORE STRA	Isochore ISOPLETH MAPS	Isochore ISOPLETHE; CARTE STRATIGRAPHIQUE	Stratigraph.Isochore ISOLINIE; STRATIGRAPH. KARTE	IZOKHORA	Isocora=estratigrafica ISOPLETA; MAPA=ESTRATIGRAFICO	Isocora CARTA DELLE ISOPLETE; CARTA STRATIGRAFICA
4278 STRATIGRAPHIC MAP STRA	STRATIGRAPHIC MAPS	CARTE STRATI-GRAPHIQUE UF: Isochore	STRATIGRAPH. KARTE UF: Stratigraph.Isochore	KARTA STRATI-GRAFICHESKAYA	MAPA=ESTRATIGRAFICO UF: Isocora=estratigrafica	CARTA STRATI-GRAFICA UF: Isocora
4279 STRATIGRAPHIC TRAP ECON	STRATIGRAPHIC TRAPS UF: Sedimentary trap	PIEGE STRATI-GRAPHIQUE	STRATIGRAPH. FALLE	LOVUSHKA STRA-TIGRAFI-CHESKAYA	TRAMPA=ESTRATIGRAFICA	TRAPPOLA STRATI-GRAFICA
4280 STRATIGRAPHIC UNIT STRA	STRATIGRAPHIC UNITS UF: Biostratigraphic unit; Chronostratigraphic unit; Eons; Epochs; Eras; Erathems; Formations; Lithostratigraphic unit; Members; Periods; Preimbrian; Series; Stratigraphic stage	UNITE STRATI-GRAPHIQUE UF: Unite biostratigraphique; Unite chronostratigraphique; Unite geochronologique; Unite lithostratigraphique	STRATIGRAPH. EINHEIT UF: Biostratigraph. Einheit; Chronostratigraph. Einheit; Geochronolog. Einheit; Lithostratigraph. Einheit	PODRAZDELENIE STRATIGRAFI-CHESKOE	UNIDAD=ESTRATIGRAFICA UF: Unidad=bioestratigrafica; Unidad=cronoestratigrafica; Unidad=geocronologica; Unidad=litoestratigrafica	UNITA STRATI-GRAFICA UF: Unita biostratigrafica; Unita cronostratigrafica; Unita geocronologica; Unita litostratigrafica
4281 STRATIGRAPHY STRA	STRATIGRAPHY	STRATIGRAPHIE UF: Lithostratigraphie	STRATIGRAPHIE UF: Litho=Stratigraphie	STRATIGRAFIYA	ESTRATIGRAFIA UF: Litoestratigrafia	STRATIGRAFIA UF: Litostratigrafia
4282 STRATOTYPE STRA	STRATOTYPES UF: Type localities	STRATOTYPE UF: Localite type	STRATOTYP UF: Typ=Lokalitaet	STRATOTIP UF: Razrez tipovoj	ESTRATOTIPO UF: Localidad=tipo	STRATOTIPO UF: Localita=tipo
4283 STRATOVOLCANO IGNE	STRATOVOL-CANOES	STRATOVOLCAN	SCHICHT=VULKAN	STRATOVULKAN	ESTRATOVOLCAN	STRATOVULCANO
4284 STREAM SURF	STREAMS UF: Affluent streams; Confluent streams; Consequent streams; Creeks; Entrenched streams; Hydraulic gradient; Intermittent stream; Profile of equilibrium; Thalweg	COURS EAU UF: Torrent	WASSER=LAUF UF: Wildbach	Potok rechnoj REKA	CURSO=AGUA UF: Torrente	CORSO D'ACQUA UF: Torrente

	ENGLISH	FRANCAIS	DEUTSCH	RUSSKIJ	ESPANOL	ITALIANO
4285 STREAM BED SURF	Stream beds VALLEYS	Lit cours eau VALLEE	Strombett TAL	LOZHE POTOKA	Lecho=de=curso=de= agua VALLE	Letto fluviale VALLE
4286 STREAM ORDER SURF	STREAM ORDER	Rang reseau hydro- graphique RESEAU HYDRO- GRAPHIQUE	Gewaesser=Anordnung HYDROGRAPH. NETZ	[STREAM ORDER] BASSEJN VODOS- BORNYJ	Categoria=cursos=agua RED= HIDROGRAFICA	Gerarchia fluviale RETICOLATO IDRO- GRAFICO
4287 STREAM SEDIMENT ECON	STREAM SEDI- MENTS	STREAM SEDI- MENT	[STREAM= SEDIMENT]	[STREAM SEDI- MENT] OSADOK; RUSLO	SEDIMENTO=DE= LECHO	[STREAM= SEDIMENT]
4288 STREAM TRANSPORT SEDI	STREAM TRANS- PORT	TRANSPORT FLU- VIATILE	FLUVIATIL. TRANSPORT	[STREAM TRANS- PORT] PERENOS; REKA	TRANSPORTE= FLUVIAL	TRASPORTO FLUVI- ALE
4289 STRENGTH PHCH	STRENGTH	RESISTANCE MECANIQUE *UF:* Forabilite; Resis- tance a la rupture	FESTIGKEIT *UF:* Bohrbarkeit; Ruptur=Festigkeit	PROCHNOST' *UF:* Fracture strength; Napryazhenie raz- rushayushchee; Tenaci- ty	RESISTENCIA= MECANICA *UF:* Perforabilidad; Tension=de=ruptura	RESISTENZA MEC- CANICA *UF:* Forza di rottura; Perforabilita
4290 STRESS PHCH	STRESS *UF:* Compressive stress; Stress mineral	CONTRAINTE *UF:* Mineral stress	MECHAN. SPANNUNG *UF:* Stress=Mineral	STRESS	TENSION *UF:* Mineral=tensional	SFORZO *UF:* Minerale stress
4291 STRESS MINERAL MINE	Stress mineral MINERALS; STRESS	Mineral stress MINERAL; CON- TRAINTE	Stress=Mineral MINERAL; MECHAN. SPANNUNG	STRESS=MINERAL	Mineral=tensional MINERALES; TEN- SION	Minerale stress MINERALE; SFORZO
4292 STRIKE STRU	STRIKE	DIRECTION	RICHTUNG	PROSTIRANIE	DIRECCION	DIREZIONE
4293 STRIKE=SLIP FAULT STRU	STRIKE=SLIP FAULTS	FAILLE DECRO- CHEMENT *UF:* Faille laterale	BLATT= VERSCHIEBUNG *UF:* Lateral=Stoerung	SBROSO=SDVIG *UF:* Dip=slip fault	FALLA= HORIZONTAL *UF:* Falla=lateral	FAGLIA A SCORRI- MENTO ORIZZON- TALE *UF:* Faglia a separaz.sec. direzione
4294 STRIP MINING MINI	Strip mining SURFACE MINING	Exploitation en decou- verte CIEL OUVERT	[STRIP=MINING] TAGEBAU	VYRABOTKA OTKRYTAYA	Mina=cielo=abierto CIELO=ABIERTO	Sfruttamento in super- ficie MINIERA A CIELO APERTO
4295 STROMATOLITES PALS	STROMATOLITES	STROMATOLITES	STROMATOLITHES	STROMATOLITHI	STROMATOLITES	STROMATOLITE
4296 STROMATOPOROIDEA PALS	STROMATOPOR- OIDEA	STROMATOPOR- OIDEA	STROMATOPOR- OIDEA	STROMATOPOR- OIDEA	STROMATOPOR- OIDEA	STROMATOPOR- OIDEA
4297 STROMBOLIAN TYPE ERUPTION IGNE	STROMBOLIAN= TYPE ERUPTIONS	STROMBOLIEN	STROMBOLI=TYP	IZVERZHENIE STROMBOLIAN- SKOGO TIPA	ESTROMBOLIANO	ERUZIONE STROM- BOLIANA
4298 STRONG MOTION ENGI	STRONG MOTION	SECOUSSE VIO- LENTE	HEFTIG. ERSCHUETTERUNG	[STRONG MOTION] GRUNT; KOLEBA- NIE	SACUDIDA= VIOLENTA	SCOSSA VIOLENTA

	ENGLISH	FRANCAIS	DEUTSCH	RUSSKIJ	ESPANOL	ITALIANO
4299 STRONTIUM CHEE	STRONTIUM	STRONTIUM	SR	STRONTSIJ	ESTRONCIO	STRONZIO
4300 STROPHOMENIDA PALS	STROPHOMENIDA	STROPHOMENIDA *UF:* Anisopleura	STROPHOMENIDA *UF:* Anisopleura	STROPHOMENIDA	STROPHOMENIDA *UF:* Anisopleura	STROPHOMENIDA *UF:* Anisopleura
4301 STRUCTURAL ANALYSIS STRU	STRUCTURAL ANALYSIS *UF:* Microtectonics	Analyse structurale TECTONIQUE	Struktur=Analyse TEKTONIK	ANALIZ STRUK-TURNYJ	Analisis=estructural TECTONICA	Analisi strutturale TETTONICA
4302 STRUCTURAL CONTROL ECON	STRUCTURAL CON-TROLS *UF:* Structural guide	Controle structural CONTROLE TEC-TONIQUE	Struktur=Genese TEKTON.EINFLUSS	KONTROL' STRUK-TURNYJ *UF:* Kontrol' tektoni-cheskij	Control=estructural CONTROL=TECTONICO	Controllo strutturale CONTROLLO TET-TONICO
4303 STRUCTURAL CRYSTALLOGRAPHY MINE	Structural crystallogra-phy CRYSTAL STRUC-TURE	Cristallographie struc-ture CRISTALLOGRA-PHIE	Struktur. Kristallographie KRISTALLOGRA-PHIE	KRISTALLOGRA-FIYA STRUK-TURNAYA	Cristalografia=estructural CRISTALOGRAFIA	Cristallografia struttur-ale CRISTALLOGRAFIA
4304 STRUCTURAL DIAGRAM STRU	Structural diagram TECTONIC MAPS	Diagramme structural REPRESENTATION GRAPHIQUE	Gefuege=Diagramm GRAPH. DARSTELL-UNG	Strukturnaya dia-gramma STRUKTURA TEK-TONICHESKAYA; DIAGRAMMA	Diagrama=estructural REPRESENTA-CION=GRAFICA	Diagramma strutturale RAPPRESENTA-ZIONE GRAFICA
4305 STRUCTURAL ELEMENT STRU	TECTONIC UNITS	UNITE TEC-TONIQUE	STRUKTURELE-MENT	Ehlement strukturnyj STRUKTURA TEK-TONICHESKAYA	UNIDAD=TECTONICA	ELEMENTO STRUT-TURALE
4306 STRUCTURAL GEOLOGY STRU	STRUCTURAL GEOLOGY	GEOLOGIE STRUC-TURALE	STRUKTUR=GEOLOGIE	Geologiya strukturnaya TEKTONIKA	GEOLOGIA=ESTRUCTURAL	GEOLOGIA STRUT-TURALE
4307 STRUCTURAL GUIDE ECON	Structural guide STRUCTURAL CON-TROLS	Guide tectonique CONTROLE TEC-TONIQUE; GUIDE	Tekton.Indikator TEKTON.EINFLUSS; INDIKATOR=MINERAL	KRITERIJ POISKOVYJ (STRUKTURNYJ)	Guia=estructural CONTROL=TECTONICO; GUIA	Guida strutturale CONTROLLO TET-TONICO; GUIDA
4308 STRUCTURAL MAP STRU	STRUCTURAL MAPS	CARTE STRUCTUR-ALE	STRUKTUR=KARTE	Karta strukturnaya KARTA TEKTONI-CHESKAYA	MAPA=ESTRUCTURAL	CARTA STRUTTUR-ALE
4309 STRUCTURAL PETROLOGY STRU	Structural petrology PETROFABRICS	Petrologie structurale PETROFABRIQUE	Gefuege=Petrologie GESTEINS=GEFUEGE	PETROLOGIYA STRUKTURNAYA	Petrologia=estructural PETROFABRICA	Petrologia strutturale STRUTTURA DELLA ROCCIA
4310 STRUCTURAL RELIEF SURF	Structural relief MORPHOSTRUC-TURES	RELIEF STRUC-TURAL *UF:* Relief morphostruc-tural; Revers; Vallee consequente; Vallee structurale	STRUKTUR=RELIEF *UF:* Einfalls=Hang; Konsequent.Strom; Morphostruktur; Tekton.Tal	REL'EF STRUK-TURNYJ *UF:* Inversiya rel'efa; Reka konsekventnaya	RELIEVE=ESTRUCTURAL *UF:* Ladera=estructural; Morfoestructura; Rio=consecuente; Valle=estructural	MORFOSTRUT-TURA *UF:* Corso d'acqua conseguente; Pendio strutturale; Rilievo morfostrutturale; Valle tettonica
4311 STRUCTURAL TRAP ECON	STRUCTURAL TRAPS	PIEGE STRUC-TURAL	TEKTON.FALLE	LOVUSHKA STRUK-TURNAYA	TRAMPA=ESTRUCTURAL	TRAPPOLA STRUT-TURALE
4312 STRUCTURAL VALLEY SURF	Structural valley VALLEYS	Vallee structurale RELIEF STRUC-TURAL; PROFIL EN LONG	Tekton.Tal STRUKTUR=RELIEF; LAENGS=PROFIL	DOLINA STRUK-TURNAYA	Valle=estructural RELIEVE=ESTRUCTURAL; PERFIL=LONGITUDINAL	Valle tettonica MORFOSTRUT-TURA; PROFILO LONGITUDINALE

	ENGLISH	FRANCAIS	DEUTSCH	RUSSKIJ	ESPANOL	ITALIANO
4313 STRUCTURE CONTOUR MAP STRU	STRUCTURE CONTOUR MAPS	CARTE STRUCTURALE ISOPLETHE	HOEHENLINIEN=KARTE	[STRUCTURE CONTOUR MAP] IZOLINIYA; KARTA STRUKTURNAYA	MAPA=ESTRUCTURAL=CURVA=NIVEL	CARTA STRUTTURALE A ISOPLETE
4314 STYLOLITE TEST	STYLOLITES	STYLOLITE	STYLOLITH	STILOLIT *UF:* Mikrostilolit	ESTILOLITO	STILOLITE
4315 STYLOPHORA PALS	STYLOPHORA	STYLOPHORA	STYLOPHORA	STYLOPHORA	STYLOPHORA	STYLOPHORA
4316 SUBAERIAL ENVIRONMENT SURF	Subaerial environment ATMOSPHERE	Milieu atmospherique ATMOSPHERE	Atmosphaer.Milieu ATMOSPHAERE	Sreda nazemnaya ATMOSFERA; SREDA OKRUZHAYUSH-CHAYA	Medio=subaereo ATMOSFERA	Ambiente subaereo ATMOSFERA
4317 SUBALKALIC COMPOSITION IGNE	Subalkalic composition ALKALIC COMPOSITION	Roche subalcaline COMPOSITION ALCALINE	Subalkal.Chemismus ALKALI=TYP	SUBSHCHELOCH-NOJ	Composicion=subalcalina COMPOSICION=ALCALINA	Composizione subalcalina COMPOSIZIONE ALCALINA
4318 SUBALUMINOUS COMPOSITION IGNE	SUBALUMINOUS COMPOSITION	Composition subalumineuse ALUMINIUM	Subaluminoes. Chemismus AL	GLINOZEMISTYJ *UF:* Perglinozemistyj	Composicion=subaluminosa ALUMINIO	Composizione ipoalluminifera ALLUMINIO
4319 SUBBOREAL SURF	Subboreal BOREAL	Subboreal ZONE FROIDE	Subboreal KALT.ZONE	Subboreal'nyj BOREAL'NYJ	Subboreal ZONA=FRIA	Subboreale ZONA FREDDA
4320 SUBDIVISION MISC	Subdivision(gen.)	Subdivision(gen.)	Untergliederung(gen.)	RASCHLENENIE	Subdivision(gen.)	Suddivisione(gen.)
4321 SUBDUCTION SOLI	SUBDUCTION	ENTRAINEMENT EN PROFONDEUR	SUBDUKTION	SUBDUKTSIYA *UF:* Obduktsiya	ARRASTRE=EN=PROFUNDIDAD	SUBDUZIONE
4322 SUBDUCTION ZONE STRU	SUBDUCTION ZONES	ZONE SUBDUCTION	SUBDUKTIONS=ZONE	ZONA SUBDUKTSII	ZONA=SUBDUCCION	ZONA DI SUBDUZIONE
4323 SUBLIMATE PHCH	SUBLIMATES	SUBLIMAT	SUBLIMAT	SUBLIMAT	SUBLIMADO	SUBLIMATO
4324 SUBLIMATION PHCH	SUBLIMATION	SUBLIMATION	SUBLIMATION	SUBLIMATSIYA	SUBLIMACION	SUBLIMAZIONE
4325 SUBLITTORAL ENVIRONMENT PALE	SUBLITTORAL ENVIRONMENT	MILIEU SUBLITTORAL	SUBLITORAL=MILIEU	Sreda sublitoral'naya SEDIMENTATSIYA MORSKAYA; LITORAL'	MEDIO=SUBLITORAL	AMBIENTE SUBLITORALE
4326 SUBMARINE CANYON MARI	SUBMARINE CANYONS *UF:* Submarine valleys	CANYON MARIN	SUBMARIN. CANYON	KAN'ON PODVODNYJ	CAÑON=MARINO	CANYON SOTTOMARINO
4327 SUBMARINE FAN SEDI	SUBMARINE FANS	CONE SOUS MARIN	SUBMARIN. FAECHER	KONUS VYNOSA PODVODNYJ	CONO=SUBMARINO	CONOIDE SOTTOMARINA
4328 SUBMARINE INSTALLATION ENGI	SUBMARINE INSTALLATIONS	INSTALLATION SOUS MARINE	SUBMARIN. KONSTRUKTION	[SUBMARINE INSTALLATION] SOORUZHENIE; MORSKOJ	INSTALACION=SUBMARINA	INSTALLAZIONE SOTTOMARINA

	ENGLISH	FRANCAIS	DEUTSCH	RUSSKIJ	ESPANOL	ITALIANO
4329 SUBMARINE MORPHOLOGY MARI	Submarine morphology OCEAN FLOORS	Morphologie sous marine RELIEF SOUS MARIN	Submarin.Morphologie SUBMARIN.RELIEF	Geomorfologiya pod-vodnaya GEOMOR-FOLOGIYA; MOR-SKOJ	Morfologia=submarina RELIEVE=SUBMARINO	Morfologia sottomarina RILIEVO SOTTO-MARINO
4330 SUBMARINE RIDGE MARI	Submarine ridge MID–OCEAN RIDGES	DORSALE	SUBMARIN. RUECKEN UF: Vulkanorium	KHREBET PODVOD-NYJ	DORSAL	DORSALE
4331 SUBMARINE SPRING GEOH	SUBMARINE SPRINGS	SOURCE SOUS MARINE	SUBMARIN. QUELLE	ISTOCHNIK SUB-MARINNYJ	MANANTIAL=SUBMARINO	SORGENTE SOTTO-MARINA
4332 SUBMARINE VALLEY MARI	Submarine valleys SUBMARINE CAN-YONS	Vallee sous marine VALLEE; MILIEU MARIN	Submarin.Tal TAL; MARIN. MILIEU	DOLINA PODVOD-NAYA	Valle=submarino VALLE; MEDIO=MARINO	Valle sottomarina VALLE; AMBIENTE MARINO
4333 SUBMARINE VOLCANO MARI	Submarine volcano VOLCANOES; MARINE ENVIRON-MENT	Volcan sous marin VOLCAN; MILIEU MARIN	Submarin.Vulkan VULKAN; MARIN. MILIEU	VULKAN PODVOD-NYJ	Volcan=submarino VOLCAN; MEDIO=MARINO	Vulcano sottomarino VULCANO; AMBI-ENTE MARINO
4334 SUBMERGENCE SEDI	SUBMERGENCE	ENNOIEMENT	UEBERFLUTUNG	Zatoplenie TRANSGRESSIYA	INMERSION	ANNEGAMENTO
4335 SUBMERSIBLE METH	SUBMERSIBLES	SUBMERSIBLE	UNTERWASSER=FAHRZEUG	APPARAT PODVOD-NYJ	SUMERGIBLE	SOMMERGIBILE
4336 SUBSEQUENT VALLEY SURF	Subsequent valleys VALLEYS	Vallee subsequente PROFIL EN LONG	Subsequent.Tal LAENGS–PROFIL	DOLINA SUBSEK-VENTNAYA	Valle=subsecuente PERFIL=LONGITUDINAL	Valle susseguente PROFILO LONGI-TUDINALE
4337 SUBSIDENCE STRU	SUBSIDENCE UF: Downwarping	SUBSIDENCE UF: Affaissement regional	SENKUNG UF: Absenkung	OPUSKANIE	SUBSIDENCIA UF: Subsidencia=regional	SUBSIDENZA UF: Abbassamento regionale
4338 SUBSTITUTION PHCH	SUBSTITUTION UF: Diadochy; Replace-ment	SUBSTITUTION UF: Remplacement	SUBSTITUTION UF: Ersatz	[SUBSTITUTION] ZAMESHCHENIE	SUSTITUCION UF: Reemplazamiento	SOSTITUZIONE UF: Rimpiazzo
4339 SUBSTRATE PALE	Substrates ECOLOGY	Substrat ECOLOGIE	Substrat OEKOLOGIE	SUBSTRAT	Substratum ECOLOGIA	Substrato ECOLOGIA
4340 SUBSURFACE GEOLOGY GEOL	Subsurface geology GEOLOGY	Geologie subsurface GEOLOGIE	[SUBSURFACE=GEOLOGY] GEOLOGIE	GEOLOGIYA GLU-BINNAYA	Geologia=del=subsuelo GEOLOGIA	Geologia del substrato GEOLOGIA
4341 SUBTIDAL ENVIRONMENT PALE	SUBTIDAL ENVI-RONMENT	MILIEU SUBTIDAL	SUBTIDAL.MILIEU	[SUBTIDAL ENVI-RONMENT] LITORAL'	MEDIO=SUBMAREA	AMBIENTE SUBTID-ALE
4342 SUBTROPICAL ENVIRONMENT SURF	SUBTROPICAL ENVIRONMENT	ZONE SUBTROPI-CALE	SUBTROPEN=ZONE	SUBTROPICHESKIJ	ZONA=SUBTROPICAL	ZONA SUBTROPI-CALE
4343 SUBVOLCANIC PROCESS IGNE	SUBVOLCANIC PROCESSES	CONDITION HYPOVOLCANIQUE	SUBVULKAN. VORGANG	SUBVULKAN	HIPOVOLCANICO	CONDIZIONI SUB-VULCANICHE

	ENGLISH	FRANCAIS	DEUTSCH	RUSSKIJ	ESPANOL	ITALIANO
4344 SUBWAY ENGI	SUBWAYS	METROPOLITAIN	U−BAHN	METROPOLITEN	METRO	METROPOLITANA
4345 SUCTORIA PALS	Suctoria PROTISTA	Suctoria PROTOZOA	Suctoria PROTOZOA	SUCTORIA	Suctoria PROTOZOA	Suctoria PROTOZOA
4346 SUDETIC OROGENY STRU	Sudetic orogeny HERCYNIAN OROGENY	Orogenie sudete OROGENIE HERCY- NIENNE	Sudet.Orogenese HERZYN. OROGENESE	Sudetskij GERTSINSKIJ	Orogenia=sudetica OROGENIA= HERCINICA	Orogenesi sudetica OROGENESI ERCINICA
4347 SUGARS CHES	SUGARS	GLUCIDE	GLUKOSE	SAKHAR	AZUCARES	ZUCCHERO
4348 SUIFORMES PALS	SUIFORMES	SUIFORMES	SUIFORMES	Suiformes ARTIODACTYLA	SUIFORMES	SUIFORMES
4349 SULFATES MING	SULFATES	SULFATE	SULFAT	SUL'FAT	SULFATO	SOLFATO
4350 SULFIDE ZONE ECON	Sulfide zone SULFIDES	Zone sulfure SULFURE	Sulfid=Zone SULFID	ZONA SUL'FIDNOGO OBO- GASHCHENIYA	Zona=de=sulfuros SULFURO	Zona dei solfuri SOLFURO
4351 SULFIDES MING	SULFIDES *UF:* Sulfide zone	SULFURE *UF:* Zone sulfure	SULFID *UF:* Sulfid=Zone	SUL'FID	SULFURO *UF:* Zona=de=sulfuros	SOLFURO *UF:* Zona dei solfuri
4352 SULFOSALTS MING	SULFOSALTS	SULFOSEL	SULFOSALZ	SUL'FOSOL'	SULFOSAL	SOLFOSALE
4353 SULFUR CHEE	SULFUR	SOUFRE	S	SERA *UF:* Sulphur=isotope ratio	AZUFRE	ZOLFO
4354 SULFUR DIOXIDE CHES	SULFUR DIOXIDE	GAZ ANHYDRIDE SULFUREUX	SCHWEFELDIOXID	DVUOKIS' SERY	GAS=SULFHIDRICO	ANIDRIDE SOL- FOROSA
4355 SULFUR RATIO ISOT	S=34/S=32	S 34=S 32	S34=S32	[SULFUR=ISOTOPE RATIO] IZOTOP; SERA	S=34=S=32	S=34/S=32
4356 SULFURIC ACID CHES	SULFURIC ACID	ACIDE SUL- FURIQUE	SCHWEFELSAEURE	KISLOTA SERNAYA	ACIDO- SULFURICO	ACIDO SOLFORICO
4357 SUMMARY ARTICLE MISC	REVIEW	MONOGRAPHIE SOMMAIRE	UEBERSICHTS- DARSTELLUNG	OBOBSHCHENIE	MONOGRAFIA= SUMARIA	SINTESI MONO- GRAFICA
4358 SUN EXTS	SUN *UF:* Solar radiation	SOLEIL *UF:* Rayonnement solaire	SONNE *UF:* Solar=Strahlung	SOLNTSE	SOL=ASTRO *UF:* Radiacion=solar	SOLE *UF:* Radiazione solare
4359 SUPERPOSED FOLD STRU	SUPERPOSED FOLDS *UF:* Superposition	PLI SUPERPOSE	UEBERPRAEGT. FALTE	SKLADKA NALOZ- HENNAYA	PLIEGUE= SUPERPUESTO	PIEGA SOVRAP- POSTA
4360 SUPERPOSITION STRU	Superposition SUPERPOSED FOLDS	Superposition TECTONIQUE SUPERPOSEE	Ueberlagerung UEBERPRAEGUNGS- TEKTONIK	ZALEGANIENE- NARUSHENNOE	Superposicion TECTONICA= SUPERPUESTA	Sovrapposizione TETTONICA SOVRAIMPOSTA
4361 SUPERSTRUCTURE MINE	SUPERSTRUCTURE	SURSTRUCTURE	UEBERSTRUKTUR	SVERKHSTRUK- TURA	SUPERESTRUC- TURA	SUPERSTRUTTURA

	ENGLISH	FRANCAIS	DEUTSCH	RUSSKIJ	ESPANOL	ITALIANO
4362 SUPRACRUSTAL ORIGIN SOLI	SUPRACRUSTAL ORIGIN	Origine supracrustale ORIGINE; LITHOSPHERE	Suprakrustal.Genese HERKUNFT; LITHOSPHAERE	PORODA SUPRAKRUS-TAL'NAYA	Origen=supracortical ORIGEN; LITOSF-ERA	Origine sopracrostale ORIGINE; LITOSF-ERA
4363 SURFACE DEFECT MINE	SURFACE DEFECTS	DEFAUT SURFACE	OBERFLAECHEN=FEHLER	[SURFACE DEFECT] DEFEKT KRISTAL-LICHESKOJ STRUK-TURY; STRUKTURA KRISTALLICHES-KAYA	IMPERFECCION-DE=SUPERFICIE	DIFETTO DI SUPER-FICIE
4364 SURFACE RUNOFF GEOH	Surface runoff RUNOFF	Ruissellement surface RUISSELLEMENT	Oberflaechen=Abfluss ABFLUSS	STOK POVERKH-NOSTNYJ *UF:* Sediment yield	Escorrentia=superficial ESCORRENTIA	Scorrimento superfi-ciale RUSCELLAMENTO
4365 SURFACE TENSION PHCH	SURFACE TENSION	TENSION SUPER-FICIELLE	OBERFLAECHEN=SPANNUNG	NATYAZHENIE POVERKHNOST-NOE	TENSION=SUPERFICIAL	TENSIONE SUPER-FICIALE
4366 SURFACE WATER GEOH	SURFACE WATER	EAU SURFACE	OBERFLAECHEN=WASSER	VODA POVERKH-NOSTNAYA *UF:* Surface water bud-get	AGUA=DE=SUPERFICIE	ACQUA SUPERFI-CIALE
4367 SURFACE WATER BUDGET GEOH	Surface water budget WATER BALANCE	BILAN EAU SUR-FACE	OBERFLAECH-ENWASSER=BILANZ	[SURFACE WATER BUDGET] BALANS VODNYJ; VODA POVERKH-NOSTNAYA	BALANCE=AGUA=SUPERFICIE	BILANCIO IDROLOGICO
4368 SURFACE WAVE SOLI	SURFACE WAVES	ONDE SURFACE *UF:* Onde Airy	OBERFLAECHEN=WELLE *UF:* Airy=Welle	VOLNA POVERKH-NOSTNAYA *UF:* Releya volna	ONDA=LONGITUDINAL *UF:* Ondas=Airy	ONDA DI SUPERFI-CIE *UF:* Onda di Airy
4369 SURFICIAL GEOLOGY SURF	SURFICIAL GEOL-OGY	FORMATION SUPERFICIELLE *UF:* Formation super-gene	OBERFLAECHEN=BILDUNGEN *UF:* Hypergenese	GEOLOGIYA CHET-VERTICHNAYA	FORMACION=SUPERFICIAL *UF:* Alteracion=supergenica	FORMAZIONE SUPERFICIALE *UF:* Rielaborazione esogena
4370 SURGE SURF	Surges OCEAN WAVES	Deferlante HOULE	Woge DUENUNG	SHTORMOVOJ NAGON	Batiente OLEAJE	Flutto ONDA MARINA
4371 SURVEY GEOL	Surveys CARTOGRAPHY	Lever CARTOGRAPHIE	Gelaende=Aufnahme KARTOGRAPHIE	S'EMKA	Investigacion CARTOGRAFIA	Rilevamento geologico CARTOGRAFIA
4372 SUSPENDED LOAD SURF	SUSPENDED MATE-RIALS	MATIERE EN SUS-PENSION	SUSPENDIERT. MATERIAL	Stok vzveshennyj VZVES'; STOK TVERDYJ	CARGA=EN=SUSPENSION	CARICO SOSPESO
4373 SUSPENSION PHCH	SUSPENSION	SUSPENSION	SUSPENSION	SUSPENZIYA	SUSPENSION	SOSPENSIONE
4374 SUTURE PALE	SUTURES	SUTURE	SUTUR	Sutura ANATOMIYA	SUTURA	SUTURA
4375 SVECOFENNIAN OROGENY STRU	Svecofennian orogeny OROGENY; PRE-CAMBRIAN	Orogenie svecofen-nienne OROGENIE ANTE-CAMBRIENNE	Svekofenniden=Orogenese PRAEKAMBR. OROGENESE	SVEKOFENSKIJ	Orogenia=svecofenniana OROGENIA=PRECAMBRICA	Orogenesi svecofenni-ana OROGENESI PRE-CAMBRIANA

	ENGLISH	FRANCAIS	DEUTSCH	RUSSKIJ	ESPANOL	ITALIANO
4376 SWAMP SURF	SWAMPS	MARAIS	SUMPF	[SWAMP] BOLOTO	CIENAGA	PALUDE
4377 SWAMP SEDIMENTATION SEDI	PALUDAL SEDI- MENTATION	SEDIMENTATION MARECAGE	FEUCHTGEBIETS= SEDIMENTATION	[SWAMP SEDIMEN- TATION] BOLOTO; SEDIMEN- TATSIYA	SEDIMENTACION= PANTANO	SEDIMENTAZIONE PALUSTRE
4378 SWELLING PHCH	SWELLING	GONFLEMENT	QUELLUNG	RAZBUKHANIE	HINCHAMIENTO	RIGONFIAMENTO
4379 SYENITE IGNS	SYENITES *UF:* Fenite	SYENITE *UF:* Fenite; Shonkinite	SYENIT *UF:* Fenit; Shonkinit	SIENIT *UF:* Shchelochnoj sienit	SIENITA *UF:* Fenita; Shonkinita	SIENITE *UF:* Fenite; Shonkinite
4380 SYLVINITE COMS	SYLVINITE	Sylvinite EVAPORITE; POTASSE	Sylvinit EVAPORIT; KALI	SIL'VINIT	Silvinita EVAPORITA; SALES=POTASICAS	Silvinite EVAPORITE; POTASSA
4381 SYMBIOSIS PALE	SYMBIOSIS	SYMBIOSE	SYMBIOSE	SIMBIOZ *UF:* Kommensalizm	SIMBIOSIS	SIMBIOSI
4382 SYMMETRICAL FOLD STRU	SYMMETRIC FOLDS	PLI SYMETRIQUE	SYMMETR.FALTE	SKLADKA NOR- MAL'NAYA	PLIEGUE= SIMETRICO	PIEGA SIM- METRICA
4383 SYMMETRODONTA PALS	SYMMETRODONTA	SYMMETRODONTA	SYMMETRODONTA	Symmetrodonta MAMMALIA	SYMMETRODONTA	SYMMETRODONTA
4384 SYMMETRY GEOL	SYMMETRY *UF:* Diploid minerals	SYMETRIE *UF:* Mineral diploide	SYMMETRIE *UF:* Diploid.Mineral	SIMMETRIYA	SIMETRIA *UF:* Diploide	SIMMETRIA *UF:* Diploide
4385 SYMPHYLA PALS	Symphyla MYRIAPODA	Symphyla MYRIAPODA	Symphyla MYRIAPODA	SYMPHYLA	Symphyla MYRIAPODA	Symphyla MYRIAPODA
4386 SYMPLECTIC TEXTURE TEST	Symplectic texture TEXTURES	Texture symplectique TEXTURE	Symplekt.Gefuege KORN=GEFUEGE	SIMPLEKTITOVYJ	Textura=simplectitica TEXTURA	Tessitura simplectitica TESSITURA
4387 SYMPOSIUM MISC	SYMPOSIA *UF:* Meeting	Symposium REUNION	Symposium TAGUNG	Simpozium SOVESHCHANIE	Simposio REUNION	Simposio RIUNIONE
4388 SYNAPSIDA PALS	SYNAPSIDA	SYNAPSIDA	SYNAPSIDA	SYNAPSIDA	SYNAPSIDA	SYNAPSIDA
4389 SYNAPTOSAURIA PALS	EURYAPSIDA	EURYAPSIDA	EURYAPSIDA	EURYAPSIDA	EURYAPSIDA	EURYAPSIDA
4390 SYNCLINE STRU	SYNCLINES *UF:* Centroclines	SYNCLINAL *UF:* Cuvette synclinale	SYNKLINALE *UF:* Zentroklinale	SINKLINAL' *UF:* Synform fold	SINCLINAL *UF:* Centroclinal	SINCLINALE *UF:* Struttura centrocli- nale
4391 SYNCLINORIUM STRU	SYNCLINORIA	SYNCLINORIUM	SYNKLINORIUM	SINKLINORIJ	SINCLINORIO	SINCLINORIUM
4392 SYNECLISE STRU	SYNECLISE	SYNECLISE	SYNEKLISE	SINEKLIZA	SINCLISIO	SINECLISI
4393 SYNFORM FOLD STRU	SYNFORM FOLDS	PLI SYNFORME	INVERS. MULDENFLUEGEL	[SYNFORM FOLD] SINKLINAL'	PLIEGUE= SINFORME	PIEGA SINFORME
4394 SYNGENESIS GEOL	SYNGENESIS *UF:* Synsedimentary processes; Synsedimen- tary structure	SYNGENESE *UF:* Origine synsedi- mentaire; Structure synsedimentaire	SYNGENESE *UF:* Synsedimentaer. Gefuege; Synsedimentaer. Herkunft	SINGENEZ	SINGENESIS *UF:* Estructura= sinsedimentaria; Origen=sinsedimentario	SINGENESI *UF:* Sinsedimentario; Struttura sinsedimen- taria

	ENGLISH	FRANCAIS	DEUTSCH	RUSSKIJ	ESPANOL	ITALIANO
4395 SYNGENETIC GEOL	Syngenetic(adj.)	Syngenetique(adj.)	Syngenetisch(adj.)	SINGENETI-CHESKIJ *UF:* Sinsedimentatsionnyj; Soft sediment deformation	Singenetico(adj.)	Singenetico(adj.)
4396 SYNONYM PALE	SYNONYMY	SYNONYME	SYNONYMIE	SYNONIMIKA	SINONIMIA	SINONIMIA
4397 SYNSEDIMENTARY ORIGIN SEDI	Synsedimentary processes SYNGENESIS	Origine synsedimentaire SYNGENESE	Synsedimentaer. Herkunft SYNGENESE	Sinsedimentatsionnyj SINGENETI-CHESKIJ	Origen=sinsedimentario SINGENESIS	Sinsedimentario SINGENESI
4398 SYNSEDIMENTARY STRUCTURE TEST	Synsedimentary structure SEDIMENTARY STRUCTURES; SYNGENESIS	Structure synsedimentaire STRUCTURE SEDIMENTAIRE; SYNGENESE	Synsedimentaer. Gefuege SEDIMENT=GEFUEGE; SYNGENESE	STRUKTURA KON-SEDIMENTATSION-NAYA *UF:* Planar bedding structure; Struktura kriptovulkanicheskaya	Estructura=sinsedimentaria ESTRUCTURA=SEDIMENTARIA; SINGENESIS	Struttura sinsedimentaria STRUTTURA SEDIMENTARIA; SINGENESI
4399 SYNTECTONIC PROCESS STRU	SYNTECTONIC PROCESSES	PROCESSUS SYNTECTONIQUE	SYNTEKTON. BILDUNG	SINTEKTONI-CHESKIJ	PROCESO=SINTECTONICO	PROCESSO SINTETTONICO
4400 SYNTEXIS IGNE	Syntexis ANATEXIS	Syntexie ANATEXIE	Syntexis ANATEXIS	SINTEKSIS	Sintexia ANATEXIA	Sintessi ANATESSI
4401 SYNTHESIS MINE	SYNTHESIS	SYNTHESE	SYNTHESE	SINTEZ	SINTESIS	SINTESI
4402 SYNTHETIC MATERIAL ECON	SYNTHETIC MATERIALS	MATERIAU SYNTHETIQUE	SYNTHET. MATERIAL	MINERAL SINTETI-CHESKIJ	MATERIAL=SINTETICO	MATERIALE SINTETICO
4403 SYSTEM GEOL	SYSTEMS(gen.)	Systeme(gen.)	System(gen.)	SISTEMA	Sistema(gen.)	Sistema(gen.)
4404 SYSTEM ANALYSIS MISC	System analysis MATHEMATICAL METHODS	Analyse systeme METHODE MATHEMATIQUE	System=Analyse MATHEMAT. METHODE	ANALIZ SISTEM-NYJ	Analisis=sistema METODO=MATEMATICO	Analisi dei sistemi METODO MATEMATICO
4405 SYSTEMS ANALOG GEOH	SYSTEMS ANALOGS	ANALOGIE SYSTEME HYDROGEOLOGIQUE	SYSTEM=ANALOGIE	[SYSTEMS ANALOG] MATEMATIKA	ANALOGIA=SISTEMA=HIDROGEOLOGICO	ANALOGIA SISTEMA IDROGEOLOGICO
4406 TABULATA PALS	TABULATA	TABULATA	TABULATA	TABULATA	TABULATA	TABULATA
4407 TACONIC OROGENY STRU	TACONIC OROGENY	OROGENIE TACONIQUE	TAKON. OROGENESE	Takonskij orogen KALEDONSKIJ; OROGENEZ	OROGENIA=TACONICA	OROGENESI TACONICA
4408 TACONITE IGMS	Taconite ITABIRITE	Taconite ITABIRITE	Taconit ITABIRIT	Takonit KVARTSIT	Taconita ITABIRITA	Taconite ITABIRITE
4409 TACTITE IGMS	Tactite SKARN	Tactite SKARN	Tactit SKARN	Taktit SKARN	Tactita SKARN	Tactite SKARN
4410 TAENIODONTA PALS	TAENIODONTA	TAENIODONTA	TAENIODONTA	TAENIODONTA	TAENIODONTA	TAENIODONTA

	ENGLISH	FRANCAIS	DEUTSCH	RUSSKIJ	ESPANOL	ITALIANO
4411 TAENIOLABOIDEA PALS	TAENIOLABOIDEA	TAENIOLABOIDEA	TAENIOLABOIDEA	Taeniolaboidea ALLOTHERIA	TAENIOLABOIDEA	TAENIOLABOIDEA
4412 TAIGA SURF	TAIGA ENVIRON- MENT	ZONE TAIGA	TAIGA	TAJGA	MEDIO=TAIGA	TAIGA
4413 TAILING MINI	TAILINGS	HALDES	HALDE	OTKHODY GORNYE	ESTERIL	SCARTO
4414 TALC COMS	TALC *UF:* Talc schist	TALC *UF:* Steatitisation	TALK *UF:* Steatitisierung	TAL'K	TALCO *UF:* Esteatizacion	TALCO SOSTANZA *UF:* Steatitizzazione
4415 TALC SCHIST IGMS	Talc schist TALC; SCHISTS	Talcschiste SCHISTE VERT	Talk=Schiefer GRUEN=SCHIEFER	Slanets tal'kovyj SLANETS ZELENYJ	Talcoesquisto ESQUISTO=VERDE	Talcoscisto SCISTO VERDE
4416 TALUS SURF	Talus TALUS SLOPES	EBOULIS *UF:* Cailloutis; Cuticule deflation; Talus eboulis	SCHUTT *UF:* Schotter; Schutt= Abhang; Wuesten= Pflaster	[TALUS] OSYP'	DERRUBIO *UF:* Guijarro; Pavimento=desertico; Talud=derrubia	DETRITO DI FALDA *UF:* Pavimentazione desertica; Pietrisco; Talus detrito di falda
4417 TALUS SLOPE SURF	TALUS SLOPES *UF:* Talus	Talus eboulis EBOULIS	Schutt=Abhang SCHUTT	Sklon osypeniya SKLON; OSYP'	Talud=derrubia DERRUBIO	Talus detrito di falda DETRITO DI FALDA
4418 TANTALATES MING	TANTALATES	TANTALATE	TANTALAT	Tantalat TANTALO=NIOBAT	TANTALATO	TANTALATO
4419 TANTALUM CHEE	TANTALUM	TANTALE	TA	TANTALUM	TANTALO	TANTALIO
4420 TAPHONOMY PALE	TAPHONOMY	TAPHONOMIE	TAPHONOMIE	TAFONOMIYA *UF:* Vymiranie	TAFONOMIA	TAFONOMIA
4421 TAPHROGENY STRU	TAPHROGENY	TAPHROGENIE	TAPHROGENESE	TAFROGENEZ	TAFROGENESIS	TAFROGENESI
4422 TAPHROGEOSYNCLINE STRU	Taphrogeosynclines GEOSYNCLINES	Taphrogeosynclinal GEOSYNCLINAL; GRABEN	Taphro=Geosynklinale GEOSYNKLINALE; GRABEN	TAFROGEOSINKLI- NAL'	Tafrogeosinclinal GEOSINCLINAL; FOSA=TECTONICA	Tafrogeosinclinale GEOSINCLINALE; GRABEN
4423 TARANNONIAN STRS	TARANNON	TARANNON	TARANNON	Tarannon SILUR NIZHNIJ	TARANONIENSE	TARANNONIANO
4424 TARDIGRADA PALS	Tardigrada ARTHROPODA	Tardigrada ARTHROPODA	Tardigrada ARTHROPODA	TARDIGRADA	Tardigrada ARTHROPODA	Tardigrada ARTHROPODA
4425 TASMANITES PALS	TASMANITES	TASMANITES	TASMANITES	Tasmanites VODOROSLI	TASMANITES	TASMANITES
4426 TATARIAN STRS	TATARIAN	TATARIEN	TATAR	TATARSKIJ	TATARIENSE	TATARIANO
4427 TAXITE IGNE	Taxite TEXTURES; VOL- CANIC ROCKS	Taxite TEXTURE; ROCHE VOLCANIQUE	Taxit KORN=GEFUEGE; VULKANIT	TAKSIT	Taxita TEXTURA; ROCA= VOLCANICA	Taxite TESSITURA; ROC- CIA VULCANICA

	ENGLISH	FRANCAIS	DEUTSCH	RUSSKIJ	ESPANOL	ITALIANO
4428 TAXONOMY PALE	TAXONOMY	TAXINOMIE	TAXONOMIE	TAKSONOMIYA *UF:* Gomonimiya; Takson anaehrobnyj	TAXONOMIA	TASSONOMIA
4429 TECHNETIUM CHEE	TECHNETIUM	TECHNETIUM	TC	TEKHNETSIJ	TECNECIO	TECNEZIO
4430 TECHNICAL COOPERATION MISC	Technical cooperation INTERNATIONAL COOPERATION	Cooperation technique COOPERATION INTERNATIONALE	Techn.Hilfe INTERNATIONAL. ZUSAMMENAR-BEIT	SOTRUDNI-CHESTVO TEKHNI-CHESKOE	Cooperacion=tecnica COOPERACION=INTERNACIONAL	Cooperazione technica COOPERAZIONE INTERNAZIONALE
4431 TECHNIQUE MISC	TECHNIQUES	Technique METHODOLOGIE	Technik METHODIK	TEKHNIKA	Tecnicas METODOLOGIA	Tecnica METODOLOGIA
4432 TECHNOLOGY MISC	TECHNOLOGY *UF:* Hydrogeological technology	TECHNOLOGIE *UF:* Technologie hydrogeologique	TECHNOLOGIE *UF:* Wasser=Technik	TEKHNOLOGIYA	TECNOLOGIA *UF:* Technologia=hidrogeologia	TECNOLOGIA *UF:* Tecnologia idrogeologica
4433 TECTOGENESIS STRU	Tectogenesis TECTONICS	TECTOGENESE *UF:* Diastrophisme; Structure volcanotectonique; Tectonique germanotype; Theorie oscillatoire	TEKTOGENESE *UF:* Diastrophismus; Oszillations=Theorie; Vulkano=tekton. Gefuege	TEKTOGENEZ	TECTOGENESIS *UF:* Diastrofismo; Estructura=vulcanotectonica; Teoria=oscilatoria	TETTOGENESI *UF:* Diastrofismo; Struttura vulcanotettonica; Teoria delle oscillazioni; Tettonica germanotipica
4434 TECTONIC BRECCIA GEOL	TECTONIC BRECCIA *UF:* Cataclasis; Crush breccia; Dislocation breccia; Fault breccia; Fold breccia; Injection breccia; Pseudoconglomerate; Shatter breccia	BRECHE TECTONIQUE *UF:* Breche broyage; Breche dislocation; Breche faille; Breche friction; Breche injection; Breche plissement; Cataclase; Pseudoconglomerat	TEKTON.BREKZIE *UF:* Dislokations=Brekzie; Falten=Brekzie; Injektions=Brekzie; Kataklase; Pseudo=Konglomerat; Reibungs=Brekzie; Schlag=Brekzie; Verwerfungs=Brekzie	BREKCHIYA TEKTONICHESKAYA *UF:* Brekchiya dislokatsionnaya; Brekchiya in'ektsionnaya; Brekchiya razlomnaya; Brekchiya treniya; Brekchiya vnutriskladchataya; Brekchiya drobleniya	BRECHA=TECTONICA *UF:* Brecha=catalcastica; Brecha=de=desmenuzamiento; Brecha=de=dislocacion; Brecha=de=falla; Brecha=de=inyeccion; Brecha=de=pliegue; Cataclasis; Pseudoconglomerado	BRECCIA TETTONICA *UF:* Breccia d'iniezione; Breccia di dislocazione; Breccia di faglia; Breccia di frizione; Breccia di piega; Breccia spigolosa; Cataclasite; Pseudoconglomerato
4435 TECTONIC CONTROL STRU	TECTONIC CONTROLS	CONTROLE TECTONIQUE *UF:* Controle structural; Guide tectonique	TEKTON.EINFLUSS *UF:* Struktur=Genese; Tekton.Indikator	Kontrol' tektonicheskij KONTROL' STRUKTURNYJ	CONTROL=TECTONICO *UF:* Control=estructural; Guia=estructural	CONTROLLO TETTONICO *UF:* Controllo strutturale; Guida strutturale
4436 TECTONIC CYCLE STRU	Tectonic cycle OROGENY	Cycle technique OROGENESE	Tekton.Zyklus OROGENESE	TSIKL TEKTONICHESKIJ	Ciclo=tectonico OROGENESIS	Ciclo tettonico OROGENESI
4437 TECTONIC MAP STRU	TECTONIC MAPS *UF:* Structural diagram	CARTE TECTONIQUE	TEKTON.KARTE	KARTA TEKTONICHESKAYA *UF:* Karta strukturnaya	MAPA=TECTONICO	CARTA TETTONICA
4438 TECTONIC STYLE STRU	Tectonic style TECTONICS	Style tectonique TECTONIQUE	Tekton.Baustil TEKTONIK	Stroenie tektonicheskoe TEKTONIKA; STRUKTURA TEKTONICHESKAYA	Estilo=tectonico TECTONICA	Stile tettonico TETTONICA
4439 TECTONIC SUPERSTRUCTURE STRU	Tectonic superstructure TECTONICS	Superstructure tectonique OROGENESE	Tekton.Ueberstruktur OROGENESE	SUPRASTRUKTURA	Superestructura=tectonica OROGENESIS	Sovrastruttura OROGENESI
4440 TECTONICS STRU	TECTONICS *UF:* Active tectonics; Diastrophism; Geologic structure; Geotectonics; Germanotype tectonics; Oscillation theory; Shortening; Tectogenesis; Tectonic style; Tectonic superstructure; Volcano=tectonic structure	TECTONIQUE *UF:* Analyse structurale; Geotectonique; Structure geologique; Style tectonique; Tectonique active	TEKTONIK *UF:* Aktuo=Tektonik; Geotektonik; Germanotyp.Tektonik; Struktur=Analyse; Tekton.Baustil; Tekton. Struktur	TEKTONIKA *UF:* Geologiya strukturnaya; Stroenie tektonicheskoe; Tektonika szhatiya/tang/	TECTONICA *UF:* Analisis=estructural; Estilo=tectonico; Estructura=geologica; Geotectonica; Tectonica=activa; Tectonica=germanica	TETTONICA *UF:* Analisi strutturale; Geotettonica; Stile tettonico; Struttura tettonica; Tettonica attiva

	ENGLISH	FRANCAIS	DEUTSCH	RUSSKIJ	ESPANOL	ITALIANO
4441 TECTONITE GEOL	TECTONITE	TECTONITE	TEKTONIT	TEKTONIT	TECTONITA	TETTONITE
4442 TECTONOPHYSICS SOLI	TECTONOPHYSICS	TECTONOPHY-SIQUE	TEKTONOPHYSIK	TEKTONOFISIKA	TECTONOFISICA	TETTONOFISICA
4443 TECTONOSPHERE STRU	TECTONOSPHERE	TECTONOSPHERE	TEKTONOSPHAERE	TEKTONOSFERA	TECTONOSFERA	TETTONOSFERA
4444 TEKTITE EXTR	TEKTITES *UF:* Australite; Indochinite; Microtektites; Moldavite	TECTITE *UF:* Australite; Indochinite; Microtectite; Moldavite	TEKTIT *UF:* Australit; Indochinit; Mikrotektit; Moldavit	TEKTIT *UF:* Avstralit; Indoshinit; Mikrotektit; Moldavit	TECTITA *UF:* Australita; Indochinita; Microtectita; Moldavita	TECTITE *UF:* Australite; Indochinite; Microtectite; Moldavite
4445 TELEOSTEI PALS	TELEOSTEI	TELEOSTEI	TELEOSTEI	TELEOSTEI	TELEOSTEI	TELEOSTEI
4446 TELESEISMIC SIGNAL SOLI	TELESEISMIC SIGNALS	SIGNAL TELES-ISMIQUE	TELESEISM.SIGNAL	[TELESEISMIC SIGNAL] ZEMLETRYASENIE	SEÑAL= TELESISMICA	SEGNALE TELES-ISMICO
4447 TELETHERMAL DEPOSIT ECON	TELETHERMAL PROCESSES	GITE TELETHER-MAL	TELETHERMAL. VORKOMMEN	TELETERMAL'NYJ	YACIMIENTO= TELETERMAL	GIACIMENTO TELETERMALE
4448 TELLURATES MING	TELLURATES	TELLURATE	TELLURAT	TELLURAT	TELURATO	TELLURATO
4449 TELLURIC METHOD APPL	Telluric methods EARTH=CURRENT METHODS	Methode tellurique METHODE ELEC-TRIQUE	Tellurik ELEKTR.METHODE	METOD TELLURI-CHESKIJ	Metodo=telurico METODO= ELECTRICO	Metodo tellurico METODO ELET-TRICO
4450 TELLURIUM CHEE	TELLURIUM	TELLURE	TE	TELLUR	TELURO	TELLURIO
4451 TEM DATA METH	TEM DATA	DONNEE MET	TEM=DATEN	[TEM DATA] DANNYE; METOD TRANSMISSII	DATO=MET	DATO MET
4452 TEMPERATE CLIMATE SURF	TEMPERATE ENVI-RONMENT	ZONE TEMPEREE	GEMAESSIGT.ZONE	UMERENNYJ (KLIMAT)	ZONA=TEMPLADA	AMBIENTE TEMP-ERATO
4453 TEMPERATURE PHCH	TEMPERATURE *UF:* Eurythermal taxa; Isotherms; Thermogram	TEMPERATURE *UF:* Isotherme; Organisme eurytherme; Thermocline; Thermogramme	TEMPERATUR *UF:* Eurythermal. Organismus; Isotherme; Thermogramm; Thermokline	TEMPERATURA *UF:* Temperatura nizkaya	TEMPERATURA *UF:* Euritermo; Isoterma; Termoclina; Termograma	TEMPERATURA *UF:* Isoterma; Organismo euritermico; Termoclina; Termografia
4454 TENACITY PHCH	Tenacity COHESION	Tenacite COHESION	Zaehigkeit KOHAESION	[TENACITY] PROCHNOST'	Tenacidad COHESION	Tenacita COESIONE
4455 TENSILE STRENGTH STRU	TENSILE STRENGTH	RESISTANCE TRACTION	ZUG= WIDERSTAND	SOPROTIVLENIE RASTYAZHENIYU	RESISTENCIA= TRACCION	RESISTENZA A TRAZIONE
4456 TENSION STRU	TENSION	TRACTION *UF:* Traction sedimentation	AUSLAENGUNG *UF:* Sediment= Auslaengung	RASTYAZHENIE	TRACCION *UF:* Traccion= sedimentacion	TENSIONE *UF:* Trazione di sedimenti

	ENGLISH	FRANCAIS	DEUTSCH	RUSSKIJ	ESPANOL	ITALIANO
4457 TENTACULITES PALS	TENTACULITES	TENTACULITES	TENTACULITES	TENTACULITES	TENTACULITES	TENTACULITES
4458 TEPHROCHRONOLOGY STRA	TEPHROCHRO- NOLOGY	TEPHROCHRO- NOLOGIE	TEPHROCHRO- NOLOGIE	TEFROKHRO- NOLOGIYA	TEFROCRONO- LOGIA	TEFROCRONO- LOGIA
4459 TERATOLOGY PALE	TERATOLOGY	TERATOLOGIE	TERATOLOGIE	TERATOLOGIYA	TERATOLOGIA	TERATOLOGIA
4460 TERBIUM CHEE	TERBIUM	TERBIUM	TB	TERBIJ	TERBIO	TERBIO
4461 TEREBRATULIDA PALS	TEREBRATULIDA	TEREBRATULIDA	TEREBRATULIDA	TEREBRATULIDA *UF:* Thecideidina	TEREBRATULIDA	TEREBRATULIDA
4462 TERMINOLOGY MISC	Terminology NOMENCLATURE	Terminologie NOMENCLATURE	Terminologie NOMENKLATUR	TERMINOLOGIYA	Terminologia NOMENCLATURA	Terminologia NOMENCLATURA
4463 TERNARY SYSTEM PHCH	Ternary system PHASE EQUILIBRIA	Systeme ternaire DIAGRAMME EQUILIBRE	Ternaer.System PHASEN- GLEICHGEWICHT	SISTEMA TREKH- KOMPONENTNAYA	Sistema=ternario DIAGRAMA= EQUILIBRIO	Sistema ternario DIAGRAMMA D'EQUILIBRIO
4464 TERRA ROSSA SUSS	TERRA ROSSA	Terra rossa SOL MEDITERRAN- EEN	Terra=rossa MEDITERRAN. BODEN	KRASNOZEM	Terra=rossa SUELO- MEDITERRANEO	Terra rossa SUOLO MEDITER- RANEO
4465 TERRACE SURF	TERRACES *UF:* River terraces	TERRASSE *UF:* Terrasse fleuve	TERRASSE *UF:* Fluss=Terrasse	TERRASA	TERRAZA *UF:* Terraza=de=rio	TERRAZZO ALLU- VIONALE *UF:* Terrazzo
4466 TERRAIN CORRECTION APPL	Terrain corrections TOPOGRAPHIC CORRECTION	Correction terrain CORRECTION TOPOGRAPHIQUE	Gelaende=Korrektur TOPOGRAPH. KORREKTUR	Popravka na rel'ef POPRAVKA TOPO- GRAFICHESKAYA	Correccion=tierra CORRECCION= TOPOGRAFICA	Correzione di terreno CORREZIONE TOPOGRAFICA
4467 TERRESTRIAL ENVIRONMENT PALE	TERRESTRIAL ENVIRONMENT	Milieu terrestre MILIEU CONTI- NENTAL	Terrestr.Milieu KONTINENTAL. MILIEU	Nazemnyj KONTINENT	Medio=terrestre MEDIO- CONTINENTAL	Ambiente terrestre AMBIENTE CON- TINENTALE
4468 TERRESTRIAL PLANET EXTR	TERRESTRIAL PLANETS	PLANETE TER- RESTRE	TERRESTR.PLANET	Planeta zemnoj gruppy ZEMLYA; PLANETA	PLANETA= TERRESTRE	PIANETA TER- RESTRE
4469 TERRIGENOUS DEPOSITS SEDI	TERRIGENOUS MATERIALS	MATERIAU TERRI- GENE	TERRIGEN. SUBSTANZ	TERRIGENNYJ	MATERIAL= TERRIGENO	DEPOSITO TERRI- GENO
4470 TERTIARY STRS	TERTIARY	TERTIAIRE	TERTIAER	Tretichnyj KAJNOZOJ	TERCIARIO	TERZIARIO
4471 TEST PALE	TESTS	TEST *UF:* Charniere test	AUSSEN=SKELETT	PANTSIR' *UF:* Karapaks	TEST *UF:* Charnela	GUSCIO *UF:* Cerniera (Pale)
4472 TESTUDINATA PALS	Testudinata CHELONIA	Testudinata CHELONIA	Testudinata CHELONIA	TESTUDINATA	Testudinata CHELONIA	Testudinata CHELONIA
4473 TETHYS STRA	TETHYS	TETHYS	TETHYS	TETIS	TETHYS	TETIDE
4474 TETRACORALLA PALS	RUGOSA *UF:* Columnariina; Cystiphyllina	RUGOSA *UF:* Columnariina; Cystiphyllina	RUGOSA *UF:* Columnariina; Cystiphyllina	RUGOSA *UF:* Columnariina; Cystiphyllina	RUGOSA *UF:* Columnariina; Cystiphyllina	RUGOSA *UF:* Columnariina; Cystiphyllina

	ENGLISH	FRANCAIS	DEUTSCH	RUSSKIJ	ESPANOL	ITALIANO
4475 TETRAPODA PALS	Tetrapoda VERTEBRATA	Tetrapoda VERTEBRATA	Tetrapoda VERTEBRATA	TETRAPODA	Tetrapoda VERTEBRATA	Tetrapoda VERTEBRATA
4476 TEXTULARIINA PALS	TEXTULARIINA	TEXTULARIINA	TEXTULARIINA	TEXTULARIINA	TEXTULARIINA	TEXTULARIINA
4477 TEXTURE TEST	TEXTURES *UF:* Blastic texture; Botryoidal texture; Boxwork texture; Cavernous texture; Coated grain; Colloform textures; Cryptocrystalline texture; Druse; Equigranular texture; Eucrystalline texture; Euhedral crystals; Eutaxitic texture; Flaggy texture; Fluidal texture; Glassy texture; Granoblastic texture; Granophyric texture; Granular texture; Groundmass; Heteroblastic texture; Holocrystalline texture; Hyalocrystalline texture; Hypidioblastic texture; Hypidiomorphic texture; Hypocrystalline texture; Idioblastic texture; Idiomorphic texture; Intersertal texture; Lamprophyric texture; Lattice texture; Lepidoblastic texture; Liesegang rings; Linophyric texture; Margination texture; Miarolitic texture; Microgranitic texture; Microscopic size; Monzonitic texture; Mosaic structure; Mottling; Orthophyric texture; Panidiomorphic texture; Parallel texture; Perlitic texture; Perthitic texture; Poikilitic texture; Poikiloblastic texture; Prismatic structure; Relict texture; Reticulate structure; Symplectic texture; Taxite; Vitroclastic texture; Vitrophyric texture; Xenomorphic crystals; Zonal structure	TEXTURE *UF:* Cristal idiomorphe; Cristal xenomorphe; Forme botryoidale; Grain enduit; Idioblaste; Mesostase; Structure chaotique; Structure reticulee; Tacheture; Taxite; Texture blastique; Texture caverneuse; Texture cloisonnee; Texture colloforme; Texture cryptocristalline; Texture dalle; Texture en mosaique; Texture equigranulaire; Texture eutaxique; Texture filamenteuse; Texture granoblastique; Texture granophyrique; Texture granulaire; Texture heteroblastique; Texture holocristalline; Texture hyaline; Texture hyalocrystalline; Texture hypidioblastique; Texture hypidiomorphique; Texture hypocristalline; Texture idiomorphe; Texture maillee; Texture marge cristaux; Texture miarolitique; Texture microscopique; Texture orthophyrique; Texture panidiomorphe; Texture parallele; Texture poecilitique; Texture poeciloblastique; Texture symplectique; Texture vitroclastique	KORN=GEFUEGE *UF:* Anhedr.Kristall; Blast.Gefuege; Chaot. Gefuege; Coated=grain; Flaser=Gefuege; Flecken=Gefuege; Gitter=Gefuege; Glasig.Gefuege; Glaskopf=Gefuege; Gleichkoernig.Gefuege; Granoblast.Gefuege; Granophyr=Gefuege; Grundmasse; Heteroblast.Gefuege; Holokristallin.Gefuege; Hyalokristallin. Gefuege; Hypidiomorph.Gefuege; Hypokristallin.Gefuege; Idioblast; Idiomorph. Gefuege; Idiomorph. Kristall; Kavernoes. Gefuege; Koernig; Kolloid=Gefuege; Kryptokristallin. Gefuege; Miarolith= Gefuege; Mikroskop. Gefuege; Mosaik= Gefuege; Orthophyr= Gefuege; Panidiomorph.Gefuege; Parallel=Gefuege; Poikilit.Gefuege; Poikiloblast.Gefuege; Rand=Gefuege; Retikular=Gefuege; Symplekt.Gefuege; Taxit; Taxit.Gefuege; Vitroklast.Gefuege; Wolken=Gefuege; Xenomorph.Kristall; Zellen=Gefuege	STRUKTURA *UF:* Fabric; Petrostruktura; Struktura biogennaya; Primary structure; Skladka ehntero=liticheskaya	TEXTURA *UF:* Cristal=idiomorfo; Cristal=xenomorfico; Estructura=caotica; Estructura=lenticular; Estructura=reticulada; Forma=botrioidal; Grano=revestido; Idioblasto; Mesostasis; Roca=moteada; Taxita; Textura=blastica; Textura=cavernosa; Textura=coloforme; Textura=criptocristalina; Textura=en=mosaico; Textura=eutaxitica; Textura=granoblastica; Textura=granofidica; Textura=granular; Textura=heteroblastica; Textura=hialocristalina; Textura=hipidioblastica; Textura=hipidiomorfica; Textura=hipocristalina; Textura=holocristalina; Textura=idiomorfica; Textura=isogranular; Textura=laminar; Textura=marginal; Textura=miarolitica; Textura=microscopica; Textura=ortofidica; Textura=panidiomorfa; Textura=paralela; Textura=poikilitica; Textura=poikiloblastica; Textura=reticulada; Textura=simplectitica; Textura=tabicada; Textura=vitrea; Textura=vitroclastica	TESSITURA *UF:* Botrioidale; Caotico; Criptocristallino; Cristallo euedrale; Cristallo xenomorfo; Granulo rivestito; Idioblasto; Massa di fondo; Scala microscopica; Stratificazione flaser; Struttura a maglie; Struttura reticolare; Taxite; Tessitura a chiazze; Tessitura a mosaico; Tessitura blastica; Tessitura cavernosa; Tessitura colloforme; Tessitura equigranulare; Tessitura eteroblastica; Tessitura eutassitica; Tessitura granoblastica; Tessitura granofirica; Tessitura granulare; Tessitura ialocristallina; Tessitura idiomorfa; Tessitura ipidioblastica; Tessitura ipidiomorfa; Tessitura ipocristallina; Tessitura listata; Tessitura marginale; Tessitura miarolitica; Tessitura olocristallina; Tessitura ortofirica; Tessitura panidiomorfica; Tessitura parallela; Tessitura peciloblastica; Tessitura reticolare; Tessitura simplectitica; Tessitura vetrosa; Tessitura vitroclastica
4478 THALASSOCRATON STRU	Thalassocraton CRATONS	Thalassocraton CRATON	Thalassokraton KRATON	TALASSOKRATON	Thalassocraton CRATON	Talassocratone CRATONE

	ENGLISH	FRANCAIS	DEUTSCH	RUSSKIJ	ESPANOL	ITALIANO
4479 THALASSOID EXTR	Thalassoid MARIA	Bassin lunaire MER LUNAIRE	[THALASSOID] MOND=MARE	TALASSOID	Thalassoide MAR=LUNAR	Thalassoide MARE LUNARE
4480 THALLIUM CHEE	THALLIUM	THALLIUM	TL	TALLIJ	TALIO	TALLIO
4481 THALLOPHYTA PALS	THALLOPHYTES	THALLOPHYTA	THALLOPHYTA	THALLOPHYTA	THALLOPHYTA	TALLOFITE
4482 THALWEG SURF	Thalweg STREAMS	Thalweg PROFIL EN LONG	Gefaells=Kurve LAENGS=PROFIL	Tal'veg DOLINA; PROFIL' RAVNOVESIYA	Vaguada PERFIL= LONGITUDINAL	Fondovalle PROFILO LONGI- TUDINALE
4483 THANATOCENOSE PALE	THANATOCENOSES *UF:* Oryctocoenosis	THANATO- COENOSE *UF:* Oryctocoenose	THANATO- COENOSE *UF:* Oryktocoenose	TANATOTSENOZ	TANATOCENOSIS *UF:* Orictocenosis	TANATOCENOSI *UF:* Oryctocoenosis
4484 THANETIAN STRS	THANETIAN	THANETIEN	THANET	TANET	TANETIENSE	THANETIANO
4485 THAWING ENGI	THAWING	DEGEL	TAUEN	TAYANIE	DESHIELO	DISGELO
4486 THECAMOEBA PALS	THECAMOEBA	THECAMOEBINA	THECAMOEBA	Thecamoeba RHIZOPODEA	THECAMOEBA	THECAMOEBA
4487 THECIDEIDINA PALS	THECIDEIDINA	THECIDEIDINA	THECIDEIDINA	Thecideidina TEREBRATULIDA	THECIDEIDINA	THECIDEIDINA
4488 THECODONTIA PALS	THECODONTIA	THECODONTIA	THECODONTIA	THECODONTIA	THECODONTIA	THECODONTIA
4489 THELODONTI PALS	Thelodonti AGNATHA	Thelodonti AGNATHA	Thelodonti AGNATHA	THELODONTI	Thelodonti AGNATHA	Thelodonti AGNATHA
4490 THEORY MISC	THEORETICAL STUDIES *UF:* Ascension theory; Darwinism; Descension theory; Determinism; Fixism; Hypothesis; Lamarckism; Neptun- ism; Oscillation theory; Parallelism; Transfor- mism	THEORIE *UF:* Darwinisme; Deter- minisme; Fixisme; Geoondulation; Hypo- these; Lamarckisme; Neptunisme; Parallel- isme; Theorie dynamo; Theorie oscillatoire; Transformisme	THEORIE *UF:* Darwinismus; Deter- minismus; Dynamo= Theorie; Fixismus; Geoundation; Hypo- these; Lamarckismus; Neptunismus; Oszillations=Theorie; Parallelitaet; Transfor- mismus	TEORIYA *UF:* Darvinizm; Descen- sion theory	TEORIA *UF:* Darwinismo; Deter- minismo; Fijismo; Geoondulacion; Hipote- sis; Neptunismo; Para- lelismo; Teoria=de=la= dinamo; Teoria=de= lamarck; Teoria= oscilatoria; Transfor- mismo	TEORIA *UF:* Darwinismo; Deter- minismo; Fissismo; Geoundazione; Ipotesi; Lamarckismo; Net- tunismo; Parallelismo; Teoria della dinamo; Teoria delle oscilla- zioni; Trasformismo
4491 THERALITE IGNS	Theralite IGNEOUS ROCKS	Theralite GABBRO; ROCHE FELDS- PATHOIDIQUE	Theralith GABBRO; FOID= GESTEIN	TERALIT	Teralita GABRO; ROCA= FELDESPATOIDICA	Teralite GABBRO; ROCCIA A FELDSPATOIDI
4492 THERAPSIDA PALS	THERAPSIDA	THERAPSIDA	THERAPSIDA	THERAPSIDA	THERAPSIDA	THERAPSIDA
4493 THERMAL MISC	Thermal(adj.)	Thermal(adj.)	Thermal(adj.)	TERMAL'NYJ	Termal(adj.)	Termale(adj.)
4494 THERMAL ANALYSIS METH	THERMAL ANALY- SIS	ANALYSE THER- MIQUE	THERMO= ANALYSE	ANALIZ TERMI- CHESKIJ	ANALISIS= TERMICO	ANALISI TERMICA

	ENGLISH	FRANCAIS	DEUTSCH	RUSSKIJ	ESPANOL	ITALIANO
4495 THERMAL CONDUCTIVITY PHCH	THERMAL CON-DUCTIVITY *UF:* Heat conduction	CONDUCTIVITE THERMIQUE *UF:* Conduction chaleur	WAERME=LEITFAEHIGKEIT *UF:* Waerme=Leitung	KOEHFFITSIENT TEPLOPROVOD-NOST'	CONDUCTIVIDAD=TERMICA *UF:* Conduccion=calor	CONDUTTIVITA TERMICA *UF:* Conducibilita di calore
4496 THERMAL DEMAGNETIZATION METH	THERMAL DEMAG-NETIZATION	DESAIMANTATION THERMIQUE	THERM. ENTMAGNETI-SIERUNG	Razmagnichivanie ter-micheskoe NAMAGNICHEN-NOST' OSTATOCH-NAYA; TERMO-GRAMMA	DESIMANTACION=TERMICA	SMAGNETIZZAZ-IONE TERMICA
4497 THERMAL EFFECT GEOL	THERMAL EFFECTS	ACTION CHALEUR *UF:* Fissure contraction	WAERME=WIRKUNG *UF:* Hitze=Sprung	EHFFEKT TERMI-CHESKIJ	ACCION=CALOR *UF:* Rotura=termica	AZIONE TERMICA *UF:* Fessura di dissecca-mento
4498 THERMAL METAMORPHISM IGNE	THERMAL META-MORPHISM *UF:* Geothermal meta-morphism	METAMORPHISME THERMIQUE	THERMO=METAMORPHOSE	METAMORFIZM TERMAL'NYJ	METAMORFISMO=TERMICO	METAMORFISMO TERMICO
4499 THERMAL PROPERTY PHCH	THERMAL PROPER-TIES	PROPRIETE THER-MIQUE	THERM. EIGENSCHAFT	SVOJSTVO TERMI-CHESKOE	PROPIEDAD=TERMICA	PROPRIETA TER-MICA
4500 THERMAL PROSPECTING APPL	Thermal prospecting HEAT FLOW	Prospection thermique GEOTHERMIE	Therm.Exploration GEOTHERM. METHODE	TERMORAZVEDKA	Prospeccion=termica GEOTERMIA	Prospezione termica GEOTERMIA
4501 THERMAL WATER GEOH	THERMAL WATERS *UF:* Hydrothermal waters	EAU THERMALE	THERMAL=WASSER *UF:* Hydrothermal=Wasser	VODA TER-MAL'NAYA	AGUA=TERMAL	ACQUA TERMALE
4502 THERMOCHEMICAL PROPERTY GEOC	THERMOCHEMI-CAL PROPERTIES	PROPRIETE THER-MOCHIMIQUE	THERMOCHEM. EIGENSCHAFT	SVOJSTVO TER-MOKHIMICHESKOE	PROPIEDAD=TERMOQUIMICA	PROPRIETA TER-MOCHIMICA
4503 THERMOCLINE MARI	THERMOCLINES	Thermocline TEMPERATURE; EAU MER	Thermokline TEMPERATUR; MEER=WASSER	TERMOKLIN	Termoclina TEMPERATURA; AGUA=DE=MAR	Termoclina TEMPERATURA; ACQUA DI MARE
4504 THERMODYNAMIC PROPERTY PHCH	THERMODYNAMIC PROPERTIES *UF:* Adiabatic processes; Degree of freedom; Exothermic reactions; Heat; Heat equivalent of fusion; Latent heat of fusion; Photochemi-cal reaction; Physical state; Thermodynamics	Propriete thermody-namique THERMODY-NAMIQUE	Thermodynam. Eigenschaft THERMODYNAMIK	SVOJSTVO TER-MODINAMI-CHESKOE	Propiedad=termodinamica TERMODINAMICA	Proprieta termodi-namica TERMODINAMICA
4505 THERMODYNAMICS PHCH	Thermodynamics THERMODYNAMIC PROPERTIES	THERMODY-NAMIQUE *UF:* Chaleur; Chaleur equivalente fusion; Degre liberte; Etat physique; Processus adiabatique; Propriete thermodynamique; Reaction; Reaction exothermique; Reaction photochimique; Serie reaction	THERMODYNAMIK *UF:* Adiabat.Vorgang; Exotherm.Reaktion; Freiheitsgrad; Latent. Schmelz=Waerme; Photochem.Reaktion; Physikal.Status; Reak-tion; Reaktions=Serie; Thermodynam. Eigenschaft; Waerme	TERMODINAMIKA *UF:* Reaktsiya ehkzoter-micheskaya; Reaktsiya ehndotermicheskaya	TERMODINAMICA *UF:* Calor; Calor=latente=de=fusion; Estado=fisico; Grado=de=libertad; Proceso=adiabatico; Propiedad=termodinamica; Reac-cion; Reaccion=exotermica; Reaccion=fotoquimica; Series=bowen	TERMODINAMICA *UF:* Calore; Calore latente di fusione; Condizione adiabatica; Grado di liberta; Pro-prieta termodinamica; Reazione; Reazione esotermica; Reazione fotochimica; Serie di reazione; Stato fisico

	ENGLISH	FRANCAIS	DEUTSCH	RUSSKIJ	ESPANOL	ITALIANO
4506 THERMOGRAM METH	Thermogram TEMPERATURE	Thermogramme TEMPERATURE	Thermogramm TEMPERATUR	TERMOGRAMMA *UF:* Razmagnichivanie termicheskoe	Termograma TEMPERATURA	Termografia TEMPERATURA
4507 THERMOGRAVIMETRY METH	THERMOGRAVI- METRIC ANALYSIS	ANALYSE THER- MOGRAVI- METRIQUE	THERMO- GRAVIMETR. ANALYSE	ANALIZ TERMOVE- SOVOJ	TERMO- GRAVIMETRICO	TERMOGRAVI- METRIA
4508 THERMOKARST SURF	THERMOKARST	THERMOKARST	THERMO-KARST	TERMOKARST	TERMOKARST	TERMOCARSISMO
4509 THERMOLUMINES- CENCE PHCH	THERMOLUMINES- CENCE	THERMOLUMINES- CENCE	THERMOLUMINES- ZENZ	TERMOLYUMIN- ESTSENTSIYA	TERMOLUMINIS- CENCIA	TERMOLUMINES- CENZA
4510 THERMOMAGNETIC ANALYSIS METH	THERMOMAGNE- TIC ANALYSIS	ANALYSE THER- MOMAGNETIQUE	THERMOMAGNET. ANALYSE	ANALIZ TERMO- MAGNITNYJ	ANALISIS= TERMOMAGNETICO	ANALISI TERMO- MAGNETICA
4511 THERMOREMANENT MAGNETIZATION PHCH	THERMOREMA- NENT MAGNE- TIZATION	AIMANTATION THERMOREMA- NENTE	THERMOREMANENT. MAGNETISIERUNG	NAMAGNICHEN- NOST' TER- MOOSTATOCH- NAYA	IMANTACION= TERMORREMANENTE	MAGNETIZZAZ- IONE TERMICA RESIDUA
4512 THESIS MISC	THESES	THESE	PRUEFUNGS= ARBEIT	TEZISY	TESIS	TESI
4513 THICKNESS MISC	THICKNESS	EPAISSEUR	DICKE	MOSHCHNOST'	POTENCIA	SPESSORE
4514 THIN SECTION METH	THIN SECTIONS	LAME MINCE	DUENNSCHLIFF	SHLIF	LAMINA= DELGADA	SEZIONE SOTTILE
4515 THIXOTROPY PHCH	THIXOTROPY	THIXOTROPIE	THIXOTROPIE	TIKSOTROPIYA *UF:* Razzhizhenie	TIXOTROPIA	TISSOTROPIA
4516 THOLEIITIC COMPOSITION IGNE	THOLEIITIC COM- POSITION	COMPOSITION THOLEITIQUE *UF:* Trapp Deccan	THOLEIIT. CHEMISMUS *UF:* Dekkan=Trapp	TOLEITOVYJ	COMPOSICION= TOLEITICA *UF:* Basalto=de=Deccan	COMPOSIZIONE THOLEITICA *UF:* Basalto del Deccan
4517 THORIUM CHEE	THORIUM	THORIUM	TH	TORIJ	TORIO	TORIO
4518 THORIUM-LEAD DATING ISOT	Th/Pb U/PB	Th=Pb U=PB	Th=Pb=Datierung U=PB=DATIERUNG	Metod torievo= svintsovyj METOD SVINT- SOVOURANOVYJ	Th=Pb U=PB	Th/Pb U/PB
4519 THRESHOLD GEOC	Geochemical threshold GEOCHEMICAL METHODS	Seuil anormal PROSPECTION GEOCHIMIQUE	Schwellenwert GEOCHEM. PROSPEKTION	PREDEL CHUVST- VITEL'NOSTI	Umbral=anomalia PROSPECCION= GEOQUIMICA	Soglia geochimica PROSPEZIONE GEOCHIMICA
4520 THRUST FAULT STRU	THRUST FAULTS	FAILLE CHEVAU- CHEMENT *UF:* Charriage; Faille horizontale	AUFSCHIEBUNG *UF:* Horizontal= Verwerfung; Uebers- chiebung	NADVIG	CABALGAMIENTO *UF:* Corrimiento; Falla= desplazamiento= horizontal	ACCAVALLA- MENTO *UF:* Faglia a piano orizzontale; Sovrascor- rimento
4521 THULIUM CHEE	THULIUM	THULIUM	TM	TULIJ	TULIO	TULIO

	ENGLISH	FRANCAIS	DEUTSCH	RUSSKIJ	ESPANOL	ITALIANO
4522 THURINGIAN STRS	THURINGIAN	THURINGIEN	THURINGIUM	TYURINGIJ	TURINGIENSE	TURINGIANO
4523 THYSANOPTERA PALS	THYSANOP- TEROIDA	THYSANOP- TEROIDA	THYSANOP- TEROIDA	THYSANOPTERA	THYSANOP- TEROIDA	THYSANOP- TEROIDA
4524 THYSANURA PALS	Thysanura ECTOTROPHA	Thysanura ECTOTROPHA	Thysanura ECTOTROPHA	THYSANURA	Thysanura ECTOTROPHA	Thysanura ECTOTROPHA
4525 TIDAL CHANNEL SURF	TIDAL CHANNELS	CHENAL MAREE	PRIEL	[TIDAL CHANNEL] PRILIVNYJ	CANAL−DE− MAREA	CANALE DI MAREA
4526 TIDAL ENVIRONMENT PALE	Tidal environment INTERTIDAL ENVI- RONMENT	Milieu tidal MILIEU LITTORAL	Tidal−Milieu GEZEITEN−MILIEU	PRILIVNYJ UF: Tidal channel; Tidal inlet	Medio−mareal MEDIO−LITORAL	Ambiente tidale AMBIENTE LITOR- ALE
4527 TIDAL FLAT SURF	TIDAL FLATS	SLIKKE	SCHLICK	OSUSHKA	ESTERO	PIANA TIDALE
4528 TIDAL INLET SURF	TIDAL INLETS	PASSE INTERTID- ALE	SEEGAT	[TIDAL INLET] PRILIVNYJ	PASO−ENTRE− MAREAS	PASSAGGIO INTER- TIDALE
4529 TIDE MARI	TIDES	MAREE	GEZEITEN	PRILIV	MAREA	MAREA
4530 TILL SURF	TILL	TILL	GESCHIEBE−LEHM	SUGLINOK MORENNYJ	TILL	TILL
4531 TILLITE SEDS	TILLITE	TILLITE	TILLIT	TILLIT	TILLITA	TILLITE
4532 TILLODONTIA PALS	TILLODONTIA	TILLODONTIA	TILLODONTIA	TILLODONTIA	TILLODONTIA	TILLODONTIA
4533 TILT SOLI	TILT	INCLINAISON	NEIGUNG	NAKLON	INCLINACION	INCLINAZIONE
4534 TILTMETER INST	TILTMETERS	CLINOMETRE UF: Pendagemetre	KLINOMETER UF: Neigungs−Messer	NAKLONOMER	CLINOMETRO UF: Pendientometro	INCLINOMETRO UF: Dip meter
4535 TIME MISC	TIME FACTOR	FACTEUR TEMPS UF: Annee galactique	ZEITFAKTOR UF: Galakt.Jahr	VREMYA	FACTOR=TIEMPO UF: Año-galactico	FATTORE TEMPO UF: Anno galattico
4536 TIME SCALE STRA	TIME SCALES	ECHELLE ABSOLUE	ABSOLUT. ALTERSSKALA	SHKALA VREMEN- NAYA	ESCALA= ABSOLUTA	SCALA ETA ASSO- LUTE
4537 TIN CHEE	TIN	ETAIN	SN	OLOVO	ESTAÑO	STAGNO
4538 TINTINNIDAE PALS	TINTINNIDAE	TINTINNIDAE	TINTINNIDAE	TINTINNIDAE	TINTINNIDAE	TINTINNIDAE
4539 TITANIUM CHEE	TITANIUM	TITANE	TI	TITAN	TITANIO	TITANIO
4540 TITHONIAN STRS	TITHONIAN	TITHONIQUE	TITHON	TITON	TITONICO	TITONIANO

	ENGLISH	FRANCAIS	DEUTSCH	RUSSKIJ	ESPANOL	ITALIANO
4541 TOARCIAN STRS	TOARCIAN	TOARCIEN	TOARCIUM	TOAR *UF:* Lejas verkhnij	TOARCIENSE	TOARCIANO
4542 TOMBOLO SURF	Tombolos SHORE FEATURES	Tombolo CORDON LITTO- RAL	Tombolo STRAND=WALL	TOMBOLO	Tombolo CORDON−LITORAL	Tombolo CORDONE LITOR- ALE
4543 TONSTEIN SEDS	TONSTEIN	TONSTEIN	TONSTEIN	TONSHTEJN	TONSTEIN	TONSTEIN
4544 TOOTH PALE	TEETH *UF:* Dentition; Odontol- ogy	DENT *UF:* Denture; Odontolo- gie	ZAHN *UF:* Dentition; Odontolo- gie	ZUBY	DIENTE *UF:* Denticion; Odon- tologia	DENTE *UF:* Dentizione; Odon- tologia
4545 TOPOGRAPHIC **CORRECTION** APPL	TOPOGRAPHIC CORRECTION *UF:* Terrain corrections	CORRECTION TOPOGRAPHIQUE *UF:* Correction terrain	TOPOGRAPH. KORREKTUR *UF:* Gelaende= Korrektur	POPRAVKA TOPO- GRAFICHESKAYA *UF:* Popravka na rel'ef	CORRECCION− TOPOGRAFICA *UF:* Correccion=tierra	CORREZIONE TOPOGRAFICA *UF:* Correzione di ter- reno
4546 TOPOGRAPHIC MAP MISC	TOPOGRAPHIC MAPS *UF:* Isohypse	CARTE TOPO- GRAPHIQUE *UF:* Isohypse	TOPOGRAPH. KARTE *UF:* Isohypse	KARTA TOPOGRA- FICHESKAYA *UF:* Karta konturnaya	MAPA= TOPOGRAFICO *UF:* Curva=de=nivel	CARTA TOPOGRA- FICA *UF:* Isoipsa
4547 TOPOGRAPHY SURF	TOPOGRAPHY	TOPOGRAPHIE	TOPOGRAPHIE	TOPOGRAFIYA *UF:* Gipsometriya; Izogipsy; Morfometriya	TOPOGRAFIA	TOPOGRAFIA
4548 TOPSET BED SEDI	TOPSET BEDS	Couche sommitale STRATIFICATION ENTRECROISEE; SEDIMENTATION DELTAIQUE	[TOPSET=BED] KREUZSCHICH- TUNG; DELTA= SEDIMENTATION	Sloj golovnoj SLOISTOST'	Capa=de=techo ESTRATIFIC- ACION=CRUZADA; SEDIMENTACION= DELTAICA	Strato di tetto STRATIFICAZIONE INCROCIATA; SEDI- MENTAZIONE DEL- TIZIA
4549 TORRENT SURF	Torrents RIVERS	Torrent COURS EAU	Wildbach WASSER=LAUF	POTOK STREMI- TEL'NYJ	Torrente CURSO=AGUA	Torrente CORSO D'ACQUA
4550 TORSION PHCH	TORSION	TORSION	TORSION	KRUCHENIE	TORSION	TORSIONE
4551 TORTONIAN STRS	TORTONIAN	TORTONIEN	TORTON	TORTON	TORTONIENSE	TORTONIANO
4552 TORTUOSITY MATH	TORTUOSITY	TORTUOSITE	KRUEMMUNG	IZVILISTOST'	TORTUOSIDAD	TORTUOSITA
4553 TOURNAISIAN STRS	TOURNAISIAN	TOURNAISIEN	TOURNAI	TURNE	TOURNAISIENSE	TOURNAISIANO
4554 TOXIC MATERIAL ENVI	TOXIC MATERIALS	SUBSTANCE TOX- IQUE	GIFT	Yad PESTITSID	SUSTANCIA= TOXICA	SOSTANZA TOS- SICA
4555 TOXICITY ENVI	TOXICITY	TOXICITE	TOXIZITAET	TOKSICHNOST'	TOXICIDAD	TOSSICITA
4556 TRACE ELEMENT GEOC	TRACE ELEMENTS	ANALYSE ELE- MENT TRACE *UF:* Metal trace	SPURENELEMENT- ANALYSE *UF:* Spuren=Metall	EHLEMENT RASSEYANNYJ *UF:* Analiz rasseyannykh ehlementov; Metall rasseyannyj	ANALISIS= ELEMENTOS= TRAZA *UF:* Metal=traza	ELEMENTO IN TRACCE *UF:* Metallo in tracce

	ENGLISH	FRANCAIS	DEUTSCH	RUSSKIJ	ESPANOL	ITALIANO
4557 TRACE FOSSIL PALE	LEBENSSPUREN *UF:* Vermiform structure	TRACE ORGANIQUE *UF:* Ichnologie; Organisme perforant; Scolite; Structure vermiforme	LEBENS=SPUR *UF:* Bohr=Organismen; Ichnologie; Skolithus; Wurmfoermig.Gefuege	SLED ZHIZ-NEDEYATEL'NOSTI *UF:* Ikhnofossilii	HUELLA=ORGANICA *UF:* Escolito; Estructura=vermiforme; Icnologia; Litofago	TRACCIA FOSSILE *UF:* Icnologia
4558 TRACE METAL GEOC	TRACE METALS	Metal trace ANALYSE ELEMENT TRACE	Spuren=Metall SPURENELEMENT=ANALYSE	Metall rasseyannyj EHLEMENT RASSEYANNYJ	Metal=traza ANALISIS=ELEMENTOS=TRAZA	Metallo in tracce ELEMENTO IN TRACCE
4559 TRACE=ELEMENT ANALYSIS GEOC	TRACE=ELEMENT ANALYSES	METHODE ANALYSE ELEMENT TRACE	SPURENELEMENT=ANALYSEN-METHODE	Analiz rasseyannykh ehlementov ANALIZ; EHLEMEN RASSEYANNYJ	METODO=ANALISIS=ELEMENTOS=TRAZA	ANALISI ELEMENTI IN TRACCE
4560 TRACER METH	TRACERS	TRACEUR	MARKIERUNG	TRASSER	TRAZADOR	TRACCIANTE
4561 TRACHYANDESITE IGNS	TRACHYANDESITE *UF:* Latite	TRACHYANDESITE *UF:* Latite	TRACHYANDESIT *UF:* Latit	TRAKHIANDEZIT	TRAQUIANDESITA *UF:* Latita	TRACHIANDESITE *UF:* Latite
4562 TRACHYBASALT IGNS	TRACHYBASALT	TRACHYBASALTE	TRACHYBASALT	TRAKHIBAZAL'T	TRAQUIBASALTO	TRACHIBASALTO
4563 TRACHYPSAMMIACEA PALS	Trachypsammiacea OCTOCORALLIA	Trachypsammiacea OCTOCORALLA	Trachypsammiacea OCTOCORALLIA	Trachypsammiacea OCTOCORALLIA	Trachypsammiacea OCTOCORALLA	Trachypsammiacea OCTOCORALLIA
4564 TRACHYTE IGNS	TRACHYTES	TRACHYTE	TRACHYT	TRAKHIT	TRAQUITA	TRACHITE
4565 TRACK–TRAIL PALE	TRAILS	PISTE	BEWEGUNGS=SPUR	SLED DVIZHENIYA	PISTA	PISTA
4566 TRADE ECON	TRADE	MARCHE	MARKT	TORGOVLYA	MERCADO	COMMERCIO
4567 TRANSCURRENT FAULT STRU	TRANSCURRENT FAULTS	FAILLE TRANS-COURANTE	QUER=VERSCHIEBUNG	Sbroso=sdvig regional'nyj RAZLOM GLUBINNYJ	FALLA=TRANSCURRENTE	FAGLIA TRASCOR-RENTE
4568 TRANSFER METH	TRANSFER	TRANSFERT	UEBERTRAGUNG	MASSOPERENOS	REPLICA	TRASFERIMENTO
4569 TRANSFORM FAULT STRU	TRANSFORM FAULTS	FAILLE TRANS-FORMANTE	TRANSFORM=STOERUNG	RAZLOM TRANS-FORMNYJ	FALLA=TRANSFORMANTE	FAGLIA TRAS-FORME
4570 TRANSFORMATION GEOL	TRANSFORMA-TIONS *UF:* Irreversibility	TRANSFORMATION *UF:* Irreversibilite	TRANSFORMATION *UF:* Irreversibilitaet	TRANSFORMAT-SIYA	TRANSFORMA-CION *UF:* Irreversibilidad	TRASFORMAZIONE *UF:* Irreversibilita
4571 TRANSFORMISM GEOL	Transformism THEORETICAL STUDIES; GRANIT-IZATION	Transformisme THEORIE; GRANI-TISATION	Transformismus THEORIE; GRANI-TISIERUNG	TRANSFORMIZM	Transformismo TEORIA; GRANIT-IZACION	Trasformismo TEORIA; GRANI-TIZZAZIONE
4572 TRANSGRESSION STRA	TRANSGRESSION *UF:* Ingression	TRANSGRESSION *UF:* Ingression	TRANSGRESSION *UF:* Ingression	TRANSGRESSIYA *UF:* Zatoplenie	TRANSGRESION *UF:* Ingresion	TRASGRESSIONE *UF:* Ingressione

	ENGLISH	FRANCAIS	DEUTSCH	RUSSKIJ	ESPANOL	ITALIANO
4573 TRANSIENT METHODS APPL	TRANSIENT METH-ODS	METHODE TRANSI-TOIRE	[TRANSIENT=METHOD]	METOD PEREK-HODNYKH PROTSESSOV	METODO=TRANSITORIO	METODO DEI TRANSIENTI
4574 TRANSMISSIBILITY COEFFICIENT GEOH	Transmissibility coefficient TRANSMISSIVITY	Coefficient transmissi-vite TRANSMISSIVITE	Transmiss.Koeffizient TRANSMISSIVIT-AET	[TRANSMISSIBIL-ITY COEFFICIENT] VODOPRONITSAE-MOST'; KOEHFFIT-SIENT	Coeficiente=transmisibilidad TRANSMISIVIDAD	Coefficiente di trasmis-sibilita TRASMISSIVITA
4575 TRANSMISSION METHOD METH	TRANSMISSION METHOD	METHODE MET	TRANSMISS. ELEKTR. MIKROSKOPIE	METOD TRANSMIS-SII UF: Tem data	METODO=MET	METODO MET
4576 TRANSMISSIVITY GEOH	TRANSMISSIVITY	TRANSMISSIVITE UF: Coefficient transmis-sivite	TRANSMISSIVIT-AET UF: Transmiss. Koeffizient	KOEHFFITSIENT FIL'TRATSII	TRANSMISIVIDAD UF: Coeficiente=transmisibilidad	TRASMISSIVITA UF: Coefficiente di trasmissibilita
4577 TRANSPORT SEDI	TRANSPORT UF: River load; Silt load	TRANSPORT UF: Charge riviere; Charge silt	TRANSPORT UF: Fluviatil.Fracht; Silt=Fracht	TRANSPORTI-ROVKA(GEOL) UF: Glacial transport; Ice rafting; Marine transport; Mass trans-fer; Ore transport; Volochenie; Wind trans-port	TRANSPORTE UF: Transporte=limo=suspendido; Transporte=suspension=fluvial	TRASPORTO UF: Carico fluviale; Carico solido
4578 TRANSPORTATION ECON	TRANSPORTATION	MOYEN TRANS-PORT	TRANSPORT=MITTEL	TRANSPOR-TIROVKA	MEDIO=DE=TRANSPORTE	MEZZO DI TRA-SPORTO
4579 TRANSVERSAL MISC	Transversal(adj.)	Transversal(adj.)	Transversal(adj.)	POPERECHNYJ UF: Oblique orientation	Transversal(adj.)	Trasversale(adj.)
4580 TRAP ECON	TRAPS UF: Gas cap; Lithologic traps; Oil trap; Trans-missibility coefficient	PIEGE UF: Accumulation gaz; Piege a petrole; Piege lithologique; Piege sedimentaire	FALLE UF: Erdoel=Falle; Gaskappe; Litholog. Falle; Sedimentaer. Falle	LOVUSHKA	TRAMPA UF: Acumulacion=gas; Trampa=petrolifera; Trampa=sedimentaria	TRAPPOLA UF: Cappello di gas; Trappola litologica; Trappola petrolifera; Trappola sedimentaria
4581 TRAVELTIME SOLI	TRAVELTIME	TEMPS PARCOURS	LAUFZEIT	VREMYA PROBEGA	TIEMPO=RECORRIDO	TEMPO DI PERCOR-RENZA
4582 TRAVELTIME CURVE SOLI	TRAVELTIME CURVES UF: Hodographs	COURBE TEMPS PARCOURS UF: Dromochronique	LAUFZEIT=KURVE UF: Hodograph	VREMYA PROBEGA	CURVA=TIEMPO=RECORRIDO UF: Hodografo	DROMOCRONA UF: Odografo
4583 TRAVERTINE SEDS	TRAVERTINE	TRAVERTIN	TRAVERTIN	TRAVERTIN	TRAVERTINO	TRAVERTINO
4584 TREATISE MISC	TEXTBOOKS	TRAITE	LEHRBUCH	[TREATISE] MONOGRAFIYA	TRATADO	TRATTATO
4585 TREE RING STRA	TREE RINGS UF: Dendrochronology	Anneau croissance DEN-DROCHRONOLO-GIE	Jahresring DENDRO=CHRONOLOGIE	KOL'TSO GODICH-NOE	Anillo=crecimiento DEN-DROCRONOLOGIA	Anello di accrescimento DEN-DROCRONOLOGIA
4586 TREMADOCIAN STRS	TREMADOCIAN	TREMADOC	TREMADOC	TREMADOK	TREMADOCIENSE	TREMADOCIANO

	ENGLISH	FRANCAIS	DEUTSCH	RUSSKIJ	ESPANOL	ITALIANO
4587 TRENCH MARI	TRENCHES	FOSSE ABYSSALE	TIEFSEE=GRABEN	ZHELOB GLUBOKO- VODNYJ	FOSA=ABISAL	FOSSA ABISSALE
4588 TREND SURFACE ANALYSIS MATH	TREND=SURFACE ANALYSIS	ANALYSE TEN- DANCE *UF:* Tendance	TREND=ANALYSE *UF:* Trend	TREND=ANALIZ	ANALISIS= TENDENCIA *UF:* Tendencia	ANALISI DELLE TENDENZE *UF:* Tendenza statistica
4589 TREPOSTOMATA PALS	TREPOSTOMATA	TREPOSTOMATA	TREPOSTOMATA	TREPOSTOMATA	TREPOSTOMATA	TREPOSTOMATA
4590 TRIANGULATION METH	TRIANGULATION	TRIANGULATION	TRIANGULATION	TRIANGULYAT- SIYA	TRIANGULACION	TRIANGOLAZIONE
4591 TRIASSIC STRS	TRIASSIC *UF:* Alpine triassic	TRIAS *UF:* Trias alpin	TRIAS *UF:* Alpin.Trias	TRIAS *UF:* Trias al'pijskij	TRIAS *UF:* Trias=alpino	TRIASSICO *UF:* Triassico alpino
4592 TRIAXIAL COMPRESSION TEST ENGI	TRIAXIAL TESTS	COMPRESSION TRIAXIALE	TRIAXIAL= KOMPRESSION	ISPYTANIE NA TREKHOSNOE SZHATIE	COMPRESION= TRIAXIAL	COMPRESSIONE TRIASSIALE
4593 TRICLINIC SYSTEM MINE	Triclinic system CRYSTAL SYSTEMS	Systeme triclinique SYSTEME CRISTAL- LIN	Triklin.System KRISTALL- SYSTEM	SINGONIYA TRIK- LINNAYA	Sistema=triclinico SISTEMA= CRISTALINO	Sistema triclino SISTEMA CRISTAL- LINO
4594 TRICONODONTA PALS	TRICONODONTA	TRICONODONTA	TRICONODONTA	Triconodonta MAMMALIA	TRICONODONTA	TRICONODONTA
4595 TRIGONAL SYSTEM MINE	Trigonal system CRYSTAL SYSTEMS	Systeme trigonal SYSTEME CRISTAL- LIN	Trigonal.System KRISTALL- SYSTEM	SINGONIYA TRIGO- NAL'NAYA	Sistema=trigonal SISTEMA= CRISTALINO	Sistema trigonale SISTEMA CRISTAL- LINO
4596 TRIGONIOIDA PALS	Trigoniidae BIVALVIA	Trigoniidae BIVALVIA	Trigoniidae LAMELLI- BRANCHIATA	TRIGONIOIDA	Trigoniidae BIVALVIA	Trigoniidae BIVALVIA
4597 TRILOBITA PALS	TRILOBITA	TRILOBITA	TRILOBITA	TRILOBITA	TRILOBITA	TRILOBITA
4598 TRILOBITOMORPHA PALS	TRILOBITOMOR- PHA	TRILOBITOMOR- PHA	TRILOBITOMOR- PHA	TRILOBITOMOR- PHA	TRILOBITOMOR- PHA	TRILOBITOMOR- PHA
4599 TRIPLE JUNCTION SOLI	TRIPLE JUNCTIONS	JONCTION TRIPLE LITHOSPHERE	DREI=PLATTEN= PUNKT	PERESECHENIE TROJNOE	UNION=TRIPLE= LITOSFERA	GIUNZIONE TRI- PLA
4600 TRIPOLI COMS	Tripoli ABRASIVES	Tripoli ABRASIF	Tripoli POLIERSTOFF	Tripoli ZEMLYA DIATO- MOVAYA	Tripoli ABRASIVO	Tripoli DIATOMITE; ABRA- SIVO
4601 TRITIUM DATING ISOT	TRITIUM	TRITIUM	TRITIUM= DATIERUNG	METOD TRITIEVYJ	TRITIO	DATAZIONE TRIZIO
4602 TROPICAL ZONE SURF	TROPICAL ENVI- RONMENT	ZONE TROPICALE *UF:* Ustalf; Ustult	TROPEN=ZONE *UF:* Ustalf; Ustult	TROPIKI	ZONA=TROPICAL *UF:* Ustalf; Ustult	AMBIENTE TROPI- CALE *UF:* Ustalf; Ustult
4603 TROUGH STRU	TROUGHS	FOSSE ABYSSALE ALLONGEE	MULDEN=ACHSE	TROG	FOSA=ABISAL= ALARGADA	FOSSA ABISSALE ALLUNGATA
4604 TSUNAMI MARI	TSUNAMIS	TSUNAMI	TSUNAMI	TSUNAMI	TSUNAMI	TSUNAMI

	ENGLISH	FRANCAIS	DEUTSCH	RUSSKIJ	ESPANOL	ITALIANO
4605 TUBOIDEA PALS	TUBOIDEA	TUBOIDEA	TUBOIDEA	TUBOIDEA	TUBOIDEA	TUBOIDEA
4606 TUBULIDENTATA PALS	TUBULIDENTATA	TUBULIDENTATA	TUBULIDENTATA	TUBULIDENTATA	TUBULIDENTATA	TUBULIDENTATA
4607 TUFF IGNS	TUFF	TUF VOLCANIQUE	VULKAN.TUFF	TUF	TOBA	TUFO VULCANICO
4608 TUFFITE IGNS	TUFFITE	TUFFITE	TUFFIT	TUFFIT	TUFITA	TUFITE
4609 TUFFLAVA IGNE	Tuff lava LAVA	Tufo=lave LAVE	Tuffo=Lava LAVA	TUFOLAVA	Toba=volcanica LAVA	Agglomerato di lava LAVA
4610 TUNDRA SURF	TUNDRA	TOUNDRA	TUNDRA	TUNDRA	TUNDRA	TUNDRA
4611 TUNDRA SOIL SUSS	TUNDRA SOILS	SOL TOUNDRA	TUNDRA=BODEN	POCHVA TUN- DROVAYA	SUELO=DE= TUNDRA	SUOLO DI TUNDRA
4612 TUNGSTATES MING	TUNGSTATES	TUNGSTATE	WOLFRAMAT	VOL'FRAMAT	TUNGSTATO	TUNGSTATO
4613 TUNGSTEN CHEE	TUNGSTEN	TUNGSTENE	W	VOL'FRAM	TUNGSTENO	TUNGSTENO
4614 TUNNEL ENGI	TUNNELS	TUNNEL	TUNNEL	TUNNEL'	TUNEL	GALLERIA
4615 TURBIDITE SEDI	TURBIDITE	TURBIDITES	TURBIDIT	TURBIDIT	TURBIDITA	TORBIDITE
4616 TURBIDITY SEDI	TURBIDITY	TURBIDITE	TURBIDIT= BILDUNG	MUTNOST'	TURBIDEZ	TORBIDITA
4617 TURBIDITY CURRENT SEDI	TURBIDITY CUR- RENTS	COURANT TURBI- DITE	TURBIDIT=STROM	POTOK MUT'EVOJ *UF:* Turbidity current structure	CORRIENTE= TURBIDEZ	CORRENTE DI TOR- BIDA
4618 TURBIDITY CURRENT STRUCTURE TEST	TURBIDITY CUR- RENT STRUC- TURES	STRUCTURE COU- RANT TURBIDITE	TURBIDIT= STRUKTUR	[TURBIDITY CUR- RENT STRUC- TURE] POTOK MUT'EVOJ	ESTRUCTURA= CORRIENTE= TURBIDEZ	STRUTTURE DA CORRENTE DI TOR- BIDA
4619 TURBODRILLING MINI	Turbodrilling BOREHOLES	Sondage turbo SONDAGE	Turbo=Bohrung BOHRUNG	BURENIE TURBIN- NOE	Turbosondeo POZO=SONDEO	Turboperforazione SONDAGGIO
4620 TURONIAN STRS	TURONIAN	TURONIEN	TURON	TURON	TURONIENSE	TURONIANO
4621 TWIN MINE	TWINNING *UF:* Baveno twin law; Glide twin	MACLE *UF:* Macle Baveno; Macle glissement	ZWILLING *UF:* Baveno=Gesetz; Gleitungs=Zwilling	DVOJNIK *UF:* Dvojnik Bavenskij /zakon/; Dvojniki skol'zheniya	MACLA *UF:* Macla=de=Baveno; Macla=deformada	GEMINAZIONE *UF:* Geminato per scor- rimento; Legge di Baveno
4622 TWO-DIMENSIONAL MODEL MATH	TWO- DIMENSIONAL MODELS	MODELE 2 DIMEN- SIONS	ZWEIDIMENS- IONAL.MODELL	MODEL' DVUM- ERNAYA	MODELO= BIDIMENSIONAL	MODELLO BIDI- MENSIONALE

	ENGLISH	FRANCAIS	DEUTSCH	RUSSKIJ	ESPANOL	ITALIANO
4623 TYLOPODA PALS	TYLOPODA	TYLOPODA	TYLOPODA	Tylopoda ARTIODACTYLA	TYLOPODA	TYLOPODA
4624 TYPE LOCALITY GEOL	Type localities STRATOTYPES	Localite type STRATOTYPE	Typ=Lokalitaet STRATOTYP	MESTNOST' STRA- TOTIPICHESKAYA	Localidad=tipo ESTRATOTIPO	Localita=tipo STRATOTIPO
4625 TYPE SECTION GEOL	TYPE SECTIONS	COUPE TYPE	TYP=PROFIL	Razrez tipovoj STRATOTIP	SECCION=TIPO	SEZIONE TIPO
4626 TYPE SPECIMEN PALE	TYPE SPECIMENS	SPECIMEN CARAC- TERISTIQUE	TYPUS=ART	Ehkzemplyar tipovoj GOLOTIP	ESPECIMEN= CARACTERISTICO	CAMPIONE CARAT- TERISTICO
4627 TYPOMORPHIC **MINERAL** MINE	TYPOMORPHIC MINERALS UF: Typomorphism	Mineral typomorphique TYPOMORPHISME	Typomorph.Mineral TYPOMORPHIE	MINERAL TIPO- MORFNYJ	Mineral=tipomorfo TIPOMORFISMO	Minerale tipomorfo TIPOMORFISMO
4628 TYPOMORPHISM GEOL	Typomorphism TYPOMORPHIC MINERALS	TYPOMORPHISME UF: Mineral typomor- phique	TYPOMORPHIE UF: Typomorph.Mineral	TIPOMORFIZM	TIPOMORFISMO UF: Mineral=tipomorfo	TIPOMORFISMO UF: Minerale tipomorfo
4629 TYRRHENIAN STRS	TYRRHENIAN	TYRRHENIEN	TYRRHENIUM	TIRREN	TIRRENIENSE	TIRRENIANO
4630 UDALF SUSS	Udalfs ALFISOLS	Udalf ALFISOL; MILIEU HUMIDE	Udalf ALFISOL; FEUCHT. MILIEU	Udalf AL'FISOL	Udalf ALFISOL; MEDIO- HUMEDO	Udalf ALFISOL; AMBI- ENTE UMIDO
4631 UDOLL SUSS	Udolls MOLLISOLS	Udoll MOLLISOL; SOL BRUN	Udoll MOLLISOL; BRAUN=ERDE	Udoll MOLLISOL	Udoll MOLLISOL; SUELO=PARDO	Udoll MOLLISOL; SUOLO BRUNO
4632 UDULT SUSS	Udults ULTISOLS	Udult ULTISOL; MILIEU HUMIDE	Udult ULTISOL; FEUCHT. MILIEU	Udult UL'TISOL	Udult ULTISOL; MEDIO- HUMEDO	Udult ULTISOL; AMBI- ENTE UMIDO
4633 UFIMIAN STRS	Ufimian KUNGURIAN; KAZANIAN	Ufien KOUNGOURIEN; KAZANIEN	Ufimium KUNGUR; KASAN	UFIMSKIJ	Ufimiense KUNGURIENSE; KAZANIENSE	Ufimiano KUNGURIANO; KAZANIANO
4634 ULTISOL SUSS	ULTISOLS UF: Aqults; Humults; Udults; Ustults; Xerults	ULTISOL UF: Aqult; Humult; Udult; Ustult; Xerult	ULTISOL UF: Aqult; Humult; Udult; Ustult; Xerult	UL'TISOL UF: Akvult; Gumult; Kserult; Udult; Ustult	ULTISOL UF: Aqult; Humult; Udult; Ustult; Xerult	ULTISOL UF: Aqult; Humult; Udult; Ustult; Xerult
4635 ULTRABASIC **COMPOSITION** IGNE	ULTRAMAFIC COMPOSITION	COMPOSITION ULTRABASIQUE	ULTRABAS. CHEMISMUS	UL'TRAOSNOVNOJ (SOSTAV)	COMPOSICION= ULTRABASICA	COMPOSIZIONE ULTRABASICA
4636 ULTRABASITE IGNS	ULTRAMAFICS	ULTRABASITE	ULTRABASIT	UL'TRABAZIT	ULTRABASITA	ULTRABASITE
4637 ULTRAMETAMORPHISM IGNE	ULTRAMETAMOR- PHISM	ULTRAMETAMOR- PHISME	ULTRA= METAMORPHOSE	UL'TRAMETA- MORFIZM	ULTRAMETAMOR- FISMO	ULTRAMETAMOR- FISMO
4638 ULTRAMYLONITE GEOL	Ultramylonite MYLONITES	Ultramylonite MYLONITE	Ultra=Mylonit MYLONIT	UL'TRAMILONIT	Ultramilonita MILONITA	Ultramilonite MILONITE
4639 ULTRASONIC METHOD APPL	ULTRASONIC METHODS	METHODE ULTRA- SON	ULTRASCHALL= METHODE	METOD UL'TRAZVUKOVYJ	METODO- ULTRASONIDO	METODO DEGLI ULTRASUONI

	ENGLISH	FRANCAIS	DEUTSCH	RUSSKIJ	ESPANOL	ITALIANO
4640 ULTRASTRUCTURE PALE	ULTRASTRUCTURE	MICROSTRUCTURE	MIKRO=STRUKTUR	[ULTRASTRUC-TURE]	MICROESTRUC-TURA	MICROSTRUTTURA
4641 ULTRAVIOLET RADIATION PHCH	ULTRAVIOLET RAYS	RAYON UV	UV=STRAHLUNG	IZLUCHENIE UL'TRAFIOLETOVOE	RAYOS=UV	RAGGI ULTRAVIO-LETTI
4642 UMBREPT SUSS	Umbrepts INCEPTISOLS	Umbrept INCEPTISOL; RANKER	Umbrept INCEPTISOL; RANKER	Umbrept INSEPTISOL	Umbrept INCEPTISOL; RANKER	Umbrept INCEPTISOL; RANKER
4643 UNAKITE IGMS	Unakite METAMORPHIC ROCKS; ALKALIC COMPOSITION	Unakite GRANITE; COMPO-SITION ALCALINE	Unakit GRANIT; ALKALI-TYP	UNAKIT	Unakita GRANITO; COMPOSICION=ALCALINA	Unakite GRANITO; COMPO-SIZIONE ALCALINA
4644 UNCONFINED AQUIFER GEOH	UNCONFINED AQUIFERS UF: Unconfined water	NAPPE LIBRE UF: Eau libre; Nappe phreatique	UNGESPANNT. GRUNDWASSER UF: Phreat. Grundwasser; Ungespannt.Wasser	GORIZONT BEZNA-PORNYJ	MANTO=LIBRE UF: Agua=freatica; Agua=libre	ACQUIFERO NON CONFINATO UF: Falda freatica; Falda libera
4645 UNCONFINED WATER GEOH	Unconfined water UNCONFINED AQUIFERS	Eau libre NAPPE LIBRE	Ungespannt.Wasser UNGESPANNT. GRUNDWASSER	VODA PODZEM-NAYA BEZNA-PORNAYA UF: Voda gruntovaya	Agua=libre MANTO=LIBRE	Falda libera ACQUIFERO NON CONFINATO
4646 UNCONFORMITY STRA	UNCONFORMITIES UF: Hiatus; Nonconfor-mities; Regional uncon-formity	DISCORDANCE UF: Discordance region-ale	DISKORDANZ UF: Regional=Diskordanz	NESOGLASIE	DISCORDANCIA UF: Discordancia=regional	DISCORDANZA UF: Discordanza region-ale
4647 UNCONSOLIDATED MATERIAL COMS	Unconsolidated materi-als SEDIMENTS	Materiau non consolide MATERIAU NON COHERENT	Unverfestigt.Gestein LOCKER=GESTEIN	PORODA RYKHLAYA	Material=no-consolidado MATERIAL=NO=COHERENTE	Materiale sciolto MATERIALE INCOERENTE
4648 UNDERCLAY SEDS	UNDERCLAY	CLAYAT	FLOEZ=LIEGENDTON	[UNDERCLAY] GLINA OGNEU-PORNAYA	ARCILLA=DE=CAPA=INFERIOR	UNDERCLAY
4649 UNDERGROUND CANALIZATION ENGI	UNDERGROUND CHANNELS	CANALISATION SOUTERRAINE	ROHRLEITUNG	[UNDERGROUND CANALIZATION] SOORUZHENIE	CANALIZACION=SUBTERRANEA	CANALIZZAZIONE SOTTERRANEA
4650 UNDERGROUND MINING MINI	Underground mining MINING GEOLOGY	Mine souterraine MINE	Unterird.Gewinnung BERGWERK	VYRABOTKA PODZEMNAYA	Mineria=subterranea MINA	Coltivazione sotter-ranea MINIERA
4651 UNDERGROUND SPACE ENGI	UNDERGROUND SPACE	ESPACE SOUTER-RAIN	HOHLRAUM	PROSTRANSTVO PODZEMNOE	ESPACIO SUBTER-RANEO	SPAZIO SOTTER-RANEO
4652 UNDERGROUND STORAGE ENGI	UNDERGROUND STORAGE	STOCKAGE SOUTERRAIN	UNTERIRD. SPEICHERUNG	KHRANENIE PODZEMNOE	ALMACEN=SUBTERRANEO	STOCCAGGIO SOT-TERRANEO
4653 UNDERGROUND STREAM GEOH	UNDERGROUND STREAMS	RIVIERE SOUTER-RAINE	UNTERIRD.FLUSS	REKA PODZEM-NAYA	RIO=SUBTERRANEO	FIUME SOTTER-RANEO

	ENGLISH	FRANCAIS	DEUTSCH	RUSSKIJ	ESPANOL	ITALIANO
4654 UNDERSATURATED PHCH	Undersaturated(adj.)	Soussature(adj.)	Untersaettigt(adj.)	NEDOSYSHCHEN-NYJ	Subsaturado(adj.)	Sottosaturo(adj.)
4655 UNDERSATURATED SOLUTION PHCH	Undersaturation SATURATION	Soussaturation SATURATION	Untersaettigt.Loesung SAETTIGUNG	NEDOSYSHCHENIE	No=saturado SATURACION	Sottosaturazione SATURAZIONE
4656 UNDERTHRUST FAULT STRU	UNDERTHRUST FAULTS	FAILLE SOUS=CHARRIAGE	UNTERSCHIEBUNG	PODDVIG	MANTO=CORRIMIENTO	FAGLIA DI SOTTOS-CORRIMENTO
4657 UNDULATORY EXTINCTION MINE	UNDULATORY EXTINCTION	EXTINCTION ROU-LANTE	UNDULOES. AUSLOESCHUNG	POGASANIE VOL-NISTOE	EXTINCION=GIRATORIA	ESTINZIONE ONDULATA
4658 UNIAXIAL-COMPRESSION TEST ENGI	UNIAXIAL TESTS	COMPRESSION UNIAXIALE	MONOAXIAL=KOMPRESSION	ISPYTANIE NA MONOOSNOE SZHATIE	COMPRESION=SIMPLE	COMPRESSIONE UNIASSIALE
4659 UNIFORMITARIANISM GEOL	UNIFORMITARIAN-ISM UF: Actualism	Uniformitarianisme ACTUALISME	Uniformitarismus AKTUALISMUS	Uniformizm AKTUALIZM	Uniformidad ACTUALISMO	Uniformitarianismo ATTUALISMO
4660 UNIONOIDA PALS	UNIONOIDAE	UNIONIDAE	UNIONIDAE	UNIONOIDA	UNIONOIDA	UNIONIDA
4661 UNIT CELL MINE	UNIT CELL	MAILLE ELE-MENTAIRE	ELEMENTAR=ZELLE	YACHEJKA EHLE-MENTARNAYA	CELDILLA=ELEMENTAL	CELLA ELEMEN-TARE
4662 UNIVERSAL STAGE INST	UNIVERSAL STAGE	PLATINE UNIVER-SELLE	UNIVERSAL=DREHTISCH	FEDOROVSKIJ STOLIK	PLATINA=UNIVERSAL	TAVOLINO UNIVERSALE
4663 UNIVERSE EXTR	UNIVERSE	Univers GEOLOGIE EXTRA TERRESTRE	Universum EXTRATERRESTR. GEOLOGIE	VSELENNAYA	Universo GEOLOGIA=EXTRATERRESTRE	Universo GEOLOGIA EXTRATERRESTRE
4664 UNSTEADY FLOW GEOH	UNSTEADY FLOW	REGIME TRANSI-TOIRE	INSTATIONAER. STROEMUNG	TECHENIE NESTATSIO-NARNOE	REGIMEN=TRANSITORIO	REGIME TRANSI-TORIO
4665 UPLIFT STRU	UPLIFTS UF: Uplift pressure; Upwarping	EXHAUSSEMENT UF: Isobase; Pression soulevement; Releve-ment	HEBUNG UF: Aufbeulung; Auftriebs=Druck; Isobase	STRUKTURA POLOZHITEL'NAYA UF: Antiforma	LEVANTAMIENTO UF: Elevamiento; Iso-base; Presion=hidrostatica=ascendente	SOLLEVAMENTO UF: Inarcamento; Iso-base; Pressione di sollevamento
4666 UPLIFT PRESSURE PHCH	Uplift pressure UPLIFTS	Pression soulevement EXHAUSSEMENT	Auftriebs=Druck DRUCK; HEBUNG	[UPLIFT PRES-SURE] DAVLENIE GIDROSTATICHES-KOE; DAVLENIE GRUNTA	Presion=hidrostatica=ascendente LEVANTAMIENTO	Pressione di solleva-mento SOLLEVAMENTO
4667 UPPER CAMBRIAN STRS	UPPER CAMBRIAN	CAMBRIEN SUP	O.KAMBRIUM	KEMBRIJ VERKH-NIJ	CAMBRICO=SUP	CAMBRIANO SUP.
4668 UPPER CARBONIFEROUS STRS	UPPER CARBONIF-EROUS	CARBONIFERE SUP	O.KARBON	KARBON VERKH-NIJ UF: Pensil'vanij	CARBONIFERO=SUP	CARBONIFERO SUP.

	ENGLISH	FRANCAIS	DEUTSCH	RUSSKIJ	ESPANOL	ITALIANO
4669 UPPER CRETACEOUS STRS	UPPER CRETA-CEOUS	CRETACE SUP	O.KREIDE	MEL VERKHNIJ	CRETACICO=SUP	CRETACEO SUP.
4670 UPPER CRUST SOLI	UPPER CRUST	CROUTE TER-RESTRE SUP	OBER.KRUSTE	Kora verkhnyaya SLOJ GRANITNYJ	CORTEZA=TERRESTRE=SUP	CROSTA SUPERI-ORE
4671 UPPER DEVONIAN STRS	UPPER DEVONIAN	DEVONIEN SUP	O.DEVON	DEVON VERKHNIJ	DEVONICO=SUP	DEVONIANO SUP.
4672 UPPER EOCENE STRS	UPPER EOCENE *UF:* Bartonian	EOCENE SUP *UF:* Bartonien	O.EOZAEN *UF:* Bartonium	Ehotsen verkhnij EHOTSEN	EOCENO=SUP *UF:* Bartoniense	EOCENE SUP. *UF:* Bartoniano
4673 UPPER JURASSIC STRS	UPPER JURASSIC	JURASSIQUE SUP	O.JURA	YURA VERKH-NYAYA	JURASICO=SUP	GIURASSICO SUP.
4674 UPPER LIASSIC STRS	UPPER LIASSIC	LIAS SUP	O.LIAS	Lejas verkhnij TOAR	LIAS=SUP	LIASSICO SUP.
4675 UPPER MANTLE SOLI	UPPER MANTLE *UF:* C layer; Chal-cosphere	MANTEAU SUP *UF:* Chalcosphere; Couche B; Couche C	OBER.ERD=MANTEL *UF:* B=Schicht; C=Schicht; Chalkosphaere	MANTIYA VERKH-NYAYA *UF:* C sloj	MANTO=GLOBO=SUP *UF:* Calcosfera; Capa=B; Capa=C	MANTELLO SUPERIORE *UF:* Calcosfera; Strato B; Strato C
4676 UPPER MIOCENE STRS	UPPER MIOCENE	MIOCENE SUP	O.MIOZAEN	Miotsen verkhnij MIOTSEN	MIOCENO=SUP *UF:* Pontiense	MIOCENE SUP.
4677 UPPER OLIGOCENE STRS	UPPER OLIGOCENE *UF:* Chattian	OLIGOCENE SUP *UF:* Chattien	O.OLIGOZAEN *UF:* Chattium	Oligotsen verkhnij KHATT	OLIGOCENO=SUP *UF:* Chattiense	OLIGOCENE SUP. *UF:* Cattiano
4678 UPPER ORDOVICIAN STRS	UPPER ORDOVI-CIAN	ORDOVICIEN SUP	O.ORDOVIZIUM	ORDOVIK VERKH-NIJ	ORDOVICICO=SUP	ORDOVICIANO SUP.
4679 UPPER PERMIAN STRS	UPPER PERMIAN	PERMIEN SUP	O.PERM	PERM' VERKH-NYAYA *UF:* Tsekhshtejn	PERMICO=SUP	PERMIANO SUP.
4680 UPPER PLEISTOCENE STRS	UPPER PLEISTO-CENE	PLEISTOCENE SUP	JUNG=PLEISTOZAEN	Plejstotsen verkhnij PLEJSTOTSEN	PLEISTOCENO=SUP	PLEISTOCENE SUP.
4681 UPPER PLIOCENE STRS	UPPER PLIOCENE *UF:* Kujalnikian	PLIOCENE SUP *UF:* Kuyalnik	O.PLIOZAEN *UF:* Kujalnikium	Pliotsen verkhnij PLIOTSEN	PLIOCENO=SUP *UF:* Kujalnikiense	PLIOCENE SUP. *UF:* Kujalnikiano
4682 UPPER PROTEROZOIC STRS	UPPER PROTERO-ZOIC	PROTEROZOIQUE SUP	O. PROTEROZOIKUM	PROTEROZOJ VERKHNIJ	PROTEROZOICO=SUP	PROTEROZOICO SUP.
4683 UPPER SILURIAN STRS	UPPER SILURIAN	SILURIEN SUP	O.SILUR	SILUR VERKHNIJ	SILURICO=SUP	SILURIANO SUP.
4684 UPPER TRIASSIC STRS	UPPER TRIASSIC	TRIAS SUP	O.TRIAS	TRIAS VERKHNIJ *UF:* Kejper	TRIAS=SUP	TRIASSICO SUP.
4685 UPTHROW STRU	Upthrow FAULTS; GEOME-TRY	Levre relevee FAILLE; GEOME-TRIE	Gehoben.Scholle STOERUNG; GEOMETRIE	PRIPODNYATYJ	Labio=levantado FALLA; GEO-METRIA	Lembo rialzato FAGLIA; GEO-METRIA
4686 UPWARPING STRU	Upwarping UPLIFTS	Relevement EXHAUSSEMENT	Aufbeulung HEBUNG	DVIZHENIE VOSKHODYASH-CHEE	Elevamiento LEVANTAMIENTO	Inarcamento SOLLEVAMENTO

	ENGLISH	FRANCAIS	DEUTSCH	RUSSKIJ	ESPANOL	ITALIANO
4687 UPWELLING GEOH	UPWELLING	UPWELLING	AUFTRIEB	APVELLING	REVESA=DE= FONDO	UPWELLING
4688 URALITIZATION IGNE	URALITIZATION	OURALITISATION	URALITISIERUNG	URALITIZATSIYA	URALITIZACION	URALITIZZAZIONE
4689 URANIUM CHEE	URANIUM	URANIUM	U	URAN (EHLEM)	URANIO	URANIO
4690 URANIUM MINERAL ECON	URANIUM MINER- ALS	MINERAL URA- NIUM	URAN=MINERAL	MINERAL URANOV	MINERAL=URANIO	MINERALE DI URANIO
4691 URANIUM- DISEQUILIBRIUM DATING ISOT	URANIUM DIS- EQUILIBRIUM	DESEQUILIBRE URANIUM	URAN= UNGLEICHGEWICHT	METOD NARUS- HENIYA URANOVOGO RAVNOVESIYA	DESEQUILIBRIO= URANIO	DISEQUILIBRIO URANIO
4692 URANIUM-LEAD DATING ISOT	U/PB *UF:* Lead=alpha age method; Th/Pb	U=PB *UF:* Datation plomb= alpha; Th=Pb	U=PB=DATIERUNG *UF:* Alpha=Blei= Datierung; Th=Pb= Datierung	METOD SVINTSOVO= URANOVIJ *UF:* Metod svintsovo= svintsovyj; Metod torievo=svintsovyj; Metod urano=torievo= svintsovyj	U=PB *UF:* Datacion=Pb=alfa; Th=Pb	U/PB *UF:* Pb=alpha; Th/Pb
4693 URANIUM-THORIUM- LEAD DATING ISOT	U/TH/PB	U=TH=PB	U=TH=PB= DATIERUNG	Metod urano=torievo= svintsovyj METOD SVINTSOVO= URANOVYJ	U=TH=PB	U/TH/PB
4694 URBAN GEOLOGY ENVI	Urban geology URBANIZATION	Geologie urbanisation URBANISATION	Stadt=Geologie SIEDLUNGSBAU	GEOLOGIYA URBANISTI- CHESKAYA	Urbanismo URBANIZACION	Geologia del territorio URBANIZZAZIONE
4695 URBAN PLANNING ENVI	URBAN PLANNING	PLANIFICATION URBAINE	STADT=PLANUNG	PLANIROVANIE GORODSKOE	PLANIFICACION= URBANA	PIANIFICAZIONE URBANISTICA
4696 URBANIZATION ENVI	URBANIZATION *UF:* Urban geology	URBANISATION *UF:* Geologie urbanisa- tion	SIEDLUNGSBAU *UF:* Stadt=Geologie	URBANIZATSIYA	URBANIZACION *UF:* Urbanismo	URBANIZZAZIONE *UF:* Geologia del terri- torio
4697 UREILITE EXTS	Ureilite ACHONDRITES	Ureilite ACHONDRITE	Ureilit ACHONDRIT	Ureilit AKHONDRIT	Ureilita ACONDRITA	Ureilite ACONDRITE
4698 URODELA PALS	Urodela AMPHIBIA	Urodela LEPOSPONDYLI	Urodela LEPOSPONDYLI	URODELA	Urodela LEPOSPONDYLI	Urodela LEPOSPONDYLI
4699 USTALF SUSS	Ustalfs ALFISOLS	Ustalf ALFISOL; ZONE TROPICALE	Ustalf ALFISOL; TROPEN= ZONE	Ustalf AL'FISOL	Ustalf ALFISOL; ZONA= TROPICAL	Ustalf ALFISOL; AMBI- ENTE TROPICALE
4700 USTERT SUSS	Usterts VERTISOLS	Ustert VERTISOL; MILIEU HUMIDE	Ustert VERTISOL; FEUCHT.MILIEU	Ustert VERTISOL	Ustert VERTISOL; MEDIO= HUMEDO	Ustert VERTISUOLO; AMBIENTE UMIDO

	ENGLISH	FRANCAIS	DEUTSCH	RUSSKIJ	ESPANOL	ITALIANO
4701 USTOLL SUSS	Ustolls MOLLISOLS	Ustoll MOLLISOL; ZONE STEPPIQUE	Ustoll MOLLISOL; STEPPEN=ZONE	Ustoll MOLLISOL	Ustoll MOLLISOL; ZONA= ESTEPARIA	Ustoll MOLLISOL; AMBI- ENTE DI STEPPA
4702 USTOX SUSS	Ustox OXISOLS	Ustox OXISOL; MILIEU ARIDE	Ustox OXISOL; ARID. MILIEU	Ustoks OKSISOL	Ustox OXISOL; MEDIO- ARIDO	Ustox OXISOL; ARIDITA
4703 USTULT SUSS	Ustults ULTISOLS	Ustult ULTISOL; ZONE TROPICALE	Ustult ULTISOL; TROPEN= ZONE	Ustult UL'TISOL	Ustult ULTISOL; ZONA= TROPICAL	Ustult ULTISOL; AMBI- ENTE TROPICALE
4704 UTILIZATION ECON	UTILIZATION *UF:* Industrial use	UTILISATION SUB- STANCE *UF:* Utilisation industr- ielle	VERWENDUNG *UF:* Industriell.Nutzung	ISPOL'ZOVANIE	UTILIZACION= SUSTANCIA *UF:* Utilizacion= industrial	UTILIZZAZIONE DI SOSTANZA *UF:* Utilizzazione industriale
4705 VACUUM FUSION ANALYSIS METH	VACUUM FUSION ANALYSIS	ANALYSE FUSION VIDE	VAKUUM= SCHMELZ= ANALYSE	[VACUUM FUSION ANALYSIS] METOD ANALITI- CHESKIJ	ANALISIS= FUSION=VACIO	ANALISI FUSIONE IN VUOTO
4706 VALANGINIAN STRS	VALANGINIAN	VALANGINIEN	VALANGINIUM	VALANZHIN	VALANGINIENSE	VALANGINIANO
4707 VALLEY SURF	VALLEYS *UF:* Entrenched streams; Gorges; Insequent valleys; Obsequent valleys; Resequent valleys; Stream beds; Structural valleys; Subsequent valleys; Wadi	VALLEE *UF:* Gorge; Lit cours eau; Vallee insequente; Vallee resequente; Vallee sous marine	TAL *UF:* Insequent.Tal; Rueckschreitend.Tal; Strombett; Submarin. Tal; Tobel	DOLINA *UF:* Dolina obsekvent- naya; Tal'veg	VALLE *UF:* Garganta; Insecu- ente; Lecho=de= curso=de=agua; Valle= resecuente; Valle= submarino	VALLE *UF:* Gola; Letto fluviale; Valle inseguente; Valle resequente; Valle sotto- marina
4708 VALVE PALE	VALVES	VALVE	KLAPPE	STVORKA	VALVA	VALVA
4709 VANADATES MING	VANADATES	VANADATE	VANADAT	VANADAT	VANADATO	VANADATO
4710 VANADIUM CHEE	VANADIUM	VANADIUM	V	VANADIJ	VANADIO	VANADIO
4711 VANE APPARATUS INST	Vane apparatus VANE TESTS	Equipement scissome- trie INSTRUMENTA- TION; SCISSOME- TRIE	Fluegelsonde INSTRUMEN- TIERUNG; FLUEGELDRUCK= SONDIERUNG	KRYL'CHATKA	Tensiometro=aspa INSTRUMENTA- CION; ENSAYO= CIZALLADURA	Scissometro STRUMENTAZ- IONE; SCISSO- METRIA
4712 VANE TEST ENGI	VANE TESTS *UF:* Vane apparatus	SCISSOMETRIE *UF:* Equipement scis- sometrie	FLUEGELDRUCK= SONDIERUNG *UF:* Fluegelsonde	ISPYTANIE KRYL'CHATKOJ	ENSAYO= CIZALLADURA *UF:* Tensiometro=aspa	SCISSOMETRIA *UF:* Scissometro
4713 VAPOR PHCH	WATER VAPOR	VAPEUR EAU	WASSER=DAMPF	PAR	VAPOR	VAPORE
4714 VARIABILITY MISC	Variability VARIATIONS	Variabilite VARIATION	Variabilitaet	IZMENCHIVOST'	Variabilidad VARIACION	Variabilita VARIAZIONE

	ENGLISH	FRANCAIS	DEUTSCH	RUSSKIJ	ESPANOL	ITALIANO
4715 VARIABLE MATH	Variable STATISTICAL ANALYSIS	Variable METHODE STATIS- TIQUE	Variable STATIST.METHODE	PEREMENNAYA	Variable METODO= ESTADISTICO	Variabile METODO STATIS- TICO
4716 VARIANCE ANALYSIS MATH	VARIANCE ANALY- SIS	ANALYSE VARI- ANCE	VARIANZ= ANALYSE	ANALIZ VARIAT- SIONNYJ	ANALISIS= VARIANZA	ANALISI DELLA VARIANZA
4717 VARIATION MATH	VARIATIONS *UF:* Magnetic field variation; Variability	VARIATION *UF:* Niveau lac; Niveau mer; Variabilite; Varia- tion niveau mer	VARIATION *UF:* Meeresspiegel; Meeresspiegel= Schwankung; See= Wasserspiegel	VARIATSIYA	VARIACION *UF:* Nivel=del=mar; Nivel=lago; Variabili- dad; Variacion=nivel= mar	VARIAZIONE *UF:* Livello del mare; Livello lacustre; Livello marino; Variabilita
4718 VARIOGRAM MATH	VARIOGRAMS	Variogramme METHODE STATIS- TIQUE	Variogramm STATIST.METHODE	VARIOGRAMMA	Variograma METODO= ESTADISTICO	Variogramma METODO STATIS- TICO
4719 VARISCAN OROGENY STRU	Variscan orogeny HERCYNIAN OROGENY	Orogenie varisque OROGENIE HERCY- NIENNE	Varist.Orogenese HERZYN. OROGENESE	Varijskij GERTSINSKIJ	Orogenia=varisca OROGENIA= HERCINICA	Orogenesi varisica OROGENESI ERCINICA
4720 VARVE SURF	VARVES *UF:* Varve chronology; Varved clay	VARVE *UF:* Argile varve; Chronologie varve	WARWE *UF:* Warwen= Chronologie; Warwen= Ton	SLOJ GODICHNYJ	VARVA *UF:* Arcilla=varvada; Varva=cronologia	VARVA *UF:* Argilla varvata; Varvecronologia
4721 VARVE CHRONOLOGY STRA	Varve chronology GEOCHRONOLOGY; VARVES	Chronologie varve GEOCHRONOLO- GIE; VARVE	Warwen=Chronologie GEOCHRONOLO- GIE; WARWE	VAR- VOKHRON- OLOGIYA	Varva=cronologia GEOCRONOLOGIA; VARVA	Varvecronologia GEOCRONOLOGIA; VARVA
4722 VARVED CLAY SURF	Varved clay VARVES	Argile varve VARVE	Warwen=Ton WARWE	GLINA LENTOCH- NAYA	Arcilla=varvada VARVA	Argilla varvata VARVA
4723 VECTOR MATH	Vector MATHEMATICAL METHODS	Vecteur METHODE MATHE- MATIQUE	Vektor MATHEMAT. METHODE	VEKTOR	Vector METODO= MATEMATICO	Vettore METODO MATEMA- TICO
4724 VEGETATION ENVI	VEGETATION	VEGETATION	VEGETATION	RASTITEL'NOST'	VEGETACION	VEGETAZIONE
4725 VEIN ECON	VEINS *UF:* Hanging wall; Horsetail; Linked veins; Lode; Segregated vein	FILON *UF:* Filon anastomose; Filon en queue de cheval; Filon segrega- tion; Gite filonien; Toit	GANG *UF:* Erz=Gang; Gang= System; Hangend. Gestein; Horsetail; Segregations=Gang	ZHILA RUDNAYA	FILON *UF:* Cola=de=caballo; Filon=segregado; Filones=ramificados; Techo; Yacimiento= filoniano	FILONE *UF:* Filone a coda di cavallo; Giacimento filoniano; Sistema di vene; Tetto; Vena di segregazione
4726 VELOCITY PHCH	VELOCITY *UF:* Seismic velocity	VITESSE *UF:* Velocite sismique	GESCHWINDIG- KEIT *UF:* Seism. Geschwindigkeit	SKOROST'	VELOCIDAD *UF:* Velocidad=sismica	VELOCITA *UF:* Velocita d'onda sismica
4727 VELOCITY STRUCTURE SOLI	VELOCITY STRUC- TURE	STRUCTURE VIT- ESSE	GESCHWINDI- GKEITS= STRUKTUR	RAZREZ SKOROST- NOJ	ESTRUCTURA= VELOCIDAD	STRUTTURA DI VELOCITA
4728 VENEROIDA PALS	VENERIDA	VENERIDA	VENEROIDA	VENEROIDA	VENERIDA	VENEROIDA
4729 VENITE IGMS	Venite METAMORPHIC ROCKS	Venite MIGMATITE	Venit MIGMATIT	VENIT	Venita MIGMATITA	Venite MIGMATITE

	ENGLISH	FRANCAIS	DEUTSCH	RUSSKIJ	ESPANOL	ITALIANO
4730 VERGENCE STRU	Vergence OVERTURNED FOLDS	Vergence ORIENTATION; PLI	Vergenz ORIENTIERUNG; FALTE	VERGENTSIYA	Vergencia ORIENTACION; PLIEGUE	Vergenza ORIENTAMENTO; PIEGA
4731 VERMES PALS	WORMS *UF:* Annelida; Chae- tognatha; Nemerta; Pogonophora; Scolite; Sipunculoida	VERMES *UF:* Annelida; Chae- tognatha; Nemerta; Pogonophora; Scolite; Sipunculoida	VERMES *UF:* Annelida; Chae- tognatha; Nemerta; Pogonophora; Sipunculoida; Skolithus	VERMES	VERMES *UF:* Annelida; Chae- tognatha; Escolito; Nemerta; Pogonophora; Sipunculoida	VERMES *UF:* Annelida; Chae- tognatha; Nemerta; Pogonophora; Scolite; Sipunculoida
4732 VERMICULITE COMS	VERMICULITE	VERMICULITE	VERMICULIT	VERMIKULIT	VERMICULITA	VERMICULITE
4733 VERMIFORM STRUCTURE TEST	Vermiform structure LEBENSSPUREN	Structure vermiforme TRACE ORGANIQUE	Wurmfoermig.Gefuege LEBENS=SPUR	CHERVEOBRAZNYJ	Estructura=vermiforme HUELLA= ORGANICA	Struttura vermiforme ICNITE
4734 VERTEBRATA PALS	VERTEBRATA *UF:* Tetrapoda	VERTEBRATA *UF:* Tetrapoda	VERTEBRATA *UF:* Tetrapoda	VERTEBRATA	VERTEBRATA *UF:* Tetrapoda	VERTEBRATA *UF:* Tetrapoda
4735 VERTICAL ORIENTATION MISC	VERTICAL ORIEN- TATION	ORIENTATION VERTICALE	SENKRECHT. ORIENTIERUNG	VERTIKAL'NYJ	ORIENTACION= VERTICAL	ORIENTAZIONE VERTICALE
4736 VERTICAL TECTONICS STRU	VERTICAL TEC- TONICS	MOUVEMENT VER- TICAL	VERTIKAL= BEWEGUNG	DVIZHENIE VERTI- KAL'NOE	MOVIMIENTO= VERTICAL	TETTONICA VERTI- CALE
4737 VERTISOL SUSS	VERTISOLS *UF:* Aquerts; Usterts	VERTISOL *UF:* Aquert; Ustert	VERTISOL *UF:* Aquert; Ustert	VERTISOL *UF:* Ustert	VERTISOL *UF:* Aquert; Ustert	VERTISOL *UF:* Aquert; Ustert
4738 VIBRATION ENGI	VIBRATION	VIBRATION	VIBRATION	VIBRATSIYA	VIBRACION	TREMORE
4739 VIBRATORY DRILLING MINI	Vibratory drilling BOREHOLES	Sondage vibration SONDAGE	Vibrations=Bohrung BOHRUNG	BURENIE VIBRAT- SIONNOE	Sondeo=vibracion POZO=SONDEO	Perforazione vibratoria SONDAGGIO
4740 VIBROSEIS APPL	VIBROSEIS	VIBROSEIS	VIBROSEIS	VIBROSEJS	VIBROSISMO	VIBROSEIS
4741 VIKING EXTR	VIKING PROGRAM	PROGRAMME VIKING	VIKING= PROGRAMM	VIKING (SPUTNIK)	PROGRAMA= VIKING	PROGRAMMA VIKING
4742 VILLAFRANCHIAN STRS	VILLAFRANCHIAN *UF:* Apsheronian	VILLAFRANCHIEN *UF:* Apsheronien	VILLAFRANCIUM *UF:* Apsheronium	Villafrank PLEJSTOTSEN	VILLAFRAN- QUIENSE *UF:* Apsheroniense	VILLAFRANCHI- ANO *UF:* Apsheroniano
4743 VIRGATION STRU	Virgation FOLDS	Virgation PLI	Virgation FALTE	VIRGATSIYA	Virgacion PLIEGUE	Virgazione PIEGA
4744 VISCOELASTICITY PHCH	VISCOELASTICITY	VISCOELASTICITE	VISKOELASTIZIT- AET	UPRUGOVYAZKIJ	VISCOELASTICI- DAD	VISCOELASTICITA
4745 VISCOSITY PHCH	VISCOSITY	VISCOSITE	VISKOSITAET	VYAZKOST'	VISCOSIDAD	VISCOSITA
4746 VISCOUS MATERIAL ENGI	VISCOUS MATERI- ALS	MATIERE VISQUEUSE	VISKOS.MATERIAL	VYASKIJ MATE- RIAL	MATERIA= VISCOSA	MATERIALE VIS- COSO
4747 VISCOUS REMANENT MAGNETIZATION PHCH	VISCOUS REMA- NENT MAGNE- TIZATION	AIMANTATION REMANENTE VISQUEUSE	VISKOREMANENT. MAGNETISIERUNG	Namagnichennost' vyazkaya NAMAGNICHEN- NOST' OSTATOCH- NAYA	IMANTACION= REMANENTE= VISCOSA	MAGNETIZZAZ- IONE VISCOSA RESIDUA

	ENGLISH	FRANCAIS	DEUTSCH	RUSSKIJ	ESPANOL	ITALIANO
4748 VISEAN STRS	VISEAN	VISEEN	VISE	VIZE	VISEENSE	VISEANO
4749 VITRAIN SEDS	Vitrain MACERALS	Vitrain MACERAL	Glanz=Kohle MAZERAL	VITREN	Vitreno MACERAL	Vitrina MACERALE
4750 VITREOUS TEXTURE TEST	Vitreous materials GLASSES	Texture vitreuse VERRE NATUREL	Glas=Gefuege NATUERL.GLAS	STRUKTURA STEKLYANNAYA	Vitrea VIDRIO=NATURAL	Vetroso VETRO NATURALE
4751 VITRIFICATION PHCH	Vitrification GLASSES	Vitrification VERRE NATUREL	Verglasung NATUERL.GLAS	VITRIFIKATSIYA	Vitrificacion VIDRIO=NATURAL	Vetrificazione VETRO NATURALE
4752 VITROCLASTIC **TEXTURE** TEST	Vitroclastic texture TEXTURES; PYRO- CLASTICS	Texture vitroclastique TEXTURE; PYRO- CLASTIQUE	Vitroklast.Gefuege KORN=GEFUEGE; PYROKLAST. GESTEIN	VITROKLASTI- CHESKIJ	Textura=vitroclastica TEXTURA; PIRO- CLASTICO	Tessitura vitroclastica TESSITURA; PRO- DOTTO PIROCLAS- TICO
4753 VITROPHYRIC TEXTURE TEST	Vitrophyric texture TEXTURES; GLASSES	Texture vitrophyrique TEXTURE POR- PHYRIQUE	Vitrophyr.Gefuege PORPHYR= GEFUEGE	VITROFIROVYJ	Textura=vitrofidica TEXTURA= PORFIDICA	Tessitura vitrofirica TESSITURA POR- FIRICA
4754 VOLANTES PALS	Volantes NEORNITHES	Volantes NEORNITHES	Volantes NEORNITHES	VOLANTES NEORNITHES	Volantes NEORNITHES	Volantes NEORNITHES
4755 VOLATILE PHCH	Volatile(adj.)	Volatil(adj.)	Volatil(adj.)	LETUCHIJ *UF:* Volatile element	Volatil(adj.)	Volatile(adj.)
4756 VOLATILE ELEMENT CHES	VOLATILE ELE- MENTS	ELEMENT VOLATIL	FLUECHTIG. ELEMENT	[VOLATILE ELEMENT] LETUCHIJ	ELEMENTO= VOLATIL	ELEMENTO VOLA- TILE
4757 VOLATILES CHES	VOLATILES	MATIERE VOLA- TILE	FLUECHTIG. VERBINDUNG	LETUCHIE	MATERIA= VOLATIL	SOSTANZA VOLA- TILE
4758 VOLATILIZATION IGNE	VOLATILIZATION	VOLATILISATION	VERFLUECH- TIGUNG	LETUCHEST'	VOLATILIZACION	VOLATILIZZAZ- IONE
4759 VOLCANIC ASH IGNE	VOLCANIC ASH	Cendre volcanique PYROCLASTIQUE	Vulkan.Asche PYROKLAST. GESTEIN	PEPEL VULKANI- CHESKIJ	Ceniza=volcanica PIROCLASTICO	Cenere vulcanica PRODOTTO PIRO- CLASTICO
4760 VOLCANIC BELT IGNE	VOLCANIC BELTS	CEINTURE VOL- CANIQUE	VULKAN= GUERTEL	TSEP' VULKANI- CHESKAYA	CINTURON= VOLCANICO	CINTURA VUL- CANICA
4761 VOLCANIC BRECCIA IGNE	VOLCANIC BREC- CIA *UF:* Explosion breccia	BRECHE VOL- CANIQUE *UF:* Breche explosion	VULKAN.BREKZIE *UF:* Explosions=Brekzie	BREKCHIYA VUL- KANICHESKAYA	BRECHA= VOLCANICA *UF:* Brecha=de= explosion	BRECCIA VUL- CANICA *UF:* Breccia d'esplosione
4762 VOLCANIC CONE IGNE	Volcanic cones VOLCANOLOGY	Cone volcanique VOLCAN	Vulkan=Kegel VULKAN	KONUS VULKANI- CHESKIJ *UF:* Konus peplovyj; Konus shlakovyj	Cono=volcanico VOLCAN	Cono vulcanico VULCANO
4763 VOLCANIC DOME IGNE	Volcanic domes VOLCANOES	Dome volcanique VOLCAN; DOME	Vulkan=Kuppe VULKAN; DOM	KUPOL VULKANI- CHESKIJ *UF:* Plug	Cupula=volcanica VOLCAN; DOMO	Duomo vulcanico VULCANO; DUOMO
4764 VOLCANIC GLASS IGNE	Volcanic glass GLASSES	Verre volcanique VERRE NATUREL	Vulkan.Glas NATUERL.GLAS	STEKLO VULKANI- CHESKOE *UF:* Steklo prirodnoe	Vidrio=volcanico VIDRIO=NATURAL	Vetro vulcanico VETRO NATURALE

	ENGLISH	FRANCAIS	DEUTSCH	RUSSKIJ	ESPANOL	ITALIANO
4765 VOLCANIC ORIGIN IGNE	Volcanic origin VOLCANOLOGY	Origine volcanique VOLCANOLOGIE	Vulkan.Entstehung VULKANOLOGIE	VULKANICHESKIJ	Volcanico VULCANOLOGIA	Vulcanico VULCANOLOGIA
4766 VOLCANIC ROCK IGNS	VOLCANIC ROCKS *UF:* Taxite	ROCHE VOL- CANIQUE *UF:* Taxite	VULKANIT *UF:* Taxit	PORODA EHFFUZ- IVNAYA	ROCA=VOLCANICA *UF:* Taxita	ROCCIA VUL- CANICA *UF:* Taxite
4767 VOLCANIC VENT IGNE	VENTS *UF:* Feeder	EVENT	AUSBRUCH	Zherlo NEKK	SALIDA= VOLCANICA	CAMINO VUL- CANICO
4768 VOLCANISM IGNE	VOLCANISM	MANIFESTATION VOLCANIQUE	VULKANISMUS	VULKANIZM	VULCANISMO	MANIFESTAZIONE VULCANICA
4769 VOLCANO IGNE	VOLCANOES *UF:* Active volcano; Island volcano; Plugs; Submarine volcano; Volcanic domes; Vul- canorium	VOLCAN *UF:* Aiguille; Bouchon; Cone scories; Cone volcanique; Dome volcanique; Volcan actif; Volcan insulaire; Volcan sous marin	VULKAN *UF:* Aktiv.Vulkan; Insel=Vulkan; Nadel; Pfropfen; Schlacken= Kegel; Submarin. Vulkan; Vulkan=Kegel; Vulkan=Kuppe	VULKAN	VOLCAN *UF:* Aguja; Cono=de= escorias; Cono= volcanico; Cupula= volcanica; Tapon; Volcan=activo; Volcan=insular; Volcan=submarino	VULCANO *UF:* Cono di eiezione; Cono vulcanico; Duomo vulcanico; Guglia vulcanica; Tappo; Vulcano attivo; Vul- cano insulare; Vulcano sottomarino
4770 VOLCANO-TECTONIC DEPRESSION STRU	Volcano=tectonic depression CALDERAS; VOL- CANOLOGY	Depression volcanotec- tonique CALDEIRA	Vulkano=tekton.Senke CALDERA	VULKANO- TEKTONICHESKAYA DEPRESSIYA	Depresion= vulcanotectonica CALDERA	Depressione vulcanotet- tonica CALDERA
4771 VOLCANO-TECTONIC STRUCTURE STRU	Volcano=tectonic struc- ture TECTONICS; VOL- CANOLOGY	Structure volcanotec- tonique TECTOGENESE; VOLCANOLOGIE	Vulkano=tekton. Gefuege TEKTOGENESE; VULKANOLOGIE	STRUKTURA VULKANO- TEKTONICHESK.	Estructura= vulcanotectonica TECTOGENESIS; VULCANOLOGIA	Struttura vulcanotet- tonica TETTOGENESI; VULCANOLOGIA
4772 VOLCANOCLASTIC IGNE	VOLCANICLASTICS	VOLCANO SEDI- MENTAIRE	VULKANO= SEDIMENTAER. BILDUNG	Vulkanoklasticheskij PIROKLAST	VULCANO= SEDIMENTARIO	VULCANO SEDI- MENTARIO
4773 VOLCANOGENIC GEOL	Volcanogenic(adj.)	Volcanogene(adj.)	Vulkanogen(adj.)	VULKANOGENNYJ	Vulcanogeno(adj.)	Vulcanogeno(adj.)
4774 VOLCANOGENIC PROCESS ECON	VOLCANIC PRO- CESSES	GITE VOLCANO- GENE	VULKAN. LAGERSTAETTE	VULKANOGENNYJ (MESTOR)	YACIMIENTO= VULCANOGENO	GIACIMENTO VUL- CANOGENICO
4775 VOLCANOLOGY IGNE	VOLCANOLOGY *UF:* Arctic suite; Atlan- tic suite; Circum= pacific region; Pacific suite; Volcanic cones; Volcanic origin; Volcano=tectonic depression; Volcano= tectonic structure	VOLCANOLOGIE *UF:* Origine volcanique; Serie arctique; Serie atlantique; Serie paci- fique; Structure volca- notectonique	VULKANOLOGIE *UF:* Arkt.Abfolge; Atlant.Abfolge; Pazif. Abfolge; Vulkan. Entstehung; Vulkano= tekton.Gefuege	VULKANOLOGIYA *UF:* Arkticheskaya seriya	VULCANOLOGIA *UF:* Estructura= vulcanotectonica; Serie=artica; Serie= atlantica; Serie= pacifica; Volcanico	VULCANOLOGIA *UF:* Serie artica; Serie atlantica; Serie pacif- ica; Struttura vulcano- tettonica; Suite artica; Vulcanico
4776 VOLUME MISC	VOLUME	VOLUME	VOLUMEN	OB'EM	VOLUMEN	VOLUME
4777 VULCANIAN TYPE ERUPTION IGNE	VULCANIAN=TYPE ERUPTIONS	VULCANIEN	VESUV=TYP	VULKANSKIJ	VULCANIANO	VULCANIANO

	ENGLISH	FRANCAIS	DEUTSCH	RUSSKIJ	ESPANOL	ITALIANO
4778 VULCANORIUM STRU	Vulcanorium VOLCANOES	Vulcanorium DORSALE OCEANIQUE	Vulcanorium SUBMARIN. RUECKEN	VULKANORIJ	Vulcanorio DORSAL= OCEANICA	Vulcanorium DORSALE OCEAN- ICA
4779 WADI SURF	Wadi DESERTS; VALLEYS	Oued DESERT; RIVIERE	Wadi WUESTE; FLUSS	Vadi REKA PER- ESYKHAYUSH- CHAYA; PUS- TYNYA	Rambla DESIERTO; RIO	Uadi DESERTO; FIUME
4780 WALLROCK ECON	WALLROCKS	EPONTE	NEBENGESTEIN	PORODA BOKOVAYA	HASTIAL	SALBANDA
4781 WASTE ENVI	Waste(gen.)	Dechet(gen.)	Abfall(gen.)	OTKHODY *UF:* Human waste	Residuo(gen.)	Rifiuti(gen.)
4782 WASTE DISPOSAL ENVI	WASTE DISPOSAL	DECHARGE DECHET	ABFALL= BESEITIGUNG	ZAKHORONENIE OTKHODOV *UF:* Landfill	VERTEDERO	DISCARICA
4783 WATER GEOH	WATER *UF:* Head; Hydroexplo- sion	EAU *UF:* Charge hydraulique; Explosion phreatique	WASSER *UF:* Hydraul.Belastung; Hydro=Explosion	VODA *UF:* Dekontaminatsiya; Voda zagryaznennaya; Vodokhozyastvennoe planirovanie; Water authority; Water map; Water pressure	AGUA *UF:* Carga=hidraulica; Hidroexplosion	ACQUA *UF:* Carico idraulico; Idroesplosione
4784 WATER AUTHORITY GEOH	WATER AUTHOR- ITY	AGENCE BASSIN	WASSERWIRT- SCHAFTS= BEHOERDE	[WATER AUTHOR- ITY] VODA; OKHRANA	CONFEDERACION= HIDROGRAFICA	ENTE DI BACINO IDROGRAFICO
4785 WATER BUDGET GEOH	Water budget WATER BALANCE	BILAN EAU *UF:* Bilan eau souter- raine	WASSER=BILANZ *UF:* Grundwasser= Bilanz	BALANS VODNYJ *UF:* Surface water bud- get	BALANCE=DE= AGUA *UF:* Balance=de=agua= subterranea	BILANCIO IDROGEOLOGICO *UF:* Bilancio idrico sotterraneo
4786 WATER CATCHMENT GEOH	WATER HARNESS- ING	CAPTAGE EAU	WASSER= GEWINNUNG	VODOZABOR	CAPTACION	CAPTAZIONE D'ACQUA
4787 WATER CONSUMPTION GEOH	Water consumption CONSUMPTION; WATER RESOURCES	Consommation eau CONSOMMATION	Wasser=Verbrauch VERBRAUCH	VODOPOTREBLE- NIE	Consumo=agua CONSUMO	Consumo idrico CONSUMO
4788 WATER DIVIDE GEOH	Divides DRAINAGE BASINS	Ligne partage eaux BASSIN VERSANT	Wasser=Scheide EINZUGSGEBIET	VODORAZDEL	Linea=de=separacion= de=aguas DIVISORIA=DE= AGUAS	Spartiacque BACINO DI RAC- COLTA
4789 WATER EROSION SURF	WATER EROSION *UF:* Fluvial erosion	EROSION EAU	WASSER=EROSION	EHROZIYA VOD- NAYA	EROSION=AGUA	EROSIONE IDRICA
4790 WATER HARDNESS GEOH	WATER HARDNESS	DURETE EAU	WASSER=HAERTE	ZHESTKOST' VODY	DUREZA=AGUA	DUREZZA DELL'ACQUA
4791 WATER MANAGEMENT GEOH	WATER MANAGE- MENT	GESTION RES- SOURCE EAU	WASSER- WIRTSCHAFT	EHKSPLUATAT- SIYA PODZEM- NYKH VOD	GESTION= RECURSOS=AGUA	GESTIONE DI RISORSE IDRICHE

	ENGLISH	FRANCAIS	DEUTSCH	RUSSKIJ	ESPANOL	ITALIANO
4792 WATER MAP GEOH	WATER MAPS	CARTE EAU	GEWAESSER= KARTE	[WATER MAP] VODA; KARTA	MAPA=AGUA	CARTA DELLE ACQUE
4793 WATER OF CRYSTALLIZATION MINE	WATER OF CRYS- TALLIZATION	EAU CRISTALLISA- TION	KRISTALL= WASSER	VODA KRISTALL- IZATSIONNAYA *UF*: Voda degidratat- sionnaya	AGUA= CRISTALIZACION	ACQUA DI CRIS- TALLIZZAZIONE
4794 WATER OF DEHYDRATION MINE	WATER OF DEHY- DRATION	EAU CONSTITU- TION	KONSTITUTIONS= WASSER	Voda degidratatsion- naya VODA KRISTALL- IZATSIONNAYA	AGUA= CONSTITUCION	ACQUA DI COSTI- TUZIONE
4795 WATER PRESSURE ENGI	WATER PRESSURE	PRESSION EAU	WASSER=DRUCK	[WATER PRES- SURE] VODA; DAVLENIE	PRESION=DE= AGUA	PRESSIONE DELL'ACQUA
4796 WATER QUALITY GEOH	WATER QUALITY	QUALITE EAU	WASSER= QUALITAET	KACHESTVO VODY	CALIDAD=AGUA	QUALITA DELLA'ACQUA
4797 WATER RECESSION GEOH	WATER RECESSION	TARISSEMENT	VERSIEGEN	ISTOSHCHENIE VODONOSNOGO GORIZONTA	AGOTAMIENTO	PROSCIUGAMENTO
4798 WATER RESOURCES GEOH	WATER RESOURCES *UF*: Water consumption	RESSOURCE EAU	NUTZBAR.WASSER	RESURSY VODNYE	RECURSOS=AGUA	RISORSE IDRICHE
4799 WATER STORAGE GEOH	WATER STORAGE	STOCKAGE EAU	WASSER= SPEICHERUNG	[WATER STORAGE] VODOKHRANIL- ISHCHE	ALMACENAMIENTO= AGUA	STOCCAGGIO DELL'ACQUA
4800 WATER SUPPLY GEOH	WATER SUPPLY	APPROVISIONNE- MENT EAU	WASSER= VERSORGUNG	VODOSNABZHENIE	APROVISION- AMIENTO=AGUA	RIFORNIMENTO D'ACQUA
4801 WATER WELL GEOH	WATER WELLS	PUITS EAU	BRUNNEN	SKVAZHINA NA VODU	POZO=AGUA	POZZO PER ACQUA
4802 WATER YIELD GEOH	WATER YIELD	EAU DISPONIBLE	VERFUEGBAR. WASSER	VODOOTDACHA	AGUA=DISPONIBLE	ACQUA DISPONI- BILE
4803 WATERFALL SURF	WATERFALLS	CHUTE EAU	WASSERFALL	VODOPAD	CATARATA	CASCATA
4804 WATERWAY ENGI	WATERWAYS	VOIE NAVIGABLE	WASSERSTRASSE	PUT' VODNYJ	VIA=NAVEGABLE	VIA NAVIGABILE
4805 WAVE PHCH	WAVES(gen.)	Onde(gen.)	Welle(gen.)	VOLNA *UF*: Volna Ehry; Wave dispersal	Ondas(gen.)	Onda(gen.)
4806 WAVE DISPERSAL PHCH	WAVE DISPERSION	DISPERSION ONDE	WELLEN- STREUUNG	[WAVE DIS- PERSAL] VOLNA; DISPERSIYA	DISPERSION= ONDA	DISPERSIONE D'ONDA
4807 WAVE PROPAGATION SOLI	WAVE PROPAGA- TION	PROPAGATION ONDE	WELLEN- AUSBREITUNG	RASPROSTRANE- NIE VOLNY	PROPAGACION= ONDA	PROPAGAZIONE D'ONDA
4808 WAX SEDS	WAX	CIRE	ERDWACHS	OZOKERIT	CERA	CERA

	ENGLISH	FRANCAIS	DEUTSCH	RUSSKIJ	ESPANOL	ITALIANO
4809 WEATHERING SURF	WEATHERING	ALTERATION METEORIQUE	ATMOSPHAER. VERWITTERUNG	VYVETRIVANIE *UF:* Vyvetrivanie mekhanicheskoe; Vyvetrivanie selektivnoe	ALTERACION=METEORICA	ALTERAZIONE METEORICA
4810 WEATHERING CRUST SURF	WEATHERING CRUST *UF:* Caliche	CROUTE ALTER-ATION *UF:* Caliche	VERWITTERUNGS=KRUSTE *UF:* Caliche	KORA VYVETRIVANIYA *UF:* Ferrikret; Kal'kret; Kaliche	COSTRA=ALTERACION *UF:* Caliche	CROSTA DI ALTERAZIONE *UF:* Caliche
4811 WEDGE STRA	STRATIGRAPHIC WEDGES	BISEAU *UF:* Amincissement	AUSKEILEN *UF:* Ausduennen	Klin VYKLINIVANIE	BORDE=CUENCA *UF:* Acuñado	CUNEO STRATI-GRAFICO *UF:* Terminazione laterale
4812 WEIGHTING MATH	Weighting MATHEMATICAL METHODS	Ponderation METHODE MATHE-MATIQUE	Wichtung MATHEMAT. METHODE	SREDNE=VZVESHENNYJE	Analisis=ponderado METODO=MATEMATICO	Peso statistico METODO MATEMA-TICO
4813 WELL MINI	WELLS *UF:* Filter; Gas well; Offshore well; Oil well; Producing well; Well completion	PUITS RECONNAIS-SANCE	SCHURF	SKVAZHINA BUROVAYA	POZO=DE=RECONOCIMIENTO	POZZO ESPLORA-TIVO
4814 WELL COMPLETION MINI	Well completion WELLS	Completion puits EXPLOITATION	Bohr=Vollendung GEWINNUNG	ZAKANCHIVANIE SKVAZHINY	Terminacion=pozo EXPLOTACION	Completamento di pozzo SFRUTTAMENTO
4815 WELL HEAD MINI	Well head BOREHOLES	Tete forage SONDAGE	Bohr=Kopf BOHRUNG	UST'E SKVAZHINY	Cruz=produccion POZO=SONDEO	Testa di pozzo SONDAGGIO
4816 WELL LOG APPL	WELL=LOGGING *UF:* Logs	DIAGRAPHIE	BOHRLOCH=MESSUNG	Karotazh skvazhinnyj KAROTAZH	DIAGRAFIA	DIAGRAFIA
4817 WELL SCREEN INST	WELL SCREENS	CREPINE	BRUNNEN=FILTER	[WELL SCREEN] SKVAZHINA; FIL'TR	FILTRO	FILTRO DI POZZO
4818 WENLOCKIAN STRS	WENLOCKIAN	WENLOCK	WENLOCK	VENLOK	WENLOKIENSE	WENLOCKIANO
4819 WEST MISC	West GEOGRAPHY	Ouest GEOGRAPHIE	Westen GEOGRAPHIE	ZAPAD	Oeste GEOGRAFIA	Ovest GEOGRAFIA
4820 WESTPHALIAN STRS	WESTPHALIAN	WESTPHALIEN	WESTFAL	VESTFAL	WESTFALIENSE	WESTFALIANO
4821 WET METHOD METH	WET METHODS	METHODE VOIE HUMIDE	NASS=CHEM. METHODE	KHIMIYA MOKRAYA	METODO=VIA=HUMEDA	METODO PER VIA UMIDA
4822 WETLAND SURF	WETLANDS	TERRAIN HUMIDE	FEUCHTGEBIET	[WETLAND] BOLOTO	TERRENO=HUMEDO	TERRE UMIDE
4823 WHITE CLAY COMS	White clay KAOLIN	Argile blanche KAOLIN	Porzellan=Ton KAOLIN	GLINA BELAYA	Arcilla=blanca CAOLIN	Argilla bianca CAOLINO
4824 WHOLE ROCK ISOT	WHOLE ROCK	ROCHE TOTALE	GESAMT=GESTEIN	PROBA VALOVAYA	ROCA=TOTAL	ROCCIA TOTALE

	ENGLISH	FRANCAIS	DEUTSCH	RUSSKIJ	ESPANOL	ITALIANO
4825 WILDFLYSCH SEDI	Wildflysch FLYSCH	Wildflysch FLYSCH	Wildflysch FLYSCH	FLISH DIKIJ	Wildflysch FLYSCH	Wildflysch FLYSCH
4826 WIND SURF	WINDS	VENT *UF:* Cyclone	WIND *UF:* Zyklone	VETER *UF:* Sreda eholovaya; Tsiklon	VIENTO *UF:* Ciclon	VENTO *UF:* Ciclone
4827 WIND EROSION SURF	WIND EROSION *UF:* Deflation	EROSION EOLIENNE *UF:* Cuticule deflation; Deflation	WIND=EROSION *UF:* Deflation; Wuesten=Pflaster	EHROZIYA VETROVAYA	EROSION=EOLICA *UF:* Deflacion; Pavimento=desertico	EROSIONE EOLICA *UF:* Deflazione; Pavi- mentazione desertica
4828 WIND RIPPLE TEST	Wind ripple RIPPLE MARKS	[WIND RIPPLE] RIPPLE MARK	Wind=Rippel RIPPELMARKE	ZNAKI RYABI EHOLOVYE	Ripple=eolico RIPPLE=MARK	Increspatura eolica RIPPLE MARK
4829 WIND TRANSPORT SURF	WIND TRANSPORT	Transport eolien SEDIMENTATION EOLIENNE	[Wind=Transport] AEOL.SEDIMENTA- TION	[WIND TRANS- PORT] TRANSPORTIROVKA (GEOL); SEDIMENTAT- SIYA EHOLOVAYA	Transporte=eolico SEDIMENTACION= EOLICA	Trasporto eolico SEDIMENTAZIONE EOLICA
4830 WINDOW STRU	WINDOWS *UF:* Inliers	FENETRE *UF:* Boutonniere	FENSTER *UF:* Inlier	OKNO TEKTONI- CHESKOE	VENTANA= TECTONICA *UF:* Ojal	FINESTRA TET- TONICA *UF:* Inlier
4831 WRENCH FAULT STRU	WRENCH FAULTS	FAILLE VERTICALE DECROCHEMENT	VERTIKAL= VERSCHIEBUNG	[WRENCH FAULT] SDVIG (TEKT)	FALLA= VERTICAL= DESGARRE	FAGLIA TRASCOR- RENTE VERTICALE
4832 X-RAY PHCH	X=RAYS	RAYON X	ROENTGEN= STRAHLUNG	X=LUCHI	RAYOS=X	RAGGI X
4833 X-RAY ANALYSIS METH	X=RAY ANALYSIS	ANALYSE RX	ROENTGEN= ANALYSE	ANALIZ RENT- GENOVSKIJ	ANALISIS=RX	ANALISI RAGGI X
4834 X-RAY DIFFRACTION **PATTERN** METH	X=RAY DIFFRAC- TION ANALYSIS *UF:* Bragg angle	DIFFRACTION RX *UF:* Reflection Bragg	ROENTGEN= BEUGUNG *UF:* Bragg=Winkel	RENTGENO- GRAMMA	DIFRACCION=RX *UF:* Angulo=de=Bragg	DIFFRAZIONE RAGGI X *UF:* Angolo di Bragg
4835 X-RAY FLUORESCENCE **ANALYSIS** METH	X=RAY FLUORES- CENCE	FLUORESCENCE RX	ROENTGEN= FLUORESZENZ	Flyuorestsentsiya rent- genospektral'naya ANALIZ RENTGENO= SPEKTRAL'NYJ	FLUORESCENCIA= RX	FLUORESCENZA A RAGGI X
4836 X-RAY SPECTROSCOPY METH	X=RAY SPECTROS- COPY	SPECTROMETRIE RX	ROENTGEN= SPEKTROMETRIE	ANALIZ RENTGENO= SPEKTRAL'NYJ *UF:* Flyuorestsentsiya rentgenospektral'naya; Metod poroshka	ESPECTROMETRIA= RX	SPETTROMETRIA RX
4837 XANTHOPHYTA PALS	Xanthophyta ALGAE	Xanthophyta ALGAE	Xanthophyta ALGAE	XANTHOPHYTA	Xanthophyta ALGAE	Xanthophyta ALGAE
4838 XENOLITH IGNE	XENOLITHS *UF:* Autoliths	ENCLAVE ROCHE *UF:* Autolite; Roof pendant	XENOLITH *UF:* Magmakammer= Dach; Magmat. Einschluss	KSENOLIT	ENCLAVE=ROCA *UF:* Autolito; Techo= rebajado	XENOLITE *UF:* Autolite; Pendente

	ENGLISH	FRANCAIS	DEUTSCH	RUSSKIJ	ESPANOL	ITALIANO
4839 XENOMORPHIC CRYSTAL TEST	Xenomorphic crystals TEXTURES	Cristal xenomorphe TEXTURE	Xenomorph.Kristall KORN=GEFUEGE	KSENOMORFNYJ *UF:* Angedral'nyj	Cristal=xenomorfico TEXTURA	Cristallo xenomorfo TESSITURA
4840 XENON CHEE	XENON	XENON	XE	KSENON	XENON	XENO
4841 XERALF SUSS	Xeralfs ALFISOLS	Xeralf ALFISOL; MILIEU ARIDE	Xeralf ALFISOL; ARID. MILIEU	Kseralf AL'FISOL	Xeralf ALFISOL; MEDIO= ARIDO	Xeralf ALFISOL; ARIDITA
4842 XEROLL SUSS	Xerolls MOLLISOLS	Xeroll MOLLISOL; ZONE STEPPIQUE	Xeroll MOLLISOL; STEPPEN=ZONE	Kseroll MOLLISOL	Xeroll MOLLISOL; ZONA= ESTEPARIA	Xeroll MOLLISOL; AMBI- ENTE DI STEPPA
4843 XEROPHILE PLANT PALE	Xerophile plant ARIDITY; FLORA	Plante xerophile MILIEU ARIDE; FLORE	Xerophil.Pflanze ARID.MILIEU; FLORA	KSEROFILY	Planta=xerofila MEDIO=ARIDO; FLORA	Pianta xerofila AMBIENTE ARIDO; FLORA
4844 XEROPHYTE PALE	Xerophyte ARIDITY; FLORA	Xerophyte FLORE; MILIEU ARIDE	Xerophyt ARID.MILIEU; FLORA	KSEROFIT	Xerofita MEDIO=ARIDO; FLORA	Xerofita AMBIENTE ARIDO; FLORA
4845 XEROSOL SUSS	Xerosols SOILS	Xerosol SOL PEU EVOLUE	Xerosol AZONAL.BODEN	KSEROSOL	Xerosol SUELO= INMADURO	Xerosol SUOLO IMMATURO
4846 XERULT SUSS	Xerults ULTISOLS	Xerult ULTISOL; MILIEU ARIDE	Xerult ULTISOL; ARID. MILIEU	Kserult UL'TISOL	Xerult ULTISOL; MEDIO= ARIDO	Xerult ULTISOL; ARIDITA
4847 YEARBOOK MISC	DIRECTORY	ANNUAIRE	JAHRBUCH	EZHEGODNIK	ANUARIO	ANNUARIO
4848 YERMOSOL SUSS	Yermosols SOILS	Sol desertique SOL BRUT	Yermosol AZONAL.BODEN	IERMOSOL	Suelo=desertico SUELO=BRUTO	Yermosol SUOLO AZONALE
4849 YIELD STRENGTH ENGI	Yield strength DEFORMATION	Point deformation DEFORMATION	Streck=Grenze VERFORMUNG	PREDEL TEKU- CHESTI	Limite=elastico DEFORMACION	Soglia di deformazione DEFORMAZIONE
4850 YOUNG MISC	Young(adj.)	Jeune(adj.)	Jung(adj.)	MOLODOJ	Joven(adj.)	Giovane(adj.)
4851 YOUNG'S MODULUS PHCH	YOUNG'S MODU- LUS	MODULE YOUNG	YOUNG=MODUL	YUNGA MODUL'	MODULO=YOUNG	MODULO DI YOUNG
4852 YPRESIAN STRS	Ypresian LOWER EOCENE	Ypresien EOCENE INF	Ypresium U.EOZAEN	IPR	Ipresiense EOCENO=INF	Ypresiano EOCENE INF.
4853 YTTERBIUM CHEE	YTTERBIUM	YTTERBIUM	YB	ITTERBIJ	ITERBIO	ITTERBIO
4854 YTTRIUM CHEE	YTTRIUM	YTTRIUM	Y	ITTRIJ	ITRIO	ITTRIO
4855 ZECHSTEIN STRS	ZECHSTEIN	ZECHSTEIN	ZECHSTEIN	Tsekhshtejn PERM' VERKH- NYAYA	ZECHSTEIN	ZECHSTEIN

	ENGLISH	FRANCAIS	DEUTSCH	RUSSKIJ	ESPANOL	ITALIANO
4856 ZEOLITE MING	ZEOLITE GROUP	ZEOLITE	ZEOLITH	TSEOLIT	ZEOLITA	ZEOLITE
4857 ZEOLITE FACIES IGNE	ZEOLITE FACIES	FACIES ZEOLITE	ZEOLITH=FAZIES	FATSIYA TSEOLI- TOVAYA *UF:* Fatsiya prenit= pumpelliitovaya	FACIES=ZEOLITA	FACIES DELLE ZEOLITI
4858 ZEOLITIZATION IGNE	ZEOLITIZATION	ZEOLITISATION	ZEOLITHISIERUNG	TSEOLITIZATSIYA	ZEOLITIZACION	ZEOLITIZZAZIONE
4859 ZIGZAG FOLD STRU	Zigzag fold FOLDS	Pli accordeon PLI EN CHEVRON	Zick=Zack=Falte FALTE	SKLADKA ZIGZA- GOOBRAZNAYA *UF:* Skladka strel'chataya	Pliegue=en=acordeon PLIEGUE	Piega a zigzag PIEGA A KINK
4860 ZINC CHEE	ZINC	ZINC	ZN	TSINK	CINC	ZINC
4861 ZIRCON COMS	ZIRCON	ZIRCON	ZIRKON	TSIRKON	CIRCON	ZIRCONE
4862 ZIRCONIUM CHEE	ZIRCONIUM	ZIRCONIUM	ZR	TSIRKONIJ	CIRCONIO	ZIRCONIO
4863 ZOANTHINIARIA PALS	ZOANTHINIARIA	ZOANTHINIARIA	ZOANTHINIARIA	Zoanthiniaria ANTHOZOA	ZOANTHINIARIA	ZOANTHINIARIA
4864 ZOECIUM PALE	Zooecium BRYOZOA	Zoecium BRYOZOA	Zoezium BRYOZOA	TSISTID	Zoecia BRYOZOA	Zoecium BRYOZOA
4865 ZONAL STRUCTURE TEST	Zonal structure ZONING; TEX- TURES	Structure zonee ZONAGE CRISTAL	Zonal.Gefuege	ZONAL'NYJ	Estructura=zonal ZONACION= CRISTAL	Struttura zonata ZONATURA CRIS- TALLINA
4866 ZONE OF AERATION GEOH	Zone of aeration MOISTURE; SOILS	Zone aeration ZONE NON SATUREE	Nichtgesaettigt.Zone UNGESAETTIGT. ZONE	ZONA AEHRATSII *UF:* Non=saturated zone	Zona=de=aireacion ZONA=NO= SATURADA	Zona d'areazione ZONA NON SATURATA
4867 ZONE OF CEMENTATION ECON	ZONE OF CEMEN- TATION	GITE CEMENTA- TION	ZEMENTATIONS= LAGERSTAETTE	ZONA TSE- MENTATSII	ZONA=DE= CEMENTACION	ZONA DI CEMENTAZIONE
4868 ZONE OF FLOW SOLI	Zone of flow PLASTICITY	Zone plastique PLASTICITE	Fliess=Zone PLASTIZITAET	ZONA PLASTICH- NOSTI	Zona=de=flujo PLASTICIDAD	Zona di scorrimento PLASTICITA
4869 ZONE OF SEDIMENTATION ECON	Zone of sedimentation SEDIMENTATION	Zone sedimentation SEDIMENTATION	Sedimentations=Zone SEDIMENTATION	ZONA SEDI- MENTATSII	Zona=de= sedimentacion SEDIMENTACION	Zona di sedimentazione SEDIMENTAZIONE
4870 ZONING GEOL	ZONING *UF:* Biozones; Ceno- zones; Chronozones; Climate zonation; Ecozones; Faunizones; Florizones; Range zone; Zonal structure	ZONAGE *UF:* Structure zonee	ZONARBAU	ZONAL'NOST' *UF:* Zonal'nost klimaticheskaya	ZONACION *UF:* Estructura=zonal	ZONATURA *UF:* Struttura zonata
4871 ZOOID PALE	Zooid ANATOMY	Zooide ANATOMIE	Zooid ANATOMIE	ZOOID	Zooide ANATOMIA	Zooide ANATOMIA

Linguistic and Field Indexes

English

ALLOYS, 139
ALLUVIAL, 140
ALLUVIAL FANS, 141
ALLUVIAL GROUNDWATER, 143
ALLUVIAL PLAINS, 144
ALLUVIAL PROSPECTING, 145
ALLUVIUM, 146
ALPHA RAYS, 147
ALPHA-RAY SPECTROSCOPY, 148
ALPINE ENVIRONMENT, 149
ALPINE OROGENY, 150
ALPINE STRUCTURE, 151
ALPINE TRIASSIC, 152
ALPINE-TYPE, 153
ALTERATION, 154
ALTITUDE, 155
ALUMINA, 156
ALUMINUM, 157
ALUNITIZATION, 158
ALVEOLINELLIDAE, 159
AMBER, 160
AMBLYPODA, 161
AMELIORATION, 2742
AMERICIUM, 162
AMINO ACIDS, 163
AMMODISCACEA, 164
AMMONIA, 165
AMMONITIDA, 166
AMMONITINA, 167
AMMONIUM, 168
AMMONOIDEA, 169
AMORPHOUS MATERIALS, 170
AMPHIBIA, 171
AMPHIBOLE, 172
AMPHIBOLITE FACIES, 174
AMPHIBOLITES, 173
AMPHIBOLITIZATION, 175
AMPHINEURA, 176
AMPLITUDE, 177
ANAEROBIC ENVIRONMENT, 178
ANAEROBIC TAXA, 179
ANALOG SIMULATION, 180
ANALOG TECHNIQUES, 181
ANALYSIS, 182
ANAPSIDA, 184
ANASPIDA, 185
ANATEXIS, 186
ANATEXITE, 187
ANATOMY, 188
ANCHIMETAMORPHISM, 189
ANCHORS, 190
ANCIENT ICE AGES, 191
ANDALUSITE, 192
ANDEPTS, 193
ANDESITE LINE, 195
ANDESITES, 194
ANDOSOLS, 196

ANELASTIC MATERIALS, 197
ANELASTICITY, 198
ANGIOSPERMS, 199
ANGLE OF REPOSE, 200
ANGRITE, 201
ANGULAR UNCONFORMITIES, 202
ANHEDRAL CRYSTALS, 203
ANHYDRITE, 204
ANHYDROUS MATERIALS, 205
ANHYSTERETIC REMANENT MAGNETIZATION, 206
ANIONS, 207
ANISIAN, 208
ANISOPLEURA, 209
ANISOTROPY, 210
ANKERITE, 211
ANNELIDA, 212
ANOMALIES, 214
ANOMALODESMATA, 213
ANORTHOSITE, 216
ANORTHOSITIZATION, 217
ANTECLISE, 218
ANTHOCYATHEA, 219
ANTHOZOA, 220
ANTHRACITE, 221
ANTHRAXOLITE, 222
ANTIARCHI, 223
ANTICLINES, 224
ANTICLINORIA, 225
ANTIDUNES, 226
ANTIFORM FOLDS, 227
ANTIMONATES, 228
ANTIMONY, 229
ANTIPATHARIA, 230
ANTIPERTHITE, 231
ANTISTRESS MINERALS, 232
ANTITHETIC FAULTS, 233
ANUROMORPHA, 234
APERTURES, 235
APHANITIC TEXTURE, 236
APICAL, 237
APLACOPHORA, 238
APLITE, 239
APLITIC TEXTURE, 240
APOGEE, 241
APOGRANITE, 242
APOLLO PROGRAM, 243
APOMAGMATIC DEPOSIT, 244
APOPHYSIS, 245
APPLIED GEOLOGY, 246
APPLIED GEOPHYSICS, 247
APPLIED HYDROGEOLOGY, 248
APPROXIMATION, 249
APRON, 250
APSHERONIAN, 251
APSIDOSPONDYLI, 252
APTIAN, 253
AQUALFS, 254

AQUENTS, 255
AQUEOUS SOLUTIONS, 256
AQUEPTS, 257
AQUERTS, 258
AQUICLUDES, 259
AQUIFER TEST, 261
AQUIFERS, 260
AQUIFUGES, 262
AQUITANIAN, 263
AQUITARDS, 264
AQUODS, 265
AQUOLLS, 266
AQUOX, 267
AQUULTS, 268
AR/AR, 297
ARACHNIDA, 269
ARCH DAMS, 271
ARCHAEOCOPIDA, 272
ARCHAEOCYATHA, 273
ARCHAEOCYATHEA, 274
ARCHAEOGASTROPODA, 275
ARCHAEOLOGICAL SITES, 280
ARCHAEOLOGY, 281
ARCHAEOMAGNETISM, 282
ARCHAEOPTERYGES, 276
ARCHAEORNITHES, 277
ARCHEAN, 278
ARCHEIDES, 279
ARCHES, 270
ARCHIPOLYPODA, 283
ARCHOSAURIA, 284
ARCOIDA, 285
ARCTIC, 286
ARCTIC SUITE, 287
ARCUATE, 288
ARCUATE FAULTS, 289
AREAL GEOLOGY, 3793
ARENACEOUS TEXTURE, 290
ARENIGIAN, 291
ARENITE, 292
ARENOSOLS, 293
ARGIDS, 294
ARGILLACEOUS TEXTURE, 295
ARGON, 296
ARHEISM, 298
ARID ENVIRONMENT, 299
ARIDISOLS, 300
ARIDITY, 301
ARKOSE, 302
ARMORED MUD BALLS, 303
ARMORICAN OROGENY, 304
AROMATIC HYDROCARBONS, 305
ARRAYS, 306
ARRIVAL TIME, 307
ARROYOS, 308
ARSENATES, 309
ARSENIC, 310

ARTERITE, 311
ARTESIAN, 312
ARTESIAN AQUIFER, 313
ARTESIAN WELLS, 314
ARTHRODIRA, 315
ARTHROPLEURIDA, 316
ARTHROPODA, 317
ARTICULATA, 318
ARTIFACTS, 319
ARTIFICIAL RECHARGE, 321
ARTINSKIAN, 323
ARTIODACTYLA, 324
ASBESTOS, 325
ASCENSION THEORY, 326
ASEISMIC DESIGN, 327
ASEISMIC MARGINS, 328
ASEISMIC REGIONS, 329
ASH, 320
ASH CONTENT, 330
ASH FALLS, 331
ASH FLOWS, 332
ASHGILLIAN, 333
ASPHALT, 334
ASPHALTITE, 335
ASSAY VALUE, 337
ASSAYS, 336
ASSEMBLAGE ZONES, 339
ASSEMBLAGES, 338
ASSIMILATION, 340
ASSOCIATIONS, 341
ASSYNTIC OROGENY, 342
ASTATINE, 343
ASTERISM, 344
ASTEROIDEA, 346
ASTEROIDS, 345
ASTEROZOA, 347
ASTHENOSPHERE, 348
ASTRAPOTHERIA, 349
ASTROBLEMES, 350
ASTROGEOLOGY, 351
ASTRONOMY, 352
ASYMMETRIC FOLDS, 353
ATAXITE, 354
ATECTONIC PROCESSES, 215
ATLANTIC SUITE, 355
ATLAS, 356
ATMOGENIC, 357
ATMOSPHERE, 358
ATMOSPHERIC ARGON, 359
ATMOSPHERIC PRECIPITATION, 360
ATMOSPHERIC PRESSURE, 361
ATOLLS, 362
ATOMIC ABSORPTION, 363
ATOMIC ABSORPTION SPECTRA, 15
ATOMIC BOND, 364
ATOMIC PACKING, 365
ATTENUATION, 366

ATTERBERG LIMITS, 367
ATTRITION, 368
AUBRITE, 369
AUGEN STRUCTURE, 370
AUGER DRILLING, 371
AULACOGENS, 372
AUSTRALITE, 374
AUTHIGENESIS, 375
AUTHIGENIC MINERALS, 376
AUTOCHTHONS, 377
AUTOCORRELATION, 378
AUTOLITHS, 379
AUTOLYSIS, 380
AUTOMATIC CARTOGRAPHY, 381
AUTOMATION, 382
AUTOMETAMORPHISM, 383
AUTOMETASOMATISM, 384
AUTORADIOGRAPHY, 385
AUTOTROPHIC ORGANISM, 386
AUTUNIAN, 387
AVALANCHES, 388
AVERAGE, 389
AVES, 390
AXIAL-PLANE CLEAVAGE, 392
AXIAL-PLANE STRUCTURES, 391
AXIS, 393
AZONAL SOILS, 394
B LAYER, 395
BACK ARC BASINS, 396
BACK REEF, 397
BACKGROUND, 398
BACKSET BEDS, 399
BACTERIA, 400
BACTERIOGENIC MATERIALS, 401
BACTRITOIDEA, 402
BADLANDS, 403
BAIKALIAN PHASE, 404
BAJADAS, 405
BAJOCIAN, 406
BALANCE, 407
BALLAST, 408
BANDED STRUCTURES, 409
BARCHANS, 411
BARITE, 412
BARIUM, 413
BARREMIAN, 414
BARREN DEPOSITS, 415
BARRIER BARS, 417
BARRIER BEACHES, 418
BARRIER LAGOONS, 419
BARRIER REEFS, 420
BARRIERS, 416
BARS, 410
BARTONIAN, 421
BASAL, 422
BASAL CLEAVAGE, 423
BASAL CONGLOMERATE, 424

BASALTIC LAYER, 426
BASALTS, 425
BASANITE, 427
BASE METALS, 428
BASEMENT, 429
BASEMENT TECTONICS, 430
BASHKIRIAN, 431
BASIFICATION, 432
BASIN-PLANNING, 433
BATHOLITHS, 434
BATHONIAN, 435
BATHYAL ENVIRONMENT, 436
BATHYMETRIC MAPS, 437
BATHYMETRY, 438
BATRACHOSAURIA, 439
BAUXITE, 440
BAUXITIZATION, 441
BAVENO TWIN LAW, 442
BAYS, 443
BE-10/BE-9, 472
BEACH PLACERS, 445
BEACH RIDGES, 446
BEACH SCARPS, 447
BEACHES, 444
BEACHROCK, 448
BEAN ORE, 449
BEARING CAPACITY, 450
BECKE LINE, 451
BEDDING, 453
BEDDING FAULTS, 456
BEDDING PLANES, 454
BEDDING-PLANE CLEAVAGE, 455
BEDLOAD, 452
BEDROCK, 457
BEHEADED STREAMS, 458
BELEMNOIDEA, 459
BELTIAN OROGENY, 460
BENCHES, 461
BENIOFF ZONE, 462
BENNETTITALES, 463
BENTHONIC ENVIRONMENT, 464
BENTHOS, 465
BENTONITE, 466
BERGSCHRUND, 467
BERKELIUM, 468
BERMS, 469
BERRIASIAN, 470
BERYLLIUM, 471
BETA RAYS, 473
BEYRICHICOPINA, 474
BIAXIAL, 475
BIAXIAL CRYSTALS, 476
BIBLIOGRAPHIC REVIEW, 3843
BIBLIOGRAPHY, 477
BIGHTS, 478
BINARY SYSTEM, 479
BINDERS, 2945

BINOMIAL DISTRIBUTION, 480
BIOCALCARENITE, 487
BIOCENOSES, 481
BIOCHEMICAL SEDIMENTATION, 482
BIOCHEMISTRY, 496
BIOCHORE, 483
BIOCHRON, 484
BIOCHRONOLOGY, 485
BIOCLASTIC SEDIMENTATION, 488
BIOCLASTS, 486
BIODEGRADATION, 489
BIOFACIES, 490
BIOFACIES MAP, 491
BIOGENESIS, 492
BIOGENIC EFFECTS, 493
BIOGENIC ORIGIN, 3172
BIOGENIC STRUCTURES, 494
BIOGEOCHEMICAL METHODS, 495
BIOGEOGRAPHY, 497
BIOGRAPHY, 498
BIOHERMS, 499
BIOLITH, 500
BIOLOGIC EVOLUTION, 1461
BIOLOGICAL CYCLE, 501
BIOLOGY, 502
BIOMASS, 503
BIOMETRY, 504
BIOMICRITE, 505
BIORHEXISTASY, 506
BIOSOMES, 507
BIOSPARITE, 508
BIOSPHERE, 509
BIOSTRATIGRAPHIC UNIT, 510
BIOSTRATIGRAPHY, 511
BIOSTROMES, 512
BIOTA, 513
BIOTOPES, 514
BIOTURBATION, 515
BIOTYPES, 516
BIOZONES, 517
BIRCH DISCONTINUITY, 518
BIREFRINGENCE, 519
BISMUTH, 520
BITUMENS, 521
BITUMINIZATION, 522
BITUMINOUS COAL, 523
BITUMINOUS SHALE, 524
BIVALVIA, 3332
BLASTIC TEXTURE, 525
BLASTOIDEA, 526
BLATTODEA, 527
BLEACHING CLAY, 528
BLIND DEPOSITS, 529
BLOCK DIAGRAMS, 531
BLOCK STRUCTURES, 532
BLOCKS, 530
BLOWHOLES, 533

BLUESCHIST, 534
BODY WAVES, 535
BOG IRON ORE, 537
BOGHEAD COAL, 538
BOGS, 536
BOMBS, 539
BONE BEDS, 540
BOOK REVIEWS, 541
BOOLEAN ALGEBRA, 542
BORALFS, 543
BORATES, 544
BORAX, 545
BORDER FACIES, 546
BORDERLAND, 547
BOREAL, 548
BOREHOLE SECTION, 550
BOREHOLES, 549
BOROLLS, 551
BORON, 552
BOTRYOIDAL TEXTURE, 553
BOTTOM SEDIMENTS, 554
BOTTOMSET BEDS, 555
BOUDINAGE, 556
BOUGUER ANOMALIES, 557
BOULDER CLAY, 559
BOULDER TRAINS, 560
BOULDERS, 558
BOUNDARY, 4275
BOUNDARY, 913
BOUNDARY LAYER, 561
BOUNDSTONE, 562
BOWEN'S REACTION SERIES, 563
BOXWORK TEXTURE, 564
BRACHIOPODA, 565
BRACHIOPTERYGII, 566
BRACHYFOLD, 567
BRACKISH WATER, 568
BRACKISH-WATER ENVIRONMENT, 569
BRAGG ANGLE, 570
BRAIDED CHANNELS, 571
BRANCHIOPODA, 572
BRANCHIURA, 573
BRAVAIS LATTICE, 574
BRECCIA, 575
BRICK CLAY, 576
BRIDGES, 577
BRINES, 578
BRITTLE, 579
BROMIDES, 580
BROMINE, 581
BROWN COAL, 582
BROWN SOILS, 583
BRYOPHYTES, 584
BRYOZOA, 585
BUILDING STONE, 588
BUILDINGS, 586
BULIMINACEA, 589

BULK DENSITY, 590
BULK MODULUS, 591
BULLARD DISCONTINUITY, 592
BUNTER, 593
BURDIGALIAN, 594
BUREAU OF MINES, 595
BURIAL, 596
BURIAL METAMORPHISM, 597
BURIED CHANNELS, 598
BURIED VALLEYS, 599
BURROWS, 600
BYSSUS, 601
C LAYER, 602
C-13/C-12, 645
C-14, 646
C.I.P.W. CLASSIFICATION, 603
CADMIUM, 604
CAFEMIC COMPOSITION, 605
CALABRIAN, 606
CALC-ALKALIC COMPOSITION, 607
CALCAREOUS ALGAE, 608
CALCAREOUS COMPOSITION, 609
CALCIC COMPOSITION, 610
CALCIPHYRE, 611
CALCISPONGEA, 612
CALCITE, 613
CALCITIZATION, 614
CALCIUM, 615
CALCRETE, 616
CALDERAS, 617
CALEDONIAN OROGENY, 618
CALICHE, 619
CALIFORNIUM, 620
CALIPER LOGGING, 621
CALLOVIAN, 622
CAMAROIDEA, 623
CAMBISOLS, 624
CAMBRIAN, 625
CAMPANIAN, 626
CAMPTOSTROMATOIDEA, 627
CANALS, 628
CANNEL COAL, 629
CANYONS, 630
CAP ROCKS, 631
CAPES, 632
CAPILLARITY, 633
CAPILLARY PRESSURE, 635
CAPILLARY WATER, 636
CARADOCIAN, 638
CARAPACE, 639
CARAT, 640
CARBOHYDRATES, 641
CARBON, 642
CARBON DIOXIDE, 643
CARBON MONOXIDE, 644
CARBONACEOUS CHONDRITES, 647
CARBONACEOUS COMPOSITION, 648

CARBONACEOUS ROCK, 649
CARBONATE COMPENSATION DEPTH, 1124
CARBONATE ROCKS, 650
CARBONATES, 651
CARBONATITES, 652
CARBONATIZATION, 653
CARBONIFEROUS, 654
CARNIAN, 655
CARNIVORA, 657
CARTERINACEA, 658
CARTOGRAMS, 659
CARTOGRAPHIC SCALES, 2669
CARTOGRAPHY, 660
CASCADE FOLDS, 661
CASCADIAN OROGENY, 662
CASE STUDIES, 663
CASSIDULINACEA, 664
CASTS, 665
CATACLASIS, 667
CATAGENESIS, 668
CATAGRAPHIA, 669
CATALYSIS, 670
CATASTROPHISM, 671
CATAZONAL METAMORPHISM, 2377
CATCHMENT HYDRODYNAMICS, 672
CATENAS, 673
CATIONS, 674
CAULDRONS, 675
CAUSTOBIOLITH, 676
CAVERNOUS TEXTURE, 679
CAVERNS, 678
CAVES, 677
CAVINGS, 680
CAYTONIALES, 681
CELL DIMENSIONS, 2451
CELLULAR DOLOMITE, 656
CELLULAR POROSITY, 682
CELLULOSE, 683
CEMENT, 684
CEMENT MATERIALS, 587
CEMENTATION, 685
CENOMANIAN, 686
CENOTYPAL, 687
CENOZOIC, 688
CENOZONES, 689
CENTROCLINES, 690
CEPHALOCARIDA, 691
CEPHALODISCIDA, 692
CEPHALOPODA, 693
CERAMIC MATERIALS, 694
CERATITIDA, 695
CERATOMORPHA, 696
CERIANTHARIA, 697
CERIANTIPATHARIA, 698
CERIUM, 699
CESIUM, 700
CETACEA, 701

CHAETOGNATHA, 702
CHALCOPHILE ELEMENTS, 703
CHALCOSPHERE, 704
CHALK, 705
CHANDLER WOBBLE, 706
CHANGES OF LEVEL, 3981
CHANNEL GEOMETRY, 708
CHANNELS, 707
CHAOTIC MATERIALS, 709
CHARACTERISTICS, 710
CHARNIAN OROGENY, 711
CHARNOCKITE, 712
CHAROPHYTA, 713
CHASSIGNITE, 714
CHATTIAN, 716
CHEILOSTOMATA, 717
CHELATION, 718
CHELICERATA, 719
CHELONIA, 720
CHEMICAL ANALYSIS, 183
CHEMICAL COMPOSITION, 721
CHEMICAL CONTROLS, 722
CHEMICAL EQUILIBRIUM, 724
CHEMICAL EVOLUTION, 725
CHEMICAL EXPLOSIONS, 726
CHEMICAL FOSSILS, 727
CHEMICAL FRACTIONATION, 1630
CHEMICAL INDUSTRY, 728
CHEMICAL PROPERTIES, 729
CHEMICAL REMANENT MAGNETIZATION, 732
CHEMICAL SEDIMENTATION, 723
CHEMICAL SYSTEM, 733
CHEMICAL WEATHERING, 735
CHEMOGENIC ORIGIN, 736
CHENIERS, 737
CHERNOZEMS, 738
CHERT, 739
CHERTIFICATION, 740
CHESTNUT SOILS, 666
CHILOPODA, 742
CHINA CLAY, 743
CHIROPTERA, 744
CHITIN, 745
CHITINOZOA, 746
CHLORIDES, 747
CHLORINATION, 748
CHLORINE, 749
CHLORINITY, 750
CHLORITE SCHIST, 751
CHLORITIZATION, 752
CHLOROPHYLL, 753
CHLOROPHYTA, 754
CHONDRICHTHYES, 755
CHONDRITES, 756
CHONDRULES, 757
CHORDATA, 758
CHROMATES, 759

CHROMATOGRAPHY, 760
CHROMIUM, 761
CHRONOSTRATIGRAPHIC UNIT, 762
CHRONOSTRATIGRAPHY, 763
CHRONOZONES, 764
CHRYSOPHYTA, 765
CILIATA, 766
CILIOPHORA, 767
CIMMERIAN OROGENY, 768
CINDER, 769
CINDER CONES, 770
CIPOLIN, 771
CIRCULAR POLARIZATION, 772
CIRCULATION, 773
CIRCUM-PACIFIC REGION, 774
CIRQUES, 775
CIRRIPEDIA, 776
CIVIL ENGINEERING, 777
CLADOCERA, 778
CLAIMS, 779
CLARAIN, 780
CLARKE, 781
CLASSIFICATION, 782
CLASTIC, 784
CLASTIC DIKES, 785
CLASTIC ROCKS, 786
CLASTS, 783
CLAY BALLS, 788
CLAY MINERALOGY, 789
CLAY MINERALS, 790
CLAYS, 787
CLAYSTONE, 791
CLEAVAGE, 792
CLIFFS, 793
CLIMATE, 794
CLIMATE EFFECTS, 795
CLIMATE TRACE, 796
CLIMATE ZONATION, 797
CLOSE PACKING, 798
CLUSTER ANALYSIS, 799
CLYMENIIDA, 800
COAL, 801
COAL BALLS, 802
COAL FIELDS, 803
COAL GAS, 804
COAL MEASURES, 805
COAL SEAMS, 806
COAL-DEPOSIT MAPS, 807
COALIFICATION, 808
COARSE GRAVEL, 809
COARSE-GRAINED MATERIALS, 810
COAST, 811
COASTAL DUNES, 812
COASTAL GEOMORPHOLOGY, 814
COASTAL PLAINS, 815
COASTLINES, 816
COATED GRAIN, 817

COBALT, 818
COBOL, 819
COCCOLITHOPHORACEAE, 820
CODA WAVES, 821
COEFFICIENT, 822
COELENTERATA, 823
COENOTHECALIA, 824
COERCIVITY, 825
COEXISTING MINERALS, 826
COHESION, 827
COHESIONLESS MATERIALS, 2202
COHESIONLESS SOILS, 828
COHESIVE MATERIALS, 829
COHESIVE SOILS, 830
COKE COAL, 831
COLD ENVIRONMENT, 832
COLEOPTEROIDA, 833
COLLAPSE, 834
COLLAPSE BRECCIA, 835
COLLAPSE CALDERA, 836
COLLAPSE STRUCTURES, 837
COLLECTING, 838
COLLECTIONS, 839
COLLEMBOLA, 840
COLLOFORM TEXTURES, 841
COLLOIDAL MATERIALS, 842
COLLUVIUM, 843
COLONIAL TAXA, 844
COLOR, 845
COLOR CENTERS, 846
COLOR INDEX, 847
COLORIMETRY, 849
COLUMNAR JOINTS, 850
COLUMNARIINA, 851
COMAGMATIC MATERIALS, 852
COMMENSALISM, 853
COMMON=DEPTH=POINT METHOD, 854
COMMUNITIES, 855
COMPACTIBILITY, 856
COMPACTION, 857
COMPANIES, 858
COMPETENT MATERIALS, 859
COMPLEXES, 860
COMPLEXING, 861
COMPLEXOMETRY, 862
COMPOSITION, 863
COMPRESSIBILITY, 864
COMPRESSION, 865
COMPRESSION TECTONICS, 867
COMPRESSION TEST, 868
COMPRESSIVE STRENGTH, 866
COMPRESSIVE STRESS, 869
COMPUTER PROGRAMS, 872
COMPUTERS, 870
CONCENTRATION, 873
CONCENTRIC FOLDS, 874
CONCHOIDAL MATERIALS, 875

CONCHOSTRACA, 876
CONCORDANCE, 877
CONCORDANT AGE, 878
CONCRETE, 879
CONCRETE DAMS, 880
CONCRETIONS, 881
CONDENSATES, 1688
CONDENSATION, 882
CONDUCTIVITY, 883
CONDYLARTHRA, 884
CONE OF DEPRESSION, 885
CONE=IN=CONE, 886
CONFINED AQUIFERS, 887
CONFLUENT STREAMS, 888
CONFORMITY, 889
CONGELATION, 890
CONGELIFRACTION, 891
CONGLOMERATE, 892
CONGRUENT MELTING, 893
CONIACIAN, 894
CONIFERALES, 895
CONJUGATE, 896
CONJUGATE FOLDS, 897
CONNATE WATERS, 898
CONOCARDIOIDA, 899
CONODONT APPARATUS, 900
CONODONTA, 901
CONRAD DISCONTINUITY, 902
CONSANGUINITY, 903
CONSEQUENT STREAMS, 904
CONSERVATION, 905
CONSISTENCY, 906
CONSOLIDATION, 907
CONSOLIDOMETER TESTS, 908
CONSTANT, 909
CONSTRUCTION, 910
CONSTRUCTION MATERIALS, 911
CONSUMPTION, 912
CONTACT BRECCIA, 914
CONTACT METAMORPHISM, 915
CONTACT METASOMATISM, 916
CONTAMINATION, 917
CONTINENTAL ACCRETION, 919
CONTINENTAL BORDERLAND, 920
CONTINENTAL CRUST, 921
CONTINENTAL DRIFT, 922
CONTINENTAL ENVIRONMENT, 923
CONTINENTAL GLACIATION, 924
CONTINENTAL MARGIN, 925
CONTINENTAL NUCLEUS, 926
CONTINENTAL RISE, 927
CONTINENTAL SEDIMENTATION, 928
CONTINENTAL SHELF, 929
CONTINENTAL SLOPE, 930
CONTINENTAL ZONE, 931
CONTINENTS, 918
CONTOUR MAPS, 932

CONTRACTING EARTH, 933
CONTRACTION, 934
CONTRAST, 935
CONTROL, 936
CONULARIDA, 937
CONVECTION, 938
CONVECTION CURRENTS, 939
CONVERGENCE, 940
CONVOLUTE BEDS, 941
COOLING, 942
COORDINATION, 943
COPEPODA, 944
COPERNICAN, 945
COPPER, 946
COPROLITES, 947
COQUINA, 948
CORAL REEFS, 949
CORALLIMORPHARIA, 950
CORALLINACEAE, 951
CORALLITE, 952
CORDAITALES, 953
CORDIERITE=AMPHIBOLITE=FACIES, 954
CORDILLERA, 955
CORE, 957
CORE DRILLING, 956
CORES, 958
CORIOLIS FORCE, 959
CORRECTIONS, 960
CORRELATION, 961
CORRELATION COEFFICIENT, 962
CORRESPONDENCE ANALYSIS, 963
CORROSION, 964
CORUNDUM, 965
CORYNEXOCHIDA, 966
COSMIC DUST, 967
COSMIC NEBULA, 968
COSMIC PARTICLE, 969
COSMIC RAYS, 970
COSMIC SYSTEM, 971
COSMOCHEMISTRY, 972
COSMOGENIC ELEMENTS, 973
COSMOLOGICAL PRINCIPLE, 974
COSMOLOGY, 975
COSTS, 976
COTYLOSAURIA, 977
COUNTRY ROCKS, 978
COVER BEDS, 979
CRANIUM, 980
CRATER LAKES, 982
CRATERING, 983
CRATERS, 981
CRATONS, 984
CREEKS, 985
CREEP, 986
CREODONTA, 987
CRESTS, 988
CRETACEOUS, 989

CREVASSE, 990
CRINOIDEA, 991
CRINOZOA, 992
CRITICAL ANGLE, 993
CRITICAL FLOW, 994
CRITICAL REVIEW, 995
CROCODILIA, 996
CROSS=BEDDING, 997
CROSS=STRATIFICATION, 998
CROSSOPTERYGII, 999
CRUSH BRECCIA, 1000
CRUST, 1001
CRUSTACEA, 1002
CRYOPEDOLOGY, 1003
CRYOTECTONICS, 1004
CRYOTURBATION, 1005
CRYPTOCRYSTALLINE TEXTURE, 1006
CRYPTODONTA, 1007
CRYPTOGENIC, 1008
CRYPTOSTOMATA, 1009
CRYPTOVOLCANIC STRUCTURE, 1010
CRYSTAL CHEMISTRY, 1012
CRYSTAL DISLOCATIONS, 1014
CRYSTAL FIELD, 1015
CRYSTAL FORM, 1016
CRYSTAL GROWTH, 1017
CRYSTAL OPTICS, 1019
CRYSTAL STRUCTURE, 1020
CRYSTAL SYSTEMS, 1021
CRYSTALLINE ROCKS, 1022
CRYSTALLINITY, 1023
CRYSTALLITES, 1024
CRYSTALLIZATION, 1025
CRYSTALLIZATION INDEX, 1026
CRYSTALLOBLAST, 1027
CRYSTALLOGRAPHY, 1028
CRYSTALS, 1011
CTENOSTOMATA, 1029
CUBIC SYSTEM, 1030
CUESTAS, 1031
CUMULATES, 1032
CURIE POINT, 1033
CURIUM, 1034
CURRENT MARKINGS, 1036
CURRENTS, 1035
CUSPS, 1037
CUTOFF GRADE, 1038
CUTTINGS, 1039
CYANOPHYTA, 1040
CYBERNETICS, 1041
CYCADALES, 1042
CYCADOFILICALES, 1043
CYCLE OF SEDIMENTATION, 1045
CYCLES, 1044
CYCLIC LOADING, 1046
CYCLIC PROCESSES, 1047
CYCLOCYSTOIDEA, 1048

CYCLONES, 1049
CYCLOSTOMATA, 1050
CYCLOTHEMS, 1051
CYLINDRICAL FOLDS, 1052
CYLINDRICAL STRUCTURES, 1053
CYNOMORPHA, 1054
CYSTIPHYLLINA, 1055
CYSTOIDEA, 1056
D LAYER, 1057
D/H, 1058
DAMS, 1059
DANIAN, 1060
DARCY'S LAW, 1061
DARWINISM, 1062
DASYCLADACEAE, 1063
DATA, 1064
DATA ACQUISITION, 1065
DATA ANALYSIS, 1066
DATA BASES, 1067
DATA HANDLING, 1068
DATA PROCESSING, 1069
DATA RETRIEVAL, 1070
DATA STORAGE, 1071
DATING, 1073
DAUGHTER PRODUCT, 1074
DEAD ICE, 1075
DEBRIS CONES, 1076
DEBRIS FLOWS, 1077
DECAPODA, 1078
DECAY CONSTANT, 1079
DECCAN TRAPS, 1080
DECLINATION, 1081
DECOLLEMENT, 1082
DECONTAMINATION, 1083
DECONVOLUTION, 1084
DECORATIVE STONES, 1085
DECREPITATION, 1086
DEDOLOMITIZATION, 1087
DEEP=FOCUS EARTHQUAKES, 1088
DEEP=SEA CHANNEL, 1089
DEEP=SEA ENVIRONMENT, 1090
DEEP=SEA SEDIMENTATION, 23
DEEP=SEATED STRUCTURES, 1094
DEEP SEISMIC SOUNDING, 1091
DEEP=TOW METHODS, 1092
DEEP=WATER ZONE, 1093
DEEPS, 17
DEFECTS, 1013
DEFLATION, 1095
DEFORMATION, 1096
DEGASSING, 1097
DEGLACIATION, 1098
DEGRADATION, 1099
DEGREE OF FREEDOM, 1100
DEHYDRATION, 1101
DEINOTHERIOIDEA, 1102
DELTAIC ENVIRONMENT, 1104

DELTAIC SEDIMENTATION, 1105
DELTAS, 1103
DEMAGNETIZATION, 1106
DEMOSPONGEA, 1107
DENDRITE, 1108
DENDRITIC DRAINAGE PATTERN, 1109
DENDROCHRONOLOGY, 1110
DENDROGRAMS, 1111
DENDROIDEA, 1112
DENSITY, 1113
DENTITION, 1115
DENUDATION, 1116
DEPOSITION, 1120
DEPRESSIONS, 1121
DEPTH, 1122
DEPTH INDICATORS, 1123
DERIVATIVE, 1125
DERMAL STRUCTURES, 1126
DERMAPTEROIDA, 1127
DERMOPTERA, 1128
DESALINIZATION, 1129
DESCENSION THEORY, 1130
DESERT DUNE, 1132
DESERT PAVEMENTS, 1133
DESERTIFICATION, 1134
DESERTS, 1131
DESICCATION, 1135
DESILICATION, 1136
DESMIDS, 1137
DESMOCERATIDA, 1138
DESMOSTYLIA, 1139
DETERMINISM, 1140
DETERMINISTIC MODEL, 1141
DETRITAL MATERIALS, 1142
DETRITAL REMANENT MAGNETIZATION, 1143
DETRITAL SEDIMENTATION, 1144
DEUTERIC COMPOSITION, 1145
DEUTERIUM, 1146
DEVELOPMENT, 1147
DEVELOPMENT DRILLING, 1148
DEVIATION, 1149
DEVITRIFICATION, 1150
DEVONIAN, 1151
DEW, 1152
DIABASE, 1234
DIACHRONISM, 1153
DIADOCHY, 1154
DIAGENESIS, 1155
DIAGNOSIS, 1156
DIAGRAMS, 1157
DIALYSIS, 1158
DIAMAGNETIC PROPERTY, 1159
DIAMICTITE, 1160
DIAMONDS, 1161
DIAPHTHORESIS, 1162
DIAPHTORITE, 1163
DIAPIRISM, 1165

DIAPIRS, 1164
DIASTEMS, 1166
DIASTROPHISM, 1167
DIATOMACEOUS EARTH, 1168
DIATOMITE, 1170
DIATOMS, 1169
DIATREMES, 1171
DICOTYLEDONEAE, 1172
DICTYONELLIDINA, 1173
DIELECTRIC CONSTANT, 1174
DIFFERENTIAL COMPACTION, 1175
DIFFERENTIAL THERMAL ANALYSIS, 1176
DIFFERENTIAL WEATHERING, 1177
DIFFERENTIATION, 1178
DIFFERENTIATION INDEX, 1179
DIFFRACTION, 1180
DIFFUSION, 1181
DIFFUSIVITY, 1182
DIGITAL SIMULATION, 1183
DIKE SWARMS, 1185
DIKES, 1184
DILATANCY, 1186
DILATION, 1187
DILUTION, 1188
DIMORPHISM, 1189
DINOCERATA, 1190
DINOFLAGELLATA, 1191
DINOSAURS, 1192
DIOGENITE, 1193
DIORITES, 1194
DIP, 1195
DIP FAULTS, 1196
DIP METER, 1197
DIP SLOPE, 1198
DIP-SLIP FAULTS, 1199
DIPLEUROZOA, 1200
DIPLOID MINERALS, 1201
DIPLOPODA, 1202
DIPLORHINA, 1203
DIPMETER LOGGING, 1204
DIPNOI, 1205
DIPOLE MOMENT, 1206
DIPTERA, 1207
DIRECTIONAL DRILLING, 1208
DIRECTORY, 4847
DISCHARGE, 1209
DISCOASTERS, 1210
DISCONFORMITIES, 1211
DISCONTINUITIES, 1212
DISCORBACEA, 1213
DISCORDANCE, 1214
DISCORDANT, 1215
DISCORDANT AGE, 1216
DISCOVERIES, 1217
DISCRIMINANT ANALYSIS, 1218
DISHARMONIC FOLDS, 1219
DISJUNCTIVE FOLDS, 1220

DISLOCATION BRECCIA, 1221
DISPERSION PATTERNS, 1223
DISPLACEMENT THEORY, 1225
DISPLACEMENTS, 1224
DISSEMINATED DEPOSITS, 1226
DISSOCIATION, 1227
DISSOLVED MATERIALS, 1229
DISTRIBUTION, 1230
DIURNAL VARIATIONS, 1231
DIVERGENCE, 1232
DIVIDES, 4788
DOCODONTA, 1233
DOLINES, 1235
DOLOMITIC LIMESTONE, 1236
DOLOMITIZATION, 1237
DOLOSTONE, 1238
DOMAIN STRUCTURE, 1239
DOMERIAN, 1241
DOMES, 1240
DOWNWARPING, 1242
DRAG FOLDS, 1243
DRAG ORE, 1244
DRAINAGE BASINS, 1245
DRAINAGE PATTERNS, 1246
DRAINING, 1247
DRAPE FOLDS, 1248
DRAWDOWN, 1249
DRAWINGS, 1250
DREDGING, 1251
DRESSING, 1252
DRIFT, 1792
DRILL SLUDGE, 1256
DRILLABILITY, 1253
DRILLING, 1254
DRILLING EQUIPMENT, 1255
DRILLING VESSEL, 1257
DROUGHT, 1258
DRUMLINS, 1259
DRUSE, 1260
DUNES, 1261
DURAIN, 1262
DURICRUST, 2205
DUST, 1263
DYE TRACERS, 848
DYNAMIC EQUILIBRIUM, 1264
DYNAMIC GEOLOGY, 1265
DYNAMIC METAMORPHISM, 1266
DYNAMICS, 1267
DYNAMO THEORY, 1268
DYNAMOTHERMAL METAMORPHISM, 1269
DYSPROSIUM, 1270
E LAYER, 1271
EARLY DIAGENESIS, 1272
EARLY PROTEROZOIDES, 1273
EARTH, 3472
EARTH DAMS, 1275
EARTH PRESSURE, 1277

EARTH ROTATION, 1278
EARTH SCIENCE, 1279
EARTH TIDES, 1280
EARTH-CURRENT METHODS, 1274
EARTH-MOON COUPLE, 1282
EARTHFLOWS, 1283
EARTHQUAKE ENGINEERING, 1285
EARTHQUAKES, 1284
EARTHWORKS, 1281
EAST, 1288
EBURNEAN OROGENY, 1289
ECCENTRICITY, 1290
ECHINODERMATA, 1291
ECHINOIDEA, 1292
ECHINOZOA, 1293
ECHO SOUNDING, 1294
ECLOGITE, 1295
ECLOGITE FACIES, 1296
ECOLOGY, 1297
ECONOMIC EVALUATION, 1117
ECONOMIC GEOLOGY, 1298
ECONOMIC GEOLOGY MAPS, 1299
ECONOMIC MINERAL, 1300
ECOSYSTEMS, 1301
ECOZONES, 1302
ECTOTROPHA, 1303
EDDY FLOW, 1304
EDENTATA, 1305
EDGE WATER, 1306
EDRIOASTEROIDEA, 1307
EDRIOBLASTOIDEA, 1308
EDUCATION, 1309
EFFLUENTS, 1310
EGGS, 1311
EH, 1312
EIFELIAN, 1313
EINSTEINIUM, 1314
EJECTA, 1315
ELASMOBRANCHII, 1316
ELASTIC CONSTANTS, 1317
ELASTIC LIMIT, 1318
ELASTIC MATERIALS, 1319
ELASTIC PROPERTIES, 1320
ELASTIC STRAIN, 1321
ELASTIC WAVES, 1322
ELASTICITY, 1323
ELECTRIC POLARIZATION, 1330
ELECTRIC POTENTIAL, 1325
ELECTRICAL CONDUCTIVITY, 1326
ELECTRICAL EXPLORATION, 1327
ELECTRICAL FIELD, 1328
ELECTRICAL LOGGING, 1324
ELECTRICAL METHODS, 1329
ELECTRICAL PROPERTIES, 1331
ELECTRICAL SOUNDING, 1332
ELECTRICAL SURVEYS, 1333
ELECTRICITY, 1334

ELECTRO-OSMOSIS, 1335
ELECTROCHEMICAL PROPERTIES, 1336
ELECTRODES, 1337
ELECTROLYSIS, 1338
ELECTROLYTES, 1339
ELECTROMAGNETIC FIELD, 1340
ELECTROMAGNETIC LOGGING, 1343
ELECTROMAGNETIC METHODS, 1341
ELECTROMAGNETIC SURVEYS, 1342
ELECTROMAGNETISM, 1344
ELECTRON DIFFRACTION ANALYSIS, 1346
ELECTRON MICROSCOPE, 1348
ELECTRON MICROSCOPY, 1349
ELECTRON MICROSCOPY DATA, 1350
ELECTRON PARAMAGNETIC RESONANCE, 1351
ELECTRON PROBE, 1347
ELECTRON PROBE DATA, 1352
ELECTRONS, 1345
ELEMENTARY PARTICLES, 1353
ELEPHANTOIDEA, 1354
ELONGATE MINERALS, 1355
ELUTRIATION, 1356
ELUVIUM, 1357
EMBANKMENTS, 1358
EMBRECHITE, 1359
EMBRITHOPODA, 1360
EMERGENCE, 1361
EMISSION SPECTROSCOPY, 1362
EMPLACEMENT, 1363
EMSIAN, 1364
EN ECHELON FAULTS, 1365
ENCRINITE, 1366
END PRODUCT, 1367
ENDEMIC TAXA, 1368
ENDOCERATOIDEA, 1369
ENDOGENE PROCESSES, 1370
ENDORHEISM, 1371
ENDOTHERMIC REACTIONS, 1372
ENERGY SOURCES, 1373
ENGINEERING GEOLOGY, 1374
ENGINEERING PROPERTIES, 1769
ENRICHMENT, 1376
ENSIALIC GEOSYNCLINES, 1377
ENSIMATIC GEOSYNCLINES, 1378
ENSTATITE CHONDRITES, 1379
ENTEROLITHIC FOLDS, 1380
ENTEROPNEUSTA, 1381
ENTHALPY, 1382
ENTISOLS, 1383
ENTRENCHED STREAMS, 1384
ENTROPY, 1385
ENVIRONMENT, 1386
ENVIRONMENTAL ANALYSIS, 1387
ENVIRONMENTAL GEOLOGY, 1388
EOCENE, 1389
EOCRINOIDEA, 1390
EOLIAN ENVIRONMENT, 1391

EOLIAN FEATURES, 1392
EOLIAN SEDIMENTATION, 1393
EONS, 1394
EOSUCHIA, 1395
EPEIROGENY, 1396
EPHEMERAL STREAMS, 1397
EPIBIOTISM, 1398
EPIBOLITE, 1399
EPICENTERS, 1400
EPIDOTE-AMPHIBOLITE FACIES, 1401
EPIDOTIZATION, 1402
EPIEUGEOSYNCLINES, 1403
EPIGENE PROCESSES, 1404
EPIMAGMATIC ORIGIN, 1405
EPITAXY, 1406
EPITHERMAL PROCESSES, 1407
EPIZONE, 1408
EPOCHS, 1409
EPR SPECTRA, 1410
EQUATIONS, 1411
EQUATIONS OF STATE, 1412
EQUATOR, 1413
EQUATORIAL REGION, 1414
EQUIGRANULAR TEXTURE, 1415
EQUILIBRIUM, 1416
EQUISETALES, 1417
ERAS, 1418
ERATHEMS, 1419
ERATOSTHENIAN, 1420
ERBIUM, 1421
ERGS, 1422
ERODIBILITY, 1423
EROSION, 1424
EROSION CONTROL, 1425
EROSION CYCLE, 1426
EROSION FEATURES, 1427
EROSION REMNANT, 1428
EROSION SURFACES, 1429
ERRATICS, 1430
ERRORS, 1431
ERUPTIONS, 1432
ESKERS, 1434
ESSEXITE, 1435
ESTUARIES, 1438
ESTUARINE ENVIRONMENT, 1436
ESTUARINE SEDIMENTATION, 1437
ETCHING, 1439
ETHANE, 1440
ETHERS, 1441
EUCILIATA, 1442
EUCRITE, 1443
EUCRYSTALLINE TEXTURE, 1444
EUGEOSYNCLINES, 1445
EUGLENOPHYTA, 1446
EUHEDRAL CRYSTALS, 1447
EUROPIUM, 1448
EURYAPSIDA, 4389

EURYHALINE TAXA, 1449
EURYTHERMAL TAXA, 1450
EUSELACHII, 1451
EUSTACY, 1452
EUTAXITIC TEXTURE, 1453
EUTECTIC CONDITIONS, 1454
EUTROPHICATION, 1455
EUXINIC ENVIRONMENT, 1456
EVALUATION, 1457
EVAPORATION, 1458
EVAPORITES, 1459
EVAPOTRANSPIRATION, 1460
EXCAVATIONS, 1462
EXCESS ARGON, 1463
EXCHANGE CAPACITY, 1464
EXFOLIATION, 1466
EXHALATIVE PROCESSES, 1467
EXINE, 1468
EXINITE, 1469
EXOBIOLOGY, 1470
EXOGENE PROCESSES, 1471
EXORHEISM, 1472
EXOTHERMIC REACTIONS, 1473
EXPANDING EARTH, 1475
EXPANSIVE MATERIALS, 1474
EXPECTATION, 2709
EXPEDITIONS, 1476
EXPERIMENTAL PETROLOGY, 1477
EXPERIMENTAL SEISMOLOGY, 1478
EXPERIMENTAL STUDIES, 1479
EXPLANATORY TEXT, 1480
EXPLOITABILITY, 1481
EXPLOITATION, 1482
EXPLORATION, 1483
EXPLORATORY WELL, 1484
EXPLOSION BRECCIA, 1486
EXPLOSIONS, 1485
EXPLOSIVE SOURCE, 1487
EXPORT, 1488
EXPOSURE AGE, 1489
EXSOLUTION, 1490
EXSURGENCE, 1491
EXTENSOMETERS, 1492
EXTERNIDES, 1493
EXTINCT LAKES, 1494
EXTINCT TAXA, 1495
EXTINCTION, 1496
EXTRAPOLATION, 1497
EXTRATERRESTRIAL GEOLOGY, 1498
F LAYER, 1499
FABRIC, 1500
FACIES, 2783
FACIES, 1501
FACTOR ANALYSIS, 1502
FAMENNIAN, 1503
FANGLOMERATE, 1505
FANS, 1504

FATTY ACIDS, 1506
FAULT BRECCIA, 1508
FAULT PLANES, 1510
FAULT SCARPS, 1511
FAULT SYSTEMS, 1512
FAULT ZONES, 1513
FAULT=PLANE STRIATIONS, 1514
FAULTING, 1515
FAULTS, 1507
FAUNA, 1516
FAUNAL LIST, 1517
FAUNAL PROVINCES, 1518
FAUNIZONES, 1519
FEASIBILITY STUDIES, 1520
FECAL PELLETS, 1521
FEEDER, 1522
FELDSPAR GROUP, 1524
FELDSPATHIZATION, 1525
FELDSPATHOID COMPOSITION, 1526
FELSIC COMPOSITION, 1527
FEMIC, 1528
FENITE, 1529
FENITIZATION, 1530
FERMIUM, 1531
FERRALSOLS, 1532
FERRIC IRON, 1533
FERRICRETE, 1534
FERRODS, 1535
FERROMAGNETIC PROPERTY, 1536
FERROMANGANESE COMPOSITION, 1537
FERROUS IRON, 1538
FERRUGINOUS COMPOSITION, 1539
FERRUGINOUS SOILS, 1540
FERTILIZERS, 1541
FIELD STUDIES, 1542
FIELD TRIPS, 1465
FIGURE OF EARTH, 4055
FILICOPSIDA, 1543
FILLING MATERIALS, 1544
FILTERS, 1546
FILTRATION, 1547
FINANCING, 1548
FINE=GRAINED MATERIALS, 1549
FINITE DIFFERENCE ANALYSIS, 1550
FINITE ELEMENT ANALYSIS, 1551
FIRECLAY, 1552
FIRN, 1553
FISSION, 1554
FISSION TRACKS, 1555
FISSION=TRACK DATING, 1556
FISSIPEDA, 1557
FIXATION, 1558
FIXISM, 1559
FJORDS, 1560
FLAGGY TEXTURE, 1561
FLAME PHOTOMETRY, 1562

FLAME STRUCTURES, 1563
FLANDRIAN, 1564
FLASER STRUCTURES, 1565
FLEXURE, 1566
FLINT, 1567
FLOCCULATION, 1568
FLOODPLAINS, 1570
FLOODS, 1569
FLORA, 1571
FLORAL PROVINCES, 1572
FLORIZONES, 1573
FLOTATION, 1574
FLOW, 1575
FLOW CLEAVAGE, 1576
FLOW FOLDS, 1577
FLUCTUATIONS, 1578
FLUID INCLUSIONS, 1579
FLUID INJECTION, 1580
FLUID MECHANICS, 1581
FLUID PHASE, 1582
FLUID PRESSURE, 1583
FLUIDAL TEXTURE, 1584
FLUORESCENCE, 1585
FLUORIDES, 1586
FLUORIMETRY, 1587
FLUORINATION, 1588
FLUORINE, 1589
FLUORSPAR, 1590
FLUTE CASTS, 1591
FLUVENTS, 1592
FLUVIAL ENVIRONMENT, 1593
FLUVIAL EROSION, 1594
FLUVIAL FILL, 142
FLUVIAL SEDIMENTATION, 1595
FLUVIOSOLS, 1598
FLYSCH, 1599
FOCAL MECHANISM, 1600
FOCUS, 2146
FOIDITE, 1601
FOLD AXES, 1603
FOLD BELTS, 1604
FOLD BRECCIA, 1605
FOLD SYSTEM, 1606
FOLD TECTONICS, 1607
FOLDING, 1608
FOLDS, 1602
FOLIATION, 1609
FORAMINIFERA, 1610
FOREDEEPS, 1612
FORELANDS, 1613
FORESET BEDS, 1614
FORESHOCKS, 1615
FORESHORE, 1616
FORMATIONS, 1617
FORMULA, 1618
FOSSIL FUELS, 1620
FOSSIL ICE WEDGES, 1621

FOSSIL LOCALITIES, 1625
FOSSIL MAN, 1622
FOSSIL WATERS, 1623
FOSSIL WOOD, 1624
FOSSILIZATION, 1626
FOSSILS, 1619
FOUNDATIONS, 1627
FOURIER ANALYSIS, 1628
FOYAITE, 1629
FRACTIONAL CRYSTALLIZATION, 1631
FRACTIONATION, 2353
FRACTURE STRENGTH, 1633
FRACTURE ZONES, 1634
FRACTURES, 1632
FRACTURING, 1635
FRAGMENTATION, 1637
FRAGMENTS, 1636
FRAMBOIDAL TEXTURE, 1638
FRANCIUM, 1639
FRASNIAN, 1640
FREE ENERGY, 1641
FREE OSCILLATIONS, 1642
FREE=AIR ANOMALIES, 1643
FREQUENCY, 1644
FREQUENCY DISTRIBUTION, 1645
FREQUENCY SOUNDING, 1646
FRESH WATER, 1647
FRESH=WATER ENVIRONMENT, 1648
FRESH=WATER SEDIMENTATION, 1649
FRICTION, 1650
FRINGING REEFS, 1651
FRONTS, 1652
FROST ACTION, 1653
FROST FEATURES, 1654
FROST HEAVING, 1655
FROZEN GROUND, 1656
FRUITS, 1657
FUEL RESOURCES, 1658
FUGACITY, 1659
FULGURITE, 1660
FULL HOLE DRILLING, 1661
FULLER'S EARTH, 1662
FULVIC ACIDS, 1663
FUMAROLES, 1664
FUNCTIONAL MORPHOLOGY, 1666
FUNCTIONS, 1665
FUNGI, 1667
FURROWS, 1668
FUSINITE, 1669
FUSULINIDAE, 1670
G LAYER, 1671
GABBROS, 1672
GADOLINIUM, 1673
GALACTIC YEAR, 1674
GALLIUM, 1675
GAMMA RAYS, 1676
GAMMA=GAMMA METHODS, 1678

GAMMA=RAY METHODS, 1679
GAMMA=RAY SPECTROSCOPY, 1677
GANGUE, 1680
GANISTER, 1681
GARNET GROUP, 1683
GAS CAP, 1685
GAS CHROMATOGRAPHY, 1686
GAS COAL, 1687
GAS FIELDS, 1689
GAS INCLUSION, 1690
GAS STORAGE, 1691
GAS WELL, 1692
GAS=OIL RATIO, 1693
GASEOUS PHASE, 1694
GASES, 1684
GASIFICATION, 1695
GASTROPODA, 1696
GASTROZOOID, 1697
GAUGING, 1698
GAUSS EPOCH, 1699
GEANTICLINES, 1700
GEDINNIAN, 1701
GELIVITY, 1704
GELS, 1702
GEMMATION, 1705
GEMMOLOGY, 1706
GEMS, 1707
GENERATION, 1708
GENESIS, 1709
GENOMORPHS, 1710
GENOTYPES, 1711
GEOBIOS, 1713
GEOBOTANICAL METHODS, 1714
GEOCHEMICAL CYCLE, 1715
GEOCHEMICAL FACIES, 1717
GEOCHEMICAL INDICATORS, 1718
GEOCHEMICAL MAPS, 1719
GEOCHEMICAL METHODS, 1716
GEOCHEMICAL PROFILES, 1720
GEOCHEMICAL THRESHOLD, 4519
GEOCHEMISTRY, 1722
GEOCHRONOLOGIC UNIT, 1724
GEOCHRONOLOGY, 1725
GEODES, 1726
GEODESY, 1727
GEODETIC COORDINATES, 1728
GEODYNAMICS, 1729
GEOFRACTURE, 1730
GEOGRAPHY, 1731
GEOID, 1732
GEOISOTHERM, 1733
GEOLOGIC BAROMETRY, 1712
GEOLOGIC HAZARDS, 1738
GEOLOGIC MAPS, 1735
GEOLOGIC STRUCTURE, 1737
GEOLOGIC THERMOMETRY, 1780
GEOLOGISTS, 1740

GEOLOGY, 1741
GEOMAGNETIC REVERSAL, 1743
GEOMAGNETISM, 1745
GEOMETRIC NETWORKS, 3013
GEOMETRY, 1746
GEOMORPHOLOGIC MAPS, 1747
GEOMORPHOLOGY, 1748
GEONOMY, 1749
GEOPHONES, 1750
GEOPHYSICAL ANOMALY, 1751
GEOPHYSICAL EXPLORATION, 1752
GEOPHYSICAL INSTRUMENTS, 1753
GEOPHYSICAL MAPS, 1754
GEOPHYSICAL METHODS, 1755
GEOPHYSICAL PROFILES, 1756
GEOPHYSICAL SURVEYS, 1757
GEOPHYSICISTS, 1758
GEOPHYSICS, 1759
GEOPRESSURE, 1760
GEOSECS, 1761
GEOSTATISTICS, 1762
GEOSYNCLINAL RIDGE, 1763
GEOSYNCLINAL SEA, 1764
GEOSYNCLINAL SEDIMENTATION, 1765
GEOSYNCLINAL TRENCH, 1766
GEOSYNCLINES, 1767
GEOTECHNICAL MAPS, 1768
GEOTECHNICAL SURVEY, 1770
GEOTECHNICS, 1771
GEOTECTONICS, 1772
GEOTHERMAL ENERGY, 1773
GEOTHERMAL FIELDS, 1774
GEOTHERMAL GRADIENT, 1775
GEOTHERMAL METAMORPHISM, 1776
GEOTHERMAL METHOD, 1777
GEOTHERMAL SYSTEMS, 1778
GEOTHERMICS, 1779
GERMANIUM, 1782
GERMANOTYPE TECTONICS, 1783
GEYSERITE, 1785
GEYSERS, 1784
GIANT FIELDS, 1786
GIGANTISM, 1787
GIGANTOSTRACA, 1788
GINKGOALES, 1789
GIVETIAN, 1790
GLACIAL, 1791
GLACIAL ENVIRONMENT, 1793
GLACIAL EROSION, 1794
GLACIAL FEATURES, 1795
GLACIAL GEOLOGY, 1796
GLACIAL LAKES, 1797
GLACIAL SEDIMENTATION, 1798
GLACIAL STAGES, 1799
GLACIAL TRANSPORT, 1801
GLACIAL VALLEYS, 1802
GLACIATION, 1803

GLACIER SURGES, 1805
GLACIERS, 1804
GLACIOFLUVIAL FEATURES, 1596
GLACIOFLUVIAL SEDIMENTATION, 1597
GLACIOLACUSTRINE SEDIMENTATION, 1806
GLACIOLOGY, 1807
GLACIOMARINE SEDIMENTATION, 1808
GLACIOTECTONICS, 1809
GLACIS, 1810
GLASS MATERIALS, 1813
GLASS=SAND, 1812
GLASSES, 1811
GLASSY TEXTURE, 1814
GLAUCONITE, 1815
GLAUCONITIC COMPOSITION, 1816
GLAUCONITIZATION, 1817
GLAUCOPHANE SCHIST FACIES, 1818
GLEYS, 1819
GLIDE TWIN, 1820
GLOBAL TECTONICS, 1821
GLOBIGERINACEA, 1822
GLOMERO=, 1823
GLOSSARIES, 1824
GLOSSOPTERIDALES, 1825
GNATHOSTOMI, 1826
GNATHOSTRACA, 1827
GNEISSES, 1828
GNEISSIC DOME, 1829
GNEISSIC TEXTURE, 1830
GNETALES, 1831
GNOMONIC PROJECTION, 1832
GOLD, 1833
GONDWANA, 1834
GONIATITIDA, 1835
GONIOMETRY, 1836
GORGES, 1837
GOSSAN, 1838
GOUGE, 1509
GRAB SAMPLING, 1839
GRABENS, 1840
GRADE, 2784
GRADE, 3156
GRADED BEDDING, 1841
GRADIENT, 1842
GRADIOMETERS, 1843
GRAINS, 1844
GRANITE GNEISS, 1846
GRANITES, 1845
GRANITIC, 1847
GRANITIC COMPOSITION, 1848
GRANITIC LAYER, 1849
GRANITIZATION, 1850
GRANITOID COMPOSITION, 1851
GRANOBLASTIC TEXTURE, 1852
GRANODIORITES, 1853
GRANOGABBRO, 1854
GRANOPHYRIC TEXTURE, 1855

GRANULAR TEXTURE, 1856
GRANULATION, 1857
GRANULITE FACIES, 1859
GRANULITES, 1858
GRANULITIC, 1860
GRANULOMETRY, 1861
GRAPESTONE, 1862
GRAPHIC DISPLAY, 1863
GRAPHIC METHODS, 1864
GRAPHIC TEXTURE, 1865
GRAPHITE, 1866
GRAPHITIZATION, 1867
GRAPTOLITHINA, 1868
GRAPTOLOIDEA, 1869
GRASSLANDS, 1870
GRAVEL, 1871
GRAVEL FILTERS, 1872
GRAVIMETERS, 1873
GRAVIMETRY, 1874
GRAVITATION, 1875
GRAVITATIONAL CONSTANT, 1876
GRAVITATIONAL DIFFERENTIATION, 1877
GRAVITY ANOMALIES, 1878
GRAVITY CONCENTRATION, 1114
GRAVITY EXPLORATION, 1879
GRAVITY FIELD, 1880
GRAVITY FOLDS, 1881
GRAVITY MAPS, 1882
GRAVITY METHODS, 1883
GRAVITY PLATFORMS, 1884
GRAVITY SLIDING, 1885
GRAVITY SURVEYS, 1886
GRAVITY TECTONICS, 1887
GRAYWACKE, 1888
GREASY LUSTER, 1889
GREAT=CIRCLE BELT, 1890
GREENHOUSE EFFECT, 1891
GREENSCHIST, 1892
GREENSCHIST FACIES, 1893
GREENSTONE, 1894
GREENSTONE BELTS, 1895
GREISEN, 1896
GREISENIZATION, 1897
GRENVILLIAN OROGENY, 1898
GRINDING, 1899
GRIT, 1900
GROINS, 1901
GROOVE CASTS, 1903
GROOVES, 1902
GROUND, 1904
GROUND ICE, 1905
GROUND METHODS, 1906
GROUND MORAINES, 1907
GROUND MOTION, 1908
GROUND PRESSURE, 1909
GROUND TRUTH, 1910
GROUND WATER, 1911

GROUNDMASS, 1919
GROUNDWATER DAMS, 1913
GROUNDWATER DISCHARGE, 1914
GROUNDWATER EXPLORATION, 1915
GROUNDWATER LEVEL, 1916
GROUNDWATER RUNOFF, 1918
GROUTING, 1920
GROWTH FAULTS, 1921
GROWTH LINES, 1922
GRUS, 1923
GUANO, 1924
GUIDE FOSSIL, 1925
GUIDEBOOK, 1926
GULFS, 1927
GULLIES, 1928
GUTENBERG DISCONTINUITY, 1929
GYMNOLAEMATA, 1930
GYMNOSPERMS, 1931
GYPSIFICATION, 1932
GYPSUM, 1933
GYTTJA, 1934
GZHELIAN, 1935
HABIT, 1936
HABITAT, 1937
HADE, 1938
HAECKEL'S LAW, 1939
HAFNIUM, 1940
HAIL, 1941
HALF=LIFE PERIOD, 1942
HALF=SPACE, 1943
HALIDES, 1944
HALITIC ENVIRONMENT, 1945
HALMYROLYSIS, 1946
HALOES, 373
HALOGENS, 1947
HALOPHILIC TAXA, 1948
HALOPHYTE, 1949
HAMMADAS, 1950
HAMMER SEISMICS, 1951
HANGING VALLEYS, 1952
HANGING WALL, 1953
HARBORS, 1954
HARDGROUND, 1955
HARDNESS, 1956
HARMONICS, 1957
HARPOLITHS, 1958
HAUTERIVIAN, 1959
HAUY'S LAW, 1960
HAWAIAN=TYPE ERUPTIONS, 1961
HE=4/HE=3, 1979
HEAD, 2094
HEAT, 1962
HEAT CAPACITY, 1963
HEAT CONDUCTION, 1964
HEAT CRACK, 1965
HEAT EQUIVALENT OF FUSION, 1966
HEAT FLOW, 1967

HEAT SOURCES, 1968
HEAT TRANSFER, 1969
HEAVE, 1970
HEAVY ISOTOPES, 1971
HEAVY METALS, 1972
HEAVY MINERALS, 1973
HEAVY OIL, 1974
HEKISTOTHERM PLANT, 1975
HELICOPLACOIDEA, 1976
HELIOZOA, 1977
HELIUM, 1978
HELVETIAN, 1980
HEMATITE, 1981
HEMICHORDATA, 4261
HEMICRUSTACEA, 1982
HEMIPTEROIDA, 1983
HERCYNIAN OROGENY, 1984
HERCYNIDES, 1985
HEREDITY, 1986
HERMATYPIC TAXA, 1987
HETEROBLASTIC TEXTURE, 1988
HETEROCHRONISM, 1989
HETEROCHRONOUS, 1990
HETEROCORALLIA, 1991
HETERODONTA, 1992
HETEROGENEITY, 1993
HETEROGENEOUS, 1994
HETEROMORPHIC TAXA, 1995
HETEROMORPHISM, 1996
HETEROSPOROUS TAXA, 1997
HETEROTROPHIC, 1998
HETTANGIAN, 1999
HEXACTINIARIA, 2001
HEXAGONAL SYSTEM, 2002
HEXAHEDRITE, 2003
HIATUS, 2004
HIEROGLYPHS, 2005
HIGH PRESSURE, 2007
HIGH TEMPERATURE, 2008
HIGH VALUE ANOMALY, 2006
HIGH=ENERGY ENVIRONMENT, 2009
HIGH=GRADE, 2010
HIGH=GRADE METAMORPHISM, 2011
HIGHLANDS, 2012
HILLS, 2013
HINGE (PALE), 2014
HINGE FAULTS, 2015
HIPPOMORPHA, 2016
HISTOGRAMS, 2017
HISTOLOGY, 2018
HISTORICAL GEOLOGY, 2019
HISTORY, 2020
HISTORY OF GEOLOGY, 2021
HISTOSOLS, 2022
HODOGRAPHS, 2023
HOLMES' CLASSIFICATION, 2024
HOLMIUM, 2025

HOLOCENE, 2026
HOLOCEPHALI, 2027
HOLOCRYSTALLINE TEXTURE, 2028
HOLOGRAPHY, 2029
HOLOPHYTE, 2030
HOLOSTEI, 2031
HOLOSTOMATOUS TAXA, 2032
HOLOTHUROIDEA, 2033
HOLOTYPES, 2034
HOMALOZOA, 2035
HOMEOMORPHISM, 2036
HOMEOMORPHY, 2037
HOMINIDAE, 2038
HOMO SAPIENS, 2039
HOMOCLINES, 2040
HOMOGENEITY, 2041
HOMOGENIZATION, 2042
HOMOIOSTELEA, 2043
HOMOLOGY, 2044
HOMONYMY, 2045
HOMOSTELEA, 2046
HOMOTAXY, 2047
HOOKE'S LAW, 2048
HORIZON DIFFERENTIATION, 2049
HORIZONTAL, 2050
HORIZONTAL COMPRESSION, 2051
HORIZONTAL DISPLACEMENTS, 2053
HORIZONTAL FAULT, 2054
HORIZONTAL ORIENTATION, 2052
HORNBLENDE-HORNFELS FACIES, 2055
HORNBLENDITE, 2056
HORNFELS, 2057
HORNFELS FACIES, 2058
HORSETAIL, 2059
HORSTS, 2060
HOST MATERIALS, 2061
HOST ROCKS, 2062
HOT, 2063
HOT BRINES, 2064
HOT SPOTS, 2065
HOT SPRINGS, 2066
HOWARDITE, 2067
HUDSONIAN OROGENY, 2068
HUMAN ECOLOGY, 2069
HUMAN WASTE, 2070
HUMIC ACIDS, 2071
HUMIC SOILS, 2072
HUMID CLIMATE, 2073
HUMID ENVIRONMENT, 2074
HUMIDITY, 2075
HUMIFICATION, 2076
HUMIN, 2077
HUMITE, 2078
HUMMOCKS, 2079
HUMODS, 2080
HUMOX, 2081
HUMULTS, 2082

HUMUS, 2083
HYALOCRYSTALLINE TEXTURE, 2084
HYALOSPONGEA, 2085
HYBRIDIZATION, 2086
HYDATOGENESIS, 2087
HYDRARCH, 2088
HYDRATION, 2089
HYDRATION OF GLASS, 2090
HYDRAULIC CONDUCTIVITY, 2091
HYDRAULIC FRACTURING, 2092
HYDRAULIC GRADIENT, 2093
HYDRAULIC MAPS, 2095
HYDRAULIC PRESSURE, 2096
HYDRAULIC RADIUS, 2097
HYDRAULICS, 2098
HYDROCARBONS, 2099
HYDROCHEMICAL FACIES, 2100
HYDROCHEMICAL MAPS, 2101
HYDROCHEMISTRY, 2102
HYDRODYNAMIC MODELS, 2103
HYDRODYNAMICS, 2104
HYDROEXPLOSION, 2105
HYDROGEN, 2106
HYDROGEN GAS, 2107
HYDROGEN SULFIDE, 2108
HYDROGEOCHEMICAL METHODS, 2109
HYDROGEOLOGIC MAPS, 2111
HYDROGEOLOGIC PROVINCES, 2112
HYDROGEOLOGICAL CONTROLS, 2110
HYDROGEOLOGICAL SURVEY, 2113
HYDROGEOLOGICAL TECHNOLOGY, 2114
HYDROGEOLOGY, 2115
HYDROGRAPHS, 2116
HYDROGRAPHY, 2117
HYDROIDA, 2118
HYDROLOGIC CYCLE, 2119
HYDROLOGIC MAPS, 2120
HYDROLOGY, 2121
HYDROMETALLURGY, 2122
HYDROMETERS, 2123
HYDROMETRY, 2124
HYDROPHONE, 2125
HYDROSPHERE, 2126
HYDROSPIRES, 2127
HYDROSTATIC PRESSURE, 2129
HYDROTHERMAL, 2130
HYDROTHERMAL ALTERATION, 2131
HYDROTHERMAL CONDITIONS, 2133
HYDROTHERMAL PROCESSES, 2132
HYDROTHERMAL WATERS, 2134
HYDROUS, 2135
HYDROXIDES, 2136
HYDROZOA, 2137
HYGROPHYLE, 2138
HYGROSCOPIC WATER, 2139
HYMENOPTERA, 2140
HYOLITHES, 2141

HYPABYSSAL ROCKS, 2142
HYPERGENESIS, 2143
HYPIDIOBLASTIC TEXTURE, 2144
HYPIDIOMORPHIC TEXTURE, 2145
HYPOCRYSTALLINE TEXTURE, 2147
HYPODIGMS, 2148
HYPOTHERMAL PROCESSES, 2149
HYPOTHESIS, 2150
HYPSOMETRY, 2151
HYRACOIDEA, 2152
HYSTRICHOMORPHA, 2153
IAPETUS, 2154
ICE, 2155
ICE CAPS, 2156
ICE FIELDS, 2157
ICE MOVEMENT, 2158
ICE RAFTING, 2159
ICE SHEETS, 2160
ICE SHELVES, 2161
ICE WEDGES, 2162
ICEBERGS, 2163
ICHNOFOSSILS, 2164
ICHNOLOGY, 2165
ICHTHYOPTERYGIA, 2166
ICHTHYOSAURIA, 2167
IDIOBLASTIC TEXTURE, 2168
IDIOGEOSYNCLINES, 2169
IDIOMORPHIC TEXTURE, 2170
IDIOMORPHISM, 2171
IGCP, 2278
IGNEOUS ACTIVITY, 2172
IGNEOUS COMPLEX, 2173
IGNEOUS FACIES, 2174
IGNEOUS PROCESSES, 2621
IGNEOUS ROCK GENESIS, 2175
IGNEOUS ROCKS, 2176
IGNIMBRITE, 2177
IMAGERY, 2178
IMBRIAN, 2179
IMBRICATE STRUCTURES, 2180
IMBRICATION, 2181
IMMATURE, 2182
IMMATURE SOILS, 2183
IMMERSION METHOD, 2184
IMMISCIBILITY, 2185
IMMISCIBLE MATERIALS, 2186
IMPACT CRATER, 2187
IMPACT FEATURES, 2188
IMPACT METAMORPHISM, 2189
IMPACT SEISMICS, 2190
IMPACT STATEMENTS, 2191
IMPACT STRUCTURES, 2192
IMPACTITE, 2193
IMPORT, 2194
IMPREGNATED DEPOSITS, 2195
IMPURITIES, 2196
IN SITU, 2197

INARTICULATA, 2198
INCEPTISOLS, 2199
INCLINATION, 2635
INCLUSIONS, 2200
INCOHERENT, 2201
INCOMPETENT MATERIALS, 2203
INCONGRUENT MELTING, 2204
INDEX FOSSILS, 2206
INDEX MAPS, 2207
INDEX MINERALS, 2208
INDICATIVE BOULDER, 2211
INDICATORS, 2212
INDIUM, 2213
INDOCHINITE, 2214
INDUCED MAGNETIZATION, 2215
INDUCED POLARIZATION, 2216
INDUCTION, 2217
INDURATION, 2218
INDUSTRIAL MINERALS, 2219
INDUSTRIAL USE, 2220
INDUSTRIAL WASTE, 2221
INDUSTRY, 2222
INERT COMPONENT, 2223
INERT GASES, 2224
INFERRED ORE, 2225
INFILTRATION, 2226
INFILTRATION GALLERY, 2227
INFORMATICS, 2228
INFORMATION SYSTEMS, 2229
INFRACAMBRIAN, 2230
INFRARED, 2231
INFRARED PHOTOGRAPHY, 2232
INFRARED SPECTRA, 2233
INFRARED SPECTROSCOPY, 2234
INFRASTRUCTURE, 2235
INGRESSION, 2236
INHERITED FEATURES, 2237
INHOMOGENEITY, 2238
INJECTION, 2239
INJECTION BRECCIA, 2240
INLETS, 2241
INLIERS, 2242
INNER CORE, 2243
INNER SHELF, 2244
INNER SLOPE, 2245
INOCERAMI, 2246
INORGANIC, 2247
INORGANIC ACIDS, 2248
INORGANIC MATERIALS, 2249
INORGANIC ORIGIN, 2250
INSECTA, 2251
INSECTIVORA, 2252
INSELBERGS, 2253
INSEQUENT VALLEYS, 2254
INSOLATION, 2255
INSOLUBLE RESIDUES, 2256
INSTITUTIONS, 2257

INSTRUMENTS, 2258
INSULATION MATERIALS, 2259
INTERCALATION, 2260
INTERFACES, 2261
INTERFERENCE, 2262
INTERFEROMETRY, 2263
INTERFLUVES, 2264
INTERFORMATIONAL, 2265
INTERGLACIAL STAGES, 2266
INTERGROWTHS, 2267
INTERIOR, 1276
INTERMEDIATE COMPOSITION, 2269
INTERMEDIATE-FOCUS EARTHQUAKES, 2270
INTERMITTENT SPRINGS, 2271
INTERMITTENT STREAM, 2272
INTERMONTANE BASINS, 2273
INTERNAL CAST, 2274
INTERNAL STRUCTURE (PALE), 2275
INTERNATIONAL COOPERATION, 2276
INTERNATIONAL GEOLOGICAL CONGRESS, 2277
INTERNIDES, 2279
INTERPLANETARY COMPARISON, 2280
INTERPLUVIAL STAGES, 2281
INTERPOLATION, 2282
INTERPRETATION, 2283
INTERSERTAL TEXTURE, 2284
INTERSTELLAR SPACE, 2285
INTERTIDAL ENVIRONMENT, 2286
INTERTIDAL SEDIMENTATION, 2287
INTRACLASTS, 2288
INTRACONTINENTAL BELTS, 2289
INTRACRATONIC BASINS, 2290
INTRAFOLIAL FOLDS, 3510
INTRAFORMATIONAL DEPOSITION, 2291
INTRAMAGMATIC DEPOSITS, 2622
INTRATELLURIC PROCESSES, 2292
INTRUSIONS, 2293
INTRUSIVE, 2294
INVENTORY, 2295
INVERSE PROBLEM, 2296
INVERTEBRATA, 2297
IO/TH, 2302
IO/U, 2303
IODINE, 2298
ION EXCHANGE, 2300
ION PROBE, 2301
IONIZATION, 2304
IONOSPHERE, 2305
IONS, 2299
IRIDESCENCE, 2306
IRIDIUM, 2307
IRON, 2308
IRON BACTERIA, 2310
IRON FORMATIONS, 2311
IRON METEORITES, 2767
IRONSTONE, 2312
IRREVERSIBILITY, 2313

IRRIGATION, 2314
ISLAND ARCS, 2316
ISLAND VOLCANO, 2317
ISLANDS, 2315
ISOANOMALY, 2318
ISOBARS, 2319
ISOBASE, 2320
ISOBATH, 2321
ISOCHEMICAL METAMORPHISM, 2322
ISOCHORE, 4277
ISOCHORES, 3393
ISOCHRONS, 2323
ISOCLINES, 2324
ISOCLINIC LINE, 2325
ISOCONS, 2326
ISOGON, 2327
ISOGRADS, 2328
ISOHYPSE, 2329
ISOLATION, 2330
ISOLITH, 2331
ISOMETRIC SYSTEM, 2332
ISOMORPH ORGANISM, 2333
ISOMORPHISM, 2334
ISOPACH MAPS, 2335
ISOPIESTIC LINE, 2336
ISOPLETH MAPS, 2337
ISOPLEURA, 2338
ISOPRENOIDS, 2339
ISORAD, 2340
ISORAT, 2341
ISOSEISMIC LINE, 2342
ISOSEISMIC MAPS, 2343
ISOSTASY, 2344
ISOSTRUCTURAL MINERALS, 2345
ISOTHERMAL REMANENT MAGNETIZATION, 2347
ISOTHERMS, 2346
ISOTOPE DILUTION, 2349
ISOTOPE EFFECT, 2350
ISOTOPE GEOCHEMISTRY, 2351
ISOTOPE RATIO, 2352
ISOTOPES, 2348
ISOTROPY, 2354
ITABIRITE, 2355
ITACOLUMITE, 2356
JADEITITE, 2357
JARAMILLO EVENT, 2358
JASPER, 2359
JAWS, 2360
JOINTING, 2362
JOINTS, 2361
JUPITER, 3473
JURASSIC, 2363
JUVENILE, 2364
JUVENILE WATERS, 2365
K/AR, 3580
K/CA, 3581
KAMES, 2366

KANGA, 2367
KAOLIN, 2368
KAOLINIZATION, 2369
KARELIAN OROGENY, 2370
KARREN, 2371
KARROO SYSTEM, 2372
KARST, 2373
KARST FILLING, 2374
KARST HYDROLOGY, 2375
KARST SPRING, 2376
KASIMOVIAN, 2382
KATAMORPHISM, 2377
KATANGAN OROGENY, 2378
KATATHERMAL DEPOSITS, 2379
KATAZONE, 2380
KAZANIAN, 2381
KELYPHYTIC RIM, 2383
KENORAN OROGENY, 2384
KERATOPHYRE, 2385
KEROGEN, 2386
KETTLES, 2387
KEUPER, 2388
KIBARA OROGENY, 2390
KIMBERLITE, 2391
KIMMERIDGIAN, 2392
KINEMATICS, 2393
KINK FOLDS, 741
KINK=BAND STRUCTURES, 2394
KINZIGITE, 2395
KLIPPEN, 2396
KREEP, 2397
KRYPTON, 2398
KUJALNIKIAN, 2399
KUNGURIAN, 2400
KURTOSIS, 2401
KUTORGINIDA, 2402
LABORATORY STUDIES, 2403
LABYRINTHODONTIA, 2404
LACCOLITHS, 2405
LACERTILIA, 2406
LACUSTRINE ENVIRONMENT, 2407
LACUSTRINE SEDIMENTATION, 2409
LADINIAN, 2410
LAGOONAL ENVIRONMENT, 2412
LAGOONAL SEDIMENTATION, 2413
LAGOONS, 2411
LAHARS, 2414
LAKE SEDIMENTS, 2408
LAKE WATER LEVEL, 2416
LAKES, 2415
LAMARCKISM, 2417
LAME CONSTANT, 2418
LAMELLAE, 2419
LAMINAR FLOW, 2420
LAMINATED STRUCTURE, 2421
LAMINATIONS, 2422
LAMPROPHYRES, 2423

LAMPROPHYRIC TEXTURE, 2424
LAND, 2425
LAND SUBSIDENCE, 2426
LAND USE, 2427
LANDFILLS, 2428
LANDFORMS, 2429
LANDSAT, 2430
LANDSLIDES, 2431
LANGUAGE (MATH), 871
LANTHANUM, 2432
LAPILLI, 2433
LAPLACE TRANSFORMATIONS, 2434
LARAMIDE OROGENY, 2435
LARGE=SCALE, 2436
LASER METHODS, 2437
LATE DIAGENESIS, 2438
LATE PROTEROZOIDES, 2439
LATENT HEAT OF FUSION, 2440
LATERAL, 2441
LATERAL FAULTS, 2442
LATERAL LOG, 2443
LATERAL MORAINES, 2444
LATERAL SECRETION, 2445
LATERITES, 2446
LATERITIC SOIL, 2447
LATERIZATION, 2448
LATITE, 2449
LATITUDE, 2450
LATTICE, 1018
LATTICE TEXTURE, 2452
LAURASIA, 2453
LAVA, 2454
LAVA CHANNELS, 2455
LAVA FIELDS, 2456
LAVA FLOWS, 2457
LAVA LAKES, 2458
LAVA TUBES, 2459
LAW OF CORRELATION OF FACIES, 2461
LAWS, 2460
LAYERED INTRUSIONS, 2463
LAYERED MEDIA, 2464
LAYERS, 2462
LEACHING, 2465
LEAD, 2466
LEAD=ALPHA AGE METHOD, 2467
LEAKAGE, 2470
LEAN, 2471
LEARNED SOCIETIES, 2472
LEAST=SQUARES ANALYSIS, 2473
LEAVES, 2469
LEBENSSPUREN, 4557
LECTOTYPES, 2474
LEGEND, 2475
LEGISLATION, 2476
LENGTH, 2477
LENGTH OF DAY, 2478
LENIAN, 2479

LENSES, 2480
LEPERDITOCOPIDA, 2481
LEPIDOBLASTIC TEXTURE, 2482
LEPIDOCYSTOIDEA, 2483
LEPIDOSAURIA, 2484
LEPOSPONDYLI, 2485
LEPTITE, 2486
LEPTOGEOSYNCLINE, 2487
LEPTYNITE, 2488
LEUCOCRATIC COMPOSITION, 2489
LEUCOSOME, 2490
LEVEES, 2491
LEVEL, 2493
LEVELING, 2492
LEXICONS, 2494
LICHENOMETRY, 2496
LICHENS, 2495
LICHIDA, 2497
LIESEGANG RINGS, 2498
LIFE, 2499
LIGHT ISOTOPE, 2500
LIGHT MINERALS, 2501
LIGHT OIL, 2502
LIGNIN, 2503
LIGNITE, 2504
LIMB, 2505
LIME, 2506
LIMESTONE, 2507
LIMNOLOGY, 2508
LIMONITE, 2510
LINEAMENTS, 2511
LINEATION, 2512
LINGULIDA, 2513
LINKED VEINS, 2514
LINOPHYRIC TEXTURE, 2515
LIPIDS, 2516
LIPOSTRACA, 2517
LIQUATION, 2518
LIQUEFACTION, 2519
LIQUID PHASE, 2520
LIQUID WASTE, 2521
LIQUIDUS, 2522
LISSAMPHIBIA, 2523
LITHIFICATION, 2524
LITHIUM, 2525
LITHOFACIES, 2526
LITHOFACIES MAPS, 2527
LITHOGENESIS, 2528
LITHOLOGIC, 2529
LITHOLOGIC CONTROLS, 2530
LITHOLOGIC MAPS, 3376
LITHOLOGIC TRAPS, 2531
LITHOLOGY, 2532
LITHOPHILE ELEMENTS, 2533
LITHOPHYTE, 2534
LITHOSOLS, 2535
LITHOSPHERE, 2536

LITHOSTRATIGRAPHIC UNIT, 2537
LITHOSTRATIGRAPHY, 2538
LITHOSTROMES, 2539
LITHOTOPES, 2540
LITHOTYPE, 2541
LITOPTERNA, 2542
LITTORAL, 2543
LITTORAL DRIFT, 2544
LITTORAL ENVIRONMENT, 2545
LITTORAL EROSION, 813
LITUOLACEA, 2547
LIVING, 2548
LIVING FOSSILS, 2549
LIXIVIATION, 2550
LLANDEILIAN, 2551
LLANDOVERIAN, 2552
LLANVIRNIAN, 2553
LOAD CASTS, 2555
LOAD TESTS, 2556
LOADING, 2554
LOAM, 2509
LOCAL, 2557
LOCOMOTION, 2558
LODE, 2559
LODRANITE, 2560
LOESS, 2561
LOGNORMAL DISTRIBUTION, 2563
LOGS, 2562
LONG-PERIOD WAVES, 2564
LONGITUDE, 2565
LONGSHORE BARS, 2567
LONGSHORE CURRENTS, 2568
LOPOLITHS, 2569
LOVE WAVES, 2570
LOW PRESSURE, 2571
LOW TEMPERATURE, 2572
LOW-ENERGY ENVIRONMENT, 2574
LOW-GRADE METAMORPHISM, 2575
LOW-VELOCITY ZONES, 2576
LOW-WATER LEVELS, 2573
LOWER CAMBRIAN, 2577
LOWER CARBONIFEROUS, 2578
LOWER CRETACEOUS, 2579
LOWER CRUST, 2580
LOWER DEVONIAN, 2581
LOWER EOCENE, 2582
LOWER JURASSIC, 2583
LOWER LIASSIC, 2584
LOWER MANTLE, 2585
LOWER MIOCENE, 2586
LOWER OLIGOCENE, 2587
LOWER ORDOVICIAN, 2588
LOWER PERMIAN, 2589
LOWER PLEISTOCENE, 2590
LOWER PLIOCENE, 2591
LOWER PROTEROZOIC, 2592
LOWER SILURIAN, 2593

LOWER TRIASSIC, 2594
LOWLANDS, 2595
LU/HF, 2608
LUDLOVIAN, 2596
LUMINESCENCE, 2597
LUNA PROGRAM, 2598
LUNAR, 2599
LUNAR CONSTITUTION, 2600
LUNAR CRATERS, 2601
LUNAR GEOLOGY, 2602
LUNAR MASSIFS, 2605
LUNAR RELIEF, 2938
LUNAR SOILS, 2603
LUNAR STATIONS, 2604
LUSTER, 2606
LUTETIUM, 2607
LUVISOLS, 2609
LYCOPSIDA, 2610
LYTOCERATIDA, 2611
MAARS, 2612
MACERALS, 2613
MACHAERIDIA, 2614
MAESTRICHTIAN, 2615
MAFIC COMPOSITION, 2616
MAGMA CHAMBERS, 2618
MAGMAS, 2617
MAGMATIC, 2619
MAGMATIC ASSOCIATIONS, 2620
MAGMATIC ORIGIN, 2623
MAGMATISM, 2624
MAGNAFACIES, 2625
MAGNESITE, 2626
MAGNESIUM, 2627
MAGNETIC ANOMALIES, 2628
MAGNETIC AZIMUTH, 2629
MAGNETIC COMPASS, 2630
MAGNETIC DOMAINS, 2631
MAGNETIC ELEMENT, 2632
MAGNETIC FIELD, 2633
MAGNETIC FIELD VARIATION, 2634
MAGNETIC MAP, 2636
MAGNETIC METHODS, 2637
MAGNETIC MINERALS, 2638
MAGNETIC PERMEABILITY, 2639
MAGNETIC PROPERTIES, 2640
MAGNETIC SPHERULES, 2641
MAGNETIC STORMS, 2642
MAGNETIC SURVEY MAPS, 2644
MAGNETIC SURVEYS, 2643
MAGNETIC SUSCEPTIBILITY, 2645
MAGNETISM, 2646
MAGNETITE, 2647
MAGNETIZATION, 2648
MAGNETOHYDRODYNAMICS, 2649
MAGNETOMETERS, 2650
MAGNETOSPHERE, 2651
MAGNETOSTRATIGRAPHY, 2652

MAGNETOSTRICTION, 2653
MAGNETOTELLURIC METHODS, 2654
MAGNETOTELLURIC SURVEYS, 2655
MAGNITUDE, 1286
MAJIAN, 2656
MAJOR ELEMENTS, 2657
MAJOR=ELEMENT ANALYSES, 2658
MALACOSTRACA, 2659
MALLEABLE MATERIALS, 2660
MAMMALIA, 2661
MANGANESE, 2662
MANGROVE SWAMPS, 2663
MANPOWER, 2664
MANTLE, 2665
MANTODEA, 2666
MANUALS, 2667
MAPPING, 2670
MAPS, 2668
MARBLES, 2671
MARGINAL PLATEAU, 2673
MARGINAL SEAS, 2674
MARGINATION TEXTURE, 2675
MARIA, 2672
MARINE ECOLOGY, 2676
MARINE ENVIRONMENT, 2677
MARINE EROSION, 2678
MARINE GEOLOGY, 2679
MARINE INSTALLATIONS, 2680
MARINE MARSH, 2681
MARINE METHODS, 2682
MARINE PLAINS, 2683
MARINE PLATFORMS, 2684
MARINE SEDIMENTATION, 2686
MARINE SEDIMENTS, 2685
MARINE SILLS, 2687
MARINE TERRACES, 2688
MARINE TRANSPORT, 2689
MARINER PROGRAM, 2690
MARKER BEDS, 2389
MARKOV CHAIN ANALYSIS, 2691
MARL, 2692
MARS, 3474
MARSH GAS, 2694
MARSHES, 2693
MARSUPIALIA, 2695
MASCONS, 2696
MASS, 2697
MASS BALANCE, 2698
MASS MOVEMENTS, 2699
MASS SPECTROSCOPY, 2700
MASS TRANSFER, 2701
MASS WASTING, 2702
MASSIVE, 2703
MASSIVE BEDDING, 2704
MASSIVE DEPOSITS, 2705
MASTIGOPHORA, 2706
MATERIALS, 2707

MATHEMATICAL ANALYSIS, 2708
MATHEMATICAL GEOGRAPHY, 2710
MATHEMATICAL GEOLOGY, 2711
MATHEMATICAL LAW, 2712
MATHEMATICAL MATRIX, 2713
MATHEMATICAL METHODS, 2714
MATHEMATICAL MODELS, 2715
MATHEMATICS, 2716
MATRIX ALGEBRA, 2717
MATURITY, 2718
MATURITY INDEX, 2719
MATUYAMA EPOCH, 2720
MAXILLOPODA, 2721
MAXIMUM, 2722
MAZATSAL REVOLUTION, 2723
MEANDERS, 2724
MECHANICAL CONTROLS, 2725
MECHANICAL ORIGIN, 2726
MECHANICAL PROPERTIES, 2727
MECHANICAL WEATHERING, 2728
MECHANICS, 2729
MECHANISM, 2730
MECOPTEROIDA, 2731
MEDIAN MASSES, 2732
MEDICAL GEOLOGY, 2733
MEDICINAL WATERS, 2734
MEDITERRANEAN SOILS, 2735
MEDIUM-GRAINED, 2736
MEETING, 2737
MEGAFACIES, 2738
MEGASPORES, 2739
MELANGE, 2740
MELANOCRATIC COMPOSITION, 2741
MELTING, 2744
MELTS, 2743
MELTWATER, 2745
MEMBERS, 2746
MENDELEVIUM, 2747
MEOTIAN, 2748
MERCURY, 2749
MERCURY PLANET, 3475
MEROSTOMATA, 2750
MEROSTOMOIDEA, 2751
MESOCRATIC COMPOSITION, 2752
MESOGAEA, 2753
MESOGASTROPODA, 2754
MESOGEOSYNCLINE, 2755
MESOPHYLE ORGANISM, 2756
MESOSAURIA, 2757
MESOSIDERITES, 2758
MESOTHERMAL PROCESSES, 2759
MESOZOIC, 2760
MESOZONAL METAMORPHISM, 2761
METABASALT, 2762
METACOPINA, 2763
METAGABBRO, 2764
METAGENESIS, 2765

METALLIC PROPERTIES, 2768
METALLIFEROUS, 2769
METALLOGENIC DISTRICT, 2770
METALLOGENIC EPOCHS, 2771
METALLOGENIC MAPS, 2772
METALLOGENIC PROVINCES, 2773
METALLOGENY, 2774
METALLOID, 2775
METALLOTECT, 2776
METALLURGY, 2777
METALS, 2766
METAMICT MINERALS, 2778
METAMICTIZATION, 2779
METAMORPHIC BELTS, 2780
METAMORPHIC COMPLEX, 2781
METAMORPHIC DIFFERENTIATION, 2782
METAMORPHIC PROCESSES, 2788
METAMORPHIC ROCKS, 2785
METAMORPHIC ZONE, 2786
METAMORPHISM, 2787
METAMORPHOSIS, 2789
METANAUPLIUS, 2790
METAPELITE, 3336
METAPHYTE, 2791
METASEDIMENTARY ROCKS, 2792
METASOMA, 2793
METASOMATIC ROCKS, 2795
METASOMATISM, 2794
METATEXIS, 2796
METEOR CRATERS, 2797
METEORITES, 2798
METEOROIDS, 2799
METEOROLOGY, 2800
METHANE, 2801
METHODS, 2802
MIAROLITIC TEXTURE, 2803
MICA, 2804
MICA SCHIST, 2805
MICRITE, 2806
MICROCOMPUTERS, 2807
MICROCRACKS, 2813
MICROCRATERS, 2808
MICROCRYSTALLINE LIMESTONE, 2809
MICROEARTHQUAKES, 2810
MICROFACIES, 2811
MICROFAUNA, 2812
MICROFLORA, 2814
MICROFOLDS, 2815
MICROFOSSILS, 2816
MICROGRAINED TEXTURE, 2817
MICROGRANITE, 2818
MICROGRANITIC TEXTURE, 2819
MICROHARDNESS, 2820
MICROLATEROLOG, 2821
MICROLITE, 2822
MICROLOG, 2823
MICROMETEORITES, 2824

MICROMORPHOLOGY, 2825
MICROORGANISMS, 2826
MICROPALEONTOLOGY, 2827
MICROPHYTOLITHS, 2828
MICROPLATES, 2829
MICROSCOPE METHODS, 2833
MICROSCOPES, 2831
MICROSCOPIC SIZE, 2832
MICROSEISMS, 2834
MICROSTYLOLITES, 2835
MICROTECTONICS, 2836
MICROTEKTITES, 2837
MICROWAVE METHODS, 2838
MICROWAVE SPECTROSCOPY, 2839
MID-OCEAN RIDGES, 2840
MIDDLE CAMBRIAN, 2841
MIDDLE DEVONIAN, 2842
MIDDLE EOCENE, 2843
MIDDLE JURASSIC, 2844
MIDDLE LIASSIC, 2845
MIDDLE MIOCENE, 2846
MIDDLE OLIGOCENE, 2847
MIDDLE ORDOVICIAN, 2848
MIDDLE PLEISTOCENE, 2849
MIDDLE PLIOCENE, 2850
MIDDLE PROTEROZOIC, 2851
MIDDLE TRIASSIC, 2852
MIGMATITES, 2853
MIGMATIZATION, 2854
MIGRATION, 2855
MIGRATION OF ELEMENTS, 2856
MILIOLACEA, 2857
MILIOLINA, 2858
MILITARY GEOLOGY, 2859
MILLER INDICES, 2860
MINERAL ASSEMBLAGES, 2863
MINERAL CLEAVAGE, 2864
MINERAL COMPOSITION, 2865
MINERAL DEPOSITS, 2866
MINERAL DEPOSITS GENESIS, 1118
MINERAL ECONOMICS, 2867
MINERAL EXPLORATION, 2868
MINERAL FACIES, 2869
MINERAL GEL, 1703
MINERAL INCLUSIONS, 2870
MINERAL INTERLAYERS, 2268
MINERAL PHASE, 2871
MINERAL RESOURCES, 2872
MINERAL SHOWS, 4086
MINERAL SPRING, 2873
MINERAL STRIATIONS, 2874
MINERAL WATERS, 2875
MINERALIZATION, 2876
MINERALOGICAL LOCALITY, 2878
MINERALOGICAL PHASE RULE, 2879
MINERALOGY, 2880
MINERALS, 2862

MINES, 2861
MINICOMPUTERS, 2881
MINIMUM, 2882
MINING, 2883
MINING ENGINEERING, 2884
MINING GEOLOGY, 2885
MINING LICENSE, 2886
MINOR ELEMENTS, 2887
MINOR=ELEMENT ANALYSES, 2888
MIOCENE, 2889
MIOGEOSYNCLINES, 2890
MIOMERA, 2891
MIOSPORES, 2892
MISCIBILITY, 2893
MISCIBILITY GAP, 2894
MISE=A=LA=MASSE, 2895
MISSISSIPPI VALLEY=TYPE, 2896
MISSISSIPPIAN, 2897
MIXED=LAYER MINERALS, 2898
MOBILE BELTS, 2899
MOBILITY, 2900
MOBILIZATION, 2901
MODAL ANALYSIS, 2902
MODE, 2903
MODELS, 2904
MODIFIED MERCALLI SCALE, 2905
MODIOMORPHOIDA, 2906
MODULUS, 2907
MODULUS OF RIGIDITY, 2908
MOHOLE PROJECT, 2909
MOHOROVICIC DISCONTINUITY, 2910
MOISTURE, 2911
MOLASSE, 2912
MOLDAVITE, 2913
MOLECULAR SPECTROSCOPY, 2914
MOLLISOLS, 2915
MOLLUSCA, 2916
MOLYBDATES, 2917
MOLYBDENUM, 2918
MONAZITE, 2919
MONOCLINES, 2920
MONOCLINIC SYSTEM, 2921
MONOCOTYLEDONEAE, 2922
MONOCYATHEA, 2923
MONOGRAPHS, 2924
MONOLETE TAXA, 2925
MONOLITH, 2926
MONOMORPHIC, 2927
MONOPLACOPHORA, 2928
MONORHINA, 2929
MONOTREMATA, 2930
MONSOONS, 2931
MONTE CARLO ANALYSIS, 2932
MONZONITES, 2933
MONZONITIC TEXTURE, 2934
MOON, 2935
MOONQUAKES, 2936

MORAINES, 2938
MORPHOGENESIS, 2939
MORPHOLOGY, 2940
MORPHOMETRY, 2941
MORPHOSCOPY, 2942
MORPHOSTRUCTURES, 2943
MOSAIC STRUCTURE, 2945
MOSCOVIAN, 2946
MOSSBAUER SPECTRA, 2948
MOSSBAUER SPECTROSCOPY, 2947
MOTTLING, 2949
MOUNTAINS, 2950
MOVEMENT, 2951
MUD, 2952
MUD BANKS, 2953
MUD FLATS, 2955
MUD LUMPS, 2957
MUD VOLCANOES, 2958
MUDCRACKS, 2954
MUDFLOWS, 2956
MUDSLIDES, 2959
MULL, 2960
MULLIONS, 2961
MULTILAYER SYSTEMS, 2962
MULTIPHASE, 2963
MULTIPHASE FLOW, 2964
MULTIPHASE TECTONICS, 2965
MULTISPECTRAL ANALYSIS, 2966
MULTITUBERCULATA, 2967
MULTIVARIATE ANALYSIS, 2968
MUSCHELKALK, 2969
MUSEUMS, 2970
MUTATION, 2971
MYLONITES, 2972
MYLONITIC STRUCTURE, 2973
MYLONITIZATION, 2974
MYODOCOPIDA, 2975
MYOIDA, 2976
MYOMORPHA, 2977
MYRIAPODA, 2978
MYRMEKITE, 2979
MYSTACOCARIDA, 2980
MYTILOIDA, 2981
MYXOMYCETES, 2982
NAKHLITE, 2983
NAMURIAN, 2984
NANNOFOSSILS, 2985
NAPPES, 2986
NATANTES, 2987
NATIVE ELEMENTS, 2988
NATURAL GAS, 2989
NATURAL GLASS, 2990
NATURAL RECHARGE, 1917
NATURAL REMANENT MAGNETIZATION, 2991
NATURAL RESOURCES, 2992
NAUTILOIDEA, 2993
NEANDERTHAL, 2994

NEARSHORE SEDIMENTATION, 2546
NEEDLES, 2996
NEKTON, 2997
NEMERTA, 2998
NEOCOMIAN, 2999
NEODYMIUM, 3000
NEOGASTROPODA, 3001
NEOGENE, 3002
NEON, 3003
NEORNITHES, 3004
NEOTECTONICS, 3005
NEOTENY, 3006
NEOTYPES, 3007
NEPHELINE, 3008
NEPHELOID LAYER, 3009
NEPTUNE, 3476
NEPTUNISM, 3010
NEPTUNIUM, 3011
NERVOUS SYSTEM, 3012
NEUTRON ACTIVATION ANALYSIS, 3015
NEUTRON DIFFRACTION ANALYSIS, 3016
NEUTRON LOGGING, 3017
NEUTRON PROBE, 3018
NEUTRONS, 3014
NEW, 3019
NEW DATA, 3020
NEW DESCRIPTION, 3021
NEW METHODS, 3022
NEW MINERALS, 3023
NEW NAMES, 3024
NEW TAXA, 3025
NICKEL, 3026
NIGGLI'S CLASSIFICATION, 3027
NILSSONIALES, 3028
NIOBATES, 3029
NIOBIUM, 3030
NIOBOTANTALATES, 3031
NITRATES, 3032
NITRIFICATION, 3033
NITROGEN, 3034
NITROGEN GAS, 3035
NIVATION, 3036
NODOSARIACEA, 3038
NODULES, 3039
NOEGGERATHIALES, 3040
NOISE, 3041
NOMENCLATURE, 3042
NOMINAL TAXON, 3043
NOMOGENESIS, 3044
NOMOGRAMS, 3045
NON=EXPLOSIVE SOURCES, 3046
NON=FERROUS METALS, 3047
NONCONFORMITIES, 3050
NONMETAL DEPOSITS, 3051
NONMETALLIC=DEPOSIT MAPS, 3052
NONMETALS, 3048
NORIAN, 3053

NORM, 3054
NORMAL DISTRIBUTION, 3055
NORMAL FAULTS, 3056
NORTH, 3057
NOSE (FOLD), 3058
NOTOSTRACA, 3059
NOTOUNGULATA, 3060
NOVACULITE, 3061
NUCLEAR ENERGY, 3062
NUCLEAR EXPLOSIONS, 3063
NUCLEAR FACILITIES, 3065
NUCLEAR MAGNETIC RESONANCE, 3064
NUCLEATION, 3066
NUCLEUS, 3067
NUCULOIDEA, 3068
NUEE ARDENTE, 3069
NUGGET, 3070
NUMMULITIDAE, 3071
NUTATION, 3072
NUTRITION, 3073
NUTRITION TRACES, 1523
O-18/O-16, 3227
OASIS, 3074
OBDUCTION, 3075
OBJECTIVES, 3076
OBLIQUE FAULT, 3077
OBLIQUE ORIENTATION, 3078
OBOLELLIDA, 3079
OBSEQUENT VALLEYS, 3080
OBSERVATIONS, 3081
OBSERVATORIES, 3082
OBSIDIAN, 3083
OCEAN BASINS, 3085
OCEAN CIRCULATION, 3086
OCEAN DRILLING PROGRAM, 3087
OCEAN FLOORS, 3088
OCEAN WAVES, 3979
OCEAN-BOTTOM SEISMOGRAPHS, 3089
OCEANIC, 3090
OCEANIC CRUST, 3091
OCEANIZATION, 3092
OCEANOGRAPHY, 3093
OCEANOLOGY, 3094
OCHER, 3095
OCHREPTS, 3096
OCTAHEDRITE, 3097
OCTOCORALLIA, 3098
ODONTOLOGY, 3099
ODONTOPLEURIDA, 3100
ODONTORNITHES, 3101
OFFSHORE, 3102
OFFSHORE DRILLING, 3103
OFFSHORE WELL, 3104
OIL AND GAS FIELDS, 3105
OIL FIELDS, 3107
OIL SANDS, 3108
OIL SEEPS, 3109

OIL SHALE, 3110
OIL TRAP, 3111
OIL WELL, 3112
OIL-AND-GAS MAPS, 3106
OIL-GAS INTERFACE, 3113
OIL-WATER INTERFACE, 3114
OLENEKIAN, 2210
OLIGOCENE, 3115
OLISTOLITHS, 3116
OLISTOSTROMES, 3117
OLIVINE, 3118
ONCOLITES, 3119
ONE-DIMENSIONAL MODELS, 3120
ONTOGENY, 3121
ONYCHOPHORA, 3122
OOID, 3123
OOLITE, 3124
OOLITIC LIMESTONE, 3125
OOLITIC TEXTURE, 3126
OOZE, 3127
OPAQUE, 3128
OPERATIONS, 3130
OPHIDIA, 3131
OPHIOCISTOIDEA, 3132
OPHIOLITE, 3133
OPHITIC TEXTURE, 3134
OPHIUROIDEA, 3135
OPISTHOBRANCHIA, 3136
OPOKA, 3137
OPTICAL DISPERSION, 1222
OPTICAL EXTINCTION, 3138
OPTICAL PROPERTIES, 3140
OPTICAL ROTATION, 3141
OPTICAL SPECTRA, 3143
OPTICAL SPECTROSCOPY, 3142
ORANGE MATERIAL, 3144
ORBICULAR TEXTURE, 3145
ORBIT, 3146
ORBITOIDACEA, 3147
ORBITOIDIDAE, 3148
ORDER-DISORDER, 3149
ORDOVICIAN, 3150
ORE, 3151
ORE BODIES, 3152
ORE CONTROLS, 3153
ORE DEPOSIT, 3154
ORE DRESSING PLANT, 3155
ORE GUIDES, 3157
ORE MICROSCOPY, 3158
ORE MINERALS, 3159
ORE MORPHOLOGY, 1119
ORE QUALITY, 3160
ORE SHOOTS, 3161
ORE SOURCES, 3162
ORE TRANSPORT, 3163
ORE-FORMING FLUIDS, 2877
ORGANIC, 3164

ORGANIC ACIDS, 3165
ORGANIC CARBON, 3166
ORGANIC COMPOUNDS, 3167
ORGANIC GEOCHEMISTRY, 3168
ORGANIC LIFE, 3169
ORGANIC MATERIALS, 3170
ORGANIC NITROGEN, 3171
ORGANIC RESIDUES, 3173
ORGANIC SEDIMENTS, 3174
ORGANIC STRUCTURE, 3175
ORGANISMS, 3176
ORIENTATION, 3177
ORIGIN, 3179
ORIGOFACIES, 3180
ORNAMENTAL MATERIALS, 3181
ORNAMENTATION, 3182
ORNITHISCHIA, 3183
OROGENIC BELT, 3184
OROGENY, 3185
ORTHENTS, 3186
ORTHIDA, 3188
ORTHIDS, 3187
ORTHOAMPHIBOLITE, 3189
ORTHODS, 3190
ORTHOGENESIS, 3191
ORTHOGNEISS, 3192
ORTHOPHYRIC TEXTURE, 3193
ORTHORHOMBIC SYSTEM, 3194
ORTHOX, 3195
ORYCTOCOENOSIS, 3196
OSCILLATION THEORY, 3198
OSCILLATIONS, 3197
OSMIUM, 3199
OSTEICHTHYES, 3200
OSTEOLOGY, 3201
OSTEOSTRACI, 3202
OSTRACODA, 3203
OSTREACEA, 3204
OTOLITHS, 3205
OUTCROPS, 3206
OUTER CORE, 3207
OUTER PLANETS, 3208
OUTER SHELF, 3209
OUTLIERS, 3210
OUTWASH, 3211
OUTWASH PLAINS, 3212
OVERBURDEN, 3213
OVERCONSOLIDATED MATERIALS, 3214
OVERGROWTHS, 3215
OVERPRINT, 1723
OVERSATURATED, 3216
OVERSATURATED SOLUTION, 3217
OVERTHRUST FAULTS, 3218
OVERTURNED FOLDS, 3219
OXFORDIAN, 3220
OXIDATION, 3221
OXIDATION ZONES, 3223

OXIDES, 3222
OXISOLS, 3224
OXYGEN, 3225
OXYGEN GAS, 3226
OXYLOPHYTE, 3228
P-T CONDITIONS, 3229
P-WAVES, 3230
PA/IO, 3643
PACIFIC SUITE, 3231
PACKING, 3232
PAHOEHOE, 3233
PAINT MINERALS, 3234
PALAEOHETERODONTA, 3235
PALAEOPHYTICUM, 3236
PALAEOTAXODONTA, 3237
PALAGONITIZATION, 3238
PALEO, 3239
PALEO-OCEANOGRAPHY, 3241
PALEOATMOSPHERE, 3240
PALEOBATHYMETRY, 3242
PALEOBIOLOGY, 3243
PALEOBOTANIC PROVINCE, 3244
PALEOBOTANY, 3245
PALEOCENE, 3246
PALEOCIRCULATION, 3247
PALEOCLIMATE, 3248
PALEOCLIMATOLOGY, 3249
PALEOCOPIDA, 3250
PALEOCURRENTS, 3251
PALEOECOLOGY, 3252
PALEOENVIRONMENT, 3253
PALEOGENE, 3254
PALEOGEOGRAPHIC CONTROLS, 3255
PALEOGEOGRAPHIC MAPS, 3256
PALEOGEOGRAPHY, 3257
PALEOGEOMORPHOLOGY, 3258
PALEOKARST, 3259
PALEOLATITUDE, 3260
PALEOLIMNOLOGY, 3261
PALEOMAGNETIC POLE, 3262
PALEOMAGNETISM, 3263
PALEONTOLOGICAL RECONSTRUCTION, 3264
PALEONTOLOGY, 3265
PALEORELIEF, 3266
PALEOSALINITY, 3267
PALEOSOLS, 3268
PALEOTEMPERATURE, 3269
PALEOTHERMOMETRY, 3270
PALEOZOIC, 3271
PALEOZOOLOGY, 3272
PALIMPSEST, 3273
PALINGENESIS, 3274
PALLADIUM, 3275
PALLASITES, 3276
PALSA, 3277
PALUDAL SEDIMENTATION, 4377
PALYNOLOGY, 3278

PALYNOMORPHS, 3279
PAMPAS, 3280
PANGAEA, 3281
PANIDIOMORPHIC TEXTURE, 3282
PANNING, 3283
PANTODONTA, 3284
PANTOTHERIA, 3285
PARA-, 3286
PARAAMPHIBOLITE, 3287
PARABLASTOIDEA, 3288
PARACRINOIDEA, 3289
PARAGENESIS, 3290
PARAGEOSYNCLINES, 3291
PARAGNEISS, 3292
PARALIAGEOSYNCLINES, 3293
PARALIC ENVIRONMENT, 3294
PARALLEL BEDDING, 3295
PARALLEL TEXTURE, 3296
PARALLELISM, 3297
PARAMAGNETIC PROPERTY, 3298
PARAMETERS, 3299
PARASITES, 3300
PARATYPES, 3301
PARENT MATERIALS, 3302
PARENT PRODUCT, 3303
PARENT ROCK, 3304
PARTIAL MELTING, 3305
PARTIAL PRESSURE, 3306
PARTICLE-TRACK DATING, 3309
PARTICLES, 3307
PARTING LINEATION, 3310
PARTITION COEFFICIENTS, 3311
PARTITIONING, 3312
PATCH REEFS, 3313
PATENTS, 3314
PATHOLOGY, 3315
PATTERNED GROUND, 3317
PATTERNS, 3316
PAUROPODA, 3318
PB/PB, 2468
PEARLS, 3319
PEAT, 3320
PEAT BOGS, 3321
PEBBLES, 3322
PECTINACEA, 3323
PEDIMENTS, 3324
PEDIPLAINS, 3325
PEDOGENESIS, 3326
PEGMATITE, 3327
PEGMATITIC, 3328
PEGMATOID TEXTURE, 3329
PELAGIC ENVIRONMENT, 3330
PELAGIC SEDIMENTATION, 3331
PELEAN-TYPE ERUPTIONS, 3333
PELITE, 3334
PELITIC TEXTURE, 3335
PELLETIZATION, 3338

PELLETS, 3337
PENEPLAINS, 3339
PENETRATION TESTS, 3340
PENETROMETERS, 3341
PENNSYLVANIAN, 3342
PENOKEAN OROGENY, 3343
PENTAMERIDA, 3344
PENTASTOMIDA, 3345
PENTOXYLALES, 3346
PERALKALINE COMPOSITION, 3347
PERALUMINOUS COMPOSITION, 3348
PERCHED AQUIFERS, 3349
PERCOLATION, 3350
PERICLINES, 3351
PERIDOTITES, 3352
PERIGEE, 3353
PERIGLACIAL ENVIRONMENT, 3355
PERIGLACIAL FEATURES, 3354
PERIODICITY, 3357
PERIODS, 3356
PERISPHINCTIDA, 3358
PERISSODACTYLA, 3359
PERLITE, 3360
PERLITIC TEXTURE, 3361
PERMAFROST, 3362
PERMEABILITY, 3363
PERMIAN, 3364
PERSONAL BIBLIOGRAPHY, 3365
PERSPECTIVE, 3366
PERTHITE, 3367
PERTHITIC TEXTURE, 3368
PESTICIDES, 3369
PETROCHEMICAL COEFFICIENT, 3370
PETROCHEMICAL DIAGRAM, 3371
PETROCHEMISTRY, 3372
PETROFABRICS, 3373
PETROGENESIS, 3374
PETROGRAPHIC ANALYSIS, 3375
PETROGRAPHIC PROVINCE, 3377
PETROGRAPHY, 3378
PETROLEUM, 3379
PETROLEUM ENGINEERING, 3380
PETROLEUM GEOLOGY, 3381
PETROLOGY, 3382
PETROPHYSICS, 3383
PH, 3384
PHACOLITH, 3385
PHACOPIDA, 3386
PHAEOPHYTA, 3387
PHAEOZEM, 3388
PHAGOTROPHIC ORGANISM, 3389
PHANEROZOIC, 3390
PHASE EQUILIBRIA, 3392
PHASE RULE, 3394
PHASES, 3391
PHENOCRYSTS, 3395
PHENOLS, 3396

PHI SCALE, 3397
PHILOSOPHY, 3398
PHOLADOMYIDA, 3399
PHOLIDOTA, 3400
PHONOLITES, 3401
PHOSPHATE ROCKS, 3402
PHOSPHATES, 3403
PHOSPHATIZATION, 3404
PHOSPHORESCENCE, 3405
PHOSPHORITE, 3406
PHOSPHOROUS, 3407
PHOTOCHEMICAL REACTION, 3408
PHOTOGEOLOGIC MAPS, 3409
PHOTOGEOLOGY, 3410
PHOTOGRAMMETRY, 3411
PHOTOGRAPHY, 3412
PHOTOMETRY, 3413
PHOTOSYNTHESIS, 3414
PHREATIC GAS, 3415
PHREATIC WATER, 3416
PHREATOPHYTE, 3417
PHYLACTOLAEMATA, 3418
PHYLLITES, 3419
PHYLLOCERATIDA, 3420
PHYLLONITES, 3421
PHYLLOPODA, 3422
PHYLOGENY, 3423
PHYLUM, 3424
PHYSICAL CHEMISTRY, 3425
PHYSICAL MODELS, 3426
PHYSICAL PROPERTIES, 3427
PHYSICAL STATE, 3428
PHYSICAL WEATHERING, 3429
PHYSICOCHEMICAL PROPERTIES, 3430
PHYSIOGRAPHIC PROVINCES, 3431
PHYSIOLOGY, 3432
PHYTANE, 3433
PHYTOCOENOSES, 3434
PHYTOLITHS, 3435
PHYTOPHAGOUS ORGANISM, 3436
PHYTOPLANKTON, 3437
PICRITE, 3438
PIEDMONT, 3439
PIERS, 3440
PIEZOCRYSTALLIZATION, 3441
PIEZOELECTRIC EFFECTS, 3442
PIEZOMETRY, 3443
PIGMENTS, 3444
PILES, 3445
PILLARS, 3446
PILLOW LAVA, 3447
PILLOW STRUCTURE, 3448
PINCH OUTS, 3449
PINGOS, 3450
PINNACLE REEFS, 3451
PINNIPEDIA, 3452
PIPELINES, 3454

PIPES, 3453
PIPING, 3455
PISCES, 3456
PISOLITHS, 3457
PISOLITIC TEXTURE, 3458
PIT, 3459
PIT SECTIONS, 3460
PLACANTICLINES, 3461
PLACENTALIA, 3462
PLACERS, 3463
PLACODERMI, 3464
PLAINS, 3465
PLANAR BEDDING STRUCTURES, 3466
PLANAR FEATURES, 3467
PLANATION, 3468
PLANCTIVOROUS ANIMAL, 3486
PLANCTOSPHAEROIDEA, 3469
PLANES, 3470
PLANETARY, 3482
PLANETARY INTERIORS, 3483
PLANETOLOGY, 3485
PLANETS, 3471
PLANKTON, 3487
PLANKTONIC TAXA, 3488
PLANNING, 3489
PLANOSOLS, 3490
PLANTAE, 3491
PLASTIC FLOW, 3492
PLASTIC MATERIALS, 3493
PLASTICITY, 3494
PLATE BOUNDARIES, 3497
PLATE COLLISION, 3496
PLATE ROTATION, 3498
PLATE TECTONICS, 3499
PLATE=BEARING TESTS, 3500
PLATEAU BASALT, 3502
PLATEAUS, 3501
PLATES, 3495
PLATFORMS, 3503
PLATINUM, 3504
PLATINUM MINERALS, 3505
PLATYRRHINA, 3506
PLAYAS, 3507
PLEISTOCENE, 3508
PLEOCHROISM, 3509
PLIENSBACHIAN, 3511
PLIOCENE, 3512
PLUGGING, 3514
PLUGS, 3513
PLUMES, 3515
PLUNGE, 3516
PLUTO, 3477
PLUTONIC METAMORPHISM, 3518
PLUTONIC ROCKS, 3519
PLUTONISM, 3520
PLUTONIUM, 3521
PLUTONS, 3517

PLUVIAL PROCESSES, 3522
PNEUMATOLYSIS, 3523
PNEUMATOLYTIC DEPOSIT, 3524
PODOCOPIDA, 3525
PODZOLS, 3526
PODZOLUVISOLS, 3527
POGONOPHORA, 3528
POIKILITIC TEXTURE, 3529
POIKILOBLASTIC TEXTURE, 3530
POISSON DISTRIBUTION, 3531
POISSON'S RATIO, 3532
POLAR, 3533
POLAR ICE CAP, 3534
POLAR WANDERING, 3535
POLARIZATION, 3139
POLAROGRAPHY, 3536
POLE POSITIONS, 1742
POLICY, 3537
POLISHED SECTIONS, 3538
POLLEN, 3539
POLLEN DIAGRAMS, 3540
POLLUTANTS, 3541
POLLUTED WATER, 3542
POLLUTION, 3543
POLONIUM, 3544
POLYCHAETIA, 3545
POLYGENETIC, 3546
POLYGONAL GROUND, 3547
POLYHEDRA, 3548
POLYMERA, 3549
POLYMERIZATION, 3550
POLYMETALLIC ORES, 3551
POLYMETAMORPHISM, 3552
POLYMICTIC TEXTURE, 3553
POLYMINERALIC, 3554
POLYMORPHISM, 3555
POLYPHASE PROCESSES, 3556
POLYPLACOPHORA, 3557
POLYTYPISM, 3558
PONTIAN, 3559
POPULAR GEOLOGY, 3560
POPULATIONS, 4243
PORCELLANITE, 3561
PORE PRESSURE, 3562
PORE WATER, 3563
PORIFERA, 3564
POROSITY, 3565
POROUS MATERIALS, 3566
PORPHYRITIC TEXTURE, 3567
PORPHYROBLASTIC TEXTURE, 3568
PORPHYRY COPPER, 3569
PORTLANDIAN, 3570
POSITIVE ORE, 3571
POSSIBILITIES, 3572
POSTGLACIAL, 3573
POSTMAGMATIC MATERIALS, 3574
POSTTECTONIC PROCESSES, 3575

POTABILITY, 3576
POTASH, 3577
POTASSIC COMPOSITION, 3578
POTASSIUM, 3579
POTENTIAL DEPOSITS, 3582
POTENTIOMETERS, 3584
POTENTIOMETRIC SURFACE, 3583
POTENTIOMETRIC SURFACE, 2128
POTHOLES, 3585
POTTER'S CLAY, 3586
POWER PLANTS, 3588
POZZOLAN, 3589
PRAECARDIIDA, 3590
PRASEODYMIUM, 3591
PRE-NEANDERTHAL, 3604
PRECAMBRIAN, 3593
PRECAMBRIAN OROGENY, 3594
PRECIOUS METALS, 3037
PRECIOUS STONE, 3595
PRECIPITATION, 3596
PREDATION, 3597
PREDICTION, 3598
PREDICTION MAPS, 1611
PREDICTIVE METALLOGENY, 3599
PREFERRED ORIENTATION, 3600
PREGLACIAL, 3601
PREHISTORY, 3602
PREHNITE-PUMPELLYITE FACIES, 3603
PREIMBRIAN, 3592
PRELIMINARY STUDIES, 3178
PREPARATION, 3605
PRESERVATION, 3606
PRESSURE, 3607
PRESSURE SOLUTION, 3608
PREVENTION, 3609
PRICE, 3610
PRIMARY DISPERSION, 3611
PRIMARY STRUCTURES, 3612
PRIMATES, 3613
PRIMITIVE, 3614
PRIORITY, 3615
PRISMATIC STRUCTURE, 3616
PROBABILITY, 3617
PROBE, 3618
PROBLEMATIC FOSSILS, 3619
PROBLEMATIC MICROFOSSILS, 2830
PROBOSCIDEA, 3620
PROCELLARIAN, 3621
PROCESSES, 3622
PROCHORDATA, 3623
PRODUCING HORIZON, 3626
PRODUCING WELL, 3627
PRODUCTION, 3624
PRODUCTION CONTROL, 3625
PROFILE OF EQUILIBRIUM, 3628
PROFITABILITY, 3629
PROGANOSAURIA, 3630

PROGNOSIS, 3631
PROGRADATION, 3632
PROGRADE METAMORPHISM, 3633
PROGRAMS, 3634
PROGRESS REPORT, 3635
PROJECTS, 3636
PROMETHIUM, 3637
PROPERTIES, 3638
PROPYLITIZATION, 3639
PROSIMII, 3640
PROSPECTING, 3641
PROTACTINIUM, 3642
PROTECTED WELL ZONES, 3644
PROTEINS, 3645
PROTELYTROPTERA, 3646
PROTEROZOIC, 3647
PROTEROZOIDES, 3648
PROTISTA, 3654
PROTOCILIATA, 3649
PROTOLITHS, 3650
PROTOMEDUSAE, 3651
PROTONS, 3652
PROTOTYPE, 3653
PROTURA, 3655
PROVENANCE, 3656
PROVINCIALITY, 3657
PSAMMENTS, 3658
PSAMMITE, 3659
PSAMMITIC, 3660
PSAMMOFAUNA, 3661
PSEPHITIC, 3662
PSEUDOCONGLOMERATE, 3663
PSEUDOGLEYS, 3664
PSEUDOMORPHISM, 3665
PSEUDOTILLITE, 3666
PSILOCERATIDA, 3667
PSILOPSIDA, 3668
PSYCHROPHYTE, 3669
PTERIDOPHYLLEN, 3670
PTERIDOPHYTES, 3671
PTERIDOSPERMAE, 3672
PTERIOIDA, 3673
PTEROBRANCHIA, 3674
PTEROMORPHIA, 3675
PTEROPODA, 3676
PTEROSAURIA, 3677
PTYCHOPARIIDA, 3678
PTYGMATIC FOLDS, 3679
PUBLIC WORKS, 3680
PUBLICATIONS, 3681
PULMONATA, 3682
PUMICE, 3683
PUMP TESTS, 3685
PUMPAGE, 3684
PUNCH CARDS, 3686
PURIFICATION, 3687
PURINE, 3688

PYCNOCLINES, 3689
PYCNOGONIDA, 3690
PYCNOMETER, 3691
PYCNOMETRY, 3692
PYRAMIDAL SYSTEM, 3693
PYRITIZATION, 3694
PYROCLASTICS, 3695
PYROLYSIS, 3696
PYROMETALLURGY, 3697
PYROMETAMORPHISM, 3698
PYROMETASOMATISM, 3699
PYROSPHERE, 3700
PYROTHERIA, 3701
PYROXENE, 3702
PYROXENE-HORNFELS FACIES, 3703
PYROXENITE, 3704
PYRROPHYTA, 3705
QUALITATIVE ANALYSIS, 3706
QUALITY, 3707
QUANTITATIVE ANALYSIS, 3708
QUARRIES, 3709
QUARTZ, 3710
QUARTZ DIORITES, 3711
QUARTZITES, 3712
QUATERNARY, 3713
QUATERNARY SYSTEM, 3714
QUICK CLAY, 3715
QUICKSAND, 3716
RACEMIZATION, 3717
RADAR METHODS, 3718
RADIAL, 3719
RADIATION, 3720
RADIATION DAMAGE, 3721
RADIO INTERFEROMETRY, 3722
RADIO SPECTROSCOPY, 3723
RADIO-WAVE METHODS, 3724
RADIOACTIVE DECAY, 3725
RADIOACTIVE DUST, 3726
RADIOACTIVE ISOTOPES, 3727
RADIOACTIVE SERIES, 3728
RADIOACTIVE TRACERS, 3729
RADIOACTIVE WASTE, 3730
RADIOACTIVITY, 3731
RADIOACTIVITY METHODS, 3732
RADIOACTIVITY SURVEYS, 3738
RADIOCARBON, 3733
RADIOLARIA, 3735
RADIOLARITE, 3736
RADIOMETER, 3737
RADIUM, 3739
RADIUS, 3740
RADON, 3741
RAIN, 3742
RAIN PRINTS, 3743
RAIN WASH, 3744
RAMAN SPECTROSCOPY, 3745
RANGE, 3746

RANGE ZONE, 3747
RANK, 3748
RANKER, 3749
RAPAKIVI, 3750
RARE EARTHS, 3751
RARE METALS, 3752
RATES, 3753
RATIOS, 730
RATITES, 3754
RAW MATERIALS, 3755
RAYLEIGH WAVES, 3756
RE/OS, 3850
REACTION RIMS, 3758
REACTION SERIES, 3759
REACTIONS, 3757
REACTIVATION, 3760
REAGENTS, 3761
RECENT SEDIMENTATION, 3762
RECEPTACULITACEAE, 3763
RECEPTION EQUIPMENT, 3764
RECHARGE, 3765
RECLAMATION, 3766
RECONSTRUCTION, 3767
RECOVERY, 3768
RECREATION, 3769
RECRYSTALLIZATION, 3770
RECUMBENT FOLDS, 3771
RECYCLING, 3772
RED BEDS, 3773
REDEPOSITION, 3774
REDLICHIIDA, 3775
REDUCTION, 731
REEF BUILDERS, 3779
REEF ENVIRONMENT, 3777
REEF SEDIMENTATION, 3778
REEFS, 3776
REFINEMENT, 3780
REFINING, 3781
REFLECTANCE, 3782
REFLECTION, 3783
REFLECTION METHODS, 3784
REFLECTIVITY, 3785
REFRACTION, 3786
REFRACTION METHODS, 3787
REFRACTIVE INDEX, 2209
REFRACTIVITY, 3788
REFRACTOMETERS, 3789
REFRACTOMETRY, 3790
REFRACTORY MATERIALS, 3791
REGENERATION, 3792
REGIONAL HYDROGEOLOGY, 3794
REGIONAL METAMORPHISM, 3795
REGIONAL UNCONFORMITY, 3796
REGMAGLYPT, 3797
REGOLITH, 3798
REGOSOLS, 3799
REGRESSION, 3800

REGRESSION ANALYSIS, 3801
REJUVENATION, 3802
RELATIONSHIP, 3803
RELATIVE AGE, 3804
RELAXATION, 3805
RELICT MATERIALS, 3806
RELICT TEXTURE, 3807
RELIEF, 3808
RELIEF EXPOSURE, 3809
RELIEF INVERSION, 3810
REMAGNETIZATION, 3811
REMANENT MAGNETIZATION, 3812
REMOTE SENSING, 3813
REMOTENESS, 3814
RENDOLLS, 3815
RENDZINAS, 3816
REPLACEMENT, 3817
REPORT, 3818
REPRESENTATIVE BASINS, 3819
REPRODUCTION, 3820
REPTILIA, 3821
RESEARCH, 3822
RESEQUENT VALLEYS, 3823
RESERVES, 3824
RESERVOIR ROCKS, 3826
RESERVOIRS, 3825
RESIDUAL, 3827
RESIDUAL CLAYS, 3828
RESIDUAL DEPOSITS, 3829
RESIDUAL MAGMA, 3830
RESINS, 3831
RESISTIVITY, 3832
RESISTIVITY METHOD, 3833
RESOURCES, 3834
RESPIRATION, 3835
RESURGENCE, 3836
RESURGENT WATER, 3837
RETENTION, 3838
RETICULAREA, 3839
RETICULATE STRUCTURE, 3840
RETROGRADE METAMORPHISM, 3841
REVERSALS, 3842
REVIEW, 4357
REVISION, 3844
REWORKED, 3845
REWORKING, 3846
RHABDOPLEURIDA, 3847
RHAETIAN, 3848
RHENIUM, 3849
RHEOLOGY, 3851
RHEOMORPHISM, 3852
RHIPIDISTIA, 3853
RHIZOPODEA, 3854
RHODIUM, 3855
RHODOPHYTA, 3856
RHOMBIC SYSTEM, 3857
RHYNCHOCEPHALIA, 3858

RHYNCHONELLIDA, 3859
RHYODACITES, 3860
RHYOLITES, 3861
RICHTER SCALE, 3862
RIEGEL, 3863
RIFT VALLEYS, 3865
RIFT ZONES, 3864
RIGIDITY, 3866
RILLES, 3867
RIMS, 3868
RING DIKES, 3869
RING FRACTURE, 3870
RING STRUCTURES, 3871
RIPPLE MARKS, 3872
RIVER LOAD, 3874
RIVER TERRACES, 3875
RIVERS, 3873
ROAD LOG, 3876
ROAD MATERIAL, 3877
ROAD TESTS, 3878
ROADWAY, 3879
ROBERTINACEA, 3880
ROCK BARS, 3882
ROCK BORER, 3889
ROCK BURSTS, 3883
ROCK GLACIERS, 3886
ROCK MATRIX, 3887
ROCK MECHANICS, 3888
ROCK=FORMING MINERALS, 3890
ROCKFALLS, 3884
ROCKFILL DAMS, 3885
ROCKS, 3881
ROCKSLIDES, 3891
RODENTIA, 3892
ROLL=TYPE DEPOSITS, 3893
ROOF COLLAPSE, 3894
ROOF PENDANT, 3895
ROOTS, 3896
ROSTROCONCHIA, 3897
ROTALIACEA, 3898
ROTALIINA, 3899
ROTARY DRILLING, 3900
ROTARY=PERCUSSION DRILLING, 3901
ROTATION, 3902
ROTATORY DISPERSION, 3903
ROTLIEGENDES, 3904
ROUGHNESS, 3905
ROUNDNESS, 3906
RUBBLE, 3907
RUBEFACTION, 3908
RUBIDIUM, 3909
RUDISTAE, 3911
RUDITE, 3912
RUGOSA, 4474
RUMINANTIA, 3913
RUN, 3914
RUNOFF, 3915

RUPELIAN, 3916
RUPTURE, 3917
RUPTURE STRENGTH, 3918
RUTHENIUM, 3919
S−34/S−32, 4355
S−WAVES, 3920
SAKMARIAN, 3921
SALIFEROUS, 3922
SALINITY, 3923
SALT, 3924
SALT DOMES, 3925
SALT FLATS, 3926
SALT LAKES, 3927
SALT MARSHES, 3928
SALT TECTONICS, 3929
SALT−WATER INTRUSION, 3930
SALTATION, 3931
SAMARIUM, 3932
SAMPLER, 3934
SAMPLES, 3933
SAMPLING, 3935
SAND, 3936
SAND BODIES, 3937
SAND WAVES, 3941
SAND−SHALE RATIO, 3938
SANDSTONE, 3939
SANDSTONE DIKES, 3940
SANIDINITE FACIES, 3942
SANTONIAN, 3943
SAPROPEL, 3944
SAPROPELITE, 3945
SARCODINA, 3946
SARCOPTERYGII, 3947
SARMATIAN, 3948
SATELLITE METHODS, 322
SATELLITES, 3484
SATURATED ZONE, 3949
SATURATION, 3950
SATURN, 3478
SAURISCHIA, 3951
SAURORNITHES, 3952
SAUSSURITIZATION, 3953
SAXONIAN, 3954
SCALE MODELS, 3955
SCANDIUM, 3956
SCANNING METHOD, 3957
SCAPHOPODA, 3958
SCAPOLITIZATION, 3959
SCARPS, 1433
SCHISTOSITY, 3961
SCHISTS, 3960
SCHLIEREN, 3962
SCINTILLATIONS, 3964
SCIUROMORPHA, 3965
SCLERACTINIA, 2000
SCLERITES, 3966
SCOLECODONTS, 3967

SCOLITE, 3968
SCORIA, 3969
SCOUR MARKS, 3970
SCOURING, 3971
SCYPHOMEDUSAE, 3972
SCYPHOZOA, 3973
SEA BOTTOM, 3975
SEA ICE, 3976
SEA LEVEL, 3977
SEA WATER, 3978
SEA−FLOOR SPREADING, 3980
SEALING, 3982
SEAMOUNTS, 3984
SEAMS, 3983
SEAS, 3974
SEASONAL VARIATIONS, 3985
SEAWALLS, 3986
SEBKHA ENVIRONMENT, 3987
SECONDARY DISPERSION, 3988
SECONDARY MINERALS, 3989
SECONDARY STRUCTURES, 3990
SECTIONS, 1736
SECULAR VARIATIONS, 1744
SEDIMENT TRACTION, 3992
SEDIMENT YIELD, 3993
SEDIMENT−WATER INTERFACE, 3994
SEDIMENTARY BASINS, 3995
SEDIMENTARY COVER, 3996
SEDIMENTARY ORE, 3997
SEDIMENTARY PETROLOGY, 3998
SEDIMENTARY PROCESSES, 3999
SEDIMENTARY ROCKS, 4000
SEDIMENTARY STRUCTURES, 4001
SEDIMENTARY TRAP, 4002
SEDIMENTATION, 4003
SEDIMENTATION RATES, 4004
SEDIMENTOGENESIS, 4005
SEDIMENTOLOGY, 4006
SEDIMENTS, 3991
SEEDS, 4007
SEEPAGE, 4008
SEGREGATED VEIN, 4009
SEGREGATION, 4010
SEICHES, 4011
SEISLOG, 4012
SEISMIC EQUIPMENT, 4013
SEISMIC METHODS, 4014
SEISMIC RISK, 4015
SEISMIC SOURCES, 4016
SEISMIC STATIONS, 4018
SEISMIC SURVEYS, 4019
SEISMIC VELOCITY, 4020
SEISMIC WAVES, 4021
SEISMICITY, 4022
SEISMOGRAMS, 4023
SEISMOGRAPHS, 4024
SEISMOLOGICAL MAPS, 4025

SEISMOLOGY, 4026
SEISMOMETERS, 4027
SEISMOTECTONICS, 4028
SELENATES, 4029
SELENIUM, 4030
SELENOCHRONOLOGY, 4031
SELENOLOGY, 4032
SELF−POTENTIAL METHODS, 4033
SEM DATA, 4034
SEMI−ARID ENVIRONMENT, 4035
SEMIPRECIOUS STONE, 4036
SENONIAN, 4037
SENSITIVE CLAYS, 4038
SEPARATION, 4039
SEPTA, 4041
SEPTARIA, 4040
SERICITE SCHIST, 4042
SERICITIZATION, 4043
SERIES, 4044
SERPENTINE GROUP, 4045
SERPENTINITE, 4046
SERPENTINIZATION, 4047
SETTLEMENT, 4048
SEXUAL DIMORPHISM, 4049
SH−WAVES, 4050
SHALE, 4051
SHALLOW−FOCUS EARTHQUAKES, 4052
SHALLOW−WATER ENVIRONMENT, 4053
SHAPE, 4054
SHAPE ANALYSIS, 3308
SHATTER BRECCIA, 4056
SHATTER CONES, 4057
SHEAR, 4058
SHEAR JOINTS, 4059
SHEAR MODULUS, 4060
SHEAR PLANES, 4061
SHEAR STRENGTH, 4062
SHEAR STRESS, 4063
SHEAR ZONES, 4064
SHEET EROSION, 4065
SHEETS, 4066
SHELF, 4067
SHELF ENVIRONMENT, 4068
SHELF SEA, 4069
SHELF SEDIMENTATION, 4070
SHELLS, 4071
SHIELD VOLCANOES, 4073
SHIELDS, 4072
SHOALS, 4074
SHOCK, 4075
SHOCK BRECCIA, 4076
SHOCK METAMORPHISM, 4077
SHOCK WAVES, 4078
SHOESTRING SANDS, 4079
SHONKINITE, 4080
SHORE FEATURES, 4081
SHORE PROTECTION, 4082

SHORELINES, 4083
SHORTENING, 4084
SHOT HOLES, 4085
SHRINKAGE, 4087
SHRINKAGE CRACKS, 4088
SIAL, 4089
SICILIAN, 4090
SIDE-SCANNING METHODS, 4091
SIDEROPHILE ELEMENTS, 4092
SIDEROPHYRES, 4093
SIEGENIAN, 4094
SIEROZEMS, 4095
SIEVES, 4096
SILICA, 4097
SILICATES, 4098
SILICEOUS COMPOSITION, 4099
SILICEOUS SINTER, 4118
SILICIFICATION, 4100
SILICIFIED WOOD, 4101
SILICOFLAGELLATA, 4102
SILICON, 4103
SILLIMANITE DEPOSITS, 4105
SILLS, 4104
SILT, 4106
SILT LOAD, 4107
SILTATION, 4108
SILTSTONE, 4109
SILURIAN, 4110
SILVER, 4111
SIMA, 4112
SIMIANS, 4113
SIMULATION, 4114
SINEMURIAN, 4115
SINGLE-CRYSTAL METHOD, 4116
SINKHOLES, 4117
SINKS, 1721
SINUOSITY, 4119
SIPHON, 4120
SIPHONAPTEROIDA, 4121
SIPUNCULOIDA, 4122
SIRENIA, 4123
SITE EXPLORATION, 4124
SIZE DISTRIBUTION, 4125
SKARN, 4126
SKELETONS, 4127
SKEWNESS, 4128
SLABS, 4129
SLATES, 4130
SLATY CLEAVAGE, 4131
SLICKENSIDES, 4132
SLIP CLEAVAGE, 4133
SLOPE DEFORMATION, 4135
SLOPE ENVIRONMENT, 4136
SLOPE STABILITY, 4137
SLOPES, 4134
SLUMP STRUCTURES, 4138
SLUMPING, 4139

SMALL SCALE, 4140
SNOW, 4142
SODIUM, 4143
SOFT CLAYS, 4144
SOFT SEDIMENT DEFORMATION, 4145
SOIL DEFORMATION, 4147
SOIL EROSION, 4148
SOIL HORIZONS, 4149
SOIL IMPROVEMENT, 4150
SOIL LEACHING, 4151
SOIL MANAGEMENT, 4152
SOIL MAPS, 4153
SOIL MECHANICS, 4154
SOIL PROFILES, 4155
SOIL SAMPLING, 4156
SOIL SCIENCES, 4157
SOIL STABILIZATION, 4158
SOILS, 4146
SOLAR RADIATION, 4159
SOLAR SYSTEM, 4160
SOLAR WIND, 4161
SOLE INJECTION, 4162
SOLE MARKS, 4163
SOLEMYIDA, 4164
SOLFATARAS, 4165
SOLID PHASE, 4166
SOLID SOLUTION, 4167
SOLID WASTE, 4168
SOLID-EARTH GEOPHYSICS, 4169
SOLIDUS, 4170
SOLIFLUCTION, 4171
SOLONCHAK SOILS, 4172
SOLONETZ SOILS, 4173
SOLUBILITY, 4174
SOLUTES, 4175
SOLUTION, 1228
SOLUTION CAVITIES, 4177
SOLUTION FEATURES, 4178
SOLUTION MINING, 4179
SOLUTIONS, 4176
SOMASTEROIDEA, 4180
SONAR METHODS, 4181
SONOBUOYS, 4182
SORPTION, 4183
SORTING, 4184
SOUNDING, 4185
SOURCE ROCKS, 4186
SOUTH, 4187
SPACE GROUPS, 4188
SPALLATION, 4189
SPARAGMITE, 4190
SPARKER, 4191
SPATTER CONES, 4192
SPECIES DIVERSITY, 4193
SPECIFIC FAUNA, 4194
SPECIFIC FLORA, 4195
SPECIFIC GRAVITY, 4196

SPECIFIC HEAT, 4197
SPECIFIC SURFACE, 4198
SPECTRA, 4202
SPECTRAL ANALYSIS, 4017
SPECTROMETERS, 4199
SPECTROPHOTOMETRY, 4200
SPECTROSCOPY, 4201
SPELEOLOGY, 4203
SPELEOTHEMS, 4204
SPHENOLITH, 4205
SPHERICAL HARMONIC ANALYSIS, 4206
SPHERICITY, 4207
SPHERULES, 4208
SPHERULITES, 4209
SPILITIZATION, 4210
SPINAL COLUMN, 4211
SPINIFEX TEXTURE, 4212
SPIRIFERIDA, 4213
SPIRILLINACEA, 4214
SPITS, 4215
SPODOSOLS, 4216
SPOILS, 4217
SPONGOLITE, 4218
SPORES, 4219
SPOROMORPH, 4220
SPOROZOA, 4221
SPOTTED SLATE, 4222
SPREADING CENTERS, 4223
SPRINGS, 4224
SQUAMATA, 4225
SR/RB, 3910
STABILITY, 4226
STABILIZATION, 4227
STABLE ISOTOPES, 4228
STAINING, 4230
STALACTITES, 4231
STALAGMITE, 4232
STAMPIAN, 4233
STANDARD DEVIATION, 4234
STANDARD MATERIALS, 4235
STANDARD METHOD, 4236
STANDARD ROCKS, 4237
STANDARDIZATION, 4238
STATISTICAL ANALYSIS, 4239
STATISTICAL DISTRIBUTION, 4240
STATISTICAL FUNCTIONS, 4241
STATISTICAL MODEL, 4242
STATISTICAL SMOOTHING, 4141
STATISTICS, 4245
STEADY FLOW, 4246
STEATITIZATION, 4247
STEEL INDUSTRY, 2309
STEP FAULTS, 4248
STEP FOLDS, 4249
STEPHANIAN, 4250
STEPPES, 4251
STEREOCHEMISTRY, 4252

STEROLS, 4253
STICK=SLIP, 4254
STIFF CLAYS, 4255
STOCHASTIC PROCESSES, 4256
STOCKS, 4257
STOCKWORK DEPOSITS, 4258
STOICHIOMETRY, 4259
STOLONOIDEA, 4260
STONE LINES, 4262
STONY IRONS, 4263
STONY METEORITES, 4264
STORAGE, 4265
STORMS, 4266
STRAIN, 4267
STRAINMETERS, 4268
STRAITS, 4269
STRATABOUND DEPOSITS, 4270
STRATEGY, 4271
STRATIFICATION, 4272
STRATIFIED STRUCTURE, 4273
STRATIFORM DEPOSITS, 4274
STRATIGRAPHIC COLUMNS, 1734
STRATIGRAPHIC CONTROLS, 4276
STRATIGRAPHIC GAPS, 1682
STRATIGRAPHIC MAPS, 4278
STRATIGRAPHIC STAGE, 4229
STRATIGRAPHIC TRAPS, 4279
STRATIGRAPHIC UNITS, 4280
STRATIGRAPHIC WEDGES, 4811
STRATIGRAPHY, 4281
STRATOTYPES, 4282
STRATOVOLCANOES, 4283
STREAM BEDS, 4285
STREAM CAPTURE, 637
STREAM GRADIENT, 2566
STREAM ORDER, 4286
STREAM SEDIMENTS, 4287
STREAM TRANSPORT, 4288
STREAMS, 4284
STRENGTH, 4289
STRESS, 4290
STRESS MINERAL, 4291
STRIATIONS, 1800
STRIKE, 4292
STRIKE=SLIP FAULTS, 4293
STRIP MINING, 4294
STROMATOLITES, 4295
STROMATOPOROIDEA, 4296
STROMBOLIAN=TYPE ERUPTIONS, 4297
STRONG MOTION, 4298
STRONTIUM, 4299
STROPHOMENIDA, 4300
STRUCTURAL ANALYSIS, 4301
STRUCTURAL CONTROLS, 4302
STRUCTURAL CRYSTALLOGRAPHY, 4303
STRUCTURAL DIAGRAM, 4304
STRUCTURAL GEOLOGY, 4306

STRUCTURAL GUIDE, 4307
STRUCTURAL MAPS, 4308
STRUCTURAL PETROLOGY, 4309
STRUCTURAL RELIEF, 4310
STRUCTURAL TRAPS, 4311
STRUCTURAL VALLEYS, 4312
STRUCTURE CONTOUR MAPS, 4313
STRUCTURES, 1375
STYLOLITES, 4314
STYLOPHORA, 4315
SUBAERIAL ENVIRONMENT, 4316
SUBALKALIC COMPOSITION, 4317
SUBALUMINOUS COMPOSITION, 4318
SUBBOREAL, 4319
SUBDIVISION, 4320
SUBDUCTION, 4321
SUBDUCTION ZONES, 4322
SUBLIMATES, 4323
SUBLIMATION, 4324
SUBLITTORAL ENVIRONMENT, 4325
SUBMARINE CANYONS, 4326
SUBMARINE FANS, 4327
SUBMARINE INSTALLATIONS, 4328
SUBMARINE MORPHOLOGY, 4329
SUBMARINE RIDGE, 4330
SUBMARINE SPRINGS, 4331
SUBMARINE VALLEYS, 4332
SUBMARINE VOLCANO, 4333
SUBMERGENCE, 4334
SUBMERSIBLES, 4335
SUBSEQUENT VALLEYS, 4336
SUBSIDENCE, 4337
SUBSTITUTION, 4338
SUBSTRATES, 4339
SUBSURFACE GEOLOGY, 4340
SUBTIDAL ENVIRONMENT, 4341
SUBTROPICAL ENVIRONMENT, 4342
SUBVOLCANIC PROCESSES, 4343
SUBWAYS, 4344
SUCTION, 634
SUCTORIA, 4345
SUDETIC OROGENY, 4346
SUGARS, 4347
SUIFORMES, 4348
SULFATES, 4349
SULFIDE ZONE, 4350
SULFIDES, 4351
SULFOSALTS, 4352
SULFUR, 4353
SULFUR DIOXIDE, 4354
SULFURIC ACID, 4356
SUN, 4358
SUPERPOSED FOLDS, 4359
SUPERPOSITION, 4360
SUPERSTRUCTURE, 4361
SUPRACRUSTAL ORIGIN, 4362
SURFACE DEFECTS, 4363

SURFACE MINING, 3129
SURFACE RUNOFF, 4364
SURFACE TENSION, 4365
SURFACE WATER, 4366
SURFACE WATER BUDGET, 4367
SURFACE WAVES, 4368
SURFICIAL GEOLOGY, 4369
SURGES, 4370
SURVEY ORGANIZATIONS, 1739
SURVEYS, 4371
SUSPENDED MATERIALS, 4372
SUSPENSION, 4373
SUTURES, 4374
SVECOFENNIAN OROGENY, 4375
SWAMPS, 4376
SWARMS, 1287
SWELLING, 4378
SYENITES, 4379
SYLVINITE, 4380
SYMBIOSIS, 4381
SYMMETRIC FOLDS, 4382
SYMMETRODONTA, 4383
SYMMETRY, 4384
SYMPHYLA, 4385
SYMPLECTIC TEXTURE, 4386
SYMPOSIA, 4387
SYNAPSIDA, 4388
SYNCLINES, 4390
SYNCLINORIA, 4391
SYNECLISE, 4392
SYNFORM FOLDS, 4393
SYNGENESIS, 4394
SYNGENETIC, 4395
SYNONYMY, 4396
SYNSEDIMENTARY PROCESSES, 4397
SYNSEDIMENTARY STRUCTURE, 4398
SYNTECTONIC PROCESSES, 4399
SYNTEXIS, 4400
SYNTHESIS, 4401
SYNTHETIC MATERIALS, 4402
SYSTEM ANALYSIS, 4404
SYSTEMS, 4403
SYSTEMS ANALOGS, 4405
TABLES, 1072
TABULATA, 4406
TACONIC OROGENY, 4407
TACONITE, 4408
TACTITE, 4409
TAENIODONTA, 4410
TAENIOLABOIDEA, 4411
TAIGA ENVIRONMENT, 4412
TAILINGS, 4413
TALC, 4414
TALC SCIST, 4415
TALUS, 4416
TALUS SLOPES, 4417
TANTALATES, 4418

TANTALUM, 4419
TAPHONOMY, 4420
TAPHROGENY, 4421
TAPHROGEOSYNCLINES, 4422
TARANNON, 4423
TARDIGRADA, 4424
TASMANITES, 4425
TATARIAN, 4426
TAXITE, 4427
TAXONOMY, 4428
TECHNETIUM, 4429
TECHNICAL COOPERATION, 4430
TECHNIQUES, 4431
TECHNOLOGY, 4432
TECTOGENESIS, 4433
TECTONIC BRECCIA, 4434
TECTONIC CONTROLS, 4435
TECTONIC CYCLE, 4436
TECTONIC MAPS, 4437
TECTONIC STYLE, 4438
TECTONIC SUPERSTRUCTURE, 4439
TECTONIC UNITS, 4305
TECTONIC WEDGES, 3963
TECTONICS, 4440
TECTONITE, 4441
TECTONOPHYSICS, 4442
TECTONOSPHERE, 4443
TEETH, 4544
TEKTITES, 4444
TELEOSTEI, 4445
TELESEISMIC SIGNALS, 4446
TELETHERMAL PROCESSES, 4447
TELLURATES, 4448
TELLURIC METHODS, 4449
TELLURIUM, 4450
TEM DATA, 4451
TEMPERATE ENVIRONMENT, 4452
TEMPERATURE, 4453
TENACITY, 4454
TENSILE STRENGTH, 4455
TENSION, 4456
TENTACULITES, 4457
TEPHROCHRONOLOGY, 4458
TERATOLOGY, 4459
TERBIUM, 4460
TEREBRATULIDA, 4461
TERMINOLOGY, 4462
TERNARY SYSTEM, 4463
TERRA ROSSA, 4464
TERRACES, 4465
TERRAIN CORRECTIONS, 4466
TERRESTRIAL COMPARISON, 3481
TERRESTRIAL ENVIRONMENT, 4467
TERRESTRIAL PLANETS, 4468
TERRIGENOUS MATERIALS, 4469
TERTIARY, 4470
TESTS, 4471

TESTUDINATA, 4472
TETHYS, 4473
TETRAPODA, 4475
TEXTBOOKS, 4584
TEXTULARIINA, 4476
TEXTURES, 4477
TH/PB, 4518
THALASSOCRATON, 4478
THALASSOID, 4479
THALLIUM, 4480
THALLOPHYTES, 4481
THALWEG, 4482
THANATOCENOSES, 4483
THANETIAN, 4484
THAWING, 4485
THECAMOEBA, 4486
THECIDEIDINA, 4487
THECODONTIA, 4488
THELODONTI, 4489
THEORETICAL STUDIES, 4490
THERALITE, 4491
THERAPSIDA, 4492
THERMAL, 4493
THERMAL ANALYSIS, 4494
THERMAL CONDUCTIVITY, 4495
THERMAL DEMAGNETIZATION, 4496
THERMAL EFFECTS, 4497
THERMAL METAMORPHISM, 4498
THERMAL PROPERTIES, 4499
THERMAL PROSPECTING, 4500
THERMAL WATERS, 4501
THERMOCHEMICAL PROPERTIES, 4502
THERMOCLINES, 4503
THERMODYNAMIC PROPERTIES, 4504
THERMODYNAMICS, 4505
THERMOGRAM, 4506
THERMOGRAVIMETRIC ANALYSIS, 4507
THERMOKARST, 4508
THERMOLUMINESCENCE, 4509
THERMOMAGNETIC ANALYSIS, 4510
THERMOREMANENT MAGNETIZATION, 4511
THESES, 4512
THICKNESS, 4513
THIN SECTIONS, 4514
THIXOTROPY, 4515
THOLEIITIC COMPOSITION, 4516
THORIUM, 4517
THRUST FAULTS, 4520
THULIUM, 4521
THURINGIAN, 4522
THYSANOPTEROIDA, 4523
THYSANURA, 4524
TIDAL CHANNELS, 4525
TIDAL ENVIRONMENT, 4526
TIDAL FLATS, 4527
TIDAL INLETS, 4528
TIDES, 4529

TILL, 4530
TILLITE, 4531
TILLODONTIA, 4532
TILT, 4533
TILTMETERS, 4534
TIME FACTOR, 4535
TIME SCALES, 4536
TIN, 4537
TINTINNIDAE, 4538
TITANIUM, 4539
TITHONIAN, 4540
TOARCIAN, 4541
TOMBOLOS, 4542
TONSTEIN, 4543
TOOL MARKS, 715
TOPOGRAPHIC CORRECTION, 4545
TOPOGRAPHIC MAPS, 4546
TOPOGRAPHY, 4547
TOPSET BEDS, 4548
TORRENTS, 4549
TORSION, 4550
TORTONIAN, 4551
TORTUOSITY, 4552
TOURNAISIAN, 4553
TOXIC MATERIALS, 4554
TOXICITY, 4555
TRACE ELEMENTS, 4556
TRACE METALS, 4558
TRACE-ELEMENT ANALYSES, 4559
TRACERS, 4560
TRACHYANDESITE, 4561
TRACHYBASALT, 4562
TRACHYPSAMMIACEA, 4563
TRACHYTES, 4564
TRADE, 4566
TRAILS, 4565
TRANSCURRENT FAULTS, 4567
TRANSFER, 4568
TRANSFORM FAULTS, 4569
TRANSFORMATIONS, 4570
TRANSFORMISM, 4571
TRANSGRESSION, 4572
TRANSIENT METHODS, 4573
TRANSMISSIBILITY COEFFICIENT, 4574
TRANSMISSION METHOD, 4575
TRANSMISSIVITY, 4576
TRANSPORT, 4577
TRANSPORTATION, 4578
TRANSVERSAL, 4579
TRAPS, 4580
TRAVELTIME, 4581
TRAVELTIME CURVES, 4582
TRAVERTINE, 4583
TREE RINGS, 4585
TREMADOCIAN, 4586
TRENCHES, 4587
TREND (MATH), 4244

TREND-SURFACE ANALYSIS, 4588
TREPOSTOMATA, 4589
TRIANGULATION, 4590
TRIASSIC, 4591
TRIAXIAL TESTS, 4592
TRICLINIC SYSTEM, 4593
TRICONODONTA, 4594
TRIGONAL SYSTEM, 4595
TRIGONIIDAE, 4596
TRILOBITA, 4597
TRILOBITOMORPHA, 4598
TRIPLE JUNCTIONS, 4599
TRIPOLI, 4600
TRITIUM, 4601
TROPICAL ENVIRONMENT, 4602
TROUGHS, 4603
TSUNAMIS, 4604
TUBOIDEA, 4605
TUBULIDENTATA, 4606
TUFF, 4607
TUFF LAVA, 4609
TUFFITE, 4608
TUNDRA, 4610
TUNDRA SOILS, 4611
TUNGSTATES, 4612
TUNGSTEN, 4613
TUNNELS, 4614
TURBIDITE, 4615
TURBIDITY, 4616
TURBIDITY CURRENT STRUCTURES, 4618
TURBIDITY CURRENTS, 4617
TURBODRILLING, 4619
TURONIAN, 4620
TWINNING, 4621
TWO-DIMENSIONAL MODELS, 4622
TYLOPODA, 4623
TYPE LOCALITIES, 4624
TYPE SECTIONS, 4625
TYPE SPECIMENS, 4626
TYPOMORPHIC MINERALS, 4627
TYPOMORPHISM, 4628
TYRRHENIAN, 4629
U/PB, 4692
U/TH/PB, 4693
UDALFS, 4630
UDOLLS, 4631
UDULTS, 4632
UFIMIAN, 4633
ULTISOLS, 4634
ULTRAMAFIC COMPOSITION, 4635
ULTRAMAFICS, 4636
ULTRAMETAMORPHISM, 4637
ULTRAMYLONITE, 4638
ULTRASONIC METHODS, 4639
ULTRASTRUCTURE, 4640
ULTRAVIOLET RAYS, 4641
UMBREPTS, 4642

UNAKITE, 4643
UNCONFINED AQUIFERS, 4644
UNCONFINED WATER, 4645
UNCONFORMITIES, 4646
UNCONSOLIDATED MATERIALS, 4647
UNDATION, 1781
UNDERCLAY, 4648
UNDERGROUND CHANNELS, 4649
UNDERGROUND MINING, 4650
UNDERGROUND SPACE, 4651
UNDERGROUND STORAGE, 4652
UNDERGROUND STREAMS, 4653
UNDERSATURATED, 4654
UNDERSATURATION, 4655
UNDERTHRUST FAULTS, 4656
UNDULATORY EXTINCTION, 4657
UNIAXIAL TESTS, 4658
UNIFORMITARIANISM, 4659
UNIONOIDAE, 4660
UNIT CELL, 4661
UNIVERSAL STAGE, 4662
UNIVERSE, 4663
UNSATURATED ZONES, 3049
UNSTEADY FLOW, 4664
UPLIFT PRESSURE, 4666
UPLIFTS, 4665
UPPER CAMBRIAN, 4667
UPPER CARBONIFEROUS, 4668
UPPER CRETACEOUS, 4669
UPPER CRUST, 4670
UPPER DEVONIAN, 4671
UPPER EOCENE, 4672
UPPER JURASSIC, 4673
UPPER LIASSIC, 4674
UPPER MANTLE, 4675
UPPER MIOCENE, 4676
UPPER OLIGOCENE, 4677
UPPER ORDOVICIAN, 4678
UPPER PERMIAN, 4679
UPPER PLEISTOCENE, 4680
UPPER PLIOCENE, 4681
UPPER PROTEROZOIC, 4682
UPPER SILURIAN, 4683
UPPER TRIASSIC, 4684
UPTHROW, 4685
UPWARPING, 4686
UPWELLING, 4687
URALITIZATION, 4688
URANIUM, 4689
URANIUM DISEQUILIBRIUM, 4691
URANIUM MINERALS, 4690
URANUS, 3479
URBAN GEOLOGY, 4694
URBAN PLANNING, 4695
URBANIZATION, 4696
UREILITE, 4697
URODELA, 4698

USTALFS, 4699
USTERTS, 4700
USTOLLS, 4701
USTOX, 4702
USTULTS, 4703
UTILIZATION, 4704
VACUUM FUSION ANALYSIS, 4705
VALANGINIAN, 4706
VALENCY, 734
VALLEYS, 4707
VALVES, 4708
VANADATES, 4709
VANADIUM, 4710
VANE APPARATUS, 4711
VANE TESTS, 4712
VARIABILITY, 4714
VARIABLE, 4715
VARIANCE ANALYSIS, 4716
VARIATIONS, 4717
VARIOGRAMS, 4718
VARISCAN OROGENY, 4719
VARVE CHRONOLOGY, 4721
VARVED CLAY, 4722
VARVES, 4720
VECTOR, 4723
VEGETATION, 4724
VEINS, 4725
VELOCITY, 4726
VELOCITY STRUCTURE, 4727
VENERIDA, 4728
VENITE, 4729
VENTS, 4767
VENUS, 3480
VERGENCE, 4730
VERMICULITE, 4732
VERMIFORM STRUCTURE, 4733
VERTEBRATA, 4734
VERTICAL ORIENTATION, 4735
VERTICAL TECTONICS, 4736
VERTISOLS, 4737
VIBRATION, 4738
VIBRATORY DRILLING, 4739
VIBROSEIS, 4740
VIKING PROGRAM, 4741
VILLAFRANCHIAN, 4742
VIRGATION, 4743
VISCOELASTICITY, 4744
VISCOSITY, 4745
VISCOUS MATERIALS, 4746
VISCOUS REMANENT MAGNETIZATION, 4747
VISEAN, 4748
VITRAIN, 4749
VITREOUS MATERIALS, 4750
VITRIFICATION, 4751
VITROCLASTIC TEXTURE, 4752
VITROPHYRIC TEXTURE, 4753
VOLANTES, 4754

VOLATILE, 4755
VOLATILE ELEMENTS, 4756
VOLATILES, 4757
VOLATILIZATION, 4758
VOLCANIC ASH, 4759
VOLCANIC BELTS, 4760
VOLCANIC BRECCIA, 4761
VOLCANIC CONES, 4762
VOLCANIC DOMES, 4763
VOLCANIC GLASS, 4764
VOLCANIC NECKS, 2995
VOLCANIC ORIGIN, 4765
VOLCANIC PROCESSES, 4774
VOLCANIC ROCKS, 4766
VOLCANICLASTICS, 4772
VOLCANISM, 4768
VOLCANO=TECTONIC DEPRESSION, 4770
VOLCANO=TECTONIC STRUCTURE, 4771
VOLCANOES, 4769
VOLCANOGENIC, 4773
VOLCANOLOGY, 4775
VOLUME, 4776
VULCANIAN=TYPE ERUPTIONS, 4777
VULCANORIUM, 4778
WADI, 4779
WALLROCKS, 4780
WASTE, 4781
WASTE DISPOSAL, 4782
WATER, 4783
WATER AUTHORITY, 4784
WATER BALANCE, 1912
WATER BUDGET, 4785
WATER CONSUMPTION, 4787
WATER EROSION, 4789
WATER HARDNESS, 4790
WATER HARNESSING, 4786
WATER MANAGEMENT, 4791
WATER MAPS, 4792
WATER OF CRYSTALLIZATION, 4793
WATER OF DEHYDRATION, 4794

WATER PRESSURE, 4795
WATER QUALITY, 4796
WATER RECESSION, 4797
WATER RESOURCES, 4798
WATER STORAGE, 4799
WATER SUPPLY, 4800
WATER VAPOR, 4713
WATER WELLS, 4801
WATER YIELD, 4802
WATERFALLS, 4803
WATERWAYS, 4804
WAVE DISPERSION, 4806
WAVE PROPAGATION, 4807
WAVES, 4805
WAX, 4808
WEATHERING, 4809
WEATHERING CRUST, 4810
WEIGHTING, 4812
WELL COMPLETION, 4814
WELL FILTER, 1545
WELL HEAD, 4815
WELL SCREENS, 4817
WELL=LOGGING, 4816
WELLS, 4813
WENLOCKIAN, 4818
WEST, 4819
WESTPHALIAN, 4820
WET METHODS, 4821
WETLANDS, 4822
WHITE CLAY, 4823
WHOLE ROCK, 4824
WILDFLYSCH, 4825
WIND EROSION, 4827
WIND RIPPLE, 4828
WIND TRANSPORT, 4829
WINDOWS, 4830
WINDS, 4826
WORLD OCEAN, 3084
WORMS, 4731
WRENCH FAULTS, 4831

X=RAY ANALYSIS, 4833
X=RAY DIFFRACTION ANALYSIS, 4834
X=RAY FLUORESCENCE, 4835
X=RAY POWDER DIFFRACTION, 3587
X=RAY RADIOGRAPHY, 3734
X=RAY SPECTROSCOPY, 4836
X=RAYS, 4832
XANTHOPHYTA, 4837
XENOLITHS, 4838
XENOMORPHIC CRYSTALS, 4839
XENON, 4840
XERALFS, 4841
XEROLLS, 4842
XEROPHILE PLANT, 4843
XEROPHYTE, 4844
XEROSOLS, 4845
XERULTS, 4846
YERMOSOLS, 4848
YIELD STRENGTH, 4849
YOUNG, 4850
YOUNG'S MODULUS, 4851
YPRESIAN, 4852
YTTERBIUM, 4853
YTTRIUM, 4854
ZECHSTEIN, 4855
ZEOLITE FACIES, 4857
ZEOLITE GROUP, 4856
ZEOLITIZATION, 4858
ZIGZAG FOLD, 4859
ZINC, 4860
ZIRCON, 4861
ZIRCONIUM, 4862
ZOANTHINIARIA, 4863
ZONAL STRUCTURE, 4865
ZONE OF AERATION, 4866
ZONE OF CEMENTATION, 4867
ZONE OF FLOW, 4868
ZONE OF SEDIMENTATION, 4869
ZONING, 4870
ZOOECIUM, 4864
ZOOID, 4871

Français

A GRANDE ECHELLE, 2436
A PETITE ECHELLE, 4140
AALENIEN, 3
ABAISSEMENT NIVEAU EAU, 1249
ABAQUE, 3045
ABLATION, 5
ABONDANCE, 16
ABRASIF, 9
ABRASION, 6
ABSORBANT, 12
ABSORPTION, 13
ABSORPTION ATOMIQUE, 363
ABYSSAL, 18
ABYSSE, 17
ACANTHARIA, 25
ACANTHODII, 26
ACCELERATION CORIOLIS, 959
ACCELEROGRAMME, 27
ACCELEROMETRE, 28
ACCRETION, 30
ACCUMULATION GAZ, 1685
ACETOLYSE, 35
ACHONDRITE, 36
ACIDE, 37
ACIDE AMINE, 163
ACIDE FULVIQUE, 1663
ACIDE GRAS, 1506
ACIDE HUMIQUE, 2071
ACIDE MINERAL, 2248
ACIDE ORGANIQUE, 3165
ACIDE SULFURIQUE, 4356
ACIDITE, 40
ACMOLITE, 98
ACRANIA, 46
ACRISOL, 47
ACRITARCHA, 48
ACROTRETIDA, 49
ACTINIARIA, 50
ACTINIUM, 51
ACTINOCERATOIDEA, 52
ACTION BIOGENE, 493
ACTION CHALEUR, 4497
ACTION CLIMATIQUE, 795
ACTION FROID, 1653
ACTION PREVENTIVE, 3609

ACTIVITE, 61
ACTIVITE IGNEE, 2172
ACTUALISME, 62
ADAPTATION, 63
ADIAGNOSTIQUE, 66
ADSORPTION, 68
AERATION, 69
AERODROME, 96
AEROSOL, 75
AFFAISSEMENT, 2426
AFFAISSEMENT REGIONAL, 1242
AFFINEMENT, 3780
AFFINITE, 76
AFFLEUREMENT, 3206
AFFLUENT, 77
AGE, 80
AGE ABSOLU, 10
AGE CONCORDANT, 878
AGE DE LA TERRE, 81
AGE DISCORDANT, 1216
AGE EXPOSITION, 1489
AGE GLACIAIRE ANCIEN, 191
AGE RELATIF, 3804
AGENCE BASSIN, 4784
AGGLOMERAT, 82
AGNATHA, 86
AGNOSTIDA, 87
AGONIATITIDA, 88
AGPAITIQUE, 89
AGRICULTURE, 90
AGROGEOLOGIE, 91
AIGUILLE, 2996
AIMANTATION, 2648
AIMANTATION INDUITE, 2215
AIMANTATION ISOTHERMOREMANENTE, 2347
AIMANTATION REMANENTE, 3812
AIMANTATION REMANENTE ANHYSTERETIQUE, 206
AIMANTATION REMANENTE CHIMIQUE, 732
AIMANTATION REMANENTE DETRITIQUE, 1143
AIMANTATION REMANENTE NATURELLE, 2991
AIMANTATION REMANENTE VISQUEUSE, 4747
AIMANTATION THERMOREMANENTE, 4511
AIR, 93
AKCHAGYLIEN, 99
ALBATRE, 100

ALBEDO, 101
ALBERTITE, 102
ALBIEN, 103
ALBITISATION, 106
ALBOLL, 107
ALCALINITE, 131
ALCYONACEA, 108
ALDANIEN, 109
ALFISOL, 110
ALGAE, 111
ALGEBRE BOOLEENNE, 542
ALGEBRE MATRICIELLE, 2717
ALGOL, 119
ALGORITHME, 121
ALGUE CALCAIRE, 608
ALIGNEMENT BLOC GLACIAIRE, 560
ALIMENTATION ARTIFICIELLE, 321
ALIMENTATION NATURELLE, 1917
ALKANE, 132
ALLIAGE, 139
ALLOCHEME, 134
ALLOCHTONIE, 135
ALLOGROMIINA, 137
ALLOMETRIE, 138
ALLUVIAL, 140
ALLUVION, 146
ALTERATION, 154
ALTERATION CHIMIQUE, 735
ALTERATION DIFFERENTIELLE, 1177
ALTERATION HYDROTHERMALE, 2131
ALTERATION MECANIQUE, 2728
ALTERATION METEORIQUE, 4809
ALTERATION PHYSIQUE, 3429
ALTERATION SOUS MARINE, 1946
ALTITUDE, 155
ALUMINE, 156
ALUMINIUM, 157
ALUNITISATION, 158
ALVEOLINELLIDAE, 159
AMAS MINERALISE, 2705
AMBLYPODA, 161
AMBRE, 160
AMENAGEMENT BASSIN, 433
AMENAGEMENT SOL, 4152
AMENDEMENT, 2742

AMENDEMENT SOL, 4150
AMERICIUM, 162
AMIANTE, 325
AMINCISSEMENT, 3449
AMMODISCACEA, 164
AMMONITIDA, 166
AMMONITINA, 167
AMMONIUM, 168
AMMONOIDEA, 169
AMPHIBIA, 171
AMPHIBOLE, 172
AMPHIBOLITE, 173
AMPHIBOLITISATION, 175
AMPHINEURA, 176
AMPLITUDE, 177
ANALOGIE SYSTEME HYDROGEOLOGIQUE, 4405
ANALYSE, 182
ANALYSE ACTIVATION, 54
ANALYSE ACTIVATION NEUTRONIQUE, 3015
ANALYSE CHIMIQUE, 721
ANALYSE CORRESPONDANCES, 963
ANALYSE DISCRIMINANTE, 1218
ANALYSE DONNEE, 1066
ANALYSE ELEMENT TRACE, 4556
ANALYSE FACTORIELLE, 1502
ANALYSE FUSION VIDE, 4705
ANALYSE GROUPE, 799
ANALYSE HARMONIQUE, 1628
ANALYSE HARMONIQUE SPHERIQUE, 4206
ANALYSE LIVRE, 541
ANALYSE MAJEURS, 2657
ANALYSE MATHEMATIQUE, 2708
ANALYSE MINEURS, 2887
ANALYSE MODALE, 2902
ANALYSE MULTIVARIABLE, 2968
ANALYSE QUALITATIVE, 3706
ANALYSE QUANTITATIVE, 3708
ANALYSE RX, 4833
ANALYSE SPECTRE SISMIQUE, 4017
ANALYSE STRUCTURALE, 4301
ANALYSE SYSTEME, 4404
ANALYSE TENDANCE, 4588
ANALYSE THERMIQUE, 4494
ANALYSE THERMIQUE DIFFERENTIELLE, 1176
ANALYSE THERMOGRAVIMETRIQUE, 4507
ANALYSE THERMOMAGNETIQUE, 4510
ANALYSE VARIANCE, 4716
ANAPSIDA, 184
ANASPIDA, 185
ANATEXIE, 186
ANATEXITE, 187
ANATOMIE, 188
ANATOMIE APPAREIL NERVEUX, 3012
ANATOMIE COLONNE VERTEBRALE, 4211
ANATOMIE SQUELETTE, 4127
ANCHIMETAMORPHISME, 189
ANCRAGE, 190

ANDALOUSITE, 192
ANDEPT, 193
ANDESITE, 194
ANDOSOL, 196
ANELASTICITE, 198
ANGIOSPERMAE, 199
ANGLE AVEC LA VERTICALE, 1938
ANGLE CRITIQUE, 993
ANGRITE, 201
ANHYDRITE, 204
ANION, 207
ANISIEN, 208
ANISOPLEURA, 209
ANISOTROPIE, 210
ANKERITE, 211
ANNEAU CROISSANCE, 4585
ANNEAUX LIESEGANG, 2498
ANNEE GALACTIQUE, 1674
ANNELIDA, 212
ANNUAIRE, 4847
ANOMALIE, 214
ANOMALIE AIR LIBRE, 1643
ANOMALIE BOUGUER, 557
ANOMALIE GEOPHYSIQUE, 1751
ANOMALIE GRAVIMETRIQUE, 1878
ANOMALIE MAGNETIQUE, 2628
ANOMALODESMATA, 213
ANORTHOSITE, 216
ANORTHOSITISATION, 217
ANSE, 478
ANTECAMBRIEN, 3593
ANTECLISE, 218
ANTHOCYATHEA, 219
ANTHOZOA, 220
ANTHRACITE, 221
ANTHRAXOLITE, 222
ANTIARCHI, 223
ANTICLINAL, 224
ANTICLINORIUM, 225
ANTIDUNE, 226
ANTIMOINE, 229
ANTIMONIATE, 228
ANTIPATHARIA, 230
ANTIPERTHITE, 231
ANUROMORPHA, 234
APICAL, 237
APLACOPHORA, 238
APLITE, 239
APOGEE, 241
APOGRANITE, 242
APOPHYSE, 245
APPAREIL CONODONTE, 900
APPROVISIONNEMENT EAU, 4800
APPROXIMATION, 249
APSHERONIEN, 251
APSIDOSPONDYLI, 252
APTIEN, 253

AQUALF, 254
AQUENT, 255
AQUEPT, 257
AQUERT, 258
AQUICLUDE, 259
AQUITANIEN, 263
AQUOD, 265
AQUOLL, 266
AQUOX, 267
AQUULT, 268
AR=AR, 297
ARACHNIDA, 269
ARC INSULAIRE, 2316
ARCHAEOCOPIDA, 272
ARCHAEOCYATHA, 273
ARCHAEOCYATHEA, 274
ARCHAEOGASTROPODA, 275
ARCHAEOPTERYGES, 276
ARCHAEORNITHES, 277
ARCHEEN, 278
ARCHEIDES, 279
ARCHEOLOGIE, 281
ARCHEOMAGNETISME, 282
ARCHETYPE, 3653
ARCHIPOLYPODA, 283
ARCHOSAURIA, 284
ARCOIDA, 285
ARCTIQUE, 286
ARDOISE, 4130
AREISME, 298
ARENE, 1923
ARENIG, 291
ARENITE, 292
ARENOSOL, 293
ARGENT, 4111
ARGID, 294
ARGILE, 787
ARGILE A BLOCAUX, 559
ARGILE A BRIQUE, 576
ARGILE BLANCHE, 4823
ARGILE FAILLE, 1509
ARGILE FLUIDE, 3715
ARGILE MINERAL, 790
ARGILE MOLLE, 4144
ARGILE POTERIE, 3586
ARGILE RAIDE, 4255
ARGILE REFRACTAIRE, 1552
ARGILE RESIDUELLE, 3828
ARGILE SENSIBLE, 4038
ARGILE VARVE, 4722
ARGILITE, 791
ARGON, 296
ARGON ATMOSPHERIQUE, 359
ARIDISOL, 300
ARIDITE, 301
ARKOSE, 302
ARQUE, 288

ARRANGEMENT GRAIN, 3232
ARROYO, 308
ARSENIATE, 309
ARSENIC, 310
ART DES MINES, 2884
ARTEFACT, 319
ARTERITE, 311
ARTESIEN, 312
ARTHRODIRA, 315
ARTHROPLEURIDA, 316
ARTHROPODA, 317
ARTICULATA, 318
ARTINSKIEN, 323
ARTIODACTYLA, 324
ASHGILL, 333
ASPHALTE, 334
ASPHALTITE, 335
ASSIMILATION MAGMATIQUE, 340
ASSISE BIOSTRATIGRAPHIQUE, 339
ASSOCIATION, 338
ASSOCIATION FOSSILE, 341
ASSOCIATION MAGMATIQUE, 2620
ASSOCIATION MINERALE, 2863
ASTATE, 343
ASTERISME, 344
ASTEROIDE, 345
ASTEROIDEA, 346
ASTEROZOA, 347
ASTHENOSPHERE, 348
ASTRAPOTHERIA, 349
ASTROBLEME, 350
ASTROGEOLOGIE, 351
ASTRONOMIE, 352
ATAXITE, 354
ATLAS, 356
ATMOSPHERE, 358
ATOLL, 362
ATTAQUE CHIMIQUE, 1439
ATTENUATION, 366
ATTRITION, 368
AUBRITE, 369
AULACOGENE, 372
AUREOLE DISPERSION, 1223
AUREOLE GEOCHIMIQUE, 373
AUREOLE PRIMAIRE, 3611
AUREOLE SECONDAIRE, 3988
AUSTRALITE, 374
AUTHIGENESE, 375
AUTOCHTONIE, 377
AUTOCORRELATION, 378
AUTOLITE, 379
AUTOLYSE, 380
AUTOMETASOMATOSE, 384
AUTORADIOGRAPHIE, 385
AUTUNIEN, 387
AVALANCHE, 388
AVANT-BUTTE, 3210

AVANT-FOSSE, 1612
AVANT-PAYS, 1613
AVANT-PLAGE, 1616
AVES, 390
AXE, 393
AXE PLI, 1603
AZIMUT MAGNETIQUE, 2629
AZOTE, 3034
AZOTE ORGANIQUE, 3171
BACTERIA, 400
BACTERIE FER, 2310
BACTRITOIDEA, 402
BAIE, 443
BAJADA, 405
BAJOCIEN, 406
BALAYAGE LATERAL, 4091
BALLAST, 408
BANC ALGAIRE, 112
BANC RECIF, 3313
BANC VASE, 2953
BANQUETTE, 461
BANQUETTE PLAGE, 469
BARITE, 412
BARKHANE, 411
BARRAGE, 1059
BARRAGE BETON, 880
BARRAGE ENROCHEMENT, 3885
BARRAGE SOUTERRAIN, 1913
BARRAGE TERRE, 1275
BARRAGE VOUTE, 271
BARRE, 410
BARRE LITTORALE, 417
BARREMIEN, 414
BARRIERE, 416
BARRIERE ROCHEUSE, 3882
BARTONIEN, 421
BARYUM, 413
BASAL, 422
BASALTE, 425
BASALTE ALCALIN, 123
BASALTE PLATEAU, 3502
BASANITE, 427
BASE DONNEE, 1067
BASHKIRIEN, 431
BASIFICATION, 432
BASSE PRESSION, 2571
BASSE TEMPERATURE, 2572
BASSES-TERRES, 2595
BASSIN HOUILLER, 803
BASSIN INTRACRATONIQUE, 2290
BASSIN LUNAIRE, 4479
BASSIN MARGINAL INTERNE, 396
BASSIN OCEANIQUE, 3085
BASSIN REPRESENTATIF, 3819
BASSIN SALE, 3507
BASSIN SEDIMENTAIRE, 3995
BASSIN VERSANT, 1245

BATHOLITE, 434
BATHONIEN, 435
BATHYMETRIE, 438
BATIMENT, 586
BATRACHOSAURIA, 439
BAUXITE, 440
BAUXITISATION, 441
BE 10-BE 9, 472
BEACH ROCK, 448
BEDROCK, 457
BELEMNOIDEA, 459
BENNETTITALES, 463
BENTHOS, 465
BENTONITE SUBSTANCE, 466
BERKELIUM, 468
BERRIASIEN, 470
BERYLLIUM, 471
BETON, 879
BEYRICHICOPINA, 474
BIAXIAL, 475
BIBLIOGRAPHIE, 477
BIBLIOGRAPHIE PERSONNELLE, 3365
BILAN, 407
BILAN EAU, 4785
BILAN EAU SOUTERRAINE, 1912
BILAN EAU SURFACE, 4367
BILAN MASSE, 2698
BIOCENOSE, 481
BIOCHIMIE, 496
BIOCHORE, 483
BIOCHRONOLOGIE, 485
BIODEGRADATION, 489
BIOFACIES, 490
BIOGENESE, 492
BIOGEOGRAPHIE, 497
BIOGRAPHIE, 498
BIOHERME, 499
BIOLITE, 500
BIOLOGIE, 502
BIOMASSE, 503
BIOMETRIE, 504
BIOMICRITE, 505
BIORHEXISTASIE, 506
BIOS, 513
BIOSOME, 507
BIOSPARITE, 508
BIOSPHERE, 509
BIOSTRATIGRAPHIE, 511
BIOSTROME, 512
BIOTOPE, 514
BIOTURBATION, 515
BIOTYPE, 516
BIOZONE, 517
BIREFRINGENCE, 519
BISCUIT ALGAIRE, 113
BISEAU, 4811
BISMUTH, 520

BITUME, 521
BITUMINISATION, 522
BIVALVIA, 3332
BLASTOIDEA, 526
BLATTOPTEROIDA, 527
BLOC, 530
BLOC DIAGRAMME, 531
BLOC ERRATIQUE, 1430
BLOC INDICATEUR, 2211
BLOC ISOLE, 558
BOIS FOSSILE, 1624
BOIS SILICIFIE, 4101
BOMBE VOLCANIQUE, 539
BONE BED, 540
BORALF, 543
BORATE, 544
BORAX, 545
BORDURE CONTINENTALE, 547
BORDURE DETRITIQUE, 250
BORE, 552
BOROLL, 551
BOUCHAGE, 3514
BOUCHON, 3513
BOUCLIER, 4072
BOUDINAGE, 556
BOUE FORAGE, 1256
BOUE ORGANIQUE, 3127
BOULANCE, 3716
BOULETAGE, 3338
BOUNDSTONE, 562
BOURGEONNEMENT, 1705
BOUSSOLE, 2630
BOUTONNIERE, 2242
BRACHIOPODA, 565
BRACHIOPTERYGII, 566
BRACHYPLI, 567
BRANCHIOPODA, 572
BRANCHIURA, 573
BRAS MER, 2241
BRECHE, 575
BRECHE BROYAGE, 1000
BRECHE CHOC, 4076
BRECHE CONTACT, 914
BRECHE DISLOCATION, 1221
BRECHE EFFONDREMENT, 835
BRECHE EXPLOSION, 1486
BRECHE FAILLE, 1508
BRECHE FRICTION, 4056
BRECHE INJECTION, 2240
BRECHE PLISSEMENT, 1605
BRECHE TECTONIQUE, 4434
BRECHE VOLCANIQUE, 4761
BREVET, 3314
BROME, 581
BROMURE, 580
BROWN COAL, 582
BROYAGE, 1899

BRUIT FOND, 3041
BRYOPHYTA, 584
BRYOZOA, 585
BULIMINACEA, 589
BUNTSANDSTEIN, 593
BURDIGALIEN, 594
BUTTE ALGAIRE, 116
BYSSUS, 601
C 13-C 12, 645
C 14, 646
CADMIUM, 604
CAILLOUTIS, 3907
CALABRIEN, 606
CALCAIRE, 2507
CALCAIRE ALGAIRE, 114
CALCAIRE BIOCLASTIQUE, 487
CALCAIRE DOLOMITIQUE, 1236
CALCAIRE ENTROQUES, 1366
CALCAIRE MICROCRISTALLIN, 2809
CALCAIRE OOLITIQUE, 3125
CALCIPHYRE, 611
CALCISPONGEA, 612
CALCITE, 613
CALCITISATION, 614
CALCIUM, 615
CALCUL PETROGRAPHIQUE, 3375
CALDEIRA, 617
CALDEIRA EFFONDREMENT, 836
CALICHE, 619
CALIFORNIUM, 620
CALLOVIEN, 622
CALOTTE GLACIAIRE, 2156
CALOTTE GLACIAIRE POLAIRE, 3534
CAMAROIDEA, 623
CAMBISOL, 624
CAMBRIEN, 625
CAMBRIEN INF, 2577
CAMBRIEN MOYEN, 2841
CAMBRIEN SUP, 4667
CAMPANIEN, 626
CAMPTOSTROMATOIDEA, 627
CANAL, 628
CANALISATION SOUTERRAINE, 4649
CANNEL COAL, 629
CANNELURE, 1902
CANYON, 630
CANYON MARIN, 4326
CAP, 632
CAP ROCK, 631
CAPACITE CALORIFIQUE, 1963
CAPACITE ECHANGE, 1464
CAPILLARITE, 633
CAPTAGE EAU, 4786
CAPTEUR, 3618
CAPTURE COURS EAU, 637
CARACTERISTIQUE, 710
CARADOC, 638

CARAPACE, 639
CARAT, 640
CARBONATATION, 653
CARBONATE, 651
CARBONATITE, 652
CARBONE, 642
CARBONE ORGANIQUE, 3166
CARBONIFERE, 654
CARBONIFERE INF, 2578
CARBONIFERE SUP, 4668
CARNIEN, 655
CARNIEULE, 656
CARNIVORA, 657
CAROTTE, 958
CARRIERE, 3709
CARTE, 2668
CARTE ASSEMBLAGE, 2207
CARTE BATHYMETRIQUE, 437
CARTE BIOFACIES, 491
CARTE COURBE NIVEAU, 932
CARTE EAU, 4792
CARTE GEOCHIMIQUE, 1719
CARTE GEOLOGIQUE, 1735
CARTE GEOMAGNETIQUE, 2644
CARTE GEOMORPHOLOGIQUE, 1747
CARTE GEOPHYSIQUE, 1754
CARTE GEOTECHNIQUE, 1768
CARTE GITES MINERAUX, 1299
CARTE GRAVIMETRIQUE, 1882
CARTE HOUILLERE, 807
CARTE HYDRAULIQUE, 2095
CARTE HYDROCARBURE, 3106
CARTE HYDROCHIMIQUE, 2101
CARTE HYDROGEOLOGIQUE, 2111
CARTE HYDROLOGIQUE, 2120
CARTE ISOSISMIQUE, 2343
CARTE LITHOLOGIQUE, 2527
CARTE MAGNETIQUE, 2636
CARTE METALLOGENIQUE, 2772
CARTE OMBREE, 659
CARTE PALEOGEOGRAPHIQUE, 3256
CARTE PEDOLOGIQUE, 4153
CARTE PERFOREE, 3686
CARTE PETROGRAPHIQUE, 3376
CARTE PHOTOGEOLOGIQUE, 3409
CARTE PREVISIONNELLE, 1611
CARTE SISMICITE, 4025
CARTE STRATIGRAPHIQUE, 4278
CARTE STRUCTURALE, 4308
CARTE STRUCTURALE ISOPLETHE, 4313
CARTE SUBSTANCE UTILE, 3052
CARTE TECTONIQUE, 4437
CARTE TOPOGRAPHIQUE, 4546
CARTERINACEA, 658
CARTOGRAPHIE, 660
CARTOGRAPHIE AERIENNE, 70
CARTOGRAPHIE AUTOMATIQUE, 381

CASSIDULINACEA, 664
CATACLASE, 667
CATAGENESE, 668
CATAGRAPHIA, 669
CATALYSE, 670
CATAMORPHISME, 2377
CATASTROPHE NATURELLE, 1738
CATASTROPHISME, 671
CATION, 674
CAUDA, 821
CAUSTOBIOLITE, 676
CAVERNE, 677
CAVITE DISSOLUTION, 4177
CAYTONIALES, 681
CEINTURE CIRCUM-PACIFIQUE, 774
CEINTURE GRAND CERCLE, 1890
CEINTURE METAMORPHIQUE, 2780
CEINTURE ROCHE VERTE, 1895
CEINTURE VOLCANIQUE, 4760
CELLULOSE, 683
CEMENTATION, 685
CENDRE INDUSTRIELLE, 320
CENDRE VOLCANIQUE, 4759
CENOMANIEN, 686
CENOZOIQUE, 688
CENOZONE, 689
CENTRALE ELECTRIQUE, 3588
CENTRALE NUCLEAIRE, 3065
CENTRE COLORE, 846
CENTRE EXPANSION, 4223
CEPHALOCARIDA, 691
CEPHALODISCIDA, 692
, CEPHALOPODA, 693
CERAMIQUE, 694
CERATITIDA, 695
CERATOMORPHA, 696
CERIANTHARIA, 697
CERIANTIPATHARIA, 698
CERIUM, 699
CESIUM, 700
CETACEA, 701
CHAETOGNATHA, 702
CHAINE GEOSYNCLINALE, 3184
CHAINE INTRACONTINENTALE, 2289
CHAINE MARKOV, 2691
CHAINE PLISSEE, 1604
CHAINE RADIOACTIVE, 3728
CHALCOSPHERE, 704
CHALEUR, 1962
CHALEUR EQUIVALENTE FUSION, 2440
CHALEUR SPECIFIQUE, 4197
CHAMBRE MAGMATIQUE, 2618
CHAMP CRISTALLIN, 1015
CHAMP ELECTRIQUE, 1328
CHAMP ELECTROMAGNETIQUE, 1340
CHAMP GAZ, 1689
CHAMP GEOTHERMIQUE, 1774

CHAMP GLACE, 2157
CHAMP GRAVIMETRIQUE, 1880
CHAMP HYDROCARBURE, 3105
CHAMP LAVE, 2456
CHAMP MAGNETIQUE, 2633
CHAMP PETROLE, 3107
CHAMP PETROLIER GEANT, 1786
CHAPEAU FER, 1838
CHARBON, 801
CHARBON BITUMINEUX, 523
CHARBON BOGHEAD, 538
CHARBON COKE, 831
CHARBON GAZ, 1687
CHARBON HUMIQUE, 2078
CHARBON SAPROPELIQUE, 3945
CHARGE FOND, 452
CHARGE HYDRAULIQUE, 2094
CHARGE RIVIERE, 3874
CHARGE SILT, 4107
CHARGEMENT, 2554
CHARNIERE TEST, 2014
CHARNOCKITE, 712
CHAROPHYCEAE, 713
CHARRIAGE, 3218
CHASSIGNITE, 714
CHATTIEN, 716
CHAUD, 2063
CHAUX, 2506
CHEILOSTOMATA, 717
CHELATION, 718
CHELICERATA, 719
CHELONIA, 720
CHENAL, 707
CHENAL ENFOUI, 598
CHENAL LAVE, 2455
CHENAL MAREE, 4525
CHENAL MER PROFONDE, 1089
CHENIER, 737
CHERNOZEM, 738
CHERTIFICATION, 740
CHILOPODA, 742
CHIMIE PHYSIQUE, 3425
CHIROPTERA, 744
CHITINE, 745
CHITINOZOA, 746
CHLORATION, 748
CHLORE, 749
CHLORINITE, 750
CHLORITISATION, 752
CHLOROPHYCOPHYTA, 754
CHLOROPHYLLE, 753
CHLORURE, 747
CHOC, 4075
CHOIX SITE, 4124
CHONDRICHTHYES, 755
CHONDRITE, 756
CHONDRITE A ENSTATITE, 1379

CHONDRITE CARBONEE, 647
CHONDRULE, 757
CHORDATA, 758
CHROMATE, 759
CHROMATOGRAPHIE, 760
CHROMATOGRAPHIE PHASE GAZEUSE, 1686
CHROME, 761
CHRONOLOGIE VARVE, 4721
CHRONOSTRATIGRAPHIE, 763
CHRONOZONE, 764
CHRYSOPHYCOPHYTA, 765
CHUTE EAU, 4803
CIEL OUVERT, 3129
CILIATA, 766
CILIOPHORA, 767
CIMENT INDUSTRIEL, 587
CIMENT ROCHE, 684
CINEMATIQUE, 2393
CIPOLIN, 771
CIRCULATION, 773
CIRCULATION OCEANIQUE, 3086
CIRE, 4808
CIRQUE GLACIAIRE, 775
CIRRIPEDIA, 776
CISAILLEMENT, 4058
CLADOCERA, 778
CLARAIN, 780
CLARKE, 781
CLASSEMENT GRANULOMETRIQUE, 4184
CLASSIFICATION, 782
CLASSIFICATION CIPW, 603
CLASSIFICATION HOLMES, 2024
CLASSIFICATION NIGGLI, 3027
CLASTE, 783
CLASTIQUE, 784
CLAYAT, 4648
CLIMAT, 794
CLIMAT HUMIDE, 2073
CLINOMETRE, 4534
CLIVAGE ARDOISIER, 4131
CLIVAGE ISOMETRIQUE, 423
CLIVAGE MINERAL, 2864
CLIVAGE PARALLELE, 455
CLIVAGE PLAN AXIAL, 392
CLIVAGE TECTONIQUE, 792
CLYMENIIDA, 800
COAL BALL, 802
COBALT, 818
COBOL, 819
COCCOLITHOPHORALES, 820
COEFFICIENT, 822
COEFFICIENT CORRELATION, 962
COEFFICIENT PARTAGE, 3311
COEFFICIENT PETROCHIMIQUE, 3370
COEFFICIENT POISSON, 3532
COEFFICIENT TRANSMISSIVITE, 4574
COELENTERATA, 823

COENOTHECALIA, 824
COERCIVITE, 825
COHESION, 827
COIN GLACE, 2162
COIN GLACE FOSSILE, 1621
COLEOPTEROIDA, 833
COLLECTION, 839
COLLEMBOLA, 840
COLLINE, 2013
COLLINE ABYSSALE, 21
COLLISION PLAQUE, 3496
COLLOIDE, 842
COLLUVION, 843
COLONNE, 3161
COLORANT, 3234
COLORATION ECHANTILLON, 4230
COLORIMETRIE, 849
COLUMNARIINA, 851
COMBUSTIBLE FOSSILE, 1620
COMMENSALISME, 853
COMPACITE, 856
COMPACTAGE, 857
COMPARAISON PLANETE, 2280
COMPARAISON TERRE, 3481
COMPLETION PUITS, 4814
COMPLEXATION, 861
COMPLEXE, 860
COMPLEXE IGNE, 2173
COMPLEXE METAMORPHIQUE, 2781
COMPLEXOMETRIE, 862
COMPOSE AMMONIAQUE, 165
COMPOSE HYDROCARBURE, 2099
COMPOSE INERTE, 2223
COMPOSITION, 863
COMPOSITION ACIDE, 39
COMPOSITION ALCALINE, 128
COMPOSITION ALCALINO-CALCIQUE, 124
COMPOSITION CAFEMIQUE, 605
COMPOSITION CALCIQUE, 610
COMPOSITION CALCOALCALINE, 607
COMPOSITION CARBONATEE, 609
COMPOSITION CARBONEE, 648
COMPOSITION DETRITIQUE, 1142
COMPOSITION FERROMANGANIFERE, 1537
COMPOSITION FERRUGINEUSE, 1539
COMPOSITION GLAUCONITIQUE, 1816
COMPOSITION HYPERALCALINE, 3347
COMPOSITION HYPERALUMINEUSE, 3348
COMPOSITION MAFIQUE, 2616
COMPOSITION MELANOCRATE, 2741
COMPOSITION MESOCRATE, 2752
COMPOSITION MINERALOGIQUE, 2865
COMPOSITION POTASSIQUE, 3578
COMPOSITION SILICEUSE, 4099
COMPOSITION SUBALUMINEUSE, 4318
COMPOSITION THOLEITIQUE, 4516
COMPOSITION ULTRABASIQUE, 4635

COMPRESSIBILITE, 864
COMPRESSION, 865
COMPRESSION HORIZONTALE, 2051
COMPRESSION OEDOMETRIQUE, 908
COMPRESSION TRIAXIALE, 4592
COMPRESSION UNIAXIALE, 4658
CONCENTRATION, 873
CONCENTRATION GRAVITE, 1114
CONCHOSTRACA, 876
CONCORDANCE STRATIGRAPHIQUE, 889
CONCRETION, 881
CONDENSATION, 882
CONDENSE GAZ, 1688
CONDITION HYDROTHERMALE, 2133
CONDITION HYPOVOLCANIQUE, 4343
CONDITION PNEUMATOLYTIQUE, 3523
CONDITION PRESSION TEMPERATURE, 3229
CONDUCTION CHALEUR, 1964
CONDUCTIVITE, 883
CONDUCTIVITE ELECTRIQUE, 1326
CONDUCTIVITE HYDRAULIQUE, 2091
CONDUCTIVITE THERMIQUE, 4495
CONDYLARTHRA, 884
CONE, 1504
CONE ALLUVION, 141
CONE CENDRE, 770
CONE DEJECTION, 1076
CONE DEPRESSION, 885
CONE IN CONE, 886
CONE SCORIES, 4192
CONE SOUS MARIN, 4327
CONE VOLCANIQUE, 4762
CONFLUENT, 888
CONGELATION, 890
CONGELIFRACTION, 891
CONGLOMERAT, 892
CONGLOMERAT BASE, 424
CONGRES GEOLOGIQUE INTERNATIONAL, 2277
CONIACIEN, 894
CONIFERALES, 895
CONJUGUE, 896
CONOCARDIOIDA, 899
CONODONTA, 901
CONSANGUINITE, 903
CONSERVATION, 3606
CONSISTANCE, 906
CONSOLIDATION, 907
CONSOMMATION, 912
CONSOMMATION EAU, 4787
CONSTANTE, 909
CONSTANTE DESINTEGRATION, 1079
CONSTANTE DIELECTRIQUE, 1174
CONSTANTE ELASTICITE, 1317
CONSTANTE GRAVITATION, 1876
CONSTANTE LAME, 2418
CONSTANTE RETICULAIRE, 2451
CONSTITUANT MINERAI, 3159

CONSTITUTION INTERNE, 1276
CONSTITUTION LUNE, 2600
CONSTITUTION PLANETE, 3483
CONSTRUCTEUR RECIF, 3779
CONSTRUCTION, 910
CONTACT, 913
CONTAMINATION MAGMATIQUE, 917
CONTINENT, 918
CONTRACTION, 934
CONTRACTION TERRE, 933
CONTRAINTE, 4290
CONTRAINTE CISAILLEMENT, 4063
CONTRAINTE COMPRESSION, 869
CONTRASTE, 935
CONTROLE, 936
CONTROLE EROSION, 1425
CONTROLE GEOCHIMIQUE GITE, 722
CONTROLE GITE, 3153
CONTROLE HYDROGEOLOGIQUE GITE, 2110
CONTROLE IGNE GITE, 2621
CONTROLE LITHOLOGIQUE, 2530
CONTROLE PALEOGEOGRAPHIQUE, 3255
CONTROLE SEDIMENTAIRE GITE, 3999
CONTROLE STRATIGRAPHIQUE, 4276
CONTROLE STRUCTURAL, 4302
CONTROLE TECTONIQUE, 4435
CONULARIDA, 937
CONVECTION, 938
CONVERGENCE, 940
COOPERATION INTERNATIONALE, 2276
COOPERATION TECHNIQUE, 4430
COORDINENCE, 943
COORDONNEE GEODESIQUE, 1728
COPEPODA, 944
COPERNICIEN, 945
COPROLITE, 947
COQUILLE, 4071
CORALLIMORPHARIA, 950
CORALLINACEAE, 951
CORALLITE, 952
CORDAITALES, 953
CORDILLERE, 955
CORDON LITTORAL, 2567
CORINDON, 965
CORNEENNE, 2057
CORPS MINERALISE, 3152
CORPS SABLE, 3937
CORRECTION, 960
CORRECTION TERRAIN, 4466
CORRECTION TOPOGRAPHIQUE, 4545
CORRELATION, 961
CORROSION, 964
CORYNEXOCHIDA, 966
COSMOCHIMIE, 972
COSMOLOGIE, 975
COTE, 811
COTE ABRASION, 7

COTE ACCUMULATION, 32
COTE FOSSILE, 4
COTYLOSAURIA, 977
COUCHE, 2462
COUCHE A, 1
COUCHE ACTIVE, 57
COUCHE B, 395
COUCHE BASALE, 555
COUCHE BASALTIQUE, 426
COUCHE C, 602
COUCHE CHARBON, 806
COUCHE CONCORDANTE, 877
COUCHE CONTREPENTEE, 399
COUCHE D, 1057
COUCHE E, 1271
COUCHE F, 1499
COUCHE FRONTALE, 1614
COUCHE G, 1671
COUCHE GRANITIQUE, 1849
COUCHE IMPERMEABLE, 262
COUCHE LIMITE, 561
COUCHE NEPHELOIDE, 3009
COUCHE ROUGE, 3773
COUCHE SEMIPERMEABLE, 264
COUCHE SOMMITALE, 4548
COULEE, 1575
COULEE BOUE, 2956
COULEE CENDRE, 332
COULEE CLASTIQUE, 1077
COULEE LAVE, 2457
COULEUR, 845
COUP CHARGE, 3883
COUPE GEOLOGIQUE, 1736
COUPE PUITS, 3460
COUPE SONDAGE, 550
COUPE TYPE, 4625
COUPLE TERRE-LUNE, 1282
COURANT, 1035
COURANT CONVECTION, 939
COURANT LITTORAL, 2568
COURANT TURBIDITE, 4617
COURBE TEMPS PARCOURS, 4582
COURS EAU, 4284
COURS EAU DERIVE, 1310
COURS EAU ENCAISSE, 1384
COUT, 976
COUVERTURE MULTIPLE, 854
CRAIE, 705
CRANE, 980
CRAQUELURE BOUE, 2954
CRAQUELURE RETRECISSEMENT, 4088
CRATERE, 981
CRATERE IMPACT, 2187
CRATERE LUNAIRE, 2601
CRATERE METEORIQUE, 2797
CRATON, 984
CREODONTA, 987

CREPINE, 4817
CRETACE, 989
CRETACE INF, 2579
CRETACE SUP, 4669
CRETE, 988
CRETE PLAGE, 446
CREVASSE, 990
CRINOIDEA, 991
CRINOZOA, 992
CRISTAL, 1011
CRISTAL IDIOMORPHE, 1447
CRISTAL XENOMORPHE, 4839
CRISTALLINITE, 1023
CRISTALLISATION, 1025
CRISTALLISATION FRACTIONNEE, 1631
CRISTALLITE, 1024
CRISTALLOBLASTE, 1027
CRISTALLOCHIMIE, 1012
CRISTALLOGRAPHIE, 1028
CRISTALLOGRAPHIE STRUCTURE, 4303
CROCODILIA, 996
CROISSANCE CONTINENTALE, 919
CROISSANCE CRISTALLINE, 1017
CROISSANCE ENCHEVETREE, 2267
CROISSANT, 1037
CROSSOPTERYGII, 999
CROUTE ALTERATION, 4810
CROUTE CALCAIRE, 616
CROUTE CONTINENTALE, 921
CROUTE FERRUGINEUSE, 1534
CROUTE OCEANIQUE, 3091
CROUTE TERRESTRE, 1001
CROUTE TERRESTRE INF, 2580
CROUTE TERRESTRE SUP, 4670
CRUE, 1569
CRUE GLACIAIRE, 1805
CRUSTACEA, 1002
CRYOPEDOLOGIE, 1003
CRYOTECTONIQUE, 1004
CRYOTURBATION, 1005
CRYPTODONTA, 1007
CRYPTOGENE, 1008
CRYPTOSTOMATA, 1009
CTENOSTOMATA, 1029
CUESTA, 1031
CUIVRE, 946
CUMULAT, 1032
CURIUM, 1034
CUTICULE DEFLATION, 1133
CUVETTE SYNCLINALE, 690
CYANOPHYTA, 1040
CYBERNETIQUE, 1041
CYCADALES, 1042
CYCADOFILICALES, 1043
CYCLE, 1044
CYCLE BIOLOGIQUE, 501
CYCLE EAU, 2119

CYCLE EROSION, 1426
CYCLE GEOCHIMIQUE, 1715
CYCLE SEDIMENTAIRE, 1045
CYCLE TECTONIQUE, 4436
CYCLOCYSTOIDEA, 1048
CYCLONE, 1049
CYCLOSTOMATA, 1050
CYCLOTHEME, 1051
CYNOMORPHA, 1054
CYSTIPHYLLINA, 1055
CYSTOIDEA, 1056
D-H, 1058
DALLE, 4129
DANIEN, 1060
DARWINISME, 1062
DASYCLADACEAE, 1063
DATATION, 1073
DATATION PLOMB-ALPHA, 2467
DATATION TRACE FISSION, 1556
DATATION TRACE PARTICULE, 3309
DEBIT, 1209
DEBIT EAU SOUTERRAINE, 1914
DEBLAI FORAGE, 1039
DEBLAI STERILE, 4217
DECAPAGE, 3971
DECAPODA, 1078
DECHARGE DECHET, 4782
DECHET, 4781
DECHET DOMESTIQUE, 2070
DECHET INDUSTRIEL, 2221
DECHET LIQUIDE, 2521
DECHET RADIOACTIF, 3730
DECHET SOLIDE, 4168
DECISION EXPLOITABILITE, 1481
DECLINAISON, 1081
DECOLLEMENT, 1082
DECONTAMINATION, 1083
DECONVOLUTION, 1084
DECOUVERTE, 1217
DECREPITOMETRIE, 1086
DEDOLOMITISATION, 1087
DEFAUT CRISTALLIN, 1013
DEFAUT SURFACE, 4363
DEFERLANTE, 4370
DEFLATION, 1095
DEFORMATION, 1096
DEFORMATION ELASTIQUE, 1321
DEFORMATION POLYPHASEE, 3556
DEFORMATION SEDIMENT, 4145
DEFORMATION SOL, 4147
DEFORMATION SOUS CONTRAINTE, 4267
DEFORMATION VERSANT, 4135
DEGEL, 4485
DEGLACIATION, 1098
DEGRE ARRONDI, 3906
DEGRE HOUILLIFICATION, 3748
DEGRE LIBERTE, 1100

DEINOTHERIOIDEA, 1102
DELTA, 1103
DEMI ESPACE, 1943
DEMI PERIODE, 1942
DEMOSPONGEA, 1107
DENDRITE, 1108
DENDROCHRONOLOGIE, 1110
DENDROGRAMME, 1111
DENDROIDEA, 1112
DENSITE, 1113
DENSITE BRUTE, 590
DENT, 4544
DENTURE, 1115
DENUDATION, 1116
DEPART GAZ, 1097
DEPLACEMENT FAILLE, 1224
DEPOT, 1120
DEPOT CONCRETIONNE, 4118
DEPOT GLACIAIRE, 1792
DEPRESSION, 1121
DEPRESSION GEOCHIMIQUE, 1721
DEPRESSION VOLCANOTECTONIQUE, 4770
DERIVATION, 1125
DERIVE CONTINENTALE, 922
DERIVE LITTORALE, 2544
DERIVE POLE, 3535
DERMAPTEROIDA, 1127
DERMOPTERA, 1128
DESAGREGATION, 1099
DESAIMANTATION, 1106
DESAIMANTATION THERMIQUE, 4496
DESEQUILIBRE URANIUM, 4691
DESERT, 1131
DESERTIFICATION, 1134
DESHYDRATATION, 1101
DESILICIFICATION, 1136
DESINTEGRATION RADIOACTIVE, 3725
DESMOCERATIDA, 1138
DESMOSTYLIA, 1139
DESQUAMATION, 1466
DESSALEMENT, 1129
DESSICATION, 1135
DESSIN, 1250
DESTRUCTION RADIATION, 3721
DETERMINISME, 1140
DETROIT, 4269
DEUTERIUM, 1146
DEVELOPPEMENT, 1147
DEVITRIFICATION, 1150
DEVONIEN, 1151
DEVONIEN INF, 2581
DEVONIEN MOYEN, 2842
DEVONIEN SUP, 4671
DIACHRONISME, 1153
DIACLASE, 2361
DIACLASE CISAILLEMENT, 4059
DIADOCHIE, 1154

DIAGENESE, 1155
DIAGENESE PRECOCE, 1272
DIAGENESE TARDIVE, 2438
DIAGNOSE, 1156
DIAGRAMME, 1157
DIAGRAMME EQUILIBRE, 3392
DIAGRAMME MONOCRISTAL, 4116
DIAGRAMME PETROCHIMIQUE, 3371
DIAGRAMME POUDRE, 3587
DIAGRAMME STRUCTURAL, 4304
DIAGRAPHIE, 4816
DIAGRAPHIE ELECTRIQUE, 1324
DIAGRAPHIE ELECTROMAGNETIQUE, 1343
DIAGRAPHIE GAMMA, 1679
DIAGRAPHIE GAMMA GAMMA, 1678
DIAGRAPHIE NEUTRON, 3017
DIAGRAPHIE RESISTIVITE, 2443
DIAGRAPHIE SONIQUE, 41
DIALYSE, 1158
DIAMANT, 1161
DIAMETRAGE, 621
DIAMICTITE, 1160
DIAPHTHORESE, 1162
DIAPHTHORITE, 1163
DIAPIR, 1164
DIAPIRISME, 1165
DIASTEME, 1166
DIASTROPHISME, 1167
DIATOMEAE, 1169
DIATOMITE, 1170
DIATREME, 1171
DICOTYLEDONES, 1172
DICTYONELLIDINA, 1173
DIFFERENCIATION HORIZON, 2049
DIFFERENCIATION MAGMATIQUE, 1178
DIFFERENCIATION METAMORPHIQUE, 2782
DIFFERENCIATION PAR GRAVITE, 1877
DIFFRACTION, 1180
DIFFRACTION ELECTRON, 1346
DIFFRACTION NEUTRON, 3016
DIFFRACTION RX, 4834
DIFFUSION, 1181
DIFFUSIVITE, 1182
DIGUE, 3986
DILATANCE, 1186
DILATATION, 1187
DILUTION, 1188
DILUTION ISOTOPIQUE, 2349
DIMORPHISME, 1189
DIMORPHISME SEXUEL, 4049
DINOCERATA, 1190
DINOFLAGELLATA, 1191
DINOSAURIA, 1192
DIOGENITE, 1193
DIORITE, 1194
DIORITE QUARTZIQUE, 3711
DIPLEUROZOA, 1200

DIPLOPODA, 1202
DIPLORHINA, 1203
DIPNOI, 1205
DIPTERA, 1207
DIRECTION, 4292
DIRECTION HORIZONTALE, 2052
DISCOASTER, 1210
DISCONTINUITE, 1212
DISCONTINUITE BIRCH, 518
DISCONTINUITE BULLARD, 592
DISCONTINUITE CONRAD, 902
DISCONTINUITE GUTENBERG, 1929
DISCORBACEA, 1213
DISCORDANCE, 4646
DISCORDANCE ANGULAIRE, 202
DISCORDANCE EROSION, 1211
DISCORDANCE REGIONALE, 3796
DISCORDANCE STRATIGRAPHIQUE, 3050
DISCORDANCE TECTONIQUE, 1214
DISCORDANT, 1215
DISLOCATION CRISTALLINE, 1014
DISPERSION ONDE, 4806
DISPERSION OPTIQUE, 1222
DISPERSION ROTATOIRE, 3903
DISSOCIATION IONIQUE, 1227
DISSOLUTION, 1228
DISSOLUTION SOUS PRESSION, 3608
DISTRIBUTION, 1230
DISTRIBUTION BINOMIALE, 480
DISTRIBUTION FREQUENCE, 1645
DISTRIBUTION LOGNORMALE, 2563
DISTRIBUTION NORMALE, 3055
DISTRIBUTION POISSON, 3531
DISTRIBUTION STATISTIQUE, 4240
DISTRICT METALLOGENIQUE, 2770
DIVERGENCE, 1232
DIVERSITE ESPECE, 4193
DOCODONTA, 1233
DOCUMENTATION, 2229
DOLERITE, 1234
DOLINE, 1235
DOLINE EFFONDREMENT, 4117
DOLOMIE, 1238
DOLOMITISATION, 1237
DOMAINE MAGNETIQUE, 2631
DOME, 1240
DOME GNEISSIQUE, 1829
DOME SEL, 3925
DOME VOLCANIQUE, 4763
DOMERIEN, 1241
DONNEE, 1064
DONNEE MEB, 4034
DONNEE MET, 4451
DONNEE MICROSCOPIE ELECTRONIQUE, 1350
DONNEE MICROSONDE ELECTRONIQUE, 1352
DORSALE, 4330
DORSALE OCEANIQUE, 2840

DRAGAGE, 1251
DRAINAGE RAYONNANT, 1109
DRAINAGE TERRAIN, 1247
DRAINANCE, 2470
DROMOCHRONIQUE, 2023
DRUMLIN, 1259
DRUSE, 1260
DUNE, 1261
DUNE CONTINENTALE, 1132
DUNE COTIERE, 812
DURAIN, 1262
DUREE JOUR, 2478
DUREE VIE, 484
DURETE, 1956
DURETE EAU, 4790
DYKE, 1184
DYKE GRESEUX, 3940
DYNAMIQUE, 1267
DYSPROSIUM, 1270
EAU, 4783
EAU ADSORBEE, 67
EAU CAPILLAIRE, 636
EAU CONNEE, 898
EAU CONSTITUTION, 4794
EAU CRISTALLISATION, 4793
EAU DISPONIBLE, 4802
EAU DOUCE, 1647
EAU FONTE, 2745
EAU FOSSILE, 1623
EAU GISEMENT, 1306
EAU HYDROTHERMALE, 2134
EAU HYGROSCOPIQUE, 2139
EAU INTERSTITIELLE, 3563
EAU JUVENILE, 2365
EAU LIBRE, 4645
EAU MER, 3978
EAU MINERALE, 2875
EAU POLLUEE, 3542
EAU RESURGENCE, 3837
EAU SAUMATRE, 568
EAU SOUTERRAINE, 1911
EAU SURFACE, 4366
EAU THERMALE, 4501
EAU THERMOMINERALE, 2734
EBOULEMENT, 3884
EBOULIS, 4416
ECAILLE, 3963
ECART, 1149
ECART TYPE, 4234
ECHANGE ION, 2300
ECHANTILLON, 3933
ECHANTILLONNAGE, 3935
ECHANTILLONNAGE PAR GRAPPIN, 1839
ECHANTILLONNAGE SOL, 4156
ECHANTILLONNEUR, 3934
ECHELLE ABSOLUE, 4536
ECHELLE CARTOGRAPHIQUE, 2669

ECHELLE MERCALLI MODIFIEE, 2905
ECHELLE PHI, 3397
ECHELLE RICHTER, 3862
ECHELLE STRATIGRAPHIQUE, 1734
ECHINODERMATA, 1291
ECHINOIDEA, 1292
ECHINOZOA, 1293
ECHOSONDAGE, 1294
ECLAT, 2606
ECLAT GRAS, 1889
ECLOGITE, 1295
ECOLOGIE, 1297
ECOLOGIE HUMAINE, 2069
ECOLOGIE MARINE, 2676
ECONOMIE MINIERE, 2867
ECOSYSTEME, 1301
ECOULEMENT EAU SOUTERRAINE, 1918
ECOULEMENT GRAVITE, 1885
ECOULEMENT INTERMITTENT, 2272
ECOULEMENT PLASTIQUE, 3492
ECOULEMENT POLYPHASIQUE, 2964
ECROULEMENT, 3891
ECROULEMENT TOIT, 3894
ECTOTROPHA, 1303
EDENTATA, 1305
EDRIOASTEROIDEA, 1307
EDRIOBLASTOIDEA, 1308
EFFET CHOC, 2188
EFFET DEUTERIQUE, 1145
EFFET ISOTOPIQUE, 2350
EFFET SERRE, 1891
EFFONDREMENT, 834
EH, 1312
EIFELIEN, 1313
EINSTEINIUM, 1314
ELASMOBRANCHII, 1316
ELASTICITE, 1323
ELECTRICITE, 1334
ELECTRO OSMOSE, 1335
ELECTRODE, 1337
ELECTROLYSE, 1338
ELECTROLYTE, 1339
ELECTROMAGNETISME, 1344
ELECTRON, 1345
ELEMENT CHALCOPHILE, 703
ELEMENT COSMOGENIQUE, 973
ELEMENT LITHOPHILE, 2533
ELEMENT MAGNETIQUE, 2632
ELEMENT METALLIQUE, 2766
ELEMENT NATIF, 2988
ELEMENT NON METALLIQUE, 3048
ELEMENT SIDEROPHILE, 4092
ELEMENT VOLATIL, 4756
ELEPHANTOIDEA, 1354
ELOIGNEMENT, 3814
ELUTRIATION, 1356
ELUVION, 1357

EMBRECHITE, 1359
EMBRITHOPODA, 1360
EMERSION, 1361
EMMAGASINEMENT, 31
EMPILEMENT ATOME, 365
EMPILEMENT DENSE, 798
EMPREINTE CANNELURE, 1903
EMPREINTE CHARGE, 2555
EMSIEN, 1364
ENCLAVE ROCHE, 4838
ENDOCERATOIDEA, 1369
ENDOREISME, 1371
ENERGIE, 1373
ENERGIE ACTIVATION, 55
ENERGIE GEOTHERMIQUE, 1773
ENERGIE LIBRE, 1641
ENERGIE NUCLEAIRE, 3062
ENFOUISSEMENT, 596
ENGRAIS, 1541
ENNOIEMENT, 4334
ENQUETE PRELIMINAIRE, 3178
ENRICHISSEMENT, 1376
ENSEIGNEMENT, 1309
ENSOLEILLEMENT, 2255
ENTEROPNEUSTA, 1381
ENTHALPIE, 1382
ENTISOL, 1383
ENTRAINEMENT EN PROFONDEUR, 4321
ENTREMONT, 2273
ENTROPIE, 1385
ENVASEMENT, 4108
EOCENE, 1389
EOCENE INF, 2582
EOCENE MOYEN, 2843
EOCENE SUP, 4672
EOCRINOIDEA, 1390
EON, 1394
EOSUCHIA, 1395
EPAISSEUR, 4513
EPANDAGE, 84
EPANDAGE FLUVIOGLACIAIRE, 3211
EPIBIOTISME, 1398
EPIBOLITE, 1399
EPICENTRE, 1400
EPIDOTISATION, 1402
EPIEUGEOSYNCLINAL, 1403
EPIGENESE, 1404
EPIROGENESE, 1396
EPISODE JARAMILLO, 2358
EPITAXIE, 1406
EPONTE, 4780
EPOQUE, 1409
EPOQUE GAUSS, 1699
EPOQUE MATUYAMA, 2720
EPOQUE METALLOGENIQUE, 2771
EPURATION, 3687
EQUATION ETAT, 1412

EQUATION MATHEMATIQUE, 1411
EQUILIBRE, 1416
EQUILIBRE CHIMIQUE, 724
EQUILIBRE DYNAMIQUE, 1264
EQUIPEMENT RECEPTION, 3764
EQUIPEMENT SCISSOMETRIE, 4711
EQUIPEMENT SISMIQUE, 4013
EQUISETALES, 1417
EQUIVALENCE, 2047
EQUIVALENT CALORIFIQUE FUSION, 1966
ERATHEME, 1419
ERATOSTHENIEN, 1420
ERBIUM, 1421
ERE, 1418
ERG, 1422
ERODABILITE, 1423
EROSION, 1424
EROSION EAU, 4789
EROSION EN NAPPE, 4065
EROSION EOLIENNE, 4827
EROSION FLUVIATILE, 1594
EROSION GLACIAIRE, 1794
EROSION LITTORALE, 813
EROSION MARINE, 2678
EROSION SOL, 4148
ERREUR, 1431
ERUPTION, 1432
ESCARPEMENT, 1433
ESCARPEMENT FAILLE, 1511
ESKER, 1434
ESPACE INTERSTELLAIRE, 2285
ESPACE SOUTERRAIN, 4651
ESSAI COMPRESSION, 868
ESSAI DEBIT, 261
ESSAI LABORATOIRE, 2403
ESSAI MINERAI, 336
ESSAI PLAQUE, 3500
ESSAI ROUTIER, 3878
ESSAIM DYKE, 1185
ESSAIM SEISME, 1287
ESSEXITE, 1435
EST, 1288
ESTACADE, 1901
ESTUAIRE, 1438
ETAGE, 4229
ETAIN, 4537
ETANCHEMENT, 3982
ETAT PHYSIQUE, 3428
ETHANE, 1440
ETHER, 1441
ETIAGE, 2573
ETUDE CAS, 663
ETUDE CRITIQUE, 995
ETUDE EN MER, 2682
ETUDE FAISABILITE, 1520
ETUDE IMPACT MILIEU, 2191
ETUDE MILIEU, 1387

EUCILIATA, 1442
EUCRITE ACHONDRITE, 1443
EUGEOSYNCLINAL, 1445
EUGLENOPHYCOPHYTA, 1446
EUPANTOTHERIA, 3285
EUROPIUM, 1448
EURYAPSIDA, 4389
EUSELACHII, 1451
EUSTATISME, 1452
EUTROPHISATION, 1455
EVALUATION, 1457
EVALUATION GITE, 1117
EVAPORATION, 1458
EVAPORITE, 1459
EVAPOTRANSPIRATION, 1460
EVENT, 4767
EVOLUTION BIOLOGIQUE, 1461
EVOLUTION CHIMIQUE, 725
EXCENTRICITE, 1290
EXCES ARGON, 1463
EXCURSION, 1465
EXHAURE MINE ACIDE, 38
EXHAUSSEMENT, 4665
EXINE, 1468
EXINITE, 1469
EXOBIOLOGIE, 1470
EXOREISME, 1472
EXPANSION, 1475
EXPANSION FOND OCEANIQUE, 3980
EXPEDITION CROISIERE, 1476
EXPERIENCE, 1479
EXPLOITATION, 1482
EXPLOITATION EN DECOUVERTE, 4294
EXPLOITATION MINIERE, 2883
EXPLOITATION PAR DISSOLUTION, 4179
EXPLOITATION PAR EBOULEMENT, 680
EXPLORATION, 3641
EXPLOSION, 1485
EXPLOSION CHIMIQUE, 726
EXPLOSION NUCLEAIRE, 3063
EXPLOSION PHREATIQUE, 2105
EXPORTATION, 1488
EXPOSITION RELIEF, 3809
EXSOLUTION, 1490
EXSURGENCE, 1491
EXTENSION STRATIGRAPHIQUE, 3746
EXTENSOMETRE, 1492
EXTERNIDES, 1493
EXTINCTION, 1496
EXTINCTION OPTIQUE, 3138
EXTINCTION ROULANTE, 4657
EXTRAPOLATION, 1497
FABRIQUE, 1500
FACIES, 1501
FACIES A SANIDINE, 3942
FACIES ACTINOTE, 53
FACIES AMPHIBOLITE, 174

FACIES AMPHIBOLITE ALBITE EPIDOTE, 104
FACIES AMPHIBOLITE CORDIERITE, 954
FACIES AMPHIBOLITE EPIDOTE, 1401
FACIES BORDURE, 546
FACIES CORNEENNE, 2058
FACIES CORNEENNE ALBITE EPIDOTE, 105
FACIES CORNEENNE HORNBLENDE, 2055
FACIES CORNEENNE PYROXENE, 3703
FACIES ECLOGITE, 1296
FACIES GEOCHIMIQUE, 1717
FACIES GLAUCOPHANITE, 1818
FACIES GRANULITE, 1859
FACIES HYDROCHIMIQUE, 2100
FACIES IGNE, 2174
FACIES METAMORPHISME, 2783
FACIES MINERAL, 2869
FACIES PREHNITE PUMPELLYITE, 3603
FACIES PRIMAIRE, 3180
FACIES SCHISTE VERT, 1893
FACIES ZEOLITE, 4857
FACTEUR TEMPS, 4535
FAILLE, 1507
FAILLE ACTIVE, 56
FAILLE ARQUEE, 289
FAILLE CHARNIERE, 2015
FAILLE CHEVAUCHEMENT, 4520
FAILLE CONTRAIRE, 233
FAILLE DECROCHEMENT, 4293
FAILLE EFFONDREMENT, 1199
FAILLE EN ECHELON, 1365
FAILLE ESCALIER, 4248
FAILLE HORIZONTALE, 2054
FAILLE LATERALE, 2442
FAILLE NORMALE, 3056
FAILLE OBLIQUE, 3077
FAILLE SOUS-CHARRIAGE, 4656
FAILLE STRATIFICATION, 456
FAILLE SYNGENETIQUE, 1921
FAILLE TRANSCOURANTE, 4567
FAILLE TRANSFORMANTE, 4569
FAILLE TRANSVERSALE, 1196
FAILLE VERTICALE DECROCHEMENT, 4831
FALAISE, 793
FAMENNIEN, 1503
FANGLOMERAT, 1505
FAUNE, 1516
FAUNE PLANCTIVORE, 3486
FAUNE SPECIFIQUE, 4194
FAUX CLIVAGE, 4133
FELDSPATH, 1524
FELDSPATHISATION, 1525
FENETRE, 4830
FENITE, 1529
FENITISATION, 1530
FER, 2308
FER FERREUX, 1538
FER FERRIQUE, 1533

FER MARAIS, 537
FERMIUM, 1531
FERROD, 1535
FEUILLE, 2469
FIGURE BASALE, 4163
FILET SABLE, 4079
FILICOPSIDA, 1543
FILON, 4725
FILON ANASTOMOSE, 2514
FILON CLASTIQUE, 785
FILON EN QUEUE DE CHEVAL, 2059
FILON SEGREGATION, 4009
FILTRAGE SIGNAL, 1546
FILTRATION, 1547
FILTRE, 1545
FINANCEMENT, 1548
FISSION NUCLEAIRE, 1554
FISSIPEDIA, 1557
FISSURATION PRISMATIQUE, 850
FISSURE CONTRACTION, 1965
FISSURE MAGMATIQUE, 1522
FIXATION ION, 1558
FIXISME, 1559
FJORD, 1560
FLANC PLI, 2505
FLANDRIEN, 1564
FLECHE LITTORALE, 4215
FLEXURE, 1566
FLOCULATION, 1568
FLORE, 1571
FLORE DESMIDIEE, 1137
FLORE SPECIFIQUE, 4195
FLORIZONE, 1573
FLOTTATION, 1574
FLUAGE, 986
FLUCTUATION, 1578
FLUIDALITE, 1584
FLUIDE MINERALISATEUR, 2877
FLUOR, 1589
FLUORESCENCE, 1585
FLUORESCENCE RX, 4835
FLUORITE, 1590
FLUORURATION, 1588
FLUORURE, 1586
FLUVENT, 1592
FLUVIOGLACIAIRE, 1596
FLUVIOSOL, 1598
FLUX GEOTHERMIQUE, 1967
FLYSCH, 1599
FOIDITE, 1601
FOLIATION, 1609
FONCAGE CIMENTATION, 1920
FONCTION, 1665
FOND MARIN, 3088
FONDATION, 1627
FORABILITE, 1253
FORAGE, 1254

FORAGE PERCUSSION ROTARY, 3901
FORAGE ROTARY, 3900
FORAMINIFERA, 1610
FORMATION, 1617
FORMATION CRATERE, 983
FORMATION FERRIFERE, 2311
FORMATION RECOUVREMENT, 979
FORMATION SUPERFICIELLE, 4369
FORMATION SUPERGENE, 2143
FORME, 4054
FORME BOTRYOIDALE, 553
FORME CRISTALLINE, 1016
FORME GLOBE, 4055
FORME GRAIN, 3308
FORMULE, 1618
FOSSE ABYSSALE, 4587
FOSSE ABYSSALE ALLONGEE, 4603
FOSSILE, 1619
FOSSILE CARACTERISTIQUE, 2206
FOSSILE CHIMIQUE, 727
FOSSILE SPECIFIQUE, 1925
FOSSILE VIVANT, 2549
FOSSILISATION, 1626
FOUILLE GENIE CIVIL, 1462
FOYAITE, 1629
FRACTION FINE, 1549
FRACTION GROSSIERE, 810
FRACTIONNEMENT, 1630
FRACTIONNEMENT ISOTOPIQUE, 2353
FRACTURATION, 1635
FRACTURATION HYDRAULIQUE, 2092
FRACTURE, 1632
FRACTURE ANNULAIRE, 3870
FRACTURE CONCHOIDALE, 875
FRAGMENT, 1636
FRAGMENTATION, 1637
FRANCIUM, 1639
FRANGE, 3868
FRANGE KELYPHITIQUE, 2383
FRANGE REACTIONNELLE, 3758
FRASNIEN, 1640
FREQUENCE, 1644
FRIABLE, 579
FRONT, 1652
FROTTEMENT, 1650
FRUCTIFICATION, 1657
FUGACITE, 1659
FULGURITE, 1660
FUMEROLLE, 1664
FUNGI, 1667
FUSINITE, 1669
FUSION, 2744
FUSION CONGRUENTE, 893
FUSION INCOMPLETE, 2204
FUSION PARTIELLE, 3305
FUSULINIDAE, 1670
GABBRO, 1672

GADOLINIUM, 1673
GALERIE CAPTANTE, 2227
GALET, 3322
GALET ARGILE, 788
GALET ARGILE INDURE, 303
GALLIUM, 1675
GANGUE, 1680
GANISTER, 1681
GASTROPODA, 1696
GASTROZOOIDE, 1697
GAZ, 1684
GAZ ANHYDRIDE SULFUREUX, 4354
GAZ AZOTE, 3035
GAZ CARBONIQUE, 643
GAZ CHARBON, 804
GAZ HYDROGENE, 2107
GAZ HYDROGENE SULFURE, 2108
GAZ MARAIS, 2694
GAZ NATUREL, 2989
GAZ OXYGENE, 3226
GAZ PHREATIQUE, 3415
GAZ RARE, 2224
GAZEIFICATION, 1695
GEANTICLINAL, 1700
GEDINNIEN, 1701
GEL COLLOIDAL, 1702
GELIVITE, 1704
GEMME, 1707
GEMMOLOGIE, 1706
GENERATION, 1708
GENESE, 1709
GENESE COMAGMATIQUE, 852
GENESE GITE, 1118
GENESE INTRATELLURIQUE, 2292
GENESE POSTMAGMATIQUE, 3574
GENESE ROCHE IGNEE, 2175
GENESE SEDIMENTAIRE, 4005
GENIE CIVIL, 777
GENIE PETROLIER, 3380
GENOMORPHE, 1710
GENOTYPE, 1711
GEOBAROMETRIE, 1712
GEOBIOS, 1713
GEOCHIMIE, 1722
GEOCHIMIE ISOTOPIQUE, 2351
GEOCHIMIE ORGANIQUE, 3168
GEOCHRONOLOGIE, 1725
GEODE, 1726
GEODESIE, 1727
GEODYNAMIQUE, 1729
GEOFRACTURE, 1730
GEOGRAPHIE, 1731
GEOGRAPHIE MATHEMATIQUE, 2710
GEOIDE, 1732
GEOISOTHERME, 1733
GEOLOGIE, 1741
GEOLOGIE APPLIQUEE, 246

GEOLOGIE DYNAMIQUE, 1265
GEOLOGIE ECONOMIQUE, 1298
GEOLOGIE ENVIRONNEMENT, 1388
GEOLOGIE EXTRA TERRESTRE, 1498
GEOLOGIE GLACIAIRE, 1796
GEOLOGIE HISTORIQUE, 2019
GEOLOGIE INGENIEUR, 1374
GEOLOGIE LUNAIRE, 2602
GEOLOGIE MARINE, 2679
GEOLOGIE MATHEMATIQUE, 2711
GEOLOGIE MINIERE, 2885
GEOLOGIE PETROLIERE, 3381
GEOLOGIE REGIONALE, 3793
GEOLOGIE STRUCTURALE, 4306
GEOLOGIE SUBSURFACE, 4340
GEOLOGIE URBANISATION, 4694
GEOLOGIE VULGARISATION, 3560
GEOMAGNETISME, 1745
GEOMETRIE, 1746
GEOMETRIE CHENAL, 708
GEOMORPHOLOGIE, 1748
GEONOMIE, 1749
GEOONDULATION, 1781
GEOPHONE, 1750
GEOPHYSICIEN, 1758
GEOPHYSIQUE, 1759
GEOPHYSIQUE APPLIQUEE, 247
GEOPRESSION, 1760
GEOSECS, 1761
GEOSTATISTIQUE, 1762
GEOSYNCLINAL, 1767
GEOSYNCLINAL ENSIALIQUE, 1377
GEOSYNCLINAL ENSIMATIQUE, 1378
GEOTECHNIQUE, 1771
GEOTECTONIQUE, 1772
GEOTHERMIE, 1779
GEOTHERMOMETRIE, 1780
GERMANIUM, 1782
GESTION PRODUCTION, 3625
GESTION RESSOURCE EAU, 4791
GEYSER, 1784
GEYSERITE, 1785
GIGANTISME, 1787
GIGANTOSTRACA, 1788
GINKGOALES, 1789
GISEMENT, 2866
GISEMENT FOSSILIFERE, 1625
GISEMENT POTENTIEL, 3582
GITE APOMAGMATIQUE, 244
GITE CACHE, 529
GITE CATATHERMAL, 2379
GITE CEMENTATION, 4867
GITE DETRITIQUE, 2725
GITE DISSEMINATION, 1226
GITE EPITHERMAL, 1407
GITE EXHALATIF, 1467
GITE FILONIEN, 2559

GITE HYDROTHERMAL, 2132
GITE HYPOTHERMAL, 2149
GITE IMPREGNATION, 2195
GITE INTRAMAGMATIQUE, 2622
GITE MESOTHERMAL, 2759
GITE METAMORPHOGENE, 2788
GITE MINERAL, 3154
GITE MINERALOGIQUE, 2878
GITE PNEUMATOLYTIQUE, 3524
GITE POLYMETALLIQUE, 3551
GITE RESIDUEL, 3829
GITE SEDIMENTAIRE, 3997
GITE STRATIFORME, 4274
GITE STRATOIDE, 4270
GITE TELETHERMAL, 4447
GITE TYPE MISSISSIPPI VALLEY, 2896
GITE TYPE ROULEAU, 3893
GITE VOLCANOGENE, 4774
GIVETIEN, 1790
GLACE, 2155
GLACE MARINE, 3976
GLACE MORTE, 1075
GLACE SOL, 1905
GLACIAIRE, 1791
GLACIATION, 1803
GLACIATION CONTINENTALE, 924
GLACIER, 1804
GLACIER ROCHE, 3886
GLACIOLOGIE, 1807
GLACIOTECTONIQUE, 1809
GLACIS, 1810
GLACIS CONTINENTAL, 927
GLAUCONITE, 1815
GLAUCONITISATION, 1817
GLEY, 1819
GLISSEMENT MASSE BOUEUSE, 2959
GLISSEMENT SACCADE, 4254
GLISSEMENT TERRAIN, 2431
GLISSEMENT TERRE, 1283
GLOBIGERINACEA, 1822
GLOMERO, 1823
GLOSSAIRE, 1824
GLOSSOPTERIDALES, 1825
GLUCIDE, 4347
GNATHOSTOMI, 1826
GNATHOSTRACA, 1827
GNEISS, 1828
GNETALES, 1831
GOLFE, 1927
GONDWANA, 1834
GONFLEMENT, 4378
GONIATITIDA, 1835
GONIOMETRIE, 1836
GORGE, 1837
GRABEN, 1840
GRADE METAMORPHIQUE, 2784
GRADIENT, 1842

GRADIENT GEOTHERMIQUE, 1775
GRADIENT HYDRAULIQUE, 2093
GRADIOMETRE, 1843
GRAIN, 1844
GRAIN ENDUIT, 817
GRAIN MOYEN, 2736
GRAINE, 4007
GRANITE, 1845
GRANITE ALCALIN, 125
GRANITIQUE, 1847
GRANITISATION, 1850
GRANITOGNEISS, 1846
GRANITOIDE, 1851
GRANOCLASSEMENT, 1841
GRANODIORITE, 1853
GRANOGABBRO, 1854
GRANULAT, 85
GRANULATION, 1857
GRANULITE, 1858
GRANULITIQUE, 1860
GRANULOMETRIE, 1861
GRAPESTONE, 1862
GRAPHIQUE, 1863
GRAPHITE, 1866
GRAPHITISATION, 1867
GRAPTOLITHINA, 1868
GRAPTOLOIDEA, 1869
GRAUWACKE, 1888
GRAVIER, 1871
GRAVIER GROSSIER, 809
GRAVIMETRE, 1873
GRAVIMETRIE, 1874
GRAVITATION, 1875
GREISEN, 1896
GREISENISATION, 1897
GRELE, 1941
GRENAT, 1683
GRES, 3939
GRES GROSSIER, 1900
GROTTE, 678
GROUPE SPATIAL, 4188
GUANO, 1924
GUERRE, 2859
GUIDE, 3157
GUIDE TECTONIQUE, 4307
GUYOT, 3984
GYMNOLAEMATA, 1930
GYMNOSPERMAE, 1931
GYPSE, 1933
GYPSIFICATION, 1932
GYTTJA, 1934
GZHELIEN, 1935
HABITAT, 1937
HABITUS, 1936
HAFNIUM, 1940
HALDES, 4413
HALOGENE, 1947

HALOGENURE, 1944
HALOPHYTE, 1949
HAMMADA, 1950
HARDGROUND, 1955
HARMONIQUE, 1957
HARPOLITE, 1958
HAUT FOND, 4074
HAUTE PRESSION, 2007
HAUTE TEMPERATURE, 2008
HAUTERIVIEN, 1959
HAUTES-TERRES, 2012
HAWAIEN, 1961
HE-HE, 1979
HELICOPLACOIDEA, 1976
HELIOZOA, 1977
HELIUM, 1978
HELVETIEN, 1980
HEMATITE, 1981
HEMICRUSTACEA, 1982
HEMIPTEROIDA, 1983
HERCYNIDES, 1985
HEREDITE, 1986
HERITAGE, 2237
HETEROCHRONE, 1990
HETEROCHRONISME, 1989
HETEROCORALLIA, 1991
HETERODONTA, 1992
HETEROGENE, 1994
HETEROGENEITE, 1993
HETEROMORPHISME, 1996
HETTANGIEN, 1999
HEXACORALLA, 2000
HEXACTINIARIA, 2001
HEXAHEDRITE, 2003
HIATUS, 2004
HIEROGLYPHE, 2005
HIPPOMORPHA, 2016
HISTOGRAMME, 2017
HISTOIRE DE LA GEOLOGIE, 2021
HISTOLOGIE, 2018
HISTORIQUE, 2020
HISTOSOL, 2022
HOLMIUM, 2025
HOLOCENE, 2026
HOLOCEPHALI, 2027
HOLOGRAPHIE, 2029
HOLOPHYTE, 2030
HOLOSTEI, 2031
HOLOSTOME, 2032
HOLOTHUROIDEA, 2033
HOLOTYPE, 2034
HOMALOZOA, 2035
HOMEOMORPHIE, 2037
HOMEOMORPHISME, 2036
HOMINIDAE, 2038
HOMME FOSSILE, 1622
HOMO SAPIENS, 2039

HOMOCLINAL, 2040
HOMOGENEISATION, 2042
HOMOGENEITE, 2041
HOMOIOSTELEA, 2043
HOMOLOGIE, 2044
HOMONYMIE, 2045
HOMOSTELEA, 2046
HORIZON PRODUCTIF, 3626
HORIZON SOL, 4149
HORIZONTAL, 2050
HORNBLENDITE, 2056
HORST, 2060
HOUILLER PRODUCTIF, 805
HOUILLIFICATION, 808
HOULE, 3979
HOWARDITE, 2067
HUILE LEGERE, 2502
HUMIDITE, 2075
HUMIDITE SOL, 2911
HUMIFICATION, 2076
HUMINE, 2077
HUMMOCK, 2079
HUMOD, 2080
HUMOX, 2081
HUMULT, 2082
HUMUS, 2083
HYALOSPONGEA, 2085
HYBRIDATION MAGMATIQUE, 2086
HYDATOGENESE, 2087
HYDRARCH, 2088
HYDRATATION, 2089
HYDRATATION VERRE, 2090
HYDRATE, 2135
HYDRAULIQUE, 2098
HYDROCARBURE, 1658
HYDROCARBURE ALIPHATIQUE, 122
HYDROCARBURE AROMATIQUE, 305
HYDROCHIMIE, 2102
HYDRODYNAMIQUE, 2104
HYDRODYNAMIQUE CAPTAGE, 672
HYDROGENE, 2106
HYDROGEOLOGIE, 2115
HYDROGEOLOGIE APPLIQUEE, 248
HYDROGEOLOGIE REGIONALE, 3794
HYDROGRAMME, 2116
HYDROGRAPHIE, 2117
HYDROIDA, 2118
HYDROLOGIE KARSTIQUE, 2375
HYDROLOGIE SURFACE, 2121
HYDROMETALLURGIE, 2122
HYDROMETRE, 2123
HYDROMETRIE, 2124
HYDROPHONE, 2125
HYDROSPHERE, 2126
HYDROSPIRE, 2127
HYDROTHERMAL, 2130
HYDROXYDE, 2136

HYDROZOA, 2137
HYGROPHYLE, 2138
HYMENOPTEROIDA, 2140
HYOLITHES, 2141
HYPOCENTRE, 2146
HYPODIGME, 2148
HYPOTHESE, 2150
HYPSOMETRIE, 2151
HYRACOIDEA, 2152
HYSTRICOMORPHA, 2153
ICEBERG, 2163
ICHNITES, 2164
ICHNOLOGIE, 2165
ICHTHYOPTERYGIA, 2166
ICHTHYOSAURIA, 2167
IDIOBLASTE, 2168
IDIOGEOSYNCLINAL, 2169
IDIOMORPHISME, 2171
IGNIMBRITE, 2177
ILE, 2315
IMAGERIE, 2178
IMBRICATION, 2181
IMBRIEN, 2179
IMMATURE, 2182
IMMISCIBILITE, 2185
IMPACTITE, 2193
IMPORTATION, 2194
IMPURETE, 2196
IN SITU, 2197
INARTICULATA, 2198
INCEPTISOL, 2199
INCLINAISON, 4533
INCLINAISON MAGNETIQUE, 2635
INCLUSION, 2200
INCLUSION FLUIDE, 1579
INCLUSION GAZEUSE, 1690
INCLUSION MINERALE, 2870
INCOHERENT, 2201
INCRUSTATION, 2205
INDICATEUR, 2212
INDICATEUR BATHYMETRIQUE, 1123
INDICATEUR GEOCHIMIQUE, 1718
INDICE COULEUR, 847
INDICE CRISTALLISATION, 1026
INDICE DIFFERENCIATION, 1179
INDICE MATURITE, 2719
INDICE MILLER, 2860
INDICE MINERAL, 4086
INDICE REFRACTION, 2209
INDIEN, 2210
INDIUM, 2213
INDOCHINITE, 2214
INDURATION, 2218
INDUSTRIE, 2222
INDUSTRIE CHIMIQUE, 728
INFILTRATION, 2226
INFORMATIQUE, 2228

INFORMATISATION, 382
INFRA-ROUGE, 2231
INFRACAMBRIEN, 2230
INFRASTRUCTURE, 2235
INGENIERIE SISMIQUE, 1285
INGRESSION, 2236
INHOMOGENEITE, 2238
INJECTION, 2239
INJECTION FAILLE, 4162
INJECTION FLUIDE, 1580
INLANDSIS, 2160
INOCERAMI, 2246
INORGANIQUE, 2247
INSECTA, 2251
INSECTIVORA, 2252
INSELBERG, 2253
INSTALLATION MARINE, 2680
INSTALLATION SOUS MARINE, 4328
INSTITUTION, 2257
INSTRUMENT GEOPHYSIQUE, 1753
INSTRUMENTATION, 2258
INTENSITE SISMIQUE, 1286
INTERCALATION, 2260
INTERCOUCHE, 2268
INTERFACE, 2261
INTERFACE AIR MER, 94
INTERFACE HUILE EAU, 3114
INTERFACE HUILE GAZ, 3113
INTERFACE SEDIMENT EAU, 3994
INTERFERENCE, 2262
INTERFEROMETRIE, 2263
INTERFEROMETRIE ONDE RADIO, 3722
INTERFLUVE, 2264
INTERFORMATIONNEL, 2265
INTERNIDES, 2279
INTERPLUVIAL, 2281
INTERPOLATION, 2282
INTERPRETATION, 2283
INTRACLASTE, 2288
INTRUSIF, 2294
INTRUSION, 2293
INTRUSION COUCHE, 4066
INTRUSION EAU SALEE, 3930
INVENTAIRE, 2295
INVERSION, 3842
INVERSION CHAMP, 1743
INVERSION RELIEF, 3810
INVERTEBRATA, 2297
IO-TH, 2302
IO-U, 2303
IODE, 2298
ION, 2299
IONISATION, 2304
IONOSPHERE, 2305
IRIDIUM, 2307
IRISATION, 2306
IRREVERSIBILITE, 2313

IRRIGATION, 2314
ISANOMALE, 2318
ISOBARE, 2319
ISOBASE, 2320
ISOBATHE, 2321
ISOCHORE, 4277
ISOCHRONE, 2323
ISOCLINE, 2325
ISOGONE, 2327
ISOGRADE, 2328
ISOHYPSE, 2329
ISOLANT, 2259
ISOLEMENT, 2330
ISOLITHE, 2331
ISOMORPHISME, 2334
ISOPAQUE, 2335
ISOPLETHE, 2337
ISOPLEURA, 2338
ISOPRENOIDE, 2339
ISORAD, 2340
ISORAT, 2341
ISOSEISTE, 2342
ISOSTASIE, 2344
ISOSTRUCTURE, 2345
ISOTHERME, 2346
ISOTOPE, 2348
ISOTOPE LEGER, 2500
ISOTOPE LOURD, 1971
ISOTOPE RADIOACTIF, 3727
ISOTOPE STABLE, 4228
ISOTROPIE, 2354
ITABIRITE, 2355
ITACOLUMITE, 2356
JADEITITE, 2357
JASPE, 2359
JAUGE DEFORMATION, 4268
JAUGEAGE, 1698
JEUNE, 4850
JONCTION TRIPLE LITHOSPHERE, 4599
JURASSIQUE, 2363
JURASSIQUE INF, 2583
JURASSIQUE MOYEN, 2844
JURASSIQUE SUP, 4673
JUVENILE, 2364
K-AR, 3580
K-CA, 3581
KAME, 2366
KANGA, 2367
KAOLIN, 2368
KAOLINISATION, 2369
KARROO, 2372
KARST, 2373
KASIMOVIEN, 2382
KAZANIEN, 2381
KERATOPHYRE, 2385
KEROGENE, 2386
KEUPER, 2388

KIMBERLITE, 2391
KIMMERIDGIEN, 2392
KINK BAND, 2394
KINZIGITE, 2395
KOUNGOURIEN, 2400
KREEP, 2397
KRYPTON, 2398
KURTOSIS, 2401
KUTORGINIDA, 2402
KUYALNIK, 2399
LABYRINTHODONTIA, 2404
LAC, 2415
LAC ARTIFICIEL, 3825
LAC CRATERE, 982
LAC GLACIAIRE, 1797
LAC LAVE, 2458
LAC MORT, 1494
LAC SALE, 3927
LACCOLITE, 2405
LACERTILIA, 2406
LACUNE MISCIBILITE, 2894
LACUNE STRATIGRAPHIQUE, 1682
LADINIEN, 2410
LAGON BARRIERE, 419
LAGUNE, 2411
LAHAR, 2414
LAMARCKISME, 2417
LAMBEAU, 2396
LAME MINCE, 4514
LAMINATION, 2422
LAMPROPHYRE, 2423
LANDSAT, 2430
LANGAGE INFORMATIQUE, 871
LANTHANE, 2432
LAPIEZ, 2371
LAPILLI, 2433
LATERAL, 2441
LATERITE, 2446
LATERITISATION, 2448
LATITE, 2449
LATITUDE, 2450
LAURASIE, 2453
LAVE, 2454
LAVE AA, 2
LECTOTYPE, 2474
LEGENDE, 2475
LEGISLATION, 2476
LENIEN, 2479
LENTILLE, 2480
LEPERDITOCOPIDA, 2481
LEPIDOCYSTOIDEA, 2483
LEPIDOSAURIA, 2484
LEPOSPONDYLI, 2485
LEPTITE, 2486
LEPTOGEOSYNCLINAL, 2487
LEPTYNITE, 2488
LESSIVAGE, 2465

LESSIVAGE SOL, 4151
LEUCOSOME, 2490
LEVE ACOUSTIQUE, 44
LEVE AEROMAGNETIQUE, 74
LEVE AEROPORTE, 72
LEVE CARTE, 2670
LEVE ELECTRIQUE, 1333
LEVE ELECTROMAGNETIQUE, 1342
LEVE GEOPHYSIQUE, 1757
LEVE GEOTECHNIQUE, 1770
LEVE GRAVIMETRIQUE, 1886
LEVE HYDROGEOLOGIQUE, 2113
LEVE MAGNETIQUE, 2643
LEVE MAGNETOTELLURIQUE, 2655
LEVE RADIOMETRIQUE, 3738
LEVE SISMIQUE, 4019
LEVEE, 2491
LEVER, 4371
LEVRE RELEVEE, 4685
LEXIQUE, 2494
LIAISON INTERATOMIQUE, 364
LIANT, 2944
LIAS INF, 2584
LIAS MOYEN, 2845
LIAS SUP, 4674
LICHENES, 2495
LICHENOMETRIE, 2496
LICHIDA, 2497
LIGNE COTIERE, 816
LIGNE CROISSANCE, 1922
LIGNE EQUATORIALE, 1413
LIGNE ISOCONCENTRATION, 2326
LIGNE ISOPIEZE, 2336
LIGNE MARSHALL, 195
LIGNE PARTAGE EAUX, 4788
LIGNE RIVAGE, 4083
LIGNINE, 2503
LIGNITE, 2504
LIMITE ATTERBERG, 367
LIMITE ELASTICITE, 1318
LIMITE PLAQUE, 3497
LIMITE STRATIGRAPHIQUE, 4275
LIMNOLOGIE, 2508
LIMON, 2509
LIMONITE, 2510
LINEAMENT, 2511
LINEATION, 2512
LINEATION COURANT, 3310
LINGULIDA, 2513
LIPIDE, 2516
LIPOSTRACA, 2517
LIQUEFACTION, 2519
LIQUIDUS, 2522
LISERE BECKE, 451
LISSAGE, 4141
LISSAMPHIBIA, 2523
LISTE FAUNISTIQUE, 1517

LIT COURS EAU, 4285
LITAGE, 453
LITAGE PARALLELE, 3295
LITHIFICATION, 2524
LITHIUM, 2525
LITHOFACIES, 2526
LITHOGENESE, 2528
LITHOLOGIE, 2532
LITHOLOGIQUE, 2529
LITHOPHYTE, 2534
LITHOSOL, 2535
LITHOSPHERE, 2536
LITHOSTRATIGRAPHIE, 2538
LITHOSTROME, 2539
LITHOTOPE, 2540
LITHOTYPE, 2541
LITOPTERNA, 2542
LITTORAL, 2543
LITUOLACEA, 2547
LIVRET GUIDE, 1926
LIXIVIATION, 2550
LLANDEILO, 2551
LLANDOVERY, 2552
LLANVIRN, 2553
LOCAL, 2557
LOCALITE TYPE, 4624
LOCOMOTION, 2558
LODRANITE, 2560
LOESS, 2561
LOG, 2562
LOG ROUTIER, 3876
LOI, 2460
LOI CORRELATION FACIES, 2461
LOI DARCY, 1061
LOI HAECKEL, 1939
LOI HAUY, 1960
LOI HOOKE, 2048
LOI MATHEMATIQUE, 2712
LOI PHASES MINERALOGIQUES, 2879
LOI STATISTIQUE, 4241
LOISIR, 3769
LONGITUDE, 2565
LONGUEUR, 2477
LOPOLITE, 2569
LOUPE BOUE, 2957
LU-HF, 2608
LUDLOW, 2596
LUMACHELLE, 948
LUMINESCENCE, 2597
LUNAIRE, 2599
LUNE, 2935
LUTETIUM, 2607
LUVISOL, 2609
LYCOPSIDA, 2610
LYTOCERATIDA, 2611
MAAR, 2612
MACERAL, 2613

MACHAERIDIA, 2614
MACHOIRE, 2360
MACLE, 4621
MACLE BAVENO, 442
MACLE GLISSEMENT, 1820
MAESTRICHTIEN, 2615
MAGMA, 2617
MAGMA RESIDUEL, 3830
MAGMATIQUE, 2619
MAGMATISME, 2624
MAGNAFACIES, 2625
MAGNESITE, 2626
MAGNESIUM, 2627
MAGNETISME, 2646
MAGNETITE, 2647
MAGNETOHYDRODYNAMIQUE, 2649
MAGNETOMETRE, 2650
MAGNETOSPHERE, 2651
MAGNETOSTRATIGRAPHIE, 2652
MAGNETOSTRICTION, 2653
MAILLE CRISTALLINE, 1018
MAILLE ELEMENTAIRE, 4661
MAILLE GEOMETRIQUE, 3013
MAIN D'OEUVRE, 2664
MALACOSTRACA, 2659
MAMMALIA, 2661
MANGANESE, 2662
MANIFESTATION VOLCANIQUE, 4768
MANIPULATION DONNEE, 1068
MANTEAU, 2665
MANTEAU INF, 2585
MANTEAU SUP, 4675
MANTODEA, 2666
MANUEL, 2667
MAQUETTE, 3955
MARAIS, 4376
MARAIS LITTORAL, 2693
MARAIS MANGROVE, 2663
MARAIS MARIN, 2681
MARAIS SALE, 3928
MARBRE, 2671
MARCHE, 4566
MARECAGE, 536
MAREE, 4529
MAREE TERRESTRE, 1280
MARGE CONTINENTALE, 925
MARGE CONTINENTALE ACTIVE, 58
MARGE CONTINENTALE ASISMIQUE, 328
MARGE CONTINENTALE COMPLEXE, 920
MARMITE, 2387
MARMITE GEANT, 3585
MARNE, 2692
MARSUPIALIA, 2695
MASCON, 2696
MASSE, 2697
MASSIF, 2703
MASSIF FILTRANT, 1872

MASSIF LUNAIRE, 2605
MASSIF MEDIAN, 2732
MASSIF MONTAGNEUX, 2950
MASTIGOPHORA, 2706
MATERIAU, 2707
MATERIAU COHERENT, 829
MATERIAU COMPETENT, 859
MATERIAU CONSTRUCTION, 911
MATERIAU EXPANSE, 1474
MATERIAU HOTE, 2061
MATERIAU INCOMPETENT, 2203
MATERIAU NON COHERENT, 2202
MATERIAU NON CONSOLIDE, 4647
MATERIAU ORNEMENTATION, 3181
MATERIAU PLASTIQUE, 3493
MATERIAU SURCONSOLIDE, 3214
MATERIAU SYNTHETIQUE, 4402
MATERIAU TERRIGENE, 4469
MATERIAU VIABILITE, 3877
MATERIEL SONDAGE, 1255
MATHEMATIQUE, 2716
MATIERE AMORPHE, 170
MATIERE DISSOUTE, 1229
MATIERE EN SUSPENSION, 4372
MATIERE FONDUE, 2743
MATIERE MINERALE, 2249
MATIERE ORANGE LUNE, 3144
MATIERE ORGANIQUE, 3170
MATIERE PARTICULAIRE, 3307
MATIERE PREMIERE, 3755
MATIERE SOLUTION, 4176
MATIERE VISQUEUSE, 4746
MATIERE VOLATILE, 4757
MATRICE, 3887
MATRICE MATHEMATIQUE, 2713
MATTE ALGAIRE, 115
MATURATION, 2718
MAUVAISES TERRES, 403
MAXILLOPODA, 2721
MAXIMUM, 2722
MAXIMUM ANOMALIE, 2006
MAYIEN, 2656
MEANDRE, 2724
MECANIQUE, 2729
MECANIQUE ROCHE, 3888
MECANIQUE SOL, 4154
MECANISME, 2730
MECANISME ECOULEMENT, 1581
MECANISME FOCAL, 1600
MECOPTEROIDA, 2731
MEDECINE, 2733
MEGAFACIES, 2738
MEGASPORE, 2739
MELANGE, 2740
MEMBRE, 2746
MENDELEVIUM, 2747
MEOTIEN, 2748

MER, 3974
MER BORDIERE, 2674
MER GEOSYNCLINALE, 1764
MER LUNAIRE, 2672
MER PLATEAU CONTINENTAL, 4069
MERCURE, 2749
MEROSTOMATA, 2750
MEROSTOMOIDEA, 2751
MESOGASTROPODA, 2754
MESOGEE, 2753
MESOGEOSYNCLINAL, 2755
MESOSAURIA, 2757
MESOSIDERITE, 2758
MESOSTASE, 1919
MESURE AU SOL, 1906
METABASALTE, 2762
METACOPINA, 2763
METAGABBRO, 2764
METAGENESE, 2765
METAL ALCALIN, 126
METAL ALCALINO TERREUX, 129
METAL BASE, 428
METAL LOURD, 1972
METAL NON FERREUX, 3047
METAL PRECIEUX, 3037
METAL RARE, 3752
METAL TRACE, 4558
METALLIFERE, 2769
METALLOGENIE, 2774
METALLOGENIE PREVISIONNELLE, 3599
METALLOIDE, 2775
METALLOTECTE, 2776
METALLURGIE, 2777
METAMICTE, 2778
METAMICTISATION, 2779
METAMORPHISME, 2787
METAMORPHISME AUTO, 383
METAMORPHISME CATAZONAL, 2380
METAMORPHISME CHOC, 4077
METAMORPHISME CONTACT, 915
METAMORPHISME DYNAMO, 1266
METAMORPHISME DYNAMOTHERMAL, 1269
METAMORPHISME ENFOUISSEMENT, 597
METAMORPHISME EPIZONAL, 1408
METAMORPHISME FAIBLE, 2575
METAMORPHISME FORT, 2011
METAMORPHISME GEOTHERMIQUE, 1776
METAMORPHISME IMPACT, 2189
METAMORPHISME ISOCHIMIQUE, 2322
METAMORPHISME MESOZONAL, 2761
METAMORPHISME PLUTONIQUE, 3518
METAMORPHISME POLY, 3552
METAMORPHISME PROGRESSIF, 3633
METAMORPHISME REGIONAL, 3795
METAMORPHISME RETRO, 3841
METAMORPHISME THERMIQUE, 4498
METAMORPHOSE, 2789

METANAUPLIUS, 2790
METAPELITE, 3336
METAPHYTE, 2791
METASOMATOSE, 2794
METASOMATOSE ALCALINE, 130
METASOMATOSE CONTACT, 916
METASOME, 2793
METATEXIE, 2796
METEORITE, 2798
METEORITE METALLIQUE, 2767
METEORITE PIERREUSE, 4264
METEOROIDE, 2799
METEOROLOGIE, 2800
METHANE, 2801
METHODE ACOUSTIQUE, 43
METHODE AEROPORTEE, 95
METHODE AFMAG, 78
METHODE ANALYSE ELEMENT TRACE, 4559
METHODE ANALYSE MAJEURS, 2658
METHODE ANALYSE MINEURS, 2888
METHODE COURANT TELLURIQUE, 1274
METHODE DIFFERENCE FINIE, 1550
METHODE ELECTRIQUE, 1329
METHODE ELECTROMAGNETIQUE, 1341
METHODE ELEMENT FINI, 1551
METHODE GEOPHYSIQUE, 1755
METHODE GEOTHERMIQUE, 1777
METHODE GRAVIMETRIQUE, 1883
METHODE INDUCTION, 2217
METHODE LASER, 2437
METHODE MAGNETIQUE, 2637
METHODE MAGNETOTELLURIQUE, 2654
METHODE MATHEMATIQUE, 2714
METHODE MEB, 3957
METHODE MET, 4575
METHODE MICROONDE, 2838
METHODE MONTE CARLO, 2932
METHODE NOUVELLE, 3022
METHODE PAR IMMERSION, 2184
METHODE RADAR, 3718
METHODE RADIOFREQUENCE, 3724
METHODE RADIOMETRIQUE, 3732
METHODE RESISTIVITE, 3833
METHODE SISMIQUE, 4014
METHODE STANDARD, 4236
METHODE STATISTIQUE, 4239
METHODE TELLURIQUE, 4449
METHODE TRACTION FOND, 1092
METHODE TRANSITOIRE, 4573
METHODE ULTRASON, 4639
METHODE VOIE HUMIDE, 4821
METHODOLOGIE, 2802
METHODOLOGIE ANALYSE, 183
METROPOLITAIN, 4344
MICA, 2804
MICASCHISTE, 2805
MICRITE, 2806

MICROCRATERE, 2808
MICRODURETE, 2820
MICROFACIES, 2811
MICROFALAISE PLAGE, 447
MICROFAUNE, 2812
MICROFISSURE, 2813
MICROFLORE, 2814
MICROFOSSILE, 2816
MICROGRANITE, 2818
MICROLATEROLOG, 2821
MICROLITE, 2822
MICROLOG, 2823
MICROMETEORITE, 2824
MICROMORPHOLOGIE SOL, 2825
MICROORDINATEUR, 2807
MICROORGANISME, 2826
MICROPALEONTOLOGIE, 2827
MICROPHYTOLITHI, 2828
MICROPLAQUE, 2829
MICROPLI, 2815
MICROSCOPE, 2831
MICROSCOPE ELECTRONIQUE, 1348
MICROSCOPIE, 2833
MICROSCOPIE ELECTRONIQUE, 1349
MICROSCOPIE MINERAI, 3158
MICROSEISME, 2834
MICROSONDE ELECTRONIQUE, 1347
MICROSONDE IONIQUE, 2301
MICROSTRUCTURE, 4640
MICROSTYLOLITE, 2835
MICROTECTITE, 2837
MICROTECTONIQUE, 2836
MICROTREMBLEMENT DE TERRE, 2810
MIGMATISATION, 2854
MIGMATITE, 2853
MIGRATION, 2855
MIGRATION ELEMENT, 2856
MILIEU, 1386
MILIEU ABYSSAL, 19
MILIEU AEROBIE, 73
MILIEU ALPIN, 149
MILIEU ANAEROBIE, 178
MILIEU ANELASTIQUE, 197
MILIEU ARIDE, 299
MILIEU ATMOSPHERIQUE, 4316
MILIEU BASSE ENERGIE, 2574
MILIEU BATHYAL, 436
MILIEU BENTHIQUE, 464
MILIEU CONTINENTAL, 923
MILIEU DELTAIQUE, 1104
MILIEU EAU DOUCE, 1648
MILIEU EAU PEU PROFONDE, 4053
MILIEU EAU PROFONDE, 1093
MILIEU ELASTIQUE, 1319
MILIEU EOLIEN, 1391
MILIEU ESTUAIRE, 1436
MILIEU EUXINIQUE, 1456

MILIEU FLUVIATILE, 1593
MILIEU FROID, 832
MILIEU GLACIAIRE, 1793
MILIEU HAUTE ENERGIE, 2009
MILIEU HUMIDE, 2074
MILIEU INTERGLACIAIRE, 2266
MILIEU INTERTIDAL, 2286
MILIEU LACUSTRE, 2407
MILIEU LAGUNAIRE, 2412
MILIEU LITTORAL, 2545
MILIEU MARGE CONTINENTALE, 4068
MILIEU MARIN, 2677
MILIEU MER PROFONDE, 1090
MILIEU PARALIQUE, 3294
MILIEU PELAGIQUE, 3330
MILIEU PERIGLACIAIRE, 3355
MILIEU POREUX, 3566
MILIEU RECIFAL, 3777
MILIEU SALIN, 1945
MILIEU SAUMATRE, 569
MILIEU SEMI ARIDE, 4035
MILIEU STRATIFIE, 2464
MILIEU SUBLITTORAL, 4325
MILIEU SUBTIDAL, 4341
MILIEU TALUS MARIN, 4136
MILIEU TERRESTRE, 4467
MILIEU TIDAL, 4526
MILIOLACEA, 2857
MILIOLINA, 2858
MINE, 2861
MINE DU PLATINE, 3505
MINE SOUTERRAINE, 4650
MINERAI, 3151
MINERAI DRAGAGE, 1244
MINERAI PISOLITIQUE, 449
MINERAI PROBABLE, 2225
MINERAI PROUVE, 3571
MINERAL, 2862
MINERAL ALLONGE, 1355
MINERAL ALLOTHIGENE, 136
MINERAL ANTISTRESS, 232
MINERAL AUTHIGENE, 376
MINERAL BIAXE, 476
MINERAL COEXISTANT, 826
MINERAL DIPLOIDE, 1201
MINERAL ECONOMIQUE, 1300
MINERAL ESSENTIEL, 3890
MINERAL GEL, 1703
MINERAL INTERSTRATIFIE, 2898
MINERAL LEGER, 2501
MINERAL LOURD, 1973
MINERAL MAGNETIQUE, 2638
MINERAL ORGANIQUE, 3167
MINERAL REPERE, 2208
MINERAL SECONDAIRE, 3989
MINERAL STRESS, 4291
MINERAL TYPOMORPHIQUE, 4627

MINERAL URANIUM, 4690
MINERAL XENOMORPHE, 203
MINERALISATION, 2876
MINERALISATION LITTORALE, 445
MINERALOGIE, 2880
MINERALOGIE ARGILE, 789
MINERAUX ACCESSOIRES, 29
MINIMUM, 2882
MINIORDINATEUR, 2881
MIOCENE, 2889
MIOCENE INF, 2586
MIOCENE MOYEN, 2846
MIOCENE SUP, 4676
MIOGEOSYNCLINAL, 2890
MIOMERA, 2891
MIOSPORE, 2892
MIROIR FAILLE, 4132
MISCIBILITE, 2893
MISE A LA MASSE, 2895
MISE EN CHARGE CYCLIQUE, 1046
MISE EN PLACE, 1363
MISSISSIPPIEN, 2897
MOBILISATION GEOCHIMIQUE, 2901
MOBILISME, 1225
MOBILITE, 2900
MODE PETROGRAPHIQUE, 2903
MODELE, 2904
MODELE 1 DIMENSION, 3120
MODELE 2 DIMENSIONS, 4622
MODELE DETERMINISTE, 1141
MODELE HYDRODYNAMIQUE, 2103
MODELE MATHEMATIQUE, 2715
MODELE PHYSIQUE, 3426
MODELE STATISTIQUE, 4242
MODIOMORPHOIDA, 2906
MODULE, 2907
MODULE CISAILLEMENT, 4060
MODULE ELASTICITE, 591
MODULE RIGIDITE, 2908
MODULE YOUNG, 4851
MOHOROVICIC, 2910
MOINDRES CARRES, 2473
MOLASSE, 2912
MOLDAVITE, 2913
MOLLISOL, 2915
MOLLUSCA, 2916
MOLYBDATE, 2917
MOLYBDENE, 2918
MOMENT DIPOLAIRE, 1206
MONAZITE, 2919
MONOCLINAL, 2920
MONOCOTYLEDONES, 2922
MONOCYATHEA, 2923
MONOGRAPHIE, 2924
MONOGRAPHIE SOMMAIRE, 4357
MONOLITHE, 2926
MONOMORPHE, 2927

MONOPLACOPHORA, 2928
MONORHINA, 2929
MONOTREMATA, 2930
MONZONITE, 2933
MORAINE, 2938
MORAINE FOND, 1907
MORAINE LATERALE, 2444
MORPHOGENESE, 2939
MORPHOLOGIE, 2940
MORPHOLOGIE COTE, 4081
MORPHOLOGIE COTIERE, 814
MORPHOLOGIE DISSOLUTION, 4178
MORPHOLOGIE EOLIENNE, 1392
MORPHOLOGIE EROSION, 1427
MORPHOLOGIE FONCTIONNELLE, 1666
MORPHOLOGIE GITE, 1119
MORPHOLOGIE GLACIAIRE, 1795
MORPHOLOGIE PERIGLACIAIRE, 1654
MORPHOLOGIE SOUS MARINE, 4329
MORPHOMETRIE, 2941
MORPHOSCOPIE, 2942
MORTS TERRAINS, 3213
MOSCOVIEN, 2946
MOULAGE, 665
MOULE INTERNE, 2274
MOUSSON, 2931
MOUVEMENT, 2951
MOUVEMENT GLACE, 2158
MOUVEMENT HORIZONTAL, 2053
MOUVEMENT MASSE, 2699
MOUVEMENT MATIERE, 2702
MOUVEMENT SOL, 1908
MOUVEMENT VERTICAL, 4736
MOYEN TRANSPORT, 4578
MOYENNE, 389
MULTIPHASE, 2963
MULTITUBERCULATA, 2967
MUSCHELKALK, 2969
MUSEE, 2970
MUTATION, 2971
MYCETOZOA, 2982
MYLONITE, 2972
MYLONITISATION, 2974
MYODOCOPINA, 2975
MYOIDA, 2976
MYOMORPHA, 2977
MYRIAPODA, 2978
MYRMEKITE, 2979
MYSTACOCARIDA, 2980
MYTILOIDA, 2981
NAKHLITE, 2983
NAMURIEN, 2984
NANOFOSSILE, 2985
NAPPE, 2986
NAPPE ALLUVION, 143
NAPPE ARTESIENNE, 313
NAPPE CAPTIVE, 887

NAPPE EAU, 260
NAPPE LIBRE, 4644
NAPPE PERCHEE, 3349
NAPPE PHREATIQUE, 3416
NATANTES, 2987
NAUTILOIDEA, 2993
NAVIRE SONDAGE, 1257
NEANDERTHALIEN, 2994
NEBULEUSE, 968
NECK, 2995
NECTON, 2997
NEIGE, 4142
NEMERTA, 2998
NEOCOMIEN, 2999
NEODYME, 3000
NEOGASTROPODA, 3001
NEOGENE, 3002
NEON, 3003
NEORNITHES, 3004
NEOTECTONIQUE, 3005
NEOTENIE, 3006
NEOTYPE, 3007
NEOVOLCANIQUE, 687
NEPHELINE, 3008
NEPTUNISME, 3010
NEPTUNIUM, 3011
NEUTRON, 3014
NEVE, 1553
NICKEL, 3026
NILSSONIALES, 3028
NIOBATE, 3029
NIOBIUM, 3030
NIOBOTANTALATE, 3031
NITRATE, 3032
NITRIFICATION, 3033
NIVATION, 3036
NIVEAU, 2493
NIVEAU EAU SOUTERRAINE, 1916
NIVEAU HYDROSTATIQUE, 2128
NIVEAU LAC, 2416
NIVEAU MER, 3977
NIVEAU REPERE, 2389
NIVELLEMENT, 2492
NODOSARIACEA, 3038
NODULE, 3039
NOEGGERATHIALES, 3040
NOM NOUVEAU, 3024
NOMENCLATURE, 3042
NOMOGENESE, 3044
NORD, 3057
NORIEN, 3053
NORMALISATION, 4238
NORME PETROGRAPHIQUE, 3054
NOSE, 3058
NOTICE EXPLICATIVE, 1480
NOTOSTRACA, 3059
NOTOUNGULATA, 3060

NOUVEAU, 3019
NOUVELLE DESCRIPTION, 3021
NOUVELLE DONNEE, 3020
NOUVELLE DONNEE MINERALE, 3023
NOVACULITE, 3061
NOYAU CONTINENTAL, 926
NOYAU CRISTALLISATION, 3067
NOYAU TERRESTRE, 957
NOYAU TERRESTRE EXTERNE, 3207
NOYAU TERRESTRE INTERNE, 2243
NUCLEATION, 3066
NUCULOIDEA, 3068
NUEE ARDENTE, 3069
NUMMULITIDAE, 3071
NUTATION, 3072
NUTATION CHANDLER, 706
NUTRITION, 3073
O 18=O 16, 3227
OASIS, 3074
OBDUCTION, 3075
OBJECTIF, 3076
OBLIQUITE, 4128
OBOLELLIDA, 3079
OBSERVATION GEOPHYSIQUE, 3081
OBSERVATOIRE, 3082
OBSIDIENNE, 3083
OCEAN, 3084
OCEAN DRILLING PROJECT, 3087
OCEANIQUE, 3090
OCEANISATION, 3092
OCEANOGRAPHIE, 3093
OCEANOLOGIE, 3094
OCHREPT, 3096
OCRE, 3095
OCTAHEDRITE, 3097
OCTOCORALLA, 3098
ODONTOLOGIE, 3099
ODONTOPLEURIDA, 3100
ODONTORNITHES, 3101
OEUF, 1311
OFFSHORE, 3102
OLIGOCENE, 3115
OLIGOCENE INF, 2587
OLIGOCENE MOYEN, 2847
OLIGOCENE SUP, 4677
OLISTOLITHE, 3116
OLISTOSTROME, 3117
OLIVINE, 3118
ONCOLITE, 3119
ONDE, 4805
ONDE ACOUSTIQUE, 45
ONDE AIRY, 97
ONDE CHOC, 4078
ONDE ELASTIQUE, 1322
ONDE LONGUE PERIODE, 2564
ONDE LOVE, 2570
ONDE P, 3230

ONDE PRECURSEUR, 1615
ONDE RAYLEIGH, 3756
ONDE S, 3920
ONDE SABLE, 3941
ONDE SH, 4050
ONDE SISMIQUE, 4021
ONDE SURFACE, 4368
ONDE VOLUME, 535
ONTOGENIE, 3121
ONYCHOPHORA, 3122
OOIDE, 3123
OOLITE, 3124
OPAQUE, 3128
OPHIDIA, 3131
OPHIOCISTIOIDEA, 3132
OPHIOLITE, 3133
OPHIUROIDEA, 3135
OPISTHOBRANCHIA, 3136
OPOKA, 3137
OPTIQUE CRISTAL, 1019
OR, 1833
ORAGE MAGNETIQUE, 2642
ORBITE, 3146
ORBITOIDACEA, 3147
ORBITOIDIDAE, 3148
ORDINATEUR, 870
ORDOVICIEN, 3150
ORDOVICIEN INF, 2588
ORDOVICIEN MOYEN, 2848
ORDOVICIEN SUP, 4678
ORDRE DESORDRE, 3149
ORGANIQUE, 3164
ORGANISME, 3176
ORGANISME AUTOTROPHE, 386
ORGANISME EURYHALIN, 1449
ORGANISME EURYTHERME, 1450
ORGANISME HALOPHILE, 1948
ORGANISME HETEROTROPHE, 1998
ORGANISME HOMEOMORPHE, 2333
ORGANISME MESOPHYLE, 2756
ORGANISME PERFORANT, 3889
ORGANISME PHAGOTROPHE, 3389
ORGANISME PHYTOPHAGUE, 3436
ORIENTATION, 3177
ORIENTATION OBLIQUE, 3078
ORIENTATION PREFERENTIELLE, 3600
ORIENTATION VERTICALE, 4735
ORIGINE, 3179
ORIGINE ATMOGENIQUE, 357
ORIGINE BACTERIOGENE, 401
ORIGINE BIOGENE, 3172
ORIGINE CHIMIQUE, 736
ORIGINE EPIMAGMATIQUE, 1405
ORIGINE INORGANIQUE, 2250
ORIGINE MAGMATIQUE, 2623
ORIGINE MECANIQUE, 2726
ORIGINE SUPRACRUSTALE, 4362

ORIGINE SYNSEDIMENTAIRE, 4397
ORIGINE VOLCANIQUE, 4765
ORNEMENTATION TEST, 3182
ORNITHISCHIA, 3183
OROGENESE, 3185
OROGENIE ALGOMIENNE, 120
OROGENIE ALLEGHENY, 133
OROGENIE ALPINE, 150
OROGENIE ANTECAMBRIENNE, 3594
OROGENIE ARMORICAINE, 304
OROGENIE ASSYNTIENNE, 342
OROGENIE BELTIENNE, 460
OROGENIE CALEDONIENNE, 618
OROGENIE CARELIENNE, 2370
OROGENIE CHARNIENNE, 711
OROGENIE CIMMERIENNE, 768
OROGENIE EBURNEENNE, 1289
OROGENIE GRENVILLE, 1898
OROGENIE HERCYNIENNE, 1984
OROGENIE HUDSONIENNE, 2068
OROGENIE KATANGIENNE, 2378
OROGENIE KENORIENNE, 2384
OROGENIE KIBARIENNE, 2390
OROGENIE LARAMIENNE, 2435
OROGENIE PENOKEENNE, 3343
OROGENIE SUDETE, 4346
OROGENIE SVECOFENNIENNE, 4375
OROGENIE TACONIQUE, 4407
OROGENIE VARISQUE, 4719
ORTHENT, 3186
ORTHID, 3187
ORTHIDA, 3188
ORTHO AMPHIBOLITE, 3189
ORTHO GNEISS, 3192
ORTHOD, 3190
ORTHOGENESE, 3191
ORTHOX, 3195
ORYCTOCOENOSE, 3196
OSCILLATION, 3197
OSCILLATION PROPRE, 1642
OSMIUM, 3199
OSTEICHTHYES, 3200
OSTEOLOGIE, 3201
OSTEOSTRACI, 3202
OSTRACODA, 3203
OSTREACEA, 3204
OTOLITE, 3205
OUED, 4779
OUEST, 4819
OURALITISATION, 4688
OUVERTURE, 235
OUVRAGE, 1375
OUVRAGE ASISMIQUE, 327
OXFORDIEN, 3220
OXISOL, 3224
OXYDATION, 3221
OXYDE, 3222

OXYDE CARBONE, 644
OXYGENE, 3225
OXYLOPHYTE, 3228
PA-IO, 3643
PAHOEHOE, 3233
PALAEOHETERODONTA, 3235
PALAEOTAXODONTA, 3237
PALAGONITISATION, 3238
PALEO, 3239
PALEOATMOSPHERE, 3240
PALEOBATHYMETRIE, 3242
PALEOBIOLOGIE, 3243
PALEOBOTANIQUE, 3245
PALEOCENE, 3246
PALEOCIRCULATION, 3247
PALEOCLIMAT, 3248
PALEOCLIMATOLOGIE, 3249
PALEOCOPIDA, 3250
PALEOCOURANT, 3251
PALEOECOLOGIE, 3252
PALEOENVIRONNEMENT, 3253
PALEOGENE, 3254
PALEOGEOGRAPHIE, 3257
PALEOKARST, 3259
PALEOLATITUDE, 3260
PALEOLIMNOLOGIE, 3261
PALEOMAGNETISME, 3263
PALEOMORPHOLOGIE, 3258
PALEONTOLOGIE, 3265
PALEOOCEANOGRAPHIE, 3241
PALEOPHYTIQUE, 3236
PALEORELIEF, 3266
PALEOSALINITE, 3267
PALEOSOL, 3268
PALEOTEMPERATURE, 3269
PALEOTHERMOMETRIE, 3270
PALEOZOOLOGIE, 3272
PALINGENESE, 3274
PALLADIUM, 3275
PALLASITE, 3276
PALSE, 3277
PALYNODIAGRAMME, 3540
PALYNOLOGIE, 3278
PALYNOMORPHE, 3279
PAMPA, 3280
PANACHE, 3515
PANGEE, 3281
PANTODONTA, 3284
PARA AMPHIBOLITE, 3287
PARA GNEISS, 3292
PARA-, 3286
PARABLASTOIDEA, 3288
PARACRINOIDEA, 3289
PARAGENESE, 3290
PARAGEOSYNCLINAL, 3291
PARALIAGEOSYNCLINAL, 3293
PARALLELISME, 3297

PARAMETRE, 3299
PARASITISME, 3300
PARATYPE, 3301
PARTICULE COSMIQUE, 969
PARTICULE ELEMENTAIRE, 1353
PASSE INTERTIDALE, 4528
PATHOLOGIE, 3315
PATTERN, 3316
PAUROPODA, 3318
PAUVRE, 2471
PB=PB, 2468
PECTINACEA, 3323
PEDIMENT, 3324
PEDIPLAINE, 3325
PEDOGENESE, 3326
PEDOLOGIE, 4157
PEGMATITE, 3327
PEGMATITIQUE, 3328
PEGMATOIDE, 3329
PELEEN, 3333
PELITE, 3334
PELOTE, 3337
PELOTE FECALE, 1521
PENDAGE, 1195
PENDAGEMETRE, 1197
PENDAGEMETRIE, 1204
PENEPLAINE, 3339
PENEPLANATION, 3468
PENETROMETRE, 3341
PENETROMETRIE, 3340
PENNSYLVANIEN, 3342
PENTAMERIDA, 3344
PENTASTOMIDA, 3345
PENTOXYLALES, 3346
PEPITE, 3070
PERCOLATION, 3350
PERGELISOL, 3362
PERICLINAL, 3351
PERIDOTITE, 3352
PERIGEE, 3353
PERIGLACIAIRE, 3354
PERIMETRE PROTECTION, 3644
PERIODE, 3356
PERIODICITE, 3357
PERISPHINCTIDA, 3358
PERISSODACTYLA, 3359
PERLE, 3319
PERLITE, 3360
PERMEABILITE, 3363
PERMEABILITE MAGNETIQUE, 2639
PERMIEN, 3364
PERMIEN INF, 2589
PERMIEN SUP, 4679
PERMIS EXPLORATION, 779
PERSPECTIVE, 3366
PERSPECTIVE EXPLOITATION, 3572
PERTHITE, 3367

PESANTEUR ABSOLUE, 11
PESTICIDE, 3369
PETROCHIMIE, 3372
PETROFABRIQUE, 3373
PETROGENESE, 3374
PETROGRAPHIE, 3378
PETROLE, 3379
PETROLE LOURD, 1974
PETROLOGIE, 3382
PETROLOGIE EXPERIMENTALE, 1477
PETROLOGIE ROCHE SEDIMENTAIRE, 3998
PETROLOGIE STRUCTURALE, 4309
PETROPHYSIQUE, 3383
PH, 3384
PHACOLITE, 3385
PHACOPIDA, 3386
PHAEOPHYCOPHYTA, 3387
PHAEOZEM, 3388
PHANEROZOIQUE, 3390
PHASE, 3391
PHASE ACADIENNE, 24
PHASE BAIKALIENNE, 404
PHASE FLUIDE, 1582
PHASE GAZEUSE, 1694
PHASE ISOCHORE, 3393
PHASE LIQUIDE, 2520
PHASE MINERALE, 2871
PHASE SOLIDE, 4166
PHENOCRISTAL, 3395
PHENOL, 3396
PHILOSOPHIE, 3398
PHOLADOMYIDA, 3399
PHOLIDOTA, 3400
PHONOLITE, 3401
PHOSPHATE, 3403
PHOSPHATISATION, 3404
PHOSPHORE, 3407
PHOSPHORESCENCE, 3405
PHOSPHORITE, 3406
PHOTO IR, 2232
PHOTOGEOLOGIE, 3410
PHOTOGRAMMETRIE, 3411
PHOTOGRAPHIE, 3412
PHOTOGRAPHIE AERIENNE, 71
PHOTOMETRIE, 3413
PHOTOMETRIE FLAMME, 1562
PHOTOSYNTHESE, 3414
PHREATOPHYTE, 3417
PHYLACTOLAEMATA, 3418
PHYLLITE, 3419
PHYLLOCERATIDA, 3420
PHYLLONITE, 3421
PHYLLOPODA, 3422
PHYLOGENIE, 3423
PHYLUM, 3424
PHYSIOLOGIE, 3432
PHYSIQUE GLOBE, 4169

PHYTANE, 3433
PHYTOCENOSE, 3434
PHYTOLITHES, 3435
PHYTOPLANCTON, 3437
PICRITE, 3438
PIEDMONT, 3439
PIEGE, 4580
PIEGE A PETROLE, 3111
PIEGE LITHOLOGIQUE, 2531
PIEGE SEDIMENTAIRE, 4002
PIEGE STRATIGRAPHIQUE, 4279
PIEGE STRUCTURAL, 4311
PIERRE ORNEMENTALE, 1085
PIERRE PRECIEUSE, 3595
PIERRE SEMIPRECIEUSE, 4036
PIERRE TAILLE, 588
PIEU, 3445
PIEZOCRISTALLISATION, 3441
PIEZOMETRIE, 3443
PIGMENT, 3444
PILIER, 3446
PILLOW LAVA, 3447
PINGO, 3450
PINNIPEDIA, 3452
PIPE, 3453
PIPELINE, 3454
PISCES, 3456
PISOLITE, 3457
PISTE, 4565
PLACANTICLINAL, 3461
PLACENTALIA, 3462
PLACER, 3463
PLACODERMI, 3464
PLAGE, 444
PLAGE BARRIERE, 418
PLAINE, 3465
PLAINE ABYSSALE, 22
PLAINE ACCUMULATION, 33
PLAINE ALLUVIALE, 144
PLAINE COTIERE, 815
PLAINE FLUVIOGLACIAIRE, 3212
PLAINE INONDABLE, 1570
PLAINE MARINE, 2683
PLAN, 3470
PLAN AXIAL, 391
PLAN CISAILLEMENT, 4061
PLAN FAILLE, 1510
PLAN STRATIFICATION, 454
PLANCTON, 3487
PLANCTOSPHAEROIDEA, 3469
PLANETAIRE, 3482
PLANETE, 3471
PLANETE EXTERIEURE, 3208
PLANETE JUPITER, 3473
PLANETE MARS, 3474
PLANETE MERCURE, 3475
PLANETE NEPTUNE, 3476

PLANETE PLUTON, 3477
PLANETE SATURNE, 3478
PLANETE TERRE, 3472
PLANETE TERRESTRE, 4468
PLANETE URANUS, 3479
PLANETE VENUS, 3480
PLANETOLOGIE, 3485
PLANIFICATION, 3489
PLANIFICATION URBAINE, 4695
PLANOSOL, 3490
PLANTE, 3491
PLANTE HEKISTOTHERME, 1975
PLANTE HETEROSPOREE, 1997
PLANTE XEROPHILE, 4843
PLAQUE, 3495
PLASTICITE, 3494
PLATEAU, 3501
PLATEAU CONTINENTAL, 4067
PLATEAU LITTORAL GLACE, 2161
PLATEFORME, 3503
PLATEFORME CONTINENTALE, 929
PLATEFORME CONTINENTALE EXTERNE, 3209
PLATEFORME CONTINENTALE INTERNE, 2244
PLATEFORME GRAVITAIRE, 1884
PLATEFORME MARGINAL, 2673
PLATEFORME MARINE, 2684
PLATINE, 3504
PLATINE UNIVERSELLE, 4662
PLATYRRHINA, 3506
PLEISTOCENE, 3508
PLEISTOCENE INF, 2590
PLEISTOCENE MOYEN, 2849
PLEISTOCENE SUP, 4680
PLEOCHROISME, 3509
PLI, 1602
PLI ACCORDEON, 4859
PLI ANTIFORME, 227
PLI ASYMETRIQUE, 353
PLI CHAPE, 1248
PLI CONCENTRIQUE, 874
PLI CONJUGUE, 897
PLI COUCHE, 3771
PLI CYCLINDRIQUE, 1052
PLI DEVERSE, 3219
PLI DISHARMONIQUE, 1219
PLI EN CASCADE, 661
PLI EN CHEVRON, 741
PLI ENTEROLITHIQUE, 1380
PLI ESCALIER, 4249
PLI ETIREMENT, 1243
PLI FAILLE, 1220
PLI FLUAGE, 1577
PLI GRAVITAIRE, 1881
PLI INTRAFOLIAIRE, 3510
PLI ISOCLINAL, 2324
PLI PTYGMATIQUE, 3679
PLI SUPERPOSE, 4359

PLI SYMETRIQUE, 4382
PLI SYNFORME, 4393
PLIENSBACHIEN, 3511
PLIOCENE, 3512
PLIOCENE INF, 2591
PLIOCENE MOYEN, 2850
PLIOCENE SUP, 4681
PLISSEMENT, 1608
PLOMB, 2466
PLONGEMENT, 3516
PLUIE, 3742
PLUIE CENDRE, 331
PLUTON ROCHE IGNEE, 3517
PLUTONISME, 3520
PLUTONIUM, 3521
PLUVIAL, 3522
PODOCOPIDA, 3525
PODZOL, 3526
PODZOLUVISOL, 3527
POGONOPHORA, 3528
POIDS SPECIFIQUE, 4196
POINT CHAUD, 2065
POINT CURIE, 1033
POINT DEFORMATION, 4849
POLAIRE, 3533
POLARISATION, 3139
POLARISATION ELECTRIQUE, 1330
POLARISATION PROVOQUEE, 2216
POLARISATION SPONTANEE, 4033
POLAROGRAPHIE, 3536
POLE GEOMAGNETIQUE, 1742
POLE PALEOMAGNETIQUE, 3262
POLITIQUE, 3537
POLLEN, 3539
POLLUANT, 3541
POLLUTION, 3543
POLONIUM, 3544
POLYCHAETA, 3545
POLYEDRE, 3548
POLYGENIQUE, 3546
POLYMERA, 3549
POLYMERISATION, 3550
POLYMICTE, 3553
POLYMINERAL, 3554
POLYMORPHISME, 3555
POLYPLACOPHORA, 3557
POLYTYPISME, 3558
POMPAGE, 3684
POMPAGE ESSAI, 3685
PONCE, 3683
PONDERATION, 4812
PONT, 577
PONTIEN, 3559
POPULATION, 855
POPULATION ENDEMIQUE, 1368
POPULATION STATISTIQUE, 4243
PORCELANITE, 3561

PORIFERA, 3564
POROSITE, 3565
POROSITE CELLULAIRE, 682
PORPHYRY COPPER, 3569
PORT, 1954
PORTANCE, 450
PORTLANDIEN, 3570
POSTGLACIAIRE, 3573
POTABILITE, 3576
POTASSE, 3577
POTASSIUM, 3579
POTENTIEL ELECTRIQUE, 1325
POTENTIOMETRIE, 3584
POUSSIERE, 1263
POUSSIERE COSMIQUE, 967
POUSSIERE RADIOACTIVE, 3726
POUVOIR CALORIFIQUE, 1963
POUVOIR REFLECTEUR, 3782
POUVOIR REFRACTEUR, 3788
POUVOIR ROTATOIRE, 772
POUZZOLANE, 3589
PRAECARDIIDA, 3590
PRAIRIE, 1870
PRASEODYME, 3591
PRE-IMBRIEN, 3592
PRECIPITATION ATMOSPHERIQUE, 360
PRECIPITATION CHIMIQUE, 3596
PRECISION, 34
PREDATION, 3597
PREGLACIAIRE, 3601
PREHISTOIRE, 3602
PRENEANDERTHALIEN, 3604
PREPARATION, 3605
PRESSIOMETRIE, 2556
PRESSION, 3607
PRESSION BAROMETRIQUE, 361
PRESSION CAPILLAIRE, 635
PRESSION EAU, 4795
PRESSION FLUIDE, 1583
PRESSION GEOSTATIQUE, 1909
PRESSION HYDRAULIQUE, 2096
PRESSION HYDROSTATIQUE, 2129
PRESSION PARTIELLE, 3306
PRESSION PORES, 3562
PRESSION SOULEVEMENT, 4666
PRESSION TERRAIN, 1277
PREVISION, 3598
PREVISION PROBABILITE, 2709
PRIMAIRE, 3271
PRIMATES, 3613
PRIMITIF, 3614
PRINCIPE COSMOLOGIQUE, 974
PRIORITE, 3615
PRIX, 3610
PROBABILITE, 3617
PROBLEMATICA, 3619
PROBLEMATICA MICRO, 2830

PROBLEME INVERSE, 2296
PROBOSCIDEA, 3620
PROCEDURE, 3130
PROCELLARIEN, 3621
PROCESSUS, 3622
PROCESSUS ADIABATIQUE, 65
PROCESSUS ATECTONIQUE, 215
PROCESSUS CYCLIQUE, 1047
PROCESSUS ENDOGENE, 1370
PROCESSUS EXOGENE, 1471
PROCESSUS POST TECTONIQUE, 3575
PROCESSUS STOCHASTIQUE, 4256
PROCESSUS SYNTECTONIQUE, 4399
PROCHORDATA, 3623
PRODUCTION, 3624
PRODUIT CHARGE, 1544
PRODUIT DEPART, 3303
PRODUIT FINAL, 1367
PRODUIT INTERMEDIAIRE, 1074
PROFIL EN LONG, 2566
PROFIL EQUILIBRE, 3628
PROFIL GEOCHIMIQUE, 1720
PROFIL GEOPHYSIQUE, 1756
PROFIL SOL, 4155
PROFONDEUR, 1122
PROFONDEUR COMPENSATION CARBONATE, 1124
PROGANOSAURIA, 3630
PROGRAMME, 3634
PROGRAMME APOLLO, 243
PROGRAMME CORRELATION PICG, 2278
PROGRAMME LUNA, 2598
PROGRAMME MARINER, 2690
PROGRAMME ORDINATEUR, 872
PROGRAMME VIKING, 4741
PROGRESSION RIVAGE, 3632
PROJECTION GNOMONIQUE, 1832
PROJECTION VOLCANIQUE, 1315
PROJET, 3636
PROJET MOHOLE, 2909
PROMETHIUM, 3637
PRONOSTIC, 3631
PROPAGATION ONDE, 4807
PROPRIETE, 3638
PROPRIETE ACOUSTIQUE, 42
PROPRIETE CHIMIQUE, 729
PROPRIETE DIAMAGNETIQUE, 1159
PROPRIETE ELASTIQUE, 1320
PROPRIETE ELECTRIQUE, 1331
PROPRIETE ELECTROCHIMIQUE, 1336
PROPRIETE FERROMAGNETIQUE, 1536
PROPRIETE GEOTECHNIQUE, 1769
PROPRIETE MAGNETIQUE, 2640
PROPRIETE MECANIQUE, 2727
PROPRIETE METALLIQUE, 2768
PROPRIETE OPTIQUE, 3140
PROPRIETE PARAMAGNETIQUE, 3298
PROPRIETE PHYSICOCHIMIQUE, 3430

PROPRIETE PHYSIQUE, 3427
PROPRIETE PIEZOELECTRIQUE, 3442
PROPRIETE THERMIQUE, 4499
PROPRIETE THERMOCHIMIQUE, 4502
PROPRIETE THERMODYNAMIQUE, 4504
PROPYLITISATION, 3639
PROSIMII, 3640
PROSPECTION, 1483
PROSPECTION ALLUVIONNAIRE, 145
PROSPECTION BATEE, 3283
PROSPECTION BIOGEOCHIMIQUE, 495
PROSPECTION EAU SOUTERRAINE, 1915
PROSPECTION ELECTRIQUE, 1327
PROSPECTION GEOBOTANIQUE, 1714
PROSPECTION GEOCHIMIQUE, 1716
PROSPECTION GEOPHYSIQUE, 1752
PROSPECTION GRAVIMETRIQUE, 1879
PROSPECTION HYDROGEOCHIMIQUE, 2109
PROSPECTION MINIERE, 2868
PROSPECTION THERMIQUE, 4500
PROTECTION ENVIRONNEMENT, 905
PROTECTION RIVAGE, 4082
PROTEINE, 3645
PROTELYTROPTERA, 3646
PROTEROZOIDES, 3648
PROTEROZOIDES INF, 1273
PROTEROZOIDES SUP, 2439
PROTEROZOIQUE, 3647
PROTEROZOIQUE INF, 2592
PROTEROZOIQUE MOYEN, 2851
PROTEROZOIQUE SUP, 4682
PROTOACTINIUM, 3642
PROTOATLANTIQUE, 2154
PROTOCILIATA, 3649
PROTOLITHE, 3650
PROTOMEDUSAE, 3651
PROTON, 3652
PROTOZOA, 3654
PROTURA, 3655
PROVENANCE, 3656
PROVINCE FAUNISTIQUE, 1518
PROVINCE FLORISTIQUE, 1572
PROVINCE HYDROGEOLOGIQUE, 2112
PROVINCE METALLOGENIQUE, 2773
PROVINCE PALEOBOTANIQUE, 3244
PROVINCE PETROGRAPHIQUE, 3377
PROVINCE PHYSIOGRAPHIQUE, 3431
PROVINCIALITE, 3657
PSAMMENT, 3658
PSAMMITE, 3659
PSAMMITIQUE, 3660
PSAMMOFAUNE, 3661
PSEPHITIQUE, 3662
PSEUDOCONGLOMERAT, 3663
PSEUDOGLEY, 3664
PSEUDOMORPHOSE, 3665
PSEUDOTILLITE, 3666

PSILOCERATIDA, 3667
PSILOPSIDA, 3668
PSYCHROPHYTE, 3669
PTERIDOPHYLLA, 3670
PTERIDOPHYTA, 3671
PTERIDOSPERMAE, 3672
PTERIOIDA, 3673
PTEROBRANCHIA, 3674
PTEROMORPHIA, 3675
PTEROPODA, 3676
PTEROSAURIA, 3677
PTYCHOPARIIDA, 3678
PUBLICATION, 3681
PUITS, 3459
PUITS A GAZ, 1692
PUITS ARTESIEN, 314
PUITS EAU, 4801
PUITS OFFSHORE, 3104
PUITS PETROLE, 3112
PUITS PRODUCTION, 3627
PUITS RECONNAISSANCE, 4813
PULMONATA, 3682
PURINE, 3688
PYCNOCLINE, 3689
PYCNOGONIDA, 3690
PYCNOMETRE, 3691
PYCNOMETRIE, 3692
PYRITISATION, 3694
PYROCLASTIQUE, 3695
PYROCLASTIQUE SOUDE, 83
PYROLYSE, 3696
PYROMETALLURGIE, 3697
PYROMETAMORPHISME, 3698
PYROMETASOMATOSE, 3699
PYROSPHERE, 3700
PYROTHERIA, 3701
PYROXENE, 3702
PYROXENITE, 3704
PYRROPHYCOPHYTA, 3705
QUAI, 3440
QUALITE, 3707
QUALITE EAU, 4796
QUALITE MINERAI, 3160
QUARTZ, 3710
QUARTZITE, 3712
QUATERNAIRE, 3713
RACCOURCISSEMENT, 4084
RACEMISATION, 3717
RACINE, 3896
RADIAL, 3719
RADIATION ADAPTATIVE, 64
RADIOACTIVITE, 3731
RADIOCARBONE, 3733
RADIOGRAPHIE RX, 3734
RADIOLARIA, 3735
RADIOLARITE, 3736
RADIOMETRE, 3737

RADIUM, 3739
RADIUS, 3740
RADON, 3741
RAFFINAGE, 3781
RAJEUNISSEMENT, 3802
RANG RESEAU HYDROGRAPHIQUE, 4286
RANKER, 3749
RAPAKIVI, 3750
RAPPORT, 3818
RAPPORT ACTIVITE, 3635
RAPPORT CHIMIQUE, 730
RAPPORT GAZ=PETROLE, 1693
RAPPORT ISOTOPIQUE, 2352
RAPPORT SABLE=ARGILE, 3938
RATITES, 3754
RAVIN, 1928
RAYON HYDRAULIQUE, 2097
RAYON IR, 2233
RAYON UV, 4641
RAYON X, 4832
RAYONNEMENT, 3720
RAYONNEMENT ALPHA, 147
RAYONNEMENT BETA, 473
RAYONNEMENT COSMIQUE, 970
RAYONNEMENT GAMMA, 1676
RAYONNEMENT SOLAIRE, 4159
RE=OS, 3850
REACTIF, 3761
REACTION, 3757
REACTION ENDOTHERMIQUE, 1372
REACTION EXOTHERMIQUE, 1473
REACTION PHOTOCHIMIQUE, 3408
REACTIVATION, 3760
REAIMANTATION, 3811
REAMENAGEMENT, 3766
RECEPTACULITACEAE, 3763
RECHARGE NAPPE, 3765
RECHERCHE SCIENTIFIQUE, 3822
RECIF, 3776
RECIF ALGUE, 117
RECIF BARRIERE, 420
RECIF CORALLIAIRE, 949
RECIF FRANGEANT, 1651
RECIF INTERNE, 397
RECIF PINACLE, 3451
RECOLTE, 838
RECONSTITUTION PALEOGEOGRAPHIQUE, 3767
RECONSTITUTION PALEONTOLOGIQUE, 3264
RECRISTALLISATION, 3770
RECUPERATION, 3768
RECYCLAGE, 3772
REDLICHIIDA, 3775
REDUCTION CHIMIQUE, 731
REFLECTION BRAGG, 570
REFLECTION ONDE, 3783
REFLECTIVITE, 3785
REFRACTAIRE, 3791

REFRACTION ONDE, 3786
REFRACTOMETRE, 3789
REFRACTOMETRIE, 3790
REFROIDISSEMENT, 942
REGENERATION, 3792
REGIME CRITIQUE, 994
REGIME LAMINAIRE, 2420
REGIME PERMANENT, 4246
REGIME TRANSITOIRE, 4664
REGIME TURBULENT, 1304
REGLE PHASE, 3394
REGMAGLYPTE, 3797
REGOLITHE, 3798
REGOSOL, 3799
REGRESSION, 3800
REGRESSION STATISTIQUE, 3801
RELATION, 3803
RELAXATION ENERGIE, 3805
RELEVEMENT, 4686
RELICTE, 3806
RELIEF, 3808
RELIEF CONTINENT, 2429
RELIEF LUNAIRE, 2937
RELIEF MORPHOSTRUCTURAL, 2943
RELIEF SOUS MARIN, 3975
RELIEF STRUCTURAL, 4310
REMANIE, 3845
REMANIEMENT FOSSILE, 3846
REMANIEMENT MINERAI, 3163
REMANIEMENT ROCHE, 3774
REMBLAI, 1358
REMBLAIEMENT, 2428
REMBLAIEMENT FLUVIATILE, 142
REMPLACEMENT, 3817
REMPLISSAGE KARSTIQUE, 2374
RENARD, 3455
RENDOLL, 3815
RENDZINE, 3816
RENTABILITE, 3629
REPARTITION ELEMENT, 3312
REPARTITION GRAIN, 4125
REPLIQUE SISMIQUE, 79
REPRESENTATION GRAPHIQUE, 1864
REPRODUCTION, 3820
REPTILIA, 3821
RESEAU ANASTOMOSE, 571
RESEAU BRAVAIS, 574
RESEAU DIACLASE, 2362
RESEAU FRACTURE, 1512
RESEAU HYDROGRAPHIQUE, 1246
RESEAU SISMIQUE, 306
RESERVE, 3824
RESIDU INSOLUBLE, 2256
RESIDU ORGANIQUE, 3173
RESIDUEL, 3827
RESINE, 3831
RESISTANCE A LA RUPTURE, 3918

RESISTANCE CISAILLEMENT, 4062
RESISTANCE COMPRESSION, 866
RESISTANCE MECANIQUE, 4289
RESISTANCE RUPTURE, 1633
RESISTANCE TRACTION, 4455
RESISTIVITE ELECTRIQUE, 3832
RESONANCE MAGNETIQUE NUCLEAIRE, 3064
RESPIRATION, 3835
RESSOURCE, 3834
RESSOURCE EAU, 4798
RESSOURCE MINERALE, 2872
RESSOURCE NATURELLE, 2992
RESTITUTION DONNEE, 1070
RESURGENCE, 3836
RETENTION, 3838
RETICULAREA, 3839
RETRECISSEMENT, 4087
REUNION, 2737
REVERS, 1198
REVISION, 3844
REVOLUTION CASCADIENNE, 662
REVOLUTION MAZATSAL, 2723
RHABDOPLEURIDA, 3847
RHENIUM, 3849
RHEOLOGIE, 3851
RHEOMORPHISME, 3852
RHETIEN, 3848
RHIPIDISTII, 3853
RHIZOPODEA, 3854
RHODIUM, 3855
RHODOPHYCOPHYTA, 3856
RHYNCHOCEPHALIA, 3858
RHYNCHONELLIDA, 3859
RHYODACITE, 3860
RHYOLITE, 3861
RICHE, 2010
RIDE GEOSYNCLINALE, 1763
RIFT, 3864
RIGIDITE, 3866
RIMAYE, 467
RING DYKE, 3869
RIPPLE MARK, 3872
RISQUE SISMIQUE, 4015
RIVIERE, 3873
RIVIERE DECAPITEE, 458
RIVIERE SOUTERRAINE, 4653
ROBERTINACEA, 3880
ROCHE, 3881
ROCHE ARGILEUSE, 295
ROCHE BIOCLASTIQUE, 486
ROCHE CARBONATEE, 650
ROCHE CARBONEE, 649
ROCHE CLASTIQUE, 786
ROCHE CRISTALLINE, 1022
ROCHE ENCAISSANTE, 978
ROCHE FELDSPATHOIDIQUE, 1526
ROCHE FELSIQUE, 1527

ROCHE FERRUGINEUSE, 2312
ROCHE GRANITIQUE, 1848
ROCHE GRENUE, 3519
ROCHE HOTE, 2062
ROCHE HYPABYSSALE, 2142
ROCHE IGNEE, 2176
ROCHE IGNEE STRATOIDE, 2463
ROCHE INTERMEDIAIRE, 2269
ROCHE LEUCOCRATE, 2489
ROCHE MAFIQUE, 1528
ROCHE MAGASIN, 3826
ROCHE MERE, 3304
ROCHE MERE HYDROCARBURE, 4186
ROCHE MERE SOL, 3302
ROCHE METAMORPHIQUE, 2785
ROCHE METASEDIMENTAIRE, 2792
ROCHE METASOMATIQUE, 2795
ROCHE MICROGRENUE, 2817
ROCHE PHOSPHATEE, 3402
ROCHE SEDIMENTAIRE, 4000
ROCHE STANDARD, 4237
ROCHE SUBALCALINE, 4317
ROCHE TOTALE, 4824
ROCHE VERTE, 1894
ROCHE VOLCANIQUE, 4766
RODENTIA, 3892
ROLE GEOLOGUE, 1740
ROOF PENDANT, 3895
ROSEE, 1152
ROSTROCONCHIA, 3897
ROTALIACEA, 3898
ROTALIINA, 3899
ROTATION, 3902
ROTATION OPTIQUE, 3141
ROTATION PLAQUE, 3498
ROTATION TERRE, 1278
ROTLIEGENDE, 3904
ROUTE, 3879
RPE, 1351
RUBEFACTION, 3908
RUBIDIUM, 3909
RUDISTAE, 3911
RUDITE, 3912
RUGOSA, 4474
RUGOSITE, 3905
RUISSEAU, 985
RUISSEAU TEMPORAIRE, 1397
RUISSELLEMENT, 3915
RUISSELLEMENT PLUIE, 3744
RUISSELLEMENT SURFACE, 4364
RUMINANTIA, 3913
RUN, 3914
RUPELIEN, 3916
RUPTURE, 3917
RUTHENIUM, 3919
S 34–S 32, 4355
SABLE, 3936

SABLE BITUMINEUX, 3108
SABLE VERRERIE, 1812
SAISIE DONNEE, 1065
SAKMARIEN, 3921
SALIFERE, 3922
SALINE, 3926
SALINITE, 3923
SALTATION, 3931
SAMARIUM, 3932
SANTONIEN, 3943
SAPROPELE, 3944
SARCODINA, 3946
SARCOPTERYGII, 3947
SARMATIEN, 3948
SATELLITE ARTIFICIEL, 322
SATELLITE PLANETE, 3484
SATURATION, 3950
SAUMURE, 578
SAUMURE CHAUDE, 2064
SAURISCHIA, 3951
SAURORNITHES, 3952
SAUSSURITISATION, 3953
SAXONIEN, 3954
SCANDIUM, 3956
SCAPHOPODA, 3958
SCAPOLITISATION, 3959
SCHISTE A HUILE, 3110
SCHISTE BITUMINEUX, 524
SCHISTE BLEU, 534
SCHISTE CHLORITE, 751
SCHISTE METAMORPHIQUE, 3960
SCHISTE TACHETE, 4222
SCHISTE VERT, 1892
SCHISTOSITE, 3961
SCHISTOSITE FLUX, 1576
SCHLIEREN, 3962
SCIENCE DE LA TERRE, 1279
SCINTILLATION, 3964
SCISSOMETRIE, 4712
SCIUROMORPHA, 3965
SCLERITE, 3966
SCOLECODONTE, 3967
SCOLITE, 3968
SCORIE, 3969
SCORIE VOLCANIQUE, 769
SCYPHOMEDUSAE, 3972
SCYPHOZOA, 3973
SEBKHA, 3987
SECHERESSE, 1258
SECONDAIRE, 2760
SECOUSSE VIOLENTE, 4298
SECRETION LATERAL, 2445
SECTION POLIE, 3538
SEDIMENT, 3991
SEDIMENT FOND, 554
SEDIMENT LACUSTRE, 2408
SEDIMENT MARIN, 2685

SEDIMENT ORGANIQUE, 3174
SEDIMENTATION, 4003
SEDIMENTATION ACTUELLE, 3762
SEDIMENTATION BIOCHIMIQUE, 482
SEDIMENTATION BIOCLASTIQUE, 488
SEDIMENTATION CHIMIQUE, 723
SEDIMENTATION CONTINENTALE, 928
SEDIMENTATION DELTAIQUE, 1105
SEDIMENTATION DETRITIQUE, 1144
SEDIMENTATION EAU DOUCE, 1649
SEDIMENTATION EOLIENNE, 1393
SEDIMENTATION ESTUAIRE, 1437
SEDIMENTATION FLUVIATILE, 1595
SEDIMENTATION FLUVIOGLACIAIRE, 1597
SEDIMENTATION GEOSYNCLINALE, 1765
SEDIMENTATION GLACIAIRE, 1798
SEDIMENTATION GLACIOLACUSTRE, 1806
SEDIMENTATION GLACIOMARINE, 1808
SEDIMENTATION INTERTIDALE, 2287
SEDIMENTATION INTRAFORMATIONNELLE, 2291
SEDIMENTATION LACUSTRE, 2409
SEDIMENTATION LAGUNAIRE, 2413
SEDIMENTATION LITTORALE, 2546
SEDIMENTATION MARECAGE, 4377
SEDIMENTATION MARGE CONTINENTALE, 4070
SEDIMENTATION MARINE, 2686
SEDIMENTATION MER PROFONDE, 23
SEDIMENTATION PELAGIQUE, 3331
SEDIMENTATION RECIFALE, 3778
SEDIMENTOLOGIE, 4006
SEGREGATION, 4010
SEICHE, 4011
SEISLOG, 4012
SEISME, 1284
SEISME INTERMEDIAIRE, 2270
SEISME LUNAIRE, 2936
SEISME PROFOND, 1088
SEISME SUPERFICIEL, 4052
SEISMOLOGIE, 4026
SEISMOLOGIE EXPERIMENTALE, 1478
SEL, 3924
SELENIATE, 4029
SELENIUM, 4030
SELENOCHRONOLOGIE, 4031
SELENOLOGIE, 4032
SENONIEN, 4037
SEPARATION, 4039
SEPARATION PHASE, 2518
SEPTARIUM, 4040
SEPTUM, 4041
SERICITISATION, 4043
SERICITOSCHISTE, 4042
SERIE, 4044
SERIE ARCTIQUE, 287
SERIE ATLANTIQUE, 355
SERIE PACIFIQUE, 3231
SERIE REACTION, 3759

SERPENTINE, 4045
SERPENTINISATION, 4047
SERPENTINITE, 4046
SERVICE DES MINES, 595
SERVICE GEOLOGIQUE, 1739
SEUIL ANORMAL, 4519
SEUIL MARIN, 2687
SHALE, 4051
SHATTER CONE, 4057
SHONKINITE, 4080
SIAL, 4089
SICILIEN, 4090
SIDEROLITE, 4263
SIDEROPHYRE, 4093
SIDERURGIE, 2309
SIEGENIEN, 4094
SIEROZEM, 4095
SIGNAL TELESISMIQUE, 4446
SILEX, 1567
SILEXITE, 739
SILICATE, 4098
SILICE, 4097
SILICIFICATION, 4100
SILICIUM, 4103
SILICOFLAGELLATA, 4102
SILL, 4104
SILLIMANITE, 4105
SILLON, 1766
SILLON EROSION, 1668
SILLON LUNAIRE, 3867
SILT, 4106
SILTSTONE, 4109
SILURIEN, 4110
SILURIEN INF, 2593
SILURIEN SUP, 4683
SIMA, 4112
SIMIEN, 4113
SIMULATION, 4114
SIMULATION ANALOGIQUE, 180
SIMULATION NUMERIQUE, 1183
SINEMURIEN, 4115
SINUOSITE RIVIERE, 4119
SIPHON, 4120
SIPHONAPTEROIDA, 4121
SIPUNCULOIDA, 4122
SIRENIA, 4123
SISMICITE, 4022
SISMIQUE MARTEAU, 1951
SISMIQUE POUSSEE, 2190
SISMIQUE REFLEXION, 3784
SISMIQUE REFRACTION, 3787
SISMOGRAMME, 4023
SISMOGRAPHE, 4024
SISMOGRAPHE FOND SOUS MARIN, 3089
SISMOMETRE, 4027
SISMOTECTONIQUE, 4028
SITE ARCHEOLOGIQUE, 280

SKARN, 4126
SLIKKE, 4527
SLUMPING, 4139
SOCIETE, 858
SOCIETE SAVANTE, 2472
SOCLE, 429
SODIUM, 4143
SOL, 4146
SOL A MULL, 2960
SOL BRUN, 583
SOL BRUT, 394
SOL CHATAIN, 666
SOL COHERENT, 830
SOL DESERTIQUE, 4848
SOL FERRALLITIQUE, 1532
SOL FERRUGINEUX, 1540
SOL FIGURE, 3317
SOL GELE, 1656
SOL LATERITIQUE, 2447
SOL LUNAIRE, 2603
SOL MEDITERRANEEN, 2735
SOL NON COHERENT, 828
SOL PEU EVOLUE, 2183
SOL POLYGONAL, 3547
SOL RICHE EN HUMUS, 2072
SOL TOUNDRA, 4611
SOLEIL, 4358
SOLEMYIDA, 4164
SOLFATARE, 4165
SOLIDUS, 4170
SOLIFLUXION, 4171
SOLONCHAK, 4172
SOLONETZ, 4173
SOLUBILITE, 4174
SOLUTE, 4175
SOLUTION AQUEUSE, 256
SOLUTION SOLIDE, 4167
SOMASTEROIDEA, 4180
SONAR, 4181
SONDAGE, 549
SONDAGE CAROTTAGE COMPLET, 1661
SONDAGE CAROTTE, 956
SONDAGE DEVELOPPEMENT, 1148
SONDAGE ELECTRIQUE, 1332
SONDAGE FREQUENCE, 1646
SONDAGE GEOPHYSIQUE, 4185
SONDAGE OFFSHORE, 3103
SONDAGE ORIENTE, 1208
SONDAGE RECONNAISSANCE, 1484
SONDAGE SISMIQUE PROFOND, 1091
SONDAGE TARIERE, 371
SONDAGE TURBO, 4619
SONDAGE VIBRATION, 4739
SONDE NEUTRON, 3018
SONOBUOY, 4182
SORPTION, 4183
SOUFRE, 4353

SOULEVEMENT, 1970
SOULEVEMENT GEL, 1655
SOURCE, 4224
SOURCE CHALEUR, 1968
SOURCE EAU MINERALE, 2873
SOURCE EXPLOSION, 1487
SOURCE INTERMITTENTE, 2271
SOURCE KARSTIQUE, 2376
SOURCE MINERAI, 3162
SOURCE NONEXPLOSIVE, 3046
SOURCE SISMIQUE, 4016
SOURCE SOUS MARINE, 4331
SOURCE THERMALE, 2066
SOUSSATURATION, 4655
SOUSSATURE, 4654
SPALLATION, 4189
SPARAGMITE, 4190
SPARKER, 4191
SPECIMEN CARACTERISTIQUE, 4626
SPECTRE, 4202
SPECTRE ABSORPTION, 15
SPECTRE MOESSBAUER, 2948
SPECTRE OPTIQUE, 3143
SPECTRE RPE, 1410
SPECTROMETRE, 4199
SPECTROMETRIE, 4201
SPECTROMETRIE ABSORPTION, 14
SPECTROMETRIE ALPHA, 148
SPECTROMETRIE EMISSION, 1362
SPECTROMETRIE FLUORESCENCE, 1587
SPECTROMETRIE GAMMA, 1677
SPECTROMETRIE HERTZIENNE, 3723
SPECTROMETRIE IR, 2234
SPECTROMETRIE MASSE, 2700
SPECTROMETRIE MICROONDE, 2839
SPECTROMETRIE MOESSBAUER, 2947
SPECTROMETRIE MOLECULAIRE, 2914
SPECTROMETRIE OPTIQUE, 3142
SPECTROMETRIE RAMAN, 3745
SPECTROMETRIE RX, 4836
SPECTROPHOTOMETRIE, 4200
SPELEOLOGIE, 4203
SPELEOTHEME, 4204
SPHENOLITE, 4205
SPHERICITE, 4207
SPHEROLITE, 4209
SPHERULE, 4208
SPHERULE MAGNETIQUE, 2641
SPILITISATION, 4210
SPINIFEX, 4212
SPIRIFERIDA, 4213
SPIRILLINACEA, 4214
SPODOSOL, 4216
SPONGOLITE, 4218
SPORE, 4219
SPOROMORPHE, 4220
SPOROZOA, 4221

SQUAMATA, 4225
SR=RB, 3910
STABILISATION, 4227
STABILISATION SOL, 4158
STABILITE, 4226
STABILITE VERSANT, 4137
STADE GLACIAIRE, 1799
STALACTITE, 4231
STALAGMITE, 4232
STAMPIEN, 4233
STANDARD CHIMIQUE, 4235
STATION LUNAIRE, 2604
STATION SISMIQUE, 4018
STATISTIQUE, 4245
STEATITISATION, 4247
STEPHANIEN, 4250
STEREOCHIMIE, 4252
STERILE, 415
STEROL, 4253
STOCK, 4257
STOCKAGE, 4265
STOCKAGE DONNEE, 1071
STOCKAGE EAU, 4799
STOCKAGE GAZ, 1691
STOCKAGE SOUTERRAIN, 4652
STOCKWERK, 4258
STOECHIOMETRIE, 4259
STOLONOIDEA, 4260
STOMOCHORDATA, 4261
STONELINE, 4262
STRATE CONVOLUTEE, 941
STRATEGIE, 4271
STRATIFICATION, 4272
STRATIFICATION ENTRECROISEE, 998
STRATIFICATION MASSIVE, 2704
STRATIFICATION OBLIQUE, 997
STRATIFICATION PLANE, 3466
STRATIGRAPHIE, 4281
STRATOTYPE, 4282
STRATOVOLCAN, 4283
STREAM SEDIMENT, 4287
STRIATION, 1514
STRIATION MINERAL, 2874
STRIE GLACIAIRE, 1800
STROMATOLITES, 4295
STROMATOPOROIDEA, 4296
STROMBOLIEN, 4297
STRONTIUM, 4299
STROPHOMENIDA, 4300
STRUCTURE ALGAIRE, 118
STRUCTURE ANNULAIRE, 3871
STRUCTURE ARQUEE, 270
STRUCTURE BIOGENE, 494
STRUCTURE CANNELEE, 1591
STRUCTURE CHAOTIQUE, 709
STRUCTURE COURANT TURBIDITE, 4618
STRUCTURE CRISTALLINE, 1020

STRUCTURE CRYPTOVOLCANIQUE, 1010
STRUCTURE CYLINDRIQUE, 1053
STRUCTURE DIAGENETIQUE, 3990
STRUCTURE DOMAINE, 1239
STRUCTURE EFFONDREMENT, 837
STRUCTURE EN COUSSIN, 3448
STRUCTURE EN MENEAU, 2961
STRUCTURE FANTOME, 3273
STRUCTURE FLAMME, 1563
STRUCTURE GEOLOGIQUE, 1737
STRUCTURE GNEISSIQUE, 1830
STRUCTURE IMPACT, 2192
STRUCTURE INTERNE, 2275
STRUCTURE LAMELLAIRE, 2419
STRUCTURE LAMINEE, 2421
STRUCTURE MYLONITIQUE, 2973
STRUCTURE OEILLEE, 370
STRUCTURE ORGANIQUE, 3175
STRUCTURE PLANE, 3467
STRUCTURE PRIMAIRE, 3612
STRUCTURE PRISMATIQUE, 3616
STRUCTURE RETICULEE, 3840
STRUCTURE RUBANEE, 409
STRUCTURE SEDIMENTAIRE, 4001
STRUCTURE SLUMPING, 4138
STRUCTURE STRATIFIEE, 4273
STRUCTURE SYNSEDIMENTAIRE, 4398
STRUCTURE VERMIFORME, 4733
STRUCTURE VITESSE, 4727
STRUCTURE VOLCANOTECTONIQUE, 4771
STRUCTURE ZONEE, 4865
STYLE TECTONIQUE, 4438
STYLOLITE, 4314
STYLOPHORA, 4315
SUBBOREAL, 4319
SUBDIVISION, 4320
SUBLIMAT, 4323
SUBLIMATION, 4324
SUBMERSIBLE, 4335
SUBSIDENCE, 4337
SUBSIDENCE EN CHAUDRON, 675
SUBSTANCE ANHYDRE, 205
SUBSTANCE MALLEABLE, 2660
SUBSTANCE NON METALLIQUE, 3051
SUBSTANCE NON=MISCIBLE, 2186
SUBSTANCE TOXIQUE, 4554
SUBSTANCE UTILE, 2219
SUBSTITUTION, 4338
SUBSTRAT, 4339
SUCCION, 634
SUCRE, 641
SUCTORIA, 4345
SUD, 4187
SUIFORMES, 4348
SUINTEMENT, 4008
SUINTEMENT PETROLE, 3109
SUITE REACTIONNELLE BOWEN, 563

SULFATE, 4349
SULFOSEL, 4352
SULFURE, 4351
SUPERPOSITION, 4360
SUPERPOSITION AGE, 1723
SUPERSTRUCTURE TECTONIQUE, 4439
SURCROISSANCE, 3215
SURFACE ABRASION, 8
SURFACE EROSION, 1429
SURFACE PIEZOMETRIQUE, 3583
SURFACE SPECIFIQUE, 4198
SURFACE TERRE, 1904
SURSATURATION, 3217
SURSATURE, 3216
SURSTRUCTURE, 4361
SUSCEPTIBILITE MAGNETIQUE, 2645
SUSPENSION, 4373
SUTURE, 4374
SYENITE, 4379
SYENITE ALCALINE, 127
SYLVINITE, 4380
SYMBIOSE, 4381
SYMETRIE, 4384
SYMMETRODONTA, 4383
SYMPHYLA, 4385
SYMPOSIUM, 4387
SYNAPSIDA, 4388
SYNCLINAL, 4390
SYNCLINORIUM, 4391
SYNECLISE, 4392
SYNGENESE, 4394
SYNGENETIQUE, 4395
SYNONYME, 4396
SYNTEXIE, 4400
SYNTHESE, 4401
SYNTHESE BIBLIOGRAPHIQUE, 3843
SYSTEME, 4403
SYSTEME BINAIRE, 479
SYSTEME CHIMIQUE, 733
SYSTEME COSMIQUE, 971
SYSTEME CRISTALLIN, 1021
SYSTEME CUBIQUE, 1030
SYSTEME EUTECTIQUE, 1454
SYSTEME GEOTHERMIQUE, 1778
SYSTEME HEXAGONAL, 2002
SYSTEME ISOMETRIQUE, 2332
SYSTEME MONOCLINIQUE, 2921
SYSTEME MULTICOUCHE, 2962
SYSTEME ORTHORHOMBIQUE, 3194
SYSTEME PLISSE, 1606
SYSTEME PYRAMIDAL, 3693
SYSTEME QUATERNAIRE, 3714
SYSTEME RHOMBIQUE, 3857
SYSTEME SOLAIRE, 4160
SYSTEME TERNAIRE, 4463
SYSTEME TRICLINIQUE, 4593
SYSTEME TRIGONAL, 4595

TABLE DONNEES, 1072
TABULATA, 4406
TACHETURE, 2949
TACONITE, 4408
TACTITE, 4409
TAENIODONTA, 4410
TAENIOLABOIDEA, 4411
TALC, 4414
TALCSCHISTE, 4415
TALUS EBOULIS, 4417
TALUS EQUILIBRE, 200
TALUS MARIN, 930
TALUS MARIN INTERNE, 2245
TAMIS, 4096
TANTALATE, 4418
TANTALE, 4419
TAPHONOMIE, 4420
TAPHROGENIE, 4421
TAPHROGEOSYNCLINAL, 4422
TARANNON, 4423
TARDIGRADA, 4424
TARISSEMENT, 4797
TASMANITES, 4425
TASSEMENT, 4048
TASSEMENT DIFFERENTIEL, 1175
TATARIEN, 4426
TAUX, 3753
TAUX SEDIMENTATION, 4004
TAXINOMIE, 4428
TAXITE, 4427
TAXON AHERMATYPIQUE, 92
TAXON ANAEROBIE, 179
TAXON COLONIAL, 844
TAXON ETEINT, 1495
TAXON HERMATYPIQUE, 1987
TAXON HETEROMORPHE, 1995
TAXON MONOLETE, 2925
TAXON NOMINALE, 3043
TAXON NOUVEAU, 3025
TAXON PLANCTONIQUE, 3488
TECHNETIUM, 4429
TECHNIQUE, 4431
TECHNIQUE ANALOGIQUE, 181
TECHNOLOGIE, 4432
TECHNOLOGIE HYDROGEOLOGIQUE, 2114
TECTITE, 4444
TECTOGENESE, 4433
TECTONIQUE, 4440
TECTONIQUE ACTIVE, 59
TECTONIQUE ALPINE, 151
TECTONIQUE CASSANTE, 532
TECTONIQUE COUVERTURE, 3996
TECTONIQUE FOND, 1094
TECTONIQUE FRACTURATION, 1515
TECTONIQUE GERMANOTYPE, 1783
TECTONIQUE GLOBALE, 1821
TECTONIQUE GRAVITE, 1887

TECTONIQUE IMBRIQUEE, 2180
TECTONIQUE PLAQUE, 3499
TECTONIQUE SALIFERE, 3929
TECTONIQUE SOCLE, 430
TECTONIQUE SOUPLE, 1607
TECTONIQUE SUPERPOSEE, 2965
TECTONIQUE TANGENTIELLE, 867
TECTONITE, 4441
TECTONOPHYSIQUE, 4442
TECTONOSPHERE, 4443
TEGUMENT VERTEBRE, 1126
TELEDETECTION, 3813
TELEDETECTION MULTISPECTRALE, 2966
TELEOSTEI, 4445
TELLURATE, 4448
TELLURE, 4450
TEMOIN EROSION, 1428
TEMPERATURE, 4453
TEMPETE, 4266
TEMPS ARRIVEE, 307
TEMPS PARCOURS, 4581
TENACITE, 4454
TENDANCE, 4244
TENEUR, 3156
TENEUR CENDRE, 330
TENEUR FOND, 398
TENEUR LIMITE, 1038
TENEUR METAL, 337
TENSION SUPERFICIELLE, 4365
TENTACULITES, 4457
TEPHROCHRONOLOGIE, 4458
TERATOLOGIE, 4459
TERBIUM, 4460
TEREBRATULIDA, 4461
TERMINOLOGIE, 4462
TERRA ROSSA, 4464
TERRAIN HUMIDE, 4822
TERRASSE, 4465
TERRASSE FLEUVE, 3875
TERRASSE MARINE, 2688
TERRASSEMENT, 1281
TERRE A FOULON, 1662
TERRE A PORCELAINE, 743
TERRE ADSORBANTE, 528
TERRE DIATOMITE, 1168
TERRE EMERGEE, 2425
TERRE RARE, 3751
TERRIER, 600
TERTIAIRE, 4470
TEST, 4471
TESTUDINATA, 4472
TETE FORAGE, 4815
TETHYS, 4473
TETRAPODA, 4475
TEXTULARIINA, 4476
TEXTURE, 4477
TEXTURE APHANITIQUE, 236

TEXTURE APLITIQUE, 240
TEXTURE ARENACEE, 290
TEXTURE BLASTIQUE, 525
TEXTURE CAVERNEUSE, 679
TEXTURE CLOISONNEE, 564
TEXTURE COLLOFORME, 841
TEXTURE CRYPTOCRISTALLINE, 1006
TEXTURE DALLE, 1561
TEXTURE EN MOSAIQUE, 2945
TEXTURE EQUIGRANULAIRE, 1415
TEXTURE EUTAXIQUE, 1453
TEXTURE FILAMENTEUSE, 1565
TEXTURE FRAMBOIDALE, 1638
TEXTURE GRANOBLASTIQUE, 1852
TEXTURE GRANOPHYRIQUE, 1855
TEXTURE GRANULAIRE, 1856
TEXTURE GRAPHIQUE, 1865
TEXTURE GRENUE, 1444
TEXTURE HETEROBLASTIQUE, 1988
TEXTURE HOLOCRISTALLINE, 2028
TEXTURE HYALINE, 1814
TEXTURE HYALOCRYSTALLINE, 2084
TEXTURE HYPIDIOBLASTIQUE, 2144
TEXTURE HYPIDIOMORPHIQUE, 2145
TEXTURE HYPOCRISTALLINE, 2147
TEXTURE IDIOMORPHE, 2170
TEXTURE INTERSERTALE, 2284
TEXTURE LAMPROPHYRIQUE, 2424
TEXTURE LEPIDOBLASTIQUE, 2482
TEXTURE LINOPHYRIQUE, 2515
TEXTURE MAILLEE, 2452
TEXTURE MARGE CRISTAUX, 2675
TEXTURE MIAROLITIQUE, 2803
TEXTURE MICROGRANITIQUE, 2819
TEXTURE MICROSCOPIQUE, 2832
TEXTURE MONZONITIQUE, 2934
TEXTURE OOLITIQUE, 3126
TEXTURE OPHITIQUE, 3134
TEXTURE ORBICULAIRE, 3145
TEXTURE ORTHOPHYRIQUE, 3193
TEXTURE PANIDIOMORPHE, 3282
TEXTURE PARALLELE, 3296
TEXTURE PELITIQUE, 3335
TEXTURE PERLITIQUE, 3361
TEXTURE PERTHITIQUE, 3368
TEXTURE PISOLITIQUE, 3458
TEXTURE POECILITIQUE, 3529
TEXTURE POECILOBLASTIQUE, 3530
TEXTURE PORPHYRIQUE, 3567
TEXTURE PORPHYROBLASTIQUE, 3568
TEXTURE RELICTE, 3807
TEXTURE SYMPLECTIQUE, 4386
TEXTURE VITREUSE, 4750
TEXTURE VITROCLASTIQUE, 4752
TEXTURE VITROPHYRIQUE, 4753
TH-PB, 4518
THALASSOCRATON, 4478

THALLIUM, 4480
THALLOPHYTA, 4481
THALWEG, 4482
THANATOCOENOSE, 4483
THANETIEN, 4484
THECAMOEBINA, 4486
THECIDEIDINA, 4487
THECODONTIA, 4488
THELODONTI, 4489
THEORIE, 4490
THEORIE DYNAMO, 1268
THEORIE MOUVEMENT PAR ASCENSION, 326
THEORIE MOUVEMENT PAR DESCENTE, 1130
THEORIE OSCILLATOIRE, 3198
THERALITE, 4491
THERAPSIDA, 4492
THERMAL, 4493
THERMOCLINE, 4503
THERMODYNAMIQUE, 4505
THERMOGRAMME, 4506
THERMOKARST, 4508
THERMOLUMINESCENCE, 4509
THESE, 4512
THIXOTROPIE, 4515
THORIUM, 4517
THULIUM, 4521
THURINGIEN, 4522
THYSANOPTEROIDA, 4523
THYSANURA, 4524
TILL, 4530
TILLITE, 4531
TILLODONTIA, 4532
TINTINNIDAE, 4538
TITANE, 4539
TITHONIQUE, 4540
TITRE MINIER, 2886
TOARCIEN, 4541
TOIT, 1953
TOMBOLO, 4542
TONSTEIN, 4543
TOPOGRAPHIE, 4547
TORRENT, 4549
TORSION, 4550
TORTONIEN, 4551
TORTUOSITE, 4552
TOUNDRA, 4610
TOURBE, 3320
TOURBIERE, 3321
TOURNAISIEN, 4553
TOXICITE, 4555
TRACE AFFOUILLEMENT, 3970
TRACE CLIMATIQUE, 796
TRACE COURANT, 1036
TRACE FISSION, 1555
TRACE MECANIQUE, 715
TRACE NUTRITION, 1523
TRACE ORGANIQUE, 4557

TRACE PLUIE, 3743
TRACEUR, 4560
TRACEUR COLORANT, 848
TRACEUR RADIOACTIF, 3729
TRACHYANDESITE, 4561
TRACHYBASALTE, 4562
TRACHYPSAMMIACEA, 4563
TRACHYTE, 4564
TRACTION, 4456
TRACTION SEDIMENTATION, 3992
TRAITE, 4584
TRAITEMENT DONNEE, 1069
TRANSFERT, 4568
TRANSFERT CHALEUR, 1969
TRANSFERT MASSE, 2701
TRANSFORMATION, 4570
TRANSFORMATION LAPLACE, 2434
TRANSFORMISME, 4571
TRANSGRESSION, 4572
TRANSMISSIVITE, 4576
TRANSPORT, 4577
TRANSPORT EOLIEN, 4829
TRANSPORT FLUVIATILE, 4288
TRANSPORT GLACE, 2159
TRANSPORT GLACIAIRE, 1801
TRANSPORT MARIN, 2689
TRANSVERSAL, 4579
TRAPP DECCAN, 1080
TRAVAUX PUBLICS, 3680
TRAVAUX TERRAIN, 1542
TRAVERTIN, 4583
TREMADOC, 4586
TREPOSTOMATA, 4589
TRIANGULATION, 4590
TRIAS, 4591
TRIAS ALPIN, 152
TRIAS INF, 2594
TRIAS MOYEN, 2852
TRIAS SUP, 4684
TRICONODONTA, 4594
TRIGONIIDAE, 4596
TRILOBITA, 4597
TRILOBITOMORPHA, 4598
TRIPOLI, 4600
TRITIUM, 4601
TROU EXPLOSION SISMIQUE, 4085
TROU SOUFFLEUR, 533
TSUNAMI, 4604
TUBOIDEA, 4605
TUBULIDENTATA, 4606
TUF VOLCANIQUE, 4607
TUFFITE, 4608
TUFO-LAVE, 4609
TUNGSTATE, 4612
TUNGSTENE, 4613
TUNNEL, 4614
TUNNEL LAVE, 2459

TURBIDITE, 4616
TURBIDITES, 4615
TURONIEN, 4620
TYLOPODA, 4623
TYPE ALPIN, 153
TYPOMORPHISME, 4628
TYRRHENIEN, 4629
U-PB, 4692
U-TH-PB, 4693
UDALF, 4630
UDOLL, 4631
UDULT, 4632
UFIEN, 4633
ULTISOL, 4634
ULTRABASITE, 4636
ULTRAMETAMORPHISME, 4637
ULTRAMYLONITE, 4638
UMBREPT, 4642
UNAKITE, 4643
UNIFORMITARIANISME, 4659
UNIONIDAE, 4660
UNITE BIOSTRATIGRAPHIQUE, 510
UNITE CHRONOSTRATIGRAPHIQUE, 762
UNITE ECOSTRATIGRAPHIQUE, 1302
UNITE GEOCHRONOLOGIQUE, 1724
UNITE LITHOSTRATIGRAPHIQUE, 2537
UNITE STRATIGRAPHIQUE, 4280
UNITE TECTONIQUE, 4305
UNIVERS, 4663
UPWELLING, 4687
URANIUM, 4689
URBANISATION, 4696
UREILITE, 4697
URODELA, 4698
USINE TRAITEMENT, 3155
USTALF, 4699
USTERT, 4700
USTOLL, 4701
USTOX, 4702
USTULT, 4703
UTILISATION INDUSTRIELLE, 2220
UTILISATION SUBSTANCE, 4704
UTILISATION TERRAIN, 2427
VALANGINIEN, 4706
VALENCE CHIMIQUE, 734
VALLEE, 4707
VALLEE CONSEQUENTE, 904
VALLEE ENFOUIE, 599
VALLEE GLACIAIRE, 1802
VALLEE INSEQUENTE, 2254
VALLEE MER PROFONDE, 20
VALLEE OBSEQUENTE, 3080
VALLEE RESEQUENTE, 3823
VALLEE RIFT, 3865
VALLEE SOUS MARINE, 4332
VALLEE STRUCTURALE, 4312
VALLEE SUBSEQUENTE, 4336

VALLEE SUSPENDUE, 1952
VALORISATION, 1252
VALVE, 4708
VANADATE, 4709
VANADIUM, 4710
VAPEUR EAU, 4713
VARIABILITE, 4714
VARIABLE, 4715
VARIATION, 4717
VARIATION CHAMP MAGNETIQUE, 2634
VARIATION DIURNE, 1231
VARIATION NIVEAU MER, 3981
VARIATION SAISONNIERE, 3985
VARIATION SECULAIRE, 1744
VARIOGRAMME, 4718
VARVE, 4720
VASE, 2952
VASIERE, 2955
VECTEUR, 4723
VEGETATION, 4724
VEINE CHARBON, 3983
VELOCITE SISMIQUE, 4020
VENERIDA, 4728
VENITE, 4729
VENT, 4826
VENT SOLAIRE, 4161
VERGENCE, 4730
VERITE TERRAIN, 1910
VERMES, 4731
VERMICULITE, 4732
VERRE, 1811
VERRE NATUREL, 2990
VERRE VOLCANIQUE, 4764
VERRERIE, 1813
VERROU, 3863
VERSANT, 4134
VERTEBRATA, 4734
VERTISOL, 4737
VIBRATION, 4738
VIBROSEIS, 4740

VIE, 2499
VIE ORGANIQUE, 3169
VILLAFRANCHIEN, 4742
VIRGATION, 4743
VISCOELASTICITE, 4744
VISCOSITE, 4745
VISEEN, 4748
VITESSE, 4726
VITRAIN, 4749
VITRIFICATION, 4751
VIVANT, 2548
VOIE NAVIGABLE, 4804
VOLANTES, 4754
VOLATIL, 4755
VOLATILISATION, 4758
VOLCAN, 4769
VOLCAN ACTIF, 60
VOLCAN BOUE, 2958
VOLCAN EN BOUCLIER, 4073
VOLCAN INSULAIRE, 2317
VOLCAN SOUS MARIN, 4333
VOLCANO SEDIMENTAIRE, 4772
VOLCANOGENE, 4773
VOLCANOLOGIE, 4775
VOLUME, 4776
VOLUME SEDIMENT, 3993
VULCANIEN, 4777
VULCANORIUM, 4778
WENLOCK, 4818
WESTPHALIEN, 4820
WILDFLYSCH, 4825
WIND RIPPLE, 4828
XANTHOPHYTA, 4837
XENON, 4840
XERALF, 4841
XEROLL, 4842
XEROPHYTE, 4844
XEROSOL, 4845
XERULT, 4846
YPRESIEN, 4852

YTTERBIUM, 4853
YTTRIUM, 4854
ZECHSTEIN, 4855
ZEOLITE, 4856
ZEOLITISATION, 4858
ZINC, 4860
ZIRCON, 4861
ZIRCONIUM, 4862
ZOANTHINIARIA, 4863
ZOECIUM, 4864
ZONAGE CRISTAL, 4870
ZONE AERATION, 4866
ZONE ASISMIQUE, 329
ZONE BENIOFF, 462
ZONE CISAILLEMENT, 4064
ZONE CLIMATIQUE, 797
ZONE CONTINENTALE, 931
ZONE EQUATORIALE, 1414
ZONE EXTENSION, 3747
ZONE FAIBLE VITESSE, 2576
ZONE FAILLEE, 1513
ZONE FAUNISTIQUE, 1519
ZONE FRACTUREE, 1634
ZONE FROIDE, 548
ZONE METAMORPHIQUE, 2786
ZONE MOBILE, 2899
ZONE NON SATUREE, 3049
ZONE OXYDATION, 3223
ZONE PLASTIQUE, 4868
ZONE SATUREE, 3949
ZONE SEDIMENTATION, 4869
ZONE STEPPIQUE, 4251
ZONE SUBDUCTION, 4322
ZONE SUBTROPICALE, 4342
ZONE SULFURE, 4350
ZONE TAIGA, 4412
ZONE TEMPEREE, 4452
ZONE TROPICALE, 4602
ZONEOGRAPHIE SOL, 673
ZOOIDE, 4871

Deutsch

A-SCHICHT, 1
AALENIUM, 3
ABBILD, 2178
ABDICHTUNG, 3982
ABFALL, 4781
ABFALL-BESEITIGUNG, 4782
ABFLUSS, 3915
ABFLUSS-VORGANG, 1581
ABFLUSSLOS.GEBIET, 298
ABFLUSSLOSIGKEIT, 1371
ABGUSS, 665
ABHANG, 4417
ABKUEHLUNG, 942
ABLAGERUNG, 1120
ABLATION, 5
ABRASION, 6
ABRASIONS-FLAECHE, 8
ABRASIONS-KUESTE, 7
ABRAUM, 3213
ABREIBUNG, 1899
ABSCHERUNG, 1082
ABSCHIEBUNG, 3056
ABSCHWEMMUNG, 4065
ABSENKUNG, 1242
ABSENKUNGS-TRICHTER, 885
ABSOLUT.ALTER, 10
ABSOLUT.ALTERSSKALA, 4536
ABSOLUT.SCHWERE, 11
ABSONDERUNG, 3310
ABSORBER, 12
ABSORPTION, 13
ABSORPTIONS-SPEKTROMETRIE, 14
ABSORPTIONS-SPEKTRUM, 15
ABTAUCHEN, 3516
ABWEICHUNG, 1149
ABYSSAL, 18
AC, 51
ACAD.OROGENESE, 24
ACANTHARIA, 25
ACANTHODII, 26
ACHONDRIT, 36
ACHSE, 393
ACHSENEBENE, 391
ACRANIA, 46
ACRISOL, 47

ACRITARCHA, 48
ACROTRETIDA, 49
ACTINIARIA, 50
ACTINOCERATOIDEA, 52
ADAPTIV.DIVERGENZ, 64
ADIABAT.VORGANG, 65
ADIAGNOSTISCH, 66
ADSORPTION, 68
AEOL.MILIEU, 1391
AEOL.MORPHOLOGIE, 1392
AEOL.SEDIMENTATION, 1393
AEON, 1394
AEQUATOR, 1413
AEQUATORIAL-ZONE, 1414
AEQUIPOTENTIAL-LINIE, 2336
AERA, 1418
AERATHEM, 1419
AEROB.MILIEU, 73
AEROMAGNET.VERMESSUNG, 74
AEROSOL, 75
AESTUAR, 1438
AESTUAR-MILIEU, 1436
AESTUAR-SEDIMENTATION, 1437
AETHAN, 1440
AETHER, 1441
AETZUNG, 1439
AEUSSER.ERD-KERN, 3207
AEUSSER.PLANET, 3208
AEUSSER.SCHELF, 3209
AFFENARTIGE, 4113
AFFINITAET, 76
AFMAG, 78
AG, 4111
AGGLOMERAT, 82
AGNATHA, 86
AGNOSTIDA, 87
AGONIATITIDA, 88
AGPAITISCH, 89
AGROGEOLOGIE, 91
AHERMATYP.TAXON, 92
AIR-BORNE-AUFNAHME, 95
AIRY-WELLE, 97
AKCHAGYLIUM, 99
AKKUMULATIONS-KUESTE, 32
AKMOLITH, 98

AKTINOLITH-FAZIES, 53
AKTIV.KONTINENTAL-RAND, 58
AKTIV.STOERUNG, 56
AKTIV.VULKAN, 60
AKTIVIERUNGS-ANALYSE, 54
AKTIVIERUNGS-ENERGIE, 55
AKTIVITAET, 61
AKTUALISMUS, 62
AKTUO-SEDIMENTATION, 3762
AKTUO-TEKTONIK, 59
AKUST.AUFNAHME, 44
AKUST.BOJE, 4182
AKUST.EIGENSCHAFT, 42
AKUST.METHODE, 43
AKUST.WELLE, 45
AKUSTIK-LOG, 41
AKZESSOR.MINERAL, 29
AL, 157
ALABASTER, 100
ALB, 103
ALBEDO, 101
ALBERTIT, 102
ALBIT-EPIDOT-AMPHIBOLIT-FAZIES, 104
ALBIT-EPIDOT-HORNFELS-FAZIES, 105
ALBITISIERUNG, 106
ALBOLL, 107
ALCYONACEA, 108
ALDANIUM, 109
ALFISOL, 110
ALGAE, 111
ALGEN-BANK, 112
ALGEN-DOM, 116
ALGEN-FLACHSTRUKTUR, 113
ALGEN-KALK, 114
ALGEN-MATTE, 115
ALGEN-RIFF, 117
ALGEN-STRUKTUR, 118
ALGOL, 119
ALGOMAN.OROGENESE, 120
ALGORITHMUS, 121
ALIPHAT.KOHLENWASSERSTOFF, 122
ALKALI-BASALT, 123
ALKALI-GRANIT, 125
ALKALI-KALK-TYP, 124
ALKALI-METALL, 126

ALKALI-METASOMATOSE, 130
ALKALI-SYENIT, 127
ALKALI-TYP, 128
ALKALINITAET, 131
ALKAN, 132
ALLEGHENY-OROGENESE, 133
ALLOCHEM, 134
ALLOCHTHONIE, 135
ALLOGROMIINA, 137
ALLOMETRIE, 138
ALLOTHIGEN.MINERAL, 136
ALLUVIAL, 140
ALLUVIAL-FLAECHE, 144
ALLUVIAL-PROSPEKTION, 145
ALLUVION, 146
ALPHA-BLEI-DATIERUNG, 2467
ALPHA-SPEKTROMETRIE, 148
ALPHA-STRAHLUNG, 147
ALPID.OROGENESE, 150
ALPIN.MILIEU, 149
ALPIN.TEKTONIK, 151
ALPIN.TRIAS, 152
ALPINOTYP.BAUSTIL, 153
ALT.PLEISTOZAEN, 2590
ALTER, 80
ALUNITISIERUNG, 158
ALVEOLINELLIDAE, 159
AM, 162
AMBLYPODA, 161
AMINO-SAEURE, 163
AMMODISCACEA, 164
AMMONIAK, 165
AMMONITIDA, 166
AMMONITINA, 167
AMMONIUM, 168
AMMONOIDEA, 169
AMORPH.SUBSTANZ, 170
AMPHIBIA, 171
AMPHIBOL, 172
AMPHIBOLIT, 173
AMPHIBOLIT-FAZIES, 174
AMPHIBOLITISIERUNG, 175
AMPHINEURA, 176
AMPLITUDE, 177
ANAEROB.MILIEU, 178
ANAEROB.TAXON, 179
ANALOG-SIMULATION, 180
ANALOG-TECHNIK, 181
ANALYSE, 182
ANALYSEN-METHODIK, 183
ANALYSEN-NETZ, 3013
ANAPSIDA, 184
ANASPIDA, 185
ANATEXIS, 186
ANATEXIT, 187
ANATOMIE, 188
ANCHI-METAMORPHOSE, 189

ANDALUSIT, 192
ANDEPT, 193
ANDESIT, 194
ANDESIT-LINIE, 195
ANDOSOL, 196
ANELAST.MEDIUM, 197
ANGEWANDT.GEOLOGIE, 246
ANGEWANDT.GEOPHYSIK, 247
ANGEWANDT.HYDROGEOLOGIE, 248
ANGIOSPERMAE, 199
ANGRIT, 201
ANHEDR.KRISTALL, 203
ANHYDRID-CHEMISMUS, 205
ANHYDRIT, 204
ANHYSTERET.REMAN.MAGNETISIERUNG, 206
ANION, 207
ANIS, 208
ANISOPLEURA, 209
ANISOTROPIE, 210
ANKERIT-GESTEIN, 211
ANKUNFTS-ZEIT, 307
ANLAGERUNG, 30
ANNELIDA, 212
ANOMALIE, 214
ANOMALODESMATA, 213
ANORGAN.ENTSTEHUNG, 2250
ANORGAN.SAEURE, 2248
ANORGAN.SUBSTANZ, 2249
ANORGANISCH, 2247
ANORTHOSIT, 216
ANORTHOSITISIERUNG, 217
ANPASSUNG, 63
ANREICHERUNG, 1376
ANSCHLIFF, 3538
ANSTEHENDES, 457
ANTEKLISE, 218
ANTHOCYATHEA, 219
ANTHOZOA, 220
ANTHRAXOLITH, 222
ANTHRAZIT, 221
ANTIARCHI, 223
ANTIFORM, 227
ANTIKLINALE, 224
ANTIKLINORIUM, 225
ANTIMONIAT, 228
ANTIPATHARIA, 230
ANTIPERTHIT, 231
ANTISTRESS-MINERAL, 232
ANTITHET.VERWERFUNG, 233
ANUROMORPHA, 234
ANWACHS-STREIFEN, 1922
APERTUR, 235
APHANIT.GEFUEGE, 236
APIKAL, 237
APLACOPHORA, 238
APLIT, 239
APLIT-GEFUEGE, 240

APOGAEUM, 241
APOLLO-PROGRAMM, 243
APOMAGMAT.GESTEIN, 242
APOMAGMAT.VORKOMMEN, 244
APOPHYSE, 245
APSHERONIUM, 251
APSIDOSPONDYLI, 252
APT, 253
AQUALF, 254
AQUENT, 255
AQUEPT, 257
AQUERT, 258
AQUICLUD, 259
AQUITAN, 263
AQUITARD, 264
AQUOD, 265
AQUOLL, 266
AQUOX, 267
AQUULT, 268
AR, 296
AR-AR-DATIERUNG, 297
ARACHNIDA, 269
ARBEITSWESEN, 2664
ARCHAEIDEN, 279
ARCHAEOCOPIDA, 272
ARCHAEOCYATHA, 273
ARCHAEOCYATHEA, 274
ARCHAEOGASTROPODA, 275
ARCHAEOLOG.STAETTE, 280
ARCHAEOLOGIE, 281
ARCHAEOMAGNETISMUS, 282
ARCHAEOPTERYGES, 276
ARCHAEORNITHES, 277
ARCHAIKUM, 278
ARCHIPOLYPODA, 283
ARCHOSAURIA, 284
ARCOIDA, 285
ARENIG, 291
ARENIT, 292
ARENOSOL, 293
ARGID, 294
ARID.MILIEU, 299
ARIDISOL, 300
ARIDITAET, 301
ARKOSE, 302
ARKT.ABFOLGE, 287
ARKTISCH, 286
ARM, 2471
ARMERZ, 1038
ARMORIKAN.OROGENESE, 304
AROMAT.KOHLENWASSERSTOFF, 305
ARROYO, 308
ARSENAT, 309
ART-VERSCHIEDENHEIT, 4193
ARTEFAKT, 319
ARTERIT, 311
ARTES.BRUNNEN, 314

ARTES.GRUNDWASSER, 313
ARTESISCH, 312
ARTHRODIRA, 315
ARTHROPLEURIDA, 316
ARTHROPODA, 317
ARTICULATA, 318
ARTINSK, 323
ARTIODACTYLA, 324
AS, 310
ASBEST, 325
ASCHEGEHALT, 330
ASCHEN-KEGEL, 770
ASCHEN-REGEN, 331
ASCHEN-STROM, 332
ASEISM.KONTINENTAL-RAND, 328
ASEISM.REGION, 329
ASHGILL, 333
ASPHALT, 334
ASPHALTIT, 335
ASSYNT.OROGENESE, 342
ASTENOSPHAERE, 348
ASTERISMUS, 344
ASTEROID, 345
ASTEROIDEA, 346
ASTEROZOA, 347
ASTRAPOTHERIA, 349
ASTROBLEM, 350
ASTROGEOLOGIE, 351
ASTRONOMIE, 352
ASYMMETR.FALTE, 353
ASZENDENZ-THEORIE, 326
AT, 343
ATAXIT, 354
ATEKTON.BILDUNG, 215
ATLANT.ABFOLGE, 355
ATLAS, 356
ATMOGEN.HERKUNFT, 357
ATMOSPHAER.ARGON, 359
ATMOSPHAER.MILIEU, 4316
ATMOSPHAER.NIEDERSCHLAG, 360
ATMOSPHAER.VERWITTERUNG, 4809
ATMOSPHAERE, 358
ATMUNG, 3835
ATOLL, 362
ATOM-ABSORPTION, 363
ATOM-PACKUNG, 365
ATTERBERG-GRENZEN, 367
ATTRITION, 368
AU, 1833
AUBRIT, 369
AUFBEREITUNG, 1252
AUFBEULUNG, 4686
AUFGEARBEITET, 3845
AUFGEARBEITET.FOSSIL, 3846
AUFGEARBEITET.GESTEIN, 3774
AUFGELOEST.STOFF, 4175
AUFLOESUNG, 1228

AUFLOESUNGS-MERKMAL, 4178
AUFSAMMLUNG, 838
AUFSCHIEBUNG, 4520
AUFSCHLUSS, 3206
AUFSCHLUSS-ALTER, 1489
AUFSCHLUSS-BOHRUNG, 1484
AUFSCHLUSS-PROFIL, 3460
AUFSCHOTTERUNG, 142
AUFSCHOTTERUNGS-EBENE, 33
AUFSCHUETTUNG, 2428
AUFTAU-ZONE, 57
AUFTRIEB, 4687
AUFTRIEBS-DRUCK, 4666
AUFWOELBUNG, 3515
AUGEN-GEFUEGE, 370
AULAKOGEN, 372
AUSBREITUNGS-ZENTRUM, 4223
AUSBRUCH, 4767
AUSDUENNEN, 3449
AUSFLOCKUNG, 1568
AUSGANGS-PRODUKT, 3303
AUSGESTORBEN.TAXON, 1495
AUSKEILEN, 4811
AUSLAENGUNG, 4456
AUSLAUGUNG, 2550
AUSLOESCHUNG, 3138
AUSSCHACHTUNG, 1462
AUSSCHLAEMMUNG, 1356
AUSSEN-SKELETT, 4471
AUSSICKERUNG, 4008
AUSSTERBEN, 1496
AUSTAUSCH-KAPAZITAET, 1464
AUSTRALIT, 374
AUSTROCKNUNG, 1135
AUSWASCHUNG, 2465
AUSWUERFLING, 1315
AUTHIGEN.MINERAL, 376
AUTHIGENESE, 375
AUTO-METAMORPHOSE, 383
AUTO-METASOMATOSE, 384
AUTOCHTHONIE, 377
AUTOKORRELATION, 378
AUTOLYSE, 380
AUTOMAT.KARTOGRAPHIE, 381
AUTOMATION, 382
AUTORADIOGRAPHIE, 385
AUTOTROPH.ORGANISMUS, 386
AUTUNIUM, 387
AVES, 390
AZETOLYSE, 35
AZIDITAET, 40
AZONAL.BODEN, 394
B, 552
B-SCHICHT, 395
BA, 413
BACH, 985
BACKSET-BED, 399

BACTERIA, 400
BACTRITOIDEA, 402
BADLANDS, 403
BAENDER-STRUKTUR, 409
BAGGERN, 1251
BAIKAL-OROGENESE, 404
BAJADA, 405
BAJOCIUM, 406
BAKTERIOGEN.ENTSTEHUNG, 401
BANKUNG, 2704
BARCHAN, 411
BARRE, 410
BARREME, 414
BARRIERE, 416
BARRIERE-LAGUNE, 419
BARRIERE-RIFF, 420
BARRIERE-STRAND, 418
BARTONIUM, 421
BARYT, 412
BASAL, 422
BASAL-KONGLOMERAT, 424
BASALT, 425
BASALT-SCHICHT, 426
BASANIT, 427
BASCHKIR, 431
BASEMENT, 429
BASIFIZIERUNG, 432
BASIS-METALL, 428
BASIS-SPALTBARKEIT, 423
BATHOLITH, 434
BATHONIUM, 435
BATHYAL-MILIEU, 436
BATHYMETR.KARTE, 437
BATHYMETRIE, 438
BATRACHOSAURIA, 439
BAUGRUND, 1627
BAUSTEIN, 588
BAUSTOFF, 911
BAUWESEN, 586
BAUWUERDIGKEIT, 1481
BAUXIT, 440
BAUXITISIERUNG, 441
BAVENO-GESETZ, 442
BAY, 443
BE, 471
BE 10-BE 9, 472
BEACH-ROCK, 448
BECKE-LINIE, 451
BEGRABEN.RINNE, 598
BEGRABEN.TAL, 599
BELASTUNG, 2554
BELASTUNGS-MARKE, 2555
BELEMNOIDEA, 459
BELT.OROGENESE, 460
BELUEFTUNG, 69
BENIOFF-ZONE, 462
BENNETTITALES, 463

BENTHON.MILIEU, 464
BENTHOS, 465
BENTONIT, 466
BEOBACHTUNG, 3081
BERG-KAMM, 988
BERGBAU, 2883
BERGBAU-BEHOERDE, 595
BERGBAU-TECHNIK, 2884
BERGMASSIV, 2950
BERGSCHRUND, 467
BERGWERK, 2861
BERGWIRTSCHAFT, 2867
BERICHT, 3818
BERME, 469
BERNSTEIN, 160
BERRIAS, 470
BERUFS-AUSBILDUNG, 1309
BESCHLEUNIGUNGS-AUFZEICHNUNG, 27
BESCHLEUNIGUNGS-MESSER, 28
BETA-STRAHLUNG, 473
BETON, 879
BETON-DAMM, 880
BEUGUNG, 1180
BEWAESSERUNG, 2314
BEWEGUNG, 2951
BEWEGUNGS-SPUR, 4565
BEWERTUNG, 1457
BEYRICHICOPINA, 474
BI, 520
BIAXIAL, 475
BIBLIOGRAPHIE, 477
BIEGE-FALTE, 874
BILANZ, 407
BIMS, 3683
BINAER.SYSTEM, 479
BINDEMITTEL, 684
BINDIG.BODEN, 830
BINDUNGSART, 364
BINOMIAL-VERTEILUNG, 480
BIO-FAZIES, 490
BIO-GEOCHEM.PROSPEKTION, 495
BIO-GEOCHEMIE, 496
BIOCHEM.SEDIMENTATION, 482
BIOCHORE, 483
BIOCHRONE, 484
BIOCHRONOLOGIE, 485
BIOCOENOSE, 481
BIODEGRADATION, 489
BIOFAZIES-KARTE, 491
BIOGEN.BILDUNG, 3172
BIOGEN.STRUKTUR, 494
BIOGEN.WIRKUNG, 493
BIOGENESE, 492
BIOGEOGRAPHIE, 497
BIOGRAPHIE, 498
BIOHERM, 499
BIOKLAST, 486

BIOKLAST.KALK, 487
BIOKLAST.SEDIMENTATION, 488
BIOLITH, 500
BIOLOG.EVOLUTION, 1461
BIOLOG.METAMORPHOSE, 2789
BIOLOG.ZYKLUS, 501
BIOLOGIE, 502
BIOMASSE, 503
BIOMETRIE, 504
BIOMIKRIT, 505
BIORHEXISTASIE, 506
BIOSOM, 507
BIOSPARIT, 508
BIOSPHAERE, 509
BIOSTRATIGRAPH.EINHEIT, 510
BIOSTRATIGRAPHIE, 511
BIOSTROM, 512
BIOTA, 513
BIOTOP, 514
BIOTURBATION, 515
BIOTYPUS, 516
BIOZONE, 517
BIRCH-DISKONTINUITAET, 518
BITUMEN, 521
BITUMINISIERUNG, 522
BITUMINOES.KOHLE, 523
BITUMINOES.SCHIEFER, 524
BK, 468
BLAST.GEFUEGE, 525
BLASTOIDEA, 526
BLATT, 2469
BLATT-VERSCHIEBUNG, 4293
BLATTOPTEROIDA, 527
BLEICH-ERDE, 528
BLOCK, 530
BLOCK-BILDUNG, 532
BLOCK-DIAGRAMM, 531
BLOCK-GLETSCHER, 3886
BLOCK-STROM, 560
BLOWHOLE, 533
BODEN, 4146
BODEN-AUSGANGSGESTEIN, 3302
BODEN-AUSWASCHUNG, 4151
BODEN-BEWEGUNG, 1908
BODEN-BEWIRTSCHAFTUNG, 4152
BODEN-BILDUNG, 3326
BODEN-EROSION, 4148
BODEN-FEUCHTE, 2911
BODEN-HEBUNG, 1970
BODEN-HORIZONT, 4149
BODEN-KARTE, 4153
BODEN-MECHANIK, 4154
BODEN-PROBENNAHME, 4156
BODEN-PROFIL, 4155
BODEN-SEDIMENT, 554
BODEN-STABILISIERUNG, 4158
BODEN-VERBESSERUNG, 4150

BODEN-VERFORMUNG, 4147
BODENFRACHT, 452
BODENKUNDE, 4157
BODENSCHAETZE, 2992
BODENSENKUNG, 2426
BOGEN-STAUDAMM, 271
BOGHEAD-KOHLE, 538
BOGIG, 288
BOHNERZ, 449
BOHR-GERAET, 1255
BOHR-GUT, 1039
BOHR-KERN, 958
BOHR-KOPF, 4815
BOHR-LOG, 2562
BOHR-ORGANISMEN, 3889
BOHR-PROFIL, 550
BOHR-SCHIFF, 1257
BOHR-SCHLAMM, 1256
BOHR-VOLLENDUNG, 4814
BOHRBARKEIT, 1253
BOHRLOCH-MESSUNG, 4816
BOHRUNG, 549
BOMBE, 539
BONE-BED, 540
BOOLE'SCHE-ALGEBRA, 542
BORALF, 543
BORAT, 544
BORAX, 545
BORDERLAND, 547
BOROLL, 551
BOTTOMSET-BED, 555
BOUDINAGE, 556
BOUGUER-ANOMALIE, 557
BOWEN'S-REAKTIONS-SERIE, 563
BR, 581
BRACHIOPODA, 565
BRACHIOPTERYGII, 566
BRACHY-FALTE, 567
BRACK-WASSER, 568
BRACK-WASSER-MILIEU, 569
BRAGG-WINKEL, 570
BRANCHIOPODA, 572
BRANCHIURA, 573
BRAUN-ERDE, 583
BRAUNKOHLE, 582
BRAVAIS-GITTER, 574
BRECHUNGS-INDEX, 2209
BRECHUNGS-VERMOEGEN, 3788
BREKZIE, 575
BROMID, 580
BRUCH, 1632
BRUCH-BILDUNG, 1635
BRUCH-FALTE, 1220
BRUCH-FESTIGKEIT, 1633
BRUCH-STUFE, 1511
BRUCH-SYSTEM, 1634
BRUCH-TEKTONIK, 1515

BRUCH=ZONE, 1513
BRUCHSTUECK, 783
BRUECKE, 577
BRUNNEN, 4801
BRUNNEN=FILTER, 4817
BRYOPHYTA, 584
BRYOZOA, 585
BUCH=BESPRECHUNG, 541
BUCHT, 478
BUHNE, 1901
BULIMINACEA, 589
BULLARD=DISKONTINUITAET, 592
BUNTSANDSTEIN, 593
BURDIGAL, 594
BYSSUS, 601
C, 642
C=14=DATIERUNG, 646
C=SCHICHT, 602
C13=C12, 645
CA, 615
CAFEM.ZUSAMMENSETZUNG, 605
CALABRIUM, 606
CALCISPONGEA, 612
CALDERA, 617
CALICHE, 619
CALLOVIUM, 622
CAMAROIDEA, 623
CAMBISOL, 624
CAMPAN, 626
CAMPTOSTROMATOIDEA, 627
CANYON, 630
CAP=ROCK, 631
CARADOC, 638
CARNIVORA, 657
CARTERINACEA, 658
CASSIDULINACEA, 664
CATAGRAPHIA, 669
CATENA, 673
CAVINGS, 680
CAYTONIALES, 681
CD, 604
CE, 699
CENOMAN, 686
CENOZONE, 689
CEPHALOCARIDA, 691
CEPHALODISCIDA, 692
CEPHALOPODA, 693
CERATITIDA, 695
CERATOMORPHA, 696
CERIANTHARIA, 697
CERIANTIPATHARIA, 698
CETACEA, 701
CF, 620
CHAETOGNATHA, 702
CHALKOPHIL.ELEMENT, 703
CHALKOSPHAERE, 704
CHANDLER=SCHWANKUNG, 706

CHAOT.GEFUEGE, 709
CHARAKTERISTIK, 710
CHARN.OROGENESE, 711
CHARNOCKIT, 712
CHAROPHYTA, 713
CHASSIGNIT, 714
CHATTIUM, 716
CHEILOSTOMATA, 717
CHELAT=BILDUNG, 718
CHELICERATA, 719
CHELONIA, 720
CHEM.ANALYSE, 721
CHEM.AUSFAELLUNG, 3596
CHEM.EIGENSCHAFT, 729
CHEM.EINFLUSS, 722
CHEM.ENTSTEHUNG, 736
CHEM.EXPLOSION, 726
CHEM.FOSSIL, 727
CHEM.GLEICHGEWICHT, 724
CHEM.INDUSTRIE, 728
CHEM.KOORDINATION, 943
CHEM.REDUKTION, 731
CHEM.SEDIMENTATION, 723
CHEM.SYSTEM, 733
CHEM.VERHAELTNIS, 730
CHEM.VERWITTERUNG, 735
CHEMISMUS, 725
CHEMOREMANENT.MAGNETISIERUNG, 732
CHENIER, 737
CHILOPODA, 742
CHIROPTERA, 744
CHITIN, 745
CHITINOZOA, 746
CHLORID, 747
CHLORINIERUNG, 748
CHLORINITAET, 750
CHLORIT=SCHIEFER, 751
CHLORITISIERUNG, 752
CHLOROPHYLL, 753
CHLOROPHYTA, 754
CHONDRE, 757
CHONDRICHTHYES, 755
CHONDRIT, 756
CHORDATA, 758
CHROMAT, 759
CHROMATOGRAPHIE, 760
CHRONOSTRATIGRAPH.EINHEIT, 762
CHRONOSTRATIGRAPHIE, 763
CHRONOZONE, 764
CHRYSOPHYTA, 765
CILIATA, 766
CILIOPHORA, 767
CIPOLIN, 771
CIPW=NORM, 603
CIRRIPEDIA, 776
CL, 749
CLADOCERA, 778

CLARAIN, 780
CLARKE, 781
CLUSTER=ANALYSE, 799
CLYMENIIDA, 800
CM, 1034
CO, 818
COATED=GRAIN, 817
COBOL, 819
COCCOLITHOPHORALES, 820
CODA=WELLE, 821
COELENTERATA, 823
COENOTHECALIA, 824
COLEOPTEROIDA, 833
COLLEMBOLA, 840
COLLUVIUM, 843
COLUMNARIINA, 851
COMAGMAT.GENESE, 852
CONCHOSTRACA, 876
CONDYLARTHRA, 884
CONE=IN=CONE, 886
CONIAC, 894
CONIFERALES, 895
CONJUGATE=FOLD, 897
CONNATE=WATER, 898
CONOCARDIOIDA, 899
CONODONTA, 901
CONODONTEN=APPARAT, 900
CONRAD=DISKONTINUITAET, 902
CONULARIDA, 937
COPEPODA, 944
CORALLIMORPHARIA, 950
CORALLINACEAE, 951
CORALLIT, 952
CORDAITALES, 953
CORDIERIT=AMPHIBOLIT=FAZIES, 954
CORDILLERE, 955
CORIOLIS=BESCHLEUNIGUNG, 959
CORYNEXOCHIDA, 966
COTYLOSAURIA, 977
CR, 761
CREODONTA, 987
CRINOIDEA, 991
CRINOZOA, 992
CROCODILIA, 996
CROSSOPTERYGII, 999
CRUSTACEA, 1002
CRYPTODONTA, 1007
CRYPTOSTOMATA, 1009
CS, 700
CTENOSTOMATA, 1029
CU, 946
CURIE=PUNKT, 1033
CUSP, 1037
CYANOPHYTA, 1040
CYCADALES, 1042
CYCADOFILICALES, 1043
CYCLOCYSTOIDEA, 1048

CYCLOSTOMATA, 1050
CYNOMORPHA, 1054
CYSTIPHYLLINA, 1055
CYSTOIDEA, 1056
D-SCHICHT, 1057
D/H-VERHAELTNIS, 1058
DAEMPFUNG, 366
DAN, 1060
DARCY-GESETZ, 1061
DARWINISMUS, 1062
DASYCLADACEAE, 1063
DATEN, 1064
DATEN-ANALYSE, 1066
DATEN-AUFBEREITUNG, 1068
DATEN-ERFASSUNG, 1065
DATEN-SPEICHERUNG, 1071
DATEN-VERARBEITUNG, 1069
DATEN-WIEDERGEWINNUNG, 1070
DATENBANK, 1067
DECAPODA, 1078
DECKE, 2986
DECKEN-INTRUSION, 4066
DECKGEBIRGE, 979
DECKGEBIRGS-TEKTONIK, 3996
DEFLATION, 1095
DEGRADATION, 1099
DEHNUNG, 1187
DEHYDRATISIERUNG, 1101
DEICH, 3986
DEINOTHERIOIDEA, 1102
DEKKAN-TRAPP, 1080
DEKLINATION, 1081
DEKONVOLUTION, 1084
DEKREPITATIONS-METHODE, 1086
DELTA, 1103
DELTA-MILIEU, 1104
DELTA-SEDIMENTATION, 1105
DEMOSPONGEA, 1107
DENDRIT, 1108
DENDRIT.ENTWAESSERUNG, 1109
DENDRO-CHRONOLOGIE, 1110
DENDROGRAMM, 1111
DENDROIDEA, 1112
DENTITION, 1115
DENUDATION, 1116
DEPRESSION, 1121
DERMAPTEROIDA, 1127
DERMOPTERA, 1128
DESMIDEEN-FLORA, 1137
DESMOCERATIDA, 1138
DESMOSTYLIA, 1139
DESZENDENZ-THEORIE, 1130
DETERMINISMUS, 1140
DETERMINIST.MODELL, 1141
DETRIT.LAGERSTAETTE, 2725
DETRIT.REMANENT.MAGNETISIERUNG, 1143
DETRIT.SEDIMENTATION, 1144

DETRIT.ZUSAMMENSETZUNG, 1142
DEUTERIUM, 1146
DEVON, 1151
DIACHRONISMUS, 1153
DIADOCHIE, 1154
DIAGENESE, 1155
DIAGENET.GEFUEGE, 3990
DIAGNOSE, 1156
DIAGRAMM, 1157
DIALYSE, 1158
DIAMAGNET.EIGENSCHAFT, 1159
DIAMANT, 1161
DIAMIKTIT, 1160
DIAPHTHORESE, 1162
DIAPHTHORIT, 1163
DIAPIR, 1164
DIAPIRISMUS, 1165
DIASTEM, 1166
DIASTROPHISMUS, 1167
DIATOMEAE, 1169
DIATOMEEN-ERDE, 1168
DIATOMIT, 1170
DIATREMA, 1171
DICHTE, 1113
DICHTE-GRADIENT, 3689
DICHTEST.PACKUNG, 798
DICKE, 4513
DICOTYLEDONEAE, 1172
DICTYONELLIDINA, 1173
DIELEKTR.KONSTANTE, 1174
DIFFERENTIAL-KOMPAKTION, 1175
DIFFERENTIATIONS-INDEX, 1179
DIFFUSION, 1181
DIFFUSIONS-VERMOEGEN, 1182
DIGITAL-TECHNIK, 1183
DILATANZ, 1186
DIMORPHISMUS, 1189
DINOCERATA, 1190
DINOFLAGELLATA, 1191
DINOSAURIA, 1192
DIOGENIT, 1193
DIORIT, 1194
DIPLEUROZOA, 1200
DIPLOID.MINERAL, 1201
DIPLOPODA, 1202
DIPLORHINA, 1203
DIPNOI, 1205
DIPOL-MOMENT, 1206
DIPTERA, 1207
DISCOASTEROIDEA, 1210
DISCORBACEA, 1213
DISHARMON.FALTE, 1219
DISKONTINUITAET, 1212
DISKORDANT, 1215
DISKORDANT.ALTER, 1216
DISKORDANZ, 4646
DISKRIMINANZ-ANALYSE, 1218

DISLOKATIONS-BREKZIE, 1221
DISPERSIONS-AUREOLE, 1223
DISSEMINATION, 1226
DISTRIKT, 2770
DIVERGENZ, 1232
DOCODONTA, 1233
DOKUMENTATION, 2229
DOLERIT, 1234
DOLINE, 1235
DOLOMIT-GESTEIN, 1238
DOLOMIT-KALK, 1236
DOLOMITISIERUNG, 1237
DOM, 1240
DOMAENEN-STRUKTUR, 1239
DOMERIUM, 1241
DOPPELBRECHUNG, 519
DRAINUNG, 1247
DRAPIER-FALTE, 1248
DREHUNGS-VERMOEGEN, 772
DREI-PLATTEN-PUNKT, 4599
DRUCK, 3607
DRUCK-LOESUNG, 3608
DRUCK-MESSUNG, 2556
DRUCK-SONDIERUNG, 3340
DRUCK-TEMPERATUR-BEDINGUNGEN, 3229
DRUCK-WIDERSTAND, 866
DRUMLIN, 1259
DRUSE, 1260
DTA, 1176
DUENE, 1261
DUENGEMITTEL, 1541
DUENNSCHLIFF, 4514
DUENUNG, 3979
DUERRE, 1258
DUKTIL.SUBSTANZ, 2660
DURAIN, 1262
DURCHFUEHRBARKEITS-STUDIE, 1520
DURCHSCHNITTSWERT, 389
DURCHSICKERUNG, 2470
DY, 1270
DYKE, 1184
DYKE-SCHWARM, 1185
DYNAM.GEOLOGIE, 1265
DYNAM.GLEICHGEWICHT, 1264
DYNAMIK, 1267
DYNAMO-METAMORPHOSE, 1266
DYNAMO-THEORIE, 1268
DYNAMO-THERM.METAMORPHOSE, 1269
E-SCHICHT, 1271
EBENE, 2595
EBURN.OROGENESE, 1289
ECHINODERMATA, 1291
ECHINOIDEA, 1292
ECHINOZOA, 1293
ECHO-LOTUNG, 1294
ECTOTROPHA, 1303
EDEL-METALL, 3037

EDELGAS, 2224
EDELSTEIN, 3595
EDENTATA, 1305
EDRIOASTEROIDEA, 1307
EDRIOBLASTOIDEA, 1308
EI, 1311
EIFEL, 1313
EIGEN-SCHWINGUNG, 1642
EIGENPOTENTIAL, 4033
EIGENSCHAFT, 3638
EINDIMENSIONAL.MODELL, 3120
EINEBNUNG, 3468
EINENGUNG, 4084
EINFALLEN, 1195
EINFALLS-HANG, 1198
EINKRISTALL-AUFNAHME, 4116
EINSCHLAG-KRATER, 2187
EINSCHLAG-STRUKTUR, 2192
EINSCHLUSS, 2200
EINSPRENGLING, 3395
EINSTURZ, 834
EINSTURZ-BREKZIE, 835
EINSTURZ-CALDERA, 836
EINSTURZ-STRUKTUR, 837
EINZUGSGEBIET, 1245
EIS, 2155
EIS-BARRIERE, 2161
EIS-BEWEGUNG, 2158
EIS-FELD, 2157
EIS-KEIL, 2162
EIS-TEKTONIK, 1809
EIS-TRANSPORT, 2159
EISBERG, 2163
EISEN-BAKTERIEN, 2310
EISEN-FORMATION, 2311
EISEN-GEHALT, 1539
EISEN-INDUSTRIE, 2309
EISEN-KRUSTE, 1534
EISEN-MANGAN-GEHALT, 1537
EISENHALTIG.BODEN, 1540
EISENSTEIN, 2312
EISERN.HUT, 1838
EKLOGIT, 1295
EKLOGIT-FAZIES, 1296
ELASMOBRANCHII, 1316
ELAST.EIGENSCHAFT, 1320
ELAST.GRENZE, 1318
ELAST.MEDIUM, 1319
ELAST.VERFORMUNG, 1321
ELAST.WELLE, 1322
ELASTIZITAET, 1323
ELASTIZITAETS-KONSTANTE, 1317
ELASTIZITAETS-MODUL, 591
ELEKTR.EIGENSCHAFT, 1331
ELEKTR.FELD, 1328
ELEKTR.LEITFAEHIGKEIT, 1326
ELEKTR.METHODE, 1329

ELEKTR.POLARISATION, 1330
ELEKTR.POTENTIAL, 1325
ELEKTR.PROSPEKTION, 1327
ELEKTR.VERMESSUNG, 1333
ELEKTR.WIDERSTAND, 3832
ELEKTRIZITAET, 1334
ELEKTRO-LOG, 1324
ELEKTRO-MAGNET.BOHRLOCH-MESSUNG, 1343
ELEKTRO-MAGNET.FELD, 1340
ELEKTRO-MAGNET.METHODE, 1341
ELEKTRO-MAGNET.VERMESSUNG, 1342
ELEKTRO-MAGNETIK, 1344
ELEKTRO-OSMOSE, 1335
ELEKTROCHEM.EIGENSCHAFT, 1336
ELEKTRODE, 1337
ELEKTROLYSE, 1338
ELEKTROLYT, 1339
ELEKTRON, 1345
ELEKTRONEN-BEUGUNGS-ANALYSE, 1346
ELEKTRONEN-MIKRO-SONDEN-DATEN, 1352
ELEKTRONEN-MIKROSKOP, 1348
ELEKTRONEN-MIKROSKOP.DATEN, 1350
ELEKTRONEN-MIKROSKOPIE, 1349
ELEMENT-VERTEILUNG, 3312
ELEMENTAR-TEILCHEN, 1353
ELEMENTAR-ZELLE, 4661
ELEMENTEN-MIGRATION, 2856
ELEPHANTOIDEA, 1354
ELUVIUM, 1357
EMBRECHIT, 1359
EMBRITHOPODA, 1360
EMERGENZ, 1361
EMISSIONS-SPEKTROMETRIE, 1362
EMPFANGS-GERAET, 3764
EMS, 1364
ENCRINIT, 1366
END-PRODUKT, 1367
ENDEM.POPULATION, 1368
ENDOCERATOIDEA, 1369
ENDOGEN.VORGANG, 1370
ENDOTHERM.REAKTION, 1372
ENERGIE, 1373
ENSIAL.GEOSYNKLINALE, 1377
ENSIMAT.GEOSYNKLINALE, 1378
ENSTATITHALTIG.CHONDRIT, 1379
ENTDECKUNG, 1217
ENTDOLOMITISIERUNG, 1087
ENTEROLITH.FALTE, 1380
ENTEROPNEUSTA, 1381
ENTGASUNG, 1097
ENTGIFTUNG, 1083
ENTGLASUNG, 1150
ENTHALPIE, 1382
ENTISOL, 1383
ENTKIESELUNG, 1136
ENTMAGNETISIERUNG, 1106
ENTMISCHUNG, 1490

ENTRENCHED STREAM, 1384
ENTROPIE, 1385
ENTSALZUNG, 1129
ENTWICKLUNG, 1147
ENTWICKLUNGS-BOHRUNG, 1148
EOCRINOIDEA, 1390
EOSUCHIA, 1395
EOZAEN, 1389
EPIBIOTISMUS, 1398
EPIBOLIT, 1399
EPIDOT-AMPHIBOLIT-FAZIES, 1401
EPIDOTISIERUNG, 1402
EPIEUGEOSYNKLINALE, 1403
EPIGENESE, 1404
EPIKONTINENTAL.BECKEN, 2290
EPIMAGMAT.EFFEKT, 1145
EPIMAGMAT.GENESE, 1405
EPIROGENESE, 1396
EPITAXIE, 1406
EPITHERMAL.VORKOMMEN, 1407
EPIZENTRUM, 1400
EPIZONE, 1408
EPOCHE, 1409
EQUISETALES, 1417
ER, 1421
ERATOSTHEN.ALTER, 1420
ERD-ALTER, 81
ERD-AUFBAU, 1276
ERD-DAMM, 1275
ERD-GESTALT, 4055
ERD-GEZEITEN, 1280
ERD-KERN, 957
ERD-KRUSTE, 1001
ERD-MANTEL, 2665
ERD-ROTATION, 1278
ERD-SCHRUMPFUNG, 933
ERDALKALI-METALL, 129
ERDARBEITEN, 1281
ERDBEBEN, 1284
ERDBEBEN-SCHWARM, 1287
ERDBEBEN-TECHNIK, 1285
ERDBEBENSICHER.BAU, 327
ERDDRUCK, 1277
ERDE-MOND-KOPPLUNG, 1282
ERDFLIESSEN, 1283
ERDGAS, 2989
ERDOBERFLAECHE, 1904
ERDOEL, 3379
ERDOEL-BOHRUNG, 3112
ERDOEL-FALLE, 3111
ERDOEL-GEOLOGIE, 3381
ERDOEL-SICKERUNG, 3109
ERDOEL-TECHNIK, 3380
ERDRUTSCH, 2431
ERDWACHS, 4808
ERG, 1422
ERHALTUNG, 3606

ERHEBUNG, 155
ERHOLUNG, 3769
ERNAEHRUNG, 3073
ERODIERBARKEIT, 1423
EROSION, 1424
EROSIONS-DISKORDANZ, 1211
EROSIONS-EINFLUSS, 1425
EROSIONS-MARKE, 3970
EROSIONS-MORPHOLOGIE, 1427
EROSIONS-OBERFLAECHE, 1429
EROSIONS-REST, 1428
EROSIONS-ZYKLUS, 1426
ERRAT.BLOCK, 1430
ERSATZ, 3817
ERSCHLIESSUNGS-ARBEITEN, 3680
ERUPTION, 1432
ERWARTUNG, 2709
ERZ, 3151
ERZ-ANALYSE, 336
ERZ-AUFBEREITUNGS-ANLAGE, 3155
ERZ-BESTANDTEIL, 3159
ERZ-FRACHT, 3163
ERZ-GANG, 2559
ERZ-GEHALT, 3156
ERZ-HOEFFIGKEIT, 3599
ERZ-LAGERSTAETTE, 3154
ERZ-MIKROSKOPIE, 3158
ERZ-PROSPEKTION, 2868
ERZ-QUALITAET, 3160
ERZ-QUELLE, 3162
ERZFALL, 3161
ERZHALTIG, 2769
ERZKOERPER, 3152
ERZSCHLAMM, 2064
ES, 1314
ESCARPMENT, 1433
ESKER, 1434
ESSEXIT, 1435
EU, 1448
EUCILIATA, 1442
EUCRIT, 1443
EUGEOSYNKLINALE, 1445
EUGLENOPHYTA, 1446
EUKRISTALLIN.GEFUEGE, 1444
EURYAPSIDA, 4389
EURYHALIN.ORGANISMUS, 1449
EURYTHERMAL.ORGANISMUS, 1450
EUSELACHII, 1451
EUSTATIK, 1452
EUTEKTIKUM, 1454
EUTROPHIERUNG, 1455
EUXIN.MILIEU, 1456
EVAPORATION, 1458
EVAPORIT, 1459
EVAPOTRANSPIRATION, 1460
EXFOLIATION, 1466
EXHALATIONS-LAGERSTAETTE, 1467

EXINE, 1468
EXINIT, 1469
EXKURSION, 1465
EXOBIOLOGIE, 1470
EXOGEN.VORGANG, 1471
EXOTHERM.REAKTION, 1473
EXPANDIEREND.MATERIAL, 1474
EXPANSION, 1475
EXPEDITION, 1476
EXPERIMENT, 1479
EXPERIMENTAL-SEISMIK, 1478
EXPERIMENTELL.PETROLOGIE, 1477
EXPLORATION, 3641
EXPLOSION, 1485
EXPLOSIONS-BREKZIE, 1486
EXPLOSIONS-QUELLE, 1487
EXPORT, 1488
EXPOSITIONS-RELIEF, 3809
EXSURGENZ, 1491
EXTENSOMETER, 1492
EXTERN.ENTWAESSERUNG, 1472
EXTERNIDEN, 1493
EXTRAPOLATION, 1497
EXTRATERRESTR.GEOLOGIE, 1498
EXZENTRIZITAET, 1290
F, 1589
F-SCHICHT, 1499
FAECHER, 1504
FAELTELUNG, 3510
FAERBUNG, 4230
FAKTOREN-ANALYSE, 1502
FALL-STUDIE, 663
FALLE, 4580
FALTE, 1602
FALTEN-BREKZIE, 1605
FALTEN-GUERTEL, 1604
FALTEN-STIRN, 3058
FALTEN-SYSTEM, 1606
FALTEN-TEKTONIK, 1607
FALTENACHSE, 1603
FALTUNG, 1608
FAMENNE, 1503
FANGLOMERAT, 1505
FARB-INDEX, 847
FARB-MARKIERUNG, 848
FARB-ZENTRUM, 846
FARBE, 845
FARBSTOFF, 3234
FASSUNGS-HYDRODYNAMIK, 672
FAUNA, 1516
FAUNEN-LISTE, 1517
FAUNEN-PROVINZ, 1518
FAUNEN-ZONE, 339
FAUNIZONE, 1519
FAZIES, 1501
FAZIES-ABFOLGE-GESETZ, 2461
FE, 2308

FEHLER, 1431
FEIN-SCHICHTUNG, 453
FEINFRAKTION, 1549
FELDINVERSION, 1743
FELDSPAT, 1524
FELDSPATISIERUNG, 1525
FELS-MECHANIK, 3888
FELSISCH.CHEMISMUS, 1527
FELSRUTSCH, 3891
FELSSTURZ, 3884
FEMISCH.CHEMISMUS, 1528
FENIT, 1529
FENITISIERUNG, 1530
FENSTER, 4830
FERNE, 3814
FERRALSOL, 1532
FERRI-EISEN, 1533
FERRO-EISEN, 1538
FERROD, 1535
FERROMAGNET.EIGENSCHAFT, 1536
FEST.ABFALL, 4168
FEST.LOESUNG, 4167
FEST.PHASE, 4166
FESTIGKEIT, 4289
FESTLAND, 2425
FETT-GLANZ, 1889
FETT-SAEURE, 1506
FEUCHT.MILIEU, 2074
FEUCHT.ZONE, 2073
FEUCHTGEBIET, 4822
FEUCHTGEBIETS-SEDIMENTATION, 4377
FEUCHTIGKEIT, 2075
FEUERFEST.ROHSTOFF, 3791
FEUERFEST.TON, 1552
FILICOPSIDA, 1543
FILTER, 1545
FILTERUNG, 1546
FILTRATION, 1547
FINANZIERUNG, 1548
FINIT.ELEMENTE-ANALYSE, 1551
FINITE-DIFFERENZ-ANALYSE, 1550
FIRMA, 858
FIRN, 1553
FISSIPEDA, 1557
FIXISMUS, 1559
FJORD, 1560
FLACHLAND, 3465
FLACHWASSER-MILIEU, 4053
FLAECHE, 3470
FLAECHEN-NORMALE, 1938
FLAMMEN-PHOTOMETRIE, 1562
FLAMMEN-STRUKTUR, 1563
FLANDRIUM, 1564
FLASER-GEFUEGE, 1565
FLECKEN-GEFUEGE, 2949
FLECKEN-SCHIEFER, 4222
FLEXUR, 1566

FLIESS-FALTE, 1577
FLIESS-SAND, 3716
FLIESS-SCHIEFERUNG, 1576
FLIESS-TON, 4038
FLIESS-ZONE, 4868
FLIESSEN, 1575
FLINT, 739
FLOEZ, 3983
FLOEZ.LIEGENDTON, 4648
FLORA, 1571
FLOREN-PROVINZ, 1572
FLORIZONE, 1573
FLOTATION, 1574
FLUECHTIG.ELEMENT, 4756
FLUECHTIG.VERBINDUNG, 4757
FLUEGELDRUCK-SONDIERUNG, 4712
FLUEGELSONDE, 4711
FLUESSIG.ABFALL, 2521
FLUESSIG.PHASE, 1582
FLUESSIGKEITS-DRUCK, 1583
FLUESSIGKEITS-EINSCHLUSS, 1579
FLUESSIGKEITS-INJEKTION, 1580
FLUESSIGKEITS-MESSUNG, 1698
FLUGPLATZ, 96
FLUIDAL-GEFUEGE, 1584
FLUKTUATION, 1578
FLUORESZENZ, 1585
FLUORID, 1586
FLUORIMETRIE, 1587
FLUORINIERUNG, 1588
FLUORIT, 1590
FLUSS, 3873
FLUSS-ANZAPFUNG, 637
FLUSS-DEICH, 2491
FLUSS-EROSION, 1594
FLUSS-TERRASSE, 3875
FLUVENT, 1592
FLUVIATIL.FRACHT, 3874
FLUVIATIL.MILIEU, 1593
FLUVIATIL.SEDIMENTATION, 1595
FLUVIATIL.TRANSPORT, 4288
FLUVIO-GLAZIAL, 1596
FLUVIO-GLAZIAL-SEDIMENTATION, 1597
FLUVIOSOL, 1598
FLYSCH, 1599
FM, 1531
FOERDER-SCHLOT, 1522
FOID-GESTEIN, 1526
FOIDIT, 1601
FORAMINIFERA, 1610
FORESET-BED, 1614
FORM, 4054
FORMATION, 1617
FORMEL, 1618
FORSCHUNG, 3822
FORTBEWEGUNG, 2558
FORTPFLANZUNG, 3820

FOSSIL, 1619
FOSSIL-FUNDPUNKT, 1625
FOSSIL-VERGESELLSCHAFTUNG, 341
FOSSIL.BRENNSTOFF, 1620
FOSSIL.FROSTSPALTE, 1621
FOSSIL.HARZ, 3831
FOSSIL.HOLZ, 1624
FOSSIL.KUESTENLINIE, 4
FOSSIL.MENSCH, 1622
FOSSIL.REGENTROPFEN, 3743
FOSSIL.WASSER, 1623
FOSSILISATION, 1626
FOURIER-ANALYSE, 1628
FOYAIT, 1629
FR, 1639
FRAGMENT, 1636
FRAKTIONIERT.KRISTALLISATION, 1631
FRAKTIONIERUNG, 1630
FRAMBOIDAL-GEFUEGE, 1638
FRASNE, 1640
FRASS-SPUR, 1523
FREI.ENERGIE, 1641
FREIHEITSGRAD, 1100
FREILUFT-ANOMALIE, 1643
FREQUENZ, 1644
FREQUENZ-SONDIERUNG, 1646
FRONT, 1652
FROST-BESTAENDIGKEIT, 1704
FROST-BODEN, 1656
FROST-ERSCHEINUNG, 1654
FROST-HEBUNG, 1655
FRUEH-DIAGENESE, 1272
FRUEH-PROTEROZOIDEN, 1273
FRUEHDIAGENET.VERFORMUNG, 4145
FRUKTIFIKATION, 1657
FUEHRER, 1926
FUELLSTOFF, 1544
FUGAZITAET, 1659
FULGURIT, 1660
FULVO-SAEURE, 1663
FUMAROLE, 1664
FUNGI, 1667
FUNKTION, 1665
FUNKTIONELL.ANPASSUNGS-FORM, 1666
FURCHE, 1668
FUSINIT, 1669
FUSULINIDAE, 1670
G-SCHICHT, 1671
GA, 1675
GABBRO, 1672
GALAKT.JAHR, 1674
GAMMA-GAMMA-LOG, 1678
GAMMA-LOG, 1679
GAMMA-SPEKTROMETRIE, 1677
GAMMA-STRAHLUNG, 1676
GANG, 4725
GANG-LETTE, 1509

GANG-SYSTEM, 2514
GANGART, 1680
GANISTER, 1681
GAS, 1684
GAS-BOHRUNG, 1692
GAS-CHROMATOGRAPHIE, 1686
GAS-EINSCHLUSS, 1690
GAS-FELD, 1689
GAS-KOHLE, 1687
GAS-KONDENSAT, 1688
GAS-OEL-VERHAELTNIS, 1693
GAS-PHASE, 1694
GAS-SPEICHERUNG, 1691
GASKAPPE, 1685
GASTROPODA, 1696
GASTROZOOID, 1697
GAUSS-EPOCHE, 1699
GD, 1673
GE, 1782
GEANTIKLINE, 1700
GEBOGEN.VERWERFUNG, 289
GEDIEGEN.ELEMENT, 2988
GEDINNE, 1701
GEFAELLE-MESSUNG, 1204
GEFAELLS-KURVE, 4482
GEFRIEREN, 890
GEFUEGE, 1500
GEFUEGE-DIAGRAMM, 4304
GEFUEGE-PETROLOGIE, 4309
GEGENRIPPEL, 226
GEHOBEN.SCHOLLE, 4685
GEHWEG, 461
GEKOEPFT.TAL, 458
GEL, 1702
GEL-MINERAL, 1703
GELAENDE-AUFNAHME, 4371
GELAENDE-KONTROLLE, 1910
GELAENDE-KORREKTUR, 4466
GELAENDE-UNTERSUCHUNG, 1542
GELAENDE-VERMESSUNG, 1906
GELOEST.FRACHT, 1229
GEMAESSIGT.ZONE, 4452
GEMMOLOGIE, 1706
GENERATION, 1708
GENESE, 1709
GENOMORPH, 1710
GENOTYP, 1711
GEOBAROMETRIE, 1712
GEOBIOS, 1713
GEOBOTAN.PROSPEKTION, 1714
GEOCHEM.AUREOLE, 373
GEOCHEM.EINBRUCHSKRATER, 1721
GEOCHEM.FAZIES, 1717
GEOCHEM.INDIKATOR, 1718
GEOCHEM.KARTE, 1719
GEOCHEM.PROFIL, 1720
GEOCHEM.PROSPEKTION, 1716

GEOCHEM.ZYKLUS, 1715
GEOCHEMIE, 1722
GEOCHRONOLOG.EINHEIT, 1724
GEOCHRONOLOGIE, 1725
GEODAESIE, 1727
GEODAET.KOORDINATE, 1728
GEODE, 1726
GEODYNAMIK, 1729
GEOELEKTR.SONDIERUNG, 1332
GEOFRAKTUR, 1730
GEOGRAPH.BREITE, 2450
GEOGRAPH.LAENGE, 2565
GEOGRAPHIE, 1731
GEOID, 1732
GEOISOTHERME, 1733
GEOLOG.DIENST, 1739
GEOLOG.KARTE, 1735
GEOLOG.MITWIRKUNG, 1740
GEOLOGIE, 1741
GEOMAGNET.KARTE, 2644
GEOMAGNET.POL, 1742
GEOMAGNETISMUS, 1745
GEOMETRIE, 1746
GEOMORPHOLOG.KARTE, 1747
GEOMORPHOLOGIE, 1748
GEONOMIE, 1749
GEOPHON, 1750
GEOPHYSIK, 1759
GEOPHYSIKAL.ANOMALIE, 1751
GEOPHYSIKAL.INSTRUMENT, 1753
GEOPHYSIKAL.KARTE, 1754
GEOPHYSIKAL.METHODE, 1755
GEOPHYSIKAL.PROFIL, 1756
GEOPHYSIKAL.PROSPEKTION, 1752
GEOPHYSIKAL.SONDIERUNG, 4185
GEOPHYSIKAL.VERMESSUNG, 1757
GEOPHYSIKER, 1758
GEOSECS, 1761
GEOSTAT.DRUCK, 1909
GEOSTATISTIK, 1762
GEOSYNKLINAL-FURCHE, 1766
GEOSYNKLINAL-KETTE, 3184
GEOSYNKLINAL-MEER, 1764
GEOSYNKLINAL-SCHWELLE, 1763
GEOSYNKLINAL-SEDIMENTATION, 1765
GEOSYNKLINALE, 1767
GEOTECHN.AUFNAHME, 1770
GEOTECHN.EIGENSCHAFT, 1769
GEOTECHN.KARTE, 1768
GEOTECHNIK, 1771
GEOTEKTONIK, 1772
GEOTHERM.ENERGIE, 1773
GEOTHERM.FELD, 1774
GEOTHERM.GRADIENT, 1775
GEOTHERM.METAMORPHOSE, 1776
GEOTHERM.METHODE, 1777
GEOTHERM.SYSTEM, 1778

GEOTHERMIK, 1779
GEOTHERMOMETRIE, 1780
GEOUNDATION, 1781
GEOWISSENSCHAFTEN, 1279
GERMANOTYP.TEKTONIK, 1783
GEROELL, 3322
GESAMT-GESTEIN, 4824
GESCHICHTE, 2020
GESCHICHTE-DER-GEOLOGIE, 2021
GESCHICHTET.GEFUEGE, 4273
GESCHICHTET.KOMPLEX, 2464
GESCHIEBE, 558
GESCHIEBE-LEHM, 4530
GESCHIEBE-MERGEL, 559
GESCHLEPPT.ERZ, 1244
GESCHWINDIGKEIT, 4726
GESCHWINDIGKEITS-STRUKTUR, 4727
GESETZ, 2460
GESPANNT.GRUNDWASSER, 887
GESTEIN, 3881
GESTEINS-GEFUEGE, 3373
GESTEINS-RIEGEL, 3882
GESTEINS-SPALTBARKEIT, 792
GESTEINSBILDEND.MINERAL, 3890
GEWAECHSHAUS-EFFEKT, 1891
GEWAESSER-ANORDNUNG, 4286
GEWAESSER-KARTE, 4792
GEWINNUNG, 1482
GEWINNUNGS-GRAD, 3768
GEYSER, 1784
GEYSERIT, 1785
GEZEITEN, 4529
GEZEITEN-MILIEU, 2286
GEZEITEN-SEDIMENTATION, 2287
GIFT, 4554
GIGANTISMUS, 1787
GIGANTOSTRACA, 1788
GINKGOALES, 1789
GIPS, 1933
GIPS-BILDUNG, 1932
GIRLANDEN-FORM, 270
GITTER-GEFUEGE, 2452
GITTERKONSTANTE, 2451
GIVET, 1790
GLACIS, 1810
GLAETTUNG, 4141
GLANZ, 2606
GLANZ-KOHLE, 4749
GLAS, 1811
GLAS-GEFUEGE, 4750
GLAS-HYDRATISIERUNG, 2090
GLAS-ROHSTOFF, 1813
GLASIG.GEFUEGE, 1814
GLASKOPF-GEFUEGE, 553
GLASSAND, 1812
GLAUKONIT, 1815
GLAUKONIT.CHEMISMUS, 1816

GLAUKONITISIERUNG, 1817
GLAUKOPHAN-FAZIES, 1818
GLAUKOPHAN-SCHIEFER, 534
GLAZIAL, 1791
GLAZIAL-ABLAGERUNG, 1792
GLAZIAL-AUFSCHUETTUNGS-EBENE, 3212
GLAZIAL-EROSION, 1794
GLAZIAL-GEOLOGIE, 1796
GLAZIAL-MILIEU, 1793
GLAZIAL-MORPHOLOGIE, 1795
GLAZIAL-SEDIMENTATION, 1798
GLAZIAL-STRIEMUNG, 1800
GLAZIAL-TRANSPORT, 1801
GLAZIO-LIMN.SEDIMENTATION, 1806
GLAZIOLOGIE, 1807
GLAZIOMARIN.SEDIMENTATION, 1808
GLEICHGEWICHT, 1416
GLEICHGEWICHTS-PROFIL, 3628
GLEICHKOERNIG.GEFUEGE, 1415
GLEITUNGS-ZWILLING, 1820
GLETSCHER, 1804
GLETSCHER-ABBRUCH, 1805
GLETSCHER-SEE, 1797
GLETSCHER-SPALTE, 990
GLETSCHER-STADIUM, 1799
GLETSCHER-TAL, 1802
GLEY, 1819
GLIMMER, 2804
GLIMMER-SCHIEFER, 2805
GLOBAL-TEKTONIK, 1821
GLOBIGERINACEA, 1822
GLOMERO-, 1823
GLOSSOPTERIDALES, 1825
GLUKOSE, 4347
GLUTWOLKE, 3069
GNATHOSTOMI, 1826
GNATHOSTRACA, 1827
GNEIS, 1828
GNEIS-GEFUEGE, 1830
GNEISKUPPEL, 1829
GNETALES, 1831
GNOMON.PROJEKTION, 1832
GOLDKLUMPEN, 3070
GOLDWAESCHE, 3283
GOLF, 1927
GONDWANA, 1834
GONIATITIDA, 1835
GONIOMETRIE, 1836
GRABEN, 1840
GRADIENT, 1842
GRADIERT.SCHICHTUNG, 1841
GRADIOMETER, 1843
GRANAT, 1683
GRANIT, 1845
GRANIT-GNEIS, 1846
GRANIT-SCHICHT, 1849
GRANIT.CHEMISMUS, 1848

GRANITISCH, 1847
GRANITISIERUNG, 1850
GRANITOID, 1851
GRANOBLAST.GEFUEGE, 1852
GRANODIORIT, 1853
GRANOGABBRO, 1854
GRANOPHYR=GEFUEGE, 1855
GRANULIERUNG, 1857
GRANULIT, 1858
GRANULIT=FAZIES, 1859
GRANULITISCH, 1860
GRANULOMETRIE, 1861
GRAPH.AUSGABE, 1863
GRAPH.DARSTELLUNG, 1864
GRAPH.GEFUEGE, 1865
GRAPHIT, 1866
GRAPHITISIERUNG, 1867
GRAPTOLITHINA, 1868
GRAPTOLOIDEA, 1869
GRASLAND, 1870
GRAUWACKE, 1888
GRAVIMETER, 1873
GRAVIMETR.FELD, 1880
GRAVIMETR.KARTE, 1882
GRAVIMETR.METHODE, 1883
GRAVIMETR.PROSPEKTION, 1879
GRAVIMETR.VERMESSUNG, 1886
GRAVIMETRIE, 1874
GRAVITATION, 1875
GRAVITATIONS=FALTE, 1881
GRAVITATIONS=KONSTANTE, 1876
GRAVITATIONS=TEKTONIK, 1887
GREIFER=PROBE, 1839
GREISEN, 1896
GRENVILLE=OROGENESE, 1898
GRENZFLAECHE, 2261
GRENZSCHICHT, 561
GRIT, 1900
GROB.GEROELL, 809
GROBFRAKTION, 810
GROSS.MASSSTAB, 2436
GROSSKREIS=GUERTEL, 1890
GROSSRIPPEL, 3941
GRUEN=SCHIEFER, 1892
GRUEN=SCHIEFER=FAZIES, 1893
GRUENSTEIN, 1894
GRUENSTEIN=GUERTEL, 1895
GRUND=MORAENE, 1907
GRUNDBRUCH, 3455
GRUNDEIS, 1905
GRUNDGEBIRGS=TEKTONIK, 430
GRUNDMASSE, 1919
GRUNDWASSER, 1911
GRUNDWASSER=ABFLUSS, 1918
GRUNDWASSER=BILANZ, 1912
GRUNDWASSER=ERGIEBIGKEIT, 1914
GRUNDWASSER=ERKUNDUNG, 1915

GRUNDWASSER=ERNEUERUNG, 3765
GRUNDWASSER=LEITER, 260
GRUNDWASSER=SPERRE, 1913
GRUNDWASSER=SPIEGEL, 1916
GRUNDWASSER=SPIEGEL=ABSENKUNG, 1249
GRUNDWASSER=STAUER, 262
GRUS, 1923
GSHEL, 1935
GUANO, 1924
GUTENBERG=DISKONTINUITAET, 1929
GYMNOLAEMATA, 1930
GYMNOSPERMAE, 1931
GYTTJA, 1934
H, 2106
HABITAT, 1937
HABITUS, 1936
HAECKEL'S=GESETZ, 1939
HAEMATIT, 1981
HAERTE, 1956
HAEUFIGKEIT, 16
HAEUFIGKEITS=VERTEILUNG, 1645
HAFEN, 1954
HAFENMAUER, 3440
HAFT=WASSER, 67
HAGEL, 1941
HALBEDELSTEIN, 4036
HALBRAUM, 1943
HALBWERTS=ZEIT, 1942
HALDE, 4413
HALID, 1944
HALIT.MILIEU, 1945
HALMYROLYSE, 1946
HALOGEN, 1947
HALOPHIL.ORGANISMUS, 1948
HALOPHYT, 1949
HAMMADA, 1950
HAMMER=SEISMIK, 1951
HAND=BOHRUNG, 371
HANDBUCH, 2667
HANG, 4134
HANG=STABILITAET, 4137
HANG=VERFORMUNG, 4135
HANGEND.GESTEIN, 1953
HANGEND.TAL, 1952
HARDGROUND, 1955
HARMON.ANALYSE, 1957
HARMONIKA=FALTE, 741
HARNISCH, 4132
HARPOLITH, 1958
HAUPTELEMENT=ANALYSE, 2657
HAUPTELEMENT=ANALYSEN=METHODIK, 2658
HAUT=BILDUNGEN, 1126
HAUTERIVE, 1959
HAUY=GESETZ, 1960
HAWAII=TYP, 1961
HE, 1978
HEBUNG, 4665

HEFTIG.ERSCHUETTERUNG, 4298
HEISS, 2063
HEISS.QUELLE, 2066
HEKISTOTHERM.PFLANZE, 1975
HELICOPLACOIDEA, 1976
HELIOZOA, 1977
HELIUM=DATIERUNG, 1979
HELVET, 1980
HEMICRUSTACEA, 1982
HEMIPTEROIDA, 1983
HERD=MECHANISMUS, 1600
HERGELEITET.ERZ, 2225
HERKUNFT, 3179
HERMATYP.TAXON, 1987
HERZYN.OROGENESE, 1984
HETEROBLAST.GEFUEGE, 1988
HETEROCHRON, 1990
HETEROCHRONIE, 1989
HETEROCORALLIA, 1991
HETERODONTA, 1992
HETEROGEN, 1994
HETEROGENITAET, 1993
HETEROMORPH.ORGANISMUS, 1995
HETEROMORPHISMUS, 1996
HETEROSPOR.PFLANZE, 1997
HETEROTROPH.ORGANISMUS, 1998
HETTANGIUM, 1999
HEXACORALLA, 2000
HEXACTINIARIA, 2001
HEXAGONAL.SYSTEM, 2002
HEXAHEDRIT, 2003
HF, 1940
HG, 2749
HIATUS, 2004
HIEROGLYPHE, 2005
HIPPOMORPHA, 2016
HISTOGRAMM, 2017
HISTOLOGIE, 2018
HISTOR.GEOLOGIE, 2019
HISTOSOL, 2022
HITZE=SPRUNG, 1965
HO, 2025
HOCH=DRUCK, 2007
HOCH=ENERGET.MILIEU, 2009
HOCH=TEMPERATUR, 2008
HOCH=WASSER, 1569
HOCHFREQUENZ, 3724
HOCHGRADIG, 2010
HOCHGRADIG.METAMORPHOSE, 2011
HOCHLAND, 2012
HODOGRAPH, 2023
HOEFFIGKEITS=KARTE, 1611
HOEHEN=MESSUNG, 2151
HOEHENLINIEN=KARTE, 4313
HOEHLE, 677
HOEHLENKUNDE, 4203
HOHER=ANOMALIE=WERT, 2006

HOHLRAUM, 4651
HOLMES-KLASSIFIKATION, 2024
HOLOCEPHALI, 2027
HOLOGRAPHIE, 2029
HOLOKRISTALLIN.GEFUEGE, 2028
HOLOPHYT, 2030
HOLOSTEI, 2031
HOLOSTOM, 2032
HOLOTHUROIDEA, 2033
HOLOTYP, 2034
HOLOZAEN, 2026
HOMALOZOA, 2035
HOMINIDAE, 2038
HOMO-SAPIENS, 2039
HOMOEOMORPHIE, 2037
HOMOEOMORPHISMUS, 2036
HOMOGENISIERUNG, 2042
HOMOGENITAET, 2041
HOMOIOSTELEA, 2043
HOMOKLINALE, 2040
HOMOLOGIE, 2044
HOMONYMIE, 2045
HOMOSTELEA, 2046
HOMOTAXIE, 2047
HOOKES-GESETZ, 2048
HORIZONT-DIFFERENZIERUNG, 2049
HORIZONTAL, 2050
HORIZONTAL-KOMPRESSION, 2051
HORIZONTAL-MERKMAL, 3467
HORIZONTAL-RICHTUNG, 2052
HORIZONTAL-SCHICHTUNG, 3466
HORIZONTAL-VERSCHIEBUNG, 2053
HORIZONTAL-VERWERFUNG, 2054
HORNBLENDE-HORNFELS-FAZIES, 2055
HORNBLENDIT, 2056
HORNFELS, 2057
HORNFELS-FAZIES, 2058
HORNSTEIN, 1567
HORNSTEIN-BILDUNG, 740
HORSETAIL, 2059
HORST, 2060
HOT-SPOT, 2065
HOWARDIT, 2067
HUDSON-OROGENESE, 2068
HUEGEL, 2013
HUMIFIZIERUNG, 2076
HUMIN, 2077
HUMIN-SAEURE, 2071
HUMMOCK, 2079
HUMOD, 2080
HUMOX, 2081
HUMULT, 2082
HUMUS, 2083
HUMUS-KOHLE, 2078
HUMUSREICH.BODEN, 2072
HYALOKRISTALLIN.GEFUEGE, 2084
HYALOSPONGEA, 2085

HYDATOGENESE, 2087
HYDRARCH, 2088
HYDRATISIERUNG, 2089
HYDRAUL.BELASTUNG, 2094
HYDRAUL.DRUCK, 2096
HYDRAUL.GRADIENT, 2093
HYDRAUL.KARTE, 2095
HYDRAUL.LEITFAEHIGKEIT, 2091
HYDRAUL.RADIUS, 2097
HYDRAUL.RISS-ERZEUGUNG, 2092
HYDRAULIK, 2098
HYDRO-EXPLOSION, 2105
HYDRO-METALLURGIE, 2122
HYDROCHEM.FAZIES, 2100
HYDROCHEM.KARTE, 2101
HYDROCHEMIE, 2102
HYDRODYNAM.MODELL, 2103
HYDRODYNAMIK, 2104
HYDROGEOCHEM.PROSPEKTION, 2109
HYDROGEOLOG.EINFLUSS, 2110
HYDROGEOLOG.ERKUNDUNG, 2113
HYDROGEOLOG.KARTE, 2111
HYDROGEOLOG.PROVINZ, 2112
HYDROGEOLOGIE, 2115
HYDROGRAPH.NETZ, 1246
HYDROGRAPHIE, 2117
HYDROIDA, 2118
HYDROLOG.KARTE, 2120
HYDROLOGIE, 2121
HYDROMETER, 2123
HYDROMETRIE, 2124
HYDROPHON, 2125
HYDROSPHAERE, 2126
HYDROSPIRE, 2127
HYDROSTAT.DRUCK, 2129
HYDROSTAT.NIVEAU, 2128
HYDROSTAT.OBERFLAECHE, 3583
HYDROTHERMAL, 2130
HYDROTHERMAL-BEDINGUNG, 2133
HYDROTHERMAL-UMWANDLUNG, 2131
HYDROTHERMAL-WASSER, 2134
HYDROTHERMAL.VORKOMMEN, 2132
HYDROXID, 2136
HYDROZOA, 2137
HYGROPHIL, 2138
HYGROSKOP.WASSER, 2139
HYMENOPTEROIDA, 2140
HYOLITHES, 2141
HYPABYSSAL-GENESE, 2142
HYPERGENESE, 2143
HYPIDIOBLAST.GEFUEGE, 2144
HYPIDIOMORPH.GEFUEGE, 2145
HYPOKRISTALLIN.GEFUEGE, 2147
HYPOTHERMAL.VORKOMMEN, 2149
HYPOTHESE, 2150
HYPOZENTRUM, 2146
HYRACOIDEA, 2152

HYSTRICOMORPHA, 2153
IAPETUS-OZEAN, 2154
ICHNITES, 2164
ICHNOLOGIE, 2165
ICHTHYOPTERYGIA, 2166
ICHTHYOSAURIA, 2167
IDIOBLAST, 2168
IDIOGEOSYNKLINALE, 2169
IDIOMORPH.GEFUEGE, 2170
IDIOMORPH.KRISTALL, 1447
IDIOMORPHIE, 2171
IGNIMBRIT, 2177
IMBRIUM, 2179
IMMERSIONS-MIKROSKOPIE, 2184
IMPAKTIT, 2193
IMPORT, 2194
IMPRAEGNATIONS-LAGERSTAETTE, 2195
IN, 2213
IN-SITU-UNTERSUCHUNG, 2197
INAKTIV.KOMPONENTE, 2223
INARTICULATA, 2198
INCEPTISOL, 2199
INDEX-KARTE, 2207
INDIEN-STUFE, 2210
INDIKATOR, 2212
INDIKATOR-MINERAL, 3157
INDOCHINIT, 2214
INDUKTIONS-METHODE, 2217
INDUSTRIE, 2222
INDUSTRIE-ASCHE, 320
INDUSTRIE-MUELL, 2221
INDUSTRIE-ZEMENT, 587
INDUSTRIELL.NUTZUNG, 2220
INDUZIERT.MAGNETISIERUNG, 2215
INDUZIERT.POLARISATION, 2216
INFILTRATION, 2226
INFORMATIK, 2228
INFRA-STRUKTUR, 2235
INFRAKAMBRIUM, 2230
INFRAROT, 2231
INGENIEUR-BAU, 1375
INGENIEUR-GEOLOGIE, 1374
INGRESSION, 2236
INHERITED, 2237
INHOMOGENITAET, 2238
INJEKTION, 2239
INJEKTIONS-BREKZIE, 2240
INKOHAERENT, 2201
INKOHLUNG, 808
INKOHLUNGS-GRAD, 3748
INKOMPETENT.GESTEIN, 2203
INKONGRUENT.SCHMELZEN, 2204
INKRUSTATION, 2205
INLAND-EIS, 2160
INLIER, 2242
INNER.ABDRUCK, 2274
INNER.ERD-KERN, 2243

INNER.KONTINENTAL=ABHANG, 2245
INNER.RAND=BECKEN, 396
INNER.SCHELF, 2244
INOCERAMI, 2246
INSECTA, 2251
INSECTIVORA, 2252
INSEL, 2315
INSEL=VULKAN, 2317
INSELBERG, 2253
INSELBOGEN, 2316
INSEQUENT.TAL, 2254
INSTATIONAER.STROEMUNG, 4664
INSTITUTION, 2257
INSTRUMENTIERUNG, 2258
INTERFERENZ, 2262
INTERFEROMETRIE, 2263
INTERFORMATIONELL, 2265
INTERGLAZIAL, 2266
INTERMEDIAER.CHEMISMUS, 2269
INTERMITTIEREND.QUELLE, 2271
INTERMITTIEREND.STROM, 2272
INTERNAT.GEOL.KORRELAT.PROGRAMM, 2278
INTERNATIONAL.GEOLOG.KONGRESS, 2277
INTERNATIONAL.ZUSAMMENARBEIT, 2276
INTERNIDEN, 2279
INTERNSTRUKTUR, 2275
INTERPLANETAR.VERGLEICH, 2280
INTERPLUVIAL, 2281
INTERPOLATION, 2282
INTERPRETATION, 2283
INTERSERTAL.GEFUEGE, 2284
INTRAKLAST, 2288
INTRAMONTAN.BECKEN, 2273
INTRATELLUR.GENESE, 2292
INTRUSIONS=KOERPER, 2293
INTRUSIV, 2294
INVENTAR, 2295
INVERS.MULDENFLUEGEL, 4393
INVERS.PROBLEM, 2296
INVERTEBRATA, 2297
IO=TH=DATIERUNG, 2302
IO=U=DATIERUNG, 2303
ION, 2299
IONEN=AUSTAUSCH, 2300
IONEN=BINDUNG, 1558
IONEN=DISSOZIIERUNG, 1227
IONEN=SONDE, 2301
IONISIERUNG, 2304
IONOSPHAERE, 2305
IR, 2307
IR=PHOTOGRAPHIE, 2232
IR=SPEKTROMETRIE, 2234
IR=STRAHLUNG, 2233
IRIDESZENZ, 2306
IRREVERSIBILITAET, 2313
ISANOMALE, 2318
ISOBARE, 2319

ISOBASE, 2320
ISOBATHE, 2321
ISOCHEM.METAMORPHOSE, 2322
ISOCHORE, 3393
ISOCHRONE, 2323
ISOCONE, 2326
ISOGONE, 2327
ISOGRADE, 2328
ISOHYPSE, 2329
ISOKLINALE, 2324
ISOKLINE, 2325
ISOLIERSTOFF, 2259
ISOLIERUNG, 2330
ISOLINIE, 2337
ISOLITH, 2331
ISOMETR.SYSTEM, 2332
ISOMORPH.ORGANISMUS, 2333
ISOMORPHIE, 2334
ISOPACHE, 2335
ISOPLEURA, 2338
ISOPRENOID, 2339
ISORAD, 2340
ISORAT, 2341
ISOSEISM.KARTE, 2343
ISOSEISTE, 2342
ISOSTASIE, 2344
ISOTHERME, 2346
ISOTHERMOREMANENT.MAGNETISIERUNG, 2347
ISOTOP, 2348
ISOTOPEN=EFFEKT, 2350
ISOTOPEN=FRAKTIONIERUNG, 2353
ISOTOPEN=GEOCHEMIE, 2351
ISOTOPEN=VERDUENNUNG, 2349
ISOTOPEN=VERHAELTNIS, 2352
ISOTROPIE, 2354
ITABIRIT, 2355
ITACOLUMIT, 2356
J, 2298
JADEITIT, 2357
JAHRBUCH, 4847
JAHRESRING, 4585
JAHRESZEITL.SCHWANKUNG, 3985
JARAMILLO=EPISODE, 2358
JASPIS, 2359
JUNG, 4850
JUNG=PLEISTOZAEN, 4680
JUPITER, 3473
JURA, 2363
JUVENIL, 2364
JUVENIL.WASSER, 2365
K, 3579
K=AR=DATIERUNG, 3580
K=CA=DATIERUNG, 3581
KAELTE=WIRKUNG, 1653
KAENOZOIKUM, 688
KALEDON.OROGENESE, 618
KALI, 3577

KALIBER=BOHRLOCH=MESSUNG, 621
KALIUM=GEHALT, 3578
KALK, 2507
KALK=ALGEN, 608
KALK=ALKALI=TYP, 607
KALK=CHEMISMUS, 610
KALK=ERDE, 2506
KALK=GEHALT, 609
KALK=KRUSTE, 616
KALT.MILIEU, 832
KALT.ZONE, 548
KALZIPHYR, 611
KALZIT, 613
KALZITISIERUNG, 614
KAMBRIUM, 625
KAMES, 2366
KANAL, 628
KANGA, 2367
KAOLIN, 2368
KAOLINISIERUNG, 2369
KAP, 632
KAPILLAR=DRUCK, 635
KAPILLAR=SOG, 634
KAPILLAR=WASSER, 636
KAPILLARITAET, 633
KAR, 775
KARAT, 640
KARBON, 654
KARBONAT, 651
KARBONAT=GESTEIN, 650
KARBONAT=SAETTIGUNGS=ZONE, 1124
KARBONAT.SANDSTEIN, 1862
KARBONATISIERUNG, 653
KARBONATIT, 652
KAREL.OROGENESE, 2370
KARN, 655
KARREN, 2371
KARROO, 2372
KARST, 2373
KARST=FUELLUNG, 2374
KARST=HYDROLOGIE, 2375
KARST=QUELLE, 2376
KARST=TRICHTER, 4117
KARST=WASSER, 3837
KARTE, 2668
KARTEN=ERLAEUTERUNG, 1480
KARTEN=MASSSTAB, 2669
KARTIERUNG, 2670
KARTOGRAPHIE, 660
KASAN, 2381
KASIMOV, 2382
KASKAD.UMBRUCH, 662
KASKADEN=FALTE, 661
KASTANNOZEM, 666
KATA=METAMORPHOSE, 2377
KATAGENESE, 668
KATAKLASE, 667

KATALYSE, 670
KATANGA-OROGENESE, 2378
KATASTROPHEN-THEORIE, 671
KATATHERMAL.VORKOMMEN, 2379
KATAZONE, 2380
KATION, 674
KAUSTOBIOLITH, 676
KAVERNE, 678
KAVERNOES.GEFUEGE, 679
KEIMBILDUNG, 3066
KELYPHIT-RINDE, 2383
KENNEL-KOHLE, 629
KENORA-OROGENESE, 2384
KERAM.ROHSTOFF, 694
KERATOPHYR, 2385
KERN, 3067
KERN-BOHRUNG, 956
KERN-ENERGIE, 3062
KERN-EXPLOSION, 3063
KERN-MAGNET.RESONANZ, 3064
KERNKRAFTWERK, 3065
KERNSPALTUNG, 1554
KEROGEN, 2386
KESSEL, 2387
KEUPER, 2388
KIBARA-OROGENESE, 2390
KIEFER, 2360
KIES, 1871
KIES-FILTER, 1872
KIESEL-GEHALT, 4099
KIESEL-KALK, 656
KIESELSAEURE, 4097
KIMBERLIT, 2391
KIMMER.OROGENESE, 768
KIMMERIDGE, 2392
KINEMATIK, 2393
KINK-BAND, 2394
KINZIGIT, 2395
KISSEN-GEFUEGE, 3448
KLAPPE, 4708
KLASSIFIKATION, 782
KLAST.GESTEIN, 786
KLASTISCH, 784
KLEIN.MASSSTAB, 4140
KLEINRECHNER, 2881
KLEINST.QUADRATE-METHODE, 2473
KLIFF, 793
KLIMA, 794
KLIMA-SPUR, 796
KLIMA-WIRKUNG, 795
KLIMA-ZONE, 797
KLINOMETER, 4534
KLIPPE, 2396
KLUEFTUNG, 2362
KLUFT, 2361
KNOLLE, 3039
KNOSPUNG, 1705

KOEFFIZIENT, 822
KOERNIG, 1856
KOERZITIV-KRAFT, 825
KOEXISTIEREND.MINERAL, 826
KOHAESION, 827
KOHAESIV.STOFF, 829
KOHLE, 801
KOHLE-BECKEN, 803
KOHLE-FLOEZ, 806
KOHLE-HORIZONT, 805
KOHLEHYDRAT, 641
KOHLEN-GAS, 804
KOHLEN-GEHALT, 648
KOHLEN-LAGERSTAETTEN-KARTE, 807
KOHLENDIOXID, 643
KOHLENMONOXID, 644
KOHLENSTOFFHALTIG.CHONDRIT, 647
KOHLENSTOFFHALTIG.GESTEIN, 649
KOHLENWASSERSTOFF, 1658
KOHLENWASSERSTOFF-VERBINDUNG, 2099
KOKS-KOHLE, 831
KOLLOID, 842
KOLLOID-GEFUEGE, 841
KOLONIEBILDEND.TAXON, 844
KOLORIMETRIE, 849
KOMMENSALISMUS, 853
KOMPAKTION, 857
KOMPAKTIONS-VERMOEGEN, 856
KOMPETENZ, 859
KOMPLEX, 860
KOMPLEX-BILDUNG, 861
KOMPLEXOMETRIE, 862
KOMPRESSIBILITAET, 864
KOMPRESSION, 865
KOMPRESSIONS-SPANNUNG, 869
KOMPRESSIONS-TEST, 868
KONDENSATION, 882
KONGLOMERAT, 892
KONGRUENT.SCHMELZE, 893
KONJUGIERT, 896
KONKORDANT.ALTER, 878
KONKORDANT.LAGERUNG, 889
KONKORDANZ, 877
KONKRETION, 881
KONSANGUINITAET, 903
KONSEQUENT.STROM, 904
KONSISTENZ, 906
KONSOLIDIERUNG, 907
KONSTANTE, 909
KONSTITUTIONS-WASSER, 4794
KONSTRUKTION, 910
KONTAKT, 913
KONTAKT-BREKZIE, 914
KONTAKT-METAMORPHOSE, 915
KONTAKT-METASOMATOSE, 916
KONTINENT, 918
KONTINENT-WACHSTUM, 919

KONTINENTAL-DRIFT, 922
KONTINENTAL-DUENE, 1132
KONTINENTAL-KERN, 926
KONTINENTAL-KETTE, 2289
KONTINENTAL-RAND, 920
KONTINENTAL-RELIEF, 2429
KONTINENTAL-SCHELF, 4067
KONTINENTAL-SEDIMENTATION, 928
KONTINENTAL-SOCKEL, 927
KONTINENTAL-TAFEL, 929
KONTINENTAL-VEREISUNG, 924
KONTINENTAL-ZONE, 931
KONTINENTAL.KRUSTE, 921
KONTINENTAL.MILIEU, 923
KONTRAKTION, 934
KONTRAST, 935
KONTROLLE, 936
KONVEKTION, 938
KONVEKTIONS-STROEMUNG, 939
KONVERGENZ, 940
KONZENTRATION, 873
KONZESSION, 2886
KOPERNIKAN.ALTER, 945
KOPROLITH, 947
KORALLEN-RIFF, 949
KORN, 1844
KORN-GEFUEGE, 4477
KORN-GESTALT, 3308
KORN-VERTEILUNG, 4125
KORREKTUR, 960
KORRELATION, 961
KORRELATIONS-KOEFFIZIENT, 962
KORRESPONDENZ-ANALYSE, 963
KORROSION, 964
KORUND, 965
KOSM.NEBEL, 968
KOSM.PARTIKEL, 969
KOSM.STAUB, 967
KOSM.STRAHLUNG, 970
KOSM.SYSTEM, 971
KOSMOCHEMIE, 972
KOSMOGEN.ELEMENT, 973
KOSMOLOG.PRINZIP, 974
KOSMOLOGIE, 975
KOSTEN, 976
KOT-PILLE, 1521
KR, 2398
KRAFTWERK, 3588
KRATER, 981
KRATER-BILDUNG, 983
KRATER-SEE, 982
KRATON, 984
KREEP, 2397
KREIDE, 989
KREUZSCHICHTUNG, 998
KRIECH-VORGANG, 986
KRISTALL, 1011

KRISTALL-FEHLER, 1013
KRISTALL-FELD, 1015
KRISTALL-GITTER, 1018
KRISTALL-OPTIK, 1019
KRISTALL-STRUKTUR, 1020
KRISTALL-SYSTEM, 1021
KRISTALL-VERSETZUNG, 1014
KRISTALL-WACHSTUM, 1017
KRISTALL-WASSER, 4793
KRISTALLCHEMIE, 1012
KRISTALLIN.GESTEIN, 1022
KRISTALLINITAET, 1023
KRISTALLISATION, 1025
KRISTALLISATIONS-SCHIEFERUNG, 1609
KRISTALLISIERUNGS-INDEX, 1026
KRISTALLIT, 1024
KRISTALLOBLAST, 1027
KRISTALLOGRAPHIE, 1028
KRIT.DARSTELLUNG, 995
KRIT.STROEMUNG, 994
KRIT.WINKEL, 993
KRUEMMUNG, 4552
KRUSTENRIFF, 3451
KRYOPEDOLOGIE, 1003
KRYOTEKTONIK, 1004
KRYOTURBATION, 1005
KRYPTOGEN, 1008
KRYPTOKRISTALLIN.GEFUEGE, 1006
KRYPTOVULKAN.STRUKTUR, 1010
KUB.SYSTEM, 1030
KUENSTL.GRUNDWASSER-ANREICHERUNG, 321
KUENSTL.SATELLIT, 322
KUENSTL.SEE, 3825
KUESTE, 4083
KUESTEN-ABDRIFT, 2544
KUESTEN-DUENE, 812
KUESTEN-EBENE, 815
KUESTEN-EROSION, 813
KUESTEN-LINIE, 816
KUESTEN-MERKMAL, 4081
KUESTEN-MORPHOLOGIE, 814
KUESTEN-SALZ-MARSCH, 3928
KUESTEN-SCHUTZ, 4082
KUESTEN-STROEMUNG, 2568
KUJALNIKIUM, 2399
KUMULAT-GESTEIN, 1032
KUNGUR, 2400
KURTOSIS, 2401
KURZDAUERND.FLUSS, 1397
KUTORGINIDA, 2402
KW-FELD, 3105
KW-GROSS-LAGERSTAETTE, 1786
KW-LAGERSTAETTEN-KARTE, 3106
KW-MUTTER-GESTEIN, 4186
KYBERNETIK, 1041
LA, 2432
LABOR-UNTERSUCHUNG, 2403

LABYRINTHODONTIA, 2404
LACERTILIA, 2406
LADIN, 2410
LAENGE, 2477
LAENGS-PROFIL, 2566
LAGER, 4274
LAGER-GANG, 4104
LAGERSTAETTE, 2866
LAGERSTAETTEN-BEWERTUNG, 1117
LAGERSTAETTEN-FORM, 1119
LAGERSTAETTEN-GENESE, 1118
LAGERSTAETTEN-INDIZ, 4086
LAGERSTAETTEN-MILIEU, 3153
LAGERSTAETTEN-WASSER, 1306
LAGUNAER.MILIEU, 2412
LAGUNAER.SEDIMENTATION, 2413
LAGUNE, 2411
LAHAR, 2414
LAKKOLITH, 2405
LAMARCKISMUS, 2417
LAMELLAR.GEFUEGE, 2419
LAME-KONSTANTE, 2418
LAMELLIBRANCHIATA, 3332
LAMINAR.STROEMUNG, 2420
LAMINIERT.GEFUEGE, 2421
LAMINIERUNG, 2422
LAMPROPHYR, 2423
LAMPROPHYR-GEFUEGE, 2424
LANDSAT, 2430
LANDWIRTSCHAFT, 90
LANGPERIOD.WELLE, 2564
LAPILLI, 2433
LAPLACE-TRANSFORMATION, 2434
LARAM.OROGENESE, 2435
LASER, 2437
LATENT.SCHMELZ-WAERME, 2440
LATERAL, 2441
LATERAL-LOG, 2443
LATERAL-SEKRETION, 2445
LATERAL-STOERUNG, 2442
LATERIT, 2446
LATERIT-BODEN, 2447
LATERITISIERUNG, 2448
LATIT, 2449
LAUFZEIT, 4581
LAUFZEIT-KURVE, 4582
LAURASIA, 2453
LAVA, 2454
LAVA-FELD, 2456
LAVA-KANAL, 2455
LAVA-SEE, 2458
LAVA-STROM, 2457
LAVA-TUNNEL, 2459
LAWINE, 388
LAYERED-INTRUSION, 2463
LEBEN, 2499
LEBEND, 2548

LEBEND.FOSSIL, 2549
LEBENS-SPUR, 4557
LECTOTYP, 2474
LEGENDE, 2475
LEGIERUNG, 139
LEHM, 2509
LEHRBUCH, 4584
LEICHT-ISOTOP, 2500
LEICHT-MINERAL, 2501
LEICHT-OEL, 2502
LEIT-FOSSIL, 1925
LEIT-GESCHIEBE, 2211
LEIT-MINERAL, 2208
LEITFAEHIGKEIT, 883
LEITFOSSIL, 2206
LEITHORIZONT, 2389
LENIUM, 2479
LEPERDITOCOPIDA, 2481
LEPIDOBLAST.GEFUEGE, 2482
LEPIDOCYSTOIDEA, 2483
LEPIDOSAURIA, 2484
LEPOSPONDYLI, 2485
LEPTIT, 2486
LEPTO-GEOSYNKLINALE, 2487
LEPTYNIT, 2488
LEUCOSOME, 2490
LEUKOKRAT.CHEMISMUS, 2489
LEXIKON, 2494
LI, 2525
LICHENES, 2495
LICHENOMETRIE, 2496
LICHIDA, 2497
LIEFERGEBIET, 3656
LIEGEND.FALTE, 3771
LIESEGANG-RINGE, 2498
LIGNIN, 2503
LIGNIT, 2504
LIMN.MILIEU, 2407
LIMN.SEDIMENT, 2408
LIMN.SEDIMENTATION, 2409
LIMNOLOGIE, 2508
LIMONIT, 2510
LINEAMENT, 2511
LINEAR-GEFUEGE, 2512
LINGULIDA, 2513
LINOPHYR.GEFUEGE, 2515
LINSE, 2480
LIPID, 2516
LIPOSTRACA, 2517
LIQUATION, 2518
LIQUID-MAGMAT.VORKOMMEN, 2622
LIQUIDUS, 2522
LISSAMPHIBIA, 2523
LITERATUR-BERICHT, 3843
LITHO-FAZIES, 2526
LITHO-STRATIGRAPHIE, 2538
LITHOFAZIES-KARTE, 2527

LITHOGENESE, 2528
LITHOLOG.EINFLUSS, 2530
LITHOLOG.FALLE, 2531
LITHOLOGIE, 2532
LITHOLOGISCH, 2529
LITHOPHIL.ELEMENT, 2533
LITHOPHYT, 2534
LITHOSPHAERE, 2536
LITHOSTRATIGRAPH.EINHEIT, 2537
LITHOSTROM, 2539
LITHOTOP, 2540
LITHOTYP, 2541
LITOPTERNA, 2542
LITORAL, 2543
LITORAL=MILIEU, 2545
LITORAL=SEDIMENTATION, 2546
LITUOLACEA, 2547
LLANDEILO, 2551
LLANDOVERY, 2552
LLANVIRN, 2553
LOCHKARTE, 3686
LOCKER=GESTEIN, 2202
LODRANIT, 2560
LOESLICHKEIT, 4174
LOESS, 2561
LOESUNG, 4176
LOESUNGS=BERGBAU, 4179
LOESUNGS=HOHLRAUM, 4177
LOKAL, 2557
LOPOLITH, 2569
LOVE=WELLE, 2570
LU, 2607
LU=HF=DATIERUNG, 2608
LUDLOW, 2596
LUECKE, 1682
LUFT, 93
LUFT=DRUCK, 361
LUFT=MEER=GRENZFLAECHE, 94
LUFT=VERMESSUNG, 72
LUFTBILD, 71
LUFTBILD=KARTIERUNG, 70
LUMINESZENZ, 2597
LUNA=PROGRAMM, 2598
LUNAR, 2599
LUVISOL, 2609
LYCOPODIALES, 2610
LYTOCERATIDA, 2611
M.DEVON, 2842
M.EOZAEN, 2843
M.JURA, 2844
M.KAMBRIUM, 2841
M.LIAS, 2845
M.MIOZAEN, 2846
M.OLIGOZAEN, 2847
M.ORDOVIZIUM, 2848
M.PLEISTOZAEN, 2849
M.PLIOZAEN, 2850

M.PROTEROZOIKUM, 2851
M.TRIAS, 2852
MAAR, 2612
MAASTRICHT, 2615
MACHAERIDIA, 2614
MAEANDER, 2724
MAEANDRIEREN, 4119
MAEOT, 2748
MAFISCH.CHEMISMUS, 2616
MAGMA, 2617
MAGMAKAMMER, 2618
MAGMAKAMMER=DACH, 3895
MAGMAKAMMER=EINBRUCH, 3894
MAGMAT.AKTIVITAET, 2172
MAGMAT.ASSIMILATION, 340
MAGMAT.DIFFERENTIATION, 1178
MAGMAT.EINFLUSS, 2621
MAGMAT.EINSCHLUSS, 379
MAGMAT.ENTSTEHUNG, 2623
MAGMAT.FAZIES, 2174
MAGMAT.HYBRIDISIERUNG, 2086
MAGMAT.KONTAMINATION, 917
MAGMAT.VERGESELLSCHAFTUNG, 2620
MAGMATISCH, 2619
MAGMATISMUS, 2624
MAGMATIT, 2176
MAGMATIT=GENESE, 2175
MAGMATIT=KOMPLEX, 2173
MAGNA=FAZIES, 2625
MAGNESIT, 2626
MAGNET=KOMPASS, 2630
MAGNET.ANOMALIE, 2628
MAGNET.AZIMUT, 2629
MAGNET.BEREICH, 2631
MAGNET.EIGENSCHAFT, 2640
MAGNET.ELEMENT, 2632
MAGNET.FELDVARIATION, 2634
MAGNET.INKLINATION, 2635
MAGNET.KARTE, 2636
MAGNET.KUGEL, 2641
MAGNET.METHODE, 2637
MAGNET.MINERAL, 2638
MAGNET.PERMEABILITAET, 2639
MAGNET.SAEKULARVARIATION, 1744
MAGNET.STURM, 2642
MAGNET.SUSZEPTIBILITAET, 2645
MAGNET.VERMESSUNG, 2643
MAGNETFELD, 2633
MAGNETIK, 2646
MAGNETISIERUNG, 2648
MAGNETIT, 2647
MAGNETO=HYDRODYNAMIK, 2649
MAGNETO=TELLUR.METHODE, 2654
MAGNETO=TELLUR.VERMESSUNG, 2655
MAGNETOMETER, 2650
MAGNETOSPHAERE, 2651
MAGNETOSTRATIGRAPHIE, 2652

MAGNETOSTRIKTION, 2653
MAJIUM, 2656
MALACOSTRACA, 2659
MAMMALIA, 2661
MANGROVE=SUMPF, 2663
MANTODEA, 2666
MARIN.EBENE, 2683
MARIN.EROSION, 2678
MARIN.MILIEU, 2677
MARIN.OEKOLOGIE, 2676
MARIN.PLATTFORM, 2684
MARIN.SEDIMENT, 2685
MARIN.SEDIMENTATION, 2686
MARIN.TRANSPORT, 2689
MARINER=PROGRAMM, 2690
MARKIERUNG, 4560
MARKOV=KETTE, 2691
MARKT, 4566
MARMOR, 2671
MARS, 3474
MARSCH, 2693
MARSUPIALIA, 2695
MASCON, 2696
MASSE, 2697
MASSEN=ABFALL, 2702
MASSEN=BEWEGUNG, 2699
MASSEN=BILANZ, 2698
MASSEN=SPEKTROMETRIE, 2700
MASSEN=TRANSFER, 2701
MASSIG.ERZKOERPER, 2705
MASSIV, 2703
MASSSTABS=MODELL, 3955
MASTIGOPHORA, 2706
MATERIAL, 2707
MATHEMAT.ANALYSE, 2708
MATHEMAT.DIFFERENZIERUNG, 1125
MATHEMAT.GEOGRAPHIE, 2710
MATHEMAT.GEOLOGIE, 2711
MATHEMAT.GESETZ, 2712
MATHEMAT.GLEICHUNG, 1411
MATHEMAT.MATRIX, 2713
MATHEMAT.METHODE, 2714
MATHEMAT.MODELL, 2715
MATHEMATIK, 2716
MATRIX, 3887
MATRIX=ALGEBRA, 2717
MATUYAMA=EPOCHE, 2720
MAXILLOPODA, 2721
MAXIMUM, 2722
MAZATSAL=REVOLUTION, 2723
MAZERAL, 2613
MD, 2747
MECHAN.EIGENSCHAFT, 2727
MECHAN.SPANNUNG, 4290
MECHAN.SPUR, 715
MECHAN.URSPRUNG, 2726
MECHAN.VERWITTERUNG, 2728

MECHANIK, 2729
MECHANISMUS, 2730
MECOPTEROIDA, 2731
MEDITERRAN.BODEN, 2735
MEDIZIN, 2733
MEER, 3974
MEER-WASSER, 3978
MEERENGE, 4269
MEERES-ANLAGE, 2680
MEERES-BODEN, 3088
MEERES-EIS, 3976
MEERES-FURCHE, 2687
MEERES-GEOLOGIE, 2679
MEERES-KUESTE, 811
MEERES-ZIRKULATION, 3086
MEERESARM, 2241
MEERESSPIEGEL, 3977
MEERESSPIEGEL-SCHWANKUNG, 3981
MEGA-FAZIES, 2738
MEGASPORE, 2739
MEHRFACH-UEBERTRAGUNG, 854
MEHRPHASEN, 2963
MEHRPHASEN-STROEMUNG, 2964
MELANGE, 2740
MELANOKRAT.ZUSAMMENSETZUNG, 2741
MELIORATION, 2742
MERGEL, 2692
MERKUR, 3475
MEROSTOMATA, 2750
MEROSTOMOIDEA, 2751
MESOGAEA, 2753
MESOGASTROPODA, 2754
MESOGEOSYNKLINALE, 2755
MESOKRAT.ZUSAMMENSETZUNG, 2752
MESOPHYL.ORGANISMUS, 2756
MESOSAURIA, 2757
MESOSIDERIT, 2758
MESOTHERMAL.VORKOMMEN, 2759
MESOZOIKUM, 2760
MESOZONE, 2761
MESS-GERAET, 3618
META-BASALT, 2762
META-GABBRO, 2764
META-PELIT, 3336
META-SEDIMENT, 2792
METACOPINA, 2763
METAGENESE, 2765
METALL, 2766
METALL-GEHALT, 337
METALL-METEORIT, 2767
METALL-PROVINZ, 2773
METALLISCH.EIGENSCHAFT, 2768
METALLOGENESE, 2774
METALLOGENET.EPOCHE, 2771
METALLOGENET.KARTE, 2772
METALLOID, 2775
METALLOTEKT, 2776

METALLURGIE, 2777
METAMIKT, 2778
METAMIKTISIERUNG, 2779
METAMORPH.DIFFERENTIATION, 2782
METAMORPH.GUERTEL, 2780
METAMORPH.KOMPLEX, 2781
METAMORPH.LAGERSTAETTE, 2788
METAMORPH.ZONE, 2786
METAMORPHIT, 2785
METAMORPHOSE, 2787
METAMORPHOSE-FAZIES, 2783
METAMORPHOSE-GRAD, 2784
METANAUPLIUS, 2790
METAPHYT, 2791
METASOM, 2793
METASOMAT.GESTEIN, 2795
METASOMATOSE, 2794
METATEXIS, 2796
METEORIT, 2798
METEORITEN-KRATER, 2797
METEOROID, 2799
METEOROLOGIE, 2800
METHAN, 2801
METHODIK, 2802
MG, 2627
MIAROLITH-GEFUEGE, 2803
MICROPHYTOLITHS, 2828
MIGMATISIERUNG, 2854
MIGMATIT, 2853
MIGRATION, 2855
MIKRIT, 2806
MIKRO-BEBEN, 2810
MIKRO-FALTE, 2815
MIKRO-FAUNA, 2812
MIKRO-FAZIES, 2811
MIKRO-FISSUR, 2813
MIKRO-FLORA, 2814
MIKRO-FOSSIL, 2816
MIKRO-GRANIT, 2818
MIKRO-HAERTE, 2820
MIKRO-KRATER, 2808
MIKRO-METEORIT, 2824
MIKRO-MORPHOLOGIE, 2825
MIKRO-ORGANISMEN, 2826
MIKRO-PALAEONTOLOGIE, 2827
MIKRO-PLATTE, 2829
MIKRO-PROBLEMATICA, 2830
MIKRO-PROZESSOR, 2807
MIKRO-SEISMIK, 2834
MIKRO-SONDE, 1347
MIKRO-STRUKTUR, 4640
MIKRO-STYLOLITH, 2835
MIKRO-TEKTIT, 2837
MIKRO-TEKTONIK, 2836
MIKROGRANIT.GEFUEGE, 2819
MIKROKRISTALLIN.KALK, 2809
MIKROLATEROLOG, 2821

MIKROLITH.GEFUEGE, 2822
MIKROLOG, 2823
MIKROSKOP, 2831
MIKROSKOP.GEFUEGE, 2832
MIKROSKOPIE, 2833
MIKROWELLEN-METHODE, 2838
MIKROWELLEN-SPEKTROMETRIE, 2839
MILIEU, 1386
MILIOLACEA, 2857
MILIOLINA, 2858
MILITAER-GEOLOGIE, 2859
MILLER-INDIZES, 2860
MINERAL, 2862
MINERAL-EINSCHLUSS, 2870
MINERAL-FAZIES, 2869
MINERAL-FUNDPUNKT, 2878
MINERAL-PHASE, 2871
MINERAL-VERGESELLSCHAFTUNG, 2863
MINERAL-WASSER, 2875
MINERAL-WASSER-QUELLE, 2873
MINERAL.ROHSTOFF, 1300
MINERAL.ROHSTOFFE, 2872
MINERALISATION, 2876
MINERALISIEREND.LOESUNG, 2877
MINERALISIERT.THERMAL-WASSER, 2734
MINERALOG.ANALYSE, 2865
MINERALOG.PHASEN-REGEL, 2879
MINERALOGIE, 2880
MINIMUM, 2882
MIOGEOSYNKLINALE, 2890
MIOMERA, 2891
MIOSPORE, 2892
MIOZAEN, 2889
MISCHBARKEIT, 2893
MISCHGITTER-MINERAL, 2898
MISCHUNGS-LUECKE, 2894
MISE-A-LA-MASSE, 2895
MISSISSIPPI-TAL-LAGERSTAETTENTYP, 2896
MISSISSIPPIAN, 2897
MITTELKOERNIG, 2736
MITTELTIEF.ERDBEBEN, 2270
MN, 2662
MO, 2818
MOBIL.GUERTEL, 2899
MOBILISIERUNG, 2901
MOBILITAET, 2900
MODAL-ANALYSE, 2902
MODE, 2903
MODELL, 2904
MODIFIZIERT.MERCALLI-SKALA, 2905
MODIOMORPHOIDA, 2906
MODUL, 2907
MOEGLICHKEIT, 3572
MOERTEL-MATERIAL, 2944
MOESSBAUER-SPEKTROMETRIE, 2947
MOESSBAUER-SPEKTRUM, 2948
MOHO, 2910

MOHOLE-PROJEKT, 2909
MOLASSE, 2912
MOLDAVIT, 2913
MOLEKULAR-SPEKTROSKOPIE, 2914
MOLLISOL, 2915
MOLLUSCA, 2916
MOLYBDAT, 2917
MONAZIT, 2919
MOND, 2935
MOND-AUFBAU, 2600
MOND-BEBEN, 2936
MOND-BODEN, 2603
MOND-FURCHE, 3867
MOND-GEOLOGIE, 2602
MOND-KRATER, 2601
MOND-MARE, 2672
MOND-RELIEF, 2937
MOND-STATION, 2604
MOND-TERRA, 2605
MONOAXIAL-KOMPRESSION, 4658
MONOCOTYLEDONEAE, 2922
MONOCYATHEA, 2923
MONOGRAPHIE, 2924
MONOKLIN.SYSTEM, 2921
MONOKLINALE, 2920
MONOLET.TAXON, 2925
MONOLITH, 2926
MONOMORPH, 2927
MONOPLACOPHORA, 2928
MONORHINA, 2929
MONOTREMATA, 2930
MONSUN, 2931
MONTAN-GEOLOGIE, 2885
MONTANGEOLOG.KARTE, 1299
MONTE-CARLO-METHODE, 2932
MONZONIT, 2933
MONZONIT-GEFUEGE, 2934
MOOR, 3321
MORAENE, 2938
MORAST, 536
MORPHOGENESE, 2939
MORPHOLOGIE, 2940
MORPHOMETRIE, 2941
MORPHOSKOPIE, 2942
MORPHOSTRUKTUR, 2943
MOSAIK-GEFUEGE, 2945
MOSKOV, 2946
MULDEN=ACHSE, 4603
MULL-BODEN, 2960
MULLION-STRUKTUR, 2961
MULTISPEKTRAL-ANALYSE, 2966
MULTITUBERCULATA, 2967
MULTIVARIAT.ANALYSE, 2968
MUSCHELIG.BRUCH, 875
MUSCHELKALK, 2969
MUSEUM, 2970
MUSTER, 3316

MUTATION, 2971
MUTTER=GESTEIN, 3304
MYLONIT, 2972
MYLONIT-GEFUEGE, 2973
MYLONITISIERUNG, 2974
MYODOCOPIDA, 2975
MYOIDA, 2976
MYOMORPHA, 2977
MYRIAPODA, 2978
MYRMEKIT, 2979
MYSTACOCARIDA, 2980
MYTILOIDA, 2981
MYXOMYCETES, 2982
N, 3034
NA, 4143
NACHBEBEN, 79
NADEL, 2996
NADELIG.MINERAL, 1355
NAEHERUNG, 249
NAKHLITH, 2983
NAMUR, 2984
NANNO-FOSSIL, 2985
NASS-CHEM.METHODE, 4821
NATANTES, 2987
NATUERL.GESTEINS-SPRENGUNG, 3883
NATUERL.GLAS, 2990
NATUERL.GRUNDWASSER-SPEISUNG, 1917
NATUERL.REMANENT.MAGNETISIERUNG, 2991
NATURKATASTROPHE, 1738
NATURRAEUML.NUTZUNG, 2427
NAUTILOIDEA, 2993
NB, 3030
ND, 3000
NE, 3003
NE-METALL, 3047
NEANDERTHALER, 2994
NEBENELEMENT-ANALYSE, 2887
NEBENELEMENT-ANALYSEN-METHODIK, 2888
NEBENFLUSS, 77
NEBENGESTEIN, 4780
NEHRUNGS-BOGEN, 4215
NEIGUNG, 4533
NEIGUNGS-MESSER, 1197
NEKTON, 2997
NEMERTA, 2998
NEO-TEKTONIK, 3005
NEOGASTROPODA, 3001
NEOGEN, 3002
NEOKOM, 2999
NEORNITHES, 3004
NEOTENIE, 3006
NEOTYP, 3007
NEOVULKANISCH, 687
NEPHELIN, 3008
NEPHELOID-HORIZONT, 3009
NEPTUN, 3476
NEPTUNISMUS, 3010

NERVENSYSTEM, 3012
NEU, 3019
NEU.DATEN, 3020
NEU.METHODE, 3022
NEU.MINERALDATEN, 3023
NEU.NAME, 3024
NEU.TAXON, 3025
NEUBESCHREIBUNG, 3021
NEUTRON, 3014
NEUTRON-LOG, 3017
NEUTRONEN-AKTIVIERUNGS-ANALYSE, 3015
NEUTRONEN-BEUGUNGS-ANALYSE, 3016
NEUTRONEN-SONDE, 3018
NI, 3026
NICHT-EXPLOSIV.QUELLE, 3046
NICHT-METALL, 3048
NICHT-METALL-LAGERSTAETTEN-KARTE, 3052
NICHT-METALL.ROHSTOFF, 3051
NICHT-MISCHBAR.MATERIAL, 2186
NICHT-MISCHBARKEIT, 2185
NICHT-QUARTAER.EISZEIT, 191
NICHTBINDIG.BODEN, 828
NICHTGESAETTIGT.ZONE, 4866
NIEDER-DRUCK, 2571
NIEDER-ENERGET.MILIEU, 2574
NIEDRIG.WASSER, 2573
NIEDRIGGRADIG.METAMORPHOSE, 2575
NIGGLI-KLASSIFIKATION, 3027
NILSSONIALES, 3028
NIOB-TANTALAT, 3031
NIOBAT, 3029
NITRAT, 3032
NITRIFIZIERUNG, 3033
NITROGEN-GAS, 3035
NIVEAU, 2493
NIVELLEMENT, 2492
NODOSARIACEA, 3038
NOEGGERATHIALES, 3040
NOMENKLATUR, 3042
NOMENKLATUR-WOERTERBUCH, 1824
NOMINAL-TAXON, 3043
NOMOGENESE, 3044
NONCONFORMITY, 3050
NOR, 3053
NORDEN, 3057
NORM, 3054
NORMAL-VERTEILUNG, 3055
NORMALLOG-VERTEILUNG, 2563
NORMUNG, 4238
NOTOSTRACA, 3059
NOTOUNGULATA, 3060
NOVACULITH, 3061
NP, 3011
NUCULOIDEA, 3068
NUMMULITIDAE, 3071
NUTATION, 3072
NUTZBAR.GESTEINE, 2219

NUTZBAR.WASSER, 4798
O, 3225
O.DEVON, 4671
O.EOZAEN, 4672
O.JURA, 4673
O.KAMBRIUM, 4667
O.KARBON, 4668
O.KREIDE, 4669
O.LIAS, 4674
O.MIOZAEN, 4676
O.OLIGOZAEN, 4677
O.ORDOVIZIUM, 4678
O.PERM, 4679
O.PLIOZAEN, 4681
O.PROTEROZOIKUM, 4682
O.SILUR, 4683
O.TRIAS, 4684
O18-O16, 3227
OASE, 3074
OBDUKTION, 3075
OBER.ERD-MANTEL, 4675
OBER.KRUSTE, 4670
OBERFLAECHEN-ABFLUSS, 4364
OBERFLAECHEN-BILDUNGEN, 4369
OBERFLAECHEN-FEHLER, 4363
OBERFLAECHEN-SPANNUNG, 4365
OBERFLAECHEN-WELLE, 4368
OBERFLAECHENNAH.BEBEN, 4052
OBERFLAECHENWASSER, 4366
OBERFLAECHENWASSER-BILANZ, 4367
OBOLELLIDA, 3079
OBSEQUENT.TAL, 3080
OBSERVATORIUM, 3082
OBSIDIAN, 3083
OCEAN DRILLING PROJECT, 3087
OCEANUS-PROCELLARIUS, 3621
OCHREPT, 3096
OCKER, 3095
OCTOCORALLIA, 3098
ODONTOLOGIE, 3099
ODONTOPLEURIDA, 3100
ODONTORNITHES, 3101
OEDOMETRIE, 908
OEKOLOGIE, 1297
OEKOSYSTEM, 1301
OEKOZONE, 1302
OEL-GAS-GRENZFLACHE, 3113
OEL-SAND, 3108
OEL-SCHIEFER, 3110
OEL-WASSER-GRENZE, 3114
OELFELD, 3107
OFF-SHORE, 3102
OFFSHORE-BOHRLOCH, 3104
OFFSHORE-BOHRUNG, 3103
OKTAEDRIT, 3097
OLIGOZAEN, 3115
OLISTHOLITH, 3116

OLISTHOSTROM, 3117
OLIVIN, 3118
ONKOLITH, 3119
ONTOGENESE, 3121
ONYCHOPHORA, 3122
OOID, 3123
OOLITH, 3124
OOLITH-GEFUEGE, 3126
OOLITH-KALK, 3125
OPAK, 3128
OPHIDIA, 3131
OPHIOCISTIOIDEA, 3132
OPHIOLITH, 3133
OPHIT-GEFUEGE, 3134
OPHIUROIDEA, 3135
OPISTHOBRANCHIA, 3136
OPOKA, 3137
OPT.DISPERSION, 1222
OPT.DREHUNG, 3141
OPT.EIGENSCHAFT, 3140
OPT.SPEKTROMETRIE, 3142
OPT.SPEKTRUM, 3143
ORANGES-MATERIAL, 3144
ORBIT, 3146
ORBITOIDACEA, 3147
ORBITOIDIDAE, 3148
ORDER-DISORDER, 3149
ORDOVIZIUM, 3150
ORGAN.GEFUEGE, 3175
ORGAN.GEOCHEMIE, 3168
ORGAN.KALK, 562
ORGAN.KOHLENSTOFF, 3166
ORGAN.LEBEN, 3169
ORGAN.LOESUNGSRUECKSTAND, 3173
ORGAN.SAEURE, 3165
ORGAN.SCHLAMM, 3127
ORGAN.STICKSTOFF, 3171
ORGAN.SUBSTANZ, 3170
ORGAN.VERBINDUNG, 3167
ORGANISCH, 3164
ORGANISMUS, 3176
ORGANOGEN.SEDIMENT, 3174
ORIENTIERUNG, 3177
ORIGO-FAZIES, 3180
ORNITHISCHIA, 3183
OROGENESE, 3185
ORTHENT, 3186
ORTHID, 3187
ORTHIDA, 3188
ORTHO-AMPHIBOLIT, 3189
ORTHO-GNEIS, 3192
ORTHOD, 3190
ORTHOGENESE, 3191
ORTHOPHYR-GEFUEGE, 3193
ORTHORHOMB.SYSTEM, 3194
ORTHOX, 3195
ORYKTOCOENOSE, 3196

OS, 3199
OSTEICHTHYES, 3200
OSTEN, 1288
OSTEOLOGIE, 3201
OSTEOSTRACI, 3202
OSTRACODA, 3203
OSTREACEA, 3204
OSZILLATION, 3197
OSZILLATIONS-THEORIE, 3198
OTOLITH, 3205
OUTLIER, 3210
OXFORD, 3220
OXID, 3222
OXIDATION, 3221
OXIDATIONS-ZONE, 3223
OXISOL, 3224
OXYLOPHYT, 3228
OZEAN, 3084
OZEAN-BECKEN, 3085
OZEAN.KRUSTE, 3091
OZEANISATION, 3092
OZEANISCH, 3090
OZEANOGRAPH.METHODE, 2682
OZEANOGRAPHIE, 3093
OZEANOLOGIE, 3094
P, 3407
P-WELLE, 3230
PA, 3642
PA-IO-DATIERUNG, 3643
PACKUNG, 3232
PAHOEHOE, 3233
PALAEO, 3239
PALAEOATMOSPHAERE, 3240
PALAEOBATHYMETRIE, 3242
PALAEOBIOLOGIE, 3243
PALAEOBODEN, 3268
PALAEOBOTAN.PROVINZ, 3244
PALAEOBOTANIK, 3245
PALAEOGEN, 3254
PALAEOGEOGRAPH.EINFLUSS, 3255
PALAEOGEOGRAPH.KARTE, 3256
PALAEOGEOGRAPH.REKONSTRUKTION, 3767
PALAEOGEOGRAPHIE, 3257
PALAEOHETERODONTA, 3235
PALAEOKARST, 3259
PALAEOKLIMA, 3248
PALAEOKLIMATOLOGIE, 3249
PALAEOLATITUDE, 3260
PALAEOLIMNOLOGIE, 3261
PALAEOMAGNET.POL, 3262
PALAEOMAGNETISMUS, 3263
PALAEOMILIEU, 3253
PALAEOMORPHOLOGIE, 3258
PALAEONTOLOG.REKONSTRUKTION, 3264
PALAEONTOLOGIE, 3265
PALAEOOZEANOGRAPHIE, 3241
PALAEOPHYTIKUM, 3236

PALAEORELIEF, 3266
PALAEOSALINITAET, 3267
PALAEOSTROEMUNG, 3251
PALAEOTAXODONTA, 3237
PALAEOTEMPERATUR, 3269
PALAEOTHERMOMETRIE, 3270
PALAEOZIRKULATION, 3247
PALAEOZOIKUM, 3271
PALAEOZOOLOGIE, 3272
PALAGONITISIERUNG, 3238
PALEOCOPIDA, 3250
PALEOZAEN, 3246
PALIMPSEST, 3273
PALINGENESE, 3274
PALLASIT, 3276
PALOEKOLOGIE, 3252
PALSA, 3277
PALYNOLOGIE, 3278
PALYNOMORPHA, 3279
PAMPA, 3280
PANGAEA, 3281
PANIDIOMORPH.GEFUEGE, 3282
PANTODONTA, 3284
PANTOTHERIA, 3285
PANZER, 639
PARA-, 3286
PARA-AMPHIBOLIT, 3287
PARA-GNEIS, 3292
PARABLASTOIDEA, 3288
PARACRINOIDEA, 3289
PARAGENESE, 3290
PARAGEOSYNKLINALE, 3291
PARAL.GEOSYNKLINALE, 3293
PARAL.MILIEU, 3294
PARALLEL-GEFUEGE, 3296
PARALLEL-SCHICHTUNG, 3295
PARALLELITAET, 3297
PARAMAGN.ELEKTR.RESON.SPEKTRUM, 1410
PARAMAGNET.EIGENSCHAFT, 3298
PARAMAGNET.ELEKTRONEN-RESONANZ, 1351
PARAMETER, 3299
PARASITISMUS, 3300
PARATYP, 3301
PARTIAL-DRUCK, 3306
PARTIELL.SCHMELZEN, 3305
PARTIKEL, 3307
PATENT, 3314
PATHOLOGIE, 3315
PAUROPODA, 3318
PAZIF.ABFOLGE, 3231
PB, 2466
PB-PB-DATIERUNG, 2468
PD, 3275
PECTINACEA, 3323
PEDIMENT, 3324
PEDIPLAIN, 3325
PEGEL-KURVE, 2116

PEGMATIT, 3327
PEGMATITISCH, 3328
PEGMATOID, 3329
PELAG.MILIEU, 3330
PELAG.SEDIMENTATION, 3331
PELEE-TYP, 3333
PELIT, 3334
PELIT-GEFUEGE, 3335
PELLET, 3337
PELLETIERUNG, 3338
PENEPLAIN, 3339
PENETROMETER, 3341
PENNSYLVANIAN, 3342
PENOKE-OROGENESE, 3343
PENTAMERIDA, 3344
PENTASTOMIDA, 3345
PENTOXYLALES, 3346
PERALKAL.CHEMISMUS, 3347
PERALUMINAT-GEHALT, 3348
PERIDOTIT, 3352
PERIGAEUM, 3353
PERIGLAZIAER, 3354
PERIGLAZIAL-ZONE, 3355
PERIKLINE, 3351
PERIODE, 3356
PERIODIZITAET, 3357
PERISPHINCTIDA, 3358
PERISSODACTYLA, 3359
PERKOLATION, 3350
PERLE, 3319
PERLIT, 3360
PERLIT-GEFUEGE, 3361
PERM, 3364
PERMAFROST, 3362
PERMEABILITAET, 3363
PERSONAL-BIBLIOGRAPHIE, 3365
PERSPEKTIVE, 3366
PERTHIT, 3367
PERTHIT-GEFUEGE, 3368
PESTIZID, 3369
PETROCHEM.DIAGRAMM, 3371
PETROCHEM.KOEFFIZIENT, 3370
PETROCHEMIE, 3372
PETROGENESE, 3374
PETROGRAPH.ANALYSE, 3375
PETROGRAPH.KARTE, 3376
PETROGRAPH.PROVINZ, 3377
PETROGRAPHIE, 3378
PETROLOGIE, 3382
PETROPHYSIK, 3383
PFAHL, 3445
PFEILER, 3446
PFLANZE, 3491
PFROPFEN, 3513
PFROPFEN-BILDUNG, 3514
PH-WERT, 3384
PHACOLITH, 3385

PHACOPIDA, 3386
PHAEOPHYTA, 3387
PHAEOZEM, 3388
PHAGOTROPH.ORGANISMUS, 3389
PHANEROZOIKUM, 3390
PHASE, 3391
PHASEN-GLEICHGEWICHT, 3392
PHASEN-REGEL, 3394
PHENOL, 3396
PHI-SKALA, 3397
PHILOSOPHIE, 3398
PHOLADOMYIDA, 3399
PHOLIDOTA, 3400
PHONOLITH, 3401
PHOSPHAT, 3403
PHOSPHAT-GESTEIN, 3402
PHOSPHATISIERUNG, 3404
PHOSPHORESZENZ, 3405
PHOSPHORIT, 3406
PHOTOCHEM.REAKTION, 3408
PHOTOGEOLOG.KARTE, 3409
PHOTOGEOLOGIE, 3410
PHOTOGRAMMETRIE, 3411
PHOTOGRAPHIE, 3412
PHOTOMETRIE, 3413
PHOTOSYNTHESE, 3414
PHREAT.GAS, 3415
PHREAT.GRUNDWASSER, 3416
PHREATOPHYT, 3417
PHYLACTOLAEMATA, 3418
PHYLLIT, 3419
PHYLLOCERATIDA, 3420
PHYLLONIT, 3421
PHYLLOPODA, 3422
PHYLOGENESE, 3423
PHYLUM, 3424
PHYSIK-DES-ERDKOERPERS, 4169
PHYSIKAL.ALTERSBESTIMMUNG, 1073
PHYSIKAL.EIGENSCHAFT, 3427
PHYSIKAL.MODELL, 3426
PHYSIKAL.STATUS, 3428
PHYSIKAL.VERWITTERUNG, 3429
PHYSIKOCHEM.EIGENSCHAFT, 3430
PHYSIKOCHEMIE, 3425
PHYSIOGRAPH.PROVINZ, 3431
PHYSIOLOGIE, 3432
PHYTAN, 3433
PHYTOCOENOSE, 3434
PHYTOLITHES, 3435
PHYTOPHAG.ORGANISMUS, 3436
PHYTOPLANKTON, 3437
PIEDMONT, 3439
PIEZO-ELEKTR.EIGENSCHAFT, 3442
PIEZO-KRISTALLISATION, 3441
PIEZOMETRIE, 3443
PIGMENT, 3444
PIKRIT, 3438

PILLOW=LAVA, 3447
PINGO, 3450
PINNIPEDIA, 3452
PIPE, 3453
PIPELINE, 3454
PISCES, 3456
PISOLITH, 3457
PISOLITH=GEFUEGE, 3458
PLACANTICLINE, 3461
PLACENTALIA, 3462
PLACODERMI, 3464
PLANCTOSPHAEROIDEA, 3469
PLANET, 3471
PLANET=ERDE, 3472
PLANETAR, 3482
PLANETEN=AUFBAU, 3483
PLANETEN=SATELLIT, 3484
PLANETOLOGIE, 3485
PLANKTON, 3487
PLANKTONFRESSER, 3486
PLANKTON.TAXON, 3488
PLANOSOL, 3490
PLANUNG, 3489
PLAST.FLIESSEN, 3492
PLAST.MATERIAL, 3493
PLASTIZITAET, 3494
PLATEAU, 3501
PLATEAU=BASALT, 3502
PLATIN=GRUPPE, 3505
PLATTE, 3495
PLATTEN=GRENZE, 3497
PLATTEN=KOLLISION, 3496
PLATTEN=ROTATION, 3498
PLATTEN=TEKTONIK, 3499
PLATTENDRUCK=VERSUCH, 3500
PLATTFORM, 3503
PLATYRRHINA, 3506
PLATZNAHME, 1363
PLAYA, 3507
PLEISTOZAEN, 3508
PLEOCHROISMUS, 3509
PLIENSBACHIUM, 3511
PLIOZAEN, 3512
PLUTO, 3477
PLUTON, 3517
PLUTON.METAMORPHOSE, 3518
PLUTONISMUS, 3520
PLUTONIT, 3519
PLUVIAL, 3522
PM, 3637
PNEUMATOLYT.BEDINGUNG, 3523
PNEUMATOLYT.LAGERSTAETTE, 3524
PO, 3544
PODOCOPIDA, 3525
PODSOL, 3526
PODZOLUVISOL, 3527
POGONOPHORA, 3528

POIKILIT.GEFUEGE, 3529
POIKILOBLAST.GEFUEGE, 3530
POISSON=KOEFFIZIENT, 3532
POISSON=VERTEILUNG, 3531
POL=KALOTTE, 2156
POL=WANDERUNG, 3535
POLAR, 3533
POLAR.EIS=KALOTTE, 3534
POLARISATION, 3139
POLAROGRAPHIE, 3536
POLIERSTOFF, 9
POLITIK, 3537
POLLEN, 3539
POLLEN=DIAGRAMM, 3540
POLY=METAMORPHOSE, 3552
POLYCHAETIA, 3545
POLYEDER, 3548
POLYGENETISCH, 3546
POLYGON=BODEN, 3547
POLYMERA, 3549
POLYMERISIERUNG, 3550
POLYMETALL.LAGERSTAETTE, 3551
POLYMIKT.GEFUEGE, 3553
POLYMINERALISCH, 3554
POLYMORPHIE, 3555
POLYPLACOPHORA, 3557
POLYTYPIE, 3558
PONTIUM, 3559
POPULAERWISSENSCHAFTL.ARTIKEL, 3560
POPULATION, 855
PORCELLANIT, 3561
POREN=DRUCK, 3562
POREN=WASSER, 3563
PORENWASSER=DRUCK, 1760
PORIFERA, 3564
POROES.MEDIUM, 3566
POROSITAET, 3565
PORPHYR=GEFUEGE, 3567
PORPHYROBLAST.GEFUEGE, 3568
PORPHYRY=COPPER, 3569
PORTLAND, 3570
PORZELLAN=ERDE, 743
PORZELLAN=TON, 4823
POSITIV.ERZ, 3571
POSTGLAZIAL, 3573
POSTMAGMAT.BILDUNG, 3574
POSTTEKTON.BILDUNG, 3575
POTENTIELL.LAGERSTAETTE, 3582
POTENTIOMETRIE, 3584
POTTERS=TON, 3586
PR, 3591
PRAE=IMBRIUM, 3592
PRAE=NEANDERTHALER, 3604
PRAECARDIIDA, 3590
PRAEGLAZIAL, 3601
PRAEKAMBR.OROGENESE, 3594
PRAEKAMBRIUM, 3593

PRAEPARATION, 3605
PRAEZISIONS=MESSUNG, 34
PREHNIT=PUMPELLYIT=FAZIES, 3603
PREIS, 3610
PRIEL, 4525
PRIMAER=AUREOLE, 3611
PRIMAER=GEFUEGE, 3612
PRIMATES, 3613
PRIMITIV, 3614
PRIORITAET, 3615
PRISMAT.GEFUEGE, 3616
PRISMAT.SAEULENBILDUNG, 850
PROBE, 3933
PROBEN=ZERKLEINERUNG, 1637
PROBENNAHME, 3935
PROBLEMATICA, 3619
PROBOSCIDEA, 3620
PROCHORDATA, 3623
PRODUKTION, 3624
PRODUKTIONS=KONTROLLE, 3625
PRODUKTIV.HORIZONT, 3626
PRODUKTIV.SCHACHT, 3627
PROFIL, 1736
PROFIT, 3629
PROGANOSAURIA, 3630
PROGNOSE, 3631
PROGRAD.METAMORPHOSE, 3633
PROGRAMM, 3634
PROJEKT, 3636
PROPYLITISIERUNG, 3639
PROSIMII, 3640
PROSPEKTION, 1483
PROTEIN, 3645
PROTELYTROPTERA, 3646
PROTEROZOIDEN, 3648
PROTEROZOIKUM, 3647
PROTOCILIATA, 3649
PROTOLITH, 3650
PROTOMEDUSAE, 3651
PROTON, 3652
PROTOTYP, 3653
PROTOZOA, 3654
PROTURA, 3655
PROVINZIALITAET, 3657
PROZESS, 3622
PRUEFUNGS=ARBEIT, 4512
PSAMMENT, 3658
PSAMMIT, 3659
PSAMMITISCH, 3660
PSAMMOFAUNA, 3661
PSEPHITISCH, 3662
PSEUDO=GLEY, 3664
PSEUDO=KONGLOMERAT, 3663
PSEUDO=SCHIEFERUNG, 4133
PSEUDOMORPHOSE, 3665
PSEUDOTILLIT, 3666
PSILOCERATIDA, 3667

PSILOPSIDA, 3668
PSYCHROPHYT, 3669
PT, 3504
PTERIDOPHYLLA, 3670
PTERIDOPHYTA, 3671
PTERIDOSPERMAE, 3672
PTERIOIDA, 3673
PTEROBRANCHIA, 3674
PTEROMORPHIA, 3675
PTEROPODA, 3676
PTEROSAURIA, 3677
PTYCHOPARIIDA, 3678
PTYGMAT.FALTE, 3679
PU, 3521
PUBLIKATION, 3681
PULMONATA, 3682
PULVER-AUFNAHME, 3587
PUMP-TEST, 3685
PUMP-VERSUCH, 261
PUMPEN, 3684
PURIN, 3688
PUZZOLAN, 3589
PYCNOGONIDA, 3690
PYKNOMETER, 3691
PYKNOMETRIE, 3692
PYRAMIDAL.SYSTEM, 3693
PYRITISIERUNG, 3694
PYRO-METALLURGIE, 3697
PYRO-METAMORPHOSE, 3698
PYRO-METASOMATOSE, 3699
PYROKLAST.GESTEIN, 3695
PYROLYSE, 3696
PYROSPHAERE, 3700
PYROTHERIA, 3701
PYROXEN, 3702
PYROXEN-HORNFELS-FAZIES, 3703
PYROXENIT, 3704
PYRROPHYTA, 3705
QUALITAET, 3707
QUALITATIV.METHODE, 3706
QUANTITATIV.METHODE, 3708
QUARTAER, 3713
QUARTAER.GRUNDWASSER, 143
QUARTAER.SYSTEM, 3714
QUARZ, 3710
QUARZ-DIORIT, 3711
QUARZIT, 3712
QUELLE, 4224
QUELLUNG, 4378
QUER-VERSCHIEBUNG, 4567
QUICK-TON, 3715
RA, 3739
RACEMISIERUNG, 3717
RADAR, 3718
RADIAL, 3719
RADIO-INTERFEROMETRIE, 3722
RADIO-KOHLENSTOFF, 3733

RADIOAKTIV.ABFALL, 3730
RADIOAKTIV.ISOTOP, 3727
RADIOAKTIV.KETTE, 3728
RADIOAKTIV.MARKIERUNG, 3729
RADIOAKTIV.METHODE, 3732
RADIOAKTIV.STAUB, 3726
RADIOAKTIV.ZERFALL, 3725
RADIOAKTIVITAET, 3731
RADIOFREQUENZ-SPEKTROMETRIE, 3723
RADIOLARIA, 3735
RADIOLARIT, 3736
RADIOMETER, 3737
RADIOMETR.PROSPEKTION, 3738
RADIUS, 3740
RAET, 3848
RAFFINIERUNG, 3781
RAHMEN-GESTEIN, 978
RAMAN-SPEKTROMETRIE, 3745
RAND-FAZIES, 546
RAND-GEFUEGE, 2675
RAND-MEER, 2674
RAND-PLATEAU, 2673
RANKER, 3749
RAPAKIWI, 3750
RASEN-EISENERZ, 537
RASTER-ELEKTR.MIKROSKOPIE, 3957
RASTER-ELEKTR.MIKROSKOPIE-DATEN, 4034
RATE, 3753
RATITES, 3754
RAUBTIERSPUR, 3597
RAUHIGKEIT, 3905
RAUM-WELLE, 535
RAUMGRUPPE, 4188
RAUSCHEN, 3041
RAYLEIGH-WELLE, 3756
RB, 3909
RB-SR-DATIERUNG, 3910
RE, 3849
RE-OS-DATIERUNG, 3850
REAGENZ, 3761
REAKTION, 3757
REAKTIONS-SERIE, 3759
REAKTIONSSAUM, 3758
REAKTIVIERUNG, 3760
RECEPTACULITIDA, 3763
RECHENPROGRAMM, 872
RECHENSPRACHE, 871
RECHNER, 870
RECHT, 2476
RECYCLING, 3772
RED-BEDS, 3773
REDLICHIIDA, 3775
REDOX-POTENTIAL, 1312
REFLEKTIVITAET, 3785
REFLEXION, 3783
REFLEXIONS-SEISMIK, 3784
REFLEXIONS-VERMOEGEN, 3782

REFRAKTIONS-SEISMIK, 3787
REFRAKTOMETER, 3789
REFRAKTOMETRIE, 3790
REGEN, 3742
REGEN-AUSWASCHUNG, 3744
REGENERATION, 3792
REGIONAL-DISKORDANZ, 3796
REGIONAL-HYDROGEOLOGIE, 3794
REGIONAL-METAMORPHOSE, 3795
REGIONAL.GEOLOGIE, 3793
REGMAGLYPT, 3797
REGOLITH, 3798
REGOSOL, 3799
REGRESSION, 3800
REIBUNG, 1650
REIBUNGS-BREKZIE, 1000
REIFEGRAD, 2719
REIFUNG, 2718
REINIGUNG, 3687
REJUVENATION, 3802
REKRISTALLISATION, 3770
REKULTIVIERUNG, 3766
RELATIV.ALTER, 3804
RELAXATIONS-ENERGIE, 3805
RELIEF, 3808
RELIEF-INVERSION, 3810
RELIEF-KARTE, 659
RELIKT, 3806
RELIKT-GEFUEGE, 3807
REMAGNETISIERUNG, 3811
REMANENT.MAGNETISIERUNG, 3812
REMOTE-SENSING, 3813
RENDOLL, 3815
RENDZINA, 3816
REPRAESENTATIV-BECKEN, 3819
REPTILIA, 3821
RESERVE, 3824
RESIDUAL, 3827
RESIDUAL-LAGERSTAETTE, 3829
RESIDUAL-TON, 3828
REST-MAGMA, 3830
RESURGENZ, 3836
RETICULAREA, 3839
RETIKULAR-GEFUEGE, 3840
RETRO-METAMORPHOSE, 3841
REVISION, 3844
RH, 3855
RHABDOPLEURIDA, 3847
RHEOLOGIE, 3851
RHEOMORPHOSE, 3852
RHIPIDISTII, 3853
RHIZOPODEA, 3854
RHODOPHYTA, 3856
RHOMB.SYSTEM, 3857
RHYNCHOCEPHALIA, 3858
RHYNCHONELLIDA, 3859
RHYODACIT, 3860

RHYOLITH, 3861
RICHTER=SKALA, 3862
RICHTUNG, 4292
RICHTUNGS=BOHREN, 1208
RIEFUNG, 2874
RIEGEL, 3863
RIFF, 3776
RIFF=BANK, 3313
RIFF=MILIEU, 3777
RIFF=SEDIMENTATION, 3778
RIFFBILDNER, 3779
RIFFEL=STRUKTUR, 1591
RIFT, 3864
RIFT=VALLEY, 3865
RIGIDITAET, 3866
RING=DYKE, 3869
RING=STRUKTUR, 3871
RINGFOERMIG.BRUCH, 3870
RINNE, 707
RINNEN=GEOMETRIE, 708
RIPPELMARKE, 3872
RN, 3741
ROBERTINACEA, 3880
RODENTIA, 3892
ROENTGEN=ANALYSE, 4833
ROENTGEN=AUFNAHME, 3734
ROENTGEN=BEUGUNG, 4834
ROENTGEN=FLUORESZENZ, 4835
ROENTGEN=SPEKTROMETRIE, 4836
ROENTGEN=STRAHLUNG, 4832
ROH=DICHTE, 590
ROHRLEITUNG, 4649
ROHSTOFF, 3755
ROLL=TYP=LAGERSTAETTE, 3893
ROSTROCONCHIA, 3897
ROTALIACEA, 3898
ROTALIINA, 3899
ROTARY=BOHRMETHODE, 3901
ROTARY=BOHRUNG, 3900
ROTATION, 3902
ROTIEREND.DISPERSION, 3903
ROTLIEGENDES, 3904
RU, 3919
RUBEFIZIERUNG, 3908
RUDISTAE, 3911
RUDIT, 3912
RUECK=RIFF, 397
RUECKSCHREITEND.TAL, 3823
RUGOSA, 4474
RUMINANTIA, 3913
RUN, 3914
RUNDUNG, 3906
RUNDUNGSGRAD, 4207
RUPEL=STUFE, 3916
RUPTUR, 3917
RUPTUR=FESTIGKEIT, 3918
RUTSCH=AUSLOESEND.ANSTOSS, 4254

RUTSCH=GEFUEGE, 4138
RUTSCHUNG, 4139
S, 4353
S=WELLE, 3920
S34=S32, 4355
SAETTIGUNG, 3950
SAETTIGUNGS=ZONE, 3949
SAEURE, 37
SAKMAR, 3921
SALINITAET, 3923
SALTATION, 3931
SALZ=LAUGE, 578
SALZ=MARSCH, 2681
SALZ=SEE, 3927
SALZ=TEKTONIK, 3929
SALZEBENE, 3926
SALZHALTIG, 3922
SALZSTOCK, 3925
SALZWASSER=INTRUSION, 3930
SAMEN, 4007
SAMMLUNG, 839
SAMPLER, 3934
SAND, 3936
SAND=KOERPER, 3937
SAND=SCHIEFER=VERHAELTNIS, 3938
SANDER, 250
SANDIG.GEFUEGE, 290
SANDSTEIN, 3939
SANDSTEIN=GANG, 3940
SANIDINIT=FAZIES, 3942
SANTON, 3943
SAPROPEL, 3944
SAPROPEL=KOHLE, 3945
SARCODINA, 3946
SARCOPTERYGII, 3947
SARMAT, 3948
SATURN, 3478
SAUERSTOFF, 3226
SAUM, 3868
SAUMRIFF, 1651
SAUR.CHEMISMUS, 39
SAUR.GRUBEN=WASSER, 38
SAURISCHIA, 3951
SAURORNITHES, 3952
SAUSSURITISIERUNG, 3953
SAXONIUM, 3954
SB, 229
SC, 3956
SCAPHOPODA, 3958
SCHACHT, 3459
SCHADSTOFF, 3541
SCHAEDEL, 980
SCHALE, 4071
SCHARNIER=VERWERFUNG, 2015
SCHELF, 925
SCHELF=HANG, 930
SCHELF=HANG=MILIEU, 4136

SCHELF=MEER, 4069
SCHELF=MILIEU, 4068
SCHELF=SEDIMENTATION, 4070
SCHENKEL, 2505
SCHER=FLAECHE, 4061
SCHER=KLUFT, 4059
SCHER=MODUL, 4060
SCHER=SPANNUNG, 4063
SCHER=WIDERSTAND, 4062
SCHER=ZONE, 4064
SCHERUNG, 4058
SCHICHT, 2462
SCHICHT=FLAECHE, 454
SCHICHT=VULKAN, 4283
SCHICHTGEBUNDEN.LAGERSTAETTE, 4270
SCHICHTGLIED, 2746
SCHICHTPARALLEL.STOERUNG, 456
SCHICHTSTUFE, 1031
SCHICHTUNG, 4272
SCHICHTUNGS=SCHIEFERUNG, 455
SCHIEFE, 4128
SCHIEFER, 3960
SCHIEFER=SPALTBARKEIT, 4131
SCHIEFER=TON, 4051
SCHIEFERUNG, 3961
SCHILD, 4072
SCHILD=VULKAN, 4073
SCHILL=KALK, 948
SCHLACKE, 3969
SCHLACKEN=KEGEL, 4192
SCHLACKEN=LAVA, 2
SCHLAG=BREKZIE, 4056
SCHLAMM, 2952
SCHLAMM=BANK, 2953
SCHLAMM=DIAPIR, 2957
SCHLAMM=RUTSCHUNG, 2959
SCHLAMM=SCHWUND=RISS, 2954
SCHLAMM=STROM, 2956
SCHLAMM=VULKAN, 2958
SCHLEIF=MARKE, 1902
SCHLEIF=MARKEN=AUSGUSS, 1903
SCHLEPP=FALTE, 1243
SCHLICK, 4527
SCHLICK=KUESTE, 2955
SCHLIEREN=GEFUEGE, 3962
SCHLIFF=KONSTANTE, 2418
SCHLOSS, 2014
SCHLOT, 2995
SCHLUCHT, 1928
SCHLUFF=SEDIMENTATION, 4108
SCHMELZ=PHASE, 2520
SCHMELZ=WAERME, 1966
SCHMELZ=WASSER, 2745
SCHMELZ=WASSER=ABLAGERUNG, 3211
SCHMELZE, 2743
SCHMELZEN, 2744
SCHMUCKSTEIN, 1707

SCHNEE, 4142
SCHNEE-EROSION, 3036
SCHOCK, 4075
SCHOCK-BREKZIE, 4076
SCHOCK-METAMORPHOSE, 4077
SCHOCK-WELLE, 4078
SCHOTTER, 3907
SCHRAEG-VERWERFUNG, 1196
SCHRAEG.ORIENTIERUNG, 3078
SCHRAEG.STOERUNG, 3077
SCHRAEGSCHICHTUNG, 997
SCHREIBKREIDE, 705
SCHRUMPFUNG, 4087
SCHUBMODUL, 2908
SCHUERF-RECHT, 779
SCHUETT-GUT, 408
SCHUETTUNG, 1209
SCHUETTUNGS-WINKEL, 200
SCHUPPE, 3963
SCHUPPEN-TEKTONIK, 2180
SCHURF, 4813
SCHUSSPUNKT, 4085
SCHUTT, 4416
SCHUTT-ABHANG, 4417
SCHUTT-KEGEL, 1076
SCHUTT-STROM, 1077
SCHUTZBEZIRK, 3644
SCHWEBEND.GRUNDWASSER, 3349
SCHWEFELDIOXID, 4354
SCHWEFELSAEURE, 4356
SCHWEFELWASSERSTOFF, 2108
SCHWELLENWERT, 4519
SCHWEMM-FAECHER, 141
SCHWER-ISOTOP, 1971
SCHWER-METALL, 1972
SCHWER-MINERAL, 1973
SCHWER-OEL, 1974
SCHWERE-ANOMALIE, 1878
SCHWERE-DIFFERENZIERUNG, 1877
SCHWERE-GLEITUNG, 1885
SCHWERE-PLATTFORM, 1884
SCHWERE-TRENNUNG, 1114
SCHWUNDRISS, 4088
SCIUROMORPHA, 3965
SCOLECODONTA, 3967
SCOURING, 3971
SCYPHOMEDUSAE, 3972
SCYPHOZOA, 3973
SE, 4030
SEA-MOUNT, 3984
SEAFLOOR-SPREADING, 3980
SEBKHA-MILIEU, 3987
SEDIMENT, 3991
SEDIMENT-AUSLAENGUNG, 3992
SEDIMENT-BECKEN, 3995
SEDIMENT-GANG, 785
SEDIMENT-GEFUEGE, 4001

SEDIMENT-GENESE, 4005
SEDIMENT-GESTEIN, 4000
SEDIMENT-PETROLOGIE, 3998
SEDIMENT-VOLUMEN, 3993
SEDIMENT-WASSER-GRENZFLAECHE, 3994
SEDIMENTAER.EINFLUSS, 3999
SEDIMENTAER.FALLE, 4002
SEDIMENTAER.LAGERSTAETTE, 3997
SEDIMENTATION, 4003
SEDIMENTATIONS-RATE, 4004
SEDIMENTATIONS-ZONE, 4869
SEDIMENTATIONS-ZYKLUS, 1045
SEDIMENTOLOGIE, 4006
SEE, 2415
SEE-AUSFLUSS, 1310
SEE-WASSERSPIEGEL, 2416
SEEGAT, 4528
SEEWAERTIG.KUESTENVERLAGERUNG, 3632
SEGREGATION, 4010
SEGREGATIONS-GANG, 4009
SEICHE, 4011
SEIFE, 3463
SEISM.GESCHWINDIGKEIT, 4020
SEISM.INSTRUMENT, 4013
SEISM.INTENSITAET, 1286
SEISM.LOG, 4012
SEISM.METHODE, 4014
SEISM.NETZ, 306
SEISM.OBSERVATORIUM, 4018
SEISM.QUELLE, 4016
SEISM.RISIKO, 4015
SEISM.SPEKTRAL-ANALYSE, 4017
SEISM.TIEFENSONDIERUNG, 1091
SEISM.VERMESSUNG, 4019
SEISM.WELLE, 4021
SEISMIZITAET, 4022
SEISMOGRAMM, 4023
SEISMOGRAPH, 4024
SEISMOLOG.KARTE, 4025
SEISMOLOGIE, 4026
SEISMOMETER, 4027
SEISMOTEKTONIK, 4028
SEITEN-MORAENE, 2444
SEKUNDAER-AUREOLE, 3988
SEKUNDAER-MINERAL, 3989
SELEKTIV.VERWITTERUNG, 1177
SELENAT, 4029
SELENOCHRONOLOGIE, 4031
SELENOLOGIE, 4032
SELTEN.ERDEN, 3751
SELTEN.METALL, 3752
SEMIARID.MILIEU, 4035
SENKRECHT.ORIENTIERUNG, 4735
SENKUNG, 4337
SENON, 4037
SEPTARIE, 4040
SEPTUM, 4041

SERIE, 4044
SERIZIT-SCHIEFER, 4042
SERIZITISIERUNG, 4043
SERPENTIN, 4045
SERPENTINISIERUNG, 4047
SERPENTINIT, 4046
SETZUNG, 4048
SEXUAL-DIMORPHISMUS, 4049
SH-WELLE, 4050
SHATTER-CONE, 4057
SHOESTRING-SAND, 4079
SHONKINIT, 4080
SI, 4103
SIAL, 4089
SICILIUM, 4090
SIDE-SCAN-METHODE, 4091
SIDEROPHIL.ELEMENT, 4092
SIDEROPHYR, 4093
SIEB, 4096
SIEDLUNGS-MUELL, 2070
SIEDLUNGSBAU, 4696
SIEGEN, 4094
SIEROZEM, 4095
SILICOFLAGELLATA, 4102
SILIKAT, 4098
SILLIMANIT, 4105
SILT, 4106
SILT-FRACHT, 4107
SILTSTEIN, 4109
SILUR, 4110
SIMA, 4112
SIMULATION, 4114
SINEMURIUM, 4115
SINTER, 4118
SIPHON, 4120
SIPHONAPTERA, 4121
SIPUNCULOIDA, 4122
SIRENIA, 4123
SKAPOLITISIERUNG, 3959
SKARN, 4126
SKELETT, 4127
SKELETT-BODEN, 2535
SKLERIT, 3966
SKOLITHUS, 3968
SKULPTUR, 3182
SM, 3932
SN, 4537
SOHLFLAECHEN-INJEKTION, 4162
SOHLMARKE, 4163
SOLAR-STRAHLUNG, 4159
SOLEMYIDA, 4164
SOLFATARA, 4165
SOLIDUS, 4170
SOLIFLUKTION, 4171
SOLONETZ, 4173
SOLONTSCHAK, 4172
SOMASTEROIDEA, 4180

SONAR, 4181
SONNE, 4358
SONNEN-EINSTRAHLUNG, 2255
SONNEN-SYSTEM, 4160
SONNEN-WIND, 4161
SORPTION, 4183
SORTIERUNG, 4184
SOZIAL-OEKOLOGIE, 2069
SPAET-DIAGENESE, 2438
SPAET-PROTEROZOIDEN, 2439
SPALLATION, 4189
SPALTBARKEIT, 2864
SPALTENFROST-VERWITTERUNG, 891
SPALTSPUREN, 1555
SPALTSPUREN-DATIERUNG, 1556
SPARAGMIT, 4190
SPARKER, 4191
SPEICHER-GESTEIN, 3826
SPEICHERUNG, 4265
SPEKTROMETER, 4199
SPEKTROMETRIE, 4201
SPEKTROPHOTOMETRIE, 4200
SPEKTRUM, 4202
SPELAEOTHEM, 4204
SPEZIF.FAUNA, 4194
SPEZIF.FLORA, 4195
SPEZIF.GEWICHT, 4196
SPEZIF.OBERFLAECHE, 4198
SPEZIF.WAERME, 4197
SPHAER.HARMON.ANALYSE, 4206
SPHAEROIDAL-GEFUEGE, 3145
SPHAEROLITH, 4209
SPHAERULIT, 4208
SPHENULITHUS, 4205
SPIEGELWERT, 398
SPILITISIERUNG, 4210
SPINIFEX-GEFUEGE, 4212
SPIRIFERIDA, 4213
SPIRILLINACEA, 4214
SPODOSOL, 4216
SPONGIOLITH, 4218
SPORE, 4219
SPOROMORPH, 4220
SPOROZOA, 4221
SPROEDE, 579
SPUREN-METALL, 4558
SPURENELEMENT-ANALYSE, 4556
SPURENELEMENT-ANALYSEN-METHODE, 4559
SQUAMATA, 4225
SR, 4299
STABIL.ISOTOP, 4228
STABILISATION, 4227
STABILITAET, 4226
STADT-GEOLOGIE, 4694
STADT-PLANUNG, 4695
STAFFEL-BRUCH, 1365
STAFFEL-STOERUNG, 4248

STALAGMIT, 4232
STALAKTIT, 4231
STAMPIEN, 4233
STANDARD, 4235
STANDARD-ABWEICHUNG, 4234
STANDARD-GESTEIN, 4237
STANDARD-METHODE, 4236
STANDORTWAHL, 4124
STARK-VERFESTIGT.MATERIAL, 3214
STATIONAER.STROEMUNG, 4246
STATIST.GESETZ, 4241
STATIST.HAEUFIGKEIT, 4243
STATIST.METHODE, 4239
STATIST.MODELL, 4242
STATIST.REGRESSION, 3801
STATIST.VERTEILUNG, 4240
STATISTIK, 4245
STAU-DAMM, 1059
STAUB, 1263
STEATITISIERUNG, 4247
STEFAN, 4250
STEIF.TON, 4255
STEIN-EISEN-METEORIT, 4263
STEIN-METEORIT, 4264
STEINBRUCH, 3709
STEINPLATTE, 4129
STEINSALZ, 3924
STEINSCHUETTUNGS-DAMM, 3885
STEPPEN-ZONE, 4251
STEREOCHEMIE, 4252
STEROL, 4253
STOCHAST.PROZESS, 4256
STOCK, 4257
STOCKWERK, 4258
STOCKWERKS-SYSTEM, 2962
STOCKWERKS-TEKTONIK, 3556
STOECHIOMETRIE, 4259
STOERUNG, 1507
STOERUNGS-EINFALLEN, 1199
STOERUNGS-SYSTEM, 1512
STOLLEN-FASSUNG, 2227
STOLONOIDEA, 4260
STOMOCHORDATA, 4261
STONE-LINE, 4262
STOSS-METAMORPHOSE, 2189
STOSS-SEISMIK, 2190
STOSSWELLEN-EFFEKT, 2188
STRAHLUNG, 3720
STRAHLUNGS-SCHADEN, 3721
STRAND, 444
STRAND-ABBRUCH, 447
STRAND-BARRIERE, 417
STRAND-SEIFE, 445
STRAND-TERRASSE, 2688
STRAND-WALL, 2567
STRASSE, 3879
STRASSEN-TEST, 3878

STRASSEN-VERMESSUNG, 3876
STRASSENBAU-MATERIAL, 3877
STRATEGIE, 4271
STRATIGRAPH.BEZUG, 4276
STRATIGRAPH.EINHEIT, 4280
STRATIGRAPH.FALLE, 4279
STRATIGRAPH.GRENZE, 4275
STRATIGRAPH.ISOCHORE, 4277
STRATIGRAPH.KARTE, 4278
STRATIGRAPH.SKALA, 1734
STRATIGRAPH.VERBREITUNG, 3746
STRATIGRAPHIE, 4281
STRATOTYP, 4282
STREAM-SEDIMENT, 4287
STRECK-GRENZE, 4849
STRESS-MINERAL, 4291
STRIEMUNG, 1514
STRIP-MINING, 4294
STROEMUNG, 1035
STROEMUNGS-SPUR, 1036
STROMATOLITHES, 4295
STROMATOPOROIDEA, 4296
STROMBETT, 4285
STROMBOLI-TYP, 4297
STROPHOMENIDA, 4300
STRUDELTOPF, 3585
STRUKTUR-ANALYSE, 4301
STRUKTUR-BODEN, 3317
STRUKTUR-GENESE, 4302
STRUKTUR-GEOLOGIE, 4306
STRUKTUR-GLEICH.MINERAL, 2345
STRUKTUR-KARTE, 4308
STRUKTUR-RELIEF, 4310
STRUKTUR.KRISTALLOGRAPHIE, 4303
STRUKTURELEMENT, 4305
STUFE, 4229
STURM, 4266
STYLOLITH, 4314
STYLOPHORA, 4315
SUBALKAL.CHEMISMUS, 4317
SUBALUMINOES.CHEMISMUS, 4318
SUBBOREAL, 4319
SUBDUKTION, 4321
SUBDUKTIONS-ZONE, 4322
SUBLIMAT, 4323
SUBLIMATION, 4324
SUBLITORAL-MILIEU, 4325
SUBMARIN.CANYON, 4326
SUBMARIN.FAECHER, 4327
SUBMARIN.KONSTRUKTION, 4328
SUBMARIN.MORPHOLOGIE, 4329
SUBMARIN.QUELLE, 4331
SUBMARIN.RELIEF, 3975
SUBMARIN.RUECKEN, 4330
SUBMARIN.RUECKEN, 2840
SUBMARIN.SEISMOGRAPH, 3089
SUBMARIN.TAL, 4332

SUBMARIN.VULKAN, 4333
SUBSEQUENT.TAL, 4336
SUBSTITUTION, 4338
SUBSTRAT, 4339
SUBSURFACE-GEOLOGY, 4340
SUBTIDAL.MILIEU, 4341
SUBTROPEN-ZONE, 4342
SUBVULKAN.GESTEIN, 2817
SUBVULKAN.VORGANG, 4343
SUCTORIA, 4345
SUDET.OROGENESE, 4346
SUEDEN, 4187
SUESS-WASSER, 1647
SUESS-WASSER-MILIEU, 1648
SUESS-WASSER-SEDIMENTATION, 1649
SUIFORMES, 4348
SULFAT, 4349
SULFID, 4351
SULFID-ZONE, 4350
SULFOSALZ, 4352
SUMPF, 4376
SUMPFGAS, 2694
SUPRAKRUSTAL.GENESE, 4362
SUSPENDIERT.MATERIAL, 4372
SUSPENSION, 4373
SUTUR, 4374
SVEKOFENNIDEN-OROGENESE, 4375
SYENIT, 4379
SYLVINIT, 4380
SYMBIOSE, 4381
SYMMETR.FALTE, 4382
SYMMETRIE, 4384
SYMMETRODONTA, 4383
SYMPHYLA, 4385
SYMPLEKT.GEFUEGE, 4386
SYMPOSIUM, 4387
SYNAPSIDA, 4388
SYNEKLISE, 4392
SYNGENESE, 4394
SYNGENETISCH, 4395
SYNKLINALE, 4390
SYNKLINORIUM, 4391
SYNONYMIE, 4396
SYNSEDIMENTAER.BILDUNG, 2291
SYNSEDIMENTAER.GEFUEGE, 4398
SYNSEDIMENTAER.HERKUNFT, 4397
SYNSEDIMENTAER.STOERUNG, 1921
SYNTEKTON.BILDUNG, 4399
SYNTEXIS, 4400
SYNTHESE, 4401
SYNTHET.MATERIAL, 4402
SYSTEM, 4403
SYSTEM-ANALOGIE, 4405
SYSTEM-ANALYSE, 4404
SZINTILLATION, 3964
TA, 4419
TABELLE, 1072

TABULATA, 4406
TACONIT, 4408
TACTIT, 4409
TAENIODONTA, 4410
TAENIOLABOIDEA, 4411
TAETIGKEITS-BERICHT, 3635
TAGEBAU, 3129
TAGES-SCHWANKUNG, 1231
TAGESDAUER, 2478
TAGUNG, 2737
TAIGA, 4412
TAKON.OROGENESE, 4407
TAL, 4707
TALK, 4414
TALK-SCHIEFER, 4415
TANGENTIAL-TEKTONIK, 867
TANTALAT, 4418
TAPHONOMIE, 4420
TAPHRO-GEOSYNKLINALE, 4422
TAPHROGENESE, 4421
TARANNON, 4423
TARDIGRADA, 4424
TASMANITES, 4425
TATAR, 4426
TAU, 1152
TAUB.GESTEIN, 4217
TAUB.LAGERSTAETTE, 415
TAUEN, 4485
TAXIT, 4427
TAXIT.GEFUEGE, 1453
TAXONOMIE, 4428
TB, 4460
TC, 4429
TE, 4450
TECHN.HILFE, 4430
TECHNIK, 4431
TECHNOLOGIE, 4432
TEILCHEN-SPUR-DATIERUNG, 3309
TEKTIT, 4444
TEKTOGENESE, 4433
TEKTON.BAUSTIL, 4438
TEKTON.BREKZIE, 4434
TEKTON.DISKORDANZ, 1214
TEKTON.EINFLUSS, 4435
TEKTON.FALLE, 4311
TEKTON.INDIKATOR, 4307
TEKTON.KARTE, 4437
TEKTON.SCHIEFERUNG, 392
TEKTON.STRUKTUR, 1737
TEKTON.TAL, 4312
TEKTON.UEBERSTRUKTUR, 4439
TEKTON.ZYKLUS, 4436
TEKTONIK, 4440
TEKTONIT, 4441
TEKTONOPHYSIK, 4442
TEKTONOSPHAERE, 4443
TELEOSTEI, 4445

TELESEISM.SIGNAL, 4446
TELETHERMAL.VORKOMMEN, 4447
TELLUR.STROM, 1274
TELLURAT, 4448
TELLURIK, 4449
TEM-DATEN, 4451
TEMPERATUR, 4453
TENTACULITES, 4457
TEPHROCHRONOLOGIE, 4458
TERATOLOGIE, 4459
TEREBRATULIDA, 4461
TERMINOLOGIE, 4462
TERNAER.SYSTEM, 4463
TERRA-ROSSA, 4464
TERRASSE, 4465
TERRESTR.MILIEU, 4467
TERRESTR.PLANET, 4468
TERRIGEN.SUBSTANZ, 4469
TERTIAER, 4470
TESTUDINATA, 4472
TETHYS, 4473
TETRAPODA, 4475
TEXTULARIINA, 4476
TH, 4517
TH-PB-DATIERUNG, 4518
THALASSOID, 4479
THALASSOKRATON, 4478
THALLOPHYTA, 4481
THANATOCOENOSE, 4483
THANET, 4484
THECAMOEBA, 4486
THECIDEIDINA, 4487
THECODONTIA, 4488
THELODONTI, 4489
THEORIE, 4490
THERALITH, 4491
THERAPSIDA, 4492
THERM.EIGENSCHAFT, 4499
THERM.ENTMAGNETISIERUNG, 4496
THERM.EXPLORATION, 4500
THERMAL, 4493
THERMAL-WASSER, 4501
THERMO-ANALYSE, 4494
THERMO-GRAVIMETR.ANALYSE, 4507
THERMO-KARST, 4508
THERMO-METAMORPHOSE, 4498
THERMOCHEM.EIGENSCHAFT, 4502
THERMODYNAM.EIGENSCHAFT, 4504
THERMODYNAMIK, 4505
THERMOGRAMM, 4506
THERMOKLINE, 4503
THERMOLUMINESZENZ, 4509
THERMOMAGNET.ANALYSE, 4510
THERMOREMANENT.MAGNETISIERUNG, 4511
THIXOTROPIE, 4515
THOLEIIT.CHEMISMUS, 4516
THURINGIUM, 4522

THYSANOPTEROIDA, 4523
THYSANURA, 4524
TI, 4539
TIDAL-MILIEU, 4526
TIEF-SCHLEPP-METHODE, 1092
TIEF-TEMPERATUR, 2572
TIEFBAU, 777
TIEFE, 1122
TIEFEN-BEBEN, 1088
TIEFEN-INDIKATOR, 1123
TIEFEN-TEKTONIK, 1094
TIEFSEE, 17
TIEFSEE-BERG, 21
TIEFSEE-BODEN, 22
TIEFSEE-GRABEN, 4587
TIEFSEE-MILIEU, 1090
TIEFSEE-RINNE, 1089
TIEFSEE-SEDIMENTATION, 23
TIEFSEE-TAL, 20
TIEFWASSER-MILIEU, 19
TIEFWASSER-ZONE, 1093
TILLIT, 4531
TILLODONTIA, 4532
TINTINNIDAE, 4538
TITHON, 4540
TL, 4480
TM, 4521
TOARCIUM, 4541
TOBEL, 1837
TOCHTER-ELEMENT, 1074
TOMBOLO, 4542
TON, 787
TON-GEHALT, 295
TON-GEROELL, 788
TON-SCHIEFER, 4130
TONERDE, 156
TONIG.GESTEIN, 791
TONMINERAL, 790
TONMINERALOGIE, 789
TONSTEIN, 4543
TOPOGRAPH.KARTE, 4546
TOPOGRAPH.KORREKTUR, 4545
TOPOGRAPHIE, 4547
TOPSET-BED, 4548
TORF, 3320
TORF-KOHLE-KONKRETION, 802
TORSION, 4550
TORTON, 4551
TOTEIS, 1075
TOURNAI, 4553
TOXIZITAET, 4555
TRACHT, 1016
TRACHYANDESIT, 4561
TRACHYBASALT, 4562
TRACHYPSAMMIACEA, 4563
TRACHYT, 4564
TRAGFAEHIGKEIT, 450

TRANSFORM-STOERUNG, 4569
TRANSFORMATION, 4570
TRANSFORMISMUS, 4571
TRANSGRESSION, 4572
TRANSIENT-METHOD, 4573
TRANSMISS.ELEKTR.MIKROSKOPIE, 4575
TRANSMISS.KOEFFIZIENT, 4574
TRANSMISSIVITAET, 4576
TRANSPORT, 4577
TRANSPORT-MITTEL, 4578
TRANSVERSAL, 4579
TRAVERTIN, 4583
TREMADOC, 4586
TREND, 4244
TREND-ANALYSE, 4588
TRENNUNG, 4039
TREPOSTOMATA, 4589
TREPPEN-FALTE, 4249
TRIANGULATION, 4590
TRIAS, 4591
TRIAXIAL-KOMPRESSION, 4592
TRICONODONTA, 4594
TRIGONAL.SYSTEM, 4595
TRIGONIIDAE, 4596
TRIKLIN.SYSTEM, 4593
TRILOBITA, 4597
TRILOBITOMORPHA, 4598
TRINKBARKEIT, 3576
TRIPOLI, 4600
TRITIUM-DATIERUNG, 4601
TROPEN-ZONE, 4602
TSCHERNOZEM, 738
TSUNAMI, 4604
TUBOIDEA, 4605
TUBULIDENTATA, 4606
TUFFIT, 4608
TUFFO-LAVA, 4609
TUNDRA, 4610
TUNDRA-BODEN, 4611
TUNNEL, 4614
TURBIDIT, 4615
TURBIDIT-BILDUNG, 4616
TURBIDIT-STROM, 4617
TURBIDIT-STRUKTUR, 4618
TURBO-BOHRUNG, 4619
TURBULENT.STROEMUNG, 1304
TURON, 4620
TYLOPODA, 4623
TYP-LOKALITAET, 4624
TYP-MATERIAL, 2148
TYP-PROFIL, 4625
TYPOMORPH.MINERAL, 4627
TYPOMORPHIE, 4628
TYPUS-ART, 4626
TYRRHENIUM, 4629
U, 4689
U-BAHN, 4344

U-PB-DATIERUNG, 4692
U-TH-PB-DATIERUNG, 4693
U.DEVON, 2581
U.EOZAEN, 2582
U.JURA, 2583
U.KAMBRIUM, 2577
U.KARBON, 2578
U.KREIDE, 2579
U.LIAS, 2584
U.MIOZAEN, 2586
U.OLIGOZAEN, 2587
U.ORDOVIZIUM, 2588
U.PERM, 2589
U.PLIOZAEN, 2591
U.PROTEROZOIKUM, 2592
U.SILUR, 2593
U.TRIAS, 2594
UDALF, 4630
UDOLL, 4631
UDULT, 4632
UEBERFLUTUNG, 4334
UEBERKIPPT.FALTE, 3219
UEBERLAGERUNG, 4360
UEBERPRAEGT.FALTE, 4359
UEBERPRAEGUNG, 1723
UEBERPRAEGUNGS-TEKTONIK, 2965
UEBERSAETTIGT, 3216
UEBERSAETTIGUNG, 3217
UEBERSCHIEBUNG, 3218
UEBERSCHUSS-ARGON, 1463
UEBERSCHWEMMUNGS-FLAECHE, 1570
UEBERSICHTS-DARSTELLUNG, 4357
UEBERSTRUKTUR, 4361
UEBERTRAGUNG, 4568
UEBERWACHSUNG, 3215
UFERWALL, 446
UFIMIUM, 4633
ULTISOL, 4634
ULTRA-METAMORPHOSE, 4637
ULTRA-MYLONIT, 4638
ULTRABAS.CHEMISMUS, 4635
ULTRABASIT, 4636
ULTRASCHALL-METHODE, 4639
UMBREPT, 4642
UMFELD-STUDIE, 2191
UMKEHRUNG, 3842
UMRISS-KARTE, 932
UMWELT-GEOLOGIE, 1388
UMWELT-SCHUTZ, 905
UMWELT-STUDIE, 1387
UNAKIT, 4643
UNDULOES.AUSLOESCHUNG, 4657
UNELASTIZITAET, 198
UNGESAETTIGT.ZONE, 3049
UNGESPANNT.GRUNDWASSER, 4644
UNGESPANNT.WASSER, 4645
UNIFORMITARISMUS, 4659

UNIONIDAE, 4660
UNIVERSAL-DREHTISCH, 4662
UNIVERSUM, 4663
UNLOESL.RUECKSTAND, 2256
UNREIF, 2182
UNREIF.BODEN, 2183
UNREINHEIT, 2196
UNTER.ERD-MANTEL, 2585
UNTER.ERD.KRUSTE, 2580
UNTERGLIEDERUNG, 4320
UNTERIRD.AKKUMULATION, 31
UNTERIRD.FLUSS, 4653
UNTERIRD.GEWINNUNG, 4650
UNTERIRD.SPEICHERUNG, 4652
UNTERSAETTIGT, 4654
UNTERSAETTIGT.LOESUNG, 4655
UNTERSCHIEBUNG, 4656
UNTERWASSER-FAHRZEUG, 4335
UNTIEFE, 4074
UNVERFESTIGT.GESTEIN, 4647
URALITISIERUNG, 4688
URAN-MINERAL, 4690
URAN-UNGLEICHGEWICHT, 4691
URANUS, 3479
UREILIT, 4697
URODELA, 4698
USTALF, 4699
USTERT, 4700
USTOLL, 4701
USTOX, 4702
USTULT, 4703
UV-STRAHLUNG, 4641
V, 4710
VAKUUM-SCHMELZ-ANALYSE, 4705
VALANGINIUM, 4706
VANADAT, 4709
VARIABILITAET, 4714
VARIABLE, 4715
VARIANZ-ANALYSE, 4716
VARIATION, 4717
VARIOGRAMM, 4718
VARIST.OROGENESE, 4719
VARISTIDEN, 1985
VEGETATION, 4724
VEKTOR, 4723
VENEROIDA, 4728
VENIT, 4729
VENUS, 3480
VERANKERUNG, 190
VERBLEND-STEIN, 1085
VERBRAUCH, 912
VERBREITUNGS-ZONE, 3747
VERDECKT.LAGERSTAETTE, 529
VERDUENNUNG, 1188
VEREISUNG, 1803
VEREISUNGS-RUECKGANG, 1098
VERERBUNG, 1986

VERFAHREN, 3130
VERFEINERUNG, 3780
VERFESTIGT.SCHLICKGEROELL, 303
VERFLUECHTIGUNG, 4758
VERFLUESSIGUNG, 2519
VERFORMUNG, 1096
VERFORMUNGS-MESSGERAET, 4268
VERFUEGBAR.WASSER, 4802
VERFUELL-MATERIAL, 1358
VERFUELLUNG, 84
VERGASUNG, 1695
VERGENZ, 4730
VERGESELLSCHAFTUNG, 338
VERGLASUNG, 4751
VERGLEICH-ERDE, 3481
VERGLEICHS-DIAGRAMM, 3045
VERGREISUNG, 1897
VERHAELTNIS, 3803
VERHAERTUNG, 2218
VERKIESELT.HOLZ, 4101
VERKIESELUNG, 4100
VERLANDET.SEE, 1494
VERMES, 4731
VERMICULIT, 4732
VERSCHIEBUNGS-THEORIE, 1225
VERSCHUPPUNG, 2181
VERSENKUNG, 596
VERSENKUNGS-METAMORPHOSE, 597
VERSIEGEN, 4797
VERSTEINERUNG, 2524
VERTEBRATA, 4734
VERTEILUNG, 1230
VERTEILUNGS-KOEFFIZIENT, 3311
VERTIKAL-BEWEGUNG, 4736
VERTIKAL-VERSCHIEBUNG, 4831
VERTISOL, 4737
VERUNREINIGT.WASSER, 3542
VERUNREINIGUNG, 3543
VERWACHSUNG, 2267
VERWENDUNG, 4704
VERWERFUNGS-BREKZIE, 1508
VERWERFUNGS-FLAECHE, 1510
VERWERFUNGS-VERSCHIEBUNG, 1224
VERWITTERUNG, 154
VERWITTERUNGS-KRUSTE, 4810
VESUV-TYP, 4777
VIBRATION, 4738
VIBRATIONS-BOHRUNG, 4739
VIBROSEIS, 4740
VIKING-PROGRAMM, 4741
VILLAFRANCIUM, 4742
VIRGATION, 4743
VISE, 4748
VISKOELASTIZITAET, 4744
VISKOREMANENT.MAGNETISIERUNG, 4747
VISKOS.MATERIAL, 4746
VISKOSITAET, 4745

VITROKLAST.GEFUEGE, 4752
VITROPHYR.GEFUEGE, 4753
VOLANTES, 4754
VOLATIL, 4755
VOLLKERN-BOHRUNG, 1661
VOLUMEN, 4776
VORBEBEN, 1615
VORBEUGEND.MASSNAHME, 3609
VORGESCHICHTE, 3602
VORHERSAGE, 3598
VORKOMMEN, 3834
VORLAEUFIG.UNTERSUCHUNG, 3178
VORLAND, 1613
VORSTRAND, 1616
VORTIEFE, 1612
VORZUGS-ORIENTIERUNG, 3600
VULKAN, 4769
VULKAN-GUERTEL, 4760
VULKAN-KEGEL, 4762
VULKAN-KUPPE, 4763
VULKAN.ASCHE, 4759
VULKAN.BREKZIE, 4761
VULKAN.EINBRUCH, 675
VULKAN.ENTSTEHUNG, 4765
VULKAN.GLAS, 4764
VULKAN.LAGERSTAETTE, 4774
VULKAN.SCHLACKE, 769
VULKAN.SCHWEISS-SCHLACKE, 83
VULKAN.TUFF, 4607
VULKANISMUS, 4768
VULKANIT, 4766
VULKANO-SEDIMENTAER.BILDUNG, 4772
VULKANO-TEKTON.GEFUEGE, 4771
VULKANO-TEKTON.SENKE, 4770
VULKANOGEN, 4773
VULKANOLOGIE, 4775
VULKANORIUM, 4778
W, 4613
WADI, 4779
WAERME, 1962
WAERME-KAPAZITAET, 1963
WAERME-LEITFAEHIGKEIT, 4495
WAERME-LEITUNG, 1964
WAERME-QUELLE, 1968
WAERME-TRANSFER, 1969
WAERME-WIRKUNG, 4497
WAERMESTROM, 1967
WAESSERIG, 2135
WAESSRIG.LOESUNG, 256
WAHRSCHEINLICHKEITS-RECHNUNG, 3617
WALK-ERDE, 1662
WARWE, 4720
WARWEN-CHRONOLOGIE, 4721
WARWEN-TON, 4722
WASSER, 4783
WASSER-BILANZ, 4785
WASSER-BOHRUNG, 1254

WASSER=DAMPF, 4713
WASSER=DRUCK, 4795
WASSER=EROSION, 4789
WASSER=GEWINNUNG, 4786
WASSER=HAERTE, 4790
WASSER=KREISLAUF, 2119
WASSER=LAUF, 4284
WASSER=QUALITAET, 4796
WASSER=SCHEIDE, 4788
WASSER=SPEICHERUNG, 4799
WASSER=TECHNIK, 2114
WASSER=VERBRAUCH, 4787
WASSER=VERSORGUNG, 4800
WASSERFALL, 4803
WASSERSTOFF, 2107
WASSERSTRASSE, 4804
WASSERWIRTSCHAFT, 4791
WASSERWIRTSCHAFTS=BEHOERDE, 4784
WASSERWIRTSCHAFTS=PLAN, 433
WECHSELLAGERUNG, 2260
WEICH.TON, 4144
WELLE, 4805
WELLEN=AUSBREITUNG, 4807
WELLEN=BRECHUNG, 3786
WELLEN=STREUUNG, 4806
WELTRAUM, 2285
WENLOCK, 4818
WERKSTEIN, 3181
WERTIGKEIT, 734
WESTEN, 4819
WESTTAL, 4820
WICHTUNG, 4812
WICKEL=SCHICHT, 941
WIDERSTANDS=MESSUNG, 3833
WILD=WASSER, 571
WILDBACH, 4549
WILDFLYSCH, 4825
WIND, 4826
WIND=EROSION, 4827
WIND=RIPPEL, 4828

WIND=TRANSPORT, 4829
WINKEL=DISKORDANZ, 202
WIRBELSAEULE, 4211
WIRTS=GESTEIN, 2062
WIRTS=MATERIAL, 2061
WIRTSCHAFTS=GEOLOGIE, 1298
WISSENSCHAFTL.GESELLSCHAFT, 2472
WOGE, 4370
WOHN=SPUR, 600
WOLFRAMAT, 4612
WOLKEN=GEFUEGE, 1561
WUESTE, 1131
WUESTEN=BILDUNG, 1134
WUESTEN=PFLASTER, 1133
WURMFOERMIG.GEFUEGE, 4733
WURZEL, 3896
XANTHOPHYTA, 4837
XE, 4840
XENOLITH, 4838
XENOMORPH.KRISTALL, 4839
XERALF, 4841
XEROLL, 4842
XEROPHIL.PFLANZE, 4843
XEROPHYT, 4844
XEROSOL, 4845
XERULT, 4846
Y, 4854
YB, 4853
YERMOSOL, 4848
YOUNG=MODUL, 4031
YPRESIUM, 4852
ZAEHIGKEIT, 4454
ZAHN, 4544
ZECHSTEIN, 4855
ZEICHNUNG, 1250
ZEITFAKTOR, 4535
ZELLEN=GEFUEGE, 564
ZELLULAR=POROSITAET, 682
ZELLULOSE, 683
ZEMENT=EINPRESSUNG, 1920

ZEMENTATION, 685
ZEMENTATIONS=LAGERSTAETTE, 4867
ZENTRAL.MASSIV, 2732
ZENTROKLINALE, 690
ZEOLITH, 4856
ZEOLITH=FAZIES, 4857
ZEOLITHISIERUNG, 4858
ZERFALLS=KONSTANTE, 1079
ZICK=ZACK=FALTE, 4859
ZIEGEL=TON, 576
ZIRKON, 4861
ZIRKULATION, 773
ZIRKUMPAZIF.GUERTEL, 774
ZN, 4860
ZOANTHINIARIA, 4863
ZOEZIUM, 4864
ZONAL.GEFUEGE, 4865
ZONARBAU, 4870
ZONE=NIEDRIG.GESCHWINDIGKEIT, 2576
ZOOID, 4871
ZR, 4862
ZUG, 4267
ZUG=WIDERSTAND, 4455
ZURUECKHALTUNG, 3838
ZUSAMMENFLUSS, 888
ZUSAMMENSETZUNG, 863
ZUSCHLAG=STOFF, 85
ZUSTANDS=GLEICHUNG, 1412
ZWECK, 3076
ZWEIACHSIG.KRISTALL, 476
ZWEIDIMENSIONAL.MODELL, 4622
ZWILLING, 4621
ZWISCHENSCHICHT, 2268
ZWISCHENSTROMLAND, 2264
ZYKL.BELASTUNG, 1046
ZYKL.PROZESS, 1047
ZYKLONE, 1049
ZYKLOTHEM, 1051
ZYKLUS, 1044
ZYLINDR.FALTE, 1052
ZYLINDR.GEFUEGE, 1053

Russkij

A SLOJ, 1
AA-LAVA, 2
AALEN, 3
ABISSAL', 19
ABISSAL'NAYA /DEPRESSIYA/, 18
ABLYATSIYA, 5
ABRAZIA, 6
ABRAZIV, 9
ABSORBENT, 12
ABSORBTSIYA, 13
ABSORBTSIYA ATOMNAYA, 363
ABYSS, 17
ACANTHARIA, 25
ACANTHODII, 26
ACID MINE DRAINAGE, 38
ACRANIA, 46
ACRITARCHA, 48
ACROTRETIDA, 49
ACTINIARIA, 50
ACTINOCERATOIDEA, 52
ADAPTATSIYA, 63
ADIABATICHESKIJ, 65
ADIAGNOSTICHESKIJ, 66
ADSORBTSIYA, 68
AEHRATSIYA, 69
AEHROBNYJ, 73
AEHRODROM, 96
AEHROFOTOS'EMKA, 72
AEHROFOTOSNIMOK, 71
AEHROKARTIROVANIE, 70
AEHROMETOD, 95
AEHROZOL', 75
AFANITOVYJ, 236
AFTERSHOK, 79
AGERMATIPNYJ, 92
AGGLYUTINAT, 83
AGGRADATSIYA, 84
AGLOMERAT, 82
AGNATHA, 86
AGNOSTIDA, 87
AGONIATITIDA, 88
AGPAITOVYJ, 89
AGREGAT, 85
AGROGEOLOGIYA, 91
AJSBERG, 2163

AKADSKIJ, 24
AKCHAGYL, 99
AKHONDRIT, 36
AKKRETSIYA, 30
AKKRETSIYA KONTINENTOV, 919
AKKUMULYATSIYA, 31
AKMOLIT, 98
AKRISOL, 47
AKSELEROGRAMMA, 27
AKSELEROMETR, 28
AKTINIJ, 51
AKTIVIZATSIYA, 59
AKTIVNOST', 61
AKTUALIZM, 62
AKVALF, 254
AKVENT, 255
AKVEPT, 257
AKVERT, 258
AKVITAN, 263
AKVOD, 265
AKVOKS, 267
AKVOLL, 266
AKVULT, 268
AL'B, 103
AL'BEDO, 101
AL'BERTIT, 102
AL'BITIZATSIYA, 106
AL'FA SPEKTROSKOPIYA, 148
AL'FA-LUCHI, 147
AL'FISOL, 110
AL'PIJSKIJ, 149
AL'PIJSKIJ /OROGEN/, 150
AL'PINOTIPNYJ, 153
ALBOLL, 107
ALCYONACEA, 108
ALDANSKIJ, 109
ALEBASTR, 100
ALEVRIT, 4533
ALEVROLIT, 4109
ALGAL BANK, 112
ALGAL BISCUIT, 113
ALGAL MAT, 115
ALGEBRA MATRICHNAYA, 2717
ALGOL, 119
ALGOMANSKIJ, 120

ALGORITM, 121
ALKAN, 132
ALLEGANSKIJ, 133
ALLOGROMIINA, 137
ALLOKHEM, 134
ALLOKHTON, 135
ALLOMETRIYA, 138
ALLOTHERIA, 2967
ALLOTIGENNYJ, 136
ALLYUVIAL'NYJ, 140
ALLYUVIJ, 146
ALMAZ, 1161
ALUNITIZATSIYA, 158
ALVEOLINELLIDAE, 159
ALYUMINIJ, 157
AMBLYPODA, 161
AMERITSIJ, 162
AMFIBOL, 172
AMFIBOLIT, 173
AMFIBOLIZATSIYA, 175
AMINOKISLOTA, 163
AMMIAK, 165
AMMODISCACEA, 164
AMMONIJ, 168
AMMONITIDA, 166
AMMONITINA, 167
AMMONOIDEA, 169
AMORFNYJ, 170
AMPHIBIA, 171
AMPHINEURA, 176
AMPLITUDA, 177
ANAEHROBNYJ, 178
ANALIZ, 182
ANALIZ AKTIVATSIONNYJ, 54
ANALIZ DANNYKH, 1066
ANALIZ DISKRIMINANTNYJ, 1218
ANALIZ FAKTORNYJ, 1502
ANALIZ FATSIAL'NYJ, 1387
ANALIZ GARMONICHESKIJ, 1957
ANALIZ GLAVNYKH EHLEMENTOV, 2658
ANALIZ GRANULOMETRICHESKIJ, 4125
ANALIZ KACHESTVENNYJ, 3706
ANALIZ KLASTERNYJ, 799
ANALIZ KOLICHESTVENNYJ, 3708
ANALIZ KOLICHESTVENNYJ RUDY, 336

ANALIZ MATEMATICHESKIJ, 2708
ANALIZ MIKROSKOPICHESKIJ, 2833
ANALIZ MINERALOGICHESKIJ, 2865
ANALIZ MODAL'NYJ, 2902
ANALIZ MONOKRISTALLA, 4116
ANALIZ MULTISPEKTRAL'NYJ, 2966
ANALIZ NA MALYE EHLEMENTY, 2888
ANALIZ NEJTRONNO-AKTIVATSIONNYJ, 3015
ANALIZ RASSEYANNYKH EHLEMENTOV, 4559
ANALIZ RENTGENO-SPEKTRAL'NYJ, 4836
ANALIZ RENTGENOVSKIJ, 4833
ANALIZ SISTEMNYJ, 4404
ANALIZ SOOTVESTVIJ, 963
ANALIZ SPEKTRAL'NYJ EHMISSIONNYJ, 1362
ANALIZ STATISTICHESKIJ, 4239
ANALIZ STRUKTURNYJ, 4301
ANALIZ TERMICHESKIJ, 4494
ANALIZ TERMICHESKIJ DIFFERENTS., 1176
ANALIZ TERMOMAGNITNYJ, 4510
ANALIZ TERMOVESOVOJ, 4507
ANALIZ VARIATSIONNYJ, 4716
ANAPSIDA, 184
ANASPIDA, 185
ANATEKSIS, 186
ANATEKSIT, 187
ANATOMIYA, 188
ANDALUZIT, 192
ANDEPT, 193
ANDEZIT, 194
ANDOSOL, 196
ANELASTIC MEDIUM, 197
ANGEDRAL'NYJ, 203
ANGIDRIT, 204
ANGIOSPERMAE, 199
ANGRIT, 201
ANION, 207
ANISOPLEURA, 209
ANIZIJ, 208
ANIZOTROPIYA, 210
ANKERIT, 211
ANKHIMETAMORFIZM, 189
ANNELIDA, 212
ANOMALIYA, 214
ANOMALIYA GEOFIZICHESKAYA, 1751
ANOMALIYA GRAVITATSIONNAYA, 1878
ANOMALIYA MAGNITNAYA, 2628
ANOMALIYA POLOZHITEL'NAYA, 2006
ANOMALIYA V SVOBODNOM VOZDUKHE, 1643
ANOMALODESMATA, 213
ANOROGENNYJ, 215
ANORTOZIT, 216
ANORTOZITIZATSIYA, 217
ANSHLIF, 3538
ANTEKLIZA, 218
ANTHOCYATHEA, 219
ANTHOZOA, 220
ANTIARCHI, 223

ANTIDYUNA, 226
ANTIFORMA, 227
ANTIKLINAL', 224
ANTIKLINORIJ, 225
ANTIMONAT, 228
ANTIPATHARIA, 230
ANTIPERTIT, 231
ANTISTRESS-MINERAL, 232
ANTITETICHESKIJ, 233
ANTRAKSOLIT, 222
ANTRATSIT, 221
ANUROMORPHA, 234
APERTURA, 235
APIKAL'NYJ, 237
APLACOPHORA, 238
APLIT, 239
APLITOVYJ, 240
APOFIZA, 245
APOGEJ, 241
APOGRANIT, 242
APOLLON PROGRAMMA, 243
APOMAGMATICHESKIJ, 244
APPARAT KONODONTOVYJ, 900
APPARAT PODVODNYJ, 4335
APPARAT ZUBNOJ, 1115
APPROKSIMATSIYA, 249
APSHERON, 251
APSIDOSPONDYLI, 252
APT, 253
APVELLING, 4687
ARACHNIDA, 269
ARCHAEOCOPIDA, 272
ARCHAEOCYATHA, 273
ARCHAEOCYATHEA, 274
ARCHAEOGASTROPODA, 275
ARCHAEOPTERYGES, 276
ARCHAEORNITHES, 277
ARCHIPOLYPODA, 283
ARCHOSAURIA, 284
ARCOIDA, 285
AREIZM, 298
ARENIG, 291
ARENIT, 292
ARGID, 294
ARGILLIT, 791
ARGON, 296
ARGON ATMOSFERNYJ, 359
ARGON IZBYTOCHNYJ, 1463
ARIDISOL, 300
ARIDNOST', 301
ARIDNYJ, 299
ARKHEIDY, 279
ARKHEJ, 278
ARKHEOLOGIYA, 281
ARKHEOMAGNETIZM, 282
ARKOZ, 302
ARKTICHESKAYA SERIYA, 287

ARKTICHESKIJ, 286
ARMORIKANSKIJ, 304
ARROIO, 308
ARSENAT, 309
ARTEFAKT, 319
ARTERIT, 311
ARTEZIANSKIJ, 312
ARTHRODIRA, 315
ARTHROPLEURIDA, 316
ARTHROPODA, 317
ARTHROPSIDA, 1417
ARTICULATA, 318
ARTINSKIJ, 323
ARTIODACTYLA, 324
ASBEST, 325
ASEJSMICHNOE SOORUZHENIE, 327
ASEJSMICHNYJ, 329
ASFAL'T, 334
ASFAL'TIT, 335
ASHGILL, 333
ASIMMETRIYA, 4128
ASSAY VALUE, 337
ASSIMILYATSIYA, 340
ASSINTSKIJ /OROGEN/, 342
ASSOTSIATSIYA, 341
ASSOTSIATSIYA MINERAL'NAYA, 2863
ASTATIN, 343
ASTENOSFERA, 348
ASTERIZM, 344
ASTEROID, 345
ASTEROIDEA, 346
ASTEROZOA, 347
ASTRAPOTHERIA, 349
ASTROBLEMA, 350
ASTROGEOLOGIYA, 351
ASTRONOMIYA, 352
ATAKSIT, 354
ATLANTICHESKIJ, 355
ATLAS, 356
ATMOGENNYJ, 357
ATMOSFERA, 358
ATOLL, 362
ATOMIC PACKING, 365
ATSETOLIZ, 35
ATTERBERGA PREDEL, 367
AUTIGENEZ, 375
AUTIGENNYJ, 376
AVES, 390
AVLAKOGEN, 372
AVSTRALIT, 374
AVTOKHTON, 377
AVTOKORRELYATSIYA, 378
AVTOLIT, 379
AVTOLIZ, 380
AVTOMATIZATSIYA, 382
AVTOMETAMORFIZM, 383
AVTOMETASOMATOZ, 384

AVTORADIOGRAFIYA, 385
AVTOTROFY, 386
AZIMUT MAGNITNYJ, 2629
AZOT, 3034
AZOT (GAZ), 3035
AZOT ORGANICHESKIJ, 3171
B SLOJ, 395
BACK ARC BASIN, 396
BACTERIA, 400
BACTRITOIDEA, 402
BAJKAL'SKIJ, 404
BAJOS, 406
BAKTERIOGENNYJ, 401
BALANS, 407
BALANS PODZEMNIKH VOD, 1912
BALANS VODNYJ, 4785
BALLAST, 408
BANK DANNYKH, 1067
BAR, 410
BAR BAR'ERNYJ, 417
BAR BEREGOVOJ, 2567
BAR'ERNYJ, 416
BARIJ, 413
BARIT, 412
BARKHAN, 411
BARREM, 414
BARREN DEPOSIT, 415
BARTON, 421
BASE METAL, 428
BASHKIRSKIJ, 431
BASSEJN GAZONOSNYJ, 1689
BASSEJN NEFTENOSNYJ, 3107
BASSEJN SEDIMENTATSII, 3995
BASSEJN UGOL'NYJ, 803
BASSEJN VODOSBORNYJ, 1245
BAT, 435
BATIAL'NYJ, 436
BATIMETRIYA, 438
BATOLIT, 434
BATRACHOSAURIA, 439
BAZAL'NYJ, 422
BAZAL'T, 425
BAZAL'T SHCHELOCHNOJ, 123
BAZANIT, 427
BAZIFIKATSIYA, 432
BEDLEND, 403
BEDNYJ, 2471
BELEMNOIDEA, 459
BELTSKIJ, 460
BEN'OFFA ZONA, 462
BENCH, 461
BENNETTITALES, 463
BENTONIT, 466
BENTONNYJ, 464
BENTOS, 465
BERCHA POVERKHNOST', 518
BEREG ABRAZIONNYJ, 7

BEREG AKKUMULYATIVNYJ, 32
BERGSHRUND, 467
BERILLYJ, 471
BERKELIJ, 468
BERMA(GEOMORFOLOGIYA), 469
BERRIAS, 470
BETA=LUCHI, 473
BETON, 879
BEYRICHICOPINA, 474
BEZVODNYJ, 205
BIBLIOGRAFIYA, 477
BIBLIOGRAFIYA PERSONAL'NAYA, 3365
BICH=ROK, 448
BIGHT, 478
BIODEGRADATSIYA, 489
BIOFATSIYA, 490
BIOGENEZ, 492
BIOGENNYJ PROTSESS, 493
BIOGEOGRAFIYA, 497
BIOGEOKHIMIYA, 496
BIOGERM, 499
BIOGRAFIYA, 498
BIOKHOR, 483
BIOKHRON, 484
BIOKHRONOLOGIYA, 485
BIOKLAST, 486
BIOLIT, 500
BIOLOGIYA, 502
BIOMASSA, 503
BIOMETRIYA, 504
BIOMIKRIT, 505
BIOREKSISTAZIYA, 506
BIOSFERA, 509
BIOSOM, 507
BIOSPARIT, 508
BIOSTRATIGRAFIYA, 511
BIOSTROM, 512
BIOTA, 513
BIOTIP, 516
BIOTOP, 514
BIOTSENOZ, 481
BIOTURBATSIYA, 515
BIOZONA, 517
BISSUS, 601
BITUM, 521
BITUMINIZATSIYA, 522
BIVALVIA, 3332
BLASTICHESKIJ, 525
BLASTOIDEA, 526
BLATTODEA, 527
BLESK, 2606
BLOK, 530
BLOK=DIAGRAMMA, 531
BOGATYJ, 2010
BOGKHED, 538
BOK VISYACHIJ, 1953
BOKSIT, 440

BOKSITIZATSIYA, 441
BOLOTO, 536
BOLOTO MORSKOE, 2681
BOLOTO TORFYANOE, 3321
BOMBA VULKANICHESKAYA, 539
BOR, 552
BORALF, 543
BORAT, 544
BORDERLEND, 547
BOREAL'NYJ, 548
BOROLL, 551
BOROZDA, 1668
BOROZDA SKOL'ZHENIYA, 1514
BOROZDA=TSARAPINA, 715
BOUNDSTONE, 562
BRACHIOPODA, 565
BRACHIOPTERYGII, 566
BRAKHISKLADKA, 567
BRANCHIOPODA, 572
BRANCHIURA, 573
BREKCHIYA, 575
BREKCHIYA DISLOKATSIONNAYA, 1221
BREKCHIYA DROBLENIYA, 4056
BREKCHIYA EHKSPLOSIVNAYA, 1486
BREKCHIYA IN'EKTSIONNAYA, 2240
BREKCHIYA KONTAKTOVAYA, 914
BREKCHIYA KOSTYANAYA, 540
BREKCHIYA RAZLOMNAYA, 1508
BREKCHIYA RUDNAYA, 1244
BREKCHIYA TEKTONICHESKAYA, 4434
BREKCHIYA TRENIYA, 1000
BREKCHIYA UDARA, 4076
BREKCHIYA VNUTRISKLADCHATAYA, 1605
BREKCHIYA VULKANICHESKAYA, 4761
BRIKETIROVANIE, 3338
BROM, 581
BROMID, 580
BRYOPSIDA, 584
BRYOZOA, 585
BUDINAZH, 556
BUGE ANOMALIYA, 557
BUGOR, 2079
BUGOR TORFYANNOJ, 3277
BUGOR VODOROSLEVYJ, 116
BUKHTA, 443
BUKHTA, 478
BULEVA ALGEBRA, 542
BULIMINACEA, 589
BULLARDA POVERKHNOST', 592
BUNTZANDSHTEJN, 593
BURA, 545
BURDIGAL, 594
BURENIE, 1254
BURENIE BESKERNOVOE, 1661
BURENIE EHKSPLUATATSIONNOE, 1148
BURENIE KOLONKOVOE, 956
BURENIE MORSKOE, 3103

BURENIE NAPRAVLENNOE, 1208
BURENIE SHNEKOVOE, 371
BURENIE TURBINNOE, 4619
BURENIE UDARNOVRASHCHATEL'NOE, 3901
BURENIE VIBRATSIONNOE, 4739
BURENIE VRASHCHATEL'NOE, 3900
BURIMOST', 1253
BUROZEM, 583
BURYA MAGNITNAYA, 2642
C SLOJ, 602
C13/C12, 645
CALCISPONGEA, 612
CAMAROIDEA, 623
CAMPTOSTROMATOIDEA, 627
CARBONACEOUS ROCK, 649
CARNIEULE, 656
CARNIVORA, 657
CARTERINACEA, 658
CASE STUDIES, 663
CASSIDULINACEA, 664
CATAGRAPHIA, 669
CATCHMENT HYDRODYNAMICS, 672
CAYTONIALES, 681
CENOTYPAL, 687
CEPHALOCARIDA, 691
CEPHALODISCIDA, 692
CEPHALOPODA, 693
CERATITIDA, 695
CERATOMORPHA, 696
CERIANTHARIA, 697
CERIANTIPATHARIA, 698
CETACEA, 701
CHAETOGNATHA, 702
CHARNIJSKIJ, 711
CHARNOKIT, 712
CHAROPHYTA, 713
CHASTITSA, 3307
CHASTITSA EHLEMENTARNAYA, 1353
CHASTITSY KOSMICHESKIE, 969
CHASTOTA, 1644
CHEILOSTOMATA, 717
CHEKHOL PLATFORMENNYJ, 979
CHELICERATA, 719
CHELONIA, 720
CHELOVEK ISKOPAEMYJ, 1622
CHELYUST', 2360
CHEMICAL EVOLUTION, 725
CHEMICAL EXPLOSION, 726
CHEREP, 980
CHERNOZEM, 738
CHERVEOBRAZNYJ, 4733
CHESHUJCHATOST', 2181
CHESHUJCHATYJ, 2180
CHESHUYA (TECTONICHESKAYA), 3963
CHETVERTICHNYJ, 3713
CHILOPODA, 742
CHIROPTERA, 744

CHITINOZOA, 746
CHLOROPHYTA, 754
CHONDRICHTHYES, 755
CHORDATA, 758
CHRYSOPHYTA, 765
CILIATA, 766
CILIOPHORA, 767
CIRRIPEDIA, 776
CLADOCERA, 778
CLYMENIIDA, 800
COAL BALL, 802
COARSE GRAVEL, 809
COBOL, 819
COCCOLITHOPHORACEAE, 820
COELENTERATA, 823
COENOTHECALIA, 824
COHESIVE MATERIAL, 829
COLEOPTEROIDA, 833
COLLAPSE BRECCIA, 835
COLLEMBOLA, 840
COLUMNARIINA, 851
CONCHOSTRACA, 876
CONDYLARTHRA, 884
CONIFERALES, 895
CONOCARDIOIDA, 899
CONODONTA, 901
CONTINENTAL ZONE, 931
CONTRASTNOST', 935
CONULATA, 937
COPEPODA, 944
CORALLIMORPHARIA, 950
CORALLINACEAE, 951
CORDAITALES, 953
CORDIL'ERA, 955
CORYNEXOCHIDA, 966
COTYLOSAURIA, 977
CREODONTA, 987
CRINOIDEA, 991
CRINOZOA, 992
CROCODILIA, 996
CROSSOPTERYGII, 999
CRUSTACEA, 1002
CRYPTODONTA, 1007
CRYPTOSTOMATA, 1009
CTENOSTOMATA, 1029
CYANOPHYTA, 1040
CYCADALES, 1042
CYCADOFILICALES, 1043
CYCLOCYSTOIDEA, 1048
CYCLOSTOMATA, 1050
CYNOMORPHA, 1054
CYSTIPHYLLINA, 1055
CYSTOIDEA, 1056
D SLOJ, 1057
D/H, 1058
DAIKA, 1184
DAIKA KLASTICHESKAYA, 785

DAIKA KOL'TSEVAYA, 3869
DAIKA PESCHANIKOVAYA, 3940
DAMBA, 3986
DANNYE, 1064
DARSI ZAKON, 1061
DARVINIZM, 1062
DASYCLADACEAE, 1063
DAT, 1060
DATIROVKA, 1073
DAVLENIE, 3607
DAVLENIE ATMOSFERNOE, 361
DAVLENIE GEOSTATICHESKOE, 1909
DAVLENIE GIDRODINAMICHESKOE, 2096
DAVLENIE GIDROSTATICHESKOE, 2129
DAVLENIE GRUNTA, 1277
DAVLENIE KAPILLYARNOE, 635
DAVLENIE NIZKOE, 2571
DAVLENIE PARTSIALNOE, 3306
DAVLENIE POROVOE, 3562
DAVLENIE VYSOKOE, 2007
DEBRIS CONE, 1076
DECAPODA, 1078
DEDOLOMITIZATSIYA, 1087
DEEP-TOW METHOD, 1092
DEFEKT KRISTALLICHESKOJ STRUKTURY, 1013
DEFLYATSIYA, 1095
DEFORMATSIYA, 1096
DEFORMATSIYA (TEKTONICHESKAYA), 4267
DEFORMATSIYA GRUNTOV, 4147
DEFORMATSIYA PLASTICHESKAYA, 3492
DEFORMATSIYA SKLONOV, 4135
DEFORMATSIYA UPRUGAYA, 1321
DEGAZATSIYA, 1097
DEGIDRATATSIYA, 1101
DEGLYATSIATSIYA, 1098
DEGRADATSIYA, 1099
DEINOTHERIOIDEA, 1102
DEJTERICHESKIJ EHFFEKT, 1145
DEJTERIJ, 1146
DEKANA TRAPPY, 1080
DEKONTAMINATSIYA, 1083
DEKONVOLUTSIYA, 1084
DEL'TA, 1103
DEL'TOVYJ, 1104
DELENIE, 1554
DEMOSPONGEA, 1107
DENDRIT, 1108
DENDROGRAMMA, 1111
DENDROIDEA, 1112
DENDROKHRONOLOGIYA, 1110
DENUDATSIYA, 1116
DEPRESSIYA (GEOMORF), 1121
DEPTH INDICATOR, 1123
DEREVO OKAMENELOE, 1624
DEREVO OKREMNELOE, 4101
DERMAL'NYJ, 1126
DERMAPTEROIDA, 1127

DERMOPTERA, 1128
DESCENSION THEORY, 1130
DESILIKATSIYA, 1136
DESKVAMATSIYA, 1466
DESMIDY, 1137
DESMOCERATIDA, 1138
DESMOSTYLIA, 1139
DETERMINIZM, 1140
DETRITOVYJ, 1142
DEVITRIFIKATSIYA, 1150
DEVON, 1151
DEVON NIZHNIJ, 2581
DEVON SREDNIJ, 2842
DEVON VERKHNIJ, 4671
DIADOKHIYA, 1154
DIAFTOREZ, 1162
DIAFTORIT, 1163
DIAGENES, 1155
DIAGNOZ, 1156
DIAGRAMMA, 1157
DIAGRAMMA PETROKHIMICHESKAYA, 3371
DIAGRAMMA PYL'TSEVAYA, 3540
DIAKHRONIZM, 1153
DIAKLAZA, 2361
DIALIZ, 1158
DIAMAGNETIK, 1159
DIAMIKTIT, 1160
DIAPIR, 1164
DIAPIRIZM, 1165
DIASTEMA, 1166
DIATOMEAE, 1169
DIATOMIT, 1170
DICOTYLEDONEAE, 1172
DICTYONELLIDINA, 1173
DIDODEKAEHDR, 1201
DIFFERENTSIATSIYA, 1178
DIFFERENTSIATSIYA GORIZONTAL'NAYA, 2049
DIFFERENTSIATSIYA GRAVITATSIONNAYA, 1877
DIFFERENTSIATSIYA METAMORFICHESKAYA, 2782
DIFFUZIYA, 1181
DIFRAKTSIYA, 1180
DILATANTSIYA, 1186
DILATATSIYA, 1187
DIMORFIZM, 1189
DINAMIKA, 1267
DINAMO TEORIYA, 1268
DINAMOMETAMORFIZM, 1266
DINOCERATA, 1190
DINOFLAGELLATA, 1191
DINOSAURIA, 1192
DIOGENIT, 1193
DIORIT, 1194
DIORIT KVARTSEVYJ, 3711
DIP=SLIP FAULT, 1199
DIPLEUROZOA, 1200
DIPLOPODA, 1202
DIPLORHINA, 1203

DIPNOI, 1205
DIPTERA, 1207
DISCORBACEA, 1213
DISKOASTERY, 1210
DISKORDANTNYJ, 1214
DISLOKATSIYA KRISTALLICHESKOJ RESHETKI, 1014
DISPERSION PATTERN, 1223
DISPERSIYA, 1222
DISPERSIYA SVETLA ROTUYUSHCHAYA, 3903
DISPROSIJ, 1270
DISSOTSIATSIYA, 1227
DIVERGENTSIYA, 1232
DLINA, 2477
DNO MORSKOE, 3975
DOBYCHA, 3624
DOBYCHA VYSHCHELACHIVANIEM, 4179
DOCODONTA, 1233
DOIMBRIJ, 3592
DOISTORICHESKIJ, 3602
DOKEMBRIJ, 3593
DOKEMBRIJSKIJ /OROGEN/, 3594
DOLEDNIKOVYJ, 3601
DOLERIT, 1234
DOLGOTA, 2565
DOLINA, 4707
DOLINA LEDNIKOVAYA, 1802
DOLINA OBSEKVENTNAYA, 3080
DOLINA PODVODNAYA, 4332
DOLINA POGREBENNAYA, 599
DOLINA RIFTOVAYA, 3865
DOLINA STRUKTURNAYA, 4312
DOLINA SUBSEKVENTNAYA, 4336
DOLINA VISYACHAYA, 1952
DOLOMIT, 1238
DOLOMITIZATSIYA, 1237
DOMEN, 2631
DOMER, 1241
DOSUG, 3769
DOZHD', 3742
DRAGA, 1251
DRENAZH (VNUTRENNIJ), 1371
DRENAZH ISKUSTVENNYJ, 1247
DRESVA, 1923
DREVNELEDNIKOVYJ, 191
DRIFT KONTINENTOV, 922
DROBLENIE, 1637
DRUMLIN, 1259
DRUZA, 1260
DUGA OSTROVNAYA, 2316
DUGOOBRAZNYJ, 288
DVIZHENIE, 2951
DVIZHENIE CHANDLEROVSKOE, 706
DVIZHENIE DIZ'UNKTIVNOE, 1515
DVIZHENIE GRUNTA, 1908
DVIZHENIE NISKHODYASHCHEE, 1242
DVIZHENIE SKLADCHATOE, 1608
DVIZHENIE TEKTONICHESKOE, 1167

DVIZHENIE VERTIKAL'NOE, 4736
DVIZHENIE VOSKHODYASHCHEE, 4686
DVOJNIK, 4621
DVOJNIK BAVENSKIJ /ZAKON/, 442
DVOJNIK SKOL'ZHENIYA, 1820
DVUOKIS' SERY, 4354
DVUOSNYJ, 475
DVUPRELOMLENIE, 519
DYKHANIE, 3855
DYUNA, 1261
DYUNA BEREGOVAYA, 812
DYUNA PUSTYNNAYA, 1132
DYUREN, 1262
E SLOJ, 1271
EARLY DIAGENESIS, 1272
EARTH=MOON COUPLE, 1282
EARTHFLOW, 1283
ECHINODERMATA, 1291
ECHINOIDEA, 1292
ECHINOZOA, 1293
ECTOTROPHA, 1303
EDENTATA, 1305
EDRIOASTEROIDEA, 1307
EDRIOBLASTOIDEA, 1308
EH, 1312
EHBURNEJSKIJ, 1289
EHFFEKT IZOTOPNYJ, 2350
EHFFEKT KLIMATICHESKIJ, 795
EHFFEKT PARNIKOVYJ, 1891
EHFFEKT TERMICHESKIJ, 4497
EHFIR, 1441
EHJFEL', 1313
EHJNSHTEJNIJ, 1314
EHKHOZONDIROVANIE, 1294
EHKLOGIT, 1295
EHKOLOGIYA, 1297
EHKOLOGIYA MORSKAYA, 2676
EHKONOMIKA MINERAL'NOGO SYR'YA, 2867
EHKOSISTEMA, 1301
EHKSGALYATSIYA, 1467
EHKSKAVATSIYA, 1462
EHKSKURSIYA, 1465
EHKSPEDITSIYA, 1476
EHKSPERIMENT, 1479
EHKSPERIMENTAL'NAYA SEJSMOLOGIYA, 1478
EHKSPLUATATSIYA, 1482
EHKSPLUATATSIYA PODZEMNYKH VOD, 4791
EHKSPORT, 1488
EHKSTENZOMETR, 1492
EHKSTERNIDY, 1493
EHKSTRAPOLYATSIYA, 1497
EHKSTSENTRISITET, 1290
EHKSTSESS, 2401
EHKVATOR, 1413
EHKZARATSIYA, 3971
EHKZEMPLYAR TIPOVOJ, 4626
EHKZINA, 1468

EHKZINIT, 1469
EHKZOBIOLOGIYA, 1470
EHKZOGENNYJ, 1471
EHLEKTRICHESTVO, 1334
EHLEKTROD, 1337
EHLEKTROLIT, 1339
EHLEKTROLIZ, 1338
EHLEKTROMAGNETIZM, 1344
EHLEKTROMAGNITNYJ KAROTAZH, 1343
EHLEKTROMETRIYA, 1329
EHLEKTRON, 1345
EHLEKTRONOGRAFIYA, 1346
EHLEKTROOSMOS, 1335
EHLEKTROPROVODNOST', 1326
EHLEKTRORAZVEDKA, 1327
EHLEKTROSTANTSIYA, 3588
EHLEKTROSTANTSIYA ATOMNAYA, 3065
EHLEKTROZONDIROVANIE, 1332
EHLEMENT GLAVNYJ, 2657
EHLEMENT KONECHNYJ, 1551
EHLEMENT MAGNETIZMA, 2632
EHLEMENT MALYJ, 2887
EHLEMENT RASSEYANNYJ, 4556
EHLEMENT SAMORODNYJ, 2988
EHLEMENT STRUKTURNYJ, 4305
EHLYUVIJ, 1357
EHMBREKHIT, 1359
EHMS, 1364
EHNDEMICHNYJ, 1368
EHNDEMIKI, 4194
EHNDOGENNYJ, 1370
EHNDOKONTAKT, 546
EHNERGIYA, 1373
EHNERGIYA AKTIVATSII, 55
EHNERGIYA GEOTERMAL'NAYA, 1773
EHNERGIYA SVOBODNAYA, 1641
EHNERGIYA YADERNAYA, 3062
EHNKRINIT, 1366
EHNTAL'PIYA, 1382
EHNTISOL, 1383
EHNTROPIYA, 1385
EHON, 1394
EHOTSEN, 1389
EHOTSEN NIZHNIJ, 2582
EHOTSEN SREDNIJ, 2843
EHOTSEN VERKHNIJ, 4672
EHPEJROGENEZ, 1396
EHPIBOLIT, 1399
EHPIDOTIZATSIYA, 1402
EHPIEHVGEOSINKLINAL', 1403
EHPIGENEZ, 1404
EHPIMAGMATICHESKIJ, 1405
EHPITAKSIYA, 1406
EHPITERMAL'NOE MESTOROZHDENIE, 1407
EHPITSENTR, 1400
EHPIZONA, 1408
EHPOKHA, 1409

EHPOKHA METALLOGENICHESKAYA, 2771
EHRA, 1418
EHRATEMA, 1419
EHRATOSFENSKIJ, 1420
EHRBIJ, 1421
EHRG, 1422
EHROZIYA, 1424
EHROZIYA LEDNIKOVAYA, 1794
EHROZIYA MORSKAYA, 2678
EHROZIYA PLOSKOSTNAYA, 4065
EHROZIYA POCHV, 4148
EHROZIYA RECHNAYA, 1594
EHROZIYA TUNNEL'NAYA, 3455
EHROZIYA VETROVAYA, 4827
EHROZIYA VODNAYA, 4789
EHSSEKSIT, 1435
EHSTUARIJ, 1438
EHTAN, 1440
EHVAPORIT, 1459
EHVGEDRAL'NYJ, 1447
EHVGEOSINKLINAL', 1445
EHVKRIT, 1443
EHVKSINNYJ, 1456
EHVM, 870
EHVOLYUTSIYA, 1461
EHVRIGALLINNYJ, 1449
EHVRITERMNYJ, 1450
EHVSTAZIYA, 1452
EHVTAKSITOVYJ, 1453
EHVTEKTIKA, 1454
EHVTROFIKATSIYA, 1455
EHZHEKTIT, 1315
ELASMOBRANCHII, 1316
ELASTIC MATERIAL, 1319
ELECTRON MICROSCOPY DATA, 1350
ELECTRON PROBE DATA, 1352
ELEPHANTOIDEA, 1354
ELONGATE MINERAL, 1355
EMBRITHOPODA, 1360
ENDOCERATOIDEA, 1369
ENSTATITE CHONDRITE, 1379
ENTEROPNEUSTA, 1381
EOCRINOIDEA, 1390
EOLIAN FEATURE, 1392
EOSUCHIA, 1395
EPIBIOTISM, 1398
ERODIBILITY, 1423
EROSION CONTROL, 1425
EROSION FEATURE, 1427
EUCILIATA, 1442
EUGLENOPHYTA, 1446
EURYAPSIDA, 4389
EUSELACHII, 1451
EVROPIJ, 1448
EXSURGENCE, 1491
EZHEGODNIK, 4847
F SLOJ, 1499

FABRIC, 1500
FABRIKA OBOGATITEL'NAYA, 3155
FAGOTROFNYJ, 3389
FAKOLIT, 3385
FAKTOR GAZOVYJ, 1693
FAMEN, 1503
FANEROZOJ, 3390
FANGLOMERAT, 1505
FATSIYA, 1501
FATSIYA AKTINOLITOVAYA, 53
FATSIYA AL'BIT=EHPIDOT=AMFIBOLIT., 104
FATSIYA AL'BIT=EHPIDOT=ROGOVIKOVAYA, 105
FATSIYA AMFIBOLITOVAYA, 174
FATSIYA EHKLOGITOVAYA, 1296
FATSIYA EHPIDOT=AMFIBOLITOVAYA, 1401
FATSIYA EHSTUARIEVAYA, 1436
FATSIYA GEOKHIMICHESKAYA, 1717
FATSIYA GIDROCHIMICHESKAYA, 2100
FATSIYA GRANULITOVAYA, 1859
FATSIYA KONTINENTAL'NAYA, 928
FATSIYA KORDIERIT=AMFIBOLITOVAYA, 954
FATSIYA LEDNIKOVAYA, 1793
FATSIYA MAGMATICHESKIKH POROD, 2174
FATSIYA METAMORFICHESKAYA, 2783
FATSIYA MINERAL'NAYA, 2869
FATSIYA PIROKSEN=ROGOVIKOVAYA, 3703
FATSIYA PRENIT=PUMPELLIITOVAYA, 3603
FATSIYA PRIBREZHNAYA, 2545
FATSIYA RECHNAYA, 1593
FATSIYA ROGOVIKOVAYA, 2058
FATSIYA ROGOVOOBMANKO=ROGOVIKOVAYA, 2055
FATSIYA SANIDINITOVAYA, 3942
FATSIYA SHEL'FOVAYA, 4068
FATSIYA SLANTSEV GLAUKOFANOVYKH, 1818
FATSIYA SLANTSEV ZELENYKH, 1893
FATSIYA TSEOLITOVAYA, 4857
FAUNA, 1516
FAZA, 3391
FAZA GAZOVAYA, 1694
FAZA MINERAL'NAYA, 2871
FAZA TVERDAYA, 4166
FAZA ZHIDKAYA, 2520
FAZA ZHIDKAYA (RASPLAV), 1582
FEDOROVSKIJ STOLIK, 4662
FEL'DSHPATIZATSIYA, 1525
FEL'ZICHESKIJ, 1527
FEMICHESKIJ, 1528
FENIT, 1529
FENITIZATSIYA, 1530
FENOKRISTALL, 3395
FENOL, 3396
FEOZEM, 3388
FERMIJ, 1531
FERRALIT, 1532
FERRIKRET, 1534
FERROD, 1535
FERROMAGNETIK, 1536

FERROMANGANESE COMPOSITION, 1537
FESTON, 1037
FIKSIZM, 1559
FIL'TR, 1545
FIL'TR GRAVIJNYJ, 1872
FIL'TRATSIYA, 1547
FIL'TRATSIYA STATSIONARNAYA, 4246
FIL'TROVANIE, 1546
FILLIT, 3419
FILLONIT, 3421
FILOGENIYA, 3423
FILOSOFIYA, 3398
FILUM, 3424
FINANSIROVANIE, 1548
FIORD, 1560
FIRN, 1553
FISSIPEDA, 1557
FITAN, 3433
FITOPLANKTON, 3437
FITOTSENOZ, 3434
FIXATION, 1558
FIZIOLOGIYA, 3432
FIZKHIMIYA, 3425
FLANDRIJ, 1564
FLEKSURA, 1566
FLINT, 1567
FLISH, 1599
FLISH DIKIJ, 4825
FLOKKULYATSIYA, 1568
FLORA, 1571
FLOTATSIYA, 1574
FLUID INJECTION, 1580
FLUID PRESSURE, 1583
FLUORIMETRIYA, 1587
FLYUIDAL'NYJ, 1584
FLYUORESTSENTSIYA, 1585
FLYUORESTSENTSIYA RENTGENOSPEKTRAL'NAYA, 4835
FLYUORIT, 1590
FLYUVENT, 1592
FLYUVIOGLYATSIAL'NYJ, 1596
FOIDIT, 1601
FOJYAIT, 1629
FON, 398
FONOLIT, 3401
FORAMINIFERA, 1610
FORLAND, 1613
FORMA, 4054
FORMA CHASTITS, 3308
FORMA KRISTALLOGRAFICHESKAYA, 1016
FORMA REL'EFA, 2429
FORMA REL'EFA BEREGOVAYA, 4081
FORMA REL'EFA LEDNIKOVAYA, 1795
FORMA ZEMLI, 4055
FORMATSIYA OFIOLITOVAYA, 3133
FORMATSIYA UGLENOSNAYA, 805
FORMATSIYA ZHELEZORUDNAYA, 2311
FORMULA, 1618

FORSHOK, 1615
FOSFAT, 3403
FOSFATIZATSIYA, 3404
FOSFOR, 3407
FOSFORESTSENTSIYA, 3405
FOSFORITY, 3406
FOSSILIZATSIYA, 1626
FOTOGEOLOGIYA, 3410
FOTOGRAFIROVANIE, 3412
FOTOGRAFIROVANIE INFRAKRASNOE, 2232
FOTOGRAMMETRIYA, 3411
FOTOMETRIYA, 3413
FOTOMETRIYA PLAMENNAYA, 1562
FOTOSINTEZ, 3414
FRACTURE STRENGTH, 1633
FRAKTSIONIROVANIE, 1630
FRAKTSIONIROVANIE IZOTOPOV, 2353
FRAKTSIYA LEGKAYA, 2501
FRAKTSIYA TONKAYA, 1549
FRAKTSIYA TYAZHELAYA, 1973
FRAMBOID, 1638
FRAN, 1640
FRANTSIJ, 1639
FREATOFIT, 3417
FRESHWATER SEDIMENTATION, 1649
FRONT, 1652
FTOR, 1589
FTORIDIT, 1586
FTORIROVANIE, 1588
FUGITIVNOST', 1659
FUL'GURIT, 1660
FUL'VOKISLOTA, 1663
FULLEROVA ZEMLYA, 1662
FUMAROLA, 1664
FUNDAMENT, 429
FUNGI, 1667
FUNKTSIYA, 1665
FUNKTSIYA STATISTICHESKAYA, 4241
FUR'E RYAD, 1628
FUSULINIDAE, 1670
FYUZINIT, 1669
G SLOJ, 1671
GABBRO, 1672
GABITUS, 1936
GADOLINIJ, 1673
GAFNIJ, 1940
GALEREYA VODOSBORNAYA, 2227
GAL'KA, 3322
GAL'MIROLIZ, 1946
GALITOVYJ, 1945
GALLIJ, 1675
GALOFIL'NYJ, 1948
GALOFIT, 1949
GALOGEN, 1947
GALOGENEZ, 2087
GALOID, 1944
GAMMA=LUCHI, 1676

GAMMA=SPEKTROSKOPIYA, 1677
GANISTER, 1681
GARPOLIT, 1958
GASTROPODA, 1696
GASTROZOOID, 1697
GAYUI ZAKON, 1960
GAUSSA EHPOKHA, 1699
GAZ, 1684
GAZ BOLOTNYJ, 2694
GAZ FREATICHESKIJ, 3415
GAZ INERTNYJ, 2224
GAZ PRIRODNYJ, 2989
GAZ UGOL'NYJ, 804
GAZIFIKATSIYA, 1695
GAZOKHRANILISHCHE, 1691
GAZOKONDENSAT, 1688
GEJZER, 1784
GEJZERIT, 1785
GEKISTOTERM, 1975
GEKKELYA ZAKON, 1939
GEKSAEHDRIT, 2003
GEL', 1702
GEL'VET, 1980
GELIJ, 1978
GELIVITY, 1704
GEMATIT, 1981
GEMMOLOGIYA, 1706
GENERATSIYA, 1708
GENEZIS, 1709
GENEZIS IZVERZHENNYKH POROD, 2175
GENEZIS MESTOROZHDENIJ, 1118
GENOMORFA, 1710
GENOTIP, 1711
GEOANTIKLINAL', 1700
GEOBAROMETRIYA, 1712
GEOBIOZ, 1713
GEOCHEMICAL SINK, 1721
GEODEZIYA, 1727
GEODINAMIKA, 1729
GEOFIZIK, 1758
GEOFIZIKA, 1759
GEOFIZIKA RAZVEDOCHNAYA, 247
GEOFON, 1750
GEOGRAFIYA, 1731
GEOGRAFIYA MATEMATICHESKAYA, 2710
GEOID, 1732
GEOIZOTERMA, 1733
GEOKHIMIYA, 1722
GEOKHIMIYA ORGANICHESKAYA, 3168
GEOKHRONOLOGIYA, 1725
GEOLOG, 1740
GEOLOGIYA, 1741
GEOLOGIYA CHETVERTICHNAYA, 4369
GEOLOGIYA DINAMICHESKAYA, 1265
GEOLOGIYA GLUBINNAYA, 4340
GEOLOGIYA INZHENERNAYA, 1374
GEOLOGIYA ISTORICHESKAYA, 2019

GEOLOGIYA IZOTOPNAYA, 2351
GEOLOGIYA KOSMICHESKAYA, 1498
GEOLOGIYA LEDNIKOVAYA, 1796
GEOLOGIYA LUNY, 2602
GEOLOGIYA MATEMATICHESKAYA, 2711
GEOLOGIYA MEDITSINSKAYA, 2733
GEOLOGIYA MORSKAYA, 2679
GEOLOGIYA NEFTYANAYA, 3381
GEOLOGIYA OKRUZHAYUSHCHEJ SREDY, 1388
GEOLOGIYA POLEVAYA, 1542
GEOLOGIYA POLEZNYKH ISKOPAEMYKH, 1298
GEOLOGIYA PRIKLADNAYA, 246
GEOLOGIYA REGIONAL'NAYA, 3793
GEOLOGIYA RUDNYKH MESTOROZHDENIJ, 2885
GEOLOGIYA STRUKTURNAYA, 4306
GEOLOGIYA URBANISTICHESKAYA, 4694
GEOLOGIYA VOENNAYA, 2859
GEOMAGNETIZM, 1745
GEOMETRIYA, 1746
GEOMETRIYA RUSLA, 708
GEOMORFOLOGIYA, 1748
GEOMORFOLOGIYA BEREGOV, 814
GEOMORFOLOGIYA PODVODNAYA, 4329
GEONOMIYA, 1749
GEOPRESSURE, 1760
GEOSECS, 1761
GEOSINKLINAL', 1767
GEOSINKLINAL' EHNSIALICHESKAYA, 1377
GEOSINKLINAL' EHNSIMATICHESKAYA, 1378
GEOSINKLINAL' VNUTRIKONTINENTAL'NAYA, 2289
GEOSINKLINAL' VNUTRIKRATONOVAYA, 2290
GEOSTATISTIKA, 1762
GEOSYNCLINAL RIDGE, 1763
GEOSYNCLINAL SEDIMENTATION, 1765
GEOSYNCLINAL TRENCH, 1766
GEOTEKHNIKA, 1771
GEOTEKTONIKA, 1772
GEOTERMIKA, 1779
GEOTERMOMETRIYA, 1780
GEOTHERMAL SYSTEM, 1778
GEOUNDATSIYA, 1781
GERMANIJ, 1782
GERMATIPNYJ, 1987
GERTSINIDY, 1985
GERTSINSKIJ, 1984
GETEROBLASTOVYJ, 1988
GETEROGENNOST', 1993
GETEROGENNYJ, 1994
GETEROKHRONIZM, 1989
GETEROMORFIZM, 1996
GETEROMORFNYJ, 1995
GETEROSPOROVYJ, 1997
GETEROTROFNYJ, 1998
GETTANG, 1999
GIALOKRISTALLICHESKIJ, 2084
GIBRIDIZATSIYA, 2086
GIDRATATSIYA, 2089

GIDRAVLIKA, 2098
GIDRICHESKIJ, 2088
GIDRODINAMIKA, 2104
GIDROEHKSPLOZIYA, 2105
GIDROFON, 2125
GIDROGEOKHIMIYA, 2102
GIDROGEOLOGIYA, 2115
GIDROGEOLOGIYA (PRIKLADNAYA), 248
GIDROGEOLOGIYA REGIONAL'NAYA, 3794
GIDROGRAF, 2116
GIDROGRAFIYA, 2117
GIDROIZOP'EZA, 2336
GIDROLAKKOLIT, 3450
GIDROLOGIYA, 2121
GIDROLOGIYA KARSTOVAYA, 2375
GIDROMEKHANIKA, 1581
GIDROMETALLURGIYA, 2122
GIDROMETR, 2123
GIDROMETRIYA, 2124
GIDROOKIS', 2136
GIDROSFERA, 2126
GIDROSPIRA, 2127
GIDROTERMAL'NOE MESTOROZHDENIE, 2132
GIDROTERMAL'NYJ, 2130
GIEROGLIF, 2005
GIGANTIZM, 1787
GIGANTOSTRACA, 1788
GIGROFIL'NYJ, 2138
GINKGOALES, 1789
GIPABISSAL'NYJ, 2142
GIPERGENEZ, 2143
GIPIDIOBLASTOVYJ, 2144
GIPIDIOMORFNYJ, 2145
GIPODIGM, 2148
GIPOKRISTALLICHESKIJ, 2147
GIPOTERMAL'NYJ, 2149
GIPOTEZA, 2150
GIPOTEZA KONTRAKTSIONNAYA, 933
GIPOTEZA OSTSILYATSIONNAYA, 3198
GIPOTEZA PEREMESHCHENIYA KONTINENTOV, 1225
GIPOTEZA RASSHIRYAYUSHCH.ZEMLI, 1475
GIPS, 1933
GIPSOMETRIYA, 2151
GISTOGRAMMA, 2017
GISTOLOGIYA, 2018
GITT'YA, 1934
GLACIAL SEDIMENTATION, 1798
GLACIAL TRANSPORT, 1801
GLACIER SURGE, 1805
GLASIS, 1810
GLAUKONIT, 1815
GLAUKONITOOBRAZOVANIE, 1817
GLAUKONITOVYJ, 1816
GLETCHER KAMENNYJ, 3886
GLINA, 787
GLINA BELAYA, 4823
GLINA FARFOROVAYA, 743

GLINA GONCHARNAYA, 3586
GLINA KIRPICHNAYA, 576
GLINA LENTOCHNAYA, 4722
GLINA NEUSTOJCHIVAYA, 4038
GLINA OGNEUPORNAYA, 1552
GLINA OSTATOCHNAYA, 3828
GLINA OTBELIVAYUSHCHAYA, 528
GLINA PLASTICHESKAYA, 4144
GLINA PROCHNAYA, 4255
GLINA VALUNNAYA, 559
GLINA VYSOKOCHUVSTVITEL'NAYA, 3715
GLINISTYJ, 295
GLINKA TRENIYA, 1509
GLINOZEM, 156
GLINOZEMISTYJ, 4318
GLOBIGERINACEA, 1822
GLOMERO-, 1823
GLOSSOPTERIDALES, 1825
GLUBINA, 1122
GLUBINA KOMPENSATSII KARBONATNOJ, 1124
GLUBINNYJ, 1094
GLUBOKOVOD'E, 1093
GLUBOKOVODNYJ, 1090
GLYATSIODISLOKATSIYA, 1809
GLYATSIOLOGIYA, 1807
GLYATSIOTEKTONICHESKIJ, 1004
GNATHOSTOMI, 1826
GNATHOSTRACA, 1827
GNEJS, 1828
GNEJSOVYJ, 1830
GNETALES, 1831
GNOMOGRAMMA, 1832
GOD GALAKTICHESKIJ, 1674
GODOGRAF, 2023
GOL'MIJ, 2025
GOLOFIT, 2030
GOLOGRAFIYA, 2029
GOLOTIP, 2034
GOLOTSEN, 2026
GOMEOMORFISM, 2036
GOMEOMORFIYA, 2037
GOMOGENIZATSIYA, 2042
GOMOGENNOST', 2041
GOMOKLINAL', 2040
GOMOLOGIYA, 2044
GOMONIMIYA, 2045
GOMOTAKSIAL'NOST, 2047
GONDVANA, 1834
GONIATITIDA, 1835
GONIOMETRIYA, 1836
GORA OSTROVNAYA, 2253
GORA PODVODNAYA, 3984
GORA=IGLA, 2996
GORIZONT ARTEZIANSKIJ, 313
GORIZONT BEZNAPORNYJ, 4644
GORIZONT MARKIRUYUSHCHIJ, 2389
GORIZONT NAPORNYJ, 887

GORIZONT POCHVENNYJ, 4149
GORIZONT PRODUKTIVNYJ, 3626
GORIZONT VODONOSNYJ, 260
GORIZONTAL'NYJ, 2050
GORNBLENDIT, 2056
GORNOE DELO, 2884
GORST, 2060
GORYACHIJ, 2063
GOTERIV, 1959
GOVARDIT, 2067
GRABEN, 1840
GRAD, 1941
GRADIENT, 1842
GRADIENT GEOTERMICHESKIJ, 1775
GRADIENT GIDRAVLICHESKIJ, 2093
GRADIOMETR, 1843
GRAFICHESKIJ, 1865
GRAFIK, 1863
GRAFIKA, 1250
GRAFIT, 1866
GRAFITIZATSIYA, 1867
GRANAT, 1683
GRANIT, 1845
GRANITIZATSIYA, 1850
GRANITNYJ, 1848
GRANITO-GNEJS, 1846
GRANITOID, 1851
GRANITOVYJ, 1847
GRANITSA POBEREZH'YA, 816
GRANITSA SEJSMICHESKAYA, 1212
GRANITSA STRATIGRAFICHESKAYA, 4275
GRANOBLASTOVYJ, 1852
GRANODIORIT, 1853
GRANOFIROVYJ, 1855
GRANOGABBRO, 1854
GRANULA, 1636
GRANULIT, 1858
GRANULITOVYJ, 1860
GRANULOMETRIYA, 1861
GRANULYATSIYA, 1857
GRAPHIC METHOD, 1864
GRAPTOLITHINA, 1868
GRAPTOLOIDEA, 1869
GRASSLAND, 1870
GRAUVAKKA, 1888
GRAVIJ, 1871
GRAVIMETR, 1873
GRAVIMETRIYA, 1874
GRAVIRAZVEDKA, 1879
GRAVITATSIYA, 1875
GREBEN', 988
GREJZEN, 1896
GREJZENIZATSIYA, 1897
GRENVIL'SKIJ, 1898
GROIN, 1901
GROUND METHOD, 1906
GROUND TRUTH, 1910

GROUNDWATER DAM, 1913
GROUTING, 1920
GROZD'EVIDNYJ, 553
GRUNT, 1904
GRUNT MERZLYJ, 1656
GRUNT NESVYASNYJ, 828
GRUNT STRUKTURNYJ, 3317
GRUNT SVYAZNYJ, 830
GRUPPA SIMMETRII, 4188
GRYADA VALUNNAYA, 560
GUANO, 1924
GUDZONSKIJ, 2068
GUKA ZAKON, 2048
GUMIDNYJ, 2074
GUMIFIKATSIYA, 2076
GUMIN, 2077
GUMINOKISLOTA, 2071
GUMOD, 2080
GUMOKS, 2081
GUMULT, 2082
GUMUS, 2083
GUTENBERGA POVERKHNOST', 1929
GYMNOLAEMATA, 1930
GYMNOSPERMAE, 1931
GZHEL'SKIJ, 1935
HAMMER SEISMICS, 1951
HEAT SOURCE, 1968
HELICOPLACOIDEA, 1976
HELIOZOA, 1977
HEMICHORDATA, 4261
HEMICRUSTACEA, 1982
HEMIPTEROIDA, 1983
HETEROCORALLIA, 1991
HETERODONTA, 1992
HEXACTINIARIA, 2001
HEXACORALLA, 2000
HIGH-GRADE METAMORPHISM, 2011
HIPPOMORPHA, 2016
HIPPURITOIDA, 3911
HOLOCEPHALI, 2027
HOLOSTEI, 2031
HOLOTHUROIDEA, 2033
HOMALOZOA, 2035
HOMINIDAE, 2038
HOMO SAPIENS, 2039
HOMOIOSTELEA, 2043
HOMOSTELEA, 2046
HUMAN ECOLOGY, 2069
HUMAN WASTE, 2070
HYALOSPONGEA, 2085
HYDRATION OF GLASS, 2090
HYDRAULIC FRACTURING, 2092
HYDRAULIC MAP, 2095
HYDROIDA, 2118
HYDROTHERMAL STAGE, 2133
HYDROZOA, 2137
HYMENOPTERA, 2140

HYOLITHES, 2141
HYRACOIDEA, 2152
HYSTRICHOMORPHA, 2153
IAPETUS, 2154
ICE MOVEMENT, 2158
ICE RAFTING, 2159
ICHTHYOPTERYGIA, 2166
ICHTHYOSAURIA, 2167
IDIOBLAST, 2168
IDIOGEOSINKLINAL', 2169
IDIOMORFIZM, 2171
IDIOMORFNYJ, 2170
IERMOSOL, 4848
IGNIMBRIT, 2177
IKHNOFOSSILII, 2164
IKHNOLOGIYA, 2165
IKHOR, 3830
IL, 2952
IL GLUBOKOVODNYJ, 3127
IMAGERY, 2178
IMBRIJ, 2179
IMPACT SEISMICS, 2190
IMPACT STATEMENT, 2191
IMPAKTIT, 2193
IMPORT, 2194
IMPREGNATSIYA, 2195
IN SITU, 2197
IN'EKTSIYA, 2239
INARTICULATA, 2198
IND, 2210
INDEKS DIFFERENTSIATSII, 1179
INDEKS KRISTALLIZATSII, 1026
INDEKS TSVETOVOJ, 847
INDEKS ZRELOSTI, 2719
INDEKS-MINERAL, 2208
INDIJ, 2213
INDIKATOR, 2212
INDIKATOR GEOKHIMICHESKIJ, 1718
INDIKATOR RADIOAKTIVNYJ, 3729
INDOSHINIT, 2214
INFIL'TRATSIYA, 2226
INFORMATIKA, 2228
INFRAKEMBRIJ, 2230
INFRAKRASNYJ, 2231
INFRASTRUKTURA, 2235
INGRESSIYA, 2236
INKRUSTATSIYA, 2205
INLET, 2241
INNER SHELF, 2244
INNER SLOPE, 2245
INOCERAMI, 2246
INSECTA, 2251
INSECTIVORA, 2252
INSEKVENTNYJ, 2254
INSEPTISOL, 2199
INSOLYATSIYA, 2255
INTERFERENTSIYA, 2262

INTERFEROMETRIYA, 2263
INTERNIDY, 2279
INTERPLANETARY COMPARISON, 2280
INTERPOLYATSIYA, 2282
INTERPRETATSIYA, 2283
INTERSERTAL'NYJ, 2284
INTERTIDAL SEDIMENTATION, 2287
INTRAKLAST, 2288
INTRATELLURICHESKIJ, 2292
INTRUZIYA, 2293
INTRUZIYA, 2294
INTRUZIYA PLASTOVAYA, 4066
INTRUZIYA RASSLOENNAYA, 2463
INVERSIYA MAGNITNOGO POLYA, 1743
INVERSIYA REL'EFA, 3810
INVERTEBRATA, 2297
IOD, 2298
ION, 2299
IONIZATSIYA, 2304
IONOSFERA, 2305
IPR, 4852
IRIDIJ, 2307
IRIZATSIYA, 2306
IRRIGATSIYA, 2314
ISKOPAEMOE RUKOVODYASHCHEE, 1925
ISKOPAEMOE ZHIVUSHCHEE, 2549
ISOPLEURA, 2338
ISOPRENOID, 2339
ISPARENIE, 1458
ISPARENIE SUMMARNOE, 1460
ISPOL'ZOVANIE, 4704
ISPOL'ZOVANIE PROMYSHLENNOE, 2220
ISPYTANIE KOMPRESSIONNOE, 868
ISPYTANIE KRYL'CHATKOJ, 4712
ISPYTANIE LABORATORNOE, 2403
ISPYTANIE NA MONOSNOE SZHATIE, 4658
ISPYTANIE NA TREKHOSNOE SZHATIE, 4592
ISPYTANIE SHTAMPOM, 3500
ISPYTANIE VODONOSNOGO GORIZONTA, 261
ISSLEDOVANIE, 3822
ISTIRANIE, 368
ISTOCHNIK, 4224
ISTOCHNIK GORYACHIJ, 2066
ISTOCHNIK KARSTOVYJ, 2376
ISTOCHNIK MINERAL'NYJ, 2873
ISTOCHNIK NEFTYANOJ, 3109
ISTOCHNIK NEVZRYVNOJ, 3046
ISTOCHNIK PODVODNYJ, 4331
ISTOCHNIK SEJSMICHESKIJ, 4016
ISTOCHNIK SUBMARINNYJ, 4331
ISTOCHNIK VREMENNYJ, 2271
ISTOCHNIK VZRYVNOJ, 1487
ISTORIYA, 2020
ISTORIYA GEOLOGII, 2021
ISTOSHCHENIE VODONOSNOGO GORIZONTA, 4797
ITABIRIT, 2355
ITAKOLUMIT, 2356

ITTERBIJ, 4853
ITTRIJ, 4854
IZLOM RAKOVISTYJ, 875
IZLUCHENIE INFRAKRASNOE, 2233
IZLUCHENIE UL'TRAFIOLETOVOE, 4641
IZMEL'CHENIE, 1899
IZMENCHIVOST', 4714
IZMENCHIVOST' VNUTRIVIDOVAYA, 4193
IZMENENIE, 154
IZMENENIE SEZONNOE, 3985
IZMENENIE UROVNYA MORYA, 3981
IZMENENIYA MAGNITNOGO POLYA, 1744
IZMERENIE, 1698
IZMERITEL' PADENIYA, 1197
IZOANOMALA, 2318
IZOBARA, 2319
IZOBATA, 2321
IZOBAZA, 2320
IZOBRETENIE, 2295
IZOCHORNYJ, 3393
IZOGIPSA, 2329
IZOGONA, 2327
IZOGRADA, 2328
IZOKHORA, 4277
IZOKHRONA, 2323
IZOKLINA, 2325
IZOKLINAL', 2324
IZOKONTSENTRATY, 2326
IZOLINIYA, 2337
IZOLIROVANNOST', 2330
IZOLITA, 2331
IZOMORFIYA, 2333
IZOMORFIZM, 2334
IZOPAKHITA, 2335
IZORADA, 2340
IZORATA, 2341
IZOSEJSMA, 2342
IZOSTAZIYA, 2344
IZOSTRUKTURNYJ, 2345
IZOTERMA, 2346
IZOTOP, 2348
IZOTOP LEGKIJ, 2500
IZOTOP RADIOAKTIVNYJ, 3727
IZOTOP STABIL'NYJ, 4228
IZOTOP TYAZHELYJ, 1971
IZOTROPIYA, 2354
IZVERZHENIE, 1432
IZVERZHENIE (AKTIV.), 2172
IZVERZHENIE GAVAJSKOGO TIPA, 1961
IZVERZHENIE PELEJSKOGO TIPA, 3333
IZVERZHENIE STROMBOLIANSKOGO TIPA, 4297
IZVEST', 2506
IZVESTKOVISTYJ, 609
IZVESTKOVO-SHCHELOCHNOJ, 124
IZVESTNYAK, 2507
IZVESTNYAK BIOKLASTICHESKIJ, 487
IZVESTNYAK DOLOMITOVYJ, 1236

IZVESTNYAK OOLITOVYJ, 3125
IZVILISTOST', 4552
IZVLECHENIE, 3768
IZYSKANIE DOROZHNOE, 3878
KACHESTVO, 3707
KACHESTVO RUDY, 3160
KACHESTVO VODY, 4796
KADMIJ, 604
KAEMKA REAKTSIONNAYA, 3758
KAFEMICHESKIJ, 605
KAJMA, 3868
KAJMA KELIFITOVAYA, 2383
KAJNOZOJ, 688
KAL'DERA, 617
KAL'DERA OBRUSHENIYA, 836
KAL'KRET, 616
KAL'TSEVO-SHCHELOCHNOJ, 607
KAL'TSIEVYJ, 610
KAL'TSIFIR, 611
KAL'TSIJ, 615
KAL'TSIT, 613
KAL'TSITIZATSIYA, 614
KALABRIJ, 606
KALEDONSKIJ, 618
KALICHE, 619
KALIEVYJ, 3578
KALIFORNYJ, 620
KALIJ, 3579
KAM, 2366
KAMBISOL, 624
KAMEN' DEKORATIVNYJ, 1085
KAMEN' DOROZHNYJ, 3877
KAMEN' DRAGOTSENNYJ, 1707
KAMEN' PODELOCHNYJ, 3181
KAMEN' POLUDRAGOTSENNYJ, 4036
KAMEN' YUVELIRNYJ, 3595
KAMERA, 2618
KAMNETOCHETS, 3889
KAMPAN, 626
KAN'ON, 630
KAN'ON PODVODNYJ, 4326
KANAL, 628
KANAL GLUBOKOVODNYJ, 1089
KANAL LAVOVYJ PODVODASHCHIJ, 2459
KANAL PODVODYASHCHIJ, 1522
KANAL PRODUVANIYA, 533
KANAL VULKANA, 2455
KANGA, 2367
KAOLIN, 2368
KAOLINIZATSIYA, 2369
KAPILLYARNOST', 633
KAR'ER, 3709
KARADOK, 638
KARAPAKS, 639
KARAT, 640
KARBON, 654
KARBON NIZHNIJ, 2578

KARBON VERKHNIJ, 4668
KARBONAT, 651
KARBONATIT, 652
KARBONATIZATSIYA, 653
KAREL'SKIJ, 2370
KARNIJ, 655
KAROTAZH, 2562
KAROTAZH AKUSTICHESKIJ, 41
KAROTAZH BOKOVOJ, 2443
KAROTAZH EHLEKTRICHESKIJ, 1324
KAROTAZH GAMMA, 1679
KAROTAZH GAMMA-GAMMA, 1678
KAROTAZH NEJTRONNYJ, 3017
KAROTAZH SEJSMICHESKIJ, 4012
KAROTAZH SKVAZHINNYJ, 4816
KARRU, 2372
KARRY, 2371
KARST, 2373
KARTA, 2668
KARTA BATIMETRICHESKAYA, 437
KARTA BIOFATSIAL'NAYA, 491
KARTA FOTOGEOLOGICHESKAYA, 3409
KARTA GEOFIZICHESKAYA, 1754
KARTA GEOKHIMICHESKAYA, 1719
KARTA GEOLOGICHESKAYA, 1735
KARTA GEOMAGNITNAYA, 2644
KARTA GEOMORFOLOGICHESKAYA, 1747
KARTA GIDROGEOLOGICHESKAYA, 2111
KARTA GIDROKHIMICHESKAYA, 2101
KARTA GIDROLOGICHESKAYA, 2120
KARTA GRAVITATSIONNAYA, 1882
KARTA INZHENERNO=GEOLOGICHESKAYA, 1768
KARTA IZOSEJSM, 2343
KARTA KONTURNAYA, 932
KARTA LITOLOGO=FATSIAL'NAYA, 2527
KARTA MAGNITNAYA, 2636
KARTA MESTOROZHDENIJ, 1299
KARTA METALLOGENICHESKAYA, 2772
KARTA NEFTEGAZONOSNOSTI, 3106
KARTA NERUDNYKH POLEZNYKH ISKOP, 3052
KARTA PALEOGEOGRAFICHESKAYA, 3256
KARTA PETROGRAFICHESKAYA, 3376
KARTA POCHVENNAYA, 4153
KARTA PROGNOZNAYA, 1611
KARTA REGISTRATSIONNAYA, 2207
KARTA SEJSMOLOGICHESKAYA, 4025
KARTA STRATIGRAFICHESKAYA, 4278
KARTA STRUKTURNAYA, 4308
KARTA TEKTONICHESKAYA, 4437
KARTA TOPOGRAFICHESKAYA, 4546
KARTA UGOL'NYKH MESTOROZHDENIJ, 807
KARTIROVANIE, 2670
KARTOGRAFIYA, 660
KARTOGRAFIYA AVTOMATIZIROVANNAYA, 381
KARTOGRAMMA, 659
KASIMOVSKIJ, 2382
KASKADNYJ, 662

KATAGENEZ, 668
KATAKLAZ, 667
KATALIZ, 670
KATAMORFIZM, 2377
KATANGSKIJ, 2378
KATASTROFIZM, 671
KATATERMAL'NOE MESTOROZHDENIE, 2379
KATAZONA, 2380
KATENA, 673
KATION, 674
KATUN GLINYANYJ, 303
KAUSTOBIOLIT, 676
KAVERNA, 678
KAVERNOMETRIYA, 621
KAVERNOZNOST', 679
KAZANSKIJ, 2381
KEHPROK, 631
KEJPER, 2388
KELLOVEJ, 622
KEMBRIJ, 625
KEMBRIJ NIZHNIJ, 2577
KEMBRIJ SREDNIJ, 2841
KEMBRIJ VERKHNIJ, 4667
KENNEL', 629
KENORANSKIJ, 2384
KERATOFIR, 2385
KERN, 958
KEROGEN, 2386
KHAL'KOFIL'NYJ, 703
KHAL'KOSFERA, 704
KHAMADA, 1950
KHAOTICHESKIJ, 709
KHARAKTERISTIKA, 710
KHATT, 716
KHELATOOBRAZOVANIE, 718
KHEMOFOSSILII, 727
KHEMOGENNYJ, 736
KHIMICHESKIJ SOSTAV, 721
KHIMIYA MOKRAYA, 4821
KHISHCHNYJ, 3597
KHISTOSOL, 2022
KHITIN, 745
KHLOR, 749
KHLORID, 747
KHLORIROVANIE, 748
KHLORITIZATSIYA, 752
KHLORNOST', 750
KHLOROFILL, 753
KHODY CHERVEJ, 600
KHOLM, 2013
KHOLM ABISSAL'NYJ, 21
KHOLODNYJ, 832
KHONDRA, 757
KHONDRIT, 756
KHONDRIT UGLERODISTYJ, 647
KHONDROLIT, 1862
KHRANENIE, 4265

KHRANENIE DANNYKH, 1071
KHRANENIE PODZEMNOE, 4652
KHREBET PODVODNYJ, 4330
KHREBET SREDINNO=OKEANICHESKIJ, 2840
KHROM, 761
KHROMAT, 759
KHROMATOGRAFIYA, 760
KHROMATOGRAFIYA GAZOVAYA, 1686
KHRONOSTRATIGRAFICHESKOE PODRAZDELENIE, 762
KHRONOSTRATIGRAFIYA, 763
KHRONOZONA, 764
KHRUPKOST', 579
KIBARSKIJ, 2390
KIBERNETIKA, 1041
KIMBERLIT, 2391
KIMERIDZH, 2392
KIMMERIJSKIJ, 768
KINEMATIKA, 2393
KINKBAND, 2394
KINTSIGIT, 2395
KISLOROD, 3225
KISLOROD GAZ, 3226
KISLOTA, 37
KISLOTA NEOGANICHESKAYA, 2248
KISLOTA ORGANICHESKAYA, 3165
KISLOTA SERNAYA, 4356
KISLOTA ZHIRNAYA, 1506
KISLOTNOST', 40
KISLYJ (SOSTAV), 39
KLAREN, 780
KLARK, 781
KLASSIFIKATSIYA, 782
KLASSIFIKATSIYA C.I.P.W., 603
KLASSIFIKATSIYA KHOLMSA, 2024
KLASSIFIKATSIYA NIGGLI, 3027
KLASSIFIKATSIYA UGLEY PROMYSHLENNAYA, 3748
KLASTIT, 786
KLIF, 793
KLIMAT, 794
KLIMAT GUMIDNYJ, 2073
KLIN, 4811
KLIN LEDYANOJ, 2162
KLIN LEDYANOJ OKAMENELYJ, 1621
KLIPP, 2396
KLIVAZH, 792
KLIVAZH BAZAL'NYJ, 423
KLIVAZH ISTECHENIYA, 4131
KLIVAZH OSEVOJ POVERKHNOSTI, 392
KLIVAZH POSLOJNYJ, 455
KLIVAZH SKOL'ZHENIYA, 4133
KLIVAZH TECHENIYA, 1576
KOBAL'T, 818
KODA VOLNA, 821
KOEHFFITSIENT, 822
KOEHFFITSIENT FIL'TRATSII, 4576
KOEHFFITSIENT KORRELYATSII, 962
KOEHFFITSIENT OTRAZHENIYA, 3782

KOEHFFITSIENT PESCHANISTOSTI, 3938
KOEHFFITSIENT PETROKHIMICHESKIJ, 3370
KOEHFFITSIENT RASPREDELENIYA, 3311
KOEHFFITSIENT TEPLOPROVODNOST', 4495
KOL'TSA LIZEGANGA, 2498
KOL'TSO GODICHNOE, 4585
KOLEBANIE, 3197
KOLEBANIE SVOBODNOE, 1642
KOLEBANIE UROVNYA (GIDROGEOL), 1578
KOLLEKTSIYA, 839
KOLLOFORMNYJ, 841
KOLLOID, 842
KOLLYUVIJ, 843
KOLONIAL'NYJ, 844
KOLONKA STRATIGRAFICHESKAYA, 1734
KOLONNA, 3446
KOLORIMETRIYA, 849
KOMAGMATICHNYJ, 852
KOMMENSALIZM, 853
KOMOCHKI FEKAL'NYE, 1521
KOMPAS(MAGNITNYJ), 2630
KOMPETENTNYJ, 859
KOMPLEKS, 860
KOMPLEKS MAGMATICHESKIJ, 2173
KOMPLEKS METAMORFICHESKIJ, 2781
KOMPLEKSOMETRIYA, 862
KOMPLEKSOOBRAZOVANIE, 861
KOMPONENT INERTNYJ, 2223
KON'YAK, 894
KONDENSATSIYA, 882
KONGLOMERAT, 892
KONGLOMERAT BAZAL'NYJ, 424
KONKRETSIYA, 881
KONRADA POVERKHNOST', 902
KONSISTENTSIYA, 906
KONSOLIDATSIYA, 907
KONSOLIDOMETRIYA, 908
KONSTANTA, 909
KONSTANTA UPRUGOSTI, 1317
KONTAKT, 913
KONTAKT GAZONEFTYANOJ, 3113
KONTAMINATSIYA, 917
KONTINENT, 918
KONTINENT (LUNNYJ), 2605
KONTRAKTSIYA, 934
KONTROL', 936
KONTROL' GIDROGEOLOGICHESKIJ, 2110
KONTROL' KHIMICHESKIJ, 722
KONTROL' LITOLOGICHESKIJ, 2530
KONTROL' MEKHANICHESKIJ, 2725
KONTROL' ORUDENENIYA, 3153
KONTROL' PALEOGEOGRAFICHESKIJ, 3255
KONTROL' STRATIGRAFICHESKIJ, 4276
KONTROL' STRUKTURNYJ, 4302
KONTROL' TEKTONICHESKIJ, 4435
KONTSENTRATSIYA, 873
KONTSENTRATSIYA GRAVIMETRICHESKAYA, 1114

KONUS DROBLENIYA, 4057
KONUS PEPLOVYJ, 4192
KONUS SHLAKOVYJ, 770
KONUS V KONUSE (STRUKTURA), 886
KONUS VULKANICHESKIJ, 4762
KONUS VYNOSA, 1504
KONUS VYNOSA ALLYUVIAL'NYJ, 141
KONUS VYNOSA PODVODNYJ, 4327
KONVEKTSIYA, 938
KONVEKTSIYA TEPLOVAYA, 939
KONVERGENTSIYA, 940
KOORDINATA(GEODET), 1728
KOORDINATSIYA, 943
KOPERNIKANSKIJ, 945
KOPROLITY, 947
KORA KONTINENTAL'NAYA, 921
KORA NIZHNYAYA, 2580
KORA OKEANICHESKAYA, 3091
KORA VERKHNYAYA, 4670
KORA VYVETRIVANIYA, 4810
KORA ZEMNAYA, 1001
KORALLIT, 952
KORNI GOR, 3896
KORREKTSIYA, 960
KORRELYATSIYA, 961
KORROZIYA, 964
KORUND, 965
KOSA, 4215
KOSMOGENNYJ, 973
KOSMOKHIMIYA, 972
KOSMOLOGIYA, 975
KOSMOS, 971
KOTEL, 3585
KOTLOVINA OKEANICHESKAYA, 3085
KOTLOVINA TERMOKARSTOVAYA, 2387
KOTLOVINA VULKANICHESKAYA, 675
KOVKIJ, 2660
KRAJ PLATFORMY, 920
KRASNOTSVETY, 3773
KRASNOZEM, 4464
KRATER, 981
KRATER LUNNYJ, 2601
KRATER METEORITNYJ, 2797
KRATER UDARA, 2187
KRATEROOBRAZOVANIE, 983
KRATON, 984
KREEP, 2397
KREMNEZEM, 4097
KREMNIJ, 4103
KREMNISTYJ, 4099
KRIOTURBATSIYA, 1005
KRIPTOGENNYJ, 1008
KRIPTON, 2398
KRISTALL, 1011
KRISTALL DVUOSNYJ, 476
KRISTALLICHESKIJ, 1022
KRISTALLICHNOST', 1023

KRISTALLIT, 1024
KRISTALLIZATSIYA, 1025
KRISTALLIZATSIYA FRAKTSIONNAYA, 1631
KRISTALLOBLAST, 1027
KRISTALLOGRAFIYA, 1028
KRISTALLOGRAFIYA STRUKTURNAYA, 4303
KRISTALLOKHIMIYA, 1012
KRISTALLOOPTIKA, 1019
KRITERIJ POISKOVYJ, 3157
KRITERIJ POISKOVYJ (STRUKTURNYJ), 4307
KRUCHENIE, 4550
KRUPNOZERNISTYJ, 810
KRYL'CHATKA, 4711
KRYLO, 2505
KSENOLIT, 4838
KSENOMORFNYJ, 4839
KSENON, 4840
KSERALF, 4841
KSEROFILY, 4843
KSEROFIT, 4844
KSEROLL, 4842
KSEROSOL, 4845
KSERULT, 4846
KUESTA, 1031
KUMULAT, 1032
KUNGUR, 2400
KUPOL, 1240
KUPOL GNEJSOVYJ, 1829
KUPOL SOLYANOJ, 3925
KUPOL VULKANICHESKIJ, 4763
KUTORGINIDA, 2402
KUYAL'NITSKIJ, 2399
KVARTS, 3710
KVARTSIT, 3712
KYURI TOCHKA, 1033
KYURIJ, 1034
LABYRINTHODONTIA, 2404
LACERTILIA, 2406
LADIN, 2410
LAGUNA, 2411
LAGUNA BAR'ERNAYA, 419
LAKHAR, 2414
LAKKOLIT, 2405
LAMARKIZM, 2417
LAMEH KONSTANTA, 2418
LAMPROFIR, 2423
LAMPROFIROVYJ, 2424
LAND USE, 2427
LANDFILL, 2428
LANDSAT, 2430
LANDSHAFT KARSTOVYJ, 4178
LANTAN, 2432
LAPILLI, 2433
LAPLASA PREOBRAZOVANIE, 2434
LARAMIJSKIJ, 2435
LATERAL'NYJ, 2441
LATERIT, 2446

LATERITIZATSIYA, 2448
LATIT, 2449
LAVA, 2454
LAVINA, 388
LAVRAZIYA, 2453
LAYERED MEDIUM, 2464
LAZER, 2437
LEAKAGE, 2470
LEARNED SOCIETY, 2472
LED, 2155
LED MERTVYJ, 1075
LED MORSKOJ, 3976
LED POGREBENNYJ, 1905
LEDNIK, 1804
LEDNIK SHEL'FOVYJ, 2161
LEDNIKOVO—MORSKOJ, 1808
LEDNIKOVO—OZERNYJ, 1806
LEDNIKOVYJ, 1791
LEGENDA, 2475
LEJAS NIZHNIJ, 2584
LEJAS SREDNIJ, 2845
LEJAS VERKHNIJ, 4674
LEJKOKRATOVYJ, 2489
LEJKOSOM, 2490
LEKSIKA, 2494
LEKTOTIP, 2474
LENGTH OF DAY, 2478
LENSKIJ, 2479
LEPERDITOCOPIDA, 2481
LEPIDOBLASTOVYJ, 2482
LEPIDOCYSTOIDEA, 2483
LEPIDOSAURIA, 2484
LEPOSPONDYLI, 2485
LEPTINIT, 2488
LEPTIT, 2486
LEPTOGEOSINKLINAL', 2487
LESS, 2561
LETUCHEST', 4758
LETUCHIE, 4757
LETUCHIJ, 4755
LEVEE, 2491
LEZHEN' KRASNYJ, 3904
LICHENES, 2495
LICHIDA, 2497
LIGNIN, 2503
LIGNIT, 2504
LIKHENOMETRIYA, 2496
LIKVATSIYA, 2518
LIKVIDUS, 2522
LIMNOLOGIYA, 2508
LIMONIT, 2510
LINEJNOST', 2512
LINGULIDA, 2513
LINIYA ANDEZITOVAYA, 195
LINIYA BEREGOVAYA, 4083
LINIYA BEREGOVAYA DREVNYAYA, 4
LINIYA NARASTANIYA, 1922

LINOFIROVYJ, 2515
LINZA, 2480
LIPIDY, 2516
LIPOSTRACA, 2517
LISSAMPHIBIA, 2523
LIST, 2469
LISTOVATOST', 1609
LITIFIKATSIYA, 2524
LITIJ, 2525
LITOFATSIYA, 2526
LITOFIL'NYJ, 2533
LITOFIT, 2534
LITOGENEZ, 2528
LITOLOGICHESKIJ, 2529
LITOLOGIYA, 2532
LITOPTERNA, 2542
LITORAL', 2543
LITOSFERA, 2536
LITOSOL, 2535
LITOSTRATIGRAFIYA, 2538
LITOSTROM, 2539
LITOTIP, 2541
LITOTOP, 2540
LITUOLACEA, 2547
LIXIVIATION, 2550
LLANDEJLO, 2551
LLANDOVERI, 2552
LLANVIRN, 2553
LODRANIT, 2560
LOKAL'NYJ, 2557
LOKATSIYA BOKOVOGO OBZORA, 4091
LOPOLIT, 2569
LOVUSHKA, 4580
LOVUSHKA LITOLOGICHESKAYA, 2531
LOVUSHKA NEFTI, 3111
LOVUSHKA STRATIGRAFICHESKAYA, 4279
LOVUSHKA STRUKTURNAYA, 4311
LOW—GRADE METAMORPHISM, 2575
LOZHE OKEANA, 3088
LOZHE POTOKA, 4285
LUDLOV, 2596
LUNA, 2935
LUNA (INST), 2598
LUNAR CONSTITUTION, 2600
LUNNYJ, 2599
LUNOTRYASENIE, 2936
LUVISOL, 2609
LYCOPSIDA, 2610
LYTOCERATINA, 2611
LYUMINESTSENTSIYA, 2597
LYUTETSIJ, 2607
MAAR, 2612
MAASTRIKHT, 2615
MACHAERIDIA, 2614
MAGMA, 2617
MAGMATIC ASSOCIATION, 2620
MAGMATIC CONTROLS, 2621

MAGMATIC ORIGIN, 2623
MAGMATICHESKIJ, 2619
MAGMATIZM, 2624
MAGNAFATSIYA, 2625
MAGNETIC SURVEY MAP, 2644
MAGNETIT, 2647
MAGNETIZM, 2646
MAGNEZIT, 2626
MAGNIJ, 2627
MAGNITOGIDRODINAMIKA, 2649
MAGNITOMETR, 2650
MAGNITORAZVEDKA, 2637
MAGNITOSFERA, 2651
MAGNITOSTRIKTSIYA, 2653
MAJSKIJ, 2656
MAKROFATSIYA, 2738
MAKSIMUM, 2722
MALACOSTRACA, 2659
MAMMALIA, 2661
MANGR, 2663
MANTIYA, 2665
MANTIYA NIZHNYAYA, 2585
MANTIYA VERKHNYAYA, 4675
MANTODEA, 2666
MARGANETS, 2662
MARINE INSTALLATION, 2680
MARINE METHODS, 2682
MARINE TRANSPORT, 2689
MARINER (SPUTNIK), 2690
MARKOVSKIJ PROTSESS, 2691
MARS, 3474
MARSH, 2693
MARSUPIALIA, 2695
MASKON, 2696
MASS BALANCE, 2698
MASS TRANSFER, 2701
MASS—SPEKTROSKOPIYA, 2700
MASSA, 2697
MASSA OSNOVNAYA, 1919
MASSHTAB, 2669
MASSHTAB KRUPNYJ, 2436
MASSHTAB MELKIJ, 4140
MASSIV GORNYJ, 2950
MASSIV SREDINNYJ, 2732
MASSIVE BEDDING, 2704
MASSIVNYJ, 2703
MASSOPERENOS, 4568
MASTIGOPHORA, 2706
MATEMATIKA, 2716
MATERIAL, 2707
MATERIAL IZOLYATSIONNYJ, 2259
MATERIAL NEORGANICHESKIJ, 2249
MATERIAL OSADOCHNYJ, 2554
MATERIAL RYKHLYJ, 2202
MATERIAL VSPUCHIVAYUSHCHIJSYA, 1474
MATRIKS, 3887
MATRITSA, 2713

MATSERAL, 2613
MATUYAMA EHPOKHA, 2720
MAXILLOPODA, 2721
MAZATTSAL'SKIJ, 2723
MEANDR, 2724
MECOPTEROIDA, 2731
MED', 946
MEDNOPORFIROVYJ, 3569
MEGASPORA, 2739
MEKHANIKA, 2729
MEKHANIKA GRUNTOV, 4154
MEKHANIKA POROD, 3888
MEKHANIZM, 2730
MEKHANIZM ZEMLETRYASENIYA, 1600
MEKHANOGENEZ, 2726
MEL, 989
MEL (PORODA), 705
MEL NIZHNIJ, 2579
MEL VERKHNIJ, 4669
MEL', 4074
MELANOKRATOVYJ, 2741
MELANZH, 2740
MELIORATSIYA, 2742
MELIORATSIYA POCHVY, 4150
MELKOVODNYJ, 4053
MENDELEVIJ, 2747
MEOTICHESKIJ, 2748
MERGEL', 2692
MERKURIJ, 3475
MEROPRIYATIYA, 3130
MEROSTOMATA, 2750
MEROSTOMOIDEA, 2751
MERZLOTA VECHNAYA, 3362
MERZLOTOVEDENIE, 1003
MESOGASTROPODA, 2754
MESOSAURIA, 2757
MESTNOST' STRATOTIPICHESKAYA, 4624
MESTOROZHDENIE, 2866
MESTOROZHDENIE MAGMATICHESKOE, 2622
MESTOROZHDENIE MASSIVNOE, 2705
MESTOROZHDENIE NEMETALLICHESKOE, 3051
MESTOROZHDENIE OSADOCHNOE, 3997
MESTOROZHDENIE OSTATOCHNOE, 3829
MESTOROZHDENIE STRATIFITSIROVANNOE, 4270
MESTOROZHDENIE STRATIFORMNOE, 4274
MESTOROZHDENIE TIPA MISSISSIPPI, 2896
MESTOROZHDENIE-GIGANT, 1786
METABAZAL'T, 2762
METACOPINA, 2763
METAFIT, 2791
METAGABBRO, 2764
METAGENEZ, 2765
METALL, 2766
METALL BLAGORODNYJ, 3037
METALL RASSEYANNYJ, 4558
METALL REDKIJ, 3752
METALL SHCHELOCHNOZEMEL'NYJ, 129

METALL SHCHELOCHNYJ, 126
METALL TSVETNOJ, 3047
METALL TYAZHELYJ, 1972
METALLICHESKIJ, 2768
METALLOGENIC DISTRICT, 2770
METALLOGENIYA, 2774
METALLOID, 2775
METALLOSODERZHASHCHIJ, 2769
METALLOTEKT, 2776
METALLURGIYA, 2777
METAMIKTNYJ, 2778
METAMORFIZM, 2787
METAMORFIZM DINAMOTERMAL'NYJ, 1269
METAMORFIZM GEOTERMAL'NYJ, 1776
METAMORFIZM GIDROTERMAL'NYJ, 2131
METAMORFIZM IMPAKTNYJ, 2189
METAMORFIZM IZOKHIMICHESKIJ, 2322
METAMORFIZM KONTAKTOVYJ, 915
METAMORFIZM NAGRUZKI, 597
METAMORFIZM PLUTONICHESKIJ, 3518
METAMORFIZM REGIONAL'NYJ, 3795
METAMORFIZM RETROGRADNYJ, 3841
METAMORFIZM TERMAL'NYJ, 4498
METAMORFIZM UDARNYJ, 4077
METAMORFOGENNYJ, 2788
METAMORFOZ, 2789
METAN, 2801
METANAUPLIUS, 2790
METAOSADOCHNYJ, 2792
METASOMA, 2793
METASOMATIT, 2795
METASOMATOZ, 2794
METASOMATOZ KONTAKTOVYJ, 916
METASOMATOZ SHCHELOCHNOJ, 130
METATEKSIS, 2796
METEORIT, 2798
METEORIT KAMENNYJ, 4264
METEORIT ZHELEZNYJ, 2767
METEOROID, 2799
METEOROLOGIYA, 2800
METOD AFMAG, 78
METOD AKUSTICHESKIJ, 43
METOD AL'FA-SVINTSOVYJ, 2467
METOD ANALITICHESKIJ, 183
METOD ARGONOVYJ AKTIVATSIONNYJ, 297
METOD BERILLIEVYJ, 472
METOD EHLEKTROMAGNITNYJ, 1341
METOD ESTESTVENNOGO EHLEKTRICHESKOGO POLYA, 4033
METOD GELIEVYJ, 1979
METOD GEOFIZICHESKIJ, 1755
METOD GEOTERMICHESKIJ, 1777
METOD GRAVIMETRICHESKIJ, 1883
METOD IMMERSIONNYJ, 2184
METOD INDUKTIVNYJ, 2217
METOD IONIEVO-URANOVYJ, 2303
METOD IONIEVYJ, 2302
METOD IZOTOPNOGO RAZBAVLENIYA, 2349

METOD KALIJ-ARGONOVYJ, 3580
METOD KALIJ-KAL'TSIEVYJ, 3581
METOD KONECHNYKH RAZNOSTEJ, 1550
METOD LYUTETSIEVO-GAFNIEVYJ, 2608
METOD MAGNITOTELLURICHESKIJ, 2654
METOD MATEMATICHESKIJ, 2714
METOD MIKROVOLNOVOJ, 2838
METOD MONTE-KARLO, 2932
METOD NAIMEN'SHIKH KVADRATOV, 2473
METOD NARUSHENIYA URANOVOGO RAVNOVESIYA, 4691
METOD NOVYJ, 3022
METOD OBSHCHEJ GLUBINNOJ TOCHKI, 854
METOD OKRASHIVANIYA, 4230
METOD OTRAZHENNYKH VOLN, 3784
METOD PEREKHODNYKH PROTSESSOV, 4573
METOD POROSHKA, 3587
METOD PRELOMLENNYKH VOLN, 3787
METOD PROTOAKTINIEVO-IONIEVYJ, 3643
METOD RADIOMETRICHESKIJ, 3732
METOD RADIOUGLERODNYJ, 646
METOD RADIOVOLNOVOJ, 3724
METOD RENIEVO-OSMIEVYJ, 3850
METOD RUBIDIEVO-STRONTSIEVYJ, 3910
METOD SEJSMICHESKIJ, 4014
METOD SOPROTIVLENIJ, 3833
METOD SVINTSOVO-SVINTSOVYJ, 2468
METOD SVINTSOVO-URANOVIJ, 4692
METOD TELLURICHESKIJ, 4449
METOD TORIEVO-SVINTSOVYJ, 4518
METOD TRANSMISSII, 4575
METOD TREKOV, 3309
METOD TREKOV DELENIYA, 1556
METOD TRITIEVYJ, 4601
METOD UL'TRAZVUKOVYJ, 4639
METOD URANO-TORIEVOSVINTSOVYJ, 4693
METOD VYZVANNOGO POTENTSIALA, 2895
METOD VYZVANNOJ POLYARIZATSII, 2216
METODOLOGIYA, 2802
METROPOLITEN, 4344
MEZHDUNAR. GEOL. KONGRESS, 2277
MEZHDUNAR. GEOL. KORRELYATS. PROGR., 2278
MEZHDURECH'E, 2264
MEZHEN', 2573
MEZHFORMATSIONNYJ, 2265
MEZHGORNYJ, 2273
MEZHLEDNIKOVYJ, 2266
MEZHPLYUVIAL'NYJ, 2281
MEZOFILY, 2756
MEZOGEA, 2753
MEZOGEOSINKLINAL', 2755
MEZOKRATOVYJ, 2752
MEZOSIDERIT, 2758
MEZOTERMAL'NYJ, 2759
MEZOZOJ, 2760
MEZOZONA, 2761
MIAROLITOVYJ, 2803
MICROCRYSTALLINE LIMESTONE, 2809

MIGMATIT, 2853
MIGMATIZATSIYA, 2854
MIGRATSIYA, 2855
MIGRATSIYA EHLEMENTOV, 2856
MIKRIT, 2806
MIKRO-EHVM, 2807
MIKROFATSIYA, 2811
MIKROFAUNA, 2812
MIKROFITOLITY, 2828
MIKROFLORA, 2814
MIKROFOSSILII, 2816
MIKROGRANIT, 2818
MIKROGRANITOVYJ, 2819
MIKROKAROTAZH, 2823
MIKROKAROTAZH BOKOVOJ, 2821
MIKROKRATER, 2808
MIKROLITOVYJ, 2822
MIKROMETEORIT, 2824
MIKROORGANIZM, 2826
MIKROPALEONTOLOGIYA, 2827
MIKROPLITA, 2829
MIKROPROBLEMATIKA(ISKOP), 2830
MIKROREL'EF, 2825
MIKROSEJSMY, 2834
MIKROSKLADKA, 2815
MIKROSKOP, 2831
MIKROSKOP EHLEKTRONNYJ, 1348
MIKROSKOPICHESKIJ, 2832
MIKROSKOPIYA EHLEKTRONNAYA, 1349
MIKROSKOPIYA EHLEKTRONNAYA SKANIRUYUSHCHAYA, 3957
MIKROSTILOLIT, 2835
MIKROTEKTIT, 2837
MIKROTEKTONIKA, 2836
MIKROTRESHCHINA, 2813
MIKROTVERDOST', 2820
MIKROZEMLETRYASENIE, 2810
MIKROZERNISTYJ, 2817
MIKROZOND EHLEKTRONNYJ, 1347
MILIOLACEA, 2857
MILIOLINA, 2858
MILONIT, 2972
MILONITIZATSIYA, 2974
MILONITOVYJ, 2973
MINERAGRAFIYA, 3158
MINERAL, 2862
MINERAL AKTSESSORNYJ, 29
MINERAL FERROMAGNITNYJ, 2638
MINERAL GLINISTYJ, 790
MINERAL INCLUSION, 2870
MINERAL PORODOOBRAZUYUSHCHIJ, 3890
MINERAL RUDNYJ, 1300
MINERAL SINTETICHESKIJ, 4402
MINERAL SMESHANNO-SLOJNYJ, 2898
MINERAL TIPOMORFNYJ, 4627
MINERAL URANOV, 4690
MINERAL VTORICHNYJ, 3989
MINERALIZATOR, 2877

MINERALIZATSIYA, 2876
MINERALOGICAL LOCALITY, 2878
MINERALOGICHESKIJ (SOSTAV), 2903
MINERALOGIYA, 2880
MINERALOGIYA GLIN, 789
MINERALOID, 1703
MINERALY SOSUSHCHESTVUYUSHCHIE, 826
MINI-EHVM, 2881
MINIMUM, 2882
MINING LICENSE, 2886
MIOGEOSINKLINAL', 2890
MIOMERA, 2891
MIOSPORA, 2892
MIOTSEN, 2889
MIOTSEN NIZHNIJ, 2586
MIOTSEN SREDNIJ, 2846
MIOTSEN VERKHNIJ, 4676
MIRMEKIT, 2979
MISCIBILITY GAP, 2894
MISSISIPIJ, 2897
MNOGOFAZNYJ, 2963
MOBIL'NOST', 2900
MOBILIZATSIYA, 2901
MODEL', 2904
MODEL' DETERMINIROVANNAYA, 1141
MODEL' DVUMERNAYA, 4622
MODEL' FIZICHESKAYA, 3426
MODEL' GIDRODINAMICHESKAYA, 2103
MODEL' MASSHTABNAYA, 3955
MODEL' MATEMATICHESKAYA, 2715
MODEL' ODNOMERNAYA, 3120
MODEL' STATISTICHESKAYA, 4242
MODELIROVANIE ANALOGOVOE, 180
MODIOMORPHOIDA, 2906
MODUL', 2907
MODUL' OB'EMNYJ, 591
MODUL' SDVIGA, 2908
MODUL' UPRUGOSTI, 4060
MOKHO PROEKT, 2909
MOKHOROVICHICHA POVERKHNOST', 2910
MOL, 3440
MOLASSA, 2912
MOLDAVIT, 2913
MOLIBDAT, 2917
MOLIBDEN, 2918
MOLLISOL, 2915
MOLLUSCA, 2916
MOLODOJ, 4850
MOMENT DIPOL'NYJ, 1206
MONATSIT, 2919
MONOCOTYLEDONEAE, 2922
MONOCYATHEA, 2923
MONOGRAFIYA, 2924
MONOKLINAL', 2920
MONOLETE TAXON, 2925
MONOLIT, 2926
MONOMORFNYJ, 2927

MONOPLACOPHORA, 2928
MONORHINA, 2929
MONOTREMATA, 2930
MONTSONIT, 2933
MONTSONITOVYJ, 2934
MORE, 3974
MORE GEOSINKLINAL'NOE, 1764
MORE LUNNOE, 2672
MORE OKRAINNOE, 2674
MORE SHEL'FOVOE, 4069
MORENA, 2938
MORENA BOKOVAYA, 2444
MORENA DONNAYA, 1907
MORFOGENEZ, 2939
MORFOLOGIYA, 2940
MORFOLOGIYA FUNKTSIAL'NAYE, 1666
MORFOLOGIYA MESTOROZHDENIJ, 1119
MORFOMETRIYA, 2941
MORFOSKOPIYA, 2942
MORFOSTRUKTURA, 2943
MORSKOJ, 2677
MOSHCHNOST', 4513
MOSKOVSKIJ, 2946
MOST, 577
MOSTOVAYA PUSTYNNAYA, 1133
MOZAICHNYJ, 2945
MRAMOR, 2671
MUD BANK, 2953
MUD LUMP, 2957
MUKA BUROVAYA, 1039
MULL', 2960
MULTILAYER SYSTEM, 2962
MUSHEL'KAL'K, 2969
MUSSON, 2931
MUTATSIYA, 2971
MUTNOST', 4616
MUZEJ, 2970
MYODOCOPIDA, 2975
MYOIDA, 2976
MYOMORPHA, 2977
MYRIAPODA, 2978
MYS, 632
MYSH'YAK, 310
MYSTACOCARIDA, 2980
MYTILOIDA, 2981
MYXOMYCETES, 2982
NABLYUDENIE, 3081
NADVIG, 4520
NAGOR'E, 2012
NAGRUZKA OPYTNAYA, 2556
NAGRUZKA TSIKLICHESKAYA, 1046
NAKLIT, 2983
NAKLON, 4533
NAKLONENIE MAGNITNOE, 2635
NAKLONOMER, 4534
NAKLONOMETRIYA, 1204
NALOZHENNOST', 1723

NAMAGNICHENNOST', 2648
NAMAGNICHENNOST' DETRITOVAYA, 1143
NAMAGNICHENNOST' IDEAL'NAYA, 206
NAMAGNICHENNOST' INDUTSIROVANNAYA, 2215
NAMAGNICHENNOST' IZOTERMICHESKAYA, 2347
NAMAGNICHENNOST' KHIMICHESKAYA, 732
NAMAGNICHENNOST' OSTATOCHNAYA, 3812
NAMAGNICHENNOST' OSTATOCHNAYA ESTESTVENNAYA, 2991
NAMAGNICHENNOST' TERMOOSTATOCHNAYA, 4511
NAMAGNICHENNOST' VYAZKAYA, 4747
NAMYUR, 2984
NANOFOSSILII, 2985
NAPOLNITEL', 1544
NAPOR GIDROSTATICHESKIJ, 2094
NAPRAVLENIE GORIZONTAL'NOE, 2052
NAPRYAZHENIE RAZRUSHAYUSHCHEE, 3918
NAPRYAZHENIE SKALYVAYUSHCHEE, 4063
NAPRYAZHENIE SZHATIYA, 869
NAROST, 3215
NARUSHENIE RADIATSIONNOE, 3721
NASLEDSTVENNOST', 1986
NASTUPANIE, 3632
NASTUPLENIE PUSTYN', 1134
NASTUPLENIE SUSHI, 1361
NASYP', 1358
NASYSHCHENIE, 3950
NATANTES, 2987
NATRIJ, 4143
NATYAZHENIE POVERKHNOSTNOE, 4365
NAUKI O ZEMLE, 1279
NAUTILOIDEA, 2993
NAVODNENIE, 1569
NAZEMNYJ, 4467
NAZVANIE NOVOE, 3024
NEANDERTHALIAN, 2994
NEDOSYSHCHENIE, 4655
NEDOSYSHCHENNYJ, 4654
NEFELIN, 3008
NEFT', 3379
NEFT' LEGKAYA, 2502
NEFT' TYAZHELAYA, 1974
NEJTRON, 3014
NEJTRONOGRAFIYA, 3016
NEKK, 2995
NEKOMPETENTNYJ, 2203
NEKTON, 2997
NEMERTA, 2998
NEMETALL, 3048
NEMETALLICHESKOE SYR'E, 2219
NEOBRATIMOST', 2313
NEODIM, 3000
NEODNORODNOST', 2238
NEOGASTROPODA, 3001
NEOGEN, 3002
NEOKOM, 2999
NEON, 3003
NEORGANICHESKIJ, 2247

NEORGANICHESKIJ (GENEZIS), 2250
NEORNITHES, 3004
NEOTEKTONIKA, 3005
NEOTENIYA, 3006
NEOTIP, 3007
NEPROZRACHNYJ, 3128
NEPTUN, 3476
NEPTUNIJ, 3011
NEPTUNIZM, 3010
NESMESHIVAYUSHCHIJSYA, 2186
NESMESIMOST', 2185
NESOGLASIE, 4646
NESOGLASIE REGIONAL'NOE, 3796
NESOGLASIE STRATIGRAFICHESKOE, 1211
NESOGLASIE STRUKTURNOE, 3050
NESOGLASIE UGLOVOE, 202
NESOGLASNYJ, 1215
NEUPRUGOST', 198
NEW DESCRIPTION, 3021
NEW MINERAL DATA, 3023
NEZRELYJ, 2182
NIKEL', 3026
NILSSONIALES, 3028
NIOBAT, 3029
NIOBIJ, 3030
NITRAT, 3032
NITRIFIKATSIYA, 3033
NIVATSIYA, 3036
NIVELIROVKA, 2492
NIZMENNOST', 2595
NODOSARIACEA, 3038
NODUL', 3039
NOEGGERATHIALES, 3040
NOMENKLATURA, 3042
NOMOGENEZ, 3044
NOMOGRAMMA, 3045
NON-SATURATED ZONE, 3049
NORIJ, 3053
NOS STRUKTURNYJ, 3058
NOTOSTRACA, 3059
NOTOUNGULATA, 3060
NOVAKULIT, 3061
NOVYE DANNYE, 3020
NOVYJ, 3019
NUCLEATION, 3066
NUCULOIDEA, 3068
NUMMULITIDAE, 3071
NUTATSIYA, 3072
OAZIS, 3074
OB'EKTIV, 3076
OB'EM, 4776
OBDUKTSIYA, 3075
OBEZVOZHIVANIE, 1135
OBLAST' NAKOPLENIYA, 4002
OBLAST' SNOSA, 3656
OBLIQUE ORIENTATION, 3078
OBLOMOCHNYJ, 784

OBLOMOK, 783
OBMEN IONNYJ, 2300
OBNAZHENIE, 3206
OBOBSHCHENIE, 4357
OBOGASHCHENIE, 1376
OBOGASHCHENIE (ZAVOD), 1252
OBOLELLIDA, 3079
OBORUDOVANIE, 2258
OBORUDOVANIE BUROVOE, 1255
OBOSNOVANIE TEKHNIKO=EHKONOMICHESKOE, 1520
OBRABOTKA DANNYKH, 1069
OBRABOTKA DANNYKH (INFORM), 1068
OBRABOTKA POCHV, 4152
OBRAZETS, 3933
OBRAZOVANIE, 1309
OBRIT, 369
OBRUSHENIE, 834
OBRUSHENIE (OPOLZ), 4139
OBRUSHENIE KROVLI, 3894
OBSERVATORIYA, 3082
OBSIDIAN, 3083
OBVAL, 3884
OBZOR, 3843
OCHAG ZEMLETRYA SENIYA, 2146
OCHISTKA, 3687
OCHKOVYJ, 370
OCTOCORALLIA, 3098
ODONTOLOGIYA, 3099
ODONTOPLEURIDA, 3100
ODONTORNITHES, 3101
OFITOVYJ, 3134
OGIPSOVANIE, 1932
OKAMENELOST', 1619
OKATANNOST', 3906
OKEAN, 3084
OKEANICHESKAYA TSIRKYULATSIYA, 3086
OKEANICHESKIJ, 3090
OKEANIZATSIYA, 3092
OKEANOGRAFIYA, 3093
OKEANOLOGIYA, 3094
OKHLAZHDENIE, 942
OKHRA, 3095
OKHRANA, 905
OKHREPT, 3096
OKIS' UGLERODA, 644
OKISEL, 3222
OKISLENIE, 3221
OKNO EHROZIONNOE, 2242
OKNO TEKTONICHESKOE, 4830
OKRAINA ASEJSMICHESKAYA, 328
OKRAINA MATERIKOVAYA, 925
OKREMNENIE, 4100
OKREMNENIE (CHERT), 740
OKSFORD, 3220
OKSILOFIT, 3228
OKSISOL, 3224
OKTAEHDRIT, 3097

OLEDENENIE, 1803
OLEDENENIE KONTINENTAL'NOE, 924
OLIGOTSEN, 3115
OLIGOTSEN NIZHNIJ, 2587
OLIGOTSEN SREDNIJ, 2847
OLIGOTSEN VERKHNIJ, 4677
OLISTOLIT, 3116
OLISTOSTROM, 3117
OLIVIN, 3118
OLOVO, 4537
OMOLOZHENIE, 3802
ONKOLIT, 3119
ONTOGENIYA, 3121
ONYCHOPHORA, 3122
OOID, 3123
OOLIT, 3124
OOLITOVYJ, 3126
OPASNOST' GEOLOGICHESKAYA, 1738
OPASNOST' SEJSMICHESKAYA, 4015
OPHIDIA, 3131
OPHIOCISTIOIDEA, 3132
OPHIUROIDEA, 3135
OPISANIE MARSHRUTNOE, 3876
OPISTHOBRANCHIA, 3136
OPOKA, 3137
OPOLZANIE, 986
OPOLZANIE GRAVITATSIONNOE, 1885
OPOLZEN', 2431
OPOLZEN' GRYAZEVOJ, 2959
OPROBOVANIE, 3935
OPROBOVANIE POCHVY, 4156
OPUSKANIE, 4337
ORANGE MATERIAL, 3144
ORBIKULYARNYJ, 3145
ORBITA, 3146
ORBITOIDACEA, 3147
ORBITOIDIDAE, 3148
ORDOVIK, 3150
ORDOVIK NIZHNIJ, 2588
ORDOVIK SREDNIJ, 2848
ORDOVIK VERKHNIJ, 4678
ORE SOURCE, 3162
ORE TRANSPORT, 3163
OREOL, 373
OREOL RASSEYANIYA PERVICHNYJ, 3611
OREOL VTORICHNYJ, 3988
ORGAN GENERATIVNYJ, 1657
ORGANIC RESIDUE, 3173
ORGANIC SEDIMENT, 3174
ORGANICHESKIJ, 3164
ORGANIZM, 3176
ORGANOGENNYJ, 3172
ORIENTIROVKA, 3177
ORIGOFATSIYA, 3180
ORIKTOTSENOZ, 3196
ORNITHISCHIA, 3183
OROGENEZ, 3185

ORTENT, 3186
ORTHIDA, 3188
ORTID, 3187
ORTOAMFIBOLIT, 3189
ORTOD, 3190
ORTOFIROVYJ, 3193
ORTOGENEZ, 3191
ORTOGNEJS, 3192
ORTOKS, 3195
OS', 393
OS' SKLADKI, 1603
OSADKA SOORUZHENIJ, 4048
OSADKI ATMOSFERNYE, 360
OSADKONAKOPLENIE RUSLOVOE, 4079
OSADKOOBRAZOVENIE SOVREMENNOE, 3762
OSADOCHNYJ, 3999
OSADOK, 3991
OSADOK DONNYJ, 554
OSADOK MORSKOJ, 2685
OSADOK OZERNYJ, 2408
OSAZHDENIE, 3596
OSAZHDENIE KHIMICHESKOE, 723
OSHIBKA, 1431
OSMIJ, 3199
OSNOVANIE SOORUZHENIJ, 1627
OSNOVNOJ, 2616
OSOV BLOCHNYJ, 3891
OSTANETS, 3210
OSTANETS EHROZIONNYJ, 1428
OSTANETS KROVLI, 3895
OSTATOCHNYJ, 3827
OSTATOK NERASTVORIMYJ, 2256
OSTEICHTHYES, 3200
OSTEOLOGIYA, 3201
OSTEOSTRACI, 3202
OSTRACODA, 3203
OSTREACEA, 3204
OSTROV, 2315
OSTROV VULKANICHESKIJ, 2317
OSUSHKA, 4527
OSYP', 2702
OTAL'KOVANIE, 4247
OTBOR PROB GREJFERNYJ, 1839
OTCHET, 3818
OTCHET INFORMATSIONNYJ, 3635
OTDEL'NOST', 3310
OTDEL'NOST' PLITCHATAYA, 2422
OTDEL'NOST' STOLBCHATAYA, 850
OTEHN, 387
OTKACHKA, 3684
OTKACHKA OPYTNAYA, 3685
OTKHODY, 4781
OTKHODY GORNYE, 4413
OTKHODY PROMYSHLENNYE, 2221
OTKHODY RADIOAKTIVNYE, 3730
OTKHODY TVERDYE, 4168
OTKHODY ZHIDKIE, 2521

OTKLONENIE, 1149
OTKLONENIE STANDARTNOE, 4234
OTKRYTIE, 1217
OTLOZHENIE, 1120
OTLOZHENIE ALLYUVIAL'NOE, 142
OTLOZHENIYA FLYUVIOGLYATSIAL'NYE, 3211
OTLOZHENIYA LEDNIKOVYE, 1792
OTMEL' ILISTAYA, 2955
OTMEL' SOLYANAYA, 3926
OTMUCHIVANIE, 1356
OTNOSHENIE IZOTOPNOE, 2352
OTOLIT, 3205
OTPECHATKI KAPEL' DOZHDYA, 3743
OTRAZHENIE, 3783
OTSENKA, 1457
OTSORTIROVANNOST', 4184
OTVAL, 4217
OTVERDENIE, 2218
OUTER SHELF, 3209
OVRAG, 1928
OXYGEN ISOTOPE RATIO, 3227
OZ, 1434
OZERNYJ, 2407
OZERO, 2415
OZERO KRATERNOE, 982
OZERO LAVOVOE, 2458
OZERO LEDNIKOVOE, 1797
OZERO OTMERSHEE, 1494
OZERO SOLYANOE, 3927
OZHIDANIE MATEMATICHESKOE, 2709
OZOKERIT, 4808
P'EZOEHLEKTRICHESTVO, 3442
P'EZOGLIPT, 3797
P'EZOKRISTALLIZATSIYA, 3441
P'EZOMETRIYA, 3443
P=T USLOVIYA, 3229
PACHKA, 2746
PADENIE, 1195
PAKHOEKHOE, 3233
PALAEOHETERODONTA, 3235
PALAEOTAXODONTA, 3237
PALAGONITIZATSIYA, 3238
PALEO, 3239
PALEOATMOSFERA, 3240
PALEOBATIMETRIYA, 3242
PALEOBIOLOGIYA, 3243
PALEOBOTANIKA, 3245
PALEOCOPIDA, 3250
PALEOEHKOLOGIYA, 3252
PALEOFIT, 3236
PALEOGEN, 3254
PALEOGEOGRAFIYA, 3257
PALEOGEOMORFOLOGIYA, 3258
PALEOKARST, 3259
PALEOKLIMAT, 3248
PALEOKLIMATOLOGIYA, 3249
PALEOLIMNOLOGIYA, 3261

PALEOMAGNETIZM, 3263
PALEONTOLOGICHESKI OKHARAKTERIZOVANNOE, 1625
PALEONTOLOGIYA, 3265
PALEOOKEANOGRAFIYA, 3241
PALEOPOCHVA, 3268
PALEOREL'EF, 3266
PALEOSHIROTA, 3260
PALEOSOLENOST', 3267
PALEOSREDA, 3253
PALEOTECHENIE, 3251
PALEOTEMPERATURA, 3269
PALEOTERMOMETRIYA, 3270
PALEOTSEN, 3246
PALEOTSIRKULYATSIYA, 3247
PALEOZOJ, 3271
PALEOZOOLOGIYA, 3272
PALIMPSESTOVYJ, 3273
PALINGENEZ, 3274
PALINOLOGIYA, 3278
PALLADIJ, 3275
PALLASIT, 3276
PALYNOMORPHA, 3279
PAMPA, 3280
PANGEA, 3281
PANIDIOMORFNYJ, 3282
PANTODONTA, 3284
PANTOTHERIA, 3285
PANTSIR', 4471
PAR, 4713
PARA (PORODA), 3286
PARAAMFIBOLIT, 3287
PARABLASTOIDEA, 3288
PARACRINOIDEA, 3289
PARAGENEZIS, 3290
PARAGEOSINKLINAL', 3291
PARAGNEJS, 3292
PARALIAGEOSINKLINAL', 3293
PARALICHESKIJ, 3294
PARALLEL'NYJ, 3296
PARALLELIZM, 3297
PARAMAGNETIK, 3298
PARAMETER, 3299
PARAMETRY RESHETKI, 2451
PARATIP, 3301
PARAZITIZM, 3300
PARTITIONING, 3312
PATCH REEF, 3313
PATENT, 3314
PATOLOGIYA, 3315
PATTERN, 3316
PAUROPODA, 3318
PECTINACEA, 3323
PEDIMENT, 3324
PEDIPLEN, 3325
PEDOGENEZIS, 3326
PEGMATIT, 3327
PEGMATITOVYJ, 3328

PEGMATOID, 3329
PELAGIC SEDIMENTATION, 3331
PELAGICHESKIJ, 3330
PELIT, 3334
PELITIC=METAMORPHIC SEQUENCE, 3336
PELITOVYJ, 3335
PELLETY, 3337
PEMZA, 3683
PENEPLEN, 3339
PENETRATSIYA, 3340
PENETROMETR, 3341
PENOKENSKIJ, 3343
PENSIL'VANIJ, 3342
PENTAMERIDA, 3344
PENTASTOMIDA, 3345
PENTOXYLALES, 3346
PEPEL VULKANICHESKIJ, 4759
PEPLOPAD, 331
PEREDVIZHENIE, 2558
PEREKRISTALLIZATSIYA, 3770
PEREMENNAYA, 4715
PEREMESHCHENIE GRAVITATSIONNOE, 2699
PEREMESHCHENIE POLYUSA, 3535
PEREOTLOZHENIE, 3774
PERERABOTANNYJ, 3845
PERERYV, 2004
PERERYV(STRATIGR), 1682
PERESECHENIE TROJNOE, 4599
PERESLAIVANIE, 2260
PERESYSHCHENNYJ, 3216
PEREUPLOTNENNYJ, 3214
PERFOKARTA, 3686
PERGLINOZEMISTYJ, 3348
PERIDOTIT, 3352
PERIGEJ, 3353
PERIGLYATSIAL'NYJ, 3354
PERIKLINAL', 3351
PERIOD, 3356
PERIOD POLURASPADA, 1942
PERIODICHNOST', 3357
PERISPHINCTIDA, 3358
PERISSODACTYLA, 3359
PERLIT, 3360
PERLITOVYJ, 3361
PERM', 3364
PERM' NIZHNYAYA, 2589
PERM' VERKHNYAYA, 4679
PEROTLOZHENNYJ, 3846
PERSPEKTIVNOST', 3366
PERTIT, 3367
PERTITOVYJ, 3368
PESCHANIK, 3939
PESCHANIK GRUBOZERNISTYJ, 1900
PESCHANISTYJ, 290
PESCHERA, 677
PESOK, 3936
PESOK NEFTENOSNYJ, 3108

PESOK STEKOL'NYJ, 1812
PESTITSID, 3369
PETROFIZIKA, 3383
PETROGENEZIS, 3374
PETROGRAFIYA, 3378
PETROGRAPHIC ANALYSIS, 3375
PETROKHIMIYA, 3372
PETROLEUM ENGINEERING, 3380
PETROLOGIYA, 3382
PETROLOGIYA EHKSPERIMENTAL'NAYA, 1477
PETROLOGIYA OSADOCHNAYA, 3998
PETROLOGIYA STRUKTURNAYA, 4309
PETROSTRUKTURA, 3373
PH, 3384
PHACOPIDA, 3386
PHAEOPHYTA, 3387
PHOLADOMYIDA, 3399
PHOLIDOTA, 3400
PHYLACTOLAEMATA, 3418
PHYLLOCERATINA, 3420
PHYLLOPODA, 3422
PHYTOLITHES, 3435
PIGMENT, 3444
PIGMENT MINERAL'NYJ, 3234
PIKNOKLIN, 3689
PIKNOMETR, 3691
PIKNOMETRIYA, 3692
PIKRIT, 3438
PILLOU, 3448
PILLOU=LAVA, 3447
PINNIPEDIA, 3452
PIRITIZATSIYA, 3694
PIROKLAST, 3695
PIROKSEN, 3702
PIROKSENIT, 3704
PIROLIZ, 3696
PIROMETALLURGIYA, 3697
PIROMETAMORFIZM, 3698
PIROMETASOMATIZM, 3699
PIROSFERA, 3700
PISCES, 3456
PITANIE (GIDRO), 3765
PITANIE (PALE), 3073
PITANIE PODZEMNYKH VOD, 1917
PITANIE PODZEMNYKH VOD ISKUSTVENNOE, 321
PITAYUSHCHIJSYA PLANKTONOM, 3486
PIZOLIT, 3457
PIZOLITOVYJ, 3458
PLACENTALIA, 3462
PLACODERMI, 3464
PLAJYA, 3507
PLAKANTIKLINAL', 3461
PLAMENNYJ, 1563
PLANAR BEDDING STRUCTURE, 3466
PLANATSIYA, 3468
PLANCTOSPHAEROIDEA, 3469
PLANET=EARTH COMPARISON, 3481

PLANETA, 3471
PLANETA YUPITERSKOJ GRUPPY, 3208
PLANETA ZEMNOJ GRUPPY, 4468
PLANETARNYJ, 3482
PLANETARNYJ SOSTAV, 3483
PLANETOLOGIYA, 3485
PLANIROVANIE, 3489
PLANIROVANIE GORODSKOE, 4695
PLANKTON, 3487
PLANOSOL, 3490
PLAST, 3983
PLAST UGOL'NYJ, 806
PLASTICHNOST', 3494
PLASTICHNYJ, 3493
PLASTINA, 2419
PLASTINCHATYJ, 2421
PLATFORMA, 3503
PLATFORMA BUROVAYA, 2684
PLATFORMA MORSKAYA STATSIONARNAYA, 1884
PLATINA, 3504
PLATINOID, 3505
PLATO, 3501
PLATO KRAEVOE, 2673
PLATOBAZAL'T, 3502
PLATYRRHINA, 3506
PLAVLENIE, 2744
PLAVLENIE CHASTICHNOE, 3305
PLAVLENIE INKONGRUEHNTNOE, 2204
PLAVLENIE KONGRUEHNTNOE, 893
PLEJSTOTSEN, 3508
PLEJSTOTSEN NIZHNIJ, 2590
PLEJSTOTSEN SREDNIJ, 2849
PLEJSTOTSEN VERKHNIJ, 4680
PLEOKHROIZM, 3509
PLINSBAKH, 3511
PLIOTSEN, 3512
PLIOTSEN NIZHNIJ, 2591
PLIOTSEN SREDNIJ, 2850
PLIOTSEN VERKHNIJ, 4681
PLITA LITOSFERNAYA, 3495
PLITCHATYJ, 1561
PLOJCHATOST', 3510
PLOSKO=PARALLEL'NYJ, 3467
PLOSKOST', 3470
PLOSKOST' OSEVAYA, 391
PLOTINA, 1059
PLOTINA AROCHNAYA, 271
PLOTINA BETONNAYA, 880
PLOTINA S KAMENNOJ NASYPKOJ, 3885
PLOTINA ZEMLYANAYA, 1275
PLOTNOST', 1113
PLUG, 3513
PLUME, 3515
PLUTON (MAGM.), 3517
PLUTON (PLANETA), 3477
PLUTONICHESKIJ, 3519
PLUTONIJ, 3521

PLUTONIZM, 3520
PLUVIAL'NYJ, 3522
PLYAZH, 444
PLYAZH NIZHNIJ, 1616
PLYVUN, 3716
PNEVMATOLIZ, 3523
POBEREZH'E, 811
POCHKOVANIE, 1705
POCHVA, 4146
POCHVA AZONAL'NAYA, 394
POCHVA GLEEVAYA, 1819
POCHVA GUMUSOVAYA, 2072
POCHVA KASHTANOVAYA, 666
POCHVA LATERITNAYA, 2447
POCHVA LUNNAYA, 2603
POCHVA NEPOLNORAZVITAYA, 2183
POCHVA PESCHANAYA, 293
POCHVA POJMENNAYA, 1598
POCHVA POLIGONAL'NAYA, 3547
POCHVA SREDIZEMNOMORSKAYA, 2735
POCHVA TUNDROVAYA, 4611
POCHVA ZHELEZISTAYA, 1540
POCHVOOBRAZUYUSHCHAYA PORODA MATERINSKAYA, 330
POCHVOVEDENIE, 4157
PODDVIG, 4656
PODGOTOVKA PROB, 3605
PODNOZH'E MATERIKOVOE, 927
PODOCOPIDA, 3525
PODOSHVA USTOJCHIVAYA, 1955
PODRAZDELENIE BIOSTRATIGRAF., 510
PODRAZDELENIE GEOKHRONOLOGICHESKOE, 1724
PODRAZDELENIE LITOSTRATIGRAF., 2537
PODRAZDELENIE STRATIGRAFICHESKOE, 4280
PODSCHET ZAPASOV, 1117
PODSVITA, 1617
PODZOL, 3526
PODZOLUVISOL, 3527
POGASANIE, 3138
POGASANIE VOLNISTOE, 4657
POGONOPHORA, 3528
POGREBENNYJ, 596
POGRUZHENIE (SKLADKI), 3516
POISK BIOGEOKHIMICHESKIJ, 495
POISK DANNYKH, 1070
POISK GEOBOTANICHESKIJ, 1714
POISK GEOKHIMICHESKIJ, 1716
POISK GIDROGEOKHIMICHESKIJ, 2109
POISKI, 3641
POJKILITOVYJ, 3529
POJKILOBLASTOVYJ, 3530
POJMA, 1570
POKAZATEL' PRELOMLENIYA, 2209
POKROV LEDNIKOVYJ, 2160
POKROV OSADOCHNYJ, 3996
POKROV TEKTONICHESKIJ, 2986
POLE EHLEKTRICHESKOE, 1328
POLE EHLEKTROMAGNITNOE, 1340

POLE GRAVITATSIONNOE, 1880
POLE KRISTALLICHESKOE, 1015
POLE LAVOVOE, 2456
POLE LEDOVOE, 2157
POLE MAGNITNOE, 2633
POLE TEPLOVOE, 1774
POLEZNYE ISKOPAEMYE, 2872
POLEZNYE ISKOPAEMYE GORYUCHIE, 1620
POLIEHDR, 3548
POLIGENETICHESKIJ, 3546
POLIMERIZATSIYA, 3550
POLIMETALLICHESKIJ, 3551
POLIMETAMORFIZM, 3552
POLIMIKTOVYJ, 3553
POLIMINERAL'NYJ, 3554
POLIMORFIZM, 3555
POLITIKA, 3537
POLITIPIYA, 3558
POLNOKRISTALLICHESKIJ, 2028
POLONIJ, 3544
POLOSCHATOST', 409
POLOSCHATYJ, 1565
POLOSKA BEKKE, 451
POLOVOJ DIMORFIZM, 4049
POLUPROSTRANSTVO, 1943
POLYARIZATSIYA, 3139
POLYARIZATSIYA EHLEKTRICHESKAYA, 1330
POLYARIZATSIYA KRUGOVAYA, 772
POLYARNYJ, 3533
POLYARNYJ LEDNIK POKROVNYJ, 3534
POLYAROGRAFIYA, 3536
POLYCHAETIA, 3545
POLYMERA, 3549
POLYPHASE PROCESS, 3556
POLYPLACOPHORA, 3557
POLYUS MAGNITNYJ, 1742
POLYUS PALEOMAGNITNYJ, 3262
PONIZHENIE UROVNYA, 1249
PONT, 3559
POPERECHNYJ, 4579
POPRAVKA NA REL'EF, 4466
POPRAVKA TOPOGRAFICHESKAYA, 4545
POPULAR GEOLOGY, 3560
PORFIROBLASTOVYJ, 3568
PORFIROVYJ, 3567
PORIFERA, 3564
PORISTOST', 3565
PORISTOST' YACHEISTAYA, 682
PORISTYJ, 3566
PORODA BOKOVAYA, 4780
PORODA EHFFUZIVNAYA, 4766
PORODA FOSFATNAYA, 3402
PORODA GORNAYA, 3881
PORODA KARBONATNAYA, 650
PORODA KORENNAYA, 457
PORODA MAGMATICHESKAYA, 2176
PORODA MATERINSKAYA, 4186

PORODA METAMORPHICHESKAYA, 2785
PORODA NEFTENOSNAYA, 3826
PORODA OSADOCHNAYA, 4000
PORODA PLASTINCHATAYA, 4129
PORODA PUSTAYA, 1680
PORODA RYKHLAYA, 4647
PORODA SHCHELOCHNAYA, 1526
PORODA SUPRAKRUSTAL'NAYA, 4362
PORODA VMESHCHAYUSHCHAYA, 2062
PORODA VMESHCHAYUSHCHAYA OBRAMLYENIYA, 978
PORODA ZELENOKAMENNAYA, 1894
PORODA ZHELEZISTAYA, 2312
POROG PODVODNYJ, 2687
PORT, 1954
PORTLAND, 3570
PORTSELLANIT, 3561
POSLELEDNIKOVYJ, 3573
POSTMAGMATICHESKIJ, 3574
POSTOYANNAYA DIEHLEKTRICHESKAYA, 1174
POSTOYANNAYA GRAVITATSIONNAYA, 1876
POSTOYANNAYA RASPADA, 1079
POSTROJKA ORGANOGENNAYA, 3175
POSTTEKTONICHESKIJ, 3575
POSTULAT KOSMOLOGICHESKIJ, 974
POTASH, 3577
POTENTSIAL EHLEKTRICHESKIJ, 1325
POTENTSIOMETR, 3584
POTOK, 1575
POTOK GRYAZEVOJ, 2956
POTOK KRITICHESKIJ, 994
POTOK LAMINARNYJ, 2420
POTOK LAVOVYJ, 2457
POTOK MNOGOFAZNYJ, 2964
POTOK MUT'EVOJ, 4617
POTOK NANOSOV, 2544
POTOK OBEZGLAVLENNYJ, 458
POTOK PEPLOVYJ, 332
POTOK PEREKHVACHENNYJ, 637
POTOK RECHNOJ, 4284
POTOK STREMITEL'NYJ, 4549
POTOK TEPLOVOJ, 1967
POTOK VIKHREVOJ, 1304
POTOK VREMENNYJ, 1397
POTOK VYTEKAYUSHCHIJ, 1310
POTREBLENIE, 912
POVERKHNOST' ABRAZIONNAYA, 8
POVERKHNOST' EHROZIONNAYA, 1429
POVERKHNOST' NAPLASTOVANIYA, 454
POVERKHNOST' P'EZOMETRICHESKAYA, 3583
POVERKHNOST' RAZDELA, 2261
POYAS KOL'TSEVOJ OKEANICHESKIJ, 1890
POYAS METAMORFICHESKIJ, 2780
POYAS OROGENICHESKIJ, 3184
POYAS PODVIZHNYJ, 2899
POYAS SKLADCHATYJ, 1604
POYAS ZELENOKAMENNYJ, 1895
POZDNIJ DIAGENEZ, 2438

PRAECARDIOIDA, 3590
PRAVILO FAZ, 3394
PRAVILO FAZ MINERALOGICHESKIKH, 2879
PRAZEODIM, 3591
PREDEL CHUVSTVITEL'NOSTI, 4519
PREDEL TEKUCHESTI, 4849
PREDEL UPRUGOSTI, 1318
PREDELPLITY, 3497
PREDGOR'E, 3439
PREDPRIYATIE, 858
PREDSKAZANIE, 3598
PREDVARITEL'NYJ POISK, 3178
PREFERRED ORIENTATION, 3600
PRELOMLENIE, 3788
PRENEANDERTHALIAN, 3604
PRESNOVODNYJ, 1648
PRESSURE SOLUTION, 3608
PRIBREZH'E, 3102
PRIBREZHNO—MORSKOJ, 2688
PRIBYL', 3629
PRILIV, 4529
PRILIVNO—OTLIVNYJ, 2286
PRILIVNYJ, 4526
PRILIVY ZEMLI, 1280
PRIMARY STRUCTURE, 3612
PRIMATES, 3613
PRIMES', 2196
PRIMITIVNYJ, 3614
PRIORITET, 3615
PRIPODNYATYJ, 4685
PRITOCHNIJ, 77
PRITOK, 888
PRIZMATICHESKIJ, 3616
PRIZNAK POISKOVYJ PRYAMOJ, 4086
PROBA VALOVAYA, 4824
PROBLEMATIKI, 3619
PROBOOTBORNIK, 3934
PROBOSCIDEA, 3620
PROCHNOST', 4289
PROCHNOST' NA SZHATIE, 866
PROCHORDATA, 3623
PRODUCTION CONTROL, 3625
PRODUKT DOCHERNIJ, 1074
PRODUKT KONECHNYJ, 1367
PRODUKT RODITEL'SKIJ, 3303
PROEKT, 3636
PROEKT GLUBOKOVODNOGO BURENIYA, 3087
PROFIL' GEOFIZICHESKIJ, 1756
PROFIL' GEOKHIMICHESKIJ, 1720
PROFIL' POCHVY, 4155
PROFIL' PRODOL'NYJ, 2566
PROFIL' RAVNOVESIYA, 3628
PROGANOSAURIA, 3630
PROGIB KRAEVOJ, 1612
PROGNOZ, 3631
PROGNOZ METALLOGENICHESKIJ, 3599
PROGRADE METAMORPHISM, 3633

PROGRAMMA, 3634
PROGRAMMIROVANIE, 872
PROISKHOZHDENIE, 3179
PROIZVODNAYA, 1125
PROLIV, 4269
PROMETIJ, 3637
PROMYSHLENNOST', 2222
PROMYSHLENNOST' KHIMICHESKAYA, 728
PROMYSHLENNOST' STALELITEJNAYA, 2309
PRONITSAEMOST', 3363
PRONITSAEMOST' MAGNITNAYA, 2639
PROPILITIZATSIYA, 3639
PRORASTANIE, 2267
PROSACHIVANIE, 3350
PROSADKA, 2426
PROSIMII, 3640
PROSLOJ, 2268
PROSTIRANIE, 4292
PROSTRANSTVO MEZHZVEZDNOE, 2285
PROSTRANSTVO PODZEMNOE, 4651
PROTAKTINIJ, 3642
PROTECTED ZONE, 3644
PROTEIN, 3645
PROTELYTROPTERA, 3646
PROTEROZOIDY, 3648
PROTEROZOIDY POZDNIE, 2439
PROTEROZOIDY RANNIE, 1273
PROTEROZOJ, 3647
PROTEROZOJ NIZHNIJ, 2592
PROTEROZOJ SREDNIJ, 2851
PROTEROZOJ VERKHNIJ, 4682
PROTOCILIATA, 3649
PROTOLITH, 3650
PROTOMEDUSAE, 3651
PROTON, 3652
PROTOTIP, 3653
PROTOZOA, 3654
PROTSELLYARIJ, 3621
PROTSESS, 3622
PROTSESS GIDROTERMAL'NYJ, 2133
PROTURA, 3655
PROVINTSII I ZONY, 3657
PROVINTSIYA FAUNISTICHESKAYA, 1518
PROVINTSIYA FIZIKO—GEOGRAFICHESKAYA, 3431
PROVINTSIYA FLORISTICHESKAYA, 1572
PROVINTSIYA GIDROGEOLOGICHESKAYA, 2112
PROVINTSIYA METALLOGENICHESKAYA, 2773
PROVINTSIYA PALEOBOTANICHESKAYA, 3244
PROVINTSIYA PETROGRAFICHESKAYA, 3377
PROVODIMOST', 883
PSAMMENT, 3658
PSAMMIT, 3659
PSAMMITOVYJ, 3660
PSAMMOFAUNA, 3661
PSEFITOVYJ, 3662
PSEVDOGLEJ, 3664
PSEVDOKONGLOMERAT, 3663

PSEVDOMORFOZA, 3665
PSEVDOTILLIT, 3666
PSIKHROFIT, 3669
PSILOCERATIDA, 3667
PSILOPSIDA, 3668
PTERIDOPHYLLA, 3670
PTERIDOPHYTA, 3671
PTERIDOSPERMIDAE, 3672
PTERIOIDA, 3673
PTEROBRANCHIA, 3674
PTEROMORPHIA, 3675
PTEROPODA, 3676
PTEROPSIDA, 1543
PTEROSAURIA, 3677
PTYCHOPARIIDA, 3678
PUASSONA KOEHFFITSIENT, 3532
PUASSONA RASPREDELENIE, 3531
PUBLIC WORKS, 3680
PUBLIKATSIYA, 3681
PULMONATA, 3682
PURIN, 3688
PUSTYNYA, 1131
PUT' VODNYJ, 4804
PUTEVODITEL', 1926
PUTSTSOLAN, 3589
PYATNISTOST', 2949
PYCNOGONIDA, 3690
PYL', 1263
PYL' KOSMICHESKAYA, 967
PYL' RADIOAKTIVNAYA, 3726
PYL'TSA, 3539
PYROTHERIA, 3701
PYRROPHYTA, 3705
RABOTY ZEMLYANYE, 1281
RADAR, 3718
RADIAL'NYJ, 3719
RADIATSIYA, 3720
RADIATSIYA ADAPTIVNAYA, 64
RADIATSIYA KOSMICHESKAYA, 970
RADIATSIYA SOLNECHNAYA, 4159
RADIJ, 3739
RADIOAKTIVNOST', 3731
RADIOINTERFEROMETRIYA, 3722
RADIOLARIA, 3735
RADIOLYARIT, 3736
RADIOMETR, 3737
RADIOSPEKTROMETRIYA, 3723
RADIOUGLEROD, 3733
RADIUS, 3740
RADIUS GIDRAVLICHESKIJ, 2097
RADON, 3741
RAJON NEFTEGAZONOSNYJ, 3105
RAKOVINA, 4071
RAKUSHNYAK, 948
RANKER, 3749
RAPA, 578
RAPAKIVI STRUKTURA, 3750

RASCHLENENIE, 4320
RASKHOD, 1209
RASPAD METAMIKTNYJ, 2779
RASPAD RADIOAKTIVNYJ, 3725
RASPAD TVERDOGO RASTVORA, 1490
RASPLAV, 2743
RASPREDELENIE, 1230
RASPREDELENIE BINOMIAL'NOE, 480
RASPREDELENIE LOGARIFMICH.=NORM., 2563
RASPREDELENIE NORMAL'NOE, 3055
RASPREDELENIE STATISTICHESKOE, 4240
RASPREDELENIE VEROYATNOSTEJ, 1645
RASPROSTRANENIE, 3746
RASPROSTRANENIE VOLNY, 4807
RASPROSTRANENNOST', 16
RASSHCHEPLENIE, 4189
RASSHIRENIE DNA OKEANA, 3980
RASSOL GORYACHIJ, 2064
RASSOLONENIE, 1129
RASSTANOVKA, 306
RASTENIE, 3491
RASTITEL'NOST', 4724
RASTRESKIVANIE, 1086
RASTVOR, 4176
RASTVOR GIDROTERMAL'NYJ, 2134
RASTVOR TVERDYJ, 4167
RASTVOR PERESYSHCHENNYJ, 3217
RASTVORENIE, 1228
RASTVORIMOST', 4174
RASTYAZHENIE, 4456
RATITES, 3754
RATSEMIZATSIYA, 3717
RAVNINA, 3465
RAVNINA ABISSAL'NAYA, 22
RAVNINA AKKUMULYATIVNAYA, 33
RAVNINA ALLYUVIAL'NAYA, 144
RAVNINA BEREGOVAYA, 2683
RAVNINA PREDGORNAYA, 405
RAVNINA PRIBREZHNAYA, 815
RAVNOVESIE, 1416
RAVNOVESIE DINAMICHESKOE, 1264
RAVNOVESIE FAZOVOE, 3392
RAVNOVESIE KHIMICHESKOE, 724
RAZBAVLENIE, 1188
RAZBUKHANIE, 4378
RAZDEL VODO-NEFTYANOJ, 3114
RAZGRUZKA PODZEMNYKH VOD, 1914
RAZLOM, 1507
RAZLOM DUGOOBRAZNYJ, 289
RAZLOM GLUBINNYJ, 2511
RAZLOM GORIZONTAL'NYJ, 2054
RAZLOM TRANSFORMNYJ, 4569
RAZLOM ZHIVUSHCHIJ, 56
RAZMAGNICHIVANIE, 1106
RAZMAGNICHIVANIE TERMICHESKOE, 4496
RAZMYV PLYAZHA, 813
RAZNOVREMENNYJ, 1990

RAZRABOTKA OBRUSHENIEM, 680
RAZRABOTKA OTKRYTAYA, 3129
RAZRABOTKA RUDNYKH MESTOROZHDENIJ, 2883
RAZREZ GEOLOGICHESKIJ, 1736
RAZREZ GEOLOGICHESKIJ SKVAZHINY, 550
RAZREZ GORNOJ VYRABOTKI, 3460
RAZREZ SKOROSTNOJ, 4727
RAZREZ TIPOVOJ, 4625
RAZRUSHENIE, 3917
RAZRYV KONSEDIMENTATSIONNYJ, 1921
RAZRYV SHARNIRNYJ, 2015
RAZRYV SOGLASNYJ, 456
RAZRYV STUPENCHATYJ, 4248
RAZVEDKA, 1483
RAZVEDKA ALLYUVIAL'NAYA, 145
RAZVEDKA GEOFIZICHESKAYA, 1752
RAZVEDKA MESTOROZHDENIJ, 2868
RAZVEDKA PODZEMNYKH VOD, 1915
RAZVEDKA RADIOMETRICHESKAYA, 3738
RAZVITIE, 1147
RAZZHIZHENIE, 2519
REAGENT, 3761
REAKTIVIZATSIYA, 3760
REAKTSIYA, 3757
REAKTSIYA EHKZOTERMICHESKAYA, 1473
REAKTSIYA EHNDOTERMICHESKA, 1372
REAKTSIYA FOTOKHIMICHESKAYA, 3408
RECEPTACULITIDA, 3763
RECHNOJ STOK TVERDYJ, 3874
RECLAMATION, 3766
REDLICHIIDA, 3775
REEF ENVIRONMENT, 3777
REEF SEDIMENTATION, 3778
REEFBUILDER, 3779
REFERAT /KNIGI/, 541
REFINEMENT, 3780
REFINING, 3781
REFRAKTOMETR, 3789
REFRAKTOMETRIYA, 3790
REFRAKTSIYA, 3786
REGENERATSIYA, 3792
REGOLIT, 3798
REGOSOL, 3799
REGRESSIYA, 3800
REGRESSIYA (MAT), 3801
REHLEYA VOLNA, 3756
REHT, 3848
REKA, 3873
REKA KONSEKVENTNAYA, 904
REKA PERESYKHAYUSHCHAYA, 2272
REKA PODZEMNAYA, 4653
REKA VREZANNAYA, 1384
REKONSTRUKTISIYA, 3767
REKONSTRUKTSIYA PALEONTOLOGICHESKAYA, 3264
REL'EF, 3808
REL'EF LUNNYJ, 2937
REL'EF MERZLOTNYJ, 1654

REL'EF STRUKTURNYJ, 4310
RELAKSATSIYA, 3805
RELIEF EXPOSURE, 3809
RELIKT, 3806
RELIKTOVYJ, 3807
REMAGNETIZATSIYA, 3811
RENDOLL, 3815
RENDZINA, 3816
RENIJ, 3849
RENTABEL'NOST', 1481
RENTGENOGRAMMA, 4834
REOLOGIYA, 3851
REOMORFIZM, 3852
REPLENISHMENT (SPELEO), 2374
REPRESENTATIVE BASIN, 3819
REPRODUKTSIYA, 3820
REPTILIA, 3821
RESEKVENTNYJ, 3823
RESHETCHATYJ, 2452
RESHETKA BRAVE, 574
RESHETKA KRISTALLICHESKAYA, 1018
RESURGENTSIYA, 3836
RESURSY PRIRODNYE, 2992
RESURSY VODNYE, 4798
RETENTION, 3838
RETICULAREA, 3839
RETSENZIYA, 995
RETSIRKULYATSIYA, 3772
REVERSAL, 3842
REVIZIYA, 3844
REZERV, 3824
REZONANS MAGNITNYJ YADERNYJ, 3064
REZONANS PARAMAGNITNYJ EHLEKTRONNYJ, 1351
RHABDOPHLEURIDA, 3847
RHIPIDISTII, 3853
RHIZOPODEA, 3854
RHODOPHYTA, 3856
RHYNCHOCEPHALIA, 3858
RHYNCHONELLIDA, 3859
RIF, 3776
RIF BAR'ERNYJ, 420
RIF KORALLOVYJ, 949
RIF OKAJMLYAYUSHCHIJ, 1651
RIF TYLOVOJ, 397
RIF VODOROSLEVYJ, 117
RIFOVYJ PIK, 3451
RIFT, 3864
RIGEL', 3863
RILL, 3867
RIODATSIT, 3860
RIOLIT, 3861
ROBERTINACEA, 3880
ROCK BAR, 3882
RODENTIA, 3892
RODIJ, 3855
RODSTVO, 76
RODSTVO GENETICHESKOE, 903

ROGOVIK, 2057
ROJ ZEMLETRYASENIJ, 1287
ROLL=TYPE DEPOSIT, 3893
ROSA, 1152
ROSSYP', 3463
ROSSYP' BEREGOVAYA, 445
ROST KRISTALLA, 1017
ROSTROCONCHIA, 3897
ROTALIACEA, 3898
ROTALIINA, 3899
RTUT', 2749
RUBEFACTION, 3908
RUBIDIJ, 3909
RUCHEJ, 985
RUDA, 3151
RUDA BOBOVAYA, 449
RUDA BOLOTNAYA, 537
RUDIT, 3912
RUDNYE ZHILY, 2559
RUDNYJ, 3159
RUGOSA, 4474
RUMINANTIA, 3913
RUSLO, 707
RUSLO POGREBENNOE, 598
RUSLO VETVYASHCHEESYA, 571
RUTENIJ, 3919
RYAD REAKTSIONNYJ, 3759
RYAD REAKTSIONNYJ BOUEHNA, 563
RYKHLYJ, 2201
RYUPEL', 3916
S'EMKA, 4371
S'EMKA AEHROMAGNITNAYA, 74
S'EMKA AKUSTICHESKAYA, 44
S'EMKA EHLEKTRICHESKAYA, 1333
S'EMKA EHLEKTROMAGNITNAYA, 1342
S'EMKA GEOFIZICHESKAYA, 1757
S'EMKA GIDROGEOLOGICHESKAYA, 2113
S'EMKA GRAVIMETRICHESKAYA, 1886
S'EMKA INZHENERNO=GEOLOGICHESKAYA, 1770
S'EMKA MAGNITNAYA, 2643
S'EMKA MAGNITOTELLURICHESKAYA, 2655
S'EMKA SEJSMICHESKAYA, 4019
SAKHAR, 4347
SAKMAR, 3921
SAKSONIJ, 3954
SALT MARSH, 3928
SAL'TATSIYA, 3931
SAMARIJ, 3932
SAMORODOK, 3070
SAND BODY, 3937
SANTON, 3943
SAPROPEL', 3944
SARCODINA, 3946
SARCOPTERYGII, 3947
SARMAT, 3948
SATURN, 3478
SAURISCHIA, 3951

SAURORNITHES, 3952
SBOR, 838
SBOR DANNYKH, 1065
SBRASYVATEL', 1510
SBROS, 3056
SBROS KOSOJ, 3077
SBROS POPERECHNYJ, 1196
SBROSO=SDVIG, 4293
SBROSO=SDVIG REGIONAL'NYJ, 4567
SCAPHOPODA, 3958
SCHELOCHNOJ SIENIT, 127
SCIUROMORPHA, 3965
SCYPHOMEDUSAE, 3972
SCYPHOZOA, 3973
SDVIG, 2442
SDVIG (MEKHANIKA), 4058
SEBKHA ENVIRONMENT, 3987
SECONDARY SEDIMENTARY STRUCTURE, 3990
SEDIMENT YIELD, 3993
SEDIMENT=WATER INTERFACE, 3994
SEDIMENTATSIYA, 4003
SEDIMENTATSIYA ABISSAL'NAYA, 23
SEDIMENTATSIYA BEREGOVAYA, 2546
SEDIMENTATSIYA BIOKHIMICHESKAYA, 482
SEDIMENTATSIYA BIOKLASTICHESKAYA, 488
SEDIMENTATSIYA DEL'TOVAYA, 1105
SEDIMENTATSIYA DETRITOVAYA, 1144
SEDIMENTATSIYA EHOLOVAYA, 1393
SEDIMENTATSIYA EHSTUARIEVAYA, 1437
SEDIMENTATSIYA FLYUVIOGLATSIAL'NAYA, 1597
SEDIMENTATSIYA LAGUNNAYA, 2413
SEDIMENTATSIYA MORSKAYA, 2686
SEDIMENTATSIYA OZERNAYA, 2409
SEDIMENTATSIYA RECHNAYA, 1595
SEDIMENTATSIYA SHEL'FOVAYA, 4070
SEDIMENTOGENEZ, 4005
SEDIMENTOLOGIYA, 4006
SEEPAGE, 4008
SEGREGATSIYA, 4010
SEISMIC SPECTRAL ANALYSIS, 4017
SEJSHI, 4011
SEJSMICHNOST', 4022
SEJSMOGRAF, 4024
SEJSMOGRAF OKEANICHESKIJ, 3089
SEJSMOGRAMMA, 4023
SEJSMOLOGIYA, 4026
SEJSMOLOGIYA INZHENERNAYA, 1285
SEJSMOSTANTSIYA, 4018
SEJSMOTEKTONIKA, 4028
SEKRETSIYA LATERAL'NAYA, 2445
SEL', 1077
SEL'SKOE KHOZYAJSTVO, 90
SELEN, 4030
SELENAT, 4029
SELENOKHRONOLOGIYA, 4031
SELENOLOGIYA, 4032
SEM DATA, 4034

SEMI=ARID ENVIRONMENT, 4035
SEMYA, 4007
SENOMAN, 686
SENON, 4037
SEPARATSIYA, 4039
SEPTA, 4041
SEPTARIYA, 4040
SERA, 4353
SEREBRO, 4111
SERITSITIZATSIYA, 4043
SERIYA, 4044
SEROVODOROD, 2108
SEROZEM, 4095
SERPENTIN, 4045
SERPENTINIT, 4046
SERPENTINIZATSIYA, 4047
SET', 3013
SET' RECHNAYA, 1246
SET' RECHNAYA DREVOVIDNAYA, 1109
SETCHATYJ, 3840
SEVER, 3057
SFENOLIT, 4205
SFERICHNOST', 4207
SFEROLIT, 4209
SFERULA, 4208
SGLAZHIVANIE, 4141
SH-VOLNA, 4050
SHAPKA GAZOVAYA, 1685
SHAPKA LEDYANAYA, 2156
SHAR'YAZH, 3218
SHARIK GLINISTYJ, 788
SHARIKI MAGNITNYE, 2641
SHASSIN'IT, 714
SHCHEBEN', 3907
SHCHELOCHNOJ, 128
SHCHELOCHNOJ GRANIT, 125
SHCHELOCHNOST', 131
SHCHIT, 4072
SHEL'F, 4067
SHEL'F KONTINENTAL'NYJ, 929
SHEROKHOVATOST', 3905
SHIP, 4211
SHIROTA, 2450
SHKALA FI, 3397
SHKALA RICHTERA, 3862
SHKALA VREMENNAYA, 4536
SHKALA ZEMLETRYASENIJ, 2905
SHLAK, 769
SHLAK VULKANICHESKIJ, 3969
SHLAM BUROVOJ, 1256
SHLEJF, 250
SHLIF, 4514
SHLIKHOVANIE, 3283
SHLIR, 3962
SHLYAPA ZHELEZNAYA, 1838
SHONKINIT, 4080
SHOSSE, 3879

SHPAT POLEVOJ, 1524
SHTOK, 4257
SHTOKVERK, 4258
SHTORM, 4266
SHTORMOVOJ NAGON, 4370
SHTRIKHOVKA, 3970
SHTRIKHOVKA LEDNIKOVAYA, 1800
SHTRIKHOVKA MINERALA, 2874
SHUM, 3041
SHURF, 3459
SIAL', 4089
SIDEROFIL'NYJ, 4092
SIDEROFIR, 4093
SIDEROLIT, 4263
SIENIT, 4379
SIFON (GIDROGEOL), 4120
SIL'VINIT, 4380
SILA KOEHRTSITIVNAYA, 825
SILA RABOCHAYA, 2664
SILA TYAZHESTI ABSOLYUTNAYA, 11
SILA ZEMLETRYASENIYA, 1286
SILICOFLAGELLATA, 4102
SILIKAT, 4098
SILL, 4104
SILLIMANIT, 4105
SILUR, 4110
SILUR NIZHNIJ, 2593
SILUR VERKHNIJ, 4683
SIMA, 4112
SIMBIOZ, 4381
SIMIAN, 4113
SIMMETRIYA, 4384
SIMPLEKTITOVYJ, 4386
SIMPOZIUM, 4387
SIMULATSIYA, 4114
SIMVOL KRISTALLICHESKOJ GRANI, 2860
SINEKLIZA, 4392
SINEMYUR, 4115
SINGENETICHESKIJ, 4395
SINGENEZ, 4394
SINGONIYA, 1021
SINGONIYA GEKSAGONAL'NAYA, 2002
SINGONIYA KUBICHESKAYA, 1030
SINGONIYA MONOKLINNAYA, 2921
SINGONIYA ORTOROMBICHESKAYA, 3194
SINGONIYA ROMBICHESKAYA, 3857
SINGONIYA TETRAGONAL'NAYA, 3693
SINGONIYA TRIGONAL'NAYA, 4595
SINGONIYA TRIKLINNAYA, 4593
SINKLINAL', 4390
SINKLINORIJ, 4391
SINSEDIMENTATSIONNYJ, 4397
SINTEKSIS, 4400
SINTEKTONICHESKIJ, 4399
SINTEZ, 4401
SINUOSITY, 4119
SIPHONAPTERA, 4121

SIPUNCULOIDA, 4122
SIRENIA, 4123
SISTEMA, 4403
SISTEMA BINARNAYA, 479
SISTEMA CHETYREKHKOMPONENTNAYA, 3714
SISTEMA DAEK, 1185
SISTEMA INFORMATSIONNAYA, 2229
SISTEMA IZOMETRICHESKAYA, 2332
SISTEMA KHIMICHESKAYA, 733
SISTEMA NERVNAYA, 3012
SISTEMA SKLADCHATAYA, 1606
SISTEMA SOLNECHNAYA, 4160
SISTEMA TREKHKOMPONENTNAYA, 4463
SITO, 4096
SITSILIJSKIJ, 4090
SKANDIJ, 3956
SKAPOLITIZATSIYA, 3959
SKARN, 4126
SKELET, 4127
SKLADCHATOST' SRYVA, 1082
SKLADKA, 1602
SKLADKA ASIMMETRICHNAYA, 353
SKLADKA CHEKHLA, 1248
SKLADKA DISGARMONICHNAYA, 1219
SKLADKA EHNTEROLITICHESKAYA, 1380
SKLADKA GRAVITATSIONNAYA, 1881
SKLADKA KASKADNAYA, 661
SKLADKA KONTSENTRICHESKAYA, 874
SKLADKA LEZHACHAYA, 3771
SKLADKA NALOZHENNAYA, 4359
SKLADKA NORMAL'NAYA, 4382
SKLADKA OPROKINUTAYA, 3219
SKLADKA PRIRAZLOMNAYA, 1220
SKLADKA PTIGMATITOVAYA, 3679
SKLADKA SOPRYAZHENNAYA, 897
SKLADKA STREL'CHATAYA, 741
SKLADKA STUPENCHATAYA, 4249
SKLADKA TECHENIYA, 1577
SKLADKA TSILINDRICHESKAYA, 1052
SKLADKA VOLOCHENIYA, 1243
SKLADKA ZIGZAGOOBRAZNAYA, 4859
SKLERIT, 3966
SKLON, 4134
SKLON KONSEKVENTNYJ, 1198
SKLON KONTINENTAL'NYJ, 930
SKLON OSYPENIYA, 4417
SKLONENIE, 1081
SKOLEKODONTY, 3967
SKOLIT (PALEONT), 3968
SKOROST', 4726
SKOROST' OSADKONAKOPLENIYA, 4004
SKOROST' SEJSMICHESKAYA, 4020
SKRYTOKRISTALLICHESKIJ, 1006
SKUL'PTURA, 3182
SKVAZHINA, 549
SKVAZHINA ARTEZIANSKAYA, 314
SKVAZHINA BUROVAYA, 4813

SKVAZHINA EHKSPLUATATSIONNAYA, 3627
SKVAZHINA GAZOVAYA, 1692
SKVAZHINA MORSKAYA, 3104
SKVAZHINA NA VODU, 4801
SKVAZHINA NEFTYANAYA, 3112
SKVAZHINA POISKOVAYA, 1484
SKVAZHINA VZRYVNAYA, 4085
SLANETS, 4130
SLANETS BITUMINOZNYJ, 524
SLANETS GLINISTYJ, 4051
SLANETS GOLUBOJ, 534
SLANETS GORYUCHIJ, 3110
SLANETS KHLORITOVYJ, 751
SLANETS KREMNISTYJ, 739
SLANETS KRISTALLICHESKIJ, 3960
SLANETS PYATNISTYJ, 4222
SLANETS SERITSITOVYJ, 4042
SLANETS TAL'KOVYJ, 4415
SLANETS ZELENYJ, 1892
SLANTSEVATOST', 3961
SLED DVIZHENIYA, 4565
SLED PALEOKLIMATA, 796
SLED PITANYA, 1523
SLED ZHIZNEDEYATEL'NOSTI, 4557
SLEPOJ, 529
SLEPOK, 665
SLOISTOST', 453
SLOISTOST' KOSAYA, 998
SLOISTOST' PARALLEL'NAYA, 3295
SLOISTOST' RITMICHNAYA, 1841
SLOJ, 2462
SLOJ AKTIVNYJ, 57
SLOJ BAZAL'TOVYJ, 426
SLOJ GODICHNYJ, 4720
SLOJ GOLOVNOJ, 4548
SLOJ GRANITNYJ, 1849
SLOJ NEFELOIDNYJ, 3009
SLOJ OBRATNYJ, 399
SLOJ PEREDOVOJ, 1614
SLOJ POGRANICHNYJ, 561
SLOJ PONIZHENNYKH SKOROSTEJ, 2576
SLOJ PRIDONNYJ, 555
SLOJCHATOST' KONVOLYUTNAYA, 941
SLOJCHATOST' KOSAYA, 997
SLOVAR', 1824
SLUCHAJNYJ PROTSESS, 4256
SLUZHBA GEOLOGICHESKAYA, 1739
SLYUDA, 2804
SLYUDYANOJ SLANETS, 2805
SMESHCHENIE, 1224
SMESHCHENIE GORIZONTAL'NOE, 2053
SMESHIVAEMOST', 2893
SMESTITEL', 4061
SMOLA, 3831
SMYV PLOSKOSTNOJ, 3744
SNEG, 4142
SNIMOK RENTGENOVSKIJ, 3734

SODERZHANIE, 3156
SODERZHANIE BORTOVOE, 1038
SOEDINENIE ORGANICHESKOE, 3167
SOFT SEDIMENT DEFORMATION, 4145
SOGLASHENIE MEZHDUNARODNOE, 2276
SOGLASNYJ, 877
SOKHRANNOST', 3606
SOKRASHCHENIE, 4084
SOL', 3924
SOL'FATARA, 4165
SOLE INJECTION, 4162
SOLE MARK, 4163
SOLEMYOIDA, 4164
SOLENOSNYJ, 3922
SOLENOST', 3923
SOLIDUS, 4170
SOLIFLYUKTSIYA, 4171
SOLNTSE, 4358
SOLONCHAK, 4172
SOLONETS, 4173
SOLONOVATOVODNYJ, 569
SOLUTE, 4175
SOLUTION CAVITY, 4177
SOMASTEROIDEA, 4180
SONAR, 4181
SONOBUJ, 4182
SOOBSHCHESTVO, 855
SOORUZHENIE, 1375
SOORUZHENIE INZHENERNOE, 910
SOOTNOSHENIE (KHIM), 730
SOPROTIVLENIE, 3832
SOPROTIVLENIE RASTYAZHENIYU, 4455
SOPROTIVLENIE SDVIGU, 4062
SOPRYAZHENNYJ, 896
SORBTSIYA, 4183
SOSSYURITIZATSIYA, 3953
SOSTAV, 863
SOSTAV NORMATIVNYJ, 3054
SOSTOYANIE (FIZICHESKOE), 3428
SOTRUDNICHESTVO TEKHNICHESKOE, 4430
SOVESHCHANIE, 2737
SOVOKUPNOST', 338
SOVOKUPNOST' GENERAL'NAYA, 4243
SPAJNOST', 2864
SPARAGMIT, 4190
SPARKER, 4191
SPEKTR, 4202
SPEKTR EHPR, 1410
SPEKTR MESSBAUEHROVSKIJ, 2948
SPEKTR OPTICHESKIJ, 3143
SPEKTR POGLOSHCHENIYA, 15
SPEKTROFOTOMETRIYA, 4200
SPEKTROMETR, 4199
SPEKTROMETRIYA MESSBAUEHROVSKAYA, 2947
SPEKTROMETRIYA MIKROVOLNOVAYA, 2839
SPEKTROMETRIYA OPTICHESKAYA, 3142
SPEKTROMETRIYA RAMANOVSKAYA, 3745

SPEKTROSKOPIYA, 4201
SPEKTROSKOPIYA ABSORBTSIONNAYA, 14
SPEKTROSKOPIYA INFRAKRASNAYA, 2234
SPEKTROSKOPIYA MOLEKULYARNAYA, 2914
SPELEOLOGIYA, 4203
SPELEOTHEM, 4204
SPHERICAL HARMONIC ANALYSIS, 4206
SPILITIZATSIYA, 4210
SPINIFEKS, 4212
SPIRIFERIDA, 4213
SPIRILLINACEA, 4214
SPISOK FAUNY, 1517
SPLAV, 139
SPODOSOL, 4216
SPONGOLIT, 4218
SPORA, 4219
SPOROMORFA, 4220
SPOROZOA, 4221
SPOSOBNOST' NESUSHCHAYA, 450
SPOSOBNOST' OBMENNAYA, 1464
SPOSOBNOST' OTRAZHATEL'NAYA, 3785
SPRAVOCHNIK, 2667
SPREADING CENTER, 4223
SPUTNIK, 322
SPUTNIK (ESTESTV), 3484
SQUAMATA, 4225
SREDA EHOLOVAYA, 1391
SREDA ESTESTVENNAYA, 1937
SREDA LAGUNNAYA, 2412
SREDA OKRUZHAYUSHCHAYA, 1386
SREDA NAZEMNAYA, 4316
SREDA SKLONA KONTINENTAL'NOGO, 4136
SREDA SPOKOJNAYA, 2574
SREDA SUBLITORAL'NAYA, 4325
SREDA VOLNENNAYA, 2009
SREDNE=VZVESHENNYJE, 4812
SREDNEZERNISTYJ, 2736
SREDNIJ (SOSTAV), 2269
STABIL'NOST', 4226
STABILIZATSIYA, 4227
STADIYA LEDNIKOVAYA, 1799
STALAGMIT, 4232
STALAKTIT, 4231
STAMP., 4233
STANDARD ROCK, 4237
STANDART, 4235
STANDARTA METOD, 4236
STANDARTIZATSIYA, 4238
STANTSIYA LUNNAYA, 2604
STATISTIKA, 4245
STATISTIKA MNOGOMERNAYA, 2968
STEFAN, 4250
STEKHIOMETRIYA, 4259
STEKLO, 1811
STEKLO PRIRODNOE, 2990
STEKLO VULKANICHESKOE, 4764
STEP', 4251

STEPEN', 3753
STEPEN' METAMORFIZMA, 2784
STEPEN' SVOBODY, 1100
STEREOKHIMIYA, 4252
STEROL, 4253
STICK=SLIP, 4254
STILOLIT, 4314
STOIMOST', 976
STOK, 3915
STOK PODZEMNYJ, 1918
STOK POVERKHNOSTNYJ, 4364
STOK TVERDYJ, 452
STOK VZVESHENNYJ, 4372
STOLB RUDNYJ, 3161
STOLKNOVENIE PLIT, 3496
STOLONOIDEA, 4260
STONE LINE, 4262
STOYANKA ARKHEOLOGICHESKAYA, 280
STRATEGIYA, 4271
STRATIFIKATSIYA, 4272
STRATIFITSIROVANNYJ, 4273
STRATIGRAFIYA, 4281
STRATIGRAFIYA PALEOMAGNITNAYA, 2652
STRATOTIP, 4282
STRATOVULKAN, 4283
STREAM ORDER, 4286
STREAM SEDIMENT, 4287
STREAM TRANSPORT, 4288
STRESS, 4290
STRESS=MINERAL, 4291
STROENIE, 586
STROENIE GLUBINNOE, 4169
STROENIE GLUBINNOE (ZEMLI), 1276
STROENIE TEKTONICHESKOE, 4438
STROENIE VNUTRENNEE, 2275
STROITEL'NOE SYR'E, 911
STROITEL'STVO GRAZHDANSKOE, 777
STROJMATERIALY, 588
STROMATOLITHI, 4295
STROMATOPOROIDEA, 4296
STRONG MOTION, 4298
STRONTSIJ, 4299
STROPHOMENIDA, 4300
STRUCTURE CONTOUR MAP, 4313
STRUKTURA, 4477
STRUKTURA BIOGENNAYA, 494
STRUKTURA DOMENNAYA, 1239
STRUKTURA IMPAKTNAYA, 2192
STRUKTURA KOL'TSEVAYA, 3871
STRUKTURA KONSEDIMENTATSIONNAYA, 4398
STRUKTURA KRIPTOVULKANICHESKAYA, 1010
STRUKTURA KRISTALLICHESKAYA, 1020
STRUKTURA MULLION, 2961
STRUKTURA OBRUSHENIYA, 837
STRUKTURA OSADOCHNAYA, 4001
STRUKTURA POLOZHITEL'NAYA, 4665
STRUKTURA RAVNOMERNO=ZERNISTAYA, 1415

STRUKTURA STEKLOVATAYA, 1814
STRUKTURA STEKLYANNAYA, 4750
STRUKTURA TEKTONICHESKAYA, 1737
STRUKTURA UDARA, 2188
STRUKTURA VODOROSLEVAYA, 118
STRUKTURA VULKANO=TEKTONICHESK., 4771
STRUKTURNAYA DIAGRAMMA, 4304
STSEPLENIE, 827
STSINTILLYATSIYA, 3964
STUPENCHATYJ, 1365
STVORKA, 4708
STYLOPHORA, 4315
SUBBOREAL'NYJ, 4319
SUBDUKTSIYA, 4321
SUBECONOMIC DEPOSIT, 3582
SUBLIMAT, 4323
SUBLIMATSIYA, 4324
SUBMARINE INSTALLATION, 4328
SUBSHCHELOCHNOJ, 4317
SUBSTITUTION, 4338
SUBSTRAT, 4339
SUBTIDAL ENVIRONMENT, 4341
SUBTROPICHESKIJ, 4342
SUBVULKAN, 4343
SUCTORIA, 4345
SUDETSKIJ, 4346
SUDNO BUROVOE, 1257
SUGLINOK, 2509
SUGLINOK MORENNYJ, 4530
SUIFORMES, 4348
SUL'FAT, 4349
SUL'FID, 4351
SUL'FOSOL, 4352
SULFUR=ISOTOPE RATIO, 4355
SUPRASTRUKTURA, 4439
SUR'MA, 229
SURFACE DEFECT, 4363
SURFACE WATER BUDGET, 4367
SUSHA, 2425
SUSPENZIYA, 4373
SUTURA, 4374
SVAYA, 3445
SVEKOFENSKIJ, 4375
SVERKHSTRUKTURA, 4361
SVINETS, 2466
SVOD, 270
SVOJSTVO, 3638
SVOJSTVO AKUSTICHESKOE, 42
SVOJSTVO EHLASTICHESKOE, 1320
SVOJSTVO EHLEKTRICHESKOE, 1331
SVOJSTVO EHLEKTROKHIMICHESKOE, 1336
SVOJSTVO FIZICHESKOE, 3427
SVOJSTVO FIZKHIMICHESKOE, 3430
SVOJSTVO GEOTEKHNICHESKOE, 1769
SVOJSTVO KHIMICHESKOE, 729
SVOJSTVO MAGNITNOE, 2640

SVOJSTVO MEKHANICHESKOE, 2727
SVOJSTVO OPTICHESKOE, 3140
SVOJSTVO TERMICHESKOE, 4499
SVOJSTVO TERMODINAMICHESKOE, 4504
SVOJSTVO TERMOKHIMICHESKOE, 4502
SVYAZ VOZDUKH=MORE, 94
SVYAZ', 3803
SVYAZ' ATOMNAYA, 364
SWAMP, 4376
SWAMP SEDIMENTATION, 4377
SYMMETRODONTA, 4383
SYMPHYLA, 4385
SYNAPSIDA, 4388
SYNFORM FOLD, 4393
SYNONIMIKA, 4396
SYR'E, 3755
SYR'E KERAMICHESKOE, 694
SYR'E STEKOL'NOE, 1813
SYR'E TOPLIVOE, 1658
SYR'E TSEMENTNOE, 2944
SYSTEMS ANALOG, 4405
SZHATIE, 865
SZHATIE GORIZONTAL'NOE, 2051
SZHIMAEMOST', 864
SZHIMAEMOST' GORNYKH POROD, 856
TABLITSA DANNYKH, 1072
TABULATA, 4406
TAENIODONTA, 4410
TAENIOLABOIDEA, 4411
TAFONOMIYA, 4420
TAFROGENEZ, 4421
TAFROGEOSINKLINAL', 4422
TAJGA, 4412
TAKONIT, 4408
TAKONSKIJ OROGEN, 4407
TAKSIT, 4427
TAKSON ANAEHROBNYJ, 179
TAKSON NOMINAL'NYJ, 3043
TAKSON NOVYJ, 3025
TAKSONOMIYA, 4428
TAKSONOMIYA PLANKTONA, 3488
TAKTIT, 4409
TAL'K, 4414
TAL'VEG, 4482
TALASSOID, 4479
TALASSOKRATON, 4478
TALLIJ, 4480
TALUS, 4416
TAMPONAZH, 3982
TAMPONIROVANIE SKVAZHINY, 3514
TANATOTSENOZ, 4483
TANET, 4484
TANTALAT, 4418
TANTALO=NIOBAT, 3031
TANTALUM, 4419
TARANNON, 4423
TARDIGRADA, 4424

TASMANITES, 4425
TATARSKIJ, 4426
TAYANIE, 4485
TECHENIE, 1035
TECHENIE BEREGOVOE, 2568
TECHENIE NESTATSIONARNOE, 4664
TEFROKHRONOLOGIYA, 4458
TEKHNETSIJ, 4429
TEKHNIKA, 4431
TEKHNIKA ANALOGOVAYA, 181
TEKHNIKA GEOFIZICHESKAYA, 1753
TEKHNIKA GIDROGEOLOGICHESKAYA, 2114
TEKHNIKA SEJSMICHESKAYA, 4013
TEKHNIKA VYCHISLITEL'NAYA, 1183
TEKHNOLOGICHESKIJ PEPEL, 320
TEKHNOLOGIYA, 4432
TEKSTURA MARGINATSIONNYJ, 2675
TEKSTURA PODVODNOGO OPOLZANIYA, 4138
TEKTIT, 4444
TEKTOGENEZ, 4433
TEKTONIKA, 4440
TEKTONIKA AL'PINOTIPNAYA, 151
TEKTONIKA BLOKOVAYA, 532
TEKTONIKA FUNDAMENTA, 430
TEKTONIKA GERMANOTIPNAYA, 1783
TEKTONIKA GLOBAL'NAYA, 1821
TEKTONIKA GRAVITATSIONNAYA, 1887
TEKTONIKA MNOGOFAZNAYA, 2965
TEKTONIKA PLIKATIVNAYA, 1607
TEKTONIKA PLIT, 3499
TEKTONIKA SOLYANAYA, 3929
TEKTONIKA SZHATIYA, 867
TEKTONIT, 4441
TEKTONOFISIKA, 4442
TEKTONOSFERA, 4443
TELEOSTEI, 4445
TELESEISMIC SIGNAL, 4446
TELETERMAL'NYJ, 4447
TELLUR, 4450
TELLURAT, 4448
TELO RUDNOE, 3152
TEM DATA, 4451
TEMPERATURA, 4453
TEMPERATURA NIZKAYA, 2572
TEMPERATUROPROVODNOST', 1182
TENACITY, 4454
TENTACULITES, 4457
TENZOMETR, 4268
TEORIYA, 4490
TEORIYA VEROYATNOSTEJ, 3617
TEORIYA VOSKHODYASHCHIKH RASTVOROV, 326
TEPLOEMKOST', 1963
TEPLOPERENOS, 1969
TEPLOPROVODNOST', 1964
TEPLOTA, 1962
TEPLOTA PLAVLENIYA, 1966
TEPLOTA PLAVLENIYA UDEL'NAYA, 2440

TEPLOTA UDEL'NAYA, 4197
TERALIT, 4491
TERATOLOGIYA, 4459
TERBIJ, 4460
TEREBRATULIDA, 4461
TERMAL'NYJ, 4493
TERMINOLOGIYA, 4462
TERMODINAMIKA, 4505
TERMOGRAMMA, 4506
TERMOKARST, 4508
TERMOKLIN, 4503
TERMOLYUMINESTSENTSIYA, 4509
TERMORAZVEDKA, 4500
TERRASA, 4465
TERRASA RECHNAYA, 3875
TERRIGENNYJ, 4469
TESTUDINATA, 4472
TETIS, 4473
TETRAPODA, 4475
TEXTULARIINA, 4476
TEZISY, 4512
THALLOPHYTA, 4481
THAWING, 4485
THECAMOEBA, 4486
THECIDEIDINA, 4487
THECODONTIA, 4488
THELODONTI, 4489
THERAPSIDA, 4492
THYSANOPTERA, 4523
THYSANURA, 4524
TIDAL CHANNEL, 4525
TIDAL INLET, 4528
TIKHOOKEANSKIJ, 3231
TIKHOOKEANSKIJ PODVIZHNYJ POYAS, 774
TIKSOTROPIYA, 4515
TILLIT, 4531
TILLODONTIA, 4532
TINTINNIDAE, 4538
TIP MESTOROZHDENIJ PNEVMATOLITOV, 3524
TIPOMORFIZM, 4628
TIRREN, 4629
TITAN, 4539
TITON, 4540
TOAR, 4541
TOCHKA PEREGREVA, 2065
TOCHNOST', 34
TOKI TELLURICHESKIE, 1274
TOKSICHNOST', 4555
TOLEITOVYJ, 4516
TOMBOLO, 4542
TONSHTEJN, 4543
TOPOGRAFIYA, 4547
TORF, 3320
TORGOVLYA, 4566
TORIJ, 4517
TORTON, 4551
TRACHYPSAMMIACEA, 4563

TRAKHIANDEZIT, 4561
TRAKHIBAZAL'T, 4562
TRAKHIT, 4564
TRANSFORMATSIYA, 4570
TRANSFORMIZM, 4571
TRANSGRESSIYA, 4572
TRANSMISSIBILITY COEFFICIENT, 4574
TRANSPORTIROVKA(ECON), 4578
TRANSPORTIROVKA(GEOL), 4577
TRASSER, 4560
TRASSER TSVETNOY, 848
TRAVERTIN, 4583
TRAVLENIE, 1439
TRAVOYADNYJ, 3436
TREATISE, 4584
TREK DELENIYA, 1555
TREMADOK, 4586
TREND (STATISTICHESKIJ), 4244
TREND=ANALIZ, 4588
TRENIE, 1650
TREPOSTOMATA, 4589
TRESHCHINA, 1632
TRESHCHINA KOL'TSEVAYA, 3870
TRESHCHINA KONTRAKTSIONNAYA, 1965
TRESHCHINA LEDNIKOVAYA, 990
TRESHCHINA SKALYVANIYA, 4059
TRESHCHINA USADOCHNAYA, 4088
TRESHCHINA USYKHANIYA, 2954
TRESHCHINOVATOST', 1635
TRESHCHINOVATOST' OTDEL'NOST', 2362
TRETICHNYJ, 4470
TRIANGULYATSIYA, 4590
TRIAS, 4591
TRIAS AL'PIJSKIJ, 152
TRIAS NIZHNIJ, 2594
TRIAS SREDNIJ, 2852
TRIAS VERKHNIJ, 4684
TRICONODONTA, 4594
TRIGONIOIDA, 4596
TRILOBITA, 4597
TRILOBITOMORPHA, 4598
TRIPOLI, 4600
TROG, 4603
TROPIKI, 4602
TRUBA, 3453
TRUBA KARSTOVAYA, 4117
TRUBKA VZRYVA, 1171
TRUBOPROVOD, 3454
TSEKHSHTEJN, 4855
TSELLYULOZA, 683
TSEMENT, 684
TSEMENT STROITEL'NYJ, 587
TSEMENTATSIYA, 685
TSENA, 3610
TSENOZONA, 689
TSENTR OKRASKI, 846
TSENTROKLINAL', 690

TSEOLIT, 4856
TSEOLITIZATSIYA, 4858
TSEP' VULKANICHESKAYA, 4760
TSEPOCHKA RADIOAKTIVNAYA, 3728
TSERIJ, 699
TSEZIJ, 700
TSIKL, 1044
TSIKL BIOLOGICHESKIJ, 501
TSIKL EHROZIONNYJ, 1426
TSIKL GEOKHIMICHESKIJ, 1715
TSIKL GIDROLOGICHESKIJ, 2119
TSIKL SEDIMENTATSII, 1045
TSIKL TEKTONICHESKIJ, 4436
TSIKLICHNOST', 1047
TSIKLON, 1049
TSIKLOTEMA, 1051
TSILINDRICHESKIJ, 1053
TSINK, 4860
TSIPOLIN, 771
TSIRK LEDNIKOVYJ, 775
TSIRKON, 4861
TSIRKONIJ, 4862
TSIRKULYATSIYA, 773
TSISTID, 4864
TSUNAMI, 4604
TSVET, 845
TUBOIDEA, 4605
TUBULIDENTATA, 4606
TUCHA PALYASHCHAYA, 3069
TUF, 4607
TUF OSADOCHNYJ, 4118
TUFFIT, 4608
TUFOLAVA, 4609
TUGOPLAVKIJ, 3791
TULIJ, 4521
TUMANNOST' KOSMICHESKAYA, 968
TUNDRA, 4610
TUNNEL', 4614
TURBIDIT, 4615
TURBIDITY CURRENT STRUCTURE, 4618
TURBOGLIF, 1591
TURNE, 4553
TURON, 4620
TVERDOST', 1956
TYLOPODA, 4623
TYURINGIJ, 4522
UCHREZHDENIE, 2257
UCHREZHDENIE GEOLOGICHESKOE, 595
UDALENNOST', 3814
UDALF, 4630
UDAR, 4075
UDEL'NAYA POVERKHNOST', 4198
UDOBRENIE, 1541
UDOLL, 4631
UDULT, 4632
UFIMSKIJ, 4633
UGLEFIKATSIYA, 808

UGLEKISLOTA, 643
UGLEROD, 642
UGLEROD ORGANICHESKIJ, 3166
UGLEVOD, 641
UGLEVODOROD, 2099
UGLEVODOROD ALIFATICHESKIJ, 122
UGLEVODOROD AROMATICHESKIJ, 305
UGLISTYJ, 648
UGOL ESTESTVENNOGO OTKOSA, 200
UGOL KRITICHESKIJ, 993
UGOL NAKLONA, 1938
UGOL OTRAZHENIYA, 570
UGOL', 801
UGOL' BITUMINOZNYJ, 523
UGOL' BURYJ, 582
UGOL' GAZOVYJ, 1687
UGOL' GUMUSOVYJ, 2078
UGOL' KOKSOVYJ, 831
UGOL' SAPROPELEVYJ, 3945
UKREPLENIE BEREGOV, 4082
UL'TISOL, 4634
UL'TRABAZIT, 4636
UL'TRAMETAMORFIZM, 4637
UL'TRAMILONIT, 4638
UL'TRAOSNOVNOJ (SOSTAV), 4635
UL'TRASHCHELOCHNOJ (SOSTAV), 3347
ULTRASTRUCTURE, 4640
UMBREPT, 4642
UMEN'SHENIE, 366
UMERENNYJ (KLIMAT), 4452
UNAKIT, 4643
UNASLEDOVANNYJ, 2237
UNDERCLAY, 4648
UNDERGROUND CANALIZATION, 4649
UNIFORMIZM, 4659
UNIONOIDA, 4660
UPAKOVKA, 3232
UPAKOVKA PLOTNEJSHAYA, 798
UPLIFT PRESSURE, 4666
UPLOTNENIE (OSADKOV), 857
UPLOTNENIE DIFFERENTSIROVANNOE, 1175
UPORYADOCHENNOST, 3149
UPRUGOST', 1323
UPRUGOVYAZKIJ, 4744
URALITIZATSIYA, 4688
URAN (EHLEM), 4689
URAN (PLANETA), 3479
URAVNENIE, 1411
URAVNENIE SOSTOYANIYA, 1412
URBANIZATSIYA, 4696
UREILIT, 4697
URODELA, 4698
UROVEN', 2493
UROVEN' GIDROSTATICHESKIJ, 2128
UROVEN' MORYA, 3977
UROVEN' PODZEMNYKH VOD, 1916
UROVEN' VODY OZERA, 2416

USADKA, 4087
USHCHEL'E, 1837
USHCHEL'E ABISSAL'NOE, 20
USLOVIYA KONTINENTAL'NYE, 923
UST'E GOLOSTOMNOE, 2032
UST'E SKVAZHINY, 4815
USTALF, 4699
USTERT, 4700
USTOJCHIVOST' SKLONA, 4137
USTOKS, 4702
USTOLL, 4701
USTROJSTVO PRIEMNOE, 3764
USTULT, 4703
USTUP, 1433
USTUP PLYAZHEVYJ, 447
USTUP SBROSOVYJ, 1511
VACUUM FUSION ANALYSIS, 4705
VADI, 4779
VAL BAR'ERNYJ, 418
VAL BEREGOVOJ, 446
VAL PLYAZHOVOJ, 737
VALANZHIN, 4706
VALENTNOST', 734
VALUN, 558
VALUN EHRRATICHESKIJ, 1430
VALUN RUKOVODYASHCHIJ, 2211
VANADAT, 4709
VANADIJ, 4710
VARIATSII MAGNITNYE, 2634
VARIATSII MAGNITNYE SUTOCHNYE, 1231
VARIATSIYA, 4717
VARIJSKIJ, 4719
VARIOGRAMMA, 4718
VARVOKHRONOLOGIYA, 4721
VEKTOR, 4723
VENERA, 3480
VENEROIDA, 4728
VENIT, 4729
VENLOK, 4818
VERGENTSIYA, 4730
VERKHOVODKA, 3349
VERMES, 4731
VERMIKULIT, 4732
VERTEBRATA, 4734
VERTIKAL'NYJ, 4735
VERTISOL, 4737
VES OB'EMNYJ, 590
VES UDEL'NYJ, 4196
VESHCHESTVO ORGANICHESKOE, 3170
VESHCHESTVO RASTVORENNOE, 1229
VESTFAL, 4820
VETER, 4826
VETER SOLNECHNYJ, 4161
VIBRATSIYA, 4738
VIBROSEJS, 4740
VID-INDEKS, 2206
VIKING (SPUTNIK), 4741

VILLAFRANK, 4742
VIRGATSIYA, 4743
VISMUT, 520
VITREN, 4749
VITRIFIKATSIYA, 4751
VITROFIROVYJ, 4753
VITROKLASTICHESKIJ, 4752
VIZE, 4748
VKLYUCHENIE, 2200
VKLYUCHENIE GAZOVOE, 1690
VKLYUCHENIE GAZOVOZHIDKOE, 1579
VKRAPLENIE, 1226
VLAGA POCHVENNAYA, 2911
VLAGOMER POCHVENNYJ NEJTRONNYJ, 3018
VLAZHNOST', 2075
VMESHCHAYUSHCHIJ, 2061
VNEDRENIE, 1363
VNESHNIJ DRENAZH, 1472
VNUTRIFORMATSIONNYJ, 2291
VODA, 4783
VODA ADSORBIROVANNAYA, 67
VODA DEGIDRATATSIONNAYA, 4794
VODA GIGROSKOPICHESKAYA, 2139
VODA GRUNTOVAYA, 3416
VODA GRUNTOVAYA ALLYUVIYA, 143
VODA ISKOPAEMAYA, 1623
VODA KAPILLAYARNAYA, 636
VODA KRISTALLIZATSIONNAYA, 4793
VODA LECHEBNAYA, 2734
VODA MINERAL'NAYA, 2875
VODA MORSKAYA, 3978
VODA OSVOBOZHDENNAYA, 3837
VODA PIT'EVAYA, 3576
VODA PODZEMNAYA, 1911
VODA PODZEMNAYA BEZNAPORNAYA, 4645
VODA POGREBENNAYA, 898
VODA POROVAYA, 3563
VODA POVERKHNOSTNAYA, 4366
VODA PRESNAYA, 1647
VODA SOLONOVATAYA, 568
VODA TALAYA, 2745
VODA TERMAL'NAYA, 4501
VODA YUVENIL'NAYA, 2365
VODA ZAGRYAZNENNAYA, 3542
VODOKHOZYASTVENNOE PLANIROVANIE, 433
VODOKHRANILISHCHE, 3825
VODOOTDACHA, 4802
VODOPAD, 4803
VODOPOTREBLENIE, 4787
VODOPRONITSAEMOST', 2091
VODORAZDEL, 4788
VODOROD, 2106
VODOROD GAZ, 2107
VODOROSLEVYJ IZVESTNYAK, 114
VODOROSLI, 111
VODOROSLI IZVESTKOVYE, 608
VODOSNABZHENIE, 4800

VODOSODERZHASHCHIJ, 2135
VODOUPOR, 259
VODOUPOR NEPRONITSAEMYJ, 262
VODOUPOR PRONITSAEMYJ, 264
VODOZABOR, 4786
VODY NEFTYANYE, 1306
VOL'FRAM, 4613
VOL'FRAMAT, 4612
VOLANTES, 4754
VOLATILE ELEMENT, 4756
VOLNA, 4805
VOLNA AKUSTICHESKAYA, 45
VOLNA DLINNOPERIODNAYA, 2564
VOLNA EHRY, 97
VOLNA LOVA, 2570
VOLNA MORSKAYA, 3979
VOLNA OB'EMNAYA, 535
VOLNA PESCHANAYA, 3941
VOLNA POPERECHNAYA, 3920
VOLNA POVERKHNOSTNAYA, 4368
VOLNA PRODOL'NAYA, 3230
VOLNA SEISMICHESKAYA, 4021
VOLNA UDARNAYA, 4078
VOLNA UPRUGAYA, 1322
VOLOCHENIE, 3992
VORONKA DEPRESSIONNAYA, 885
VORONKA KARSTOVAYA, 1235
VOSPRIIMCHIVOST' MAGNITNAYA, 2645
VOSSTANOVLENIE (KHIM), 731
VOSTOK, 1288
VOZDEJSTVIE MOROZNOE, 1653
VOZDUKH, 93
VOZMOZHNOST', 3572
VOZRAST ABSOLYUTNYJ, 10
VOZRAST DISKORDANTNYJ, 1216
VOZRAST GEOLOGICHESKIJ, 80
VOZRAST KONKORDANTNYJ, 878
VOZRAST KOSMICHESKIJ, 1489
VOZRAST OTNOSITEL'NYJ, 3804
VOZRAST ZEMLI, 81
VPITYVANIE, 634
VRASHCHENIE, 3902
VRASHCHENIE OPTICHESKOE, 3141
VRASHCHENIE PLIT, 3498
VRASHCHENIE ZEMLI, 1278
VREMYA, 4535
VREMYA PROBEGA, 4582
VREMYA PROBEGA, 4581
VREMYA VSTUPLENIYA, 307
VSELENNAYA, 4663
VSKRYSHA, 3213
VSPUCHIVANIE, 1970
VSPUCHIVANIE MOROZNOE, 1655
VTORZHENIE SOLENYKH VOD, 3930
VULKAN, 4769
VULKAN DEJSTVUYUSHCHIJ, 60
VULKAN GRYAZEVOJ, 2958

VULKAN PODVODNYJ, 4333
VULKAN SHCHITOVOJ, 4073
VULKANICHESKIJ, 4765
VULKANIZM, 4768
VULKANO-TEKTONICHESKAYA DEPRESSIYA, 4770
VULKANOGENNYJ, 4773
VULKANOGENNYJ (MESTOR), 4774
VULKANOKLASTICHESKIJ, 4772
VULKANOLOGIYA, 4775
VULKANORIJ, 4778
VULKANSKIJ, 4777
VYAZKIJ MATERIAL, 4746
VYAZKOST', 4745
VYBOR MESTA ZALOZHENIYA, 4124
VYKLINIVANIE, 3449
VYMERSHIJ, 1495
VYMIRANIE, 1496
VYRABOTKA GORNAYA, 2861
VYRABOTKA OTKRYTAYA, 4294
VYRABOTKA PODZEMNAYA, 4650
VYSHCHELACHIVANIE, 2465
VYSHCHELACHIVANIE POCHVY, 4150
VYSOKOTEMPERATURNYJ, 2008
VYSOTA, 155
VYVETRIVANIE, 4809
VYVETRIVANIE FIZICHESKOE, 3429
VYVETRIVANIE KHIMICHESKOE, 735
VYVETRIVANIE MEKHANICHESKOE, 2728
VYVETRIVANIE MOROZNOE, 891
VYVETRIVANIE SELEKTIVNOE, 1177
VZRYV, 1485
VZRYV GORNYJ, 3883
VZRYV YADERNYJ, 3063
VZVES', 4107
WATER AUTHORITY, 4784
WATER MAP, 4792
WATER PRESSURE, 4795
WATER STORAGE, 4799
WAVE DISPERSAL, 4806
WELL SCREEN, 4817
WETLAND, 4822
WIND TRANSPORT, 4829
WRENCH FAULT, 4831
X-LUCHI, 4832
XANTHOPHYTA, 4837
YACHEISTYJ, 564
YACHEJKA EHLEMENTARNAYA, 4661
YAD, 4554
YADRO, 3067
YADRO (PALE), 2274
YADRO KONTINENTAL'NOE, 926
YADRO VNESHNEE ZEMLI, 3207
YADRO VNUTRENNEE, 2243
YADRO ZEMLI, 957
YAJTSO, 1311
YANTAR', 160
YARAMILLO EHPOKHA, 2358

YARUS, 4229
YASHMA, 2359
YAVNOKRISTALLICHESKIJ, 1444
YAZYK EHVM, 871
YUG, 4187
YUNGA MODUL', 4851
YUPITER, 3473
YURA, 2363
YURA NIZHNYAYA, 2583
YURA SREDNYAYA, 2844
YURA VERKHNYAYA, 4673
YUVENIL'NYJ, 2364
ZADACHA OBRATNAYA, 2296
ZAGRYAZNENIE, 3543
ZAGRYAZNITEL', 3541
ZAILENIE, 4108
ZAKANCHIVANIE SKVAZHINY, 4814
ZAKHORONENIE OTKHODOV, 4782
ZAKON, 2460
ZAKON KORIOLISA, 959
ZAKON KORRELYATSSII FATSIJ, 2461
ZAKON MATEMATICHESKIJ, 2712
ZAKONODATEL'STVO, 2476
ZAKRAINA AKTIVNAYA, 58
ZAKREPLENIE GRUNTOV, 4158
ZALEGANIE NENARUSHENNOE, 4360
ZALEGANIE SOGLASNOE, 889
ZALEZH', 3914
ZALEZH' RUDNAYA, 3154
ZALIV, 1927
ZAMESHCHENIE, 3817
ZAMOK, 2014
ZANDR, 3212
ZAPAD, 4819
ZAPASY, 3834
ZAPASY PROGNOZNYE, 2225
ZAPASY PROMYSHLENNYE, 3571
ZAPISKA OB'YASNITEL'NAYA, 1480
ZASHCHITA, 3609
ZASUKHA, 1258
ZATOPLENIE, 4334

ZATVERDENIE, 890
ZAYAKORIVANIE, 190
ZAYAVKA, 779
ZEMLETRYASENIE, 1284
ZEMLETRYASENIE GLUBOKOFOKUSNOE, 1088
ZEMLETRYASENIE MELKOFOKUSNOE, 4052
ZEMLETRYASENIE SREDNEFOKUSNOE, 2270
ZEMLI REDKIE, 3751
ZEMLYA, 3472
ZEMLYA DIATOMOVAYA, 1168
ZERKALO SKOL'ZHENIYA, 4132
ZERNISTYJ, 1856
ZERNO, 1844
ZERNO OKUTANNOE, 817
ZHADEITIT, 2357
ZHEDIN, 1701
ZHELEZISTYJ, 1539
ZHELEZO, 2308
ZHELEZO OKISNOE, 1538
ZHELEZO ZAKISNOE, 1533
ZHELEZOBAKTERIYA, 2310
ZHELOB GLUBOKOVODNYJ, 4587
ZHELOB LEDNIKOVYJ, 1902
ZHEMCHUG, 3319
ZHEODA, 1726
ZHERLO, 4767
ZHESTKOST', 3866
ZHESTKOST' VODY, 4790
ZHILA RUDNAYA, 4725
ZHILA SEGREGATSIONNAYA, 4009
ZHILA STUPENCHATAYA, 2514
ZHILA TIPA KONSKOGO KHVOSTA, 2059
ZHIRNYJ BLESK, 1889
ZHIVET, 1790
ZHIVUSHCHIJ, 2548
ZHIZN', 2499
ZHIZN' ORGANICHESKAYA, 3169
ZIGEN, 4094
ZNACHENIE SREDNEE, 389
ZNAK VOLOCHENIYA, 1903
ZNAKI RYABI, 3872

ZNAKI RYABI EHOLOVYE, 4828
ZNAKI TECHENIYA, 1036
ZNAKI VNEDRENIYA, 2555
ZOANTHINIARIA, 4863
ZOL'NOST', 330
ZOLOTO, 1833
ZONA AEHRATSII, 4866
ZONA DROBLENIYA, 4064
ZONA EHKOLOGICHESKYA, 1302
ZONA EHKVATORIAL'NAYA, 1414
ZONA FAUNISTICHESKAYA, 1519
ZONA FLORISTICHESKAYA, 1573
ZONA GLUBINNAYA, 2786
ZONA KOMPLEKSNAYA, 339
ZONA NARUSHENIYA, 1512
ZONA NASYSHCHENIYA, 3949
ZONA OKISLENIYA, 3223
ZONA OLEDENENIYA KRAEVAYA, 3355
ZONA PLASTICHNOSTI, 4868
ZONA RANGOVAYA, 3747
ZONA RAZLOMA, 1513
ZONA SEDIMENTATSII, 4869
ZONA SOCHLENENIYA, 1730
ZONA SUBDUKTSII, 4322
ZONA SUL'FIDNOGO OBOGASHCHENIYA, 4350
ZONA TRESHCHINOVATOSTI, 1634
ZONA TSEMENTATSII, 4867
ZONAL'NOST', 4870
ZONAL'NOST' KLIMATICHESKAYA, 797
ZONAL'NYJ, 4865
ZOND, 3618
ZOND IONNYJ, 2301
ZONDIROVANIE, 4185
ZONDIROVANIE CHASTOTNOE EHLEKTRICHESKOE, 1646
ZONDIROVANIE DISTANTSIONNOE, 3813
ZONDIROVANIE SEJSMICHESKOE GLUBINNOE, 1091
ZOOID, 4871
ZRELOST', 2718
ZUBY, 4544

Español

AALIENSE, 3
ABACO, 3045
ABANICO=FLUVIAL, 141
ABANICO=FLUVIOGLACIAR, 250
ABISAL, 18
ABLACION, 5
ABOMBAMIENTO, 1970
ABOMBAMIENTO=POR=HIELO, 1655
ABONO, 1541
ABRASION, 6
ABRASIVO, 9
ABSORBENTE, 12
ABSORCION, 13
ABSORCION=ATOMICA, 363
ABUNDANCIA, 16
ACANTHARIA, 25
ACANTHODII, 26
ACANTILADO, 793
ACCION=BIOGENICA, 493
ACCION=CALOR, 4497
ACCION=CLIMATICA, 795
ACCION=FRIO, 1653
ACCION=PREVENTIVA, 3609
ACEITE=PESADO, 1974
ACELERACION=CORIOLIS, 959
ACELEROGRAMA, 27
ACELEROMETRO, 28
ACETOLISIS, 35
ACIDEZ, 40
ACIDO, 37
ACIDO=FULVICO, 1663
ACIDO=GRASO, 1506
ACIDO=HUMICO, 2071
ACIDO=MINERAL, 2248
ACIDO=ORGANICO, 3165
ACIDO=SULFURICO, 4356
ACMOLITO, 98
ACONDRITA, 36
ACORTAMIENTO, 4084
ACRANIA, 46
ACRECENTAMIENTO, 84
ACRECION, 30
ACRISOL, 47
ACRITARCHA, 48
ACROTRETIDA, 49

ACROZONA, 3747
ACTINIARIA, 50
ACTINIO, 51
ACTINOCERATOIDEA, 52
ACTIVACION=NEUTRONICA, 3015
ACTIVIDAD, 61
ACTIVIDAD=IGNEA, 2172
ACTUALISMO, 62
ACUIFERO, 260
ACUIFUGO, 262
ACUMULACION=GAS, 1685
ACUÑADO, 3449
ACUOSO, 2135
ADAPTACION, 63
ADIAGNOSTICO, 66
ADSORBCION, 68
AERODROMO, 96
AEROSOL, 75
AFININIDAD, 76
AFINO, 3780
AFLORAMIENTO, 3206
AFLUENTE, 77
AFORO, 1698
AGLOMERADO, 82
AGLOMERANTE, 2944
AGNATHA, 86
AGNOSTIDA, 87
AGONIATITIDA, 88
AGOTAMIENTO, 4797
AGPAITICO, 89
AGREGADO, 85
AGRICULTURA, 90
AGROGEOLOGIA, 91
AGUA, 4783
AGUA=CAPILAR, 636
AGUA=CONGENITA, 898
AGUA=CONSTITUCION, 4794
AGUA=CONTAMINADA, 3542
AGUA=CRISTALIZACION, 4793
AGUA=DE=ADSORCION, 67
AGUA=DE=FUNDICION, 2745
AGUA=DE=MAR, 3978
AGUA=DE=SUPERFICIE, 4366
AGUA=DISPONIBLE, 4802
AGUA=DULCE, 1647

AGUA=FOSIL, 1623
AGUA=FREATICA, 3416
AGUA=HIDROTERMAL, 2134
AGUA=HIGROSCOPICA, 2139
AGUA=INTERSTICIAL, 3563
AGUA=JUVENIL, 2365
AGUA=LIBRE, 4645
AGUA=MARGINAL, 1306
AGUA=MINERAL, 2875
AGUA=RESURGENTE, 3837
AGUA=SALOBRE, 568
AGUA=SUBTERRANEA, 1911
AGUA=TERMAL, 4501
AGUA=TERMOMINERAL, 2734
AGUJA, 2996
AIRE=ATMOSFERICO, 93
AIREACION, 69
AISLAMIENTO, 2330
AISLANTE, 2259
AKTCHAGYLIENSE, 99
ALABASTRO, 100
ALBEDO, 101
ALBERTITA, 102
ALBIENSE, 103
ALBITIZACION, 106
ALBOLL, 107
ALBUFERA, 2411
ALBUFERA=DE=ATOLON, 419
ALCALINIDAD, 131
ALCANO, 132
ALCYONACEA, 108
ALDANIENSE, 109
ALEACION, 139
ALFISOL, 110
ALGA=CALCAREA, 608
ALGAE, 111
ALGEBRA=BOOLEANA, 542
ALGEBRA=MATRICIAL, 2717
ALGOL, 119
ALGORITMO, 121
ALIMENTACION=NATURAL, 1917
ALINEACION=BLOQUE=GLACIAR, 560
ALLOGROMIINA, 137
ALMACEN=SUBTERRANEO, 4652
ALMACENAMIENTO, 4265

ALMACENAMIENTO-AGUA, 4799
ALMACENAMIENTO-DATOS, 1071
ALMACENAMIENTO-GAS, 1691
ALOCTONIA, 135
ALOMETRIA, 138
ALOQUIMICO, 134
ALOTRIOMORFO, 203
ALPINO, 151
ALTERACION, 154
ALTERACION-DIFERENCIAL, 1177
ALTERACION-FISICA, 3429
ALTERACION-HIDROTERMAL, 2131
ALTERACION-MECANICA, 2728
ALTERACION-METEORICA, 4809
ALTERACION-QUIMICA, 735
ALTERACION-SUBMARINA, 1946
ALTERACION-SUPERGENICA, 2143
ALTITUD, 155
ALTO-GRADO, 2010
ALUMINA, 156
ALUMINIO, 157
ALUNITIZACION, 158
ALUVIAL, 140
ALUVION, 146
ALVEOLINELLIDAE, 159
AMBAR, 160
AMBLYPODA, 161
AMERICIO, 162
AMIANTO, 325
AMINOACIDO, 163
AMMODISCACEA, 164
AMMONITIDA, 166
AMMONITINA, 167
AMMONOIDEA, 169
AMONIACO, 165
AMONIO, 168
AMORFO, 170
AMPHIBIA, 171
AMPHINEURA, 176
AMPLITUD, 177
ANALISIS, 182
ANALISIS-ACTIVACION, 54
ANALISIS-ARMONICO, 1628
ANALISIS-ARMONICO-ESFERICO, 4206
ANALISIS-CONTENIDO, 337
ANALISIS-CORRESPONDENCIA, 963
ANALISIS-CUALITATIVO, 3706
ANALISIS-CUANTITATIVO, 3708
ANALISIS-DE-DATOS, 1066
ANALISIS-DIFRACCION-ELECTRON, 1346
ANALISIS-DIFRACCION-NEUTRON, 3016
ANALISIS-DISCRIMINANTE, 1218
ANALISIS-ELEMENTOS-TRAZA, 4556
ANALISIS-ESPECTRO-SISMICO, 4017
ANALISIS-ESTRUCTURAL, 4301
ANALISIS-FACTORIAL, 1502
ANALISIS-FUSION-VACIO, 4705

ANALISIS-GRUPO, 799
ANALISIS-LIBRO, 541
ANALISIS-MATEMATICO, 2708
ANALISIS-MAYORES, 2657
ANALISIS-MODAL, 2902
ANALISIS-MULTIVARIABLE, 2968
ANALISIS-PONDERADO, 4812
ANALISIS-QUIMICO, 721
ANALISIS-RX, 4833
ANALISIS-SISTEMA, 4404
ANALISIS-TENDENCIA, 4588
ANALISIS-TERMICO, 4494
ANALISIS-TERMOMAGNETICO, 4510
ANALISIS-VARIANZA, 4716
ANALOGIA-SISTEMA-HIDROGEOLOGICO, 4405
ANAPSIDA, 184
ANASPIDA, 185
ANATEXIA, 186
ANATEXITA, 187
ANATOMIA, 188
ANATOMIA-APARATO-NERVIOSO, 3012
ANATOMIA-COLUMNA-VERTEBRAL, 4211
ANATOMIA-ESQUELETO, 4127
ANATOMIA-LOCOMOCION, 2558
ANCHIMETAMORFISMO, 189
ANCLAJE, 190
ANDALUCITA, 192
ANDEPT, 193
ANDESITA, 194
ANDOSOL, 196
ANFIBOL, 172
ANFIBOLITA, 173
ANFIBOLITIZACION, 175
ANGIOSPERMAE, 199
ANGREITA, 201
ANGULO-CON-LA-VERTICAL, 1938
ANGULO-CRITICO, 993
ANGULO-DE-BRAGG, 570
ANGULO-DE-REPOSO, 200
ANHIDRITA, 204
ANHIDRO, 205
ANILLO-CRECIMIENTO, 4585
ANILLOS-LIESEGANG, 2498
ANIMAL-PLANCTIVORO, 3486
ANION, 207
ANISIENSE, 208
ANISOPLEURA, 209
ANISOTROPIA, 210
ANNELIDA, 212
AÑO-GALACTICO, 1674
ANOMALIA, 214
ANOMALIA-AIRE-LIBRE, 1643
ANOMALIA-BOUGUER, 557
ANOMALIA-GEOFISICA, 1751
ANOMALIA-GRAVIMETRICA, 1878
ANOMALIA-MAGNETICA, 2628
ANOMALODESMATA, 213

ANORTOSITA, 216
ANORTOSITIZACION, 217
ANTEFOSA, 1612
ANTEPAIS, 1613
ANTEPLAYA, 1616
ANTHOCYATHEA, 219
ANTHOZOA, 220
ANTIARCHI, 223
ANTICLINAL, 224
ANTICLINORIO, 225
ANTICLISIO, 218
ANTIDUNA, 226
ANTIMONIATO, 228
ANTIMONIO, 229
ANTIPATHARIA, 230
ANTIPERTITA, 231
ANTRACITA, 221
ANTRAXOLITA, 222
ANUARIO, 4847
ANUROMORPHA, 234
APARATO-MASTICADOR-CONODONTOS, 900
APERTURA, 235
APICAL, 237
APLACOPHORA, 238
APLITA, 239
APOFISIS, 245
APOGEO, 241
APOGRANITO, 242
APROVISIONAMIENTO-AGUA, 4800
APROXIMACION, 249
APSHERONIENSE, 251
APSIDOSPONDYLI, 252
APTIENSE, 253
AQUALF, 254
AQUENT, 255
AQUEPT, 257
AQUERT, 258
AQUITANIENSE, 263
AQUOD, 265
AQUOLL, 266
AQUOX, 267
AQUULT, 268
ARACHNIDA, 269
ARCAICO, 278
ARCHAEOCOPIDA, 272
ARCHAEOCYATHA, 273
ARCHAEOCYATHEA, 274
ARCHAEOGASTROPODA, 275
ARCHAEOPTERYGES, 276
ARCHAEORNITHES, 277
ARCHEIDES, 279
ARCHIPOLYPODA, 283
ARCHOSAURIA, 284
ARCILLA, 787
ARCILLA-BLANCA, 4823
ARCILLA-CON-BLOQUES, 559
ARCILLA-DE-CAPA-INFERIOR, 4648

ARCILLA-DE-CERAMICA, 3586
ARCILLA-DE-LADRILLO, 576
ARCILLA-ESMECTICA, 528
ARCILLA-FLUIDA, 3715
ARCILLA-FRIABLE, 4144
ARCILLA-INTERESTRATIFICADA, 2898
ARCILLA-MINERAL, 790
ARCILLA-REFRACTARIA, 1552
ARCILLA-RESIDUAL, 3828
ARCILLA-SENSIBLE, 4038
ARCILLA-TENAZ, 4255
ARCILLA-VARVADA, 4722
ARCILLITA, 791
ARCO-INSULAR, 2316
ARCOIDA, 285
ARCOSA, 302
AREA-CONTINENTAL, 2425
ARENA, 3936
ARENA-BITUMINOSA, 3108
ARENA-DE-VIDRIO, 1812
ARENA-FOSILIFERA, 3661
ARENA-MOVEDIZA, 3716
ARENIGIENSE, 291
ARENISCA, 3939
ARENISCA-GROSERA, 1900
ARENITA, 292
ARENOSOL, 293
ARGID, 294
ARGON, 296
ARGON-ATMOSFERICO, 359
ARIDEZ, 301
ARIDISOL, 300
ARMONICO, 1957
ARQUEADO, 288
ARQUEOLOGIA, 281
ARQUEOMAGNETISMO, 282
ARRASTRE-EN-PROFUNDIDAD, 4321
ARRECIFE, 3776
ARRECIFE-CORAL, 949
ARRECIFE-DE-ALGAS, 117
ARRECIFE-DE-BARRERA, 1651
ARRECIFE-INTERNO, 397
ARROYO, 308
ARSENIATO, 309
ARSENICO, 310
ARTERITA, 311
ARTESIANO, 312
ARTHRODIRA, 315
ARTHROPLEURIDA, 316
ARTHROPODA, 317
ARTICO, 286
ARTICULATA, 318
ARTINSKIENSE, 323
ARTIODACTYLA, 324
ASFALTITA, 335
ASFALTO, 334
ASGILLIENSE, 333

ASIENTO, 4048
ASIMILACION-MAGMATICA, 340
ASISMICO, 329
ASOCIACION, 338
ASOCIACION-DIQUES, 1185
ASOCIACION-MAGMATICA, 2620
ASOCIACION-MINERAL, 2863
ASOCIACIONES-FOSILES, 341
ASTATO, 343
ASTENOSFERA, 348
ASTEROIDE, 345
ASTEROIDEA, 346
ASTEROZOA, 347
ASTRAPOTHERIA, 349
ASTROBLEMA, 350
ASTROGEOLOGIA, 351
ASTRONOMIA, 352
ATAQUE-QUIMICO, 1439
ATAXITA, 354
ATENUACION-ONDA, 366
ATLAS, 356
ATMOSFERA, 358
ATOLON, 362
ATRICION, 368
AUBRITA, 369
AULACOGENO, 372
AUREOLA-DISPERSION, 1223
AUREOLA-GEOQUIMICA, 373
AUREOLA-PRIMARIA, 3611
AUREOLA-SECUNDARIA, 3988
AUSTRALITA, 374
AUTIGENESIS, 375
AUTOCORRELACION, 378
AUTOCTONIA, 377
AUTOLISIS, 380
AUTOLITO, 379
AUTOMATIZACION, 382
AUTOMETAMORFISMO, 383
AUTOMETASOMATISMO, 384
AUTOPISTA, 3879
AUTORADIOGRAFIA, 385
AUTUNIENSE, 387
AVALANCHA, 388
AVENAMIENTO-DENDRITICO, 1109
AVES, 390
AZIMUT-MAGNETICO, 2629
AZUCARES, 4347
AZUFRE, 4353
BACKSET-BED, 399
BACTERIA, 400
BACTERIA-HIERRO, 2310
BACTERIOGENICO, 401
BACTRITOIDEA, 402
BAHAMITA, 1862
BAHIA, 443
BAJADA, 405
BAJO-GRADO, 2471

BAJOCIENSE, 406
BALANCE, 407
BALANCE-AGUA-SUPERFICIE, 4367
BALANCE-DE-AGUA, 4785
BALANCE-DE-AGUA-SUBTERRANEA, 1912
BALANCE-MASA, 2698
BALASTO, 408
BANCO-ARRECIFAL, 3313
BANCO-DE-ALGAS, 112
BANCO-DE-DATOS, 1067
BANCO-TERRAZA, 461
BANDAS-DE-BARRO, 2953
BARCO-SONDEO, 1257
BARIO, 413
BARITA, 412
BARRA, 410
BARRA-BARRERA, 417
BARRANCO, 1928
BARREMIENSE, 414
BARRERA, 416
BARRERA-ARRECIFAL, 420
BARRERA-PLAYA, 418
BARRO-ORGANICO, 3127
BARTONIENSE, 421
BASAL, 422
BASALTO, 425
BASALTO-ALCALINO, 123
BASALTO-DE-DECCAN, 1080
BASALTO-DE-MESETA, 3502
BASANITA, 427
BASHKIRIENSE, 431
BASIFICACION, 432
BATEA, 3283
BATIENTE, 4370
BATIENTE-GLACIAR, 1805
BATIMETRIA, 438
BATOLITO, 434
BATONIENSE, 435
BATRACHOSAURIA, 439
BAUXITA, 440
BAUXITIZACION, 441
BE-10-BE-9, 472
BEACH-ROCK, 448
BELEMNOIDEA, 459
BENEFICIACION, 1252
BENNETTITALES, 463
BENTONITA, 466
BENTOS, 465
BERILIO, 471
BERKELIO, 468
BERRIASIENSE, 470
BETUN, 521
BEYRICHICOPINA, 474
BIAXIAL, 475
BIBLIOGRAFIA, 477
BIBLIOGRAFIA-PERSONAL, 3365
BIOCENOSIS, 481

BIOCLASTO, 486
BIOCORO, 483
BIOCRONA, 484
BIOCRONOLOGIA, 485
BIODEGRADACION, 489
BIOESTRATIGRAFIA, 511
BIOFACIES, 490
BIOGENESIS, 492
BIOGEOGRAFIA, 497
BIOGEOQUIMICA, 496
BIOGRAFIA, 498
BIOHERMES, 499
BIOLITO, 500
BIOLOGIA, 502
BIOMASA, 503
BIOMETRIA, 504
BIOMICRITA, 505
BIORRESISTASIA, 506
BIOSFERA, 509
BIOSOMA, 507
BIOSPARITA, 508
BIOSTROMA, 512
BIOTA, 513
BIOTIPO, 516
BIOTOPO, 514
BIOTURBACION, 515
BIOZONA, 517
BIRREFRINGENCIA, 519
BISMUTO, 520
BISO, 601
BITUMINIZACION, 522
BIVALVIA, 3332
BIZCOCHO-ALGAS, 113
BLASTOIDEA, 526
BLATTOPTEROIDA, 527
BLOQUE, 530
BLOQUE-AISLADO, 558
BLOQUE-CENTRAL, 2732
BLOQUE-DIAGRAMA, 531
BLOQUE-ERRATICO, 1430
BLOQUE-INDICADOR, 2211
BOLA, 3337
BOLA-DE-ARCILLA, 788
BOLA-DE-CARBON, 802
BOMBA-VOLCANICA, 539
BOMBEO, 3684
BOMBEO-DE-ENSAYO, 3685
BONE-BED, 540
BORALF, 543
BORATO, 544
BORAX, 545
BORDE-CONTINENTAL, 547
BORDE-CUENCA, 4811
BORO, 552
BOROLL, 551
BOUDINAGE, 556
BOUNDSTONE, 562

BOYA-ACUSTICA, 4182
BRACHIOPODA, 565
BRACHIOPTERYGII, 566
BRANCHIOPODA, 572
BRANCHIURA, 573
BRAQUIPLIEGUE, 567
BRAZO-DE-MAR, 2241
BRECHA, 575
BRECHA-CATACLASTICA, 1000
BRECHA-DE-CONTACTO, 914
BRECHA-DE-DESMENUZAMIENTO, 4056
BRECHA-DE-DESPLOME, 835
BRECHA-DE-DISLOCACION, 1221
BRECHA-DE-EXPLOSION, 1486
BRECHA-DE-FALLA, 1508
BRECHA-DE-IMPACTO, 4076
BRECHA-DE-INYECCION, 2240
BRECHA-DE-PLIEGUE, 1605
BRECHA-TECTONICA, 4434
BRECHA-VOLCANICA, 4761
BRILLO, 2606
BRILLO-GRASO, 1889
BROMO, 581
BROMURO, 580
BRUJULA, 2630
BRYOPHYTA, 584
BRYOZOA, 585
BUFADERO, 533
BULIMINACEA, 589
BUNTSANDSTEIN, 593
BURDIGALIENSE, 594
BURILADO, 3971
BUSQUEDA-DE-DATOS, 1070
BUZAMIENTO, 1195
C-14, 646
CABALGAMIENTO, 4520
CABO, 632
CADENA-DE-MARKOV, 2691
CADENA-GEOSINCLINAL, 1763
CADMIO, 604
CAL, 2506
CALA, 985
CALABRIENSE, 606
CALCIFIDO, 611
CALCIO, 615
CALCISPONGEA, 612
CALCITA, 613
CALCITIZACION, 614
CALCOSFERA, 704
CALCULO-PETROGRAFICO, 3375
CALCULO-PROBABILIDAD, 3617
CALDERA, 617
CALDERA-DE-HUNDIMIENTO, 836
CALICHE, 619
CALIDAD, 3707
CALIDAD-AGUA, 4796
CALIDAD-MINERAL, 3160

CALIENTE, 2063
CALIFORNIO, 620
CALIZA, 2507
CALIZA-BIOCLASTICA, 487
CALIZA-DE-ALGAS, 114
CALIZA-DOLOMITICA, 1236
CALIZA-MICROCRISTALINA, 2809
CALIZA-OOLITICA, 3125
CALLOVIENSE, 622
CALOR, 1962
CALOR-ESPECIFICO, 4197
CALOR-LATENTE-DE-FUSION, 2440
CAMARA-MAGMATICA, 2618
CAMAROIDEA, 623
CAMBIO-IONICO, 2300
CAMBISOL, 624
CAMBRICO, 625
CAMBRICO-INF, 2577
CAMBRICO-MEDIO, 2841
CAMBRICO-SUP, 4667
CAMPAÑA-PROSPECCION-ACUSTICA, 44
CAMPAÑA-PROSPECCION-AEROPORTADA, 72
CAMPANIENSE, 626
CAMPO-CRISTALINO, 1015
CAMPO-DE-HIELO, 2157
CAMPO-DE-LAVA, 2456
CAMPO-DE-NIEVE, 1553
CAMPO-ELECTRICO, 1328
CAMPO-ELECTROMAGNETICO, 1340
CAMPO-GAS-HIDROCARBURO, 1689
CAMPO-GEOTERMICO, 1774
CAMPO-GRAVITATORIO, 1880
CAMPO-HIDROCARBURO, 3105
CAMPO-MAGNETICO, 2633
CAMPO-PETROLIFERO, 3107
CAMPO-PETROLIFERO-GIGANTE, 1786
CAMPTOSTROMATOIDEA, 627
CANAL, 628
CANAL-DE-LAVA, 2455
CANAL-DE-MAREA, 4525
CANAL-FOSIL, 598
CANAL-MARINO-PROFUNDO, 1089
CANALIZACION-SUBTERRANEA, 4649
CAÑON, 630
CAÑON-MARINO, 4326
CANTERA, 3709
CANTO-ARCILLA-ENDURECIDA, 303
CANTO-RODADO, 3322
CAOLIN, 2368
CAOLIN-CHINA, 743
CAOLINIZACION, 2369
CAP-ROCK, 631
CAPA-A, 1
CAPA-ACTIVA, 57
CAPA-B, 395
CAPA-BASAL, 555
CAPA-BASALTICA, 426

CAPA=C, 602
CAPA=CARBON, 806
CAPA=CONCORDANTE, 877
CAPA=CONFINANTE, 259
CAPA=D, 1057
CAPA=DE=TECHO, 4548
CAPA=E, 1271
CAPA=F, 1499
CAPA=FRONTAL, 1614
CAPA=G, 1671
CAPA=GRANITICA, 1849
CAPA=LIMITE, 561
CAPA=NEFELOIDE, 3009
CAPA=PRODUCTIVA, 3983
CAPA=ROJA, 3773
CAPA=SEMIPERMEABLE, 264
CAPACIDAD=ALMACENAMIENTO, 31
CAPACIDAD CALORIFICA, 1963
CAPACIDAD=DE=CAMBIO, 1464
CAPACIDAD=DE=CARGA, 450
CAPARAZON, 639
CAPAS=DE=CARBON, 805
CAPILARIDAD, 633
CAPTACION, 4786
CAPTURA, 637
CARACTERISTICA, 710
CARBON, 801
CARBON=BITUMINOSO, 523
CARBON=BOGHEAD, 538
CARBON=CANNEL, 629
CARBON=COQUE, 831
CARBON=PARDO, 582
CARBON=SAPROPELICO, 3945
CARBONATITA, 652
CARBONATIZACION, 653
CARBONATO, 651
CARBONIFERO, 654
CARBONIFERO=INF, 2578
CARBONIFERO=SUP, 4668
CARBONIFICACION, 808
CARBONO, 642
CARBONO=ORGANICO, 3166
CARBONO=RADIOACTIVO, 3733
CARGA, 2554
CARGA=CICLICA, 1046
CARGA=EN=SUSPENSION, 4372
CARGA=FONDO, 452
CARGA=HIDRAULICA, 2094
CARNIENSE, 655
CARNIOLA, 656
CARNIVORA, 657
CARODICIENSE, 638
CARTERINACEA, 658
CARTOGRAFIA, 660
CARTOGRAFIA=AEREA, 70
CARTOGRAFIA=AUTOMATICA, 381
CARTOGRAFIAR, 2670

CASIGNITA, 714
CASQUETE=GLACIAR, 2156
CASQUETE=POLAR, 3534
CASSIDULINACEA, 664
CATACLASIS, 667
CATAGENESIS, 668
CATAGRAFIA, 669
CATALISIS, 670
CATAMORFISMO, 2377
CATARATA, 4803
CATASTROFE=NATURAL, 1738
CATASTROFISMO, 671
CATEGORIA=CURSOS=AGUA, 4286
CATION, 674
CAUCE, 707
CAUDAL, 1209
CAUSTOBIOLITA, 676
CAVERNA, 677
CAVIDAD=DISOLUCION, 4177
CAYTONIALES, 681
CELDILLA=ELEMENTAL, 4661
CELULOSA, 683
CEMENTACION, 685
CEMENTO, 879
CEMENTO=INDUSTRIAL, 587
CEMENTO=ROCA, 684
CENIZA=ARTIFICIAL, 320
CENIZA=VOLCANICA, 4759
CENOMANIENSE, 686
CENOTIPICO, 687
CENOZOICO, 688
CENOZONA, 689
CENTELLEO, 3964
CENTRAL=ELECTRICA, 3588
CENTRAL=NUCLEAR, 3065
CENTRO=COLOREADO, 846
CENTRO=EXPANSION, 4223
CENTROCLINAL, 690
CEPHALOCARIDA, 691
CEPHALODISCIDA, 692
CEPHALOPODA, 693
CERA, 4808
CERAMICA, 694
CERATITIDA, 695
CERATOMORPHA, 696
CERIANTHARIA, 697
CERIANTIPATHARIA, 698
CERIO, 699
CERRO=DE=ALGAS, 116
CESIO, 700
CETACEA, 701
CHAETOGNATHA, 702
CHARNELA, 2014
CHARNOCKITA, 712
CHAROPHYTA, 713
CHATTIENSE, 716
CHEILOSTOMATA, 717

CHELICERATA, 719
CHELONIA, 720
CHERNOZEM, 738
CHERTIFICACION, 740
CHILOPODA, 742
CHIMENEA, 2995
CHIROPTERA, 744
CHITINOZOA, 746
CHLOROPHYTA, 754
CHONDRICHTHYES, 755
CHORDATA, 758
CHRYSOPHYCOPHYTA, 765
CIANOFITA, 1040
CIBERNETICA, 1041
CICLO, 1044
CICLO=AGUA, 2119
CICLO=BIOLOGICO, 501
CICLO=DE=EROSION, 1426
CICLO=GEOQUIMICO, 1715
CICLO=SEDIMENTARIO, 1045
CICLO=TECTONICO, 4436
CICLON, 1049
CICLOTEMA, 1051
CIELO=ABIERTO, 3129
CIENAGA, 4376
CIENCIA=DE=LA=TIERRA, 1279
CILIATA, 766
CILIOPHORA, 767
CIMENTACION, 1627
CINC, 4860
CINEMATICA, 2393
CINTURON=CIRCUMPACIFICO, 774
CINTURON=METAMORFICO, 2780
CINTURON=OFIOLITICO, 1895
CINTURON=PLEGADO, 1604
CINTURON=TECTONICO, 1890
CINTURON=VOLCANICO, 4760
CIPOLINO, 771
CIRCO=GLACIAR, 775
CIRCON, 4861
CIRCONIO, 4862
CIRCULACION, 773
CIRCULACION=OCEANICA, 3086
CIRRIPEDIA, 776
CIZALLAMIENTO, 4058
CLADOCERA, 778
CLARENO, 780
CLARKE, 781
CLASIFICACION, 782
CLASIFICACION=C.I.P.W., 603
CLASIFICACION=DE=HOLMES, 2024
CLASIFICACION=DE=NIGGLI, 3027
CLASIFICACION=GRANULOMETRICA, 4184
CLASTICO, 784
CLASTO, 783
CLIMA, 794
CLIMA=HUMEDO, 2073

CLIMA=POLAR, 3533
CLINOMETRO, 4534
CLIVAGE=TECTONICO, 792
CLORACION, 748
CLORINIDAD, 750
CLORITIZACION, 752
CLORITOESQUISTO, 751
CLORO, 749
CLOROFILA, 753
CLORURO, 747
CLYMENIDA, 800
COBALTO, 818
COBOL, 819
COBRE, 946
COBRE=PORFIDICO, 3569
COCCOLITHOPHORALES, 820
COEFICIENTE, 822
COEFICIENTE=CORRELACION, 962
COEFICIENTE=PETROQUIMICO, 3370
COEFICIENTE=POISSON, 3532
COEFICIENTE=REPARTICION, 3311
COEFICIENTE=TRANSMISIBILIDAD, 4574
COELENTERATA, 823
COENOTHECALIA, 824
COERCITIVIDAD, 825
COHESION, 827
COLA=DE=CABALLO, 2059
COLADA, 1575
COLADA=CENIZA, 332
COLADA=DE=BARRO, 2956
COLADA=LAVA, 2457
COLECCION, 839
COLEOPTEROIDA, 833
COLINA, 2013
COLINA=ABISAL, 21
COLISION=PLACA, 3496
COLLEMBOLA, 840
COLMATACION, 3982
COLOIDE, 842
COLOR, 845
COLORACION=MUESTRA, 4230
COLORANTE, 3234
COLORIMETRIA, 849
COLUMNA, 3161
COLUMNARIINA, 851
COLUVION, 843
COMAGMATICO, 852
COMBA, 478
COMENSALISMO, 853
COMPACIDAD, 856
COMPACTACION, 857
COMPACTACION=DIFERENCIAL, 1175
COMPARACION=PLANETA, 2280
COMPARACION=TIERRA, 3481
COMPLEJO, 860
COMPLEJO=IGNEO, 2173
COMPLEJO=METAMORFICO, 2781

COMPLEXACION, 861
COMPLEXOMETRIA, 862
COMPONENTE=INERTE, 2223
COMPOSICION, 863
COMPOSICION=ACIDA, 39
COMPOSICION=ALCALINA, 128
COMPOSICION=ALCALINO=CALCICA, 124
COMPOSICION=CAFEMICA, 605
COMPOSICION=CALCICA, 610
COMPOSICION=CALCOALCALINA, 607
COMPOSICION=CARBONATADA, 609
COMPOSICION=CARBONOSA, 648
COMPOSICION=DETRITICA, 1142
COMPOSICION=FELSICA, 1527
COMPOSICION=FEMICA, 1528
COMPOSICION=FERROMANGANIFERA, 1537
COMPOSICION=FERRUGINOSA, 1539
COMPOSICION=GLAUCONITICA, 1816
COMPOSICION=GRANITICA, 1848
COMPOSICION=INTERMEDIA, 2269
COMPOSICION=LEUCOCRATICA, 2489
COMPOSICION=MAFICA, 2616
COMPOSICION=MELANOCRATICA, 2741
COMPOSICION=MESOCRATICA, 2752
COMPOSICION=PERALCALINA, 3347
COMPOSICION=PERALUMINOSA, 3348
COMPOSICION=POTASICA, 3578
COMPOSICION=SILICEA, 4099
COMPOSICION=SUBALCALINA, 4317
COMPOSICION=SUBALUMINOSA, 4318
COMPOSICION=TOLEITICA, 4516
COMPOSICION=ULTRABASICA, 4635
COMPRESIBILIDAD, 864
COMPRESION, 865
COMPRESION=EDOMETRICA, 908
COMPRESION=HORIZONTAL, 2051
COMPRESION=SIMPLE, 4658
COMPRESION=TRIAXIAL, 4592
COMPUESTO=HIDROCARBURO, 2099
CONCENTRACION, 873
CONCENTRACION=ALGAS, 115
CONCENTRACION=POR=GRAVEDAD, 1114
CONCESION=MINERA, 2886
CONCHA, 4071
CONCHOSTRACA, 876
CONCORDANCIA=ESTRATIGRAFICA, 889
CONCRECION, 881
CONDENSACION, 882
CONDENSADO, 1688
CONDICION=NEUMATOLITICA, 3523
CONDICION=PRESION=TEMPERATURA, 3229
CONDRITA, 756
CONDRITA=CARBONOSA, 647
CONDRITA=ENSTATITA, 1379
CONDRULO, 757
CONDUCCION=CALOR, 1964
CONDUCTIVIDAD, 883

CONDUCTIVIDAD=ELECTRICA, 1326
CONDUCTIVIDAD=HIDRAULICA, 2091
CONDUCTIVIDAD=TERMICA, 4495
CONDUCTO=ALIMENTADOR, 1522
CONDYLARTHRA, 884
CONE=IN=CONE, 886
CONFEDERACION=HIDROGRAFICA, 4784
CONFLUENTE, 888
CONGELACION, 890
CONGLOMERADO, 892
CONGLOMERADO=BASE, 424
CONGRESO=GEOL=INTER, 2277
CONIACIENSE, 894
CONIFERALES, 895
CONJUGADO, 896
CONO, 1504
CONO=DE=CENIZAS, 770
CONO=DE=DEPRESION, 885
CONO=DE=DEYECCION, 1076
CONO=DE=ESCORIAS, 4192
CONO=SUBMARINO, 4327
CONO=VOLCANICO, 4762
CONOCARDIOIDA, 899
CONODONTA, 901
CONSANGUINEIDAD, 903
CONSERVACION, 3606
CONSISTENCIA, 906
CONSOLIDACION, 907
CONSTANTE, 909
CONSTANTE=DE=DESINTEGRACION, 1079
CONSTANTE=DIELECTRICA, 1174
CONSTANTE=ELASTICIDAD, 1317
CONSTANTE=GRAVITACIONAL, 1876
CONSTANTE=LAME, 2418
CONSTANTE=RETICULAR, 2451
CONSTITUCION=GLOBO, 1276
CONSTITUCION=LUNA, 2600
CONSTITUCION=PLANETA, 3483
CONSTITUYENTE=MINERAL, 3159
CONSTRUCCION, 910
CONSTRUCCION=ANTISISMICA, 327
CONSTRUCTOR=DE=ARRECIFES, 3779
CONSUMO, 912
CONSUMO=AGUA, 4787
CONTACTO, 913
CONTAMINACION, 3543
CONTAMINACION=MAGMATICA, 917
CONTAMINANTE, 3541
CONTENIDO=CENIZA, 330
CONTENIDO=LIMITE, 1038
CONTINENTE, 918
CONTRACCION, 934
CONTRACCION=TIERRA, 933
CONTRASTE, 935
CONTROL, 936
CONTROL=EROSION, 1425
CONTROL=ESTRATIGRAFICO, 4276

CONTROL=ESTRUCTURAL, 4302
CONTROL=HIDROGEOLOGICO, 2110
CONTROL=LITOLOGICO, 2530
CONTROL=PALEOGEOGRAFICO, 3255
CONTROL=PRODUCCION, 3625
CONTROL=TECTONICO, 4435
CONTROL=YACIMIENTO, 3153
CONULARIDA, 937
CONVECCION=TERMICA, 938
CONVERGENCIA, 940
COOPERACION=INTERNACIONAL, 2276
COOPERACION=TECNICA, 4430
COORDENADAS=GEODESICAS, 1728
COORDINACION, 943
COPEPODA, 944
COPERNICIENSE, 945
COPROLITO, 947
CORALICO, 952
CORALLIMORPHARIA, 950
CORALLINACEAE, 951
CORDAITALES, 953
CORDILLERA, 955
CORDILLERA=GEOSINCLINAL, 3184
CORDILLERA=INTRACONTINENTAL, 2289
CORDON=LITORAL, 2567
CORINDON, 965
CORNEANA, 2057
CORONA=QUELIFITICA, 2383
CORRECCION, 960
CORRECCION=TIERRA, 4466
CORRECCION=TOPOGRAFICA, 4545
CORRELACION, 961
CORRIDA, 3914
CORRIENTE, 1035
CORRIENTE=CONVECCION, 939
CORRIENTE=DE=BARRO=VOLCANICO, 2414
CORRIENTE=DECAPITADA, 458
CORRIENTE=INTERMITENTE=AGUA, 2272
CORRIENTE=LITORAL, 2568
CORRIENTE=TELURICA, 1274
CORRIENTE=TEMPORAL, 1397
CORRIENTE=TURBIDEZ, 4617
CORRIMIENTO, 3218
CORRIMIENTO=GRAVEDAD, 1885
CORRIMIENTO=PLASTICO, 3492
CORROSION, 964
CORTE=GEOL=POZO, 3460
CORTE=GEOL=SONDEO, 550
CORTE=GEOLOGICO, 1736
CORTEZA=CONTINENTAL, 921
CORTEZA=OCEANICA, 3091
CORTEZA=TERRESTRE, 1001
CORTEZA=TERRESTRE=INF, 2580
CORTEZA=TERRESTRE=SUP, 4670
CORYNEXOCHIDA, 966
COSMOLOGIA, 975
COSMOQUIMICA, 972

COSTA, 811
COSTA=ABANDONADA, 4
COSTA=ABRASION, 7
COSTA=ACUMULACION, 32
COSTE, 976
COSTRA=ALTERACION, 4810
COSTRA=CALCAREA, 616
COSTRA=FERRUGINOSA, 1534
COTYLOSAURIA, 977
COVERTURA=MULTIPLE, 854
CRANEO, 980
CRATER, 981
CRATER=DE=IMPACTO, 2187
CRATER=LUNAR, 2601
CRATER=METEORICO, 2797
CRATERES=IMPACTO, 983
CRATON, 984
CRECIDA=RIO, 1569
CRECIMIENTO=CONTINENTAL, 919
CRECIMIENTO=CRISTALINO, 1017
CRENULACION, 3510
CREODONTA, 987
CRESTA, 988
CRESTA=PLAYA, 446
CRETA, 705
CRETACICO, 989
CRETACICO=INF, 2579
CRETACICO=SUP, 4669
CRINOIDEA, 991
CRINOZOA, 992
CRIOPEDOLOGIA, 1003
CRIOTECTONICO, 1004
CRIOTURBACION, 1005
CRIPTOGENO, 1008
CRIPTON, 2398
CRISTAL, 1011
CRISTAL=IDIOMORFO, 1447
CRISTAL=XENOMORFICO, 4839
CRISTALINIDAD, 1023
CRISTALITICO, 1024
CRISTALIZACION, 1025
CRISTALIZACION=FRACCIONADA, 1631
CRISTALOBLASTO, 1027
CRISTALOGRAFIA, 1028
CRISTALOGRAFIA=ESTRUCTURAL, 4303
CRISTALOQUIMICA, 1012
CROCODILIA, 996
CROMATO, 759
CROMATOGRAFIA, 760
CROMATOGRAFIA=FASE=GASEOSA, 1686
CROMO, 761
CRONOESTRATIGRAFIA, 763
CRONOZONA, 764
CROSSOPTERYGII, 999
CRUSTACEA, 1002
CRUZ=PRODUCCION, 4815
CRYPTODONTA, 1007

CRYPTOSTOMATA, 1009
CTENOSTOMATA, 1029
CUARCITA, 3712
CUARZO, 3710
CUARZO=DIORITA, 3711
CUATERNARIO, 3713
CUENCA=ENDORREICA, 1371
CUENCA=ENTRE=MONTA#AS, 2273
CUENCA=HULLERA, 803
CUENCA=INTRACRATONICA, 2290
CUENCA=MARGINAL=INTERNA, 396
CUENCA=OCEANICA, 3085
CUENCA=REPRESENTATIVA, 3819
CUENCA=SEDIMENTARIA=FOSIL, 3995
CUERPO=ARENA, 3937
CUERPO=MINERALIZADO, 3152
CUESTA, 1031
CUEVA, 678
CUMULO, 1032
CUÑA=DE=HIELO, 2162
CUÑA=DE=HIELO=FOSIL, 1621
CUPULA=VOLCANICA, 4763
CURIO, 1034
CURSO=AGUA, 4284
CURSO=ENCAJADO, 1384
CURTOSIS, 2401
CURVA=DE=NIVEL, 2329
CURVA=TIEMPO=RECORRIDO, 4582
CUSPIDE, 1037
CYCADALES, 1042
CYCADOFILICALES, 1043
CYCLOCYSTOIDEA, 1048
CYCLOSTOMATA, 1050
CYNOMORPHA, 1054
CYSTIPHYLLINA, 1055
CYSTOIDEA, 1056
D/H, 1058
DANIENSE, 1060
DARWINISMO, 1062
DASYCLADACEAE, 1063
DATACION, 1073
DATACION=AR, 297
DATACION=PB=ALFA, 2467
DATACION=TRAZA=FISION, 1556
DATACION=TRAZA=PARTICULA, 3309
DATO=MEB, 4034
DATO=MET, 4451
DATO=MICROSCOPIA=ELECTRONICA, 1350
DATO=MICROSONDA=ELECTRONICA, 1352
DATOS, 1064
DECAPODA, 1078
DECISION=EXPLOTABILIDAD, 1481
DECLINACION, 1081
DECONVOLUCION, 1084
DECREPITACION, 1086
DEDOLOMITIZACION, 1087
DEFLACION, 1095

DEFORMACION, 1096
DEFORMACION=BAJO-TENSION, 4267
DEFORMACION=ELASTICA, 1321
DEFORMACION=MEDIDA, 4268
DEFORMACION=POLIFASICA, 3556
DEFORMACION=SEDIMENTO-BLANDO, 4145
DEFORMACION=SUELO, 4147
DEFORMACION=TALUD, 4135
DEGRADACION, 1099
DEINOTHERIOIDEA, 1102
DELTA, 1103
DEMOSPONGEA, 1107
DENDRITA, 1108
DENDROCRONOLOGIA, 1110
DENDROGRAMA, 1111
DENDROIDEA, 1112
DENSIDAD, 1113
DENSIDAD=APARENTE, 590
DENTICION, 1115
DENUDACION, 1116
DEPOSITO=APOMAGMATICO, 244
DEPOSITO=CATATERMAL, 2379
DEPOSITO=ESTERIL, 415
DEPOSITO=FLUVIAL, 142
DEPOSITO=GLACIAR, 1792
DEPOSITO=MINERAL, 2866
DEPOSITO=RESIDUAL, 3829
DEPOSITO=SEDIMENTARIO, 3997
DEPREDACION, 3597
DEPRESION, 1121
DEPRESION=VULCANOTECTONICA, 4770
DEPURACION, 3687
DERIVA=CONTINENTAL, 922
DERIVA=LITORAL, 2544
DERIVA=POLAR, 3535
DERIVACION, 1125
DERMAPTEROIDA, 1127
DERMOPTERA, 1128
DERRUBIO, 4416
DESAGUE=ACIDO-DE-MINA, 38
DESALINIZACION, 1129
DESARROLLO, 1147
DESCARGA=AGUA-SUBTERRANEA, 1914
DESCENSO=NIVEL-DE-AGUA, 1249
DESCONTAMINACION, 1083
DESCUBRIMIENTO, 1217
DESECACION, 1135
DESEQUILIBRIO=URANIO, 4691
DESERTIZACION, 1134
DESGLACIACION, 1098
DESHIDRATACION, 1101
DESHIELO, 4485
DESIERTO, 1131
DESILIZACION, 1136
DESIMANTACION, 1106
DESIMANTACION=TERMICA, 4496
DESINTEGRACION=RADIOACTIVA, 3725

DESLIZAMIENTO=DE-BARRO, 2959
DESLIZAMIENTO=DE-MASAS-ROCOSAS, 3891
DESLIZAMIENTO=LADERA, 2426
DESLIZAMIENTO=TERRENO, 2431
DESMIDS, 1137
DESMOCERATIDA, 1138
DESMOSTYLIA, 1139
DESPEGUE, 1082
DESPLAZAMIENTO, 1224
DESPLAZAMIENTO=HORIZONTAL, 2053
DESPRENDIMIENTO, 3884
DESVIACION, 1149
DESVIACION=TIPICA, 4234
DESVITRIFICACION, 1150
DETERIORO=POR-RADIACION, 3721
DETERMINACION=MINERAL, 2865
DETERMINISMO, 1140
DEUTERIO, 1146
DEVONICO, 1151
DEVONICO=INF, 2581
DEVONICO=MEDIO, 2842
DEVONICO=SUP, 4671
DIACLASA, 2361
DIACLASAMIENTO, 2362
DIACRONISMO, 1153
DIADOQUIA, 1154
DIAFTORESIS, 1162
DIAFTORITA, 1163
DIAGENESIS, 1155
DIAGENESIS=PRECOZ, 1272
DIAGENESIS=TARDIA, 2438
DIAGNOSIS, 1156
DIAGRAFIA, 4816
DIAGRAFIA=ELECTRICA, 1324
DIAGRAFIA=ELECTROMAGNETICA, 1343
DIAGRAFIA=NEUTRON, 3017
DIAGRAFIA=SONICA, 41
DIAGRAMA, 1157
DIAGRAMA=DE-POLVO, 3587
DIAGRAMA=EQUILIBRIO, 3392
DIAGRAMA=ESTRUCTURAL, 4304
DIAGRAMA=LAUE, 344
DIAGRAMA=MONOCRISTAL, 4116
DIAGRAMA=PETROQUIMICO, 3371
DIALISIS, 1158
DIAMANTE, 1161
DIAMETRO=SONDEO, 621
DIAMICTITA, 1160
DIAPIRISMO, 1165
DIAPIRO, 1164
DIASTEMA, 1166
DIASTROFISMO, 1167
DIATOMEAE, 1169
DIATOMITA, 1170
DIATREMA, 1171
DIBUJO, 1250
DICOTYLEDONEAE, 1172

DICTYONELLIDINA, 1173
DIENTE, 4544
DIFERENCIACION=GRAVITACIONAL, 1877
DIFERENCIACION=HORIZONTE, 2049
DIFERENCIACION=MAGMATICA, 1178
DIFERENCIACION=METAMORFICA, 2782
DIFRACCION, 1180
DIFRACCION=RX, 4834
DIFUSION, 1181
DIFUSIVIDAD, 1182
DILATACION, 1187
DILATANCIA, 1186
DILUCION, 1188
DIMORFISMO, 1189
DIMORFISMO=SEXUAL, 4049
DINAMICA, 1267
DINAMOMETAMORFISMO, 1266
DINOCERATA, 1190
DINOFLAGELATA, 1191
DINOSAURIA, 1192
DIOGENITA, 1193
DIORITA, 1194
DIPLEUROZOA, 1200
DIPLOIDE, 1201
DIPLOPODA, 1202
DIPLORHINA, 1203
DIPNOI, 1205
DIPTERA, 1207
DIQUE, 1184
DIQUE=ANULAR, 3869
DIQUE=DE-ARENISCAS, 3940
DIQUE=DE-CONTENCION, 2491
DIRECCION, 4292
DIRECCION=HORIZONTAL, 2052
DISCOASTER, 1210
DISCONTINUIDAD, 1212
DISCONTINUIDAD=DE-BIRCH, 518
DISCONTINUIDAD=DE-BULLARD, 592
DISCONTINUIDAD=DE-CONRAD, 902
DISCONTINUIDAD=DE-GUTENBERG, 1929
DISCORBACEA, 1213
DISCORDANCIA, 4646
DISCORDANCIA=ANGULAR, 202
DISCORDANCIA=EROSION, 1211
DISCORDANCIA=IGNEO-SEDIMENTARIA, 3050
DISCORDANCIA=REGIONAL, 3796
DISCORDANCIA=TECTONICA, 1214
DISCORDANTE, 1215
DISLOCACION=CRISTALINA, 1014
DISOCIACION=IONICA, 1227
DISOLUCION, 1228
DISOLUCION=BAJO-PRESION, 3608
DISPERSION=ONDA, 4806
DISPERSION=OPTICA, 1222
DISPERSION=ROTATORIA, 3903
DISPROSIO, 1270
DISTANCIA, 3814

DISTRIBUCION, 1230
DISTRIBUCION-BINOMIAL, 480
DISTRIBUCION-DE-FRECUENCIA, 1645
DISTRIBUCION-ESTADISTICA, 4240
DISTRIBUCION-LOGNORMAL, 2563
DISTRIBUCION-NORMAL, 3055
DISTRIBUCION-POISSON, 3531
DISTRITO-METALOGENICO, 2770
DIVERGENCIA, 1232
DIVERSIDAD-ESPECIES, 4193
DIVISORIA-DE-AGUAS, 1245
DOCODONTA, 1233
DOCUMENTACION, 2229
DOLERITA, 1234
DOLINA, 1235
DOLINA-DE-HUNDIMIENTO, 4117
DOLOMIA, 1238
DOLOMITIZACION, 1237
DOMERIENSE, 1241
DOMINIO-ESTRUCTURAL, 1239
DOMINIO-MAGNETICO, 2631
DOMO, 1240
DOMO-DE-SAL, 3925
DOMO-GNEISICO, 1829
DORSAL, 4330
DORSAL-OCEANICA, 2840
DRAGADO, 1251
DRENAJE-TERRENO, 1247
DRUMLING, 1259
DRUSA, 1260
DUNA, 1261
DUNA-CONTINENTAL, 1132
DUNA-COSTERA, 812
DURACION-DEL-DIA, 2478
DURENO, 1262
DUREZA, 1956
DUREZA-AGUA, 4790
ECHINODERMATA, 1291
ECHINOIDEA, 1292
ECHINOZOA, 1293
ECLOGITA, 1295
ECOLOGIA, 1297
ECOLOGIA-HUMANA, 2069
ECOLOGIA-MARINA, 2676
ECONOMIA-MINERA, 2867
ECOSISTEMA, 1301
ECOSONDEO, 1294
ECOZONA, 1302
ECTOTROPHA, 1303
ECUACION-MATEMATICA, 1411
EDAD, 80
EDAD-ABSOLUTA, 10
EDAD-ACORDE, 878
EDAD-DE-LA-TIERRA, 81
EDAD-EXPOSICION, 1489
EDAD-GLACIACION, 191
EDAD-NO-ACORDE, 1216

EDAD-RELATIVA, 3804
EDAFOGENESIS, 3326
EDAFOLOGIA, 4157
EDENTATA, 1305
EDIFICIO, 586
EDRIOASTEROIDEA, 1307
EDRIOBLASTOIDEA, 1308
EFECTO-DE-CHOQUE, 2188
EFECTO-INVERNADERO, 1891
EFECTO-ISOTOPICO, 2350
EH, 1312
EIFELIENSE, 1313
EINSTEINIO, 1314
EJE, 393
EJE-DE-PLEGAMIENTO, 1603
ELASMOBRANCHII, 1316
ELASTICIDAD, 1323
ELECCION-DE-LUGAR, 4124
ELECTRICIDAD, 1334
ELECTRODO, 1337
ELECTROLISIS, 1338
ELECTROLITO, 1339
ELECTROMAGNETISMO, 1344
ELECTRON, 1345
ELECTROOSMOSIS, 1335
ELEMENTO-CALCOFILO, 703
ELEMENTO-COSMOGENICO, 973
ELEMENTO-FINITO, 1551
ELEMENTO-LITOFILO, 2533
ELEMENTO-MAGNETICO, 2632
ELEMENTO-NATIVO, 2988
ELEMENTO-NO-METALICO, 3048
ELEMENTO-SIDEROFILO, 4092
ELEMENTO-VOLATIL, 4756
ELEMENTOS-MENORES, 2887
ELEPHANTOIDEA, 1354
ELEVAMIENTO, 4686
ELUTRIACION, 1356
ELUVION, 1357
ELUVION-GRANITICO, 1923
EMBALSE-SUBTERRANEO, 1913
EMBRECHITA, 1359
EMBRITHOPODA, 1360
EMERSION, 1361
EMISARIO, 1310
EMPAQUETADO-ATOMICO-CERRADO, 798
EMPAQUETADURA-GRAVA, 1872
EMPLAZAMIENTO, 1363
EMSIENSE, 1364
ENCLAVE-ROCA, 4838
ENCRINITA, 1366
ENDOCERATOIDEA, 1369
ENDORREISMO, 298
ENDURECIMIENTO, 2218
ENERGIA, 1373
ENERGIA-ACTIVACION, 55
ENERGIA-GEOTERMICA, 1773

ENERGIA-LIBRE, 1641
ENERGIA-NUCLEAR, 3062
ENERGIA-RELAJACION, 3805
ENFRIAMIENTO, 942
ENLACE-INTERATOMICO, 364
ENMIENDA-SUELO, 4150
ENRIQUECIMIENTO, 1376
ENSAYO, 336
ENSAYO-BOMBEO, 261
ENSAYO-CARRETERA, 3878
ENSAYO-CIZALLADURA, 4712
ENSAYO-COMPRENSION, 868
ENSAYO-DE-PLACA, 3500
ENSAYO-LABORATORIO, 2403
ENSEÑANZA, 1309
ENTALPIA, 1382
ENTEROPNEUSTA, 1381
ENTIERRO, 596
ENTISOL, 1383
ENTROPIA, 1385
EOCENO, 1389
EOCENO-INF, 2582
EOCENO-MEDIO, 2843
EOCENO-SUP, 4672
EOCRINOIDEA, 1390
EON, 1394
EOSUCHIA, 1395
EPIBIOTISMO, 1398
EPIBOLITA, 1399
EPICENTRO, 1400
EPIDOTIZACION, 1402
EPIEUGEOSINCLINAL, 1403
EPIGENESIS, 1404
EPIMAGMATICO, 1145
EPIROGENESIS, 1396
EPISODIO-JARAMILLO, 2358
EPITAXIA, 1406
EPOCA, 1409
EPOCA-GAUSS, 1699
EPOCA-MATUYAMA, 2720
EPOCA-METALOGENICA, 2771
EQUILIBRIO, 1416
EQUILIBRIO-DINAMICO, 1264
EQUILIBRIO-QUIMICO, 724
EQUIPO-SISMICO, 4013
EQUISETALES, 1417
EQUIVALENTE-CALOR-DE-FUSION, 1966
ERA, 1418
ERATEMA, 1419
ERATOSTENSIENSE, 1420
ERBIO, 1421
ERG, 1422
EROSION, 1424
EROSION-AGUA, 4789
EROSION-EOLICA, 4827
EROSION-FLUVIAL, 1594
EROSION-GLACIAR, 1794

EROSION=LITORAL, 813
EROSION=MARINA, 2678
EROSION=POR=ARROYADA, 4065
EROSION=SUELO, 4148
ERROR, 1431
ERTS=LANDSAT, 2430
ERUPCION, 1432
ESCALA, 2669
ESCALA=ABSOLUTA, 4536
ESCALA=ESTRATIGRAFICA, 1734
ESCALA=FI, 3397
ESCALA=MERCALLI=MODIFICADA, 2905
ESCALA=RICHTER, 3862
ESCAMA, 3963
ESCANDIO, 3956
ESCAPE, 1244
ESCAPOLITIZACION, 3959
ESCARPE, 1433
ESCARPE=DE=FALLA, 1511
ESCARPE=DE=PLAYA, 447
ESCLERITES, 3966
ESCOLECODONTO, 3967
ESCOLITO, 3968
ESCOMBRERA, 4217
ESCOMBRERA=DE=TESTIGOS, 1039
ESCORIA, 3969
ESCORIA VULCANICA, 769
ESCORRENTIA, 3915
ESCORRENTIA=AGUA=SUBTERRANEA, 1918
ESCORRENTIA=LLUVIA, 3744
ESCORRENTIA=POLIFASICA, 2964
ESCORRENTIA=SUPERFICIAL, 4364
ESCUDO, 4072
ESFENOLITO, 4205
ESFERICIDAD, 4207
ESFERULA, 4208
ESFERULITO, 4209
ESKER, 1434
ESPACIO SUBTERRANEO, 4651
ESPACIO=INTERESTELAR, 2285
ESPACIO=INTERPLANAR, 2268
ESPALACION, 4189
ESPARAGMITA, 4190
ESPATO=FLUOR, 1590
ESPECIMEN=CARACTERISTICO, 4626
ESPECTRO, 4202
ESPECTRO=ABSORCION, 15
ESPECTRO=MOSSBAUER, 2948
ESPECTRO=OPTICO, 3143
ESPECTRO=RPE, 1410
ESPECTROFOTOMETRIA, 4200
ESPECTROMETRIA, 4201
ESPECTROMETRIA=ALFA, 148
ESPECTROMETRIA=DE=ABSORCION, 14
ESPECTROMETRIA=DE=EMISION, 1362
ESPECTROMETRIA=DE=MASA, 2700
ESPECTROMETRIA=GAMMA, 1677

ESPECTROMETRIA=HERTZIANA, 3723
ESPECTROMETRIA=IR, 2234
ESPECTROMETRIA=MICROONDA, 2839
ESPECTROMETRIA=MOLECULAR, 2914
ESPECTROMETRIA=MOSSBAUER, 2947
ESPECTROMETRIA=OPTICA, 3142
ESPECTROMETRIA=RAMAN, 3745
ESPECTROMETRIA=RX, 4836
ESPECTROMETRO, 4199
ESPEJO=FALLA, 4132
ESPELEOLOGIA, 4203
ESPELEOTEMA, 4204
ESPILITIZACION, 4210
ESPONJOLITA, 4218
ESPORA, 4219
ESPOROMORFO, 4220
ESQUISTO, 3960
ESQUISTO=AZUL, 534
ESQUISTO=BITUMINOSO, 524
ESQUISTO=MOTEADO, 4222
ESQUISTO=PETROLIFERO, 3110
ESQUISTO=SERICITICO, 4042
ESQUISTO=VERDE, 1892
ESQUISTOSIDAD, 3961
ESSEXITA, 1435
ESTABILIDAD, 4226
ESTABILIDAD=TALUD, 4137
ESTABILIZACION, 4227
ESTABILIZACION=TERRENO, 4158
ESTACADA, 1901
ESTACION=LUNAR, 2604
ESTACION=SISMICA, 4018
ESTADISTICA, 4245
ESTADO=FISICO, 3428
ESTADO=HIDROTERMAL, 2133
ESTALACTITA, 4231
ESTALAGMITA, 4232
ESTAÑO, 4537
ESTE, 1288
ESTEATIZACION, 4247
ESTEFANIENSE, 4250
ESTEQUIOMETRIA, 4259
ESTEREOQUIMICA, 4252
ESTERIL, 4413
ESTERO, 4527
ESTEROLES, 4253
ESTIAJE, 2573
ESTILO=TECTONICO, 4438
ESTILOLITO, 4314
ESTRATEGIA, 4271
ESTRATIFICACION, 4272
ESTRATIFICACION=CONCORDANTE, 3295
ESTRATIFICACION=CRUZADA, 998
ESTRATIFICACION=FINA, 453
ESTRATIFICACION=MASIVA, 2704
ESTRATIFICACION=OBLICUA, 997
ESTRATIFICACION=PLANA, 3466

ESTRATIGRAFIA, 4281
ESTRATO, 2462
ESTRATO=CONVOLUTO, 941
ESTRATOTIPO, 4282
ESTRATOVOLCAN, 4283
ESTRECHO, 4269
ESTRIA, 1668
ESTRIA=GLACIAR, 1800
ESTRIACION, 1514
ESTRIACION=MINERAL, 2874
ESTROMBOLIANO, 4297
ESTRONCIO, 4299
ESTRUCTURA=ACANALADA, 1591
ESTRUCTURA=ALGAREA, 118
ESTRUCTURA=ALMOHADILLADA, 2961
ESTRUCTURA=ANULAR, 3871
ESTRUCTURA=ARQUEADA, 270
ESTRUCTURA=ATOMO, 365
ESTRUCTURA=BIOGENA, 494
ESTRUCTURA=CAOTICA, 709
ESTRUCTURA=CILINDRICA, 1053
ESTRUCTURA=CORRIENTE=TURBIDEZ, 4618
ESTRUCTURA=CRIPTOVOLCANICA, 1010
ESTRUCTURA=CRISTALINA, 1020
ESTRUCTURA=DE=DESPLOME, 837
ESTRUCTURA=DIAGENETICA, 3990
ESTRUCTURA=EN=BANDAS, 409
ESTRUCTURA=EN=LLAMA, 1563
ESTRUCTURA=ESTRATIFICADA, 4273
ESTRUCTURA=GEOLOGICA, 1737
ESTRUCTURA=GLANDULAR, 370
ESTRUCTURA=IMPACTO, 2192
ESTRUCTURA=INTERNA, 2275
ESTRUCTURA=LAMINAR, 2421
ESTRUCTURA=LENTICULAR, 1565
ESTRUCTURA=MILONITICA, 2973
ESTRUCTURA=ORGANICA, 3175
ESTRUCTURA=PILLOW, 3448
ESTRUCTURA=PRIMARIA, 3612
ESTRUCTURA=RETICULADA, 3840
ESTRUCTURA=SEDIMENTARIA, 4001
ESTRUCTURA=SINSEDIMENTARIA, 4398
ESTRUCTURA=SLUMPING, 4138
ESTRUCTURA=VELOCIDAD, 4727
ESTRUCTURA=VERMIFORME, 4733
ESTRUCTURA=VULCANOTECTONICA, 4771
ESTRUCTURA=ZONAL, 4865
ESTUARIO, 1438
ESTUDIO=CRITICO, 995
ESTUDIO=DE=CAMPO, 1542
ESTUDIO=DE=UN=CASO, 663
ESTUDIO=EN=MAR, 2682
ESTUDIO=FACTIBILIDAD, 1520
ESTUDIO=GEOTECNICO, 1770
ESTUDIO=HIDROGEOLOGICO, 2113
ESTUDIO=IMPACTO=MEDIO, 2191
ESTUDIO=MEDIO, 1387

ESTUDIO=PRELIMINAR, 3178
ETANO, 1440
ETAPA=GLACIAR, 1799
ETER, 1441
EUCILIATA, 1442
EUCRITA, 1443
EUGEOSINCLINAL, 1445
EUGLENOPHYTA, 1446
EURIHALINO, 1449
EURITERMO, 1450
EUROPIO, 1448
EURYAPSIDA, 4389
EUSELACHII, 1451
EUSTATISMO, 1452
EUTECTICA, 1454
EUTROFIZACION, 1455
EVALUACION, 1457
EVALUACION=YACIMIENTO, 1117
EVAPORACION, 1458
EVAPORITA, 1459
EVAPOTRANSPIRACION, 1460
EVOLUCION=BIOLOGICA, 1461
EXAEDRITA, 2003
EXCAVACION, 1462
EXCENTRICIDAD, 1290
EXCESO=ARGON, 1463
EXCURSION, 1465
EXFOLIACION, 1466
EXFOLIACION=BASAL, 423
EXFOLIACION=CONCORDANTE, 455
EXFOLIACION=FLUIDAL, 1576
EXFOLIACION=MINERAL, 2864
EXFOLIACION=PIZARROSA, 4131
EXFOLIACION=SEGUN=PLANO=AXIAL, 392
EXINA, 1468
EXINITA, 1469
EXOBIOLOGIA, 1470
EXORREISMO, 1472
EXPANSION, 1475
EXPANSION=FLUVIOGLACIAR, 3211
EXPANSION=FONDO=OCEANICO, 3980
EXPEDICION=CRUCERO, 1476
EXPERIMENTO, 1479
EXPLORACION, 3641
EXPLORACION=AGUA=SUBTERRANEA, 1915
EXPLOSION, 1485
EXPLOSION=NUCLEAR, 3063
EXPLOSION=QUIMICA, 726
EXPLOTACION, 1482
EXPLOTACION=MINERA, 2883
EXPLOTACION=POR=DISOLUCION, 4179
EXPLOTACION=POR=HUNDIMIENTO, 680
EXPORTACION, 1488
EXPOSICION=RELIEVE, 3809
EXSOLUCION, 1490
EXTENSION=ESTRATIGRAFICA, 3746
EXTENSOMETRO, 1492

EXTERNIDES, 1493
EXTINCION, 1496
EXTINCION=GIRATORIA, 4657
EXTINCION=OPTICA, 3138
EXTRAPOLACION, 1497
EXUDACION, 4008
EXUDACION=PETROLEO, 3109
FABRICA, 1500
FACIES, 1501
FACIES=ACTINOTA, 53
FACIES=ANFIBOLITA, 174
FACIES=ANFIBOLITA=ALBITA=EPIDOTA, 104
FACIES=ANFIBOLITA=CORDIERITA, 954
FACIES=ANFIBOLITA=EPIDOTA, 1401
FACIES=BORDE, 546
FACIES=CORNEANA, 2058
FACIES=CORNEANA=ALBITA=EPIDOTA, 105
FACIES=CORNEANA=HORNEBLENDA, 2055
FACIES=CORNEANA=PIROXENICA, 3703
FACIES=ECLOGITA, 1296
FACIES=ESQUISTO=VERDE, 1893
FACIES=GEOQUIMICA, 1717
FACIES=GLAUCOFANA, 1818
FACIES=GRANULITA, 1859
FACIES=HIDROQUIMICA, 2100
FACIES=IGNEA, 2174
FACIES=METAMORFISMO, 2783
FACIES=MINERAL, 2869
FACIES=PREHNITA=PUMPELLYITA, 3603
FACIES=SANIDINA, 3942
FACIES=ZEOLITA, 4857
FACOLITO, 3385
FACTOR=TIEMPO, 4535
FAEOZEM, 3388
FALLA, 1507
FALLA=ACTIVA, 56
FALLA=ANTITETICA, 233
FALLA=ARQUEADA, 289
FALLA=DE=CHARNELA, 2015
FALLA=DESPLAZAMIENTO=HORIZONTAL, 2054
FALLA=EN=ESCALERA, 4248
FALLA=EN=ESCALON, 1365
FALLA=ESTRATIFICADA, 456
FALLA=HORIZONTAL, 4293
FALLA=HUNDIMIENTO, 1199
FALLA=LATERAL, 2442
FALLA=NORMAL, 3056
FALLA=OBLICUA, 3077
FALLA=SINGENETICA, 1921
FALLA=TRANSCURRENTE, 4567
FALLA=TRANSFORMANTE, 4569
FALLA=TRANSVERSAL, 1196
FALLA=VERTICAL=DESGARRE, 4831
FALLADO, 1515
FALSO=CRUCERO, 4133
FAMENIENSE, 1503
FANEROZOICO, 3390

FANGLOMERADO, 1505
FANGO, 2952
FASE, 3391
FASE=BAIKALIANA, 404
FASE=FLUIDA, 1582
FASE=GASEOSA, 1694
FASE=LIQUIDA, 2520
FASE=MINERAL, 2871
FASE=OROGENICA=CASCADIENSE, 662
FASE=SOLIDA, 4166
FAUNA, 1516
FAUNA=ESPECIFICA, 4194
FELDESPATIZACION, 1525
FELDESPATO, 1524
FENITA, 1529
FENITIZACION, 1530
FENOCRISTAL, 3395
FENOL, 3396
FERMIO, 1531
FERROD, 1535
FIGURA=BASAL, 4163
FIJACION=ION, 1558
FIJISMO, 1559
FILICOPSIDA, 1543
FILITA, 3419
FILOGENIA, 3423
FILON, 4725
FILON=CLASTICO, 785
FILON=SEGREGADO, 4009
FILONES=RAMIFICADOS, 2514
FILONITA, 3421
FILOSOFIA, 3398
FILTRACION, 1547
FILTRADO, 1546
FILTRO, 4817
FILTRO=TAMIZ, 1545
FILUM, 3424
FINANCIACION, 1548
FIORDO, 1560
FISICA=DEL=GLOBO, 4169
FISICOQUIMICA, 3425
FISIOLOGIA, 3432
FISION=NUCLEAR, 1554
FISSIPEDIA, 1557
FISURACION=PRISMATICA, 850
FITAN, 3433
FITOCENOSIS, 3434
FITOPLANCTON, 3437
FLANCO, 2505
FLANDRIENSE, 1564
FLECHA=LITORAL, 4215
FLEXURA, 1566
FLOCULACION, 1568
FLORA, 1571
FLORA=ESPECIFICA, 4195
FLORIZONA, 1573
FLOTACION, 1574

FLUCTUACION, 1578
FLUIDO-MINERALIZADOR, 2877
FLUJO-DE-TIERRA, 1283
FLUJO-GEOTERMICO, 1967
FLUJO-MATERIAL-DETRITICO, 1077
FLUOR, 1589
FLUORATO, 1586
FLUORESCENCIA, 1585
FLUORESCENCIA-RX, 4835
FLUORIMETRIA, 1587
FLUORURACION, 1588
FLUVENT, 1592
FLUVIO-GLACIAR, 1596
FLUVIOSOL, 1598
FLYSCH, 1599
FOCO-EXPLOSIONES, 1487
FOIDITA, 1601
FOLIACION, 1609
FONDO-MARINO, 3088
FONDO-REGIONAL, 398
FONDO-SOMERO, 4074
FONOLITA, 3401
FORAMINIFERA, 1610
FORMA, 4054
FORMA-BOTRIOIDAL, 553
FORMA-CRISTALINA, 1016
FORMA-GLOBO, 4055
FORMA-GRANO, 3308
FORMACION, 1617
FORMACION-FERRIFERA, 2311
FORMACION-RECUBRIMIENTO, 979
FORMACION-SUPERFICIAL, 4369
FORMULA, 1618
FOSA-ABISAL, 4587
FOSA-ABISAL-ALARGADA, 4603
FOSA-DE-HUNDIMIENTO, 3865
FOSA-TECTONICA, 1840
FOSFATIZACION, 3404
FOSFATO, 3403
FOSFORESCENCIA, 3405
FOSFORITA, 3406
FOSFORO, 3407
FOSIL, 1619
FOSIL-CARACTERISTICO, 2206
FOSIL-ESPECIFICO, 1925
FOSIL-QUIMICO, 727
FOSIL-VIVIENTE, 2549
FOSILIZACION, 1626
FOTO-IR, 2232
FOTOGEOLOGIA, 3410
FOTOGRAFIA, 3412
FOTOGRAFIA-AEREA, 71
FOTOGRAMETRIA, 3411
FOTOMETRIA, 3413
FOTOMETRIA-LLAMA, 1562
FOTOSINTESIS, 3414
FOYAITA, 1629

FRACCION-FINA, 1549
FRACCION-GRUESA, 810
FRACCIONAMIENTO, 1630
FRACCIONAMIENTO-ISOTOPICO, 2353
FRACTURA, 1632
FRACTURA-ANULAR, 3870
FRACTURA-CONCOIDEA, 875
FRACTURA-DE-CIZALLAMIENTO, 4059
FRACTURACION, 1635
FRACTURACION-HIDRAULICA, 2092
FRAGIL, 579
FRAGMENTACION, 1637
FRAGMENTO, 1636
FRANCIO, 1639
FRANJA-DE-REACCION, 3758
FRANJA-DE-SEGREGACION, 3962
FRASNIENSE, 1640
FREATOFITA, 3417
FRECUENCIA, 1644
FRENTE, 1652
FRICCION, 1650
FRUCTIFICACION, 1657
FUEL-FOSIL, 1620
FUENTE-AGUA-MINERAL, 2873
FUENTE-INTERMITENTE, 2271
FUENTE-KARSTICA, 2376
FUENTE-NO-EXPLOSIVA, 3046
FUENTE-SISMICA, 4016
FUENTE-TERMAL, 2066
FUENTES-DE-CALOR, 1968
FUGA, 2470
FUGACIDAD, 1659
FULGURITA, 1660
FUMAROLA, 1664
FUNCION, 1665
FUNGI, 1667
FUSINITA, 1669
FUSION, 2744
FUSION-INCONGRUENTE, 2204
FUSION-PARCIAL, 3305
FUSULINIDAE, 1670
GABRO, 1672
GADOLINIO, 1673
GALERIA-DE-CAPTACION, 2227
GALIO, 1675
GANGA, 1680
GANISTER, 1681
GARGANTA, 1837
GAS, 1684
GAS-CARBON, 804
GAS-CARBONICO, 643
GAS-FREATICO, 3415
GAS-HIDROGENO, 2107
GAS-NATURAL, 2989
GAS-NITROGENO, 3035
GAS-OXIGENO, 3226
GAS-PANTANOS, 2694

GAS-RARO, 2224
GAS-SULFHIDRICO, 4354
GAS-SULFUROSO, 2108
GASIFICACION, 1695
GASTROPODA, 1696
GASTROZOIDE, 1697
GEANTICLINAL, 1700
GEDINIENSE, 1701
GEL-COLOIDAL, 1702
GELIVACION, 1704
GEMA, 1707
GEMACION, 1705
GEMOLOGIA, 1706
GENERACION, 1708
GENERADOR-CHISPA, 4191
GENESIS, 1709
GENESIS-ROCA-IGNEA, 2175
GENESIS-YACIMIENTO, 1118
GENOMORFO, 1710
GENOTIPO, 1711
GEOBAROMETRIA, 1712
GEOBIOS, 1713
GEOCRONOLOGIA, 1725
GEODA, 1726
GEODESIA, 1727
GEODINAMICA, 1729
GEOESTADISTICA, 1762
GEOFISICA, 1759
GEOFISICA-APLICADA, 247
GEOFISICO, 1758
GEOFONO, 1750
GEOFRACTURA, 1730
GEOGRAFIA, 1731
GEOGRAFIA-MATEMATICA, 2710
GEOIDE, 1732
GEOISOTERMA, 1733
GEOLOGIA, 1741
GEOLOGIA-APLICADA, 246
GEOLOGIA-DEL-INGENIERO, 1374
GEOLOGIA-DEL-MEDIO-AMBIENTE, 1388
GEOLOGIA-DEL-PETROLEO, 3381
GEOLOGIA-DEL-SUBSUELO, 4340
GEOLOGIA-DINAMICA, 1265
GEOLOGIA-DIVULGACION, 3560
GEOLOGIA-ECONOMICA, 1298
GEOLOGIA-ESTRUCTURAL, 4306
GEOLOGIA-EXTRATERRESTRE, 1498
GEOLOGIA-GLACIAR, 1796
GEOLOGIA-HISTORICA, 2019
GEOLOGIA-LUNAR, 2602
GEOLOGIA-MARINA, 2679
GEOLOGIA-MATEMATICA, 2711
GEOLOGIA-MINERA, 2885
GEOLOGIA-REGIONAL, 3793
GEOMAGNETISMO, 1745
GEOMETRIA, 1746
GEOMETRIA-CANAL, 708

GEOMORFOLOGIA, 1748
GEONOMIA, 1749
GEOONDULACION, 1781
GEOPRESION, 1760
GEOQUIMICA, 1722
GEOQUIMICA=ISOTOPICA, 2351
GEOQUIMICA=ORGANICA, 3168
GEOSECS, 1761
GEOSINCLINAL, 1767
GEOSINCLINAL=SIALICO, 1377
GEOSINCLINAL=SIMATICO, 1378
GEOTECNIA, 1771
GEOTECTONICA, 1772
GEOTERMIA, 1779
GEOTERMOMETRIA, 1780
GERMANIO, 1782
GESTION=RECURSOS=AGUA, 4791
GEYSER, 1784
GEYSERITA, 1785
GIGANTISMO, 1787
GIGANTOSTRACA, 1788
GINKGOALES, 1789
GIVETIENSE, 1790
GLACIACION, 1803
GLACIACION=CONTINENTAL, 924
GLACIAR, 1791
GLACIAR, 1804
GLACIAR=COLGADO, 1075
GLACIAR=CONTINENTAL, 2160
GLACIAR=ROCAS, 3886
GLACIOLOGIA, 1807
GLACIOTECTONICO, 1809
GLACIS, 1810
GLACIS=LLANURA=ABISAL, 927
GLAUCONITA, 1815
GLAUCONITIZACION, 1817
GLEY, 1819
GLOBIGERINACEA, 1822
GLOMERO, 1823
GLOSARIO, 1824
GLOSSOPTERIDALES, 1825
GNATHOSTOMI, 1826
GNATHOSTRACA, 1827
GNEIS, 1828
GNEIS=GRANITICO, 1846
GNETALES, 1831
GOLFO, 1927
GONDWANA, 1834
GONIATITIDA, 1835
GONIOMETRIA, 1836
GRADIENTE, 1842
GRADIENTE=GEOTERMICO, 1775
GRADIENTE=HIDRAULICO, 2093
GRADIOMETRO, 1843
GRADO=CARBONIFICACION, 3748
GRADO=DE=LIBERTAD, 1100
GRADO=METAMORFICO, 2784

GRADO=REDONDEZ, 3906
GRAFICO, 1863
GRAFITIZACION, 1867
GRAFITO, 1866
GRAN=ESCALA, 2436
GRANATE, 1683
GRANITICO, 1847
GRANITIZACION, 1850
GRANITO, 1845
GRANITO=ALCALINO, 125
GRANITOIDE, 1851
GRANIZO, 1941
GRANO, 1844
GRANO=MEDIO, 2736
GRANO=REVESTIDO, 817
GRANODIORITA, 1853
GRANOGABRO, 1854
GRANOSELECCION, 1841
GRANULACION, 1857
GRANULITA, 1858
GRANULITICO, 1860
GRANULOMETRIA, 1861
GRAPTOLITHINA, 1868
GRAPTOLOIDEA, 1869
GRAUWACA, 1888
GRAVA, 1871
GRAVA=GRUESA, 809
GRAVEDAD=ABSOLUTA, 11
GRAVIMETRIA, 1874
GRAVIMETRO, 1873
GRAVITACION, 1875
GREISEN, 1896
GREISENIZACION, 1897
GRIETA=DE=HIELO, 990
GRIETA=DESECACION, 2954
GRIETA=RETRACCION, 4088
GRUPO=DEL=PLATINO, 3505
GRUPO=ESPACIAL, 4188
GUANO, 1924
GUERRA, 2859
GUIA, 3157
GUIA=ESTRUCTURAL, 4307
GUIJARRO, 3907
GUYOT, 3984
GYMNOSPERMAE, 1931
GYMOLAEMATA, 1930
GYTJA, 1934
GZHELIENSE, 1935
HABITAT, 1937
HABITO, 1936
HAFNIO, 1940
HALITICO, 1945
HALOFILO, 1948
HALOFITA, 1949
HALOGENO, 1947
HALURO, 1944
HAMMADA, 1950

HARDGROUND, 1955
HARPOLITO, 1958
HASTIAL, 4780
HAUTERIVIENSE, 1959
HAWAIANO, 1961
HE=HE, 1979
HELICOPLACOIDEA, 1976
HELIO, 1978
HELIOZOA, 1977
HELVETIENSE, 1980
HEMATITES, 1981
HEMICRUSTACEA, 1982
HEMIPTEROIDA, 1983
HERCINIDES, 1985
HEREDADO, 2237
HERENCIA, 1986
HETEROCORALLIA, 1991
HETEROCRONISMO, 1989
HETEROCRONO, 1990
HETERODONTA, 1992
HETEROGENEIDAD, 1993
HETEROGENEO, 1994
HETEROMORFISMO, 1996
HETEROMORFO, 1995
HETTANGIENSE, 1999
HEXACORALLA, 2000
HEXACTINIARIA, 2001
HIATO, 2004
HIBRIDO, 2086
HIDATOGENESIS, 2087
HIDRATACION, 2089
HIDRATACION=VIDRIO, 2090
HIDRATO=DE=CARBONO, 641
HIDRAULICA, 2098
HIDROCARBURO, 1658
HIDROCARBURO=ALIFATICO, 122
HIDROCARBURO=AROMATICO, 305
HIDRODINAMICA, 2104
HIDRODINAMICA=CAPTACION, 672
HIDROEXPLOSION, 2105
HIDROFONO, 2125
HIDROGENO, 2106
HIDROGEOLOGIA, 2115
HIDROGEOLOGIA=KARST, 2375
HIDROGEOLOGIA=REGIONAL, 3794
HIDROGRAFIA, 2117
HIDROGRAMA, 2116
HIDROLACOLITO, 3450
HIDROLOGIA=SUPERFICIE, 2121
HIDROMETALURGIA, 2122
HIDROMETRIA, 2124
HIDROMETRO, 2123
HIDROQUIMICA, 2102
HIDROSFERA, 2126
HIDROSPIRA, 2127
HIDROTERMAL, 2130
HIDROXIDO, 2136

HIELO, 2155
HIELO-MARINO, 3976
HIELO-SUELO, 1905
HIERRO, 2308
HIERRO-DE-LOS-PANTANOS, 537
HIERRO-FERRICO, 1533
HIERRO-FERROSO, 1538
HIGROFILO, 2138
HINCHAMIENTO, 4378
HIPOCENTRO, 2146
HIPODIGMA, 2148
HIPOTESIS, 2150
HIPOVOLCANICO, 4343
HIPPOMORPHA, 2016
HIPSOMETRIA, 2151
HISTOGRAMA, 2017
HISTOLOGIA, 2018
HISTORIA-DE-LA-GEOLOGIA, 2021
HISTORICA, 2020
HISTOSOL, 2022
HODOGRAFO, 2023
HOJA, 2469
HOLMIO, 2025
HOLOCENO, 2026
HOLOCEPHALI, 2027
HOLOFITA, 2030
HOLOGRAFIA, 2029
HOLOSTEI, 2031
HOLOSTOMADO, 2032
HOLOTHUROIDEA, 2033
HOLOTIPO, 2034
HOMALOZOA, 2035
HOMBRE-FOSIL, 1622
HOMINIDAE, 2038
HOMO-SAPIENS, 2039
HOMOCLINAL, 2040
HOMOGENEIDAD, 2041
HOMOGENEIZACION, 2042
HOMOIOSTELEA, 2043
HOMOLOGIA, 2044
HOMOMORFISMO, 2036
HOMOMORFO, 2037
HOMONIMO, 2045
HOMOSTELEA, 2046
HOMOTAXIA, 2047
HORIZONTAL, 2050
HORIZONTE-PRODUCTIVO, 3626
HORIZONTE-SUELO, 4149
HORNBLENDITA, 2056
HORST, 2060
HOWARDITA, 2067
HUELLA-CLIMATICA, 796
HUELLA-CORRIENTE, 1036
HUELLA-DE-LLUVIA, 3743
HUELLA-DERRUBIO, 3970
HUELLA-MECANICA, 715
HUELLA-NUTRICION, 1523

HUELLA-ORGANICA, 4557
HUEVO, 1311
HULLA-GRASA, 1687
HUMEDAD, 2075
HUMEDAD-SUELO, 2911
HUMIFICACION, 2076
HUMIN, 2077
HUMITA, 2078
HUMMOCK, 2079
HUMOD, 2080
HUMOX, 2081
HUMULT, 2082
HUMUS, 2083
HUNDIMIENTO, 834
HUNDIMIENTO-GEOQUIMICO, 1721
HUNDIMIENTO-TECHO, 3894
HYALOSPONGEA, 2085
HYDRARCH, 2088
HYDROIDA, 2118
HYDROZOA, 2137
HYMENOPTERA, 2140
HYOLITHES, 2141
HYRACOIDEA, 2152
HYSTRICHOMORPHA, 2153
ICEBERG, 2163
ICHNITES, 2164
ICHTHYOPTERYGIA, 2166
ICHTHYOSAURIA, 2167
ICNOLOGIA, 2165
IDIOBLASTO, 2168
IDIOGEOSINCLINAL, 2169
IDIOMORFISMO, 2171
IGNIMBRITA, 2177
IMAGEN, 2178
IMANTACION, 2648
IMANTACION-INDUCIDA, 2215
IMANTACION-ISOTERMORREMANENTE, 2347
IMANTACION-REMANENTE, 3812
IMANTACION-REMANENTE-ANHISTERETICA, 206
IMANTACION-REMANENTE-DETRITICA, 1143
IMANTACION-REMANENTE-NATURAL, 2991
IMANTACION-REMANENTE-QUIMICA, 732
IMANTACION-REMANENTE-VISCOSA, 4747
IMANTACION-TERMORREMANENTE, 4511
IMBRICACION, 2181
IMBRIENSE, 2179
IMPACTITA, 2193
IMPACTO, 4075
IMPERFECCION-CRISTALINA, 1013
IMPERFECCION-DE-SUPERFICIE, 4363
IMPORTACION, 2194
IMPRESION-FOSIL, 1903
IMPUREZAS-AGUA, 2196
IN-SITU, 2197
INARTICULATA, 2198
INCEPTISOL, 2199
INCLINACION, 4533

INCLINACION-MAGNETICA, 2635
INCLUSION, 2200
INCLUSION-FLUIDA, 1579
INCLUSION-GAS, 1690
INCLUSION-MINERAL, 2870
INCOHERENTE, 2201
INCRUSTACION, 2205
INDICADOR, 2212
INDICADOR-BATIMETRICO, 1123
INDICADOR-GEOQUIMICO, 1718
INDICE-COLOR, 847
INDICE-CRISTALIZACION, 1026
INDICE-DE-MADUREZ, 2719
INDICE-DE-MILLER, 2860
INDICE-DE-REFRACCION, 2209
INDICE-DIFERENCIACION, 1179
INDICIO-MINERAL, 4086
INDIENSE, 2210
INDIO, 2213
INDOCHINITA, 2214
INDUSTRIA, 2222
INDUSTRIA-QUIMICA, 728
INELASTICIDAD, 198
INFILTRACION, 2226
INFORMATICA-GEOLOGICA, 2228
INFORME, 3818
INFORME-ACTIVIDAD, 3635
INFRACAMBRICO, 2230
INFRAESTRUCTURA, 2235
INFRARROJO, 2231
INGENIERIA-CIVIL, 777
INGENIERIA-MINERA, 2884
INGENIERIA-SISMICA, 1285
INGRESION, 2236
INHOMOGENEIDAD, 2238
INMADURO, 2182
INMERSION, 4334
INMISCIBILIDAD, 2185
INMISCIBLE, 2186
INOCERAMI, 2246
INORGANICO, 2247
INSECTA, 2251
INSECTIVORA, 2252
INSECUENTE, 2254
INSOLACION, 2255
INSTALACION-MARINA, 2680
INSTALACION-SUBMARINA, 4328
INSTITUCION, 2257
INSTRUMENTACION, 2258
INSTRUMENTOS-GEOFISICOS, 1753
INTERCALACION, 2260
INTERCRECIMIENTO, 2267
INTERFASE, 2261
INTERFASE-ACEITE-GAS, 3113
INTERFASE-AGUA-PETROLEO, 3114
INTERFASE-AIRE-MAR, 94
INTERFASE-SEDIMENTO-AGUA, 3994

INTERFERENCIA, 2262
INTERFEROMETRIA, 2263
INTERFLUVIO, 2264
INTERFORMACIONAL, 2265
INTERNIDAS, 2279
INTERPLUVIAL, 2281
INTERPOLACION, 2282
INTERPRETACION, 2283
INTERSERTAL, 2284
INTRACLASTO, 2288
INTRATELURICO, 2292
INTRUSION, 2293
INTRUSION=AGUA=SALADA, 3930
INTRUSION=PLANO=FALLA, 4162
INTRUSIVO, 2294
INVENTARIO, 2295
INVERSION, 3842
INVERSION=DE=CAMPO, 1743
INVERSION=RELIEVE, 3810
INVERTEBRATA, 2297
INVESTIGACION, 4371
INVESTIGACION=CIENTIFICA, 3822
INYECCION, 2239
INYECCION=FLUIDA, 1580
INYECCION=LECHADA, 1920
IO=TH, 2302
IO=U, 2303
ION, 2299
IONIZACION, 2304
IONOSFERA, 2305
IPRESIENSE, 4852
IRIDIO, 2307
IRIDISCENCIA, 2306
IRREVERSIBILIDAD, 2313
IRRIGACION, 2314
ISLA, 2315
ISLA=DE=DELTA, 2957
ISOANOMALIA, 2318
ISOBARA, 2319
ISOBASE, 2320
ISOBATA, 2321
ISOCLINAL, 2324
ISOCORA, 3393
ISOCORA=ESTRATIGRAFICA, 4277
ISOCRONA, 2323
ISOESTRUCTURAL, 2345
ISOGONA, 2327
ISOGRADA, 2328
ISOLITO, 2331
ISOMORFISMO, 2334
ISOMORFO, 2333
ISOPACA, 2335
ISOPIEZA, 2336
ISOPLETA, 2337
ISOPLEURA, 2338
ISOPRENOIDE, 2339
ISORAD, 2340

ISORAT, 2341
ISOSALINIDAD, 2326
ISOSTASIA, 2344
ISOTERMA, 2346
ISOTOPO, 2348
ISOTOPO=ESTABLE, 4228
ISOTOPO=LIGERO, 2500
ISOTOPO=PESADO, 1971
ISOTOPO=RADIOACTIVO, 3727
ISOTROPIA, 2354
ITABIRITA, 2355
ITACOLUMITA, 2356
ITERBIO, 4853
ITINERARIO=EXCURSION, 3876
ITRIO, 4854
JADEITITA, 2357
JASPE, 2359
JEROGLIFICO, 2005
JOVEN, 4850
JURASICO, 2363
JURASICO=INF, 2583
JURASICO=MEDIO, 2844
JURASICO=SUP, 4673
JUVENIL, 2364
K=AR, 3580
K=CA, 3581
KANGA, 2367
KARROO, 2372
KARST, 2373
KASIMOVIENSE, 2382
KAZANIENSE, 2381
KEROGENO, 2386
KEUPER, 2388
KIMBERLITA, 2391
KIMMERIDGIENSE, 2392
KINK=BAND, 2394
KINZIGITA, 2395
KLIPPE, 2396
KREEP, 2397
KUJALNIKIENSE, 2399
KUNGURIENSE, 2400
KUTORGINIDA, 2402
LABIO=LEVANTADO, 4685
LABYRINTHODONTIA, 2404
LACERTILIA, 2406
LACOLITO, 2405
LADERA, 4134
LADERA=ESTRUCTURAL, 1198
LADINIENSE, 2410
LAGO, 2415
LAGO=ARTIFICAL, 3825
LAGO=CERRADO, 1494
LAGO=DE=CRATER, 982
LAGO=GLACIAR, 1797
LAGO=LAVA, 2458
LAGO=LITORAL, 2693
LAGO=SALADO, 3927

LAGUNA, 1682
LAGUNA=DE=MISCIBILIDAD, 2894
LAJA=TECTONICA, 4129
LAMINA=DELGADA, 4514
LAMINA=MAGMATICA, 4066
LAMINACION, 2422
LAMINAR, 2419
LAMPROFIDO, 2423
LANTANO, 2432
LAPIAZ, 2371
LAPILLI, 2433
LASER, 2437
LATERAL, 2441
LATERITA, 2446
LATERITIZACION, 2448
LATITA, 2449
LATITUD, 2450
LAURASIA, 2453
LAVA, 2454
LAVA=AA, 2
LAVADO, 2465
LAVADO=SUELO, 4151
LECHO=DE=CURSO=DE=AGUA, 4285
LECHO=MAYOR, 1570
LECTOTIPO, 2474
LEGISLACION, 2476
LENGUAJE=INFORMATICO, 871
LENIENSE, 2479
LENTEJON, 2480
LENTEJON=DE=ARENA, 4079
LEPERDITOCOPIDA, 2481
LEPIDOBLASTICA, 2482
LEPIDOCYSTOIDEA, 2483
LEPIDOSAURIA, 2484
LEPOSPONDYLI, 2485
LEPTINITA, 2488
LEPTITA, 2486
LEPTOGEOSINCLINAL, 2487
LEUCOSOMA, 2490
LEVANTAMIENTO, 4665
LEVANTAMIENTO=CARTOGRAFICO, 2670
LEVANTAMIENTO=ELECTRICO, 1333
LEVANTAMIENTO=ELECTROMAGNETICO, 1342
LEVANTAMIENTO=GEOFISICO, 1757
LEVANTAMIENTO=GRAVIMETRICO, 1886
LEVANTAMIENTO=MAGNETICO, 2643
LEVANTAMIENTO=SISMICO, 4019
LEXICO, 2494
LEY, 2460
LEY=DE=CORRELACION=DE=FACIES, 2461
LEY=DE=DARCY, 1061
LEY=DE=HAECKEL, 1939
LEY=DE=HAUY, 1960
LEY=DE=HOOKE, 2048
LEY=ESTADISTICA, 4241
LEY=MATEMATICA, 2712
LEY=MINERAL, 3156

LEYENDA, 2475
LIAS-INF, 2584
LIAS-MEDIO, 2845
LIAS-SUP, 4674
LIBRO-GUIA, 1926
LICHIDA, 2497
LICUACION, 2518
LICUEFACCION, 2519
LIGNINA, 2503
LIGNITO, 2504
LIMITE-ATTERBERG, 367
LIMITE-ELASTICIDAD, 1318
LIMITE-ELASTICO, 4849
LIMITE-ESTRATIGRAFICO, 4275
LIMITE-PLACA, 3497
LIMNOLOGIA, 2508
LIMO, 4106
LIMOLITA, 4109
LIMON, 2509
LIMONITA, 2510
LINEA-COSTA, 4083
LINEA-COSTERA, 816
LINEA-DE-BECKE, 451
LINEA-DE-CRECIMIENTO, 1922
LINEA-DE-LA-ANDESITA, 195
LINEA-DE-SEPARACION-DE-AGUAS, 4788
LINEA-ECUATORIAL, 1413
LINEA-ISOCLINA, 2325
LINEA-ISOSISMICA, 2342
LINEACION, 2512
LINEACION-CORRIENTE, 3310
LINEAMIENTO, 2511
LINGULIDA, 2513
LINOFIDICA, 2515
LIPIDO, 2516
LIPOSTRACA, 2517
LIQUENES, 2495
LIQUENOMETRIA, 2496
LIQUIDUS, 2522
LISSAMPHIBIA, 2523
LISTA-FAUNISTICA, 1517
LITIFICACION, 2524
LITIO, 2525
LITOESTRATIGRAFIA, 2538
LITOFACIES, 2526
LITOFAGO, 3889
LITOFITA, 2534
LITOGENESIS, 2528
LITOLOGIA, 2532
LITOLOGICO, 2529
LITOPTERNA, 2542
LITORAL, 2543
LITOSFERA, 2536
LITOSOL, 2535
LITOSTROMA, 2539
LITOTIPO, 2541
LITOTOPO, 2540

LITUOLACEA, 2547
LIXIVIACION, 2550
LLANDEILOIENSE, 2551
LLANDOVERIENSE, 2552
LLANURA, 3465
LLANURA-ABISAL, 22
LLANURA-ALUVIAL, 144
LLANURA-CONTINENTAL, 4067
LLANURA-COSTERA, 815
LLANURA-DE-ABRASION, 2683
LLANURA-FLUVIOGLACIAR, 3212
LLANURA-SALINA, 3926
LLANVIRNIENSE, 2553
LLUVIA, 3742
LLUVIA-CENIZA, 331
LOCAL, 2557
LOCALIDAD-TIPO, 4624
LODO-DE-SONDEO, 1256
LODRANITO, 2560
LOES, 2561
LONGITUD, 2477
LONGITUD-GEOGRAFICA, 2565
LOPOLITO, 2569
LU-HF, 2608
LUDLOWIENSE, 2596
LUGAR-ARQUEOLOGICO, 280
LUMAQUELA, 948
LUMINISCENCIA, 2597
LUNA, 2935
LUNAR, 2599
LUTECIO, 2607
LUVISOL, 2609
LYCOPSIDA, 2610
LYTOCERATIDA, 2611
MAAR, 2612
MACERAL, 2613
MACHAERIDIA, 2614
MACIZO-LUNAR, 2605
MACIZO-MONTAÑOSO, 2950
MACLA, 4621
MACLA-DE-BAVENO, 442
MACLA-DEFORMADA, 1820
MADERA-FOSIL, 1624
MADERA-SILIFICADA, 4101
MADRIGUERA, 600
MADURACION, 2718
MAESTRICHTIENSE, 2615
MAGMA, 2617
MAGMA-RESIDUAL, 3830
MAGMATICO, 2619
MAGMATISMO, 2624
MAGNAFACIES, 2625
MAGNESIO, 2627
MAGNESITA, 2626
MAGNETISMO, 2646
MAGNETITA, 2647
MAGNETOESTRATIGRAFIA, 2652

MAGNETOESTRICCION, 2653
MAGNETOHIDRODINAMICA, 2649
MAGNETOMETRO, 2650
MAGNETOSFERA, 2651
MAJIENSE, 2656
MALACOSTRACA, 2659
MALEABLE, 2660
MALECON, 3986
MALLA-CRISTALINA, 1018
MALLA-GEOMETRICA, 3013
MALPAIS, 403
MAMMALIA, 2661
MANANTIAL, 4224
MANANTIAL-SUBMARINO, 4331
MANDIBULA, 2360
MANGANESO, 2662
MANGLAR, 2663
MANIPULACION-DATOS, 1068
MANO-DE-OBRA, 2664
MANTO, 2986
MANTO-ALUVIAL, 143
MANTO-ARTESIANO, 313
MANTO-CAUTIVO, 887
MANTO-COLGADO, 3349
MANTO-CORRIMIENTO, 4656
MANTO-GLOBO, 2665
MANTO-GLOBO-INF, 2585
MANTO-GLOBO-SUP, 4675
MANTO-LIBRE, 4644
MANTODEA, 2666
MANUAL, 2667
MAPA, 2668
MAPA-AGUA, 4792
MAPA-BATIMETRICO, 437
MAPA-BIOFACIES, 491
MAPA-DE-CURVAS-DE-NIVEL, 932
MAPA-EDAFOLOGICO, 4153
MAPA-ESTRATIGRAFICO, 4278
MAPA-ESTRUCTURAL, 4308
MAPA-ESTRUCTURAL-CURVA-NIVEL, 4313
MAPA-FOTOGEOLOGICO, 3409
MAPA-GEOFISICO, 1754
MAPA-GEOLOGICO, 1735
MAPA-GEOLOGICO-ECONOMICO, 1299
MAPA-GEOMAGNETICO, 2644
MAPA-GEOMORFOLOGICO, 1747
MAPA-GEOQUIMICO, 1719
MAPA-GEOTECNICO, 1768
MAPA-GRAVIMETRICO, 1882
MAPA-HIDRAULICO, 2095
MAPA-HIDROCARBURO, 3106
MAPA-HIDROGEOLOGICO, 2111
MAPA-HIDROLOGICO, 2120
MAPA-HIDROQUIMICO, 2101
MAPA-HULLERO, 807
MAPA-INDICE, 2207
MAPA-ISOSISMICO, 2343

MAPA-LITOFACIES, 2527
MAPA-MAGNETICO, 2636
MAPA-METALOGENETICO, 2772
MAPA-PALEOGEOGRAFICO, 3256
MAPA-PETROGRAFICO, 3376
MAPA-PREVISORIO, 1611
MAPA-RELIEVE-SOMBREADO, 659
MAPA-SISMICO, 4025
MAPA-SUSTANCIAS-UTILES, 3052
MAPA-TECTONICO, 4437
MAPA-TOPOGRAFICO, 4546
MAQUETA, 3955
MAQUINARIA-SONDEO, 1255
MAR, 3974
MAR-GEOSINCLINAL, 1764
MAR-LUNAR, 2672
MAR-MARGINAL, 2674
MAR-PLATAFORMA-CONTINENTAL, 4069
MAREA, 4529
MAREA-TERRESTRE, 1280
MARGA, 2692
MARGEN-CONTINENTAL, 925
MARGEN-CONTINENTAL-ACTIVO, 58
MARGEN-CONTINENTAL-ASISMICA, 328
MARGEN-CONTINENTAL-COMPLEJO, 920
MARINER, 2690
MARISMA, 2955
MARISMA-SALADA, 3928
MARMITA-DE-GIGANTE, 3585
MARMITA-GLACIAR, 2387
MARMOL, 2671
MARSUPIALIA, 2695
MASA, 2697
MASA-MINERALIZADA, 2705
MASCON, 2696
MASIVO, 2703
MASTIGOPHORA, 2706
MATEMATICAS, 2716
MATERIA-DISUELTA, 1229
MATERIA-ELEMENTAL, 1353
MATERIA-LUNAR-NARANJA, 3144
MATERIA-ORGANICA, 3170
MATERIA-PRIMA, 3755
MATERIA-VISCOSA, 4746
MATERIA-VOLATIL, 4757
MATERIAL, 2707
MATERIAL-CARRETERA, 3877
MATERIAL-COHERENTE, 829
MATERIAL-COMPETENTE, 859
MATERIAL-CONSTRUCCION, 911
MATERIAL-DILATABLE, 1474
MATERIAL-HUESPED, 2061
MATERIAL-INORGANICO, 2249
MATERIAL-NO-COHERENTE, 2202
MATERIAL-NO-CONSOLIDADO, 4647
MATERIAL-ORNAMENTACION, 3181
MATERIAL-PLASTICO, 3493

MATERIAL-SINTETICO, 4402
MATERIAL-SOBRECONSOLIDADO, 3214
MATERIAL-TERRIGENO, 4469
MATERIAL-VIDRIO, 1813
MATRIZ-MATEMATICA, 2713
MATRIZ-ROCA, 3887
MAXILLOPODA, 2721
MAXIMO, 2722
MAXIMO-ANOMALIA, 2006
MEANDRO, 2724
MECANICA, 2729
MECANICA-ROCAS, 3888
MECANICA-SUELO, 4154
MECANISMO, 2730
MECANISMO-ESCORRENTIA, 1581
MECANISMO-FOCAL, 1600
MECOPTEROIDA, 2731
MEDANO-SEMILUNAR, 411
MEDIA, 389
MEDICINA, 2733
MEDIDA-EN-EL-SUELO, 1906
MEDIDA-ISOTOPICA, 2349
MEDIO-ABISAL, 19
MEDIO-AEROBIO, 73
MEDIO-AGUA-DULCE, 1648
MEDIO-AGUA-POCO-PROFUNDA, 4053
MEDIO-ALBUFERA, 2412
MEDIO-ALPINO, 149
MEDIO-ALTA-ENERGIA, 2009
MEDIO-AMBIENTE, 1386
MEDIO-ANAEROBIO, 178
MEDIO-ARIDO, 299
MEDIO-ARRECIFAL, 3777
MEDIO-BAJA-ENERGIA, 2574
MEDIO-BATIAL, 436
MEDIO-BENTONICO, 464
MEDIO-CONTINENTAL, 923
MEDIO-DE-TRANSPORTE, 4578
MEDIO-DELTAICO, 1104
MEDIO-ELASTICO, 1319
MEDIO-EOLICO, 1391
MEDIO-ESTUARIO, 1436
MEDIO-EUXINICO, 1456
MEDIO-FLUVIAL, 1593
MEDIO-FRIO, 832
MEDIO-GLACIAR, 1793
MEDIO-HUMEDO, 2074
MEDIO-INELASTICO, 197
MEDIO-INTERGLACIAR, 2266
MEDIO-LACUSTRE, 2407
MEDIO-LITORAL, 2545
MEDIO-MAR-PROFUNDO, 1090
MEDIO-MAREA, 2286
MEDIO-MAREAL, 4526
MEDIO-MARGEN-CONTINENTAL, 4068
MEDIO-MARINO, 2677
MEDIO-MATERIAL-DEPOSITADO, 2464

MEDIO-PARALICO, 3294
MEDIO-PELAGICO, 3330
MEDIO-PERIGLACIAR, 3355
MEDIO-POROSO, 3566
MEDIO-SALOBRE, 569
MEDIO-SEBKHA, 3987
MEDIO-SEMIARIDO, 4035
MEDIO-SUBAEREO, 4316
MEDIO-SUBLITORAL, 4325
MEDIO-SUBMAREA, 4341
MEDIO-TAIGA, 4412
MEDIO-TALUD-MARINO, 4136
MEDIO-TERRESTRE, 4467
MEGAESPORA, 2739
MEGAFACIES, 2738
MEGATECTONICA, 1821
MEJORA-SUELO, 2742
MENA, 3151
MENDELEVIO, 2747
MEOTIENSE, 2748
MERCADO, 4566
MERCURIO, 2749
MEROSTOMATA, 2750
MEROSTOMOIDEA, 2751
MESETA, 3501
MESETA-LITORAL-HELADA, 2161
MESOGASTROPODA, 2754
MESOGEA, 2753
MESOGEOSINCLINAL, 2755
MESOSAURIA, 2757
MESOSIDERITA, 2758
MESOSTASIS, 1919
MESOZOICO, 2760
METABASALTO, 2762
METACOPINA, 2763
METAFITA, 2791
METAGABRO, 2764
METAGENESIS, 2765
METAL, 2766
METAL-ALCALINO, 126
METAL-ALCALINOTERREO, 129
METAL-BASE, 428
METAL-PRECIOSO, 3037
METAL-TRAZA, 4558
METALES-NO-FERREOS, 3047
METALES-PESADOS, 1972
METALES-RAROS, 3752
METALICO, 2768
METALIFERO, 2769
METALOGENIA, 2774
METALOGENIA-PREVISORIA, 3599
METALOIDE, 2775
METALOTECTO, 2776
METALURGIA, 2777
METAMICTIZACION, 2779
METAMICTO, 2778
METAMORFISMO, 2787

METAMORFISMO-BAJO-GRADO, 2575
METAMORFISMO-CATAZONAL, 2380
METAMORFISMO-CHOQUE, 4077
METAMORFISMO-CONTACTO, 915
METAMORFISMO-DE-SUBSIDENCIA, 597
METAMORFISMO-DINAMOTERMAL, 1269
METAMORFISMO-EPIZONAL, 1408
METAMORFISMO-FUERTE, 2011
METAMORFISMO-GEOTERMICO, 1776
METAMORFISMO-IMPACTO, 2189
METAMORFISMO-ISOQUIMICO, 2322
METAMORFISMO-MESOZONAL, 2761
METAMORFISMO-PLUTONICO, 3518
METAMORFISMO-PROGRESIVO, 3633
METAMORFISMO-REGIONAL, 3795
METAMORFISMO-TERMICO, 4498
METAMORFOSIS, 2789
METANAUPLIUS, 2790
METANO, 2801
METASOMA, 2793
METASOMATISMO, 2794
METASOMATISMO-ALCALINO, 130
METASOMATISMO-DE-CONTACTO, 916
METASOMATITA, 2795
METATEXIA, 2796
METEORITO, 2798
METEORITO-LITICO, 4264
METEORITO-METALICO, 2767
METEORIZACION-POR-HELADA, 891
METEOROIDE, 2799
METEOROLOGIA, 2800
METODO-ACUSTICO, 43
METODO-AEROPORTADO, 95
METODO-AFMAG, 78
METODO-ANALISIS-ELEMENTOS-TRAZA, 4559
METODO-ANALISIS-MAYORES, 2658
METODO-ANALISIS-MENORES, 2888
METODO-DE-BARRIDO-LATERAL, 4091
METODO-DE-INDUCCION, 2217
METODO-DE-INMERSION, 2184
METODO-DIFERENCIA-FINITA, 1550
METODO-ELECTRICO, 1329
METODO-ELECTROMAGNETICO, 1341
METODO-ESTADISTICO, 4239
METODO-GEOFISICO, 1755
METODO-GEOTERMICO, 1777
METODO-GRAVIMETRICO, 1883
METODO-MAGNETICO, 2637
METODO-MAGNETOTELURICO, 2654
METODO-MATEMATICO, 2714
METODO-MEB, 3957
METODO-MET, 4575
METODO-MICROONDAS, 2838
METODO-MONTE-CARLO, 2932
METODO-NUEVO, 3022
METODO-RADIOFRECUENCIA, 3724
METODO-RADIOMETRICO, 3732

METODO-RESISTIVIDAD, 3833
METODO-SISMICO, 4014
METODO-STANDARD, 4236
METODO-TELURICO, 4449
METODO-TRACCION-FONDO, 1092
METODO-TRANSITORIO, 4573
METODO-ULTRASONIDO, 4639
METODO-VIA-HUMEDA, 4821
METODOLOGIA, 2802
METODOLOGIA-ANALISIS, 183
METRO, 4344
MEZCLA-CONGRUENTE, 893
MEZCLA-FUNDIDA, 2743
MICA, 2804
MICAESQUISTO, 2805
MICRITA, 2806
MICRO-ESTILOLITO, 2835
MICROCRATER, 2808
MICRODUREZA, 2820
MICROESTRUCTURA, 4640
MICROFACIES, 2811
MICROFAUNA, 2812
MICROFISURA, 2813
MICROFLORA, 2814
MICROFOSIL, 2816
MICROGRANITO, 2818
MICROLOG, 2823
MICROMETEORITO, 2824
MICROMORFOLOGIA-SUELO, 2825
MICROORDENADOR, 2807
MICROORGANISMO, 2826
MICROPALEONTOLOGIA, 2827
MICROPHYTOLITHI, 2828
MICROPLACA, 2829
MICROPLANOS-IMPACTO, 3467
MICROPLIEGUE, 2815
MICROSCOPIA, 2833
MICROSCOPIA-ELECTRONICA, 1349
MICROSCOPIA-MINERAL, 3158
MICROSCOPIO, 2831
MICROSCOPIO-ELECTRONICO, 1348
MICROSISMO, 2834
MICROSONDA-ELECTRONICA, 1347
MICROSONDA-IONICA, 2301
MICROTECTITA, 2837
MICROTECTONICA, 2836
MICROTERREMOTO, 2810
MIEMBRO, 2746
MIGMATITA, 2853
MIGMATIZACION, 2854
MIGRACION, 2855
MIGRACION-DE-ELEMENTOS, 2856
MILIOLACEA, 2857
MILIOLINA, 2858
MILONITA, 2972
MILONITIZACION, 2974
MINA, 2861

MINA-CIELO-ABIERTO, 4294
MINERAL-ALARGADO, 1355
MINERAL-ALOTIGENO, 136
MINERAL-ANTITENSIONAL, 232
MINERAL-AUTIGENO, 376
MINERAL-BIAXIAL, 476
MINERAL-GEL, 1703
MINERAL-INDICE, 2208
MINERAL-INDUSTRIAL, 1300
MINERAL-MAGNETICO, 2638
MINERAL-ORGANICO, 3167
MINERAL-PISOLITICO, 449
MINERAL-PROBABLE, 2225
MINERAL-SECUNDARIO, 3989
MINERAL-TENSIONAL, 4291
MINERAL-TIPOMORFO, 4627
MINERAL-URANIO, 4690
MINERALES, 2862
MINERALES-ACCESORIOS, 29
MINERALES-COEXISTENTES, 826
MINERALES-CONSTITUYENTES, 3890
MINERALES-LIGEROS, 2501
MINERALES-PESADOS, 1973
MINERALIZACION, 2876
MINERALIZACION-LITORAL, 445
MINERALOGIA, 2880
MINERALOGIA-ARCILLAS, 789
MINERIA-SUBTERRANEA, 4650
MINIMO, 2882
MINIMOS-CUADRADOS, 2473
MINIORDENADOR, 2881
MIOCENO, 2889
MIOCENO-INF, 2586
MIOCENO-MEDIO, 2846
MIOCENO-SUP, 4676
MIOGEOSINCLINAL, 2890
MIOMERA, 2891
MIOSPORA, 2892
MIRMEKITA, 2979
MISCIBILIDAD, 2893
MISSISSIPPIENSE, 2897
MODA, 2903
MODELO, 2904
MODELO-BIDIMENSIONAL, 4622
MODELO-DETERMINATIVO, 1141
MODELO-ESTADISTICO, 4242
MODELO-FISICO, 3426
MODELO-HIDRODINAMICO, 2103
MODELO-MATEMATICO, 2715
MODELO-UNIDIMENSIONAL, 3120
MODIOMORPHOIDA, 2906
MODULO, 2907
MODULO-CIZALLAMIENTO, 4060
MODULO-ELASTICIDAD, 591
MODULO-RIGIDEZ, 2908
MODULO-YOUNG, 4851
MOHOROVICIC, 2910

MOLASA, 2912
MOLDAVITA, 2913
MOLDE=DE=CARGA, 2555
MOLDE=INTERNO, 2274
MOLIBDATO, 2917
MOLIBDENO, 2918
MOLIENDA, 1899
MOLLISOL, 2915
MOLLUSCA, 2916
MOMENTO=DIPOLAR, 1206
MONACITA, 2919
MONOCLINAL, 2920
MONOCOTYLEDONES, 2922
MONOCYATHEA, 2923
MONOGRAFIA, 2924
MONOGRAFIA=SUMARIA, 4357
MONOLITO, 2926
MONOMORFICO, 2927
MONOPLACOPHORA, 2928
MONORHINA, 2929
MONOTREMATA, 2930
MONTE=ISLA, 2253
MONTERA=DE=HIERRO, 1838
MONTICULO=MATERIALES=GLACIARES, 2366
MONZONITA, 2933
MORFOESTRUCTURA, 2943
MORFOGENESIS, 2939
MORFOLOGIA, 2940
MORFOLOGIA=COSTA, 4081
MORFOLOGIA=COSTERA, 814
MORFOLOGIA=DE=DISOLUCION, 4178
MORFOLOGIA=EOLICA, 1392
MORFOLOGIA=EROSION, 1427
MORFOLOGIA=FUNCIONAL, 1666
MORFOLOGIA=GLACIAR, 1795
MORFOLOGIA=PERIGLACIAR, 1654
MORFOLOGIA=SUBMARINA, 4329
MORFOLOGIA=YACIMIENTO, 1119
MORFOMETRIA, 2941
MORFOSCOPIA, 2942
MORRENA, 2938
MORRENA=DE=FONDO, 1907
MORRENA=LATERAL, 2444
MOSCOVIENSE, 2946
MOVILIDAD, 2900
MOVILIZACION, 2901
MOVIMIENTO=A=TIRONES, 4254
MOVIMIENTO=DE=TIERRAS, 1281
MOVIMIENTO=HIELO, 2158
MOVIMIENTO=MASA, 2699
MOVIMIENTO=SUELO, 1908
MOVIMIENTO=VERTICAL, 4736
MOVIMIENTOS, 2951
MUELLE, 3440
MUESTRA, 3933
MUESTREO, 3935
MUESTREO=ALEATORIO, 1839

MUESTREO=SUELO, 4156
MULL=SOIL, 2960
MULTITUBERCULATA, 2967
MUSCHELKALK, 2969
MUSEO, 2970
MUTACION, 2971
MYODOCOPIDA, 2975
MYOIDA, 2976
MYOMORPHA, 2977
MYRIAPODA, 2978
MYSTACOCARIDA, 2980
MYTILOIDA, 2981
MYXOMYCETES, 2982
NAKHLITA, 2983
NAMURIENSE, 2984
NANOFOSIL, 2985
NATANTES, 2987
NAUTILOIDEA, 2993
NEANDERTHALIENSE, 2994
NEBULOSA, 968
NECTON, 2997
NEFELINA, 3008
NEMERTA, 2998
NEOCOMIENSE, 2999
NEODIMIO, 3000
NEOGASTROPODA, 3001
NEOGENO, 3002
NEON, 3003
NEORNITHES, 3004
NEOTECTONICA, 3005
NEOTENIA, 3006
NEOTIPO, 3007
NEPTUNIO, 3011
NEPTUNISMO, 3010
NEUTRON, 3014
NIEVE, 4142
NILSSONIALES, 3028
NIOBATO, 3029
NIOBIO, 3030
NIOBOTANTALATO, 3031
NIQUEL, 3026
NITRATO, 3032
NITRIFICACION, 3033
NITROGENO, 3034
NITROGENO=ORGANICO, 3171
NIVACION, 3036
NIVEL, 2493
NIVEL=AGUA=SUBTERRANEA, 1916
NIVEL=DEL=MAR, 3977
NIVEL=GUIA, 2389
NIVEL=HIDROSTATICO, 2128
NIVEL=LAGO, 2416
NIVELACION, 2492
NO=METAL, 3051
NO=SATURADO, 4655
NODOSARIACEA, 3038
NODULO, 3039

NOEGGERATHIALES, 3040
NOMBRE=NUEVO, 3024
NOMENCLATURA, 3042
NOMOGENESIS, 3044
NORIENSE, 3053
NORMA, 3054
NORMALIZACION, 4238
NORTE, 3057
NOTA=EXPLICATIVA, 1480
NOTOSTRACA, 3059
NOTOUNGULATA, 3060
NOVACULITA, 3061
NOVEDAD, 3021
NUBE=ARDIENTE, 3069
NUCLEACION, 3066
NUCLEO, 3067
NUCLEO=CONTINENTAL, 926
NUCLEO=EXTERNO, 3207
NUCLEO=GLOBO, 957
NUCLEO=INTERNO, 2243
NUCULOIDEA, 3068
NUEVO, 3019
NUEVO=DATO, 3020
NUEVO=DATO=MINERAL, 3023
NUEVO=TAXON, 3025
NUMMULITIDAE, 3071
NUTACION, 3072
NUTACION=CHANDLER, 706
NUTRICION, 3073
O=18=O=16, 3227
OASIS, 3074
OBDUCCION, 3075
OBJETIVO=HIDROGEOLOGICO, 248
OBJETIVOS, 3076
OBLICUIDAD, 4128
OBOLELLIDA, 3079
OBRA=PUBLICA, 3680
OBRAS, 1375
OBRAS=DEL=PETROLEO, 3380
OBSERVACION=GEOFISICA, 3081
OBSERVATORIO, 3082
OBSIDIANA, 3083
OCEAN=DRILLING=PROJECT, 3087
OCEANICO, 3090
OCEANIZACION, 3092
OCEANO, 3084
OCEANOGRAFIA, 3093
OCEANOLOGIA, 3094
OCHREPT, 3096
OCRE, 3095
OCTAEDRITA, 3097
OCTOCORALLA, 3098
ODONTOLOGIA, 3099
ODONTOPLEURIDA, 3100
ODONTORNITHES, 3101
OESTE, 4819
OFF=SHORE, 3102

OFIOLITA, 3133
OJAL, 2242
OLEAJE, 3979
OLEODUCTO, 3454
OLIGOCENO, 3115
OLIGOCENO-INF, 2587
OLIGOCENO-MEDIO, 2847
OLIGOCENO-SUP, 4677
OLISTOLITO, 3116
OLISTOSTROMA, 3117
OLIVINO, 3118
ONCOLITO, 3119
ONDA-ACUSTICA, 45
ONDA-ARENA, 3941
ONDA-CHOQUE, 4078
ONDA-ELASTICA, 1322
ONDA-LARGO-PERIODO, 2564
ONDA-LONGITUDINAL, 4368
ONDA-LOVE, 2570
ONDA-P, 3230
ONDA-PRECURSORA, 1615
ONDA-RAYLEIGH, 3756
ONDA-S, 3920
ONDA-SH, 4050
ONDA-SISMICA, 4021
ONDA-TERMINAL, 821
ONDA-TRANSVERSAL, 535
ONDAS, 4805
ONDAS-AIRY, 97
ONTOGENIA, 3121
ONYCHOPHORA, 3122
OOIDE, 3123
OOLITO, 3124
OPACO, 3128
OPHIDIA, 3131
OPHIOCISTIOIDEA, 3132
OPHIUROIDEA, 3135
OPISTHOBRANCHIA, 3136
OPOKA, 3137
OPTICA-CRISTALINA, 1019
ORBITA, 3146
ORBITOIDACEA, 3147
ORBITOIDIDAE, 3148
ORDEN-DESORDEN, 3149
ORDENACION-GRANOS, 3232
ORDENADOR, 870
ORDOVICICO, 3150
ORDOVICICO-INF, 2588
ORDOVICICO-MEDIO, 2848
ORDOVICICO-SUP, 4678
ORGANICO, 3164
ORGANISMO, 3176
ORGANISMO-AUTOTROFICO, 386
ORGANISMO-FAGOTROFICO, 3389
ORGANISMO-FITOFAGO, 3436
ORGANISMO HETEROTROFICO, 1998
ORGANISMO-MESOFILO, 2756

ORICTOCENOSIS, 3196
ORIENTACION, 3177
ORIENTACION-OBLIGUA, 3078
ORIENTACION-PREFERENTE, 3600
ORIENTACION-VERTICAL, 4735
ORIGEN, 3179
ORIGEN-ATMOGENICO, 357
ORIGEN-BIOGENICO, 3172
ORIGEN-EPIMAGMATICO, 1405
ORIGEN-INORGANICO, 2250
ORIGEN-MAGMATICO, 2623
ORIGEN-MECANICO, 2726
ORIGEN-QUIMICO, 736
ORIGEN-SINSEDIMENTARIO, 4397
ORIGEN-SUPRACORTICAL, 4362
ORIGOFACIES, 3180
ORLA, 3868
ORNAMENTACION-EXTERIOR, 3182
ORNITHISCHIA, 3183
ORO, 1833
OROGENESIS, 3185
OROGENIA-ACADIENSE, 24
OROGENIA-ALGOMANA, 120
OROGENIA-ALLEGHENY, 133
OROGENIA-ALPINA, 150
OROGENIA-ARMORICANA, 304
OROGENIA-ASINTICA, 342
OROGENIA-BELTICA, 460
OROGENIA-CALEDONIANA, 618
OROGENIA-CARELIANA, 2370
OROGENIA-CHARNIA, 711
OROGENIA-CIMMERICA, 768
OROGENIA-EBURNEA, 1289
OROGENIA-GRENVILLE, 1898
OROGENIA-HERCINICA, 1984
OROGENIA-HUDSONIANA, 2068
OROGENIA-KATANGIANA, 2378
OROGENIA-KENORAN, 2384
OROGENIA-KIBARA, 2390
OROGENIA-LARAMICA, 2435
OROGENIA-PENOKEENSE, 3343
OROGENIA-PRECAMBRICA, 3594
OROGENIA-SUDETICA, 4346
OROGENIA-SVECOFENNIANA, 4375
OROGENIA-TACONICA, 4407
OROGENIA-VARISCA, 4719
ORTHENT, 3186
ORTHID, 3187
ORTHIDA, 3188
ORTHOD, 3190
ORTHOX, 3195
ORTOANFIBOLITA, 3189
ORTOGENESIS, 3191
ORTOGNEIS, 3192
OSCILACION, 3197
OSCILACION-PROPIA, 1642
OSMIO, 3199

OSTEICHTHYES, 3200
OSTEOLOGIA, 3201
OSTEOSTRACI, 3202
OSTRACODA, 3203
OSTREACEA, 3204
OTOLITO, 3205
OXFORDIENSE, 3220
OXIDACION, 3221
OXIDO, 3222
OXIDO-DE-CARBONO, 644
OXIGENO, 3225
OXILOFITA, 3228
OXISOL, 3224
PA-IO, 3643
PAHOEHOE, 3233
PALADIO, 3275
PALAEOHETERODONTA, 3235
PALAEOTAXODONTA, 3237
PALAGONITIZACION, 3238
PALEO, 3239
PALEO-OCEANOGRAFIA, 3241
PALEOAMBIENTE, 3253
PALEOATMOSFERA, 3240
PALEOBATIMETRIA, 3242
PALEOBIOLOGIA, 3243
PALEOBOTANICA, 3245
PALEOCENO, 3246
PALEOCIRCULACION, 3247
PALEOCLIMA, 3248
PALEOCLIMATOLOGIA, 3249
PALEOCOPIDA, 3250
PALEOCORRIENTE, 3251
PALEOECOLOGIA, 3252
PALEOFITICO, 3236
PALEOGENO, 3254
PALEOGEOGRAFIA, 3257
PALEOKARST, 3259
PALEOLATITUD, 3260
PALEOLIMNOLOGIA, 3261
PALEOMAGNETISMO, 3263
PALEOMORFOLOGIA, 3258
PALEONTOLOGIA, 3265
PALEORELIEVE, 3266
PALEOSALINIDAD, 3267
PALEOSUELO, 3268
PALEOTEMPERATURA, 3269
PALEOTERMOMETRIA, 3270
PALEOZOOLOGIA, 3272
PALIMPSESTICA, 3273
PALINGENESIS, 3274
PALINODIAGRAMA, 3540
PALINOLOGIA, 3278
PALINOMORFO, 3279
PALLASITA, 3276
PALSA, 3277
PAMPA, 3280
PANGEA, 3281

PANTAÑO, 536
PANTAÑO MARINO, 2681
PANTODONTA, 3284
PANTOTHERIA, 3285
PAPEL-DEL-GEOLOGO, 1740
PAR-TIERRA-LUNA, 1282
PARA-, 3286
PARA-ANFIBOLITA, 3287
PARABLASTOIDEA, 3288
PARACRINOIDEA, 3289
PARAGENESIS, 3290
PARAGEOSINCLINAL, 3291
PARAGNEIS, 3292
PARALELISMO, 3297
PARALIAGEOSINCLINAL, 3293
PARAMETRO, 3299
PARASITISMO, 3300
PARATIPO, 3301
PARTICULA-COSMICA, 969
PARTICULA-ELEMENTAL, 3307
PASO-ENTRE-MAREAS, 4528
PATENTE, 3314
PATOLOGIA, 3315
PATRON, 3316
PATRON-QUIMICO, 4235
PAUROPODA, 3318
PAVIMENTO-DESERTICO, 1133
PB-PB, 2468
PECTINACEA, 3323
PEDIMENTO, 3324
PEDIPLANO, 3325
PEGMATITA, 3327
PEGMATITICO, 3328
PEGMATOIDE, 3329
PELEANO, 3333
PELET-FECAL, 1521
PELETIZACION, 3338
PELITA, 3334
PENDIENTE-DE-TALUD-MARINO, 4417
PENDIENTOMETRIA, 1204
PENDIENTOMETRO, 1197
PENEPLANACION, 3468
PENETROMETRIA, 3340
PENETROMETRO, 3341
PENILLANURA, 3339
PENSYLVANIENSE, 3342
PENTAMERIDA, 3344
PENTASTOMIDA, 3345
PENTOXYLALES, 3346
PEPITA, 3070
PEQUEÑA-ESCALA, 4140
PERCOLACION, 3350
PERFIL-DE-EQUILIBRIO, 3628
PERFIL-GEOFISICO, 1756
PERFIL-GEOQUIMICO, 1720
PERFIL-LONGITUDINAL, 2566
PERFIL-SUELO, 4155

PERFORABILIDAD, 1253
PERFORACION-OFF-SHORE, 3104
PERICLINAL, 3351
PERIDOTITA, 3352
PERIGEO, 3353
PERIGLACIAR, 3354
PERIMETRO-PROTECCION, 3644
PERIODICIDAD, 3357
PERIODO, 3356
PERISPHINCTIDA, 3358
PERISSODACTYLA, 3359
PERLA, 3319
PERLITA, 3360
PERMAFROST, 3362
PERMEABILIDAD, 3363
PERMEABILIDAD-MAGNETICA, 2639
PERMICO, 3364
PERMICO-INF, 2589
PERMICO-SUP, 4679
PERMISO-INVESTIGACION-MINA, 779
PERSPECTIVA, 3366
PERSPECTIVA-EXPLOTACION, 3572
PERTITA, 3367
PESO-ESPECIFICO, 4196
PESTICIDA, 3369
PETROFABRICA, 3373
PETROFISICA, 3383
PETROGENESIS, 3374
PETROGRAFIA, 3378
PETROLEO, 3379
PETROLEO-LIGERO, 2502
PETROLOGIA, 3382
PETROLOGIA-ESTRUCTURAL, 4309
PETROLOGIA-EXPERIMENTAL, 1477
PETROLOGIA-ROCA-SEDIMENTARIA, 3998
PETROQUIMICA, 3372
PH, 3384
PHACOPIDA, 3386
PHAEOPHYCOPHYTA, 3387
PHOLADOMYIDA, 3399
PHOLIDOTA, 3400
PHYLACTOLAEMATA, 3418
PHYLLOCERATIDA, 3420
PHYLLOPODA, 3422
PHYTOLITHES, 3435
PICNOCLINA, 3689
PICNOMETRIA, 3692
PICNOMETRO, 3691
PICRITA, 3438
PIE-DE-MONTE, 3439
PIEDRA-DE-CONSTRUCCION, 588
PIEDRA-ORNAMENTACION, 1085
PIEDRA-POMEZ, 3683
PIEDRA-PRECIOSA, 3595
PIEDRA-SEMIPRECIOSA, 4036
PIEZOCRISTALIZACION, 3441
PIEZOMETRIA, 3443

PIGMENTO, 3444
PILASTRA, 3446
PILLOW-LAVA, 3447
PILOTE-DE-CIMENTACION, 3445
PINACULO-DE-ARRECIFES, 3451
PINCHAMIENTO, 3516
PINNIPEDIA, 3452
PIPA, 3453
PIRITIZACION, 3694
PIROCLASTICO, 3695
PIROCLASTICO-CEMENTADO, 83
PIROLISIS, 3696
PIROMETALURGIA, 3697
PIROMETAMORFISMO, 3698
PIROMETASOMATISMO, 3699
PIROSFERA, 3700
PIROXENITA, 3704
PIROXENO, 3702
PISCES, 3456
PISO-ESTRATIGRAFICO, 4229
PISOLITO, 3457
PISTA, 4565
PIZARRA, 4130
PIZARRA-NO-METAMORFICA, 4051
PLACA, 3495
PLACA-ANTICLINAL, 3461
PLACENTALIA, 3462
PLACER, 3463
PLACODERMI, 3464
PLANCTON, 3487
PLANCTOSPHAEROIDEA, 3469
PLANETA, 3471
PLANETA-EXTERIOR, 3208
PLANETA-JUPITER, 3473
PLANETA-MARTE, 3474
PLANETA-MERCURIO, 3475
PLANETA-NEPTUNO, 3476
PLANETA-PLUTON, 3477
PLANETA-SATURNO, 3478
PLANETA-TERRESTRE, 4468
PLANETA-TIERRA, 3472
PLANETA-URANO, 3479
PLANETA-VENUS, 3480
PLANETARIO, 3482
PLANETOLOGIA, 3485
PLANIFICACION, 3489
PLANIFICACION-CUENCA-HIDROGEOLOGICA, 433
PLANIFICACION-SUELO, 4152
PLANIFICACION-URBANA, 4695
PLANO, 3470
PLANO-ACUMULACION, 33
PLANO-AXIAL, 391
PLANO-DE-CIZALLAMIENTO, 4061
PLANO-ESTRATIFICACION, 454
PLANO-FALLA, 1510
PLANOSOL, 3490
PLANTA-FOSIL, 3491

PLANTA-HEKISTOTERMICA, 1975
PLANTA-HETEROSPORA, 1997
PLANTA-PREPARACION-MENA, 3155
PLANTA-XEROFILA, 4843
PLASTICIDAD, 3494
PLATA, 4111
PLATAFORMA, 3503
PLATAFORMA-CONTINENTAL, 929
PLATAFORMA-CONTINENTAL-EXTERNA, 3209
PLATAFORMA-CONTINENTAL-INTERNA, 2244
PLATAFORMA-GRAVIMETRICA, 1884
PLATAFORMA-MARGINAL, 2673
PLATAFORMA-MARINA, 2684
PLATINA-UNIVERSAL, 4662
PLATINO, 3504
PLATYRRHINA, 3506
PLAYA, 444
PLEGAMIENTO, 1608
PLEISTOCENO, 3508
PLEISTOCENO-INF, 2590
PLEISTOCENO-MEDIO, 2849
PLEISTOCENO-SUP, 4680
PLEOCROISMO, 3509
PLIEGUE, 1602
PLIEGUE-ANTIFORME, 227
PLIEGUE-ASIMETRICO, 353
PLIEGUE-CAPA, 1248
PLIEGUE-CILINDRICO, 1052
PLIEGUE-CONCENTRICO, 874
PLIEGUE-CONJUGADO, 897
PLIEGUE-DE-ARRASTRE, 1243
PLIEGUE-DE-FLUENCIA, 1577
PLIEGUE-DE-GRAVEDAD, 1881
PLIEGUE-DISARMONICO, 1219
PLIEGUE-EN-ACORDEON, 4859
PLIEGUE-EN-CASCADA, 661
PLIEGUE-EN-ZIG-ZAG, 741
PLIEGUE-ENTEROLITICO, 1380
PLIEGUE-FALLA, 1220
PLIEGUE-INVERTIDO, 3219
PLIEGUE-MONOCLINAL, 4249
PLIEGUE-PTIGMATICO, 3679
PLIEGUE-SIMETRICO, 4382
PLIEGUE-SINFORME, 4393
PLIEGUE-SUPERPUESTO, 4359
PLIEGUE-TUMBADO, 3771
PLIENSBACHIENSE, 3511
PLIOCENO, 3512
PLIOCENO-INF, 2591
PLIOCENO-MEDIO, 2850
PLIOCENO-SUP, 4681
PLOMO, 2466
PLUMA, 3515
PLUTON, 3517
PLUTONIO, 3521
PLUTONISMO, 3520
PLUVIAL, 3522

POBLACION, 855
POBLACION-ENDEMICA, 1368
POBLACION-ESTADISTICA, 4243
PODER-REFLECTOR, 3782
PODER-ROTATORIO, 3141
PODOCOPIDA, 3525
PODZOL, 3526
PODZOLUVISOL, 3527
POGONOPHORA, 3528
POLAR, 3533
POLARIZACION, 3139
POLARIZACION-CIRCULAR, 772
POLARIZACION-ELECTRICA, 1330
POLARIZACION-ESPONTANEA, 4033
POLARIZACION-PROVOCADA, 2216
POLAROGRAFIA, 3536
POLEN, 3539
POLIEDRO, 3548
POLIFASICA, 2963
POLIGENETICO, 3546
POLIMERIZACION, 3550
POLIMETAMORFISMO, 3552
POLIMICTO, 3553
POLIMINERAL, 3554
POLIMORFISMO, 3555
POLITICA, 3537
POLITIPISMO, 3558
POLO-GEOMAGNETICO, 1742
POLO-PALEOMAGNETICO, 3262
POLONIO, 3544
POLVO, 1263
POLVO-COSMICO, 967
POLVO-MAGNETICO, 2641
POLVO-RADIOACTIVO, 3726
POLYCHAETA, 3545
POLYMERA, 3549
POLYPLACOPHORA, 3557
PONTIENSE, 3559
PORCELANITA, 3561
PORIFERA, 3564
POROSIDAD, 3565
POROSIDAD-CELULAR, 682
PORTLANDIENSE, 3570
POSTECTONICO, 3575
POSTGLACIAR, 3573
POSTMAGMATICO, 3574
POTABILIDAD, 3576
POTASIO, 3579
POTENCIA, 4513
POTENCIAL-ELECTRICO, 1325
POTENCIOMETRIA, 3584
POZO-ARTESIANO, 314
POZO-DE-AGUA, 4801
POZO-DE-DISPARO, 4085
POZO-DE-EXPLORACION, 1484
POZO-DE-GAS, 1692
POZO-DE-PETROLEO, 3112

POZO-DE-RECONOCIMIENTO, 4813
POZO-MINERO, 3459
POZO-PRODUCTIVO, 3627
POZO-SONDEO, 549
PRADERA, 1870
PRAECARDIIDA, 3590
PRASEODIMIO, 3591
PRECAMBRICO, 3593
PRECIO, 3610
PRECIPITACION-ATMOSFERICA, 360
PRECIPITACION-QUIMICA, 3596
PRECISION-DE-MEDIDA, 34
PREGLACIAR, 3601
PREHISTORIA, 3602
PREIMBRIENSE, 3592
PRENEANDERTHALIENENSE, 3604
PREPARACION, 3605
PRESA, 1059
PRESA-ARQUEADA, 271
PRESA-ESCOLLERA, 3885
PRESA-HORMIGON, 880
PRESA-TIERRA, 1275
PRESIOMETRIA, 2556
PRESION, 3607
PRESION-ALTA, 2007
PRESION-ATMOSFERICA, 361
PRESION-BAJA, 2571
PRESION-CAPILAR, 635
PRESION-DE-AGUA, 4795
PRESION-FLUIDO, 1583
PRESION-GEOSTATICA, 1909
PRESION-HIDRAULICA, 2096
PRESION-HIDROSTATICA, 2129
PRESION-HIDROSTATICA-ASCENDENTE, 4666
PRESION-PARCIAL, 3306
PRESION-POROS, 3562
PRESION-TERRENO, 1277
PREVISION, 3598
PREVISION-PROBABILIDAD, 2709
PRIMARIO, 3271
PRIMATES, 3613
PRIMITIVO, 3614
PRINCIPIO-COSMOLOGICO, 974
PRIORIDAD, 3615
PRISMATICA, 3616
PROBLEMA-INVERSO, 2296
PROBLEMATICA, 3619
PROBLEMATICA-MICRO, 2830
PROBOSCIDEA, 3620
PROCEDENCIA, 3656
PROCEDIMIENTO, 3130
PROCELLARIENSE, 3621
PROCESO, 3622
PROCESO-ADIABATICO, 65
PROCESO-ATECTONICO, 215
PROCESO-ENDOGENO, 1370
PROCESO-ESTOCASTICO, 4256

PROCESO=EXOGENO, 1471
PROCESO=SINTECTONICO, 4399
PROCESOS=CICLICOS, 1047
PROCHORDATA, 3623
PRODUCCION, 3624
PRODUCTO=CARGA, 1544
PRODUCTO=FINAL, 1367
PRODUCTO=HIJO, 1074
PRODUCTO=PADRE, 3303
PROFUNDIDAD, 1122
PROFUNDIDAD=COMPENSACION=CARBONATO, 1124
PROGANOSAURIA, 3630
PROGRAMA, 3634
PROGRAMA=APOLO, 243
PROGRAMA=CORRELACION=PICG, 2278
PROGRAMA=LUNA, 2598
PROGRAMA=ORDENADOR, 872
PROGRAMA=VIKING, 4741
PROGRESION=COSTA, 3632
PROMETIO, 3637
PROMONTORIO=ANTICLINAL, 3058
PRONOSTICO, 3631
PROPAGACION=ONDA, 4807
PROPIEDAD, 3638
PROPIEDAD=ACUSTICA, 42
PROPIEDAD=DIAMAGNETICA, 1159
PROPIEDAD=ELASTICA, 1320
PROPIEDAD=ELECTRICA, 1331
PROPIEDAD=ELECTROQUIMICA, 1336
PROPIEDAD=FERROMAGNETICA, 1536
PROPIEDAD=FISICA, 3427
PROPIEDAD=FISICO=QUIMICA, 3430
PROPIEDAD=GEOTECNICA, 1769
PROPIEDAD=MAGNETICA, 2640
PROPIEDAD=MECANICA, 2727
PROPIEDAD=OPTICA, 3140
PROPIEDAD=PARAMAGNETICA, 3298
PROPIEDAD=PIEZOELECTRICA, 3442
PROPIEDAD=QUIMICA, 729
PROPIEDAD=TERMICA, 4499
PROPIEDAD=TERMODINAMICA, 4504
PROPIEDAD=TERMOQUIMICA, 4502
PROPILITIZACION, 3639
PROPORCION, 3753
PROPORCION=MATERIAL=SEDIMENTADO, 4004
PROSIMII, 3640
PROSPECCION, 1483
PROSPECCION=AEROMAGNETICA, 74
PROSPECCION=ALUVIONAR, 145
PROSPECCION=BIOGEOQUIMICA, 495
PROSPECCION=ELECTRICA, 1327
PROSPECCION=GEOBOTANICA, 1714
PROSPECCION=GEOFISICA, 1752
PROSPECCION=GEOQUIMICA, 1716
PROSPECCION=GRAVIMETRICA, 1879
PROSPECCION=HIDROGEOQUIMICA, 2109
PROSPECCION=MAGNETOTELURICA, 2655

PROSPECCION=MINERA, 2868
PROSPECCION=RADIOMETRICA, 3738
PROSPECCION=TERMICA, 4500
PROTACTINIO, 3642
PROTECCION=COSTERA, 4082
PROTECCION=MEDIO=AMBIENTE, 905
PROTEINA, 3645
PROTELYTROPTERA, 3646
PROTEROZOICO, 3647
PROTEROZOICO=INF, 2592
PROTEROZOICO=MEDIO, 2851
PROTEROZOICO=SUP, 4682
PROTEROZOIDES, 3648
PROTEROZOIDES=INF, 1273
PROTEROZOIDES=SUP, 2439
PROTOATLANTICO, 2154
PROTOCILIATA, 3649
PROTOLITO, 3650
PROTOMEDUSAE, 3651
PROTON, 3652
PROTOTIPO, 3653
PROTOZOA, 3654
PROTURA, 3655
PROVINCIA=FAUNISTICA, 1518
PROVINCIA=FISOGRAFICA, 3431
PROVINCIA=FLORISTICA, 1572
PROVINCIA=HIDROGEOLOGICA, 2112
PROVINCIA=METALOGENICA, 2773
PROVINCIA=PALEOBOTANICA, 3244
PROVINCIA=PETROGRAFIA, 3377
PROVINCIALIDAD, 3657
PROYECCION=GNOMONICA, 1832
PROYECCION=VOLCANICA, 1315
PROYECTO, 3636
PROYECTO=MOHOLE, 2909
PRUEBA, 3618
PSAMITA, 3659
PSAMITICO, 3660
PSAMMENT, 3658
PSEFITICO, 3662
PSEUDOCONGLOMERADO, 3663
PSEUDOGLEY, 3664
PSEUDOMORFOSIS, 3665
PSEUDOTILLITA, 3666
PSICROFITA, 3669
PSILOCERATIDA, 3667
PSILOPSIDA, 3668
PTERIDOPHYLLA, 3670
PTERIDOPHYTA, 3671
PTERIDOSPERMAE, 3672
PTERIOIDA, 3673
PTEROBRANCHIA, 3674
PTEROMORPHIA, 3675
PTEROPODA, 3676
PTEROSAURIA, 3677
PTYCHOPARIIDA, 3678
PUBLICACION, 3681

PUENTE, 577
PUERTO, 1954
PUESTA=A=MASA, 2895
PULMONATA, 3682
PULVERIZADO=FALLA, 1509
PUNTO=CALIENTE, 2065
PUNTO=DE=CURIE, 1033
PURINA, 3688
PUZOLANA, 3589
PYCNOGONIDA, 3690
PYROTHERIA, 3701
PYRROPHYCOPHYTA, 3705
QUELACION, 718
QUERATOFIDO, 2385
QUILATE, 640
QUIMISMO, 725
QUITINA, 745
RACEMISACION, 3717
RADAR, 3718
RADIACION, 3720
RADIACION=ADAPTABLE, 64
RADIACION=SOLAR, 4159
RADIAL, 3719
RADIO, 3739
RADIO=HIDRAULICO, 2097
RADIO=INTERFEROMETRIA, 3722
RADIOACTIVIDAD, 3731
RADIOGRAFIA=RX, 3734
RADIOLARIA, 3735
RADIOLARITA, 3736
RADIOMETRO, 3737
RADIUS, 3740
RADON, 3741
RAIZ=DE=MANTO, 3896
RAMBLA, 4779
RANKER, 3749
RAPAKIVI, 3750
RATITES, 3754
RAYOS=ALFA, 147
RAYOS=BETA, 473
RAYOS=COSMICOS, 970
RAYOS=GAMMA, 1676
RAYOS=IR, 2233
RAYOS=UV, 4641
RAYOS=X, 4832
RAZON=DE=ESTADO, 1412
RB=SR, 3910
RE=OS, 3850
REACCION, 3757
REACCION=ENDOTERMICA, 1372
REACCION=EXOTERMICA, 1473
REACCION=FOTOQUIMICA, 3408
REACONDICIONAMIENTO, 3766
REACTIVACION, 3760
REACTIVO, 3761
RECARGA, 3765
RECARGA=ARTIFICIAL, 321

RECEPCION–SEÑAL, 3764
RECEPTACULITIDA, 3763
RECICLAJE, 3772
RECOGIDA, 838
RECONSTRUCCION–PALEOGEOGRAFICA, 3767
RECONSTRUCCION–PALEONTOLOGICA, 3264
RECRISTALIZACION, 3770
RECUBRIMIENTO–ESTERIL, 3213
RECUPERACION, 3768
RECURSO–MINERO, 2872
RECURSO–NATURAL, 2992
RECURSOS, 3834
RECURSOS–AGUA, 4798
RED–DE–BRAVAIS, 574
RED–DE–CANALES, 571
RED–FRACTURA, 1512
RED–HIDROGRAFICA, 1246
RED–SISMICA, 306
REDLICHIIDA, 3775
REDUCCION–QUIMICA, 731
REEMPLAZAMIENTO, 3817
REFINO, 3781
REFLECTIVIDAD, 3785
REFLEXION–ONDA, 3783
REFRACCION–ONDA, 3786
REFRACTARIO, 3791
REFRACTIVIDAD, 3788
REFRACTOMETRIA, 3790
REFRACTOMETRO, 3789
REGENERACION, 3792
REGIMEN–CRITICO, 994
REGIMEN–LAMINAR, 2420
REGIMEN–PERMANENTE, 4246
REGIMEN–TRANSITORIO, 4664
REGIMEN–TURBULENTO, 1304
REGISTRO–SISMICO, 4012
REGLA–FASES, 3394
REGLA–MINERALOGICA–DE–LAS–FASES, 2879
REGMAGLYPTO, 3797
REGOLITO, 3798
REGOSOL, 3799
REGRESION, 3800
REGRESION–ESTADISTICA, 3801
REIMANTACION, 3811
REJUVENECIMIENTO, 3802
RELACION, 3803
RELACION–ARENITA–LUTITA, 3938
RELACION–CARBONO, 645
RELACION–GAS–PETROLEO, 1693
RELACION–ISOTOPICA, 2352
RELACION–QUIMICA, 730
RELICTA, 3806
RELICTO–EROSION, 1428
RELIEVE, 3808
RELIEVE–CONTINENTAL, 2429
RELIEVE–ESTRUCTURAL, 4310
RELIEVE–LUNAR, 2937

RELIEVE–SUBMARINO, 3975
RELLENO, 1120
RELLENO–KARSTICO, 2374
RELLENO–TERRAPLEN, 1358
REMOCION–FOSIL, 3846
REMOCION–SEDIMENTARIA, 3774
RENDOLL, 3815
RENDZINA, 3816
RENIO, 3849
RENTABILIDAD, 3629
REOLOGIA, 3851
REOMORFISMO, 3852
REPARTICION–FASE, 3312
REPARTICION–GRANOS, 4125
REPLICA, 4568
REPRESENTACION–GRAFICA, 1864
REPRODUCCION, 3820
REPTACION, 986
REPTILIA, 3821
RESERVA–MINERAL, 3824
RESIDUAL, 3827
RESIDUO, 4781
RESIDUO–INDUSTRIAL, 2221
RESIDUO–INSOLUBLE, 2256
RESIDUO–LIQUIDO, 2521
RESIDUO–ORGANICO, 3173
RESIDUO–RADIOACTIVO, 3730
RESIDUO–SOLIDO, 4168
RESIDUO–URBANO, 2070
RESINA, 3831
RESISTENCIA–A–LA–COMPRESION, 866
RESISTENCIA–A–LA–EROSION, 1423
RESISTENCIA–A–LA–RUPTURA, 1633
RESISTENCIA–CIZALLAMIENTO, 4062
RESISTENCIA–MECANICA, 4289
RESISTENCIA–TRACCION, 4455
RESISTIVIDAD–ELECTRICA, 3832
RESONANCIA–MAGNETICA–NUCLEAR, 3064
RESPIRACION, 3835
RESPUESTA–SISMICA, 79
RESTINGA, 737
RESURGENCIA, 3836
RETENCION, 3838
RETICULAREA, 3839
RETRABAJADO, 3845
RETRACCION, 4087
RETROMETAMORFISMO, 3841
REUNION, 2737
REVESA–DE–FONDO, 4687
REVISION, 3844
REVOLUCION–MAZATSAL, 2723
RHABDOPLEURIDA, 3847
RHETIENSE, 3848
RHIPIDISTII, 3853
RHIZOPODEA, 3854
RHODOPHYCOPHYTA, 3856
RHYNCHOCEPHALIA, 3858

RHYNCHONELLIDA, 3859
RIESGO–SISMICO, 4015
RIFT, 3864
RIGIDEZ, 3866
RIMAYA, 467
RIO, 3873
RIO–CONSECUENTE, 904
RIO–SUBTERRANEO, 4653
RIODACITA, 3860
RIOLITA, 3861
RIPPLE–EOLICO, 4828
RIPPLE–MARK, 3872
ROBERTINACEA, 3880
ROCA, 3881
ROCA–ALMACEN, 3826
ROCA–ARCILLOSA, 295
ROCA–CARBONATADA, 650
ROCA–CARBONOSA, 649
ROCA–CLASTICA, 786
ROCA–CRISTALINA, 1022
ROCA–DE–ANKERITA, 211
ROCA–DE–IMPACTO, 4057
ROCA–DE–MEZCLA, 2740
ROCA–ENCAJANTE, 978
ROCA–FELDESPATOIDICA, 1526
ROCA–FERRUGINOSA, 2312
ROCA–FIRME, 457
ROCA–FOSFATADA, 3402
ROCA–GRANUDA, 3519
ROCA–HIPOABISAL, 2142
ROCA–HUESPED, 2062
ROCA–IGNEA, 2176
ROCA–IGNEA–ESTRATOIDE, 2463
ROCA–INCOMPETENTE, 2203
ROCA–MADRE, 3304
ROCA–MADRE–HIDROCARBURO, 4186
ROCA–MADRE–SUELO, 3302
ROCA–METAMORFICA, 2785
ROCA–MICROGRANUDA, 2817
ROCA–MOTEADA, 2949
ROCA–PATRON, 4237
ROCA–SEDIMENTARIA, 4000
ROCA–TOTAL, 4824
ROCA–VERDE, 1894
ROCA–VOLCANICA, 4766
ROCIO, 1152
RODENTIA, 3892
RODIO, 3855
ROSTROCONCHIA, 3897
ROTACION, 3902
ROTACION–PLACA, 3498
ROTACION–TIERRA, 1278
ROTALIACEA, 3898
ROTALIINA, 3899
ROTLIEGENDE, 3904
ROTURA–ROCA, 3883
ROTURA–TERMICA, 1965

RPE, 1351
RUBEFACCION, 3908
RUBIDIO, 3909
RUDISTAE, 3911
RUDITA, 3912
RUGOSA, 4474
RUGOSIDAD, 3905
RUIDO-DE-FONDO, 3041
RUMINANTIA, 3913
RUPELIENSE, 3916
RUPTURA, 3917
RUTENIO, 3919
S-34=S-32, 4355
SACA=MUESTRAS, 3934
SACUDIDA-VIOLENTA, 4298
SAKMARIENSE, 3921
SAL=GEMA, 3924
SALES-POTASICAS, 3577
SALIDA=GAS, 1097
SALIDA-VOLCANICA, 4767
SALINA, 3507
SALINIDAD, 3923
SALINO, 3922
SALMUERA, 578
SALMUERA=CALIENTE, 2064
SALTACION, 3931
SAMARIO, 3932
SANTONIENSE, 3943
SAPROPEL, 3944
SARCODINA, 3946
SARCOPTERYGII, 3947
SARMATIENSE, 3948
SATELITE, 3484
SATELITE=ARTIFICIAL, 322
SATURACION, 3950
SAURISCHIA, 3951
SAURORNITHES, 3952
SAUSURITIZACION, 3953
SAXONIENSE, 3954
SCAPHOPODA, 3958
SCIUROMORPHA, 3965
SCYPHOMEDUSAE, 3972
SCYPHOZOA, 3973
SECCION-PULIDA, 3538
SECCION-TIPO, 4625
SECRECION-LATERAL, 2445
SECUENCIA-METASEDIMENTARIA, 2792
SECUENCIA-PELITICO-METAMORFICA, 3336
SEDIMENTACION, 4003
SEDIMENTACION=ACTUAL, 3762
SEDIMENTACION=AGUA-DULCE, 1649
SEDIMENTACION=AGUA-PROFUNDA, 23
SEDIMENTACION=ALBUFERA, 2413
SEDIMENTACION=ARRECIFAL, 3778
SEDIMENTACION=BIOCLASTICA, 488
SEDIMENTACION=BIOQUIMICA, 482
SEDIMENTACION=CONTINENTAL, 928

SEDIMENTACION-DE-LIMOS, 4108
SEDIMENTACION-DELTAICA, 1105
SEDIMENTACION-DETRITICA, 1144
SEDIMENTACION-EOLICA, 1393
SEDIMENTACION-ESTUARIO, 1437
SEDIMENTACION-FLUVIAL, 1595
SEDIMENTACION-FLUVIOGLACIAR, 1597
SEDIMENTACION-GEOSINCLINAL, 1765
SEDIMENTACION-GLACIAR, 1798
SEDIMENTACION-GLACIOLACUSTRE, 1806
SEDIMENTACION-GLACIOMARINA, 1808
SEDIMENTACION-INTRAFORMACIONAL, 2291
SEDIMENTACION-LACUSTRE, 2409
SEDIMENTACION-LITORAL, 2546
SEDIMENTACION-MAREA, 2287
SEDIMENTACION-MARGEN-CONTINENTAL, 4070
SEDIMENTACION-MARINA, 2686
SEDIMENTACION-PANTANO, 4377
SEDIMENTACION-PELAGICA, 3331
SEDIMENTACION-QUIMICA, 723
SEDIMENTO, 3991
SEDIMENTO-BASE, 554
SEDIMENTO-DE-LECHO, 1512
SEDIMENTO-LACUSTRE, 2408
SEDIMENTO-MARINO, 2685
SEDIMENTO-ORGANICO, 3174
SEDIMENTOGENESIS, 4005
SEDIMENTOLOGIA, 4006
SEGREGACION, 4010
SEICHE, 4011
SELENIATO, 4029
SELENIO, 4030
SELENOCRONOLOGIA, 4031
SELENOLOGIA, 4032
SEMIESPACIO, 1943
SEMILLA, 4007
SEÑAL-TELESISMICA, 4446
SENONIENSE, 4037
SEPARACION, 4039
SEPTA, 4041
SEPTARIA, 4040
SEQUIA, 1258
SERICITIZACION, 4043
SERIE, 4044
SERIE-ARTICA, 287
SERIE-ATLANTICA, 355
SERIE-PACIFICA, 3231
SERIE-SISMOS, 1287
SERIES-BOWEN, 3759
SERIES-DE-REACCION-DE-BOWEN, 563
SERIES-RADIOACTIVAS, 3728
SERPENTINA, 4045
SERPENTINITA, 4046
SERPENTINIZACION, 4047
SERVICIO-DE-MINAS, 595
SERVICIO-GEOLOGICO, 1739
SHONKINITA, 4080

SIAL, 4089
SICILIENSE, 4090
SIDEROFIRO, 4093
SIDEROLITO, 4263
SIDERURGIA, 2309
SIEGENINSE, 4094
SIENITA, 4379
SIENITA-ALCALINA, 127
SIEROZEM, 4095
SIFON, 4120
SILEX, 1567
SILEXITA, 739
SILICATO, 4098
SILICE, 4097
SILICIFICACION, 4100
SILICIO, 4103
SILICOFLAGELLATA, 4102
SILL, 4104
SILLIMANITA, 4105
SILURICO, 4110
SILURICO-INF, 2593
SILURICO-SUP, 4683
SILVINITA, 4380
SIMA, 4112
SIMA-MARINA, 17
SIMBIOSIS, 4381
SIMETRIA, 4384
SIMIEN, 4113
SIMPOSIO, 4387
SIMULACION, 4114
SIMULACION-ANALOGICA, 180
SIMULACION-NUMERICA, 1183
SINCLINAL, 4390
SINCLINORIO, 4391
SINCLISIO, 4392
SINEMURIENSE, 4115
SINGENESIS, 4394
SINGENETICO, 4395
SINONIMIA, 4396
SINTER, 4118
SINTESIS, 4401
SINTESIS-BIBLIOGRAFICA, 3843
SINTEXIA, 4400
SINUOSIDAD-RIO, 4119
SIPHONAPTEROIDA, 4121
SIPUNCULOIDA, 4122
SIRENIA, 4123
SISMICA-IMPACTO, 2190
SISMICA-MARTILLO, 1951
SISMICA-REFLEXION, 3784
SISMICA-REFRACCION, 3787
SISMICIDAD, 4022
SISMO-INTERMEDIO, 2270
SISMO-LUNAR, 2936
SISMO-PROFUNDO, 1088
SISMO-SUPERFICIAL, 4052
SISMODINAMOMETRIA, 1286

SISMOGRAFO, 4024
SISMOGRAFO-FONDO-SUBMARINO, 3089
SISMOGRAMA, 4023
SISMOLOGIA, 4026
SISMOLOGIA-EXPERIMENTAL, 1478
SISMOMETRO, 4027
SISMOTECTONICA, 4028
SISTEMA, 4403
SISTEMA-BINARIO, 479
SISTEMA-COSMICO, 971
SISTEMA-CRISTALINO, 1021
SISTEMA-CUATERNARIO, 3714
SISTEMA-CUBICO, 1030
SISTEMA-DE-PLIEGUES, 1606
SISTEMA-GEOTERMICO, 1778
SISTEMA-HEXAGONAL, 2002
SISTEMA-MONOCLINICO, 2921
SISTEMA-MULTICAPA, 2962
SISTEMA-ORTOROMBICO, 3194
SISTEMA-PIRAMIDAL, 3693
SISTEMA-QUIMICO, 733
SISTEMA-REGULAR, 2332
SISTEMA-ROMBICO, 3857
SISTEMA-SOLAR, 4160
SISTEMA-TERNARIO, 4463
SISTEMA-TRICLINICO, 4593
SISTEMA-TRIGONAL, 4595
SKARN, 4126
SLUMPING, 4139
SOBRESATURACION, 3217
SOBRESATURADO, 3216
SOCAVON, 3455
SOCIEDAD, 858
SOCIEDAD-CIENTIFICA, 2472
SODIO, 4143
SOL-ASTRO, 4358
SOLEMYIDA, 4164
SOLFATARA, 4165
SOLIDUS, 4170
SOLIFLUXION, 4171
SOLONCHAK, 4172
SOLONEZ, 4173
SOLUBILIDAD, 4174
SOLUCION, 4176
SOLUCION-ACUOSA, 256
SOLUCION-SOLIDA, 4167
SOLUTO, 4175
SOMASTEROIDEA, 4180
SONAR, 4181
SONDA-NEUTRON, 3018
SONDEO-AUGER, 371
SONDEO-DIRECCION, 1208
SONDEO-ELECTRICO, 1332
SONDEO-EXPLOTACION, 1148
SONDEO-FRECUENCIA-VARIABLE, 1646
SONDEO-GEOFISICO, 4185
SONDEO-MECANICO, 1254

SONDEO-OFF-SHORE, 3103
SONDEO-ROTACION-PERCUSION, 3901
SONDEO-ROTARY, 3900
SONDEO-SISMICO-PROFUNDO, 1091
SONDEO-TESTIGO-CONTINUO, 1661
SONDEO-VIBRACION, 4739
SORBCION, 4183
SPINIFEX, 4212
SPIRIFERIDA, 4213
SPIRILLINACEA, 4214
SPODOSOL, 4216
SPOROZOA, 4221
SQUAMATA, 4225
STAMPIENSE, 4233
STOCK, 4257
STOCKWERK, 4258
STOLONOIDEA, 4260
STOMOCHORDATA, 4261
STONELINE, 4262
STROMATOLITES, 4295
STROMATOPOROIDEA, 4296
STROPHOMENIDA, 4300
STYLOPHORA, 4315
SUAVIZADO, 4141
SUBBOREAL, 4319
SUBDIVISION, 4320
SUBLIMACION, 4324
SUBLIMADO, 4323
SUBSATURADO, 4654
SUBSIDENCIA, 4337
SUBSIDENCIA-REGIONAL, 1242
SUBSIDENCIA-VOLCANICA, 675
SUBSTRATUM, 4339
SUCCION, 634
SUCTORIA, 4345
SUELO, 4146
SUELO-BRUTO, 394
SUELO-CASTAÑO, 666
SUELO-COHESIVO, 830
SUELO-CON-FIGURAS, 3317
SUELO-DE-TUNDRA, 4611
SUELO-DESERTICO, 4848
SUELO-FERRALITICO, 1532
SUELO-FERRUGINOSO, 1540
SUELO-HELADO, 1656
SUELO-INMADURO, 2183
SUELO-LATERITICO, 2447
SUELO-LUNAR, 2603
SUELO-MEDITERRANEO, 2735
SUELO-PARDO, 583
SUELO-POLIGONAL, 3547
SUELO-RECREATIVO, 3769
SUELO-RICO-EN-HUMUS, 2072
SUELO-SIN-COHESION, 828
SUIFORMES, 4348
SULFATO, 4349
SULFOSAL, 4352

SULFURO, 4351
SUMERGIBLE, 4335
SUPERCRECIMIENTO, 3215
SUPERESTRUCTURA, 4361
SUPERESTRUCTURA-TECTONICA, 4439
SUPERFICIE-ACANALADURA, 1902
SUPERFICIE-DE-ABRASION, 8
SUPERFICIE-DE-EROSION, 1429
SUPERFICIE-ESPECIFICA, 4198
SUPERFICIE-PIEZOMETRICA, 3583
SUPERPOSICION, 4360
SUPERPOSICION-EDAD, 1723
SUR, 4187
SURCO, 1766
SURCO-ABISAL, 20
SURGENCIA, 1491
SUSCEPTIBILIDAD-MAGNETICA, 2645
SUSPENSION, 4373
SUSTANCIA-TOXICA, 4554
SUSTANCIA-UTIL, 2219
SUSTITUCION, 4338
SUTURA, 4374
SYMMETRODONTA, 4383
SYMPHYLA, 4385
SYNAPSIDA, 4388
TABLA-DE-DATOS, 1072
TABULATA, 4406
TACONITA, 4408
TACTITA, 4409
TAENIODONTA, 4410
TAENIOLABOIDEA, 4411
TAFONOMIA, 4420
TAFROGENESIS, 4421
TAFROGEOSINCLINAL, 4422
TALCO, 4414
TALCOESQUISTO, 4415
TALIO, 4480
TALUD-DERRUBIA, 4417
TALUD-MARINO, 930
TALUD-MARINO-INTERNO, 2245
TAMIZ, 4096
TANATOCENOSIS, 4483
TANETIENSE, 4484
TANTALATO, 4418
TANTALO, 4419
TAPON, 3513
TAPONADO, 3514
TARANONIENSE, 4423
TARDIGRADA, 4424
TARJETA-PERFORADA, 3686
TASMANITES, 4425
TATARIENSE, 4426
TAXITA, 4427
TAXON-AHERMATIPICO, 92
TAXON-ANAEROBIO, 179
TAXON-COLONIAL, 844
TAXON-EXTINGUIDO, 1495

TAXON=HERMATIPICO, 1987
TAXON=MONOLETE, 2925
TAXON=NOMINAL, 3043
TAXON=PLANCTONICO, 3488
TAXONOMIA, 4428
TECHO, 1953
TECHO=REBAJADO, 3895
TECNECIO, 4429
TECNICA=ANALOGICA, 181
TECNICAS, 4431
TECNOLOGIA, 4432
TECNOLOGIA=HIDROGEOLOGICA, 2114
TECTITA, 4444
TECTOGENESIS, 4433
TECTONICA, 4440
TECTONICA=ACTIVA, 59
TECTONICA=DE=COBERTERA, 3996
TECTONICA=DE=FONDO, 1094
TECTONICA=DE=FRACTURA, 532
TECTONICA=DE=PLACAS, 3499
TECTONICA=DE=PLEGAMIENTO, 1607
TECTONICA=DE=ZOCALO, 430
TECTONICA=GERMANICA, 1783
TECTONICA=GRAVEDAD, 1887
TECTONICA=IMBRICADA, 2180
TECTONICA=SALIFERA, 3929
TECTONICA=SUPERPUESTA, 2965
TECTONICA=TANGENCIAL, 867
TECTONITA, 4441
TECTONOFISICA, 4442
TECTONOSFERA, 4443
TEFROCRONOLOGIA, 4458
TEGUMENTO=VERTEBRADO, 1126
TELEDETECCION, 3813
TELEDETECCION=MULTIESPECTRAL, 2966
TELEOSTEI, 4445
TELURATO, 4448
TELURO, 4450
TEMPERATURA, 4453
TEMPERATURA=ALTA, 2008
TEMPERATURA=BAJA, 2572
TEMPESTAD, 4266
TENACIDAD, 4454
TENDENCIA, 4244
TENSIOMETRO=ASPA, 4711
TENSION, 4290
TENSION=CIZALLAMIENTO, 4063
TENSION=DE=COMPRESION, 869
TENSION=DE=RUPTURA, 3918
TENSION=SUPERFICIAL, 4365
TENTACULITES, 4457
TEORIA, 4490
TEORIA=DE=LA=DINAMO, 1268
TEORIA=DE=LAMARCK, 2417
TEORIA=DEL=DESPLAZAMIENTO, 1225
TEORIA=MOVIMIENTO=ASCENDENTE, 326
TEORIA=MOVIMIENTO=DESCENDENTE, 1130

TEORIA=OSCILATORIA, 3198
TERALITA, 4491
TERATOLOGIA, 4459
TERBIO, 4460
TERCIARIO, 4470
TEREBRATULIDA, 4461
TERMAL, 4493
TERMINACION=POZO, 4814
TERMINOLOGIA, 4462
TERMO=DIFERENCIAL, 1176
TERMO=GRAVIMETRICO, 4507
TERMOCLINA, 4503
TERMODINAMICA, 4505
TERMOGRAMA, 4506
TERMOKARST, 4508
TERMOLUMINISCENCIA, 4509
TERRA=ROSSA, 4464
TERRAPLENAMIENTO, 2428
TERRAZA, 4465
TERRAZA=COSTERA, 469
TERRAZA=DE=RIO, 3875
TERRAZA=MARINA, 2688
TERREMOTO, 1284
TERRENO, 1904
TERRENO=HUMEDO, 4822
TERRENO=VERDADERO, 1910
TESIS, 4512
TEST, 4471
TESTIFICACION, 2562
TESTIFICACION=DE=MICRORRESISTIVIDAD, 2821
TESTIFICACION=DE=RESISTIVIDAD, 2443
TESTIFICACION=GAMMA, 1679
TESTIFICACION=GAMMA=GAMMA, 1678
TESTIGO, 958
TESTIGO=DE=EROSION, 3210
TESTIGO=SONDEO, 956
TESTUDINATA, 4472
TETHYS, 4473
TETRAPODA, 4475
TEXTULARIINA, 4476
TEXTURA, 4477
TEXTURA=AFANITICA, 236
TEXTURA=APLITICA, 240
TEXTURA=ARENOSA, 290
TEXTURA=BLASTICA, 525
TEXTURA=CAVERNOSA, 679
TEXTURA=COLOFORME, 841
TEXTURA=CRIPTOCRISTALINA, 1006
TEXTURA=EN=MOSAICO, 2945
TEXTURA=EUTAXITICA, 1453
TEXTURA=FLUIDAL, 1584
TEXTURA=FRAMBOIDAL, 1638
TEXTURA=GNEISICA, 1830
TEXTURA=GRAFICA, 1865
TEXTURA=GRANOBLASTICA, 1852
TEXTURA=GRANOFIDICA, 1855
TEXTURA=GRANUDA, 1444

TEXTURA=GRANULAR, 1856
TEXTURA=HETEROBLASTICA, 1988
TEXTURA=HIALOCRISTALINA, 2084
TEXTURA=HIPIDIOBLASTICA, 2144
TEXTURA=HIPIDIOMORFICA, 2145
TEXTURA=HIPOCRISTALINA, 2147
TEXTURA=HOLOCRISTALINA, 2028
TEXTURA=IDIOMORFICA, 2170
TEXTURA=ISOGRANULAR, 1415
TEXTURA=LAMINAR, 1561
TEXTURA=LAMPROFIDICA, 2424
TEXTURA=MARGINAL, 2675
TEXTURA=MIAROLITICA, 2803
TEXTURA=MICROGRANITICA, 2819
TEXTURA=MICROLITICA, 2822
TEXTURA=MICROSCOPICA, 2832
TEXTURA=MONZONITICA, 2934
TEXTURA=OFITICA, 3134
TEXTURA=OOLITICA, 3126
TEXTURA=ORBICULAR, 3145
TEXTURA=ORTOFIDICA, 3193
TEXTURA=PANIDIOMORFA, 3282
TEXTURA=PARALELA, 3296
TEXTURA=PELITICA, 3335
TEXTURA=PERLITICA, 3361
TEXTURA=PERTITICA, 3368
TEXTURA=PISOLITICA, 3458
TEXTURA=POIKILITICA, 3529
TEXTURA=POIKILOBLASTICA, 3530
TEXTURA=PORFIDICA, 3567
TEXTURA=PORFIDOBLASTICA, 3568
TEXTURA=RELICTA, 3807
TEXTURA=RETICULADA, 2452
TEXTURA=SIMPLECTITICA, 4386
TEXTURA=TABICADA, 564
TEXTURA=VITREA, 1814
TEXTURA=VITROCLASTICA, 4752
TEXTURA=VITROFIDICA, 4753
TH=PB, 4518
THALASSOCRATON, 4478
THALASSOIDE, 4479
THALLOPHYTA, 4481
THECAMOEBA, 4486
THECIDEIDINA, 4487
THECODONTIA, 4488
THELODONTI, 4489
THERAPSIDA, 4492
THYSANOPTEROIDA, 4523
THYSANURA, 4524
TIEMPO=DE=LLEGADA, 307
TIEMPO=RECORRIDO, 4581
TIERRA=DE=BATAN, 1662
TIERRA=DE=DIATOMEAS, 1168
TIERRAS=ALTAS, 2012
TIERRAS=BAJAS, 2595
TIERRAS=RARAS, 3751
TILL, 4530

TILLITA, 4531
TILLODONTIA, 4532
TINTINNIDAE, 4538
TIPO-ALPINO, 153
TIPOMORFISMO, 4628
TIRRENIENSE, 4629
TITANIO, 4539
TITONICO, 4540
TIXOTROPIA, 4515
TOARCIENSE, 4541
TOBA, 4607
TOBA-VOLCANICA, 4609
TOMA-DE-DATOS, 1065
TOMBOLO, 4542
TONSTEIN, 4543
TOPOGRAFIA, 4547
TORIO, 4517
TORMENTA-MAGNETICA, 2642
TORRENTE, 4549
TORSION, 4550
TORTONIENSE, 4551
TORTUOSIDAD, 4552
TOURNAISIENSE, 4553
TOXICIDAD, 4555
TRACCION, 4456
TRACCION-SEDIMENTACION, 3992
TRACHYPSAMMIACEA, 4563
TRAMPA, 4580
TRAMPA-ESTRATIGRAFICA, 4279
TRAMPA-ESTRUCTURAL, 4311
TRAMPA-LITOLOGICA, 2531
TRAMPA-PETROLIFERA, 3111
TRAMPA-SEDIMENTARIA, 4002
TRANSFERENCIA-DE-MASA, 2701
TRANSFORMACION, 4570
TRANSFORMACION-LAPLACE, 2434
TRANSFORMISMO, 4571
TRANSGRESION, 4572
TRANSMISIVIDAD, 4576
TRANSPORTE, 4577
TRANSPORTE-DE-MINERAL, 3163
TRANSPORTE-EN-MASAS, 2702
TRANSPORTE-EOLICO, 4829
TRANSPORTE-FLUVIAL, 4288
TRANSPORTE-GLACIAR, 1801
TRANSPORTE-HIELO, 2159
TRANSPORTE-LIMO-SUSPENDIDO, 4107
TRANSPORTE-MARINO, 2689
TRANSPORTE-SUSPENSION-FLUVIAL, 3874
TRANSVERSAL, 4579
TRAQUIANDESITA, 4561
TRAQUIBASALTO, 4562
TRAQUITA, 4564
TRASMISION, 1969
TRATADO, 4584
TRATAMIENTO-DATOS, 1069
TRAVERTINO, 4583

TRAZA-FISION, 1555
TRAZADOR, 4560
TRAZADOR-COLORANTE, 848
TRAZADOR-RADIOACTIVO, 3729
TREMADOCIENSE, 4586
TREPOSTOMATA, 4589
TRIANGULACION, 4590
TRIAS, 4591
TRIAS-ALPINO, 152
TRIAS-INF, 2594
TRIAS-MEDIO, 2852
TRIAS-SUP, 4684
TRICONODONTA, 4594
TRIGONIIDAE, 4596
TRILOBITA, 4597
TRILOBITOMORPHA, 4598
TRIPOLI, 4600
TRITIO, 4601
TSUNAMI, 4604
TUBOIDEA, 4605
TUBULIDENTATA, 4606
TUFITA, 4608
TULIO, 4521
TUNDRA, 4610
TUNEL, 4614
TUNEL-DE-LAVA, 2459
TUNGSTATO, 4612
TUNGSTENO, 4613
TURBA, 3320
TURBERA, 3321
TURBIDEZ, 4616
TURBIDITA, 4615
TURBOSONDEO, 4619
TURINGIENSE, 4522
TURONIENSE, 4620
TYLOPODA, 4623
U-PB, 4692
U-TH-PB, 4693
UDALF, 4630
UDOLL, 4631
UDULT, 4632
UFIMIENSE, 4633
ULTISOL, 4634
ULTRABASITA, 4636
ULTRAMETAMORFISMO, 4637
ULTRAMILONITA, 4638
UMBRAL, 2687
UMBRAL-ANOMALIA, 4519
UMBRAL-GLACIAR, 3863
UMBRAL-ROCOSO, 3882
UMBREPT, 4642
UNAKITA, 4643
UNIDAD-BIOESTRATIGRAFICA, 510
UNIDAD-CRONOESTRATIGRAFICA, 762
UNIDAD-ESTRATIGRAFICA, 4280
UNIDAD-GEOCRONOLOGICA, 1724
UNIDAD-LITOESTRATIGRAFICA, 2537

UNIDAD-TECTONICA, 4305
UNIFORMIDAD, 4659
UNION-TRIPLE-LITOSFERA, 4599
UNIONOIDA, 4660
UNIVERSO, 4663
URALITIZACION, 4688
URANIO, 4689
URBANISMO, 4694
URBANIZACION, 4696
UREILITA, 4697
URODELA, 4698
USTALF, 4699
USTERT, 4700
USTOLL, 4701
USTOX, 4702
USTULT, 4703
UTIL-PREHISTORICO, 319
UTILIZACION-INDUSTRIAL, 2220
UTILIZACION-SUSTANCIA, 4704
UTILIZACION-TERRENO, 2427
VACIADO, 665
VAGUADA, 4482
VALANGINIENSE, 4706
VALENCIA-QUIMICA, 734
VALLE, 4707
VALLE-COLGADO, 1952
VALLE-ESTRUCTURAL, 4312
VALLE-FOSILIZADO, 599
VALLE-GLACIAR, 1802
VALLE-LUNAR, 3867
VALLE-OBSECUENTE, 3080
VALLE-RESECUENTE, 3823
VALLE-SUBMARINO, 4332
VALLE-SUBSECUENTE, 4336
VALVA, 4708
VANADATO, 4709
VANADIO, 4710
VAPOR, 4713
VARIABILIDAD, 4714
VARIABLE, 4715
VARIACION, 4717
VARIACION-CAMPO-MAGNETICO, 2634
VARIACION-DIURNA, 1231
VARIACION-ESTACIONARIA, 3985
VARIACION-NIVEL-MAR, 3981
VARIACION-SECULAR, 1744
VARIOGRAMA, 4718
VARVA, 4720
VARVA-CRONOLOGIA, 4721
VECTOR, 4723
VEGETACION, 4724
VELOCIDAD, 4726
VELOCIDAD-SISMICA, 4020
VENERIDA, 4728
VENITA, 4729
VENTANA-TECTONICA, 4830
VERGENCIA, 4730

VERMES, 4731
VERMICULITA, 4732
VERTEBRATA, 4734
VERTEDERO, 4782
VERTISOL, 4737
VIA−NAVEGABLE, 4804
VIBRACION, 4738
VIBROSISMO, 4740
VIDA, 2499
VIDA−MEDIA−RADIOACTIVA, 1942
VIDA−ORGANICA, 3169
VIDRIO, 1811
VIDRIO−NATURAL, 2990
VIDRIO−VOLCANICO, 4764
VIENTO, 4826
VIENTO−MONZON, 2931
VIENTO−SOLAR, 4161
VILLAFRANQUIENSE, 4742
VIRGACION, 4743
VISCOELASTICIDAD, 4744
VISCOSIDAD, 4745
VISEENSE, 4748
VITREA, 4750
VITRENO, 4749
VITRIFICACION, 4751
VIVIENTE, 2548
VOLANTES, 4754
VOLATIL, 4755
VOLATILIZACION, 4758
VOLCAN, 4769
VOLCAN−ACTIVO, 60
VOLCAN−DE−BARRO, 2958
VOLCAN−EN−ESCUDO, 4073
VOLCAN−INSULAR, 2317
VOLCAN−SUBMARINO, 4333
VOLCANICO, 4765
VOLUMEN, 4776
VOLUMEN−DE−SEDIMENTOS, 3993
VULCANIANO, 4777
VULCANISMO, 4768
VULCANO−SEDIMENTARIO, 4772

VULCANOGENO, 4773
VULCANOLOGIA, 4775
VULCANORIO, 4778
WENLOKIENSE, 4818
WESTFALIENSE, 4820
WILDFLYSCH, 4825
XANTHOPHYTA, 4837
XENON, 4840
XERALF, 4841
XEROFITA, 4844
XEROLL, 4842
XEROSOL, 4845
XERULT, 4846
YACIMIENTO, 3162
YACIMIENTO−DETRITICO, 2725
YACIMIENTO−DISEMINACION, 1226
YACIMIENTO−EPITERMAL, 1407
YACIMIENTO−ESTRATIFORME, 4274
YACIMIENTO−ESTRATOIDE, 4270
YACIMIENTO−EXHALATIVO, 1467
YACIMIENTO−FILONIANO, 2559
YACIMIENTO−FOSILIFERO, 1625
YACIMIENTO−HIDROTERMAL, 2132
YACIMIENTO−HIPOTERMAL, 2149
YACIMIENTO−IGNEO, 2621
YACIMIENTO−IMPREGNACION, 2195
YACIMIENTO−INTRAMAGMATICO, 2622
YACIMIENTO−MESOTERMAL, 2759
YACIMIENTO−METAMORFOGENO, 2788
YACIMIENTO−MINERAL, 3154
YACIMIENTO−MINERALOGICO, 2878
YACIMIENTO−NEUMATOLITICO, 3524
YACIMIENTO−NO−AFLORANTE, 529
YACIMIENTO−POLIMETALICO, 3551
YACIMIENTO−POSITIVO, 3571
YACIMIENTO−POTENCIAL, 3582
YACIMIENTO−QUIMICO, 722
YACIMIENTO−SEDIMENTARIO, 3999
YACIMIENTO−TELETERMAL, 4447
YACIMIENTO−TIPO−MISSISSIPPI−VALLEY, 2896
YACIMIENTO−TIPO−ROLL, 3893

YACIMIENTO−VULCANOGENO, 4774
YESIFICACION, 1932
YESO, 1933
YODO, 2298
ZECHSTEIN, 4855
ZEOLITA, 4856
ZEOLITIZACION, 4858
ZOANTHINIARIA, 4863
ZOCALO, 429
ZOECIA, 4864
ZONA−BENIOFF, 462
ZONA−BIOESTRATIGRAFICA, 339
ZONA−CIZALLA, 4064
ZONA−CLIMATICA, 797
ZONA−CONTINENTAL, 931
ZONA−DE−AIREACION, 4866
ZONA−DE−BAJA−VELOCIDAD, 2576
ZONA−DE−CEMENTACION, 4867
ZONA−DE−FLUJO, 4868
ZONA−DE−SEDIMENTACION, 4869
ZONA−DE−SULFUROS, 4350
ZONA−ECUATORIAL, 1414
ZONA−ESTEPARIA, 4251
ZONA−FALLA, 1513
ZONA−FAUNISTICA, 1519
ZONA−FRACTURADA, 1634
ZONA−FRIA, 548
ZONA−MAR−PROFUNDO, 1093
ZONA−METAMORFICA, 2786
ZONA−MOVIL, 2899
ZONA−NO−SATURADA, 3049
ZONA−OXIDACION, 3223
ZONA−SATURADA, 3949
ZONA−SUBDUCCION, 4322
ZONA−SUBTROPICAL, 4342
ZONA−TEMPLADA, 4452
ZONA−TROPICAL, 4602
ZONACION−CRISTAL, 4870
ZONOGRAFIA−SUELO, 673
ZOOIDE, 4871

Italiano

A GRANA MEDIA, 2736
A GRANDE SCALA, 2436
A PICCOLA SCALA, 4140
AALENIANO, 3
ABACO, 3045
ABBASSAMENTO DELLA FALDA, 1249
ABBASSAMENTO REGIONALE, 1242
ABBATTIMENTO GEOCHIMICO, 1721
ABBONDANZA, 16
ABISSALE, 18
ABISSO, 17
ABITO CRISTALLINO, 1936
ABLAZIONE, 5
ABRASIONE, 6
ABRASIVO, 9
ACANTHARIA, 25
ACANTHODII, 26
ACCAVALLAMENTO, 4520
ACCELEROGRAMMA, 27
ACCELEROMETRO, 28
ACCRESCIMENTO, 30
ACCRESCIMENTO CONTINENTALE, 919
ACCRESCIMENTO CRISTALLINO, 1017
ACETOLISI, 35
ACIDITA, 40
ACIDO, 37
ACIDO FULVICO, 1663
ACIDO GRASSO, 1506
ACIDO INORGANICO, 2248
ACIDO ORGANICO, 3165
ACIDO SOLFORICO, 4356
ACIDO UMICO, 2071
ACMOLITE, 98
ACONDRITE, 36
ACQUA, 4783
ACQUA CAPILLARE, 636
ACQUA CONNATA, 898
ACQUA CURATIVA, 2734
ACQUA DI ADSORBIMENTO, 67
ACQUA DI COSTITUZIONE, 4794
ACQUA DI CRISTALLIZZAZIONE, 4793
ACQUA DI FUSIONE, 2745
ACQUA DI GIACIMENTO, 1306
ACQUA DI MARE, 3978
ACQUA DI RISORGIVA, 3837

ACQUA DISPONIBILE, 4802
ACQUA DOLCE, 1647
ACQUA FOSSILE, 1623
ACQUA GIOVANILE, 2365
ACQUA IGROSCOPICA, 2139
ACQUA INQUINATA, 3542
ACQUA INTERSTIZIALE, 3563
ACQUA MINERALE, 2875
ACQUA SALMASTRA, 568
ACQUA SOTTERRANEA, 1911
ACQUA SUPERFICIALE, 4366
ACQUA TERMALE, 4501
ACQUA TERMOMINERALE, 2134
ACQUIFERO, 260
ACQUIFERO NON CONFINATO, 4644
ACQUISIZIONE DEI DATI, 1065
ACQUITRINO, 536
ACRANIA, 46
ACRISOL, 47
ACRITARCHA, 48
ACROTRETIDA, 49
ACTINIARIA, 50
ACTINOCERATOIDEA, 52
ADATTAMENTO, 63
ADIAGNOSTICO, 66
ADSORBIMENTO, 68
AERAZIONE, 69
AERODROMO, 96
AEROFOTOGRAFIA, 71
AEROSOL, 75
AFFINAMENTO, 3780
AFFINITA, 76
AFFIORAMENTO, 3206
AFFLUENTE, 77
AFFOSSAMENTO, 2426
AFNIO, 1940
AGGLOMERATO, 82
AGGLOMERATO DI LAVA, 4609
AGGLUTINATO, 83
AGGRADATION, 84
AGGREGATO, 85
AGNATHA, 86
AGNOSTIDA, 87
AGONIATITIDA, 88
AGPAITICO, 89

AGRICOLTURA, 90
AGROGEOLOGIA, 91
AKTCHAGYLIANO, 99
ALABASTRO, 100
ALBEDO, 101
ALBERTITE, 102
ALBIANO, 103
ALBITIZZAZIONE, 106
ALBOLL, 107
ALCALINITA, 131
ALCANI, 132
ALCYONACEA, 108
ALDANIANO, 109
ALFISOL, 110
ALGAE, 111
ALGEBRA BOOLEANA, 542
ALGEBRA MATRICIALE, 2717
ALGHE CALCAREE, 608
ALGOL, 119
ALGORITMO, 121
ALIMENTAZIONE ARTIFICIALE, 321
ALIMENTAZIONE NATURALE, 1917
ALLINEAMENTO DI ERRATICI, 560
ALLOCHIMICO, 134
ALLOCTONO, 135
ALLOGROMIINA, 137
ALLOMETRIA, 138
ALLUMINA, 156
ALLUMINIO, 157
ALLUVIONALE, 140
ALLUVIONE, 146
ALOFITA, 1949
ALOGENO, 1947
ALOGENURO, 1944
ALPINOTIPO, 153
ALTA PRESSIONE, 2007
ALTA TEMPERATURA, 2008
ALTERAZIONE, 154
ALTERAZIONE CHIMICA, 735
ALTERAZIONE DIFFERENZIALE, 1177
ALTERAZIONE FISICA, 3429
ALTERAZIONE IDROTERMALE, 2131
ALTERAZIONE MECCANICA, 2728
ALTERAZIONE METEORICA, 4809
ALTERAZIONE SOTTOMARINA, 1946

ALTETERRE, 2012
ALTITUDINE, 155
ALTOFONDO, 4074
ALTOPIANO, 3501
ALUNITIZZAZIONE, 158
ALVEOLINELLIDAE, 159
AMBIENTE, 1386
AMBIENTE A BASSA ENERGIA, 2574
AMBIENTE ABISSALE, 19
AMBIENTE AD ALTA ENERGIA, 2009
AMBIENTE AEROBICO, 73
AMBIENTE ALPINO, 149
AMBIENTE ANAEROBICO, 178
AMBIENTE ARIDO, 299
AMBIENTE BATIALE, 436
AMBIENTE BENTONICO, 464
AMBIENTE CONTINENTALE, 923
AMBIENTE D'ACQUA DOLCE, 1648
AMBIENTE D'ESTUARIO, 1436
AMBIENTE DELTIZIO, 1104
AMBIENTE DI ACQUA POCO PROFONDA, 4053
AMBIENTE DI MARE PROFONDO, 1090
AMBIENTE DI SCOGLIERA, 3777
AMBIENTE DI STEPPA, 4251
AMBIENTE DI TALUS MARINO, 4136
AMBIENTE EOLICO, 1391
AMBIENTE EQUATORIALE, 1414
AMBIENTE EUXINICO, 1456
AMBIENTE FLUVIALE, 1593
AMBIENTE FREDDO, 832
AMBIENTE GLACIALE, 1793
AMBIENTE INTERTIDALE, 2286
AMBIENTE LACUSTRE, 2407
AMBIENTE LAGUNARE, 2412
AMBIENTE LITORALE, 2545
AMBIENTE MARGINE CONTINENTALE, 4068
AMBIENTE MARINO, 2677
AMBIENTE PARALICO, 3294
AMBIENTE PELAGICO, 3330
AMBIENTE SALATO, 1945
AMBIENTE SALMASTRO, 569
AMBIENTE SEMIARIDO, 4035
AMBIENTE SUBAEREO, 4316
AMBIENTE SUBLITORALE, 4325
AMBIENTE SUBTIDALE, 4341
AMBIENTE TEMPERATO, 4452
AMBIENTE TERRESTRE, 4467
AMBIENTE TIDALE, 4526
AMBIENTE TROPICALE, 4602
AMBIENTE UMIDO, 2074
AMBLYPODA, 161
AMBRA, 160
AMERICIO, 162
AMIANTO, 325
AMMASSO MINERALIZZATO, 2705
AMMENDAMENTO, 2742
AMMENDAMENTO SUOLO, 4150

AMMINOACIDO, 163
AMMODISCACEA, 164
AMMONIACA, 165
AMMONIO, 168
AMMONITIDA, 166
AMMONITINA, 167
AMMONOIDEA, 169
AMPHIBIA, 171
AMPHINEURA, 176
AMPIEZZA, 177
ANALISI, 182
ANALISI ARMONICA SFERICA, 4206
ANALISI ATTIVAZIONE NEUTRONICA, 3015
ANALISI DEI DATI, 1066
ANALISI DEI SISTEMI, 4404
ANALISI DELLA VARIANZA, 4716
ANALISI DELLE TENDENZE, 4588
ANALISI DI CORRISPONDENZA, 963
ANALISI DI FOURIER, 1628
ANALISI DIFFRAZIONE ELETTRONICA, 1346
ANALISI DIFFRAZIONE NEUTRONICA, 3016
ANALISI DISCRIMINANTE, 1218
ANALISI ELEMENTI IN TRACCE, 4559
ANALISI ELEMENTI MAGGIORI, 2658
ANALISI ELEMENTI MINORI, 2888
ANALISI FATTORIALE, 1502
ANALISI FUSIONE IN VUOTO, 4705
ANALISI MATEMATICA, 2708
ANALISI MODALE, 2902
ANALISI MULTIVARIATA, 2968
ANALISI PER ATTIVAZIONE, 54
ANALISI QUALITATIVA, 3706
ANALISI QUANTITATIVA, 3708
ANALISI RAGGI X, 4833
ANALISI SISMICA SPETTRALE, 4017
ANALISI STRUTTURALE, 4301
ANALISI TERMICA, 4494
ANALISI TERMODIFFERENZIALE, 1176
ANALISI TERMOMAGNETICA, 4510
ANALOGIA SISTEMA IDROGEOLOGICO, 4405
ANAPSIDA, 184
ANASPIDA, 185
ANATESSI, 186
ANATESSITE, 187
ANATOMIA, 188
ANATOMIA APPARATO RESPIRATORIO, 3835
ANATOMIA DELLA NUTRIZIONE, 3073
ANATOMIA DELLO SCHELETRO, 4127
ANATOMIA SISTEMA NERVOSO, 3012
ANCHIMETAMORFISMO, 189
ANCORAGGIO, 190
ANDALUSITE, 192
ANDEPT, 193
ANDESITE, 194
ANDOSOL, 196
ANELASTICITA, 198
ANELLI DI LIESEGANG, 2498

ANELLO CHELIFITICO, 2383
ANELLO DI ACCRESCIMENTO, 4585
ANELLO DI REAZIONE, 3758
ANFIBOLI, 172
ANFIBOLITE, 173
ANFIBOLITIZZAZIONE, 175
ANGIOSPERMAE, 199
ANGOLO CON LA VERTICALE, 1938
ANGOLO CRITICO, 993
ANGOLO DI BRAGG, 570
ANGOLO DI RIPOSO, 200
ANGRITE, 201
ANIDRIDE CARBONICA, 643
ANIDRIDE SOLFOROSA, 4354
ANIDRITE, 204
ANIMALE PLANTIVORO, 3486
ANIONE, 207
ANISICO, 208
ANISOPLEURA, 209
ANISOTROPIA, 210
ANKERITE, 211
ANNEGAMENTO, 4334
ANNELIDA, 212
ANNO GALATTICO, 1674
ANNUARIO, 4847
ANOMALIA, 214
ANOMALIA DI BOUGUER, 557
ANOMALIA GEOFISICA, 1751
ANOMALIA GRAVIMETRICA, 1878
ANOMALIA IN ARIA LIBERA, 1643
ANOMALIA MAGNETICA, 2628
ANOMALODESMATA, 213
ANORTOSITE, 216
ANORTOSITIZZAZIONE, 217
ANSA, 478
ANTECLISI, 218
ANTHOCYATHEA, 219
ANTHOZOA, 220
ANTIARCHI, 223
ANTICLINALE, 224
ANTICLINORIO, 225
ANTIDUNA, 226
ANTIFORME, 227
ANTIMONIATO, 228
ANTIMONIO, 229
ANTIPATHARIA, 230
ANTIPERTITE, 231
ANTISPIAGGIA, 1616
ANTRACITE, 221
ANTRAXOLITE, 222
ANUROMORPHA, 234
APERTURA, 235
APICALE, 237
APLACOPHORA, 238
APLITE, 239
APOFISI, 245
APOGEO, 241

APOGRANITO, 242
APPARATO CONODONTE, 900
APPARECCHI PER PERFORAZIONE, 1255
APPARECCHIATURA SISMICA, 4013
APPROSSIMAZIONE, 249
APSHERONIANO, 251
APSIDOSPONDYLI, 252
APTIANO, 253
AQUALF, 254
AQUENT, 255
AQUEPT, 257
AQUERT, 258
AQUITANIANO, 263
AQUITARDO, 264
AQUOD, 265
AQUOLL, 266
AQUOX, 267
AQUULT, 268
AR/AR, 297
ARACHNIDA, 269
ARCHAEOCYATA, 273
ARCHAEOCYATHEA, 274
ARCHAEOGASTROPODA, 275
ARCHAEOPTERYGES, 276
ARCHAEORNITHES, 277
ARCHEANO, 278
ARCHEIDES, 279
ARCHEOCOPIDA, 272
ARCHEOLOGIA, 281
ARCHEOMAGNETISMO, 282
ARCHIPOLYPODA, 283
ARCHIVIAZIONE DATI, 1071
ARCHOSAURIA, 284
ARCO INSULARE, 2316
ARCOIDA, 285
ARCOSE, 302
ARCUATO, 288
ARDESIA, 4130
AREA MARGINALE, 547
AREISMO, 298
ARENARIA, 3939
ARENARIA GROSSOLANA, 1900
ARENIGIANO, 291
ARENITE, 292
ARENOSOL, 293
ARGENTO, 4111
ARGID, 294
ARGILLA, 787
ARGILLA A BLOCCHI, 559
ARGILLA BIANCA, 4823
ARGILLA DA CERAMICA, 3586
ARGILLA DI FAGLIA, 1509
ARGILLA DURA, 4255
ARGILLA FLUIDA, 3715
ARGILLA MOLLE, 4144
ARGILLA PER LATERIZI, 576
ARGILLA PER PORCELLANA, 743

ARGILLA REFRATTARIA, 1552
ARGILLA RESIDUALE, 3828
ARGILLA SBIANCANTE, 528
ARGILLA SENSIBILE, 4038
ARGILLA VARVATA, 4722
ARGILLITE, 791
ARGILLITE LAMINATA, 4051
ARGINE, 1358
ARGO, 296
ARGON ATMOSFERICO, 359
ARGON IN ECCESSO, 1463
ARIA, 93
ARIDISOL, 300
ARIDITA, 301
ARMONICA, 1957
ARPOLITE, 1958
ARRICCHIMENTO, 1376
ARROTONDAMENTO, 3906
ARROYO, 308
ARSENIATO, 309
ARSENICO, 310
ARTEFATTO, 319
ARTERITE, 311
ARTESIANO, 312
ARTHRODIRA, 315
ARTHROPLEURIDA, 316
ARTHROPODA, 317
ARTICO, 286
ARTICULATA, 318
ARTINSKIANO, 323
ARTIODACTYLA, 324
ASFALTITE, 335
ASFALTO, 334
ASHGILLIANO, 333
ASSE, 393
ASSE DI PIEGA, 1603
ASSESTAMENTO, 4048
ASSIMILAZIONE MAGMATICA, 340
ASSOCIAZIONE, 341
ASSOCIAZIONE CULTURALE, 2472
ASSOCIAZIONE DI FOSSILI, 338
ASSOCIAZIONE MAGMATICA, 2620
ASSOCIAZIONE MINERALOGICA, 2863
ASSORBENTE, 12
ASSORBIMENTO, 13
ASSORBIMENTO ATOMICO, 363
ASTATO, 343
ASTENOSFERA, 348
ASTERISMO, 344
ASTEROIDE, 345
ASTEROIDEA, 346
ASTEROZOA, 347
ASTRAPOTHERIA, 349
ASTROBLEMA, 350
ASTROGEOLOGIA, 351
ASTRONOMIA, 352
ATAXITE, 354

ATLANTE, 356
ATMOSFERA, 358
ATOLLO, 362
ATTACCO CHIMICO, 1439
ATTENUAZIONE, 366
ATTINIO, 51
ATTIVITA, 61
ATTIVITA MAGMATICA, 2172
ATTUALISMO, 62
AUBRITE, 369
AULACOGENO, 372
AUREOLA DI DISPERSIONE, 1223
AUREOLA GEOCHIMICA, 373
AUREOLA PRIMARIA, 3611
AUREOLA SECONDARIA, 3988
AUSTRALITE, 374
AUTIGENESI, 375
AUTOCORRELAZIONE, 378
AUTOCTONIA, 377
AUTOLISI, 380
AUTOLITE, 379
AUTOMAZIONE, 382
AUTOMETAMORFISMO, 383
AUTOMETASOMATISMO, 384
AUTORADIOGRAFIA, 385
AUTUNIANO, 387
AVAMPAESE, 1613
AVANFOSSA, 1612
AVANZAMENTO GLACIALE, 1805
AVES, 390
AZIMUT MAGNETICO, 2629
AZIONE BIOGENICA, 493
AZIONE CLIMATICA, 795
AZIONE DEL GELO, 1653
AZIONE PREVENTIVA, 3609
AZIONE TERMICA, 4497
AZOTO, 3034
AZOTO ORGANICO, 3171
BACINO CARBONIFERO, 803
BACINO DI RETROARCO, 396
BACINO IDROGRAFICO, 1245
BACINO INTERMONTANO, 2273
BACINO INTRACRATONICO, 2290
BACINO OCEANICO, 3085
BACINO RAPPRESENTATIVO, 3819
BACINO SALINO, 3507
BACINO SEDIMENTARIO, 3995
BACKSET BED, 399
BACTERIA, 400
BACTRITIDA, 402
BADLANDS, 403
BAIA, 443
BAJADA, 405
BAJOCIANO, 406
BALLAST, 408
BANCA DI DATI, 1067
BANCHINA, 461

BANCO ALGALE, 112
BANCO DI FANGO, 2953
BARCANA, 411
BARIO, 413
BARITE, 412
BARRA, 410
BARRA LITTORALE, 417
BARREMIANO, 414
BARRIERA, 416
BARRIERA CORALLINA, 420
BARRIERA ROCCIOSA, 3882
BARTONIANO, 421
BASALE, 422
BASALTO, 425
BASALTO ALCALINO, 123
BASALTO DEL DECCAN, 1080
BASALTO DI PLATEAU, 3502
BASAMENTO, 429
BASANITE, 427
BASHKIRIANO, 431
BASIFICAZIONE, 432
BASSA PRESSIONE, 2571
BASSA TEMPERATURA, 2572
BATIMETRIA, 438
BATOLITE, 434
BATONIANO, 435
BATRACHOSAURIA, 439
BAUXITE, 440
BAUXITIZZAZIONE, 441
BE-10/BE-9, 472
BEACHROCK, 448
BEDROCK, 457
BELEMNOIDEA, 459
BENNATA, 1839
BENNETTITALES, 463
BENTHOS, 465
BENTONITE, 466
BERILLIO, 471
BERKELIO, 468
BERMA, 469
BERRIASIANO, 470
BEYRICHICOPINA, 474
BIASSIALE, 475
BIBLIOGRAFIA, 477
BIBLIOGRAFIA PERSONALE, 3365
BILANCIO, 407
BILANCIO DI MASSA, 2698
BILANCIO IDRICO SOTTERRANEO, 1912
BILANCIO IDROGEOLOGICO, 4785
BILANCIO IDROLOGICO, 4367
BIOCALCARENITE, 487
BIOCENOSI, 481
BIOCLASTO, 486
BIOCORA, 483
BIOCRONA, 484
BIOCRONOLOGIA, 485
BIODEGRADAZIONE, 489

BIOFACIES, 490
BIOGENESI, 492
BIOGEOCHIMICA, 496
BIOGEOGRAFIA, 497
BIOGRAFIA, 498
BIOHERMA, 499
BIOLITE, 500
BIOLOGIA, 502
BIOMASSA, 503
BIOMETRIA, 504
BIOMICRITE, 505
BIORESISTASI, 506
BIOSFERA, 509
BIOSOMA, 507
BIOSPARITE, 508
BIOSTRATIGRAFIA, 511
BIOSTROMA, 512
BIOTA, 513
BIOTIPO, 516
BIOTOPO, 514
BIOTURBAZIONE, 515
BIOZONA, 517
BIRIFRANGENZA, 519
BISCOTTO ALGALE, 113
BISMUTO, 520
BISSO, 601
BITUME, 521
BITUMINIZZAZIONE, 522
BIVALVIA, 3332
BLASTOIDEA, 526
BLATTOPTEROIDA, 527
BLOCCO, 530
BLOCCO ISOLATO, 558
BLOCK-DIAGRAMMA, 531
BOA SONORA, 4182
BOMBA, 539
BONE BED, 540
BORACE, 545
BORALF, 543
BORATO, 544
BORO, 552
BOROLL, 551
BOTRIOIDALE, 553
BOUDINAGE, 556
BOUNDSTONE, 562
BRACCIO DI MARE, 2241
BRACHIOPODA, 565
BRACHIOPTERYGII, 566
BRACHIPIEGA, 567
BRANCHIOPODA, 572
BRANCHIURA, 573
BRECCIA, 575
BRECCIA D'ESPLOSIONE, 1486
BRECCIA D'IMPATTO, 4076
BRECCIA D'INIEZIONE, 2240
BRECCIA DI COLLASSO, 835
BRECCIA DI CONTATTO, 914

BRECCIA DI DISLOCAZIONE, 1221
BRECCIA DI FAGLIA, 1508
BRECCIA DI FRIZIONE, 1000
BRECCIA DI PIEGA, 1605
BRECCIA SPIGOLOSA, 4056
BRECCIA TETTONICA, 4434
BRECCIA VULCANICA, 4761
BREVETTO, 3314
BRINA, 578
BRINA CALDA, 2064
BROMO, 581
BROMURO, 580
BROWN COAL, 582
BRYOPHYTA, 584
BRYOZOA, 585
BULIMINACEA, 589
BUNTSANDSTEIN, 593
BURDIGALIANO, 594
BURROW, 600
BUSSOLA MAGNETICA, 2630
C-14, 646
C13/C12, 645
CADMIO, 604
CALABRIANO, 606
CALCARE, 2507
CALCARE ALGALE, 114
CALCARE DOLOMITICO, 1236
CALCARE MICROCRISTALLINO, 2809
CALCARE OOLITICO, 3125
CALCE, 2506
CALCEFIRO, 611
CALCESTRUZZO, 879
CALCIO, 615
CALCISPONGEA, 612
CALCITE, 613
CALCITIZZAZIONE, 614
CALCOLATORE, 870
CALCOLO PETROGRAFICO, 3375
CALCOSFERA, 704
CALDERA, 617
CALDERA DI SPROFONDAMENTO, 836
CALDO, 2063
CALIBRAZIONE, 1698
CALICHE, 619
CALIFORNIO, 620
CALLOVIANO, 622
CALORE, 1962
CALORE LATENTE DI FUSIONE, 2440
CALORE SPECIFICO, 4197
CALOTTA GLACIALE, 2156
CALOTTA POLARE, 3534
CAMAROIDEA, 623
CAMBISOL, 624
CAMBRIANO, 625
CAMBRIANO INF., 2577
CAMBRIANO MED., 2841
CAMBRIANO SUP., 4667

CAMERA MAGMATICA, 2618
CAMINO VULCANICO, 4767
CAMPANIANO, 626
CAMPIONATORE, 3934
CAMPIONATURA, 3935
CAMPIONATURA DEL SUOLO, 4156
CAMPIONE, 3933
CAMPIONE CARATTERISTICO, 4626
CAMPO CRISTALLINO, 1015
CAMPO D'IDROCARBURI, 3105
CAMPO DI LAVA, 2456
CAMPO DI METANO, 1689
CAMPO ELETTRICO, 1328
CAMPO ELETTROMAGNETICO, 1340
CAMPO GEOTERMICO, 1774
CAMPO GRAVIMETRICO, 1880
CAMPO MAGNETICO, 2633
CAMPO PETROLIFERO, 3107
CAMPO PETROLIFERO GIGANTE, 1786
CAMPO SOLCATO, 2371
CAMPTOSTROMATOIDEA, 627
CANALE, 707
CANALE ARTIFICIALE, 628
CANALE DI LAVA, 2455
CANALE DI MAREA, 4525
CANALE SEPOLTO, 598
CANALE SOTTOMARINO, 1089
CANALIZZAZIONE SOTTERRANEA, 4649
CANYON, 630
CANYON SOTTOMARINO, 4326
CAOLINIZZAZIONE, 2369
CAOLINO, 2368
CAOTICO, 709
CAP ROCK, 631
CAPACITA DI SCAMBIO, 1464
CAPACITA TERMICA, 1963
CAPILLARITA, 633
CAPO, 632
CAPPELLO DI FERRO, 1838
CAPPELLO DI GAS, 1685
CAPTAZIONE D'ACQUA, 4786
CARADOCIANO, 638
CARAPACE, 639
CARATO, 640
CARATTERE, 710
CARBOIDRATO, 641
CARBONATAZIONE, 653
CARBONATITE, 652
CARBONATO, 651
CARBONE, 801
CARBONE A FIAMMA LUNGA, 629
CARBONE BITUMINOSO, 523
CARBONE BOGHEAD, 538
CARBONE COKE, 831
CARBONE DA GAS, 1687
CARBONE SAPROPELITICO, 3945
CARBONIFERO, 654

CARBONIFERO INF., 2578
CARBONIFERO SUP., 4668
CARBONIO, 642
CARBONIO ORGANICO, 3166
CARBONIZZAZIONE, 808
CARICAMENTO, 2554
CARICAMENTO CICLICO, 1046
CARICO DI FONDO, 452
CARICO FLUVIALE, 3874
CARICO IDRAULICO, 2094
CARICO IN SOLUZIONE, 1229
CARICO SOLIDO, 4107
CARICO SOSPESO, 4372
CARNICO, 655
CARNIOLA, 656
CARNIVORA, 657
CAROTA, 958
CAROTAGGIO, 956
CARSISMO, 2373
CARTA, 2668
CARTA A ISOLINEE, 932
CARTA A OMBREGGIO, 659
CARTA BATIMETRICA, 437
CARTA D'INSIEME, 2207
CARTA DELLA SISMICITA, 4025
CARTA DELLE ACQUE, 4792
CARTA DELLE BIOFACIES, 491
CARTA DELLE ISOPACHE, 2335
CARTA DELLE ISOPLETE, 2337
CARTA DELLE LITOFACIES, 2527
CARTA FOTOGEOLOGICA, 3409
CARTA GEOCHIMICA, 1719
CARTA GEOFISICA, 1754
CARTA GEOLOGICA, 1735
CARTA GEOMAGNETICA, 2644
CARTA GEOMORFOLOGICA, 1747
CARTA GEOTECNICA, 1768
CARTA GIACIMENTI D'IDROCARBURI, 3106
CARTA GIACIMENTI DI CARBONE, 807
CARTA GRAVIMETRICA, 1882
CARTA IDRAULICA, 2095
CARTA IDROCHIMICA, 2101
CARTA IDROGEOLOGICA, 2111
CARTA IDROLOGICA, 2120
CARTA ISOSISMICA, 2343
CARTA LITOLOGICA, 3376
CARTA MAGNETICA, 2636
CARTA MATERIE PRIME, 1299
CARTA METALLOGENICA, 2772
CARTA MINERALI NON METALLIFERI, 3052
CARTA PALEOGEOGRAFICA, 3256
CARTA PEDOLOGICA, 4153
CARTA PREVISIONALE, 1611
CARTA STRATIGRAFICA, 4278
CARTA STRUTTURALE, 4308
CARTA STRUTTURALE A ISOPLETE, 4313
CARTA TETTONICA, 4437

CARTA TOPOGRAFICA, 4546
CARTERINACEA, 658
CARTOGRAFIA, 660
CARTOGRAFIA AEREA, 70
CARTOGRAFIA AUTOMATIZZATA, 381
CASCATA, 4803
CASSIDULINACEA, 664
CATACLASITE, 667
CATAGENESI, 668
CATAGRAPHIA, 669
CATALISI, 670
CATAMETAMORFISMO, 2377
CATASTROFISMO, 671
CATATERMALE, 2149
CATAZONA, 2380
CATENA A PIEGHE, 1604
CATENA DI MARKOV, 2691
CATENA GEOSINCLINALE, 3184
CATENA INTRACONTINENTALE, 2289
CATENA RADIOATTIVA, 3728
CATIONE, 674
CATTIANO, 716
CATTURA DI CORSO D'ACQUA, 637
CAUSTOBIOLITE, 676
CAVA, 3709
CAVERNA, 677
CAVITA DI DISSOLUZIONE, 4177
CAYTONIALES, 681
CELLA CRISTALLINA, 1018
CELLA ELEMENTARE, 4661
CELLULOSA, 683
CEMENTAZIONE, 685
CEMENTAZIONE PENETRANTE, 1920
CEMENTO INDUSTRIALE, 587
CEMENTO SEDIMENTARIO, 684
CENERE ARTIFICIALE, 320
CENERE VULCANICA, 4759
CENOMANIANO, 686
CENOTIPICO, 687
CENOZOICO, 688
CENOZONA, 689
CENTRALE ELETTRICA, 3588
CENTRALE NUCLEARE, 3065
CENTRO CROMATICO, 846
CENTRO DI ESPANSIONE, 4223
CEPHALOCARIDA, 691
CEPHALODISCIDA, 692
CEPHALOPODA, 693
CERA, 4808
CERAMICA, 694
CERATITIDA, 695
CERATOMORPHA, 696
CERIANTHARIA, 697
CERIANTIPATHARIA, 698
CERIO, 699
CERNIERA (PALE), 2014
CERNOZEM, 738

CESIO, 700
CETACEA, 701
CHAETOGNATHA, 702
CHARNOCKITE, 712
CHAROPHYTA, 713
CHASSIGNITE, 714
CHEILOSTOMATA, 717
CHELAZIONE, 718
CHELICERATA, 719
CHELONIA, 720
CHENIER, 737
CHERATOFIRO, 2385
CHILOPODA, 742
CHIMICA FISICA, 3425
CHIMISMO, 725
CHIROPTERA, 744
CHITINA, 745
CHITINOZOA, 746
CHLOROPHYTA, 754
CHONDRICHTHYES, 755
CHORDATA, 758
CHRYSOPHYTA, 765
CIBERNETICA, 1041
CICLO, 1044
CICLO BIOLOGICO, 501
CICLO D'EROSIONE, 1426
CICLO DELL'ACQUA, 2119
CICLO GEOCHIMICO, 1715
CICLO SEDIMENTARIO, 1045
CICLO TETTONICO, 4436
CICLONE, 1049
CICLOTEMA, 1051
CILIATA, 766
CILIOPHORA, 767
CINEMATICA, 2393
CINTURA CIRCUMPACIFICA, 774
CINTURA DI ROCCE VERDI, 1895
CINTURA MASSIMA, 1890
CINTURA METAMORFICA, 2780
CINTURA VULCANICA, 4760
CIOTTOLO, 3322
CIOTTOLO D'ARGILLA INDURITO, 303
CIPOLLINO, 771
CIRCO, 775
CIRCOLAZIONE, 773
CIRCOLAZIONE OCEANICA, 3086
CIRRIPEDIA, 776
CLADOCERA, 778
CLARAIN, 780
CLARKE, 781
CLASSAZIONE GRANULOMETRICA, 4184
CLASSIFICAZIONE, 782
CLASSIFICAZIONE C.I.P.W., 603
CLASSIFICAZIONE DI HOLMES, 2024
CLASSIFICAZIONE DI NIGGLI, 3027
CLASTICO, 784
CLASTO, 783

CLIMA, 794
CLIMA UMIDO, 2073
CLIVAGGIO DI PIANO ASSIALE, 392
CLIVAGGIO PARALLELO, 455
CLIVAGGIO TETTONICO, 792
CLORINITA, 750
CLORITIZZAZIONE, 752
CLORO, 749
CLOROFILLA, 753
CLORURAZIONE, 748
CLORURO, 747
CLUSTER ANALYSIS, 799
CLYMENIIDA, 800
COAL BALL, 802
COBALTO, 818
COBOL, 819
COCCOLITHOPHORACEAE, 820
COEFFICIENTE, 822
COEFFICIENTE DI CORRELAZIONE, 962
COEFFICIENTE DI PARTIZIONE, 3311
COEFFICIENTE DI POISSON, 3532
COEFFICIENTE DI TRASMISSIBILITA, 4574
COEFFICIENTE PETROCHIMICO, 3370
COELENTERATA, 823
COENOTHECALIA, 824
COERCITIVITA, 825
COESIONE, 827
COLATA, 1575
COLATA DI CENERE, 332
COLATA DI DETRITO, 1077
COLATA DI FANGO, 2956
COLATA LAVICA, 2457
COLATA PLASTICA, 3492
COLATA POLIFASICA, 2964
COLEOPTEROIDA, 833
COLLAUDO STRADALE, 3878
COLLEMBOLA, 840
COLLEZIONE, 839
COLLINA, 2013
COLLINA ABISSALE, 21
COLLISIONE DI PLACCHE, 3496
COLLOIDE, 842
COLLUVIUM, 843
COLMATAZIONE, 2428
COLONNA, 3161
COLONNA VERTEBRALE, 4211
COLORANTE, 3234
COLORAZIONE, 4230
COLORE, 845
COLORIMETRIA, 849
COLPO DI TENSIONE, 3883
COLTIVAZIONE MINERARIA, 2883
COLTIVAZIONE SOTTERRANEA, 4650
COLUMNARIINA, 851
COMAGMATICO, 852
COMBUSTIBILE FOSSILE, 1620
COMMENSALISMO, 853

COMMERCIO, 4566
COMPARAZIONE INTERPLANETARIA, 2280
COMPARAZIONE TERRESTRE, 3481
COMPATTABILITA, 856
COMPATTAZIONE, 857
COMPLESSAZIONE, 861
COMPLESSO, 860
COMPLESSO ERUTTIVO, 2173
COMPLESSO METAMORFICO, 2781
COMPLESSOMETRIA, 862
COMPLETAMENTO DI POZZO, 4814
COMPONENTE INERTE, 2223
COMPOSIZIONE, 863
COMPOSIZIONE ACIDA, 39
COMPOSIZIONE ALCALICALCICA, 124
COMPOSIZIONE ALCALINA, 128
COMPOSIZIONE ANIDRA, 205
COMPOSIZIONE ARGILLOSA, 295
COMPOSIZIONE CAFEMICA, 605
COMPOSIZIONE CALCALCALINA, 607
COMPOSIZIONE CALCAREA, 609
COMPOSIZIONE CALCICA, 610
COMPOSIZIONE CARBONIOSA, 648
COMPOSIZIONE CHIMICA, 721
COMPOSIZIONE DETRITICA, 1142
COMPOSIZIONE FELSICA, 1527
COMPOSIZIONE FEMICA, 1528
COMPOSIZIONE FERROMANGANESIFERA, 1537
COMPOSIZIONE FERRUGGINOSA, 1539
COMPOSIZIONE GLAUCONITICA, 1816
COMPOSIZIONE GRANITICA, 1848
COMPOSIZIONE INTERMEDIA, 2269
COMPOSIZIONE IPERALLUMINIFERA, 3348
COMPOSIZIONE IPOALLUMINIFERA, 4318
COMPOSIZIONE LEUCOCRATICA, 2489
COMPOSIZIONE MAFICA, 2616
COMPOSIZIONE MELANOCRATICA, 2741
COMPOSIZIONE MESOCRATICA, 2752
COMPOSIZIONE MINERALOGICA, 2865
COMPOSIZIONE PERALCALINA, 3347
COMPOSIZIONE POTASSICA, 3578
COMPOSIZIONE SILICEA, 4099
COMPOSIZIONE SUBALCALINA, 4317
COMPOSIZIONE THOLEITICA, 4516
COMPOSIZIONE ULTRABASICA, 4635
COMPOSTO ORGANICO, 3167
COMPRESSIBILITA, 864
COMPRESSIONE, 865
COMPRESSIONE ORIZZONTALE, 2051
COMPRESSIONE TRIASSICA, 4592
COMPRESSIONE UNIASSIALE, 4658
COMUNITA, 855
CONCENTRAZIONE, 873
CONCENTRAZIONE PER GRAVITA, 1114
CONCHIGLIA, 4071
CONCHOSTRACA, 876
CONCORDANZA, 889

CONCORDANZA DI STRATIFICAZIONE, 877
CONCRESCIMENTO, 2267
CONCREZIONE, 881
CONDENSATO DI GAS, 1688
CONDENSAZIONE, 882
CONDIZIONE ADIABATICA, 65
CONDIZIONI P-T, 3229
CONDIZIONI PNEUMATOLITICHE, 3523
CONDIZIONI SUBVULCANICHE, 4343
CONDOTTA, 3454
CONDOTTO ALIMENTATORE, 1522
CONDRITE, 756
CONDRITE CARBONIOSA, 647
CONDRITE ENSTATITICA, 1379
CONDRULA, 757
CONDUCIBILITA, 883
CONDUCIBILITA DI CALORE, 1964
CONDUTTIVITA ELETTRICA, 1326
CONDUTTIVITA IDRAULICA, 2091
CONDUTTIVITA TERMICA, 4495
CONDYLARTHRA, 884
CONE-IN-CONE, 886
CONFLUENZA, 888
CONGELAMENTO, 890
CONGELIFRAZIONE, 891
CONGLOMERATO, 892
CONGLOMERATO BASALE, 424
CONGRESSO GEOL.INTERNAZIONALE, 2277
CONIACIANO, 894
CONIFERALES, 895
CONIUGATO, 896
CONO D'ESPLOSIONE, 4057
CONO DI CENERE, 770
CONO DI DEPRESSIONE, 885
CONO DI DETRITO, 1076
CONO DI EIEZIONE, 4192
CONO VULCANICO, 4762
CONOCARDIOIDA, 899
CONODONTA, 901
CONOIDE, 1504
CONOIDE ALLUVIONALE, 141
CONOIDE SOTTOMARINA, 4327
CONSANGUINEITA, 903
CONSERVAZIONE DI FOSSILI, 3606
CONSISTENZA, 906
CONSOLIDAMENTO, 907
CONSUMO, 912
CONSUMO IDRICO, 4787
CONTAMINAZIONE, 917
CONTATTO, 913
CONTENUTO IN CENERE, 330
CONTINENTE, 918
CONTRASTO, 935
CONTRAZIONE, 934
CONTRAZIONE TERRESTRE, 933
CONTROIMPRONTA LOBATA, 1591
CONTROLLO, 936

CONTROLLO DELL'EROSIONE, 1425
CONTROLLO DI MINERALIZZAZIONE, 3153
CONTROLLO GEOCHIMICO GIACIMENTO, 722
CONTROLLO IDROGEOLOGICO GIACIMENTO, 2110
CONTROLLO IGNEO GIACIMENTO, 2621
CONTROLLO LITOLOGICO, 2530
CONTROLLO PALEOGEOGRAFICO, 3255
CONTROLLO SEDIMENTARIO GIACIMENTO, 3999
CONTROLLO STRATIGRAFICO, 4276
CONTROLLO STRUTTURALE, 4302
CONTROLLO TETTONICO, 4435
CONULARIDA, 937
CONVERGENZA, 940
CONVEZIONE, 938
COOPERAZIONE INTERNAZIONALE, 2276
COOPERAZIONE TECNICA, 4430
COORDINATE GEODETICHE, 1728
COORDINAZIONE, 943
COPEPODA, 944
COPERNICANO, 945
COPERTURA MULTIPLA, 854
COPPIA TERRA-LUNA, 1282
COPROLITE, 947
CORALLIMORPHARIA, 950
CORALLINACEAE, 951
CORALLITE, 952
CORDAITALES, 953
CORDIGLIERA, 955
CORDIGLIERA INTRAGEOSINCLINALE, 1763
CORDONE LITTORALE, 2567
CORINDONE SOSTANZA, 965
CORNUBIANITE, 2057
CORPO DELLE MINIERE, 595
CORPO MINERALIZZATO, 3152
CORPO SABBIOSO, 3937
CORRELAZIONE, 961
CORRENTE, 1035
CORRENTE DI CONVEZIONE, 939
CORRENTE DI TORBIDA, 4617
CORRENTE LITTORALE, 2568
CORRENTE TELLURICA, 1274
CORREZIONE, 960
CORREZIONE DI TERRENO, 4466
CORREZIONE TOPOGRAFICA, 4545
CORROSIONE, 964
CORSO D'ACQUA, 4284
CORSO D'ACQUA CONSEGUENTE, 904
CORSO D'ACQUA INCASSATO, 1384
CORSO D'ACQUA INTERMITTENTE, 2272
CORYNEXOCHIDA, 966
COSMOCHIMICA, 972
COSMOLOGIA, 975
COSTA, 811
COSTA D'ABRASIONE, 7
COSTA D'ACCUMULAZIONE, 32
COSTANTE, 909
COSTANTE DI DECADIMENTO, 1079

COSTANTE DI ELASTICITA, 1317
COSTANTE DI LAME, 2418
COSTANTE DIELETTRICA, 1174
COSTANTE GRAVITAZIONALE, 1876
COSTANTE RETICOLARE, 2451
COSTIPAMENTO DIFFERENZIALE, 1175
COSTIPAZIONE, 3232
COSTITUZIONE DELLA LUNA, 2600
COSTITUZIONE DI PIANETA, 3483
COSTO, 976
COSTRUTTORE DI SCOGLIERA, 3779
COSTRUZIONE, 910
COSTRUZIONE ALGALE, 116
COSTRUZIONE ASISMICA, 327
COTYLOSAURIA, 977
CRANIO, 980
CRATERE, 981
CRATERE D'IMPATTO, 2187
CRATERE LUNARE, 2601
CRATERE METEORITICO, 2797
CRATONE, 984
CREEP, 986
CREODONTA, 987
CREPACCIO, 990
CREPACCIO TERMINALE, 467
CRESTA, 988
CRETA, 1662
CRETACEO, 989
CRETACEO INF., 2579
CRETACEO SUP., 4669
CRINOIDEA, 991
CRINOZOA, 992
CRIOPEDOLOGIA, 1003
CRIOTETTONICA, 1004
CRIOTURBAZIONE, 1005
CRIPTOCRISTALLINO, 1006
CRIPTOGENICO, 1008
CRISTALLINITA, 1023
CRISTALLITE, 1024
CRISTALLIZZAZIONE, 1025
CRISTALLIZZAZIONE FRAZIONATA, 1631
CRISTALLO, 1011
CRISTALLO ANEDRALE, 203
CRISTALLO BIASSIALE, 476
CRISTALLO EUEDRALE, 1447
CRISTALLO XENOMORFO, 4839
CRISTALLOBLASTO, 1027
CRISTALLOCHIMICA, 1012
CRISTALLOGRAFIA, 1028
CRISTALLOGRAFIA STRUTTURALE, 4303
CROCODILIA, 996
CROLLO DI TETTO, 3894
CROMATO, 759
CROMATOGRAFIA, 760
CROMO, 761
CRONOSTRATIGRAFIA, 763
CRONOZONA, 764

CROSSOPTERYGII, 999
CROSTA CALCAREA, 616
CROSTA CONTINENTALE, 921
CROSTA DI ALTERAZIONE, 4810
CROSTA FERRUGGINOSA, 1534
CROSTA OCEANICA, 3091
CROSTA PROFONDA, 2580
CROSTA SUPERIORE, 4670
CROSTA TERRESTRE, 1001
CRUSTACEA, 1002
CRYPTODONTA, 1007
CRYPTOSTOMATA, 1009
CTENOSTOMATA, 1029
CUESTA, 1031
CUNEO DI GHIACCIO, 2162
CUNEO DI GHIACCIO FOSSILE, 1621
CUNEO STRATIGRAFICO, 4811
CUPOLA DI GNEISS, 1829
CURIO, 1034
CUSPIDE, 1037
CUTTINGS, 1039
CYANOPHYTA, 1040
CYCADALES, 1042
CYCADOFILICALES, 1043
CYCLOCYSTOIDEA, 1048
CYCLOSTOMATA, 1050
CYNOMORPHA, 1054
CYSTIPHYLLINA, 1055
CYSTOIDEA, 1056
D/H, 1058
DANIANO, 1060
DANNO DA RADIAZIONE, 3721
DARWINISMO, 1062
DASYCLADACEAE, 1063
DATAZIONE, 1073
DATAZIONE TRIZIO, 4601
DATO, 1064
DATO MES, 4034
DATO MET, 4451
DATO MICROSCOPIO ELETTRONICO, 1350
DATO MICROSONDA ELETTRONICA, 1352
DATO NUOVO, 3020
DECADIMENTO RADIOATTIVO, 3725
DECAPITAZIONE FLUVIALE, 458
DECAPODA, 1078
DECLINAZIONE, 1081
DECONTAMINAZIONE, 1083
DECONVOLUZIONE, 1084
DECREPITOMETRIA, 1086
DEDOLOMITIZZAZIONE, 1087
DEFLAZIONE, 1095
DEFLUSSO DI FALDA IDRICA, 1918
DEFORMAZIONE, 1096
DEFORMAZIONE DEL SUOLO, 4147
DEFORMAZIONE DI SEDIMENTI MOLLI, 4145
DEFORMAZIONE DI VERSANTE, 4135
DEFORMAZIONE ELASTICA, 1321

DEFORMAZIONE POLIFASICA, 3556
DEFORMAZIONE SOTTO SFORZO, 4267
DEGASSAZIONE, 1097
DEGLACIAZIONE, 1098
DEINOTHERIOIDEA, 1102
DELTA, 1103
DEMOSPONGEA, 1107
DENDRITE, 1108
DENDROCRONOLOGIA, 1110
DENDROGRAMMA, 1111
DENDROIDEA, 1112
DENSITA, 1113
DENSITA APPARENTE, 590
DENTE, 4544
DENTIZIONE, 1115
DENUDAMENTO, 1116
DEPOSITO, 1120
DEPOSITO CONCREZIONATO, 4118
DEPOSITO TERRIGENO, 4469
DEPRESSIONE, 1121
DEPRESSIONE VULCANOTETTONICA, 4770
DEPURAZIONE, 3687
DERIVA CONTINENTALE, 922
DERIVA DEL POLO, 3535
DERIVA LITORALE, 2544
DERIVATA, 1125
DERMAPTEROIDA, 1127
DERMOPTERA, 1128
DESERTIFICAZIONE, 1134
DESERTO, 1131
DESILICIZZAZIONE, 1136
DESMOCERATIDA, 1138
DESMOSTYLIA, 1139
DESQUAMAZIONE, 1466
DETERMINISMO, 1140
DETRITO DI FALDA, 4416
DETRITO GLACIALE, 1792
DEUTERIO, 1146
DEVETRIFICAZIONE, 1150
DEVIAZIONE, 1149
DEVIAZIONE STANDARD, 4234
DEVONIANO, 1151
DEVONIANO INF., 2581
DEVONIANO MED., 2842
DEVONIANO SUP., 4671
DI ALTO GRADO, 2010
DIABASE, 1234
DIACLASI, 2361
DIACLASI DI TAGLIO, 4059
DIACRONISMO, 1153
DIADOCHIA, 1154
DIAFTORESI, 1162
DIAFTORITE, 1163
DIAGENESI, 1155
DIAGENESI PRECOCE, 1272
DIAGENESI TARDIVA, 2438
DIAGNOSI, 1156

DIAGRAFIA, 4816
DIAGRAFIA ACUSTICA, 41
DIAGRAFIA ELETTRICA, 1324
DIAGRAFIA ELETTROMAGNETICA, 1343
DIAGRAFIA GAMMA, 1679
DIAGRAFIA GAMMA=GAMMA, 1678
DIAGRAFIA NEUTRONICA, 3017
DIAGRAMMA, 1157
DIAGRAMMA D'EQUILIBRIO, 3392
DIAGRAMMA PETROCHIMICO, 3371
DIAGRAMMA POLLINICO, 3540
DIAGRAMMA STRUTTURALE, 4304
DIALISI, 1158
DIAMANTE, 1161
DIAMETRAGGIO, 621
DIAMICTITE, 1160
DIAPIRISMO, 1165
DIAPIRO, 1164
DIASPRO, 2359
DIASTEMA, 1166
DIASTROFISMO, 1167
DIATOMEAE, 1169
DIATOMITE, 1170
DIATREMA, 1171
DICCO, 1184
DICCO DI ARENARIA, 3940
DICOTILEDONI, 1172
DICTYONELLIDINA, 1173
DIFETTO CRISTALLINO, 1013
DIFETTO DI SUPERFICIE, 4363
DIFFERENZIAZIONE GRAVITATIVA, 1877
DIFFERENZIAZIONE MAGMATICA, 1178
DIFFERENZIAZIONE METAMORFICA, 2782
DIFFRAZIONE, 1180
DIFFRAZIONE CRISTALLO SINGOLO, 4116
DIFFRAZIONE DI POLVERI, 3587
DIFFRAZIONE RAGGI X, 4834
DIFFUSIONE, 1181
DIFFUSIVITA, 1182
DIGA, 1059
DIGA AD ARCO, 271
DIGA IN CALCESTRUZZO, 880
DIGA IN TERRA, 1275
DIGA ROCK FILL, 3885
DILATANZA, 1186
DILATAZIONE, 1187
DILAVAMENTO, 4065
DILUIZIONE, 1188
DILUIZIONE ISOTOPICA, 2349
DIMORFISMO, 1189
DIMORFISMO SESSUALE, 4049
DINAMICA, 1267
DINOCERATA, 1190
DINOFLAGELLATA, 1191
DINOSAURIA, 1192
DIOGENITE, 1193
DIORITE, 1194

DIORITE QUARZIFERA, 3711
DIP METER, 1197
DIPLEUROZOA, 1200
DIPLOIDE, 1201
DIPLOPODA, 1202
DIPLORINA, 1203
DIPNOI, 1205
DIPTERA, 1207
DIREZIONE, 4292
DIREZIONE ORIZZONTALE, 2052
DISAGGREGAZIONE, 1099
DISCARICA, 4782
DISCOASTER, 1210
DISCONTINUITA, 1212
DISCONTINUITA DI BIRCH, 518
DISCONTINUITA DI BULLARD, 592
DISCONTINUITA DI CONRAD, 902
DISCONTINUITA DI GUTENBERG, 1929
DISCONTINUITA DI MOHOROVICIC, 2910
DISCORBACEA, 1213
DISCORDANTE, 1215
DISCORDANZA, 4646
DISCORDANZA ANGOLARE, 202
DISCORDANZA EROSIVA, 1211
DISCORDANZA REGIONALE, 3796
DISCORDANZA SEMPLICE, 3050
DISCORDANZA TETTONICA, 1214
DISEGNO, 1250
DISEQUILIBRIO URANIO, 4691
DISGELO, 4485
DISGREGAZIONE IN MASSA, 2702
DISIDRATAZIONE, 1101
DISLOCAZIONE CRISTALLINA, 1014
DISPERSIONE D'ONDA, 4806
DISPERSIONE OTTICA, 1222
DISPERSIONE ROTATORIA, 3903
DISPROSIO, 1270
DISSALAZIONE, 1129
DISSIMMETRIA, 4128
DISSOCIAZIONE IONICA, 1227
DISSOLUZIONE, 1228
DISSOLUZIONE SOTTO PRESSIONE, 3608
DISTANZA, 3814
DISTRETTO METALLOGENICO, 2770
DISTRIBUZIONE, 1230
DISTRIBUZIONE BINOMIALE, 480
DISTRIBUZIONE DI FREQUENZA, 1645
DISTRIBUZIONE DI POISSON, 3531
DISTRIBUZIONE LOGNORMALE, 2563
DISTRIBUZIONE NORMALE, 3055
DISTRIBUZIONE STATISTICA, 4240
DIVERGENZA, 1232
DIVERSITA DI SPECIE, 4193
DIVULGAZIONE GEOLOGICA, 3560
DOCODONTA, 1233
DOCUMENTAZIONE, 2229
DOLINA, 1235

DOLINA DI SPROFONDAMENTO, 4117
DOLOMIA, 1238
DOLOMITIZZAZIONE, 1237
DOMERIANO, 1241
DOMINIO MAGNETICO, 2631
DORSALE, 4330
DORSALE OCEANICA, 2840
DRAG ORE, 1244
DRAGAGGIO, 1251
DRENAGGIO, 1247
DRENAGGIO ACIDO, 38
DRENAGGIO ANASTOMIZZATO, 571
DROMOCRONA, 4582
DRUMLIN, 1259
DRUSA, 1260
DUNA, 1261
DUNA CONTINENTALE, 1132
DUNA COSTIERA, 812
DUOMO, 1240
DUOMO SALINO, 3925
DUOMO VULCANICO, 4763
DURATA DEL GIORNO, 2478
DUREZZA, 1956
DUREZZA DELL'ACQUA, 4790
DURITE, 1262
ECCENTRICITA, 1290
ECHINODERMATA, 1291
ECHINOIDEA, 1292
ECHINOZOA, 1293
ECLOGITE, 1295
ECOLOGIA, 1297
ECOLOGIA MARINA, 2676
ECOLOGIA UMANA, 2069
ECONOMIA MINERARIA, 2867
ECOSISTEMA, 1301
ECOSONDAGGIO, 1294
ECOZONA, 1302
ECTOTROPHA, 1303
EDENTATA, 1305
EDIFICIO, 586
EDRIOASTEROIDEA, 1307
EDRIOBLASTOIDEA, 1308
EFFETTO DEUTERICO, 1145
EFFETTO DI CORIOLIS, 959
EFFETTO DI SERRA, 1891
EFFETTO DI SHOCK, 2188
EFFETTO ISOTOPICO, 2350
EH, 1312
EIFELIANO, 1313
EINSTEINIO, 1314
ELASMOBRANCHII, 1316
ELASTICITA, 1323
ELEMENTI MAGGIORI, 2657
ELEMENTI MINORI, 2887
ELEMENTO CALCOFILO, 703
ELEMENTO COSMOGENICO, 973
ELEMENTO FINITO, 1551

ELEMENTO IN TRACCE, 4556
ELEMENTO LITOFILO, 2533
ELEMENTO MAGNETICO, 2632
ELEMENTO METALLICO, 2766
ELEMENTO NATIVO, 2988
ELEMENTO NON METALLICO, 3048
ELEMENTO SIDEROFILO, 4092
ELEMENTO STRUTTURALE, 4305
ELEMENTO VOLATILE, 4756
ELEPHANTOIDEA, 1354
ELETTRICITA, 1334
ELETTRODO, 1337
ELETTROLISI, 1338
ELETTROLITA, 1339
ELETTROMAGNETISMO, 1344
ELETTRONE, 1345
ELETTROOSMOSI, 1335
ELIO, 1978
ELUTRIAZIONE, 1356
ELUVIUM, 1357
ELVEZIANO, 1980
EMATITE, 1981
EMBRECHITE, 1359
EMBRITHOPODA, 1360
EMERGENZA, 1491
EMERSIONE, 1361
EMISSARIO, 1310
EMSIANO, 1364
ENCRINITE, 1366
ENDOCERATOIDEA, 1369
ENDOREISMO, 1371
ENERGIA, 1373
ENERGIA DI ATTIVAZIONE, 55
ENERGIA GEOTERMICA, 1773
ENERGIA LIBERA, 1641
ENERGIA NUCLEARE, 3062
ENTALPIA, 1382
ENTE DI BACINO IDROGRAFICO, 4784
ENTEROPNEUSTA, 1381
ENTISOL, 1383
ENTROPIA, 1385
EOCENE, 1389
EOCENE INF., 2582
EOCENE MED., 2843
EOCENE SUP., 4672
EOCRINOIDEA, 1390
EONE, 1394
EOSUCHIA, 1395
EPIBIOSI, 1398
EPIBOLITE, 1399
EPICENTRO, 1400
EPIDOTIZZAZIONE, 1402
EPIEUGEOSINCLINALE, 1403
EPIGENESI, 1404
EPIROGENESI, 1396
EPITASSIA, 1406
EPIZONA, 1408

EPOCA, 1409
EPOCA GAUSS, 1699
EPOCA MATUYAMA, 2720
EPOCA METALLOGENICA, 2771
EQUATORE, 1413
EQUAZIONE, 1411
EQUAZIONE DI STATO, 1412
EQUILIBRIO, 1416
EQUILIBRIO CHIMICO, 724
EQUILIBRIO DINAMICO, 1264
EQUISETALES, 1417
EQUIVALENTE CALORICO DI FUSIONE, 1966
ERA, 1418
ERATEMA, 1419
ERATOSTENIANO, 1420
ERBIO, 1421
EREDITARIETA, 1986
ERG, 1422
ERODIBILITA, 1423
EROSIONE, 1424
EROSIONE COSTIERA, 813
EROSIONE DEL SUOLO, 4148
EROSIONE EOLICA, 4827
EROSIONE FLUVIALE, 1594
EROSIONE GLACIALE, 1794
EROSIONE IDRICA, 4789
EROSIONE MARINA, 2678
ERRATICO INDICATORE, 2211
ERRORE, 1431
ERUZIONE, 1432
ERUZIONE HAWAIANA, 1961
ERUZIONE STROMBOLIANA, 4297
ESAEDRITE, 2003
ESCURSIONE, 1465
ESKER, 1434
ESOBIOLOGIA, 1470
ESOREISMO, 1472
ESPANDIMENTO FLUVIOGLACIALE, 3211
ESPANSIONE DI FONDO OCEANICO, 3980
ESPANSIONE TERRESTRE, 1475
ESPERIMENTO, 1479
ESPLORAZIONE, 3641
ESPLORAZIONE DI FALDE IDRICHE, 1915
ESPLOSIONE, 1485
ESPLOSIONE CHIMICA, 726
ESPLOSIONE NUCLEARE, 3063
ESPORTAZIONE, 1488
ESPOSIZIONE, 3809
ESSEXITE, 1435
ESSICCAZIONE, 1135
ESSOLUZIONE, 1490
EST, 1288
ESTENSIONE, 3746
ESTENSOMETRO, 1492
ESTERNIDI, 1493
ESTINZIONE, 1496
ESTINZIONE ONDULATA, 4657

ESTINZIONE OTTICA, 3138
ESTUARIO, 1438
ETANO, 1440
ETA, 80
ETA ASSOLUTA, 10
ETA CONCORDANTE, 878
ETA D'ESPOSIZIONE, 1489
ETA DELLA TERRA, 81
ETA DISCORDANTE, 1216
ETA RELATIVA, 3804
ETA TRACCE DI FISSIONE, 1556
ETERE, 1441
ETEROCRONISMO, 1989
ETEROCRONO, 1990
ETEROGENEITA, 1993
ETEROGENEO, 1994
ETEROMORFISMO, 1996
EUCILIATA, 1442
EUCRITE (METEORITE), 1443
EUGEOSINCLINALE, 1445
EUGLENOPHYTHA, 1446
EUROPIO, 1448
EURYAPSIDA, 4389
EUSELACHII, 1451
EUSTATISMO, 1452
EUTETTICO, 1454
EUTROFIZZAZIONE, 1455
EVAPORAZIONE, 1458
EVAPORITE, 1459
EVAPOTRASPIRAZIONE, 1460
EVENTO DI JARAMILLO, 2358
EVIDENZA DI TERRENO, 1910
EVOLUZIONE BIOLOGICA, 1461
EVOLUZIONE IPOGENICA, 1370
EVOLUZIONE SUPERGENICA, 1471
EXINE, 1468
EXINITE, 1469
EXTRAPOLAZIONE, 1497
FACIES, 1501
FACIES ANFIBOLITE A EPIDOTO, 1401
FACIES ANFIBOLITI A CORDIERITE, 954
FACIES ANFIBOLITI ALBITE EPIDOTO, 104
FACIES ATTINOTO, 53
FACIES CORNUBIAN. ALBITE EPIDOTO, 105
FACIES CORNUBIANITI A ORNEBLENDA, 2055
FACIES CORNUBIANITI PIROSSENICHE, 3703
FACIES DEGLI SCISTI VERDI, 1893
FACIES DELLE ANFIBOLITI, 174
FACIES DELLE CORNUBIANITI, 2058
FACIES DELLE ECLOGITI, 1296
FACIES DELLE GRANULITI, 1859
FACIES DELLE SANIDINITI, 3942
FACIES DELLE ZEOLITI, 4857
FACIES DI CONTATTO, 546
FACIES GEOCHIMICA, 1717
FACIES IDROCHIMICA, 2100
FACIES MAGMATICA, 2174

FACIES METAMORFICA, 2783
FACIES MINERALE, 2869
FACIES PREHNITE=PUMPELLYITE, 3603
FACIES SCISTI BLU, 1818
FACOLITE, 3385
FAGLIA, 1507
FAGLIA A CERNIERA, 2015
FAGLIA A GRADINI, 4248
FAGLIA A PIANO ORIZZONTALE, 2054
FAGLIA A SCORRIMENTO ORIZZONTALE, 4293
FAGLIA A SEPARAZ.SEC.DIREZIONE, 2442
FAGLIA ANTITETICA, 233
FAGLIA ARCUATA, 289
FAGLIA ATTIVA, 56
FAGLIA D'IMMERSIONE, 1199
FAGLIA DI SOTTOSCORRIMENTO, 4656
FAGLIA DI STRATO, 456
FAGLIA NORMALE, 3056
FAGLIA OBLIQUA, 3077
FAGLIA SINGENETICA, 1921
FAGLIA TRASCORRENTE, 4567
FAGLIA TRASCORRENTE VERTICALE, 4831
FAGLIA TRASFORME, 4569
FAGLIA TRASVERSALE, 1196
FAGLIAMENTO, 1515
FAGLIE A GRADINATA, 1365
FALDA, 2986
FALDA ARTESIANA, 313
FALDA CIRCOSCRITTA, 887
FALDA FREATICA, 3416
FALDA IN ALLUVIONI, 143
FALDA LIBERA, 4645
FALDA SOSPESA, 3349
FALESIA, 793
FALSO CLIVAGGIO, 4133
FAMENNIANO, 1503
FANEROZOICO, 3390
FANGLOMERATO, 1505
FANGO, 2952
FANGO DI PERFORAZIONE, 1256
FANGO ORGANICO, 3127
FASCIA DETRITICA, 250
FASE, 3391
FASE ACADIANA, 24
FASE FLUIDA, 1582
FASE GASSOSA, 1694
FASE IDROTERMALE, 2133
FASE LIQUIDA, 2520
FASE MINERALOGICA, 2871
FASE SOLIDA, 4166
FATTORE TEMPO, 4535
FAUNA, 1516
FAUNA SPECIFICA, 4194
FAUNIZONA, 1519
FELDSPATIZZAZIONE, 1525
FELDSPATO, 1524
FELTRO ALGALE, 115

FENITE, 1529
FENITIZZAZIONE, 1530
FENOCRISTALLO, 3395
FENOLO, 3396
FENOMENO DI TAGLIO, 4058
FERMIO, 1531
FERRALSOL, 1532
FERRO, 2308
FERRO DELLE PALUDI, 537
FERRO FERRICO, 1533
FERRO FERROSO, 1538
FERROBATTERI, 2310
FERROD, 1535
FERTILIZZANTE, 1541
FESSURA DI CONTRAZIONE, 4088
FESSURA DI DISSECCAMENTO, 1965
FESSURAZIONE DA DISSECCAMENTO, 2954
FESSURAZIONE PRISMATICA, 850
FIANCO, 2505
FIGURA BASALE, 4163
FILICALES, 1543
FILLITE, 3419
FILLONITE, 3421
FILOGENESI, 3423
FILONE, 4725
FILONE A CODA DI CAVALLO, 2059
FILONE AD ANELLO, 3869
FILONE CLASTICO, 785
FILOSOFIA, 3398
FILTRAGGIO DI SEGNALE, 1546
FILTRAZIONE, 1547
FILTRO DI POZZO, 1545
FILTRO DI POZZO, 4817
FINANZIAMENTO, 1548
FINESTRA TETTONICA, 4830
FIORDO, 1560
FISIOLOGIA, 3432
FISSAZIONE IONICA, 1558
FISSILITA, 4131
FISSIONE, 1554
FISSIPEDA, 1557
FISSISMO, 1559
FITANO, 3433
FITOCENOSI, 3434
FITOPLANCTON, 3437
FIUME, 3873
FIUME SOTTERRANEO, 4653
FLANDRIANO, 1564
FLESSURA, 1566
FLOCCULAZIONE, 1568
FLORA, 1571
FLORA A DESMIDI, 1137
FLORA SPECIFICA, 4195
FLORIZONA, 1573
FLOTTAZIONE, 1574
FLUIDO MINERALIZZANTE, 2877
FLUORESCENZA, 1585

FLUORESCENZA A RAGGI X, 4835
FLUORITE, 1590
FLUORO, 1589
FLUORURAZIONE, 1588
FLUORURO, 1586
FLUSSO DI CALORE, 1967
FLUTTO, 4370
FLUTTUAZIONE, 1578
FLUVENT, 1592
FLUVIOGLACIALE, 1596
FLUVIOSOL, 1598
FLYSCH, 1599
FOGLIA, 2469
FOIDITE, 1601
FOLIAZIONE, 1609
FONDAZIONE, 1627
FONDO MARINO, 3088
FONDOVALLE, 4482
FONOLITE, 3401
FONTE MINERARIA, 3162
FONTE NON D'ESPLOSIONE, 3046
FORAMINIFERA, 1610
FORMA, 4054
FORMA CRISTALLINA, 1016
FORMA DEL GRANULO, 3308
FORMA DEL RILIEVO, 2429
FORMA DELLA TERRA, 4055
FORMAZIONE, 1617
FORMAZIONE DI CRATERE, 983
FORMAZIONE FERRIFERA, 2311
FORMAZIONE SUPERFICIALE, 4369
FORMULA, 1618
FORRA, 1928
FORZA DI COMPRESSIONE, 869
FORZA DI ROTTURA, 3918
FOSFATIZZAZIONE, 3404
FOSFATO, 3403
FOSFORESCENZA, 3405
FOSFORITE, 3406
FOSFORO, 3407
FOSSA ABISSALE, 4587
FOSSA ABISSALE ALLUNGATA, 4603
FOSSA INTRAGEOSINCLINALE, 1766
FOSSILE, 1619
FOSSILE CHIMICO, 727
FOSSILE GUIDA, 2206
FOSSILE SPECIFICO, 1925
FOSSILE VIVENTE, 2549
FOSSILIZZAZIONE, 1626
FOTOGEOLOGIA, 3410
FOTOGRAFIA, 3412
FOTOGRAFIA ALL'INFRAROSSO, 2232
FOTOGRAMMETRIA, 3411
FOTOMETRIA, 3413
FOTOMETRIA DI FIAMMA, 1562
FOTOSINTESI, 3414
FOYAITE, 1629

FRAGILE, 579
FRAMMENTAZIONE, 1637
FRAMMENTO, 1636
FRANA DI CROLLO, 3884
FRANA DI SCIVOLAMENTO IN ROCCIA, 3891
FRANCIO, 1639
FRANTUMAZIONE, 1899
FRASNIANO, 1640
FRATTURA, 1632
FRATTURA AD ANELLO, 3870
FRATTURA CONCOIDE, 875
FRATTURAZIONE, 1635
FRATTURAZIONE A GIUNTI, 2362
FRATTURAZIONE IDRAULICA, 2092
FRAZIONAMENTO, 1630
FRAZIONAMENTO ISOTOPICO, 2353
FRAZIONE FINE, 1549
FRAZIONE GROSSOLANA, 810
FREQUENZA, 1644
FRIZIONE, 1650
FRONTE, 1652
FRUTTIFICAZIONE, 1657
FUGACITA, 1659
FULGURITE, 1660
FUMAROLA, 1664
FUNGI, 1667
FUNZIONE, 1665
FUNZIONE STATISTICA, 4241
FUSINITE, 1669
FUSIONE, 2744
FUSIONE CONGRUENTE, 893
FUSIONE INCONGRUENTE, 2204
FUSIONE PARZIALE, 3305
FUSULINIDAE, 1670
GABBRO, 1672
GADOLINIO, 1673
GALESTRO, 1681
GALLERIA, 4614
GALLERIA DI CAPTAZIONE, 2227
GALLERIA LAVICA, 2459
GALLIO, 1675
GANGA, 1680
GAS, 1684
GAS AZOTO, 3035
GAS DI CARBONE, 804
GAS DI PALUDE, 2694
GAS FREATICO, 3415
GAS IDROGENO, 2107
GAS INERTE, 2224
GAS NATURALE, 2989
GAS OSSIGENO, 3226
GASCROMATOGRAFIA, 1686
GASSIFICAZIONE, 1695
GASTROPODA, 1696
GASTROZOOIDE, 1697
GEDINNIANO, 1701
GEL, 1702

GEL MINERALE, 1703
GELIVITA, 1704
GEMINATO PER SCORRIMENTO, 1820
GEMINAZIONE, 4621
GEMMA, 1707
GEMMAZIONE, 1705
GEMMOLOGIA, 1706
GENERAZIONE, 1708
GENESI, 1709
GENESI DI GIACIMENTO, 1118
GENESI DI ROCCIA INTRUSIVA, 2175
GENESI DI SEDIMENTO, 4005
GENESI INTRATELLURICA, 2292
GENOMORFO, 1710
GENOTIPO, 1711
GEOANTICLINALE, 1700
GEOBAROMETRIA, 1712
GEOBIOS, 1713
GEOCHIMICA, 1722
GEOCHIMICA ISOTOPICA, 2351
GEOCHIMICA ORGANICA, 3168
GEOCRONOLOGIA, 1725
GEODE, 1726
GEODESIA, 1727
GEODINAMICA, 1729
GEOFISICA, 1759
GEOFISICA APPLICATA, 247
GEOFISICA DELLA TERRA SOLIDA, 4169
GEOFISICO, 1758
GEOFONO, 1750
GEOGRAFIA, 1731
GEOGRAFIA MATEMATICA, 2710
GEOIDE, 1732
GEOISOTERMA, 1733
GEOLOGIA, 1741
GEOLOGIA AMBIENTALE, 1388
GEOLOGIA APPLICATA, 246
GEOLOGIA DEL PETROLIO, 3381
GEOLOGIA DEL SUBSTRATO, 4340
GEOLOGIA DEL TERRITORIO, 4694
GEOLOGIA DINAMICA, 1265
GEOLOGIA ECONOMICA, 1298
GEOLOGIA EXTRATERRESTRE, 1498
GEOLOGIA GLACIALE, 1796
GEOLOGIA LUNARE, 2602
GEOLOGIA MARINA, 2679
GEOLOGIA MATEMATICA, 2711
GEOLOGIA MILITARE, 2859
GEOLOGIA MINERARIA, 2885
GEOLOGIA REGIONALE, 3793
GEOLOGIA SANITARIA, 2733
GEOLOGIA STORICA, 2019
GEOLOGIA STRUTTURALE, 4306
GEOLOGIA TECNICA, 1374
GEOMAGNETISMO, 1745
GEOMETRIA, 1746
GEOMETRIA DEL CANALE, 708

GEOMORFOLOGIA, 1748
GEONOMIA, 1749
GEOPRESSIONE, 1760
GEOSECS, 1761
GEOSINCLINALE, 1767
GEOSINCLINALE INTERCRATONICO, 1378
GEOSINCLINALE INTRACRATONICO, 1377
GEOSTATISTICA, 1762
GEOTECNICA, 1771
GEOTERMIA, 1779
GEOTERMOMETRIA, 1780
GEOTETTONICA, 1772
GEOUNDAZIONE, 1781
GERARCHIA FLUVIALE, 4286
GERMANIO, 1782
GERME CRISTALLINO, 3067
GEROGLIFICO, 2005
GESSIFICAZIONE, 1932
GESSO, 705
GESSO SOSTANZA, 1933
GESTIONE DELLA PRODUZIONE, 3625
GESTIONE DI BACINO, 433
GESTIONE DI RISORSE IDRICHE, 4791
GEYSER, 1784
GEYSERITE, 1785
GHIACCIAIO, 1804
GHIACCIO, 2155
GHIACCIO DEL SUOLO, 1905
GHIACCIO MARINO, 3976
GHIACCIO MORTO, 1075
GHIAIA, 1871
GHIAIA GROSSOLANA, 809
GIACIMENTO, 2866
GIACIMENTO ACCERTATO, 3571
GIACIMENTO APOMAGMATICO, 244
GIACIMENTO CATATERMALE, 2379
GIACIMENTO CHIUSO, 529
GIACIMENTO D'IMPREGNAZIONE, 2195
GIACIMENTO DETRITICO, 2725
GIACIMENTO DI ESALAZIONE, 1467
GIACIMENTO DISSEMINATO, 1226
GIACIMENTO EPITERMALE, 1407
GIACIMENTO FILONIANO, 2559
GIACIMENTO IDROTERMALE, 2132
GIACIMENTO INTERSTRATO, 4270
GIACIMENTO INTRAMAGMATICO, 2622
GIACIMENTO LITTORALE, 445
GIACIMENTO MESOTERMALE, 2759
GIACIMENTO METALLIFERO, 3154
GIACIMENTO METAMORFOGENO, 2788
GIACIMENTO MINERARIO, 3151
GIACIMENTO PNEUMATOLITICO, 3524
GIACIMENTO POLIMETALLICO, 3551
GIACIMENTO POTENZIALE, 3582
GIACIMENTO RESIDUALE, 3829
GIACIMENTO ROLL-TYPE, 3893
GIACIMENTO SEDIMENTOGENO, 3997

GIACIMENTO STRATIFORME, 4274
GIACIMENTO TELETERMALE, 4447
GIACIMENTO TIPO VALLE DEL MISSISSIPPI, 2896
GIACIMENTO VULCANOGENICO, 4774
GIADEITITE, 2357
GIGANTISMO, 1787
GIGANTOSTRACA, 1788
GIMNOSPERME, 1931
GINKGOALES, 1789
GIOVANE, 4850
GIOVANILE, 2364
GIUNZIONE TRIPLA, 4599
GIURASSICO, 2363
GIURASSICO INF., 2583
GIURASSICO MED., 2844
GIURASSICO SUP., 4673
GIVETIANO, 1790
GLACIALE, 1791
GLACIALE ANTICO, 191
GLACIAZIONE, 1803
GLACIAZIONE CONTINENTALE, 924
GLACIOLOGIA, 1807
GLACIOTETTONICA, 1809
GLAUCONITE, 1815
GLAUCONITIZZAZIONE, 1817
GLEY, 1819
GLOBIGERINACEA, 1822
GLOMERO-, 1823
GLOSSARIO, 1824
GLOSSOPTERIDALES, 1825
GNATHOSTRACA, 1827
GNATOSTOMI, 1826
GNEISS, 1828
GNETALES, 1831
GOLA, 1837
GOLFO, 1927
GONDWANA, 1834
GONIATITIDA, 1835
GONIOMETRIA, 1836
GRABEN, 1840
GRADIENTE, 1842
GRADIENTE GEOTERMICO, 1775
GRADIENTE IDRAULICO, 2093
GRADIMETRO, 1843
GRADO DI CARBONIZZAZIONE, 3748
GRADO DI LIBERTA, 1100
GRADO METAMORFICO, 2784
GRAFICO, 1863
GRAFITE, 1866
GRAFITIZZAZIONE, 1867
GRANA, 1844
GRANATO, 1683
GRANDINE, 1941
GRANITICO, 1847
GRANITIZZAZIONE, 1850
GRANITO, 1845
GRANITO ALCALINO, 125

GRANITO GNEISSICO, 1846
GRANITOIDE, 1851
GRANOCLASSAZIONE, 1841
GRANODIORITE, 1853
GRANOGABBRO, 1854
GRANULAZIONE, 1857
GRANULITE, 1858
GRANULITICO, 1860
GRANULO RIVESTITO, 817
GRANULOMETRIA, 1861
GRAPESTONE, 1862
GRAPTOLITHINA, 1868
GRAPTOLOIDEA, 1869
GRAVIMETRIA, 1874
GRAVIMETRO, 1873
GRAVITAZIONE, 1875
GRAVITA ASSOLUTA, 11
GREISEN, 1896
GREISENIZZAZIONE, 1897
GROTTA, 678
GROVACCA, 1888
GRUMO DI FANGO, 2957
GRUMO FECALE, 1521
GRUPPO IDROCARBURI, 2099
GRUPPO SPAZIALE, 4188
GUANO, 1924
GUGLIA VULCANICA, 2996
GUIDA, 3157
GUIDA STRUTTURALE, 4307
GUSCIO, 4471
GUYOT, 3984
GYMNOLAEMATA, 1930
GYTTJA, 1934
GZHELIANO, 1935
HABITAT, 1937
HAMMADA, 1950
HARDGROUND, 1955
HAUTERIVIANO, 1959
HE/HE, 1979
HELICOPLACOIDEA, 1976
HELIOZOA, 1977
HEMICHORDATA, 4261
HEMICRUSTACEA, 1982
HEMIPTEROIDEA, 1983
HETEROCORALLIA, 1991
HETERODONTA, 1992
HETTANGIANO, 1999
HEXACORALLA, 2000
HEXACTINIARIA, 2001
HIATUS, 2004
HIPPOMORPHA, 2016
HISTOSOL, 2022
HOLOCEPHALI, 2027
HOLOPHYTE, 2030
HOLOSTEI, 2031
HOLOTHURIOIDEA, 2033
HOMALOZOA, 2035

HOMINIDAE, 2038
HOMO SAPIENS, 2039
HOMOIOSTELEA, 2043
HOMOSTELEA, 2046
HORST, 2060
HOWARDITE, 2067
HUMITE, 2078
HUMMOCK, 2079
HUMOD, 2080
HUMOX, 2081
HUMULT, 2082
HUMUS, 2083
HYALOSPONGEA, 2085
HYDRARCH, 2088
HYDROIDA, 2118
HYDROZOA, 2137
HYMENOPTERA, 2140
HYOLITHES, 2141
HYRACOIDEA, 2152
HYSTRICHOMORPHA, 2153
IBRIDAZIONE MAGMATICA, 2086
ICEBERG, 2163
ICHTHYOPTERYGIA, 2166
ICHTYOSAURIA, 2167
ICNITE, 2164
ICNOLOGIA, 2165
IDATOGENESI, 2087
IDIOBLASTO, 2168
IDIOGEOSINCLINALE, 2169
IDIOMORFISMO, 2171
IDRATATO, 2135
IDRATAZIONE, 2089
IDRATAZIONE DEL VETRO, 2090
IDRAULICA, 2098
IDROCARBURO ALIFATICO, 122
IDROCARBURO AROMATICO, 305
IDROCHIMICA, 2102
IDRODINAMICA, 2104
IDRODINAMICA DI CAPTAZIONE, 672
IDROESPLOSIONE, 2105
IDROFONO, 2125
IDROGENO, 2106
IDROGEOLOGIA, 2115
IDROGEOLOGIA APPLICATA, 248
IDROGEOLOGIA REGIONALE, 3794
IDROGRAFIA, 2117
IDROGRAMMA, 2116
IDROLOGIA, 2121
IDROLOGIA CARSICA, 2375
IDROMETALLURGIA, 2122
IDROMETRIA, 2124
IDROMETRO, 2123
IDROSFERA, 2126
IDROSPIRA, 2127
IDROSSIDO, 2136
IDROTERMALE, 2130
IGCP, 2278

IGNIMBRITE, 2177
IGROFILO, 2138
IMBRIANO, 2179
IMBRICAZIONE, 2181
IMMAGAZZINAMENTO, 31
IMMAGINE, 2178
IMMATURO, 2182
IMMERSIONE, 1195
IMMERSIONE ASSIALE DI PIEGA, 3516
IMMISCIBILITA, 2185
IMPACCHETTAMENTO FITTO, 798
IMPACTITE, 2193
IMPATTO, 4075
IMPATTO AMBIENTALE, 2191
IMPIANTO DI TRATTAMENTO, 3155
IMPILAMENTO ATOMICO, 365
IMPORTAZIONE, 2194
IMPRONTA, 665
IMPRONTA DI CARICO, 2555
IMPRONTA DI PIOGGIA, 3743
IMPRONTA MECCANICA, 715
IMPUREZZA, 2196
INARCAMENTO, 4686
INARTICULATA, 2198
INCEPTISOL, 2199
INCLINAZIONE, 4533
INCLINAZIONE MAGNETICA, 2635
INCLINOMETRO, 4534
INCLUSIONE, 2200
INCLUSIONE FLUIDA, 1579
INCLUSIONE GASSOSA, 1690
INCLUSIONE MINERALE, 2870
INCOERENTE, 2201
INCRESPATURA EOLICA, 4828
INCROSTAZIONE, 2205
INDIANO, 2210
INDICATORE, 2212
INDICATORE BATIMETRICO, 1123
INDICATORE GEOCHIMICO, 1718
INDICE DI COLORE, 847
INDICE DI CRISTALLIZZAZIONE, 1026
INDICE DI DIFFERENZIAZIONE, 1179
INDICE DI MATURITA, 2719
INDICE DI RIFRAZIONE, 2209
INDICI DI MILLER, 2860
INDIO, 2213
INDIZIO MINERARIO, 4086
INDOCHINITE, 2214
INDURIMENTO, 2218
INDUSTRIA, 2222
INDUSTRIA CHIMICA, 728
INFANGAMENTO, 4108
INFILTRAZIONE, 2226
INFORMATICA, 2228
INFRACAMBRIANO, 2230
INFRAROSSO, 2231
INFRASTRUTTURA, 2235

INGEGNERIA CIVILE, 777
INGEGNERIA DEL PETROLIO, 3380
INGEGNERIA MINERARIA, 2884
INGEGNERIA SISMICA, 1285
INGRESSIONE, 2236
INIEZIONE, 2239
INIEZIONE BASALE, 4162
INIEZIONE DI FLUIDO, 1580
INLANDSIS, 2160
INLIER, 2242
INOCERAMI, 2246
INOMOGENEITA, 2238
INORGANICO, 2247
INQUINAMENTO, 3543
INQUINANTE, 3541
INSECTA, 2251
INSECTIVORA, 2252
INSEGNAMENTO, 1309
INSELBERG, 2253
INSOLAZIONE, 2255
INSTALLAZIONE MARINA, 2680
INSTALLAZIONE SOTTOMARINA, 4328
INTERCALAZIONE, 2260
INTERFACCIA, 2261
INTERFACCIA ACQUA=SEDIMENTO, 3994
INTERFACCIA ARIA=MARE, 94
INTERFACCIA OLIO=ACQUA, 3114
INTERFACCIA OLIO=GAS, 3113
INTERFERENZA, 2262
INTERFEROMETRIA, 2263
INTERFLUVIO, 2264
INTERFORMAZIONALE, 2265
INTERGLACIALE, 2266
INTERNIDI, 2279
INTERNO DELLA TERRA, 1276
INTERPLUVIALE, 2281
INTERPOLAZIONE, 2282
INTERPRETAZIONE, 2283
INTERSTRATO, 2268
INTRACLASTO, 2288
INTRAFORMAZIONALE, 2291
INTRUSIONE, 2293
INTRUSIONE DI ACQUA SALATA, 3930
INTRUSIONE STRATIFICATA, 4066
INTRUSIVO, 2294
INVENTARIO, 2295
INVERSIONE, 3842
INVERSIONE DEL CAMPO, 1743
INVERSIONE DEL RILIEVO, 3810
INVERTEBRATA, 2297
IO/TH, 2302
IO/U, 2303
IODIO, 2298
IONE, 2299
IONIZZAZIONE, 2304
IONOSFERA, 2305
IPOCENTRO, 2146

IPODIGMA, 2148
IPOTESI, 2150
IPSOMETRIA, 2151
IRIDESCENZA, 2306
IRIDIO, 2307
IRREVERSIBILITA, 2313
IRRIGAZIONE, 2314
ISOANOMALA, 2318
ISOBARA, 2319
ISOBASE, 2320
ISOBATA, 2321
ISOCLINALE, 2324
ISOCORA, 4277
ISOCORA DI FASE, 3393
ISOCRONA, 2323
ISOGONA, 2327
ISOGRADA, 2328
ISOIPSA, 2329
ISOLA, 2315
ISOLAMENTO, 2330
ISOLANTE, 2259
ISOLITO, 2331
ISOMORFISMO, 2334
ISOPLEURA, 2338
ISOPRENOIDE, 2339
ISORAD, 2340
ISORAT, 2341
ISOSISTA, 2342
ISOSTASIA, 2344
ISOTERMA, 2346
ISOTOPO, 2348
ISOTOPO LEGGERO, 2500
ISOTOPO PESANTE, 1971
ISOTOPO STABILE, 4228
ISOTROPIA, 2354
ISTITUZIONE, 2257
ISTOGRAMMA, 2017
ISTOLOGIA, 2018
ITABIRITE, 2355
ITACOLUMITE, 2356
ITTERBIO, 4853
ITTRIO, 4854
K/AR, 3580
K/CA, 3581
KAME, 2366
KANGA, 2367
KARROO, 2372
KASIMOVIANO, 2382
KAZANIANO, 2381
KEROGENE, 2386
KEUPER, 2388
KIMBERLITE, 2391
KIMMERIDGIANO, 2392
KINK BAND, 2394
KINZIGITE, 2395
KLIPPE, 2396
KREEP, 2397

KRIPTON, 2398
KUJALNIKIANO, 2399
KUNGURIANO, 2400
KURTOSI, 2401
KUTORGINIDA, 2402
LABYRINTHODONTIA, 2404
LACCIO DA SCARPA, 4079
LACCOLITE, 2405
LACERTILIA, 2406
LACUNA, 1682
LACUNA DI MISCIBILITA, 2894
LADINICO, 2410
LAGO, 2415
LAGO ARTIFICIALE, 3825
LAGO CRATERICO, 982
LAGO DI LAVA, 2458
LAGO ESTINTO, 1494
LAGO GLACIALE, 1797
LAGO SALATO, 3927
LAGUNA, 2411
LAGUNA CHIUSA, 419
LAHAR, 2414
LAMARCKISMO, 2417
LAMINA, 2419
LAMINAZIONE, 2422
LAMPROFIRO, 2423
LANDSAT, 2430
LANTANIO, 2432
LAPILLI, 2433
LASER, 2437
LASTRA, 4129
LATERALE, 2441
LATERITE, 2446
LATERITIZZAZIONE, 2448
LATITE, 2449
LATITUDINE, 2450
LAURASIA, 2453
LAVA, 2454
LAVA AA, 2
LAVAGGIO A BATEA, 3283
LAVORI PUBBLICI, 3680
LEACHING, 2465
LECTOTIPO, 2474
LEGA, 139
LEGAME INTERATOMICO, 364
LEGANTE, 2944
LEGENDA, 2475
LEGGE, 2460
LEGGE DELLA CORRELAZ.DI FACIES, 2461
LEGGE DI BAVENO, 442
LEGGE DI DARCY, 1061
LEGGE DI HAECKEL, 1939
LEGGE DI HAUY, 1960
LEGGE DI HOOKE, 2048
LEGGE MATEMATICA, 2712
LEGISLAZIONE, 2476
LEGNO FOSSILE, 1624

LEGNO SILICIZZATO, 4101
LEMBO RIALZATO, 4685
LENIANO, 2479
LENTE, 2480
LEPERDITOCOPIDA, 2481
LEPIDOCYSTOIDEA, 2483
LEPIDOSAURIA, 2484
LEPOSPONDYLI, 2485
LEPTINITE, 2488
LEPTITE, 2486
LEPTOGEOSINCLINALE, 2487
LESSICO, 2494
LETTO FLUVIALE, 4285
LETTO PRODUTTIVO, 3983
LEUCOSOMA, 2490
LIASSICO INF., 2584
LIASSICO MED., 2845
LIASSICO SUP., 4674
LIBRETTO GUIDA, 1926
LICHENES, 2495
LICHENOMETRIA, 2496
LICHIDA, 2497
LIDO, 418
LIGNINA, 2503
LIGNITE, 2504
LIMITE DI ELASTICITA, 1318
LIMITE DI PLACCA, 3497
LIMITE STRATIGRAFICO, 4275
LIMITI DI ATTERBERG, 367
LIMNOLOGIA, 2508
LIMONITE, 2510
LIMONITE PISOLITICA, 449
LINEA D'ISOSALINITA, 2326
LINEA DELLE ANDESITI, 195
LINEA DI BECKE, 451
LINEA DI COSTA, 816
LINEA DI RIVA, 4083
LINEA DI RIVA ABBANDONATA, 4
LINEA ISOCLINA, 2325
LINEA ISOPIESTICA, 2336
LINEAMENTO, 2511
LINEAZIONE, 2512
LINGUA LITORALE, 4215
LINGUAGGIO INFORMATICO, 871
LINGULIDA, 2513
LIPIDE, 2516
LIPOSTRACA, 2517
LIQUEFAZIONE, 2519
LIQUIDUS, 2522
LISCIAMENTO, 4141
LISCIVIAZIONE, 2550
LISCIVIAZIONE DEL SUOLO, 4151
LISSAMPHIBIA, 2523
LISTA FAUNISTICA, 1517
LITHOPHYTE, 2534
LITIFICAZIONE, 2524
LITIO, 2525

LITOFACIES, 2526
LITOGENESI, 2528
LITOLOGIA, 2532
LITOLOGICO, 2529
LITOPTERNA, 2542
LITOSFERA, 2536
LITOSTRATIGRAFIA, 2538
LITOSTROMA, 2539
LITOTIPO, 2541
LITOTOPO, 2540
LITTORALE, 2543
LITUOLACEA, 2547
LIVELLAMENTO, 2492
LIVELLI CARBONIFERI, 805
LIVELLO, 2493
LIVELLO, 2462
LIVELLO BASALTICO, 426
LIVELLO DEL MARE, 3981
LIVELLO DELLA FALDA IDRICA, 1916
LIVELLO DI CARBONE, 806
LIVELLO DI COPERTURA, 979
LIVELLO DI MAGRA, 2573
LIVELLO FOSSILIFERO, 1625
LIVELLO GRANITICO, 1849
LIVELLO IDROSTATICO, 2128
LIVELLO IMPERMEABILE, 262
LIVELLO LACUSTRE, 2416
LIVELLO MARINO, 3977
LIVELLO NEFELOIDE, 3009
LLANDEILIANO, 2551
LLANDOVERIANO, 2552
LLANVIRNIANO, 2553
LOCALE, 2557
LOCALITA MINERALOGICA, 2878
LOCALITA-TIPO, 4624
LOCOMOZIONE, 2558
LODRANITE, 2560
LOESS, 2561
LOG, 2562
LOG LATERALE, 2443
LONGITUDINE, 2565
LOPOLITE, 2569
LU/HF, 2608
LUCENTEZZA, 2606
LUCENTEZZA GRASSA, 1889
LUDLOWIANO, 2596
LUMACHELLA, 948
LUMINESCENZA, 2597
LUNA, 2935
LUNARE, 2599
LUNGHEZZA, 2477
LUTEZIO, 2607
LUVISOL, 2609
LYCOPODIALES, 2610
LYTOCERATIDA, 2611
MAAR, 2612
MAASTRICHTIANO, 2615

MACERALE, 2613
MACHAERIDIA, 2614
MAGLIA GEOMETRICA, 3013
MAGMA, 2617
MAGMA RESIDUO, 3830
MAGMATICO, 2619
MAGMATISMO, 2624
MAGNAFACIES, 2625
MAGNESIO, 2627
MAGNESITE, 2626
MAGNETISMO, 2646
MAGNETITE, 2647
MAGNETIZZAZIONE, 2648
MAGNETIZZAZIONE ANISTERETICA RESIDUA, 206
MAGNETIZZAZIONE CHIMICA RESIDUA, 732
MAGNETIZZAZIONE DETRITICA RESIDUA, 1143
MAGNETIZZAZIONE INDOTTA, 2215
MAGNETIZZAZIONE ISOTERMICA RESIDUA, 2347
MAGNETIZZAZIONE NATURALE RESIDUA, 2991
MAGNETIZZAZIONE RESIDUA, 3812
MAGNETIZZAZIONE TERMICA RESIDUA, 4511
MAGNETIZZAZIONE VISCOSA RESIDUA, 4747
MAGNETOIDRODINAMICA, 2649
MAGNETOMETRO, 2650
MAGNETOSFERA, 2651
MAGNETOSTRIZIONE, 2653
MAGNITUDO SISMICA, 1286
MAGRO, 2471
MAJIANO, 2656
MALACOSTRACA, 2659
MALLEABILITA, 2660
MAMMALIA, 2661
MANGANESE, 2662
MANIFESTAZIONE VULCANICA, 4768
MANIPOLAZIONE DI DATI, 1068
MANO D'OPERA, 2664
MANTELLO, 2665
MANTELLO INFERIORE, 2585
MANTELLO SUPERIORE, 4675
MANTODEA, 2666
MANUALE, 2667
MARE, 3974
MARE DI PIATTAFORMA, 4069
MARE GEOSINCLINALE, 1764
MARE LUNARE, 2672
MARE MARGINALE, 2674
MARE PROFONDO, 1093
MAREA, 4529
MAREA TERRESTRE, 1280
MARGINE CONTINENTALE, 925
MARGINE CONTINENTALE ASISMICO, 328
MARGINE CONTINENTALE ATTIVO, 58
MARGINE CONTINENTALE COMPLESSO, 920
MARINER, 2690
MARMITTA, 2387
MARMITTA DEI GIGANTI, 3585
MARMO, 2671

MARNA, 2692
MARSUPIALIA, 2695
MASCELLA, 2360
MASCON, 2696
MASSA, 2697
MASSA DI FONDO, 1919
MASSA FILTRANTE, 1872
MASSICCIO CENTRALE, 2732
MASSICCIO LUNARE, 2605
MASSICCIO MONTUOSO, 2950
MASSIMO, 2722
MASSIMO DI ANOMALIA, 2006
MASSIVO, 2703
MASSO ERRATICO, 1430
MASTIGOPHORA, 2706
MATEMATICA, 2716
MATERIA AMORFA, 170
MATERIA FUSA, 2743
MATERIA INORGANICA, 2249
MATERIA ORGANICA, 3170
MATERIA PRIMA, 3755
MATERIALE ARANCIONE, 3144
MATERIALE COERENTE, 829
MATERIALE COMPETENTE, 859
MATERIALE DA COSTRUZIONE, 911
MATERIALE DA ORNAMENTAZIONE, 3181
MATERIALE DI RIEMPIMENTO, 1544
MATERIALE ESPANDIBILE, 1474
MATERIALE IMMISCIBILE, 2186
MATERIALE INCASSANTE, 2061
MATERIALE INCOERENTE, 2202
MATERIALE PER MASSICCIATA, 3877
MATERIALE PLASTICO, 3493
MATERIALE SCIOLTO, 4647
MATERIALE SINTETICO, 4402
MATERIALE SOVRACONSOLIDATO, 3214
MATERIALE VISCOSO, 4746
MATERIALI, 2707
MATRICE, 3887
MATRICE MATEMATICA, 2713
MATURITA, 2718
MAXILLOPODA, 2721
MEANDRO, 2724
MECCANICA, 2729
MECCANICA DEI FLUIDI, 1581
MECCANICA DEL SUOLO, 4154
MECCANICA DELLE ROCCE, 3888
MECCANISMO, 2730
MECCANISMO FOCALE, 1600
MECOPTEROIDA, 2731
MEDIA, 389
MEGAFACIES, 2738
MEGASPORA, 2739
MELANGE, 2740
MEMBRO, 2746
MENDELEVIO, 2747
MEOZIANO, 2748

MERCURIO, 2749
MEROSTOMATA, 2750
MEROSTOMOIDEA, 2751
MESOGASTROPODA, 2754
MESOGEA, 2753
MESOGEOSINCLINALE, 2755
MESOSAURIA, 2757
MESOSIDERITE, 2758
MESSA A TERRA, 2895
MESSA IN POSTO, 1363
METABASALTO, 2762
METACOPINA, 2763
METAGABBRO, 2764
METAGENESI, 2765
METALLI ALCALINO TERROSI, 129
METALLIFERO, 2769
METALLO ALCALINO, 126
METALLO BASE, 428
METALLO IN TRACCE, 4558
METALLO NON FERROSO, 3047
METALLO PESANTE, 1972
METALLO PREZIOSO, 3037
METALLO RARO, 3752
METALLOGENESI, 2774
METALLOGENESI PREVISIONALE, 3599
METALLOIDE, 2775
METALLOTECTO, 2776
METALLURGIA, 2777
METAMITTIZZAZIONE, 2779
METAMORFISMO D'IMPATTO, 2189
METAMORFISMO DA IMPATTO, 4077
METAMORFISMO DEBOLE, 2575
METAMORFISMO DI AFFOSSAMENTO, 597
METAMORFISMO DI CONTATTO, 915
METAMORFISMO DINAMICO, 1266
METAMORFISMO FORTE, 2011
METAMORFISMO GEOTERMICO, 1776
METAMORFISMO ISOCHIMICO, 2322
METAMORFISMO MESOZONALE, 2761
METAMORFISMO PLUTONICO, 3518
METAMORFISMO PROGRESSIVO, 3633
METAMORFISMO REGIONALE, 3795
METAMORFISMO TERMICO, 4498
METAMORFOSI, 2789
METANAUPLIUS, 2790
METANO, 2801
METAPELITE, 3336
METAPHYTE, 2791
METASOMA, 2793
METASOMATISMO ALCALINO, 130
METASOMATISMO DI CONTATTO, 916
METASOMATOSI, 2794
METATESSI, 2796
METEORITE, 2798
METEORITE METALLICA, 2767
METEORITE ROCCIOSA, 4264
METEOROIDE, 2799

METEOROLOGIA, 2800
METODO A IMMERSIONE, 2184
METODO A SCANSIONE LATERALE, 4091
METODO ACUSTICO, 43
METODO AFMAG, 78
METODO ANALITICO, 183
METODO D'INDUZIONE, 2217
METODO DEGLI ULTRASUONI, 4639
METODO DEI TRANSIENTI, 4573
METODO DELLA RESISTIVITA, 3833
METODO DI MONTE CARLO, 2932
METODO DIFFERENZE FINITE, 1550
METODO ELETTRICO, 1329
METODO ELETTROMAGNETICO, 1341
METODO GEOFISICO, 1755
METODO GEOTERMICO, 1777
METODO GRAVIMETRICO, 1883
METODO MAGNETICO, 2637
METODO MAGNETOTELLURICO, 2654
METODO MATEMATICO, 2714
METODO MET, 4575
METODO MICROONDE, 2838
METODO NUOVO, 3022
METODO PER VIA UMIDA, 4821
METODO RADIOMETRICO, 3732
METODO SISMICO, 4014
METODO SONAR, 4181
METODO STANDARD, 4236
METODO STATISTICO, 4239
METODO TELLURICO, 4449
METODO TRAZIONE DI FONDO, 1092
METODOLOGIA, 2802
METROPOLITANA, 4344
MEZZO ANELASTICO, 197
MEZZO DI TRASPORTO, 4578
MEZZO ELASTICO, 1319
MEZZO POROSO, 3566
MEZZO STRATIFICATO, 2464
MICA, 2804
MICASCISTO, 2805
MICRITE, 2806
MICROCALCOLATORE, 2807
MICROCRATERE, 2808
MICRODUREZZA, 2820
MICROFACIES, 2811
MICROFAUNA, 2812
MICROFESSURA, 2813
MICROFLORA, 2814
MICROFOSSILE, 2816
MICROFOSSILE PROBLEMATICO, 2830
MICROGRANITO, 2818
MICROLATEROLOG, 2821
MICROLOG, 2823
MICROMETEORITE, 2824
MICROMORFOLOGIA DEL SUOLO, 2825
MICROORGANISMO, 2826
MICROPALEONTOLOGIA, 2827

MICROPHYTOLITHI, 2828
MICROPIEGA, 2815
MICROPLACCA, 2829
MICROSCOPIA, 2833
MICROSCOPIA ELETTR.SCANSIONE, 3957
MICROSCOPIA ELETTRONICA, 1349
MICROSCOPIA MINERARIA, 3158
MICROSCOPIO, 2831
MICROSCOPIO ELETTRONICO, 1348
MICROSCOSSA, 2810
MICROSISMA, 2834
MICROSONDA A IONI, 2301
MICROSONDA ELETTRONICA, 1347
MICROSTILOLITE, 2835
MICROSTRUTTURA, 4640
MICROTECTITE, 2837
MICROTETTONICA, 2836
MIGMATITE, 2853
MIGMATIZZAZIONE, 2854
MIGRAZIONE, 2855
MIGRAZIONE DI ELEMENTI, 2856
MILIOLACEA, 2857
MILIOLINA, 2858
MILONITE, 2972
MILONITIZZAZIONE, 2974
MINERALE, 2862
MINERALE A STRATI MISTI, 2898
MINERALE ACCESSORIO, 29
MINERALE ALLOTIGENO, 136
MINERALE ALLUNGATO, 1355
MINERALE ANTISTRESS, 232
MINERALE AUTIGENO, 376
MINERALE COESISTENTE, 826
MINERALE DI URANIO, 4690
MINERALE ESSENZIALE, 3890
MINERALE INDICE, 2208
MINERALE ISOSTRUTTURALE, 2345
MINERALE MAGNETICO, 2638
MINERALE METAMITTICO, 2778
MINERALE NUOVO, 3023
MINERALE PISOLTICA, 449
MINERALE PRESUNTO, 2225
MINERALE SECONDARIO, 3989
MINERALE SFRUTTABILE, 1300
MINERALE STRESS, 4291
MINERALE TIPOMORFO, 4627
MINERALE UTILE, 3159
MINERALI DELLE ARGILLE, 790
MINERALI LEGGERI, 2501
MINERALI PESANTI, 1973
MINERALIZZAZIONE, 2876
MINERALOGIA, 2880
MINERALOGIA DELLE ARGILLE, 789
MINICALCOLATORE, 2881
MINIERA, 2861
MINIERA A CIELO APERTO, 3129
MINIMI QUADRATI, 2473
MINIMO, 2882

MIOCENE, 2889
MIOCENE INF., 2586
MIOCENE MED., 2846
MIOCENE SUP., 4676
MIOGEOSINCLINALE, 2890
MIOMERA, 2891
MIOSPORA, 2892
MIRMECHITE, 2979
MISCIBILITA, 2893
MISSISSIPPIANO, 2897
MISURA AL SUOLO, 1906
MISURA DI DEFORMAZIONE, 4268
MISURAZIONE DI PENDENZA, 1204
MOBILITA, 2900
MOBILIZZAZIONE GEOCHIMICA, 2901
MODA, 2903
MODELLO, 2904
MODELLO BIDIMENSIONALE, 4622
MODELLO DETERMINISTICO, 1141
MODELLO FISICO, 3426
MODELLO IDRODINAMICO, 2103
MODELLO IN SCALA, 3955
MODELLO INTERNO, 2274
MODELLO MATEMATICO, 2715
MODELLO MONODIMENSIONALE, 3120
MODELLO STATISTICO, 4242
MODIOMORPHOIDA, 2906
MODULO, 2907
MODULO D'ELASTICITA, 591
MODULO DI RIGIDITA, 2908
MODULO DI TAGLIO, 4060
MODULO DI YOUNG, 4851
MOLASSA, 2912
MOLDAVITE, 2913
MOLIBDATO, 2917
MOLIBDENO, 2918
MOLLISOL, 2915
MOLLUSCA, 2916
MOMENTO DIPOLARE, 1206
MONAZITE, 2919
MONOCLINALE, 2920
MONOCOTILEDONI, 2922
MONOCYATHEA, 2923
MONOGRAFIA, 2924
MONOLITO, 2926
MONOMORFO, 2927
MONOPLACOPHORA, 2928
MONORHINA, 2929
MONOSSIDO DI CARBONIO, 644
MONOTREMATA, 2930
MONSONE, 2931
MONZONITE, 2933
MORENA, 2938
MORENA DI FONDO, 1907
MORENA LATERALE, 2444
MORFOGENESI, 2939
MORFOLOGIA, 2940

MORFOLOGIA COSTIERA, 4081
MORFOLOGIA D'EROSIONE, 1427
MORFOLOGIA DI COSTA, 814
MORFOLOGIA DI DISSOLUZIONE, 4178
MORFOLOGIA DI GIACIMENTO, 1119
MORFOLOGIA EOLICA, 1392
MORFOLOGIA EREDITATA, 2237
MORFOLOGIA FUNZIONALE, 1666
MORFOLOGIA GLACIALE, 1795
MORFOLOGIA SOTTOMARINA, 4329
MORFOMETRIA, 2941
MORFOSCOPIA, 2942
MORFOSTRUTTURA, 4310
MOSCOVIANO, 2946
MOVIMENTO, 2951
MOVIMENTO DEL SUOLO, 1908
MOVIMENTO DI GHIACCIO, 2158
MOVIMENTO DI MASSA, 2699
MOVIMENTO ORIZZONTALE, 2053
MULTITUBERCULATA, 2967
MUSCHELKALK, 2969
MUSEO, 2970
MUTAZIONE, 2971
MYODOCOPIDA, 2975
MYOIDA, 2976
MYOMORPHA, 2977
MYRIAPODA, 2978
MYSTACOCARIDA, 2980
MYTILOIDA, 2981
MYXOMYCETES, 2982
NAKHLITE, 2983
NAMURIANO, 2984
NANNOFOSSILE, 2985
NATANTES, 2987
NAUTILOIDEA, 2993
NAVE PER TRIVELLAZIONE, 1257
NEANDERTHALIANO, 2994
NEBULOSA, 968
NECK, 2995
NECTON, 2997
NEFELINA, 3008
NEMERTA, 2998
NEOCOMIANO, 2999
NEODIMIO, 3000
NEOGASTROPODA, 3001
NEOGENE, 3002
NEON, 3003
NEORNITHES, 3004
NEOTENIA, 3006
NEOTETTONICA, 3005
NEOTIPO, 3007
NETTUNIO, 3011
NETTUNISMO, 3010
NEUTRONE, 3014
NEVATO, 1553
NEVE, 4142
NICKEL, 3026

NILSSONIALES, 3028
NIOBATO, 3029
NIOBATO-TANTALATO, 3031
NIOBIO, 3030
NITRATO, 3032
NITRIFICAZIONE, 3033
NIVAZIONE, 3036
NODOSARIACEA, 3038
NODULO, 3039
NOEGGERATHIALES, 3040
NOME NUOVO, 3024
NOMENCLATURA, 3042
NOMOGENESI, 3044
NORD, 3057
NORICO, 3053
NORMA PETROGRAFICA, 3054
NORMALIZZAZIONE, 4238
NOSE, 3058
NOTA ILLUSTRATIVA, 1480
NOTOSTRACA, 3059
NOTOUNGULATA, 3060
NOVACULITE, 3061
NOVITA, 3021
NUBE ARDENTE, 3069
NUCLEAZIONE, 3066
NUCLEO CONTINENTALE, 926
NUCLEO ESTERNO, 3207
NUCLEO INTERNO, 2243
NUCLEO TERRESTRE, 957
NUCULOIDEA, 3068
NUMMULITIDAE, 3071
NUOVO, 3019
NUTAZIONE, 3072
NUTAZIONE CHANDLER, 706
O-18/O-16, 3227
OASI, 3074
OBBIETTIVO, 3076
OBDUZIONE, 3075
OBOLELLIDA, 3079
OCEAN DRILLING PROJECT, 3087
OCEANICO, 3090
OCEANIZZAZIONE, 3092
OCEANO, 3084
OCEANOGRAFIA, 3093
OCEANOLOGIA, 3094
OCHREPT, 3096
OCRA, 3095
OCTOCORALLIA, 3098
ODOGRAFO, 2023
ODONTOLOGIA, 3099
ODONTOPLEURIDA, 3100
ODONTORNITHES, 3101
OFFSHORE, 3102
OFIOLITE, 3133
OLIGOCENE, 3115
OLIGOCENE INF., 2587
OLIGOCENE MED., 2847

OLIGOCENE SUP., 4677
OLISTOLITE, 3116
OLISTOSTROMA, 3117
OLIVINA, 3118
OLMIO, 2025
OLOCENE, 2026
OLOGRAFIA, 2029
OLOSTOMATO, 2032
OLOTIPO, 2034
OMEOMORFIA, 2037
OMEOMORFISMO, 2036
OMOCLINALE, 2040
OMOGENEITA, 2041
OMOGENEIZZAZIONE, 2042
OMOLOGIA, 2044
OMONIMIA, 2045
OMOTASSIA, 2047
ONCOLITE, 3119
ONDA, 4805
ONDA A LUNGO PERIODO, 2564
ONDA ACUSTICA, 45
ONDA D'URTO, 4078
ONDA DI AIRY, 97
ONDA DI CODA, 821
ONDA DI RAYLEIGH, 3756
ONDA DI SABBIA, 3941
ONDA DI SUPERFICIE, 4368
ONDA DI VOLUME, 535
ONDA ELASTICA, 1322
ONDA LOVE, 2570
ONDA MARINA, 3979
ONDA P, 3230
ONDA PREMONITRICE, 1615
ONDA S, 3920
ONDA SH, 4050
ONDA SISMICA, 4021
ONTOGENESI, 3121
ONYCHOPHORA, 3122
OOIDE, 3123
OOLITE, 3124
OPACO, 3128
OPERA INGEGNERISTICA, 1375
OPERAZIONI, 3130
OPHIDIA, 3131
OPHIOCISTIOIDEA, 3132
OPHIUROIDEA, 3135
OPISTHOBRANCHIA, 3136
OPOKA, 3137
ORBITA, 3146
ORBITOIDACEA, 3147
ORBITOIDIDAE, 3148
ORDINE-DISORDINE, 3149
ORDOVICIANO, 3150
ORDOVICIANO INF., 2588
ORDOVICIANO MED., 2848
ORDOVICIANO SUP., 4678
ORGANICO, 3164

ORGANISMO, 3176
ORGANISMO ALOFILO, 1948
ORGANISMO AUTOTROFICO, 386
ORGANISMO ETEROTROFO, 1998
ORGANISMO EURIALINO, 1449
ORGANISMO EURITERMICO, 1450
ORGANISMO FAGOTROPICO, 3389
ORGANISMO FITOFAGO, 3436
ORGANISMO ISOMORFO, 2333
ORGANISMO LITOFAGO, 3889
ORGANISMO MESOFILO, 2756
ORIENTAZIONE, 3177
ORIENTAZIONE OBLIQUA, 3078
ORIENTAZIONE PREFERENZIALE, 3600
ORIENTAZIONE VERTICALE, 4735
ORIGINE, 3179
ORIGINE ATMOGENA, 357
ORIGINE BATTERIOGENA, 401
ORIGINE CHIMICA, 736
ORIGINE EPIMAGMATICA, 1405
ORIGINE INORGANICA, 2250
ORIGINE IPOABISSALE, 2142
ORIGINE MAGMATICA, 2623
ORIGINE MECCANICA, 2726
ORIGINE ORGANOGENA, 3172
ORIGINE POSTMAGMATICA, 3574
ORIGINE SOPRACROSTALE, 4362
ORIGOFACIES, 3180
ORIZZONTALE, 2050
ORIZZONTE DEL SUOLO, 4149
ORIZZONTE DIFFERENZIATO, 2049
ORIZZONTE PRODUTTIVO, 3626
ORLO, 3868
ORNAMENTAZIONE, 3182
ORNEBLENDITE, 2056
ORNITHISCHIA, 3183
ORO, 1833
OROGENESI, 3185
OROGENESI ALGOMANIANA, 120
OROGENESI ALLEGHENIANA, 133
OROGENESI ALPINA, 150
OROGENESI ARMORICANA, 304
OROGENESI ASSINTICA, 342
OROGENESI BAIKALIANA, 404
OROGENESI BELTIANA, 460
OROGENESI CALEDONIANA, 618
OROGENESI CARELIANA, 2370
OROGENESI CHARNIANA, 711
OROGENESI CIMMERICA, 768
OROGENESI EBURNEA, 1289
OROGENESI ERCINICA, 1984
OROGENESI GRENVILLIANA, 1898
OROGENESI HUDSONIANA, 2068
OROGENESI KATANGHIANA, 2378
OROGENESI KENORIANA, 2384
OROGENESI KIBARIANA, 2390
OROGENESI LARAMICA, 2435

OROGENESI PENOKEANA, 3343
OROGENESI PRECAMBRIANA, 3594
OROGENESI SUDETICA, 4346
OROGENESI SVECOFENNIANA, 4375
OROGENESI TACONICA, 4407
OROGENESI VARISICA, 4719
ORTHENT, 3186
ORTHID, 3187
ORTHIDA, 3188
ORTHOD, 3190
ORTHOX, 3195
ORTOANFIBOLITE, 3189
ORTOGENESI, 3191
ORTOGNEISS, 3192
ORYCTOCOENOSIS, 3196
OSCILLAZIONE, 3197
OSCILLAZIONE PROPRIA, 1642
OSMIO, 3199
OSSERVATORIO, 3082
OSSERVAZIONE GEOFISICA, 3081
OSSIDAZIONE, 3221
OSSIDIANA, 3083
OSSIDO, 3222
OSSIGENO, 3225
OSTEICHTHYES, 3200
OSTEOLOGIA, 3201
OSTEOSTRACI, 3202
OSTRACEA, 3204
OSTRACODA, 3203
OTOLITE, 3205
OTTAEDRITE, 3097
OTTICA MINERALOGICA, 1019
OUTLIER, 3210
OVEST, 4819
OXFORDIANO, 3220
OXISOL, 3224
OXYLOPHYTE, 3228
PA/IO, 3643
PAHOEHOE, 3233
PALAEOHETERODONTA, 3235
PALAEOTAXODONTA, 3237
PALAGONITIZZAZIONE, 3238
PALEO, 3239
PALEOAMBIENTE, 3253
PALEOATMOSFERA, 3240
PALEOBATIMETRIA, 3242
PALEOBIOLOGIA, 3243
PALEOBOTANICA, 3245
PALEOCARSISMO, 3259
PALEOCENE, 3246
PALEOCIRCOLAZIONE, 3247
PALEOCLIMA, 3248
PALEOCLIMATOLOGIA, 3249
PALEOCOPIDA, 3250
PALEOCORRENTE, 3251
PALEOECOLOGIA, 3252
PALEOFITICO, 3236

PALEOGENE, 3254
PALEOGEOGRAFIA, 3257
PALEOLATITUDINE, 3260
PALEOLIMNOLOGIA, 3261
PALEOMAGNETISMO, 3263
PALEOMORFOLOGIA, 3258
PALEONTOLOGIA, 3265
PALEOOCEANOGRAFIA, 3241
PALEORILIEVO, 3266
PALEOSALINITA, 3267
PALEOSUOLO, 3268
PALEOTEMPERATURA, 3269
PALEOTERMOMETRIA, 3270
PALEOZOICO, 3271
PALEOZOOLOGIA, 3272
PALINGENESI, 3274
PALINOLOGIA, 3278
PALINOMORFO, 3279
PALINSESTO, 3273
PALLA D'ARGILLA, 788
PALLADIO, 3275
PALLASITE, 3276
PALO, 3445
PALO-PILA, 3440
PALSA, 3277
PALUDE, 4376
PALUDE A MANGROVIA, 2663
PALUDE LITORALE, 2693
PALUDE MARINA, 2681
PALUDE SALMASTRA, 3928
PAMPA, 3280
PANGEA, 3281
PANTODONTA, 3284
PANTOTHERIA, 3285
PARA-, 3286
PARAANFIBOLITE, 3287
PARABLASTOIDEA, 3288
PARACRINOIDEA, 3289
PARAFORA, 1730
PARAGENESI, 3290
PARAGEOSINCLINALE, 3291
PARAGNEISS, 3292
PARALIAGEOSINCLINALE, 3293
PARALLELISMO, 3297
PARAMETRO, 3299
PARASSITISMO, 3300
PARATIPO, 3301
PARTICELLA, 3307
PARTICELLA COSMICA, 969
PARTICELLA ELEMENTARE, 1353
PASSAGGIO INTERTIDALE, 4528
PATCH REEF, 3313
PATOLOGIA, 3315
PATTERN, 3316
PAUROPODA, 3318
PAVIMENTAZIONE DESERTICA, 1133
PB-ALPHA, 2467

PB/PB, 2468
PB/TH, 4518
PECTINACEA, 3323
PEDEMONTE, 3439
PEDEPIANO, 3325
PEDIMENTO, 3324
PEDOGENESI, 3326
PEGMATITE, 3327
PEGMATITICO, 3228
PEGMATOIDE, 3329
PELEEANO, 3333
PELITE, 3334
PELLETIZZAZIONE, 3338
PELLETS, 3337
PENDENTE, 3895
PENDIO, 1810
PENDIO STRUTTURALE, 1198
PENEPIANO, 3339
PENEPLANAZIONE, 3468
PENETROMETRIA, 3340
PENETROMETRO, 3341
PENNACCHIO, 3515
PENNELLO, 1901
PENNSILVANIANO, 3342
PENTAMERIDA, 3344
PENTASTOMIDA, 3345
PENTOXYLALES, 3346
PEPITA, 3070
PERCOLAZIONE, 3350
PERDITA, 2470
PERFORABILITA, 1253
PERFORAZIONE, 1661
PERFORAZIONE AUGER, 371
PERFORAZIONE DIREZIONATA, 1208
PERFORAZIONE OFFSHORE, 3103
PERFORAZIONE ROTARY, 3900
PERFORAZIONE ROTOPERCUSSIONE, 3901
PERFORAZIONE VIBRATORIA, 4739
PERICLINALE, 3351
PERIDOTITE, 3352
PERIGEO, 3353
PERIGLACIALE, 3354
PERIMETRO DI PROTEZIONE, 3644
PERIODICITA, 3357
PERIODO, 3356
PERIODO DI DIMEZZAMENTO, 1942
PERISPHINCTIDA, 3358
PERISSODACTYLA, 3359
PERLA, 3319
PERLITE, 3360
PERMAFROST, 3362
PERMEABILITA, 3363
PERMEABILITA MAGNETICA, 2639
PERMESSO DI ESPLORAZONE, 779
PERMIANO, 3364
PERMIANO INF., 2589
PERMIANO SUP., 4679

PERTITE, 3367
PESO SPECIFICO, 4196
PESO STATISTICO, 4812
PESTICIDA, 3369
PETROCHIMICA, 3372
PETROFISICA, 3383
PETROGENESI, 3374
PETROGRAFIA, 3378
PETROLIO, 3379
PETROLIO LEGGERO, 2502
PETROLIO PESANTE, 1974
PETROLOGIA, 3382
PETROLOGIA DEL SEDIMENTARIO, 3998
PETROLOGIA SPERIMENTALE, 1477
PETROLOGIA STRUTTURALE, 4309
PH, 3384
PHACOPIDA, 3386
PHAEOPHYTA, 3387
PHAEOZEM, 3388
PHOLADOMYIDA, 3399
PHOLIDOTA, 3400
PHREATOPHYTE, 3417
PHYLACTOLAEMATA, 3418
PHYLLOCERATIDA, 3420
PHYLLOPODA, 3422
PHYLUM, 3424
PHYTOLITHES, 3435
PIANA, 3465
PIANA ABISSALE, 22
PIANA ALLUVIONALE, 144
PIANA COSTIERA, 815
PIANA D'ACCUMULAZIONE, 33
PIANA D'ORIGINE MARINA, 2683
PIANA DI FANGO, 2955
PIANA FLUVIOGLACIALE, 3212
PIANA INONDABILE, 1570
PIANA LITORALE GLACIALE, 2161
PIANA SALINA, 3926
PIANA TIDALE, 4527
PIANETA, 3471
PIANETA ESTERNO, 3208
PIANETA GIOVE, 3473
PIANETA MARTE, 3474
PIANETA MERCURIO, 3475
PIANETA NETTUNO, 3476
PIANETA PLUTONE, 3477
PIANETA SATURNO, 3478
PIANETA TERRA, 3472
PIANETA TERRESTRE, 4468
PIANETA URANO, 3479
PIANETA VENERE, 3480
PIANIFICAZIONE, 3489
PIANIFICAZIONE URBANISTICA, 4695
PIANO, 3470
PIANO ASSIALE, 391
PIANO DI FAGLIA, 1510
PIANO DI STRATIFICAZIONE, 454

PIANO DI TAGLIO, 4061
PIANO STRATIGRAFICO, 4229
PIANTA, 3491
PIANTA ECHISTOTERMA, 1975
PIANTA ETEROSPORICA, 1997
PIANTA XEROFILA, 4843
PIANURA, 2595
PIATTAFORMA, 4067
PIATTAFORMA CONTINENTALE, 929
PIATTAFORMA CONTINENTALE ESTERNA, 3209
PIATTAFORMA CONTINENTALE INTERNA, 2244
PIATTAFORMA GRAVIMETRICA, 1884
PIATTAFORMA INTRACRATONICA, 3503
PIATTAFORMA MARGINALE, 2673
PIATTAFORMA MARINA, 2684
PICNOCLINO, 3689
PICNOMETRIA, 3692
PICNOMETRO, 3691
PICRITE, 3438
PIEGA, 1602
PIEGA A CASCATA, 661
PIEGA A GRADINO, 4249
PIEGA A KINK, 741
PIEGA A ZIGZAG, 4859
PIEGA ASIMMETRICA, 353
PIEGA CILINDRICA, 1052
PIEGA CONCENTRICA, 874
PIEGA CONIUGATA, 897
PIEGA CORICATA, 3771
PIEGA DI FLUSSO, 1577
PIEGA DI RIVESTIMENTO, 1248
PIEGA DI TRASCINAMENTO, 1243
PIEGA DISARMONICA, 1219
PIEGA ENTEROLITICA, 1380
PIEGA GRAVITATIVA, 1881
PIEGA INTRAFOLIARE, 3510
PIEGA PTIGMATICA, 3679
PIEGA ROVESCIATA, 3219
PIEGA SIMMETRICA, 4382
PIEGA SINFORME, 4393
PIEGA SOVRAPPOSTA, 4359
PIEGA-FAGLIA, 1220
PIEGAMENTO, 1608
PIENA, 1569
PIETRA DA TAGLIO, 588
PIETRA DECORATIVA, 1085
PIETRA PREZIOSA, 3595
PIETRA SEMIPREZIOSA, 4036
PIETRISCO, 3907
PIEZOCRISTALLIZZAZIONE, 3441
PIEZOMETRIA, 3443
PIGMENTO, 3444
PILASTRO, 3446
PILLOW LAVA, 3447
PINGO, 3450
PINNIPEDIA, 3452
PIOGGIA, 3742

PIOGGIA DI CENERE, 331
PIOMBO, 2466
PIPING, 3455
PIRITIZZAZIONE, 3694
PIROLISI, 3696
PIROMETALLURGIA, 3697
PIROMETAMORFISMO, 3698
PIROMETASOMATOSI, 3699
PIROSFERA, 3700
PIROSSENITE, 3704
PIROSSENO, 3702
PISCES, 3456
PISOLITE, 3457
PISTA, 4565
PISTE DI NUTRIZIONE, 1523
PLACANTICLINE, 3461
PLACCA, 3495
PLACENTALIA, 3462
PLACER, 3463
PLACODERMI, 3464
PLANCTON, 3487
PLANCTOSPHAEROIDEA, 3469
PLANETARIO, 3482
PLANETOLOGIA, 3485
PLANOSOL, 3490
PLASTICITA, 3494
PLATINO, 3504
PLATINOIDI, 3505
PLATYRRHINA, 3506
PLEISTOCENE, 3508
PLEISTOCENE INF., 2590
PLEISTOCENE MED., 2849
PLEISTOCENE SUP., 4680
PLEOCROISMO, 3509
PLIENSBACHIANO, 3511
PLIOCENE, 3512
PLIOCENE INF., 2591
PLIOCENE MED., 2850
PLIOCENE SUP., 4681
PLUTONE, 3517
PLUTONIO, 3521
PLUTONISMO, 3520
PLUVIALE, 3522
PODOCOPIDA, 3525
PODSOL, 3526
PODZOLUVISOL, 3527
POGONOPHORA, 3528
POLARE, 3533
POLARIZZAZIONE, 3139
POLARIZZAZIONE ELETTRICA, 1330
POLARIZZAZIONE INDOTTA, 2216
POLARIZZAZIONE SPONTANEA, 4033
POLAROGRAFIA, 3536
POLIEDRO, 3548
POLIFASICO, 2963
POLIGENICO, 3546
POLIMERIZZAZIONE, 3550

POLIMETAMORFISMO, 3552
POLIMINERALE, 3554
POLIMITTICO, 3553
POLIMORFISMO, 3555
POLITICA, 3537
POLITIPISMO, 3558
POLLINE, 3539
POLO GEOMAGNETICO, 1742
POLO PALEOMAGNETICO, 3262
POLONIO, 3544
POLVERE, 1263
POLVERE COSMICA, 967
POLVERE RADIOATTIVA, 3726
POLYCHAETIA, 3545
POLYMERA, 3549
POLYPLACOPHORA, 3557
POMICE, 3683
POMPAGGIO, 3684
PONTE, 577
PONTIANO, 3559
POPOLAZIONE ENDEMICA, 1368
POPOLAZIONE STATISTICA, 4243
PORCELLANITE, 3561
PORIFERA, 3564
POROSITA, 3565
POROSITA CELLULARE, 682
PORPHYRY COPPER, 3569
PORTANZA, 450
PORTATA, 1209
PORTATA D'ACQUA SOTTERRANEA, 1914
PORTLANDIANO, 3570
PORTO, 1954
POSSIBILITA, 3572
POSTGLACIALE, 3573
POSTTETTONICO, 3575
POTABILITA, 3576
POTASSA, 3577
POTASSIO, 3579
POTENZIALE ELETTRICO, 1325
POTENZIOMETRO, 3584
POTERE RIFLETTENTE, 3782
POTERE RIFRANGENTE, 3788
POTERE ROTATORIO, 772
POZZO ARTESIANO, 314
POZZO DI METANO, 1692
POZZO DI MINIERA, 3459
POZZO ESPLORATIVO, 4813
POZZO OFFSHORE, 3104
POZZO PER ACQUA, 4801
POZZO PETROLIFERO, 3112
POZZO PRODUTTIVO, 3627
POZZOLANA, 3589
PRAECARDIIDA, 3590
PRASEODIMIO, 3591
PRATERIA, 1870
PRECAMBRIANO, 3593
PRECIPITAZIONE ATMOSFERICA, 360

PRECIPITAZIONE CHIMICA, 3596
PRECISIONE, 34
PREDAZIONE, 3597
PREGLACIALE, 3601
PREIMBRIANO, 3592
PREISTORIA, 3602
PRENEANDERTHALIANO, 3604
PREPARAZIONE, 3605
PREPARAZIONE DEI MINERALI, 1252
PRESSIONE, 3607
PRESSIONE BAROMETRICA, 361
PRESSIONE CAPILLARE, 635
PRESSIONE DEL TERRENO, 1277
PRESSIONE DELL'ACQUA, 4795
PRESSIONE DI FLUIDO, 1583
PRESSIONE DI SOLLEVAMENTO, 4666
PRESSIONE GEOSTATICA, 1909
PRESSIONE IDRAULICA, 2096
PRESSIONE IDROSTATICA, 2129
PRESSIONE NEI PORI, 3562
PRESSIONE PARZIALE, 3306
PRESSOMETRIA, 2556
PREVISIONE, 3598
PREZZO, 3610
PRIMATES, 3613
PRIMITIVO, 3614
PRINCIPIO COSMOLOGICO, 974
PRIORITA, 3615
PROBABILITA, 3617
PROBLEMA INVERSO, 2296
PROBLEMATICA, 3619
PROBOSCIDEA, 3620
PROCELLARIANO, 3621
PROCESSO, 3622
PROCESSO ATETTONICO, 215
PROCESSO CICLICO, 1047
PROCESSO SINTETTONICO, 4399
PROCESSO STOCASTICO, 4256
PROCHORDATA, 3623
PRODOTTO DI DECADIMENTO, 1074
PRODOTTO FINALE, 1367
PRODOTTO PIROCLASTICO, 3695
PRODUZIONE, 3624
PROFILO D'EQUILIBRIO, 3628
PROFILO DEL SUOLO, 4155
PROFILO GEOCHIMICO, 1720
PROFILO GEOFISICO, 1756
PROFILO GEOLOGICO, 1736
PROFILO LONGITUDINALE, 2566
PROFILO STRADALE, 3876
PROFONDITA, 1122
PROFONDITA DI COMPENSAZIONE, 1124
PROGANOSAURIA, 3630
PROGETTO, 3636
PROGETTO MOHOLE, 2909
PROGNOSI, 3631
PROGRADAZIONE DI RIVA, 3632

PROGRAMMA, 3634
PROGRAMMA APOLLO, 243
PROGRAMMA DI CALCOLO, 872
PROGRAMMA LUNA, 2598
PROGRAMMA VIKING, 4741
PROIEZIONE GNOMONICA, 1832
PROIEZIONE VULCANICA, 1315
PROMEZIO, 3637
PROPAGAZIONE D'ONDA, 4807
PROPILITIZZAZIONE, 3639
PROPRIETA, 3638
PROPRIETA ACUSTICA, 42
PROPRIETA CHIMICA, 729
PROPRIETA DIAMAGNETICA, 1159
PROPRIETA ELASTICA, 1320
PROPRIETA ELETTRICA, 1331
PROPRIETA ELETTROCHIMICA, 1336
PROPRIETA FERROMAGNETICA, 1536
PROPRIETA FISICA, 3427
PROPRIETA FISICO-CHIMICA, 3430
PROPRIETA GEOTECNICA, 1769
PROPRIETA MAGNETICA, 2640
PROPRIETA MECCANICA, 2727
PROPRIETA METALLICA, 2768
PROPRIETA OTTICA, 3140
PROPRIETA PARAMAGNETICA, 3298
PROPRIETA PIEZOELETTRICA, 3442
PROPRIETA TERMICA, 4499
PROPRIETA TERMOCHIMICA, 4502
PROPRIETA TERMODINAMICA, 4504
PROSCIUGAMENTO, 4797
PROSIMII, 3640
PROSPETTIVA, 3366
PROSPEZIONE, 1483
PROSPEZIONE AEREA, 72
PROSPEZIONE BIOGEOCHIMICA, 495
PROSPEZIONE GEOBOTANICA, 1714
PROSPEZIONE GEOCHIMICA, 1716
PROSPEZIONE GEOELETTRICA, 1327
PROSPEZIONE GEOFISICA, 1752
PROSPEZIONE GRAVIMETRICA, 1879
PROSPEZIONE IDROGEOCHIMICA, 2109
PROSPEZIONE IN ALLUVIONI, 145
PROSPEZIONE MINERARIA, 2868
PROSPEZIONE RADIOMETRICA, 3738
PROSPEZIONE TERMICA, 4500
PROTEINA, 3645
PROTELYTROPTERA, 3646
PROTEROZOICO, 3647
PROTEROZOICO INF., 2592
PROTEROZOICO MED., 2851
PROTEROZOICO SUP., 4682
PROTEROZOIDI, 3648
PROTEROZOIDI INF., 1273
PROTEROZOIDI SUP., 2439
PROTEZIONE AMBIENTALE, 905
PROTEZIONE COSTIERA, 4082

PROTOATLANTICO, 2154
PROTOATTINIO, 3642
PROTOCILIATA, 3649
PROTOLITO, 3650
PROTOMEDUSAE, 3651
PROTONE, 3652
PROTOTIPO, 3653
PROTOZOA, 3654
PROTURA, 3655
PROVA ALLA PIASTRA, 3500
PROVA DI COMPRESSIONE, 868
PROVA DI LABORATORIO, 2403
PROVA DI POMPAGGIO, 3685
PROVA DI PORTATA, 261
PROVA EDOMETRICA, 908
PROVA IN SITO, 2197
PROVENIENZA, 3656
PROVINCIA FAUNISTICA, 1518
PROVINCIA FISIOGRAFICA, 3431
PROVINCIA FLORISTICA, 1572
PROVINCIA IDROGEOLOGICA, 2112
PROVINCIA METALLOGENICA, 2773
PROVINCIA PALEOBOTANICA, 3244
PROVINCIA PETROGRAFICA, 3377
PROVINCIALITA, 3657
PSAMMENT, 3658
PSAMMITE, 3659
PSAMMITICO, 3660
PSAMMOFAUNA, 3661
PSEFITICO, 3662
PSEUDOCONGLOMERATO, 3663
PSEUDOGLEY, 3664
PSEUDOMORFOSI, 3665
PSEUDOTILLITE, 3666
PSILOCERATIDA, 3667
PSILOPSIDA, 3668
PSYCHROPHYTE, 3669
PTERIDOFITE, 3671
PTERIDOPHYLLA, 3670
PTERIDOSPERMAE, 3672
PTERIOIDA, 3673
PTEROBRANCHIA, 3674
PTEROMORPHIA, 3675
PTEROPODA, 3676
PTEROSAURIA, 3677
PTYCHOPARIIDA, 3678
PUBBLICAZIONE, 3681
PULMONATA, 3682
PUNTO CALDO, 2065
PUNTO DI CURIE, 1033
PUNTO DI SCOPPIO, 4085
PURINA, 3688
PYCNOGONIDA, 3690
PYROTHERIA, 3701
PYRROPHYTA, 3705
QUALITA, 3707
QUALITA DEL MINERALE, 3160

QUALITA DELLACQUA, 4796
QUARZITE, 3712
QUARZO, 3710
QUATERNARIO, 3713
RACCOLTA, 838
RACCORCIAMENTO CROSTALE, 4084
RACEMIZZAZIONE, 3717
RADAR, 3718
RADIALE, 3719
RADIAZIONE, 3720
RADIAZIONE ADATTATIVA, 64
RADIAZIONE SOLARE, 4159
RADICE, 3896
RADIO, 3739
RADIOATTIVITA, 3731
RADIOCARBONIO, 3733
RADIOFREQUENZA, 3724
RADIOGRAFIA R-X, 3734
RADIOINTERFEROMETRIA, 3722
RADIOISOTOPO, 3727
RADIOLARIA, 3735
RADIOLARITE, 3736
RADIOMETRO, 3737
RADIONUCLIDE INIZIALE, 3303
RADON, 3741
RAFFINAZIONE, 3781
RAFFREDDAMENTO, 942
RAGGI ALFA, 147
RAGGI BETA, 473
RAGGI COSMICI, 970
RAGGI GAMMA, 1676
RAGGI INFRAROSSI, 2233
RAGGI ULTRAVIOLETTI, 4641
RAGGI X, 4832
RAGGIO, 3740
RAGGIO IDRAULICO, 2097
RAME, 946
RANGE ZONE, 3747
RANKER, 3749
RAPAKIVI, 3750
RAPPORTO, 3818
RAPPORTO CHIMICO, 730
RAPPORTO D'ATTIVITA, 3635
RAPPORTO GAS-PETROLIO, 1693
RAPPORTO ISOTOPICO, 2352
RAPPORTO SABBIA=ARGILLA, 3938
RAPPRESENTAZIONE GRAFICA, 1864
RATITES, 3754
RE/OS, 3850
REAGENTE, 3761
REAZIONE, 3757
REAZIONE ENDOTERMICA, 1372
REAZIONE ESOTERMICA, 1473
REAZIONE FOTOCHIMICA, 3408
RECENSIONE, 541
RECEPTACULITIDA, 3763
RECUPERO, 3768

RECUPERO AMBIENTALE, 3766
RED BEDS, 3773
REDDITIVITA, 3629
REDLICHIIDA, 3775
REFRATTARIO, 3791
REGIME CRITICO, 994
REGIME LAMINARE, 2420
REGIME STAZIONARIO, 4246
REGIME TRANSITORIO, 4664
REGIME TURBOLENTO, 1304
REGIONE ASISMICA, 329
REGMAGLIPTO, 3797
REGOLA DELLE FASI, 3394
REGOLA DELLE FASI MINERALOGICHE, 2879
REGOLITE, 3798
REGOSUOLO, 3799
REGRESSIONE, 3800
REGRESSIONE STATISTICA, 3801
RELAZIONE, 3803
RELITTO, 3806
RELITTO D'EROSIONE, 1428
RENDOLL, 3815
RENDZINA, 3816
RENIO, 3849
REOLOGIA, 3851
REOMORFISMO, 3852
REPERE SISMICO, 2389
REPERIMENTO DATI, 1070
REPLICA SISMICA, 79
REPTILIA, 3821
RESIDUALE, 3827
RESIDUO INSOLUBILE, 2256
RESIDUO ORGANICO, 3173
RESINA, 3831
RESISTENZA A ROTTURA, 1633
RESISTENZA A TRAZIONE, 4455
RESISTENZA AL TAGLIO, 4062
RESISTENZA ALLA COMPRESSIONE, 866
RESISTENZA MECCANICA, 4289
RESISTIVITA ELETTRICA, 3832
RETE IDROGRAFICA DENDRITICA, 1109
RETE SISMICA, 306
RETICO, 3848
RETICOLATO IDROGRAFICO, 1246
RETICOLO DI BRAVAIS, 574
RETICOLO DI FRATTURE, 1512
RETICULAREA, 3839
RETROMETAMORFISMO, 3841
RETROSCOGLIERA, 397
REVISIONE, 3844
RHABDOPHLEURIDA, 3847
RHIPIDISTII, 3853
RHIZOPODEA, 3854
RHODOPHYTA, 3856
RHYNCHOCEPHALIA, 3858
RHYNCHONELLIDA, 3859
RIATTIVAZIONE, 3760

RICARICA DI FALDA, 3765
RICERCA SCIENTIFICA, 3822
RICEZIONE, 3764
RICICLAGGIO, 3772
RICOSTRUZIONE PALEOGEOGRAFICA, 3767
RICOSTRUZIONE PALEONTOLOGICA, 3264
RICRISTALLIZZAZIONE, 3770
RIDEPOSIZIONE, 3774
RIDUZIONE CHIMICA, 731
RIELABORAZIONE ESOGENA, 2143
RIEMPIMENTO CARSICO, 2374
RIEMPIMENTO FLUVIALE, 142
RIFIUTI, 4781
RIFIUTI DOMESTICI, 2070
RIFIUTI INDUSTRIALI, 2221
RIFIUTI LIQUIDI, 2521
RIFIUTI RADIOATTIVI, 3730
RIFIUTI SOLIDI, 4168
RIFIUTO INERTE, 4217
RIFLESSIONE D'ONDA, 3783
RIFLETTIVITA, 3785
RIFORNIMENTO D'ACQUA, 4800
RIFRATTOMETRIA, 3790
RIFRATTOMETRO, 3789
RIFRAZIONE, 3786
RIFT, 3864
RIGENERAZIONE, 3792
RIGETTO, 1224
RIGIDITA, 3866
RIGONFIAMENTO, 4378
RILASSAMENTO, 3805
RILEVAMENTO, 2491
RILEVAMENTO ACUSTICO, 44
RILEVAMENTO AEROMAGNETICO, 74
RILEVAMENTO AEROPORTATO, 95
RILEVAMENTO CARTOGRAFICO, 2670
RILEVAMENTO ELETTRICO, 1333
RILEVAMENTO ELETTROMAGNETICO, 1342
RILEVAMENTO GEOFISICO, 1757
RILEVAMENTO GEOLOGICO, 4371
RILEVAMENTO GEOTECNICO, 1770
RILEVAMENTO GRAVIMETRICO, 1886
RILEVAMENTO IDROGEOLOGICO, 2113
RILEVAMENTO MAGNETICO, 2643
RILEVAMENTO MAGNETOTELLURICO, 2655
RILEVAMENTO SISMICO, 4019
RILIEVO, 3808
RILIEVO LUNARE, 2937
RILIEVO MORFOSTRUTTURALE, 2943
RILIEVO SOTTOMARINO, 3975
RIMAGNETIZZAZIONE, 3811
RIMANEGGIAMENTO DI FOSSILE, 3846
RIMANEGGIAMENTO DI MINERALE, 3163
RIMANEGGIATO, 3845
RIMPIAZZO, 3817
RINGIOVANIMENTO, 3802
RIODACITE, 3860

RIOLITE, 3861
RIPARTIZIONE DI ELEMENTI, 3312
RIPARTIZIONE DI GRANA, 4125
RIPPLE MARK, 3872
RIPRODUZIONE, 3820
RISCHIO GEOLOGICO, 1738
RISCHIO SISMICO, 4015
RISERVA, 3824
RISERVE COMBUSTIBILI, 1658
RISONANZA MAGNETICO-NUCLEARE, 3064
RISORGIVA, 3836
RISORSE, 3834
RISORSE IDRICHE, 4798
RISORSE MINERARIE, 2872
RISORSE NATURALI, 2992
RITENZIONE, 3838
RITIRO, 4087
RIUNIONE, 2737
RIVOLUZIONE CASCADIANA, 662
RIVOLUZIONE MAZATSALIANA, 2723
ROBERTINACEA, 3880
ROCCE VERDI, 1894
ROCCIA, 3881
ROCCIA A FELDSPATOIDI, 1526
ROCCIA CARBONATICA, 650
ROCCIA CARBONIOSA, 649
ROCCIA CLASTICA, 786
ROCCIA CRISTALLINA, 1022
ROCCIA DI CUMULO, 1032
ROCCIA FERRIFERA, 2312
ROCCIA FOSFATICA, 3402
ROCCIA IGNEA, 2176
ROCCIA IGNEA STRATOIDE, 2463
ROCCIA INCASSANTE, 978
ROCCIA INCOMPETENTE, 2203
ROCCIA MADRE, 3304
ROCCIA MADRE IDROCARBURI, 4186
ROCCIA MADRE SUOLO, 3302
ROCCIA METAMORFICA, 2785
ROCCIA METASEDIMENTARIA, 2792
ROCCIA METASOMATICA, 2795
ROCCIA OSPITE, 2062
ROCCIA PLUTONICA, 3519
ROCCIA SEDIMENTARIA, 4000
ROCCIA SERBATOIO, 3826
ROCCIA STANDARD, 4237
ROCCIA STERILE, 415
ROCCIA TOTALE, 4824
ROCCIA VULCANICA, 4766
ROCK GLACIER, 3886
RODENTIA, 3892
RODIO, 3855
ROSTROCONCHIA, 3897
ROTALIACEA, 3898
ROTALIINA, 3899
ROTAZIONE, 3902
ROTAZIONE DI PLACCA, 3498

ROTAZIONE OTTICA, 3141
ROTAZIONE TERRESTRE, 1278
ROTLIEGENDES, 3904
ROTTURA, 3917
RPE, 1351
RUBEFAZIONE, 3908
RUBIDIO, 3909
RUDISTAE, 3911
RUDITE, 3912
RUGA DI SPIAGGIA, 446
RUGIADA, 1152
RUGOSA, 4474
RUGOSITA, 3905
RUMINANTIA, 3913
RUMORE DI FONDO, 3041
RUN, 3914
RUOLO DEL GEOLOGO, 1740
RUPELIANO, 3916
RUSCELLAMENTO, 3915
RUSCELLAMENTO PLUVIALE, 3744
RUSCELLO, 985
RUSCELLO TEMPORANEO, 1397
RUTENIO, 3919
S-34/S-32, 4355
SABBIA, 3936
SABBIA BITUMINOSA, 3108
SABBIA DA VETRO, 1812
SABBIA MOBILE, 3716
SABBIONE, 1923
SAGGIO, 336
SAKMARIANO, 3921
SALBANDA, 4780
SALGEMMA, 3924
SALIFERO, 3922
SALINITA, 3923
SALTAZIONE, 3931
SAMARIO, 3932
SANTONIANO, 3943
SAPROPEL, 3944
SARCODINA, 3946
SARCOPTERYGII, 3947
SARMATIANO, 3948
SASSONIANO, 3954
SATELLITE, 3484
SATELLITE ARTIFICIALE, 322
SATURAZIONE, 3950
SAURISCHIA, 3951
SAURORNITHES, 3952
SAUSSURITIZZAZIONE, 3953
SBANCAMENTO, 680
SBARRAMENTO, 3986
SBARRAMENTO IDROGEOLOGICO, 1913
SCAGLIA, 3963
SCALA, 2669
SCALA ETA ASSOLUTE, 4536
SCALA MERCALLI MODIFICATA, 2905
SCALA MICROSCOPICA, 2832

SCALA PHI, 3397
SCALA RICHTER, 3862
SCALA STRATIGRAFICA, 1734
SCAMBIO IONICO, 2300
SCANALATURA, 1902
SCANDIO, 3956
SCAPHOPODA, 3958
SCAPOLITIZZAZIONE, 3959
SCARPATA CONTINENTALE, 927
SCARPATA D'EROSIONE, 1433
SCARPATA DI FAGLIA, 1511
SCARPATA LITORALE, 447
SCARTO, 4413
SCAVO, 1462
SCELTA DEL SITO, 4124
SCHEDA PERFORATA, 3686
SCHLIEREN, 3962
SCIAME DI FILONI, 1185
SCIAME DI TERREMOTI, 1287
SCIENZA DEI SUOLI, 4157
SCIENZE DELLA TERRA, 1279
SCINTILLAZIONE, 3964
SCISSOMETRIA, 4712
SCISSOMETRO, 4711
SCISTI BLU, 534
SCISTO, 3960
SCISTO A CLORITE, 751
SCISTO A OLIO, 3110
SCISTO BITUMINOSO, 524
SCISTO MACULATO, 4222
SCISTO SERICITICO, 4042
SCISTO VERDE, 1892
SCISTOSITA, 3961
SCISTOSITA FLUIDALE, 1576
SCIUROMORPHA, 3965
SCIVOLAMENTO DA SCOSSA, 4254
SCIVOLAMENTO DI TERRA, 1283
SCIVOLAMENTO DI TERRENO, 2431
SCIVOLAMENTO GRAVITATIVO, 1885
SCLERITE, 3966
SCOGLIERA, 3776
SCOGLIERA A PINNACOLO, 3451
SCOGLIERA CORALLINA, 949
SCOGLIERA D'ALGHE, 117
SCOGLIERA FRANGENTE, 1651
SCOLECODONTI, 3967
SCOLITE, 3968
SCOLLAMENTO, 1082
SCOPERTA, 1217
SCORIA, 769
SCORIE, 3969
SCORRIMENTO SUPERFICIALE, 4364
SCOSSA VIOLENTA, 4298
SCUDO, 4072
SCYPHOMEDUSAE, 3972
SCYPHOZOA, 3973
SEBKHA, 3987

SECONDARIO, 2760
SECREZIONE LATERALE, 2445
SEDIMENTAZIONE, 4003
SEDIMENTAZIONE ATTUALE, 3762
SEDIMENTAZIONE BIOCHIMICA, 482
SEDIMENTAZIONE BIOCLASTICA, 488
SEDIMENTAZIONE CHIMICA, 723
SEDIMENTAZIONE CONTINENTALE, 928
SEDIMENTAZIONE D'ESTUARIO, 1437
SEDIMENTAZIONE DELTIZIA, 1105
SEDIMENTAZIONE DETRITICA, 1144
SEDIMENTAZIONE DI LAGUNA, 2413
SEDIMENTAZIONE DI MARE PROFONDO, 23
SEDIMENTAZIONE DI SCOGLIERA, 3778
SEDIMENTAZIONE EOLICA, 1393
SEDIMENTAZIONE FLUVIALE, 1595
SEDIMENTAZIONE FLUVIOGLACIALE, 1597
SEDIMENTAZIONE GEOSINCLINALE, 1765
SEDIMENTAZIONE GLACIALE, 1798
SEDIMENTAZIONE GLACIOLACUSTRE, 1806
SEDIMENTAZIONE GLACIOMARINA, 1808
SEDIMENTAZIONE IN ACQUA DOLCE, 1649
SEDIMENTAZIONE INTERTIDALE, 2287
SEDIMENTAZIONE LACUSTRE, 2409
SEDIMENTAZIONE LITORALE, 2546
SEDIMENTAZIONE MARGINE CONTINENTALE, 4070
SEDIMENTAZIONE MARINA, 2686
SEDIMENTAZIONE PALUSTRE, 4377
SEDIMENTAZIONE PELAGICA, 3331
SEDIMENTO, 3991
SEDIMENTO DI FONDO, 554
SEDIMENTO LACUSTRE, 2408
SEDIMENTO MARINO, 2685
SEDIMENTO ORGANICO, 3174
SEDIMENTOLOGIA, 4006
SEGNALE TELESISMICO, 4446
SEGREGAZIONE, 4010
SEISLOG, 4012
SELCE, 1567
SELCIFICAZIONE, 740
SELENIATO, 4029
SELENIO, 4030
SELENOCRONOLOGIA, 4031
SELENOLOGIA, 4032
SEME, 4007
SEMISPAZIO, 1943
SENONIANO, 4037
SEPARAZIONE, 4039
SEPARAZIONE DEL LIQUIDO, 2518
SEPPELLIMENTO, 596
SEPTARIA, 4040
SERICITIZZAZIONE, 4043
SERIE, 4044
SERIE ARTICA, 287
SERIE ATLANTICA, 355
SERIE DI REAZIONE, 3759
SERIE PACIFICA, 3231

SERIE REAZIONALE DI BOWEN, 563
SERPENTINITE, 4046
SERPENTINIZZAZIONE, 4047
SERPENTINO, 4045
SERVIZIO GEOLOGICO, 1739
SESSA, 4011
SETACCIO, 4096
SETTO, 4041
SEZIONE DI POZZO, 3460
SEZIONE DI SONDAGGIO, 550
SEZIONE LUCIDA, 3538
SEZIONE SOTTILE, 4514
SEZIONE TIPO, 4625
SFALDATURA, 2864
SFALDATURA BASALE, 423
SFENOLITE, 4205
SFERICITA, 4207
SFERULA, 4208
SFERULA MAGNETICA, 2641
SFERULITE, 4209
SFIATATOIO, 533
SFORZO, 4290
SFORZO DI TAGLIO, 4063
SFREGAMENTO, 368
SFRUTTABILITA, 1481
SFRUTTAMENTO, 1482
SFRUTTAMENTO IN SUPERFICIE, 4294
SFRUTTAMENTO PER DISSOLUZIONE, 4179
SHONKINITE, 4080
SIAL, 4089
SICCITA, 1258
SICILIANO, 4090
SIDEROFIRO, 4093
SIDEROLITE, 4263
SIDERURGIA, 2309
SIEGENIANO, 4094
SIENITE, 4379
SIENITE A FELDSPATOIDI, 127
SIEROZEM, 4095
SIFONE, 4120
SIGILLATURA, 3982
SILEXITE, 739
SILICATO, 4098
SILICE, 4097
SILICIO, 4103
SILICIZZAZIONE, 4100
SILICOFLAGELLATA, 4102
SILL, 4104
SILLIMANITE, 4105
SILT, 4106
SILTITE, 4109
SILURIANO, 4110
SILURIANO INF., 2593
SILURIANO SUP., 4683
SILVINITE, 4380
SIMA, 4112
SIMBIOSI, 4381

SIMIANI, 4113
SIMMETRIA, 4384
SIMPOSIO, 4387
SIMULAZIONE, 4114
SIMULAZIONE ANALOGICA, 180
SINCLINALE, 4390
SINCLINORIUM, 4391
SINECLISI, 4392
SINEMURIANO, 4115
SINGENESI, 4394
SINGENETICO, 4395
SINONIMIA, 4396
SINSEDIMENTARIO, 4397
SINTESI, 4401
SINTESI BIBLIOGRAFICA, 3843
SINTESI MONOGRAFICA, 4357
SINTESSI, 4400
SINTETTONICO, 4399
SINUOSITA, 4119
SIPHONAPTERA, 4121
SIPUNCULOIDA, 4122
SIRENIA, 4123
SISMA A FUOCO INTERMEDIO, 2270
SISMA LUNARE, 2936
SISMA PROFONDO, 1088
SISMA SUPERFICIALE, 4052
SISMICA A PERCUSSIONE, 1951
SISMICA A RIFLESSIONE, 3784
SISMICA A RIFRAZIONE, 3787
SISMICA AVANZATA, 2190
SISMICITA, 4022
SISMOGRAFIA, 4024
SISMOGRAFO SOTTOMARINO, 3089
SISMOGRAMMA, 4023
SISMOLOGIA, 4026
SISMOLOGIA SPERIMENTALE, 1478
SISMOMETRO, 4027
SISMOTETTONICA, 4028
SISTEMA, 4403
SISTEMA BINARIO, 479
SISTEMA CHIMICO, 733
SISTEMA COSMICO, 971
SISTEMA CRISTALLINO, 1021
SISTEMA DI PIEGHE, 1606
SISTEMA DI VENE, 2514
SISTEMA ESAGONALE, 2002
SISTEMA GEOTERMICO, 1778
SISTEMA ISOMETRICO, 2332
SISTEMA MONOCLINO, 2921
SISTEMA MONOMETRICO, 1030
SISTEMA MULTISTRATO, 2962
SISTEMA ORTOROMBICO, 3194
SISTEMA PIRAMIDALE, 3693
SISTEMA QUATERNARIO, 3714
SISTEMA ROMBICO, 3857
SISTEMA SOLARE, 4160
SISTEMA TERNARIO, 4463

SISTEMA TRICLINO, 4593
SISTEMA TRIGONALE, 4595
SITO ARCHEOLOGICO, 280
SKARN, 4126
SLUMPING, 4139
SMAGNETIZZAZIONE, 1106
SMAGNETIZZAZIONE TERMICA, 4496
SMOTTAMENTO, 2959
SOCIETA, 858
SODIO, 4143
SOGLIA, 3863
SOGLIA DI DEFORMAZIONE, 4849
SOGLIA GEOCHIMICA, 4519
SOLCO D'EROSIONE, 1668
SOLCO DA TRASCINAMENTO, 1903
SOLCO SOTTOMARINO, 2687
SOLE, 4358
SOLEMYIDA, 4164
SOLFATARA, 4165
SOLFATO, 4349
SOLFOSALE, 4352
SOLFURO, 4351
SOLFURO DI IDROGENO, 2108
SOLIDUS, 4170
SOLIFLUSSO, 4171
SOLLEVAMENTO, 4665
SOLLEVAMENTO DA GELO, 1655
SOLLEVAMENTO DEL TERRENO, 1970
SOLUBILITA, 4174
SOLUTO, 4175
SOLUZIONE, 4176
SOLUZIONE ACQUOSA, 256
SOLUZIONE SOLIDA, 4167
SOMASTEROIDEA, 4180
SOMMERGIBILE, 4335
SONDA, 3618
SONDA A NEUTRONI, 3018
SONDAGGIO, 549
SONDAGGIO DI FREQUENZA, 1646
SONDAGGIO ELETTRICO, 1332
SONDAGGIO ESPLORATIVO, 1484
SONDAGGIO GEOFISICO, 4185
SONDAGGIO SISMICO PROFONDO, 1091
SOPRASATURO, 3216
SORGENTE, 4224
SORGENTE CARSICA, 2376
SORGENTE D'ESPLOSIONE, 1487
SORGENTE DI CALORE, 1968
SORGENTE INTERMITTENTE, 2271
SORGENTE MINERALE, 2873
SORGENTE SISMICA, 4016
SORGENTE SOTTOMARINA, 4331
SORGENTE TERMALE, 2066
SORPTION, 4183
SOSPENSIONE, 4373
SOSTANZA NON METALLICA, 3051
SOSTANZA TOSSICA, 4554

SOSTANZA UTILE, 2219
SOSTANZA VOLATILE, 4757
SOSTITUZIONE, 4338
SOTTOSATURAZIONE, 4655
SOTTOSATURO, 4654
SOVRACCRESCITA, 3215
SOVRAPPOSIZIONE, 4360
SOVRAPPOSIZIONE DI ETA, 1723
SOVRASATURAZIONE, 3217
SOVRASCORRIMENTO, 3218
SOVRASTRUTTURA, 4439
SPALLAZIONE, 4189
SPARAGMITE, 4190
SPARKER, 4191
SPARTIACQUE, 4788
SPAZIO INTERSTELLARE, 2285
SPAZIO SOTTERRANEO, 4651
SPECCHIO DI FAGLIA, 4132
SPEDIZIONE=CROCIERA, 1476
SPELEOLOGIA, 4203
SPELEOTEMA, 4204
SPERANZA MATEMATICA, 2709
SPESSORE, 4513
SPETTRO, 4202
SPETTRO D'ASSORBIMENTO, 15
SPETTRO MOESSBAUER, 2948
SPETTRO OTTICO, 3143
SPETTRO RPE, 1410
SPETTROFOTOMETRIA, 4200
SPETTROMETRIA, 4201
SPETTROMETRIA ALFA, 148
SPETTROMETRIA ALL'INFRAROSSO, 2234
SPETTROMETRIA D'ASSORBIMENTO, 14
SPETTROMETRIA D'EMISSIONE, 1362
SPETTROMETRIA DI FLUORESCENZA, 1587
SPETTROMETRIA DI MASSA, 2700
SPETTROMETRIA GAMMA, 1677
SPETTROMETRIA MICROONDE, 2839
SPETTROMETRIA MOESSBAUER, 2947
SPETTROMETRIA MOLECOLARE, 2914
SPETTROMETRIA OTTICA, 3142
SPETTROMETRIA RADIOFREQUENZE, 3723
SPETTROMETRIA RAMAN, 3745
SPETTROMETRIA RX, 4836
SPETTROMETRO, 4199
SPIAGGIA, 444
SPILITIZZAZIONE, 4210
SPINIFEX, 4212
SPIRIFERIDA, 4213
SPIRILLINACEA, 4214
SPODOSOL, 4216
SPONGOLITE, 4218
SPORA, 4219
SPOROMORFO, 4220
SPOROZOA, 4221
SPROFONDAMENTO, 834
SPROFONDAMENTO VULCANICO, 675

SQUAMATA, 4225
SR/RB, 3910
STABILITA, 4226
STABILITA DI VERSANTE, 4137
STABILIZZAZIONE, 4227
STABILIZZAZIONE DEL SUOLO, 4158
STADIO GLACIALE, 1799
STAGNO, 4537
STALAGMITE, 4232
STALATTITE, 4231
STAMPIANO, 4233
STANDARD CHIMICO, 4235
STATISTICA, 4245
STATO FISICO, 3428
STAZIONE LUNARE, 2604
STAZIONE SISMICA, 4018
STEATITIZZAZIONE, 4247
STECHIOMETRIA, 4259
STEFANIANO, 4250
STEREOCHIMICA, 4252
STEROLO, 4253
STILE TETTONICO, 4438
STILLICIDIO, 4008
STILLICIDIO DI OLIO, 3109
STILOLITE, 4314
STIMA, 1457
STOCCAGGIO, 4265
STOCCAGGIO DEL GAS, 1691
STOCCAGGIO DELL'ACQUA, 4799
STOCCAGGIO SOTTERRANEO, 4652
STOCK, 4257
STOCKWORK, 4258
STOLONOIDEA, 4260
STONELINE, 4262
STORIA, 2020
STORIA DELLA GEOLOGIA, 2021
STRATEGIA, 4271
STRATI CONVOLUTI, 941
STRATIFICAZIONE, 4272
STRATIFICAZIONE FLASER, 1565
STRATIFICAZIONE INCROCIATA, 998
STRATIFICAZIONE MASSICCIA, 2704
STRATIFICAZIONE OBLIQUA, 997
STRATIFICAZIONE PARALLELA, 3295
STRATIFICAZIONE PIANA, 3466
STRATIFICAZIONE SOTTILE, 453
STRATIGRAFIA, 4281
STRATIGRAFIA MAGNETICA, 2652
STRATO A, 1
STRATO ATTIVO, 57
STRATO B, 395
STRATO C, 602
STRATO D, 1057
STRATO DI LETTO, 555
STRATO DI TETTO, 4548
STRATO E, 1271
STRATO F, 1499

STRATO FRONTALE, 1614
STRATO G, 1671
STRATO IMPERMEABILE, 259
STRATO LIMITE, 561
STRATOTIPO, 4282
STRATOVULCANO, 4283
STREAM-SEDIMENT, 4287
STRETTO, 4269
STRIA DI ACCRESCIMENTO, 1922
STRIAMENTO, 3971
STRIATURA, 1514
STRIATURA DI CRISTALLO, 2874
STRIATURA GLACIALE, 1800
STROMATOLITE, 4295
STROMATOPOROIDEA, 4296
STRONZIO, 4299
STROPHOMENIDA, 4300
STRUMENTAZIONE, 2258
STRUMENTO GEOFISICO, 1753
STRUTTURA, 1500
STRUTTURA A BANDE, 409
STRUTTURA A CUSCINI, 3448
STRUTTURA A DOMINII, 1239
STRUTTURA A FIAMME, 1563
STRUTTURA A MAGLIE, 2452
STRUTTURA A MULLION, 2961
STRUTTURA A SLUMPING, 4138
STRUTTURA AD ARCO, 270
STRUTTURA ALGALE, 118
STRUTTURA ANULARE, 3871
STRUTTURA BIOGENICA, 494
STRUTTURA CENTROCLINALE, 690
STRUTTURA CILINDRICA, 1053
STRUTTURA CRIPTOVULCANICA, 1010
STRUTTURA CRISTALLINA, 1020
STRUTTURA D' IMPATTO, 2192
STRUTTURA DA GELO, 1654
STRUTTURA DELLA ROCCIA, 3373
STRUTTURA DI COLLASSO, 837
STRUTTURA DI VELOCITA, 4727
STRUTTURA DIAGENETICA, 3990
STRUTTURA INTERNA, 2275
STRUTTURA LAMINATA, 2421
STRUTTURA MILONITICA, 2973
STRUTTURA ORGANICA, 3175
STRUTTURA PIANA, 3467
STRUTTURA PRIMARIA, 3612
STRUTTURA PRISMATICA, 3616
STRUTTURA RETICOLARE, 3840
STRUTTURA SEDIMENTARIA, 4001
STRUTTURA SINSEDIMENTARIA, 4398
STRUTTURA STRATIFICATA, 4273
STRUTTURA TETTONICA, 1737
STRUTTURA VERMIFORME, 4733
STRUTTURA VULCANOTETTONICA, 4771
STRUTTURA ZONATA, 4865
STRUTTURE DA CORRENTE DI TORBIDA, 4618

STUDIO AMBIENTALE, 1387
STUDIO CRITICO, 995
STUDIO DEL CASO, 663
STUDIO DI CAMPAGNA, 1542
STUDIO DI FATTIBILITA, 1520
STUDIO IN MARE, 2682
STUDIO PRELIMINARE, 3178
STYLOPHORA, 4315
SUBBOREALE, 4319
SUBDUZIONE, 4321
SUBLIMATO, 4323
SUBLIMAZIONE, 4324
SUBSIDENZA, 4337
SUBSTRATO, 4339
SUCTORIA, 4345
SUD, 4187
SUDDIVISIBILITA, 3310
SUDDIVISIONE, 4320
SUIFORMES, 4348
SUITE ARTICA, 287
SUOLO, 4146
SUOLO A MULL, 2960
SUOLO AZONALE, 394
SUOLO BRUNO, 583
SUOLO CASTANO, 666
SUOLO COERENTE, 830
SUOLO DI TUNDRA, 4611
SUOLO FERRUGINOSO, 1540
SUOLO FIGURATO, 3317
SUOLO GELATO, 1656
SUOLO IMMATURO, 2183
SUOLO INCOERENTE, 828
SUOLO LATERITICO, 2447
SUOLO LIMOSO, 2509
SUOLO LUNARE, 2603
SUOLO MEDITERRANEO, 2735
SUOLO POLIGONALE, 3547
SUOLO SCHELETRICO, 2535
SUOLO SOLONCHAK, 4172
SUOLO SOLONETZ, 4173
SUOLO UMICO, 2072
SUPERFICIE D'ABRASIONE, 8
SUPERFICIE D'EROSIONE, 1429
SUPERFICIE PIEZOMETRICA, 3583
SUPERFICIE SPECIFICA, 4198
SUPERSTRUTTURA, 4361
SUSCETTIVITA MAGNETICA, 2645
SUTURA, 4374
SUZIONE, 634
SVILUPPO, 1147
SVILUPPO DI POZZO, 1148
SYMMETRODONTA, 4383
SYMPHYLA, 4385
SYNAPSIDA, 4388
TABELLA, 1072
TABULATA, 4406
TACONITE, 4408

TACTITE, 4409
TAENIODONTA, 4410
TAENIOLABOIDEA, 4411
TAFONOMIA, 4420
TAFROGENESI, 4421
TAFROGEOSINCLINALE, 4422
TAIGA, 4412
TALASSOCRATONE, 4478
TALCO SOSTANZA, 4414
TALCOSCISTO, 4415
TALLIO, 4480
TALLOFITE, 4481
TALUS DETRITO DI FALDA, 4417
TALUS MARINO, 930
TALUS MARINO INTERNO, 2245
TAMPONATURA, 3514
TANATOCENOSI, 4483
TANTALATO, 4418
TANTALIO, 4419
TAPPO, 3513
TARANNONIANO, 4423
TARDIGRADA, 4424
TASMANITES, 4425
TASSO, 3753
TASSO DI SEDIMENTAZIONE, 4004
TASSONOMIA, 4428
TATARIANO, 4426
TAVOLINO UNIVERSALE, 4662
TAXITE, 4427
TAXON AHERMATIPICO, 92
TAXON ANAEROBICO, 179
TAXON COLONIALE, 844
TAXON ESTINTO, 1495
TAXON ETEROMORFO, 1995
TAXON HERMATIPICO, 1987
TAXON MONOLETO, 2925
TAXON NOMINALE, 3043
TAXON NUOVO, 3025
TAXON PLANCTONICO, 3488
TECNEZIO, 4429
TECNICA, 4431
TECNICA ANALOGICA, 181
TECNICA DIGITALE, 1183
TECNOLOGIA, 4432
TECNOLOGIA IDROGEOLOGICA, 2114
TECTITE, 4444
TEFROCRONOLOGIA, 4458
TEGUMENTO, 1126
TELEOSSERVAZIONE, 3813
TELEOSSERVAZIONE MULTISPETTRALE, 2966
TELEOSTEI, 4445
TELLURATO, 4448
TELLURIO, 4450
TEMPERATURA, 4453
TEMPESTA, 4266
TEMPESTA MAGNETICA, 2642
TEMPO D'ARRIVO, 307

TEMPO DI PERCORRENZA, 4581
TEMPO LIBERO, 3769
TENACITA, 4454
TENDENZA STATISTICA, 4244
TENORE, 3156
TENORE DI FONDO, 398
TENORE IN METALLO, 337
TENORE LIMITE, 1038
TENSIONE, 4456
TENSIONE SUPERFICIALE, 4365
TENTACULITES, 4457
TEORIA, 4490
TEORIA ASCENSIONALE, 326
TEORIA DELLA DINAMO, 1268
TEORIA DELLE OSCILLAZIONI, 3198
TEORIA DERIVA CONTINENTALE, 1225
TEORIA DISCENDENTE, 1130
TERALITE, 4491
TERATOLOGIA, 4459
TERBIO, 4460
TEREBRATULIDA, 4461
TERMALE, 4493
TERMINAZIONE LATERALE, 3449
TERMINOLOGIA, 4462
TERMOCARSISMO, 4508
TERMOCLINA, 4503
TERMODINAMICA, 4505
TERMODINAMOMETAMORFISMO, 1269
TERMOGRAFIA, 4506
TERMOGRAVIMETRIA, 4507
TERMOLUMINESCENZA, 4509
TERRA A DIATOMEE, 1168
TERRA EMERSA, 2425
TERRA ROSSA, 4464
TERRAZZAMENTO ARTIFICIALE, 1281
TERRAZZO, 3875
TERRAZZO ALLUVIONALE, 4465
TERRAZZO MARINO, 2688
TERRE RARE, 3751
TERRE UMIDE, 4822
TERREMOTO, 1284
TERRENO, 1904
TERRENO STERILE, 3213
TERZIARIO, 4470
TESI, 4512
TESSITURA, 4477
TESSITURA A CHIAZZE, 2949
TESSITURA A MOSAICO, 2945
TESSITURA AFANITICA, 236
TESSITURA APLITICA, 240
TESSITURA ARENACEA, 290
TESSITURA BLASTICA, 525
TESSITURA CAVERNOSA, 679
TESSITURA COLLOFORME, 841
TESSITURA EQUIGRANULARE, 1415
TESSITURA ETEROBLASTICA, 1988
TESSITURA EUCRISTALLINA, 1444

TESSITURA EUTASSITICA, 1453
TESSITURA FLUIDALE, 1584
TESSITURA FRAMBOIDALE, 1638
TESSITURA GNEISSICA, 1830
TESSITURA GRAFICA, 1865
TESSITURA GRANOBLASTICA, 1852
TESSITURA GRANOFIRICA, 1855
TESSITURA GRANULARE, 1856
TESSITURA IALOCRISTALLINA, 2084
TESSITURA IDIOMORFA, 2170
TESSITURA INTERSERTALE, 2284
TESSITURA IPIDIOBLASTICA, 2144
TESSITURA IPIDIOMORFA, 2145
TESSITURA IPOCRISTALLINA, 2147
TESSITURA LAMPROFIRICA, 2424
TESSITURA LEPIDOBLASTICA, 2482
TESSITURA LINOFIRICA, 2515
TESSITURA LISTATA, 1561
TESSITURA MARGINALE, 2675
TESSITURA MIAROLITICA, 2803
TESSITURA MICROGRANITICA, 2819
TESSITURA MICROGRANULARE, 2817
TESSITURA MICROLITICA, 2822
TESSITURA MONZONITICA, 2934
TESSITURA OCCHIADINA, 370
TESSITURA OFITICA, 3134
TESSITURA OLOCRISTALLINA, 2028
TESSITURA OOLITICA, 3126
TESSITURA ORBICOLARE, 3145
TESSITURA ORTOFIRICA, 3193
TESSITURA PANIDIOMORFICA, 3282
TESSITURA PARALLELA, 3296
TESSITURA PECILITICA, 3529
TESSITURA PECILOBLASTICA, 3530
TESSITURA PELITICA, 3335
TESSITURA PERLITICA, 3361
TESSITURA PERTITICA, 3368
TESSITURA PISOLITICA, 3458
TESSITURA PORFIRICA, 3567
TESSITURA PORFIROBLASTICA, 3568
TESSITURA RELITTA, 3807
TESSITURA RETICOLARE, 564
TESSITURA SIMPLECTITICA, 4386
TESSITURA VETROSA, 1814
TESSITURA VITROCLASTICA, 4752
TESSITURA VITROFIRICA, 4753
TESTA DI POZZO, 4815
TESTUDINATA, 4472
TETIDE, 4473
TETRAPODA, 4475
TETTO, 1953
TETTOGENESI, 4433
TETTONICA, 4440
TETTONICA A PIEGHE, 1607
TETTONICA A PLACCHE, 3499
TETTONICA ALPINA, 151
TETTONICA ATTIVA, 59

TETTONICA COMPRESSIVA, 867
TETTONICA DEL BASAMENTO, 430
TETTONICA DEL SALE, 3929
TETTONICA DELLA COPERTURA, 3996
TETTONICA DISGIUNTIVA, 532
TETTONICA GERMANOTIPICA, 1783
TETTONICA GLOBALE, 1821
TETTONICA GRAVITATIVA, 1887
TETTONICA IMBRICATA, 2180
TETTONICA PROFONDA, 1094
TETTONICA SOVRAIMPOSTA, 2965
TETTONICA VERTICALE, 4736
TETTONITE, 4441
TETTONOFISICA, 4442
TETTONOSFERA, 4443
TEXTULARIINA, 4476
THALASSOIDE, 4479
THANETIANO, 4484
THECAMOEBA, 4486
THECIDEIDINA, 4487
THECODONTIA, 4488
THELODONTI, 4489
THERAPSIDA, 4492
THYSANOPTEROIDA, 4523
THYSANURA, 4524
TILL, 4530
TILLITE, 4531
TILLODONTIA, 4532
TINTINNIDAE, 4538
TIPO DI METAMORFISMO, 2787
TIPOMORFISMO, 4628
TIRRENIANO, 4629
TISSOTROPIA, 4515
TITANIO, 4539
TITOLO MINERARIO, 2886
TITONIANO, 4540
TOARCIANO, 4541
TOMBOLO, 4542
TONSTEIN, 4543
TOPOGRAFIA, 4547
TORBA, 3320
TORBIDITA, 4616
TORBIDITE, 4615
TORBIERA, 3321
TORIO, 4517
TORRENTE, 4549
TORSIONE, 4550
TORTONIANO, 4551
TORTUOSITA, 4552
TOSSICITA, 4555
TOURNAISIANO, 4553
TRACCE DI FISSIONE, 1555
TRACCIA CLIMATICA, 796
TRACCIA DI CORRENTE, 1036
TRACCIA DI ESCAVAZIONE, 3970
TRACCIA DI PARTICELLA, 3309
TRACCIA FOSSILE, 4557

TRACCIANTE, 4560
TRACCIANTE COLORANTE, 848
TRACCIANTE RADIOATTIVO, 3729
TRACCIATO, 3879
TRACHIANDESITE, 4561
TRACHIBASALTO, 4562
TRACHITE, 4564
TRACHYPSAMMIACEA, 4563
TRAPPOLA, 4580
TRAPPOLA LITOLOGICA, 2531
TRAPPOLA PETROLIFERA, 3111
TRAPPOLA SEDIMENTARIA, 4002
TRAPPOLA STRATIGRAFICA, 4279
TRAPPOLA STRUTTURALE, 4311
TRASFERIMENTO, 4568
TRASFERIMENTO DI CALORE, 1969
TRASFERIMENTO DI MASSA, 2701
TRASFORMATA DI LAPLACE, 2434
TRASFORMAZIONE, 4570
TRASFORMISMO, 4571
TRASGRESSIONE, 4572
TRASMISSIVITA, 4576
TRASPORTO, 4577
TRASPORTO DI GHIACCIO, 2159
TRASPORTO EOLICO, 4829
TRASPORTO FLUVIALE, 4288
TRASPORTO GLACIALE, 1801
TRASPORTO MARINO, 2689
TRASVERSALE, 4579
TRATTAMENTO DEI DATI, 1069
TRATTAMENTO DEL SUOLO, 4152
TRATTATO, 4584
TRAVERTINO, 4583
TRAZIONE DI SEDIMENTI, 3992
TREMADOCIANO, 4586
TREMORE, 4738
TREPOSTOMATA, 4589
TRIANGOLAZIONE, 4590
TRIASSICO, 4591
TRIASSICO ALPINO, 152
TRIASSICO INF., 2594
TRIASSICO MED., 2852
TRIASSICO SUP., 4684
TRICONODONTA, 4594
TRIGONIIDAE, 4596
TRILOBITA, 4597
TRILOBITOMORPHA, 4598
TRIPOLI, 4600
TRIVELLAZIONE, 1254
TSUNAMI, 4604
TUBO, 3453
TUBOIDEA, 4605
TUBULIDENTATA, 4606
TUFITE, 4608
TUFO VULCANICO, 4607
TULIO, 4521
TUNDRA, 4610

TUNGSTATO, 4612
TUNGSTENO, 4613
TURBOPERFORAZIONE, 4619
TURINGIANO, 4522
TURONIANO, 4620
TYLOPODA, 4623
U/PB, 4692
U/TH/PB, 4693
UADI, 4779
UDALF, 4630
UDOLL, 4631
UDULT, 4632
UFIMIANO, 4633
ULTISOL, 4634
ULTRABASITE, 4636
ULTRAMETAMORFISMO, 4637
ULTRAMILONITE, 4638
UMBREPT, 4642
UMIDITA, 2075
UMIDITA DEL SUOLO, 2911
UMIFICAZIONE, 2076
UMINA, 2077
UNAKITE, 4643
UNDERCLAY, 4648
UNIFORMITARIANISMO, 4659
UNIONIDA, 4660
UNITA BIOSTRATIGRAFICA, 510
UNITA CRONOSTRATIGRAFICA, 762
UNITA GEOCRONOLOGICA, 1724
UNITA LITOSTRATIGRAFICA, 2537
UNITA STRATIGRAFICA, 4280
UNIVERSO, 4663
UOMO FOSSILE, 1622
UOVO, 1311
UPWELLING, 4687
URALITIZZAZIONE, 4688
URANIO, 4689
URBANIZZAZIONE, 4696
UREILITE, 4697
URODELA, 4698
USTALF, 4699
USTERT, 4700
USTOLL, 4701
USTOX, 4702
USTULT, 4703
UTILIZZAZIONE DEL TERRITORIO, 2427
UTILIZZAZIONE DI SOSTANZA, 4704
UTILIZZAZIONE INDUSTRIALE, 2220
VALANGA, 388
VALANGINIANO, 4706
VALENZA CHIMICA, 734
VALLE, 4707
VALLE ABISSALE, 20
VALLE DI RIFT, 3865
VALLE GLACIALE, 1802
VALLE INSEGUENTE, 2254
VALLE LUNARE, 3867

VALLE OBSEQUENTE, 3080
VALLE RESEQUENTE, 3823
VALLE SEPOLTA, 599
VALLE SOSPESA, 1952
VALLE SOTTOMARINA, 4332
VALLE SUSSEGUENTE, 4336
VALLE TETTONICA, 4312
VALUTAZIONE DI GIACIMENTO, 1117
VALVA, 4708
VANADATO, 4709
VANADIO, 4710
VAPORE, 4713
VARIABILE, 4715
VARIABILITA, 4714
VARIAZIONE, 4717
VARIAZIONE DI CAMPO MAGNETICO, 2634
VARIAZIONE DIURNA, 1231
VARIAZIONE SECOLARE, 1744
VARIAZIONE STAGIONALE, 3985
VARIOGRAMMA, 4718
VARISCIDI, 1985
VARVA, 4720
VARVECRONOLOGIA, 4721
VEDRETTA, 2157
VEGETAZIONE, 4724
VELOCITA, 4726
VELOCITA D'ONDA SISMICA, 4020
VENA DI SEGREGAZIONE, 4009
VENEROIDA, 4728
VENITE, 4729
VENTO, 4826
VENTO SOLARE, 4161
VERGENZA, 4730
VERMES, 4731
VERMICULITE, 4732
VERSANTE, 4134
VERTEBRATA, 4734
VERTISOL, 4737
VETRERIA, 1813
VETRIFICAZIONE, 4751
VETRO, 1811

VETRO NATURALE, 2990
VETRO VULCANICO, 4764
VETROSO, 4750
VETTORE, 4723
VIA NAVIGABILE, 4804
VIBROSEIS, 4740
VILLAFRANCHIANO, 4742
VIRGAZIONE, 4743
VISCOELASTICITA, 4744
VISCOSITA, 4745
VISEANO, 4748
VITA, 2499
VITA ORGANICA, 3169
VITRINA, 4749
VIVENTE, 2548
VOLANTES, 4754
VOLATILE, 4755
VOLATILIZZAZIONE, 4758
VOLUME, 4776
VOLUME DI SEDIMENTI, 3993
VULCANIANO, 4777
VULCANICO, 4765
VULCANO, 4769
VULCANO A SCUDO, 4073
VULCANO ATTIVO, 60
VULCANO DI FANGO, 2958
VULCANO INSULARE, 2317
VULCANO SEDIMENTARIO, 4772
VULCANO SOTTOMARINO, 4333
VULCANOGENO, 4773
VULCANOLOGIA, 4775
VULCANORIUM, 4778
WENLOCKIANO, 4818
WESTFALIANO, 4820
WILDFLYSCH, 4825
XANTHOPHYTA, 4837
XENO, 4840
XENOLITE, 4838
XERALF, 4841
XEROFITA, 4844

XEROLL, 4842
XEROSOL, 4845
XERULT, 4846
YERMOSOL, 4848
YPRESIANO, 4852
ZECHSTEIN, 4855
ZEOLITE, 4856
ZEOLITIZZAZIONE, 4858
ZINC, 4860
ZIRCONE, 4861
ZIRCONIO, 4862
ZOANTHINIARIA, 4863
ZOECIUM, 4864
ZOLFO, 4353
ZONA A BASSA VELOCITA, 2576
ZONA CONTINENTALE, 931
ZONA D'AREAZIONE, 4866
ZONA D'ASSOCIAZIONE, 339
ZONA DEI SOLFURI, 4350
ZONA DI BENIOFF, 462
ZONA DI CEMENTAZIONE, 4867
ZONA DI OSSIDAZIONE, 3223
ZONA DI SATURAZIONE, 3949
ZONA DI SCORRIMENTO, 4868
ZONA DI SEDIMENTAZIONE, 4869
ZONA DI SUBDUZIONE, 4322
ZONA DI TAGLIO, 4064
ZONA FAGLIATA, 1513
ZONA FRATTURATA, 1634
ZONA FREDDA, 548
ZONA METAMORFICA, 2786
ZONA MOBILE, 2899
ZONA NON SATURATA, 3049
ZONA PERIGLACIALE, 3355
ZONA SUBTROPICALE, 4342
ZONALITA CLIMATICA, 797
ZONATURA, 4870
ZONEOGRAFIA DEL SUOLO, 673
ZOOIDE, 4871
ZUCCHERO, 4347

Field Index

KRYPTON, 2398
LANTHANUM, 2432
LEAD, 2466
LITHIUM, 2525
LUTETIUM, 2607
MAGNESIUM, 2627
MANGANESE, 2662
MENDELEVIUM, 2747
MERCURY, 2749
MOLYBDENUM, 2918
NEODYMIUM, 3000
NEON, 3003
NEPTUNIUM, 3011
NICKEL, 3026
NIOBIUM, 3030
NITROGEN, 3034
OSMIUM, 3199
OXYGEN, 3225
PALLADIUM, 3275
PHOSPHOROUS, 3407
PLATINUM, 3504
PLUTONIUM, 3521
POLONIUM, 3544
POTASSIUM, 3579
PRASEODYMIUM, 3591
PROMETHIUM, 3637
PROTACTINIUM, 3642
RADIUM, 3739
RADON, 3741
RHENIUM, 3849
RHODIUM, 3855
RUBIDIUM, 3909
RUTHENIUM, 3919
SAMARIUM, 3932
SCANDIUM, 3956
SELENIUM, 4030
SILICON, 4103
SILVER, 4111
SODIUM, 4143
STRONTIUM, 4299
SULFUR, 4353
TANTALUM, 4419
TECHNETIUM, 4429
TELLURIUM, 4450
TERBIUM, 4460
THALLIUM, 4480
THORIUM, 4517
THULIUM, 4521
TIN, 4537
TITANIUM, 4539
TUNGSTEN, 4613
URANIUM, 4689
VANADIUM, 4710
XENON, 4840
YTTERBIUM, 4853
YTTRIUM, 4854
ZINC, 4860

ZIRCONIUM, 4862

CHES CHEMICAL COMPOSITION, CHEMICAL COMPOUNDS

ACID, 37
ALIPHATIC HYDROCARBON, 122
ALKALI=METALS, 126
ALKALINE EARTH METALS, 129
ALKANES, 132
ALLOY, 139
ALUMINA, 156
AMINO ACID, 163
AMMONIA, 165
AMMONIUM, 168
AROMATIC HYDROCARBON, 305
CARBOHYDRATE, 641
CARBON DIOXIDE, 643
CARBON MONOXIDE, 644
CARBONACEOUS COMPOSITION, 648
CELLULOSE, 683
CHALCOPHILE ELEMENT, 703
CHITIN, 745
CHLOROPHYLL, 753
COSMOGENIC ELEMENT, 973
ETHANE, 1440
ETHERS, 1441
FATTY ACID, 1506
FERRIC IRON, 1533
FERROUS IRON, 1538
FULVIC ACID, 1663
GAS, 1684
HALOGEN, 1947
HEAVY METALS, 1972
HUMIC ACID, 2071
HUMIN, 2077
HYDROCARBON, 2099
HYDROGEN GAS, 2107
HYDROGEN SULFIDE, 2108
INERT GAS, 2224
INORGANIC ACID, 2248
INORGANIC MATERIAL, 2249
INSOLUBLE RESIDUE, 2256
ISOPRENOID, 2339
KEROGEN, 2386
LIGNIN, 2503
LIPID, 2516
LITHOPHILE ELEMENT, 2533
MAJOR ELEMENT, 2657
MARSH GAS, 2694
METAL, 2766
METALLOID, 2775
METHANE, 2801
MINOR ELEMENT, 2887
NITROGEN GAS, 3035
NOBLE METAL, 3037
NON=FERROUS METALS, 3047

NON=METALS, 3048
ORGANIC ACID, 3165
ORGANIC CARBON, 3166
ORGANIC COMPOUND, 3167
ORGANIC MATERIAL, 3170
ORGANIC NITROGEN, 3171
OXYGEN GAS, 3226
PHENOLS, 3396
PHYTANE, 3433
PIGMENTS, 3444
PROTEINS, 3645
PURINE, 3688
RARE EARTHS, 3751
RARE METAL, 3752
SIDEROPHILE ELEMENT, 4092
STEROLS, 4253
SUGARS, 4347
SULFUR DIOXIDE, 4354
SULFURIC ACID, 4356
VOLATILE ELEMENT, 4756
VOLATILES, 4757

COMS MINERAL COMMODITIES

ABRASIVE, 9
ABSORBENT, 12
AGGREGATE, 85
ANDALUSITE, 192
ANHYDRITE, 204
ANTHRACITE, 221
ARTIFICIAL ASH, 320
ASBESTOS, 325
ASPHALT, 334
ASPHALTITE, 335
BARITE, 412
BAUXITE, 440
BEAN ORE, 449
BENTONITE, 466
BITUMEN, 521
BITUMINOUS COAL, 523
BITUMINOUS SHALE, 524
BLEACHING CLAY, 528
BOG IRON ORE, 537
BOGHEAD COAL, 538
BORAX, 545
BRICK CLAY, 576
BROWN COAL, 582
BUILDING CEMENT, 587
BUILDING STONE, 588
CALCITE, 613
CANNEL COAL, 629
CERAMICS, 694
CHINA CLAY, 743
COAL, 801
COAL GAS, 804
COKING COAL, 831

CONSTRUCTION MATERIALS, 911
CORUNDUM, 965
DECORATIVE STONE, 1085
DIAMOND, 1161
DIATOMACEOUS EARTH, 1168
EXPANDABLE MATERIAL, 1474
FERTILIZER, 1541
FILLING MATERIAL, 1544
FIRECLAY, 1552
FLUORSPAR, 1590
FOSSIL FUEL, 1620
FUEL RESOURCES, 1658
FULLER'S EARTH, 1662
GANISTER, 1681
GAS COAL, 1687
GAS CONDENSATE, 1688
GEMSTONE, 1707
GLASS-MAKING MATERIAL, 1813
GLASS-SAND, 1812
GLAUCONITE, 1815
GYPSUM, 1933
HEAVY OIL, 1974
HEMATITE, 1981
INCOHERENT MATERIAL, 2202
INDUSTRIAL MINERAL, 2219
INSULANT, 2259
KAOLIN, 2368
LIGHT OIL, 2502
LIGNITE, 2504
LIME, 2506
LIMONITE, 2510
MAGNESITE, 2626
MAGNETITE, 2647
MONAZITE, 2919
MORTAR MATERIAL, 2944
NATURAL GAS, 2989
NATURAL RESOURCE, 2992
NUGGET, 3070
OCHER, 3095
OIL SAND, 3108
OIL SHALE, 3110
ORNAMENTAL STONE, 3181
PAINT MATERIAL, 3234
PEAT, 3320
PETROLEUM, 3379
PHOSPHATE ROCK, 3402
POTASH, 3577
POTTER'S CLAY, 3586
POZZOLAN, 3589
PRECIOUS STONE, 3595
RAW MATERIAL, 3755
REFRACTORY, 3791
RESOURCE, 3834
ROAD MATERIAL, 3877
SALT, 3924
SAPROPELITIC COAL, 3945
SEMIPRECIOUS STONE, 4036

SILLIMANITE, 4105
SYLVINITE, 4380
TALC, 4414
TRIPOLI, 4600
UNCONSOLIDATED MATERIAL, 4647
VERMICULITE, 4732
WHITE CLAY, 4823
ZIRCON, 4861

ECON   ECONOMIC GEOLOGY

ALLUVIAL PROSPECTING, 145
APOMAGMATIC DEPOSIT, 244
ASCENSION THEORY, 326
ASSAY, 336
ASSAY VALUE, 337
BARREN DEPOSIT, 415
BASE METAL, 428
BAUXITIZATION, 441
BEACH PLACER, 445
BIOGEOCHEMICAL EXPLORATION, 495
BLIND DEPOSIT, 529
CAP ROCK, 631
CARAT, 640
CHEMICAL CONTROLS, 722
CHEMICAL INDUSTRY, 728
CLAIM, 779
COAL FIELD, 803
COAL MEASURES, 805
COAL SEAM, 806
COAL-DEPOSIT MAP, 807
COMPANY, 858
CONSUMPTION, 912
CUTOFF GRADE, 1038
CUTTINGS, 1039
DENSITY CONCENTRATION, 1114
DEPOSIT EVALUATION, 1117
DEPOSIT GENESIS, 1118
DEPOSIT MORPHOLOGY, 1119
DESCENSION THEORY, 1130
DISSEMINATION, 1226
DRAG ORE, 1244
ECONOMIC GEOLOGY, 1298
ECONOMIC GEOLOGY MAP, 1299
ECONOMIC MINERAL, 1300
EDGE WATER, 1306
ENDOGENE PROCESS, 1370
ENERGY SOURCE, 1373
ENRICHMENT, 1376
EPITHERMAL DEPOSIT, 1407
EXHALATIVE PROCESS, 1467
EXOGENE PROCESS, 1471
EXPLOITABILITY, 1481
EXPLORATION, 1483
EXPLORATORY WELL, 1484
EXPORT, 1488

GAS CAP, 1685
GAS FIELD, 1689
GAS WELL, 1692
GAS-OIL RATIO, 1693
GEOBOTANICAL EXPLORATION, 1714
GEOCHEMICAL EXPLORATION, 1716
GEOCHEMICAL INDICATOR, 1718
GEOCHEMICAL SINK, 1721
GEOTHERMAL ENERGY, 1773
GEOTHERMAL FIELD, 1774
GEOTHERMAL SYSTEM, 1778
GIANT FIELD, 1786
GOSSAN, 1838
HANGING WALL, 1953
HIGH-GRADE, 2010
HORSETAIL, 2059
HOT BRINE, 2064
HYDROGEOCHEMICAL PROSPECTING, 2109
HYDROGEOLOGICAL CONTROL, 2110
HYDROTHERMAL DEPOSIT, 2132
HYPOTHERMAL DEPOSIT, 2149
IMPORT, 2194
IMPREGNATION, 2195
INDUSTRIAL USE, 2220
INDUSTRY, 2222
INFERRED ORE, 2225
IRON AND STEEL INDUSTRY, 2309
KANGA, 2367
KATATHERMAL DEPOSIT, 2379
LATERAL SECRETION, 2445
LEAN, 2471
LINKED VEINS, 2514
LITHOLOGIC CONTROLS, 2530
LITHOLOGIC TRAP, 2531
LODE, 2559
MAGMATIC CONTROLS, 2621
MAGMATIC ORE DEPOSIT, 2622
MASSIVE DEPOSIT, 2705
MECHANICAL CONTROL, 2725
MESOTHERMAL DEPOSIT, 2759
METALLIC PROPERTY, 2768
METALLIFEROUS, 2769
METALLOGENIC DISTRICT, 2770
METALLOGENIC EPOCH, 2771
METALLOGENIC MAP, 2772
METALLOGENIC PROVINCE, 2773
METALLOGENY, 2774
METALLOTECT, 2776
METAMORPHOGENIC DEPOSIT, 2788
MINERAL DEPOSIT, 2866
MINERAL ECONOMICS, 2867
MINERAL EXPLORATION, 2868
MINERAL RESOURCE, 2872
MINERALIZATION, 2876
MINERALIZER, 2877
MINING LICENSE, 2886
MISSISSIPPI VALLEY-TYPE, 2896

NONMETAL DEPOSIT, 3051
NONMETALLIC=DEPOSIT MAP, 3052
NUCLEAR ENERGY, 3062
OIL AND GAS FIELD, 3105
OIL AND GAS MAP, 3106
OIL FIELD, 3107
OIL SEEP, 3109
OIL TRAP, 3111
OIL=GAS INTERFACE, 3113
OIL=WATER INTERFACE, 3114
ORE, 3151
ORE BODY, 3152
ORE CONTROL, 3153
ORE DEPOSIT, 3154
ORE GRADE, 3156
ORE GUIDE, 3157
ORE MINERAL, 3159
ORE QUALITY, 3160
ORE SHOOT, 3161
ORE SOURCE, 3162
ORE TRANSPORT, 3163
ORIENTATION SURVEY, 3178
OXIDIZED ZONE, 3223
PALEOGEOGRAPHIC CONTROLS, 3255
PERSPECTIVE, 3366
PETROLEUM GEOLOGY, 3381
PLACER, 3463
PNEUMATOLYTIC DEPOSIT, 3524
POLYMETALLIC DEPOSIT, 3551
PORPHYRY COPPER, 3569
POSITIVE ORE, 3571
POSSIBILITIES, 3572
POTENTIAL ORE, 3582
PREDICTIVE METALLOGENY, 3599
PRICE, 3610
PRODUCTION, 3624
PRODUCTION CONTROL, 3625
PRODUCTION HORIZON, 3626
PROFIT, 3629
PROSPECTING, 3641
QUALITY, 3707
RANK OF COAL, 3748
RECYCLING, 3772
RESERVE, 3824
RESERVOIR ROCK, 3826
RESIDUAL DEPOSIT, 3829
ROLL=TYPE DEPOSIT, 3893
RUN, 3914
SEAM, 3983
SEDIMENTARY ORE, 3997
SEDIMENTARY PROCESS, 3999
SEGREGATED VEIN, 4009
SHOWING, 4086
STORAGE, 4265
STRATABOUND DEPOSIT, 4270
STRATEGY, 4271
STRATIFORM DEPOSIT, 4274

STRATIGRAPHIC CONTROL, 4276
STRATIGRAPHIC TRAP, 4279
STREAM SEDIMENT, 4287
STRUCTURAL CONTROL, 4302
STRUCTURAL GUIDE, 4307
STRUCTURAL TRAP, 4311
SULFIDE ZONE, 4350
SYNTHETIC MATERIAL, 4402
TELETHERMAL DEPOSIT, 4447
TRADE, 4566
TRANSPORTATION, 4578
TRAP, 4580
URANIUM MINERAL, 4690
UTILIZATION, 4704
VEIN, 4725
VOLCANOGENIC PROCESS, 4774
WALLROCK, 4780
ZONE OF CEMENTATION, 4867
ZONE OF SEDIMENTATION, 4869

ENGI ENGINEERING GEOLOGY

AIRFIELD, 96
ANCHORING, 190
ANGLE OF REPOSE, 200
ARCH DAM, 271
ASEISMIC DESIGN, 327
ATTERBERG LIMIT, 367
BALLAST, 408
BEARING CAPACITY, 450
BRIDGE, 577
BUILDING, 586
CANAL, 628
CHEMICAL EXPLOSION, 726
CIVIL ENGINEERING, 777
COHESIONLESS SOIL, 828
COHESIVE MATERIAL, 829
COHESIVE SOIL, 830
COMPETENT MATERIAL, 859
COMPRESSIBILITY, 864
COMPRESSION STRENGTH, 866
COMPRESSION TEST, 868
CONCRETE, 879
CONCRETE DAM, 880
CONGELIFRACTION, 891
CONSTRUCTION, 910
CRITICAL ANGLE, 993
CYCLIC LOADING, 1046
DAM, 1059
DRAINING, 1247
DRILLING SLUDGE, 1256
EARTH DAM, 1275
EARTH PRESSURE, 1277
EARTH WORK, 1281
EARTHFLOW, 1283
EARTHQUAKE ENGINEERING, 1285

ELASTIC MATERIAL, 1319
EMBANKMENT, 1358
ENGINEERING GEOLOGY, 1374
ENGINEERING WORKS, 1375
EXCAVATION, 1462
FLUID INJECTION, 1580
FLUID PRESSURE, 1583
FOUNDATION, 1627
FRACTURE STRENGTH, 1633
GAS STORAGE, 1691
GEOPRESSURE, 1760
GEOTECHNICAL MAP, 1768
GEOTECHNICAL PROPERTY, 1769
GEOTECHNICAL SURVEY, 1770
GEOTECHNICS, 1771
GRAVITY PLATFORM, 1884
GROIN, 1901
GROUTING, 1920
HARBOR, 1954
HYDRAULIC FRACTURING, 2092
INCOMPETENT ROCK, 2203
INJECTION, 2239
LABORATORY TEST, 2403
LAND SUBSIDENCE, 2426
LANDFILL, 2428
LANDSLIDE, 2431
LEVEE, 2491
LOAD, 2554
LOAD TEST, 2556
MARINE INSTALLATION, 2680
MARINE PLATFORM, 2684
NUCLEAR POWER PLANT, 3065
OVERCONSOLIDATED MATERIAL, 3214
PENETRATION TEST, 3340
PETROLEUM ENGINEERING, 3380
PETROPHYSICS, 3383
PIER, 3440
PILE, 3445
PILLAR, 3446
PIPELINE, 3454
PIPING, 3455
PLASTIC FLOW, 3492
PLASTIC MATERIAL, 3493
PLATE=BEARING TEST, 3500
POROUS MEDIUM, 3566
POWER PLANT, 3588
PUBLIC WORKS, 3680
QUICK CLAY, 3715
QUICKSAND, 3716
ROAD TEST, 3878
ROADWAY, 3879
ROCK BURST, 3883
ROCK FILL DAM, 3885
ROCK MECHANICS, 3888
SEALING, 3982
SEAWALL, 3986
SEEPAGE, 4008

SEISMIC RISK, 4015
SENSITIVE CLAY, 4038
SETTLEMENT, 4048
SHEAR MODULUS, 4060
SHORE PROTECTION, 4082
SILTING, 4108
SITE SELECTION, 4124
SLOPE STABILITY, 4137
SOFT CLAY, 4144
SOIL DEFORMATION, 4147
SOIL MECHANICS, 4154
SOIL STABILIZATION, 4158
STABILIZATION, 4227
STIFF CLAY, 4255
STRONG MOTION, 4298
SUBMARINE INSTALLATION, 4328
SUBWAY, 4344
THAWING, 4485
TRIAXIAL COMPRESSION TEST, 4592
TUNNEL, 4614
UNDERGROUND CANALIZATION, 4649
UNDERGROUND SPACE, 4651
UNDERGROUND STORAGE, 4652
UNIAXIAL=COMPRESSION TEST, 4658
VANE TEST, 4712
VIBRATION, 4738
VISCOUS MATERIAL, 4746
WATER PRESSURE, 4795
WATERWAY, 4804
YIELD STRENGTH, 4849

ENVI ENVIRONMENTAL GEOLOGY

ACID MINE DRAINAGE, 38
ALPINE ENVIRONMENT, 149
ATMOSPHERE, 358
CONSERVATION, 905
DECONTAMINATION, 1083
ENVIRONMENT, 1386
ENVIRONMENTAL GEOLOGY, 1388
EUTROPHICATION, 1455
GRASSLAND, 1870
HUMAN ECOLOGY, 2069
HUMAN WASTE, 2070
IMPACT STATEMENT, 2191
INDUSTRIAL WASTE, 2221
LAND USE, 2427
LIQUID WASTE, 2521
NUCLEAR EXPLOSION, 3063
PESTICIDE, 3369
POLLUTANT, 3541
POLLUTED WATER, 3542
POLLUTION, 3543
PURIFICATION, 3687
RADIOACTIVE WASTE, 3730
RECLAMATION, 3766

RECREATION, 3769
SOIL MANAGEMENT, 4152
SOLID WASTE, 4168
TOXIC MATERIAL, 4554
TOXICITY, 4555
URBAN GEOLOGY, 4694
URBAN PLANNING, 4695
URBANIZATION, 4696
VEGETATION, 4724
WASTE, 4781
WASTE DISPOSAL, 4782

EXTR EXTRATERRESTRIAL GEOLOGY

APOGEE, 241
APOLLO PROGRAM, 243
ASTEROID, 345
ASTROBLEME, 350
ASTROGEOLOGY, 351
ASTRONOMY, 352
CHONDRULE, 757
COPERNICAN, 945
COSMIC DUST, 967
COSMIC NEBULA, 968
COSMIC PARTICLE, 969
COSMIC RADIATION, 970
COSMIC SYSTEM, 971
COSMOCHEMISTRY, 972
COSMOLOGICAL PRINCIPLE, 974
COSMOLOGY, 975
EARTH=MOON COUPLE, 1282
ERATOSTHENIAN, 1420
EXOBIOLOGY, 1470
EXTRATERRESTRIAL GEOLOGY, 1498
IMBRIAN, 2179
IMPACT CRATER, 2187
IMPACT FEATURE, 2188
IMPACTITE, 2193
INTERPLANETARY COMPARISON, 2280
INTERSTELLAR SPACE, 2285
KREEP, 2397
LUNA, 2598
LUNAR, 2599
LUNAR CONSTITUTION, 2600
LUNAR CRATER, 2601
LUNAR GEOLOGY, 2602
LUNAR SOIL, 2603
LUNAR STATION, 2604
LUNAR TERRA, 2605
MAGNETIC SPHERULE, 2641
MARE, 2672
MARINER, 2690
MASCON, 2696
METEOR CRATER, 2797
MOONQUAKE, 2936
MOONSCAPE, 2937

ORANGE MATERIAL, 3144
ORBIT, 3146
OUTER PLANET, 3208
PERIGEE, 3353
PLANET=EARTH COMPARISON, 3481
PLANETARY, 3482
PLANETARY CONSTITUTION, 3483
PLANETARY SATELLITE, 3484
PLANETOLOGY, 3485
PRE=IMBRIAN, 3592
PROCELLARIAN, 3621
RADIOACTIVE DUST, 3726
REGMAGLYPT, 3797
RILLES, 3867
SELENOCHRONOLOGY, 4031
SELENOLOGY, 4032
SOLAR SYSTEM, 4160
SOLAR WIND, 4161
TEKTITE, 4444
TERRESTRIAL PLANET, 4468
THALASSOID, 4479
UNIVERSE, 4663
VIKING, 4741

EXTS EXTRATERRESTRIAL GEOLOGY, METEORITES, PLANETS

ACHONDRITE, 36
ANGRITE, 201
ATAXITE, 354
AUBRITE, 369
AUSTRALITE, 374
CARBONACEOUS CHONDRITE, 647
CHASSIGNITE, 714
CHONDRITE, 756
DIOGENITE, 1193
ENSTATITE CHONDRITE, 1379
EUCRITE, 1443
HEXAHEDRITE, 2003
HOWARDITE, 2067
INDOCHINITE, 2214
LODRANITE, 2560
MESOSIDERITE, 2758
METALLIC METEORITE, 2767
METEORITE, 2798
METEOROIDS, 2799
MICROMETEORITE, 2824
MICROTEKTITE, 2837
MOLDAVITE, 2913
MOON, 2935
NAKHLITE, 2983
OCTAHEDRITE, 3097
PALLASITE, 3276
PLANET, 3471
PLANET JUPITER, 3473
PLANET MARS, 3474

PLANET MERCURY, 3475
PLANET NEPTUNE, 3476
PLANET PLUTO, 3477
PLANET SATURN, 3478
PLANET URANUS, 3479
PLANET VENUS, 3480
SIDEROPHYRE, 4093
STONY IRON METEORITE, 4263
STONY METEORITE, 4264
SUN, 4358
UREILITE, 4697

GEOC GEOCHEMISTRY

ANHYDROUS COMPOSITION, 205
AUREOLE, 373
BACKGROUND, 398
BIOGEOCHEMISTRY, 496
BOUNDARY LAYER, 561
CHELATION, 718
CHEMICAL COMPOSITION, 721
CHEMICAL EQUILIBRIUM, 724
CHEMICAL RATIO, 730
CHLORINATION, 748
CLARKE, 781
COMPOSITION, 863
DISPERSION PATTERN, 1223
FERRUGINOUS COMPOSITION, 1539
FLUORINATION, 1588
GEOCHEMICAL CYCLE, 1715
GEOCHEMICAL FACIES, 1717
GEOCHEMICAL MAP, 1719
GEOCHEMICAL PROFILE, 1720
GEOCHEMISTRY, 1722
HYDROCHEMICAL FACIES, 2100
MIGRATION OF ELEMENT, 2856
ORGANIC GEOCHEMISTRY, 3168
PARTITION COEFFICIENT, 3311
PRIMARY DISPERSION PATTERN, 3611
SECONDARY AUREOLE, 3988
SILICEOUS COMPOSITION, 4099
THERMOCHEMICAL PROPERTY, 4502
THRESHOLD, 4519
TRACE ELEMENT, 4556
TRACE=ELEMENT ANALYSIS, 4559
TRACE METAL, 4558

GEOH HYDROGEOLOGY

ADSORBED WATER, 67
ALLUVIAL GROUNDWATER, 143
APPLIED HYDROGEOLOGY, 248
AQUICLUDE, 259
AQUIFER, 260

AQUIFER TEST, 261
AQUIFUGE, 262
AQUITARD, 264
ARHEISM, 298
ARTESIAN, 312
ARTESIAN AQUIFER, 313
ARTESIAN WELL, 314
ARTIFICIAL RECHARGE, 321
ATMOSPHERIC PRECIPITATION, 360
BASIN PLANNING, 433
BRACKISH WATER, 568
CAPILLARY WATER, 636
CATCHMENT HYDRODYNAMICS, 672
CONE OF DEPRESSION, 885
CONFINED AQUIFER, 887
CRITICAL FLOW, 994
DARCY'S LAW, 1061
DENDRITIC DRAINAGE PATTERN, 1109
DISCHARGE, 1209
DRAINAGE BASIN, 1245
DRAWDOWN, 1249
EFFLUENT, 1310
ENDORHEISM, 1371
EVAPORATION, 1458
EVAPOTRANSPIRATION, 1460
EXORHEISM, 1472
EXSURGENCE, 1491
FILTRATION, 1547
FLOOD, 1569
FLUCTUATION, 1578
FOSSIL WATER, 1623
FRESH WATER, 1647
GAUGING, 1698
GRAVEL FILTER, 1872
GROUND WATER, 1911
GROUND=WATER BUDGET, 1912
GROUND=WATER DAM, 1913
GROUND=WATER DISCHARGE, 1914
GROUND=WATER EXPLORATION, 1915
GROUND=WATER LEVEL, 1916
GROUND=WATER RECHARGE, 1917
GROUND=WATER RUNOFF, 1918
HOT SPRING, 2066
HYDRAULIC CONDUCTIVITY, 2091
HYDRAULIC GRADIENT, 2093
HYDRAULIC HEAD, 2094
HYDRAULIC MAP, 2095
HYDRAULIC PRESSURE, 2096
HYDRAULIC RADIUS, 2097
HYDRAULICS, 2098
HYDROCHEMICAL MAP, 2101
HYDROCHEMISTRY, 2102
HYDRODYNAMIC MODEL, 2103
HYDRODYNAMICS, 2104
HYDROGEOLOGICAL MAP, 2111
HYDROGEOLOGICAL PROVINCE, 2112
HYDROGEOLOGICAL SURVEY, 2113

HYDROGEOLOGICAL TECHNOLOGY, 2114
HYDROGEOLOGY, 2115
HYDROGRAPH, 2116
HYDROGRAPHY, 2117
HYDROLOGIC CYCLE, 2119
HYDROLOGICAL MAP, 2120
HYDROLOGY, 2121
HYDROMETRY, 2124
HYDROSTATIC LEVEL, 2128
HYDROSTATIC PRESSURE, 2129
HYDROTHERMAL WATER, 2134
INFILTRATION, 2226
INFILTRATION GALLERY, 2227
INTERMITTENT SPRING, 2271
INTERMITTENT STREAM, 2272
IRRIGATION, 2314
JUVENILE WATER, 2365
KARST HYDROLOGY, 2375
KARST SPRING, 2376
LAKE WATER LEVEL, 2416
LAMINAR FLOW, 2420
LEAKAGE, 2470
LOW WATER, 2573
MEDICINAL WATER, 2734
MELTWATER, 2745
MINERAL SPRING, 2873
MINERAL WATER, 2875
MOISTURE, 2911
MULTILAYER SYSTEM, 2962
MULTIPHASE FLOW, 2964
NON=SATURATED ZONE, 3049
PERCHED AQUIFER, 3349
PERCOLATION, 3350
PHREATIC WATER, 3416
PIEZOMETRY, 3443
PORE PRESSURE, 3562
POTABLE WATER, 3576
POTENTIOMETRIC SURFACE, 3583
PROTECTED ZONE, 3644
PUMPAGE, 3684
RAIN, 3742
RAIN WASH, 3744
RECHARGE, 3765
REGIONAL HYDROGEOLOGY, 3794
REPRESENTATIVE BASIN, 3819
RESERVOIR, 3825
RESURGENCE, 3836
RESURGENT WATER, 3837
RETENTION, 3838
RUNOFF, 3915
SALT=WATER INTRUSION, 3930
SATURATED ZONE, 3949
SEICHE, 4011
SIPHON, 4120
SPRING, 4224
STEADY FLOW, 4246
SUBMARINE SPRING, 4331

SURFACE RUNOFF, 4364
SURFACE WATER, 4366
SURFACE WATER BUDGET, 4367
SYSTEMS ANALOG, 4405
THERMAL WATER, 4501
TRANSMISSIBILITY COEFFICIENT, 4574
TRANSMISSIVITY, 4576
UNCONFINED AQUIFER, 4644
UNCONFINED WATER, 4645
UNDERGROUND STREAM, 4653
UNSTEADY FLOW, 4664
UPWELLING, 4687
WATER, 4783
WATER AUTHORITY, 4784
WATER BUDGET, 4785
WATER CATCHMENT, 4786
WATER CONSUMPTION, 4787
WATER DIVIDE, 4788
WATER HARDNESS, 4790
WATER MANAGEMENT, 4791
WATER MAP, 4792
WATER QUALITY, 4796
WATER RECESSION, 4797
WATER RESOURCES, 4798
WATER STORAGE, 4799
WATER SUPPLY, 4800
WATER WELL, 4801
WATER YIELD, 4802
ZONE OF AERATION, 4866

GEOL   GENERAL GEOLOGY

ABYSSAL, 18
ACCRETION, 30
ACCUMULATION, 31
ACTUALISM, 62
AERIAL MAPPING, 70
AERIAL PHOTOGRAPH, 71
AERIAL SURVEY, 72
AGE, 80
AGE OF THE EARTH, 81
AIR, 93
ALLOCHTHON, 135
ANOROGENIC PROCESS, 215
APPLIED GEOLOGY, 246
AUTOCHTHON, 377
BACTERIOGENIC ORIGIN, 401
BASAL, 422
BEDROCK, 457
BIOGENESIS, 492
BIOGENIC EFFECT, 493
BIOSPHERE, 509
BLOCK, 530
BLOCK DIAGRAM, 531
BOREHOLE SECTION, 550
BRECCIA, 575

BURIAL, 596
CALCITIZATION, 614
CATASTROPHISM, 671
CHEMICAL EVOLUTION, 725
CIRCUM PACIFIC BELT, 774
CLIMATIC EFFECT, 795
CLIMATIC TRACE, 796
COLLAPSE BRECCIA, 835
COLLECTING, 838
COLLECTION, 839
COMPLEX, 860
CONJUGATE, 896
CONSOLIDATION, 907
CONTACT, 913
CONTACT BRECCIA, 914
CONTINENT, 918
CONVERGENCE, 940
CORE SAMPLE, 958
COUNTRY ROCK, 978
CRUSH BRECCIA, 1000
CRYPTOGENIC, 1008
CURRENT, 1035
CYCLE, 1044
CYCLIC PROCESS, 1047
DEPRESSION, 1121
DESILICATION, 1136
DETERMINISM, 1140
DISCORDANT, 1215
DISLOCATION BRECCIA, 1221
DIVERGENCE, 1232
DOME, 1240
DYNAMIC GEOLOGY, 1265
EARTH SCIENCE, 1279
EPIGENESIS, 1404
EXCURSION, 1465
EXPEDITION=CRUISE, 1476
EXPERIMENTAL STUDIES, 1479
EXPLOSION BRECCIA, 1486
FACIES, 1501
FAULT BRECCIA, 1508
FIELD STUDY, 1542
FIXISM, 1559
FOLD BRECCIA, 1605
GENESIS, 1709
GEOBAROMETRY, 1712
GEOGRAPHY, 1731
GEOLOGIC MAP, 1735
GEOLOGIC SECTION, 1736
GEOLOGICAL HAZARDS, 1738
GEOLOGY, 1741
GEONOMY, 1749
GEOTHERMOMETRY, 1780
GREAT=CIRCLE BELT, 1890
GUIDEBOOK, 1926
HISTORY OF GEOLOGY, 2021
HORIZONTAL DIRECTION, 2052
HYDATOGENESIS, 2087

HYDROUS, 2135
IMMATURE, 2182
INJECTION BRECCIA, 2240
INORGANIC ORIGIN, 2250
IRON BACTERIA, 2310
ISOBATH, 2321
ISOMORPHISM, 2334
LANDSAT, 2430
LATERAL, 2441
LENS, 2480
LEVELS, 2493
LITHOLOGIC, 2529
LITHOLOGY, 2532
LITHOTYPE, 2541
LOG, 2562
MAP, 2668
MASS TRANSFER, 2701
MATERIAL, 2707
MATURITY, 2718
MECHANISM, 2730
MEDICAL GEOLOGY, 2733
MELANGE, 2740
MIGRATION, 2855
MILITARY GEOLOGY, 2859
MORPHOLOGY, 2940
MULTIPHASE, 2963
MYLONITE, 2972
NEPTUNISM, 3010
ORGANIC ORIGIN, 3172
ORIENTATION, 3177
ORIGIN, 3179
OUTCROP, 3206
PALEO, 3239
PARENT ROCK, 3304
PERIODICITY, 3357
PETROGENESIS, 3374
PETROGRAPHY, 3378
PETROLOGY, 3382
PHASE, 3391
PHOTOGEOLOGIC MAP, 3409
PHOTOGEOLOGY, 3410
PHREATIC GAS, 3415
PHYSICAL MODEL, 3426
PIT SECTION, 3460
PLUTONISM, 3520
POLYGENETIC, 3546
POLYPHASE PROCESS, 3556
POPULAR GEOLOGY, 3560
PREVENTION, 3609
PROCESS, 3622
RECONSTRUCTION, 3767
REGENERATION, 3792
REGIONAL GEOLOGY, 3793
REJUVENATION, 3802
RELICT, 3806
REWORKED, 3845
RIM, 3868

RING STRUCTURE, 3871
ROCK, 3881
ROTATION, 3902
SALIFEROUS, 3922
SCALE MODEL, 3955
SEASONAL VARIATION, 3985
SHATTER BRECCIA, 4056
SHOCK BRECCIA, 4076
SILICIFICATION, 4100
STABILITY, 4226
STANDARD ROCK, 4237
SUBSURFACE GEOLOGY, 4340
SURVEY, 4371
SYMMETRY, 4384
SYNGENESIS, 4394
SYNGENETIC, 4395
SYSTEM, 4403
TECTONIC BRECCIA, 4434
TECTONITE, 4441
THERMAL EFFECT, 4497
TRANSFORMATION, 4570
TRANSFORMISM, 4571
TYPE LOCALITY, 4624
TYPE SECTION, 4625
TYPOMORPHISM, 4628
ULTRAMYLONITE, 4638
UNIFORMITARIANISM, 4659
VOLCANOGENIC, 4773
ZONING, 4870

IGMS METAMORPHIC ROCKS

AMPHIBOLITE, 173
ANATEXITE, 187
ARTERITE, 311
BLUESCHIST, 534
CALCIPHYRE, 611
CARBONATITE, 652
CHARNOCKITE, 712
CHLORITE SCHIST, 751
CIPOLIN, 771
DIAPHTHORITE, 1163
ECLOGITE, 1295
EMBRECHITE, 1359
EPIBOLITE, 1399
FENITE, 1529
GNEISS, 1828
GRANITE GNEISS, 1846
GRANULITE, 1858
GREENSCHIST, 1892
GREENSTONE, 1894
HORNFELS, 2057
ITABIRITE, 2355
ITACOLUMITE, 2356
JADEITITE, 2357
KERATOPHYRE, 2385

KINZIGITE, 2395
LEPTITE, 2486
LEPTYNITE, 2488
MARBLE, 2671
METABASALT, 2762
METAGABBRO, 2764
METAMORPHIC ROCK, 2785
METASEDIMENTARY ROCK, 2792
METASOMATITE, 2795
MICASCHIST, 2805
MIGMATITE, 2853
ORTHOAMPHIBOLITE, 3189
ORTHOGNEISS, 3192
PARAAMPHIBOLITE, 3287
PARAGNEISS, 3292
PELITIC-METAMORPHIC SEQUENCE, 3336
PHYLLITE, 3419
PHYLLONITE, 3421
QUARTZITE, 3712
SCHIST, 3960
SERICITE SCHIST, 4042
SERPENTINITE, 4046
SKARN, 4126
SLATE, 4130
SPOTTED SLATE, 4222
TACONITE, 4408
TACTITE, 4409
TALC SCHIST, 4415
UNAKITE, 4643
VENITE, 4729

IGNE PETROLOGY, CRYSTALLINE ROCKS

AA LAVA, 2
ACCESSORY MINERAL, 29
ACIDIC COMPOSITION, 39
ACTINOLITE FACIES, 53
ACTIVE VOLCANO, 60
AGGLUTINATE, 83
AGPAITIC, 89
AKMOLITH, 98
ALBITE-EPIDOTE-AMPHIBOL.FACIES, 104
ALBITE-EPIDOTE-HORNFELS FACIES, 105
ALBITIZATION, 106
ALKALI-CALCIC COMPOSITION, 124
ALKALIC COMPOSITION, 128
ALKALINE METASOMATISM, 130
ALUNITIZATION, 158
AMPHIBOLITE FACIES, 174
AMPHIBOLIZATION, 175
ANATEXIS, 186
ANCHIMETAMORPHISM, 189
ANDESITE LINE, 195
ANORTHOSITIZATION, 217
APOGRANITE, 242
APOPHYSIS, 245

ARCTIC SUITE, 287
ASH FALL, 331
ASH FLOW, 332
ASSIMILATION, 340
ATLANTIC SUITE, 355
AUTOLITH, 379
AUTOLYSIS, 380
AUTOMETAMORPHISM, 383
AUTOMETASOMATISM, 384
BASIFICATION, 432
BATHOLITH, 434
BOMB, 539
BORDER FACIES, 546
BOWEN'S REACTION SERIES, 563
BURIAL METAMORPHISM, 597
C.I.P.W. CLASSIFICATION, 603
CAFEMIC COMPOSITION, 605
CALC-ALKALIC COMPOSITION, 607
CALCIC COMPOSITION, 610
CENOTYPAL, 687
CHLORITIZATION, 752
CINDER, 769
COLOR INDEX, 847
COMAGMATIC GENESIS, 852
CONGRUENT MELTING, 893
CONSANGUINITY, 903
CONTACT METAMORPHISM, 915
CONTACT METASOMATISM, 916
CONTAMINATION, 917
COOLING, 942
CORDIERITE-AMPHIBOLITE FACIES, 954
CRYSTALLINE ROCK, 1022
CRYSTALLIZATION, 1025
CRYSTALLIZATION INDEX, 1026
CRYSTALLOBLAST, 1027
CUMULATE, 1032
DECCAN BASALT, 1080
DEUTERIC EFFECT, 1145
DIAPHTHORESIS, 1162
DIATREME, 1171
DIFFERENTIATION, 1178
DIFFERENTIATION INDEX, 1179
DIKE, 1184
DIKE SWARM, 1185
DYNAMIC METAMORPHISM, 1266
DYNAMOTHERMAL METAMORPHISM, 1269
ECLOGITE FACIES, 1296
EJECTA, 1315
EMPLACEMENT, 1363
EPIDOTE-AMPHIBOLITE FACIES, 1401
EPIDOTIZATION, 1402
EPIMAGMATIC ORIGIN, 1405
EPIZONE, 1408
ERUPTION, 1432
EXPERIMENTAL PETROLOGY, 1477
FEEDER, 1522
FELDSPATHIZATION, 1525

STRATOVOLCANO, 4283
STROMBOLIAN TYPE ERUPTION, 4297
SUBALKALIC COMPOSITION, 4317
SUBALUMINOUS COMPOSITION, 4318
SUBVOLCANIC PROCESS, 4343
SYNTEXIS, 4400
TAXITE, 4427
THERMAL METAMORPHISM, 4498
THOLEIITIC COMPOSITION, 4516
TUFFLAVA, 4609
ULTRABASIC COMPOSITION, 4625
ULTRAMETAMORPHISM, 4637
URALITIZATION, 4688
VOLATILIZATION, 4758
VOLCANIC ASH, 4759
VOLCANIC BELT, 4760
VOLCANIC BRECCIA, 4761
VOLCANIC CONE, 4762
VOLCANIC DOME, 4763
VOLCANIC GLASS, 4764
VOLCANIC ORIGIN, 4765
VOLCANIC VENT, 4767
VOLCANISM, 4768
VOLCANO, 4769
VOLCANOCLASTIC, 4772
VOLCANOLOGY, 4775
VULCANIAN TYPE ERUPTION, 4777
XENOLITH, 4838
ZEOLITE FACIES, 4857
ZEOLITIZATION, 4858

IGNS   IGNEOUS ROCKS

ALKALI=BASALT, 123
ALKALI=GRANITE, 125
ALKALI=SYENITE, 127
ANDESITE, 194
ANORTHOSITE, 216
APLITE, 239
BASALT, 425
BASANITE, 427
DIORITE, 1194
DOLERITE, 1234
ESSEXITE, 1435
FELDSPATHOIDIC ROCK, 1526
FOIDITE, 1601
FOYAITE, 1629
GABBRO, 1672
GRANITE, 1845
GRANITOID, 1851
GRANODIORITE, 1853
GRANOGABBRO, 1854
HORNBLENDITE, 2056
IGNEOUS ROCKS, 2176
IGNIMBRITE, 2177
KIMBERLITE, 2391

LAMPROPHYRE, 2423
LATITE, 2449
MICROGRANITE, 2818
MONZONITE, 2933
OBSIDIAN, 3083
PEGMATITE, 3327
PERIDOTITE, 3352
PHONOLITE, 3401
PICRITE, 3438
PLUTONIC ROCK, 3519
PYROXENITE, 3704
QUARTZ DIORITE, 3711
RHYODACITE, 3860
RHYOLITE, 3861
SHONKINITE, 4080
SYENITE, 4379
THERALITE, 4491
TRACHYANDESITE, 4561
TRACHYBASALT, 4562
TRACHYTE, 4564
TUFF, 4607
TUFFITE, 4608
ULTRABASITE, 4636
VOLCANIC ROCK, 4766

INST   INSTRUMENTS, EQUIPMENT

ACCELEROMETER, 28
ARTIFICIAL SATELLITE, 322
DIP METER, 1197
DRILLING EQUIPMENT, 1255
DRILLING VESSEL, 1257
ELECTRON MICROSCOPE, 1348
EXTENSOMETER, 1492
FILTER, 1545
GEOPHONE, 1750
GEOPHYSICAL INSTRUMENTS, 1753
GRADIOMETER, 1843
GRAVIMETER, 1873
HYDROMETER, 2123
HYDROPHONE, 2125
INSTRUMENTATION, 2258
MAGNETIC COMPASS, 2630
MAGNETOMETER, 2650
MICROSCOPE, 2831
NEUTRON=SOIL=MOISTURE METER, 3018
OCEAN=BOTTOM SEISMOGRAPH, 3089
PENETROMETER, 3341
PYCNOMETER, 3691
RADIOMETER, 3737
RECEPTION EQUIPMENT, 3764
REFRACTOMETER, 3789
SAMPLER, 3934
SEISMIC EQUIPMENT, 4013
SEISMOGRAPH, 4024
SEISMOMETER, 4027

SIEVE, 4096
SONOBUOY, 4182
SPARKER, 4191
SPECTROMETER, 4199
STRAINMETER, 4268
TILTMETER, 4534
UNIVERSAL STAGE, 4662
VANE APPARATUS, 4711
WELL SCREEN, 4817

ISOT   GEOCHRONOLOGY, ISOTOPE GEOLOGY

ABSOLUTE AGE, 10
ARGON=ARGON DATING, 297
ATMOSPHERIC ARGON, 359
BERYLLIUM DATING, 472
CARBON RATIO, 645
CARBON=14 DATING, 646
CONCORDANT AGE, 878
D/H RATIO, 1058
DATING, 1073
DAUGHTER PRODUCT, 1074
DECREPITATION, 1086
DEUTERIUM, 1146
DISCORDANT AGE, 1216
END PRODUCT, 1367
EXCESS ARGON, 1463
EXPOSURE AGE, 1489
FISSION TRACKS, 1555
FISSION=TRACK DATING, 1556
GEOCHRONOLOGIC OVERPRINT, 1723
GEOCHRONOLOGY, 1725
HALF=LIFE PERIOD, 1942
HEAVY ISOTOPE, 1971
HELIUM AGE METHOD, 1979
HYDRATION OF GLASS, 2090
IONIUM=THORIUM DATING, 2302
IONIUM=URANIUM DATING, 2303
ISOCHRON, 2323
ISORAD, 2340
ISORAT, 2341
ISOTOPE, 2348
ISOTOPE DILUTION, 2349
ISOTOPE EFFECT, 2350
ISOTOPE GEOLOGY, 2351
ISOTOPE RATIO, 2352
ISOTOPIC FRACTIONATION, 2353
LEAD=ALPHA AGE METHOD, 2467
LEAD=LEAD DATING, 2468
LIGHT ISOTOPE, 2500
LUTETIUM=HAFNIUM DATING, 2608
OXYGEN RATIO, 3227
PARENT PRODUCT, 3303
PARTICLE=TRACK DATING, 3309
POTASSIUM=ARGON DATING, 3580
POTASSIUM=CALCIUM DATING, 3581

PROTACTINIUM—IONIUM DATING, 3643
RACEMIZATION, 3717
RADIATION DAMAGE, 3721
RADIOACTIVE DECAY, 3725
RADIOACTIVE ISOTOPE, 3727
RADIOACTIVE SERIES, 3728
RADIOCARBON, 3733
RHENIUM—OSMIUM DATING, 3850
RUBIDIUM—STRONTIUM DATING, 3910
STABLE ISOTOPE, 4228
SULFUR RATIO, 4355
THORIUM—LEAD DATING, 4518
TRITIUM DATING, 4601
URANIUM—DISEQUILIBRIUM DATING, 4691
URANIUM—LEAD DATING, 4692
URANIUM—THORIUM—LEAD DATING, 4693
WHOLE ROCK, 4824

MARI   MARINE GEOLOGY

ABYSS, 17
ABYSSAL GAP, 20
ABYSSAL HILL, 21
ABYSSAL PLAIN, 22
AIR—SEA INTERFACE, 94
ATOLL, 362
BARRIER BAR, 417
BATHYMETRIC CHART, 437
BATHYMETRY, 438
CIRCULATION, 773
CONTINENTAL MARGIN, 925
CONTINENTAL RISE, 927
CONTINENTAL SHELF, 929
CONTINENTAL SLOPE, 930
CORAL REEF, 949
DEEP—SEA CHANNEL, 1089
DEEP—WATER ZONE, 1093
DEPTH INDICATOR, 1123
DEPTH OF COMPENSATION, 1124
FORESHORE, 1616
GEOSECS, 1761
GEOSYNCLINAL SEA, 1764
HALMYROLYSIS, 1946
ICEBERG, 2163
INNER SHELF, 2244
INNER SLOPE, 2245
LONGSHORE CURRENT, 2568
MARGINAL PLATEAU, 2673
MARGINAL SEA, 2674
MARINE GEOLOGY, 2679
MARINE SILL, 2687
MID—OCEANIC RIDGE, 2840
NEPHELOID LAYER, 3009
OCEAN, 3084
OCEAN BASIN, 3085
OCEAN CIRCULATION, 3086

OCEAN DRILLING PROGRAM, 3087
OCEAN FLOOR, 3088
OCEANIC, 3090
OCEANOGRAPHY, 3093
OCEANOLOGY, 3094
OFFSHORE, 3102
OUTER SHELF, 3209
PYCNOCLINE, 3689
SANDWAVE, 3941
SEA, 3974
SEA BOTTOM, 3975
SEA ICE, 3976
SEA LEVEL, 3977
SEA WATER, 3978
SEA WAVE, 3979
SEA—LEVEL CHANGE, 3981
SEAMOUNT, 3984
SHELF, 4067
SHELF SEA, 4069
SHOAL, 4074
STRAIT, 4269
SUBMARINE CANYON, 4326
SUBMARINE MORPHOLOGY, 4329
SUBMARINE RIDGE, 4330
SUBMARINE VALLEY, 4332
SUBMARINE VOLCANO, 4333
THERMOCLINE, 4503
TIDE, 4529
TRENCH, 4587
TSUNAMI, 4604

MATH   MATHEMATICAL GEOLOGY

ABUNDANCE, 16
ACCURACY, 34
ALGOL, 119
ALGORITHM, 121
ANALOG SIMULATION, 180
ANALOG TECHNIQUE, 181
APPROXIMATION, 249
AUTOCORRELATION, 378
AVERAGE, 389
AXIS, 393
BINOMIAL DISTRIBUTION, 480
BOOLEAN ALGEBRA, 542
CLUSTER ANALYSIS, 799
COBOL, 819
COMPUTER, 870
COMPUTER LANGUAGE, 871
COMPUTER PROGRAM, 872
CORRECTION, 960
CORRELATION COEFFICIENT, 962
CORRESPONDENCE ANALYSIS, 963
CYBERNETICS, 1041
DATA ACQUISITION, 1065
DATA ANALYSIS, 1066

DATA BANK, 1067
DATA HANDLING, 1068
DATA PROCESSING, 1069
DATA RETRIEVAL, 1070
DATA STORAGE, 1071
DATA TABLE, 1072
DECONVOLUTION, 1084
DENDROGRAM, 1111
DERIVATIVE, 1125
DETERMINISTIC MODEL, 1141
DEVIATION, 1149
DIGITAL TECHNIQUE, 1183
DISCRIMINANT ANALYSIS, 1218
DISTRIBUTION, 1230
EQUATION, 1411
ERROR, 1431
FACTOR ANALYSIS, 1502
FILTERING, 1546
FINITE DIFFERENCE ANALYSIS, 1550
FINITE ELEMENTS, 1551
FOURIER ANALYSIS, 1628
FREQUENCY, 1644
FREQUENCY DISTRIBUTION, 1645
FUNCTION, 1665
GEOSTATISTICS, 1762
GRAPHIC, 1863
HARMONICS, 1957
HISTOGRAM, 2017
INFORMATICS, 2228
INTERPOLATION, 2282
KURTOSIS, 2401
LAPLACE TRANSFORMATION, 2434
LEAST SQUARES, 2473
LOGNORMAL DISTRIBUTION, 2563
MARKOV PROCESS, 2691
MATHEMATICAL ANALYSIS, 2708
MATHEMATICAL EXPECTATION, 2709
MATHEMATICAL GEOGRAPHY, 2710
MATHEMATICAL GEOLOGY, 2711
MATHEMATICAL LAW, 2712
MATHEMATICAL MATRIX, 2713
MATHEMATICAL METHODS, 2714
MATHEMATICAL MODEL, 2715
MATHEMATICS, 2716
MATRIX ALGEBRA, 2717
MAXIMUM, 2722
MICROCOMPUTER, 2807
MINICOMPUTER, 2881
MINIMUM, 2882
MONTE CARLO METHOD, 2932
MULTIVARIATE STATISTICS, 2968
NOMOGRAM, 3045
NORMAL DISTRIBUTION, 3055
ONE—DIMENSIONAL MODEL, 3120
POISSON DISTRIBUTION, 3531
PROBABILITY, 3617
PUNCH CARD, 3686

RADIUS, 3740
REGRESSION STATISTICS, 3801
SIMULATION, 4114
SKEWNESS, 4128
SMOOTHING, 4141
SPHERICAL HARMONIC ANALYSIS, 4206
STANDARD DEVIATION, 4234
STATISTICAL ANALYSIS, 4239
STATISTICAL DISTRIBUTION, 4240
STATISTICAL FUNCTION, 4241
STATISTICAL MODEL, 4242
STATISTICAL POPULATION, 4243
STATISTICAL TREND, 4244
STATISTICS, 4245
STOCHASTIC PROCESS, 4256
TORTUOSITY, 4552
TREND SURFACE ANALYSIS, 4588
TWO=DIMENSIONAL MODEL, 4622
VARIABLE, 4715
VARIANCE ANALYSIS, 4716
VARIATION, 4717
VARIOGRAM, 4718
VECTOR, 4723
WEIGHTING, 4812

METH METHODOLOGY

ABSORPTION SPECTROSCOPY, 14
ACETOLYSIS, 35
ACTIVATION ANALYSIS, 54
ALPHA=RAY SPECTROSCOPY, 148
ANALYSIS, 182
ANALYTICAL METHOD, 183
ARRAY, 306
ATOMIC ABSORPTION, 363
AUTOMATED CARTOGRAPHY, 381
AUTOMATION, 382
BOREHOLE, 549
CARTOGRAPHY, 660
CHROMATOGRAPHY, 760
COLOR TRACER, 848
COLORIMETRY, 849
COMPLEXOMETRY, 862
CONSOLIDOMETER TEST, 908
DESALINATION, 1129
DIFFERENTIAL THERMAL ANALYSIS, 1176
DREDGING, 1251
DRILLING, 1254
ELECTRON DIFFRACTION ANALYSIS, 1346
ELECTRON MICROPROBE, 1347
ELECTRON MICROSCOPY, 1349
ELECTRON MICROSCOPY DATA, 1350
ELECTRON=PROBE DATA, 1352
EMISSION SPECTROSCOPY, 1362
ETCHING, 1439
EXTRAPOLATION, 1497

FLAME PHOTOMETRY, 1562
FLOTATION, 1574
FLUORIMETRY, 1587
FRAGMENTATION, 1637
GAMMA RAY SPECTROSCOPY, 1677
GAS CHROMATOGRAPHY, 1686
GASIFICATION, 1695
GONIOMETRY, 1836
GRAB SAMPLING, 1839
GRINDING, 1899
HOLOGRAPHY, 2029
HYDROMETALLURGY, 2122
IMMERSION METHOD, 2184
INFRARED PHOTOGRAPHY, 2232
INFRARED SPECTROSCOPY, 2234
INTERFEROMETRY, 2263
ION PROBE, 2301
LEVELING, 2492
MAJOR=ELEMENT ANALYSIS, 2658
MAPPING, 2670
MASS SPECTROSCOPY, 2700
METALLURGY, 2777
METHODOLOGY, 2802
MICROSCOPY, 2833
MICROWAVE SPECTROSCOPY, 2839
MINOR=ELEMENT ANALYSIS, 2888
MODAL ANALYSIS, 2902
MOLECULAR SPECTROSCOPY, 2914
MORPHOMETRY, 2941
MORPHOSCOPY, 2942
MOSSBAUER SPECTROSCOPY, 2947
MULTISPECTRAL ANALYSIS, 2966
NETWORK, 3013
NEUTRON ACTIVATION, 3015
NEUTRON DIFFRACTION ANALYSIS, 3016
NUCLEAR MAGNETIC RESONANCE, 3064
OPTICAL SPECTROSCOPY, 3142
ORE MICROSCOPY, 3158
PALEOTHERMOMETRY, 3270
PANNING, 3283
PELLETIZING, 3338
PHI SCALE, 3397
PHOTOGRAMMETRY, 3411
PHOTOGRAPHY, 3412
PHOTOMETRY, 3413
POLAROGRAPHY, 3536
POLISHED SECTION, 3538
POTENTIOMETRY, 3584
POWDER DIFFRACTION, 3587
PREPARATION, 3605
PROBE, 3618
PUMPING TEST, 3685
PYCNOMETRY, 3692
PYROLYSIS, 3696
PYROMETALLURGY, 3697
QUALITATIVE METHOD, 3706
QUANTITATIVE METHOD, 3708

RADAR, 3718
RADIO SPECTROSCOPY, 3723
RADIOACTIVE TRACER, 3729
RADIOGRAPHY, 3734
RAMAN SPECTROSCOPY, 3745
REFINING, 3781
REFRACTOMETRY, 3790
REMOTE SENSING, 3813
ROAD LOG, 3876
ROTARY DRILLING, 3900
SAMPLE, 3933
SAMPLING, 3935
SCANNING ELECTRON MICROSCOPY, 3957
SEISMIC SPECTRAL=ANALYSIS, 4017
SEM DATA, 4034
SEPARATION, 4039
SINGLE=CRYSTAL ANALYSIS, 4116
SOIL SAMPLING, 4156
SPECTROPHOTOMETRY, 4200
SPECTROSCOPY, 4201
STAINING, 4230
STANDARD METHOD, 4236
SUBMERSIBLE, 4335
TEM DATA, 4451
THERMAL ANALYSIS, 4494
THERMAL DEMAGNETIZATION, 4496
THERMOGRAM, 4506
THERMOGRAVIMETRY, 4507
THERMOMAGNETIC ANALYSIS, 4510
THIN SECTION, 4514
TRACER, 4560
TRANSFER, 4568
TRANSMISSION METHOD, 4575
TRIANGULATION, 4590
VACUUM FUSION ANALYSIS, 4705
WET METHOD, 4821
X=RAY ANALYSIS, 4833
X=RAY DIFFRACTION PATTERN, 4834
X=RAY FLUORESCENCE ANALYSIS, 4835
X=RAY SPECTROSCOPY, 4836

MINE MINERALOGY

ALLOGENIC MINERAL, 136
ANHEDRAL CRYSTAL, 203
ANTISTRESS MINERAL, 232
ASTERISM, 344
ATOMIC PACKING, 365
AUTHIGENESIS, 375
AUTHIGENIC MINERAL, 376
BASAL CLEAVAGE, 423
BAVENO TWIN LAW, 442
BECKE LINE, 451
BIAXIAL, 475
BIAXIAL CRYSTAL, 476
BIREFRINGENCE, 519

BRAGG ANGLE, 570
BRAVAIS LATTICE, 574
CIRCULAR POLARIZATION, 772
CLAY MINERALOGY, 789
CLOSE PACKING, 798
COEXISTING MINERALS, 826
COLOR CENTER, 846
CRYSTAL, 1011
CRYSTAL CHEMISTRY, 1012
CRYSTAL DEFECT, 1013
CRYSTAL DISLOCATION, 1014
CRYSTAL FIELD, 1015
CRYSTAL FORM, 1016
CRYSTAL GROWTH, 1017
CRYSTAL LATTICE, 1018
CRYSTAL OPTICS, 1019
CRYSTAL STRUCTURE, 1020
CRYSTAL SYSTEM, 1021
CRYSTALLINITY, 1023
CRYSTALLITE, 1024
CRYSTALLOGRAPHY, 1028
CUBIC SYSTEM, 1030
DIADOCHY, 1154
DIPLOID, 1201
DOMAIN STRUCTURE, 1239
DRUSE, 1260
EPITAXY, 1406
EUHEDRAL CRYSTAL, 1447
FLUID INCLUSION, 1579
GAS INCLUSION, 1690
GEL MINERAL, 1703
GEMMOLOGY, 1706
GLIDE TWIN, 1820
HABIT, 1936
HAUY'S LAW, 1960
HEXAGONAL SYSTEM, 2002
HOMEOMORPHISM, 2036
IDIOMORPHISM, 2171
INCLUSION, 2200
INTERLAYER, 2268
ISOMETRIC SYSTEM, 2332
ISOSTRUCTURAL MINERAL, 2345
LATTICE CONSTANT, 2451
MAGNETIC DOMAIN, 2631
MAGNETIC MINERALS, 2638
METAMICT, 2778
MILLER INDICES, 2860
MINERAL, 2862
MINERAL ASSEMBLAGE, 2863
MINERAL CLEAVAGE, 2864
MINERAL COMPOSITION, 2865
MINERAL STRIATION, 2874
MINERALOGICAL LOCALITY, 2878
MINERALOGY, 2880
MISCIBILITY GAP, 2894
MIXED=LAYER MINERAL, 2898
MONOCLINIC SYSTEM, 2921

NEW MINERAL DATA, 3023
NUCLEUS, 3067
ORDER=DISORDER, 3149
ORTHORHOMBIC SYSTEM, 3194
OVERGROWTH, 3215
PARAGENESIS, 3290
POLYHEDRA, 3548
POLYMORPHISM, 3555
POLYTYPISM, 3558
PSEUDOMORPHISM, 3665
PYRAMIDAL SYSTEM, 3693
REACTION RIM, 3758
RECRYSTALLIZATION, 3770
REFINEMENT, 3780
REPLACEMENT, 3817
RHOMBIC SYSTEM, 3857
SECONDARY MINERAL, 3989
SPACE GROUP, 4188
STRESS MINERAL, 4291
STRUCTURAL CRYSTALLOGRAPHY, 4303
SUPERSTRUCTURE, 4361
SURFACE DEFECT, 4363
SYNTHESIS, 4401
TRICLINIC SYSTEM, 4593
TRIGONAL SYSTEM, 4595
TWIN, 4621
TYPOMORPHIC MINERAL, 4627
UNDULATORY EXTINCTION, 4657
UNIT CELL, 4661
WATER OF CRYSTALLIZATION, 4793
WATER OF DEHYDRATION, 4794

MING   MINERAL GROUPS

AMPHIBOLE, 172
ANTIMONATES, 228
ARSENATES, 309
BORATES, 544
BROMIDES, 580
CARBONATES, 651
CHLORIDES, 747
CHROMATES, 759
CLAY MINERALS, 790
FELDSPAR, 1524
FLUORIDES, 1586
GARNET, 1683
GRAPHITE, 1866
HALIDES, 1944
HYDROXIDES, 2136
MICA, 2804
MOLYBDATES, 2917
NATIVE ELEMENTS, 2988
NEPHELINE, 3008
NIOBATES, 3029
NIOBOTANTALATES, 3031
NITRATES, 3032

OLIVINE, 3118
OXIDES, 3222
PHOSPHATES, 3403
PLATINUM GROUP, 3505
PYROXENE, 3702
QUARTZ, 3710
SELENATES, 4029
SERPENTINE, 4045
SILICA, 4097
SILICATES, 4098
SULFATES, 4349
SULFIDES, 4351
SULFOSALTS, 4352
TANTALATES, 4418
TELLURATES, 4448
TUNGSTATES, 4612
VANADATES, 4709
ZEOLITE, 4856

MINI   MINING GEOLOGY

AUGER DRILLING, 371
CAVINGS, 680
CORE DRILLING, 956
DEVELOPMENT DRILLING, 1148
DIRECTIONAL DRILLING, 1208
DRESSING, 1252
DRILLABILITY, 1253
EXPLOITATION, 1482
FULL HOLE DRILLING, 1661
MINE, 2861
MINING, 2883
MINING ENGINEERING, 2884
MINING GEOLOGY, 2885
OFFSHORE DRILLING, 3103
OFFSHORE WELL, 3104
OIL WELL, 3112
OPEN=PIT MINING, 3129
ORE DRESSING PLANT, 3155
OVERBURDEN, 3213
PIT, 3459
PLUGGING, 3514
PRODUCTION WELL, 3627
QUARRY, 3709
RECOVERY, 3768
ROTARY=PERCUSSION DRILLING, 3901
SOLUTION MINING, 4179
SPOIL, 4217
STRIP MINING, 4294
TAILING, 4413
TURBODRILLING, 4619
UNDERGROUND MINING, 4650
VIBRATORY DRILLING, 4739
WELL, 4813
WELL COMPLETION, 4814
WELL HEAD, 4815

MISC   GENERAL (TIME, SPACE, ORGANIZATION, INFORMATION)

ADIAGNOSTIC, 66
AFFINITY, 76
ALTITUDE, 155
APICAL, 237
ARCHEOLOGICAL SITE, 280
ARCHEOLOGY, 281
ARCUATE, 288
ATLAS, 356
BALANCE, 407
BIBLIOGRAPHY, 477
BIOGRAPHY, 498
BIOLOGY, 502
BOOK REVIEW, 541
BUREAU OF MINES, 595
CARTOGRAM, 659
CASE STUDIES, 663
CHARACTERISTICS, 710
CLASSIFICATION, 782
CONSTANT, 909
CONTOUR MAP, 932
CONTROL, 936
COSTS, 976
CRITICAL REVIEW, 995
DATA, 1064
DEPTH, 1122
DEVELOPMENT, 1147
DIAGNOSIS, 1156
DIAGRAM, 1157
DISCOVERY, 1217
DRAWINGS, 1250
EAST, 1288
ECCENTRICITY, 1290
EDUCATION, 1309
EQUATOR, 1413
EVALUATION, 1457
EXPLANATORY NOTE, 1480
FEASIBILITY STUDY, 1520
FINANCING, 1548
FORECASTING MAP, 1611
GALACTIC YEAR, 1674
GENERATION, 1708
GEODETIC COORDINATE, 1728
GEOLOGICAL SURVEY, 1739
GEOLOGIST, 1740
GEOMETRY, 1746
GLASS, 1811
GLOSSARY, 1824
GRADIENT, 1842
GRAPHIC METHOD, 1864
HETEROGENEITY, 1993
HETEROGENEOUS, 1994
HISTORY, 2020
HOMOGENEITY, 2041
HOMOGENIZATION, 2042

HORIZONTAL, 2050
HYPOTHESIS, 2150
IN SITU, 2197
INCOHERENT, 2201
INDEX MAPS, 2207
INDICATOR, 2212
INFORMATION SYSTEM, 2229
INORGANIC, 2247
INSTITUTION, 2257
INTERNAT. AGREEMENT, 2276
INTERNAT. GEOL. CONGRESS, 2277
INTERNAT. GEOL. CORREL. PROGR., 2278
INTERPRETATION, 2283
INVENTORY, 2295
IRREVERSIBILITY, 2313
ISOPLETH, 2337
LARGE=SCALE, 2436
LATITUDE, 2450
LAW, 2460
LEARNED SOCIETY, 2472
LEGEND, 2475
LEGISLATION, 2476
LENGTH, 2477
LEXICON, 2494
LOCAL, 2557
LONGITUDE, 2565
MANPOWER, 2664
MANUAL, 2667
MAP SCALE, 2669
MEETING, 2737
MODEL, 2904
MONOGRAPH, 2924
MOVEMENT, 2951
MUSEUM, 2970
NEW, 3019
NEW DATA, 3020
NEW DESCRIPTION, 3021
NEW METHODS, 3022
NEW NAME, 3024
NOMENCLATURE, 3042
NORTH, 3057
OBJECTIVE, 3076
OBSERVATION, 3081
OBSERVATORY, 3082
OPERATIONS, 3130
ORGANIC, 3164
PARAMETER, 3299
PATENT, 3314
PATTERN, 3316
PERSONAL BIBLIOGRAPHY, 3365
PHILOSOPHY, 3398
PLANE, 3470
PLANNING, 3489
POLICY, 3537
PREDICTION, 3598
PRIMITIVE, 3614
PRIORITY, 3615

PROGNOSIS, 3631
PROGRAM, 3634
PROGRESS REPORT, 3635
PROJECT, 3636
PUBLICATION, 3681
RATE, 3753
RELATIONSHIP, 3803
REMOTENESS, 3814
REPORT, 3818
RESEARCH, 3822
REVIEW ARTICLE, 3843
REVISION, 3844
SHAPE, 4054
SMALL=SCALE, 4140
SOUTH, 4187
STANDARD MATERIALS, 4235
STANDARDIZATION, 4238
SUBDIVISION, 4320
SUMMARY ARTICLE, 4357
SYMPOSIUM, 4387
SYSTEM ANALYSIS, 4404
TECHNICAL COOPERATION, 4430
TECHNIQUE, 4431
TECHNOLOGY, 4432
TERMINOLOGY, 4462
THEORY, 4490
THERMAL, 4493
THESIS, 4512
THICKNESS, 4513
TIME, 4535
TOPOGRAPHIC MAP, 4546
TRANSVERSAL, 4579
TREATISE, 4584
VARIABILITY, 4714
VERTICAL ORIENTATION, 4735
VOLUME, 4776
WEST, 4819
YEARBOOK, 4847
YOUNG, 4850

PALE   PALEONTOLOGY, PALEOECOLOGY

ABYSSAL ENVIRONMENT, 19
ADAPTATION, 63
ADAPTIVE RADIATION, 64
AEROBIC CONDITION, 73
AHERMATYPIC TAXON, 92
ALLOMETRY, 138
ANAEROBIC CONDITION, 178
ANAEROBIC TAXON, 179
ANATOMY, 188
APERTURE, 235
ARTIFACT, 319
ASSEMBLAGE, 338
ASSOCIATION, 341
AUTOTROPHIC ORGANISM, 386
BATHYAL ENVIRONMENT, 436

BENTHONIC ENVIRONMENT, 464
BENTHOS, 465
BIOCENOSIS, 481
BIOCHORE, 483
BIOGEOGRAPHY, 497
BIOLOGICAL CYCLE, 501
BIOMASS, 503
BIOMETRY, 504
BIOTA, 513
BIOTOPE, 514
BIOTYPE, 516
BRACKISH=WATER ENVIRONMENT, 569
BURROW, 600
BYSSUS, 601
CARAPACE, 639
CAST, 665
CHEMICAL FOSSILS, 727
COLONIAL TAXON, 844
COMMENSALISM, 853
COMMUNITY, 855
CONODONT APPARATUS, 900
CONTINENTAL ENVIRONMENT, 923
COPROLITE, 947
CORALLITE, 952
CRANIUM, 980
DARWINISM, 1062
DEEP=SEA ENVIRONMENT, 1090
DELTAIC ENVIRONMENT, 1104
DENTITION, 1115
DERMAL STRUCTURE, 1126
DIMORPHISM, 1189
ECOLOGY, 1297
ECOSYSTEM, 1301
EGG, 1311
ENDEMIC POPULATION, 1368
EOLIAN ENVIRONMENT, 1391
EPIBIOTISM, 1398
ESTUARINE ENVIRONMENT, 1436
EURYHALINE ORGANISM, 1449
EURYTHERMAL ORGANISM, 1450
EUXINIC ENVIRONMENT, 1456
EVOLUTION, 1461
EXINE, 1468
EXTINCT TAXON, 1495
EXTINCTION, 1496
FAUNA, 1516
FAUNAL LIST, 1517
FAUNAL PROVINCE, 1518
FECAL PELLET, 1521
FEEDING TRACK, 1523
FLORA, 1571
FLORAL PROVINCE, 1572
FLUVIAL ENVIRONMENT, 1593
FOSSIL, 1619
FOSSIL MAN, 1622
FOSSIL WOOD, 1624
FOSSILIFEROUS DEPOSIT, 1625

FOSSILIZATION, 1626
FRESHWATER ENVIRONMENT, 1648
FRUCTIFICATION, 1657
FUNCTIONAL MORPHOLOGY, 1666
GASTROZOOID, 1697
GEMMATION, 1705
GENOMORPH, 1710
GENOTYPE, 1711
GEOBIOS, 1713
GIGANTISM, 1787
GROWTH LINE, 1922
HABITAT, 1937
HAECKEL'S LAW, 1939
HALOPHILIC ORGANISM, 1948
HALOPHYTE, 1949
HEKISTOTHERM PLANT, 1975
HEREDITY, 1986
HERMATYPIC TAXON, 1987
HETEROMORPH ORGANISM, 1995
HETEROSPOROUS PLANT, 1997
HETEROTROPHIC ORGANISM, 1998
HINGE, 2014
HISTOLOGY, 2018
HOLOPHYTE, 2030
HOLOSTOMATOUS, 2032
HOLOTYPE, 2034
HOMEOMORPHY, 2037
HOMOLOGY, 2044
HOMONYMY, 2045
HYDRARCH, 2088
HYDROSPIRE, 2127
HYGROPHYLE ORGANISM, 2138
HYPODIGM, 2148
ICHNOFOSSIL, 2164
ICHNOLOGY, 2165
INTERNAL CAST, 2274
INTERNAL STRUCTURE, 2275
INTERTIDAL ENVIRONMENT, 2286
ISOLATION, 2330
ISOMORPH ORGANISM, 2333
JAW, 2360
LACUSTRINE ENVIRONMENT, 2407
LAGOONAL ENVIRONMENT, 2412
LAMARCKISM, 2417
LEAF, 2469
LECTOTYPE, 2474
LIFE, 2499
LITHOPHYTE, 2534
LITTORAL ENVIRONMENT, 2545
LIVING, 2548
LIVING FOSSIL, 2549
LOCOMOTION, 2558
MARINE ECOLOGY, 2676
MARINE ENVIRONMENT, 2677
MEGASPORES, 2739
MESOPHYLE ORGANISM, 2756
METAMORPHOSIS, 2789

METANAUPLIUS, 2790
METAPHYTE, 2791
MICROFAUNA, 2812
MICROFLORA, 2814
MICROFOSSIL, 2816
MICROORGANISM, 2826
MICROPALEONTOLOGY, 2827
MIOSPORE, 2892
MONOLETE TAXON, 2925
MONOMORPHIC, 2927
MUTATION, 2971
NANNOFOSSIL, 2985
NEKTON, 2997
NEOTENY, 3006
NEOTYPE, 3007
NERVOUS SYSTEM, 3012
NEW TAXON, 3025
NOMINAL TAXON, 3043
NOMOGENESIS, 3044
NUTRITION, 3073
ODONTOLOGY, 3099
ONTOGENY, 3121
ORGANIC LIFE, 3169
ORGANISM, 3176
ORNAMENTATION, 3182
ORTHOGENESIS, 3191
ORYCTOCOENOSIS, 3196
OSTEOLOGY, 3201
OTOLITH, 3205
OXYLOPHYTE, 3228
PALEOBIOLOGY, 3243
PALEOBOTANIC PROVINCE, 3244
PALEOBOTANY, 3245
PALEOECOLOGY, 3252
PALEOENVIRONMENT, 3253
PALEONTOLOGICAL RECONSTRUCTION, 3264
PALEONTOLOGY, 3265
PALEOZOOLOGY, 3272
PALYNOLOGY, 3278
PARALIC ENVIRONMENT, 3294
PARALLELISM, 3297
PARASITISM, 3300
PARATYPE, 3301
PATHOLOGY, 3315
PELAGIC ENVIRONMENT, 3330
PHAGOTROPHIC ORGANISM, 3389
PHREATOPHYTE, 3417
PHYLOGENY, 3423
PHYLUM, 3424
PHYSIOLOGY, 3432
PHYTOCOENOSIS, 3434
PHYTOPHAGOUS ORGANISM, 3436
PHYTOPLANKTON, 3437
PLANKTIVOROUS ANIMAL, 3486
PLANKTON, 3487
PLANKTONIC TAXON, 3488
PLANTAE, 3491

POLLEN, 3539
PREDATION, 3597
PRESERVATION, 3606
PROTOTYPE, 3653
PROVINCIALITY, 3657
PSAMMOFAUNA, 3661
PSYCHROPHYTE, 3669
REEF ENVIRONMENT, 3777
REEFBUILDER, 3779
REPRODUCTION, 3820
RESPIRATION, 3835
REWORKED FOSSIL, 3846
ROCK-BORING ORGANISM, 3889
SCLERITE, 3966
SCOLECODONT, 3967
SCOLITE, 3968
SEED, 4007
SEPTUM, 4041
SEXUAL DIMORPHISM, 4049
SHALLOW-WATER ENVIRONMENT, 4053
SHELF ENVIRONMENT, 4068
SHELL, 4071
SILICIFIED WOOD, 4101
SKELETON, 4127
SLOPE ENVIRONMENT, 4136
SPECIES DIVERSITY, 4193
SPECIFIC FAUNA, 4194
SPECIFIC FLORA, 4195
SPINE, 4211
SPORE, 4219
SPOROMORPH, 4220
SUBLITTORAL ENVIRONMENT, 4325
SUBSTRATE, 4339
SUBTIDAL ENVIRONMENT, 4341
SUTURE, 4374
SYMBIOSIS, 4381
SYNONYM, 4396
TAPHONOMY, 4420
TAXONOMY, 4428
TERATOLOGY, 4459
TERRESTRIAL ENVIRONMENT, 4467
TEST, 4471
THANATOCENOSE, 4483
TIDAL ENVIRONMENT, 4526
TOOTH, 4544
TRACE FOSSIL, 4557
TRACK-TRAIL, 4565
TYPE SPECIMEN, 4626
ULTRASTRUCTURE, 4640
VALVE, 4708
XEROPHILE PLANT, 4843
XEROPHYTE, 4844
ZOECIUM, 4864
ZOOID, 4871

ACANTHARIA, 25
ACANTHODII, 26
ACRANIA, 46
ACRITARCHA, 48
ACROTRETIDA, 49
ACTINIARIA, 50
ACTINOCERATOIDEA, 52
AGNATHA, 86
AGNOSTIDA, 87
AGONIATITIDA, 88
ALCYONACEA, 108
ALGAE, 111
ALLOGROMIINA, 137
ALVEOLINELLIDAE, 159
AMBLYPODA, 161
AMMODISCACEA, 164
AMMONITIDA, 166
AMMONITINA, 167
AMMONOIDEA, 169
AMPHIBIA, 171
AMPHINEURA, 176
ANAPSIDA, 184
ANASPIDA, 185
ANGIOSPERMAE, 199
ANISOPLEURA, 209
ANNELIDA, 212
ANOMALODESMATA, 213
ANTHOCYATHEA, 219
ANTHOZOA, 220
ANTIARCHI, 223
ANTIPATHARIA, 230
ANUROMORPHA, 234
APLACOPHORA, 238
APSIDOSPONDYLI, 252
ARACHNIDA, 269
ARCHAEOCOPIDA, 272
ARCHAEOCYATHA, 273
ARCHAEOCYATHEA, 274
ARCHAEOGASTROPODA, 275
ARCHAEOPTERYGES, 276
ARCHAEORNITHES, 277
ARCHIPOLYPODA, 283
ARCHOSAURIA, 284
ARCOIDA, 285
ARTHRODIRA, 315
ARTHROPLEURIDA, 316
ARTHROPODA, 317
ARTICULATA, 318
ARTIODACTYLA, 324
ASTEROIDEA, 346
ASTEROZOA, 347
ASTRAPOTHERIA, 349
AVES, 390
BACTERIA, 400
BACTRITOIDEA, 402

BATRACHOSAURIA, 439
BELEMNOIDEA, 459
BENNETTITALES, 463
BEYRICHICOPINA, 474
BLASTOIDEA, 526
BLATTOIDEA, 527
BRACHIOPODA, 565
BRACHIOPTERYGII, 566
BRANCHIOPODA, 572
BRANCHIURA, 573
BRYOPHYTA, 584
BRYOZOA, 585
BULIMINACEA, 589
CALCAREOUS ALGAE, 608
CALCISPONGEA, 612
CAMAROIDEA, 623
CAMPTOSTROMATOIDEA, 627
CARNIVORA, 657
CARTERINACEA, 658
CASSIDULINACEA, 664
CATAGRAPHIA, 669
CAYTONIALES, 681
CEPHALOCARIDA, 691
CEPHALODISCIDA, 692
CEPHALOPODA, 693
CERATITIDA, 695
CERATOMORPHA, 696
CERIANTHARIA, 697
CERIANTIPATHARIA, 698
CETACEA, 701
CHAETOGNATHA, 702
CHAROPHYTA, 713
CHEILOSTOMATA, 717
CHELICERATA, 719
CHELONIA, 720
CHILOPODA, 742
CHIROPTERA, 744
CHITINOZOA, 746
CHLOROPHYTA, 754
CHONDRICHTHYES, 755
CHORDATA, 758
CHRYSOPHYTA, 765
CILIATA, 766
CILIOPHORA, 767
CIRRIPEDIA, 776
CLADOCERA, 778
CLYMENIIDA, 800
COCCOLITHOPHORACEAE, 820
COELENTERATA, 823
COENOTHECALIA, 824
COLEOPTEROIDA, 833
COLLEMBOLA, 840
COLUMNARIINA, 851
CONCHOSTRACA, 876
CONDYLARTHRA, 884
CONIFERALES, 895
CONOCARDIOIDA, 899

CONODONTA, 901
CONULARIDA, 937
COPEPODA, 944
CORALLIMORPHARIA, 950
CORALLINACEAE, 951
CORDAITALES, 953
CORYNEXOCHIDA, 966
COTYLOSAURIA, 977
CREODONTA, 987
CRINOIDEA, 991
CRINOZOA, 992
CROCODILIA, 996
CROSSOPTERYGII, 999
CRUSTACEA, 1002
CRYPTODONTA, 1007
CRYPTOSTOMATA, 1009
CTENOSTOMATA, 1029
CYANOPHYTA, 1040
CYCADALES, 1042
CYCADOFILICALES, 1043
CYCLOCYSTOIDEA, 1048
CYCLOSTOMATA, 1050
CYNOMORPHA, 1054
CYSTIPHYLLINA, 1055
CYSTOIDEA, 1056
DASYCLADACEAE, 1063
DECAPODA, 1078
DEINOTHERIOIDEA, 1102
DEMOSPONGEA, 1107
DENDROIDEA, 1112
DERMAPTERA, 1127
DERMOPTERA, 1128
DESMIDS, 1137
DESMOCERATIDA, 1138
DESMOSTYLIA, 1139
DIATOMEAE, 1169
DICOTYLEDONEAE, 1172
DICTYONELLIDINA, 1173
DINOCERATA, 1190
DINOFLAGELLATA, 1191
DINOSAURIA, 1192
DIPLEUROZOA, 1200
DIPLOPODA, 1202
DIPLORHINA, 1203
DIPNOI, 1205
DIPTERA, 1207
DISCOASTERS, 1210
DISCORBACEA, 1213
DOCODONTA, 1233
ECHINODERMATA, 1291
ECHINOIDEA, 1292
ECHINOZOA, 1293
ECTOTROPHA, 1303
EDENTATA, 1305
EDRIOASTEROIDEA, 1307
EDRIOBLASTOIDEA, 1308
ELASMOBRANCHII, 1316

ELEPHANTOIDEA, 1354
EMBRITHOPODA, 1360
ENDOCERATOIDEA, 1369
ENTEROPNEUSTA, 1381
EOCRINOIDEA, 1390
EOSUCHIA, 1395
EQUISETALES, 1417
EUCILIATA, 1442
EUGLENOPHYTA, 1446
EUSELACHII, 1451
FILICALES, 1543
FISSIPEDA, 1557
FORAMINIFERA, 1610
FUNGI, 1667
FUSULINIDAE, 1670
GASTROPODA, 1696
GIGANTOSTRACA, 1788
GINKGOALES, 1789
GLOBIGERINACEA, 1822
GLOSSOPTERIDALES, 1825
GNATHOSTOMI, 1826
GNATHOSTRACA, 1827
GNETALES, 1831
GONIATITIDA, 1835
GRAPTOLITHINA, 1868
GRAPTOLOIDEA, 1869
GYMNOLAEMATA, 1930
GYMNOSPERMAE, 1931
HELICOPLACOIDEA, 1976
HELIOZOA, 1977
HEMICRUSTACEA, 1982
HEMIPTEROIDA, 1983
HETEROCORALLIA, 1991
HETERODONTA, 1992
HEXACORALLA, 2000
HEXACTINIARIA, 2001
HIPPOMORPHA, 2016
HOLOCEPHALI, 2027
HOLOSTEI, 2031
HOLOTHURIOIDEA, 2033
HOMALOZOA, 2035
HOMINIDAE, 2038
HOMO SAPIENS, 2039
HOMOIOSTELEA, 2043
HOMOSTELEA, 2046
HYALOSPONGEA, 2085
HYDROIDA, 2118
HYDROZOA, 2137
HYMENOPTEROIDA, 2140
HYOLITHES, 2141
HYRACOIDEA, 2152
HYSTRICHOMORPHA, 2153
ICHTHYOPTERYGIA, 2166
ICHTHYOSAURIA, 2167
INARTICULATA, 2198
INOCERAMI, 2246
INSECTA, 2251

INSECTIVORA, 2252
INVERTEBRATA, 2297
ISOPLEURA, 2338
KUTORGINIDA, 2402
LABYRINTHODONTIA, 2404
LACERTILIA, 2406
LEPERDITOCOPIDA, 2481
LEPIDOCYSTOIDEA, 2483
LEPIDOSAURIA, 2484
LEPOSPONDYLI, 2485
LICHENES, 2495
LICHIDA, 2497
LINGULIDA, 2513
LIPOSTRACA, 2517
LISSAMPHIBIA, 2523
LITOPTERNA, 2542
LITUOLACEA, 2547
LYCOPODIALES, 2610
LYTOCERATIDA, 2611
MACHAERIDIA, 2614
MALACOSTRACA, 2659
MAMMALIA, 2661
MANTODEA, 2666
MARSUPIALIA, 2695
MASTIGOPHORA, 2706
MAXILLOPODA, 2721
MECOPTEROIDA, 2731
MEROSTOMATA, 2750
MEROSTOMOIDEA, 2751
MESOGASTROPODA, 2754
MESOSAURIA, 2757
METACOPINA, 2763
MICROPHYTOLITH, 2828
MICROPROBLEMATICA, 2830
MILIOLACEA, 2857
MILIOLINA, 2858
MIOMERA, 2891
MODIOMORPHOIDA, 2906
MOLLUSCA, 2916
MONOCOTYLEDONEAE, 2922
MONOCYATHEA, 2923
MONOPLACOPHORA, 2928
MONORHINA, 2929
MONOTREMATA, 2930
MULTITUBERCULATA, 2967
MYODOCOPIDA, 2975
MYOIDA, 2976
MYOMORPHA, 2977
MYRIAPODA, 2980
MYSTACOCARIDA, 2980
MYTILOIDA, 2981
MYXOMYCETES, 2982
NATANTES, 2987
NAUTILOIDEA, 2993
NEANDERTHALIAN, 2994
NEMERTA, 2998
NEOGASTROPODA, 3001

NEORNITHES, 3004
NILSSONIALES, 3028
NODOSARIACEA, 3038
NOEGGERATHIALES, 3040
NOTOSTRACA, 3059
NOTOUNGULATA, 3060
NUCULOIDEA, 3068
NUMMULITIDAE, 3071
OBOLELLIDA, 3079
OCTOCORALLA, 3098
ODONTOPLEURIDA, 3100
ODONTORNITHES, 3101
ONYCHOPHORA, 3122
OPHIDIA, 3131
OPHIOCISTIOIDEA, 3132
OPHIUROIDEA, 3135
OPISTHOBRANCHIA, 3136
ORBITOIDACEA, 3147
ORBITOIDIDAE, 3148
ORNITHISCHIA, 3183
ORTHIDA, 3188
OSTEICHTHYES, 3200
OSTEOSTRACI, 3202
OSTRACODA, 3203
OSTREACEA, 3204
PALAEOHETERODONTA, 3235
PALAEOTAXODONTA, 3237
PALEOCOPIDA, 3250
PALYNOMORPH, 3279
PANTODONTA, 3284
PANTOTHERIA, 3285
PARABLASTOIDEA, 3288
PARACRINOIDEA, 3289
PAUROPODA, 3318
PECTINACEA, 3323
PELECYPODA, 3332
PENTAMERIDA, 3344
PENTASTOMIDA, 3345
PENTOXYLALES, 3346
PERISPHINCTIDA, 3358
PERISSODACTYLA, 3359
PHACOPIDA, 3386
PHAEOPHYTA, 3387
PHOLADOMYIDA, 3399
PHOLIDOTA, 3400
PHYLACTOLAEMATA, 3418
PHYLLOCERATIDA, 3420
PHYLLOPODA, 3422
PHYTOLITHES, 3435
PINNIPEDIA, 3452
PISCES, 3456
PLACENTALIA, 3462
PLACODERMI, 3464
PLANCTOSPHAEROIDEA, 3469
PLATYRRHINA, 3506
PODOCOPIDA, 3525
POGONOPHORA, 3528

POLYCHAETA, 3545
POLYMERA, 3549
POLYPLACOPHORA, 3557
PORIFERA, 3564
PRAECARDIOIDA, 3590
PRENEANDERTHALIAN, 3604
PRIMATES, 3613
PROBLEMATICA, 3619
PROBOSCIDEA, 3620
PROCHORDATA, 3623
PROGANOSAURIA, 3630
PROSIMII, 3640
PROTELYTROPTERA, 3646
PROTOCILIATA, 3649
PROTOMEDUSAE, 3651
PROTOZOA, 3654
PROTURA, 3655
PSILOCERATIDA, 3667
PSILOPSIDA, 3668
PTERIDOPHYLLA, 3670
PTERIDOPHYTA, 3671
PTERIDOSPERMAE, 3672
PTERIOIDA, 3673
PTEROBRANCHIA, 3674
PTEROMORPHIA, 3675
PTEROPODA, 3676
PTEROSAURIA, 3677
PTYCHOPARIIDA, 3678
PULMONATA, 3682
PYCNOGONIDA, 3690
PYROTHERIA, 3701
PYRROPHYTA, 3705
RADIOLARIA, 3735
RATITES, 3754
RECEPTACULITIDA, 3763
REDLICHIIDA, 3775
REPTILIA, 3821
RETICULAREA, 3839
RHABDOPLEURIDA, 3847
RHIPIDISTII, 3853
RHIZOPODEA, 3854
RHODOPHYTA, 3856
RHYNCHOCEPHALIA, 3858
RHYNCHONELLIDA, 3859
ROBERTINACEA, 3880
RODENTIA, 3892
ROSTROCONCHIA, 3897
ROTALIACEA, 3898
ROTALIINA, 3899
RUDISTAE, 3911
RUMINANTIA, 3913
SARCODINA, 3946
SARCOPTERYGII, 3947
SAURISCHIA, 3951
SAURORNITHES, 3952
SCAPHOPODA, 3958
SCIUROMORPHA, 3965

SCYPHOMEDUSAE, 3972
SCYPHOZOA, 3973
SILICOFLAGELLATA, 4102
SIMIAN, 4113
SIPHONAPTERA, 4121
SIPUNCULOIDA, 4122
SIRENIA, 4123
SOLEMYIDA, 4164
SOMASTEROIDEA, 4180
SPIRIFERIDA, 4213
SPIRILLINACEA, 4214
SPOROZOA, 4221
SQUAMATA, 4225
STOLONOIDEA, 4260
STOMOCHORDATA, 4261
STROMATOLITES, 4295
STROMATOPOROIDEA, 4296
STROPHOMENIDA, 4300
STYLOPHORA, 4315
SUCTORIA, 4345
SUIFORMES, 4348
SYMMETRODONTA, 4383
SYMPHYLA, 4385
SYNAPSIDA, 4388
SYNAPTOSAURIA, 4389
TABULATA, 4406
TAENIODONTA, 4410
TAENIOLABOIDEA, 4411
TARDIGRADA, 4424
TASMANITES, 4425
TELEOSTEI, 4445
TENTACULITES, 4457
TEREBRATULIDA, 4461
TESTUDINATA, 4472
TETRACORALLA, 4474
TETRAPODA, 4475
TEXTULARIINA, 4476
THALLOPHYTA, 4481
THECAMOEBA, 4486
THECIDEIDINA, 4487
THECODONTIA, 4488
THELODONTI, 4489
THERAPSIDA, 4492
THYSANOPTERA, 4523
THYSANURA, 4524
TILLODONTIA, 4532
TINTINNIDAE, 4538
TRACHYPSAMMIACEA, 4563
TREPOSTOMATA, 4589
TRICONODONTA, 4594
TRIGONIOIDA, 4596
TRILOBITA, 4597
TRILOBITOMORPHA, 4598
TUBOIDEA, 4605
TUBULIDENTATA, 4606
TYLOPODA, 4623
UNIONOIDA, 4660

URODELA, 4698
VENEROIDA, 4728
VERMES, 4731
VERTEBRATA, 4734
VOLANTES, 4754
XANTHOPHYTA, 4837
ZOANTHINIARIA, 4863

PHCH PHYSICAL AND CHEMICAL PROPERTIES AND PHENOMENA

ABSORPTION, 13
ABSORPTION SPECTRUM, 15
ACIDITY, 40
ACOUSTIC PROPERTY, 42
ACTIVATION ENERGY, 55
ACTIVITY, 61
ADIABATIC CONDITION, 65
ADSORPTION, 68
ALBEDO, 101
ALKALINITY, 131
ALPHA RAY, 147
AMPLITUDE, 177
ANELASTICITY, 198
ANHYSTERETIC REMANENT MAGNETIZATION, 206
ANION, 207
ANISOTROPY, 210
ANOMALY, 214
AQUEOUS SOLUTION, 256
ATOMIC BOND, 364
ATTENUATION, 366
BETA RAY, 473
BINARY SYSTEM, 479
BRITTLE, 579
BULK DENSITY, 590
BULK MODULUS, 591
CAPILLARITY, 633
CAPILLARY PERCOLATION, 634
CAPILLARY PRESSURE, 635
CATALYSIS, 670
CATION, 674
CELLULAR POROSITY, 682
CHEMICAL PROPERTY, 729
CHEMICAL REDUCTION, 731
CHEMICAL REMANENT MAGNETIZATION, 732
CHEMICAL SYSTEM, 733
CHEMICAL VALENCY, 734
CHLORINITY, 750
COEFFICIENT, 822
COHESION, 827
COLLOID, 842
COLOR, 845
COMPACTIBILITY, 856
COMPLEXING, 861
COMPRESSION, 865
COMPRESSIVE STRESS, 869
CONCENTRATION, 873

CONDENSATION, 882
CONDUCTIVITY, 883
CONGELATION, 890
CONSISTENCY, 906
CONTRACTION, 934
CONTRAST, 935
CONVECTION, 938
COORDINATION, 943
CURIE POINT, 1033
DECAY CONSTANT, 1079
DEGASSING, 1097
DEGREE OF FREEDOM, 1100
DEHYDRATION, 1101
DEMAGNETIZATION, 1106
DENSITY, 1113
DESICCATION, 1135
DETRITAL REMANENT MAGNETIZATION, 1143
DEVITRIFICATION, 1150
DIALYSIS, 1158
DIAMAGNETIC PROPERTY, 1159
DIELECTRIC CONSTANT, 1174
DIFFRACTION, 1180
DIFFUSION, 1181
DIFFUSIVITY, 1182
DILATANCY, 1186
DILATION, 1187
DILUTION, 1188
DISPERSION, 1222
DISSOCIATION, 1227
DISSOLUTION, 1228
DYNAMIC EQUILIBRIUM, 1264
DYNAMICS, 1267
EDDY FLUX, 1304
EH, 1312
ELASTIC CONSTANT, 1317
ELASTIC PROPERTY, 1320
ELASTIC STRAIN, 1321
ELASTICITY, 1323
ELECTRIC POTENTIAL, 1325
ELECTRICAL CONDUCTIVITY, 1326
ELECTRICAL FIELD, 1328
ELECTRICAL POLARIZATION, 1330
ELECTRICAL PROPERTY, 1331
ELECTRICITY, 1334
ELECTRO-OSMOSIS, 1335
ELECTROCHEMICAL PROPERTY, 1336
ELECTROLYSIS, 1338
ELECTROLYTE, 1339
ELECTROMAGNETISM, 1344
ELECTRON, 1345
ELECTRON PARAMAGNETIC RESONANCE, 1351
ELEMENTARY PARTICLE, 1353
ENDOTHERMIC REACTION, 1372
ENTHALPY, 1382
ENTROPY, 1385
EPR SPECTRUM, 1410
EQUATION OF STATE, 1412

EQUILIBRIUM, 1416
EUTECTIC CONDITION, 1454
EXCHANGE CAPACITY, 1464
EXOTHERMIC REACTION, 1473
EXSOLUTION, 1490
FERROMAGNETIC PROPERTY, 1536
FISSION, 1554
FIXATION, 1558
FLOCCULATION, 1568
FLOW, 1575
FLUID MECHANICS, 1581
FLUID PHASE, 1582
FLUORESCENCE, 1585
FORMULA, 1618
FRACTIONATION, 1630
FREE ENERGY, 1641
FRICTION, 1650
FUGACITY, 1659
GAMMA RAY, 1676
GASEOUS PHASE, 1694
GEL, 1702
GRAVITATIONAL CONSTANT, 1876
GREASY LUSTER, 1889
HARDNESS, 1956
HEAT, 1962
HEAT CAPACITY, 1963
HEAT CONDUCTION, 1964
HEAT EQUIVALENT OF FUSION, 1966
HIGH PRESSURE, 2007
HIGH TEMPERATURE, 2008
HOOKE'S LAW, 2048
HOT, 2063
HUMIDITY, 2075
HYDRATION, 2089
IMMISCIBILITY, 2185
IMMISCIBLE MATERIAL, 2186
IMPURITY, 2196
INDEX OF REFRACTION, 2209
INDUCED MAGNETIZATION, 2215
INERT COMPONENT, 2223
INFRARED, 2231
INFRARED RADIATION, 2233
INHOMOGENEITY, 2238
INTERFACE, 2261
INTERFERENCE, 2262
ION, 2299
ION EXCHANGE, 2300
IONIZATION, 2304
IRIDESCENCE, 2306
ISOANOMALY, 2318
ISOBAR, 2319
ISOCON, 2326
ISOPIESTIC LINE, 2336
ISOTHERM, 2346
ISOTHERMAL REMANENT MAGNETIZATION, 2347
ISOTROPY, 2354
KINEMATICS, 2393

LAME CONSTANT, 2418
LASER, 2437
LATENT HEAT OF FUSION, 2440
LEACHING, 2465
LIQUEFACTION, 2519
LIQUID PHASE, 2520
LIXIVIATION, 2550
LOW PRESSURE, 2571
LOW TEMPERATURE, 2572
LUMINESCENCE, 2597
LUSTER, 2606
MAGNETIC ELEMENT, 2632
MAGNETIC PERMEABILITY, 2639
MAGNETIC PROPERTY, 2640
MAGNETIC SUSCEPTIBILITY, 2645
MAGNETIZATION, 2648
MAGNETOSTRICTION, 2653
MALLEABLE CONSISTENCE, 2660
MASS, 2697
MECHANICAL PROPERTY, 2727
MECHANICS, 2729
MELTING, 2744
MICROHARDNESS, 2820
MINERAL PHASE, 2871
MINERALOGICAL PHASE RULE, 2879
MISCIBILITY, 2893
MOBILITY, 2900
MOBILIZATION, 2901
MODULUS, 2907
MODULUS OF RIGIDITY, 2908
MOSSBAUER SPECTRUM, 2948
NATURAL REMANENT MAGNETIZATION, 2991
NEUTRON, 3014
NUCLEATION, 3066
OPAQUE, 3128
OPTICAL EXTINCTION, 3138
OPTICAL POLARIZATION, 3139
OPTICAL PROPERTY, 3140
OPTICAL ROTATION, 3141
OPTICAL SPECTRUM, 3143
OVERSATURATED, 3216
OVERSATURATED SOLUTION, 3217
OXIDATION, 3221
P=T CONDITIONS, 3229
PARAMAGNETIC PROPERTY, 3298
PARTIAL PRESSURE, 3306
PARTITIONING, 3312
PERMEABILITY, 3363
PH, 3384
PHASE EQUILIBRIA, 3392
PHASE ISOCHORE, 3393
PHASE RULE, 3394
PHOSPHORESCENCE, 3405
PHOTOCHEMICAL REACTION, 3408
PHOTOSYNTHESIS, 3414
PHYSICAL CHEMISTRY, 3425
PHYSICAL PROPERTY, 3427

PHYSICAL STATE, 3428
PHYSICOCHEMICAL PROPERTY, 3430
PIEZOELECTRIC EFFECT, 3442
PLASTICITY, 3494
PLEOCHROISM, 3509
POLYMERIZATION, 3550
POROSITY, 3565
PRECIPITATION, 3596
PRESSURE, 3607
PROPERTY, 3638
PROTON, 3652
QUATERNARY SYSTEM, 3714
RADIATION, 3720
RADIOACTIVITY, 3731
REACTION, 3757
REAGENT, 3761
REFLECTANCE, 3782
REFLECTION, 3783
REFLECTIVITY, 3785
REFRACTION, 3786
REFRACTIVITY, 3788
RELAXATION ENERGY, 3805
REMAGNETIZATION, 3811
REMANENT MAGNETIZATION, 3812
RESISTIVITY, 3832
REVERSAL, 3842
RHEOLOGY, 3851
RIGIDITY, 3866
ROTATORY DISPERSION, 3903
ROUGHNESS, 3905
RUPTURE STRENGTH, 3918
SALINITY, 3923
SATURATION, 3950
SCINTILLATION, 3964
SHEAR, 4058
SHEAR STRENGTH, 4062
SHEAR STRESS, 4063
SHOCK, 4075
SHOCK WAVE, 4078
SOLAR RADIATION, 4159
SOLID PHASE, 4166
SOLID SOLUTION, 4167
SOLUBILITY, 4174
SOLUTE, 4175
SOLUTION, 4176
SORPTION, 4183
SPALLATION, 4189
SPECIFIC GRAVITY, 4196
SPECIFIC HEAT, 4197
SPECIFIC SURFACE, 4198
SPECTRUM, 4202
STEREOCHEMISTRY, 4252
STOICHIOMETRY, 4259
STRAIN, 4267
STRENGTH, 4289
STRESS, 4290
SUBLIMATE, 4323

SUBLIMATION, 4324
SUBSTITUTION, 4338
SURFACE TENSION, 4365
SUSPENSION, 4373
SWELLING, 4378
TEMPERATURE, 4453
TENACITY, 4454
TERNARY SYSTEM, 4463
THERMAL CONDUCTIVITY, 4495
THERMAL PROPERTY, 4499
THERMODYNAMIC PROPERTY, 4504
THERMODYNAMICS, 4505
THERMOLUMINESCENCE, 4509
THERMOREMANENT MAGNETIZATION, 4511
THIXOTROPY, 4515
TORSION, 4550
ULTRAVIOLET RADIATION, 4641
UNDERSATURATED, 4654
UNDERSATURATED SOLUTION, 4655
UPLIFT PRESSURE, 4666
VAPOR, 4713
VELOCITY, 4726
VISCOELASTICITY, 4744
VISCOSITY, 4745
VISCOUS REMANENT MAGNETIZATION, 4747
VITRIFICATION, 4751
VOLATILE, 4755
WAVE, 4805
WAVE DISPERSAL, 4806
X=RAY, 4832
YOUNG'S MODULUS, 4851

SEDI   SEDIMENTATION

ABYSSAL SEDIMENTATION, 23
ALGAL BANK, 112
ALGAL BISCUIT, 113
ALGAL MAT, 115
ALGAL MOUND, 116
ALGAL REEF, 117
ALGAL STRUCTURE, 118
ALLOCHEM, 134
ARGILLACEOUS COMPOSITION, 295
ARMORED MUD BALL, 303
ASH CONTENT, 330
ATMOGENIC ORIGIN, 357
BACK REEF, 397
BACKSET BED, 399
BARRIER REEF, 420
BASAL CONGLOMERATE, 424
BEACHROCK, 448
BIOCHEMICAL SEDIMENTATION, 482
BIOCLAST, 486
BIOCLASTIC SEDIMENTATION, 488
BIOFACIES, 490
BIOFACIES MAP, 491

BIOHERM, 499
BIOLITH, 500
BIOSTROME, 512
BIOTURBATION, 515
BITUMINIZATION, 522
BONE BED, 540
BOTTOM SEDIMENT, 554
BOTTOMSET BED, 555
BOULDER, 558
CALCAREOUS COMPOSITION, 609
CARBONATIZATION, 653
CATAGENESIS, 668
CEMENT, 684
CEMENTATION, 685
CHEMICAL DEPOSITION, 723
CHEMOGENIC ORIGIN, 736
CHERTIFICATION, 740
CLAST, 783
CLASTIC, 784
CLASTIC DIKE, 785
CLAY BALL, 788
COAL BALL, 802
COALIFICATION, 808
COARSE-GRAINED FRACTION, 810
COMPACTION, 857
CONCRETION, 881
CONNATE WATER, 898
CONTINENTAL SEDIMENTATION, 928
CROSS-BEDDING, 997
CURRENT MARK, 1036
CYCLE OF SEDIMENTATION, 1045
CYCLOTHEM, 1051
DEDOLOMITIZATION, 1087
DELTAIC SEDIMENTATION, 1105
DENDRITE, 1108
DEPOSITION, 1120
DETRITAL COMPOSITION, 1142
DETRITAL SEDIMENTATION, 1144
DIAGENESIS, 1155
DIFFERENTIAL COMPACTION, 1175
DISSOLVED LOAD, 1229
DOLOMITIZATION, 1237
DUST, 1263
EARLY DIAGENESIS, 1272
ELUTRIATION, 1356
ENVIRONMENTAL ANALYSIS, 1387
EOLIAN SEDIMENTATION, 1393
ERRATIC BLOCK, 1430
ESTUARINE SEDIMENTATION, 1437
FANGLOMERATE, 1505
FERROMANGANESE COMPOSITION, 1537
FINE-GRAINED FRACTION, 1549
FLUVIAL SEDIMENTATION, 1595
FLUVIOGLACIAL SEDIMENTATION, 1597
FLYSCH, 1599
FORESET BED, 1614
FRAGMENT, 1636

FRESHWATER SEDIMENTATION, 1649
FRINGING REEF, 1651
GEODE, 1726
GEOSYNCLINAL SEDIMENTATION, 1765
GLACIAL SEDIMENTATION, 1798
GLACIOLACUSTRINE SEDIMENTATION, 1806
GLACIOMARINE SEDIMENTATION, 1808
GLAUCONITIC COMPOSITION, 1816
GLAUCONITIZATION, 1817
GRANULOMETRY, 1861
GROOVE CAST, 1903
GYPSIFICATION, 1932
HALITIC ENVIRONMENT, 1945
HARDGROUND, 1955
HEAVY MINERALS, 1973
HIGH-ENERGY ENVIRONMENT, 2009
IMBRICATION, 2181
INDURATION, 2218
INTERCALATION, 2260
INTERFORMATIONAL, 2265
INTERTIDAL SEDIMENTATION, 2287
INTRACLAST, 2288
INTRAFORMATIONAL DEPOSITION, 2291
IRON FORMATION, 2311
ISOLITH, 2331
ISOPACH, 2335
KAOLINIZATION, 2369
LACUSTRINE SEDIMENT, 2408
LACUSTRINE SEDIMENTATION, 2409
LAGOONAL SEDIMENTATION, 2413
LATE DIAGENESIS, 2438
LIGHT MINERALS, 2501
LITHIFICATION, 2524
LITHOFACIES, 2526
LITHOFACIES MAP, 2527
LITHOGENESIS, 2528
LITHOSTROME, 2539
LITHOTOPE, 2540
LITTORAL SEDIMENTATION, 2546
LOAD CAST, 2555
LOW-ENERGY ENVIRONMENT, 2574
MACERAL, 2613
MAGNAFACIES, 2625
MARINE SEDIMENT, 2685
MARINE SEDIMENTATION, 2686
MARINE TRANSPORT, 2689
MASSIVE BEDDING, 2704
MATURITY INDEX, 2719
MECHANICAL ORIGIN, 2726
MEGAFACIES, 2738
METAGENESIS, 2765
MICROFACIES, 2811
MOLASSE, 2912
MUD BANK, 2953
MUD FLOW, 2956
MUD LUMP, 2957
NITRIFICATION, 3033

NODULE, 3039
OLISTOLITH, 3116
OLISTOSTROME, 3117
ONCOLITE, 3119
OOLITE, 3124
ORGANIC RESIDUE, 3173
ORIGOFACIES, 3180
PALEOCURRENT, 3251
PALEOSALINITY, 3267
PARTICLE, 3307
PARTICLE SHAPE, 3308
PATCH REEF, 3313
PEARL, 3319
PEBBLE, 3322
PELAGIC SEDIMENTATION, 3331
PELLET, 3337
PHOSPHATIZATION, 3404
PINCH OUT, 3449
PINNACLE REEF, 3451
PISOLITE, 3457
POLYMICTIC TEXTURE, 3553
PORE WATER, 3563
PRESSURE SOLUTION, 3608
PROVENANCE, 3656
PYRITIZATION, 3694
RAIN PRINT, 3743
RECENT SEDIMENTATION, 3762
RED BEDS, 3773
REDEPOSITION, 3774
REEF, 3776
REEF SEDIMENTATION, 3778
ROUNDNESS, 3906
SALTATION, 3931
SAND BODY, 3937
SAND-SHALE RATIO, 3938
SANDSTONE DIKE, 3940
SEBKHA ENVIRONMENT, 3987
SEDIMENT, 3991
SEDIMENT TRACTION, 3992
SEDIMENT YIELD, 3993
SEDIMENT-WATER INTERFACE, 3994
SEDIMENTARY BASIN, 3995
SEDIMENTARY PETROLOGY, 3998
SEDIMENTARY TRAP, 4002
SEDIMENTATION, 4003
SEDIMENTATION RATE, 4004
SEDIMENTOGENESIS, 4005
SEDIMENTOLOGY, 4006
SEPTARIUM, 4040
SHELF SEDIMENTATION, 4070
SHOESTRING SAND, 4079
SINTER, 4118
SLUMPING, 4139
SOFT SEDIMENT DEFORMATION, 4145
SOURCE ROCK, 4186
STRATIFICATION, 4272
STREAM TRANSPORT, 4288

SUBMARINE FAN, 4327
SUBMERGENCE, 4334
SWAMP SEDIMENTATION, 4377
SYNSEDIMENTARY ORIGIN, 4397
TERRIGENOUS DEPOSITS, 4469
TOPSET BED, 4548
TRANSPORT, 4577
TURBIDITE, 4615
TURBIDITY, 4616
TURBIDITY CURRENT, 4617
WILDFLYSCH, 4825

SEDS SEDIMENTARY ROCKS

AGGLOMERATE, 82
ALABASTER, 100
ALBERTITE, 102
ALGAL LIMESTONE, 114
AMBER, 160
ANKERITE, 211
ANTHRAXOLITE, 222
ARENITE, 292
ARKOSE, 302
BIOCLASTIC LIMESTONE, 487
BIOMICRITE, 505
BIOSPARITE, 508
BOUNDSTONE, 562
BRINE, 578
CARBONACEOUS ROCK, 649
CARBONATE ROCKS, 650
CARNIEULE, 656
CAUSTOBIOLITH, 676
CHALK, 705
CHERT, 739
CLARAIN, 780
CLASTICS, 786
CLAY, 787
CLAYSTONE, 791
COARSE GRAVEL, 809
CONGLOMERATE, 892
COQUINA, 948
DIAMICTITE, 1160
DIATOMITE, 1170
DOLOMITIC LIMESTONE, 1236
DOLOSTONE, 1238
DURAIN, 1262
ENCRINITE, 1366
EVAPORITE, 1459
EXINITE, 1469
FLINT, 1567
FUSINITE, 1669
GEYSERITE, 1785
GRAPESTONE, 1862
GRAVEL, 1871
GRAYWACKE, 1888
GRIT, 1900

GRUS, 1923
GUANO, 1924
GYTTJA, 1934
HUMITE COAL, 2078
IRONSTONE, 2312
JASPER, 2359
LATERITE, 2446
LIMESTONE, 2507
LIMON, 2509
LOESS, 2561
MARL, 2692
MICRITE, 2806
MICROCRYSTALLINE LIMESTONE, 2809
MUD, 2952
NOVACULITE, 3061
OOLITIC LIMESTONE, 3125
OOZE, 3127
OPOKA, 3137
ORGANIC SEDIMENT, 3174
PELITE, 3334
PHOSPHORITE, 3406
PORCELLANITE, 3561
PSAMMITE, 3659
PSEUDOTILLITE, 3666
RADIOLARITE, 3736
RESIDUAL CLAY, 3828
RESIN, 3831
RUDITE, 3912
SAND, 3936
SANDSTONE, 3939
SAPROPEL, 3944
SEDIMENTARY ROCK, 4000
SHALE, 4051
SILT, 4106
SILTSTONE, 4109
SPARAGMITE, 4190
SPONGOLITE, 4218
TILLITE, 4531
TONSTEIN, 4543
TRAVERTINE, 4583
UNDERCLAY, 4648
VITRAIN, 4749
WAX, 4808

SOLI SOLID—EARTH GEOPHYSICS

A LAYER, 1
ABSOLUTE GRAVITY, 11
ACOUSTICAL WAVE, 45
ACTIVE MARGIN, 58
AFTERSHOCK, 79
AIRY WAVE, 97
ANELASTIC MEDIUM, 197
ARCHEOMAGNETISM, 282
ASEISMIC MARGINS, 328
ASEISMIC REGION, 329

ASTHENOSPHERE, 348
B LAYER, 395
BASALTIC LAYER, 426
BENIOFF ZONE, 462
BIRCH DISCONTINUITY, 518
BODY WAVE, 535
BULLARD DISCONTINUITY, 592
C LAYER, 602
CHALCOSPHERE, 704
CHANDLER WOBBLE, 706
CODA WAVE, 821
COERCIVITY, 825
CONRAD DISCONTINUITY, 902
CONTINENTAL ACCRETION, 919
CONTINENTAL CRUST, 921
CONTINENTAL DRIFT, 922
CONTRACTING EARTH, 933
CONVECTION CURRENT, 939
CORE OF THE EARTH, 957
CORIOLIS EFFECT, 959
CRUST, 1001
D LAYER, 1057
DECLINATION, 1081
DEEP—FOCUS EARTHQUAKE, 1088
DISCONTINUITY, 1212
DIURNAL VARIATION, 1231
DYNAMO THEORY, 1268
E LAYER, 1271
EARTH CURRENT, 1274
EARTH INTERIOR, 1276
EARTH ROTATION, 1278
EARTH TIDE, 1280
EARTHQUAKE, 1284
EARTHQUAKE MAGNITUDE, 1286
EARTHQUAKE SWARM, 1287
ELASTIC WAVE, 1322
ELECTROMAGNETIC FIELD, 1340
EPICENTER, 1400
EXPANDING EARTH, 1475
EXPERIMENTAL SEISMOLOGY, 1478
F LAYER, 1499
FOCAL MECHANISM, 1600
FORESHOCK, 1615
FREE OSCILLATION, 1642
G LAYER, 1671
GEODESY, 1727
GEODYNAMICS, 1729
GEOID, 1732
GEOISOTHERM, 1733
GEOMAGNETIC POLE, 1742
GEOMAGNETIC REVERSAL, 1743
GEOMAGNETIC SECULAR VARIATION, 1744
GEOMAGNETISM, 1745
GEOPHYSICAL ANOMALY, 1751
GEOPHYSICAL MAP, 1754
GEOPHYSICAL PROFILE, 1756
GEOPHYSICIST, 1758

GEOPHYSICS, 1759
GEOTHERMAL GRADIENT, 1775
GNOMONIC PROJECTION, 1832
GRANITIC LAYER, 1849
GRAVIMETRY, 1874
GRAVITATION, 1875
GRAVITY FIELD, 1880
GRAVITY MAP, 1882
GROUND MOTION, 1908
GROUND PRESSURE, 1909
GUTENBERG DISCONTINUITY, 1929
HALF-SPACE, 1943
HEAT FLOW, 1967
HEAT SOURCE, 1968
HEAT TRANSFER, 1969
HIGH ANOMALY VALUE, 2006
HODOGRAPH, 2023
HOT SPOT, 2065
HYDROSPHERE, 2126
HYPOCENTER, 2146
IAPETUS, 2154
INNER CORE, 2243
INTERMEDIATE-FOCUS EARTHQUAKE, 2270
IONOSPHERE, 2305
ISOCLINIC LINE, 2325
ISOGON, 2327
ISOSEISM, 2342
ISOSEISMIC MAP, 2343
ISOSTASY, 2344
LAYERED MEDIUM, 2464
LENGTH OF DAY, 2478
LITHOSPHERE, 2536
LONG-PERIOD WAVE, 2564
LOVE WAVE, 2570
LOW-VELOCITY LAYER, 2576
LOWER CRUST, 2580
LOWER MANTLE, 2585
MAGNETIC AZIMUTH, 2629
MAGNETIC FIELD, 2633
MAGNETIC FIELD VARIATION, 2634
MAGNETIC INCLINATION, 2635
MAGNETIC MAP, 2636
MAGNETIC STORM, 2642
MAGNETIC SURVEY MAP, 2644
MAGNETISM, 2646
MAGNETOHYDRODYNAMICS, 2649
MAGNETOSPHERE, 2651
MANTLE, 2665
MICROEARTHQUAKE, 2810
MICROPLATE, 2829
MICROSEISM, 2834
MODIFIED MERCALLI SCALE, 2905
MOHOLE PROJECT, 2909
MOHOROVICIC DISCONTINUITY, 2910
NUTATION, 3072
OBDUCTION, 3075
OCEANIC CRUST, 3091

OCEANIZATION, 3092
OSCILLATION, 3197
OUTER CORE, 3207
P-WAVES, 3230
PALEOMAGNETIC POLE, 3262
PALEOMAGNETISM, 3263
PLANET EARTH, 3472
PLATE, 3495
PLATE COLLISION, 3496
PLATE LIMIT, 3497
PLUME, 3515
POLAR WANDERING, 3535
PYROSPHERE, 3700
RAYLEIGH WAVE, 3756
RICHTER SCALE, 3862
S WAVE, 3920
SEISMIC SOURCE, 4016
SEISMIC VELOCITY, 4020
SEISMIC WAVE, 4021
SEISMICITY, 4022
SEISMOGRAM, 4023
SEISMOLOGICAL MAP, 4025
SEISMOLOGY, 4026
SEISMOTECTONICS, 4028
SH-WAVE, 4050
SHALLOW FOCUS EARTHQUAKE, 4052
SHAPE OF THE EARTH, 4055
SIAL, 4089
SIMA, 4112
SLAB, 4129
SOLID-EARTH GEOPHYSICS, 4169
SPREADING CENTER, 4223
STICK SLIP, 4254
SUBDUCTION, 4321
SUPRACRUSTAL ORIGIN, 4362
SURFACE WAVE, 4368
TECTONOPHYSICS, 4442
TELESEISMIC SIGNAL, 4446
TILT, 4533
TRAVELTIME, 4581
TRAVELTIME CURVE, 4582
TRIPLE JUNCTION, 4599
UPPER CRUST, 4670
UPPER MANTLE, 4675
VELOCITY STRUCTURE, 4727
WAVE PROPAGATION, 4807
ZONE OF FLOW, 4868

STRA   STRATIGRAPHY, HISTORICAL GEOLOGY

ANCIENT ICE AGE, 191
ASSEMBLAGE ZONE, 339
BIOCHRON, 484
BIOCHRONOLOGY, 485
BIOSOME, 507
BIOSTRATIGRAPHIC UNIT, 510

BIOSTRATIGRAPHY, 511
BIOZONE, 517
CENOZONE, 689
CHRONOSTRATIGRAPHIC UNIT, 762
CHRONOSTRATIGRAPHY, 763
CHRONOZONE, 764
CONCORDANCE, 877
CONFORMITY, 889
CORRELATION, 961
DENDROCHRONOLOGY, 1110
DIACHRONISM, 1153
DIASTEM, 1166
DISCONFORMITY, 1211
DISCORDANCE, 1214
ECOZONE, 1302
EON, 1394
EPOCH, 1409
ERA, 1418
ERATHEM, 1419
FAUNIZONE, 1519
FLORIZONE, 1573
FORMATION, 1617
GAP, 1682
GAUSS EPOCH, 1699
GEOCHRONOLOGIC UNIT, 1724
GEOLOGIC COLUMN, 1734
GLACIAL STAGE, 1799
GONDWANA, 1834
GUIDE FOSSIL, 1925
HETEROCHRONISM, 1989
HETEROCHRONOUS, 1990
HIATUS, 2004
HISTORICAL GEOLOGY, 2019
HOMOTAXY, 2047
INDEX FOSSIL, 2206
INTERGLACIAL, 2266
INTERPLUVIAL, 2281
JARAMILLO EVENT, 2358
KARROO, 2372
KEY BED, 2389
LAURASIA, 2453
LAW OF CORRELATION OF FACIES, 2461
LICHENOMETRY, 2496
LITHOSTRATIGRAPHIC UNIT, 2537
LITHOSTRATIGRAPHY, 2538
MAGNETOSTRATIGRAPHY, 2652
MATUYAMA EPOCH, 2720
MEMBER, 2746
MESOGAEA, 2753
NONCONFORMITY, 3050
PALEO-ATMOSPHERE, 3240
PALEO-OCEANOGRAPHY, 3241
PALEOBATHYMETRY, 3242
PALEOCIRCULATION, 3247
PALEOCLIMATE, 3248
PALEOCLIMATOLOGY, 3249
PALEOGEOGRAPHIC MAP, 3256

PALEOGEOGRAPHY, 3257
PALEOGEOMORPHOLOGY, 3258
PALEOKARST, 3259
PALEOLATITUDE, 3260
PALEOLIMNOLOGY, 3261
PALEORELIEF, 3266
PALEOSOL, 3268
PALEOTEMPERATURE, 3269
PANGEA, 3281
PERIOD, 3356
PLUVIAL, 3522
POLLEN DIAGRAM, 3540
PREHISTORIC AGE, 3602
RANGE, 3746
RANGE ZONE, 3747
REGIONAL UNCONFORMITY, 3796
REGRESSION, 3800
RELATIVE AGE, 3804
SERIES, 4044
STAGE, 4229
STRATIGRAPHIC BOUNDARY, 4275
STRATIGRAPHIC ISOCHORE, 4277
STRATIGRAPHIC MAP, 4278
STRATIGRAPHIC UNIT, 4280
STRATIGRAPHY, 4281
STRATOTYPE, 4282
TEPHROCHRONOLOGY, 4458
TETHYS, 4473
TIME SCALE, 4536
TRANSGRESSION, 4572
TREE RING, 4585
UNCONFORMITY, 4646
VARVE CHRONOLOGY, 4721
WEDGE, 4811

STRS STRATIGRAPHY—SYSTEMATICS

AALENIAN, 3
AKTCHAGYLIAN, 99
ALBIAN, 103
ALDANIAN, 109
ALPINE TRIASSIC, 152
ANISIAN, 208
APSHERONIAN, 251
APTIAN, 253
AQUITANIAN, 263
ARCHEAN, 278
ARENIGIAN, 291
ARTINSKIAN, 323
ASHGILLIAN, 333
AUTUNIAN, 387
BAJOCIAN, 406
BARREMIAN, 414
BARTONIAN, 421
BASHKIRIAN, 431
BATHONIAN, 435

BERRIASIAN, 470
BUNTSANDSTEIN, 593
BURDIGALIAN, 594
CALABRIAN, 606
CALLOVIAN, 622
CAMBRIAN, 625
CAMPANIAN, 626
CARADOCIAN, 638
CARBONIFEROUS, 654
CARNIAN, 655
CENOMANIAN, 686
CENOZOIC, 688
CHATTIAN, 716
CONIACIAN, 894
CRETACEOUS, 989
DANIAN, 1060
DEVONIAN, 1151
DOMERIAN, 1241
EIFELIAN, 1313
EMSIAN, 1364
EOCENE, 1389
FAMENNIAN, 1503
FLANDRIAN, 1564
FRASNIAN, 1640
GEDINNIAN, 1701
GIVETIAN, 1790
GZHELIAN, 1935
HAUTERIVIAN, 1959
HELVETIAN, 1980
HETTANGIAN, 1999
HOLOCENE, 2026
INDIAN, 2210
INFRACAMBRIAN, 2230
JURASSIC, 2363
KAZANIAN, 2381
KAZIMOVIAN, 2382
KEUPER, 2388
KIMMERIDGIAN, 2392
KUJALNIKIAN, 2399
KUNGURIAN, 2400
LADINIAN, 2410
LENIAN, 2479
LLANDEILIAN, 2551
LLANDOVERIAN, 2552
LLANVIRNIAN, 2553
LOWER CAMBRIAN, 2577
LOWER CARBONIFEROUS, 2578
LOWER CRETACEOUS, 2579
LOWER DEVONIAN, 2581
LOWER EOCENE, 2582
LOWER JURASSIC, 2583
LOWER LIASSIC, 2584
LOWER MIOCENE, 2586
LOWER OLIGOCENE, 2587
LOWER ORDOVICIAN, 2588
LOWER PERMIAN, 2589
LOWER PLEISTOCENE, 2590

LOWER PLIOCENE, 2591
LOWER PROTEROZOIC, 2592
LOWER SILURIAN, 2593
LOWER TRIASSIC, 2594
LUDLOVIAN, 2596
MAESTRICHTIAN, 2615
MAJIAN, 2656
MEOTIAN, 2748
MESOZOIC, 2760
MIDDLE CAMBRIAN, 2841
MIDDLE DEVONIAN, 2842
MIDDLE EOCENE, 2843
MIDDLE JURASSIC, 2844
MIDDLE LIASSIC, 2845
MIDDLE MIOCENE, 2846
MIDDLE OLIGOCENE, 2847
MIDDLE ORDOVICIAN, 2848
MIDDLE PLEISTOCENE, 2849
MIDDLE PLIOCENE, 2850
MIDDLE PROTEROZOIC, 2851
MIDDLE TRIASSIC, 2852
MIOCENE, 2889
MISSISSIPPIAN, 2897
MOSCOVIAN, 2946
MUSCHELKALK, 2969
NAMURIAN, 2984
NEOCOMIAN, 2999
NEOGENE, 3002
NORIAN, 3053
OLIGOCENE, 3115
ORDOVICIAN, 3150
OXFORDIAN, 3220
PALAEOPHYTICUM, 3236
PALEOCENE, 3246
PALEOGENE, 3254
PALEOZOIC, 3271
PENNSYLVANIAN, 3342
PERMIAN, 3364
PHANEROZOIC, 3390
PLEISTOCENE, 3508
PLIENSBACHIAN, 3511
PLIOCENE, 3512
PONTIAN, 3559
PORTLANDIAN, 3570
PRECAMBRIAN, 3593
PROTEROZOIC, 3647
QUATERNARY, 3713
RHAETIAN, 3848
ROTLIEGENDES, 3904
RUPELIAN, 3916
SAKMARIAN, 3921
SANTONIAN, 3943
SARMATIAN, 3948
SAXONIAN, 3954
SENONIAN, 4037
SICILIAN, 4090
SIEGENIAN, 4094

SILURIAN, 4110
SINEMURIAN, 4115
STAMPIAN, 4233
STEPHANIAN, 4250
TARANNONIAN, 4423
TATARIAN, 4426
TERTIARY, 4470
THANETIAN, 4484
THURINGIAN, 4522
TITHONIAN, 4540
TOARCIAN, 4541
TORTONIAN, 4551
TOURNAISIAN, 4553
TREMADOCIAN, 4586
TRIASSIC, 4591
TURONIAN, 4620
TYRRHENIAN, 4629
UFIMIAN, 4633
UPPER CAMBRIAN, 4667
UPPER CARBONIFEROUS, 4668
UPPER CRETACEOUS, 4669
UPPER DEVONIAN, 4671
UPPER EOCENE, 4672
UPPER JURASSIC, 4673
UPPER LIASSIC, 4674
UPPER MIOCENE, 4676
UPPER OLIGOCENE, 4677
UPPER ORDOVICIAN, 4678
UPPER PERMIAN, 4679
UPPER PLEISTOCENE, 4680
UPPER PLIOCENE, 4681
UPPER PROTEROZOIC, 4682
UPPER SILURIAN, 4683
UPPER TRIASSIC, 4684
VALANGINIAN, 4706
VILLAFRANCHIAN, 4742
VISEAN, 4748
WENLOCKIAN, 4818
WESTPHALIAN, 4820
YPRESIAN, 4852
ZECHSTEIN, 4855

STRU TECTONICS

ACADIAN OROGENY, 24
ACTIVE FAULT, 56
ACTIVE TECTONICS, 59
ALGOMAN OROGENY, 120
ALLEGHENY OROGENY, 133
ALPINE OROGENY, 150
ALPINE TECTONICS, 151
ALPINE=TYPE, 153
ANGULAR UNCONFORMITY, 202
ANTECLISE, 218
ANTICLINE, 224
ANTICLINORIUM, 225

ANTIFORM, 227
ANTITHETIC FAULT, 233
ARCH, 270
ARCHEIDES, 279
ARCUATE FAULT, 289
ARMORICAN OROGENY, 304
ASSYNTIAN OROGENY, 342
ASYMMETRIC FOLD, 353
AULACOGEN, 372
AXIAL PLANE, 391
AXIAL=PLANE CLEAVAGE, 392
BACK ARC BASIN, 396
BAIKALIAN OROGENY, 404
BASEMENT, 429
BASEMENT TECTONICS, 430
BEDDING, 453
BEDDING PLANE, 454
BEDDING=PLANE CLEAVAGE, 455
BEDDING=PLANE FAULT, 456
BELTIAN OROGENY, 460
BLOCK FAULTING, 532
BORDERLAND, 547
BOUDINAGE, 556
BRACHYFOLD, 567
CALEDONIAN OROGENY, 618
CASCADE FOLD, 661
CASCADIAN REVOLUTION, 662
CATACLASIS, 667
CENTROCLINE, 690
CHARNIAN OROGENY, 711
CHEVRON FOLD, 741
CIMMERIAN OROGENY, 768
CLEAVAGE, 792
COLLAPSE STRUCTURE, 837
COMPRESSION TECTONICS, 867
CONCENTRIC FOLD, 874
CONJUGATE FOLD, 897
CONTINENTAL BORDERLAND, 920
CONTINENTAL NUCLEUS, 926
CORDILLERA, 955
COVER, 979
CRATON, 984
CYLINDRICAL FOLD, 1052
DECOLLEMENT, 1082
DEEPSEATED TECTONICS, 1094
DEFORMATION, 1096
DIAPIR, 1164
DIAPIRISM, 1165
DIASTROPHISM, 1167
DIP, 1195
DIP FAULT, 1196
DIP=SLIP FAULT, 1199
DISHARMONIC FOLD, 1219
DISJUNCTIVE FOLD, 1220
DISPLACEMENT, 1224
DISPLACEMENT THEORY, 1225
DOWNWARPING, 1242

DRAG FOLD, 1243
DRAPE FOLD, 1248
EARLY PROTEROZOIDES, 1273
EBURNEAN OROGENY, 1289
ELASTIC LIMIT, 1318
EN ECHELON, 1365
ENSIALIC GEOSYNCLINE, 1377
ENSIMATIC GEOSYNCLINE, 1378
ENTEROLITHIC FOLD, 1380
EPEIROGENY, 1396
EPIEUGEOSYNCLINE, 1403
EUGEOSYNCLINE, 1445
EUSTATISM, 1452
EXTERNIDES, 1493
FAULT, 1507
FAULT GOUGE, 1509
FAULT PLANE, 1510
FAULT SYSTEM, 1512
FAULT ZONE, 1513
FAULT=PLANE STRIATION, 1514
FAULTING, 1515
FLEXURE, 1566
FLOW CLEAVAGE, 1576
FLOW FOLD, 1577
FOLD, 1602
FOLD AXIS, 1603
FOLD BELT, 1604
FOLD SYSTEM, 1606
FOLD TECTONICS, 1607
FOLDING, 1608
FOREDEEP, 1612
FORELAND, 1613
FRACTURE, 1632
FRACTURE ZONE, 1634
FRACTURING, 1635
GEANTICLINE, 1700
GEOFRACTURE, 1730
GEOLOGIC STRUCTURE, 1737
GEOSYNCLINAL RIDGE, 1763
GEOSYNCLINAL TRENCH, 1766
GEOSYNCLINE, 1767
GEOTECTONICS, 1772
GEOUNDATION, 1781
GERMANOTYPE TECTONICS, 1783
GLACIOTECTONICS, 1809
GLOBAL TECTONICS, 1821
GNEISSIC DOME, 1829
GRABEN, 1840
GRAVITY FOLD, 1881
GRAVITY SLIDING, 1885
GRAVITY TECTONICS, 1887
GREENSTONE BELT, 1895
GRENVILLE OROGENY, 1898
GROWTH FAULT, 1921
HADE, 1938
HERCYNIAN OROGENY, 1984
HERCYNIDES, 1985

HINGE FAULT, 2015
HOMOCLINE, 2040
HORIZONTAL COMPRESSION, 2051
HORIZONTAL DISPLACEMENT, 2053
HORIZONTAL FAULT, 2054
HORST, 2060
HUDSONIAN OROGENY, 2068
IDIOGEOSYNCLINE, 2169
IMBRICATE STRUCTURE, 2180
INFRASTRUCTURE, 2235
INLIER, 2242
INTERNIDES, 2279
INTRACONTINENTAL BELT, 2289
INTRACRATONIC BASIN, 2290
ISLAND ARC, 2316
ISOBASE, 2320
ISOCLINE, 2324
JOINT, 2361
JOINTING, 2362
KARELIAN OROGENY, 2370
KATANGA OROGENY, 2378
KENORAN OROGENY, 2384
KIBARA OROGENY, 2390
KINK BAND, 2394
KLIPPE, 2396
LARAMIDE OROGENY, 2435
LATE PROTEROZOIDES, 2439
LATERAL FAULT, 2442
LEPTOGEOSYNCLINE, 2487
LIMB, 2505
LINEAMENT, 2511
LINEATION, 2512
MAZATSAL REVOLUTION, 2723
MEDIAN MASS, 2732
MESOGEOSYNCLINE, 2755
MICROFOLD, 2815
MICROTECTONICS, 2836
MIOGEOSYNCLINE, 2890
MOBILE BELT, 2899
MONOCLINE, 2920
MULTIPHASE TECTONICS, 2965
MYLONITIZATION, 2974
NAPPE, 2986
NEOTECTONICS, 3005
NORMAL FAULT, 3056
NOSE, 3058
OBLIQUE FAULT, 3077
OBLIQUE ORIENTATION, 3078
OROGENIC BELT, 3184
OROGENY, 3185
OSCILLATION THEORY, 3198
OUTLIER, 3210
OVERTHRUST, 3218
OVERTURNED FOLD, 3219
PARAGEOSYNCLINE, 3291
PARALIAGEOSYNCLINE, 3293
PENOKEAN OROGENY, 3343

PERICLINE, 3351
PLACANTICLINE, 3461
PLATE ROTATION, 3498
PLATE TECTONICS, 3499
PLATFORM, 3503
PLICATION, 3510
PLUNGE, 3516
POISSON'S RATIO, 3532
POSTTECTONIC PROCESS, 3575
PRECAMBRIAN OROGENY, 3594
PROTEROZOIDES, 3648
PSEUDOCONGLOMERATE, 3663
REACTIVATION, 3760
RECUMBENT FOLD, 3771
RIFT, 3864
RIFT VALLEY, 3865
RING FRACTURE, 3870
ROOT, 3896
RUPTURE, 3917
SALT DOME, 3925
SALT TECTONICS, 3929
SCHISTOSITY, 3961
SCHUPPE, 3963
SEA=FLOOR SPREADING, 3980
SEDIMENTARY COVER, 3996
SHEAR JOINT, 4059
SHEAR PLANE, 4061
SHEAR ZONE, 4064
SHIELD, 4072
SHORTENING, 4084
SLATY CLEAVAGE, 4131
SLIP CLEAVAGE, 4133
STEP FAULT, 4248
STEP FOLD, 4249
STRIKE, 4292
STRIKE=SLIP FAULT, 4293
STRUCTURAL ANALYSIS, 4301
STRUCTURAL DIAGRAM, 4304
STRUCTURAL ELEMENT, 4305
STRUCTURAL GEOLOGY, 4306
STRUCTURAL MAP, 4308
STRUCTURAL PETROLOGY, 4309
STRUCTURE CONTOUR MAP, 4313
SUBDUCTION ZONE, 4322
SUBSIDENCE, 4337
SUDETIC OROGENY, 4346
SUPERPOSED FOLD, 4359
SUPERPOSITION, 4360
SVECOFENNIAN OROGENY, 4375
SYMMETRICAL FOLD, 4382
SYNCLINE, 4390
SYNCLINORIUM, 4391
SYNECLISE, 4392
SYNFORM FOLD, 4393
SYNTECTONIC PROCESS, 4399
TACONIC OROGENY, 4407
TAPHROGENY, 4421

TAPHROGEOSYNCLINE, 4422
TECTOGENESIS, 4433
TECTONIC CONTROL, 4435
TECTONIC CYCLE, 4436
TECTONIC MAP, 4437
TECTONIC STYLE, 4438
TECTONIC SUPERSTRUCTURE, 4439
TECTONICS, 4440
TECTONOSPHERE, 4443
TENSILE STRENGTH, 4455
TENSION, 4456
THALASSOCRATON, 4478
THRUST FAULT, 4520
TRANSCURRENT FAULT, 4567
TRANSFORM FAULT, 4569
TROUGH, 4603
UNDERTHRUST FAULT, 4656
UPLIFT, 4665
UPTHROW, 4685
UPWARPING, 4686
VARISCAN OROGENY, 4719
VERGENCE, 4730
VERTICAL TECTONICS, 4736
VIRGATION, 4743
VOLCANO=TECTONIC DEPRESSION, 4770
VOLCANO=TECTONIC STRUCTURE, 4771
VULCANORIUM, 4778
WINDOW, 4830
WRENCH FAULT, 4831
ZIGZAG FOLD, 4859

SURF SURFICIAL GEOLOGY, GLACIAL GEOLOGY, GEOMORPHOLOGY

ABANDONED SHORELINE, 4
ABLATION, 5
ABRASION, 6
ABRASION COAST, 7
ABRASION SURFACE, 8
ACCUMULATIVE COAST, 32
ACCUMULATIVE PLAIN, 33
ACTIVE LAYER, 57
AERATION, 69
AEROSOL, 75
AFFLUENT, 77
AGGRADATION, 84
AGRICULTURE, 90
AGROGEOLOGY, 91
ALLUVIAL, 140
ALLUVIAL FAN, 141
ALLUVIAL FILL, 142
ALLUVIAL PLAIN, 144
ALLUVIUM, 146
ALTERATION, 154
ANTIDUNE, 226
APRON, 250
ARCTIC, 286

ARID ENVIRONMENT, 299
ARIDITY, 301
ARROYO, 308
ATMOSPHERIC PRESSURE, 361
ATTRITION, 368
AVALANCHE, 388
BADLANDS, 403
BAJADA, 405
BAR, 410
BARCHAN, 411
BARRIER, 416
BARRIER BEACH, 418
BARRIER LAGOON, 419
BAY, 443
BEACH, 444
BEACH RIDGE, 446
BEACH SCARP, 447
BED LOAD, 452
BEHEADED STREAM, 458
BENCH, 461
BERGSCHRUND, 467
BERM, 469
BIGHT, 478
BIODEGRADATION, 489
BIORHEXISTASY, 506
BLOWHOLE, 533
BOG, 536
BOREAL, 548
BOULDER CLAY, 559
BOULDER TRAIN, 560
BRAIDED CHANNEL, 571
BURIED CHANNEL, 598
BURIED VALLEY, 599
CALCRETE, 616
CALDERA, 617
CALICHE, 619
CANYON, 630
CAPE, 632
CAPTURED STREAM, 637
CATENA, 673
CAULDRON, 675
CAVE, 677
CAVERN, 678
CHANNEL, 707
CHANNEL GEOMETRY, 708
CHEMICAL WEATHERING, 735
CHENIER, 737
CINDER CONE, 770
CIRQUE, 775
CLIFF, 793
CLIMATE, 794
CLIMATIC ZONATION, 797
COAST, 811
COASTAL DUNE, 812
COASTAL EROSION, 813
COASTAL GEOMORPHOLOGY, 814
COASTAL PLAIN, 815

COASTLINE, 816
COLD ENVIRONMENT, 832
COLLAPSE, 834
COLLAPSE CALDERA, 836
COLLUVIUM, 843
CONFLUENCE, 888
CONSEQUENT STREAM, 904
CONTINENTAL GLACIATION, 924
CONTINENTAL ZONE, 931
CORROSION, 964
CRATER, 981
CRATER LAKE, 982
CRATERING, 983
CREEK, 985
CREEP, 986
CREST, 988
CREVASSE, 990
CRYOPEDOLOGY, 1003
CRYOTECTONICS, 1004
CRYPTOVOLCANIC STRUCTURE, 1010
CUESTA, 1031
CUSP, 1037
CYCLONE, 1049
DEAD ICE, 1075
DEBRIS CONE, 1076
DEBRIS FLOW, 1077
DEFLATION, 1095
DEGLACIATION, 1098
DEGRADATION, 1099
DELTA, 1103
DENUDATION, 1116
DESERT, 1131
DESERT DUNE, 1132
DESERT PAVEMENT, 1133
DESERTIFICATION, 1134
DEW, 1152
DIFFERENTIAL WEATHERING, 1177
DIP SLOPE, 1198
DOLINE, 1235
DRAINAGE SYSTEM, 1246
DROUGHT, 1258
DRUMLIN, 1259
DUNE, 1261
ELUVIUM, 1357
EMERGENCE, 1361
ENTRENCHED STREAM, 1384
EOLIAN FEATURE, 1392
EPHEMERAL STREAM, 1397
EQUATORIAL ZONE, 1414
ERG, 1422
ERODIBILITY, 1423
EROSION, 1424
EROSION CONTROL, 1425
EROSION CYCLE, 1426
EROSION FEATURE, 1427
EROSION REMNANT, 1428
EROSION SURFACE, 1429

ESCARPMENT, 1433
ESKER, 1434
ESTUARY, 1438
EXFOLIATION, 1466
EXTINCT LAKE, 1494
FAN, 1504
FAULT SCARP, 1511
FERRICRETE, 1534
FIRN, 1553
FJORD, 1560
FLOOD PLAIN, 1570
FLUVIAL EROSION, 1594
FLUVIOGLACIAL, 1596
FOSSIL ICE WEDGE, 1621
FROST ACTION, 1653
FROST FEATURE, 1654
FROST HEAVING, 1655
FROZEN GROUND, 1656
FURROW, 1668
GELIVITY, 1704
GEOMORPHOLOGICAL MAP, 1747
GEOMORPHOLOGY, 1748
GLACIAL, 1791
GLACIAL DRIFT, 1792
GLACIAL ENVIRONMENT, 1793
GLACIAL EROSION, 1794
GLACIAL FEATURE, 1795
GLACIAL GEOLOGY, 1796
GLACIAL LAKE, 1797
GLACIAL STRIATION, 1800
GLACIAL TRANSPORT, 1801
GLACIAL VALLEY, 1802
GLACIATION, 1803
GLACIER, 1804
GLACIER SURGE, 1805
GLACIOLOGY, 1807
GLACIS, 1810
GORGE, 1837
GREENHOUSE EFFECT, 1891
GROOVE, 1902
GROUND, 1904
GROUND ICE, 1905
GROUND MORAINE, 1907
GULF, 1927
GULLY, 1928
HAIL, 1941
HAMMADA, 1950
HANGING VALLEY, 1952
HEAVE, 1970
HIGHLAND, 2012
HILL, 2013
HORIZON DIFFERENTIATION, 2049
HUMID CLIMATE, 2073
HUMID ENVIRONMENT, 2074
HUMIFICATION, 2076
HUMMOCK, 2079
HUMUS, 2083

HYGROSCOPIC WATER, 2139
HYPERGENESIS, 2143
HYPSOMETRY, 2151
ICE, 2155
ICE CAP, 2156
ICE FIELD, 2157
ICE MOVEMENT, 2158
ICE RAFTING, 2159
ICE SHEET, 2160
ICE SHELF, 2161
ICE WEDGE, 2162
INCRUSTATION, 2205
INDICATIVE BOULDER, 2211
INGRESSION, 2236
INHERITED FEATURES, 2237
INLET, 2241
INSELBERG, 2253
INSEQUENT VALLEY, 2254
INSOLATION, 2255
INTERFLUVE, 2264
INTERMONTANE UNIT, 2273
ISLAND, 2315
ISOHYPSE, 2329
KAME, 2366
KARREN, 2371
KARST, 2373
KARST FILLING, 2374
KETTLE, 2387
LAGOON, 2411
LAKE, 2415
LAND, 2425
LANDFORM, 2429
LATERAL MORAINE, 2444
LATERIZATION, 2448
LIMNOLOGY, 2508
LITTORAL, 2543
LITTORAL DRIFT, 2544
LONGITUDINAL PROFILE, 2566
LONGSHORE BAR, 2567
LOWLAND, 2595
MAAR, 2612
MANGROVE SWAMP, 2663
MARINE EROSION, 2678
MARINE MARSH, 2681
MARINE PLAIN, 2683
MARINE TERRACE, 2688
MARSH, 2693
MASS BALANCE, 2698
MASS MOVEMENT, 2699
MASS WASTING, 2702
MEANDER, 2724
MECHANICAL WEATHERING, 2728
MELIORATION, 2742
METEOROLOGY, 2800
MICROMORPHOLOGY, 2825
MONOLITH, 2926
MONSOON, 2931

MORAINE, 2938
MORPHOGENESIS, 2939
MORPHOSTRUCTURE, 2943
MOUNTAINS, 2950
MUD CRACK, 2954
MUD FLAT, 2955
MUDSLIDE, 2959
NEEDLE, 2996
NIVATION, 3036
OASIS, 3074
OBSEQUENT VALLEY, 3080
OUTWASH, 3211
OUTWASH PLAIN, 3212
PALSA, 3277
PAMPA, 3280
PARENT MATERIAL SOILS, 3302
PATTERNED GROUND, 3317
PEAT BOG, 3321
PEDIMENT, 3324
PEDIPLAIN, 3325
PEDOGENESIS, 3326
PENEPLAIN, 3339
PERIGLACIAL FEATURES, 3354
PERIGLACIAL ZONE, 3355
PERMAFROST, 3362
PHYSICAL WEATHERING, 3429
PHYSIOGRAPHIC PROVINCE, 3431
PIEDMONT, 3439
PINGO, 3450
PLAIN, 3465
PLANATION, 3468
PLATEAU, 3501
PLAYA, 3507
POLAR, 3533
POLAR ICECAP, 3534
POLYGONAL GROUND, 3547
POSTGLACIAL, 3573
POTHOLE, 3585
PREGLACIAL, 3601
PROFILE OF EQUILIBRIUM, 3628
PROGRADATION, 3632
REGOLITH, 3798
RELIEF, 3808
RELIEF EXPOSURE, 3809
RELIEF INVERSION, 3810
RESEQUENT VALLEY, 3823
RESIDUAL, 3827
RIEGEL, 3863
RIVER, 3873
RIVER LOAD, 3874
RIVER TERRACE, 3875
ROCK BAR, 3882
ROCK FALL, 3884
ROCK GLACIER, 3886
ROCKSLIDE, 3891
RUBBLE, 3907
RUBEFACTION, 3908

SALT FLAT, 3926
SALT LAKE, 3927
SALT MARSH, 3928
SCOURING, 3971
SEMI−ARID ENVIRONMENT, 4035
SHEET EROSION, 4065
SHORE FEATURE, 4081
SHORELINE, 4083
SHRINKAGE, 4087
SILT LOAD, 4107
SINKHOLE, 4117
SINUOSITY, 4119
SLOPE, 4134
SLOPE DEFORMATION, 4135
SNOW, 4142
SOIL, 4146
SOIL EROSION, 4148
SOIL HORIZON, 4149
SOIL IMPROVEMENT, 4150
SOIL LEACHING, 4151
SOIL MAP, 4153
SOIL PROFILES, 4155
SOIL SCIENCE, 4157
SOLIFLUCTION, 4171
SOLUTION CAVITY, 4177
SOLUTION FEATURE, 4178
SPELEOLOGY, 4203
SPELEOTHEM, 4204
SPIT, 4215
STALACTITE, 4231
STALAGMITE, 4232
STEPPE ZONE, 4251
STONE LINE, 4262
STORM, 4266
STREAM, 4284
STREAM BED, 4285
STREAM ORDER, 4286
STRUCTURAL RELIEF, 4310
STRUCTURAL VALLEY, 4312
SUBAERIAL ENVIRONMENT, 4316
SUBBOREAL, 4319
SUBSEQUENT VALLEY, 4336
SUBTROPICAL ENVIRONMENT, 4342
SURFICIAL GEOLOGY, 4369
SURGE, 4370
SUSPENDED LOAD, 4372
SWAMP, 4376
TAIGA, 4412
TALUS, 4416
TALUS SLOPE, 4417
TEMPERATE CLIMATE, 4452
TERRACE, 4465
THALWEG, 4482
THERMOKARST, 4508
TIDAL CHANNEL, 4525
TIDAL FLAT, 4527
TIDAL INLET, 4528

TILL, 4530
TOMBOLO, 4542
TOPOGRAPHY, 4547
TORRENT, 4549
TROPICAL ZONE, 4602
TUNDRA, 4610
VALLEY, 4707
VARVE, 4720
VARVED CLAY, 4722
WADI, 4779
WATER EROSION, 4789
WATERFALL, 4803
WEATHERING, 4809
WEATHERING CRUST, 4810
WETLAND, 4822
WIND, 4826
WIND EROSION, 4827
WIND TRANSPORT, 4829

SUSS   SOILS, SYSTEMATICS

ACRISOL, 47
ALBOLL, 107
ALFISOL, 110
ANDEPT, 193
ANDOSOL, 196
AQUALF, 254
AQUENT, 255
AQUEPT, 257
AQUERT, 258
AQUOD, 265
AQUOLL, 266
AQUOX, 267
AQUULT, 268
ARENOSOL, 293
ARGID, 294
ARIDISOL, 300
AZONAL SOIL, 394
BORALF, 543
BOROLL, 551
BROWN SOIL, 583
CAMBISOL, 624
CASTANOZEM, 666
CHERNOZEM, 738
ENTISOL, 1383
FERRALSOL, 1532
FERROD, 1535
FERRUGINOUS SOIL, 1540
FLUVENT, 1592
FLUVIOSOL, 1598
GLEYSOL, 1819
HISTOSOL, 2022
HUMIC SOIL, 2072
HUMOD, 2080
HUMOX, 2081
HUMULT, 2082

IMMATURE SOIL, 2183
INCEPTISOL, 2199
LATERITIC SOIL, 2447
LITHOSOL, 2535
LUVISOL, 2609
MEDITERRANEAN SOIL, 2735
MOLLISOL, 2915
MULL SOIL, 2960
OCHREPT, 3096
ORTHENT, 3186
ORTHID, 3187
ORTHOD, 3190
ORTHOX, 3195
OXISOL, 3224
PHAEOZEM, 3388
PLANOSOL, 3490
PODZOL, 3526
PODZOLUVISOL, 3527
PSAMMENT, 3658
PSEUDOGLEY, 3664
RANKER, 3749
REGOSOL, 3799
RENDOLL, 3815
RENDZINA, 3816
SIEROZEM, 4095
SOLONCHAK, 4172
SOLONETZ, 4173
SPODOSOL, 4216
TERRA ROSSA, 4464
TUNDRA SOIL, 4611
UDALF, 4630
UDOLL, 4631
UDULT, 4632
ULTISOL, 4634
UMBREPT, 4642
USTALF, 4699
USTERT, 4700
USTOLL, 4701
USTOX, 4702
USTULT, 4703
VERTISOL, 4737
XERALF, 4841
XEROLL, 4842
XEROSOL, 4845
XERULT, 4846
YERMOSOL, 4848

TEST   TEXTURES, STRUCTURES

AMORPHOUS MINERAL, 170
ANTIPERTHITE, 231
APHANITIC TEXTURE, 236
APLITIC TEXTURE, 240
ARENACEOUS, 290
AUGEN STRUCTURE, 370
BANDING, 409

BIOGENIC STRUCTURE, 494
BLASTIC TEXTURE, 525
BOTRYOIDAL, 553
BOXWORK, 564
CAVERNOUS TEXTURE, 679
CHAOTIC STRUCTURE, 709
CHATTERMARK, 715
COATED GRAIN, 817
COLLOFORM TEXTURE, 841
COLUMNAR JOINTING, 850
CONCHOIDAL FRACTURE, 875
CONE=IN=CONE, 886
CONVOLUTE BEDDING, 941
CROSS=STRATIFICATION, 998
CRYOTURBATION, 1005
CRYPTOCRYSTALLINE TEXTURE, 1006
CYLINDRICAL STRUCTURE, 1053
ELONGATE MINERAL, 1355
EQUIGRANULAR TEXTURE, 1415
EUCRYSTALLINE TEXTURE, 1444
EUTAXITIC TEXTURE, 1453
FABRIC, 1500
FLAGGY TEXTURE, 1561
FLAME STRUCTURE, 1563
FLASER STRUCTURE, 1565
FLUIDAL TEXTURE, 1584
FLUTE CAST, 1591
FOLIATION, 1609
FRAMBOIDAL TEXTURE, 1638
FULGURITE, 1660
GLASSY TEXTURE, 1814
GLOMERO=, 1823
GNEISSIC TEXTURE, 1830
GRADED BEDDING, 1841
GRAIN, 1844
GRANOBLASTIC TEXTURE, 1852
GRANOPHYRIC TEXTURE, 1855
GRANULAR, 1856
GRANULATION, 1857
GRANULITIC, 1860
GRAPHIC TEXTURE, 1865
GROUNDMASS, 1919
HEAT CRACK, 1965
HETEROBLASTIC TEXTURE, 1988
HIEROGLYPH, 2005
HOLOCRYSTALLINE TEXTURE, 2028
HYALOCRYSTALLINE TEXTURE, 2084
HYPIDIOBLASTIC TEXTURE, 2144
HYPIDIOMORPHIC TEXTURE, 2145
HYPOCRYSTALLINE TEXTURE, 2147
IDIOBLAST, 2168
IDIOMORPHIC TEXTURE, 2170
IMPACT STRUCTURE, 2192
INTERGROWTH, 2267
INTERSERTAL TEXTURE, 2284
LAMELLAE, 2419
LAMINATED STRUCTURE, 2421

LAMINATION, 2422
LAMPROPHYRIC TEXTURE, 2424
LATTICE TEXTURE, 2452
LAYER, 2462
LEPIDOBLASTIC TEXTURE, 2482
LIESEGANG RINGS, 2498
LINOPHYRIC TEXTURE, 2515
MARGINATION TEXTURE, 2675
MASSIVE, 2703
MEDIUM-GRAINED, 2736
MIAROLITIC TEXTURE, 2803
MICROCRATER, 2808
MICROFISSURE, 2813
MICROGRAINED TEXTURE, 2817
MICROGRANITIC TEXTURE, 2819
MICROLITIC TEXTURE, 2822
MICROSCOPIC SIZE, 2832
MICROSTYLOLITE, 2835
MONZONITIC TEXTURE, 2934
MOSAIC STRUCTURE, 2945
MOTTLING, 2949
MULLION STRUCTURE, 2961
MYLONITIC STRUCTURE, 2973
MYRMEKITE, 2979
OOID, 3123
OOLITIC TEXTURE, 3126
OPHITIC TEXTURE, 3134
ORBICULAR TEXTURE, 3145
ORGANIC STRUCTURE, 3175
ORTHOPHYRIC TEXTURE, 3193

PACKING, 3232
PANIDIOMORPHIC TEXTURE, 3282
PARALLEL BEDDING, 3295
PARALLEL TEXTURE, 3296
PARTING, 3310
PEGMATITIC, 3328
PELITIC TEXTURE, 3335
PERLITIC TEXTURE, 3361
PERTHITE, 3367
PERTHITIC TEXTURE, 3368
PETROFABRIC, 3373
PHENOCRYST, 3395
PILLOW STRUCTURE, 3448
PISOLITIC TEXTURE, 3458
PLANAR BEDDING STRUCTURE, 3466
PLANAR FEATURES, 3467
POIKILITIC TEXTURE, 3529
POIKILOBLASTIC TEXTURE, 3530
PORPHYRITIC TEXTURE, 3567
PORPHYROBLASTIC TEXTURE, 3568
PREFERRED ORIENTATION, 3600
PRIMARY STRUCTURE, 3612
PRISMATIC STRUCTURE, 3616
PSAMMITIC, 3660
PSEPHITIC, 3662
RADIAL, 3719
RAPAKIVI, 3750
RELICT TEXTURE, 3807
RETICULATE STRUCTURE, 3840
RIPPLE MARK, 3872

ROCK MATRIX, 3887
SCHLIEREN, 3962
SCOUR MARK, 3970
SECONDARY SEDIMENTARY STRUCTURE, 3990
SEDIMENTARY STRUCTURE, 4001
SHATTER CONE, 4057
SHRINKAGE CRACK, 4088
SIZE DISTRIBUTION, 4125
SLICKENSIDE, 4132
SLUMP STRUCTURE, 4138
SOLE MARK, 4163
SORTING, 4184
SPHERICITY, 4207
SPHERULE, 4208
SPHERULITE, 4209
SPINIFEX, 4212
STRATIFIED STRUCTURE, 4273
STYLOLITE, 4314
SYMPLECTIC TEXTURE, 4386
SYNSEDIMENTARY STRUCTURE, 4398
TEXTURE, 4477
TURBIDITY CURRENT STRUCTURE, 4618
VERMIFORM STRUCTURE, 4733
VITREOUS TEXTURE, 4750
VITROCLASTIC TEXTURE, 4752
VITROPHYRIC TEXTURE, 4753
WIND RIPPLE, 4828
XENOMORPHIC CRYSTAL, 4839
ZONAL STRUCTURE, 4865